EULER: $y_{n+1} = y_n + f(x_n, y_n)h$

SECOND-ORDER RUNGE–KUTTA: $y_{n+1} = y_n + \frac{1}{2}(k_1 + k_2)$

$$k_1 = hf(x_n, y_n) \qquad k_2 = hf(x_{n+1}, y_n + k_1)$$

FOURTH-ORDER RUNGE–KUTTA: $y_{n+1} = y_n + \frac{1}{6}(k_1 + 2k_2 + 2k_3 + k_4)$

$$k_1 = hf(x_n, y_n) \qquad\qquad k_2 = hf\left(x_n + \frac{h}{2}, y_n + \frac{1}{2}k_1\right)$$
$$k_3 = hf\left(x_n + \frac{h}{2}, y_n + \frac{1}{2}k_2\right) \qquad k_4 = hf(x_{n+1}, y_n + k_3)$$

GAMMA FUNCTION: (In exercises, Section 13.2)

$$\Gamma(p) = \int_0^\infty x^{p-1}e^{-x}\,dx \qquad (p > 0)$$

$$\Gamma\left(\frac{1}{2}\right) = \sqrt{\pi}, \quad \Gamma(n) = (n-1)!, \quad \Gamma(p) = (p-1)\Gamma(p-1) \text{ for } p > 1$$

HEAVISIDE FUNCTION: $H(t) = \begin{cases} 1, & t > 0 \\ 0, & t < 0 \end{cases}$

$$\int_0^t f(\tau)H(\tau - a)\,d\tau = H(t - a)\int_a^t f(\tau)\,d\tau$$

MAPLE INDEX

Differential Equations
&Linear Algebra

Differential Equations & Linear Algebra

MICHAEL D. GREENBERG

University of Delaware

PRENTICE HALL
Upper Saddle River, New Jersey 07458

Library of Congress Cataloging-in-Publication Data
Greenberg, Michael D.
    Differential equations and linear algebra/Michael D. Greenberg.
      p. cm.
    Includes bibliographical references index.
    ISBN 0-13-011118-X
    1. Algebras, Linear. 2. Differential equations. I. Title.

QA184.G72   2001
512'–dc21   00-055053
                                 CIP

Acquisitions Editor: *George Lobell*
Editor-in-Chief: *Sally Yagan*
Assistant Vice President of Production and Manufacturing: *David W. Riccardi*
Executive Managing Editor: *Kathleen Schiaparelli*
Senior Managing Editor: *Linda Mihatov Behrens*
Production Editor: *Bob Walters*
Manufacturing Buyer: *Alan Fischer*
Manufacturing Manager: *Trudy Pisciotti*
Marketing Manager: *Angela Battle*
Marketing Assistant: *Vince Jansen*
Director of Marketing: *John Tweeddale*
Editorial Assistant: *Gale Epps*
Art Director: *Maureen Eide*
Interior Designer: *Donna Wickes*
Cover Designer: *Daniel Conte*
Cover Photo: *Shuji Yamada/ITSUKO HASEGAWA ATELIER*

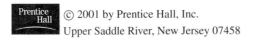 © 2001 by Prentice Hall, Inc.
Upper Saddle River, New Jersey 07458

Printed in the United States of America
10 9 8 7 6 5 4 3 2 1

ISBN 0-13-011118-X

Prentice-Hall International (UK) Limited, *London*
Prentice-Hall of Australia Pty. Limited, *Sydney*
Prentice-Hall Canada, Inc., *Toronto*
Prentice-Hall HispanoAmericana, S.A., *Mexico*
Prentice-Hall of India Private Limited, *New Delhi*
Prentice-Hall of Japan, Inc., *Tokyo*
Pearson Education Asia Pte. Ltd.
Editora Prentice-Hall do Brasil, Ltda., *Rio de Janeiro*

Contents

6 LINEAR DIFFERENTIAL EQUATIONS OF SECOND ORDER AND HIGHER   251

7 APPLICATIONS OF LINEAR CONSTANT-COEFFICIENT EQUATIONS   317

8 POWER SERIES SOLUTIONS   353

9 THE EIGENVALUE PROBLEM   395

10 SYSTEMS OF LINEAR DIFFERENTIAL EQUATIONS   433

Preface

PURPOSE AND PREREQUISITES

This book is intended as a textbook for a course in differential equations with linear algebra, to follow the differential and integral calculus. Since the syllabus of such a course is by no means standard, we have included more material than can be covered in a single course—possibly enough material for a two-semester course. This additional material is included to broaden the menu for the instructor and to increase the text's subsequent usefulness as a reference book for the student.

Written for engineering, science, and computer science students, the approach is aimed at the applications oriented student but is also intended to be rigorous and to reveal the beauty and elegance of the subject.

Why blend linear algebra with the differential equations? Since mid-twentieth century, the traditional course in differential equations has been offered in the first or second semester of the sophomore year and has relied on only a minimum of linear algebra, most notably the use of determinants. More recently, beginning with the advent of digital computers on campuses and in industry around the 1960s, a course or part of a course in linear algebra has become a part of most engineering science curricula. Given the current interest in introducing linear algebra earlier in curricula, the growing importance of systems of differential equations, and the natural use of linear algebra concepts in the study of differential equations, it seems best to move toward an integrated approach.

FLEXIBILITY

The text is organized so as to be flexible. For instance, it is generally considered desirable to include some nonlinear phase plane analysis in a course on differential equations since the qualitative topological approach complements the traditional analytical approach and also powerfully emphasizes the differences between linear and nonlinear systems. However, that topic usually proves to be a "luxury" to which one can devote one or two classes at best. Thus, we have arranged the phase plane material to allow anywhere from a one-class introduction to a moderately detailed discussion: We introduce the phase plane in only four pages in Section 7.3 in support of our discussion of the harmonic oscillator and we return to it in Chapter 11. There, Section 11.2 affords a more detailed overview of the method and provides another possible stopping point.

To assist the instructor in the syllabus design we list some sections and subsections as optional but emphasize that these designations are subjective and intended only as a rule of thumb. (To the student we note that "optional" is not intended to mean unimportant, but only as a guide as to which material can be omitted by virtue of not being a prerequisite for the material that follows.)

SPECIFIC PEDAGOGICAL DECISIONS
Several pedagogical decisions made in writing this text deserve explanation.

1. *Chapter sequence*: Some instructors prefer to discuss numerical solution early, even within the study of first-order equations. Placement of the material on numerical solution near the end of this text does not rule out such an approach for one could cover Sections 12.1-12.2 on Euler's method, say, at any point in Chapters 2 or 3. Here, it seemed preferable to group Chapters 11 (on the phase plane) and 12 (on numerical solution) together since they complement the analytical approach, the former being qualitative and the latter being quantitative. As such, these two chapters might well have been made the final chapters, with the Laplace transform chapter moving up to precede or to follow Chapter 8 on series solution. Such movement is possible in a course syllabus since other chapters do not depend on series solution or on the Laplace transform. Also along these lines, it might seem awkward that Chapters 4 and 5 on vectors and matrices are separated from Chapter 9 on the eigenvalue problem. This separation may not be as great as it appears since in a one-semester course Chapter 8 might well be omitted. In any case in a combined approach to differential equations and linear algebra it seems logical to intersperse these two topics as naturally as possible rather than presenting them end-to-end. It may even be true that for optimal student retention it is good to have a gap between first meeting the linear algebra in Chapters 4 and 5 and returning to it in Chapter 9, so that it feels more like one is studying the subject twice.

2. *Placing Gauss elimination in Chapter 4 on vectors rather than in Chapter 5 on matrices and linear algebraic equations:* Just as one studies the real number axis before studying functions (mappings from one such axis to another), it seems appropriate to study vector spaces before studying matrices (which provide mappings from one vector space to another). In that case we find—in discussing span, linear dependence, bases, and expansions in Chapter 4—that we need to solve systems of coupled linear algebraic equations. Hence, we devote Section 4.5, which precedes that discussion, to Gauss elimination.

3. *Introducing the Heaviside function in the chapter on first-order differential equations rather than in the chapter on the Laplace transform*: If the forcing function is given piecewise, solution of the differential equation by a computer algebra system (*Maple* in this text) requires us to give a single expression for that function, and that can be accomplished using the Heaviside function. Further, including the Heaviside function in Chapter 2 makes it possible to include that topic even if the chapter on the Laplace transform is not covered.

Computer Algebra System
As a representative computer algebra system this text uses *Maple*, but does not as-

sume prior knowledge of that system. The *Maple* discussion is confined to subsections at the end of most sections, immediately preceding the exercises; see, for example, Sections 2.2 and 2.3. The reader can bypass those discussions entirely since they are supplemental and intended to show the student how to carry out various *Maple* calculations relevant to the material in that section. In some cases they explain how text figures were generated. The view represented here is that it would be foolish not to use the powerful computer algebra systems that are now available, but that primary emphasis should continue to rest firmly on fundamentals and understanding of the theory and methods. See also the section on supplements, below.

EXERCISES

End-of-section exercises are of different kinds and are arranged, typically, as follows. First, and usually near the beginning of the exercise group, are exercises that follow up on small gaps in the reading, thus engaging the student more fully in the reading (e.g., Exercises 1 and 2 of Section 3.5). Second, there are usually numerous "drill" type exercises that ask the student to mimic steps or calculations that are essentially similar to those demonstrated in the text (e.g., there are 19 matrices to invert by hand in Exercise 1 of Section 5.6). Third, there are exercises that call for the use of *Maple* (e.g., Exercise 3 of Section 5.6 and Exercise 4 of Section 10.4). Fourth, some exercises involve physical applications (e.g., Exercise 22 of Section 2.4 on the distribution of a pollutant in a river, Exercises 17 and 18 of Section 5.6 on electrical circuits, and Exercise 14 of Section 5.8 on computer graphics). And, fifth, there are exercises intended to extend the text and increase its value as a reference book (e.g., Exercises 7-12 of Section 2.3 on the Bernoulli, Riccati, Alembert-Lagrange, and Clairaut equations, and Exercise 2 of Section 3.3 on envelopes). Answers to selected exercises (which are denoted in the text by underlining the equation number) are given at the end of the book.

SYLLABUS DESIGN

Designing a two-semester course is simple in the sense that one would probably cover virtually everything in the text. Thus, let us restrict our comments to the design of a one-semester course. As a general comment we note that sections and subsections are arranged with an eye toward flexibility. In Chapter 10, for instance, one could limit the coverage to Sections 10.1–10.3 or one could cover Sections 10.1, 10.2, and 10.4. As a specific example, at the University of Delaware mechanical engineers are currently required to take a three-course sequence in their sophomore year as follows. In the fall they take a three-credit course on differential equations and linear algebra following a syllabus somewhat as follows: Chapters 1–7 and 9–10 with these sections omitted—2.3.3, 2.4.2, 3.4, 4.4.2, 4.4.3, 4.5.6, 4.5.7, 4.8.3, 4.9.4, 4.9.5, 5.6.5, 5.7.2, 5.8, 6.6.2, 6.7.3, 6.7.4, 9.4.2, 9.4.3, 10.3.3, and 10.5–10.7.

In the Spring they take two more courses, one covering Laplace transforms, field theory, and partial differential equations, and the other covering numerical methods, including the numerical solution of ordinary and partial differential equations.

SUPPLEMENTS

For information regarding the Instructor's Solution Manual and other supplements, see the publisher's website, available 1/1/01 at www.prenhall.com/greenberg. The site will contain quizzes and other text related activities that will be free to all text users. Suggestions, comments, and errata will be gratefully received at the author's e-mail address given below.

ACKNOWLEDGMENTS

I am grateful for extensive support in the preparation of this text and thank my mathematics editor at Prentice Hall, George Lobell, for his insight and support. I am also pleased to thank Professors Idris Assani (University of North Carolina, Chapel Hill), Michael Kirby (Colorado State University), Dan Knopf (University of Wisconsin, Madison), Paul Milewski (University of California, Santa Barbara), Gustavo Ponce (University of California, Santa Barbara), Christopher Raymond (University of Wisconsin, Madison), Dragan Skropanic (Western Wyoming Community College), Marshall Slemrod (University of Wisconsin, Madison), and especially Dr. David Weidner from the University of Delaware.

I dedicate this book, with love, to my wife Yisraela. Finally and most important: "From whence cometh my help? My help cometh from the Lord, who made heaven and earth." (Psalm 121).

MICHAEL D. GREENBERG
greenberg@me.udel.edu

Differential Equations & Linear Algebra

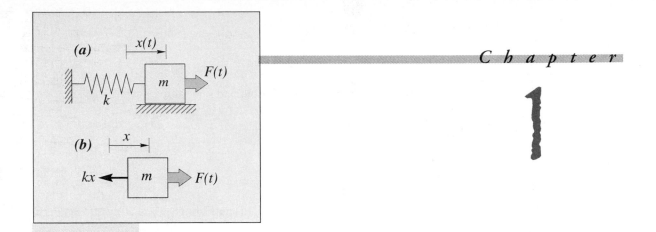

INTRODUCTION TO DIFFERENTIAL EQUATIONS

1.1 INTRODUCTION

Most phenomena in science and engineering are governed by equations involving derivatives of one or more unknown functions.

To illustrate, consider the motion of a body of mass m along a straight line, which we designate as an x axis. Let the mass be subjected to a force $F(t)$ along that axis, where t is the time. Then according to Newton's second law of motion,

$$m\frac{d^2x}{dt^2} = F(t), \tag{1}$$

where $x(t)$ is the mass's displacement measured from the origin. If we know the displacement $x(t)$ and we wish to determine the force $F(t)$ required to produce that displacement, then the solution is simple: According to (1), we merely differentiate the given $x(t)$ twice and multiply the result by m.

However, if we know the applied force $F(t)$ and wish to determine the displacement $x(t)$ that results, then we say that (1) is a "differential equation" on $x(t)$ since it involves the derivative, more precisely the second derivative in this example, of the unknown function $x(t)$ with respect to t. The question is: What function $x(t)$, when differentiated twice with respect to t and then multiplied by m (which is a constant), gives the prescribed function $F(t)$? To solve (1) for $x(t)$, we need to undo the differentiations; that is, we need to integrate (1), twice in fact.

For definiteness and simplicity, suppose that $F(t) = F_0$ is a constant, so that (1) becomes

$$m\frac{d^2x}{dt^2} = F_0. \tag{2}$$

Integrating (2) once with respect to t gives

$$m\frac{dx}{dt} = F_0 t + A,\qquad (3)$$

where A is an arbitrary constant of integration, and integrating again gives

$$mx = \frac{F_0}{2}t^2 + At + B,$$

so

$$x(t) = \frac{1}{m}\left(\frac{F_0}{2}t^2 + At + B\right).\qquad (4)$$

The constants of integration, A and B, can be found from (3) and (4) if the displacement and the velocity are prescribed at the initial time, which we take to be $t = 0$. That is, we can solve for A and B if $x(0)$ and $\frac{dx}{dt}(0)$ are known. For instance, suppose we know that

$$x(0) = 0 \quad\text{and that}\quad \frac{dx}{dt}(0) = 0.\qquad (5)$$

Then, by setting $t = 0$ in (4), we find that $B = 0$, and by setting $t = 0$ in (3) we find that $A = 0$. Thus, (4) gives the solution

$$x(t) = \frac{1}{m}\frac{F_0}{2}t^2,\qquad (6)$$

which evidently holds for all $t \geq 0$, that is, for $0 \leq t < \infty$. It is easily verified that (6) does satisfy both the differential equation (2) and the initial conditions (5).

Unfortunately, most differential equations cannot be solved this readily, that is, by merely undoing the derivatives by integration. For instance, suppose that the mass is restrained by a coil spring that supplies a restoring force (i.e., in the direction opposite to the displacement) proportional to the displacement x, with constant of proportionality k (Fig. 1a). Then (Fig. 1b) the total force on the mass is $-kx + F(t)$, so in place of (1), the differential equation governing the motion is

$$m\frac{d^2x}{dt^2} = -kx + F(t)$$

or,

$$m\frac{d^2x}{dt^2} + kx = F(t).\qquad (7)$$

Let us try to solve (7) for $x(t)$ as we did before, by integrating twice with respect to t. After one integration (7) becomes

$$m\frac{dx}{dt} + k\int x(t)\,dt = \int F(t)\,dt + A,\qquad (8)$$

where A is the constant of integration. Since $F(t)$ is a known function, the integral of $F(t)$ in (8) can be evaluated. However, since the solution $x(t)$ is not yet known, the integral $\int x(t)\,dt$ cannot be evaluated, and we cannot proceed with our

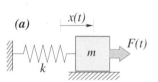

(a)

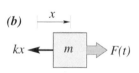

(b)

FIGURE 1
Mass/spring system.

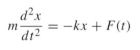

intended technique of solution-by-repeated-integration. Be clear that it would be incorrect to state that the integral $\int x(t)\, dt$ in (8) is $x(t)t$. The latter would be correct if $x(t)$ were a constant, but $x(t)$ is a thus-far-unknown function that is probably not merely a constant. If we were to distinguish small mistakes from large ones, writing "$\int x(t)\, dt = x\, t$" would be a large one, so be sure to understand this point.

Thus, we see that solving differential equations is not merely a matter of undoing the derivatives by direct integration. The theory and technique involved is considerable and will occupy us throughout this book.

In order to develop the theory, it is convenient, and even necessary, to use concepts and methods from the mathematical domain known as Linear Algebra. Our plan is to get started in our study of differential equations until linear algebra concepts are about to force their way upon the discussion. At that point we take a "break," in Chapters 4 and 5, to consider as much linear algebra as needed to proceed with our discussion of differential equations. Additional linear algebra material, on the so-called "eigenvalue problem," is introduced in Chapters 9 and 10.

1.2   DEFINITIONS AND TERMINOLOGY

To begin, we introduce several fundamental definitions.

Differential equation. By a **differential equation** we mean an equation containing one or more derivatives of the function under consideration. Some examples, which we put forward without derivation, are as follows:

$$\frac{dN}{dt} = \kappa N, \tag{1}$$

$$m\frac{d^2x}{dt^2} + kx = F(t), \tag{2}$$

$$L\frac{d^2i}{dt^2} + \frac{1}{C}i = \frac{dE}{dt}, \tag{3}$$

$$\frac{d^2y}{dx^2} = C\sqrt{1 + \left(\frac{dy}{dx}\right)^2}, \tag{4}$$

$$\frac{d^2\theta}{dt^2} + \frac{g}{l}\sin\theta = 0, \tag{5}$$

$$EI\frac{d^4y}{dx^4} = -w(x). \tag{6}$$

Equation (1) is the differential equation governing the population $N(t)$ of a particular species, where κ is a net birth/death rate and t is the time.

Equation (2) governs the linear displacement $x(t)$ of a body of mass m, subjected to an applied force $F(t)$, and a restraining spring of stiffness k, as was discussed in Section 1.1.

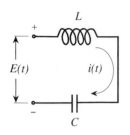

FIGURE 1
Electrical circuit,
equation (3)

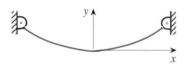

FIGURE 2
Hanging cable, equation (4).

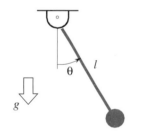

FIGURE 3
Pendulum,
equation (5).

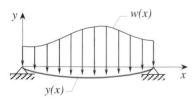

FIGURE 4
Loaded beam, equation (6).

Equation (3) governs the current $i(t)$ in an electrical circuit containing an inductor with inductance L, a capacitor with capacitance C, and an applied voltage source of strength $E(t)$ (Fig. 1), where t is the time.

Equation (4) governs the deflection $y(x)$ of a flexible cable or string, hanging under the action of gravity, where C is a constant (Fig. 2).

Equation (5) governs the angular motion $\theta(t)$ of a pendulum of length l under the action of gravity, where g is the acceleration of gravity and t is the time (Fig. 3).

Finally, equation (6) governs the deflection $y(x)$ of a beam subjected to a load distribution $w(x)$ (Fig. 4), where E and I are physical constants that involve the beam material and cross sectional shape, respectively.

Ordinary and partial differential equations.   We classify a differential equation as an **ordinary differential equation** if it contains ordinary derivatives with respect to a single independent variable. Thus, equations (1)–(6) are ordinary differential equations (traditionally abbreviated as **ODE**'s). The independent variable is t in (1), (2), (3), and (5), and x in (4) and (6), but the mathematics that we develop will be insensitive to the physical nature of the independent and dependent variables.

If the dependent variable is a function of more than one independent variable then we can expect the governing differential equation to contain derivatives with respect to those various independent variables, partial derivatives this time, in which case we call the differential equation a **partial differential equation** (abbreviated as **PDE**). To illustrate, three of the most important PDE's in science and engineering are these:

$$k\left(\frac{\partial^2 T}{\partial x^2} + \frac{\partial^2 T}{\partial y^2} + \frac{\partial^2 T}{\partial z^2}\right) = \rho c \frac{\partial T}{\partial t}, \tag{7}$$

$$\frac{\partial^2 T}{\partial x^2} + \frac{\partial^2 T}{\partial y^2} + \frac{\partial^2 T}{\partial z^2} = 0, \tag{8}$$

$$\tau\left(\frac{\partial^2 w}{\partial x^2} + \frac{\partial^2 w}{\partial y^2}\right) = \rho \frac{\partial^2 w}{\partial t^2}, \tag{9}$$

which are examples of the **diffusion equation**, the **Laplace equation**, and the **wave equation**, respectively. The first governs the temperature distribution $T(x, y, z, t)$ in some domain of three-dimensional x, y, z space, such as the interior of a hot ingot that is cooling down under the action of the heat transfer mechanism known as conduction; the physical constants k, ρ, c are the conductivity, mass density, and specific heat of the medium, respectively. In physical terms, (7) expresses the thermodynamic law that the rate of change of the heat contained in any volume element of the material (given by the right-hand side of the equation) is equal to the net rate of heat flowing into the element through its surface by the mechanism of conduction (given by the left-hand side of the equation). The second, equation (8), governs the same phenomenon but in the event that the temperature distribution is in "steady state"—that is, it is not changing with time. And the third, equation (9), governs the deflection $w(x, y, t)$ of a stretched membrane such as a drumhead, normal to the x, y

plane; the physical constants τ and ρ are the tension per unit length in the membrane and the membrane's mass per unit area, respectively.

However, at this point we limit our subsequent attention to *ordinary* differential equations. Thus, we will omit the adjective "ordinary," for brevity, and will henceforth speak only of "differential equations." Besides the possibility of having more than one independent variable, as in (7)–(9), there could be more than one dependent variable. For instance,

$$
\begin{aligned}
\frac{dx_1}{dt} &= -(k_{21} + k_{31})x_1 + k_{12}x_2 + k_{13}x_3, \\
\frac{dx_2}{dt} &= k_{21}x_1 - (k_{12} + k_{32})x_2 + k_{32}x_3, \\
\frac{dx_3}{dt} &= k_{31}x_1 + k_{32}x_2 - (k_{13} + k_{23})x_3
\end{aligned}
\tag{10}
$$

is a set, or system, of three ODE's governing the three unknowns $x_1(t)$, $x_2(t)$, $x_3(t)$; (10) arises in the study of chemical kinetics, where x_1, x_2, x_3 are concentrations of three reacting chemical species, such as in a combustion chamber, where the k_{ij}'s are reaction rate constants, and where the reactions are, in chemical jargon, first-order reactions. Our study in this text will include such *systems* of ODE's.

Order. We define the **order** of a differential equation as the order of the highest derivative (of the unknown function or functions) therein. Thus, (1) is of first order, (2)–(5) are of second order, (6) is of fourth order, and (10) is a system of first-order equations.

More generally,

$$
F\left(x, u(x), u'(x), u''(x), \ldots, u^{(n)}(x)\right) = 0
\tag{11}
$$

is said to be an nth-order differential equation on the unknown $u(x)$, where n is the order of the highest derivative of u present in (11).

For compactness, we generally use the standard prime notation for derivatives: $u'(x)$ denotes du/dx, $u''(x)$ denotes the second derivative, and $u^{(n)}(x)$ denotes the nth derivative. In the fourth-order differential equation (6), for instance, in which the dependent variable is y rather than u, $F(x, y, y', y'', y''', y'''')$ is $EIy'''' + w(x)$, which happens not to contain y, y', y'', or y'''.

Solution. We've already used the term "solution" in Section 1.1, but have not yet defined it. A function is said to be a **solution** of a differential equation, over a particular domain of the independent variable, if its substitution into the equation reduces that equation to an identity everywhere within the domain.

EXAMPLE 1

The function $y(x) = 4\sin x - x\cos x$ is a solution of the differential equation

$$
y'' + y = 2\sin x
\tag{12}
$$

on the entire x axis because its substitution into (12) yields

$$(-4\sin x + 2\sin x + x\cos x) + (4\sin x - x\cos x) = 2\sin x,$$

which is an identity for all x. Note that we said "a" solution rather than "the" solution since there happen to be many solutions of (12):

$$y(x) = A\sin x + B\cos x - x\cos x \tag{13}$$

is a solution for *any* values of the constants A and B, as can be verified by substituting (13) into (12). [In Chapter 6, we will be in a position to derive the solution (13), and to show that it is the most general possible solution, that is, that every solution of (12) can be expressed in the form (13).] ∎

EXAMPLE 2
The function $y(x) = 1/x$ is a solution of the differential equation

$$y' + y^2 = 0 \tag{14}$$

over any interval that does not contain the origin since its substitution into (14) gives $-1/x^2 + 1/x^2 = 0$, which relation is an identity, provided that $x \neq 0$ since $-1/0 + 1/0$ is simply undefined. ∎

EXAMPLE 3
Whereas (12) admits an *infinity* of solutions [one for each choice of A and B in (13)], the equation

$$\left|y'\right| + |y| + 3 = 0 \tag{15}$$

illustrates that it is possible for an equation to have *no solution*, since the sum of the two nonnegative terms and the one positive term cannot be zero for any choice of $y(x)$. ∎

In applications, however, one normally expects that if a problem is formulated carefully then it should indeed have a solution, and that the solution should be unique, that is, there should be one and only one. Thus, the issues of *existence* (Does the equation have a solution?) and *uniqueness* (If it does have a solution, is that solution unique?) will be of important interest.

Initial value and boundary value problems. Besides the differential equation to be satisfied, the unknown function is often subjected to conditions at one or more points on the interval under consideration. Conditions specified at a single point (often the left end point of the interval), are called **initial conditions**, and the differential equation together with those initial conditions is called an **initial value problem**. In contrast, conditions specified at both ends are called **boundary conditions**, and the differential equation together with the boundary conditions is called a **boundary value problem**. For initial value problems the independent variable is often the time, and for boundary value problems the independent variable is often a space variable.

E X A M P L E   4   *Straight-Line Motion of a Mass*.
Consider once again the problem of predicting the straight-line motion of a body of mass m subjected to a time-varying force $F(t)$. According to Newton's second law of motion, the differential equation governing the displacement $x(t)$ is $mx'' = F(t)$. Besides the differential equation, suppose that we wish to impose the conditions $x(0) = 0$ and $x'(0) = V$; that is, the initial displacement and the initial velocity are 0 and V, respectively. Then the complete problem statement is the initial value problem

$$mx''(t) = F(t), \qquad (0 \le t < \infty)$$
$$x(0) = 0, \quad x'(0) = V. \tag{16}$$

That is, $x(t)$ is to satisfy the differential equation $mx'' = F(t)$ on the interval $0 \le t < \infty$ *and* the initial conditions $x(0) = 0$ and $x'(0) = V$.   ∎

E X A M P L E   5   *Deflection of a Loaded Cantilever Beam*.
Consider the deflection $y(x)$ of a cantilever beam of length L, under a distributed load $w(x)$ newtons per meter (Fig. 5). According to the so-called Euler beam theory, the governing problem is as follows:

$$EIy'''' = -w(x) \qquad (0 \le x \le L)$$
$$y(0) = 0, \quad y'(0) = 0, \quad y''(L) = 0, \quad y'''(L) = 0, \tag{17}$$

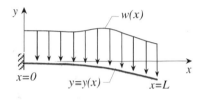

FIGURE 5
Loaded cantilever beam.

where E and I are known physical constants. The appended conditions are boundary conditions because at least one is specified at one end, and at least one at the other end; therefore, we classify (17) as a boundary value problem. The physical significance of the boundary conditions is as follows: $y(0) = 0$ is true simply by virtue of our chosen placement of the origin of the x, y coordinate system; $y'(0) = 0$ follows since the beam is cantilevered out of the wall, so that its slope at $x = 0$ is zero; $y''(L) = 0$ and $y'''(L) = 0$ because the "moment" and "shear force," respectively, are zero at the end of the beam, as learned in traditional courses in civil and mechanical engineering.   ∎

Linear and nonlinear differential equations. Finally, there is a distinction to be made between "linear" and "nonlinear" differential equations. An nth-order differential equation is said to be **linear** if it is expressible in the form

$$\boxed{a_0(x)y^{(n)}(x) + a_1(x)y^{(n-1)}(x) + \cdots + a_n(x)y(x) = f(x),} \tag{18}$$

where $a_0(x), \dots, a_n(x)$ are functions of the independent variable x, and it is **nonlinear** if it is not expressible in that form. To illustrate, equations (1), (2), (3), and (6) are linear. In (1), for instance, where the independent and dependent variables are t and N instead of x and y, respectively, we see that $n = 1$, $a_0(t) = 1$, $a_1(t) = -\kappa$, and $f(t) = 0$; in (6), $a_0(x) = EI$, $a_1(x) = a_2(x) = a_3(x) = a_4(x) = 0$, and $f(x) = -w(x)$; and so on.

   In both of these cases the a_j coefficients happen to be constants, but we are interested in linear equations with nonconstant coefficients as well, such as the equation

$$xy''' - 3(\cos x)\,y'' + 2y = e^x, \tag{19}$$

which is seen from (18) to be a linear third-order ODE, with $a_0(x) = x$, $a_1(x) = -3\cos x$, $a_2(x) = 0$, $a_3(x) = 2$, and $f(x) = e^x$.

If $f(x) = 0$ we say that (18) is **homogeneous**; if not it is **nonhomogeneous**. For instance, (1) is homogeneous because if the κN term is moved to the left-hand side, to put (1) into the form (18), then the right-hand side is zero; however, (2) is nonhomogeneous because of the $F(t)$ term on the right-hand side.

In contrast with the linear equation (18), (4) and (5) are not linear; they are nonlinear because they cannot be expressed in the form (18).

We will find that the theory of linear differential equations is quite comprehensive insofar as our major concerns — the existence and uniqueness of solutions and how to *find* solutions, especially if the $a_j(x)$ coefficients are constants. Even in the nonconstant coefficient case the theory often provides substantial guidance.

However, nonlinear equations prove to be far more difficult in general, and the available theory and solution technique is not nearly as comprehensive as for linear equations. Whereas for linear equations solutions can generally be found either in closed form or as integrals or infinite series, for nonlinear equations one might need to settle for qualitative information about the solution or numerical solution by computer simulation, or both.

The strategy, in science and engineering, until around 1960 when high-speed digital computers became widely available, was to try to "get by" almost exclusively with linear theory. As a classic example that will be studied in this text, consider the nonlinear equation (5), namely,

$$\frac{d^2\theta}{dt^2} + \frac{g}{l}\sin\theta = 0, \tag{20}$$

governing the motion of a pendulum, where $\theta(t)$ is the angular displacement from the vertical and t is the time. If one is willing to limit one's attention to small motions, that is, where θ is sufficiently small, then one can use the Taylor series approximation

$$\sin\theta = \theta - \frac{1}{3!}\theta^3 + \frac{1}{5!}\theta^5 - \cdots \approx \theta \tag{21}$$

to replace the nonlinear equation (20) by the approximate "linearized" equation

$$\frac{d^2\theta}{dt^2} + \frac{g}{l}\theta = 0, \tag{22}$$

which, as we shall see in a later chapter, is readily solved for $\theta(t)$.

Unfortunately, the linearized version (22) is not only less and less accurate as larger motions are considered, it may even be incorrect in a qualitative sense as well. For instance, whereas (22) predicts only motions that are oscillatory, (20) shows (as we shall see) that nonoscillatory motions are possible as well. That is, from a phenomenological standpoint, replacing a nonlinear differential equation by an approximate linear one may amount to "throwing out the baby with the bath water." Whereas we will be able to solve the linear equation (22) quickly, easily, and completely by hand (i.e., by means of a brief analysis), we will have to rely on qualitative and computer methods to study the nonlinear equation (20).

Thus, it will be important for us to keep coming back to this distinction between linear and nonlinear equations as we proceed with our study of differential equations.

Closure. Notice that we have begun, in this section, to classify differential equations, that is, to categorize them by types. Thus far we have distinguished ODE's (ordinary differential equations) from PDE's (partial differential equations), established the idea of the order of a differential equation, distinguished initial value problems from boundary value problems, linear equations from nonlinear ones, and homogeneous equations from nonhomogeneous ones.

Why do we classify so extensively? Because the most general differential equation is far too difficult for us to deal with. The most reasonable program, then, is to break the set of all possible differential equations into various categories and to try to develop theory and solution strategies that are tailored to a given category. Historically, however, the early work on differential equations—by such well known mathematicians as *Leonhard Euler* (1707–1783), *Jakob* (or James) *Bernoulli* (1654–1705) and his brother *Johann* (or John) (1667–1748), *Joseph-Louis Lagrange* (1736–1813), *Alexis-Claude Clairaut* (1713–1765), and *Jean le Rond d'Alembert* (1717–1783)— generally involved attempts at solving specific equations rather than the development of a general theory.

We shall find that in many cases diverse physical phenomena are governed by the same differential equation. For example, consider equations (2) and (3) and observe that they are actually the same equation, to within a change in the names of the various quantities: $m \to L$, $k \to 1/C$, $F(t) \to dE(t)/dt$, and $x(t) \to i(t)$. Thus, to within these correspondences, their solutions are identical. We speak of the mechanical system and the electrical circuit as **analogs** of each other; for instance, the mechanical system in Fig. 1 of Section 1.1 is the "mechanical analog" of the electrical circuit in Fig. 1 of this section. This idea is deeper and more general than can be seen from this one example, and the result is that if one knows a lot about mechanical systems then one thereby knows a lot about electrical, biological, and social systems to whatever extent they are governed by differential equations of the same form. The significance of this fact can hardly be overstated as a justification for a careful study of the mathematical field of differential equations, or as a cause for marvel at the underlying design of the physical universe.

EXERCISES 1.2

1. State the order of each differential equation, and show whether or not the given functions are solutions of that equation.

(a) $(y')^2 = 4y$; $y_1(x) = x^2$, $y_2(x) = 2x^2$,
$y_3(x) = e^{-x}$

(b) $2yy' = 9 \sin 2x$; $y_1(x) = \sin x$, $y_2(x) = 3 \sin x$,
$y_3(x) = e^x$

(c) $y'' - 9y = 0$; $y_1(x) = e^{3x} - e^x$, $y_2(x) = 3 \sinh 3x$,

$y_3(x) = 2e^{3x} - e^{-3x}$

(d) $(y')^2 - 4xy' + 4y = 0$; $y_1(x) = x^2 - x$,
$y_2(x) = 2x - 1$

(e) $y'' + 9y = 0$; $y_1(x) = 4 \sin 3x + 3 \cos 3x$,
$y_2(x) = 6 \sin(3x + 2)$

(f) $y'' - y' - 2y = 6$; $y_1(x) = 5e^{2x} - 3$,
$y_2(x) = -2e^{-x} - 3$

(g) $y''' - 6y'' + 12y' - 8y = 32 - 10x$;

$$y_1(x) = 2x - 1 + (3 + x + x^2)e^{2x}$$

(h) $x^6 y''' = 6y^2;$ $y_1(x) = x^3,$ $y_2(x) = x^2,$ $y_3(x) = 0$

(i) $y'' + y' = y^2 - 4;$ $y_1(x) = x, y_2(x) = 1, y_3(x) = 2$

(j) $y' + 2xy = 1;$ $y_1(x) = 4e^{-x^2} \int_0^x e^{t^2}\, dt,$

$$y_2(x) = e^{-x^2}\left(\int_0^x e^{t^2}\, dt + A\right) \text{ for any value of } A.$$

HINT: Recall the *fundamental theorem of the integral calculus*, that if $F(x) = \int_a^x f(t)\, dt$ and $f(t)$ is continuous on $a \le x \le b$, then $F'(x) = f(x)$ on $a \le x \le b$.

2. First, verify that the given function $y(x)$ is a solution of the given differential equation, for any value of A. Then, solve for A, so that $y(x)$ satisfies the given initial condition.

(a) $y' + y = 1;$ $y(x) = 1 + Ae^{-x};$ $y(0) = 3$

(b) $y' - y = x;$ $y(x) = Ae^x - x - 1;$ $y(2) = 5$

(c) $y' + 6y = 0;$ $y(x) = Ae^{-6x};$ $y(4) = -1$

(d) $y' = 2xy^2;$ $y(x) = -1/(x^2 + A);$ $y(0) = 5$

(e) $yy' = x;$ $y(x) = \sqrt{x^2 + A};$ $y(1) = 10$

3. First, verify that the given function is a solution of the given differential equation, for any constants A, B. Then, solve for A, B so that y satisfies the given initial or boundary conditions.

(a) $y'' + 4y = 8x^2;$ $y(x) = 2x^2 - 1 + A\sin 2x + B\cos 2x;$
$y(0) = 1,$ $y'(0) = 0$

(b) $y'' - y = x^2;$ $y(x) = -x^2 - 2 + A\sinh x + B\cosh x;$
$y(0) = -2,$ $y'(0) = 0$

(c) $y'' - 2y' + y = 0;$ $y(x) = (A + Bx)e^x;$
$y(0) = 1,$ $y(2) = 0$

(d) $y'' - y' = 0;$ $y(x) = A + Be^x;$ $y'(0) = 1,$
$y(3) = 0$

(e) $y'' + 2y' = 4x;$ $y(x) = A + Be^{-2x} + x^2 - x;$
$y(0) = 0,$ $y'(0) = 0$

4. Classify each equation as linear or nonlinear:

(a) $y' + e^x y = 4$ **(b)** $yy' = x + y$

(c) $e^x y' = x - 2y$ **(d)** $y' - \exp y = \sin x$

(e) $y'' + (\sin x)y = x^2$ **(f)** $y'' - y = \exp x$

(g) $yy''' + 4y = 3x$ **(h)** $y''' = y$

(i) $y''' + y^2 + 6y = x$ **(j)** $\dfrac{y'' - y}{y' + y} = 4$

(k) $y'' = x^3 y'$ **(l)** $y''' + y''y' = 3x$

NOTE: We sometimes use the notation exp( ) in place of $e^{(\,)}$ since the former takes up less vertical space.

5. For what value(s) of the constant r will $y = \exp(rx)$ be a solution of the given differential equation? If there are no such r's state that and explain your reasoning. See the note in Exercise 4.

(a) $y' + 3y = 0$ **(b)** $y' + 3y^2 = 0$

(c) $y'' - 3y' + 4y^3 = 0$ **(d)** $y'' - 2y' + y = 0$

(e) $y''' - y' = 0$ **(f)** $y''' - 2y'' - y' + 2y = 0$

(g) $y'''' - 6y'' + 5y = 0$ **(h)** $y'' + 5yy' + y = 0$

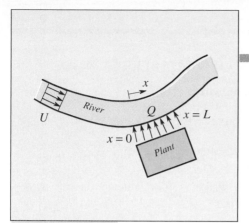

LINEAR FIRST-ORDER EQUATIONS

2.1 INTRODUCTION

In the study of differential equations it is traditional and sensible to begin with the simplest case, first-order equations, and that is the subject of this chapter and the next. We will separate our discussions of the linear case (Chapter 2) and of the general (primarily nonlinear) case (Chapter 3) since one of the aims of this text is to highlight the distinction between linear and nonlinear equations and the important ways in which their theories and solutions differ.

Thus, whereas the general first-order ODE is of the form

$$F(x, y, y') = 0, \tag{1}$$

in this chapter we consider only the linear case,

$$a_0(x)y'(x) + a_1(x)y(x) = f(x), \tag{2}$$

where the functions $a_0(x)$, $a_1(x)$ and $f(x)$ are prescribed; that is, they are known functions of x. If we assume that $a_0(x)$ is nonzero over the x interval of interest, then we can divide (2) by $a_0(x)$ and obtain the slightly simpler looking form

$$\boxed{y'(x) + p(x)y(x) = q(x),} \tag{3}$$

which equation is the focus of this chapter. We will consider the functions $p(x)$ and $q(x)$ to be continuous on the x interval of interest.

2.2 HOMOGENEOUS CASE;
SOLUTION BY SEPARATION OF VARIABLES

To begin, we consider the homogeneous case, where $q(x) = 0$:

$$y'(x) + p(x)y(x) = 0. \tag{1}$$

To solve (1), we first write it in the more suggestive form

$$\boxed{\frac{dy}{dx} + p(x)y = 0.} \tag{2}$$

Let us first explore a method that seems natural but does *not* work, so we can better understand a revised version of that method that *does* work. If we multiply (1) through by dx and integrate we have[1]

$$\int dy + \int p(x)\,y(x)\,dx = 0, \tag{3}$$

or,

$$y(x) + \int p(x)\,y(x)\,dx = C, \tag{4}$$

where C is an arbitrary integration constant. Unfortunately, the integral in (4) cannot be evaluated because the $y(x)$ in the integrand is not yet known. Thus, we come to a dead end and conclude that we cannot solve (2) by merely integrating it with respect to x.

However, suppose we separate the x and y variables in (2) by multiplying through by dx (as above) and also dividing by y (which is permissible if $y \neq 0$), obtaining

$$\frac{dy}{y} + p(x)dx = 0. \tag{5}$$

Now we can integrate because the first integral involves only y and the second involves only x:

$$\int \frac{dy}{y} + \int p(x)\,dx = 0. \tag{6}$$

Thus,

$$\ln|y| + \int p(x)\,dx = C, \tag{7}$$

---

[1]Observe that in multiplying (2) by dx we have treated the derivative dy/dx as a fraction, the ratio of two numbers, whereas it is actually a single quantity defined in the calculus as the limit of a difference quotient. Nevertheless, we can justify the result (4) by showing that the derivative of (4) does give us (2). To do so we recall the fundamental theorem of the integral calculus, that $(d/dx) \int F(x)\,dx = F(x)$ if $F(x)$ is continuous. In the present case $p(x)$ is continuous by assumption and so is $y(x)$ [after all, the differentiability of $y(x)$, assumed in (2), requires that $y(x)$ be continuous]. Hence, $(d/dx) \int p(x)y(x)\,dx = p(x)y(x)$ so the derivative of (4) does give us (2).

where C is an arbitrary integration constant and where $\int p(x)\,dx$ can be evaluated once a specific function $p(x)$ is prescribed. Equivalent to (7), we have

$$\ln|y| = -\int p(x)\,dx + C, \tag{8}$$

$$|y| = e^{-\int p(x)\,dx+C} = e^{C} e^{-\int p(x)\,dx}, \tag{9}$$

so

$$y(x) = \pm e^{C} e^{-\int p(x)\,dx}. \tag{10}$$

Recall that in dividing (2) by y, to obtain (5), we needed to restrict our search to functions y that are nonzero. The solution (10) is indeed nonzero because it is a product of exponentials, and an exponential function is not zero for *any* finite value of its exponent. Replacing the $\pm e^{C}$ in (10) by A, say, for simplicity, we have the solution

$$\boxed{y(x) = A e^{-\int p(x)\,dx}} \tag{11}$$

of (2), for any value of A other than zero. In addition, the function $y(x) = 0$ obtained by setting $A = 0$ in (11) is seen to satisfy (2), for it gives $0 + 0 = 0$, so we can now remove our restriction that A be nonzero.

Thus, we have the solution (11) where A is arbitrary — positive, negative, or zero. Actually we should say "solutions" since the scale factor A is arbitrary; (11) gives a different solution of (2) for each different choice of A. We say that (11) is a **general solution** of (2) in that it contains all possible solutions of (2). It is a whole "family" of solutions, with each specific choice of A giving a specific member of that family, which member we call a **particular solution**. That is, a particular solution is merely one specific solution among the set of all possible solutions.

Since the key was in separating the x and y variables so as to obtain one integral involving only y and one integral involving only x, we call the method **separation of variables**, which method will be used later on as well. Be sure to understand the steps from (5) to (11).

In fact, we can directly verify that (11) does satisfy the ODE (2), as we have claimed. [However, we don't *need* to since each step in the derivation of (11) is reversible.] Using chain differentiation, we obtain from (11)

$$y'(x) = A e^{-\int p(x)\,dx} \frac{d}{dx}\left[-\int p(x)\,dx\right] = -A p(x) e^{-\int p(x)\,dx}, \tag{12}$$

where the last step in (12) follows from the **Fundamental Theorem of the Calculus** [that the derivative of the integral equals the integrand if the integrand is a continuous function, the use of which theorem is justified since we have assumed $p(x)$ to be continuous]. Thus, (2) becomes

$$y' + py = -A p e^{-\int p\,dx} + p A e^{-\int p\,dx} = 0, \tag{13}$$

so (11) is indeed a solution of (2).

EXAMPLE 1 *Population Dynamics*.

Let $N(t)$ be the population of a certain species, where t is the time, and let β be its birth rate (births per individual per unit time) and δ be its death rate (deaths per individual per unit time), with both β and δ assumed to be constant. Then, for any time interval Δt we have

$$N(t + \Delta t) = N(t) + \beta N(t)\Delta t - \delta N(t)\Delta t, \tag{14}$$

or,

$$\frac{N(t + \Delta t) - N(t)}{\Delta t} = (\beta - \delta)N(t). \tag{15}$$

[(14) is not so much a physical law as it is simple bookkeeping: the number of individuals at time $t + \Delta t$ equals the number that we started with at time t plus the number that have been born minus the number that have died over that time interval.] If we let $\Delta t \to 0$ and define the net birth/death rate as $\kappa = \beta - \delta$, then (15) gives

$$\boxed{N'(t) = \kappa N(t),} \tag{16}$$

which is a linear first-order homogeneous differential equation on $N(t)$, homogeneous because it is equivalent to $N' - \kappa N = 0$. By comparing (16) with (2), we see that $p(t) = -\kappa$, where in this case the independent and dependent variables are t and N instead of x and y, respectively. Thus, (11) gives the solution as

$$N(t) = Ae^{-\int (-\kappa)\,dt} = Ae^{\kappa t}, \tag{17}$$

where A is an arbitrary constant.

Or, instead of plugging into the "off-the-shelf" formula (11) we could have solved (16) by carrying out the steps of the same separation of variables method that we used to derive the general formula (11):

$$\frac{dN}{N} = \kappa\,dt, \quad \ln N = \kappa t + C, \quad N(t) = e^{\kappa t + C} = e^C e^{\kappa t} = Ae^{\kappa t}, \tag{18}$$

where we simply wrote $\ln N$ rather than $\ln |N|$ because surely the population $N(t)$ cannot be negative.

If we impose an initial condition

$$N(0) = N_0 \tag{19}$$

on $N(t)$ then (17) gives $N(0) = N_0 = Ae^{(\kappa)(0)} = A$ so $A = N_0$ and the infinite set of solutions given by (17) (i.e., for all possible choices of A) narrows down to the single solution

$$N(t) = N_0 e^{\kappa t}, \tag{20}$$

which we have plotted in Fig. 1 for representative values of κ: one negative (i.e., where the death rate exceeds the birth rate), one zero (where the death rate equals the birth rate), and one positive (where the birth rate exceeds the death rate).

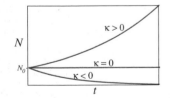

FIGURE 1
Exponential growth and decay.

COMMENT 1.   Strictly speaking, $N(t)$ is integer-valued since one cannot have a population of 28.37. Thus, its graph develops in a stepwise manner. Since $N(t)$ is therefore a discontinuous function of t it is not a differentiable function of t, so the derivative $N'(t)$ in (16) does not even exist. However, if N is large then the steps will be small compared to N and we can regard N, approximately, as a continuous function of t and we can tolerate the dN/dt derivative in (16).

COMMENT 2.   We see that if the death rate exceeds the birth rate ($\kappa < 0$) then (20) expresses exponential decrease, with $N(t)$ tending to zero as $t \to \infty$. That result seems reasonable (although when N becomes small enough our approximation of N as a continuous function of t comes into question, as noted in Comment 1). But if the birth rate exceeds the death rate ($\kappa > 0$) then (20) indicates exponential growth, with $N(t)$ tending to infinity as $t \to \infty$. That result eventually becomes unrealistic because if N becomes sufficiently large, then undoubtedly other factors will come into play, such as insufficient food or other necessary resources, factors that have not been accounted for in our simple mathematical model. Specifically, κ will not really be a constant but will vary with N. In particular, we expect it to decrease as N increases; for instance, if food becomes scarce as N increases then the death rate will increase (so κ will decrease) due to an increasing prevalence of death by starvation. As a simple model of such behavior, suppose that κ varies linearly with N: $\kappa = a - bN$, with a and b positive so that κ diminishes as N increases. Then (16) is to be replaced by the equation,

$$\frac{dN}{dt} = (a - bN)N, \tag{21}$$

which is known as the **logistic equation**, or the **Verhulst equation**, after the Belgian mathematician *P. F. Verhulst* (1804–1849) who introduced it in his work on population dynamics. We could solve (21) by the separation of variables method that we used above, but let us defer this equation for the moment since (21) is not of the type being considered in the present chapter. That is, (21) is not a homogeneous linear first-order equation; it is *nonlinear* because of the N^2 term.   ■

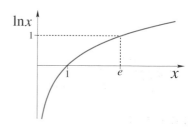

FIGURE 2
Exponential function.

TIME OUT: Before continuing, let us give a brief list of some of the important properties of the exponential and logarithmic functions, which will continue to figure prominently in these chapters: The exponential function e^x [sometimes written as $\exp(x)$ because the latter takes up less vertical space] is defined on $-\infty < x < \infty$ and its graph is shown in Fig. 2 (together with the graph of e^{-x}). The logarithmic function $\ln x$ is defined on $0 < x < \infty$ and its graph is shown in Fig. 3. We see that e^x tends to 0 as $x \to -\infty$, equals 1 when $x = 0$, and tends to ∞ as $x \to +\infty$; $\ln x$ tends to $-\infty$ as $x \to 0$, is 0 when $x = 1$, is 1 when $x = e$, and tends to $+\infty$ as $x \to +\infty$. Some key useful properties are as follows:

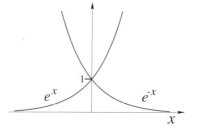

FIGURE 3
Logarithmic function.

1. $e^a e^b = e^{a+b}$

2. $(e^a)^b = e^{ab}$

3. $\ln(ab) = \ln a + \ln b$

4. $\ln(a/b) = \ln a - \ln b$

5. $\ln a^b = b \ln a$

6. If $a = e^b$ then (by taking the logarithm of both sides and recalling that $\ln e = 1$) $b = \ln a$; that is, the exponential and logarithmic functions are inverses of each other.

7. $a = e^{\ln a}$ for all $a > 0$

8. $\dfrac{d}{dx} e^x = e^x$ and $\dfrac{d}{dx} e^{ax} = ae^{ax}$

9. $\displaystyle\int e^x \, dx = e^x$ and $\displaystyle\int e^{ax} \, dx = \dfrac{e^{ax}}{a}$ (plus arbitrary constants)

Continuing, consider another example.

EXAMPLE 2

Solve the differential equation

$$y' + 2xy = 0. \tag{22}$$

Using separation of variables,

$$\frac{dy}{y} + 2x \, dx = 0, \tag{23}$$

$$\int \frac{dy}{y} + 2 \int x \, dx = 0, \tag{24}$$

$$\ln|y| + x^2 = C, \tag{25}$$

$$|y| = e^{C - x^2} = e^C e^{-x^2}, \tag{26}$$

$$y(x) = \pm e^C e^{-x^2} = Ae^{-x^2}, \tag{27}$$

where the constant A is arbitrary. The graphs of the solutions, (27), are called the solution curves or integral curves corresponding to the given differential equation (22), and these are displayed for several values of A in Fig. 4. From the bottom curve upward, $A = -2, -1, 0, 1, 2,$ and 3, respectively. For instance, $A = 0$ gives the solution curve $y(x) = 0$.

Alternative to our separation-of-variable solution steps, we could have plugged into the "off-the-shelf" formula (11). Doing so, with $p(x) = 2x$, gives $y(x) = A\exp(-x^2)$, which is the same result as (27). ∎

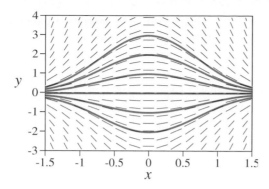

FIGURE 4 Representative solution curves and direction field for
$$y' + 2xy = 0.$$

In Example 2 we used the term "solution curve." A **solution curve**, or **integral curve**, corresponding to a given differential equation, is simply the graph of a solution to that equation.

Besides several solution curves, Fig. 4 also contains a field of lineal elements through a discrete set of points, or grid. By a **lineal element** at a point (x_0, y_0), corresponding to a differential equation of the form

$$y' = f(x, y),$$

we mean a short straight-line segment through the point (x_0, y_0), centered at that point and having the slope $f(x_0, y_0)$ [since $y' = f(x, y)$ is the slope]. That is, each lineal element has the same slope as the solution curve passing through that point and is therefore a small part of the tangent line to that solution curve. The set of all the lineal elements is called the **direction field** corresponding to the given differential equation.[1]

In intuitive language, the direction field shows the "flow" of solution curves. Given a sufficiently dense plot of the direction field, presumably done by computer, one can begin to visualize the various solution curves, or sketch the solution curve through a given initial point. The plots shown here were generated using the *Maple* computer algebra system. Application of the *Maple* software to the material in this section is discussed at the end of this section under the heading *Maple*.

Inclusion of an initial condition. The presence of the arbitrary constant A in (11) permits the solution (11) to satisfy an initial condition, if one is specified. Thus, suppose that we seek a solution $y(x)$ of our differential equation $y' + p(x)y = 0$ that satisfies an initial condition $y(a) = b$, for specified values of a and b. For this purpose it is convenient to first re-express (11) as

$$y(x) = Ae^{-\int_a^x p(s)\,ds}, \tag{28}$$

---

[1] Some authors call the linear element field a **slope field**, reserving the term **direction field** for the case where barbed arrows are used in place of line elements. For simplicity, we use the term direction field for both cases.

which form is equivalent to (11) because the indefinite integral $\int p(x)\,dx$ and the definite integral $\int_a^x p(s)\,ds$ differ by at most an additive constant, say D, and the resulting e^D factor in (28) can be absorbed by the arbitrary constant A.

Before continuing, is it clear why we change the integration variable from x to s in (28)? The idea is that it is improper notation to write $\int_a^x p(x)\,dx$ because we would be using the letter x for two different purposes in the same expression: The upper limit x is a *fixed endpoint* of the integration interval, whereas the x in $p(x)\,dx$ is the *integration variable* that runs from the fixed point a to the fixed point x. To avoid this confusion it is standard practice to use a different variable for the dummy integration variable; we chose the letter s, but the letter used is immaterial. For instance, $\int_0^x s\,ds$, $\int_0^x t\,dt$, and $\int_a^x u\,du$ are all the same, namely, $x^2/2$; thus, we call the integration variable a "dummy" variable.

The advantage of (28) over (11) is that if we impose an initial condition

$$y(a) = b, \tag{29}$$

then (28) readily gives

$$y(a) = b = Ae^{\int_a^a p(s)\,ds} = Ae^0 = A, \tag{30}$$

because an integral from a given point to itself is simply zero. Thus, $A = b$ and the particular solution of the initial value problem

$$\boxed{y' + p(x)y = 0; \quad y(a) = b} \tag{31}$$

is

$$\boxed{y(x) = be^{-\int_a^x p(s)\,ds}.} \tag{32}$$

EXAMPLE 3

Consider the initial value problem

$$(x + 2)y' - xy = 0, \tag{33a}$$

$$y(0) = 3. \tag{33b}$$

Since an initial value is prescribed, let us use (32) rather than (11), with $p(x) = -x/(x + 2)$:

$$y(x) = 3e^{\int_0^x s\,ds/(s+2)} = 3e^{[s + 2 - 2\ln|s + 2|]\big|_0^x}$$

$$= 3e^{[x + 2 - 2\ln|x + 2| - (2 - 2\ln 2)]} = 3e^x e^{\ln|x + 2|^{-2}} e^{\ln 2^2}$$

$$= 12\frac{e^x}{|x + 2|^2} = 12\frac{e^x}{x^2 + 4x + 4}, \tag{34}$$

where we have used some of the nine identities listed before Example 2, and the change of variables $s + 2 = t$ to evaluate the integral in the exponent.

Alternatively, we could have used (11) and then applied the initial condition to that result:

$$y(x) = Ae^{\int x\,dx/(x+2)} = Ae^{x+2-2\ln|x+2|} = A\frac{e^2 e^x}{|x+2|^2}, \tag{35}$$

where $y(0) = 3 = Ae^2/4$ gives $A = 12e^{-2}$, which once again gives the particular solution

$$y(x) = 12\frac{e^x}{x^2 + 4x + 4}. \tag{36}$$

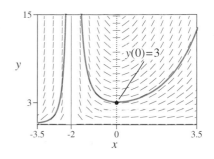

FIGURE 5
Solution to $(x + 2)y' - xy = 0$; $y(0) = 3$, together with the direction field.

The graph of that solution is shown over the nominal x interval $-3.5 < x < 3.5$ in Fig. 5, in which we also display the direction field of the equation (33a), namely, the equation $y' = \dfrac{xy}{x+2}$.

There is a problem associated with (36): Both the right-hand side and its first derivative are undefined at $x = -2$ because $x^2 + 4x + 4 = 0$ there. Hence, we cannot say that (36) satisfies (33a) *at* $x = -2$. [It is tempting to argue that (33a) gives $-\infty + \infty = 0$ at $x = -2$ and that the latter is indeed a numerical identity. However, we cannot say that $-\infty$ and $+\infty$ "cancel to zero."] Thus we conclude that (36) is a solution to (33) only on $-2 < x < \infty$. In contrast, the solutions of (22) that were given by (27), for any value of the constant A, were valid without restriction, that is, on $-\infty < x < \infty$. ∎

Existence and uniqueness of solutions. Thus far we have been preoccupied with how to find solutions of equations of the form $y' + p(x)y = 0$, but let us now address the following fundamental questions: *Is* there a solution (or solutions)? If so, is the solution *unique*; that is, is there precisely one solution or are there more than one? These are the questions of **existence** and **uniqueness**, respectively. We have answered the question of existence by actually *finding* solutions, namely, the infinite set of solutions given by (11), infinite because A is arbitrary so each different choice of A gives another solution. We verified, by substitution of (11) into (2), that (11) is indeed a solution for any A, subject to the assumption that $p(x)$ is continuous on the interval in question. (We call such a proof of existence a *constructive* proof; that is, when we show that a solution exists by actually finding one.)

For the initial value problem (31), however, we found that the initial condition $y(a) = b$ forced a specific choice for the constant A, so in that case we found a *unique* solution.

We have the following theorem:

THEOREM 2.2.1

Existence and Uniqueness for the Homogeneous Linear First-Order Equation
The homogeneous linear equation $y' + p(x)y = 0$ does admit a solution through an initial point $y(a) = b$ if $p(x)$ is continuous at $x = a$. That solution,

$$y(x) = be^{-\int_a^x p(s)\,ds}, \tag{37}$$

is unique, and it exists at least on the largest x interval, containing $x = a$, over which $p(x)$ is continuous.

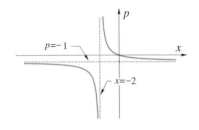

FIGURE 6
The graph of $p(x) = -x/(x+2)$.

To illustrate this result, let us reconsider Example 3. The unique solution to the initial value problem (33) is given by (36). On what x interval is that solution valid? Theorem 2.2.1 tells us that it is valid on the broadest interval, containing the initial point $x = 0$, over which

$$p(x) = -\frac{x}{x+2}$$

is continuous, namely, on $-2 < x < \infty$ because (Fig. 6) $p(x)$ is infinite (undefined) at $x = -2$. Thus, the theorem assures us that the solution likewise will be valid on the interval $-2 < x < \infty$. Notice that not only does $p(x)$ "blow up" at $x = -2$ (it tends to infinity as $x \to -2$), but so does the solution (35). That is, we can think of the solution as "inheriting" a singular behavior at $x = -2$ from the singular behavior of $p(x)$ at that point.

Closure. In this section we have considered the homogeneous linear first-order differential equation

$$y' + p(x)y = 0. \tag{38}$$

Using a method of separation of variables, we derived the general solution

$$y(x) = Ae^{-\int p(x)\,dx}, \tag{39}$$

namely, the whole family of solutions (39) for arbitrary values of the constant A, which constant is essentially the constant that arises from integrating (38). If an initial condition is prescribed then the problem statement takes the form

$$y' + p(x)y = 0; \quad y(a) = b. \tag{40}$$

In this case it is more convenient to work with the general solution form

$$y(x) = Ae^{-\int_a^x p(s)\,ds}, \tag{41}$$

which is equivalent to (39). Applying the initial condition $y(a) = b$ to (41) gives $A = b$, and hence the particular solution

$$y(x) = be^{-\int_a^x p(s)\,ds}, \tag{42}$$

to (40), which solution is unique and which is valid at least over the largest x interval containing the initial point a, over which $p(x)$ is continuous.

The example

$$N' = \kappa N \tag{43}$$

studied in connection with population dynamics, is especially important. If one were to take a poll among scientists and engineers, of the single most important differential equation, the "winning equation" might very well be (43) (or, in terms of our original x, y variables, $y' = \kappa y$, where κ is a constant). The general solution

$$N(t) = Ae^{\kappa t} \tag{44}$$

expresses exponential change, decay if $\kappa < 0$ and growth if $\kappa > 0$. We will discuss this equation in other connections as well, in exercises and subsequent sections—for instance, in connection with radioactive decay and carbon dating.

Maple. There are now several powerful computer-algebra systems, such as *Mathematica*, *MATLAB*, and *Maple*, that can be used to implement much of the mathematics presented in this text—numerically, symbolically, and graphically. Consider the application of *Maple* (specifically, *Maple* V, release 5), as a representative software, to the material in this section.

There are two types of applications involved in this section. One entails finding the general solution of a given first-order differential equation, or a particular solution satisfying a given initial condition. These can be carried out on *Maple* using the dsolve command ("function," in *Maple* terminology).

For example, to solve the equation $(x + 2)y' - xy = 0$ of Example 3, for $y(x)$, enter

```
dsolve((x+2)*diff(y(x),x)-x*y(x)=0, y(x));
```

(including the semicolon) and return; **dsolve** is the differential equation solver, and **diff** is the derivative command for y'. [The command for y'' would be diff($y(x), x, x$), and so on for higher derivatives.] The output is the general solution

$$y(x) = \frac{\_C1\, e^x}{(x + 2)^2}$$

where $\_C1$ is *Maple* notation for an arbitrary constant.

To solve the same equation, but this time with the initial condition $y(0) = 3$, enter

```
dsolve({(x+2)*diff(y(x),x)-x*y(x)=0, y(0)=3}, y(x));
```

and return. The output is the particular solution

$$y(x) = 12\frac{e^x}{(x + 2)^2}$$

which agrees with our result in Example 3.

The dsolve command can cope with differential equations that contain unspecified parameters or functions. For example, to solve $y' + p(x)y = 0$, where $p(x)$ is not specified, enter

```
dsolve(diff(y(x),x)+p(x)*y(x)=0, y(x));
```

and return. The output is the general solution

$$y(x) = \_C1 \exp\left(-\int p(x)\,dx\right)$$

The second type of application entails generating the graphical display of various solution curves and the direction field for a given differential equation. Both of these tasks can be carried out on *Maple* using the **phaseportrait** command. For example, to obtain the plot shown as Fig. 4, enter

```
with(DEtools):
```

to access the phaseportrait command; then return and enter

```
phaseportrait (diff(y(x),x)+2*x*y(x)=0, y(x), x=-1.5..1.5,
{[0,-2],[0,-1],[0,0],[0,1],[0,2],[0,3]}, arrows = LINE);
```

and return. The items within the outer parentheses are as follows:

```
phaseportrait(differential equation, dependent
(independent) variable, range of independent variable,
{initial points}, optional specifications);
```

All items, if any, following the {initial points} are optional. For instance, the dependent variable range is set automatically, but it can be specified as an optional item if you wish. Similarly, the default direction field arrows are small but can be selected as medium or large or as lines (i.e., without barbs, as in Fig. 4), or suppressed altogether, by including an option arrows = VALUE, where VALUE is MEDIUM, LARGE, LINE, or NONE, respectively. For instance, if we want the y range to be $-2 < y < \pi$ and we want the direction field arrows to be omitted, end the command either as ... [0,3]},y=-2..Pi,arrows=NONE); or, since the order of the options is immaterial, as ... [0,3]}, arrows=NONE, y=-2..Pi);

To run phaseportrait repeatedly, one needs to enter "with(DEtools):" only at the beginning of the session.

If we desire only the direction field, without any solution curves, we can use the **dfieldplot** command instead of phaseportrait. To obtain the direction field shown in Fig. 4, for instance, enter with(DEtools): as for phaseportrait, followed by the command

```
dfieldplot (diff(y(x),x)+2*x*y(x)=0, y(x), x=-1.5..1.5,
y=-3..4, arrows=LINE);
```

Incidentally, to obtain a list of all commands supported by DEtools follow the entry with(DEtools) with a semicolon instead of a colon.

To obtain a listing of the mathematical commands enter **?lib** and return. Within that list one would find such commands as dsolve and phaseportrait. To learn how to use a command enter a question mark, then the command name, then return. For example, type **?dsolve** and return. This is important because as newer versions of *Maple* are developed the comments might differ slightly from those listed in this text.

In the exercises that follow, and those in subsequent sections, problems are included that require the use of a computer-algebra system such as *Maple*. It will be important to develop skill in the use of at least one such system in parallel with, but not in place of, developing understanding of the underlying mathematics presented in this text. As we progress through this text we will cite various *Maple* commands.

A list of these commands and their location within the text is given in the end-papers.

NOTE: In working the *Maple* exercises in this text it is expected not only that you generate the necessary *Maple* output but also add whatever comments are needed to explain your results.

EXERCISES 2.2

1. Find the general solution two different ways: First, use (11); second, derive the solution using the separation of variables method.

   (<u>a</u>) $y' = 6x^2 y$
   (**b**) $y' + (\sin x)y = 0$
   (<u>c</u>) $y' - (\cos x)y = 0$
   (**d**) $y' - (1 + 6x^2)y = 0$
   (**e**) $xy' - y = 0$
   (**f**) $xy' + 3y = 0$
   (**g**) $(1 + x)y' - 2y = 0$
   (**h**) $(1 + x)y' + 4y = 0$
   (**i**) $e^x y' - 2y = 0$
   (**j**) $(2 + x)^2 y' + y = 0$
   (<u>k</u>) $x^2 y' - y = 0$
   (**l**) $x^2 y' + y = 0$
   (**m**) $xy' + (1 + x)y = 0$
   (**n**) $y' + (3 + 2x - 6x^2)y = 0$
   (**o**) $y' = (\tan x)y$
   (**p**) $y' = -(\tan x)y$

2. (**a**)–(**p**) Solve the differential equation given in the corresponding part of Exercise 1 and find the particular solution corresponding to the initial condition $y(1) = 2$. Over what x interval does that unique solution exist? Sketch its graph over that interval.

3. (**a**)–(**p**) Same as Exercise 2, but with the initial condition $y(-3) = 5$ [in place of the initial condition $y(1) = 2$].

4. (**a**)–(**p**) Same as Exercise 2, but with the initial condition $y(2) = 0$ [in place of the initial condition $y(1) = 2$].

5. (**a**)–(**p**) Use *Maple* to obtain the general solution to the differential equation given in the corresponding part of Ex-

ercise 1. Then use *Maple* to obtain the particular solution of that differential equation subject to the initial condition $y(5) = 4$.

6. (**a**)–(**p**) Use *Maple* to obtain the direction field for the differential equation given in the corresponding part of Exercise 1, over $-2 \le x \le 2$ and $-2 \le y \le 8$, say. On that direction field, sketch (by hand) the graph of the solution through the initial point $y(1) = 1$.

7. (**a**) Obtain the general solution of $xy' = 2y$.

   (**b**) Obtain the particular solution satisfying the initial condition $y(2) = 12$. On what x interval (if any) is that solution unique, and is that result consistent with Theorem 2.2.1? Explain.

   (**c**) Obtain the particular solution satisfying the initial condition $y(0) = 0$. On what interval (if any) is that solution unique, and is that result consistent with Theorem 2.2.1? Explain.

   (**d**) Give an initial condition such that there is *no* solution and, in graphical terms, explain your choice.

8. Show, from (44), that if $\kappa > 0$ then the population is doubled every $(\ln 2)/\kappa$ years, and that if $\kappa < 0$ then the population is halved every $(\ln 2)/|\kappa|$ years.

2.3 NONHOMOGENEOUS CASE

In Section 2.2 we studied the homogeneous linear first-order differential equation $y' + p(x)y = 0$. In this section we consider the general linear first-order equation

$$\boxed{y'(x) + p(x)y(x) = q(x),} \tag{1}$$

that is, where $q(x)$ is not necessarily zero so that (1) is **nonhomogeneous**. In this case our solution by separation of variables fails because when we try to separate variables in (1) by re-expressing it as

$$\frac{1}{y}dy + p(x)dx = \frac{q(x)}{y}dx \tag{2}$$

we see that the term on the right-hand side is a "mixed" term, involving both x and y. Thus, we will need a different line of approach to solve (1).

2.3.1 Solution by integrating factor method.   We will develop two different solution techniques for (1), the integrating factor method and the method of variation of parameters. To introduce the integrating factor method, which was invented by the great mathematician *Leonhard Euler* (1707–1783)[1] let us illustrate its use in a simple example, and then generalize the idea.

E X A M P L E 1   *Introduction to Integrating Factor Method.*
To solve the equation

$$y' + \frac{1}{x}y = 12x^2, \tag{3}$$

notice that if we multiply through by x, we obtain

$$xy' + y = 12x^3. \tag{4}$$

The key point is that, unlike the left-hand side of (3), the left-hand side of (4) is the derivative of a single quantity, namely, xy; that is, $xy' + y$ is the same as $(xy)'$, where we remember that primes denote d/dx and that y is a function of x. Thus, (4) can be expressed as

$$\frac{d}{dx}(xy) = 12x^3 \tag{5}$$

so we can solve by integration:

$$d(xy) = 12x^3dx, \tag{6a}$$

---

[1] Euler, pronounced "oiler" rather than "yuler," was surely one of the greatest and most prolific mathematicians of all time. He contributed to virtually every branch of mathematics and to the application of mathematics to the science of mechanics. Though totally blind during the last 17 years of his life he produced several books and some 400 research papers during that time. He knew, by heart, the first six powers of the first 100 prime numbers and the entire Aeneid.

$$\int d(xy) = \int 12x^3 \, dx, \tag{6b}$$

$$xy = 12\frac{x^4}{4} + C, \tag{6c}$$

where C is an arbitrary constant. Thus, we have the general solution

$$y(x) = 3x^3 + \frac{C}{x} \tag{7}$$

of (3), which solution can be verified, by substituting it into (3) and seeing that the result is an identity for all x (except for $x = 0$, where the $+1/x^2$ and $-1/x^2$ terms that arise are undefined, and cannot be considered to cancel each other).   ∎

One might say that the integrating factor method is similar to the familiar method of solving a quadratic equation by completing the square. In completing the square, we add a suitable quantity to both sides so that the left-hand side becomes a "perfect square," and the equation can then be solved by taking square roots. In the integrating factor method we multiply both sides by a suitable quantity so that the left-hand side becomes a "perfect derivative," and the equation can then be solved by integration.

To apply this method to the general equation (1), multiply (1) by a not-yet-determined function $\sigma(x)$:

$$\sigma y' + \sigma p y = \sigma q. \tag{8}$$

[In Example 1, $\sigma(x)$ was simply x.] By a suitable choice of σ, the left-hand side of (8) can be made to be the derivative $(\sigma y)'$, for if we expand $(\sigma y)'$ as $\sigma y' + \sigma' y$ and compare the latter with the terms $\sigma y' + \sigma p y$ in (8), we see that they will be identical if we choose $\sigma(x)$ so that

$$\sigma' = \sigma p. \tag{9}$$

But the latter, rewritten in the form

$$\sigma' - p(x)\sigma = 0, \tag{10}$$

is of the same form as the equation $y' + p(x)y = 0$ that we solved in the preceding section [if we change $y(x)$ to $\sigma(x)$ and $p(x)$ to $-p(x)$], so we can solve it by separation of variables [or by using (11) in Section 2.2], obtaining

$$\sigma(x) = Ae^{\int p(x) \, dx}.$$

We don't need the most general integrating factor, we simply need *an* integrating factor, so let us choose $A = 1$. Then

$$\boxed{\sigma(x) = e^{\int p(x) \, dx}.} \tag{11}$$

With $\sigma(x)$ so chosen, (8) becomes

$$(\sigma y)' = \sigma q \quad \text{or} \quad \frac{d(\sigma y)}{dx} = \sigma q, \tag{12}$$

which can be integrated to give

$$\int d(\sigma y) = \int \sigma(x)q(x)\,dx, \tag{13a}$$

$$\sigma y = \int \sigma(x)q(x)\,dx + C, \tag{13b}$$

$$y(x) = \sigma(x)^{-1}\left(\int \sigma(x)q(x)\,dx + C\right). \tag{13c}$$

Finally, putting the integrating factor (11) into (13c) gives the desired general solution

$$\boxed{y(x) = e^{-\int p(x)\,dx}\left(\int e^{\int p(x)\,dx}q(x)\,dx + C\right)} \tag{14}$$

of (1).

EXAMPLE 2   *Application of (14).*
Solve

$$y' + 3y = x. \tag{15}$$

With $p(x) = 3$ and $q(x) = x$, we have

$$e^{\int p(x)\,dx} = e^{\int 3\,dx} = e^{3x},$$

so (14) gives

$$y(x) = e^{-3x}\left(\int e^{3x}x\,dx + C\right) = \frac{x}{3} - \frac{1}{9} + Ce^{-3x} \tag{16}$$

as the general solution of (15).

COMMENT.   Rather than "plugging into" the formula (14), it is a good idea to simply use the integrating factor method—at least until the method is thoroughly understood. That procedure involves the following steps.

Step 1. Multiply the differential equation (15) through by $\sigma(x)$:

$$\sigma y' + 3\sigma y = \sigma x.$$

Step 2. Choose $\sigma(x)$ so that the left-hand side is the derivative $(\sigma y)'$:

$$\sigma y' + 3\sigma y = (\sigma y)'.$$

Expanding $(\sigma y)'$ as $\sigma y' + \sigma' y$ we see that to match those two terms with $\sigma y' + 3\sigma y$ we need

$$\sigma' = 3\sigma, \quad \text{so } \sigma(x) = e^{3x}.$$

Step 3. Rewrite the differential equation accordingly:

$$(e^{3x}y)' = xe^{3x}.$$

Step 4. Now we can solve by integration:

$$e^{3x}y = \int e^{3x}x\,dx + C$$

or, solving for y,

$$y(x) = e^{-3x}\left(\int e^{3x}x\,dx + C\right) = \frac{x}{3} - \frac{1}{9} + Ce^{-3x},$$

as in (16).   ∎

Next, suppose that we have not only the differential equation (1), but also an initial condition $y(a) = b$ to satisfy. That is, we have the initial value problem

$$y' + p(x)y = q(x), \tag{17a}$$
$$y(a) = b. \tag{17b}$$

Then we need to impose (17b) upon the general solution (14) and solve for C, thereby obtaining the desired particular solution. For that purpose, it is more convenient to work not with (14) but with an equivalent version that contains definite rather than indefinite integrals. Specifically, return to (11) and write

$$\sigma(x) = e^{\int_a^x p(s)\,ds} \tag{18}$$

instead. The integrals in the exponents in (11) and (18) differ by at most an arbitrary additive constant, say B, and the scale factor e^B that results, in $\sigma(x)$, can be discarded without loss. Accordingly, let us change both p integrals in (14) to definite integrals, with the initial point a as the lower limit. Likewise, the q integral in (14) can be changed to a definite integral with a as lower limit since this step changes the integral by at most an arbitrary additive constant. This constant can then be absorbed by the arbitrary constant C. Thus, equivalent to (14), let us write

$$y(x) = e^{-\int_a^x p(s)\,ds}\left(\int_a^x e^{\int_a^s p(u)\,du}q(s)\,ds + C\right), \tag{19}$$

where s and u are dummy integration variables.

If we impose the initial condition $y(a) = b$ on (19) we obtain $y(a) = b = e^{-0}(0 + C) = C$, where each s integral is zero because it is an integral from a to a. Thus, $C = b$ and

$$\boxed{y(x) = e^{-\int_a^x p(s)\,ds}\left(\int_a^x e^{\int_a^s p(u)\,du}q(s)\,ds + b\right).} \tag{20}$$

As a partial check on (20), notice that (20) does reduce, as it should, to equation (32) of Section 2.2 in the event that $q(x) = 0$.

Whereas (14) was a general solution of (1), (20) is a particular solution of (1) since it corresponds to one particular solution curve, the solution curve through the initial point (a, b). It may look intimidating because of the two dummy integration variables (s and u) but, as the following example illustrates, it is easy to use. Further, it will be an especially convenient solution form for cases in which the function $q(x)$ is discontinuous, as we will see in Section 2.4.2.

E X A M P L E 3   *Application of (20).*
Use (20) to solve the initial value problem

$$y' - 2xy = \sin x, \tag{21a}$$
$$y(0) = 3. \tag{21b}$$

With $p(x) = -2x$, $q(x) = \sin x$, $a = 0$, and $b = 3$, we have

$$e^{\int_0^x p(s)\,ds} = e^{\int_0^x (-2s)\,ds} = e^{-s^2 \big|_0^x} = e^{-(x^2 - 0)} = e^{-x^2},$$

so that (20) gives the desired solution as

$$y(x) = e^{x^2}\left(\int_0^x e^{-s^2} \sin s \, ds + 3\right). \tag{22}$$

COMMENT.   The integral in (22) is said to be **nonelementary** in that it cannot be evaluated in closed form in terms of the so-called **elementary functions**: powers of x, trigonometric functions, exponentials, and logarithms. Thus, we will leave it as it is. It can be evaluated in terms of *non*elementary functions or integrated numerically, for any desired value(s) of x, but such discussion is beyond the purpose of this example and we will be content to accept (22) as our answer. ∎

This concludes our discussion of the integrating factor method of solution of (1). Let us postpone our discussion of the second method, the method of variation of parameters, until after we have discussed the questions of existence and uniqueness.

2.3.2 Existence and uniqueness for the linear equation.   As mentioned earlier, a fundamental issue in the theory of differential equations is whether a given differential equation $F(x, y, y') = 0$ *has* a solution through a given initial point $y(a) = b$ in the x, y plane and, if so, over what x interval it is valid. That is the question of existence. If a solution does exist, then our next question is that of uniqueness: Is that solution unique, or is there more than one such solution?

For the *linear* initial value problem

$$y' + p(x)y = q(x), \quad y(a) = b, \tag{23}$$

these questions are readily answered. Our proof of existence is said to be constructive because we actually found a solution, (20). To verify (20), we can see by inspection that it does satisfy the initial condition $y(a) = b$. To verify that it also satisfies the differential equation $y' + p(x)y = q(x)$, we can put (20) into that equation and see if we obtain an identity. Differentiating (20) gives

$$y' = e^{-\int_a^x p(s)\,ds}\frac{d}{dx}\left(-\int_a^x p(s)\,ds\right)\left(\int_a^x e^{\int_a^s p(u)\,du}q(s)\,ds + b\right)$$
$$+ e^{-\int_a^x p(s)\,ds}\frac{d}{dx}\left(\int_a^x e^{\int_a^s p(u)\,du}q(s)\,ds + b\right)$$

$$= e^{-\int_a^x p(s)\,ds}[-p(x)]\left(\int_a^x e^{\int_a^s p(u)\,du} q(s)\,ds + b\right) \tag{24}$$

$$+ e^{-\int_a^x p(s)\,ds}\, e^{\int_a^x p(u)\,du} q(x)$$

$$= -p(x)y + q(x),$$

which does show that the $y(x)$ given by (20) satisfies the equation $y' + p(x)y = q(x)$. In (24) we used chain differentiation and the Fundamental Theorem of the Calculus (that the derivative of the integral equals the integrand if the integrand is continuous). The latter holds over any x interval, containing the initial point a, over which $p(x)$ and $q(x)$ are continuous. Thus, we can say that (20) holds over the broadest x interval, containing $x = a$, over which both $p(x)$ and $q(x)$ are continuous.

We are also assured that the solution (20), of the initial value problem (23), is unique since each step in its derivation leaves no room for a different or alternative result, as can be seen by reviewing those steps. Nonetheless, let us show how to prove that the solution is unique using a different line of approach.

Suppose that we have two solutions of (23), say $y_1(x)$ and $y_2(x)$, on any x interval I containing the initial point a. That is,

$$y_1' + p(x)y_1 = q(x), \quad y_1(a) = b, \tag{25a}$$

and

$$y_2' + p(x)y_2 = q(x), \quad y_2(a) = b. \tag{25b}$$

Next, denote the difference $y_1(x) - y_2(x)$ as $u(x)$, say. If we subtract (25b) from (25a), and use the fact that $(f - g)' = f' - g'$, known from the calculus, we obtain the "homogenized" problem

$$u' + p(x)u = 0, \quad u(a) = 0 \tag{26}$$

on $u(x)$. Equivalent to $u' + p(x)u = 0$ is

$$e^{\int p\,dx} u' + p e^{\int p\,dx} u = 0$$

since $\exp\left(\int p\,dx\right) \neq 0$. Thus,

$$\left(u e^{\int p\,dx}\right)' = 0,$$

so

$$u e^{\int p\,dx} = C,$$

$$u(x) = C e^{-\int p\,dx}.$$

Now $\exp\left(-\int p\,dx\right)$ is nonzero, so the only way that $u(x)$ can satisfy the condition $u(a) = 0$, in (26), is if $C = 0$. But then $u(x)$ is identically zero and, since $u(x) = y_1(x) - y_2(x)$, it follows that $y_1(x)$ and $y_2(x)$ must be identical on I. Since $y_1(x)$ and $y_2(x)$ are *any* two solutions of (23), it follows that the solution of (23) must be unique.

The approach that we just used, in proving uniqueness, is somewhat standard. Namely, we suppose that we have two solutions to the given problem, we let their difference be u, say, obtain a homogenized problem on u, and show that the solution u of that problem must be identically zero. It follows that the two solutions must be identical, so that the solution (if it exists in the first place) must be unique. In summary, we have the following result.

THEOREM 2.3.1

Existence and Uniqueness for the Linear First-Order Equation
The linear equation $y' + p(x)y = q(x)$ does admit a solution

$$y(x) = e^{-\int_a^x p(s)\,ds}\left(\int_a^x e^{\int_a^s p(u)\,du}q(s)\,ds + b\right) \tag{27}$$

through an initial point $y(a) = b$ if $p(x)$ and $q(x)$ are continuous at a. That solution exists at least on the broadest x interval containing $x = a$, over which $p(x)$ and $q(x)$ are continuous, and is unique.

EXAMPLE 4   *Interval of Existence.*
Using (27), the initial value problem

$$y' + \frac{1}{x}y = \frac{1}{(x-5)^2}, \quad y(1) = 3 \tag{28}$$

gives the solution

$$y(x) = \frac{1}{x}\left(\ln\left|\frac{x-5}{4}\right| - \frac{5}{x-5} + \frac{7}{4}\right), \tag{29}$$

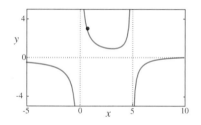

FIGURE 1
Graph of the function (29).

shown in Fig. 1. Observe that $p(x) = 1/x$ is discontinuous at $x = 0$ and that $q(x) = 1/(x-5)^2$ is discontinuous at $x = 5$, so Theorem 2.3.1 guarantees that a unique solution through the initial point $(1, 3)$ will exist at least on the interval $0 < x < 5$. From Fig. 1 we see that the solution (29) does indeed exist on that interval, and that it can be extended no further since the $1/x$ factor in (29) becomes undefined at $x = 0$ and the $\ln|(x-5)/4|$ and $5/(x-5)$ terms become undefined at $x = 5$. In picturesque language, we say that the solution "blows up" at $x = 0$ and at $x = 5$.   ∎

EXAMPLE 5   *A More Subtle Case.*
The condition of continuity of $p(x)$ and $q(x)$ is sufficient to imply the conclusions stated in the theorem, but is not necessary, as illustrated by the problem

$$xy' + 3y = 6x^3. \tag{30}$$

The general solution of (30) is readily found from (14) to be

$$y(x) = x^3 + \frac{C}{x^3}. \tag{31}$$

The graphs of the solution for several values of C (i.e., the solution curves) are shown in Fig. 2. In this case $p(x) = 3/x$ is continuous for all x except $x = 0$, and $q(x) = 6x^2$ is continuous for all x, so if we append to (30) an initial condition $y(a) = b$, for some $a > 0$, then Theorem 2.3.1 tells us that the solution (31) that passes through that initial point will be valid over the interval $0 < x < \infty$ at least. For instance, if $a = 1$ and $b = 2.5$ (the point P_1), then $C = 1.5$ and the solution (31) is valid only over $0 < x < \infty$ since it is undefined at $x = 0$ because of the $1/x^3$ term, as can also be seen from the figure. However, if $a = 1$ and $b = 1$ (the point P_2) then $C = 0$ and the solution (31) is valid over the broader interval $-\infty < x < \infty$ because $C = 0$ removes the singular $1/x^3$ term in (31). That is, if the initial point happens to lie on the solution curve $y(x) = x^3$ through the origin, then the solution $y(x) = x^3$ is valid on $-\infty < x < \infty$; if not, then the solution (30) is valid on $0 < x < \infty$ if $a > 0$ and on $-\infty < x < 0$ if $a < 0$.   ∎

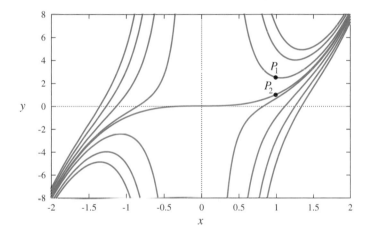

FIGURE 2   Representative integral curves (31) for the equation (30).

2.3.3 Solution by variation of parameters. (Optional).   A second method for the solution of the general first-order linear equation

$$y' + p(x)y = q(x) \tag{32}$$

is the method of **variation of parameters**, due to the great French mathematician *Joseph-Louis Lagrange* (1736–1813) who, like Euler, also worked on the applications of mathematics to mechanics, especially celestial mechanics.

Lagrange's method is as follows. We begin by considering the homogeneous version of (32),

$$y' + p(x)y = 0, \tag{33}$$

which is more readily solved. Recall that we solved it by integrating

$$\int \frac{dy}{y} + \int p(x)\,dx = 0$$

and obtaining the general solution

$$y_h(x) = Ae^{-\int p(x)\,dx}. \tag{34}$$

We use the subscript h here because $y_h(x)$ is called the **homogeneous solution** of (32). That is, it is the solution of the homogeneous version (33) of the original non-homogeneous equation (32). [In place of $y_h(x)$, some authors write $y_c(x)$ and call it the **complementary solution**.] Lagrange's idea is to try varying the "parameter" A, the arbitrary constant in (34). Thus, we seek a solution $y(x)$ of the *non*homogeneous equation in the form

$$y(x) = A(x)e^{-\int p(x)\,dx}. \tag{35}$$

(The general idea of seeking a solution of a differential equation in a particular form is important and is developed further in subsequent chapters.)

Putting (35) into (32) gives

$$\left(A'e^{-\int p\,dx} + A(-p)e^{-\int p\,dx} \right) + pAe^{-\int p\,dx} = q. \tag{36}$$

Canceling the two A terms and solving for A' gives

$$A'(x) = q(x)e^{\int p(x)\,dx}, \tag{37}$$

which can be integrated to give

$$A(x) = \int e^{\int p(x)\,dx}q(x)\,dx + C$$

and hence the general solution

$$y(x) = A(x)e^{-\int p(x)\,dx} = e^{-\int p(x)\,dx}\left(\int e^{\int p(x)\,dx}q(x)\,dx + C \right), \tag{38}$$

which is identical to our previous result (20).

NOTE: It is easy to miss how remarkable is Lagrange's idea. To explain, let us put Lagrange's idea aside for a moment and consider the second-order linear equation $y'' + y' - 2y = 0$, for instance. In Chapter 6 we will learn to seek solutions of such equations in the exponential form $y = e^{rx}$, where r is a constant that needs to be determined. Putting that form into $y'' + y' - 2y = 0$ gives the equation $(r^2 + r - 2)e^{rx} = 0$, which implies that r needs to satisfy the quadratic equation $r^2 + r - 2 = 0$, with roots $r = 1$ and $r = -2$. Thus, we are successful, in this example, in finding two solutions of the assumed form, $y(x) = e^x$ and $y(x) = e^{-2x}$. Notice how easily this idea works. It is readily implemented because most of the work has been done in deciding to look for solutions in the correct form, that is exponentials, rather than looking within the vast set of all possible functions. Similarly, if we lose our eyeglasses, the task of finding them is much easier if we know that they are somewhere on our desk, than if we know that they are somewhere in the universe.

Returning to Lagrange's idea, observe that the form (35), however, is completely *non*specific. That is, *every* function can be expressed in that form by a suitable choice of $A(x)$. Thus it is difficult to see why the idea seemed attractive to Lagrange. In fact, the equation (36), governing $A(x)$, is itself a first-order nonhomogeneous equation of the same form as the original equation (32), and looks even harder than (32)—except for the fact that the two A terms cancel, so that we obtain the simple equation (37) that can be solved by direct integration. Thus, the key to the success of Lagrange's method is in the cancelation of the two A terms, and that cancelation was not an accident. To explain, suppose that $A(x)$ is a constant. Then the A' term in (36) drops out and the two A terms *must* cancel to zero because if A is a constant then (35) is a solution of the homogeneous equation $y' + p(x)y = 0$!

Thus, Lagrange's idea was quite clever. In Chapter 6 we will adapt his variation-of-parameter method to higher-order differential equations.

Closure. Whereas in Section 2.2 we studied the homogeneous equation $y'+p(x)y = 0$, in this section we have studied the nonhomogeneous equation $y' + p(x)y = q(x)$, which is the general linear first-order differential equation. Using an integrating factor method we derived its general solution

$$y(x) = e^{-\int p(x)\,dx} \left(\int e^{\int p(x)\,dx} q(x)\,dx + C \right), \tag{39}$$

as well as the particular solution

$$y(x) = e^{-\int_a^x p(s)\,ds} \left(\int_a^x e^{\int_a^s p(u)\,du} q(s)\,ds + b \right) \tag{40}$$

satisfying an initial condition $y(a) = b$. Finally, we also derived (39) by a different method, Lagrange's method of variation of parameters.

Note that even if we have an initial value problem we do not *have* to use (40); we can use (39) to get the general solution and then evaluate the constant C by applying the initial condition to that solution. We mention this approach as an alternative in case you feel intimidated by the dummy variables s and μ in (40).

The solution (40) of the initial value problem

$$y' + p(x)y = q(x), \quad y(a) = b$$

exists and is unique over the broadest interval, containing the initial point a, over which both $p(x)$ and $q(x)$ are continuous, although that continuity is sufficient, not necessary.

In closing, we call attention to the exercises, below, that introduce additional important special cases: the **Bernoulli**, **Riccati**, **d'Alembert-Lagrange** and **Clairaut** equations. In the sequel we occasionally refer to those well known equations and those exercises.

Maple. Consider representative applications of *Maple* to the material covered in this section. For instance, consider the integral in (22), which we claimed was nonelementary. Let us see if it can be evaluated by *Maple*. Using the **int** integration command, enter

```
int(exp(-s^2)*sin(s),s=0..x);
```

and return. Here, the first item within the outer parentheses gives the integrand and the second gives the integration variable and limits. (If we desire the indefinite integral instead, then in place of the $s = 0..x$ we type only an s, to indicate the integration variable.) The output is the integral itself,

$$\int_0^x e^{-s^2} \sin s \, ds,$$

which indicates that *Maple* was not able to evaluate the integral analytically.

If we wish, we can nevertheless obtain a series expression for the integral by following the unsuccessful int command with the **series** command

```
series(%,x=0);
```

where the % is essentially a ditto mark that inputs the preceding output, in this case the integral, and where the $x = 0$ calls for the expansion to be about the point $x = 0$ (or any other point of our choosing). The result is the Taylor series

$$\frac{1}{2}x^2 - \frac{7}{24}x^4 + O(x^6),$$

where the $O(x^6)$ denotes that terms of order 6 and higher have been omitted, 6 being the default. If we want the series carried out to a higher order, we could indicate that by adding another (optional) argument. For instance,

```
series(%,x=0,12);
```

would give the desired expansion up to terms of order $O(x^{12})$. Unfortunately, an infinite series solution is in what we call **open form**, but sometimes that is the best we can do. In contrast, the solution (29) of the initial value problem (28) is in **closed form** since it involves only a finite number of calculations.

As one more example of the series command,

```
series(sin(x),x=0,9);
```

would give the output

$$x - \frac{1}{6}x^3 + \frac{1}{120}x^5 - \frac{1}{5040}x^7 + O(x^9)$$

Next, consider our generation of the graph in Fig. 1. If, to solve the initial value problem (29) and plot the solution, over the interval $-5 \leq x \leq 10$, say, we use the **phaseportrait** command

```
with(DEtools):
phaseportrait (diff(y(x),x) + (1/x)*y(x) = 1/(x-5)^2,y(x),
x=- 5..10, {[1,3]},arrows=NONE);
```

the result is the error message: "Error, (in phaseportrait) Stopping integration due to, division by zero." That is, the command cannot cope with the singularity in $p(x) = 1/x$ at $x = 0$ and in $q(x) = 1/(x - 5)^2$ at $x = 5$. To successfully run the phaseportrait command we need to keep the x interval within $0 < x < 5$ by writing $x = 0.0001..4.9999$ say, in place of the $x = -5..10$. Even so, we find that the output graph is plotted on a vertical scale that is so broad (namely, from $y = 0$ to $y = 4e + 06 = 4e^6$) that the details of the solution curve cannot be seen. If we include the option $y = 0..10$ say, then we do obtain a useful graph of the solution curve. Nevertheless, the branches of the solution curve to the left of $x = 0$ and to the right of $x = 5$ are not accessible to us using phaseportrait.

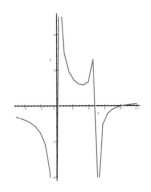

FIGURE 3
Maple plot of the solution (29).

Thus, to obtain the plot shown in Fig. 1 we used a plotting command instead, namely, the **implicitplot** command. Entering with(plots) first, to access the plotting commands, we used the command sequence

```
with(plots):
implicitplot(y=(1/x)*(log(abs((x-5)/4))-5/(x-5)+7/4),
x=-5..10,y=-4..5);
```

and obtained the result shown in Fig. 3. Implicitplot plots the graph of a relation $f(x, y) = 0$ on a specified rectangle in the x, y plane. The arguments within the outer parentheses are the relation and then the x and y limits, respectively. The default grid is only 25 by 25 (hence, 625 points), so the graph may be too coarse, as in Fig. 3. Further, implicitplot interpolates across discontinuities, as occurs at $x = 0$ and at $x = 5$ in the figure. To improve the quality of the plot we can refine the grid by means of an option such as *grid*. For instance, following the $y = -4..5$ we could type a comma then the option grid = [75, 100], if we desire a 75×100 grid. The resulting graph, which is indeed an improvement over that in Fig. 3, is given in Fig. 4.

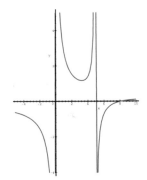

FIGURE 4
Improved *Maple* plot of the solution (29).

In our use of the implicitplot command, above, the x, y relation happened to give y explicitly in terms of x. More generally, as its name implies, implicitplot can be used when the relation does not give y explicitly in terms of x. Furthermore, it can be used to superimpose more than one such plot. For instance, to plot the graphs of the (implicit) relation $x^2 + y^2 = 5$ and the (explicit) relation $y = \sqrt{x + 5}\sin x$ over the rectangle $-4 \leq x \leq 9, -8 \leq y \leq 10$ use the command

```
implicitplot({x^2+y^2=5,y=sqrt(x+5)*sin(x)},x=-4..9,y=-8..10);
```

EXERCISES 2.3

1. In each case find the general solution, both using (14) and then again by using the integrating factor method, as we did in the COMMENT part of Example 2. The x interval under consideration will be understood to be the broadest interval on which both $p(x)$ and $q(x)$ are continuous. For example, in part (a) the x interval is $-\infty < x < \infty$, in part (e) it is any interval on which $\tan x$ is continuous (such as $\pi/2 < x < 3\pi/2$), and in part (f) it is either $-\infty < x < 0$ or $0 < x < \infty$ [to ensure the continuity of $p(x) = 2/x$].

(a) $y' - y = 3e^x$ (b) $y' + 4y = 8$

(c) $y' + y = x^2$ (d) $y' = y - 5\sin 2x$

(e) $y' - (\tan x)y = 6$ (f) $xy' + 2y = x^3$

(g) $xy' - 2y = x^3$ (h) $y' + (\cot x)y = 2\cos x$

(i) $(x - 5)(xy' + 3y) = 2$ (j) $\dfrac{dx}{dy} - 6x = 10e^y$

(k) $y\dfrac{dx}{dy} - y^5 + 3x = 0$ (l) $y^2\dfrac{dx}{dy} + xy - 4y^2 = 1$

(m) $t\dfrac{dx}{dt} - 4t^5 = x$ (n) $\dfrac{dr}{d\theta} + 2(\cot 2\theta)r = 1$

2. (a)–(n) For the equation given in the corresponding part of Exercise 1 find the general solution using *Maple*. Verify your result by showing that it does satisfy the given differential equation.

3. Solve $xy' + y = 6x^2$ subject to the given initial condition using any method of this section, and state the (broadest) interval of validity of the solution. Also, sketch the graph of the solution, by hand, and label any key values.

(a) $y(1) = 0$ (b) $y(2) = 2$ (c) $y(1) = 2$

(d) $y(-3) = 0$ (e) $y(-3) = 0$ (f) $y(1) = 8$

4. Solve $xy' + 2y = x + 2$ subject to the given initial condition using any method of this section, and state the (broadest) interval of validity of the solution. Also, sketch the graph of the solution, by hand, and label any key values.

(a) $y(2) = 0$ (b) $y(0) = 1$ (c) $y(-1) = 1$

(d) $y(1) = 1$ (e) $y(-2) = 0$ (f) $y(-3) = 1$

5. (*Interchange of independent and dependent variables*) The following equations on $y(x)$ are not linear first-order equations. Nevertheless, you should find that if you change your view point by regarding y as the independent variable and x as the dependent variable, rather than vice versa, and recall from the calculus that $dy/dx = 1/(dx/dy)$, then the revised equation will be a linear first-order equation on $x(y)$. Thus, use any method in this section to solve for x as a function of y, say $x = f(y)$. If, finally, you can solve the equation $x = f(y)$ for y as a function of x (to return to our original point of view) then do so; if not, leave it in

that form.

(a) $\dfrac{dy}{dx} = \dfrac{1}{x + 3e^y}$ (b) $\dfrac{dy}{dx} = \dfrac{1}{6x + y^2}$

(c) $(6y^2 - x)\dfrac{dy}{dx} - y = 0$ (d) $(y^2\sin y + x)\dfrac{dy}{dx} = y$

6. (*Direction fields*) The direction field concept was discussed in Section 2.2. For the differential equation given, use computer software to plot the direction field over the specified rectangular region in the x, y plane, as well as the integral curve through the specified point P. Also, if you can identify any integral curves exactly, from an inspection of the direction field, then give the equations of those curves, and verify that they do satisfy the given differential equation.

(a) $y' = 2 + (2x - y)^3$ on $0 \le x \le 3$, $|y| \le 4$; $P = (2, 1)$

(b) $y' = y(y^2 - 4)$ on $|x| \le 4$, $|y| \le 4$; $P = (1, 1)$

(c) $y' = (3 - y^2)^2$ on $|x| \le 2$, $|y| \le 3$; $P = (0, 0)$

(d) $y' + 2y = e^{-x}$ on $|x| \le 3$, $|y| \le 2$; $P = (0, 0.5)$

(e) $y' = x^2/(y^2 - 1)$ on $|x| \le 3$, $|y| \le 3$; $P = (-3, -3)$

(f) $y' + x = y$ on $|x| \le 20$, $-10 \le y \le 20$; $P = (0, 1)$

(g) $y' = e^{-x}y$ on $0 \le x \le 20$, $0 \le y \le 40$; $P = (0, 10)$

(h) $y' = x\sin y$ on $0 \le x \le 10$, $0 \le y \le 10$; $P = (2, 2)$

7. (*Bernoulli equation*) The equation

$$\boxed{y' + p(x)y = q(x)y^n,} \qquad (7.1)$$

where n is a constant (not necessarily an integer), is called **Bernoulli's equation**, after the Swiss mathematician *Jakob Bernoulli*. Jakob (1654–1705), his brother Johann (1667–1748), and Johann's son Daniel (1700–1782), are the best known of the eight members of the Bernoulli family who were prominent mathematicians and scientists.

(a) Give the general solution of (7.1) for the special cases $n = 0$ and $n = 1$.

(b) If n is neither 0 nor 1, then (7.1) is nonlinear. Nevertheless, show that by transforming the dependent variable from $y(x)$ to $v(x)$ according to $v = y^{1-n}$ (for $n \ne 0, 1$), (7.1) can be converted to the equation

$$v' + (1 - n)p(x)v = (1 - n)q(x), \qquad (7.2)$$

which is linear and can be solved by the methods developed in this section. This method of solution was discovered by *Gottfried Wilhelm Leibniz* (1646–1716) in 1696.

8. Use the method suggested in Exercise 7(b) to find the general solution to each of the following:

(**a**) $y' - 4y = 4y^2$ (**b**) $xy' - 2y = x^3 y^2$

(**c**) $2xyy' + y^2 = 2x$ (**d**) $\sqrt{y}(3y' + y) = x$

(**e**) $y' = y^2$ (**f**) $y' = xy^3$

(**g**) $y'' = (y')^2$ HINT: First, let $y'(x) = u(x)$.

9. (*Riccati equation*) The equation

$$y' = p(x)y^2 + q(x)y + r(x) \qquad (9.1)$$

is called **Riccati's equation**, after the Italian mathematician *Jacopo Francesco Riccati* (1676–1754). The Riccati equation is nonlinear if $p(x)$ is not identically zero. Recall from Exercise 7 that the Bernoulli equation can always be reduced to a linear equation by a suitable change of variables. Likewise, the Riccati equation can be converted to a linear equation, provided that any one particular solution can be found.

Let $Y(x)$ be any one particular solution of (9.1), as found by inspection, trial and error, or any other means. [Depending on $p(x)$, $q(x)$, and $r(x)$, finding such a $Y(x)$ may be easy, or it may prove too great a task.] Show that by changing the dependent variable from $y(x)$ to $u(x)$ according to

$$y = Y(x) + \frac{1}{u} \qquad (9.2)$$

the Riccati equation (9.1) can be converted to the equation

$$u' + [2p(x)Y(x) + q(x)]u = -p(x), \qquad (9.3)$$

which is linear and can be solved by the methods developed in this section. This method of solution was discovered by *Leonhard Euler* (1707–1783) in 1760.

10. Use the method suggested in Exercise 9 to find the general solution to each of the following. Nonelementary integrals, such as $\int \exp(ax^2)\,dx$, may be left as is.

(**a**) $y' - 4y = y^2$ HINT: $Y(x) = -4$

(**b**) $y' = y^2 - xy + 1$ HINT: $Y(x) = x$

(**c**) $(\cos x)y' = 1 - y^2$ HINT: $Y(x) = \sin x$

(**d**) $y' = e^{-x}y^2 - y$ HINT: $Y(x) = 2e^x$

(**e**) $y' = (2 - y)y$

(**f**) $y' = (1 - y)(2 - y)$

(**g**) $y' = y^2 - 4$

11. (*d'Alembert-Lagrange equation*) The *nonlinear* first-order differential equation

$$y = xf(p) + g(p) \qquad (11.1)$$

on $y(x)$, in which p denotes y', and where f and g are given functions of p, is known as a **d'Alembert-Lagrange equation** after the French mathematicians *Jean le Rond d'Alembert* (1717–1783) and *Joseph-Louis Lagrange* (1736–1813).

(**a**) Differentiating (11.1) with respect to x, show that

$$p - f(p) = \left[xf'(p) + g'(p) \right] \frac{dp}{dx}. \qquad (11.2)$$

Observe that this nonlinear equation on $p(x)$ can be converted to a linear equation if we interchange the roles of x and p by now regarding x as the independent variable and p as the dependent variable. Thus, obtain from (11.2) the linear equation

$$\frac{dx}{dp} - \frac{f'(p)}{p - f(p)}x = \frac{g'(p)}{p - f(p)} \qquad (11.3)$$

on $x(p)$. Since we have divided by $p - f(p)$ we must restrict $f(p)$ so that $f(p) \neq p$. Solving the simpler equation (11.3) for $x(p)$, the solution of (11.1) is thereby obtained in *parametric form*: $x = x(p)$ from solving (11.3), and $y = x(p)f(p) + g(p)$ from (11.1). This result is the key idea of this exercise, and is illustrated in parts (b)–(c). In parts (d)–(k) we consider a more specialized result, namely, for the case where $f(p)$ happens to have a "fixed point."

(**b**) To illustrate part (a), consider the equation $y = 2xy' + 3y'$ [i.e., where $f(p) = 2p$ and $g(p) = 3p$], and derive a parametric solution as discussed in part (a).

(**c**) To illustrate the approach outlined in part (a), consider the equation $y = x(y' + (y')^2)$ [i.e., $f(p) = p + p^2$ and $g(p) = 0$], and derive the parametric solution discussed in part (a).

(**d**) Suppose that $f(p)$ has a **fixed point** P_0, that is, such that $f(P_0) = P_0$. [A given function f may have none, one, or any number of fixed points. They are found as the solutions of the equation $f(p) = p$.] Show that (11.1) then has the straight line

$$y = P_0 x + g(P_0) \qquad (11.4)$$

as a particular solution, by verifying that the latter satisfies the differential equation (11.1). [This fact is of special interest in cases where the integrals that occur in the general solution of (11.3) are difficult to evaluate.]

(**e**) Show that $f(p) = 3p^2$ has two fixed points, $p = 0$ and $p = 1/3$, and hence show that the equation $y = 3xp^2 + g(p)$ has straight-line solutions $y = g(0)$ and $y = \frac{1}{3}x + g\left(\frac{1}{3}\right)$ for any given function g.

(**f**) Determine all particular solutions of the form (11.4), if any, for the equation $y = x\left((y')^2 - 2y' + 2\right) + e^{y'}$.

(**g**) Same as (f), for $y = xe^{y'} - 5\cos y'$.

(**h**) Same as (f), for $y = x\left((y')^2 - 2y'\right) + 6(y')^3$.

(**i**) Same as (f), for $y = x\left((y')^3 - 3y'\right) - 2\sin y'$.

12. (*Clairaut equation*) For the special case $f(p) = p$, the d'Alembert-Lagrange equation (11.1) in the preceding exercise becomes

$$\boxed{y = xp + g(p),} \tag{12.1}$$

which is known as the **Clairaut equation**, after the French mathematician *Alexis Claude Clairaut* (1713–1765). (Recall that p denotes y'.)

(**a**) Verify, by direct substitution into (12.1), that (12.1) admits the family of solutions

$$y = Cx + g(C), \tag{12.2}$$

where C is an arbitrary constant.

(**b**) Recall that (11.3) does not hold if $f(p) = p$, but (11.2) does. Letting $f(p) = p$ in (11.2), derive the family of solutions (12.2), as well as the additional particular solution given parametrically by

$$x = -g'(p), \tag{12.3a}$$
$$y = -pg'(p) + g(p). \tag{12.3b}$$

(**c**) To illustrate, find the parametric solution (12.3) for the equation $y = xy' - (y')^2$. Show that in this example (12.3) can be gotten into the explicit form $y = x^2/4$ by eliminating the parameter p between (12.3a) and (12.3b). Plot, by hand, the family (12.2), for $C = 0, \pm 1/2, \pm 1, \pm 2$, together with the solution $y = x^2/4$. (Observe, from that plot, that the particular solution $y = x^2/4$ forms an "envelope" of the family of straight-line solutions. Such a solution is called a **singular solution** of the differential equation.)

(**d**) Instead of a hand plot, do a computer plot of $y = x^2/4$ and the family (12.2), for $C = 0, \pm 0.25, \pm 0.5, \pm 0.75, \ldots, \pm 3$, on $-8 \le x \le 8$, $-10 \le y \le 12$.

2.4 APPLICATIONS OF LINEAR FIRST-ORDER EQUATIONS

In this section we consider several representative physical applications that are governed by linear first-order equations: mechanical systems, electrical circuits, radioactivity, and mixing problems. Additional applications are contained in the exercises.

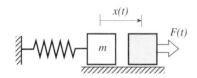

FIGURE 1
Mechanical system.

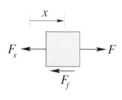

FIGURE 2
The forces, if $x > 0$ and $x' > 0$.

2.4.1 Mechanical systems.   Consider a block of mass m lying on a table and restrained laterally by an ordinary coil spring (Fig. 1), and denote by x the displacement of the mass (measured as positive to the right, say) from its equilibrium position. That is, $x = 0$ when the spring is neither stretched nor compressed. We imagine the mass to be disturbed from its rest position by an applied force $F(t)$, where t is the time, and we seek the equation (which will turn out to be a differential equation) governing the resulting time history of the displacement, $x(t)$.

Our first step is to identify the relevant physics which, in this case, is Newton's second law of motion. Since the motion is to be along a straight line we need consider only the forces in the x direction, and these are shown in Fig. 2: F_s is the force exerted on the mass by the spring, F_f is the force exerted on the bottom of the mass due to its sliding friction, and F is the applied force, the driving force. How do we know if F_s and F_f act to the left or to the right? The idea is to make assumptions on the signs of the displacement $x(t)$ and the velocity $x'(t)$ at the instant under consideration. For definiteness, suppose that $x > 0$ and $x' > 0$. Then it follows that each of the forces F_s and F_f are directed to the left, as shown in Fig. 2. (The equation of motion that we obtain will be insensitive to those assumptions, as we shall see.) Newton's second law then gives

$$(\text{mass})(x \text{ acceleration}) = \text{sum of } x \text{ forces}$$

or,

$$mx'' = F - F_s - F_f, \tag{1}$$

and we now need to express each of the forces F_s and F_f, in terms of the dependent and independent variables x and t.

Consider F_s first. If one knows enough about the geometry of the spring and the material of which it is made, one can derive an expression for F_s as a function of the extension x, as might be discussed in a course in Strength of Materials or Theoretical Elasticity. In practice, however, one can proceed empirically, and more simply, by actually applying various positive forces (i.e., to the right, in the positive x direction) and negative forces (to the left) to the spring and measuring the resulting displacement x (Fig. 3). For a typical coil spring, the resulting graph will be somewhat as sketched in Fig. 4, where its steepening at A and B is due to the coils becoming completely compressed and completely extended, respectively. Thus, F_s in (1) is the function the graph of which is shown as the curve AB. (Ignore the dashed line L for the moment.)

Next, consider the friction force F_f. The modeling of F_f will depend upon the nature of the contact between the mass and the table—in particular, upon whether it is dry or lubricated. Let us suppose it is lubricated, so that the mass rides on a thin film of lubricant such as oil. To model F_f, then, we must consider the fluid mechanics of the lubricating film. The essential idea is that the stress τ (force per unit area) on the bottom of the mass is proportional to the gradient du/dy of the fluid velocity u (Fig. 5), where the constant of proportionality is the coefficient of viscosity μ: $\tau = \mu\, du/dy$. But $u(y)$ is found, in a course in Fluid Mechanics, to be a linear function (Fig. 5), namely,

$$u(y) = \frac{u(h) - u(0)}{h} y = \frac{x'(t) - 0}{h} y = \frac{x'(t)}{h} y,$$

so

$$\tau = \mu \frac{du}{dy} = \frac{\mu}{h} x'(t).$$

Thus,

$$F_f = (\text{stress } \tau)\,(\text{area } A \text{ of bottom of block})$$

$$= \left(\frac{\mu\, x'(t)}{h} \right)(A).$$

That is, it is of the form

$$F_f = cx'(t), \tag{2}$$

for some constant c that we may consider as known. The upshot is that the friction force is proportional to the velocity. We will call c the *damping coefficient* because,

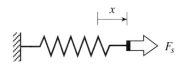

FIGURE 3
Spring force and displacement.

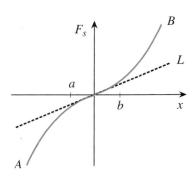

FIGURE 4
Force-displacement graph.

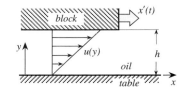

FIGURE 5
Lubricating film.

as we will see in Chapter 7, the effect of the cx' term in the governing differential equation is to cause the motion to "damp out," that is, to die out as $t \to \infty$.

Likewise, one can model an aerodynamic drag force on the block, but let us neglect that force on the tentative assumption that it can be shown to be small compared to the other two forces. Then (1) becomes

$$mx'' + cx' + F_s(x) = F(t). \tag{3}$$

It is important to realize that the governing differential equation (3) is nonlinear, because $F_s(x)$ is not a linear function of x (i.e., a constant times x), as can be seen from the graph AB shown in Fig. 4. With the exception of certain types of first-order equations, nonlinear differential equations are, in general, anywhere from extremely difficult to virtually impossible to solve analytically, although numerical solution is a viable option.

Since linear equations are much simpler than nonlinear ones, let us see if we can *approximate* the nonlinear equation (3) by a linear one. With that objective in mind, suppose that the motion is small enough, say between a and b in Fig. 4, so that we can approximate the actual nonlinear graph AB by the linear tangent line L (Fig. 4). If we denote the slope of L as k, then we can express

$$F_s(x) \approx kx. \tag{4}$$

We call k the *spring stiffness*.

Thus, the final form of our governing differential equation, or equation of motion, is the linearized approximation

$$\boxed{mx'' + cx' + kx = F(t),} \tag{5}$$

on $0 \leq t < \infty$, where the constants m, c, k, and the applied force $F(t)$ are known. Equation (5) is important, and we will return to it repeatedly. It is a nonhomogeneous linear second-order differential equation with constant coefficients (namely, m, c, and k).

Before continuing, let us focus our attention on the linearization process whereby we approximated the nonlinear equation (3) by the linear equation (5). By limiting our attention to small motions, namely, in the neighborhood of the equilibrium position $x = 0$, we were able to approximate the nonlinear graph [i.e., the nonlinear function $F_s(x)$] by the linear tangent line L (i.e., the linear function kx). Aside from this graphical approach, observe that we can proceed alternatively as follows:

1. Identify the equilibrium configuration, namely, the point in the neighborhood of which the motion takes place: $x = 0$.

2. Expand $F_s(x)$ in a Taylor series about that point:

$$F_s(x) = F_s(0) + F_s'(0)x + \frac{1}{2!}F_s''(0)x^2 + \cdots . \tag{6}$$

3. "Linearize" that series by cutting it off after the linear term:

$$F_s(x) \approx F_s(0) + F_s'(0)x = 0 + kx = kx. \tag{7}$$

Let us continue. We wish to append suitable initial or boundary conditions to (5). This particular problem is most naturally of initial value type since we envision initiating the motion in some manner at the initial time, say $t = 0$, and then solving for the motion that results. Thus, to (5) we append initial conditions

$$x(0) = x_0 \quad \text{and} \quad x'(0) = x_0', \tag{8}$$

for some (positive, negative, or zero) specified constants x_0 and x_0'. It should be plausible intuitively that we do need to specify both the initial displacement $x(0)$ and the initial velocity $x'(0)$ if we are to determine the resulting motion.

The differential equation (5) and initial conditions (8) comprise our resulting mathematical model of the physical system. By no means is there an exact correspondence between the model and the system since approximations were made in modeling the forces. Indeed, even our use of Newtonian mechanics, rather than relativistic mechanics, was an approximation.

This completes the *modeling* phase. The next step would be to solve the differential equation (5) subject to the initial conditions (8), for the motion $x(t)$.

COMMENT 1. Let us examine our claim that the resulting differential equation is insensitive to the assumptions made as to the signs of x and x'. In place of our assumption that $x > 0$ and $x' > 0$ at the instant under consideration, suppose we assume that $x > 0$ and $x' < 0$. Since $x > 0$, it follows that F_s acts to the left, and since $x' < 0$, it follows that F_f acts to the right. Then (Fig. 6a)

$$mx'' = F - F_s + F_f, \tag{9}$$

the signs of the force terms being $+$ if they act to the right (i.e., in the positive x direction) and $-$ if they act to the left. The sign of the F_f term is different in (9), compared with (1), because F_f now acts to the right, but notice that F_f now needs to be written as $F_f = c\left(-x'(t)\right)$, rather than $cx'(t)$ since x' is negative. Further, F_s is still kx, so (9) becomes

$$mx'' = F(t) - kx + (-cx'), \tag{10}$$

which is indeed equivalent to (5), as claimed.

Next, what if $x < 0$ and $x' > 0$? This time (Fig. 6b)

$$mx'' = F + F_s - F_f, \tag{11}$$

which differs from (1) in the sign of the F_s term. But now F_s needs to be written as $F_s = k\left(-x(t)\right)$ since x is negative. Further, F_f is cx', so (11) becomes

$$mx'' = F + k(-x) - cx',$$

which, again, is equivalent to (5). The remaining case, where $x < 0$ and $x' < 0$, is left for the exercises.

(a) $x > 0, \ x' < 0$

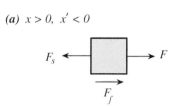

(b) $x < 0, \ x' > 0$

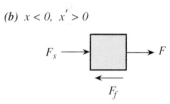

FIGURE 6
Other assumptions
on the signs of x and x'.

COMMENT 2. The final equation for F_s, $F_s = kx$, is well known as **Hooke's law**, after *Robert Hooke* (1635–1703). Hooke published his law of elastic behavior in 1676 as the anagram *ceiiinosssttuv* and, two years later, the solution *ut tensio sic vis*: roughly, "as the force, so is the displacement." That is, the force is proportional to the displacement. In view of the complexity with which we can now deal, Hooke's law must seem quite modest, but one must appreciate it within its historical context. In spirit, it followed *Galileo Galilei* (1564–1642) who, in breaking with lines established by the ancient Greeks, sought to establish a quantitative science, expressed in formulas and mathematical terms. For example, where Aristotle explained the increasing speed of a falling body in terms of the body moving more and more jubilantly as it approached its natural place (the center of the earth, which was believed to coincide with the center of the universe), Galileo sidestepped the question of cause entirely, and instead put forward the formula $v = 9.8t$, where v is the speed (in meters per second) and t is the time (in seconds). It may be argued that such departure from the less productive ancient Greek tradition marked the beginning of modern science.

Finally, let us turn to the solution for $x(t)$. We are not yet prepared to solve (5) because it is a second-order equation, not a first-order equation. However, if we consider the case where the mass m is sufficiently small for us to neglect the mx'' term in (5), compared to the cx' and kx terms, then we have the approximate *first-order* equation

$$cx' + kx = F(t) \tag{12}$$

or, putting it in our standard form "$y' + p(x)y = q(x)$" and appending an initial condition, we have the initial value problem[1]

$$x' + \frac{k}{c}x = \frac{1}{c}F(t); \quad x(0) = x_0. \tag{13}$$

The solution of (13) is, according to (20) of Section 2.3,

$$x(t) = e^{-\int_0^t \frac{k}{c}\,ds} \left(\int_0^t e^{\int_0^s \frac{k}{c}\,du} \, \frac{F(s)}{c} ds + x_0 \right)$$

$$= e^{-kt/c} \left(\int_0^t e^{ks/c} \, \frac{F(s)}{c} ds + x_0 \right). \tag{14}$$

Let us use (14) to work out the response $x(t)$ for several important types of the forcing function $F(t)$.

EXAMPLE 1   *No Force*.

If there is no force [i.e., $F(t) = 0$] then (14) simply gives

$$x(t) = x_0 e^{-kt/c}. \tag{15}$$

Thus, beginning with $x(0) = x_0$, x tends monotonely to zero as $t \to \infty$, as seen

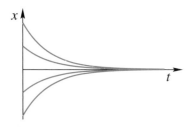

FIGURE 7
Response $x(t)$ when
$F(t) = 0$, for various x_0's.

---

[1] Actually, even if m is very small it is still possible that x'' is a large enough for the product mx'' *not* to be negligible compared to other terms in (5). For out present purpose let us assume that that situation is not present and that (13) models the physics sufficiently well.

in Fig. 7, in which we plot $x(t)$ for several different values of x_0, both positive and negative, for a fixed value of k/c. Alternatively, let us keep x_0 fixed and plot $x(t)$ for different values of k/c. As seen in Fig. 8, the approach to zero as $t \to \infty$ is more and more rapid for larger and larger values of k/c. That result seems quite reasonable: The return to the equilibrium position $x = 0$ is increasingly rapid as the spring stiffness k is increased and as the damping coefficient c is decreased. ∎

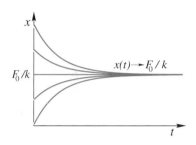

FIGURE 8
Effect of k/c on decay rate.

EXAMPLE 2   *Constant Force.*

Next, consider the case where the force is a constant, $F(t) = \text{constant} = F_0$. Then (14) gives

$$x(t) = \underbrace{\left(x_0 - \frac{F_0}{k}\right) e^{-kt/c}}_{\text{transient}} + \underbrace{\frac{F_0}{k}}_{\text{steady state.}} \qquad (16)$$

As $t \to \infty$, the exponential term in (16) tends to zero and $x(t) \to F_0/k$. Thus we call the $(x_0 - F_0/k)e^{-kt/c}$ term in (16) the **transient** part of the solution, and we call the F_0/k term the **steady-state** solution. Graphs of $x(t)$, in Fig. 9, show the approach to steady state for several different initial conditions. As in Example 1, the transient part of the response dies out faster for larger values of k/c.

The upshot, then, is that under the action of a steady force F_0 there will be a static deflection F_0/k. If the initial deflection x_0 is other than F_0/k then a dynamical process takes place, governed by equation (12), whereby $x(t) \to F_0/k$ as $t \to \infty$. ∎

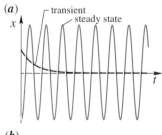

FIGURE 9
Response $x(t)$ for the case $F(t) = \text{constant} = F_0$, for various x_0's.

EXAMPLE 3   *Sinusoidal Force.*

As another important case, let the force be oscillatory, say $F(t) = F_0 \sin \omega t$, with amplitude F_0 and frequency ω. Then, with the help of the integral table given in the endpapers, (14) gives

$$x(t) = \underbrace{\left(x_0 + \frac{F_0 \omega c}{k^2 + (\omega c)^2}\right) e^{-kt/c}}_{\text{transient}} + \underbrace{\frac{F_0 k}{k^2 + (\omega c)^2}\left(\sin \omega t - \frac{\omega c}{k}\cos \omega t\right)}_{\text{steady state}}. \qquad (17)$$

Once again we have a transient response, transient in that it tends to zero as $t \to \infty$, as well as a steady-state response, namely, that which is left after the transients have died out. Note that steady-state does not necessarily mean constant; in Example 2 the applied force was held constant so the steady state was, likewise, a constant or "static" displacement, but in this example the applied force is oscillatory and the steady state is oscillatory as well. The response is plotted, for representative values of various parameters, in Fig. 10; the transient and steady-state parts are shown in (a) and the total response $x(t)$ is shown in (b). For the choice of parameters used in Fig. 10 we see that the transient part of $x(t)$ has virtually died out after three or four cycles. ∎

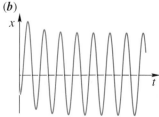

FIGURE 10
The response (17).

Observe, from the foregoing, that we not only derive the solution to a particular problem, such as the solution (17) to problem (13), we also try to interpret and understand the results, which process normally involves suitable graphical displays. We urge you not only to follow the step-by-step solutions, but also to appreciate that the physical interpretation of results and their presentation (graphical or otherwise) is an integral part of the overall process of analysis.

EXAMPLE 4   *Piecewise-Constant Force*.

Besides the cases of constant and oscillatory forcing functions, a third particularly important case is the case where $F(t)$ is piecewise constant, that is, where it is a constant F_0 over $0 < t < t_1$, then a different constant F_1 over $t_1 < t < t_2$, and so on. To illustrate, consider the case shown in Fig. 11.

To obtain the solution from (14) we need to distinguish two cases: $t < t_1$ and $t > t_1$. If $t < t_1$ then the $F(s)$ in the integrand is simply F_0 and we obtain

$$x(t) = e^{-kt/c} \left(\int_0^t \frac{F_0}{c} e^{ks/c} \, ds + x_0 \right)$$

$$= \left(x_0 - \frac{F_0}{k} \right) e^{-kt/c} + \frac{F_0}{k}, \tag{18}$$

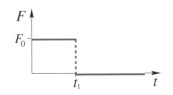

FIGURE 11
A piecewise-constant forcing function.

the same result as found in Example 2. But if $t > t_1$ then, since $F(t)$ is defined piecewise, we need to break the integral into two parts:

$$x(t) = e^{-kt/c} \left(\int_0^{t_1} \frac{F_0}{c} e^{ks/c} \, ds + \int_{t_1}^t \frac{0}{c} e^{ks/c} \, ds + x_0 \right)$$

$$= \left[x_0 + \frac{F_0}{k} \left(e^{kt_1/c} - 1 \right) \right] e^{-kt/c}. \tag{19}$$

Summarizing,

$$x(t) = \begin{cases} \left(x_0 - \dfrac{F_0}{k} \right) e^{-kt/c} + \dfrac{F_0}{k}, & 0 < t < t_1 \\[2ex] \left[x_0 + \dfrac{F_0}{k} \left(e^{kt_1/c} - 1 \right) \right] e^{-kt/c}, & t_1 < t < \infty, \end{cases} \tag{20}$$

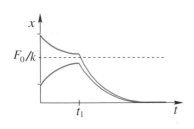

FIGURE 12
Response to piecewise-constant force.

as shown in Fig. 12. One curve is a typical response curve (i.e., solution curve) for $x_0 > F_0/k$, and the other is a typical response curve for $x_0 < F_0/k$. We see that $x(t)$ begins an exponential approach to the steady-state value F_0/k, as in Fig. 9, until time t_1 when the force is removed, then it decreases exponentially to zero.

COMMENT.   Remember, from the existence and uniqueness theorem 2.3.1 that existence and uniqueness are guaranteed only as far as "$p(x)$" and "$q(x)$" remain continuous. In the present example "$p(x)$"$= k/c$ is a constant so it is continuous for *all* t, but "$q(x)$"$= F(t)/c$ is continuous only as far as $t = t_1$, where it suffers a step discontinuity. Sure enough, the solution $x(t)$ suffers a kink at $t = t_1$, so the derivative $x'(t)$

does not exist at that point;[1] hence, the differential equation is not satisfied at t_1 so we can claim existence and uniqueness only over $0 < t < t_1$. However, we suggest that this breakdown is not serious. One way to view the situation is to think of two separate problems, one over $0 < t < t_1$ and the other over $t_1 < t < \infty$, where the final value of $x(t)$ from the first problem is taken as the initial condition of the second problem, with the two solutions "fitting together" at t_1.   ∎

Observe, in Example 4, that even if the forcing function $F(t)$ in

$$x' + \frac{k}{c}x = \frac{1}{c}F(t) \tag{21}$$

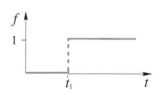

is discontinuous, the response $x(t)$ is nevertheless continuous; that is, even though the *input* $F(t)$ is discontinuous at t_1, the *output* $x(t)$ is continuous there, though its graph suffers a "kink" (i.e., a discontinuity in the slope) at that point (Fig. 12). Essentially, the idea is that the solution of (21) involves integration, and the integral of a function with a finite discontinuity at a point t_1 is continuous at t_1. That is, the integration is a smoothing process. To illustrate, consider the simple function $f(t)$ shown in Fig. 13. The integral function

FIGURE 13
A function f with finite discontinuity.

$$f_1(t) = \int_0^t f(s)\,ds$$

$$= \begin{cases} \int_0^{t_1} 0\,ds, & t < t_1 \\ \int_0^{t_1} 0\,ds + \int_{t_1}^t 1\,ds, & t > t_1 \end{cases}$$

$$= \begin{cases} 0, & t < t_1 \\ t - t_1, & t > t_1 \end{cases} \tag{22}$$

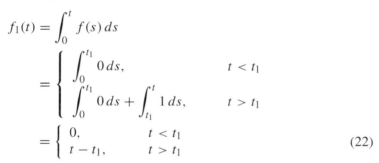

shown in Fig. 14 is continuous but with a kink at t_1, just as $x(t)$ in Fig. 12 is continuous but with a kink at the point t_1 at which the forcing function $F(t)$ is discontinuous (Fig. 11). In fact, if we integrate again, then

$$f_2(t) = \int_0^t f_1(s)\,ds = \begin{cases} 0, & t < t_1 \\ \frac{1}{2}(t - t_1)^2, & t > t_1 \end{cases} \tag{23}$$

and we see from Fig. 14 that not only is f_2 continuous at t_1, it is smooth there (i.e., it has no kink) as well. Again, the essential idea is that *integration is a smoothing process.*

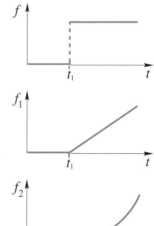

FIGURE 14
Integration is a smoothing process.

---

[1]Recall from the calculus that

$$x'(t_1) = \lim_{\Delta t \to 0} \frac{x(t_1 + \Delta t) - x(t_1)}{\Delta t}$$

where $\Delta t \to 0$ in any manner. For the functions $x(t)$ shown in Fig. 12, the limit as $\Delta t \to 0$ through positive values is not the same as the limit as $\Delta t \to 0$ through negative values, so a limit does not exist, therefore $x'(t_1)$ does not exist. Put more simply, $x'(t_1)$ does not exist because the graph of x does not have a unique and finite slope at t_1.

2.4.2 Electrical circuits.   In the case of electrical circuits the relevant underlying physics, analogous to Newton's second law for mechanical systems, is provided by Kirchhoff's laws. Instead of forces and displacements in a mechanical system comprised of various elements such as masses and springs, we are interested now in voltages and currents in an electrical system comprised of various elements such as resistors, inductors, and capacitors.

First, by a current we mean a flow of charges: The *current* through a given control surface, such as the cross section of a wire, is the charge per unit time crossing that surface. Each electron carries a negative charge of 1.6×10^{-19} *coulombs*, and each proton carries an equal positive charge. Current is measured in *amperes*, with one ampere being a flow of one coulomb per second. By convention, a current is counted as positive in a given direction if it is the flow of positive charge in that direction. While, in general, currents can involve the flow of positive or negative charges, in an electrical circuit the flow is of negative charges, free electrons. Thus, when one speaks of a current of one ampere in a given direction in an electrical circuit one really means the flow of one coulomb per second of negative charges (electrons) in the opposite direction.

Just as heat flows due to a temperature difference, from one point to another, an electric current flows due to a difference in the electric potential, or *voltage*, measured in *volts*.

We will need to know the relationship between the voltage difference across a given circuit element and the corresponding current flow. The circuit elements of interest here are resistors, inductors, and capacitors.

For a **resistor**, the voltage drop $E(t)$, where t is the time (in seconds), is proportional to the current $i(t)$ through it:

$$E(t) = Ri(t), \tag{24}$$

where the constant of proportionality R is called the *resistance* and is measured in *ohms*; (24) is called **Ohm's law**. By a resistor we usually mean an "off-the-shelf" electrical device, often made of carbon, that offers a specified resistance—such as 100 ohms, 500 ohms, and so on. But even the current-carrying wire in a circuit is itself a resistor, with its resistance directly proportional to its length and inversely proportional to its cross-sectional area, though that resistance is probably negligible compared to that of other resistors in the circuit. The standard symbolic representation of a resistor is shown in Fig. 15.

For an **inductor**, the voltage drop is proportional to the time rate of change of current through it:

$$E(t) = L\frac{di(t)}{dt}, \tag{25}$$

where the constant of proportionality L is called the **inductance** and is measured in *henrys*. Physically, most inductors are coils of wire; hence, the symbolic representation shown in Fig. 15.

For a **capacitor**, the voltage drop is proportional to the charge $Q(t)$ on the capacitor:

$$E(t) = \frac{1}{C}Q(t), \tag{26}$$

Resistor :

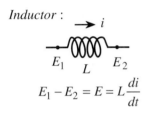

$$E_1 - E_2 = E = Ri$$

Inductor :

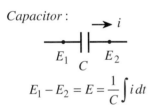

$$E_1 - E_2 = E = L\frac{di}{dt}$$

Capacitor :

$$E_1 - E_2 = E = \frac{1}{C}\int i\,dt$$

FIGURE 15
The circuit elements.

where C is called the **capacitance** and is measured in *farads*. Physically, a capacitor is normally comprised of two plates separated by a gap across which no current flows, and $Q(t)$ is the charge on one plate relative to the other. Though no current flows across the gap, there will be a current $i(t)$ that flows through the circuit that links the two plates and is equal to the time rate of change of charge on the capacitor:

$$i(t) = \frac{dQ(t)}{dt}. \tag{27}$$

From (26) and (27) it follows that the desired voltage/current relation for a capacitor can be expressed as

$$E(t) = \frac{1}{C} \int i(t)\, dt. \tag{28}$$

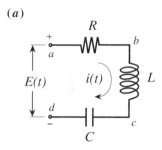

(a)

Now that we have equations (24), (25), and (28) relating the voltage drop to the current, for our various circuit elements, how do we deal with a grouping of such elements within a circuit? The relevant physics that we need, for that purpose, is given by Kirchhoff's laws, named after the German physicist *Gustav Robert Kirchhoff* (1824–1887):

Kirchhoff's current law states that the algebraic sum of the currents approaching (or leaving) any point of a circuit is zero.

Kirchhoff's voltage law states that the algebraic sum of the voltage drops around any loop of a circuit is zero.

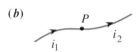

(b)

To apply these ideas, consider the circuit shown in Fig. 16a, consisting of a single loop containing a resistor, an inductor, a capacitor, a voltage source (such as a battery or generator), and the necessary wiring. Let us consider the current $i(t)$ to be positive clockwise; if it actually flows counterclockwise then its numerical value will be negative. In this case Kirchhoff's current law simply says that the current i is a constant from point to point within the circuit and therefore varies only with time. That is, the current law states that at any given point P in the circuit (Fig. 16b), $i_1 + (-i_2) = 0$ or, $i_1 = i_2$. Kirchhoff's voltage law, which is really the self-evident algebraic identity

FIGURE 16
RLC circuit.

$$(E_a - E_d) + (E_b - E_a) + (E_c - E_b) + (E_d - E_c) = 0, \tag{29}$$

gives

$$E(t) - Ri - L\frac{di}{dt} - \frac{1}{C} \int i\, dt = 0. \tag{30}$$

The latter is called an **integrodifferential equation** because it contains both derivatives and integrals of the unknown function, but we can convert it to a differential equation in either of two ways.

First, we could differentiate (30) with respect to t to eliminate the integral sign, thereby obtaining

$$\boxed{L\frac{d^2 i}{dt^2} + R\frac{di}{dt} + \frac{1}{C}i = \frac{dE(t)}{dt}.} \tag{31}$$

Alternatively, we could use $Q(t)$ instead of $i(t)$ as our dependent variable, for then $\int i \, dt = Q(t)$, so $i(t) = dQ/dt$ and (30) becomes

$$L\frac{d^2Q}{dt^2} + R\frac{dQ}{dt} + \frac{1}{C}Q = E(t). \tag{32}$$

Either way, we obtain a linear second-order differential equation.

In this chapter our interest is in *first*-order equations. Nevertheless, in place of the second-order equations (31) and (32), we do obtain first-order equations in the following special cases.

EXAMPLE 5   *RC Circuit.*

If $L = 0$, then (31) reduces to the linear first-order equation

$$R\frac{di}{dt} + \frac{1}{C}i = \frac{dE(t)}{dt}, \tag{33}$$

where we regard $E(t)$ as prescribed (Fig. 17a).   ∎

EXAMPLE 6   *RL Circuit.*

If, instead of removing the inductor from the circuit shown in Fig. 16, we remove the capacitor (Fig. 17b), then (30) gives the first-order equation

$$L\frac{di}{dt} + Ri = E(t). \tag{34}$$

COMMENT.   It would be natural to expect that deleting the capacitor is equivalent to setting $C = 0$, yet in that case the capacitor term in (30) becomes infinite rather than zero. However, realize that if we wish to effectively remove the capacitor we can do that by moving its plates together until they touch. Since the capacitance C varies as the inverse of the gap dimension, then as the gap diminishes to zero $C \to \infty$ and the capacitor term in (30) does indeed drop out because of its $1/C$ factor.   ∎

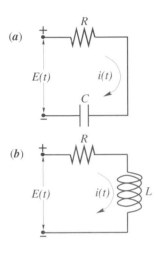

FIGURE 17
Two special cases:
RC and *RL* circuits.

We will not solve equations (33) and (34) since they are of the same form as equation (12) for the mechanical system, the solution of which was covered in Section 2.4.1.

Before closing this discussion of electrical circuits, we wish to emphasize the correspondence, or **analogy**, between the RLC electrical circuit and the mechanical oscillator studied in Section 2.4.1, and governed by the equation

$$m\frac{d^2x}{dt^2} + c\frac{dx}{dt} + kx = F(t). \tag{35}$$

For we see that equations (31) (the current formulation) and (32) (the charge formulation) are of exactly the same form as (35). Thus, their mathematical solutions are identical, and hence their physical behaviors are identical too. Consider (31), for instance. Comparing it with (35), we note the correspondence

$$L \leftrightarrow m, \quad R \leftrightarrow c, \quad 1/C \leftrightarrow k, \quad i(t) \leftrightarrow x(t), \quad \frac{dE(t)}{dt} \leftrightarrow F(t). \tag{36}$$

Thus, given the values of m, c, k, and the function $F(t)$, we can construct an electrical *analog circuit* by setting $L = m$, $R = c$, $C = 1/k$, and $E(t) = \int F(t)\, dt$. If we also match the initial conditions by setting $i(0) = x(0)$ and $\dfrac{di}{dt}(0) = \dfrac{dx}{dt}(0)$, then the resulting current $i(t)$ will be identical to the motion $x(t)$.

Or, we could use (32) to create a different analog, namely,

$$L \leftrightarrow m, \quad R \leftrightarrow c, \quad 1/C \leftrightarrow k, \quad Q(t) \leftrightarrow x(t), \quad E(t) \leftrightarrow F(t). \tag{37}$$

In either case we see that, in mechanical terminology, the inductor provides "inertia" (as does the mass), the resistor provides energy dissipation (as does the friction force), and the capacitor provides a means of energy storage (as does the spring).

Our interest in such analogs is at least twofold. First, to whatever extent we understand the mechanical oscillator, we thereby also understand its electrical analog circuit, and vice versa. Second, if the system is too complex to solve analytically, we may wish to study it experimentally. If so, by virtue of the analogy we have the option of studying whichever is more convenient. For instance, it would no doubt be simpler, experimentally, to study the RLC circuit than the mechanical oscillator.

2.4.3 Radioactive decay; carbon dating.

Another important application of linear first-order linear equations involves radioactive decay and carbon dating.

Radioactive materials, such as carbon-14, einsteinium-253, plutonium-241, radium-226, and thorium-234, are found to decay at a rate that is proportional to the amount of mass present. This observation is consistent with the supposition that the disintegration of a given nucleus, within the mass, is independent of the past or future disintegrations of the other nuclei, for then the number of nuclei disintegrating, per unit time, will be proportional to the total number of nuclei present:

$$\frac{dN}{dt} = -kN, \tag{38}$$

where k is known as the disintegration constant, or decay rate.[1] Actually, the graph of $N(t)$ proceeds in unit steps since $N(t)$ is integer-valued, so $N(t)$ is discontinuous and hence nondifferentiable. However, if N is very large, then the steps are very small compared to N. Thus, we can regard N, approximately, as a continuous function of t and can tolerate the dN/dt derivative in (38). However, it is inconvenient to work with N since one cannot count the number of atoms in a given mass. Thus, we multiply both sides of (38) by the atomic mass, in which case (38) becomes the simple linear first-order equation

$$\frac{dm}{dt} = -km, \tag{39}$$

where $m(t)$ is the total mass, a quantity which is more readily measured. Solving, we obtain

$$m(t) = m_0 e^{-kt}, \tag{40}$$

---

[1] We could write $dN/dt = kN$ where k is negative, but it seems best to put the negativeness of k in evidence by including a minus sign explicitly and writing $dN/dt = -kN$ where k is positive.

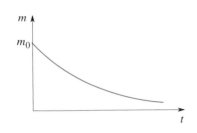

FIGURE 18
Exponential decay.

where $m(0) = m_0$ is the initial amount of mass (Fig. 18). This result is indeed the exponential decay that is observed experimentally.

Since k gives the rate of decay, it can be expressed in terms of the **half-life** T of the material, the time required for any initial amount of mass m_0 to be reduced by half, to $m_0/2$. Then (40) gives

$$\frac{m_0}{2} = m_0 e^{-kT},$$

so $k = (\ln 2)/T$, and (40) can be re-expressed in terms of T, if we prefer, as

$$m(t) = m_0 2^{-t/T}. \tag{41}$$

Thus, if $t = T, 2T, 3T, 4T, \ldots$, then $m(t) = m_0, m_0/2, m_0/4, m_0/8$, and so on.

Radioactivity has had an important archaeological application in connection with **dating**. The basic idea behind any dating technique is to identify a physical process that proceeds at a known rate. If we measure the state of the system now, and we know its state at the initial time, then from these two quantities together with the known rate of the process, we can infer how much time has elapsed; the mathematics enables us to "travel back in time as easily as a wanderer walks up a frozen river."[1]

In particular, consider carbon dating, developed by the American chemist *Willard Libby* in the 1950s. The essential idea is as follows. Cosmic rays consisting of high-velocity nuclei penetrate the earth's lower atmosphere. Collisions of these nuclei with atmospheric gases produce free neutrons. These, in turn, collide with nitrogen, thus changing some of the nitrogen to carbon-14, which is radioactive, and which decays to nitrogen-14 with a half-life of around 5,570 years. Thus, some of the carbon dioxide which is formed in the atmosphere contains this radioactive C-14. Plants absorb both radioactive and nonradioactive CO_2, and humans and animals inhale both and also eat the plants. Consequently, the countless plants and animals living today contain both C-12 and, to a much lesser extent, its radioactive isotope C-14, in a ratio that is essentially the same from one plant or animal to another.

E X A M P L E   7   *Carbon Dating*.

Consider a wood sample that we wish to date. Since C-14 emits approximately 15 beta particles per minute per gram, we can determine how many grams of C-14 are contained in the sample by measuring the rate of beta particle emission. Suppose that we find that the sample contains 0.002 gram of C-14, whereas if it were alive today it would, based upon its weight, contain around 0.0045 grams. Thus, we assume that it contained 0.0045 grams of C-14 when it died. That mass of C-14 will have decayed, over the subsequent time span t, to 0.002 gram. Then (41) gives

$$0.002 = (0.0045)\, 2^{-t/5570},$$

and, solving for t, we determine the sample to be around $t = 6{,}516$ years old.

However, it must be emphasized that this method (and the various others that are based upon radioactive decay) depend critically upon assumptions of uniformity.

---

[1] Ivar Ekeland, *Mathematics and the Unexpected* (Chicago: University of Chicago Press, 1988).

To date the wood sample studied in this example, for instance, we need to know the amount of C-14 present in the sample when the tree died, and what the decay rate was over the time period in question. To apply the method, we assume, first, that the decay rate has remained constant over the time period in question and, second, that the ratio of the amounts of C-14 to C-12 was the same when the tree died as it is today. Observe that although these assumptions are usually stated as fact they can never be proved, since it is too late for direct observation and the only evidence available now is necessarily circumstantial. ■

2.4.4 Mixing problems; one-compartment models. In this final application we consider a mixing tank with an inflow of $Q(t)$ gallons per minute and an equal outflow, where t is the time; see Fig. 19. The inflow is at a constant concentration c_1 of a particular solute (pounds per gallon), and the tank is constantly stirred, so that the concentration $c(t)$ within the tank is uniform. Hence, the outflow is at concentration $c(t)$. Let v denote the volume within the tank, in gallons; v is a constant because the inflow and outflow rates are equal. To keep track of the instantaneous mass of solute $x(t)$ within the tank, let us carry out a mass balance for the "control volume" V (dashed lines in the figure):

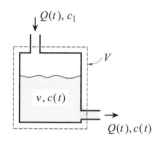

FIGURE 19
Mixing tank.

$$
\begin{array}{c}
\text{Rate of increase} \\
\text{of mass of solute} \\
\text{within } V
\end{array}
= \text{Rate in} - \text{Rate out}, \tag{42}
$$

$$
\frac{dx}{dt}\frac{\text{lb}}{\text{min}} = \left(Q(t)\frac{\text{gal}}{\text{min}}\right)\left(c_1\frac{\text{lb}}{\text{gal}}\right) - \left(Q(t)\frac{\text{gal}}{\text{min}}\right)\left(c(t)\frac{\text{lb}}{\text{gal}}\right), \tag{43}
$$

or, since $c(t) = x(t)/v$,

$$
\frac{dx(t)}{dt} + \frac{Q(t)}{v}x(t) = c_1 Q(t), \tag{44}
$$

which is a linear first-order equation on $x(t)$. Alternatively, we have the linear first-order equation

$$
\frac{dc(t)}{dt} + \frac{Q(t)}{v}c(t) = \frac{c_1 Q(t)}{v} \tag{45}
$$

on the concentration $c(t)$.

Recall that in modeling a physical system one needs to incorporate the relevant physics such as Newton's second law or Kirchhoff's laws. In the present application, the relevant physics is provided entirely by (42). To better understand (42), suppose we rewrite it with one more term included on the right-hand side:

$$
\begin{array}{c}
\text{Rate of increase} \\
\text{of mass of solute} \\
\text{within } V
\end{array}
=
\begin{array}{c}
\text{Rate} \\
\text{into} \\
V
\end{array}
-
\begin{array}{c}
\text{Rate} \\
\text{out of} \\
V
\end{array}
+
\begin{array}{c}
\text{Rate of creation} \\
\text{of mass} \\
\text{within } V.
\end{array}
\tag{46}
$$

The equation (46) is merely a matter of logic or bookkeeping, not physics. Since (42) follows from (46) only if there is no creation (or destruction) of mass, we can now understand (42) to be a statement of the physical principle of *conservation of mass*, namely, that matter can neither be created nor destroyed (except under exceptional circumstances that are not present in this situation).

The tank shown in Fig. 19 could, literally, be a mixing tank in a chemical plant, but in some applications the figure could be intended only schematically. For instance, in biological applications it is common to represent the interacting parts of the biological system, that are under consideration, as one or more interconnected **compartments**, with inflows, outflows, and exchanges between compartments. One component could be an organ such as the liver; another could be all of the blood in the circulatory system. Thus, the system represented in Fig. 19 is a one-compartment system.

In our mixing tank formulation the parameters c_1 and v are considered as known and our purpose is to determine the time history of $c(t)$ by solving (45) subject to some appropriate initial condition. For instance, if $Q(t) = \text{constant} = Q$, $c_1 = 0$, and $c(0) = c_0$, then we obtain

$$c(t) = c_0 e^{-\frac{Q}{v}t}. \tag{47}$$

In a biological application, on the other hand, we might monitor $c(t)$ and infer, by comparing the monitored response with the analytical solution (47), the system parameter Q/v.[1]

2.4.5 Heaviside step function.   Recall that in Example 4 the forcing function was defined "piecewise" as

$$F(t) = \begin{cases} F_0, & 0 < t < t_1 \\ 0, & t_1 < t < \infty. \end{cases} \tag{48}$$

Such functions are common in engineering. For instance, a residential heating system is either "on" or "off"; it does not vary continuously between these two extremes. To handle such functions in analysis it is convenient to introduce the **Heaviside step function** $H(t)$,[2] defined as

$$H(t) = \begin{cases} 1, & t > 0 \\ 0, & t < 0. \end{cases} \tag{49}$$

The value assigned to $H(t)$ at the point of discontinuity is not standard; some authors use $H(0) = 0$ while others use $H(0) = 1/2$ or $H(0) = 1$. The value assigned to $H(0)$ will not be consequential in this discussion so we have simply omitted its specification in (49) and similarly for $F(t_1)$ in (48).[3]

Since $H(t)$ is a unit step at $t = 0$ (Fig. 20a), $H(t - a)$ is a unit step shifted to $t = a$, as shown in Fig. 20b.

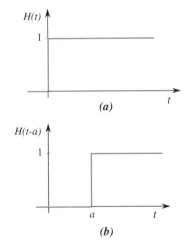

FIGURE 20
Unit step function.

---

[1] For a discussion of *compartmental analysis* in biology see, for example, L. Edelstein-Keshet, *Mathematical Models in Biology* (New York: Random House, 1988).

[2] *Oliver Heaviside* (1850–1925), initially a telegraph and telephone engineer, is best known for

In fact, the Heaviside function can be used as a basic building block to build up more complicated cases. For instance, the unit **rectangular pulse** $P(t; a, b)$, which is unity over $a < t < b$ and zero outside that interval (Fig. 21), is expressible in terms of Heaviside functions as

$$\boxed{P(t; a, b) = H(t - a) - H(t - b).}$$ (50)

Such pulses can be used to build up functions that are defined piecewise.

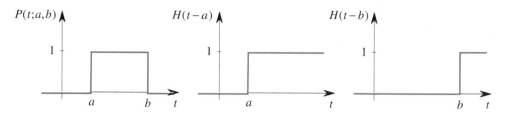

FIGURE 21   Rectangular pulse: $P(t; a, b) = H(t - a) - H(t - b)$.

E X A M P L E   8   *Piecewise-Defined Function.*

For instance, the function

$$f(t) = \begin{cases} 2 + t^2, & 0 < t < 2 \\ 6, & 2 < t < 3 \\ 2/(2t - 5), & 3 < t < \infty \end{cases}$$ (51)

shown in Fig. 22 can be expressed as a unit pulse over $0 < t < 2$ times $2 + t^2$, plus a unit pulse over $2 < t < 3$ times 6, plus a unit pulse over $3 < t < \infty$ times $2/(2t - 5)$:

$$f(t) = (2 + t^2)[H(t - 0) - H(t - 2)] + 6[H(t - 2) - H(t - 3)]$$

$$+ \frac{2}{2t - 5}[H(t - 3) - H(t - \infty)],$$ (52)

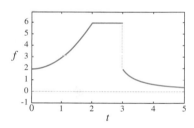

FIGURE 22
$f(t)$ of Example 8.

or, since we cannot distinguish $H(t - 0)$ [i.e., $H(t)$] from 1 over the interval $0 < t < \infty$ and since $H(t - \infty)$ is simply zero for all finite t,

$$\begin{aligned} f(t) &= (2 + t^2)[1 - H(t - 2)] + 6[H(t - 2) - H(t - 3)] + \frac{2}{2t - 5}H(t - 3) \\ &= 2 + t^2 + (4 - t^2)H(t - 2) + \left(\frac{2}{2t - 5} - 6\right)H(t - 3), \end{aligned}$$ (53)

which form is more compact than the three-tier definition (51).   ∎

his contributions to vector field theory and to the development of a systematic Laplace transform methodology for the solution of differential equations. Note the spelling: Heaviside, not Heavyside.

[3]The reason the value $F(t_1)$ is inconsequential is that the response $x(t)$ due to a forcing function $F(t)$ is given by an integral that contains $F(t)$ in its integrand as, for example, in (14). Recall the area significance of a definite integral and note that the value of the integral is insensitive to the value of the integrand at any given point because there is no area under a single point of finite height.

EXAMPLE 9

Finally, returning to the piecewise-constant function $F(t)$ in Example 4, observe that we can express that function as F_0 times a unit pulse over $0 < t < t_1$:

$$F(t) = F_0 [H(t) - H(t - t_1)] = F_0 [1 - H(t - t_1)]. \tag{54}$$

■

Besides the Heaviside notation being more compact than multi-tier expressions, we *must* use the Heaviside form when using *Maple*, as discussed in the *Maple* section below. In addition, when we solve differential equations by the Laplace transform section (Chapter 13) we will have to use the Heaviside form rather than the multi-tier form.

Closure. In this section we have studied applications of linear first-order equations to mechanical systems, RLC circuits, radioactivity, and mixing problems. These applications are representative but by no means exhaustive. Although the mechanical system gives rise to a linear *second*-order equation we reduced that equation to one of first order by neglecting the mass (i.e., by neglecting the inertial mx'' term relative to the damping force term cx' and the spring force term kx). We prescribed several different types of forcing functions $F(t)$, in each case solving for the resulting displacement $x(t)$, plotting solution curves, and discussing both the transient and steady-state parts of the solution.

Note that, normally, the terms "transient" and "steady state" are used when the independent variable is the time t, as in the examples in this section (though not in all of the exercises). In (16) and (17), for instance, the transient and steady-state parts are clearly discerned, and we labeled them accordingly. In other cases the solution might not tend to a constant or a steady oscillation, as $t \to \infty$, so in those cases we say that there is no steady-state solution.

Like the mck mechanical system, an RLC circuit gives rise to a linear second-order equation. We found that we could work with a first-order equation by omitting either the inductor or the capacitor.

Additional applications are introduced in the exercises.

Maple. Consider, as representative, the generation of Fig. 12 by *Maple*. The problem is

$$x' + \frac{k}{c}x = \frac{1}{c}F(t); \qquad x(0) = x_0. \tag{55}$$

If we wish to use the phaseportrait command we need numerical values for k, c, and x_0, so let us (arbitrarily) choose $k = c = 1$, and let us use two different initial conditions $x_0 = 1$ and $x_0 = 5$. The two-tier definition (48) of $F(t)$ is not usable for *Maple*, so let us re-express it as $F_0 [1 - H(t - t_1)]$, as in (54), and let us choose $F_0 = 3$, say, and $t_1 = 2$. Since the *Maple* designation of $H(t)$ is

```
Heaviside(t)
```

we can use these commands:

```
with(DEtools):
phaseportrait(diff(x(t),t)+x(t)=3*(1-Heaviside(t-2)),
x(t), t=0..10, {[0,1],[0,5]}, arrows=NONE);
```

CAUTION: Within the *Maple* software Heaviside(t) is 0 for $t < 0$ and 1 for $t > 0$, but it is undefined for $t = 0$, that is, at the point at which the function is discontinuous. Thus, a *Maple* calculation that confronts the quantity $H(0)$ will give an error message about an undefined quantity. In particular, the default phaseportrait command breaks the independent variable interval into 20 equal parts which are called the **stepsize** of the calculation. In the calculation shown above, the interval is $0 < t < 10$, so calculations are carried out at 0, 0.5, 1.0, 1.5, 2.0, 2.5, ..., 10 (and at the midpoints 0.25, 0.75, 1.25, ... of those subintervals as well). Since $t = 2$, where the Heaviside function $H(t - 2)$ is discontinuous, is among the calculation points, *Maple* will give an error message. To avoid this difficulty we can use the stepsize option to change the stepsize so as to avoid the point $t = 2$. (Also, the smaller the stepsize the greater the accuracy.) To change the stepsize from $(10 - 0)/20 = 0.5$ to 0.3, say, insert a comma then stepsize $= 0.3$ after arrows = NONE. Note that the sequencing of options is immaterial, so we could insert the stepsize option before the arrows option if we wish.

EXERCISES 2.4

1. In Example 1, show that the time required to reduce $x(t)$ from x_0 to $x_0/2$ is $(\ln 2)(c/k)$.

2. In Example 3, show that the steady-state part of (17) can be expressed more compactly as

$$A \sin(\omega t - \phi),$$

where the **amplitude** A is

$$A = \frac{F_0}{\sqrt{k^2 + (\omega c)^2}}$$

and the **phase angle** ϕ is the angle between 0 and $\pi/2$ such that $\tan\phi = \omega c/k$. HINT: Use the trigonometric identity $\sin(a - b) = \sin a \cos b - \sin b \cos a$. NOTE: Isn't it striking that the linear combination of $\sin \omega t$ and $\cos \omega t$ in (17) is equivalent to a single sinusoid with a phase shift?

3. For the mechanical system governed by (13), where $F(t)$ is given below, determine $x(t)$ and identify the steady-state solution. Also, sketch the graph of $x(t)$ and label key values, as we have in Fig. 12.

(a) $F(t) = \begin{cases} F_0, & 0 < t < t_1 \\ 0, & t_1 < t < \infty \end{cases}$

(b) $F(t) = \begin{cases} 0, & 0 < t < t_1 \\ F_0, & t_1 < t < t_2 \\ 0, & t_2 < t < \infty \end{cases}$

4. (*Maple*) Consider the mechanical system governed by (13), where $k = 2$, $c = 1$, and x_0 and $F(t)$ are defined below. Use dsolve to solve for $x(t)$. Also, use phaseportrait to obtain a graph of $x(t)$ over the the indicated t interval. NOTE: Remember that if $F(t)$ is defined piecewise then you will first need to re-express it in terms of Heaviside functions.

(a) $F(t) = 5e^{-2t}$; $x_0 = 10$; $0 \leq t \leq 20$

(b) $F(t) = te^{-t}$; $x_0 = 3, 6$; $0 \leq t \leq 5$

(c) $F(t) = 10(1 - e^{-2t})$; $x_0 = 3, 4, 5, 6, 7$; $0 \leq t \leq 5$

(d) $F(t) = 3(1 - e^{-t})$; $x_0 = 1.0, 1.5, 2.0$; $0 \leq t \leq 10$

(e) $F(t) = 50$ for $0 < t < 5$ and 0 for $t > 5$; $x_0 = 15, 20, 25, 30, 35$; $0 \leq t \leq 10$

(f) $F(t) = 0$ for $t < 5$ and 40 for $t > 5$; $x_0 = 0$; $0 \leq t \leq 15$

(g) $F(t) = 0$ for $t < 5$ and for $t > 10$, and 25 for $5 < t < 10$; $x_0 = 2, 6$; $0 \leq t \leq 20$

(h) $F(t) = 10t$ for $0 < t < 10$ and 0 for $t > 10$; $x_0 = 10, 40$; $0 \leq t \leq 20$

(i) $F(t) = 10t$ for $0 < t < 5$, $100 - 10t$ for $5 < t < 10$ and 0 for $t > 10$; $x_0 = 0$; $0 \leq t \leq 20$

(j) $F(t)$, x_0, and t interval supplied by your instructor

5. For the RC circuit of Example 5, let $R = 10$, $C = 0.5$, $E(t) = te^{-t}$, and $i(0) = 0$. Determine $i(t)$ at $t = 1$ and at $t = 10$.

6. Same as Exercise 5, with $E(t) = 10t$.

7. For the RL circuit of Example 7, let $R = 5$, $L = 1$, $E(t) = e^{-0.3t}$, and $i(0) = 0.5$. Determine $i(t)$ at $t = 1, 5$, and 10.

8. Same as Exercise 7, with $E(t) = 10$ for $0 < t < 5$ and $E(t) = 0$ for $5 < t < \infty$.

9. A seashell contains 90% as much C-14 as a living shell of the same size. How old is it? Approximately how many years did it take for its C-14 content to diminish from its initial value to 99% of its initial value?

10. If 10 grams of some radioactive substance will be reduced to 8 grams in 60 years, in how many years will 2 grams be left? In how many years will 0.1 gram be left?

11. If 20% of a radioactive substance disappears in 70 days, what is its half-life?

12. Show that if m_1 and m_2 grams of a radioactive substance are present at times t_1 and t_2, respectively, then its half-life is

$$T = (t_2 - t_1) \frac{\ln 2}{\ln (m_1/m_2)}.$$

13. For the mixing tank shown in Fig. 19, let $Q(t) = $ constant $= Q$ and $c(0) = 0$. Solve (45) for the concentration $c(t)$, in terms of v, c_1, Q and t, and sketch the graph of $c(t)$ over $0 < t < \infty$, labeling any key values.

14. Same as Exercise 13, with $c(0) = 2c_1$.

15. In (45), let $v = c_1 = 1$, let $Q(t) = 4$ for $0 < t < 2$ and 2 for $t > 2$, and let $c(0) = 0$. Solve for $c(t)$ at $t = 0.5, 2, 3, \infty$.

16. (*Mass on an inclined plane*) The equation $mx'' + cx' = mg \sin \alpha$ governs the straight-line displacement $x(t)$ of a mass m along a plane that is inclined at an angle α with respect to the horizontal, if it slides under the action of gravity and friction. If $x(0) = 0$ and $x'(0) = 0$, solve for $x(t)$. HINT: First, integrate the equation once with respect to t to reduce it to a first-order equation. Then solve that first-order equation.

17. (*Free fall; terminal velocity*) The equation of motion of a body of mass m falling vertically under the action of a downward gravitational force mg and an upward aerodynamic drag force $f(v)$ is

$$mv' = mg - f(v), \qquad (17.1)$$

where $v(t)$ is the velocity [so that $v'(t)$ is the acceleration]. The determination of the form of $f(v)$, for the given body shape, would require either careful wind tunnel measurements, or sophisticated theoretical and/or numerical analy-

sis. All we need to know here is that for a variety of body shapes, the result of such an analysis is that $f(v)$ can be approximated (over some limited range of velocities) in the form cv^β, for suitable constants c and β. For low velocities (in fluid mechanics terminology, for low "Reynolds numbers") $\beta \approx 1$, and for high velocities (i.e., for high "Reynolds numbers") $\beta \approx 2$.

(**a**) Solve (17.1), together with the initial condition $v(0) = 0$, for the case where $f(v) \approx cv$. What is the terminal (i.e., steady-state) velocity?

(**b**) Same as (a), for $f(v) \approx cv^2$. HINT: Read Exercise 9 in Section 2.3.

18. (*Light extinction*) As light passes through window glass some of it is absorbed. If x is a coordinate normal to the glass (with $x = 0$ at the incident face) and $I(x)$ is the light intensity at x, then the fractional loss in intensity, $-dI/I$ (with the minus sign included because dI will be negative), will be proportional to dx: $-dI/I = k\, dx$, where k is a positive constant. Thus, $I(x)$ satisfies the differential equation

$$I'(x) = -kI(x). \qquad (18.1)$$

(**a**) If 80% of the light penetrates a 1-inch thick slab of this glass, how thin must the glass be to let 95% penetrate? NOTE: Your answer should be numerical, not in terms of an unknown k. You will need to solve (18.1).

(**b**) If 50% of the light penetrates three inches, how far does 25% penetrate? How far does 1% penetrate?

19. (*Newton's law of cooling*) Newton's law of cooling states that a body that is hotter than its environment will cool at a rate that is proportional to the temperature difference between the body and its environment, so that the temperature $u(t)$ of the body is governed by the differential equation

$$\frac{du}{dt} = k(U - u), \qquad (19.1)$$

where U is the temperature of the environment (assumed here to be constant), t is the time, and k is a constant of proportionality.

(**a**) Derive the general solution of (19.1),

$$u(t) = U + Ae^{-kt}. \qquad (19.2)$$

(**b**) A cup of coffee in a room that is at 70° F is initially at 200° F. After 10 minutes it has cooled to 180° F. How long will it take to cool to 100° F? NOTE: You may use (19.2).

(**c**) An interesting application of (19.1) and its solution (19.2) is in connection with the determination of the time of death in a homicide. For instance, suppose that a body is discovered at a time T after death and its temperature is measured to be 90° F. We wish to solve for T. Suppose that the ambient temperature is $U = 70°$ F and assume that the temperature of the body at the time of death was $u_0 = 98.6°$ F. Putting this information into (19.2) we can solve for T, provided that we know k, but we don't. Proceeding indirectly, we can infer the value of k by taking one more temperature reading. Thus, suppose that we wait an hour and again measure the temperature of the body, and find that $u(T + 1) = 87°$ F. Use this information to solve for T (in hours).

20. (*Compound interest*) Suppose that a sum of money earns interest at a rate k, compounded yearly, monthly, weekly, or even daily. If it is compounded continuously, then $dS/dt = kS$, where $S(t)$ denotes the sum at time t. If $S(0) = S_0$, then the solution is

$$S(t) = S_0 e^{kt}. \qquad (20.1)$$

Instead, suppose that interest is compounded yearly. Then after t years

$$S(t) = S_0(1 + k)^t,$$

and if the compounding is done n times per year, then

$$S(t) = S_0 \left(1 + \frac{k}{n}\right)^{nt}. \qquad (20.2)$$

(a) Show that if we let $n \to \infty$ in (20.2), then we do recover the continuous compounding result (20.1). HINT: Recall, from the calculus, that

$$\lim_{m \to \infty} \left(1 + \frac{1}{m}\right)^m = e.$$

(b) Let $k = 0.05$ (i.e., 5% interest) and compare $S(t)/S_0$ after 1 year ($t = 1$) if interest is compounded yearly, monthly, weekly, daily, and continuously.

21. (*Drug delivery*) A certain drug is known to be helpful if its concentration C in the blood is above some minimum level C_1, but harmful if it exceeds some upper level C_2. Further, the concentration $C(t)$ is known to decrease at a rate that is proportional to $C(t)$, where t is the time, so

$$\frac{dC}{dt} = -kC, \qquad (21.1)$$

where C is an empirical constant. As an approximation, we can suppose that if the drug is delivered intravenously then it diffuses through the circulatory system so rapidly that the increase of C to its new level is immediate. The idea, then, is to increase C, initially, to $C(0) = C_2$ and, when it diminishes to C_1, to increase it again, intravenously, to C_2, and so on. Determine the time interval between intravenous doses, in terms of k, C_1, and C_2.

22. (*Pollution in a river*) Suppose that a pollutant is discharged into a river at a steady rate Q (grams/second) over a distance L, as sketched in the figure, and we wish to determine the distribution of pollutant in the river—that is, its concentration c (grams/meter^3).

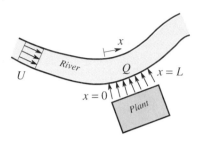

Measure x as arc length along the river, positive downstream. The river flows with velocity U (meters/second) and has a cross-sectional area A (meters^2), both of which, for simplicity, we assume to be constant. Also for simplicity, suppose that c is a function of x only. That is, it is a constant over each cross section of the stream. This is evidently a poor approximation near the interval $0 < x < L$, where we expect appreciable across-stream and vertical variations in c, but it should suffice if we are concerned mostly with the far field, that is, more than several river widths upstream or downstream of the plant. Then it can be shown that $c(x)$ is governed by the differential equation

$$kc'' - Uc' - \beta c = -\frac{Q(x)}{A}, \quad (-\infty < x < \infty) \quad (22.1)$$

where k (meters^2/second) is a diffusion constant, β (grams per second per gram) is a chemical decay constant, and $Q(x)$ is the constant Q over $0 < x < L$ and 0 outside that interval. [Physically, (22.1) expresses a mass balance between the *input* $-Q(x)/A$, the transport of pollutant by *diffusion*, kc'', the transport of pollutant by *convection* with the moving stream, Uc', and by disappearance through *chemical decay*, βc.] We assume that the river is clear upstream; that is, we have the initial condition $c(-\infty) = 0$.

(a) Let $L = \infty$. Suppose that k is sufficiently small so that we can neglect the diffusion term. Then (22.1) reduces to the linear first-order equation

$$Uc' + \beta c = \frac{Q(x)}{A}, \qquad (22.2)$$

where $Q(x) = 0$ for $x < 0$ and $Q(x) = Q$ for $x > 0$. Solve the latter for $c(x)$ and sketch its graph over $-\infty < x < \infty$, labeling any key values.

(b) Repeat part (a) for the case where L is finite.

CHAPTER 2 REVIEW

To get started in our study of differential equations we began in this chapter by limiting our attention to the **linear first-order equation**

$$y' + p(x)y = q(x). \tag{1}$$

Considering first the **homogeneous** version of (1), where $q(x) = 0$,

$$y' + p(x)y = 0, \tag{2}$$

we solved by **separation of variables** and obtained the **general solution**

$$y(x) = Ae^{-\int p(x)\,dx}, \tag{3}$$

where the constant A is arbitrary. Or, appending an initial condition $y(a) = b$ to (2), we obtained, for the **initial value problem**

$$y' + p(x)y = 0; \quad y(a) = b, \tag{4}$$

the solution

$$y(x) = be^{-\int_a^x p(s)\,ds}, \tag{5}$$

which solution exists and is unique on the broadest x interval, containing the initial point a, on which $p(x)$ is continuous.

Turning to the more difficult nonhomogeneous equation (1), we solved by two alternate methods, the **integrating factor method** and Lagrange's method of **variation of parameters**, obtaining the general solution

$$y(x) = e^{-\int p(x)\,dx} \left(\int e^{\int p(x)\,dx} q(x)\,dx + C \right). \tag{6}$$

Or, appending an initial condition $y(a) = b$ to (1) we obtained, for the initial value problem

$$y' + p(x)y = q(x); \quad y(a) = b, \tag{7}$$

the solution

$$y(x) = e^{-\int_a^x p(s)\,ds} \left(\int_a^x e^{\int_a^s p(u)\,du} q(s)\,ds + b \right), \tag{8}$$

which solution exists and is unique on the broadest x interval, containing the initial point a, on which $p(x)$ and $q(x)$ are continuous.

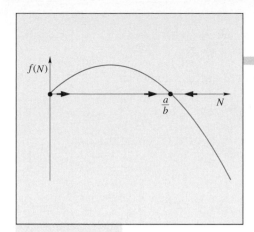

GENERAL FIRST-ORDER EQUATIONS

3.1 INTRODUCTION

In this chapter we complete the discussion of first-order equations that was begun in Chapter 2. There we studied only the linear case

$$y' + p(x)y = q(x); \tag{1}$$

here we consider the general case

$$F(x, y, y') = 0. \tag{2}$$

If we can solve (2) for y' then it is convenient to re-express (2) in the form

$$\boxed{y' = f(x, y),} \tag{3}$$

which form we take as our starting point in this chapter.

E X A M P L E 1
The equation

$$e^x y' = x + y^3 y' + 5, \tag{4}$$

or, equivalently

$$F(x, y, y') \equiv e^x y' - x - y^3 y' - 5 = 0, \tag{5}$$

can be solved by algebra for y', giving

$$y' = \frac{x+5}{e^x - y^3} \equiv f(x, y). \tag{6}$$

Evidently, (4) cannot be gotten into the form (1), so it is nonlinear.   ■

EXAMPLE 2
In contrast with (4), the equation

$$xy' - y = e^{y'} + 4 \tag{7}$$

or, equivalently,

$$F(x, y, y') = xy' - y - e^{y'} - 4 \tag{8}$$

cannot be solved by algebra for y' in terms of x and y. However, such examples are rarely encountered in applications, so we sacrifice little when we limit our attention to differential equations of the form (3).   ■

Our objectives in this chapter, as in Chapter 2, are to study

1. Methods of solution;

2. The questions of the existence and uniqueness of solutions;

3. Applications.

In addition, we will be able to clarify

4. The important consequences of a differential equation being linear or nonlinear.

As to the methods of solution, we will rely chiefly on the method of separation of variables that was used in Section 2.2 for the linear homogeneous equation $y' + p(x)y = 0$, and on a generalization of the integrating factor method that was used in Section 2.3 for the linear nonhomogeneous equation $y' + p(x)y = q(x)$.

3.2 SEPARABLE EQUATIONS

If $f(x, y)$ in

$$\frac{dy}{dx} = f(x, y) \tag{1}$$

can be factored as a function of x times a function of y, so that (1) can be written as

$$\boxed{\frac{dy}{dx} = X(x)Y(y),} \tag{2}$$

then we say that the differentiable equation (1) is **separable**. For instance, $y' = xe^{x+2y}$ is separable because it can be written as $y' = (xe^x)(e^{2y})$, whereas it is evident that $y' = 3x + y$ is not.

To solve (2), separate the variables x and y by multiplying (2) by dx and dividing through by $Y(y)$ [tentatively assuming that $Y(y) \neq 0$ on the y domain of interest], and then integrate both sides. Those steps give[1]

$$\boxed{\int \frac{dy}{Y(y)} = \int X(x)\,dx.}$$ (3)

Hence the solution method is known as **separation of variables**.

EXAMPLE 1

Solve

$$y' = 2(x - 1)e^{-y}.$$ (4)

Separating the variables and integrating gives

$$\int e^y\,dy = 2\int (x - 1)\,dx,$$ (5)

so

$$e^y + C_1 = x^2 - 2x + C_2,$$ (6)

where C_1, C_2 are the integration constants. Equivalently and more simply,

$$e^y = x^2 - 2x + C$$ (7)

where $C = C_2 - C_1$, so we obtain the general solution

$$y(x) = \ln(x^2 - 2x + C),$$ (8)

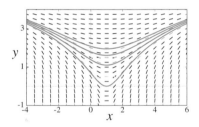

FIGURE 1
Representative members of the family of solutions given by (8).

where C is an arbitrary constant. The latter is a whole "family" of solutions, so it is of interest to plot at least a few of them, as we have done in Fig. 1 for the representative values $C = 2, 4, 6, 8$; we have included the direction field as well, to suggest the overall "flow."

If, in addition to (4), we have an initial condition such as $y(1) = 2$, then we have an initial value problem

$$y' = 2(x - 1)e^{-y}; \quad y(1) = 2.$$ (9)

---

[1] Observe that in moving the dx in (2) to the right-hand side we have treated dy/dx as a fraction, the ratio of two numbers, whereas it is actually a single quantity defined in the calculus as the limit of a difference quotient. Nevertheless, we can justify the result (3) by showing that the derivative of (3) does give us (2). The chain rule and the fundamental theorem of the integral calculus give

$$\frac{d}{dx}\left(\int \frac{dy}{Y(y)}\right) = \frac{d}{dy}\left(\int \frac{dy}{Y(y)}\right)\frac{dy}{dx} = \frac{1}{Y(y)}\frac{dy}{dx}$$

and

$$\frac{d}{dx}\left(\int X(x)\,dx\right) = X(x)$$

if $1/Y(y)$ and $X(x)$ are continuous, so differentiating (3) does give (2). Hence (3) and (2) differ by at most an arbitrary additive constant, but the integrals in (3) give arbitrary constants anyway.

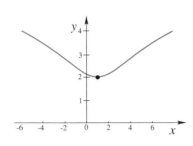

FIGURE 2
Particular solution (11) of the
initial value problem (9).

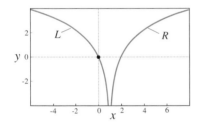

FIGURE 3
Particular solution (13)
corresponding to the initial
condition $y(0) = 0$.

Since the general solution of the differential equation in (9) is given by (8), we impose the initial condition on that solution:

$$y(1) = 2 = \ln(1 - 2 + C), \tag{10}$$

which gives $C = 1 + e^2$. Hence the particular solution of (9) is

$$
\begin{aligned}
y(x) &= \ln(x^2 - 2x + 1 + e^2) \\
&= \ln\left[(x-1)^2 + e^2\right], \tag{11}
\end{aligned}
$$

the graph of which is shown in Fig. 2.

Over what x interval is (11) a valid solution[1] of (9)? Recalling that the logarithm tends to $-\infty$ (i.e., it "blows up") as its argument tends to zero, we need to see if the argument in (11) is zero anywhere. Indeed it is not, its smallest value being e^2 at $x = 1$, so the solution (11) is valid on the entire axis, $-\infty < x < \infty$.

However, a different initial condition might lead to a particular solution that is valid only on a part of the x axis. For instance, the initial condition

$$y(0) = 0 \tag{12}$$

yields $C = 1$ and hence the particular solution

$$y(x) = \ln(x^2 - 2x + 1) = \ln(x - 1)^2 = 2\ln|x - 1|, \tag{13}$$

displayed in Fig. 3. Since $2\ln|x - 1| \to -\infty$ as $x \to 1$, we can think of the graph as consisting of two branches, one to the left of $x = 1$ (labeled L in Fig. 3) and one to the right of $x = 1$ (labeled R). The solution through $(0, 0)$ can be extended arbitrarily far to the left, along L, but cannot be extended to the right beyond $x = 1$ since the solution (13) fails to exist at $x = 1$. Thus, the solution through $(0, 0)$ exists only on $-\infty < x < 1$. If, on the other hand, we specified an initial point anywhere on the right-hand branch R, then the solution (13) would exist only on $1 < x < \infty$ and its graph would be given by R.

Finally, recall the tentative assumption that $Y(y) \neq 0$, made between (2) and (3). In the present example $Y(y) = e^{-y}$, which is indeed nonzero for all finite values of y. ∎

Thus, we are interested not only in solution technique but also in the questions of the existence and uniqueness of solutions. The solution technique considered here, separation of variables, is extremely simple, merely amounting to the steps given in equations (1)–(3). Of course, in a given application one or both of the integrals in (3) might be difficult to evaluate; in Example 1 both were readily evaluated. The questions of the existence and uniqueness of solutions will be the subject of Section 3.3, but since that section is optional we will give a brief and pragmatic discussion in the present section, which discussion we began in Example 1.

---

[1] By including the adjective "valid" we do not mean to make a technical distinction between a "solution" and a "valid solution;" it is used only for emphasis.

Observe a striking difference between linear and nonlinear first-order equations. For the linear equation

$$y' + p(x)y = q(x) \tag{14}$$

we can predict, even before we solve for $y(x)$, that the solution through a given initial point can be extended to the left and right over the broadest interval over which $p(x)$ and $q(x)$ are continuous. As seen in Example 1, the issue is more subtle for nonlinear equations. Think of it this way: If we re-express the nonlinear equation (4) in the form

$$y' + \left(\frac{2(1-x)e^{-y}}{y}\right)y = 0 \tag{15}$$

then it is of the form (14) where

$$p(x) = \frac{2(1-x)e^{-y(x)}}{y(x)} \tag{16}$$

and $q(x) = 0$. The problem is that $p(x)$ contains the not-yet-known solution $y(x)$, so we can not tell in advance the extent of the x interval, containing the initial point, over which $p(x)$ is continuous.

EXAMPLE 2
As another illustration of these ideas consider the equation

$$y' = -y^2. \tag{17}$$

Separating the variables gives

$$\int \frac{dy}{y^2} = -\int dx$$

and if we integrate, combine the two integration constants into one, and solve for y, we obtain the general solution

$$y(x) = \frac{1}{x+C}. \tag{18}$$

If we impose an initial condition $y(0) = y_0$ then we can solve for C and obtain the particular solution

$$y(x) = \frac{1}{x+1/y_0}, \tag{19}$$

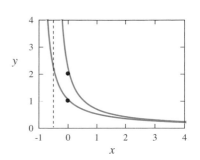

FIGURE 4
Particular solutions given by (19), for $y_0 = 1$ and $y_0 = 2$.

which is plotted in Fig. 4 for the representative values $y_0 = 1$ and $y_0 = 2$. The solution through the initial point $(0, 1)$ exists over $-1 < x < \infty$, and the one through $(0, 2)$ exists over $-1/2 < x < \infty$. More generally, the one through $(0, y_0)$ exists over $-1/y_0 < x < \infty$ because the denominator in (19) becomes zero at $x = -1/y_0$. We could plot (19) to the left of that point as well, but such extension of the graph would be meaningless because the point $x = -1/y_0$ serves as a "barrier;" y and y' fail to exist there, so the solution cannot be continued to the left of that point. ∎

EXAMPLE 3

Solve the initial value problem

$$y' = \frac{4x}{1 + 2e^y}; \qquad y(0) = 1. \tag{20}$$

Separating variables and integrating gives

$$\int \left(1 + 2e^y\right) dy = \int 4x \, dx, \tag{21}$$

$$y + 2e^y = 2x^2 + C. \tag{22}$$

Unfortunately, the latter is a transcendental equation[1] in y and, evidently, we cannot solve it for y explicitly (i.e., in closed form) as a function of x, as we were able to solve (7), for instance. Nevertheless, we can impose the initial condition on (22) to evaluate C: $1 + 2e = 0 + C$, so $C = 1 + 2e$ and the solution is given by

$$y + 2e^y = 2x^2 + 1 + 2e. \tag{23}$$

Since we are not able to solve the latter for y, we accept it as it is, a **relation** on x and y that defines $y(x)$ only implicitly rather than explicitly. Thus, we say that (23) is the solution in **implicit** form rather than in (the more desirable) explicit form.

In spite of its implicit form, (23) can be used to obtain a computer plot of y versus x, as discussed in the *Maple* subsection at the end of this section. The result is shown in Fig. 5, along with the direction field.

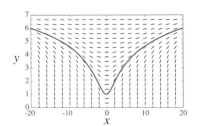

FIGURE 5
The solution (23) of Problem (20).

COMMENT. What is the interval of existence of the solution? In Examples 1 and 2 we were able to ascertain that interval by direct examination of the solution. Here, such examination is not possible because the solution (23) is not in explicit form. It appears, from Fig. 5, that the solution exists for all x, but we can't be sure since Fig. 5 covers only $-20 < x < 20$. However, (23) reveals the asymptotic behavior $2e^y \sim 2x^2$, or $y \sim 2\ln|x|$, as $x \to \pm\infty$, so it seems clear that the solution continues to vary smoothly as $|x|$ increases. Of course this approach has been informal, not rigorous. ∎

Closure. If the equation $y' = f(x, y)$ is in the separable form

$$y' = X(x)Y(y), \tag{24}$$

then solution by the method of separation of variables amounts to dividing both sides by $Y(y)$, multiplying both sides by dx, and integrating:

$$\int \frac{dy}{Y(y)} = \int X(x) \, dx. \tag{25}$$

---

[1] An equation $f(y) = 0$ is said to be **algebraic** if f is a finite-degree polynomial and is **transcendental** otherwise. Equation (22) is transcendental in y (though it is algebraic in x) because of the e^y term, which is the infinite-degree polynomial $1 + y + y^2/2! + y^3/3! + \cdots$. Some algebraic equations can be solved explicitly and some cannot; likewise, some transcendental equations (such as $e^y = 1$) can be solved and some [such as (22)] cannot.

There are two potential difficulties. First, one or both integrals might be too difficult to evaluate. Second, (25) gives the solution in implicit form and we might not be able to solve for y explicitly.

Maple.   The *Maple* commands used to produce Fig. 1 were

```
with(DEtools):
phaseportrait(diff(y(x),x)=2*(x-1)*exp(-y),y(x),x=-4..6,
{[0,ln(2)],[0,ln(4)],[0,ln(6)],[0,ln(8)]},stepsize=.05,
y=-1..4,arrows=LINE);
```

For Figs. 2 and 3 we modified the phaseportrait part as follows: for Fig. 2 we used

```
phaseportrait(diff(y(x),x)=2*(x-1)*exp(-y),y(x),x=-6..8,
{[1,2]}, stepsize=.05,y=0..4,arrows=NONE);
```

and for Fig. 3 we used

```
phaseportrait(diff(y(x),x)=2*(x-1)*exp(-y),y(x),x=-4..6,
{[0,0],[2,0]},stepsize=.05,y=-4..4,arrows=NONE);
```

Regarding the latter command, observe that corresponding to the initial point $[0, 0]$ the solution is given by (13), the graph of which is given in Fig. 3. However, if we use only the initial point $[0, 0]$ in the phaseportrait command we obtain only the left-hand branch of that graph (labeled L in Fig. 3). To pick up the right-hand branch R as well, we need to specify an initial point on that branch; we used the point $[2, 0]$. Hence, the two initial points, $[0, 0]$ and $[2, 0]$, are listed in the phaseportrait command.

To generate Fig. 5, we could have used the implicitplot command on the implicit solution (23) [in which case the e at the end of (23) would be entered as $\exp(1)$], but chose to use phaseportrait again, so we could display the direction field.

Besides using the dsolve command to obtain analytical solutions and phaseportrait to obtain plots of the solution we can use the dsolve command with a numeric option to obtain a table of values of the solution. To illustrate, let us solve (20) for the values of $y(x)$ at $x = -20, -10, 0,$ and 15, $x = 0$ being included just as a check because $y(0)$ should equal 1 because of the initial condition $y(0) = 1$. Enter

```
dsolve({diff(y(x),x)=4*x/(1+2*exp(y(x))),y(0)=1},y(x),
type=numeric,value=array([-20,-10,0,15]));
```

and return. The printed result is

$$
\left[\begin{array}{c}
\left[\begin{array}{cc}
 & [x, y(x)] \\
-20 & 5.992020060 \\
-10 & 4.614240533 \\
0 & 1 \\
15 & 5.418360508
\end{array} \right]
\end{array} \right]
$$

which values are, as far as our eyes can discern, in agreement with the plot in Fig. 5.

EXERCISES 3.2

NOTE: Solutions should be expressed in explicit form if possible.

1. Use separation of variables to find the general solution. Then, obtain the particular solution satisfying the given initial condition. Sketch the graph of the solution, showing the key features, and label any key values.

   (a) $y' - 3x^2 e^{-y} = 0$; $y(0) = 0$

   (b) $y' = 6x^2 + 5$; $y(0) = 0$

   (c) $y' + 4y = 0$; $y(-1) = 0$

   (d) $y' = 1 + y^2$; $y(0) = 1$

   (e) $y' = (y^2 - y)e^x$; $y(0) = 2$

   (f) $y' = y^2 + y - 6$; $y(1) = 1$

   (g) $y' = y(y + 3)$; $y(0) = -4$

   (h) $xy' = 6y \ln y$ $y(1) = e$

   (i) $y' = e^{x+2y}$; $y(0) = 1$

   (j) $2xy' = y$; $y(3) = -1$

   (k) $y' + 3y(y + 1)\sin 2x = 0$; $y(0) = 0$

   (l) $y = \ln y'$; $y(0) = 5$

   (m) $(1 + e^y)y' = 2x$; $y(0) = 0$

   (n) $y' + 2y = y^2 + 1$; $y(-3) = 0$

   (o) $y' = \dfrac{y+1}{x+1}$; $y(0) = 1$

   (p) $(1 + y)y' = e^{x-y}$; $y(0) = 0$

   (q) $y' = y(y - 2)$; $y(0) = 4$

   (r) $xy' = 1 + y$; $y(5) = 3$

2. (a)–(r) For the problem given in the corresponding part of Exercise 1, use *Maple* to solve for $y(x)$ and to plot the graph of $y(x)$ over a reasonably chosen x interval. Verify, by hand, that the *Maple* solution does satisfy the differential equation and initial condition.

3. In general, the Bernoulli equation $y' + p(x)y = q(x)y^n$ is not variable separable, but it is if p and q are constants, if one of the functions $p(x)$ and $q(x)$ is zero, if one is a constant times the other, or if $n = 1$. Obtain the general solution for the case where each of $p(x)$ and $q(x)$ is a nonzero constant, for any real number n other than 0 or 1 (because in those cases the equation is merely first-order linear, which case was solved in Chapter 2). HINT: A difficult integral will occur. Our discussion of the Bernoulli equation in the exercises for Section 2.3 should help you to find a change of variables that will simplify that integration.

4. (*Homogeneous functions*) A function $f(x_1, \ldots, x_n)$ is said to be **homogeneous of degree k** if $f(\lambda x_1, \ldots, \lambda x_n) = $

$\lambda^k f(x_1, \ldots, x_n)$ for any λ. For example,

$$f(x, y, z) = \frac{4x^5}{y^2 + 3z^2}\sin\left(\frac{y}{z}\right)$$

is homogeneous of degree 3 because

$$f(\lambda x, \lambda y, \lambda z) = \frac{4(\lambda x)^5}{(\lambda y)^2 + 3(\lambda z)^2}\sin\left(\frac{\lambda y}{\lambda z}\right)$$

$$= \lambda^3 f(x, y, z).$$

State whether f is homogeneous or not. If it is, determine its degree.

   (a) $f(x, y) = x^2 + 4y^2 - 7$

   (b) $f(x, y, z) = \cos\left(\dfrac{x - y}{5z}\right)$

   (c) $f(x, y, z) = x^2 - y^2 + 7xz - 3xy$

   (d) $f(x, y) = \sin\left(x^2 + y^2\right)$

5. (*Homogeneous equation*) The equation

$$\boxed{y' = f\left(\frac{y}{x}\right)} \tag{5.1}$$

is said to be **homogeneous** because $f(y/x)$ is homogeneous (of degree zero); see the preceding exercise. CAUTION: The term homogeneous is also used to describe a linear differential equation that has zero as its "forcing function" on the right-hand side, as defined in Section 1.2. Thus, one needs to use the context to determine which meaning is intended.

   (a) Show, by examples, that (5.1) may, but need not, be separable.

   (b) In any case, show that the change of dependent variable $w = y/x$, from $y(x)$ to $w(x)$, reduces (5.1) to the separable form

$$w' = \frac{f(w) - w}{x}. \tag{5.2}$$

6. Use the idea contained in the preceding exercise, to find the general solution to each of the following equations.

   (a) $y' = \dfrac{2x + 2y}{3x + y}$

   (b) $y' = \dfrac{4x^2 + 3y^2}{2xy}$

   (c) $y' = \dfrac{xy + 2y^2}{x^2}$

   (d) $y' = -\dfrac{2x + y}{x}$

   (e) $y' = \sqrt{1 + (y/x)^2} + (y/x)$

7. (*Almost-homogeneous equation*)

    (a) Show that

$$y' = \frac{ax + by + c}{dx + ey + f} \quad (a, b, \dots, f \text{ constants}) \quad (7.1)$$

    can be reduced to homogeneous form by the change of variables $x = u + h$, $y = v + k$, where h and k are suitably chosen constants, provided that $ae - bd \neq 0$.

    (b) Thus, find the general solution of $y' = (2x + 2y - 6)/(3x + y - 7)$.

    (c) Similarly, for $y' = (x - y - 4)/(x + y - 4)$.

    (d) Devise a method of solution that will work in the exceptional case where $ae - bd = 0$, and apply it to the case $y' = (x + 2y - 3)/(2x + 4y - 1)$.

8. (*Algebraic, exponential, and explosive growth*) We saw, in Example 1 of Section 2.2, that the population model

$$\frac{dN}{dt} = \kappa N \qquad (\kappa > 0) \qquad (8.1)$$

gives the exponential growth $N(t) = Ae^{kt}$, whereby $N \to \infty$ as $t \to \infty$ (if $A > 0$). More generally, consider the model

$$\frac{dN}{dt} = \kappa N^p, \qquad (\kappa > 0) \qquad (8.2)$$

where p is a positive constant. Our purpose in this exercise is to examine how the rate of growth of $N(t)$ varies with the exponent p in (8.2).

    (a) Solve (8.2) and show that if $0 < p < 1$ then the solution exhibits **algebraic growth** [i.e., $N(t) \sim \alpha t^\beta$ as $t \to \infty$].

    (b) Show that as $p \to 0$ the exponent β tends to unity, and as $p \to 1$ the exponent β tends to infinity. (Of course, when $p = 1$ we then have **exponential growth**, as mentioned above, so we can think—crudely—of exponential growth as a limiting case of algebraic growth, in the limit as the exponent β becomes infinite. Thus, exponential growth is powerful indeed.)

    (c) If p is increased beyond 1 then we expect the growth to be even more spectacular. Show that if $p > 1$ then the solution exhibits **explosive growth**, explosive in the sense that not only does $N \to \infty$, but it does so in *finite* time, as $t \to T$, where

$$T = \frac{1}{\kappa(p - 1)N_0^{p-1}}; \qquad (8.3)$$

N_0 denotes the initial value $N(0)$. Observe that T diminishes as p increases.

3.3 EXISTENCE AND UNIQUENESS (OPTIONAL)

Recall, from Theorem 2.3.1, that the linear equation $y' + p(x)y = q(x)$ does admit a solution through an initial point $y(a) = b$ if $p(x)$ and $q(x)$ are continuous at a; that solution exists at least on the broadest x interval containing $x = a$, on which p and q are continuous, and is unique. What can be said about existence and uniqueness for the more general equation

$$y' = f(x, y)$$

which could be linear but in general is not? We have the following theorem.

THEOREM 3.3.1

Existence and Uniqueness

If $f(x, y)$ is a continuous function of x and y on some rectangle (Fig. 1)

$$R: \qquad |x - a| < A, \quad |y - b| < B,$$

then the initial value problem

$$y' = f(x, y); \qquad y(a) = b \qquad (1)$$

has at least one solution defined on some open x interval[1] containing $x = a$. If, in addition, $\partial f/\partial y$ is continuous on R, then the solution to (1) is unique on some open x interval containing $x = a$.

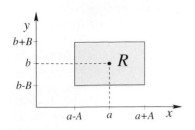

FIGURE 1
The rectangle R in Theorem 3.3.1.

Notice that whereas Theorem 2.3.1 predicts a minimum interval of existence and uniqueness for the linear equation, Theorem 3.3.1 merely ensures existence and uniqueness for the general equation $y' = f(x, y)$ over *some* interval; it gives no clue as to how broad that interval will be. Thus, for the general equation $y' = f(x, y)$ we say that Theorem 3.3.1 is a *local* result; it tells us that under the stipulated conditions all is well locally, in some neighborhood of $x = a$. Stronger (i.e., more informative) theorems could be cited, but this one will suffice here.

Let us illustrate Theorem 3.3.1 with three examples.

E X A M P L E   1
Consider the equation

$$y' = \frac{y - 1}{x}. \tag{2}$$

Solving by separation of variables,

$$\int \frac{dy}{y - 1} = \int \frac{dx}{x}, \tag{3a}$$

$$\ln |y - 1| = \ln |x| + A, \tag{3b}$$

$$\ln \left| \frac{y - 1}{x} \right| = A, \tag{3c}$$

$$\left| \frac{y - 1}{x} \right| = e^A, \tag{3d}$$

$$\frac{y - 1}{x} = \pm e^A \equiv C, \tag{3e}$$

$$y = 1 + Cx, \tag{3f}$$

where A is an arbitrary integration constant and where, in (3e), we set $C = \pm e^A$, for brevity. Thus, the general solution of (2) is given by (3f), where C is an arbitrary constant, so the solution curves are the set of all straight lines through the point $(0, 1)$ (Fig. 2), except that the y axis is not an acceptable solution curve since the slope y' is infinite all along that line.

Applying Theorem 3.3.1 to (2), we see that both $f(x, y) = (y - 1)/x$ and $\partial f/\partial y = 1/x$ are continuous everywhere in the plane except on the line $x = 0$ (the y axis). Thus, through any initial point *not* on the y axis there exists a solution and that solution is unique. For instance, through the initial point P there is the unique solution the graph of which is the line L.

What does the theorem tell us about the existence and uniqueness of solutions through initial points *on* the y axis? It provides no information since the conditions of the theorem are not met there. Thus, through a point on the y axis there might be no solution, a unique solution, or a nonunique solution. From Fig. 2 we can see that

---

[1]By an **open interval** we mean $x_1 < x < x_2$, and by a **closed interval** we mean $x_1 \leq x \leq x_2$. Thus, a closed interval includes its endpoints, an open x interval does not. It is common to use the notation (x_1, x_2) and $[x_1, x_2]$ for open and closed intervals, respectively. Further, $(x_1, x_2]$ means $x_1 < x \leq x_2$, and $[x_1, x_2)$ means $x_1 \leq x < x_2$.

through the point (0, 1) there is a nonunique solution, namely, the family of straight lines other than the y axis itself, but through any other point on the y axis there is no solution.

COMMENT 1. Equation (2) happens to be linear, with "$p(x)$"$= -1/x$ and "$q(x)$"$= -1/x$, so we could, alternatively, have used Theorem 2.3.1.

COMMENT 2. Example 5 in Section 2.3 was similar to the present example since both $f(x, y) = (6x^3 - 3y)/x$ and $\partial f/\partial y = -3/x$, in that example, were continuous everywhere in the plane except along the line $x = 0$ (the y axis). But in that example the result was different: Through any point on the y axis, other than the origin, there was no solution, and through the origin there was a unique solution. ■

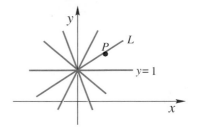

FIGURE 2

The solution curves (3f) for equation (2).

EXAMPLE 2

The equation

$$y' = -\frac{x}{4y} \tag{4}$$

is readily solved by separation of variables. We obtain the solution in implicit form as

$$x^2 + 4y^2 = C \tag{5}$$

or, solving for y,

$$y(x) = \pm\frac{1}{2}\sqrt{C - x^2}. \tag{6}$$

From (5) we see that the integral curves are a family of ellipses, several of which are shown in Fig. 3.

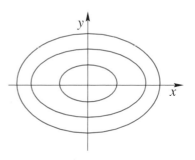

FIGURE 3

The solution curves (5) for equation (4).

Applying Theorem 3.3.1 to (4), observe that both $f(x, y) = -x/(4y)$ and $\partial f/\partial y = x/(4y^2)$ are continuous everywhere in the plane except on the line $y = 0$ (the x axis). Thus, through any initial point not on the x axis there exists a solution and that solution is unique. For instance, through the point (3, 1) there is the unique solution

$$y = +\frac{1}{2}\sqrt{13 - x^2}, \tag{7}$$

where we have solved for C by putting (3,1) into (5). That is, the solution curve is the upper branch of the ellipse $x^2 + 4y^2 = 13$ on $-\sqrt{13} < x < \sqrt{13}$ (Fig. 4). Similarly, through (3, −1) there is the unique solution

$$y = -\frac{1}{2}\sqrt{13 - x^2}$$

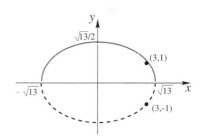

FIGURE 4

The solution curve (7) through (3, 1).

on $-\sqrt{13} < x < \sqrt{13}$, corresponding to the lower branch of the ellipse.

However, if the initial point is *on* the x axis then the theorem does not guarantee the existence or uniqueness of a solution. In fact, we can see from Fig. 3 that if the initial point is on the x axis then there is no solution through that point because the slope is infinite there; that is, y' is undefined there, so the differential equation cannot be satisfied there. ■

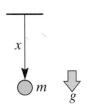

FIGURE 5
Free fall.

E X A M P L E 3 *Free Fall*.

This time consider a physical application. Suppose that a body of mass m is dropped, from rest, at time $t = 0$. With its displacement $x(t)$ measured downward from the point of release, the equation of motion is $mx'' = mg$, where g is the acceleration due to gravity and t is the time (Fig. 5). Thus,

$$x'' = g, \qquad (0 \le t < \infty) \tag{8a}$$
$$x(0) = 0, \tag{8b}$$
$$x'(0) = 0. \tag{8c}$$

Equation (8a) is of second order, whereas this chapter is about first-order equations, but it is readily integrated twice with respect to t. Doing so, and invoking (8b) and (8c) gives the solution

$$x(t) = \frac{g}{2}t^2, \tag{9}$$

which result is probably familiar to you from a first course in physics.

However, instead of multiplying (8a) through by dt and integrating on t, let us multiply it by dx and integrate on x. Then $x'' dx = g\,dx$ and since, from the calculus,

$$x'' dx = \frac{dx'}{dt}dx = \frac{dx'}{dt}\frac{dx}{dt}dt = x'\frac{dx'}{dt}dt = x' dx', \tag{10}$$

$x'' dx = g\,dx$ becomes

$$x' dx' = g\,dx. \tag{11}$$

Integrating (11) gives

$$\frac{1}{2}(x')^2 = gx + A, \tag{12}$$

and $x(0) = x'(0) = 0$ imply that $A = 0$. Thus, we have reduced (8) to the *first*-order problem

$$x' = \sqrt{2g}\,x^{1/2}, \qquad (0 \le t < \infty) \tag{13a}$$
$$x(0) = 0, \tag{13b}$$

which shall be our starting point in this example. Equation (13a) is separable and readily solved. The result is the general solution

$$x(t) = \frac{1}{4}\left(\sqrt{2g}\,t + C\right)^2, \tag{14}$$

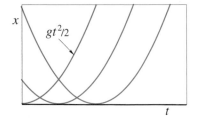

FIGURE 6

The parabolic solution curves given by (14).

which gives, as the solution curves in the x, t plane, a family of parabolas, several of which are shown in Fig. 6, with their "noses" on the t axis. Finally, applying the initial condition (13b) to (14) gives $C = 0$, so (14) gives $x(t) = gt^2/2$, which corresponds to the half-parabola through the origin in Fig. 6 and which is identical to (9).

By inspection, however, we see that another solution of (13) is simply $x(t) = 0$, because the latter reduces (13a) to the identity $0 = 0$ and also satisfies (13b); the graph of that solution is the positive t axis. The result $x(t) = 0$ seems unrealistic because it amounts to the mass levitating, yet it does satisfy (13). Further, there is no violation of Theorem 3.3.1 because (with x, y changed to t, x, respectively, in this example) $\partial f/\partial x = \partial(\sqrt{2g}x^{1/2})/\partial x = \sqrt{2g}/(2\sqrt{x})$ is not continuous in any neighborhood of the initial point $(0, 0)$ because it "blows up" to infinity at $x = 0$. Thus, the theorem gives no guarantee of the uniqueness of a solution to (13).

To understand the mysterious levitation solution, consider the physics. When we multiply force by distance we get work, and work shows up (in a system without dissipation, as in this example) as energy. Thus, multiplying (8a) by dx and integrating converted the original force equation (Newton's second law) to an energy equation. That is, (12) tells us that the total energy (kinetic plus potential) is conserved; it is constant for all time: $(x')^2/2 + (-gx) = $ constant or, equivalently,

$$\frac{1}{2}m(x')^2 + (-mgx) = A, \tag{15}$$

$$\text{Kinetic energy} + \text{Potential energy} = \text{Constant}.$$

Since $x(0) = x'(0) = 0$, the total energy A is zero. When the mass falls, its kinetic energy becomes positive and its potential energy becomes negative such that their total remains zero for all $t > 0$. However, the energy equation is also satisfied if the released mass levitates for all $t \geq 0$ (or, indeed, if it levitates for any amount of time and then falls, as considered in the exercises). Thus, our additional solution $x(t) = 0$ is indeed physically meaningful in that it does satisfy the requirement of conservation of energy. Observe, however, that it does not satisfy the equation of motion (8a) since the insertion of $x(t) = 0$ into that equation gives $0 = g$. Thus, the spurious additional solution $x(t) = 0$, of (13), must have entered somewhere between (8) and (13). In fact, we introduced it inadvertently when we multiplied (8a) by dx because $x''dx = g\,dx$ is satisfied not only by $x'' = g$, but also by $dx = 0$ and hence by $x(t) = $ constant.

The upshot is that although the solution to (13) is nonunique, a look at our derivation of (13) shows that we should discount the solution $x(t) = 0$ of (13) since it does not also satisfy the original equation of motion $x'' = g$. In that case we are indeed left with the unique solution $x(t) = gt^2/2$. ∎

Closure.   Regarding the existence and uniqueness of solutions, our chief result was Theorem 3.3.1, which states that the initial value problem

$$y' = f(x, y); \qquad y(a) = b \tag{16}$$

has at least one solution if $f(x, y)$ is continuous in some rectangular neighborhood of (a, b) in the x, y plane, and has a unique solution if both $f(x, y)$ and $\partial f(x, y)/\partial y$ are continuous in some rectangular neighborhood of (a, b). The latter is a "local" result in that it does not predict how broad the interval will be over which the solution exists, but only that it will exist over *some* x interval containing the point $x = a$.

EXERCISES 3.3

1. Use Theorem 3.3.1 in each case, if applicable, to predict the existence and uniqueness of a solution to the given differential equation through each given initial point. (That is, does there exist a solution curve through the first initial point, and is it unique? Through the second initial point? And so on.) Further, find the general solution of the differential equation and the particular solution(s), if any, through the given initial points. Give the interval of existence of those particular solutions. Provide either a labeled sketch or a computer plot of those solutions.

(a) $y' = 2xy$; $y(0) = 0$, $y(0) = 2$, $y(0) = -1$, $y(1) = -5$

(b) $y' = x + y$; $y(0) = 3$, $y(0) = 0$, $y(0) = -2$, $y(-1) = 0$

(c) $yy' = x$; $y(0) = 2$, $y(0) = 0$, $y(0) = -2$, $y(5) = 3$, $y(4) = 0$, $y(-5) = -3$

(d) $yy' + x = 0$; $y(0) = 4$, $y(3) = 0$, $y(-3) = -4$

(e) $2yy' = 1$; $y(0) = 2$, $y(0) = 0$, $y(0) = -3$, $y(2) = 5$

(f) $y' = 1/(x + y)$; $y(0) = 0$, $y(1) = -1$, $y(2) = 0$

(g) $y' = \tan y$; $y(3) = 2$, $y(3) = 0$, $y(0) = 2$

(h) $y' = -\tan y$; $y(3) = 2$, $y(3) = 0$

2. (*Envelopes*) In this exercise we develop the geometrical idea of an "envelope" of a family of curves; in the next exercise we relate that idea to certain solutions of differential equations. Consider a one-parameter family of curves

$$g(x, y, c) = 0; \qquad (2.1)$$

for instance, $x^2 + y^2 - c^2 = 0$ is the family of concentric circles centered at the origin. Such a family of curves may, but need not, have an envelope, such as the curve Γ in the first figure. (A curve Γ is an **envelope** of a family of curves if every member of the family is tangent to Γ and if Γ is tangent, at each of its points, to some member of the family.) If we are given $g(x, y, c)$, how can we find any such envelopes? The coordinates x, y of point P (in the second figure) must satisfy both $g(x, y, c) = 0$ and $g(x, y, c + \Delta c) = 0$ or, equivalently,

$$g(x, y, c) = 0 \qquad (2.2)$$

and

$$\frac{g(x, y, c + \Delta c) - g(x, y, c)}{\Delta c} = 0. \qquad (2.3)$$

Equation (2.3) is valid for Δc arbitrarily small, so it must hold in the limit as $\Delta c \to 0$, in which limit P approaches Γ, if there is such a curve. Thus (2.2) and (2.3) become

$$g(x, y, c) = 0, \qquad (2.4)$$

$$\frac{\partial g}{\partial c}(x, y, c) = 0. \qquad (2.5)$$

Eliminating c between (2.4) and (2.5) gives the desired equation of Γ, if there is such a curve. To illustrate, consider the family of lines $y = x + c$ (which surely has no envelope). Then (2.4) and (2.5) give $g(x, y, c) = y - x - c = 0$ and $\frac{\partial g}{\partial c}(x, y, c) = -1 = 0$, which give no envelope. On the other hand, the family of circles $(x - c)^2 + y^2 = 9$ gives $g(x, y, c) = (x - c)^2 + y^2 - 9 = 0$ and $\frac{\partial g}{\partial c}(x, y, c) = -2(x - c) = 0$, and eliminating c between these gives the two straight line envelopes $y = +3$ and $y = -3$ which, from a sketch of the family of circles, is seen to be correct. Here is the problem: In each case determine all envelopes, if any, of the given family of curves, and illustrate with a sketch.

(a) $y = cx + \dfrac{1}{c}$

(b) $(x - c)^2 + y^2 = \dfrac{c^2}{2}$

(c) $y^2 = cx - c^{3/2}$

(d) $y^2 - (x - c)^3 = 0$

(e) the family of circles of radius 1, with their centers on the circle $x^2 + y^2 = 1$

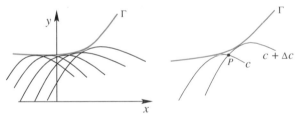

3. (*Singular solutions of differential equations*) Let a general solution of the differential equation

$$y' = f(x, y) \qquad (3.1)$$

be given by

$$g(x, y, c) = 0, \qquad (3.2)$$

where c is a constant of integration, and suppose that the family of solution curves (3.2) has an envelope Γ (as in the first figure in Exercise 2). Note carefully that at each point on Γ the values of x, y and the slope y' are such that (3.1) is satisfied, so Γ *itself must be a solution curve*, a so-called **singular solution** of (3.1). In that case a "general solution" does not really contain *all* solutions of (3.1) since Γ is not itself a member of the curves (3.2)! [Even so, one still calls (3.2) a "general solution."] Here is the problem:

In each case determine all singular solutions of the given differential equation, if any, and illustrate with a sketch.

(a) $y' + y = 0$

(b) $y' = \dfrac{3}{2} y^{1/3}$

(c) $y' = 8y^{3/4}$

NOTE: The concept of the envelope of a family of curves is of importance not only in connection with singular solutions of differential equations but also in optics (in connection with "caustics") and in acoustics.

4. (*No singular solutions for linear equations*) First, read Exercises 2 and 3, above. Show that a *linear* first-order differential equation cannot have a singular solution.

5. (*Example 3, on free fall*) First, read Exercises 2 and 3, above. We stated, in Example 3, that the solution of (13) is nonunique. Specifically, besides the solution given by (14) there is the solution $x(t) = 0$, which corresponds to the mass levitating for all $t \geq 0$, the mass can levitate for any amount of time T and then fall, namely,

$$x(t) = \begin{cases} 0, & 0 \leq t \leq T \\ \dfrac{g}{2}(t - T)^2, & T \leq t < \infty. \end{cases}$$

Discuss this result, using a suitable labeled sketch, in terms of singular solutions.

3.4 EXACT EQUATIONS AND ONES THAT CAN BE MADE EXACT

Thus far we have developed solution techniques for first-order differential equations that are linear or separable. In addition, Bernoulli, Riccati, Clairaut, homogeneous, and almost-homogeneous equations were discussed in the exercises. In this section we consider one more important case, equations that are "exact," and ones that are not exact but can be made exact.

First, let us review some information, from the calculus, about partial derivatives. Specifically, recall that the symbol $\dfrac{\partial^2 f}{\partial x \partial y}$ is understood to mean $\dfrac{\partial}{\partial x}\left(\dfrac{\partial f}{\partial y}\right)$. If we use the standard subscript notation instead, then this quantity would be expressed as f_{yx}, that is, $(f_y)_x$. Does the order of differentiation matter? That is, is $f_{yx} = f_{xy}$?[1] It is shown in the calculus that a *sufficient condition for f_{xy} to equal f_{yx} is that f_x, f_y, f_{yx}, and f_{xy} all be continuous within the region in question.* These conditions are met so typically in applications, that in textbooks on engineering and science f_{xy} and f_{yx} are generally treated as indistinguishable. Here, however, we assume that $f_{xy} = f_{yx}$ only if we explicitly assume the continuity of f_x, f_y, f_{yx}, and f_{xy}.

3.4.1 Exact differential equations.
To motivate the idea of exact equations, consider the equation

$$\frac{dy}{dx} = \frac{\sin y}{2y - x \cos y} \tag{1}$$

---

[1] Sometimes the sequencing of the two operations does matter. For instance, note that you are careful to first apply the toothpaste and then brush your teeth; you do not first brush your teeth and then apply the toothpaste. Likewise, the results in mathematics may depend upon the sequence of the operations.

or, rewritten in differential form,

$$\sin y \, dx + (x \cos y - 2y)dy = 0. \tag{2}$$

Notice that the left-hand side is the differential of $F(x, y) = x \sin y - y^2$ because, by the chain rule,

$$dF = \frac{\partial F}{\partial x}dx + \frac{\partial F}{\partial y}dy = \sin y \, dx + (x \cos y - 2y)dy, \tag{3}$$

so (2) is simply $dF = 0$, which can be integrated to give $F = $ constant; that is,

$$F(x, y) = x \sin y - y^2 = C, \tag{4}$$

where C is an arbitrary integration constant. Equation (4) gives the general solution to (1), in implicit form.[1]

To generalize the method outlined above, we consider the differential equation

$$\frac{dy}{dx} = -\frac{M(x, y)}{N(x, y)}, \tag{5}$$

where the minus sign is included for convenience, so that when we re-express (5) in the differential form

$$M(x, y)dx + N(x, y)dy = 0, \tag{6}$$

then both signs on the left will be positive. It is important to be aware that in equation (5) the variable y is regarded as a function of x, as is clear from the presence of the derivative dy/dx. That is, there is a hierarchy whereby x is the independent variable and y is the dependent variable. But upon re-expressing (5) in the form (6) we change our viewpoint and now consider x and y as having the same status; now they are both independent variables.

We have seen that integration of (6) is simple if $M dx + N dy$ happens to be the differential of some function $F(x, y)$, for if there does exist a function $F(x, y)$ such that

$$dF(x, y) = M(x, y)dx + N(x, y)dy, \tag{7}$$

---

[1]Really, our use of the differential form (2) begs justification since we seem to have thereby treated dy/dx as a fraction of computable quantities dy and dx, whereas it is actually the limit of a difference quotient. In fact the use of differentials is a matter of convenience and is not essential to the method. For instance, observe that if we write (1) as

$$\sin y + (x \cos y - 2y)\frac{dy}{dx} = 0 \tag{A}$$

in place of (2), to avoid any questionable use of differentials, then the left-hand side of the latter equation is the x derivative (total, not partial) of $F(x, y) = x \sin y - y^2$:

$$\frac{d}{dx}F(x, y(x)) = \frac{d}{dx}\left(x \sin y - y^2\right) = \sin y + (x \cos y - 2y)\frac{dy}{dx},$$

so (A) amounts to $dF/dx = 0$. Integrating the latter gives $F(x, y) = x \sin y - y^2 = C$, just as before. Thus, let us continue, without concern about manipulating dy/dx as though it were a fraction.

then (6) is simply

$$dF(x, y) = 0, \tag{8}$$

which can be integrated to give the general solution

$$F(x, y) = C, \tag{9}$$

where C is an arbitrary constant.

Given $M(x, y)$ and $N(x, y)$, suppose that there does exist an $F(x, y)$ such that $Mdx + Ndy = dF$. Then we say that $Mdx + Ndy$ is an **exact differential**, and that (6) is an **exact differential equation**. That case is of great interest because its general solution is given immediately, in implicit form, by (9).

Two questions arise. *How do we determine if such an F exists* and, *if it does, then how do we find it?* The first is addressed by the following theorem.

THEOREM 3.4.1

Test for Exactness
Let $M(x, y)$, $N(x, y)$, $\partial M/\partial y$, and $\partial N/\partial x$ be continuous within a rectangle R in the x, y plane. Then $Mdx + Ndy$ is an exact differential in R if and only if

$$\boxed{\frac{\partial M}{\partial y} = \frac{\partial N}{\partial x}} \tag{10}$$

everywhere in R.

Partial Proof.   Let us suppose that $Mdx + Ndy$ is exact, so that there is an F satisfying (7). Then it must be true, according to the chain rule of the calculus, that

$$M - \frac{\partial F}{\partial x} \tag{11a}$$

and

$$N = \frac{\partial F}{\partial y}. \tag{11b}$$

Differentiating (11a) partially with respect to y, and (11b) partially with respect to x, gives

$$M_y = F_{xy}, \tag{12a}$$

and

$$N_x = F_{yx}. \tag{12b}$$

Since M, N, M_y, and N_x have been assumed continuous in R, it follows from (11) and (12) that F_x, F_y, F_{xy}, and F_{yx} are too, so $F_{xy} = F_{yx}$. Then it follows from (12) that $M_y = N_x$, which is equation (10). Because of the "if and only if" wording in the theorem, we also need to prove the reverse, that the truth of (10) implies the existence of F, but we will omit that part of the proof. ∎

Assuming that the conditions of the theorem are met, so that we are assured that such an F exists, how do we *find F*? We can find it by integrating (11a) with respect to x, and (11b) with respect to y. Let us illustrate the method by reconsidering the example given above.

EXAMPLE 1
Consider equation (1) once again, or, in differential form,

$$\sin y \, dx + (x \cos y - 2y)dy = 0. \tag{13}$$

First, we identify $M = \sin y$, and $N = x \cos y - 2y$. Clearly, M, N, M_y, and N_x are continuous in the whole plane, so we turn to the exactness condition (10): $M_y = \cos y$, and $N_x = \cos y$, so (10) is satisfied, and it follows from Theorem 3.4.1 that there does exist an $F(x, y)$ such that the left-hand side of (13) is dF. Next, we find F from (11):

$$\frac{\partial F}{\partial x} = \sin y, \tag{14a}$$

$$\frac{\partial F}{\partial y} = x \cos y - 2y. \tag{14b}$$

Integrating (14a) partially, with respect to x, gives

$$F(x, y) = \int \sin y \, \partial x = x \sin y + A(y), \tag{15}$$

where the $\sin y$ integrand was treated as a constant in the integration since we performed a "partial integration" on x, holding y fixed [just as y was held fixed in computing $\partial F / \partial x$ in (14a)]. The constant of integration A must therefore be allowed to depend upon y since y was held fixed and was therefore constant. If you are not convinced of this point, observe that taking a partial x-derivative of (15) does indeed recover (14a).

Observe that initially $F(x, y)$ was unknown. The integration of (14a) reduced the problem from an unknown function F of x and y to an unknown function A of y alone. In turn, $A(y)$ can now be determined from (14b). Specifically, we put the right-hand side of (15) into the left-hand side of (14b) and obtain

$$x \cos y + A'(y) = x \cos y - 2y, \tag{16}$$

where the prime denotes d/dy. Canceling terms gives $A'(y) = -2y$, so

$$A(y) = - \int 2y \, dy = -y^2 + B, \tag{17}$$

where this integration was not a "partial integration," it was an ordinary integration on y since $A'(y)$ was an ordinary derivative of $A(y)$. Combining (17) and (15) gives

$$F(x, y) = x \sin y - y^2 + B = \text{constant}. \tag{18}$$

Finally, absorbing B into the constant, and calling the result C, gives the general solution

$$x \sin y - y^2 = C \tag{19}$$

of (1), in implicit form.

COMMENT 1.   Be aware that the partial integration notation $\int (\ )\partial x$ and $\int (\ )\partial y$ is not standard; we use it here because we find it reasonable and helpful in reminding us that any y's in the integrand of $\int (\ )\partial x$ are to be treated as constants, and likewise any for any x's in $\int (\ )\partial y$.

COMMENT 2.   From (13) all the way through (19), x and y have been regarded as independent variables. With (19) in hand, we can now return to our original viewpoint of y being a function of x. We can, if possible, solve (19) by algebra for $y(x)$ [in this case it is not possible because (19) is transcendental], plot the result, and so on. Even failing to solve (19) for $y(x)$, we can nevertheless verify that $x \sin y - y^2 = C$ satisfies (1) by differentiating with respect to x. That step gives $\sin y + x(\cos y)y' - 2yy' = 0$ or $y' = (\sin y)/(2y - x \cos y)$, which does agree with (1).

COMMENT 3.   It would be natural to wonder how this method can *fail* to work. That is, whether or not $M_y = N_x$, why can't we always successfully integrate (11) to find F? The answer, in this example, can be found in (16). Suppose (16) were $2x \cos y + A'(y) = x \cos y - 2y$ instead. Then the $x \cos y$ terms would not cancel, as they did in (16), and we would have $A'(y) = -x \cos y - 2y$, which is impossible because it expresses a relationship between x and y, whereas x and y are regarded here as independent variables. In other words, $A'(y)$ is a function of y only, so it cannot be a function of x as well. Thus, the cancelation of the $x \cos y$ terms in (16) was crucial and was not an accident; it was a consequence of the fact that M and N satisfied the exactness condition (10).

COMMENT 4.   Though we used (14a) first, then (14b), the order is immaterial and could have been reversed.   ∎

3.4.2 Making an equation exact; integrating factors.

It may be discouraging to realize that for any given pair of functions M and N it is unlikely that the exactness condition (10) will be satisfied. However, there is power available to us that we have not yet tapped, for even if M and N fail to satisfy (10), so that the equation

$$M(x, y)dx + N(x, y)dy = 0 \qquad (20)$$

is not exact, it *may* be possible to find a multiplicative factor $\sigma(x, y)$ so that

$$\sigma(x, y)M(x, y)dx + \sigma(x, y)N(x, y)dy = 0 \qquad (21)$$

is exact. That is, we seek a function $\sigma(x, y)$ so that the revised exactness condition

$$\boxed{\frac{\partial}{\partial y}(\sigma M) = \frac{\partial}{\partial x}(\sigma N)} \qquad (22)$$

is satisfied for (21) even though the exactness condition $\partial M/\partial y = \partial N/\partial x$ was not satisfied for (20). Of course, we need $\sigma(x, y) \neq 0$ for (21) to be equivalent to (20).

If we can find a $\sigma(x, y)$ satisfying (22), then we call it an **integrating factor** of (20) because then (21) is equivalent to $dF = 0$, for some $F(x, y)$, and $dF = 0$ can be integrated immediately to give the solution of the original differential equation as $F(x, y) = $ constant.

How do we find such a σ? It is any (nonzero) solution of (22), that is, of

$$\sigma_y M + \sigma M_y = \sigma_x N + \sigma N_x. \tag{23}$$

Of course, (23) is a first-order partial differential equation on σ, so we have made dubious headway: In order to solve our original first-order ordinary differential equation on $y(x)$, we must now solve the first-order partial differential equation (23) on $\sigma(x, y)$!

However, perhaps an integrating factor σ can be found that is a function of x alone: $\sigma(x)$. Then (23) reduces to the ordinary differential equation

$$\sigma M_y = \frac{d\sigma}{dx} N + \sigma N_x$$

or

$$\frac{d\sigma}{dx} = \sigma \left(\frac{M_y - N_x}{N} \right). \tag{24}$$

This idea succeeds if and only if the $(M_y - N_x)/N$ ratio on the right-hand side of (24) is a function of x only, for if it did contain any y dependence then (24) would amount to the impossible situation of a function of x equaling a function of x and y, where x and y are independent variables. Thus, if

$$\boxed{\frac{M_y - N_x}{N} = \text{function of } x \text{ alone,}} \tag{25}$$

then integration of (24) by separation of variables gives

$$\boxed{\sigma(x) = e^{\int \frac{M_y - N_x}{N} \, dx}.} \tag{26}$$

Actually, the general solution of (24) for $\sigma(x)$ includes an arbitrary constant factor, but that factor is inconsequential and can be taken to be 1. Also, remember that we need σ to be nonzero and we are pleased to see, a posteriori, that the σ given by (26) cannot equal zero because it is an exponential function.[1]

If $(M_y - N_x)/N$ is *not* a function of x alone, then an integrating factor $\sigma(x)$ does not exist, but perhaps we can find σ as a function of y alone: $\sigma(y)$. Then (23) reduces to

$$\frac{d\sigma}{dy} M + \sigma M_y = \sigma N_x$$

or

$$\frac{d\sigma}{dy} = -\sigma \left(\frac{M_y - N_x}{M} \right).$$

---

[1] Sketch the graph of e^u versus its argument u, over $-\infty < u < \infty$, and you will see that e^u has no zeros—no values of u for which $e^u = 0$.

If

$$\boxed{\frac{M_y - N_x}{M} = \text{function of } y \text{ alone},}$$

(27)

then

$$\boxed{\sigma(y) = e^{-\int \frac{M_y - N_x}{M} dy}.}$$

(28)

EXAMPLE 2

Consider the equation (already expressed in differential form)

$$dx + \left(3x - e^{-2y}\right) dy = 0.$$

(29)

Then $M(x, y) = 1$ and $N(x, y) = 3x - e^{-2y}$, so (10) is not satisfied and (29) is not exact. Seeking an integrating factor that is a function of x alone, we find that

$$\frac{M_y - N_x}{N} = \frac{0 - 3}{3x - e^{-2y}} \neq \text{function of } x \text{ alone},$$

(30)

and conclude that $\sigma(x)$ is not possible. Seeking instead an integrating factor that is a function of y alone,

$$\frac{M_y - N_x}{M} = \frac{0 - 3}{1} = -3 = \text{function of } y \text{ alone},$$

(31)

so that $\sigma(y)$ *is* possible, and is given by

$$\sigma(y) = e^{-\int \frac{M_y - N_x}{M} dy} = e^{\int 3 \, dy} = e^{3y}.$$

(32)

Thus, we multiply (29) through by the integrating factor $\sigma = e^{3y}$ and obtain

$$e^{3y} dx + e^{3y} \left(3x - e^{-2y}\right) dy = 0,$$

which is now guaranteed to be exact. Thus,

$$\frac{\partial F}{\partial x} = e^{3y}$$

(33a)

and

$$\frac{\partial F}{\partial y} = e^{3y} \left(3x - e^{-2y}\right)$$

(33b)

so (33a) gives

$$F(x, y) = \int e^{3y} \, \partial x = xe^{3y} + A(y).$$

(34)

Next, put the right-hand side of (34) into the left-hand side of (33b):

$$\frac{\partial F}{\partial y} = 3xe^{3y} + A'(y) = e^{3y} \left(3x - e^{-2y}\right).$$

(35)

The latter gives

$$A'(y) = -e^y$$

so

$$A(y) = -e^y + B.$$

Thus,

$$F(x, y) = xe^{3y} + A(y) = xe^{3y} + \left(-e^y + B\right) = \text{constant}$$

or

$$xe^{3y} - e^y = C, \tag{36}$$

where C is an arbitrary constant; (36) is a general solution of (29), in implicit form.

COMMENT.   Can we solve (36) for y? If we let $e^y \equiv z$, then (36) is the cubic equation $xz^3 - z = C$ in z, and there is a known solution to cubic equations. If we can solve for z, then we have y as $y = \ln z$. However, the solution of that cubic equation (as can be obtained using the *Maple* **solve** command) is quite a messy expression.   ■

E X A M P L E   3   *First-Order Linear Equation*.
We've already solved the general first-order linear equation

$$\frac{dy}{dx} + p(x)y = q(x) \tag{37}$$

in Section 2.3, but let us see if we can solve it again, using the ideas of this section. First, express (37) in the form

$$[p(x)y - q(x)]\,dx + dy = 0. \tag{38}$$

Thus, $M = p(x)y - q(x)$ and $N = 1$, so $M_y = p(x)$ and $N_x = 0$. Hence $M_y \neq N_x$, so (38) is not exact [except in the trivial case when $p(x) = 0$]. Since

$$\frac{M_y - N_x}{N} = \frac{p(x) - 0}{1} = \text{function of } x \text{ alone,}$$

$$\frac{M_y - N_x}{M} = \frac{p(x) - 0}{p(x)y - q(x)} \neq \text{function of } y \text{ alone,}$$

we can find an integrating factor that is a function of x alone, but not one that is a function of y alone. We leave it for the exercises to show that the integrating factor is

$$\sigma(x) = e^{\int p(x)\,dx},$$

and that the final solution (this time obtainable in explicit form) is

$$y(x) = e^{-\int p\,dx}\left(\int e^{\int p\,dx} q\,dx + C\right), \tag{39}$$

as found earlier, in Section 2.3.   ■

Closure. Let us summarize the main results. Given a differential equation $dy/dx = f(x, y)$, the first step in using the method of exact differentials is to re-express it in the differential form $M(x, y)dx + N(x, y)dy = 0$. If M, N, M_y, and N_x are all continuous in the region of interest, check to see if the exactness condition $M_y = N_x$ is satisfied. If it is, then the equation is exact, and its general solution is $F(x, y) = C$, where F is found by integrating (11a) and (11b). As a check on your work, a differential of $F(x, y) = C$ should give you back the original equation $Mdx + Ndy = 0$.

If the equation is not exact, see if $(M_y - N_x)/N$ is a function of x alone. If it is, then an integrating factor $\sigma(x)$ can be found from (26). Multiply the given equation $Mdx + Ndy = 0$ through by that $\sigma(x)$ so the new equation will be exact and you can proceed as outlined above for an exact equation.

If $(M_y - N_x)/N$ is not a function of x alone, check to see if $(M_y - N_x)/M$ is a function of y alone. If it is, then an integrating factor $\sigma(y)$ can be found from (28). Multiply $Mdx + Ndy = 0$ through by that $\sigma(y)$ so the new equation will be exact and you can proceed as outlined above for an exact equation.

If $M_y \neq N_x$, $(M_y - N_x)/N$ is not a function of x alone, and $(M_y - N_x)/M$ is not a function of y alone, then the method is of no help unless an integrating factor σ can be found that is a function of both x and y.

Finally, recall that we have studied three types of first-order equations:

1. Linear, $y' + p(x)y = q(x)$, Chapter 2.

2. Separable, $y' = X(x)Y(y)$, Section 3.2.

3. Exact or can be made exact (by an integrating factor), present section.

Are these three cases mutually exclusive? No. For instance, a subset of linear equations is also separable, namely, if $q(x)$ is zero or if $p(x)$ is a constant time $q(x)$. Further, *every* separable equation is exact if we write it in the form

$$X(x)dx - \frac{1}{Y(y)}dy = 0,$$

and every linear equation can be made exact (as we did in Example 3). In summary, the result is as shown schematically in Fig. 1.

We see that, in principle, it would suffice to study only equations that are exact (or can be made exact) since that case includes the cases of linear and separable equations. However, it is important and traditional to study each of these cases explicitly, the linear equation in order to begin to mark the distinction between linear and non-linear equations, and separable equations because the separation-of-variables method is so simple. In fact, given a specific first-order differential equation, we recommend that you first look to see if it is separable. If so, solve it by separation of variables; if not, see if it is linear or exact, whichever method you prefer. If it is neither, see if you can make it exact.

If none of the aforementioned methods work you might see if it is a special case covered in the exercises, such as a Bernoulli equation or a Riccati equation.

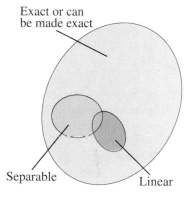

Exact or can be made exact

Separable   Linear

FIGURE 1
Schematic of the sets of first-order equations that are exact (or can be made exact), separable, or linear.

EXERCISES 3.4

NOTE: Solutions should be expressed in explicit form if possible.

1. Show that the equation is exact, and obtain its general solution. Also, find the particular solution corresponding to the given initial condition.

   (a) $3dx - dy = 0$; $y(0) = 6$

   (b) $x^2 dx + y^2 dy = 0$; $y(9) = -1$

   (c) $x\,dx + 2y\,dy = 0$; $y(1) = 2$

   (d) $4\cos 2u\,du - e^{-5v}dv = 0$; $v(0) = -2$

   (e) $e^y dx + (xe^y - 1)dy = 0$; $y(-5) = 6$

   (f) $(e^y + z)dy - (\sin z - y)dz = 0$; $z(0) = 2$

   (g) $(x - 2z)dx - (2x - z)dz = 0$; $z(-3) = 5$

   (h) $(\sin y + y\cos x)dx + (\sin x + x\cos y)dy = 0$; $y(2) = 3$

   (i) $(\sin xy + xy\cos xy)dx + x^2\cos xy\,dy = 0$; $y(2) = -1$

   (j) $(3x^2\sin 2y - 2xy)dx + (2x^3\cos 2y - x^2)dy = 0$; $y(0.5) = 3.1$

   (k) $(4x^3y^5\sin 3x + 3x^4y^5\cos 3x)dx + 5x^4y^4\sin 3x\,dy = 0$; $y(1) = 1$

   (l) $3x^2y\ln y\,dx + (x^3\ln y + x^3 - 2y)dy = 0$; $y(8) = 3$

2. (a)–(m) Find the general solution of the equation given in the corresponding part of Exercise 1 using *Maple*, and also the particular solution corresponding to the given initial condition.

3. Make up three different examples of exact equations.

4. Determine whatever conditions, if any, are needed on the constants $a, b, \ldots, f, A, B, \ldots, F$ for the equation to be exact.

   (a) $(ax + by + c)dx + (Ax + By + C)dy = 0$

   (b) $(ax^2 + by^2 + cxy + dx + ey + f)dx + (Ax^2 + By^2 + Cxy + Dx + Ey + F)dy = 0$

5. Find a suitable integrating factor $\sigma(x)$ or $\sigma(y)$, if possible, and use it to find a general solution of a differential equation. If neither is possible, state that.

   (a) $3y\,dx + dy = 0$

   (b) $y\,dx + x\ln x\,dy = 0$

   (c) $y\ln y\,dx + (x + y)dy = 0$

   (d) $dx + (x - e^{-y})dy = 0$

   (e) $dx + x\,dy = 0$

   (f) $(ye^{-x} + 1)dx + (xe^{-x})dy = 0$

   (g) $\cos y\,dx - [2(x - y)\sin y + \cos y]dy = 0$

   (h) $(1 - x - z)dx + dz = 0$

   (i) $(2 + \tan^2 x)(1 + e^{-y})dx - e^{-y}\tan x\,dy = 0$

(j) $2xy\,dx + (y^2 - x^2)dy = 0$

(k) $\cos x\,dx + (3\sin x + 3\cos y - \sin y)dy = 0$

(l) $(y\ln y + 2xy^2)dx + (x + x^2y)dy = 0$

(m) $(3x - 2p)dx - x\,dp = 0$

(n) $y\,dx + (x^2 - x)dy = 0$

6. (*First-order linear equation*) Verify that $\sigma(x) = e^{\int p(x)\,dx}$ is an integrating factor for the general linear first-order equation (37), and use it to derive the general solution (39).

7. Show that the given equation is not exact and that an integrating factor depending on x alone or y alone does not exist. If possible, find an integrating factor in the form $\sigma(x, y) = x^a y^b$, where a and b are suitably chosen constants. If such a σ can be found, then use it to obtain the general solution of the differential equation; if not, state that.

   (a) $(3xy - 2y^2)dx + (2x^2 - 3xy)dy = 0$

   (b) $(3xy + 2y^2)dx + (3x^2 + 4xy)dy = 0$

   (c) $(x + y^2)dx + (x - y)dy = 0$

   (d) $y\,dx - (x^2y - x)dy = 0$

8. Show that the equation is not exact and that an integrating factor depending on x alone or y alone does not exist. Nevertheless, find a suitable integrating factor by inspection, and use it to obtain the general solution.

   (a) $e^y dx + e^x dy = 0$ (b) $y^2 dx - e^{3x}dy = 0$

   (c) $e^{2y}dx - \tan x\,dy = 0$

9. Obtain the general solution, using the methods of this section.

   (a) $\dfrac{dy}{dx} = \dfrac{x - y}{x + y}$ (b) $\dfrac{dr}{d\theta} = -\dfrac{r^2\cos\theta}{2r\sin\theta + 1}$

   (c) $\dfrac{dy}{dx} = \dfrac{2xy - e^y}{x(e^y - x)}$ (d) $\dfrac{dy}{dx} = -\dfrac{\sin y + y\cos x}{\sin x + x\cos y}$

10. What do the integrating factors defined by (26) and (28) turn out to be if the equation is exact to begin with?

11. (a) Show that $(x^3 + y)dx + (y^3 + x)dy = 0$ is exact.

    (b) More generally, is $M(x, y)dx + M(y, x)dy = 0$ exact? Explain.

12. If $F(x, y) = C$ is the general solution (in implicit form) of a given first-order equation, then what is the particular solution (in implicit form) satisfying the initial condition $y(a) = b$?

13. If $M\,dx + N\,dy = 0$ and $P\,dx + Q\,dy = 0$ are exact, is $(M + P)dx + (N + Q)dy = 0$ exact? Explain.

14. Show that for $[p(x) + q(y)]dx + [r(x) + s(y)]dy = 0$ to be exact, it is necessary and sufficient that $q(y)dx + r(x)dy$ be an exact differential.

15. We solved (1) by using the fact that (2) is exact. Observe that although (1) it is neither separable nor first-order linear, it *is* first-order linear if we change our viewpoint and regard x as a function of y. Use that idea to solve for $x(y)$ and verify that your solution agrees with (4).

3.5 ADDITIONAL APPLICATIONS

3.5.1 Verhulst population model.   Let $N(t)$ be the population of a certain species as a function of the time t, and let β be its birth rate (births per individual per unit time) and δ be its death rate (deaths per individual per unit time), with both β and δ assumed to be constant. Then $N(t)$ is governed by the differential equation

$$N'(t) = (\beta - \delta)N(t) \tag{1}$$

or, if we define the net birth/death rate as $\kappa = \beta - \delta$,

$$N'(t) = \kappa N(t), \tag{2}$$

with general solution

$$N(t) = Ae^{\kappa t}, \tag{3}$$

where A is an arbitrary constant.

This simple model was discussed in Example 1 of Section 2.2. There, we pointed out that if $\kappa > 0$ (i.e., the birth rate β exceeds the death rate δ) then the resulting exponential growth of $N(t)$ surely becomes unrealistic because if N becomes sufficiently large we can expect other factors to come into play, such as insufficient food or other necessary resources, that have not been accounted for in our simple model (1). Specifically, κ will not really be a constant but will vary with N. We expect it to decrease as N increases; for instance, if food becomes scarce as N increases then the death rate δ will increase (hence κ will decrease) due to an increasing prevalence of death by starvation. As a simple model of such behavior, let us suppose that κ varies linearly with N so $\kappa = a - bN$, with a and b positive so that κ diminishes as N increases. Then (2) is to be replaced by the equation

$$N' = (a - bN)N, \tag{4}$$

which is known as the **logistic equation**, or the **Verhulst equation**, after the Belgian mathematician *P. F. Verhulst* (1804–1849) who used it in his work on population dynamics. Appending an initial condition $N(0) = N_0$, we consider the initial value problem

$$N'(t) = (a - bN)N; \qquad N(0) = N_0. \tag{5}$$

The differential equation is not linear; it is nonlinear because of the N^2 term on the right-hand side. But it is separable and can be solved accordingly:

$$\int \frac{dN}{(a - bN)N} = \int dt. \tag{6}$$

We expand the integrand by partial fractions (this technique is reviewed in the Appendix),

$$\frac{1}{(a - bN)N} = -\frac{1}{b}\frac{1}{(N - \frac{a}{b})N} = -\frac{1}{a}\frac{1}{N - \frac{a}{b}} + \frac{1}{a}\frac{1}{N} \tag{7}$$

so (6) gives

$$-\frac{1}{a}\ln\left|N - \frac{a}{b}\right| + \frac{1}{a}\ln N = t + C, \tag{8}$$

where C is an arbitrary constant ($-\infty < C < \infty$). [Whether we write $\ln|N|$ or $\ln N$ in (8) is immaterial since $N > 0$. However, $N - a/b$ can be positive *or* negative so we keep the absolute value signs in $\ln|N - a/b|$.] Whereas (8) gives the general solution of (4) in implicit form, we can get it into explicit form by solving (8) for N as follows. First, re-express (8), equivalently, as

$$\ln\left|N - \frac{a}{b}\right|^{-1/a} + \ln N^{1/a} = t + C,$$

$$\left|\frac{N}{N - \frac{a}{b}}\right|^{1/a} = e^{t+C},$$

$$\left|\frac{N}{N - \frac{a}{b}}\right| = e^{at+aC} = Be^{at},$$

where we have replaced $\exp(aC)$ by B, so $0 \leq B < \infty$ [since the exponential function is nonnegative for all values of its argument, $B = 0$ being permissible because it gives $N(t) = 0$, which does satisfy (4)]. Thus

$$\frac{N}{N - \frac{a}{b}} = \pm Be^{at} \equiv Ae^{at}, \tag{9}$$

where A is arbitrary. Finally, imposing the initial condition $N(0) = N_0$ on (9) gives $A = N_0/(N_0 - a/b)$, and putting that expression into (9) and solving for N gives

$$N(t) = \frac{a}{b}\frac{N_0}{N_0 + \left(\frac{a}{b} - N_0\right)e^{-at}}. \tag{10}$$

To ascertain the behavior of $N(t)$, from (10), we can examine its asymptotic behavior as $t \to \infty$ and we can also obtain representative computer plots of N versus t. Since the e^{-at} tends to zero as $t \to \infty$, (10) reveals the asymptotic behavior

$$N(t) \sim \frac{a}{b} \quad \text{as} \quad t \to \infty, \tag{11}$$

unless $N_0 = 0$, in which case (10) gives $N(t) = 0$ for *all t*; let us assume that $N_0 \neq 0$ since $N_0 = 0$ gives only the trivial case where there is no population! According to the terminology introduced in Section 2.4, we say that a/b is the *steady-state* value of N, the value approached asymptotically as $t \to \infty$.

To plot N versus t, using (10), we need to assign representative numerical values to the parameters a, b, and N_0. Recalling that our differential equation model assumes that $N(t)$ is large enough so that $N(t)$ can be smoothed out into a continuous function (rather than being an integer-valued function) of t, let the steady-state population a/b be 1000, say; for instance, let $a = 1$ and $b = 0.001$. Choosing values of N_0 both below and above the steady-state value, let $N_0 = 200, 400, 600, 800, 1000, 1200,$ and 1400. The results are shown in Fig. 1, where the approach to the steady-state value $a/b = 1000$ is evident.

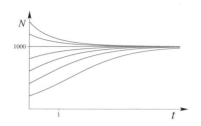

FIGURE 1
The solution (10) for $a = 1$, $b = 0.001$, $N_0 = 200, 400, \ldots, 1400$.

COMMENT 1.   Besides being separable, (4) happens to be both a Bernoulli and a Riccati equation so it could, alternatively, be solved by techniques specific to those equations, as outlined in the exercises for Section 2.3. We leave those alternative solutions of (4) to the exercises.

COMMENT 2.   It is also illuminating to plot the right-hand side of (4), $f(N) = (a - bN)N$, versus N, as we have in Fig. 2. The values $N = 0$ and $N = a/b$ correspond to equilibrium points in that they give $N'(t) = f(N) = 0$, so $N(t) = $ constant. If we start at $N(0) = 0$ then $N(t) = 0$ for all t, and if we start at $N(0) = a/b$ then $N(t) = a/b$ for all t. The arrows on the N axis indicate the "flow" direction: For $0 < N < a/b$ we have $f(N) > 0$ so $N(t)$ is increasing [because $N'(t) = f(N) > 0$], and for $a/b < N < \infty$ we have $f(N) < 0$ so $N(t)$ is decreasing. Notice how the plot in Fig. 2 [namely, N versus $f(N)$, i.e., N versus N'] complements the plot of $N(t)$ versus t in Fig. 1. Fig. 2 is an example of a "phase plane" plot, which idea is developed in Section 7.3 and in Chapter 11.

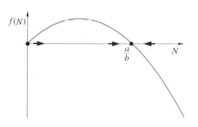

FIGURE 2
Equilibrium points and the "flow."

Generalizing this example, we say that N_{eq} is an **equilibrium point** of $N'(t) = f(N)$ if $f(N_{eq}) = 0$. The latter is **stable** if $N(t)$ can be kept arbitrarily close to N_{eq} for all $t \geq 0$ by taking it to be sufficiently close initially; otherwise it is **unstable**. For the Verhulst model (4), we can see from Fig. 2 that the equilibrium points $N = 0$ and $N = a/b$ are unstable and stable, respectively. The former is unstable because we *cannot* keep $|N(t) - N_{eq}| = |N(t) - 0| < \epsilon$ for all $t \geq 0$, where ϵ is arbitrarily small, no matter how close to $N_{eq} = 0$ we start out. And the latter is stable because we *can* keep $|N(t) = N_{eq}| = |N(t) - a/b| < \epsilon$ for all t, where ϵ is arbitrarily small, by starting out closer to a/b than ϵ.

Further, we classify a stable equilibrium point as **asymptotically stable** if $N(t)$ not only stays close to N_{eq} for all $t \geq 0$ but if it actually *tends to* N_{eq}, that is to say if $N(t) \to N_{eq}$ as $t \to \infty$. From Fig. 2 we see that the equilibrium point at a/b is not only stable, it is asymptotically stable.

Equilibrium points and their stability will be discussed further in Chapter 11 on the phase plane.

3.5.2 Projectile dynamics subject to gravitational attraction.   *Newton's law of gravitation* states that the force of attraction F exerted by any one point mass M on

any other point mass m is[1]

$$F = G\frac{Mm}{d^2},\tag{12}$$

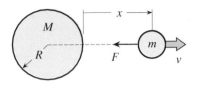

FIGURE 3
Projectile launch.

where d is the distance between them and $G\ (=6.67\times10^{-8}\ \mathrm{cm}^3/\mathrm{gram\ sec}^2)$ is called the *universal gravitational constant*; (12) is said to be an *inverse-square law* since the force F varies as the inverse square of the distance d. (By M and m being point masses, we mean that their sizes are negligible compared with the distance between them.)

Consider the linear motion of a projectile of mass m that is launched from the surface of the earth, as sketched in Fig. 3, where M and R are the mass and radius of the earth, respectively. From Newton's second law of motion and his law of gravitation (12), it follows that the equation of motion of the projectile is

$$m\frac{d^2x}{dt^2} = -G\frac{Mm}{(x+R)^2},\tag{13}$$

the minus sign because the force is in the negative x direction, and the $x + R$ being the distance between mass centers. Although (13) is a second-order equation, we can reduce it to one of first order by noting that

$$\frac{d^2x}{dt^2} = \frac{d}{dt}\left(\frac{dx}{dt}\right) = \frac{dv}{dt} = \frac{dv}{dx}\frac{dx}{dt} = v\frac{dv}{dx},\tag{14}$$

---

[1]Newton derived (12) from *Kepler's laws* of planetary motion which, in turn, were inferred empirically from the voluminous measurements recorded by the Danish astronomer *Tycho Brahe* (1546–1601). Usually, in applications (not to mention homework assignments in mechanics), one is given the force exerted on a mass and asked to determine the motion by twice integrating Newton's second law of motion. In deriving (12), however, Newton worked "backwards": The motion of the planets was supplied in sufficient detail by Kepler's laws, and Newton used those laws to infer the force needed to sustain that motion. It turned out to be an inverse-square force directed toward the sun. Being aware of other such forces between masses, for example, the force that kept his shoes on the floor, Newton then proposed the bold generalization that (12) holds not just between planets and the sun, but between any two bodies in the universe; hence the name *universal law of gravitation*.

Just as it is difficult to overestimate the importance of Newton's law of gravitation and its impact upon science, it is also difficult to overestimate how the idea of a force acting at a distance, rather than through physical contact, must have been incredible when first proposed. In fact, such eminent scientists and mathematicians as Huygens, Leibniz, and John Bernoulli referred to Newton's idea of gravitation as absurd and revolting. Imagine Newton's willingness to stand nonetheless upon the results of his mathematics, in inferring the concept of gravitation, even in the absence of any physical mechanism or physical plausibility, and in the face of such opposition.

Remarkably, *Coulomb's law* subsequently stated an inverse-square type of electrical attraction or repulsion between two charges. Why these two types of force field turn out to be of the same mathematical form is not known. Equally remarkable is the fact that although the forms of the two laws are identical, the magnitudes of the forces are staggeringly different. Specifically, the ratio of the electrical force of repulsion to the gravitational force exerted on each other by two electrons (which is independent of the distance of separation) is 4.17×10^{42}.

where $v = dx/dt$ is the velocity, and where the third equality follows from the chain rule. Thus (13) becomes the first-order equation

$$v\frac{dv}{dx} = -\frac{GM}{(x+R)^2},\tag{15}$$

which is separable and which gives

$$\int v\,dv = -GM \int \frac{dx}{(x+R)^2},\tag{16}$$

$$\frac{v^2}{2} = \frac{GM}{x+R} + C\tag{17}$$

for $v(x)$. If the *launch velocity* (when $x = 0$) is $v(0) = V$, then (17) gives

$$C = \frac{V^2}{2} - \frac{GM}{R},$$

and putting the latter back in (17) and solving for v gives

$$v = \sqrt{V^2 - \frac{2GM}{R}\frac{x}{x+R}}\tag{18}$$

as the desired expression of v as a function of x.

Before we proceed, let us simplify (18) by noting that when $x = 0$ the right-hand side of (13) must be the weight force $-mg$, where g is the familiar gravitational acccleration at the earth's surface. Thus, $-mg = -GMm/R^2$, so $GM = R^2g$ and (18) becomes

$$v = \sqrt{V^2 - 2gR\frac{x}{x+R}}.\tag{19}$$

Remember that a key step in any analysis is the interpretation of the results. In the present case we derived the formula (19) for $v(x)$, but how do we use (19) to generate understanding of the projectile motion beyond using it to evaluate the velocity v at a particular x location? We encountered the same problem in Section 3.5.1, where the solution was given by (10). There, we used (10) to obtain the asymptotic behavior (11) as $t \to \infty$ and to generate the representative plots shown in Fig. 1. Similarly in this problem, we see from (19) that $v(x)$ diminishes from its initial value $v(0) = V$ to the asymptotic constant value

$$v(x) \sim \sqrt{V^2 - 2gR} \quad \text{as} \quad x \to \infty.\tag{20}$$

However, the latter is meaningful only if the launch velocity V is sufficiently great so that $V^2 - 2gR \geq 0$, that is, if $V \geq \sqrt{2gR}$; otherwise, the square root does not give a real number. The idea is that if $V < \sqrt{2gR}$, then (19) does not hold on $0 \leq x < \infty$ but only as far as the argument of the square root remains positive, namely, over

$$0 \leq x \leq \frac{V^2R}{2gR - V^2},\tag{21}$$

where the right-hand member is obtained by setting the argument of the square root in (19) equal to zero. That is, the projectile does not escape the gravitational pull of the earth; it travels only as far as

$$x_{\max} = \frac{V^2 R}{2gR - V^2},$$
(22)

stops, and is then drawn back toward the earth. For the return trip (19) still holds, provided that this time we choose the *negative* square root. Finally, the projectile will strike the earth with a velocity of $-V$.

The critical launch velocity

$$V_e = \sqrt{2gR}$$
(23)

mentioned in connection with (20) is called the *escape velocity* for we see from (22) that as $V \to V_e$, $x_{\max} \to \infty$; if V is equal to or greater than the minimum escape velocity the projectile travels to "infinity," with the velocity approaching the asymptotic value given by (20) as $x \to \infty$. Numerically, $V_e \approx 11.1$ km/s.

In addition to these useful results we can use (19) to plot v versus x for various values of V [recall that V is the launch velocity $v(0)$] less than and greater than the critical velocity V_e, as we have done in Fig. 4. The uppermost curve corresponds to a launch velocity of $v(0) = V = 1.5V_e$; that curve reveals that the velocity $v(x)$ diminishes as x increases, and approaches the asymptotic value $\sqrt{(1.5V_e)^2 - V_e^2} \approx 1.12V_e$ predicted by (20). The next curve corresponds to a launch velocity of $v(0) = V = 1.3V_e$; the velocity diminishes as x increases, and approaches the asymptotic value $\sqrt{(1.3V_e)^2 - V_e^2} \approx 0.83V_e$ predicted by (20). The same is true for the third curve. The fourth curve corresponds to a subcritical launch velocity $v(0) = V = 0.9V_e$, in which case the projectile does not continue to $x = \infty$ but gets only as far as the $x_{\max}$ given by (22) before reversing its motion. The effect is even more pronounced for the fifth curve.

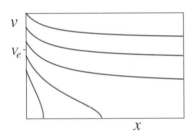

FIGURE 4
Velocity as a function of x from (19), for various values of the launch velocity V.

COMMENT 1. Recall that whereas the law of gravitation (12) applies to two point masses separated by a distance d, the earth is hardly a "point mass;" it is quite large. Thus, it is appropriate to question the use of (12) in the right-hand side of (13). It turns out that (13) is indeed correct, but verification of that claim is outside the scope of this discussion.

COMMENT 2. If we wish, we can recall that $v = dx/dt$ so that (19) is really the first-order differential equation

$$\frac{dx}{dt} = \sqrt{V^2 - 2gR\frac{x}{x+R}}$$
(24)

on $x(t)$, and we can solve the latter for $x(t)$ because it is variable separable. However, the x integration turns out to be quite tedious. We will leave that step to be done by *Maple*, in the exercises.

This is a good time to call attention to our various changes in viewpoint. That is, in the first two equalities in (14) we regarded the displacement x and the velocity v as dependent variables, functions of the independent variable t. But beginning with the right-hand side of that equality, we began to regard v as a function of x. Then, once we solved (15) for $v(x)$ we replaced v, in (19), by dx/dt, and x changed from the independent variable to the dependent variable once again. The point to notice is that which variable is regarded as the independent variable and which is the dependent variable is not so much an absolute truth as it is a decision that we make, and that decision, or viewpoint, is sometimes changed over the course of the analysis.

COMMENT 3.   The steps in (14), whereby we were able to reduce our second-order equation (13) to the first-order equation (15), were not limited to this specific application. They apply whenever the force is a function of x alone, for if we apply (14) to the equation

$$m\frac{d^2x}{dt^2} = f(x), \tag{25}$$

we get the separable first-order equation

$$mv\frac{dv}{dx} = f(x) \tag{26}$$

with solution

$$\frac{mv^2}{2} = \int f(x)\,dx + C \tag{27}$$

or, equivalently,

$$\left.\frac{mv^2}{2}\right|_{x_1}^{x_2} = \int_{x_1}^{x_2} f(x)\,dx. \tag{28}$$

In the language of mechanics, the right-hand side of (28) is the work done by the force $f(x)$ as the body moves from x_1 to x_2, and $mv^2/2$ is the kinetic energy. Thus, the physical significance of (28) is that it is a work-energy statement: The change in the kinetic energy of the body is equal to the work done on it.

Closure.   In this section we did not introduce any new mathematics. Rather, we used already-developed solution techniques to solve two specific problems, one in population dynamics and one in Newtonian mechanics.

Maple.   The *Maple* commands used to generate Fig. 1, for instance, were

```
with(plots):
implicitplot({N=1000*200/(200+800*exp(-t)),
N=1000*400/(400+600*exp(-t)),
N=1000*600/(600+400*exp(-t)),
N=1000*800/(800+200*exp(-t)), N=1000,
N=1000*1200/(1200-200*exp(-t)),
N=1000*1400/(1400-400*exp(-t))}, t=0..5, N=0..2000);
```

EXERCISES 3.5

1. We solved (4) by separation of variables but it also happens to be a Riccati equation and a Bernoulli equation, which equations were discussed in the exercises for Section 2.3.

   (a) Solve (4) by treating it as a Riccati equation and verify that your general solution agrees with that given by (9), which was obtained by separation of variables.

   (b) Solve (4) by treating it as a Bernoulli equation and verify that your general solution agrees with (9).

2. Show that every solution curve of (4) changes its concavity where it crosses the horizontal line $N = a/2b$ (which is $N = 500$ in Fig. 1). HINT: You can deduce that result from the solution (10), but it is simpler, and more elegant, to deduce it directly from (4).

3. Determine all equilibrium points, if any, on $0 \le N < \infty$, and classify each as stable or unstable.

   (a) $N' = (1 - N)(2 - N)N$

   (b) $N' = (1 + 4N)N$

   (c) $N' = (1 - N^2)N$

   (d) $N' = (1 - N^2)(5 - N)(24 - 3N)N$

   (e) $N' = 50 + N^2$

4. In Fig. 4 we did not include a solution curve corresponding to an initial velocity of V_e. Sketch that solution curve and indicate any key features; for example, whether the curve intersects the x axis, whether it has an asymptote as $x \to \infty$, and so on.

5. If $V < \sqrt{2gR}$, then (19) indicates that v becomes imaginary as $x \to \infty$. Can you interpret this seemingly strange result?

6. (a) The solution of (24) is simple for the special case where $V = V_e = \sqrt{2gR}$. For that case, solve (24) subject to the initial condition $x(0) = 0$. Show that $x(t)$ has the behavior

   $$x(t) \sim V_e t \quad \text{as} \quad t \to 0 \tag{6.1}$$

   and

   $$x(t) \sim \left(\frac{3}{2} V_e \sqrt{R} \right)^{2/3} t^{2/3} \quad \text{as} \quad t \to \infty, \tag{6.2}$$

   and sketch the graph of $x(t)$.

   (b) If, on the other hand, $V > V_e$ then show that $x(t)$ has the asymptotic behavior

   $$x(t) \sim V t \quad \text{as} \quad t \to 0 \tag{6.3}$$

   and

   $$x(t) \sim \sqrt{V^2 - V_e^2} \, t \quad \text{as} \quad t \to \infty. \tag{6.4}$$

   HINT: You should be able to obtain (6.3) and (6.4) directly from (24).

   (c) In contrast with (6.2) and (6.4), which show that $x(t) \to \infty$ as $t \to \infty$, show, from (24), that if $V < V_e$ then $x(t)$ does not tend to infinity as $t \to \infty$ but that it attains a finite maximum value

   $$x_{\max} = V^2 R / \left(V_e^2 - V^2 \right). \tag{6.5}$$

   (d) From the results outlined in parts (a)–(c), sketch the graphs of $x(t)$ versus t for the representative cases where the launch velocity V is $V_e/3$, $2V_e/3$, V_e, $2V_e$, and $4V_e$.

7. We saw, in Section 1.1, that the displacement $x(t)$ of a body of mass m restrained by a coil spring of stiffness k and driven by an applied force $F(t)$ is governed by the differential equation

   $$m x'' + k x = F(t). \tag{7.1}$$

   Let $F(t) = 0$, and suppose we have the initial conditions $x(0) = x_0$, $x'(0) = 0$. That is, we displace the mass a distance x_0, hold it still [since the velocity $x'(0)$ is zero] and, at $t = 0$, release it. We are interested in the ensuing motion $x(t)$ that is governed by the initial value problem

   $$m x'' + k x = 0; \quad x(0) = x_0, \quad x'(0) = 0. \tag{7.2}$$

   Of course, the differential equation is of second order whereas we have not yet studied second-order equations, but it can be solved using the idea presented in Comment 3 of Section 3.5.2.

   (a) Thus, denoting x' as the velocity v, show that

   $$\frac{1}{2} m v^2 + \frac{1}{2} k x^2 = \text{constant} \tag{7.3}$$

   or, using the initial conditions given in (7.2),

   $$\frac{1}{2} m v^2 + \frac{1}{2} k x^2 = \frac{1}{2} k x_0^2. \tag{7.4}$$

   [Whereas $m x'' + k x = 0$ is a statement of Newton's second law of motion, (7.3) is a statement of the conservation of energy for this system: kinetic energy + potential energy = constant.]

   (b) Next, solve (7.4) for v and replace v by $x'(t)$ so that you now have a first-order differential equation on $x(t)$. Solve for $x(t)$ and show that the result is

   $$x(t) = x_0 \cos \left(\sqrt{\frac{k}{m}} \, t \right). \tag{7.5}$$

NOTE: Thus, the result is an oscillation (as we might well have anticipated), of amplitude x_0 and frequency $\sqrt{k/m}$. It seems quite reasonable that the amplitude equals the initial displacement x_0 and that the frequency increases with the spring stiffness k and decreases with the mass m.

3.6 LINEAR AND NONLINEAR EQUATIONS CONTRASTED

Recall that the equation

$$y' + p(x)y = q(x) \tag{1}$$

is linear, and that

$$y' = f(x, y) \tag{2}$$

is nonlinear if it is not expressible in the form (1). For instance, each of the equations $y' + 2xy^3 = \sin x$, $y' = \sin y$, $y' = x + e^y$, and $yy' = x - 3y$ is nonlinear.

The distinction between the linear and nonlinear cases is so important and fundamental that it is worthwhile pausing to review our study of first-order equations, in Chapters 2 and 3, in that light.

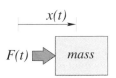

FIGURE 1
Cause and effect.

We are familiar with the idea of "cause and effect." For instance, if we push on a mass with a force $F(t)$ (Fig. 1), then there results a displacement $x(t)$; $F(t)$ is the cause and $x(t)$ is the effect. In place of the terms cause and effect, engineers usually use the terms **input** and **output** (or **response**), respectively: The output is a result, or response, to whatever inputs are present. For instance, for the electrical circuit shown in Fig. 2 and governed by the initial value problem

$$L\frac{di}{dt} + Ri = E(t); \qquad i(0) = i_0 \tag{3}$$

it is natural to think of the resulting current $i(t)$ as the output. There is more than one input: both the applied voltage $E(t)$ and the initial condition i_0. The solution to (3) is

$$i(t) = i_0 e^{-Rt/L} + \frac{1}{L}e^{-Rt/L}\int_0^t e^{Rs/L}E(s)\,ds. \tag{4}$$

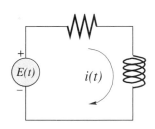

FIGURE 2
RL circuit.

If $E(t) = 0$, so only the i_0 input is present, then (4) gives

$$i(t) = i_0 e^{-Rt/L} \tag{5}$$

as the response to the i_0 input, and if $i_0 = 0$, so only the $E(t)$ input is present, then (4) gives

$$i(t) = \frac{1}{L}e^{-Rt/L}\int_0^t e^{Rs/L}E(s)\,ds \tag{6}$$

as the response to the $E(t)$ input.

We see, from (4), that the total response (i.e., to both inputs acting together) is simply the sum of the responses to each input acting alone. The same is true if there is more than one voltage source in the circuit, as in the circuit shown in Fig. 3

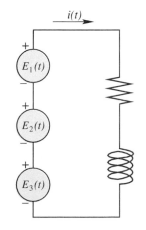

FIGURE 3
Multiple voltage sources.

where there are three such sources. For instance, if there are k such sources then (3) becomes

$$L\frac{di}{dt} + Ri = E_1(t) + E_2(t) + \cdots + E_k(t); \qquad i(0) = i_0 \tag{7}$$

and the solution to (7) is

$$
\begin{aligned}
i(t) &= i_0 e^{-Rt/L} + \frac{1}{L} e^{-Rt/L} \int_0^t e^{Rs/L}[E_1(s) + \cdots + E_k(s)]\, ds \\
&= i_0 e^{-Rt/L} + \frac{1}{L} e^{-Rt/L} \int_0^t e^{Rs/L} E_1(s)\, ds \\
&\quad + \cdots + \frac{1}{L} e^{-Rt/L} \int_0^t e^{Rs/L} E_k(s)\, ds.
\end{aligned}
\tag{8}
$$

From (8) we see that the same result holds:

(i) *The total response is the sum of the responses to each individual input.* [In the present example, the inputs are i_0, $E_1(t), \ldots, E_k(t)$.] This result is known as a superposition principle.

(ii) Also, *each response is proportional to its corresponding input.*

That is, the response $i_0 e^{-Rt/L}$ to the i_0 input is proportional to i_0, the response $\frac{1}{L}e^{-Rt/L}\int_0^t e^{Rs/L}E_1(s)\,ds$ to the $E_1(t)$ input is proportional to $E_1(t)$ (the s merely being a dummy integration variable in place of t), $\ldots$, and finally the response $\frac{1}{L}e^{-Rt/L}\int_0^t e^{Rs/L}E_k(s)\,ds$ to the $E_k(t)$ input is proportional to $E_k(t)$. Thus, if we double i_0 we double its response, if we triple i_0 we triple its response, and so on, and similarly for the inputs $E_k(t), \ldots, E_k(t)$.

The two italicized results (i) and (ii), above, were in regard to the specific circuit problem (7), but the same results hold for the general linear first-order initial value problem

$$y' + p(x)y = q_1(x) + \cdots + q_k(x); \qquad y(a) = b, \tag{9}$$

which has the solution

$$
\begin{aligned}
y(x) &= b e^{-\int_a^x p(s)\, ds} + e^{-\int_a^x p(s)\, ds} \int_a^x e^{\int_a^s p(u)\, du} q_1(s)\, ds \\
&\quad + \cdots + e^{-\int_a^x p(s)\, ds} \int_a^x e^{\int_a^s p(u)\, du} q_k(s)\, ds.
\end{aligned}
\tag{10}
$$

[Of course it doesn't matter whether the variables are t and $x(t)$, x and $y(x)$, or whatever.]

To solidify understanding of (i) and (ii) let us offer a more graphical illustration. Let our system be a cantilever beam such as a diving board, let the inputs be downward forces at certain points along the beam as if due to divers standing at those points, and let the output (i.e., the response) be regarded as the vertical deflection of the beam (Fig. 4), namely, the function $y(x)$. Without even writing down any differential equation governing $y(x)$ we claim, and ask you to accept, that the system is linear and has the properties (i) and (ii). Fig. 4 illustrates property (i): The deflection $y(x)$ due to the collective action of F_1 and F_2 is the sum of the deflection $y_1(x)$ due to the action of F_1 alone, plus the deflection $y_2(x)$ due to the action of F_2 alone. According to property (ii), if we double the force F_1 then we double the deflection $y_1(x)$, if we triple F_1 then we triple $y_1(x)$, and so on.

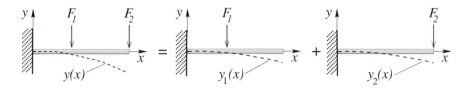

FIGURE 4   Property (i) for a diving board: $y(x) = y_1(x) + y_2(x)$.

In contrast, *nonlinear equations do not have the properties (i) and (ii).* To illustrate, consider the following comparison.

E X A M P L E 1
The *linear* problem

$$y' + y = 0; \qquad y(0) = y_0 \tag{11}$$

admits the solution

$$y(x) = y_0 e^{-x}, \tag{12}$$

which solution, in accordance with property (ii), is proportional to the initial condition input $y(0) = y_0$. But let us modify (11) as

$$y' + y^2 = 0; \qquad y(0) = y_0, \tag{13}$$

which is now *nonlinear* because of the y^2 term. Its solution is found, by separation of variables, to be

$$y(x) = \frac{y_0}{y_0 x + 1} \tag{14}$$

and we see that the latter response to the initial condition input is *not* proportional to y_0; if we double y_0, say, we do not double $y(x)$ because

$$\frac{(2y_0)}{(2y_0)x + 1} \neq 2\frac{y_0}{y_0 x + 1}. \tag{15}$$

COMMENT.   Observe that if we plot (12) for various values of y_0 the result is not so interesting, in the sense that all of the curves are merely scalings of the single curve defined by e^{-x} (Fig. 5). In contrast, there is more variety present in (14) since the curves obtained by assigning different values to y_0 are not merely scalings of a single curve (Fig. 6).

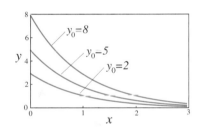

FIGURE 5
The solution curves (12) for the linear problem (11).

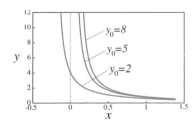

FIGURE 6
The solution curves (14) for the nonlinear problem (13).

For instance, the solution $y(x)$ given by (14) "blows up" (i.e., $y \to \infty$) as $x \to -0.5$ if $y_0 = 2$, as $x \to -0.2$ if $y_0 = 5$, and as $x \to -0.125$ if $y_0 = 8$. Qualitatively put, we can say that the solution to a nonlinear problem is "richer" and contains more variety than the solution to a similar linear problem, which idea will be developed more fully in Chapter 11.   ■

Thus far we have stressed the properties (i) and (ii) for linear problems, and their absence for nonlinear problems, but there are other interesting comparisons to be made between linear and nonlinear differential equations:

Their solutions. The *linear* equation $y' + p(x)y = q(x)$ can be solved once and for all, its general solution being

$$ y(x) = e^{-\int p\, dx} \left(\int e^{\int p\, dx} q\, dx + C \right), \tag{16} $$

and the latter is in explicit form, rather than the less convenient implicit form.

For the *nonlinear* equation $y' = f(x, y)$, however, there is no known solution in general. For a specific $f(x, y)$ one might be able to solve by one or more of the methods that we have studied, such as separation of variables, but even if we are successful the solution might be in implicit rather than explicit form.

The existence and uniqueness of their solutions. For the *linear* equation we have the existence and uniqueness Theorem 2.3.1: "The linear equation $y' + p(x)y = q(x)$ does admit a solution

$$ y(x) = e^{-\int_a^x p(s)\, ds} \left(\int_a^x e^{\int_a^s p(u)\, du} q(s)\, ds + b \right) \tag{17} $$

through an initial point $y(a) = b$ if $p(x)$ and $q(x)$ are continuous at a. That solution exists at least on the broadest x interval, containing $x = a$, over which $p(x)$ and $q(x)$ are continuous, and is unique." In Example 1, for instance, both $p(x) = 1$ and $q(x) = 0$ are continuous for all x so we know, in advance, that there will exist a unique solution of (11) on $-\infty < x < \infty$.

For the *nonlinear* case we have the existence and uniqueness Theorem 3.3.1: "If $f(x, y)$ is a continuous function of x and y on some rectangle R ($|x - a| < A$, $|y - b| < B$) in the x, y plane, then the initial value problem

$$ y' = f(x, y); \qquad y(a) = b \tag{18} $$

has at least one solution defined on some open x interval containing $x = a$. If, in addition, $\partial f / \partial y$ is continuous on R, then the solution to (18) is unique on some open interval containing $x = a$." Observe that this theorem is not as strong as Theorem 2.3.1 for it does not give us the solution or tell us how to find it, and it is not informative about the x interval on which the solution exists. More informative theorems could be cited, but none is as informative as Theorem 2.3.1 for the linear case. In Example 1, for instance, Theorem 3.3.1 does guarantee the existence of a unique solution of $y' = -y^2$ (through *any* initial point in fact), but it does not give us the solution. Nor does it predict the interval over which that solution exists, it simply says over "some" interval containing the initial point. For the initial condition $y(0) = y_0$ we find the solution (14) by separation of variables. With that solution in hand we can see that it blows up ($y \to \infty$) at $x = -1/y_0$. Thus, if $y_0 = 2$, say, then the solution (14) through $y(0) = 2$ exists over $-1/2 < x < \infty$; if $y_0 = 5$, then the solution (14) through $y(0) = 5$ exists over $-1/5 < x < \infty$; and so on.

Closure.   In summary, we have compared the following features of linear and nonlinear equations:

Linear: The general solution is known, for the general linear first-order equation $y' + p(x)y = q(x)$, and is given in explicit form by (16). If there is an initial condition $y(a) = b$, then the solution is given by (17). The latter exists and is unique over the broadest x interval, containing the initial point $x = a$, on which $p(x)$ and $q(x)$ are continuous. Further, the total response equals the sum of the responses to the individual inputs, and the response to each input is proportional to that input.

Nonlinear: The results stated above for linear equations do not hold for nonlinear equations. In a specific case we might be able to solve a nonlinear equation but, even then, the solution might be in implicit rather than explicit form. Theorem 3.3.1 can be used to predict the existence and uniqueness of a solution through an initial point $y(a) = b$, but it is only a "local" theorem in that it does not predict the extent of the x interval on which that solution exists and is unique.

EXERCISES 3.6

1. For the nonlinear initial value problem

$$yy' = f(x); \qquad y(0) = y_0$$

the response is $y(x)$ and there are two inputs, $f(x)$ and y_0. For simplicity, let $f(x) = $ constant $= f_0$. Solve for $y(x)$ and show whether or not the total response $y(x)$ is the sum of the individual responses to y_0 and f_0.

2. For the nonlinear initial value problem

$$yy' = f(x); \qquad y(0) = 0,$$

where $f(x) = f_0$ is a constant, for simplicity, show whether or not the response $y(x)$ is proportional to f_0.

3. A certain first-order differential equation, with initial condition $y(0) = y_0$, admits the following solution. From the form of that solution infer whether the differential equation is linear or nonlinear. Explain your reasoning.

(**a**) $y(x) = y_0(1 + y_0 x^3)$

(**b**) $y(x) = y_0^2/(y_0 + \sin x)$

(**c**) $y(x) = 3y_0^2 \cos(2x)/(3y_0 + 5x)$

(**d**) $y(x) = y_0 e^{(y_0+2)x}$

(**e**) $y(x) = (y_0 + 1 - e^x)^2/(y_0 + 3x)$

(**f**) $y(x) = y_0 \cos[x/(y_0 + x)]$

CHAPTER 3 REVIEW

Whereas in Chapter 2 we studied the *linear* first-order equation $y' + p(x)y = q(x)$, in Chapter 3 we have considered the first-order equation $y' = f(x, y)$ which, in general, is *nonlinear*.

Two solution techniques were put forward, **separation of variables** in Section 3.2 and the method of **exact equations** in Section 3.4. Separation of variables is readily implemented: if $f(x, y)$ can be factored as $X(x)Y(y)$, then the solution of $y'(x) = f(x, y)$ can be found by evaluating the integrals in

$$\int \frac{dy}{Y(y)} = \int X(x)\, dx. \tag{1}$$

According to the method of exact equations, it is convenient (but not essential) to first re-express the differential equation in the differential form

$$M(x, y)dx + N(x, y)dy = 0. \tag{2}$$

If

$$M_y = N_x, \tag{3}$$

with subscripts denoting partial derivatives, then $M\,dx + N\,dy$ is said to be an *exact differential* and the equation (2) is said to be an *exact differential equation*. In that case (see Theorem 3.4.1 for the precise conditions) there exists a function $F(x, y)$ such that

$$M(x, y)\, dx + N(x, y)\, dy = dF(x, y), \tag{4}$$

so (2) becomes $dF = 0$, with solution $F(x, y) = C$. The latter is a general solution of the original differential equation, in implicit form. To find $F(x, y)$ we integrate the equations

$$\frac{\partial F}{\partial x} = M(x, y) \quad \text{and} \quad \frac{\partial F}{\partial y} = N(x, y) \tag{5}$$

with respect to x and y, respectively. If (2) is not exact it is possible that it can be made exact by multiplying it through by a suitable function $\sigma(x, y)$ which is an *integrating factor* of (2). Hopefully, an integrating factor can be found that is a function of x or y alone.

Of these two methods, separation of variables is the simpler, but the method of exact equations is more powerful. Thus, it is suggested that you solve by separation of variables if that method applies; if not, re-express the differential equation in the form (2) and see if the latter is exact or can be made exact by use of an integrating factor.

In Section 3.5 we explored two applications, to population dynamics and to Newtonian mechanics. In the final section, 3.6, we contrasted linear and nonlinear equations. In general, linear problems are much simpler than nonlinear ones, and nonlinear ones contain greater richness and diversity. We can summarize the main results as follows:

Linear: For the linear equation we have the existence and uniqueness Theorem 2.3.1: "The linear equation $y' + p(x)y = q(x)$ does admit an explicit solution

$$y(x) = e^{-\int_a^x p(s)\, ds} \left(\int_a^x e^{\int_a^s p(u)\, du} q(s)\, ds + b \right) \tag{6}$$

through an initial point $y(a) = b$ if $p(x)$ and $q(x)$ are continuous at a. That solution exists at least on the broadest x interval, containing $x = a$, over which $p(x)$ and $q(x)$ are continuous, and is unique." Further, the total response $y(x)$ is the sum of the responses to each individual input, and each response is proportional to its corresponding input.

Nonlinear: For the nonlinear case we have the existence and uniqueness Theorem 3.3.1: "If $f(x, y)$ is a continuous function of x and y on some rectangle R in the x, y plane, then the initial value problem

$$y' = f(x, y); \qquad y(a) = b \tag{7}$$

has at least one solution defined on some open x interval containing $x = a$. If, in addition, $\partial f / \partial y$ is continuous on R, then the solution to (7) is unique on some open interval containing $x = a$." This theorem is not as strong as Theorem 2.3.1 for it does not give us the solution or tell us how to find it, and it is not informative about the x interval on which the solution exists. More informative theorems could be cited, but none is as informative as Theorem 2.3.1 for the linear case. Finally, we found (in the exercises of Section 3.3) that a nonlinear equation can, in exceptional cases, admit additional solutions, called **singular solutions**, that are not contained within the general solution, if the family of solution curves has one or more **envelopes**. Linear equations cannot have singular solutions.

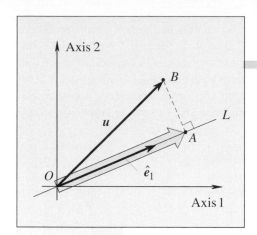

VECTORS AND *n*-SPACE

OVERVIEW OF CHAPTERS 4, 5, AND 9 ON LINEAR ALGEBRA

Chapters 4 and 5 amount to an introduction to a branch of mathematics known as LINEAR ALGEBRA. It would indeed be possible to omit these chapters and to continue, uninterrupted, with our study of differential equations, turning next to equations of second and higher order. But numerous aspects of linear algebra would be needed, and we would need to introduce them on an as-needed basis. Pedagogically, it seems better to develop the linear algebra concepts first. Thus, these two chapters, together with Chapter 9 (which can be studied following Chapter 5 if you wish to keep the linear algebra material together) can be thought of as a mini course in linear algebra, which is included in order to serve our more detailed study of differential equations.

To put Chapters 4 and 5, on vectors and matrices, in perspective, recall our study, in precalculus, of the concept of a function. A **function** f of a single real variable x is a **mapping**, from an x axis to an f axis, as illustrated in Fig. 1a; we can think of x as the input and $f(x)$ as the output. For instance, if $f(x) = x^2$ then the mapping f sends the point $x = 3$ on the x axis into the point $f(x) = 9$ on the f axis. To obtain a graphical display of the mapping we could draw arrows from a number of x points to their image points on the f axis, as we have done in Fig. 1b for the function $f(x) = x^2$.

However, *Descartes* (1596–1650) had a better idea: place the x and f axes at right angles to each other and plot all points $(x, f(x))$ in the "Cartesian" x, f plane. That procedure gives the familiar **graph** of f, as in Fig. 1c.

Thus, logically, one first studies the real axis (integers, rational numbers, and irrational numbers) and then mappings (functions) from one real axis to another.

Similarly in linear algebra, but here the inputs will be "n-dimensional vectors" rather than points on a real axis, and the mappings will be "matrices," which send one vector into another. Thus, just as we studied the real number axis first, in grade school, before we could study mappings from one axis to another, in linear algebra we begin by studying vectors. We do so in Chapter 4. In Chapter 5 we introduce matrices, the principal purpose of which will be to map vectors in one vector space into vectors in another vector space. When we return to differential equations the following will be important to us: the vector concepts of *linear dependence* and *independence*, the compact *matrix notation*, *determinants*, and *systems of linear algebraic equations*.

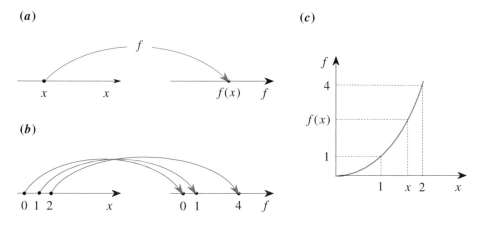

FIGURE 1   Functions as mappings.

Finally, in Chapter 9 we will study the "eigenvalue problem," which will be used to study *systems* of differential equations in Chapter 10.

4.1   INTRODUCTION

Normally, one meets vectors for the first time within some physical context, such as mechanics or the study of electric and magnetic fields. There, the vectors exist within two-dimensional or three-dimensional space and correspond to such entities as force, velocity, position, and magnetic field. They have both *magnitude* and *direction*; they can be *scaled* by multiplicative factors and *added* according to the parallelogram law; *dot* and *cross product* operations are defined between vectors; the *angle* between two vectors is defined; and so on. Alternatively, there is a more formal approach to vectors that generalizes the primitive concept of "arrow vectors" in two-dimensional and three-dimensional space to *n-dimensional space* where n can be greater than three and even infinite. It is this vector concept that we will need when we continue our study of differential equations.

4.2 GEOMETRICAL REPRESENTATION OF "ARROW" VECTORS

4.2.1 Arrow vectors.   Some quantities that we encounter can be completely defined by a single real number, or magnitude, such as the mass or kinetic energy of a given particle, and the temperature or salinity at some point in the ocean. Others are not defined solely by a magnitude but by a magnitude and a direction, examples being force, velocity, momentum, and acceleration. Such quantities are called **vectors**.

The defining features of a vector being magnitude and direction suggests the geometric representation of a vector as a directed line segment, or "arrow," where the length of the arrow is scaled according to the magnitude of the vector. For example, if the wind is blowing at 8 meters/sec from the northeast, that defines a wind-velocity vector **v**, where we adopt **boldface type** to signify that the quantity is a vector; alternative notations include the use of an overhead arrow as in $\vec{v}$. Choosing, according to convenience, a scale of 5 meters/sec per centimeter, say, the geometric representation of **v** is as shown in Fig. 1. Denoting the magnitude, or **norm**, of any vector **v** as $\|\mathbf{v}\|$, we have $\|\mathbf{v}\| = 8$ for the **v** vector in Fig. 1.

Observe that the *location* of a vector is not specified, only its magnitude and direction. Thus, the two unlabeled arrows in Fig. 1 are equally valid alternative representations of **v**. That is not to say that the physical *effect* of the vector will be entirely independent of its position. For example, it should be apparent that the motion of the body $\mathcal{B}$ induced by a force **F** (Fig. 2) will certainly depend on the point of application of **F** as will the stress field induced in $\mathcal{B}$.[1] Nevertheless, the two vectors in Fig. 2 are still regarded as equal, as are the three in Fig. 1.

Like numbers, vectors do not become useful until we introduce rules for their manipulation, that is, a vector algebra. Having elected the arrow representation of vectors, the vector algebra that we now introduce will, likewise, be geometric.

First, we say that two vectors are **equal** if and only if their lengths are identical and if their directions are identical as well.

Next, we define a process of **addition** between any two vectors **u** and **v**. The first step is to move **v** (if necessary), parallel to itself, so that its tail coincides with the head of **u**. Then the sum, or *resultant*, **u** + **v** is defined as the arrow from the tail of **u** to the head of **v**, as in Fig. 3a. Reversing the order, **v** + **u** is as shown in Fig. 3b. Equivalently, we may place **u** and **v** tail to tail, as in Fig. 3c. Comparing Fig. 3c with Fig. 3a and Fig. 3b, we see that the diagonal of the parallelogram (Fig. 3c), is both **u** + **v** and **v** + **u**. Thus,

$$\mathbf{u} + \mathbf{v} = \mathbf{v} + \mathbf{u} \tag{1}$$

so addition is commutative. One can show (Exercise 1) that it is associative as well,

$$(\mathbf{u} + \mathbf{v}) + \mathbf{w} = \mathbf{u} + (\mathbf{v} + \mathbf{w}). \tag{2}$$

---

[1] Students of mechanics know that the point of application of **F** affects the *rotational* part of the motion but not the *translational* part.

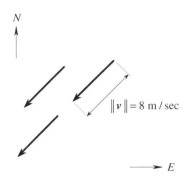

Scale: 5 m/sec/cm

FIGURE 1
Geometric representation of **v**.

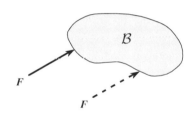

FIGURE 2
Position of a vector.

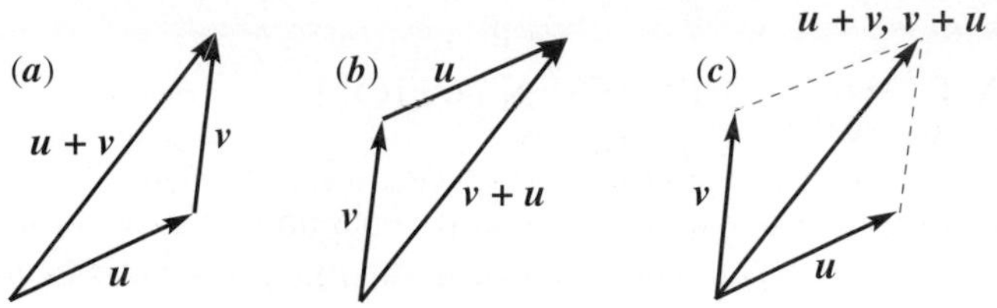

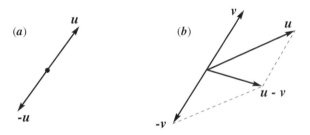

FIGURE 3   Vector addition.

Next, we define any vector of zero length to be a **zero vector**, denoted as **0**. Its length being zero, its direction is immaterial; any direction may be assigned if desired. From the graphical definition of addition above, it follows that

$$\mathbf{u} + \mathbf{0} = \mathbf{0} + \mathbf{u} = \mathbf{u} \tag{3}$$

for each vector **u**.

Corresponding to **u** we define a **negative inverse** "−**u**" such that if **u** is any nonzero vector, then −**u** is determined uniquely, as shown in Fig. 4a; that is, it is of the same length as **u** but is directed in the opposite direction (again, **u** and −**u** have the same length, the length of −**u** is *not* negative). For the zero vector we have −**0** = **0**.

To define vector **subtraction**, we denote **u** + (−**v**) as **u** − **v** ("**u** minus **v**") but emphasize that it is really the addition of **u** and −**v**, as in Fig. 4b.

FIGURE 4   −**u** and vector subtraction.

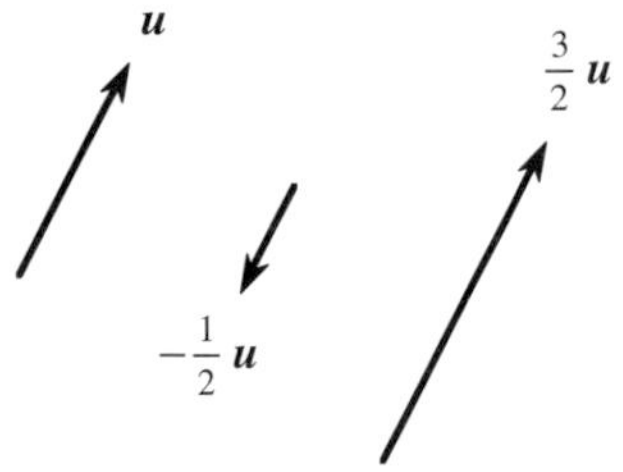

FIGURE 5
Scalar multiplication.

Finally, we introduce another operation, called **scalar multiplication**, between any vector **u** and any scalar (i.e., a real number) α: If $\alpha \neq 0$ and $\mathbf{u} \neq \mathbf{0}$, then $\alpha\mathbf{u}$ is a vector whose length is $|\alpha|$ times the length of **u** and whose direction is the same as that of **u** if $\alpha > 0$, and the opposite if $\alpha < 0$; if $\alpha = 0$ and/or $\mathbf{u} = \mathbf{0}$, then $\alpha\mathbf{u} = \mathbf{0}$. This definition is illustrated in Fig. 5. It follows from this definition that scalar multiplication has the following algebraic properties:

$$\alpha(\beta\mathbf{u}) = (\alpha\beta)\mathbf{u}, \tag{4a}$$

$$(\alpha + \beta)\mathbf{u} = \alpha\mathbf{u} + \beta\mathbf{u}, \tag{4b}$$

$$\alpha(\mathbf{u} + \mathbf{v}) = \alpha\mathbf{u} + \alpha\mathbf{v}, \tag{4c}$$

$$1\mathbf{u} = \mathbf{u}, \tag{4d}$$

where α, β are any scalars and $\mathbf{u}$, $\mathbf{v}$ are any vectors.

Observe that the parallelogram rule of vector addition is a definition so it does not need to be proved. Nevertheless, definitions are not necessarily fruitful so it is worthwhile to reflect for a moment on why the parallelogram rule has turned out to be important and useful. Basically, if we say that "the sum of $\mathbf{u}$ and $\mathbf{v}$ is $\mathbf{w}$," and thereby pass from the two vectors $\mathbf{u}$, $\mathbf{v}$ to the single vector $\mathbf{w}$, it seems fair to expect some sort of equivalence to exist between the action of $\mathbf{w}$ and the joint action of $\mathbf{u}$ and $\mathbf{v}$. For example, if $\mathbf{F}_1$ and $\mathbf{F}_2$ are two forces acting on a body $\mathcal{B}$, as shown in Fig. 6, it is known from fundamental principles of mechanics that their combined effect will be the same as that due to the single force $\mathbf{F}$, so it seems reasonable and natural to say that $\mathbf{F}$ is the sum of $\mathbf{F}_1$ and $\mathbf{F}_2$. This concept goes back at least as far as *Aristotle* (384–322 B.C.). Thus, while the algebra of vectors is developed here as an essentially mathematical matter, it is important to appreciate the role of physics and physical motivation.

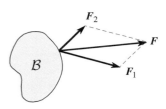

FIGURE 6
Physical motivation for parallelogram addition.

4.2.2 Angle and dot product.   Continuing our discussion, we now define the angle between two vectors and a "dot product" operation between two vectors. The **angle** θ between two nonzero vectors $\mathbf{u}$ and $\mathbf{v}$ will be understood to mean the ordinary angle between the two vectors when they are arranged tail to tail as in Fig. 7. (We will not attempt to define θ if one or both of the vectors is $\mathbf{0}$.) Of course, this definition of θ is ambiguous in that there are *two* such angles, an interior angle ($\leq \pi$) and an exterior angle ($\geq \pi$); for definiteness, we choose θ to be the interior angle,

$$0 \leq \theta \leq \pi, \tag{5}$$

as in Fig. 7. Unless explicitly stated otherwise, angular measure will be understood to be in *radians*, where 1 radian $= 360/2\pi \approx 57.3°$.

Next, we define the so-called **dot product**, $\mathbf{u \cdot v}$, between two vectors $\mathbf{u}$ and $\mathbf{v}$ as

$$\boxed{\mathbf{u \cdot v} \equiv \|\mathbf{u}\| \, \|\mathbf{v}\| \cos\theta;} \tag{6}$$

FIGURE 7
The angle θ between $\mathbf{u}$ and $\mathbf{v}$.

$\|\mathbf{u}\|$, $\|\mathbf{v}\|$, and $\cos\theta$ are scalars so $\mathbf{u \cdot v}$ is a scalar, too. However, the right-hand side of (6) is undefined if $\mathbf{u}$ and/or $\mathbf{v}$ are $\mathbf{0}$, because if one or both are $\mathbf{0}$ then the angle θ in (6) is undefined. To cover this case we *define* $\mathbf{u \cdot v} \equiv \mathbf{0}$ if $\mathbf{u}$ and/or $\mathbf{v}$ are $\mathbf{0}$.

By way of geometrical interpretation, observe that $\|\mathbf{u}\| \cos\theta$ is the length of the orthogonal projection of $\mathbf{u}$ onto the line of action of $\mathbf{v}$ so that $\mathbf{u \cdot v} = \|\mathbf{u}\| \, \|\mathbf{v}\| \cos\theta = (\|\mathbf{v}\|)\,(\|\mathbf{u}\| \cos\theta)$ is the length of $\mathbf{v}$ times the length of the orthogonal projection of $\mathbf{u}$ onto the line of action of $\mathbf{v}$.[1] Alternatively, we could express $\mathbf{u \cdot v} = \|\mathbf{u}\| \, \|\mathbf{v}\| \cos\theta = (\|\mathbf{u}\|)\,(\|\mathbf{v}\| \cos\theta)$; that is, as the length of $\mathbf{u}$ times the length of the orthogonal projection of $\mathbf{v}$ onto the line of action of $\mathbf{u}$ (Fig. 8).

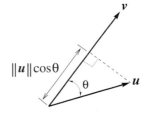

FIGURE 8
Projection of $\mathbf{u}$ onto $\mathbf{v}$.

---

[1]Actually, that statement holds if $0 \leq \theta \leq \pi/2$; if $\pi/2 < \theta \leq \pi$, the cosine is negative, and $\mathbf{u \cdot v} = \|\mathbf{u}\| \, \|\mathbf{v}\| \cos\theta$ is the negative of the length of $\mathbf{v}$ times the length of the orthogonal projection of $\mathbf{u}$ on the line of action of $\mathbf{v}$.

EXAMPLE 1   ***Work Done by a Force.***

In mechanics the *work* W done when a body undergoes a linear displacement from an initial point A to a final point B, under the action of a constant force $\mathbf{F}$ (Fig. 9), is defined as the length of the orthogonal projection of $\mathbf{F}$ onto the line of displacement, positive if $\mathbf{F}$ is "assisting" the motion (i.e., if $0 \le \theta < \pi/2$, as in Fig. 9a) and negative if $\mathbf{F}$ is "opposing" the motion (i.e., if $\pi/2 < \theta \le \pi$, as in Fig. 9b), times the displacement. By the displacement we mean the length of the vector $\mathbf{AB}$ with head at B and tail at A. But that product is precisely the dot product of $\mathbf{F}$ with $\mathbf{AB}$,

$$W = \mathbf{F} \cdot \mathbf{AB}. \tag{7}$$

∎

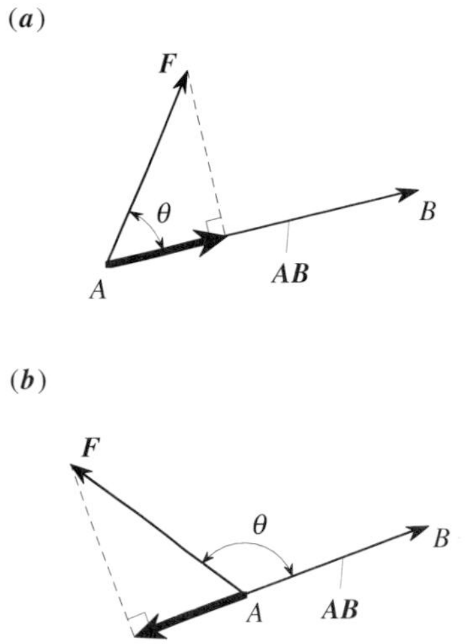

(*a*)

(*b*)

FIGURE 9
Work done by $\mathbf{F}$.

An important special case of the dot product occurs when $\theta = \pi/2$. Then $\mathbf{u}$ and $\mathbf{v}$ are perpendicular, and

$$\mathbf{u} \cdot \mathbf{v} = \|\mathbf{u}\| \, \|\mathbf{v}\| \cos \frac{\pi}{2} = 0. \tag{8}$$

Also of importance is the case where $\mathbf{u} = \mathbf{v}$. Then, according to (6),

$$\mathbf{u} \cdot \mathbf{u} \equiv \|\mathbf{u}\| \, \|\mathbf{u}\| \cos 0 = \|\mathbf{u}\|^2, \tag{9}$$

so that we have

$$\boxed{\|\mathbf{u}\| = \sqrt{\mathbf{u} \cdot \mathbf{u}}.} \tag{10}$$

The relationship (10) between the dot product and the norm will be useful in subsequent sections.

Closure.   For most readers this section will have been a review of ideas met elsewhere, such as in a course in physics. It will be needed as the foundation on which we build "*n*-space" in Section 4.3.

EXERCISES 4.2

1. Show that the associative property (2) follows from the graphical definition of vector addition.

2. Derive the following from the definitions of vector addition and scalar multiplication:

(**a**) property (4a)
(**b**) property (4b)
(**c**) property (4c)
(**d**) property (4d)

3. **(a)** If $\|\mathbf{A}\| = 1$, $\|\mathbf{B}\| = 2$, and $\|\mathbf{C}\| = 5$, can $\mathbf{A} + \mathbf{B} + \mathbf{C} = \mathbf{0}$? HINT: Use the *law of cosines* $s^2 = q^2 + r^2 - 2qr \cos\theta$ (see the accompanying figure) or the Euclidean proposition that the length of any one side of a triangle cannot exceed the sum of the lengths of the other two sides.

   (b) Repeat part (a), with "$\|\mathbf{A}\| = 1$" changed to $\|\mathbf{A}\| = 4$.

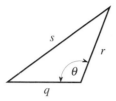

4. Use the definitions and properties given in the reading to show that $\mathbf{A} + \mathbf{B} = \mathbf{C}$ implies that $\mathbf{A} = \mathbf{C} - \mathbf{B}$.

5. **(a)** Show that if $\mathbf{A} + \mathbf{B} = \mathbf{0}$ and $\mathbf{A}$ and $\mathbf{B}$ are not parallel, then each of $\mathbf{A}$ and $\mathbf{B}$ must be $\mathbf{0}$.

   (b) Vectors can be of help in deriving geometrical relationships. For example, to show that the *diagonals of a parallelogram bisect each other* one may proceed as follows. From the accompanying figure $\mathbf{A} + \mathbf{B} = \mathbf{C}$, $\mathbf{A} - \alpha\mathbf{D} = \beta\mathbf{C}$, and $\mathbf{A} = \mathbf{B} + \mathbf{D}$.

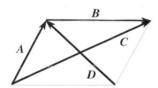

   Eliminating $\mathbf{A}$ and $\mathbf{B}$, we obtain $(2\beta - 1)\mathbf{C} = (1 - 2\alpha)\mathbf{D}$, and since $\mathbf{C}$ and $\mathbf{D}$ are not parallel, it must be true [per part (a)] that $2\beta - 1 = 1 - 2\alpha = 0$ (i.e., $\alpha = \beta = \frac{1}{2}$), which completes the proof. We now state the problem: Use this sort of procedure to show that *a line from one vertex of a parallelogram to the midpoint of a nonadjacent side trisects a diagonal*.

6. If (see the accompanying figure) the vector $\mathbf{A} + \alpha\mathbf{B}$ is placed with its tail at point P, show the line generated by its head as α varies between $-\infty$ and $+\infty$.

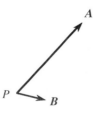

7. If (see the accompanying figure) $\|\mathbf{AB}\| / \|\mathbf{AC}\| = \alpha$, show that $\mathbf{OB} = \alpha\,\mathbf{OC} + (1 - \alpha)\mathbf{OA}$.

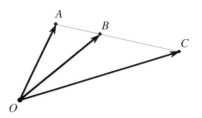

8. Evaluate $\mathbf{u}\cdot\mathbf{v}$ in each case. In (a) $\|\mathbf{u}\| = 5$, in (b) $\|\mathbf{u}\| = 3$, and in (c) $\|\mathbf{u}\| = 6$.

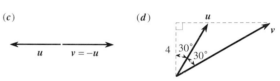

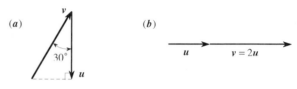

9. (*Properties of the dot product*) Prove each of the following properties of the dot product, where α, β are any scalars, and $\mathbf{u}, \mathbf{v}, \mathbf{w}$ are any vectors.

   (a) $\mathbf{u}\cdot\mathbf{v} = \mathbf{v}\cdot\mathbf{u}$ (commutativity)

   (b) $\mathbf{u}\cdot\mathbf{u} > 0$ for $\mathbf{u} \neq \mathbf{0}$ (nonnegativeness)

   $\quad\quad\;\; = 0$ for $\mathbf{u} = \mathbf{0}$

   (c) $(\alpha\mathbf{u} + \beta\mathbf{v})\cdot\mathbf{w} = \alpha(\mathbf{u}\cdot\mathbf{w}) + \beta(\mathbf{v}\cdot\mathbf{w})$ (linearity)

   HINT: In proving part (c), you may wish to show, first, that part (c) is equivalent to the two conditions $(\mathbf{u} + \mathbf{v})\cdot\mathbf{w} = \mathbf{u}\cdot\mathbf{w} + \mathbf{v}\cdot\mathbf{w}$ and $(\alpha\mathbf{u})\cdot\mathbf{v} = \alpha(\mathbf{u}\cdot\mathbf{v})$.

10. Using the properties given in Exercise 9, show that

$$(\mathbf{u} + \mathbf{v})\cdot(\mathbf{w} + \mathbf{x}) = \mathbf{u}\cdot\mathbf{w} + \mathbf{u}\cdot\mathbf{x} + \mathbf{v}\cdot\mathbf{w} + \mathbf{v}\cdot\mathbf{x}. \quad (10.1)$$

11. If $\mathbf{u}$ and $\mathbf{v}$ are nonzero, show that $\mathbf{w} = \|\mathbf{v}\|\,\mathbf{u} + \|\mathbf{u}\|\,\mathbf{v}$ bisects the angle between $\mathbf{u}$ and $\mathbf{v}$. (You may use any of the properties given in Exercise 9.)

4.3 *n*-SPACE

Here we move away from our dependence on the arrow representation of vectors, in two-dimensional space and three-dimensional space, by introducing an alternative representation in terms of 2-tuples and 3-tuples. This step will lead us to a more general notion of vectors in "*n*-space."

The idea is simple and is based on the familiar representation of *points* in Cartesian 1-space , 2-space , and 3-space as 1-tuples, 2-tuples, and 3-tuples of real numbers. For example, the 2-tuple (a_1, a_2) denotes the point P indicated in Fig. 1a, where a_1, a_2 are the x, y coordinates, respectively. But it can also serve to denote the vector **OP** in Fig. 1b or, indeed, any equivalent vector **QR**.

Thus the vector is now represented as the 2-tuple (a_1, a_2) rather than as an arrow, and while pictures may still be drawn, as in Fig. lb, they are no longer essential and can be discarded if we wish—at least once the algebra of 2-tuples is established (in the next paragraph). The set of all such real 2-tuple vectors will be called **2-space** and will be denoted by the symbol $\mathbb{R}^2$; that is,

$$\mathbb{R}^2 = \{(a_1, a_2) \mid a_1, a_2 \text{ real numbers}\}. \tag{1}$$

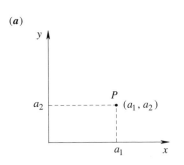

(a)

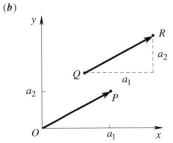

(b)

FIGURE 1
2-tuple representation.

Vectors $\mathbf{u} = (u_1, u_2)$ and $\mathbf{v} = (v_1, v_2)$ in $\mathbb{R}^2$ are *equal* if $u_1 = v_1$ and $u_2 = v_2$; their *sum* is

$$\mathbf{u} + \mathbf{v} = (u_1 + v_1, u_2 + v_2) \tag{2}$$

as can be seen from Fig. 2; the *scalar multiple* $\alpha\mathbf{u}$ is, for any scalar α,

$$\alpha\mathbf{u} = (\alpha u_1, \alpha u_2); \tag{3}$$

the *zero vector* is

$$\mathbf{0} = (0, 0); \tag{4}$$

and the *negative of* $\mathbf{u}$ is

$$-\mathbf{u} = (-u_1, -u_2). \tag{5}$$

Similarly, for $\mathbb{R}^3$:

$$\mathbb{R}^3 = \{(a_1, a_2, a_3) \mid a_1, a_2, a_3 \text{ real numbers}\}. \tag{6}$$

$$\mathbf{u} + \mathbf{v} \equiv (u_1 + v_1, u_2 + v_2, u_3 + v_3), \tag{7}$$

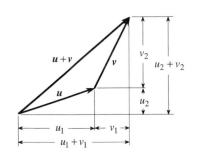

FIGURE 2
Establishing (2).

and so on.[1]

Indeed, why stop at 3-tuples? One may introduce the set of all ordered real **n-tuple** vectors, even if n is greater than 3. We call this **n-space**, and denote it as $\mathbb{R}^n$, that is,

$$\mathbb{R}^n = \{(a_1, \ldots, a_n) \mid a_1, \ldots, a_n \text{ real numbers }\}. \tag{8}$$

Consider two vectors, $\mathbf{u} = (u_1, \ldots, u_n)$ and $\mathbf{v} = (v_1, \ldots, v_n)$, in $\mathbb{R}^n$. The scalars $u_1, \ldots, u_n$ and $v_1, \ldots, v_n$ are called the **components** of $\mathbf{u}$ and $\mathbf{v}$. As you may well

---

[1]The space $\mathbb{R}^1$ of 1-tuples will not be of interest here.

expect, based on our foregoing discussion of $\mathbb{R}^2$ and $\mathbb{R}^3$, **u** and **v** are said to be *equal* if $u_1 = v_1, \ldots, u_n = v_n$, and we define[1]

$$\mathbf{u} + \mathbf{v} \equiv (u_1 + v_1, \ldots, u_n + v_n), \qquad \text{(addition)} \qquad (9a)$$

$$\alpha\mathbf{u} \equiv (\alpha u_1, \ldots, \alpha u_n), \qquad \text{(scalar multiplication)} \qquad (9b)$$

$$\mathbf{0} \equiv (0, \ldots, 0), \qquad \text{(zero vector)} \qquad (9c)$$

$$-\mathbf{u} \equiv (-1)\mathbf{u}, \qquad \text{(negative inverse)} \qquad (9d)$$

$$\mathbf{u} - \mathbf{v} \equiv \mathbf{u} + (-\mathbf{v}). \qquad (9e)$$

From these definitions we may deduce the following properties:

$$\mathbf{u} + \mathbf{v} = \mathbf{v} + \mathbf{u}, \qquad \text{(commutative)} \qquad (10a)$$

$$(\mathbf{u} + \mathbf{v}) + \mathbf{w} = \mathbf{u} + (\mathbf{v} + \mathbf{w}), \qquad \text{(associative)} \qquad (10b)$$

$$\mathbf{u} + \mathbf{0} = \mathbf{u}, \qquad (10c)$$

$$\mathbf{u} + (-\mathbf{u}) = \mathbf{0}, \qquad (10d)$$

$$\alpha(\beta\mathbf{u}) = (\alpha\beta)\mathbf{u}, \qquad \text{(associative)} \qquad (10e)$$

$$(\alpha + \beta)\mathbf{u} = \alpha\mathbf{u} + \beta\mathbf{u}, \qquad \text{(distributive)} \qquad (10f)$$

$$\alpha(\mathbf{u} + \mathbf{v}) = \alpha\mathbf{u} + \alpha\mathbf{v}, \qquad \text{(distributive)} \qquad (10g)$$

$$1\mathbf{u} = \mathbf{u}, \qquad (10h)$$

$$0\mathbf{u} = \mathbf{0}, \qquad (10i)$$

$$(-1)\mathbf{u} = -\mathbf{u}, \qquad (10j)$$

$$\alpha\mathbf{0} = \mathbf{0}. \qquad (10k)$$

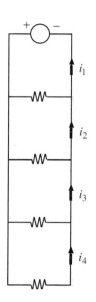

FIGURE 3

Relevance of *n*-space to an electrical circuit.

From property (10b) we see that we can write $\mathbf{u} + \mathbf{v} + \mathbf{w}$ without ambiguity. That is, whether we first add **u** and **v** and then add **w** to their sum, or whether we first add **v** and **w** and then add **u** to their sum, doesn't matter for, according to (10b), the results are the same. Similarly, from (10e) we see that we can write $\alpha\beta\mathbf{u}$ without ambiguity.

To illustrate how such *n*-tuples arise, observe that the state of the electrical system shown in Fig. 3 may be defined by the four currents i_1, i_2, i_3, i_4. These may be regarded as the components of a single vector $\mathbf{i} = (i_1, i_2, i_3, i_4)$ in $\mathbb{R}^4$.

Of course, the notation of $(u_1, \ldots, u_n)$ as a *point* or *arrow* in "*n*-dimensional space" can be realized graphically only if $n \leq 3$; if $n > 3$, the interpretation is valid only in an abstract, or schematic, sense. However, our inability to carry out traditional Cartesian graphical constructions for $n > 3$ will be no hindrance. Indeed, part of the idea here is to move *away* from a dependence on graphical constructions.

Closure. We have used the familiar arrow vector concept as a ladder on which to stand, to introduce a more general kind of vector, an *n*-tuple $\mathbf{u} = (u_1, \ldots, u_n)$. Just as arrow vectors "live" in ordinary physical 2-space and 3-space, an *n*-tuple vector lives in *n*-space. (We refrain from calling it "*n*-dimensional space," for now, because the *dimension* of a space is a concept that we have not yet introduced.) Beginning in 2-space, the key idea was to introduce a Cartesian coordinate system in terms of

[1]We use $\equiv$ to denote *equal to by definition*.

which we could re-express arrow vectors $\mathbf{u}$ and $\mathbf{v}$ as $\mathbf{u} = (u_1, u_2)$ and $\mathbf{v} = (v_1, v_2)$, where the scalars u_1, u_2, v_1, v_2 are the components of $\mathbf{u}$ and $\mathbf{v}$, respectively. In terms of this 2-tuple notation, the definitions of vector addition, scalar multiplication, and so on, can be expressed as $\mathbf{u} + \mathbf{v} = (u_1 + v_1, u_2 + v_2)$, $\alpha\mathbf{u} = (\alpha u_1, \alpha u_2)$, and so on. The same is true for 3-space. To build on these results, we defined n-space as the set of all vectors of the form $\mathbf{u} = (u_1, \ldots, u_n)$, we defined vector addition by $\mathbf{u} + \mathbf{v} = (u_1 + v_1, \ldots, u_n + v_n)$, we defined scalar multiplication by $\alpha\mathbf{u} = (\alpha u_1, \ldots \alpha u_n)$, and so on.

This transition, from arrow vectors in 2-space and 3-space to abstract vectors in n-space illustrates the power of mathematical *notation*, for it was the 2-tuple and 3-tuple notation itself that suggested the idea of n-space.

EXERCISES 4.3

1. If $\mathbf{t} = (5, 0, 1, 2)$, $\mathbf{u} = (2, -1, 3, 4)$, $\mathbf{v} = (4, -5, 1)$, $\mathbf{w} = (-1, -2, 5, 6)$, evaluate each of the following (as a single vector); if the operation is undefined (i.e., if it has not been defined here), state that. At each step cite the equation number of the definition or property being used.

  (**a**) $2\mathbf{t} + 7\mathbf{u}$ (**b**) $3\mathbf{t} - 5\mathbf{u}$
  (**c**) $4[\mathbf{u} + 5(\mathbf{w} - 2\mathbf{u})]$ (**d**) $4\mathbf{t}\mathbf{u} + \mathbf{w}$
  (**e**) $-\mathbf{w} + \mathbf{t}$ (**f**) $2\mathbf{t}/\mathbf{u}$
  (**g**) $\mathbf{t} + 2\mathbf{u} + 3\mathbf{v}$ (**h**) $\mathbf{t} \quad 2\mathbf{u} - 4\mathbf{v}$
  (**i**) $\mathbf{u}(3\mathbf{t} + \mathbf{w})$ (**j**) $\mathbf{u}^2 + 2\mathbf{t}$
  (**k**) $2\mathbf{t} + 7\mathbf{u} - 4$ (**l**) $\mathbf{u} + \mathbf{w}\mathbf{t}$
  (**m**) $\sin\mathbf{u}$ (**n**) $\mathbf{w} + \mathbf{t} - 2\mathbf{u}$

2. Let $\mathbf{u} = (1, 3, 0, -2)$, $\mathbf{v} = (2, 0, -5, 0)$, and $\mathbf{w} = (4, 3, 2, -1)$.

  (**a**) If $3\mathbf{u} - \mathbf{x} = 4(\mathbf{v} + 2\mathbf{x})$, solve for $\mathbf{x}$.
  (**b**) If $\mathbf{x} + \mathbf{u} + \mathbf{v} + \mathbf{w} = \mathbf{0}$, solve for $\mathbf{x}$.

3. Let $\mathbf{u}$, $\mathbf{v}$, and $\mathbf{w}$ be as given in Exercise 2. Citing the definition or property used, at each step, solve each of the following for $\mathbf{x}$. NOTE: Besides the definitions and properties stated in this section, it should be clear that if $\mathbf{x} = \mathbf{y}$, then $\mathbf{x} + \mathbf{z} = \mathbf{y} + \mathbf{z}$ for any $\mathbf{z}$, and $\alpha\mathbf{x} = \alpha\mathbf{y}$ for any α (adding and multiplying equals by equals).

  (**a**) $3\mathbf{x} + 2(\mathbf{u} - 5\mathbf{v}) = \mathbf{w}$
  (**b**) $3\mathbf{x} = 40 + (1, 0, 0, 0)$
  (**c**) $\mathbf{u} - 4\mathbf{x} = \mathbf{0}$
  (**d**) $\mathbf{u} + \mathbf{v} - 2\mathbf{x} = \mathbf{w}$

4. If $\mathbf{t} = (2, 1, 3)$, $\mathbf{u} = (1, 2, -4)$, $\mathbf{v} = (0, 1, 1)$, $\mathbf{w} = (-2, 1, -1)$, solve each of the following for the scalars α_1, α_2. If no such scalars exist, state that.

  (**a**) $\alpha_1\mathbf{t} + \alpha_2\mathbf{u} = \mathbf{0}$
  (**b**) $\alpha_1\mathbf{t} + \alpha_2\mathbf{v} = \mathbf{0}$
  (**c**) $\alpha_1\mathbf{t} + \alpha_2\mathbf{u} = (5, 4, 2)$
  (**d**) $\alpha_1\mathbf{t} + \alpha_2\mathbf{v} = (4, -1, 3)$
  (**e**) $\alpha_1\mathbf{u} + \alpha_2\mathbf{v} = (0, 0, 1)$
  (**f**) $\alpha_1\mathbf{u} + \alpha_2\mathbf{v} = 3\mathbf{w} - (7, 3, -3)$
  (**g**) $\alpha_1\mathbf{u} - \alpha_2\mathbf{w} = 2\mathbf{t} - (3, -5, 19)$
  (**h**) $\alpha_1\mathbf{u} - \alpha_2\mathbf{t} = (1, 6, -14) - \mathbf{v}$

4.4   DOT PRODUCT, NORM, AND ANGLE FOR n-SPACE

4.4.1 Dot product, norm, and angle. We wish to define the norm of an n-tuple vector, and the dot product and angle between two n-tuple vectors, just as we did for "arrow vectors." These definitions should be expressed in terms of the *components* of the n-tuples since the graphical and geometrical arguments used for arrow vectors will not be possible here for $n > 3$. Thus, if $\mathbf{u} = (u_1, \ldots, u_n)$, we wish to define the *norm* or "length" of $\mathbf{u}$, denoted as $\|\mathbf{u}\|$, in terms of the components $u_1, \ldots, u_n$ of $\mathbf{u}$;

and given another vector $\mathbf{v} = (v_1, \ldots, v_n)$, we wish to define the *angle* θ between $\mathbf{u}$ and $\mathbf{v}$, and the *dot product* $\mathbf{u} \cdot \mathbf{v}$, in terms of $u_1, \ldots, u_n$ and $v_1, \ldots, v_n$.

Furthermore, we would like these definitions to reduce to the definitions given in Section 4.2 in the event that $n = 2$ or 3.

Let us begin with the dot product. Our plan is to return to the arrow vector formula

$$\mathbf{u} \cdot \mathbf{v} = \|\mathbf{u}\| \, \|\mathbf{v}\| \cos \theta, \tag{1}$$

to re-express it in terms of vector components for $\mathbb{R}^2$ and $\mathbb{R}^3$, and then to generalize those forms to $\mathbb{R}^n$.

If $\mathbf{u}$ and $\mathbf{v}$ are vectors in $\mathbb{R}^2$ as shown in Fig. 1, formula (1) may be expressed in terms of the components of $\mathbf{u}$ and $\mathbf{v}$ as follows:

$$\begin{aligned}
\mathbf{u} \cdot \mathbf{v} &= \|\mathbf{u}\| \, \|\mathbf{v}\| \cos \theta \\
&= \|\mathbf{u}\| \, \|\mathbf{v}\| \cos (\beta - \alpha) \\
&= \|\mathbf{u}\| \, \|\mathbf{v}\| (\cos \beta \cos \alpha + \sin \beta \sin \alpha) \\
&= (\|\mathbf{u}\| \cos \alpha) (\|\mathbf{v}\| \cos \beta) + (\|\mathbf{u}\| \sin \alpha) (\|\mathbf{v}\| \sin \beta) \\
&= u_1 v_1 + u_2 v_2. \tag{2}
\end{aligned}$$

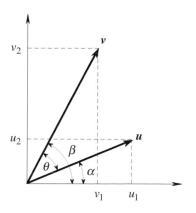

FIGURE 1
$\mathbf{u} \cdot \mathbf{v}$ in terms of
the components of $\mathbf{u}$ and $\mathbf{v}$.

We state, without derivation, that the analogous result for $\mathbb{R}^3$ is

$$\mathbf{u} \cdot \mathbf{v} = u_1 v_1 + u_2 v_2 + u_3 v_3. \tag{3}$$

Generalizing (2) and (3) to $\mathbb{R}^n$, it is eminently reasonable to define the (scalar-valued) dot product of two *n*-tuple vectors $\mathbf{u} = (u_1, \ldots, u_n)$ and $\mathbf{v} = (v_1, \ldots, v_n)$ as

$$\mathbf{u} \cdot \mathbf{v} \equiv u_1 v_1 + u_2 v_2 + \cdots + u_n v_n = \sum_{j=1}^{n} u_j v_j. \tag{4}$$

Understand that we have not proved (4); it is a definition.

Defining the dot product is the key, for now $\|\mathbf{u}\|$ and θ follow readily. Do you recall the arrow vector formulas $\|\mathbf{u}\| = \sqrt{\mathbf{u} \cdot \mathbf{u}}$ and $\mathbf{u} \cdot \mathbf{v} = \|\mathbf{u}\| \, \|\mathbf{v}\| \cos \theta$ in Section 4.2.2? Having now defined a dot product for *n*-space, according to (4), we can use those formulas to define $\|\mathbf{u}\|$ and θ for *n*-space as follows:

$$\|\mathbf{u}\| \equiv \sqrt{\mathbf{u} \cdot \mathbf{u}} = \sqrt{\sum_{j=1}^{n} u_j^2} \tag{5}$$

and

$$\theta \equiv \cos^{-1} \left(\frac{\mathbf{u} \cdot \mathbf{v}}{\|\mathbf{u}\| \, \|\mathbf{v}\|} \right), \tag{6}$$

where the inverse cosine is understood to be in the interval $[0, \pi]$.[1] Again, we emphasize that we have not derived or proved (5) and (6), they are *definitions* based upon the arrow vector formulas $\|\mathbf{u}\| = \sqrt{\mathbf{u} \cdot \mathbf{u}}$ and $\mathbf{u} \cdot \mathbf{v} = \|\mathbf{u}\| \|\mathbf{v}\| \cos \theta$.

Notice that (5) amounts to a generalization of the familiar Pythagorean theorem, which is

$$\|\mathbf{u}\| = \sqrt{u_1^2 + u_2^2}$$

in 2-space, and

$$\|\mathbf{u}\| = \sqrt{u_1^2 + u_2^2 + u_3^2}$$

in 3-space.

Other dot products and norms are sometimes defined for n-space, but we choose to use (4) and (5), which are known as the **Euclidean dot product** and **Euclidean norm**, respectively. To signify that the Euclidean dot product and norm have been adopted, we henceforth refer to the space as **Euclidean n-space**, rather than just n-space. We will still denote it by the symbol $\mathbb{R}^n$ (although some authors prefer the notation $\mathbb{E}^n$).

EXAMPLE 1
Let $\mathbf{u} = (1, 0)$ and $\mathbf{v} = (2, -2)$. Then

$$
\begin{aligned}
\mathbf{u} \cdot \mathbf{v} &= (1)(2) + (0)(-2) = 2, \\
\|\mathbf{u}\| &= \sqrt{(1)^2 + (0)^2} = 1, \\
\|\mathbf{v}\| &= \sqrt{(2)^2 + (-2)^2} = 2\sqrt{2}, \\
\theta &= \cos^{-1}\left(\frac{2}{2\sqrt{2}}\right) = \frac{\pi}{4} \quad \text{(or } 45°\text{)},
\end{aligned}
$$

as is readily verified if we sketch $\mathbf{u}$ and $\mathbf{v}$ as arrow vectors in a Cartesian plane.   ∎

EXAMPLE 2
Let $\mathbf{u} = (2, -2, 4, -1)$ and $\mathbf{v} = (5, 9, -1, 0)$. Then,

$$\mathbf{u} \cdot \mathbf{v} = (2)(5) + (-2)(9) + (4)(-1) + (-1)(0) = -12, \tag{7}$$

$$\|\mathbf{u}\| = \sqrt{(2)^2 + (-2)^2 + (4)^2 + (-1)^2} = 5, \tag{8}$$

$$\|\mathbf{v}\| = \sqrt{(5)^2 + (9)^2 + (-1)^2 + (0)^2} = \sqrt{107}, \tag{9}$$

$$\theta = \cos^{-1}\left(\frac{-12}{5\sqrt{107}}\right) = \cos^{-1}(-0.232) = 1.805 \quad \text{(or } 103.4°\text{)}. \tag{10}$$

In this case, $n \,(= 4)$ is greater than 3 so (7) through (10) are not to be understood in any physical or graphical sense, but in terms of the definitions (4) to (6).

---

[1] By the "interval $[a, b]$ on a real x axis," we mean the points $a \leq x \leq b$. Such an interval is said to be **closed** since it includes the two endpoints. To denote the **open** interval $a < x < b$, we would write (a, b). Similarly, $[a, b)$ means $a \leq x < b$, and $(a, b]$ means $a < x \leq b$. Implicit in the closed-interval notation $[a, b]$ is the finiteness of a and b.

COMMENT.   The dot product of $\mathbf{u} = (2, -2, 4)$ and $\mathbf{v} = (5, 9, -1, 0)$, on the other hand, is *not defined* since here $\mathbf{u}$ and $\mathbf{v}$ are members of different spaces, $\mathbb{R}^3$ and $\mathbb{R}^4$, respectively. It is *not* legitimate to augment $\mathbf{u}$ to the form $(2, -2, 4, 0)$ on the grounds that "surely adding a zero can't hurt." ∎

There is one catch that you may have noticed: (6) serves to define a (real) angle θ only if the argument of the inverse cosine is less than or equal to unity in magnitude, that is, only if

$$-1 \leq \frac{\mathbf{u}\cdot\mathbf{v}}{\|\mathbf{u}\|\|\mathbf{v}\|} \leq 1$$

or, equivalently,

$$\boxed{|\mathbf{u}\cdot\mathbf{v}| \leq \|\mathbf{u}\|\|\mathbf{v}\|.} \tag{11}$$

That is, if for given vectors $\mathbf{u}$ and $\mathbf{v}$ the argument of the $\cos^{-1}$ in (6) is greater than unity in magnitude, then the definition (6) fails. That did not happen in Examples 1 or 2, but we wonder if it might happen in other cases. In fact, (11) holds for *any* two vectors $\mathbf{u}$ and $\mathbf{v}$ in $\mathbb{R}^n$, proof of which is deferred to the exercises, and is known as the **Schwarz inequality**.[1] Whereas double braces denote vector norms, the single braces in (11) denote the absolute value of the scalar $\mathbf{u}\cdot\mathbf{v}$; for instance, $|5| = 5$ and $|-5| = 5$. As a result of the Schwarz inequality (11), we are assured that the argument of the inverse cosine in (6) is indeed less than or equal to unity in magnitude, so (6) does define a (real) angle θ for any given vectors $\mathbf{u}$ and $\mathbf{v}$ in $\mathbb{R}^n$.

4.4.2 Properties of the dot product.   The dot product defined by (4) possesses the following important properties:

$$\begin{array}{llr} \textit{Commutative}: & \mathbf{u}\cdot\mathbf{v} = \mathbf{v}\cdot\mathbf{u}, & (12\mathrm{a}) \\[6pt] \textit{Nonnegative}: & \mathbf{u}\cdot\mathbf{u} > 0 \quad \text{for all } \mathbf{u} \neq \mathbf{0} & \\[4pt] & \quad\quad\quad = 0 \quad \text{for } \mathbf{u} = \mathbf{0}, & (12\mathrm{b}) \\[6pt] \textit{Linear}: & (\alpha\mathbf{u} + \beta\mathbf{v})\cdot\mathbf{w} = \alpha(\mathbf{u}\cdot\mathbf{w}) + \beta(\mathbf{v}\cdot\mathbf{w}), & (12\mathrm{c}) \end{array}$$

for any scalars α, β and any vectors $\mathbf{u}$, $\mathbf{v}$, $\mathbf{w}$. The linearity condition (12c) is equivalent to the two conditions $(\mathbf{u} + \mathbf{v})\cdot\mathbf{w} = (\mathbf{u}\cdot\mathbf{w}) + (\mathbf{v}\cdot\mathbf{w})$ and $(\alpha\mathbf{u})\cdot\mathbf{v} = \alpha(\mathbf{u}\cdot\mathbf{v})$. Verification of these claims is left for the exercises.

EXAMPLE 3
Expand the dot product $(\alpha\mathbf{p} + \beta\mathbf{q})\cdot(\gamma\mathbf{r} + \delta\mathbf{s})$. Using (12), we obtain

$$\begin{aligned} (\alpha\mathbf{p} + \beta\mathbf{q})\cdot(\gamma\mathbf{r} + \delta\mathbf{s}) &= \alpha[\mathbf{p}\cdot(\gamma\mathbf{r} + \delta\mathbf{s})] + \beta[\mathbf{q}\cdot(\gamma\mathbf{r} + \delta\mathbf{s})] && \text{by (12c)} \\ &= \alpha[(\gamma\mathbf{r} + \delta\mathbf{s})\cdot\mathbf{p}] + \beta[(\gamma\mathbf{r} + \delta\mathbf{s})\cdot\mathbf{q}] && \text{by (12a)} \\ &= \alpha[\gamma(\mathbf{r}\cdot\mathbf{p}) + \delta(\mathbf{s}\cdot\mathbf{p})] + \beta[\gamma(\mathbf{r}\cdot\mathbf{q}) + \delta(\mathbf{s}\cdot\mathbf{q})] && \text{by (12c)} \end{aligned}$$

---

[1] After *Hermann Amandus Schwarz* (1843–1921). The names *Cauchy* and *Bunyakovsky* are also associated with this inequality.

$$= \alpha\gamma\,(\mathbf{r}\cdot\mathbf{p}) + \alpha\delta\,(\mathbf{s}\cdot\mathbf{p}) + \beta\gamma\,(\mathbf{r}\cdot\mathbf{q}) + \beta\delta\,(\mathbf{s}\cdot\mathbf{q})$$
$$= \alpha\gamma\,(\mathbf{p}\cdot\mathbf{r}) + \alpha\delta\,(\mathbf{p}\cdot\mathbf{s}) + \beta\gamma\,(\mathbf{q}\cdot\mathbf{r}) + \beta\delta\,(\mathbf{q}\cdot\mathbf{s}) \qquad \text{by (12a)}$$

in much the same way that we obtain $(a+b)(c+d) = ac+ad+bc+bd$ in scalar arithmetic. ■

4.4.3 Properties of the norm.   Since the norm is related to the dot product according to

$$\|\mathbf{u}\| = \sqrt{\mathbf{u}\cdot\mathbf{u}}, \tag{13}$$

the properties (12) of the dot product imply certain corresponding properties of the norm. These properties are as follows:

$$\text{\textit{Scaling}:} \qquad \|\alpha\mathbf{u}\| = |\alpha|\,\|\mathbf{u}\|, \tag{14a}$$
$$\text{\textit{Nonnegative}:} \qquad \|\mathbf{u}\| > 0 \quad \text{for all } \mathbf{u} \neq \mathbf{0} \tag{14b}$$
$$= 0 \quad \text{for } \mathbf{u} = \mathbf{0},$$
$$\text{\textit{Triangle Inequality}:} \qquad \|\mathbf{u}+\mathbf{v}\| \leq \|\mathbf{u}\| + \|\mathbf{v}\|. \tag{14c}$$

Equation (14a) simply says that $\alpha\mathbf{u}$ is $|\alpha|$ times as long as $\mathbf{u}$. For arrow representations of 2-tuples or 3-tuples the triangle inequality (14c) amounts to the Euclidean proposition that the length of any one side of a triangle cannot exceed the sum of the lengths of the other two sides (Fig. 2). Less obvious, however, is the fact that (14c) holds for n-tuples for n's > 3.

Let us prove only (14a) and (14c) since (14b) follows readily from (5). First, (14a):

$$\|\alpha\mathbf{u}\| = \sqrt{(\alpha\mathbf{u})\cdot(\alpha\mathbf{u})} \qquad\qquad \text{by (13)}$$
$$= \sqrt{\alpha\mathbf{u}\cdot(\alpha\mathbf{u})} \qquad\qquad \text{by (12c) with } \beta = 0 \text{ and } \mathbf{w} = \alpha\mathbf{u}$$
$$= \sqrt{\alpha(\alpha\mathbf{u})\cdot\mathbf{u}} \qquad\qquad \text{by (12a)}$$
$$= \sqrt{\alpha^2\mathbf{u}\cdot\mathbf{u}} \qquad\qquad \text{by (12c) with } \beta = 0 \text{ and } \mathbf{w} = \mathbf{u}$$
$$= |\alpha|\,\sqrt{\mathbf{u}\cdot\mathbf{u}} = |\alpha|\,\|\mathbf{u}\|.$$

Turning to (14c), we find that

$$\|\mathbf{u}+\mathbf{v}\|^2 = (\mathbf{u}+\mathbf{v})\cdot(\mathbf{u}+\mathbf{v}) \qquad\qquad \text{by (13)}$$
$$= \mathbf{u}\cdot\mathbf{u} + \mathbf{v}\cdot\mathbf{u} + \mathbf{u}\cdot\mathbf{v} + \mathbf{v}\cdot\mathbf{v} \qquad\qquad \text{as in Example 3}$$
$$= \|\mathbf{u}\|^2 + 2\mathbf{u}\cdot\mathbf{v} + \|\mathbf{v}\|^2 \qquad\qquad \text{by (13) and (12a)}$$
$$\leq \|\mathbf{u}\|^2 + 2\,|\mathbf{u}\cdot\mathbf{v}| + \|\mathbf{v}\|^2$$
$$\leq \|\mathbf{u}\|^2 + 2\,\|\mathbf{u}\|\,\|\mathbf{v}\| + \|\mathbf{v}\|^2 \qquad\qquad \text{by (11)}$$
$$= (\|\mathbf{u}\| + \|\mathbf{v}\|)^2$$

so that

$$\|\mathbf{u}+\mathbf{v}\| \leq \|\mathbf{u}\| + \|\mathbf{v}\|,$$

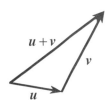

FIGURE 2
Triangle inequality.

as claimed. A key step was the use of the Schwarz inequality (11), but we also used the inequality $\mathbf{u}\cdot\mathbf{v} \leq |\mathbf{u}\cdot\mathbf{v}|$, which is simply the inequality $a \leq |a|$ that holds for *any* real number a.

EXAMPLE 4

Let us illustrate the triangle inequality (14c) for a specific example. Let $\mathbf{u} = (2, 1, 3, -1)$ and $\mathbf{v} = (0, 4, 2, 1)$. Then $\mathbf{u} + \mathbf{v} = (2, 5, 5, 0)$ so (14c) becomes

$$\sqrt{54} \leq \sqrt{15} + \sqrt{21}$$

or $7.348 \leq 3.873 + 4.583$, which is indeed true.   ∎

4.4.4 Orthogonality.   If $\mathbf{u}$ and $\mathbf{v}$ are nonzero vectors such that $\mathbf{u}\cdot\mathbf{v} = 0$, then

$$\theta = \cos^{-1}\left(\frac{\mathbf{u}\cdot\mathbf{v}}{\|\mathbf{u}\|\,\|\mathbf{v}\|}\right) = \cos^{-1}\left(\frac{0}{\|\mathbf{u}\|\,\|\mathbf{v}\|}\right) = \cos^{-1}(0) = \frac{\pi}{2}, \qquad (15)$$

and we say that $\mathbf{u}$ and $\mathbf{v}$ are *perpendicular*. But to say that the condition $\mathbf{u}\cdot\mathbf{v} = 0$ implies perpendicularity ($\theta = \pi/2$) would not be correct, since $\mathbf{u}\cdot\mathbf{v}$ will be zero also in the event that $\mathbf{u}$ and/or $\mathbf{v}$ are $\mathbf{0}$, in which case θ is not defined. Let us therefore make a distinction between perpendicularity and "orthogonality." We will say that $\mathbf{u}$ and $\mathbf{v}$ are **orthogonal** if

$$\mathbf{u}\cdot\mathbf{v} = 0. \qquad (16)$$

Only if $\mathbf{u}$ and $\mathbf{v}$ are both nonzero does their orthogonality imply their perpendicularity (i.e., $\theta = \pi/2$). This definition implies that the zero vector $\mathbf{0}$ is orthogonal to *every* vector including itself.

Finally, we say that a *set* of vectors, say $\{\mathbf{u}_1, \ldots, \mathbf{u}_k\}$, is an **orthogonal set** if every vector in the set is orthogonal to every other one:

$$\mathbf{u}_i\cdot\mathbf{u}_j = 0 \quad \text{if } i \neq j. \qquad (17)$$

EXAMPLE 5

$\mathbf{u}_1 = (2, 3, -1, 0)$, $\mathbf{u}_2 = (1, 2, 8, 3)$, $\mathbf{u}_3 = (9, -6, 0, 1)$ is an orthogonal set because $\mathbf{u}_1\cdot\mathbf{u}_2 = \mathbf{u}_1\cdot\mathbf{u}_3 = \mathbf{u}_2\cdot\mathbf{u}_3 = 0$.   ∎

EXAMPLE 6

$\mathbf{u}_1 = (1, 3)$, $\mathbf{u}_2 = (0, 0)$ is an orthogonal set because $\mathbf{u}_1\cdot\mathbf{u}_2 = 0$.   ∎

4.4.5 Normalization.   Any nonzero vector $\mathbf{u}$ can be scaled to have unit length by multiplying it by $1/\|\mathbf{u}\|$ so we say that the vector

$$\boxed{\hat{\mathbf{u}} = \frac{1}{\|\mathbf{u}\|}\mathbf{u}} \qquad (18)$$

has been "normalized." That $\hat{\mathbf{u}}$ has unit length is readily verified:

$$\|\hat{\mathbf{u}}\| = \left\| \frac{1}{\|\mathbf{u}\|}\mathbf{u} \right\| = \left| \frac{1}{\|\mathbf{u}\|} \right| \|\mathbf{u}\| \qquad \text{by (14a)}$$

$$= \frac{1}{\|\mathbf{u}\|}\|\mathbf{u}\| \qquad \text{by (14b)}$$

$$= 1.$$

A vector of unit length is called a **unit vector**. We will often use the caret notation $\hat{\mathbf{u}}$ for unit vectors. Be careful: In this book the term "normalized" does not mean "made perpendicular to"; it means scaled to have unit length.

EXAMPLE 7
Normalize $\mathbf{u} = (1, -1, 0, 2)$. Since $\|\mathbf{u}\| = \sqrt{\mathbf{u}\cdot\mathbf{u}} = \sqrt{6}$, we have

$$\hat{\mathbf{u}} = \frac{1}{\|\mathbf{u}\|}\mathbf{u} = \frac{1}{\sqrt{6}}(1, -1, 0, 2) = \left(\frac{1}{\sqrt{6}}, -\frac{1}{\sqrt{6}}, 0, \frac{2}{\sqrt{6}} \right). \qquad \blacksquare$$

A set of vectors is said to be **orthonormal** if it is orthogonal and if each vector is normalized (i.e., is a unit vector). We will use that term so frequently that it will be useful to abbreviate it as ON, but be cautioned that *that abbreviation is not standard outside of this book*. Thus, $\{\mathbf{u}_1, \ldots, \mathbf{u}_k\}$ is ON if and only if $\mathbf{u}_i\cdot\mathbf{u}_j = 0$ whenever $i \neq j$ (for orthogonality), and $\mathbf{u}_j\cdot\mathbf{u}_j = 1$ for each j (so $\|\mathbf{u}_j\| = 1$, so the set is normalized).

The symbol

$$\delta_{ij} = \left\{ \begin{array}{ll} 1, & i = j \\ 0, & i \neq j \end{array} \right. \tag{19}$$

will be useful, and is known as the **Kronecker delta**, after *Leopold Kronecker* (1823–1891). Thus, a vector set $\{\mathbf{u}_1, \ldots, \mathbf{u}_k\}$ being ON means that

$$\boxed{\mathbf{u}_i\cdot\mathbf{u}_j = \delta_{ij}} \tag{20}$$

for $i = 1, 2, \ldots, k$ and $j = 1, 2, \ldots, k$.

EXAMPLE 8
Let

$$\mathbf{u}_1 = (1, 0, 0, 0), \quad \mathbf{u}_2 = \left(0, \frac{1}{\sqrt{2}}, 0, \frac{1}{\sqrt{2}} \right), \quad \mathbf{u}_3 = \left(0, \frac{1}{\sqrt{2}}, 0, -\frac{1}{\sqrt{2}} \right).$$

Then $\{\mathbf{u}_1, \mathbf{u}_2, \mathbf{u}_3\}$ is ON because $\|\mathbf{u}_1\| = \|\mathbf{u}_2\| = \|\mathbf{u}_3\| = 1$ and $\mathbf{u}_1\cdot\mathbf{u}_2 = \mathbf{u}_1\cdot\mathbf{u}_3 = \mathbf{u}_2\cdot\mathbf{u}_3 = 0$. $\qquad \blacksquare$

Closure. In this section we introduced a dot product $\mathbf{u}\cdot\mathbf{v}$, a norm $\|\mathbf{u}\|$, and an angle θ between two vectors in n-space. Their introduction was not a matter of derivation but, rather, a matter of definition. The definitions were designed as extensions of the definitions for the familiar "arrow" vectors of 2-space and 3-space, somewhat as the upper floors of a home are built upon the foundation rather than being placed on an adjacent lot. Those extensions became apparent once we expressed the dot product, norm, and angle for arrow vectors in *n*-tuple notation. The key was the definition

$$\mathbf{u} \cdot \mathbf{v} = u_1 v_1 + u_2 v_2 + \cdots + u_n v_n,$$

because then the arrow vector formula $\|\mathbf{u}\| = \sqrt{\mathbf{u} \cdot \mathbf{u}}$ gave us a norm, and the arrow vector formula $\theta = \cos^{-1}(\mathbf{u} \cdot \mathbf{v}/\|\mathbf{u}\|\,\|\mathbf{v}\|)$ gave us an angle θ between $\mathbf{u}$ and $\mathbf{v}$.

From these definitions we then derived the properties (12a,b,c) of the dot product and (14a,b,c) of the norm. In 2-space and 3-space the triangle inequality amounts to the familiar Euclidean proposition that the length of any one side of a triangle cannot exceed the sum of the lengths of the other two sides, but in *n*-space, for $n > 3$, it amounts to an abstract generalization of that notion and does not have such a realizable physical or geometrical interpretation.

EXERCISES 4.4

1. Given the following vectors $\mathbf{u}$ and $\mathbf{v}$, determine $\|\mathbf{u}\|$, $\|\mathbf{v}\|$, $\mathbf{u} \cdot \mathbf{v}$, and θ (in radians and degrees). If $\mathbf{u}$ and $\mathbf{v}$ are orthogonal, state that.

(a) $\mathbf{u} = (4, 3)$, $\mathbf{v} = (1, 2)$
(b) $\mathbf{u} = (1, 2, 3, 4)$, $\mathbf{v} = (-4, -3, -2, -1)$
(c) $\mathbf{u} = (3, 0, 1)$, $\mathbf{v} = (-2, 3, 6)$
(d) $\mathbf{u} = (2, 2, 2)$, $\mathbf{v} = (-4, -5, -6)$
(e) $\mathbf{u} = (2, 5)$, $\mathbf{v} = (10, -4)$
(f) $\mathbf{u} = (1, 2, 3, 4)$, $\mathbf{v} = (4, 3, 2, 1)$
(g) $\mathbf{u} = (3, 2, 0, -1, 1)$, $\mathbf{v} = (-5, 0, 0, 2, 4)$
(h) $\mathbf{u} = (1, 2, 0, 0, 0)$, $\mathbf{v} = (0, 1, 1, 1, 1)$

2. State whether or not each of the following expressions is defined.

(a) $\|\mathbf{u}\|\,\mathbf{u}$
(b) $\mathbf{u} \cdot (\mathbf{v} \cdot \mathbf{w})$
(c) $\|(\mathbf{u} \cdot \mathbf{v})\mathbf{v}\|$
(d) $(\mathbf{u} + \mathbf{v}) \cdot \mathbf{w}$
(e) $(\mathbf{u} + \mathbf{v}) \cdot (\mathbf{u} - \mathbf{v})$
(f) $\mathbf{u} + 6(\mathbf{v} \cdot \mathbf{w})$
(g) $\cos^{-1}(2\mathbf{u} + \mathbf{v})$
(h) $\mathbf{u}/\|\mathbf{u}\|^2$
(i) $(7\mathbf{u}) \cdot (2\mathbf{v})$
(j) $\|\mathbf{u} + 3\mathbf{u}^2\|$

3. Let us denote, as points in 2-space and 3-space, $A = (2, 0)$, $B = (3, -1)$, $C = (5, 0)$, $D = (4, 2)$, $E = (2, 2)$, $F = (0, 0, 4)$, $G = (1, 0, 3)$, $H = (2, 1, 1)$, $I = (2, 2, 0)$, $J = (0, 1, 3)$. Determine, by vector methods, all interior angles and their sum, in degrees, for each of the following polygons.

(a) $ABCA$
(b) $ABCDA$
(c) $ABCDEA$
(d) $BCDB$
(e) $BCDEB$
(f) $FGHF$
(g) $FGIF$
(h) $GHIG$
(i) $FGJF$
(j) $GHJG$
(k) $HIJH$
(l) $FIJF$

4. (a)–(g) Normalize each pair of $\mathbf{u}$, $\mathbf{v}$ vectors in Exercise 1; that is, obtain $\hat{\mathbf{u}}$ and $\hat{\mathbf{v}}$.

5. If vectors $\mathbf{A}$, $\mathbf{B}$, $\mathbf{C}$, represented as arrows, form a triangle such that $\mathbf{A} = \mathbf{B} + \mathbf{C}$, derive the *law of cosines* $C^2 = A^2 + B^2 - 2AB\cos\alpha$, where α is the interior angle between $\mathbf{A}$ and $\mathbf{B}$, and where A, B, C are the lengths of $\mathbf{A}$, $\mathbf{B}$, $\mathbf{C}$, respectively, by starting with the identity $\mathbf{C} \cdot \mathbf{C} = (\mathbf{A} - \mathbf{B}) \cdot (\mathbf{A} - \mathbf{B})$.

6. If $\mathbf{u} = (1, 3, -4, 2)$ and $\mathbf{v} = (2, 0, 0, 3)$, evaluate the following.

(a) $\|\mathbf{u} - \mathbf{v}\|$
(b) $\|3\mathbf{u} - 2\mathbf{v}\| + \|-\mathbf{v}\|$
(c) $\left\|\dfrac{1}{\|\mathbf{u}\|}\mathbf{u}\right\|$
(d) $\|\mathbf{u}\| + \|\mathbf{v}\|$

7. Derive the following identities.

(a) $\|\mathbf{u} + \mathbf{v}\|^2 + \|\mathbf{u} - \mathbf{v}\|^2 = 2\|\mathbf{u}\|^2 + 2\|\mathbf{v}\|^2$
(b) $\|\mathbf{u} + \mathbf{v}\|^2 - \|\mathbf{u} - \mathbf{v}\|^2 = 4\mathbf{u} \cdot \mathbf{v}$
(c) Verify parts (a) and (b) for the case $\mathbf{u} = (2, 0, 1, 1)$ and $\mathbf{v} = (1, -3, 0, 2)$.

8. (*Orthogonal separation*) Sometimes it is desired to separate a given nonzero vector $\mathbf{u}$ into the sum of two orthogonal vectors, one parallel to and one perpendicular to some other nonzero vector $\mathbf{v}$, as sketched in part (a) of the accompanying figure. That is, $\mathbf{u} = \mathbf{u}_\parallel + \mathbf{u}_\perp$, where $\mathbf{u}_\parallel$ is of the form $\alpha\mathbf{v}$, and $\mathbf{u}_\perp \cdot \mathbf{u}_\parallel = 0$. We call $\mathbf{u}_\parallel$ the *orthogonal projection of $\mathbf{u}$ on $\mathbf{v}$*, and we call $\mathbf{u}_\perp$ the *component of $\mathbf{u}$ orthogonal to $\mathbf{v}$*.

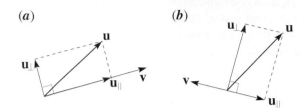

(*a*)   (*b*)

(a) Show that $\mathbf{u}_\parallel$ and $\mathbf{u}_\perp$ can be found, in terms of $\mathbf{u}$ and $\mathbf{v}$, as

$$\mathbf{u}_\parallel = (\mathbf{u}\cdot\hat{\mathbf{v}})\hat{\mathbf{v}} \qquad \text{where } \hat{\mathbf{v}} = \mathbf{v}/\|\mathbf{v}\|,$$
$$\mathbf{u}_\perp = \mathbf{u} - \mathbf{u}_\parallel. \tag{8.1}$$

Is (8.1) valid only for 2-space and 3-space, or does it hold, without modification, for n-space as well? Explain.

(b) Use (8.1) to determine $\mathbf{u}_\parallel$ and $\mathbf{u}_\perp$ if $\mathbf{u} = (1, 3)$ and $\mathbf{v} = (4, 3)$. Interpret your results graphically.

(c) Same as part (b), with $\mathbf{u} = (1, 3)$ and $\mathbf{v} = (-4, 0)$.

(d) Use (8.1) to determine $\mathbf{u}_\parallel$ and $\mathbf{u}_\perp$ if $\mathbf{u} = (2, 3, 1)$ and $\mathbf{v} = (0, 2, 3)$.

(e) Use (8.1) to determine $\mathbf{u}_\parallel$ and $\mathbf{u}_\perp$ if $\mathbf{u} = (3, 0, 5, 6)$ and $\mathbf{v} = (1, -2, 0, 4)$.

(f) Use (8.1) to determine $\mathbf{u}_\parallel$ and $\mathbf{u}_\perp$ if $\mathbf{u} = (2, 1, 0, 0, 3)$ and $\mathbf{v} = (0, 0, 1, -2, 1)$.

9. (a) Prove the associative property $(\alpha\mathbf{u})\cdot\mathbf{v} = \alpha(\mathbf{u}\cdot\mathbf{v})$.

(b) Prove the distributive property $(\mathbf{u} + \mathbf{v})\cdot\mathbf{w} = \mathbf{u}\cdot\mathbf{w} + \mathbf{v}\cdot\mathbf{w}$.

(c) Prove that the linearity property (12c) is equivalent to the two properties given in parts (a) and (b) of this exercise.

10. If $\mathbf{u}\cdot\mathbf{v} = 0$ and $\mathbf{v}\cdot\mathbf{w} = 0$, does that imply that $\mathbf{u}\cdot\mathbf{w} = 0$? Prove or disprove. HINT: If a claim is true, it needs to be proved in general, that is, for all possible cases. But if it is false, it can be disproved merely by putting forward a single counterexample.

11. Determine whether or not the vector set is orthogonal.

(a) $(1, 3)$, $(-6, 2)$, $(0, 0)$

(b) $(2, 3, 0)$, $(-3, 2, 1)$, $(1, 1, 1)$, $(1, -3, 1)$

(c) $(1, 0, 0, 0)$, $(0, 1, 0, 0)$, $(0, 0, 1, 0)$, $(0, 0, 0, 1)$

(d) $(1, 1, 1, 1)$, $(1, -1, 1, -1)$, $(0, 1, 0, -1)$, $(2, 0, -2, 0)$

(e) $(2, 1, -1, 1)$, $(1, 1, 3, 0)$, $(1, -1, 0, -1)$, $(2, 1, 1, 1)$

12. (*Derivation of Schwarz inequality*) Derive the Schwarz inequality

$$|\mathbf{u}\cdot\mathbf{v}| \le \|\mathbf{u}\|\,\|\mathbf{v}\|. \tag{12.1}$$

HINT: Begin with the inequality

$$(\mathbf{u} + \alpha\mathbf{v})\cdot(\mathbf{u} + \alpha\mathbf{v}) \ge 0 \tag{12.2}$$

which is guaranteed by (12b) for any scalar α and any vectors $\mathbf{u}$ and $\mathbf{v}$ in $\mathbb{R}^n$. Expanding the terms on the left, obtain

$$\|\mathbf{u}\|^2 + 2\alpha\,\mathbf{u}\cdot\mathbf{v} + \alpha^2\|\mathbf{v}\|^2 \ge 0. \tag{12.3}$$

Regarding $\mathbf{u}$ and $\mathbf{v}$ as fixed and α as variable, the left-hand side is then a quadratic function of α. If we choose α so as to minimize the left-hand side, then (12.3) will be as close to an equality as possible and hence as informative as possible. Thus, setting $d(\text{left-hand side})/d\alpha = 0$, obtain

$$\alpha = -\mathbf{u}\cdot\mathbf{v}/\|\mathbf{v}\|^2. \tag{12.4}$$

Putting this optimal value of α into (12.3), and rearranging terms, derive the Schwarz inequality (12.1). Thus, we see that the Schwarz inequality can be thought of as an optimal version of (12.2).

4.5 GAUSS ELIMINATION

4.5.1 Systems of linear algebraic equations. In the sections to follow, on span, linear dependence, bases, and expansions, we will need to know how to solve **systems of linear algebraic equations**, namely, sets of equations of the form

$$
\begin{aligned}
a_{11}x_1 + a_{12}x_2 + \cdots + a_{1n}x_n &= c_1, \quad &\text{(eq.1)}\\
a_{21}x_1 + a_{22}x_2 + \cdots + a_{2n}x_n &= c_2, \quad &\text{(eq.2)}\\
&\;\;\vdots\\
a_{m1}x_1 + a_{m2}x_2 + \cdots + a_{mn}x_n &= c_m. \quad &\text{(eq.}m\text{)}
\end{aligned}
\tag{1}
$$

The purpose of this section is to present a systematic solution method known as Gauss elimination. We will study (1) in Chapter 5 in greater detail using matrix methods, but the present section on the method of Gauss elimination will suffice until then.

In (1), $x_1, \ldots, x_n$ are the unknowns and the a_{ij} coefficients and the c_j's are prescribed numbers. Often, the number of equations (m) is equal to the number of unknowns (n) so that $m = n$, but that is *not* necessarily true; we could have $m < n$, $m = n$, or $m > n$. We will assume that both m and n are finite so that we have only a finite number of equations in a finite number of unknowns. If all of the c_j's are zero then (1) is **homogeneous**; if at least one of them is nonzero then (1) is **nonhomogeneous**.

The subscript notation adopted in (1) is not essential but it is rather natural, it renders inherent patterns more apparent, and it will permit a natural transition to the matrix notation introduced in Chapter 5. The first subscript in a_{ij} indicates the equation and the second indicates the x_j variable that it multiplies. For instance, a_{21} appears in the second equation and multiplies the x_1 variable. To avoid ambiguity we should write $a_{2,1}$ rather than a_{21} so that one does not mistakenly read the subscripts as twenty-one, but we will omit commas except when such ambiguity is not easily resolved from the context.

We say that a set of numbers $s_1, s_2, \ldots, s_n$ is a **solution** of (1) if and only if each of the m equations is satisfied numerically when we substitute s_1 for x_1, s_2 for x_2, and so on. If there exist *one or more* solutions to (1), we say that the system is **consistent**; if there is precisely *one* solution, that solution is **unique**; and if there is more than one, the solution is **nonunique**. If, on the other hand, there are *no* solutions to (1), the system is said to be **inconsistent**. The collection of all solutions to (1) is called its **solution set** so, by "solving (1)" we mean finding its solution set; if the system is inconsistent, we say that its solution set is **empty**.

Let us illustrate how such systems might arise in applications.

EXAMPLE 1   *Curve Fitting*.

Suppose we are interested in the functional dependence of a certain dependent variable y on a certain independent variable x, and suppose that k experimental runs indicate that to the x values $x_1, \ldots, x_k$ there correspond y values $y_1, \ldots, y_k$. Through these k datum points in the x, y plane we wish to fit a polynomial $y(x)$ of as low a degree as possible (Fig. 1).

Evidently, that polynomial is of degree $k - 1$;[1] for instance, if $k = 2$ then we need a first-degree polynomial (i.e., a straight line). Thus, we seek $y(x)$ in the form

$$y(x) = a_0 + a_1 x + a_2 x^2 + \cdots + a_{k-1} x^{k-1}, \qquad (2)$$

and the problem is to determine $a_0, a_1, \ldots, a_{k-1}$ so that the graph of (2) passes through the k datum points. Imposing those k conditions on (2) gives

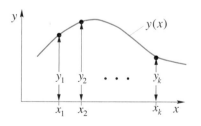

FIGURE 1
Curve fitting by polynomial.

---

[1]If, for example, $k = 4$ and (2) is $y(x) = 5 - 2x + 0x^2 + 0x^3$, we still consider the latter as a second-degree polynomial even though it can be simplified as $y(x) = 5 - 2x$.

$$y(x_1) = y_1 = a_0 + a_1 x_1 + a_2 x_1^2 + \cdots + a_{k-1} x_1^{k-1},$$
$$y(x_2) = y_2 = a_0 + a_1 x_2 + a_2 x_2^2 + \cdots + a_{k-1} x_2^{k-1},$$
$$\vdots \qquad\qquad\qquad\qquad\qquad\qquad (3)$$
$$y(x_k) = y_k = a_0 + a_1 x_k + a_2 x_k^2 + \cdots + a_{k-1} x_k^{k-1},$$

or putting (3) into the form (1),

$$a_0 + x_1 a_1 + x_1^2 a_2 + \cdots + x_1^{k-1} a_{k-1} = y_1,$$
$$a_0 + x_2 a_1 + x_2^2 a_2 + \cdots + x_2^{k-1} a_{k-1} = y_2,$$
$$\vdots \qquad\qquad\qquad\qquad\qquad\qquad (4)$$
$$a_0 + x_k a_1 + x_k^2 a_2 + \cdots + x_k^{k-1} a_{k-1} = y_k.$$

Here the x_j's and y_j's are known, and (4) is a system of k linear algebraic equations in the k unknowns $a_0, a_1, \ldots, a_{k-1}$. Of course, the unknown a_j's in (4) are playing the role of the x_j's in (1), the y_j's in (4) are playing the role of the c_j's in (1), and so on. The idea is to solve (4) for the a_j's; putting these into (2) gives the desired polynomial. Here, our purpose was merely to formulate the problem (4), not to solve it.   ∎

EXAMPLE 2   A DC Circuit.

Consider the DC (direct current) circuit shown in Fig. 2, where i_1, i_2, i_3 are the three currents (measured as positive in the directions assumed in the figure), R_1, R_2, R_3 are the resistances of the three resistors, and E is the voltage rise from e to a induced by the battery or other source.

Kirchhoff's current and voltage laws were discussed in Section 2.4.2. If we apply his current law to junctions a and c, and his voltage law to loops $abcda$, $abcea$, and $adcea$, we obtain the following system of linear algebraic equations,

$$i_1 - i_2 - i_3 = 0, \quad \text{(junction } a\text{)}$$
$$i_1 - i_2 - i_3 = 0, \quad \text{(junction } c\text{)}$$
$$R_2 i_2 - R_3 i_3 = 0, \quad \text{(loop } abcda\text{)} \qquad (5)$$
$$R_1 i_1 + R_2 i_2 = E, \quad \text{(loop } abcea\text{)}$$
$$R_1 i_1 + R_3 i_3 = E, \quad \text{(loop } adcea\text{)}$$

on the unknown currents i_1, i_2, i_3. Of these equations, the first two are identical so we can discard one and reduce (5) to four equations on the three unknowns.   ∎

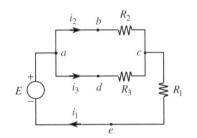

FIGURE 2
DC circuit.

4.5.2 Preliminary ideas and geometrical approach.   Actually, we are interested not only in the practical matter of how to solve systems of linear algebraic equations, we are also interested in the questions of existence and uniqueness: Does a solution of (1) *exist*; that is, is (1) consistent? If so, is the solution *unique*? And, of course, how do we *find* the solution(s)?

In this subsection we use a geometrical discussion to develop some basic understanding of the problem. A great advantage of geometric reasoning is that it brings our visual system into play. It is estimated that at least a third of the neurons in our brain are devoted to vision, hence our visual sense is extremely powerful. No wonder we say "Now I see what you mean; now I get the picture." The more geometry, pictures, and visual images to assist our thinking, the better.

(a)

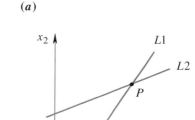

To start, it makes sense to begin with the simplest case, where $m = n = 1$: then (1) is simply

$$a_{11}x_1 = c_1. \tag{6}$$

In the generic case, $a_{11} \neq 0$ and (6) admits the unique solution $x_1 = c_1/a_{11}$, but if $a_{11} = 0$ there are two possibilities: If $c_1 \neq 0$ then there are no values of x_1 such that $0x_1 = c_1$ so (6) is inconsistent, but if $c_1 = 0$ then (6) becomes $0x_1 = 0$ and $x_1 = \alpha$ is a solution for *any* value of α; that is, the solution is nonunique.

(b)

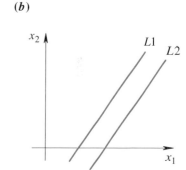

Far from being too simple to be of interest, the latter case where $m = n = 1$ establishes a pattern that will hold in general, for any values of m and n, as we shall see. Specifically, the system (1) will admit a unique solution, no solution, or an infinity of solutions. For instance, it will never admit 4 solutions, 12 solutions, or 137 solutions.

Next, consider the case where $m = n = 2$:

$$a_{11}x_1 + a_{12}x_2 = c_1, \quad \text{(eq.1)} \tag{7a}$$
$$a_{21}x_1 + a_{22}x_2 = c_2. \quad \text{(eq.2)} \tag{7b}$$

If a_{11} and a_{12} are not both zero, then (eq.1) defines a straight line, say $L1$, in a Cartesian x_1, x_2 plane; that is, the solution set of (eq.1) is the set of all points on that line. Similarly, if a_{21} and a_{22} are not both zero then the solution set of (eq.2) is the set of all points on a straight line $L2$. There are exactly three possibilities, and these are illustrated in Fig. 3. First, the lines may intersect at a point, say P, in which case (7) admits the unique solution given by the coordinate pair x_1, x_2 of the point P (Fig. 3a). That is, any solution pair x_1, x_2 of (7) needs to be in the solution set of (eq.1) *and* in the solution set of (eq.2), hence at an intersection of $L1$ and $L2$. This is the generic case, and it occurs (Exercise 1) if and only if

(c)

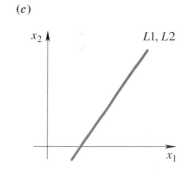

$$a_{11}a_{22} - a_{12}a_{21} \neq 0; \tag{8}$$

(8) is the analog of the $a_{11} \neq 0$ condition for the $m = n = 1$ case discussed above.

Second, the lines may be parallel and therefore nonintersecting (Fig. 3b), in which case there is no solution. Then (7) is inconsistent, the solution set is empty.

FIGURE 3
Existence and uniqueness for the system (7).

Third, the lines may coincide (Fig. 3c), in which case the coordinate pair of each point on the line is a solution. Then (7) is consistent and there is an infinite number of solutions.

EXAMPLE 3

These three cases (Fig. 3a,b,c) are illustrated by the systems

$$\begin{aligned} 2x_1 - x_2 &= 5, \\ x_1 + 3x_2 &= -1, \end{aligned} \tag{9}$$

and

$$x_1 + 3x_2 = 1,$$
$$x_1 + 3x_2 = 0, \tag{10}$$

and

$$x_1 + 3x_2 = 1,$$
$$2x_1 + 6x_2 = 2, \tag{11}$$

respectively: (9) admits the unique solution $x_1 = 2$, $x_2 = -1$; (10) has no solution; (11) admits a nonunique solution, namely, the infinity of solutions consisting of all x_1, x_2 pairs along the line $x_1 + 3x_2 = 1$.   ∎

Of course, only if a_{11} and a_{12} are not both zero does (eq.1) define a line $L1$ in the x_1, x_2 plane, and only if a_{21} and a_{22} are not both zero does (eq.2) define a line $L2$; if $a_{11} = a_{12} = 0$ and/or $a_{21} = a_{22} = 0$, then Fig. 3 does not apply.

What if $a_{11} = a_{12} = 0$? Then if $c_1 \neq 0$ there is no solution of (7a), and hence there is no solution to the system (7). But if $c_1 = 0$, then (7a) reduces to $0 = 0$ and can be discarded, leaving just (7b). If a_{21} and a_{22} are not both zero, then (7b) gives a line of solutions, but if they are both zero then everything hinges on c_2. If $c_2 \neq 0$ there is no solution and (7) is inconsistent, but if $c_2 = 0$, so both (7a) and (7b) are simply $0 = 0$, then both x_1 and x_2 are arbitrary, and every point in the plane is a solution!

Let us summarize the case where $m = n = 2$: (7) will admit no solution, a unique solution, or an infinity of solutions, the latter comprising either a line in the x_1, x_2 plane or (if $a_{11} = a_{12} = c_1 = a_{21} = a_{22} - c_2 = 0$) the entire plane.

Next, consider the case where $m = n = 3$:

$$a_{11}x_1 + a_{12}x_2 + a_{13}x_3 = c_1, \qquad \text{(eq.1)} \tag{12a}$$
$$a_{21}x_1 + a_{22}x_2 + a_{23}x_3 = c_2, \qquad \text{(eq.2)} \tag{12b}$$
$$a_{31}x_1 + a_{32}x_2 + a_{33}x_3 = c_3. \qquad \text{(eq.3)} \tag{12c}$$

Continuing the geometric approach employed in Fig. 3, observe that if a_{11}, a_{12}, a_{13} are not all zero then (eq.1) defines a plane, say $P1$, in Cartesian x_1, x_2, x_3 space, and similarly for (eq.2) and (eq.3). In the generic case, $P1$ and $P2$ intersect along a line L, and L pierces $P3$ at a point P. Then the x_1, x_2, x_3 coordinates of P give the unique solution of (12).

In the nongeneric case we can have no solution or an infinity of solutions in the following ways. There will be no solution if L (the intersection of $P1$ and $P2$) is parallel to $P3$ and hence fails to pierce it, or if any two of the planes are parallel and not coincident. There will be an infinity of solutions if L lies in $P3$ (i.e., a line of solutions), if two planes are coincident and intersect the third (again, a line of solutions), or if all three planes are coincident (this time an entire plane of solutions).

Further, if all three a_{ij}'s in any of the three equations (12) are zero then there is no solution if the corresponding c_i is nonzero. In the extreme case where all of the nine a_{ij}'s and all three c_i's are zero, every point x_1, x_2, x_3 is a solution; the solution set is all of 3-space.

An abstract extension of such geometrical reasoning can be continued even if $m = n \geq 4$. For instance, one speaks of $a_{11}x_1 + a_{12}x_2 + a_{13}x_3 + a_{14}x_4 = c_1$ as defining a *hyperplane* in an abstract four-dimensional space. In fact, perhaps we should mention that even the familiar x_1, x_2 plane and x_1, x_2, x_3 space discussed here could be abstract as well. For instance, if x_1 and x_2 are unknown currents in loops of an electrical circuit, then what physical meaning is there to an x_1, x_2 plane? None, we introduced it—created it— to make available to us the geometry and vector mathematics of 2-space.

4.5.3 Solution by Gauss elimination.   To proceed further with our study of the linear system (1) we need to develop a systematic method of *finding* solutions. We do that next, by putting forward a solution technique known as **Gauss elimination**. To motivate the idea we begin with an example.

E X A M P L E 4
Determine the solution set of the system

$$\begin{aligned} x_1 + x_2 - x_3 &= 1, \\ 3x_1 + x_2 + x_3 &= 9, \\ x_1 - x_2 + 4x_3 &= 8. \end{aligned} \tag{13}$$

Keep the first equation intact, and add -3 times the first equation to the second (as a replacement for the second equation), and add -1 times the first equation to the third (as a replacement for the third equation). These steps yield the new "indented" system

$$\begin{aligned} x_1 + x_2 - x_3 &= 1, \\ -2x_2 + 4x_3 &= 6, \\ -2x_2 + 5x_3 &= 7. \end{aligned} \tag{14}$$

Next, keep the first two of these intact, and add -1 times the second equation to the third, and obtain

$$\begin{aligned} x_1 + x_2 - x_3 &= 1, \\ -2x_2 + 4x_3 &= 6, \\ x_3 &= 1. \end{aligned} \tag{15}$$

Finally, multiplying the second of these by $-1/2$ to normalize the leading coefficient (to unity), gives

$$\begin{aligned} x_1 + x_2 - x_3 &= 1, \quad \text{(eq.1)} \\ x_2 - 2x_3 &= -3, \quad \text{(eq.2)} \\ x_3 &= 1. \quad \text{(eq.3)} \end{aligned} \tag{16}$$

It is helpful to think of the original system (13) as a tangle of string that we wish to unravel. The first step is to find a loose end and that is, in effect, what the foregoing process of successive indentations has done for us. Specifically, (eq.3) in (16) is the "loose end," and with that in hand we may unravel (16) just as we would unravel a tangle: putting $x_3 = 1$ into (eq.2) gives $x_2 = -1$, and then putting $x_3 = 1$ and $x_2 = -1$ into (eq.1) gives $x_1 = 3$. Thus, we obtain the unique solution

$$x_3 = 1, \quad x_2 = -1, \quad x_1 = 3. \tag{17}$$

COMMENT 1.   From a mathematical point of view, the system (16) was a "tangle" because the equations were **coupled**; that is, each equation contained more than one unknown. Actually, the final system (16) is coupled too, since (eq.1) contains all three unknowns and (eq.2) contains two of them. However, the coupling in (16) is not as debilitating because (16) is in what we call **triangular form**. Thus, we were able to solve (eq.3) for x_3, put that value into (eq.2) and solve for x_2, and then put these values into (eq.1) and solve for x_1, which steps are known as **back substitution**.

COMMENT 2.   However, the process begs this question: Is it obvious that the systems (13)–(16) all have the same solution sets so that when we solve (16) we are actually solving (13)? That is, is it not conceivable that in applying the arithmetic steps that carried us from (13) to (16) we might, inadvertently, have altered the solution set? For example, $x - 1 = 4$ has the unique solution $x = 5$, but if we innocently square both sides, the resulting equation $(x - 1)^2 = 16$ admits the *two* solutions $x = 5$ and $x = -3$.   ■

The question just raised applies to the general linear system (1). It is answered in Theorem 4.5.1 that follows, but first we define two terms: "elementary equation operations" and "equivalent systems."

The following operations on linear systems are known as **elementary equation operations**:

1. Addition of a multiple of one equation to another
   Symbolically: $(\text{eq.}j) \rightarrow (\text{eq.}j) + \alpha\,(\text{eq.}k)$

2. Multiplication of an equation by a nonzero constant
   Symbolically: $(\text{eq.}j) \rightarrow \alpha\,(\text{eq.}j)$

3. Interchange of two equations
   Symbolically: $(\text{eq.}j) \leftrightarrow (\text{eq.}k)$

We say that two systems of m linear algebraic equations in n unknowns are **equivalent** if one can be obtained from the other by a finite number of elementary equation operations.

THEOREM 4.5.1
Equivalent Systems
If two linear algebraic systems of m equations in n unknowns are equivalent, then their solution sets are identical.

Outline of Proof.   The truth of this claim for elementary equation operations of types 2 and 3 should be evident, so we confine our remarks to operations of type 1. It suffices to look at the effect of one such operation. Thus, suppose that a given linear system A is altered by replacing its jth equation by its jth plus α times its kth, its other equations being kept intact. Let us call the new system A'. Surely, every solution of A will also be a solution of A' since we have merely added equal quantities to equal quantities. That is, *if A' results from A by the application of an* *elementary equation operation of type 1, then every solution of A is also a solution of A'.* Further, we can convert A' back to A by an elementary equation operation of type 1, namely, by replacing the jth equation of A' by the jth equation of A' plus $-\alpha$ times the kth equation of A'. Consequently, it follows from the italicized result (two sentences back) that every solution of A' is also a solution of A. Thus, is A and A' are equivalent, then their solution sets are identical.　　■

In Example 4 we saw that each step is an elementary equation operation: Three elementary equation operations of type 1 took us from (13) to (15), and one of type 2 took us from (15) to (16); finally, the back substitution amounted to several operations of type 1. Thus, according to Theorem 4.5.1, equivalence was maintained throughout so we can be sure that (17) is the solution set of the original system (13) (as can be verified by direct substitution).

The system in Example 4 admitted a unique solution. To see how the method of successive elimination works out when there is no solution, or a nonunique solution, let us work two more examples.

E X A M P L E　5　*Inconsistent System.*
Consider the system

$$
\begin{aligned}
2x_1 + 3x_2 - 2x_3 &= 4, \\
x_1 - 2x_2 + x_3 &= 3, \\
7x_1 \qquad\ - x_3 &= 2.
\end{aligned} \tag{18}
$$

Keep the first equation intact, add $-\frac{1}{2}$ times the first equation to the second (eq.2 $\to$ eq.2 $-\frac{1}{2}$ eq.1), and add $-\frac{7}{2}$ times the first to the third (eq.3 $\to$ eq.3 $-\frac{7}{2}$ eq.1):

$$
\begin{aligned}
2x_1 + 3x_2 - 2x_3 &= 4, \\
-\tfrac{7}{2}x_2 + 2x_3 &= 1, \\
-\tfrac{21}{2}x_2 + 6x_3 &= -12.
\end{aligned} \tag{19}
$$

Keep the first two equations intact, and add -3 times the second equation to the third (eq.3 $\to$ eq.3 -3 eq.2):

$$
\begin{aligned}
2x_1 + 3x_2 - 2x_3 &= 4, \\
-\tfrac{7}{2}x_2 + 2x_3 &= 1, \\
0 &= -15.
\end{aligned} \tag{20}
$$

Any solution x_1, x_2, x_3 of (20) must satisfy each of the three equations, but there are no values of x_1, x_2, x_3 that can satisfy $0 = -15$. Thus, (20) is inconsistent (its solution set is "empty"), and therefore (18) is as well.

COMMENT.　The source of the inconsistency is that whereas the left-hand side of the third equation, in (18), is 2 times the left-hand side of the first equation plus 3 times the left-hand side of the second, the right-hand sides do not bear that relationship: $2(4) + 3(3) = 17 \neq 2$. [While that built-in contradiction is not obvious from (18), it eventually comes to light in the third equation in (20).] If we modify the system (18) by changing the final 2 in (18) to 17, then the final -12 in (21) becomes a 3, and the final -15 in (20) becomes a zero:

$$\begin{aligned}
2x_1 + 3x_2 - 2x_3 &= 4, \\
-\tfrac{7}{2}x_2 + 2x_3 &= 1, \\
0 &= 0
\end{aligned} \tag{21}$$

or, multiplying the first by $\tfrac{1}{2}$ and the second by $-\tfrac{2}{7}$,

$$\begin{aligned}
x_1 + \tfrac{3}{2}x_2 - x_3 &= 2, \\
x_2 - \tfrac{4}{7}x_3 &= -\tfrac{2}{7},
\end{aligned} \tag{22a,b}$$

where we have discarded the identity $0 = 0$. Thus, by changing the c_j's so as to be "compatible," the system now admits an infinity of solutions rather than none. Specifically, we can let x_3 (or x_2, it doesn't matter which) in (22b) be *any* value, say α, where α is arbitrary. Then (22b) gives $x_2 = -\tfrac{2}{7} + \tfrac{4}{7}\alpha$, and putting these into (22a) gives $x_1 = \tfrac{17}{7} + \tfrac{1}{7}\alpha$. Thus, we have the infinity of solutions

$$x_3 = \alpha, \quad x_2 = -\frac{2}{7} + \frac{4}{7}\alpha, \quad x_1 = \frac{17}{7} + \frac{1}{7}\alpha \tag{23}$$

for any α. Evidently, two of the three planes defined by (18) intersect, giving a line that lies in the third plane, and equations (23) are parametric equations of that line![1]   ■

E X A M P L E   6   *Nonunique Solution.*
Consider the system of four equations in six unknowns ($m = 4, n = 6$)

$$\begin{aligned}
2x_2 + x_3 + 4x_4 + 3x_5 + x_6 &= 2, \\
x_1 - x_2 + x_3 \qquad\qquad\quad + 2x_6 &= 0, \\
x_1 + x_2 + 2x_3 + 4x_4 + x_5 + 2x_6 &= 3, \\
x_1 - 3x_2 \qquad\quad - 4x_4 - 2x_5 + x_6 &= 0.
\end{aligned} \tag{24}$$

Wanting the top equation to begin with x_1 and subsequent equations to indent at the left, let us first move the top equation to the bottom (eq.1 $\leftrightarrow$ eq.4):

$$\begin{aligned}
x_1 - 3x_2 \qquad\quad - 4x_4 - 2x_5 + x_6 &= 0, \\
x_1 - x_2 + x_3 \qquad\qquad\quad + 2x_6 &= 0, \\
x_1 + x_2 + 2x_3 + 4x_4 + x_5 + 2x_6 &= 3, \\
2x_2 + x_3 + 4x_4 + 3x_5 + x_6 &= 2.
\end{aligned} \tag{25}$$

---

[1]Recall, from a course in calculus and analytic geometry, that a space curve is defined parametrically by three equations $x = x(\tau)$, $y = y(\tau)$, $z = z(\tau)$, where the parameter τ (which is α in the present case) varies between specified limits τ_1 and τ_2. In particular, equations of the form $x = a + b\tau$, $y = c + d\tau$, $z = e + f\tau$, for $-\infty < \tau < \infty$, define a straight line of infinite length in x, y, z space, which result is consistent with our geometrical interpretation of (23) as defining a line in x_1, x_2, x_3 space.

Add -1 times the first equation to the second (eq.2 $\rightarrow$ eq.2 -1 eq.1) and third (eq.3 $\rightarrow$ eq.3 -1 eq.1) equations:

$$\begin{aligned}
x_1 - 3x_2 \quad\quad - 4x_4 - 2x_5 + x_6 &= 0, \\
2x_2 + \; x_3 + 4x_4 + 2x_5 + x_6 &= 0, \\
4x_2 + 2x_3 + 8x_4 + 3x_5 + x_6 &= 3, \\
2x_2 + \; x_3 + 4x_4 + 3x_5 + x_6 &= 2.
\end{aligned} \tag{26}$$

Add -2 times the second to the third (eq.3 $\rightarrow$ eq.3 -2 eq.2) and -1 times the second to the fourth (eq.4 $\rightarrow$ eq.4 -1 eq.2):

$$\begin{aligned}
x_1 - 3x_2 \quad\quad - 4x_4 - 2x_5 + x_6 &= 0, \\
2x_2 + x_3 + 4x_4 + 2x_5 + x_6 &= 0, \\
-x_5 - x_6 &= 3, \\
x_5 \quad\quad &= 2.
\end{aligned} \tag{27}$$

Add the third to the fourth (eq.4 $\rightarrow$ eq.4 $+$ eq.3):

$$\begin{aligned}
x_1 - 3x_2 \quad\quad - 4x_4 - 2x_5 + \; x_6 &= 0, \\
2x_2 + x_3 + 4x_4 + 2x_5 + \; x_6 &= 0, \\
-x_5 - \; x_6 &= 3, \\
-x_6 &= 5.
\end{aligned} \tag{28}$$

Finally, multiply the second, third, and fourth by $\frac{1}{2}, -1$, and -1, respectively, to normalize the leading coefficients (eq.2 $\rightarrow$ $\frac{1}{2}$ eq.2, eq.3 $\rightarrow$ -1 eq.3, eq.4 $\rightarrow$ -1 eq.4):

$$\begin{aligned}
x_1 - 3x_2 \quad\quad - 4x_4 - 2x_5 + \; x_6 &= \; 0, \\
x_2 + \tfrac{1}{2}x_3 + 2x_4 + \; x_5 + \tfrac{1}{2}x_6 &= \; 0, \\
x_5 + \; x_6 &= -3, \\
x_6 &= -5.
\end{aligned} \tag{29}$$

The last two equations give $x_6 = -5$ and $x_5 = 2$, and these values can be substituted back into the second equation. In that equation we can let x_4 be arbitrary, say α_1, and we can also let x_3 be arbitrary, say α_2. Then that equation gives x_2 and, again by back substitution, the first equation gives x_1. The result is the infinity of solutions

$$\begin{aligned}
x_6 = -5, \quad x_5 = 2, \quad x_4 = \alpha_1, \quad x_3 = \alpha_2, \\
x_2 = \frac{1}{2} - 2\alpha_1 - \frac{1}{2}\alpha_2, \quad x_1 = \frac{21}{2} - 2\alpha_1 - \frac{3}{2}\alpha_2,
\end{aligned} \tag{30}$$

where α_1 and α_2 are arbitrary.   ∎

If a solution set contains N independent arbitrary parameters $(\alpha_1, \ldots, \alpha_N)$, we call it (in this text) an **N-parameter family of solutions**. Thus, (23) and (30) are 1-parameter and 2-parameter families of solutions, respectively. Each choice of values for $\alpha_1, \ldots, \alpha_N$ yields a **particular solution**. In (30), for instance, the choice $\alpha_1 = 1$ and $\alpha_2 = 0$ yields the particular solution $x_1 = \frac{17}{2}$, $x_2 = -\frac{3}{2}$, $x_3 = 0$, $x_4 = 1$, $x_5 = 2$, and $x_6 = -5$.

The method of Gauss elimination,[1] illustrated in Examples 4–6, can be applied to *any* linear system (1), whether $m < n$, $m = n$, or $m > n$. Though not obvious from the foregoing examples, the method is efficient and is commonly incorporated in computer systems and software.

Observe that the end result of the Gauss elimination process enables us to determine, merely from the pattern of the final equations, whether or not a solution exists and is unique. For instance, we can see from the pattern of (16) that there is a unique solution, from the bottom equation in (20) that there is no solution, and from the extra double indentation in (29) that there is a 2-parameter family of solutions.

4.5.4 Matrix notation.   In applying Gauss elimination, we quickly discover that writing the variables $x_1, \ldots, x_n$ over and over is inefficient, and even tends to upstage the more central role of the a_{ij}'s and c_j's. It is therefore preferable to omit the x_j's altogether and to work directly with the rectangular array

$$\begin{bmatrix} a_{11} & a_{12} & \cdots & a_{1n} & c_1 \\ a_{21} & a_{22} & \cdots & a_{2n} & c_2 \\ \vdots & \vdots & & \vdots & \vdots \\ a_{m1} & a_{m2} & \cdots & a_{mn} & c_m \end{bmatrix}, \tag{31}$$

known as the **augmented matrix** of the system (1), that is, the **coefficient matrix**

$$\begin{bmatrix} a_{11} & a_{12} & \cdots & a_{1n} \\ a_{21} & a_{22} & \cdots & a_{2n} \\ \vdots & \vdots & & \vdots \\ a_{m1} & a_{m2} & \cdots & a_{mn} \end{bmatrix}, \tag{32}$$

augmented by the column of c_j's. By **matrix** we simply mean a rectangular array of numbers, called **elements**; it is customary to enclose the elements between brackets or parentheses to emphasize that the entire matrix is regarded as a single entity. A horizontal line of elements is called a **row**, and a vertical line is called a **column**. Counting rows from the top, and columns from the left,

$$a_{21} \quad a_{22} \quad \cdots \quad a_{2n} \quad c_2 \qquad \text{and} \qquad \begin{matrix} c_1 \\ c_2 \\ \vdots \\ c_m \end{matrix},$$

for instance, are the second row and $(n+1)$th column, respectively, of the augmented matrix (31).

In terms of this abbreviated matrix notation, the calculation in Example 4 would look like this.

---

[1]The method is attributed to *Karl Friedrich Gauss* (1777–1855), who is generally regarded as the foremost mathematician of the nineteenth century and is often referred to as the "prince of mathematicians."

Original system:

$$\begin{bmatrix} 1 & 1 & -1 & 1 \\ 3 & 1 & 1 & 9 \\ 1 & -1 & 4 & 8 \end{bmatrix}.$$

Add -3 times first row to second row and add -1 times first row to third row:

$$\begin{bmatrix} 1 & 1 & -1 & 1 \\ 0 & -2 & 4 & 6 \\ 0 & -2 & 5 & 7 \end{bmatrix}.$$

Add -1 times second row to third row and multiply second row by $-\frac{1}{2}$:

$$\begin{bmatrix} 1 & 1 & -1 & 1 \\ 0 & 1 & -2 & -3 \\ 0 & 0 & 1 & 1 \end{bmatrix}. \tag{33}$$

Thus, corresponding to the so-called elementary equation operations on members of a system of linear equations there are **elementary row operations** on the augmented matrix, as follows:

1. Addition of a multiple of one row to another
   Symbolically: (jth row) $\rightarrow$ (jth row) $+ \alpha(k$th row)

2. Multiplication of a row by a nonzero constant
   Symbolically: (jth row) $\rightarrow \alpha(j$th row)

3. Interchange of two rows
   Symbolically: (jth row) $\leftrightarrow$ (kth row)

And we say that two matrices are **row equivalent** if one can be obtained from the other by finitely many elementary row operations.

Recall that the main purpose of this Section 4.5 is to provide a systematic solution procedure for systems of linear algebraic equations, namely, Gauss elimination, which will be relied on extensively in subsequent sections and chapters. With that objective now accomplished, Sections 4.5.5–4.5.7 are optional.

4.5.5 Gauss-Jordan reduction. (Optional). With the Gauss elimination completed, the remaining steps consist of back substitution. In fact, those steps amount to elementary row operations as well. The difference is that whereas in the Gauss elimination we proceed from the top down, in the back substitution we proceed from the bottom up.

E X A M P L E   7

To illustrate, let us return to Example 4 and pick up at the end of the Gauss elimination, with (16), and complete the back substitution steps using elementary row operations. In matrix format, we begin with

$$\begin{bmatrix} 1 & 1 & -1 & 1 \\ 0 & 1 & -2 & -3 \\ 0 & 0 & 1 & 1 \end{bmatrix}. \tag{34}$$

Keeping the bottom row intact, add 2 times that row to the second and add 1 times that row to the first:

$$\begin{bmatrix} 1 & 1 & 0 & 2 \\ 0 & 1 & 0 & -1 \\ 0 & 0 & 1 & 1 \end{bmatrix}. \tag{35}$$

Now keeping the bottom two rows intact add -1 times the second row to the first:

$$\begin{bmatrix} 1 & 0 & 0 & 3 \\ 0 & 1 & 0 & -1 \\ 0 & 0 & 1 & 1 \end{bmatrix}, \tag{36}$$

which is the solution: $x_1 = 3$, $x_2 = -1$, $x_3 = 1$ as obtained in Example 4.   ∎

The entire process of Gauss elimination plus back substitution is known as **Gauss-Jordan reduction**, after Gauss and *Wilhelm Jordan* (1842–1899). The final result is an augmented matrix in **reduced row-echelon form**. That is:

1. In each row not made up entirely of zeros, the first nonzero element is a 1, a so-called **leading 1**.

2. In any two consecutive rows not made up entirely of zeros, the leading 1 in the lower row is to the right of the leading 1 in the upper row.

3. If a column contains a leading 1, every other element in that column is a zero.

4. All rows made up entirely of zeros are grouped at the bottom of the matrix.

For instance, (36) is in reduced row-echelon form, as is the final matrix in the next example.

E X A M P L E 8

Let us return to Example 6 and finish the Gauss-Jordan reduction, beginning with (29):

$$
\begin{bmatrix}
1 & -3 & 0 & -4 & -2 & 1 & 0 \\
0 & 1 & \frac{1}{2} & 2 & 1 & \frac{1}{2} & 0 \\
0 & 0 & 0 & 0 & 1 & 1 & -3 \\
0 & 0 & 0 & 0 & 0 & 1 & -5
\end{bmatrix}
\rightarrow
\begin{bmatrix}
1 & -3 & 0 & -4 & -2 & 0 & 5 \\
0 & 1 & \frac{1}{2} & 2 & 1 & 0 & \frac{5}{2} \\
0 & 0 & 0 & 0 & 1 & 0 & 2 \\
0 & 0 & 0 & 0 & 0 & 1 & -5
\end{bmatrix}
\rightarrow
$$

$$
\begin{bmatrix}
1 & -3 & 0 & -4 & 0 & 0 & 9 \\
0 & 1 & \frac{1}{2} & 2 & 0 & 0 & \frac{1}{2} \\
0 & 0 & 0 & 0 & 1 & 0 & 2 \\
0 & 0 & 0 & 0 & 0 & 1 & -5
\end{bmatrix}
\rightarrow
\begin{bmatrix}
\mathbf{1} & 0 & \frac{3}{2} & 2 & 0 & 0 & \frac{21}{2} \\
0 & \mathbf{1} & \frac{1}{2} & 2 & 0 & 0 & \frac{1}{2} \\
0 & 0 & 0 & 0 & \mathbf{1} & 0 & 2 \\
0 & 0 & 0 & 0 & 0 & \mathbf{1} & -5
\end{bmatrix}.
$$

The last augmented matrix is in reduced row-echelon form. The four leading 1's are displayed in **bold type**, and we see that, as a result of the back substitution steps, only 0's are to be found above each leading 1. The final augmented matrix once again gives the solution (30). ■

Think of the Jordan part as an option; we can omit it and do the back substitution by hand, if we prefer, as we did in Examples 4–6. If we *do* the Jordan part, that amounts to obtaining zeros above each leading 1 (except the one in the first row, of course).

4.5.6 Results regarding existence and uniqueness. (Optional). As we have noted, we can see directly from the Gauss eliminated form of (1) whether there is no solution, a unique solution, or an N-parameter family of solutions. For instance, (16) reveals that there is a unique solution, (20) that there is no solution, (21) that there is a 1-parameter family of solutions, and (29) that there is a 2-parameter family of solutions. Generalizing this idea, one can draw a number of conclusions regarding the existence and uniqueness of solutions of the linear system (1):

THEOREM 4.5.2
Existence / Uniqueness for Linear Systems
If $m < n$, the system (1) can be consistent or inconsistent. If it is consistent it cannot have a unique solution; it will have an N-parameter family of solutions where $n - m \leq N \leq n$. If $m \geq n$, then (1) can be consistent or inconsistent. If it is consistent it can have a unique solution or an N-parameter family of solutions, where $1 \leq N \leq n$.

The next theorem follows immediately from Theorem 4.5.2, but we state it separately for emphasis. It confirms the claim that we made in the paragraph below (6).

THEOREM 4.5.3
Existence / Uniqueness for Linear Systems
Every system (1) admits no solution, a unique solution, or an infinity of solutions.

Observe that a system (1) is inconsistent only if, in its Gauss-eliminated form, one or more of the equations is of the form zero equal to a nonzero number. But that can never happen if every c_j in (1) is zero, that is, if (1) is **homogeneous**.

THEOREM 4.5.4
Existence / Uniqueness for Homogeneous Systems
Every homogeneous linear system of m equations in n unknowns is consistent. Either it admits the unique trivial solution or else it admits an infinity of nontrivial solutions in addition to the trivial solution. If $m < n$, then there *is* an infinity of nontrivial solutions in addition to the trivial solution.

In summary, not only did the method of Gauss elimination provide us with an efficient and systematic solution procedure, it also led us to important results regarding the existence and uniqueness of solutions.

4.5.7 Pivoting. (Optional).   Recall that the first step in the Gauss elimination of the system

$$
\begin{aligned}
a_{11}x_1 + a_{12}x_2 + \cdots + a_{1n}x_n &= c_1, \\
a_{21}x_1 + a_{22}x_2 + \cdots + a_{2n}x_n &= c_2, \\
&\vdots \\
a_{m1}x_1 + a_{m2}x_2 + \cdots + a_{mn}x_n &= c_m,
\end{aligned}
\tag{37}
$$

is to subtract a_{21}/a_{11} times the first equation from the second, a_{31}/a_{11} times the first equation from the third, and so on, while keeping the first equation intact. The first equation is called the **pivot equation** (or, the first row is the pivot row if one is using the matrix format), and a_{11} is called the **pivot**. That step produces an indented system of the form

$$
\begin{aligned}
a_{11}x_1 + a_{12}x_2 + \cdots + a_{1n}x_n &= c_1, \\
a'_{22}x_2 + \cdots + a'_{2n}x_n &= c'_2, \\
&\vdots \\
a'_{m2}x_2 + \cdots + a'_{mn}x_n &= c'_m.
\end{aligned}
\tag{38}
$$

Next, we keep the first *two* equations intact and use the second equation as the new pivot equation to indent the third through mth equations, and so on.

Naturally, we need each pivot to be nonzero. For instance, we need $a_{11} \neq 0$ for $a_{21}/a_{11}, a_{31}/a_{11}, \ldots$ to be defined. If a pivot is zero, interchange that equation with any one below it, such as the next equation or last equation (as we did in Example 6), until a nonzero pivot is available. Such interchange of equations is called **partial pivoting**. If a pivot is zero we have no choice but to use partial pivoting, but in practice even a nonzero pivot should be rejected if it is "very small," since the smaller it is the more susceptible is the calculation to the adverse effect of machine roundoff error (see Exercise 18). To be as safe as possible, one can choose the pivot equation as the one with the largest leading coefficient (relative to the other coefficients in that equation).

Closure. Beginning with a system of coupled linear algebraic equations, one can use a sequence of elementary operations to minimize the coupling between the equations while leaving the solution set intact. Besides putting forward the important method of Gauss elimination, which is used heavily in the sequel, we used that method to establish several important theoretical results regarding the existence and uniqueness of solutions. For instance, we found that (1) necessarily admits no solution (inconsistent), one solution (unique), or an infinity of solutions (N-parameter family of solutions), and that a homogeneous system is necessarily consistent since, at the least, it has the solution $x_1 = x_2 = \cdots = x_n = 0$ (which we call the *trivial solution*).

Maple. We will see that many calculations in linear algebra can be carried out using computer algebra systems such as *Maple*. Using *Maple*, a great many commands ("functions") are contained within the **linalg** package. A listing of these commands can be obtained by entering ?linalg. That list includes the **linsolve** command, which can be used to solve a system of m linear algebraic equations in n unknowns. To access linsolve (or any other command within the linalg package), first enter with(linalg). Then, linsolve(A,b) solves (1) for $x_1, \ldots, x_n$, where A is the coefficient matrix and b is the column of c_j's. For instance, Gauss elimination reveals that the system

$$\begin{aligned} x_1 - x_2 + 2x_3 - 3x_4 &= 4, \\ x_1 + 2x_2 - x_3 + 3x_4 &= 1 \end{aligned} \tag{39}$$

admits the 2-parameter family of solutions

$$x_4 = \alpha_1, \quad x_3 = \alpha_2, \quad x_2 = -1 - 2\alpha_1 + \alpha_2, \quad x_1 = 3 + \alpha_1 - \alpha_2, \tag{40}$$

where α_1, α_2 are arbitrary. To solve (39) using *Maple*, enter

```
with(linalg):
```

then return and enter

```
linsolve(array([[1,-1,2,-3], [1,2,-1,3]]), array([4,1]));
```

and return. The output is

$$[3 - \_t_1 + \_t_2, \quad -1 + \_t_1 - 2\_t_2, \quad \_t_1, \quad \_t_2]$$

for $x_1, \ldots, x_4$, where $\_t_1$ and $\_t_2$ are arbitrary constants. With $\_t_1 = \alpha_2$ and $\_t_2 = \alpha_1$, this result is the same as (40). If you prefer, you could use the sequence

```
with(linalg):
A:=array([[1,-1,2,-3], [1,2,-1,3]]):
b:=array([4,1]):
linsolve(A,b);
```

instead. If the system is inconsistent, then either the output will be NULL, or there will be no output.

EXERCISES 4.5

1. For $m = n = 2$, the case of a unique solution of (1) is illustrated in Fig. 3b. We stated that that case will occur if and only if $a_{11}a_{22} - a_{12}a_{21} \neq 0$. Prove that claim using ideas or methods contained in this section.

2. Derive the solution set for each of the following systems using Gauss elimination. Document each step (e.g., 2nd row→ 2nd row + 5 times 1st row), and classify the result (e.g., unique solution, inconsistent, 3-parameter family of solutions, etc.). Use either the equation format (as in Example 4) or the matrix format (as in Section 4.5.4), whichever you prefer.

(a) $\quad x - 3y = 1$
$\quad\ 5x + \ y = 2$

(b) $\ 2x + \ y = 0$
$\quad 3x - 2y = 0$

(c) $\ x + 2y = 4$

(d) $\ 2x_1 - x_2 - x_3 - 5x_4 = 6$

(e) $\quad x - y + z = 1$
$\quad 2x - y - z = 8$

(f) $\ 2x_1 - x_2 - \ x_3 - 3x_4 = 0$
$\quad x_1 - x_2 + 4x_3 \qquad\ = 2$

(g) $\quad x + \ 2y + \ 3z = \ 4$
$\quad 5x + \ 6y + \ 7z = \ 8$
$\quad 9x + 10y + 11z = 12$

(h) $\ x_1 + \ x_2 - 2x_3 = \ \ 3$
$\quad x_1 - \ x_2 - 3x_3 = \ \ 1$
$\quad x_1 - 3x_2 - 4x_3 = -1$

(i) $\ 2x_1 - \ x_2 = \ \ 6$
$\quad 3x_1 + 2x_2 = \ \ 4$
$\quad x_1 + 10x_2 = -12$
$\quad 6x_1 + 11x_2 = \ \ -2$

(j) $\quad x_1 - \ x_2 + 2x_3 + \ x_4 = -1$
$\quad 2x_1 + \ x_2 + \ x_3 - \ x_4 = \ \ 4$
$\quad x_1 + 2x_2 - \ x_3 - 2x_4 = \ \ 5$
$\quad x_1 \qquad\quad + \ x_3 \qquad\quad = \ \ 1$

(k) $\ x_1 + 2x_2 = 1$
$\quad 2x_1 - \ x_2 = 3$
$\quad 3x_1 + \ x_2 = 4$
$\quad 4x_1 + 3x_2 = 5$

(l) $\qquad\qquad\ x_3 + x_4 = 2$
$\qquad 4x_2 - \ x_3 + x_4 = 0$
$\quad x_1 - \ x_2 + 2x_3 + x_4 = 4$

(m) $\quad x + 2y + 3z = 5$
$\quad 2x + 3y + 4z = 8$
$\quad 3x + 4y + 5z = c$
$\quad x + \ y \qquad\quad = 2$

for $c = 10$, and again, for $c = 11$

(n) $\ 2x + \ y + \ z = \ \ 10$
$\quad 3x + \ y - \ z = \ \ 6$
$\quad x - 2y - 4z = -10$

(o) $\ 2x_1 + \ x_2 \qquad\qquad = 5$
$\quad x_1 + 2x_2 + \ x_3 \qquad = 5$
$\qquad\quad x_2 + 2x_3 + \ x_4 = 5$
$\qquad\qquad\quad x_3 + 2x_4 = 5$

(p) $\ 2x_1 + \ x_2 \qquad\quad = \ \ 0$
$\quad x_1 + 2x_2 + \ x_3 = -1$
$\qquad\quad x_2 + 2x_3 = -4$

(q) $\ 2x_1 + \ x_2 \qquad\quad + \ x_4 + 2x_5 = 0$
$\quad x_1 + \ x_2 - x_3 \qquad\qquad\quad = 0$
$\quad x_1 + \ x_2 + x_3 - 3x_4 + 2x_5 = 0$
$\quad 2x_1 + 2x_2 - x_3 \qquad\quad + \ x_5 = 0$

3. (a)–(q) Same as Exercise 2 but using Gauss-Jordan reduction instead of Gauss elimination.

4. (a)–(q) Same as Exercise 2 but using computer software such as the *Maple* linsolve command.

5. Let $m = 1$ and $n = 4$ in (1). If possible, make up an example as follows. If it is not possible, say so.

(a) no solution

(b) a unique solution

(c) a 1-parameter family of solutions

(d) a 2-parameter family of solutions

(e) a 3-parameter family of solutions

(f) a 4-parameter family of solutions

6. (a)–(f) Same as Exercise 5, but with $m = 2, n = 4$.

7. (a)–(f) Same as Exercise 5, but with $m = 3, n = 4$.

8. (a)–(f) Same as Exercise 5, but with $m = 4, n = 4$.

9. (a)–(f) Same as Exercise 5, but with $m = 5, n = 4$.

10. Can 20 linear algebraic equations in 14 unknowns have a unique solution? Be inconsistent? Have a 2-parameter family of solutions? Have a 14-parameter family of solutions? Have a 16-parameter family of solutions? Explain.

11. Let
$$a_1x_1 + a_2x_2 + a_3x_3 = 0,$$
$$b_1x_1 + b_2x_2 + b_3x_3 = 0$$

represent any two planes through the origin in a Cartesian x_1, x_2, x_3 space. For the case where the planes intersect along a line, show whether or not that line necessarily passes through the origin.

12. If possible, adapt the methods of this section to solve the

following *nonlinear* systems. If it is *not* possible, say so.

(a) $x_1^2 + 2x_2^2 - x_3^2 = 29$

$\quad\ x_1^2 + \ x_2^2 + x_3^2 = 19$

$\quad 3x_1^2 + 4x_2^2 \qquad = 67$

(b) $\quad\ x + 3y = 13$

$\quad \sin x + 2y = \ 5$

(c) $\ \sin x + \sin y \qquad\quad = 1$

$\quad \sin x - \sin y + 4\cos z = 1.2$

$\quad \sin x + \sin y + 2\cos z = 1.6$

where $-\pi/2 \le x \le \pi/2$, $-\pi/2 \le y \le \pi/2$, and $0 \le z \le 2\pi$.

13. For what values of the parameter λ do the following homogeneous (do you agree that they are homogeneous?) systems admit *non*trivial solutions? Find the nontrivial solutions corresponding to each such λ.

(a) $\quad 2x + \ y = \lambda x$

$\qquad\ x + 2y = \lambda y$

(b) $\quad 2x - \ y = \lambda x$

$\qquad -x + 2y = \lambda y$

(c) $\quad\ x - 2y = \lambda x$

$\qquad 4x - 8y = \lambda y$

(d) $\qquad\qquad\ z = \lambda x$

$\qquad\qquad\quad\ z = \lambda y$

$\qquad x + y + z = \lambda z$

(e) $\ x + y + \ z = \lambda x$

$\qquad\ y + \ z = \lambda y$

$\qquad\qquad 2z = \lambda z$

(f) $\quad 2x + \ y + \ z = \lambda x$

$\qquad\ x + 2y + \ z = \lambda y$

$\qquad\ x + \ y + 2z = \lambda z$

14. Give a critical evaluation of these excerpts from examination papers.

(a) "Given the system

$$x_1 - 2x_2 = 0,$$
$$2x_1 - 4x_2 = 0,$$

add -2 times the first equation to the second and add $-\frac{1}{2}$ times the second equation to the first. By these Gauss elimination steps we obtain the equivalent system $0 = 0$ and $0 = 0$, and hence the 2-parameter family of solutions $x_1 = \alpha_1$ (arbitrary), $x_2 = \alpha_2$ (arbitrary)."

(b) "Given the system

$$x_1 + x_2 - 4x_3 = 0,$$
$$2x_1 - x_2 + \ x_3 = 0,$$

since both left-hand sides equal zero, they must equal each other. Hence we have the equation

$$x_1 + x_2 - 4x_3 = 2x_1 - x_2 + x_3,$$

which equation is equivalent to the original system."

15. Make up an example of an inconsistent linear algebraic system of equations, with

(a) $m = 2, n = 4$
(b) $m = 1, n = 4$
(c) $m = 3, n = 2$
(d) $m = 4, n = 2$
(e) $m = 5, n = 3$
(f) $m = 3, n = 5$

16. (*Physical example of nonexistence and nonuniqueness; DC circuit*) Equations (5) govern the *DC* circuit in Fig. 2. Solve for i_1, i_2, i_3 by Gauss elimination. If there is no solution or if there is a nonunique solution explain that result in physical terms. In each case let $E = 20$ and take

(a) $R_1 = R_2 = R_3 = 5$
(b) $R_1 = R_2 = 5$, $R_3 = 10$

(c) $R_1 = 5, R_2 = R_3 = 10$
(d) $R_1 = 5, R_2 = 20$, $R_3 = 30$

(e) $R_1 = R_3 = 0, R_2 = 5$
(f) $R_2 = R_3 = 0, R_1 = 5$

(g) $R_1 = R_2 = R_3 = 0$

17. (*Physical example of nonexistence and nonuniqueness; statically indeterminate structures*)

(a) Consider the static equilibrium of the system shown, consisting of two weightless cables connected at P, at which point a vertical load F is applied. Use the equilibrium of vertical force components, and horizontal force components too, to derive two linear algebraic equations on the unknown tensions T_1 and T_2. Are there any combinations of angles θ_1 and θ_2 (where $0 \le \theta_1 \le \frac{\pi}{2}$ and $0 \le \theta_2 \le \frac{\pi}{2}$) such that there is either no solution or a nonunique solution? Explain.

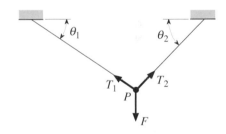

(b) This time let there be three cables at angles of $45°$, $60°$, and $30°$ as shown. Again, use the equilibrium of vertical and horizontal forces at P to derive two linear algebraic equations on the unknown tensions T_1, T_2, T_3. Show that the equations are consistent so there is a nonunique solution. NOTE: We say that such a structure is **statically indeterminate** because the forces in it cannot be determined from the laws of statics alone. What information needs to be added if we are to complete the evaluation of T_1, T_2, T_3? What is needed is information about the relative stiffness of the cables.

We pursue this to a conclusion in (c), below.

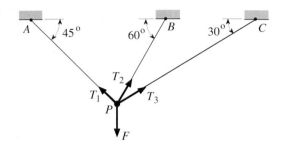

(c) (*Completion of part (b).*) Before the load F is applied, locate an x, y Cartesian coordinate system at P. Let P be one foot below the "ceiling" so the coordinates of A, B, C are $(-1, 1)$, $(1/\sqrt{3}, 1)$, and $(\sqrt{3}, 1)$, respectively. Now apply the load F. The point P will move to a point (x, y), and we assume that the cables are stiff enough so that x and y are very small: $|x| \ll 1$ and $|y| \ll 1$. Let the cables obey *Hooke's law*: $T_1 = k_1 \delta_1$, $T_2 = k_2 \delta_2$, and $T_3 = k_3 \delta_3$, where δ_j is the increase in length of the jth cable due to the tension T_j. Since P moves to (x, y), it follows that

$$\delta_1 = \sqrt{(x+1)^2 + (y-1)^2} - \sqrt{2}$$

$$= \sqrt{2 + 2(x - y) + (x^2 + y^2)} - \sqrt{2}$$

$$\approx \sqrt{2 + 2(x - y)} - \sqrt{2}$$

$$= \sqrt{2}\,[1 + (x - y)]^{1/2} - \sqrt{2}$$

$$\approx \sqrt{2}\left[1 + \frac{1}{2}(x - y)\right] - \sqrt{2} = \frac{1}{\sqrt{2}}(x - y).$$
(17.1)

Explain each step in (17.1), and show, similarly, that

$$\delta_2 \approx -\frac{1}{2}x - \frac{\sqrt{3}}{2}y, \tag{17.2}$$

$$\delta_3 \approx -\frac{\sqrt{3}}{2}x - \frac{1}{2}y. \tag{17.3}$$

Thus,

$$T_1 = k_1 \delta_1 \approx \frac{k_1}{\sqrt{2}}(x - y),$$

$$T_2 = k_2 \delta_2 \approx -\frac{k_2}{2}(x + \sqrt{3}y), \tag{17.4}$$

$$T_3 = k_3 \delta_3 \approx -\frac{k_3}{2}(\sqrt{3}x + y).$$

Putting (17.4) into the two equilibrium equations obtained in (b) then gives two equations in the unknown

displacements x, y. With $k_1 = k_2 = k_3 = 100$ for definiteness, solve that system for x and y, and thus complete the solution for T_1, T_2, T_3.

18. (*Roundoff error difficulty due to small pivots*) To illustrate how small pivots can accentuate the effects of roundoff error, consider the system

$$\begin{aligned} 0.005x_1 + 1.47x_2 &= 1.49, \\ 0.975x_1 + 2.32x_2 &= 6.22 \end{aligned} \tag{18.1}$$

with exact solution $x_1 = 4$ and $x_2 = 1$. Suppose that our computer carries three significant figures and then rounds off. Using the first equation as our pivot equation, Gauss elimination gives

$$\begin{bmatrix} 0.005 & 1.47 & 1.49 \\ 0.975 & 2.32 & 6.22 \end{bmatrix} \rightarrow \begin{bmatrix} 0.005 & 1.47 & 1.49 \\ 0 & -285 & -284 \end{bmatrix}$$

so $x_2 = 284/285 = 0.996$ and $x_1 = [1.49 - (1.47)(0.996)]/0.005 = (1.49 - 1.46)/0.005 = 6$. Show that if we use partial pivoting and then use the first equation of the system

$$\begin{aligned} 0.975x_1 + 2.32x_2 &= 6.22, \\ 0.005x_1 + 1.47x_2 &= 1.49 \end{aligned} \tag{18.2}$$

as our pivot equation, we obtain the result $x_1 = 4.00$ and $x_2 = 1.00$ (which happens to be exactly correct).

19. (*Ill-conditioned systems*) Practically speaking, our numerical calculations are normally carried out on computers, be they hand-held calculators or large digital computers. Such machines carry only a finite number of significant figures and thus introduce *roundoff error* into most calculations. One might expect (or hope) that such slight deviations will lead to answers that are only slightly in error. For example, the solution of

$$\begin{aligned} x + y &= 2, \\ x - 1.014y &= 0 \end{aligned} \tag{19.1}$$

is $x \approx 1.007$, $y \approx 0.993$, whereas the solution of the rounded-off version

$$\begin{aligned} x + y &= 2, \\ x - 1.01y &= 0 \end{aligned}$$

is very much the same, namely $x \approx 1.005$, $y \approx 0.995$. In sharp contrast, the solutions of

$$\begin{aligned} x + y &= 2, \\ x + 1.014y &= 0 \end{aligned} \tag{19.2}$$

and the rounded-off version

$$x + \quad y = 2,$$
$$x + 1.01y = 0,$$

$x \approx 144.9$, $y \approx -142.9$ and $x = 202$, $y = -200$, respectively, are quite different; (19.2) is an example of a so-called **ill-conditioned** system (ill-conditioned in the sense that small changes in the coefficients lead to large changes in the solution). Here, we ask the following: Explain why (19.2) is much more sensitive to roundoff than (19.1) by exploring the two cases graphically, that is, in the x, y plane.

4.6   SPAN

In Sections 4.6, 4.7, and 4.9 we develop a sequence of closely related vector concepts: *span*, *linear dependence*, *bases*, and *expansions*. To motivate where we are headed, consider by analogy an artist who wishes to be able to paint with all the colors of the rainbow but who cannot stock that many colors because their number is infinite! Instead, he (or she) maintains a small set of colors and obtains others by suitable mixing, that is, by taking suitable combinations of his basic set. Even without being artists, we know that he can get by with a minimum set of three basic colors, such as red, blue, and yellow. To form green he can use 50% blue and 50% yellow, to form blue-green he can use 70% blue and 30% yellow, say, and so on. The strategy works because red, blue, and yellow "span" the set of colors of the rainbow in that any given color can be obtained by a suitable combination of them. Further, we say that red, blue, and yellow are "linearly independent" in that no one of them can be obtained by a combination of the others. For instance, surely no combination of red and blue can yield yellow or, indeed, any color in the yellow/green part of the spectrum. Thus we need all three. Continuing to adapt vector terminology to this discussion, if only by rough analogy, we say that red, blue, and yellow form a "basis" of colors for the set of colors of the rainbow.

Similarly, we will have great interest in finding a minimal set of "base vectors" that, by suitable linear combinations, can produce any vector in $\mathbb{R}^n$. To understand how this works we need first to establish the concepts of span (in this section) and linear dependence (in the next).

We begin with the idea of the "span" of a set of vectors.

Definition 4.6.1   *Span*

If $\mathbf{u}_1, \ldots, \mathbf{u}_k$ are vectors, then the set of all linear combinations of these vectors, that is, all vectors of the form

$$\alpha_1 \mathbf{u}_1 + \cdots + \alpha_k \mathbf{u}_k \tag{1}$$

where $\alpha_1, \ldots, \alpha_k$ are scalars, is called the **span** of $\mathbf{u}_1, \ldots, \mathbf{u}_k$ and is denoted as span $\{\mathbf{u}_1, \ldots, \mathbf{u}_k\}$.

The set $\{\mathbf{u}_1, \ldots, \mathbf{u}_k\}$ is called a **generating set** of span $\{\mathbf{u}_1, \ldots, \mathbf{u}_k\}$.

Let us illustrate with some vector sets in $\mathbb{R}^2$ and $\mathbb{R}^3$ so we can support the discussion with diagrams.

EXAMPLE 1
Determine the span of the single vector (Fig. 1)

$$\mathbf{u}_1 = (4, 2) \tag{2}$$

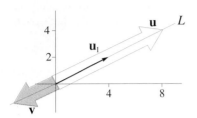

FIGURE 1
Span $\{\mathbf{u}_1\}$.

in $\mathbb{R}^2$. From (1), span $\{\mathbf{u}_1\}$ is the set of all vectors $\alpha_1\mathbf{u}_1$, that is, the set of all vectors that are scalar multiples of $\mathbf{u}_1$. Hence, span $\{\mathbf{u}_1\}$ is the set of all vectors on the line L, such as $\mathbf{u} = 2\mathbf{u}_1 = (8, 4)$, $\mathbf{v} = -\frac{1}{2}\mathbf{u}_1 = (-2, -1)$, and $\mathbf{0} = 0\mathbf{u}_1 = (0, 0)$. We say that $\mathbf{u}_1$ *spans* or *generates* the line L. ∎

EXAMPLE 2
Determine the span of the two vectors

$$\mathbf{u}_1 = (4, 2), \quad \mathbf{u}_2 = (-8, -4). \tag{3}$$

Span $\{\mathbf{u}_1, \mathbf{u}_2\}$ is, once again, the line L in Fig. 1 (i.e., the set of all vectors on L), for both $\mathbf{u}_1$ and $\mathbf{u}_2$ lie along L, so any linear combination of them does too. Similarly, span $\{(4, 2), (-8, -4), (18, 9), (0, 0)\}$ is the line L. ∎

EXAMPLE 3
Determine the span of

$$\mathbf{u}_1 = (4. 2), \quad \mathbf{u}_2 = (1, 5). \tag{4}$$

This time we can see, from the parallelogram law of addition, that span $\{\mathbf{u}_1, \mathbf{u}_2\}$ is all of $\mathbb{R}^2$; that is, $\mathbf{u}_1$ and $\mathbf{u}_2$ span the entire plane. Consider, for instance, the vector $\mathbf{v} = (6, -3)$ shown in Fig. 2. Drawing lines parallel to $\mathbf{u}_1$ and $\mathbf{u}_2$ through both the head and tail of the $\mathbf{v}$ vector gives the dashed parallelogram. Thus, $\mathbf{v} = \mathbf{OA} + \mathbf{OB}$ where (with the aid of a scale) $\mathbf{OA} \approx 1.8\mathbf{u}_1$ and $\mathbf{OB} \approx -1.3\mathbf{u}_2$, so

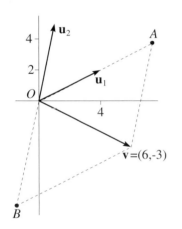

FIGURE 2
Span $\{\mathbf{u}_1, \mathbf{u}_2\}$.

$$\mathbf{v} \approx 1.8\mathbf{u}_1 - 1.3\mathbf{u}_2. \tag{5}$$

Evidently, we can use this parallelogram idea to express *any* vector $\mathbf{v}$ in $\mathbb{R}^2$ as a linear combination of $\mathbf{u}_1$ and $\mathbf{u}_2$, so $\mathbf{u}_1$ and $\mathbf{u}_2$ do indeed span all of $\mathbb{R}^2$.

While the parallelogram graphical construction approach provides understanding, that method is inexact and becomes unwieldy for $\mathbb{R}^3$ and impossible for $\mathbb{R}^4$, $\mathbb{R}^5, \ldots$. Thus, let us reconsider this example from an analytical point of view.

Letting $\mathbf{v} = (v_1, v_2)$ be any given vector in $\mathbb{R}^2$, try to express

$$\mathbf{v} = \alpha_1\mathbf{u}_1 + \alpha_2\mathbf{u}_2. \tag{6}$$

That is, recalling the definition of scalar multiplication and vector addition in n-space,

$$\begin{aligned}
(v_1, v_2) &= \alpha_1(4, 2) + \alpha_2(1, 5) \\
&= (4\alpha_1, 2\alpha_1) + (\alpha_2, 5\alpha_2) \\
&= (4\alpha_1 + \alpha_2, 2\alpha_1 + 5\alpha_2).
\end{aligned} \tag{7}$$

Equating components gives the linear algebraic equations

$$\begin{aligned}
4\alpha_1 + \alpha_2 &= v_1, \\
2\alpha_1 + 5\alpha_2 &= v_2
\end{aligned} \tag{8}$$

in the unknowns α_1, α_2. (Remember, we are considering $\mathbf{v}$ as given, so v_1 and v_2 are given; α_1 and α_2 are the unknowns.) Applying Gauss elimination, (8) becomes

$$\alpha_1 + \tfrac{1}{4}\alpha_2 = \tfrac{1}{4}v_1,$$
$$\alpha_2 = \tfrac{2}{9}v_2 - \tfrac{1}{9}v_1. \tag{9}$$

We see from (9) that the system is *consistent for every given vector $\mathbf{v}$ in $\mathbb{R}^2$.* Thus, every $\mathbf{v}$ in $\mathbb{R}^2$ can be expressed as a linear combination of $\mathbf{u}_1$ and $\mathbf{u}_2$. For instance, let $\mathbf{v} = (6, -3)$, as above. Then $v_1 = 6$, $v_2 = -3$, and (9) gives $\alpha_2 = -\tfrac{4}{3}$ and $\alpha_1 = \tfrac{11}{6}$, so

$$\mathbf{v} = \frac{11}{6}\mathbf{u}_1 - \frac{4}{3}\mathbf{u}_2 \approx 1.8\mathbf{u}_1 - 1.3\mathbf{u}_2, \tag{10}$$

in agreement with the result (5) of our graphical approach.   ■

EXAMPLE 4

Suppose that we add $\mathbf{u}_3 = (-1, -2)$ to the set, so we have

$$\mathbf{u}_1 = (4, 2), \quad \mathbf{u}_2 = (1, 5), \quad \mathbf{u}_3 = (-1, -2). \tag{11}$$

It should be evident that span$\{\mathbf{u}_1, \mathbf{u}_2, \mathbf{u}_3\}$ is all of $\mathbb{R}^2$ again, since $\{\mathbf{u}_1, \mathbf{u}_2\}$ spanned $\mathbb{R}^2$ even "without any help" from $\mathbf{u}_3$. But in case this is not clear, let us go through steps analogous to (6) to (9):

$$\mathbf{v} = \alpha_1\mathbf{u}_1 + \alpha_2\mathbf{u}_2 + \alpha_3\mathbf{u}_3, \tag{12}$$

so $(v_1, v_2) = (4\alpha_1 + \alpha_2 - \alpha_3, 2\alpha_1 + 5\alpha_2 - 2\alpha_3)$. Thus,

$$4\alpha_1 + \alpha_2 - \alpha_3 = v_1,$$
$$2\alpha_1 + 5\alpha_2 - 2\alpha_3 = v_2, \tag{13}$$

or, after Gauss elimination,

$$\alpha_1 + \tfrac{1}{4}\alpha_2 - \tfrac{1}{4}\alpha_3 = \tfrac{1}{4}v_1,$$
$$\alpha_2 - \tfrac{1}{3}\alpha_3 = \tfrac{2}{9}v_2 - \tfrac{1}{9}v_1. \tag{14}$$

Like (9), (14) is consistent for every $\mathbf{v}$ in $\mathbb{R}^2$, so $\{\mathbf{u}_1, \mathbf{u}_2, \mathbf{u}_3\}$ does span $\mathbb{R}^2$, as claimed. Whereas (9) had a unique solution so that the representation (6) was unique, (14) has a 1-parameter family of solutions so that the representation (12) is *not* unique. For instance, we can choose α_3 arbitrarily and then solve (14) for α_1 and α_2.   ■

EXAMPLE 5

What is the span of the two vectors

$$\mathbf{u}_1 = (1, 2, 2), \quad \mathbf{u}_2 = (-1, 0, 2) \tag{15}$$

in $\mathbb{R}^3$? Setting

$$\mathbf{v} = \alpha_1\mathbf{u}_1 + \alpha_2\mathbf{u}_2, \tag{16}$$

we have

$$\begin{aligned} \alpha_1 - \alpha_2 &= v_1, \\ 2\alpha_1 &= v_2, \\ 2\alpha_1 + 2\alpha_2 &= v_3, \end{aligned} \tag{17}$$

or, after Gauss elimination,

$$\begin{aligned} \alpha_1 - \alpha_2 &= v_1, \\ \alpha_2 &= \tfrac{1}{2}v_2 - v_1, \\ 0 &= v_3 - 2v_2 + 2v_1. \end{aligned} \tag{18}$$

Now, span$\{\mathbf{u}_1, \mathbf{u}_2\}$ is the set of all vectors $\mathbf{v}$ such that (18) is consistent, namely, all vectors $\mathbf{v} = (v_1, v_2, v_3)$ such that

$$2v_1 - 2v_2 + v_3 = 0, \tag{19}$$

so that the last equation in (18) is $0 = 0$ rather than a contradiction.

In geometrical terms, on the other hand, span$\{\mathbf{u}_1, \mathbf{u}_2\}$ should be the subset of $\mathbb{R}^3$ consisting of the *plane* that passes through $\mathbf{u}_1$ and $\mathbf{u}_2$ ($\mathbf{u}_1$ and $\mathbf{u}_2$ are shown in Fig. 3). How does that fact correlate with (19)? As a matter of fact, (19) *is* the equation of a plane in 3-space,[1] and that plane does pass through the origin, through the tip of $\mathbf{u}_1$ [i.e., the point $(1, 2, 2)$], and through the tip of $\mathbf{u}_2$ [the point $(-1, 0, 2)$]. Hence, it *is* the plane through $\mathbf{u}_1$ and $\mathbf{u}_2$ so the analytical approach in (16) to (19) and our geometrical interpretation are in agreement.

We conclude that span$\{\mathbf{u}_1, \mathbf{u}_2\}$ is not all of $\mathbb{R}^3$; it is only that part of $\mathbb{R}^3$ consisting of the plane (i.e., all vectors in the plane) containing the given vectors $\mathbf{u}_1$ and $\mathbf{u}_2$.

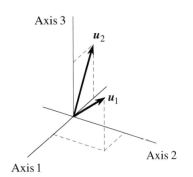

FIGURE 3
$\mathbf{u}_1$ and $\mathbf{u}_2$.

COMMENT. Since span$\{\mathbf{u}_1, \mathbf{u}_2\}$ is a plane, would it be correct to say that span$\{\mathbf{u}_1, \mathbf{u}_2\}$ is $\mathbb{R}^2$? No, that would be incorrect; $\mathbb{R}^2$ is made up of 2-tuples, while the vectors in the above-mentioned plane are 3-tuples. Thus, $\mathbb{R}^2$ space is not relevant in this problem. All that can be said here is that span$\{\mathbf{u}_1, \mathbf{u}_2\}$ is that part of $\mathbb{R}^3$ consisting of the plane containing the vectors $\mathbf{u}_1$ and $\mathbf{u}_2$, that is, the plane defined by (19). ∎

EXAMPLE 6

As our final example, what is the span of

$$\mathbf{u}_1 = (1, 2, 3, 4), \quad \mathbf{u}_2 = (2, 4, 6, 8) \tag{20}$$

in $\mathbb{R}^4$? This example is very much like Example 2 because $\mathbf{u}_1$ and $\mathbf{u}_2$ in (20) are *colinear*; that is, they lie along a common line, say L, which is an infinite straight line

---

[1] Recall that the general equation of a plane in x, y, z space is $ax + by + cz = d$ and that that plane passes through the origin if and only if $d = 0$.

passing through the origin, the point $(1, 2, 3, 4)$, and the point $(2, 4, 6, 8)$. Span$\{\mathbf{u}_1, \mathbf{u}_2\}$ is the line L, namely, the set of vectors $\beta(1, 2, 3, 4)$ for all values of β such that $-\infty < \beta < \infty$. Understand that this is a "line" in $\underline{4}$-space so we cannot draw a picture of it the way we did for the line L in 2-space (Fig. 1).

Although we can see this result by inspection of (20), let us also employ our analytical approach as one last illustration of the method. Setting

$$\mathbf{v} = \alpha_1 \mathbf{u}_1 + \alpha_2 \mathbf{u}_2, \tag{21}$$

we have

$$\alpha_1 + 2\alpha_2 = v_1,$$
$$2\alpha_1 + 4\alpha_2 = v_2,$$
$$3\alpha_1 + 6\alpha_2 = v_3, \tag{22}$$
$$4\alpha_1 + 8\alpha_2 = v_4,$$

or, after Gauss elimination,

$$\alpha_1 + 2\alpha_2 = v_1, \tag{23a}$$
$$0 = v_2 - 2v_1, \tag{23b}$$
$$0 = v_3 - 3v_1, \tag{23c}$$
$$0 = v_4 - 4v_1. \tag{23d}$$

For the latter system [and hence the original system (22)] to be consistent we need $\mathbf{v} = (v_1, v_2, v_3, v_4)$ to be such that (23b,c,d) are satisfied. Thus we need $v_2 = 2v_1$, $v_3 = 3v_1$, and $v_4 = 4v_1$, so $\mathbf{v}$ must be of the form

$$\mathbf{v} = (v_1, v_2, v_3, v_4) = (v_1, 2v_1, 3v_1, 4v_1)$$
$$= v_1(1, 2, 3, 4), \tag{24}$$

where v_1 is arbitrary. Thus, the set of $\mathbf{v}$'s that can be formed by linear combinations of $\mathbf{u}_1$ and $\mathbf{u}_2$, as in (21), is the set of vectors that are scalar multiples of the vector $(1, 2, 3, 4)$. That result is the same as that obtained, in the first paragraph of this example, by inspection. ∎

Closure. If $\mathbf{u}_1, \ldots, \mathbf{u}_k$ are vectors in $\mathbb{R}^n$, then span$\{\mathbf{u}_1, \ldots, \mathbf{u}_k\}$, namely, the set of all linear combinations of $\mathbf{u}_1, \ldots, \mathbf{u}_k$, is all or part of $\mathbb{R}^n$. To determine their span, write

$$\mathbf{v} = \alpha_1 \mathbf{u}_1 + \alpha_2 \mathbf{u}_2 + \cdots + \alpha_k \mathbf{u}_k. \tag{25}$$

Expressing (25) in component form gives n linear algebraic equations in the unknowns $\alpha_1, \ldots, \alpha_k$. Solving the latter system by Gauss elimination will reveal the set of $\mathbf{v}$'s for which the system is *consistent*. That set of $\mathbf{v}$'s is the span of $\mathbf{u}_1, \ldots, \mathbf{u}_k$. In 3-space, for instance, the span of a set of 3-tuple vectors will be the point at the origin (which case occurs if and only if each of the $\mathbf{u}_j$'s is $\mathbf{0}$), a line through the origin, a plane through the origin, or all of 3-space.

EXERCISES 4.6

1. (a) Sketch any two vectors that span the space of all vectors in the plane of the paper.
   (b) Sketch any three such vectors.
   (c) Sketch any four such vectors.

2. Using the analytical (rather than graphical) approach, show whether the vectors
   (a) $(1, 2)$, $(2, 1)$ span $\mathbb{R}^2$
   (b) $(1, 2)$, $(2, 1)$, $(4, 5)$ span $\mathbb{R}^2$
   (c) $(1, 2)$, $(2, 1)$, $(2, 3)$, $(2, -4)$ span $\mathbb{R}^2$
   (d) $(0, 0, 0, 1)$, $(0, 0, 1, 1)$, $(0, 1, 1, 1)$, $(1, 1, 1, 1)$ span $\mathbb{R}^4$
   (e) $(1, 2, 0, 4)$, $(2, 3, 1, -1)$, $(0, 1, 0, 1)$, $(0, 0, 0, 0)$, $(1, 1, 2, 3)$ span $\mathbb{R}^4$
   (f) $(1, 3, 2, 2)$, $(5, 7, 1, 0)$, $(-1, -2, -4, 3)$ span $\mathbb{R}^4$
   (g) $(1, 0, 1)$, $(2, 1, -1)$, $(1, 2, -5)$ span $\mathbb{R}^3$
   (h) $(1, 1, 2)$, $(0, 0, 0)$, $(2, 1, 0)$, $(-1, 0, 3)$ span $\mathbb{R}^3$
   (i) $(2, 0, 3)$, $(-1, 2, 4)$, $(-5, 2, -2)$ span $\mathbb{R}^3$
   (j) $(1, 3, 0)$, $(2, -1, 1)$, $(1, 1, 4)$ span $\mathbb{R}^3$
   (k) $(-1, 2, 4)$, $(-5, 2, -2)$, $(2, 0, 3)$, $(1, 2, 3)$ span $\mathbb{R}^3$
   (l) $(0, 0, 0)$, $(2, 1, 4)$, $(-1, 3, 5)$ span $\mathbb{R}^3$
   (m) $(2, 1, 3)$, $(1, -1, 2)$ span $\mathbb{R}^3$
   (n) $(2, 1, -1)$, $(1, 3, 1)$, $(5, 5, -1)$, $(0, 5, 3)$ span $\mathbb{R}^3$
   (o) $(-4, 1, 0)$, $(2, 2, 2)$, $(1, 2, 3)$ span $\mathbb{R}^3$
   (p) $(-3, 1, 0)$, $(1, 1, 1)$, $(-1, 7, 5)$ span $\mathbb{R}^3$
   (q) $(1, 0, \ldots, 0)$, $(0, 1, 0, \ldots, 0)$, $\ldots$, $(0, \ldots, 0, 1)$ span $\mathbb{R}^n$

3. Find any two vectors in $\mathbb{R}^3$ that span the plane
   (a) $x_1 - 2x_2 + 4x_3 = 0$
   (b) $2x_1 + x_2 - 6x_3 = 0$
   (c) $x_1 + 5x_3 = 0$
   (d) $x_1 + 4x_2 + x_3 = 0$
   (e) $x_2 + 2x_3 = 0$
   (f) $3x_1 - x_2 - x_3 = 0$
   (g) $x_1 + x_2 - x_3 = 0$
   (h) $x_1 + 2x_2 + 3x_3 = 0$

4. Show whether the given sets are identical. Explain.
   (a) span $\{(1, 4), (1, 1)\}$ and span $\{(1, 0), (1, 1), (2, 1)\}$
   (b) span $\{(2, 3), (1, 0)\}$ and span $\{(1, -1), (-4, 4)\}$
   (c) span $\{(2, -1, -1), (3, 1, 0)\}$ and span $\{(2, -1, -1), (5, 5, 1)\}$
   (d) span $\{(1, 2, 3), (2, -1, 1)\}$ and span $\{(1, 2, 3), (3, 1, 5)\}$
   (e) span $\{(4, 1, 0), (1, 1, 1)\}$ and span $\{(1, 1, 1), (2, -1, -2)\}$
   (f) span $\{(1, 2, -1), (-3, 0, 0), (2, -2, 1)\}$ and span $\{(1, 0, 0), (1, 3, 0)\}$
   (g) span $\{(1, 0, 1, 2), (-1, 1, 1, 0)\}$ and span $\{(0, 1, 2, 2), (1, 1, 3, 4), (1, -2, -3, -2)\}$
   (h) span $\{(1, 0, 1, 2), (1, 1, 1, 1), (1, 2, 3, 4)\}$ and span $\{(2, 0, -1, 0), (0, -1, 2, 3), (4, 3, 2, 1)\}$
   (i) span $\{(1, 0, 1, 1), (2, 1, 1, 0), (1, 2, 2, 2)\}$ and span $\{(2, -1, 0, 0), (1, -2, 0, 1), (3, 5, 4, 1)\}$
   (j) span $\{(1, 2, 3, 0), (0, 1, 0, 2), (2, 3, 0, 1)\}$ and span $\{(1, 0, -3, -1), (-1, 1, 3, 3), (1, 2, 1, 1)\}$

5. Find any two ON (orthonormal) vectors in
   (a) span $\{(1, 2), (6, -1)\}$
   (b) span $\{(1, 2, 4), (2, -1, 3)\}$
   (c) span $\{(1, -1, 0), (1, 2, 3)\}$
   (d) span $\{(2, 1, 0), (0, 1, 2)\}$
   (e) span $\{(1, 1, 0, 1), (0, 2, -1, 1)\}$
   (f) span $\{(-2, 3, 1, 1), (0, 2, -1, 1)\}$

4.7 LINEAR DEPENDENCE AND INDEPENDENCE

To recall our plan for Sections 4.6, 4.7, and 4.9, we urge you to re-read the first paragraph of Section 4.6 before continuing. We stated, there, that red, blue, and yellow are "linearly independent" in that no one of them can be obtained by a linear combination of the others. Turning from that rough paint analogy to vectors, we have the following fundamental definition.

Definition 4.7.1 *Linear Dependence and Linear Independence*

A set of vectors $\{\mathbf{u}_1, \ldots, \mathbf{u}_k\}$ is said to be **linearly dependent** if at least one of them can be expressed as a linear combination of the others. If none can be so expressed, then the set is **linearly independent**.[1]

For brevity, we will usually use the abbreviations **LD** and **LI** for linearly dependent and linearly independent, respectively, but be aware that this notation is *not* standard among other authors.

EXAMPLE 1

Let $\mathbf{u}_1 = (1, 0)$, $\mathbf{u}_2 = (1, 1)$, and $\mathbf{u}_3 = (5, 4)$. These are LD since, by inspection, we can express $\mathbf{u}_3$ as a linear combination of $\mathbf{u}_1$ and $\mathbf{u}_2$: $\mathbf{u}_3 = \mathbf{u}_1 + 4\mathbf{u}_2$. (Alternatively, we could express $\mathbf{u}_2 = \frac{1}{4}\mathbf{u}_3 - \frac{1}{4}\mathbf{u}_1$, or $\mathbf{u}_1 = -4\mathbf{u}_2 + \mathbf{u}_3$). ∎

EXAMPLE 2

Let $\mathbf{u}_1 = (1, 0)$ and $\mathbf{u}_2 = (1, 1)$. These are LI since $\mathbf{u}_1$ cannot be expressed as a "linear combination of the others," namely, as a scalar multiple of $\mathbf{u}_2$, nor can $\mathbf{u}_2$ be expressed as a scalar multiple of $\mathbf{u}_1$. ∎

EXAMPLE 3

Let $\mathbf{u}_1 = (2, -1)$, $\mathbf{u}_2 = (0, 0)$, and $\mathbf{u}_3 = (0, 1)$. These are LD since we can express $\mathbf{u}_2 = 0\mathbf{u}_1 + 0\mathbf{u}_3$. (The fact that we *cannot* express $\mathbf{u}_1$ as a linear combination of $\mathbf{u}_2$ and $\mathbf{u}_3$, nor $\mathbf{u}_3$ as a linear combination of $\mathbf{u}_1$ and $\mathbf{u}_2$ does not alter our conclusion, for recall the words "at least one" in the definition.) ∎

The preceding examples were simple enough to be worked by inspection. In more complicated cases, the following theorem provides a systematic approach for determining whether a given vector set is linearly dependent or linearly independent.

THEOREM 4.7.1

Test for Linear Dependence / Independence
A finite set of vectors $\{\mathbf{u}_1, \ldots, \mathbf{u}_k\}$ is LD if and only if there exist scalars α_j, not all zero, such that

$$\alpha_1\mathbf{u}_1 + \cdots + \alpha_k\mathbf{u}_k = \mathbf{0}; \tag{1}$$

if (1) holds only if all the α's are zero, then the set is LI.

Proof. Because of the "if and only if" we must prove the statement in both directions. First, suppose the set is LD. Then, according to the definition of linear dependence one of the vectors, say $\mathbf{u}_j$, can be expressed as a linear combination of the others:

$$\mathbf{u}_j = \alpha_1\mathbf{u}_1 + \cdots + \alpha_{j-1}\mathbf{u}_{j-1} + \alpha_{j+1}\mathbf{u}_{j+1} + \cdots + \alpha_k\mathbf{u}_k, \tag{2}$$

which equation can be rearranged as

$$\alpha_1\mathbf{u}_1 + \cdots + \alpha_{j-1}\mathbf{u}_{j-1} + (-1)\mathbf{u}_j + \alpha_{j+1}\mathbf{u}_{j+1} + \cdots + \alpha_k\mathbf{u}_k = \mathbf{0}. \tag{3}$$

Comparing (3) with (1), we see that $\alpha_j = -1$ is nonzero, so there exist scalars $\alpha_1, \ldots, \alpha_k$ not all zero such that (1) holds.

---

[1]Usually, $k > 1$. If $k = 1$ the set $\{\mathbf{u}_1\}$ is defined to be linearly dependent if $\mathbf{u}_1 = \mathbf{0}$ and linearly independent if $\mathbf{u}_1 \neq \mathbf{0}$.

Conversely, suppose that (1) holds with the α's not all zero. If α_p, for instance, is nonzero, then (1) can be divided by α_p and solved for $\mathbf{u}_p$ as a linear combination of the other $\mathbf{u}$'s, in which case $\{\mathbf{u}_1, \ldots, \mathbf{u}_k\}$ is LD.   ∎

It may be helpful to think of it this way. If all the α's in (1) are zero, then (1) becomes $0\mathbf{u}_1 + \cdots + 0\mathbf{u}_k = \mathbf{0}$. The latter is hardly a relation on $\mathbf{u}_1, \ldots, \mathbf{u}_k$, it is simply the identity $\mathbf{0} = \mathbf{0}$. However, if (1) holds with the α's not all zero, that *is* a relation on $\mathbf{u}_1, \ldots, \mathbf{u}_k$; they are "related," hence LD.

EXAMPLE 4

Consider the 4-tuples

$$\mathbf{u}_1 = (2, 0, 1, -3), \quad \mathbf{u}_2 = (0, 1, 1, 1), \quad \mathbf{u}_3 = (2, 2, 3, 0). \tag{4}$$

To see if these vectors are LD or LI, appeal directly to (1):

$$\alpha_1(2, 0, 1, -3) + \alpha_2(0, 1, 1, 1) + \alpha_3(2, 2, 3, 0) = (0, 0, 0, 0), \tag{5}$$

or $(2\alpha_1 + 2\alpha_3, \ \alpha_2 + 2\alpha_3, \ \alpha_1 + \alpha_2 + 3\alpha_3, \ -3\alpha_1 + \alpha_2) = (0, 0, 0, 0)$. Thus,

$$\begin{aligned}
2\alpha_1 \qquad + 2\alpha_3 &= 0, \\
\alpha_2 + 2\alpha_3 &= 0, \\
\alpha_1 + \alpha_2 + 3\alpha_3 &= 0, \\
-3\alpha_1 + \alpha_2 \qquad &= 0.
\end{aligned} \tag{6}$$

Applying Gauss elimination yields

$$\begin{aligned}
2\alpha_1 \qquad + 2\alpha_3 &= 0, \\
\alpha_2 + 2\alpha_3 &= 0, \\
\alpha_3 &= 0, \\
0 &= 0.
\end{aligned} \tag{7}$$

This system admits only the trivial solution, $\alpha_1 = \alpha_2 = \alpha_3 = 0$ so $\mathbf{u}_1, \mathbf{u}_2, \mathbf{u}_3$ are LI.   ∎

EXAMPLE 5

Consider the 3-tuples

$$\mathbf{u}_1 = (1, 0, 1), \quad \mathbf{u}_2 = (1, 1, 1), \quad \mathbf{u}_3 = (1, 1, 2), \quad \mathbf{u}_4 = (1, 2, 1). \tag{8}$$

Working from (1), as in Example 4, we have

$$\begin{aligned}
\alpha_1 + \alpha_2 + \ \alpha_3 + \ \alpha_4 &= 0, \\
\alpha_2 + \ \alpha_3 + 2\alpha_4 &= 0, \\
\alpha_1 + \alpha_2 + 2\alpha_3 + \ \alpha_4 &= 0,
\end{aligned} \tag{9}$$

or, after Gauss elimination,

$$\begin{aligned}
\alpha_1 + \alpha_2 + \alpha_3 + \ \alpha_4 &= 0, \\
\alpha_2 + \alpha_3 + 2\alpha_4 &= 0, \\
\alpha_3 \qquad &= 0.
\end{aligned} \tag{10}$$

This time, there exist nontrivial solutions for the α_j's so the vectors $\mathbf{u}_1, \mathbf{u}_2, \mathbf{u}_3$ are LD. Specifically, (10) gives $\alpha_3 = 0$, $\alpha_4 = \beta$, $\alpha_2 = -2\beta$, $\alpha_1 = \beta$, where β is arbitrary. With $\beta = 1$, say, (1) becomes $\mathbf{u}_1 - 2\mathbf{u}_2 + 0\mathbf{u}_3 + \mathbf{u}_4 = \mathbf{0}$.   ∎

We conclude this section with four "little" theorems.

THEOREM 4.7.2

Linear Dependence / Independence of Two Vectors
A set of *two* vectors $\{\mathbf{u}_1, \mathbf{u}_2\}$ is LD if and only if one is expressible as a scalar multiple of the other.

THEOREM 4.7.3

Linear Dependence of Sets Containing the Zero Vector
A set containing the zero vector is LD.

THEOREM 4.7.4

Equating Coefficients
Let $\{\mathbf{u}_1, \ldots, \mathbf{u}_k\}$ be LI. Then, for

$$a_1\mathbf{u}_1 + \cdots + a_k\mathbf{u}_k = b_1\mathbf{u}_1 + \cdots + b_k\mathbf{u}_k$$

to hold, it is necessary and sufficient that $a_j = b_j$ for each $j = 1, \ldots, k$. That is, the coefficients of corresponding vectors on the left-hand and right-hand sides must match.

THEOREM 4.7.5

Orthogonal Sets
Every finite orthogonal set of (nonzero) vectors is LI.

Proof of Theorem 4.7.5.   Dot $\mathbf{u}_1$ into both sides of

$$\alpha_1\mathbf{u}_1 + \alpha_2\mathbf{u}_2 + \cdots + \alpha_k\mathbf{u}_k = \mathbf{0}. \tag{11}$$

In other words,

$$
\begin{aligned}
\mathbf{u}_1 \cdot (\alpha_1\mathbf{u}_1 + \alpha_2\mathbf{u}_2 + \cdots + \alpha_k\mathbf{u}_k) &= \mathbf{u}_1 \cdot \mathbf{0}, \\
\alpha_1\mathbf{u}_1 \cdot \mathbf{u}_1 + \alpha_2\mathbf{u}_1 \cdot \mathbf{u}_2 + \cdots + \alpha_k\mathbf{u}_1 \cdot \mathbf{u}_k &= 0, \\
\alpha_1 \|\mathbf{u}_1\|^2 + 0 + \cdots + 0 &= 0.
\end{aligned}
\tag{12}
$$

Now $\mathbf{u}_1 \neq \mathbf{0}$ implies that $\|\mathbf{u}_1\| \neq 0$ so it follows from (12) that $\alpha_1 = 0$. Similarly, dotting $\mathbf{u}_2$ into (11) gives $\alpha_2 = 0$, and so on. Since $\alpha_1 = \alpha_2 = \cdots = \alpha_k = 0$, the $\mathbf{u}_j$'s must be LI, as claimed. ∎

Understand that we do not have degrees of linear dependence; two vectors are either LD or not. Nevertheless, it might be helpful to think of orthogonality as the extreme case of linear independence. For suppose the angle between two vectors is $\theta = 0$; then they are LD. If we rotate one of them, then when $\theta = \pi$ they are once again LD. It is when $\theta = \pi/2$, when the vectors are orthogonal, that they are as far from being LD as possible.

We close with illustrations of Theorems 4.7.2 and 4.7.3.

EXAMPLE 6

The set $\{(2, 1), (1, 5)\}$ in $\mathbb{R}^2$ is LI because neither vector can be expressed as a scalar multiple of the other. ∎

EXAMPLE 7

Let
$$\mathbf{u}_1 = (4, -1, 1, 2), \quad \mathbf{u}_2 = (3, 0, 2, 5), \quad \mathbf{u}_3 = (0, 0, 0, 0)$$

in $\mathbb{R}^4$. The set is LD, according to Theorem 4.7.3 because it contains the zero vector $\mathbf{u}_3 = \mathbf{0}$. That is, $\mathbf{u}_3$ can be expressed as a linear combination of $\mathbf{u}_1$ and $\mathbf{u}_2$: $\mathbf{u}_3 = 0\mathbf{u}_1 + 0\mathbf{u}_2$. If the preceding sentence is not clear, rewrite the equation as $0\mathbf{u}_1 + 0\mathbf{u}_2 - 1\mathbf{u}_3 = \mathbf{0}$ and observe that the coefficients $\alpha_1 = 0, \alpha_2 = 0, \alpha_3 = -1$ are not all zero. ∎

Closure. This section is of special importance because when we return to our study of differential equations in Chapter 6 we will introduce the same concept of linear dependence and linear independence for *functions*, and that concept will be essential to our further development of the subject of differential equations.

If we cannot establish the linear dependence or independence of a given vector set $\mathbf{u}_1, \ldots, \mathbf{u}_k$ by appealing to Definition 4.7.1 and using inspection, as we did in Examples 1–3, we can use the test provided by Theorem 4.7.1. Accordingly, write

$$\alpha_1 \mathbf{u}_1 + \cdots + \alpha_k \mathbf{u}_k = \mathbf{0}$$

and re-express the latter, equivalently, as a homogeneous system of linear algebraic equation on the α's. [In Example 4, for instance, that system is given by (6).] Then use Gauss elimination to solve for the α's. If we obtain the unique trivial solution $\alpha_1 = \cdots = \alpha_k = 0$ then $\mathbf{u}_1, \ldots, \mathbf{u}_k$ are LI; if, besides the trivial solution, there are nontrivial solutions as well, then $\mathbf{u}_1, \ldots, \mathbf{u}_k$ are LD.

EXERCISES 4.7

1. **(a)** Can a set be neither LD nor LI? Explain.
   (b) Can a set be both LD and LI? Explain.

2. Show that the following sets are LD by expressing one of the vectors as a linear combination of the others.
   (a) $\{(1, 1), (1, 2), (3, 4)\}$
   (b) $\{(1, 4), (2, 8), (3, -1)\}$
   (c) $\{(1, -1), (4, 2), (-3, 3)\}$
   (d) $\{(1, 2, 3), (3, 2, 1), (5, 5, 5)\}$
   (e) $\{(1, 0, 0), (0, 1, 0), (3, 3, 0), (2, -7, 9)\}$

3. Determine whether the following set is LI or LD. If it is LD, then give a linear relation among the vectors.
   (a) $(1, 3), (2, 0), (-1, 3), (7, 3)$
   (b) $(1, 3), (2, 0), (1, 2), (-1, 5)$
   (c) $(2, 3, 0), (1, -2, 3)$
   (d) $(2, 3, 0), (1, -2, 4), (1, 1, 0), (1, 1, 1)$
   (e) $(0, 0, 2), (0, 0, 3), (2, -1, 5), (1, 2, 4)$
   (f) $(2, 3, 0, 0), (1, -5, 0, 2), (3, 1, 2, 2)$
   (g) $(1, 3, 2, 0), (4, 1, -2, -2), (0, 2, 0, 3), (4, 7, 1, 2)$
   (h) $(2, 0, 1, -1, 0), (1, 2, 0, 3, 1), (4, -4, 3, -9, -2)$
   (i) $(1, 3, 0), (0, 1, -1), (0, 0, 0)$
   (j) $(1, 1, 0, 0), (1, -1, 0, 0), (0, 0, -2, 2), (0, 0, 1, 1)$
   (k) $(1, -3, 0, 2, 1), (-2, 6, 0, -4, -2)$
   (l) $(5, 4, 1, 1), (0, 0, 0, 0), (1, 9, -7, 2)$
   (m) $(1, 2, 3, 4), (2, 3, 4, 5)$
   (n) $(2, 1, -1), (1, 4, 2), (3, -2, -4)$
   (o) $(7, 1, 0), (-1, 1, 4), (2, 3, 5)$

(p) $(1, 2, -1), (1, 0, 1), (3, -2, 5)$

(q) $(3, 1, 0, 0), (1, -2, 4, 1), (2, 1, 6, 5)$

(r) $(2, 4, 0, 1), (1, 0, 1, 2), (1, -3, 1, 2),$
$(1, 1, -1, -1)$

4. Show, by graphical means, that the vector sets shown below, and lying in the plane of the paper, are LD. (The emphasis here is on the method and ideas, not on graphical precision.)

(a)

(b)

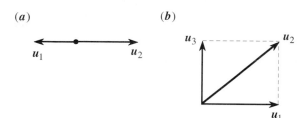

(c)

(d)

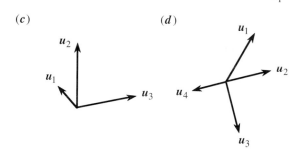

(e)

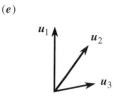

5. If $\mathbf{u}_1$ and $\mathbf{u}_2$ are LI, $\mathbf{u}_1$ and $\mathbf{u}_3$ are LI and $\mathbf{u}_2$ and $\mathbf{u}_3$ are LI, does it follow that $\{\mathbf{u}_1, \mathbf{u}_2, \mathbf{u}_3\}$ is LI? Prove or disprove.

6. Prove or disprove:

    (a) $\mathbf{v}$ is in span $\{\mathbf{u}_1, \ldots, \mathbf{u}_k\}$ if $\{\mathbf{v}, \mathbf{u}_1, \ldots, \mathbf{u}_k\}$ is LD.

    (b) $\mathbf{v}$ is not in span $\{\mathbf{u}_1, \ldots, \mathbf{u}_k\}$ if $\{\mathbf{v}, \mathbf{u}_1, \ldots, \mathbf{u}_k\}$ is LI.

    (c) $\mathbf{v}$ is not in span $\{\mathbf{u}_1, \ldots, \mathbf{u}_k\}$ if and only if $\{\mathbf{v}, \mathbf{u}_1, \ldots, \mathbf{u}_k\}$ is LI.

7. **(a)** Prove Theorem 4.7.2.

    (b) Prove Theorem 4.7.3.

    (c) Prove Theorem 4.7.4.

4.8 VECTOR SPACE

4.8.1 Vector space.   Recall that we began with "arrow vectors" and, building upon that concept, we defined n-tuple vectors in n-space. In this section we generalize still further and provide a mathematical framework known as vector space.

The idea is as follows. In preceding sections we introduced the vectors and arithmetic rules for their manipulation, and then derived the various properties, such as $\mathbf{u} + \mathbf{v} = \mathbf{v} + \mathbf{u}$, $\mathbf{u} + \mathbf{0} = \mathbf{u}$, $\alpha(\beta\mathbf{u}) = (\alpha\beta)\mathbf{u}$, and so on. In generalizing, the essential idea is to reverse the cart and the horse. Specifically, we elevate the derived properties to axioms, or requirements, and regard the vectors as "objects," the nature of which is not restricted in advance. They may be chosen to be arrow vectors, n-tuples or whatever; all that we ask is that a plus (+) operation, a zero vector, a negative inverse, and scalar multiplication be defined such that all of the vector space axioms are satisfied. Thus:

Definition 4.8.1   *Vector Space*

We call a (nonempty) set $\mathcal{S}$ of elements, which are denoted by **boldface type** and referred to as *vectors*, a **vector space** if the following requirements are met:

(i) An operation, which will be called vector *addition* and denoted as $+$, is defined between any two vectors in $\mathcal{S}$ in such a way that if $\mathbf{u}$ and $\mathbf{v}$ are in $\mathcal{S}$, then $\mathbf{u} + \mathbf{v}$ is too (i.e., $\mathcal{S}$ is *closed under addition*). Furthermore,

$$\mathbf{u} + \mathbf{v} = \mathbf{v} + \mathbf{u}, \qquad \text{(commutative)} \qquad (1)$$

$$(\mathbf{u} + \mathbf{v}) + \mathbf{w} = \mathbf{u} + (\mathbf{v} + \mathbf{w}). \qquad \text{(associative)} \qquad (2)$$

(ii) $\mathcal{S}$ contains a unique *zero vector* $\mathbf{0}$ such that

$$\mathbf{u} + \mathbf{0} = \mathbf{u} \qquad (3)$$

for each $\mathbf{u}$ in $\mathcal{S}$.

(iii) For each $\mathbf{u}$ in $\mathcal{S}$ there is a unique vector "$-\mathbf{u}$" in $\mathcal{S}$, called the *negative inverse* of $\mathbf{u}$, such that

$$\mathbf{u} + (-\mathbf{u}) = \mathbf{0}. \qquad (4)$$

We denote $\mathbf{u} + (-\mathbf{v})$ as $\mathbf{u} - \mathbf{v}$ for brevity, but emphasize that it is actually the $+$ operation between $\mathbf{u}$ and $-\mathbf{v}$.

(iv) Another operation, called *scalar multiplication*, is defined such that if $\mathbf{u}$ is any vector in $\mathcal{S}$ and α is any scalar, then the scalar multiple $\alpha\mathbf{u}$ is in $\mathcal{S}$, too (i.e., $\mathcal{S}$ is *closed under scalar multiplication*). Further, we require that

$$\alpha(\beta\mathbf{u}) = (\alpha\beta)\mathbf{u}, \qquad \text{(associative)} \qquad (5)$$

$$(\alpha + \beta)\mathbf{u} = \alpha\mathbf{u} + \beta\mathbf{u}, \qquad \text{(distributive)} \qquad (6)$$

$$\alpha(\mathbf{u} + \mathbf{v}) = \alpha\mathbf{u} + \alpha\mathbf{v}, \qquad \text{(distributive)} \qquad (7)$$

$$1\mathbf{u} = \mathbf{u}, \qquad (8)$$

if the vectors $\mathbf{u}, \mathbf{v}$ are in $\mathcal{S}$, and α, β are scalars.

As in earlier sections of this chapter, all scalar quantities will be understood to be real, not complex. For instance, if $\mathbf{u}$ is a 2-tuple vector (u_1, u_2) then both u_1 and u_2 are to be real, and if we multiply $\mathbf{u}$ by a scalar α then α is to be real.

Observe that if we write $\mathbf{u} + \mathbf{v} + \mathbf{w}$, it is not clear whether we mean $(\mathbf{u} + \mathbf{v}) + \mathbf{w}$ (i.e., first add $\mathbf{u}$ and $\mathbf{v}$, and then add the result to $\mathbf{w}$) or $\mathbf{u} + (\mathbf{v} + \mathbf{w})$. However, the associative property (2) guarantees that it does not matter, so the parentheses can be omitted without ambiguity. Similarly, $\alpha\beta\mathbf{u}$ is unambiguous by virtue of (5).

EXAMPLE 1   *Arrow Vector Space*.

Let the vectors be arrow vectors in physical three-dimensional space, with vector addition, the scalar multiplication of vectors, the zero vector, and the negative inverse of a vector defined as in Section 4.2.1. Then the axioms listed in Definition 4.8.1 are identical to the properties listed in Section 4.2.1, so the arrow vector system is indeed a vector space.   ∎

E X A M P L E   2 $\mathbb{R}^n$-*Space*.
Similarly, the n-space $\mathbb{R}^n$ defined earlier constitutes a vector space. After all, the axioms listed in Definition 4.8.1 were inspired by the properties of n-space listed in Section 4.3.

COMMENT.   However, to further illustrate the use of the axioms, suppose we modify the space by changing the addition operation from

$$\mathbf{u} + \mathbf{v} \equiv (u_1 + v_1, \dots, u_n + v_n) \tag{9}$$

to

$$\mathbf{u} + \mathbf{v} \equiv (u_1 + 2v_1, \dots, u_n + 2v_n). \tag{10}$$

With that change, do we still have a vector space? That is, are the axioms still satisfied? Let us see. According to (10),

$$\mathbf{v} + \mathbf{u} \equiv (v_1 + 2u_1, \dots, v_n + 2u_n) \tag{11}$$

so a comparison of (10) and (11) shows that the commutativity axiom (1) is satisfied only if $u_j + 2v_j = v_j + 2u_j$ $(j = 1, \dots, n)$, hence only if $v_j = u_j$, hence only if $\mathbf{v} = \mathbf{u}$. Since (1) does not hold for all possible chosen vectors $\mathbf{u}$ and $\mathbf{v}$, but only for vectors $\mathbf{u}$ and $\mathbf{v}$ that are equal, we conclude that if $\mathbf{u} + \mathbf{v}$ is defined by (10), then we do *not* have a vector space. Of course, it is possible that (10) violates other axioms besides (1), but one failure is sufficient to show that the set is not a legitimate vector space.

Understand that we have not shown that vector addition *must* be defined as in (9); all we've shown is that (9) does work [i.e., it satisfies the requirements listed under (i)] and that (10) does not. It is still conceivable that some other definition might work.

It is important to appreciate that the plus signs on the left-hand and right-hand sides of (9) are not the same. The ones on the right denote the usual addition of real numbers (e.g., $2 + 5 = 7$), whereas the one on the left denotes a certain operation between vectors $\mathbf{u}$ and $\mathbf{v}$, which is being defined by (9). To emphasize that point we could use a different symbol such as $\mathbf{u} * \mathbf{v}$, in place of $\mathbf{u} + \mathbf{v}$, as some authors do. However, having made this point let us continue to use $\mathbf{u} + \mathbf{v}$.   ∎

Arrow vector space and $\mathbb{R}^n$ are but two examples of a vector space. Many other useful spaces can be introduced. For example, the vectors may be functions, matrices, or whatever, provided that vector addition, a zero vector, a negative inverse, and scalar multiplication are defined such that all of the vector space axioms are satisfied, for nowhere in Definition 4.8.1 is the *nature* of the vectors specified or in any way restricted.

EXAMPLE 3   *A Function Space.*

This time, let the vectors be functions. Specifically, let $\mathbf{u} = u(x)$ be any continuous function defined on $0 \leq x \leq 1$, say. For the addition operation let

$$\mathbf{u} + \mathbf{v} \equiv u(x) + v(x); \tag{12}$$

that is, let $\mathbf{u} + \mathbf{v}$ be the function whose values are the ordinary sum $u(x) + v(x)$. For scalar multiplication let

$$\alpha \mathbf{u} \equiv \alpha u(x); \tag{13}$$

for the zero vector choose the zero function

$$\mathbf{0} \equiv 0; \tag{14}$$

and for the negative inverse of $\mathbf{u}$ define

$$-\mathbf{u} \equiv -u(x); \tag{15}$$

that is, the function whose values are $-u(x)$.

With these definitions, we can verify that all of the vector space requirements are satisfied, so that the set $\mathcal{S}$ of such vectors is a bona fide vector space. For instance, if $\mathbf{u} = u(x)$ and $\mathbf{v} = v(x)$ are continuous on $0 \leq x \leq 1$, then so is $\mathbf{u} + \mathbf{v} = u(x) + v(x)$ so $\mathcal{S}$ is closed under addition. Further, $\mathbf{v} + \mathbf{u} = v(x) + u(x) = u(x) + v(x) = \mathbf{u} + \mathbf{v}$,[1] so addition satisfies the commutative property (1), and so on.

This $\mathcal{S}$ is but one example of a **function space**, a space in which the vectors are functions.   ∎

The following theorem is useful, and its proof further illustrates the axiomatic approach.

THEOREM 4.8.1

Properties of Scalar Multiplication

If $\mathbf{u}$ is any vector in a vector space $\mathcal{S}$ and α is any scalar, then

$$0\,\mathbf{u} = \mathbf{0}, \tag{16}$$

$$(-1)\mathbf{u} = -\mathbf{u}, \tag{17}$$

$$\alpha\,\mathbf{0} = \mathbf{0}. \tag{18}$$

Proof.   As obvious as these results seem, we need to show that they follow from our definition of vector space. To prove (16), one line of approach is as follows:

$$
\begin{aligned}
0\mathbf{u} + \mathbf{u} &= 0\mathbf{u} + 1\mathbf{u} && \text{by (8)} \\
&= (0 + 1)\mathbf{u} && \text{by (6)} \\
&= 1\mathbf{u} && \\
&= \mathbf{u} && \text{by (8).}
\end{aligned}
$$

---

[1] The second equality holds because $v(x) + u(x)$ is the ordinary sum of two real numbers; e.g., $4 + 3 = 3 + 4$.

Then

$$0\mathbf{u} + \mathbf{u} + (-\mathbf{u}) = \mathbf{u} + (-\mathbf{u})$$
$$0\mathbf{u} + \mathbf{0} = \mathbf{0} \qquad \text{by (4),}$$
$$0\mathbf{u} = \mathbf{0} \qquad \text{by (3).}$$

The remaining two, (17) and (18), are left for the exercises. ∎

E X A M P L E   4   *Zero Vector Space*.
Let S consist of a single vector, $\mathbf{0}$, and define vector addition by $\mathbf{0} + \mathbf{0} = \mathbf{0}$ and scalar multiplication by $\alpha\mathbf{0} = \mathbf{0}$ for all scalars α. We find that all of the vector space requirements are satisfied and we call this the **zero vector space**. ∎

4.8.2 Subspace.   If a subset $\mathcal{T}$ of a vector space $\mathcal{S}$ is itself a vector space (with the same definitions of vector addition $\mathbf{u} + \mathbf{v}$, scalar multiplication $\alpha\mathbf{u}$, zero vector $\mathbf{0}$, and negative inverse of a vector $-\mathbf{u}$) then we call $\mathcal{T}$ a **subspace** of $\mathcal{S}$.

E X A M P L E   5
Consider possible subspaces of the vector space $\mathbb{R}^2$. If $\mathbf{p}$ is any nonzero vector in $\mathbb{R}^2$, then the line L (Fig. 1) generated by $\mathbf{p}$ (namely, span$\{\mathbf{p}\}$) is itself a vector space, a subspace of $\mathbb{R}^2$. To verify that claim we don't need to verify (1)–(8) in Definition 4.8.1 since we already know that they hold for $\mathbb{R}^2$, and the line L is within $\mathbb{R}^2$. But we do need to verify that L is closed under vector addition, that it does contain a unique zero vector, that it does contain a unique negative inverse of each vector, and that it is closed under scalar multiplication. It is evident from Fig. 1 that each of these requirements is satisfied, for if $\mathbf{u}$ and $\mathbf{v}$ are any vectors in L then $\mathbf{u} + \mathbf{v}$ is in L too; L does contain the zero vector, which is the point at the origin, and $-\mathbf{u}$, and $\alpha\mathbf{u}$. Thus, the line L is a subspace of $\mathbb{R}^2$.

How about the subset of $\mathbb{R}^2$ consisting of the entire first quadrant (Fig. 2)? Is that a subspace of $\mathbb{R}^2$? No, because it does not contain a negative inverse for each $\mathbf{u}$ nor is it closed under scalar multiplication, for if $\mathbf{u}$ is in the first quadrant then $-\mathbf{u}$ is in the third quadrant, as is $\alpha\mathbf{u}$ if α is negative. ∎

E X A M P L E   6
Similarly for $\mathbb{R}^3$. If $\mathbf{p}$ is any nonzero vector then the line that it generates (namely, span$\{\mathbf{p}\}$) is a subspace of $\mathbb{R}^3$, and if $\mathbf{p}$ and $\mathbf{q}$ are any two LI vectors in $\mathbb{R}^3$ then the plane that they generate (namely, span$\{\mathbf{p}, \mathbf{q}\}$) is a subspace of $\mathbb{R}^3$. ∎

Thus, the idea is that subspaces of a given vector space S can be obtained as the span of one or more vectors in S. We have the following theorem, proof of which is left for the exercises.

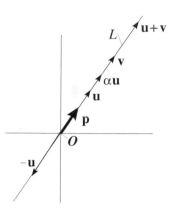

FIGURE 1
The subspace L of $\mathbb{R}^2$.

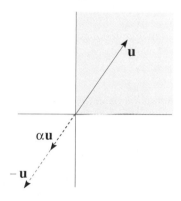

FIGURE 2
First quadrant not a subspace of $\mathbb{R}^2$; $\alpha = -0.5$, say.

THEOREM 4.8.2

Span as Subspace

If $\mathbf{u}_1, \ldots, \mathbf{u}_k$ are vectors in a vector space $\mathcal{S}$, then span$\{\mathbf{u}_1, \ldots, \mathbf{u}_k\}$ is itself a vector space, a subspace of $\mathcal{S}$.

An important example of subspaces arises in connection with homogeneous systems of linear algebraic equations,

$$
\begin{aligned}
a_{11}x_1 + \cdots + a_{1n}x_1 &= 0, \\
&\;\;\vdots \\
a_{m1}x_1 + \cdots + a_{mn}x_n &= 0.
\end{aligned}
\tag{19}
$$

EXAMPLE 7

For instance, consider the homogeneous linear system

$$
\begin{aligned}
x_1 + x_2 + x_3 + x_4 &= 0, \\
x_1 - x_2 + 3x_3 + x_4 &= 0, \\
x_1 \phantom{{}- x_2} + 2x_3 + x_4 &= 0, \\
2x_1 - x_2 + 5x_3 + 2x_4 &= 0,
\end{aligned}
\tag{20}
$$

Gauss elimination of which gives $x_4 = \alpha_1$, $x_3 = \alpha_2$, $x_2 = \alpha_2$, $x_1 = -\alpha_1 - 2\alpha_2$ or, in vector form,

$$
\begin{aligned}
\mathbf{x} &= (-\alpha_1 - 2\alpha_2, \alpha_2, \alpha_2, \alpha_1) \\
&= \alpha_1 (-1, 0, 0, 1) + \alpha_2 (-2, 1, 1, 0) \\
&\equiv \alpha_1 \mathbf{x}_1 + \alpha_2 \mathbf{x}_2.
\end{aligned}
\tag{21}
$$

That is, the set of solutions of (20) (expressed in vector form) is the set span $\{\mathbf{x}_1, \mathbf{x}_2\}$. According to Theorem 4.8.2 the solution set is a vector space, a subspace of $\mathbb{R}^4$. ∎

In fact it is *always* true that the general solution of (19) is a vector space, a subspace of $\mathbb{R}^n$, for (19) has either a nonunique solution or the unique trivial solution $\mathbf{x} = \mathbf{0}$. In the former case the general solution is of the form $\mathbf{x} = \alpha_1 \mathbf{x}_1 + \cdots + \alpha_N \mathbf{x}_N$ and is the span of $\mathbf{x}_1, \ldots, \mathbf{x}_N$ and hence, by Theorem 4.8.2, a subspace of $\mathbb{R}^n$. Even in the latter case, $\mathbf{x} = \mathbf{0}$ is the zero vector space and is a subspace of $\mathbb{R}^n$. Since the set of all solutions of (19) is a vector space it is called the **solution space** of (19). Thus, we have the following theorem.

THEOREM 4.8.3

Solution Space as Subspace

The set of all vectors $\mathbf{x} = (x_1, \ldots, x_n)$ that are solutions of the homogenous linear system (19) is a vector space, a subspace of $\mathbb{R}^n$.

4.8.3 Inclusion of inner product and/or norm. (Optional).   Observe that there is no mention of a dot product or a norm in Definition 4.8.1. Indeed, a vector space $\mathcal{S}$ need not *have* a dot product (also called an **inner product**) or a norm defined for it. If it does have an inner product it is called an **inner product space**; if it has a norm it is called a **normed vector space**; and if it has both it is called a **normed inner product space**.

If we do choose to introduce an inner product for $\mathcal{S}$, how is it to be defined? Do you remember the idea of reversing the cart and the horse? That is how we do it. Equations (12a,b,c) in Section 4.4.2 were shown to be properties of the inner product $\mathbf{u} \cdot \mathbf{v} = u_1 v_1 + \cdots + u_n v_n$. We now take those properties and elevate them to axioms, or requirements, that are to be satisfied by any inner product of any vector space.

Similarly, we take the properties (14a,b,c) of the norm, in Section 4.4.3, and elevate them to axioms, or requirements, that are to be satisfied by any norm of any vector space.

Let us illustrate.

E X A M P L E   8 $\mathbb{R}^n$-*Space*.
If we wish to add an inner product to the vector space $\mathbb{R}^n$, we can use the choice

$$\mathbf{u} \cdot \mathbf{v} \equiv u_1 v_1 + \cdots + u_n v_n = \sum_{j=1}^{n} u_j v_j. \tag{22}$$

We know that (22) satisfies the inner product axioms because the latter were deduced, in Section 4.4.2, as properties that *follow* from (22). A variation of (22) that still satisfies the inner product axioms is (Exercise 12)

$$\mathbf{u} \cdot \mathbf{v} \equiv w_1 u_1 v_1 + \cdots + w_n u_n v_n = \sum_{j=1}^{n} w_j u_j v_j, \tag{23}$$

where the w_j's are fixed positive constants known as "weights" because they attach more or less weight to the different components of $\mathbf{u}$ and $\mathbf{v}$. For instance, consider $\mathbb{R}^2$ and let $w_1 = 5$ and $w_2 = 3$. Then if $\mathbf{u} = (2, -4)$ and $\mathbf{v} = (1, 6)$ we have $\mathbf{u} \cdot \mathbf{v} = 5(2)(1) + 3(-4)(6) = -62$.

Note that for (23) to be a legitimate inner product we must have $w_j > 0$ for each j. For suppose, still in $\mathbb{R}^2$, that $w_1 = 3$ and $w_2 = -2$. Then, for $\mathbf{u} = (1, 5)$, say, we have $\mathbf{u} \cdot \mathbf{u} = 3(1)(1) - 2(5)(5) = -47 < 0$, in violation of the axiom that $\mathbf{u} \cdot \mathbf{u} > 0$ for all $\mathbf{u} \neq \mathbf{0}$. Or, suppose that $w_1 = 3$ and $w_2 = 0$. Then, for $\mathbf{u} = (0, 4)$, say, we have $\mathbf{u} \cdot \mathbf{u} = 3(0)(0) + 0(4)(4) = 0$ even though $\mathbf{u} \neq \mathbf{0}$, in violation of the axiom that $\mathbf{u} \cdot \mathbf{u} = 0$ only if $\mathbf{u} = \mathbf{0}$.

Now, suppose that we wish to add a norm. If for any vector space $\mathcal{S}$ we already have an inner product, then a legitimate norm can always be obtained from that inner product as $\|\mathbf{u}\| = \sqrt{\mathbf{u} \cdot \mathbf{u}}$, and that choice is called the **natural norm**. Thus, the natural norms corresponding to the dot products (22) and (23) are, respectively,

$$\|\mathbf{u}\| \equiv \sqrt{\sum_{1}^{n} u_j^2} \quad \text{and} \quad \|\mathbf{u}\| \equiv \sqrt{\sum_{1}^{n} w_j u_j^2}, \tag{24a,b}$$

However, we do not *have* to choose the natural norm. For instance, we could use (22) as our inner product, and choose

$$\|\mathbf{u}\| \equiv |u_1| + \cdots + |u_n| = \sum_{j=1}^{n} |u_j| \tag{25}$$

as our norm (Exercise 14). The latter is used by Struble in his book on differential equations,[1] probably because it is algebraically simpler than the *Euclidean norm* (24a) or the *modified Euclidean norm* (24b). Furthermore, he defines no inner product whatsoever. Struble calls (25) the *taxicab norm* since a taxicab driver judges the distance from the corner of 5th Avenue and 34th Street to the corner of 2nd Avenue and 49th Street as 18 blocks, not $\sqrt{234}$ blocks.   ∎

EXAMPLE 9   *The Function Space of Example 3.*

How might we choose an inner product for the *function* space $\mathcal{S}$ defined in Example 3? To motivate our choice, let us imagine approximating any given function (i.e., vector) $u(x)$ in $\mathcal{S}$ in a piecewise-constant manner as depicted in Fig. 3. That is, divide the x interval ($0 \leq x \leq 1$) into n equal parts and define the approximating piecewise-constant function, over each subinterval as the value of $u(x)$ at the left endpoint of that subinterval. If we represent the piecewise-constant function as the n-tuple $(u_1, \ldots, u_n)$, then we have, in a heuristic sense,

$$u(x) \approx (u_1, \ldots, u_n). \tag{26}$$

Similarly, for any other function $v(x)$ in $\mathcal{S}$,

$$v(x) \approx (v_1, \ldots, v_n). \tag{27}$$

The n-tuple vectors on the right-hand sides of (26) and (27) are members of $\mathbb{R}^n$. For that space, let us adopt the inner product

$$(u_1, \ldots, u_n)\cdot(v_1, \ldots, v_n) = \sum_{j=1}^{n} u_j v_j \Delta x, \tag{28}$$

that is, (23) with all of the w_j weights equal to the subinterval width Δx. If we let $n \to \infty$, the "staircase approximations" approach $u(x)$ and $v(x)$, and the sum in (28) tends to the integral $\int_0^1 u(x)v(x)\,dx$.

This heuristic reasoning suggests the inner product

$$\left\langle u(x), v(x) \right\rangle \equiv \int_0^1 u(x)v(x)\,dx. \tag{29}$$

We can denote it as $\mathbf{u}\cdot\mathbf{v}$ and call it the dot product, or we can denote it as $\langle u(x), v(x) \rangle$ and call it the inner product. For function spaces, the latter notation is somewhat standard, and is our choice here.

---

[1] R. A. Struble, *Nonlinear Differential Equations* (New York: McGraw-Hill, 1962).

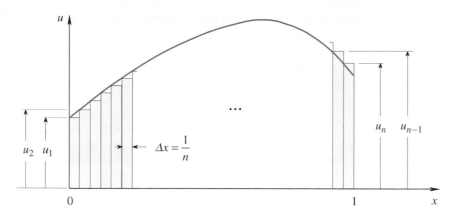

FIGURE 3   Staircase approximation of $u(x)$.

Recall that we took all the weights to be the same in (28), namely, Δx. If, instead, we have them vary with j as $w_j \Delta x$ then, in place of (29), we are led to the definition

$$\Big\langle u(x), v(x) \Big\rangle \equiv \int_0^1 u(x)v(x)w(x)\,dx, \tag{30}$$

where $w(x) > 0$ for $0 \leq x \leq 1$.

Understand that we have not *derived* (29) and (30), we have used our "staircase" idea to *motivate* them as *definitions*. One can then verify that they do satisfy the inner product axioms, which verification we leave for the exercises.

If, besides an inner product, we wish to define a norm as well, we could use a natural norm $\|\mathbf{u}\| = \sqrt{\langle u(x), u(x) \rangle}$ in conjunction with (29) or (30). Using (29), for instance, gives

$$\|\mathbf{u}\| \equiv \sqrt{\int_0^1 u^2(x)\,dx}. \tag{31}$$

COMMENT 1.   Note that the concept of the *dimension* of a vector space has not yet been introduced; it will be in Section 4.9. There, we define dimension and find that $\mathbb{R}^n$ is n-dimensional (which claim is probably not a great shock). Since the staircase approximation (26) becomes exact only as $n \to \infty$, it appears that our function space $\mathcal{S}$ is infinite dimensional! That suspicion turns out to be true.

COMMENT 2.   A bit of standard notation: the set of functions that are defined and continuous on [0, 1] (i.e., $0 \leq x \leq 1$) is usually denoted as $C^0[0, 1]$. If the functions are continuous and all their derivatives through order k are continuous as well, then the set is denoted as $C^k[0, 1]$.   ∎

Closure.   In this section we've accomplished our final generalization of the vector concept by taking the various properties that were derived for arrow vectors and *n*-tuple vectors (such as $\mathbf{u} + \mathbf{v} = \mathbf{v} + \mathbf{u}$) and elevating them to be axioms, or requirements, to be met by any *vector space*, in which there is no restriction on the nature of the vectors, which can be arrows, *n*-tuples, functions, matrices—or whatever.

In this book we are concerned primarily with the vector space $\mathbb{R}^n$ and with subspaces thereof, but when we return to our study of differential equations in Chapter 6 we will be concerned with function spaces as well.

EXERCISES 4.8

1. Recall that $\mathbb{R}^n$ is the vector space ("real" vector space since all scalars are to be real numbers) in which the vectors are *n*-tuples $\mathbf{u} = (u_1, \ldots, u_n)$, with the definitions

$$\mathbf{u} + \mathbf{v} = (u_1, \ldots, u_n) + (v_1, \ldots, v_n)$$
$$\equiv (u_1 + v_1, \ldots, u_n + v_n), \tag{1.1}$$
$$\mathbf{0} \equiv (0, \ldots, 0), \tag{1.2}$$
$$-\mathbf{u} \equiv (-u_1, \ldots, -u_n), \tag{1.3}$$
$$\alpha\mathbf{u} \equiv (\alpha u_1, \ldots, \alpha u_n). \tag{1.4}$$

If we make the following modifications, do we still have a vector space? If not, specify requirements within Definition 4.8.1 that are not satisfied.

(a) only vectors of the form $\mathbf{u} = (u, u, \ldots, u)$ admitted, where $-\infty < u < \infty$

(b) only vectors of the form $\mathbf{u} = (u, 2u, 3u, \ldots, nu)$ admitted, where $-\infty < u < \infty$

(c) $\mathbf{u} + \mathbf{v} \equiv (u_1 - v_1, \ldots, u_n - v_n)$, in place of (1.1)

(d) $\mathbf{u} + \mathbf{v} \equiv (0, \ldots, 0)$ for all $\mathbf{u}$'s and $\mathbf{v}$'s, in place of (1.1)

(e) $\alpha\mathbf{u} \equiv (\alpha^2 u_1, \ldots, \alpha^2 u_n)$, in place of (1.4)

2. We noted in Example 2 that the definition (10) of vector addition violates axiom (1). Does it violate any others as well? Explain.

3. Let S be the set of real-valued polynomial functions, of degree n, defined on $a \le x \le b$. If $\mathbf{u} = a_0 + a_1 x + \cdots + a_n x^n$ and $\mathbf{v} = b_0 + b_1 x + \cdots + b_n x^n$ are any two such functions, and α is any (real) scalar, define the sum $\mathbf{u} + \mathbf{v}$ and the scalar multiple $\alpha\mathbf{u}$ as

$$(\mathbf{u} + \mathbf{v})(x) = (a_0 + b_0) + (a_1 + b_1)x + \cdots + (a_n + b_n)x^n,$$
$$(\alpha\mathbf{u})(x) = \alpha a_0 + \alpha a_1 x + \cdots + \alpha a_n x^n,$$

respectively. Further, let $\mathbf{0}$ be the function $0 + 0x + \cdots + 0x^n$, and let $-\mathbf{u}$ be the function $-a_0 - a_1 x + \cdots - a_n x^n$. Show that S is a vector space.

4. If the system (19) were *non*homogeneous, would the set of solutions still be a vector space? Explain.

5. (*Solution space*) Using Gauss elimination, determine the solution space as the span of one or more vectors (or simply as $\mathbf{x} = \mathbf{0}$ if the system admits only the trivial solution), as we did in Example 7 for the system (20).

(a) $x_1 - x_2 + 4x_3 = 0$ in $\mathbb{R}^3$

(b) $x_1 + x_2 + x_3 - x_4 = 0$ in $\mathbb{R}^4$

(c) $\begin{aligned} x_1 - x_2 + x_3 &= 0 \\ x_1 + x_2 + x_3 &= 0 \end{aligned}$ in $\mathbb{R}^3$

(d) $\begin{aligned} x_1 + 3x_2 - x_3 + x_4 &= 0 \\ x_1 + 2x_3 + x_4 &= 0 \end{aligned}$ in $\mathbb{R}^4$

(e) $\begin{aligned} x_1 - x_2 + x_3 - 2x_4 &= 0 \\ x_1 - x_2 + x_4 + 2x_5 &= 0 \end{aligned}$ in $\mathbb{R}^5$

(f) $\begin{aligned} x_1 + x_2 - x_3 + x_4 &= 0 \\ x_1 + 2x_2 - x_4 &= 0 \\ x_2 + x_3 &= 0 \end{aligned}$ in $\mathbb{R}^4$

(g) $\begin{aligned} x_1 + x_2 + 2x_3 - 2x_4 &= 0 \\ x_1 + x_2 + 2x_3 + x_5 &= 0 \\ 2x_4 + x_5 &= 0 \end{aligned}$ in $\mathbb{R}^5$

6. Are the following vector sets subspaces of $\mathbb{R}^2$? (See accompanying figure.) Explain.

(a) the straight line L that extends from the origin to infinity

(b) the wedge-shaped region (including its boundary lines) that extends to infinity in both directions

(c) the upper half plane $x_2 \ge 0$

(a)

(b)

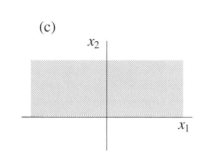

(c)

7. We saw, in Example 5, that the set of all vectors lying along the line L (Fig. 1) constitute a subspace of $\mathbb{R}^2$. Suppose, instead, that L does not pass through the origin. Does the set of all vectors with their tails at the origin and their tips on L constitute a vector space? How about a plane P in $\mathbb{R}^3$ that does not pass through the origin? Explain.

8. Prove (16).

9. Prove (17).

10. Prove (18).

11. Prove that if $\alpha\mathbf{u} = \mathbf{0}$ then $\alpha = 0$ and/or $\mathbf{u} = \mathbf{0}$.

12. Show that the inner product (23) does satisfy the inner product axioms.

13. We stated in Example 8 that if for any vector space $\mathcal{S}$ we al-

ready have an inner product, then a legitimate norm can always be obtained from that inner product as $\|\mathbf{u}\| = \sqrt{\mathbf{u}\cdot\mathbf{u}}$, which choice is called the natural norm. Prove that claim.

14. Show that the "taxicab norm" (25) is a legitimate norm—that is, that it satisfies the requirements (14a,b,c) in Section 4.4.

15. (a) Does the choice $\|\mathbf{u}\| = \max\limits_{1\le j\le n} |u_j|$, for $\mathbb{R}^n$, satisfy the required axioms? Explain.

(b) How about $\|\mathbf{u}\| = \min\limits_{1\le j\le n} |u_j|$, for $\mathbb{R}^n$?

16. Show that the inner product (30) does satisfy the inner product axioms.

17. (Schwarz inequality) We discussed the Schwarz inequality

$$|\mathbf{u}\cdot\mathbf{v}| \le \|\mathbf{u}\|\,\|\mathbf{v}\| \qquad (17.1)$$

for $\mathbb{R}^n$ space in Section 4.4.1. The latter holds not only for $\mathbb{R}^n$ but for any normed inner product space with the natural norm $\|\mathbf{u}\| = \sqrt{\mathbf{u}\cdot\mathbf{u}}$. Verify (17.1) by working out the left-hand and right-hand sides for these specific cases:

(a) $\mathbf{u} = (3, 1, -1, 0)$ and $\mathbf{v} = (1, 2, 5, -4)$ in $\mathbb{R}^4$, with the inner product (19)

(b) $\mathbf{u} = (1, 2, 4, -3)$ and $\mathbf{v} = (0, 4, 1, 1)$ in $\mathbb{R}^4$, with $\mathbf{u}\cdot\mathbf{v} = u_1v_1 + 5u_2v_2 + 3u_3v_3 + 2u_4v_4$

(c) $\mathbf{u} = (1, 1, 1, 1, 1)$ and $\mathbf{v} = (2, 2, 2, 2, 2)$ in $\mathbb{R}^5$, with $\mathbf{u}\cdot\mathbf{v} = u_1v_1 + 2u_2v_2 + 3u_3v_3 + 4u_4v_4 + 5u_5v_5$

(d) $\mathbf{u} = 2 + x$ and $\mathbf{v} = 3x^2$ in the function space of Example 9, with the inner product $\mathbf{u}\cdot\mathbf{v} = \langle u(x), v(x)\rangle = \int_0^1 u(x)v(x)\,dx$

(e) Same as (d), but with $\langle u(x), v(x)\rangle = \int_0^1 u(x)v(x)(2 + 5x)\,dx$

18. Prove Theorem 4.8.2.

4.9 BASES AND EXPANSIONS

This section is important and relies heavily upon the material in Sections 4.2–4.8. It is a bit long and contains several theorems and proofs so, to see the "forest" as expeditiously as possible without getting lost in the "trees," we suggest that you omit the proofs on a first reading and save them for a second, more detailed, reading.

4.9.1 Bases and expansions.   In the calculus we learn that some functions $f(x)$ can be "expanded" as a linear combination of powers of x (namely $1, x, x^2, \ldots$),

$$f(x) = a_0 + a_1 x + a_2 x^2 + \cdots . \qquad (1)$$

We call $a_0, a_1, a_2, \ldots$ the "expansion coefficients," and (from our knowledge of Taylor series) these can be computed from $f(x)$ as $a_j = f^{(j)}(0)/j!$. Such representation of a given function is important, and examples such as $e^x = 1 + x + \frac{1}{2!}x^2 + \frac{1}{3!}x^3 + \cdots$ and $\sin x = x - \frac{1}{3!}x^3 + \frac{1}{5!}x^5 - \cdots$ are familiar to us.

Similarly, it will be important for us to be able to expand a given vector $\mathbf{u}$ in terms of a set of "base vectors" $\mathbf{e}_1, \ldots, \mathbf{e}_k$:

$$\mathbf{u} = \alpha_1 \mathbf{e}_1 + \cdots + \alpha_k \mathbf{e}_k. \tag{2}$$

How do we come up with such sets of base vectors and, once we know the $\mathbf{e}_j$'s and the given $\mathbf{u}$, how do we compute the expansion coefficients α_j? The story is simpler than for the power series of functions because whereas (1) is an infinite series and one needs to deal with the issue of convergence, vector expansions in n-space entail only a *finite* number of terms.

Let us begin simply. Consider the vector space $\mathbb{R}^2$, the set of all vectors in the plane of the paper. In particular, consider the vectors $\mathbf{e}_1$ and $\mathbf{e}_2$ shown in Fig. 1a. It should be evident (Theorem 4.7.2) that $\mathbf{e}_1$ and $\mathbf{e}_2$ are LI and that they span the space so that any given vector, such as $\mathbf{u}$ in Fig. 1b and $\mathbf{v}$ in Fig. 1c, can be expressed as a linear combination of them.

For the vector $\mathbf{u}$, for example, $\mathbf{u} = \mathbf{OA} + \mathbf{OB}$. With the aid of a scale, $\mathbf{OA} \approx 1.6\mathbf{e}_1$ and $\mathbf{OB} \approx 2\mathbf{e}_2$, so that

$$\mathbf{u} \approx 1.6\mathbf{e}_1 + 2\mathbf{e}_2. \tag{3}$$

Similarly (Fig. 1c),

$$\mathbf{v} \approx 2\mathbf{e}_1 - 2.5\mathbf{e}_2, \tag{4}$$

and so on, for any given vector in the plane. Of course, the zero vector is simply $\mathbf{0} = 0\mathbf{e}_1 + 0\mathbf{e}_2$.

The formulas (3) and (4) are examples of the *expansion* of a given vector [$\mathbf{u}$ in (3), $\mathbf{v}$ in (4)] in terms of a set of *base vectors* [the set $\{\mathbf{e}_1, \mathbf{e}_2\}$ in this case].

Definition 4.9.1   *Basis*
A finite set of vectors $\{\mathbf{e}_1, \ldots, \mathbf{e}_k\}$ in a vector space $\mathcal{S}$ is a **basis** for $\mathcal{S}$ if each vector $\mathbf{u}$ in $\mathcal{S}$ can be expressed (i.e., "expanded") uniquely in the form

$$\mathbf{u} = \alpha_1 \mathbf{e}_1 + \cdots + \alpha_k \mathbf{e}_k = \sum_{j=1}^{k} \alpha_j \mathbf{e}_j. \tag{5}$$

The vectors $\mathbf{e}_1, \ldots, \mathbf{e}_k$ are called **base vectors**.

By the expansion (5) being unique, we mean that the α_j *expansion coefficients are uniquely determined*. We need to be careful about the zero vector space (Example 4 of Section 4.8). It is tempting to think that the single vector $\mathbf{e}_1 = \mathbf{0}$ is a basis for that space because surely the only vector in the space (namely, $\mathbf{0}$) can be expressed as $\mathbf{0} = \alpha_1 \mathbf{e}_1 = \alpha_1 \mathbf{0}$. However, α_1 is not uniquely determined in the latter expansion, so we conclude that *the zero vector does not have a basis*.

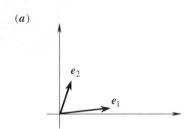

(a)

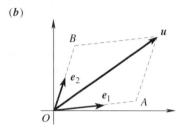

(b)

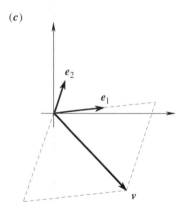

(c)

FIGURE 1
Vector expansion in $\mathbb{R}^2$.

THEOREM 4.9.1

Test for Basis

A finite set $\{\mathbf{e}_1, \ldots, \mathbf{e}_k\}$ in a vector space S is a basis for S if and only if it spans S and is LI.

Proof. First, it follows from the definition of the verb span that every vector $\mathbf{u}$ in S can be expanded as in (5) if and only if the set $\{\mathbf{e}_1, \ldots, \mathbf{e}_k\}$ spans S. Turning to the question of the uniqueness of the expansion, suppose that both expansions

$$\mathbf{u} \;=\; \alpha_1\mathbf{e}_1 + \cdots + \alpha_k\mathbf{e}_k, \tag{6}$$

$$\mathbf{u} \;=\; \beta_1\mathbf{e}_1 + \cdots + \beta_k\mathbf{e}_k \tag{7}$$

hold for a given vector $\mathbf{u}$ in S. Subtracting (7) from (6) gives

$$(\alpha_1 - \beta_1)\mathbf{e}_1 + \cdots + (\alpha_k - \beta_k)\mathbf{e}_k = \mathbf{0}. \tag{8}$$

Each of the coefficients $(\alpha_1 - \beta_1), \ldots, (\alpha_k - \beta_k)$ in (8) must be zero if and only if the set $\{\mathbf{e}_1, \ldots, \mathbf{e}_k\}$ is LI. In that case $\alpha_1 = \beta_1, \ldots, \alpha_k = \beta_k$. Hence, the expansion (5) is unique if and only if the set is LI, and this completes the proof. ∎

The key idea revealed in the foregoing proof is that *a basis needs to contain* **enough** *vectors but* **not too many**: *enough so that the set spans the space and can therefore be used to expand any given vector in the space, but not so many that the set is LD for then expansions may be nonunique.*

Let us illustrate. First, consider the vector $\mathbf{e}_1$ alone, in Fig. 1a. The set $\{\mathbf{e}_1\}$ falls short of being a basis for $\mathbb{R}^2$ because it does not span $\mathbb{R}^2$, it spans only a line. Any vector $\mathbf{v}$ not lying on that line cannot be expanded as $\mathbf{v} = \alpha_1\mathbf{e}_1$.

Second, consider the set $\{\mathbf{e}_1, \mathbf{e}_2\}$ shown in Fig. 1a. That set does span $\mathbb{R}^2$ and is LI so it is a basis for $\mathbb{R}^2$.

Third, consider the set $\{\mathbf{e}_1, \mathbf{e}_2, \mathbf{e}_3\}$, where $\mathbf{e}_1$ and $\mathbf{e}_2$ are shown in Fig. 1a and $\mathbf{e}_3$ is any third vector in $\mathbb{R}^2$. Surely the set spans $\mathbb{R}^2$, because $\mathbf{e}_1$ and $\mathbf{e}_2$ do so even without help from $\mathbf{e}_3$. However, the set is not LI because $\mathbf{e}_1$ and $\mathbf{e}_2$ themselves are a basis for $\mathbb{R}^2$; hence $\mathbf{e}_3$ must be expressible in terms of them, so $\{\mathbf{e}_1, \mathbf{e}_2, \mathbf{e}_3\}$ is LD. Thus, $\{\mathbf{e}_1, \mathbf{e}_2, \mathbf{e}_3\}$ is not a basis for $\mathbb{R}^2$. The problem is that although we can expand any vector in $\mathbb{R}^2$ in terms of them, the expansion coefficients may not be uniquely determined.

EXAMPLE 1

Determine whether or not the vectors

$$\mathbf{e}_1 = (1, 1, 2), \quad \mathbf{e}_2 = (1, 2, 1), \quad \mathbf{e}_3 = (2, 0, 3) \tag{9}$$

are a basis for $\mathbb{R}^3$. From Theorem 4.9.1 we see that we can proceed by testing the set to see if it spans $\mathbb{R}^2$ (as discussed in Section 4.6) and if it is LI (as discussed in Section 4.7). That approach would require two assessments, one for span and one for linear independence. It is more direct to use Definition 4.9.1, namely, to see if, for any given $\mathbf{u}$ in $\mathbb{R}^3$, (5) can be solved uniquely for the α's. To do so we write out (5) as

$$(u_1, u_2, u_3) = \alpha_1(1, 1, 2) + \alpha_2(1, 2, 1) + \alpha_3(2, 0, 3) \tag{10}$$

or, equivalently,

$$\begin{aligned} \alpha_1 + \alpha_2 + 2\alpha_3 &= u_1, \\ \alpha_1 + 2\alpha_2 &= u_2, \\ 2\alpha_1 + \alpha_2 + 3\alpha_3 &= u_3. \end{aligned} \tag{11}$$

Then, Gauss elimination gives

$$\begin{aligned} \alpha_1 + \alpha_2 + 2\alpha_3 &= u_1, \\ \alpha_2 - 2\alpha_3 &= u_2 - u_1, \\ \alpha_3 &= -\tfrac{1}{3}(u_3 + u_2 - 3u_1). \end{aligned} \tag{12}$$

Clearly, the latter has a unique solution for $\alpha_1, \alpha_2, \alpha_3$ for any given $\mathbf{u}$ vector in $\mathbb{R}^3$ (i.e., for any values of u_1, u_2, u_3). Thus, (9) is indeed a basis for $\mathbb{R}^3$.

COMMENT. How could we modify the set (9) to illustrate the case where the set is *not* a basis for $\mathbb{R}^3$? In various ways. For instance, if we delete the $\mathbf{e}_3$ vector then $\mathbf{e}_1$ and $\mathbf{e}_2$ will fall short of spanning $\mathbb{R}^3$; they will span only a plane. How will this failure show up in (12)? Deleting $\mathbf{e}_3$ is equivalent to setting $\alpha_3 = 0$ in (10)–(12). Thus, in place of (12) we would obtain

$$\begin{aligned} \alpha_1 + \alpha_2 &= u_1, \\ \alpha_2 &= u_2 - u_1, \\ 0 &= -\tfrac{1}{3}(u_3 + u_2 - 3u_1), \end{aligned} \tag{13}$$

which does *not* give a unique solution for the α's unless $u_3 + u_2 - 3u_1 = 0$, that is, unless $\mathbf{u}$ happens to lie in the plane spanned by $\mathbf{e}_1$ and $\mathbf{e}_2$. ∎

EXAMPLE 2
This problem is a bit different: Are the vectors

$$\mathbf{e}_1 = (1, 2, 4), \quad \mathbf{e}_2 = (3, 1, -1) \tag{14}$$

a basis for span$\{\mathbf{e}_1, \mathbf{e}_2\}$? Of course, span$\{\mathbf{e}_1, \mathbf{e}_2\}$ is not all of $\mathbb{R}^3$, it is only the subspace of $\mathbb{R}^3$ consisting of the plane spanned by $\mathbf{e}_1$ and $\mathbf{e}_2$. Surely $\mathbf{e}_1$ and $\mathbf{e}_2$ span the set span$\{\mathbf{e}_1, \mathbf{e}_2\}$, by the definition of span. Further, $\mathbf{e}_1$ and $\mathbf{e}_2$ are LI, by Theorem 4.7.2, since neither one is a multiple of the other. Thus, yes, $\mathbf{e}_1$ and $\mathbf{e}_2$ are a basis for span$\{\mathbf{e}_1, \mathbf{e}_2\}$. ∎

4.9.2 Dimension. If we always worked in 2-space or 3-space, the concept of dimension would hardly need elaboration; for example, 3-space is three-dimensional, a plane within it is two-dimensional, and a line within it is one-dimensional. However, having generalized our vector concept beyond 3-space, we need to formally define the dimension of a vector space.

Definition 4.9.2   *Dimension*

If the greatest number of LI vectors that can be found in a vector space S is k, where $1 \le k < \infty$, then S is **k-dimensional**, and we write

$$\dim S = k. \tag{15}$$

If S is the zero vector space (i.e., if it contains only the zero vector), we define $\dim S = 0$. If an arbitrarily large number of LI vectors can be found in S, we say that S is **infinite-dimensional**.

To determine the dimension of a given vector space, it may be more convenient to use the following theorem than to work directly from Definition 4.9.2.

THEOREM 4.9.2

Number of Vectors in a Basis

A vector space S admits a basis consisting of k vectors (with $k \ge 1$) if and only if S is k-dimensional.[1]

Proof.   Let $\{\mathbf{e}_1, \ldots, \mathbf{e}_k\}$ be a basis for S. Because these vectors form a basis, they must be LI. Hence, we have *at least k* LI vectors in S. To show that *no more than k* LI vectors can be found in S, suppose that vectors $\mathbf{e}'_1, \ldots, \mathbf{e}'_{k+1}$ in S are LI. Each of these can be expanded in terms of the given base vectors, as

$$\mathbf{e}'_1 = a_{11}\mathbf{e}_1 + \cdots + a_{1k}\mathbf{e}_k,$$

$$\vdots \tag{16}$$

$$\mathbf{e}'_{k+1} = a_{k+1,1}\mathbf{e}_1 + \cdots + a_{k+1,k}\mathbf{e}_k,$$

say. Putting these expressions into the equation

$$\alpha_1 \mathbf{e}'_1 + \alpha_2 \mathbf{e}'_2 + \cdots + \alpha_{k+1}\mathbf{e}'_{k+1} = \mathbf{0} \tag{17}$$

and grouping terms gives

$$\left(\alpha_1 a_{11} + \cdots + \alpha_{k+1}a_{k+1,1}\right)\mathbf{e}_1 + \cdots + \left(\alpha_1 a_{1k} + \cdots + \alpha_{k+1}a_{k+1,k}\right)\mathbf{e}_k = \mathbf{0}.$$

But the set $\{\mathbf{e}_1, \ldots, \mathbf{e}_k\}$ is LI since it is a basis, so each coefficient in the preceding equation must be zero:

$$a_{11}\alpha_1 + \cdots + a_{k+1,1}\alpha_{k+1} = 0,$$

$$\vdots \tag{18}$$

$$a_{1k}\alpha_1 + \cdots + a_{k+1,k}\alpha_{k+1} = 0.$$

---

[1]We include the stipulation $k \ge 1$ because of the exceptional zero vector space, which is zero-dimensional and has no basis.

These are k linear homogeneous equations in the $k + 1$ unknowns α_1 through α_{k+1}, and such a system necessarily admits nontrivial solutions (Theorem 4.5.4). Thus, the α's in (18) are not all necessarily zero so the vectors $\mathbf{e}'_1, \ldots, \mathbf{e}'_{k+1}$ could not have been LI after all. Hence, it is not possible to find more than k LI vectors in $\mathcal{S}$, so S is k-dimensional. Proof of the converse is left for the exercises. ∎

The spaces of chief concern in this text are the n-tuple spaces $\mathbb{R}^n$ and subspaces thereof. For $\mathbb{R}^n$ we can say the following.

THEOREM 4.9.3
Dimension of $\mathbb{R}^n$
The dimension of $\mathbb{R}^n$ is n: $\dim \mathbb{R}^n = n$.

Proof.   The vectors

$$
\begin{aligned}
\mathbf{e}_1 &= (1, 0, 0, \ldots, 0), \\
\mathbf{e}_2 &= (0, 1, 0, \ldots, 0), \\
&\vdots \\
\mathbf{e}_n &= (0, \ldots, 0, 0, 1)
\end{aligned}
\tag{19}
$$

constitute a basis for $\mathbb{R}^n$ because any vector $\mathbf{u} = (u_1, \ldots, u_n)$ in $\mathbb{R}^n$ can be expanded uniquely as $\mathbf{u} = u_1 \mathbf{e}_1 + \cdots + u_n \mathbf{e}_n$. Since this basis contains n vectors, it follows from Theorem 4.9.2 that $\mathbb{R}^n$ is n-dimensional. ∎

Indeed, we might well have questioned the reasonableness of our definition of dimension if $\mathbb{R}^n$ had turned out to be other than n-dimensional! The ON basis (19) is called the **standard basis** for $\mathbb{R}^n$ (and is the n-space generalization of the "$\hat{\mathbf{i}}, \hat{\mathbf{j}}, \hat{\mathbf{k}}$" ON basis that might be known to you from a course in physics or mechanics).

Finally, what about the dimension of a *subspace*, such as the subspace of $\mathbb{R}^3$ that is spanned by two given vectors?

THEOREM 4.9.4
Dimension of Span $\{\mathbf{u}_1, \ldots, \mathbf{u}_k\}$
The dimension of span $\{\mathbf{u}_1, \ldots, \mathbf{u}_k\}$, where the $\mathbf{u}_j$'s are not all zero, denoted as $\dim [\operatorname{span} \{\mathbf{u}_1, \ldots, \mathbf{u}_k\}]$, is equal to the greatest number of LI vectors within the generating set $\{\mathbf{u}_1, \ldots, \mathbf{u}_k\}$.

Proof.   Denote the generating set $\{\mathbf{u}_1, \ldots, \mathbf{u}_k\}$ as U. Let the greatest number of LI vectors in U be N, so $1 \le N \le k$. It may be assumed, without loss of generality, that the members of U have been numbered so that $\mathbf{u}_1, \ldots, \mathbf{u}_N$ are LI. Then each of the remaining members of U, namely $\mathbf{u}_{N+1}, \ldots, \mathbf{u}_k$, can be expressed as a linear combination of $\mathbf{u}_1, \ldots, \mathbf{u}_N$. Surely, then, each vector in span U can similarly be expressed as a linear combination of $\mathbf{u}_1, \ldots, \mathbf{u}_N$. Now $\{\mathbf{u}_1, \ldots, \mathbf{u}_N\}$ is LI and spans span U. According to Theorem 4.9.3, then, the dimension of span U is N; that is, it is the same as the greatest number of LI vectors in U, as was to be proved. ∎

E X A M P L E 3
Let
$$\mathbf{u}_1 = (3, -1, 2, 1), \quad \mathbf{u}_2 = (1, 1, 0, -1), \quad \mathbf{u}_3 = (4, 0, 2, 0). \quad (20)$$

These vectors are, of course, members of $\mathbb{R}^4$. But since $\mathbf{u}_1, \mathbf{u}_2, \mathbf{u}_3$ are only three vectors, dim [span $\{\mathbf{u}_1, \mathbf{u}_2, \mathbf{u}_3\}$] is *at most* three. In fact, it is *not* three since we see that $\mathbf{u}_3 = \mathbf{u}_1 + \mathbf{u}_2$. But $\mathbf{u}_1$ and $\mathbf{u}_2$, say, are LI since neither is a scalar multiple of the other. Thus, there are only two LI vectors within the generating set so dim [span $\{\mathbf{u}_1, \mathbf{u}_2, \mathbf{u}_3\}$] = 2. ∎

In Example 3 we were able to determine that the greatest number of LI vectors in the generating set was 2, merely by inspection. What if we wish to determine dim [span $\{\mathbf{u}_1, \ldots, \mathbf{u}_k\}$] where the $\mathbf{u}_j$'s are members of $\mathbb{R}^8$, and $k = 6$, say? For such a large problem we cannot expect "inspection" to work. Yet, what are we to do, test the $\mathbf{u}_j$'s for linear independence one at a time, two at a time, three at a time, and so on, until we determine the greatest number of LI vectors in $\{\mathbf{u}_1, \ldots, \mathbf{u}_k\}$? That would be quite tedious. No, we will see later, in Chapter 5, that the best way to determine the greatest number of LI vectors in a given set is to determine the "rank" of a certain matrix, and that can be done by the extremely efficient method of elementary row operations. Meanwhile, in the present section, we will "get by" by keeping the examples and exercises simple enough so that we can rely on inspection.

Now let us return to our discussion of bases and expansions.

4.9.3 Orthogonal basis.   Note carefully that any given vector space having a basis[1] has an *infinite* number of bases. For instance, *any* two nonzero and noncolinear vectors in $\mathbb{R}^2$ constitute a basis for $\mathbb{R}^2$. How then do we decide which basis to select in a given application? We will find that the most convenient basis to use is suggested by the context, so we will not not worry about that now. That point is addressed in Chapter 9 on the eigenvalue problem.

Here, we wish to show that *orthogonal* bases are especially convenient and are therefore to be preferred. Let us return to Example 1 to see why. If we wish to expand a given vector $\mathbf{u}$ in terms of the basis (9), then we need to solve (11) for $\alpha_1, \alpha_2, \alpha_3$; that is, we need to solve three equations in three unknowns. Similarly, in $\mathbb{R}^{100}$ we need to solve 100 equations in 100 unknowns. Thus, expanding vectors can be quite a computational chore!

On the other hand, suppose that $\{\mathbf{e}_1, \ldots, \mathbf{e}_k\}$ is an *orthogonal* basis for our vector space $\mathcal{S}$; that is, it is not only a basis but also happens to be an orthogonal set:

$$\mathbf{e}_i \cdot \mathbf{e}_j = 0 \quad \text{if} \quad i \neq j. \quad (21)$$

Suppose that we wish to expand a given vector $\mathbf{u}$ in $\mathcal{S}$ in terms of that basis; that is, we wish to determine the coefficients $\alpha_1, \ldots, \alpha_k$ in the expansion

$$\mathbf{u} = \alpha_1 \mathbf{e}_1 + \alpha_2 \mathbf{e}_2 + \cdots + \alpha_k \mathbf{e}_k. \quad (22)$$

---

[1]Recall that the zero vector space has no basis.

To accomplish this, we can proceed as in Example 1 and obtain k coupled linear algebraic equations on $\alpha_1, \ldots, \alpha_k$. Alternatively, we can use the orthogonality of the basis to our advantage. Specifically, dot (22) with $\mathbf{e}_1, \mathbf{e}_2, \ldots, \mathbf{e}_k$, in turn. Doing so, and using (21), we obtain the linear system

$$
\begin{aligned}
\mathbf{u}\cdot\mathbf{e}_1 &= (\mathbf{e}_1\cdot\mathbf{e}_1)\alpha_1 + 0\alpha_2 + \cdots + 0\alpha_k, \\
\mathbf{u}\cdot\mathbf{e}_2 &= 0\alpha_1 + (\mathbf{e}_2\cdot\mathbf{e}_2)\alpha_2 + 0\alpha_3 + \cdots + 0\alpha_k, \\
&\;\;\vdots \\
\mathbf{u}\cdot\mathbf{e}_k &= 0\alpha_1 + \cdots + 0\alpha_{k-1} + (\mathbf{e}_k\cdot\mathbf{e}_k)\alpha_k,
\end{aligned}
\tag{23}
$$

where all the scalar quantities $\mathbf{u}\cdot\mathbf{e}_1, \ldots, \mathbf{u}\cdot\mathbf{e}_k, \mathbf{e}_1\cdot\mathbf{e}_1, \ldots, \mathbf{e}_k\cdot\mathbf{e}_k$ are computable since $\mathbf{u}, \mathbf{e}_1, \ldots, \mathbf{e}_k$ are known. The crucial point is that even though (23) is still a system of k equations in the k unknown α_j's, the system (23) is *uncoupled* (i.e., the only unknown in the first equation is α_1, the only one in the second is α_2, and so on) and readily gives

$$
\alpha_1 = \frac{\mathbf{u}\cdot\mathbf{e}_1}{\mathbf{e}_1\cdot\mathbf{e}_1}, \quad \alpha_2 = \frac{\mathbf{u}\cdot\mathbf{e}_2}{\mathbf{e}_2\cdot\mathbf{e}_2}, \quad \ldots, \quad \alpha_k = \frac{\mathbf{u}\cdot\mathbf{e}_k}{\mathbf{e}_k\cdot\mathbf{e}_k},
\tag{24}
$$

provided, of course, that none of the denominators vanish. But these quantities cannot vanish because $\mathbf{e}_j\cdot\mathbf{e}_j = \|\mathbf{e}_j\|^2$, which is zero if and only if $\mathbf{e}_j = \mathbf{0}$, and this cannot be because if any $\mathbf{e}_j$ were $\mathbf{0}$, then the set $\{\mathbf{e}_1, \ldots, \mathbf{e}_k\}$ would be LD (Theorem 4.7.3), and hence not a basis.

Thus, *if the $\{\mathbf{e}_1, \ldots, \mathbf{e}_k\}$ basis is orthogonal, then the expansion of any given $\mathbf{u}$ is simply*

$$
\boxed{\mathbf{u} = \left(\frac{\mathbf{u}\cdot\mathbf{e}_1}{\mathbf{e}_1\cdot\mathbf{e}_1}\right)\mathbf{e}_1 + \cdots + \left(\frac{\mathbf{u}\cdot\mathbf{e}_k}{\mathbf{e}_k\cdot\mathbf{e}_k}\right)\mathbf{e}_k = \sum_{j=1}^{k}\left(\frac{\mathbf{u}\cdot\mathbf{e}_j}{\mathbf{e}_j\cdot\mathbf{e}_j}\right)\mathbf{e}_j.}
\tag{25}
$$

To use (25), all we need to do is to evaluate the dot products $\mathbf{u}\cdot\mathbf{e}_j$ and $\mathbf{e}_j\cdot\mathbf{e}_j$ for $j = 1$ through k. If our base vectors are not only orthogonal but are also scaled to have unit length (i.e., they are *normalized*) then we have the ON (orthonormal) basis $\{\hat{\mathbf{e}}_1, \ldots, \hat{\mathbf{e}}_k\}$, and (25) simplifies slightly to

$$
\boxed{\mathbf{u} = \left(\mathbf{u}\cdot\hat{\mathbf{e}}_1\right)\hat{\mathbf{e}}_1 + \cdots + \left(\mathbf{u}\cdot\hat{\mathbf{e}}_k\right)\hat{\mathbf{e}}_k = \sum_{j=1}^{k}\left(\mathbf{u}\cdot\hat{\mathbf{e}}_j\right)\hat{\mathbf{e}}_j}
\tag{26}
$$

because $\hat{\mathbf{e}}_j\cdot\hat{\mathbf{e}}_j = \|\hat{\mathbf{e}}_j\|^2 = 1$ in (25). We see that there is a great benefit in using an orthogonal basis, but whether or not the basis is normalized is not as significant.

EXAMPLE 4

Expand $\mathbf{u} = (4, 3, -3, 6)$ in terms of the orthogonal base vectors $\mathbf{e}_1 = (1, 0, 2, 0)$, $\mathbf{e}_2 = (0, 1, 0, 0)$, $\mathbf{e}_3 = (-2, 0, 1, 5)$, $\mathbf{e}_4 = (-2, 0, 1, -1)$ of $\mathbb{R}^4$. This basis is orthogonal but not ON so we use (25) rather than (26). Computing $\mathbf{u}\cdot\mathbf{e}_1 = -2$, $\mathbf{e}_1\cdot\mathbf{e}_1 = 5$, and so on, (25) gives

$$\mathbf{u} = -\frac{2}{5}\mathbf{e}_1 + 3\mathbf{e}_2 + \frac{19}{30}\mathbf{e}_3 - \frac{17}{6}\mathbf{e}_4. \tag{27}$$

Alternatively, we could have inferred, from $\mathbf{u} = \alpha_1\mathbf{e}_1 + \cdots + \alpha_4\mathbf{e}_4$, the four equations

$$\begin{aligned}
\alpha_1 \quad\quad - 2\alpha_3 - 2\alpha_4 &= \quad 4, \\
\alpha_2 \quad\quad\quad\quad &= \quad 3, \\
2\alpha_1 \quad\quad + \alpha_3 + \alpha_4 &= -3, \\
5\alpha_3 - \alpha_4 &= \quad 6
\end{aligned} \tag{28}$$

on the four unknown α_j's, and solved these by Gauss elimination, but it is much easier to "cash in" on the orthogonality of the basis and to use (25). If we prefer to work with an ON basis, we can scale the $\mathbf{e}_j$'s as $\hat{\mathbf{e}}_1 = \frac{1}{\sqrt{5}}(1, 0, 2, 0)$, $\hat{\mathbf{e}}_2 = (0, 1, 0, 0)$, $\hat{\mathbf{e}}_3 = \frac{1}{\sqrt{30}}(-2, 0, 1, 5)$, $\hat{\mathbf{e}}_4 = \frac{1}{\sqrt{6}}(-2, 0, 1, -1)$. Then (26) gives

$$\mathbf{u} = -\frac{2}{\sqrt{5}}\hat{\mathbf{e}}_1 + 3\hat{\mathbf{e}}_2 + \frac{19}{\sqrt{30}}\hat{\mathbf{e}}_3 - \frac{17}{\sqrt{6}}\hat{\mathbf{e}}_4, \tag{29}$$

which result is equivalent to (27) because $\hat{\mathbf{e}}_1 = \mathbf{e}_1/\sqrt{5}$, $\hat{\mathbf{e}}_2 = \mathbf{e}_2$, $\hat{\mathbf{e}}_3 = \mathbf{e}_3/\sqrt{30}$, and $\hat{\mathbf{e}}_4 = \mathbf{e}_4/\sqrt{6}$. ∎

4.9.4 Gram-Schmidt orthogonalization. (Optional).

In view of the advantage of orthogonal vector sets over ones that are merely LI, there is interest in being able to convert a given LI vector set to one that is orthogonal.

Given k LI vectors $\mathbf{v}_1, \ldots, \mathbf{v}_k$, let us denote span$\{\mathbf{v}_1, \ldots, \mathbf{v}_k\}$ as $\mathcal{S}$. We wish to obtain, by suitable linear combinations of the $\mathbf{v}_j$'s, k *orthogonal* vectors $\mathbf{w}_1, \ldots, \mathbf{w}_k$. Then $\mathbf{w}_1, \ldots, \mathbf{w}_k$ will, likewise, be in $\mathcal{S}$. In fact, they will constitute an orthogonal basis for $\mathcal{S}$. For suppose there is a vector in $\mathcal{S}$ that cannot be expressed as a linear combination of the $\mathbf{w}_j$'s. Then $\mathcal{S}$ must be at least $(k + 1)$-dimensional. But we know that $\mathcal{S} = $ span$\{\mathbf{v}_1, \ldots, \mathbf{v}_k\}$ is k-dimensional (Theorem 4.9.4), so every vector in $\mathcal{S}$ must be expressible as a linear combination of the $\mathbf{w}_j$'s, so the $\mathbf{w}_j$'s span $\mathcal{S}$. Further, they must be LI because they are orthogonal (Theorem 4.7.5). Hence they constitute an orthogonal basis for $\mathcal{S}$.

There is a systematic procedure for obtaining the $\mathbf{w}_j$'s, known as the **Gram-Schmidt orthogonalization procedure**, after *Jürgen P. Gram* (1850–1916) and *Erhardt Schmidt* (1876–1959). Our plan is to explain this procedure using two examples and to leave additional discussion to the exercises.

EXAMPLE 5
Consider

$$\mathbf{v}_1 = (2, 1, 1), \quad \mathbf{v}_2 = (1, 0, 3)$$

which are LI but not orthogonal. These vectors span a plane $\mathcal{S}$ within $\mathbb{R}^3$, and we wish to find two orthogonal vectors $\mathbf{w}_1$ and $\mathbf{w}_2$ within $\mathcal{S}$. We will take

$$\mathbf{w}_1 \equiv \mathbf{v}_1 = (2, 1, 1), \tag{30}$$

say, and will seek $\mathbf{w}_2$ in the form

$$\mathbf{w}_2 = \mathbf{v}_2 + \alpha \mathbf{v}_1, \tag{31}$$

with α determined so that $\mathbf{w}_2 \cdot \mathbf{w}_1 = 0$. That is, $(\mathbf{v}_2 + \alpha \mathbf{v}_1) \cdot \mathbf{v}_1 = \mathbf{v}_2 \cdot \mathbf{v}_1 + \alpha \mathbf{v}_1 \cdot \mathbf{v}_1 = 0$ so

$$\alpha = -\frac{\mathbf{v}_2 \cdot \mathbf{v}_1}{\mathbf{v}_1 \cdot \mathbf{v}_1} = -\frac{5}{6}. \tag{32}$$

Then (30)–(32) give, as the result,

$$\mathbf{w}_1 = (2, 1, 1), \quad \mathbf{w}_2 = \left(-\frac{4}{6}, -\frac{5}{6}, \frac{13}{6} \right). \tag{33}$$

If we wish, we could scale $\mathbf{w}_2$ as $(4, 5, -13)$, to clear it of fractions, or we could normalize the $\mathbf{w}_j$'s as

$$\hat{\mathbf{w}}_1 = \frac{1}{\sqrt{6}}(2, 1, 1), \quad \hat{\mathbf{w}}_2 = \frac{1}{\sqrt{186}}(4, 5, -13). \tag{34}$$

Then (33) and (34) are orthogonal and ON bases for $\mathcal{S}$, respectively.   ∎

EXAMPLE 6
Consider

$$\mathbf{v}_1 = (1, 1, 0), \quad \mathbf{v}_2 = (2, 1, -1), \quad \mathbf{v}_3 = (1, 2, 3),$$

which are LI but not orthogonal. We take

$$\mathbf{w}_1 \equiv \mathbf{v}_1 = (1, 1, 0) \tag{35}$$

and seek

$$\mathbf{w}_2 = \mathbf{v}_2 + \alpha \mathbf{v}_1, \tag{36}$$

$$\mathbf{w}_3 = \mathbf{v}_3 + \beta \mathbf{v}_2 + \gamma \mathbf{v}_1, \tag{37}$$

with α determined so that $\mathbf{w}_2 \cdot \mathbf{w}_1 = 0$. Then, with α known, we determine β and δ so that $\mathbf{w}_3 \cdot \mathbf{w}_1 = 0$ and $\mathbf{w}_3 \cdot \mathbf{w}_2 = 0$. The first step gives

$$\alpha = -\frac{\mathbf{v}_2 \cdot \mathbf{v}_1}{\mathbf{v}_1 \cdot \mathbf{v}_1} = -\frac{3}{2} \tag{38}$$

so that

$$\mathbf{w}_2 = (2, 1, -1) - \frac{3}{2}(1, 1, 0) = \left(\frac{1}{2}, -\frac{1}{2}, -1 \right). \tag{39}$$

Then, using (35)–(37) and the definitions of $\mathbf{v}_1$, $\mathbf{v}_2$, and $\mathbf{v}_3$, $\mathbf{w}_3 \cdot \mathbf{w}_1 = 0$ and $\mathbf{w}_3 \cdot \mathbf{w}_2 = 0$ give

$$
\begin{aligned}
3\beta + 2\gamma &= -3, \\
\tfrac{3}{2}\beta + 0\gamma &= \tfrac{7}{2},
\end{aligned}
\tag{40}
$$

so $\beta = 7/3$ and $\gamma = -5$. With these values (37) gives

$$
\mathbf{w}_3 = (1, 2, 3) + \frac{7}{3}(2, 1, -1) - 5(1, 1, 0) = \left(\frac{2}{3}, -\frac{2}{3}, \frac{2}{3} \right).
\tag{41}
$$

The result is given by (35), (39), and (41). Or, scaling (39) by 2 and (41) by $3/2$,

$$
\mathbf{w}_1 = (1, 1, 0), \quad \mathbf{w}_2 = (1, -1, -2), \quad \mathbf{w}_3 = (1, -1, 1),
\tag{42}
$$

which vectors constitute an orthogonal basis for $\mathcal{S} = \text{span}\{\mathbf{v}_1, \mathbf{v}_2, \mathbf{v}_3\} = \mathbb{R}^3$. ∎

It should not be difficult to visualize the generalization of this "boot-strapping" procedure. In fact, explicit formulas for $\mathbf{w}_1, \ldots, \mathbf{w}_k$ are given in the exercises.

4.9.5 Best approximation and orthogonal projection. (Optional).

We know that if $\{\mathbf{e}_1, \ldots, \mathbf{e}_N\}$ is a basis for a vector space $\mathcal{S}$, then any vector $\mathbf{u}$ in $\mathcal{S}$ can be (uniquely) expanded in the form $\mathbf{u} = \sum_{j=1}^{N} c_j \mathbf{e}_j$. If the basis is orthogonal, then the expansion process is simple, with the c_j's computed from the given vector $\mathbf{u}$ and the base vectors $\mathbf{e}_j$ as $c_j = (\mathbf{u} \cdot \mathbf{e}_j)/(\mathbf{e}_j \cdot \mathbf{e}_j)$. And if the basis is not only orthogonal but ON, then $\mathbf{u} = \sum_{j=1}^{N} c_j \hat{\mathbf{e}}_j$, where $c_j = \mathbf{u} \cdot \hat{\mathbf{e}}_j$.

However, what if we do not have a "full deck?" That is, what if $\{\hat{\mathbf{e}}_1, \ldots, \hat{\mathbf{e}}_N\}$ is ON, but falls short of being a basis for $\mathcal{S}$ (i.e., $N < \dim \mathcal{S}$)? If $\mathbf{u}$ happens to fall within span $\{\hat{\mathbf{e}}_1, \ldots, \hat{\mathbf{e}}_N\}$, which subspace of $\mathcal{S}$ we denote as $\mathcal{T}$, then it can nevertheless be expanded in terms of $\hat{\mathbf{e}}_1, \ldots, \hat{\mathbf{e}}_N$, but if it is not in $\mathcal{T}$, then it cannot be so expanded. In the latter case the question arises, what is the best *approximation* of $\mathbf{u}$ in terms of $\hat{\mathbf{e}}_1, \ldots, \hat{\mathbf{e}}_N$? In this section we answer that question in general, and illustrate the results for the case where $\mathcal{S}$ is $\mathbb{R}^n$. The concepts and results are important in applied mathematics, though we will not have occasion to use them in this book. Hence this section is listed as optional.

The *best approximation problem* which we address is this: Given a vector $\mathbf{u}$ in $\mathcal{S}$, and an ON set $\{\hat{\mathbf{e}}_1, \ldots, \hat{\mathbf{e}}_N\}$ in $\mathcal{S}$, what is the best approximation

$$
\mathbf{u} \approx c_1 \hat{\mathbf{e}}_1 + \cdots + c_N \hat{\mathbf{e}}_N = \sum_{j=1}^{N} c_j \hat{\mathbf{e}}_j ?
\tag{43}
$$

That is, how do we compute the c_j coefficients so as to render the error vector $\mathbf{E} = \mathbf{u} - \sum_{j=1}^{N} c_j \hat{\mathbf{e}}_j$ as small as possible? In other words, how do we choose the c_j's so as to minimize the norm of the error vector $\|\mathbf{E}\|$? If $\|\mathbf{E}\|$ is a minimum, then so is $\|\mathbf{E}\|^2$, so let us minimize $\|\mathbf{E}\|^2$ (to avoid square roots), where

$$
\|\mathbf{E}\|^2 = \mathbf{E} \cdot \mathbf{E} = \left(\mathbf{u} - \sum_{j=1}^{N} c_j \hat{\mathbf{e}}_j \right) \cdot \left(\mathbf{u} - \sum_{j=1}^{N} c_j \hat{\mathbf{e}}_j \right) = \mathbf{u} \cdot \mathbf{u} - 2 \sum_{j=1}^{N} c_j \left(\mathbf{u} \cdot \hat{\mathbf{e}}_j \right) + \sum_{j=1}^{N} c_j^2,
\tag{44}
$$

the last term following from

$$\left(\sum_1^N c_j \hat{\mathbf{e}}_j\right) \cdot \left(\sum_1^N c_j \hat{\mathbf{e}}_j\right) = (c_1 \hat{\mathbf{e}}_1 + \cdots + c_N \hat{\mathbf{e}}_N) \cdot (c_1 \hat{\mathbf{e}}_1 + \cdots + c_N \hat{\mathbf{e}}_N)$$

$$= c_1^2 + \cdots + c_N^2 = \sum_1^N c_j^2 \tag{45}$$

because of the orthonormality of the $\hat{\mathbf{e}}_j$'s.

Defining $\mathbf{u} \cdot \hat{\mathbf{e}}_j \equiv \alpha_j$ and noting that $\mathbf{u} \cdot \mathbf{u} = \|\mathbf{u}\|^2$, we may express (44) as

$$\|\mathbf{E}\|^2 = \sum_{j=1}^N c_j^2 - 2 \sum_{j=1}^N \alpha_j c_j + \|\mathbf{u}\|^2,$$

or, completing the square, as

$$\|\mathbf{E}\|^2 = \sum_{j=1}^N \left(c_j - \alpha_j\right)^2 + \|\mathbf{u}\|^2 - \sum_{j=1}^N \alpha_j^2. \tag{46}$$

Note that $\mathbf{u}$ and the ON set $\{\hat{\mathbf{e}}_1, \ldots, \hat{\mathbf{e}}_N\}$ are given, so $\|\mathbf{u}\|$ and the α_j's in (46) are fixed computable quantities: $\|\mathbf{u}\| = \sqrt{\mathbf{u} \cdot \mathbf{u}}$ and $\alpha_j = \mathbf{u} \cdot \hat{\mathbf{e}}_j$ for $j = 1, 2, \ldots, N$. Thus, in seeking to minimize the right-hand side of (46), the only control we exercise is in our choice of the c_j's. The right-hand side of (46) is greater than or equal to zero,[1] and so is the $\sum_{j=1}^N \left(c_j - \alpha_j\right)^2$ term containing the c_j's. Thus, the best that we can do is to set $c_j = \alpha_j$ $(j = 1, 2, \ldots, N)$. With that choice, our best approximation (43) becomes

$$\boxed{\mathbf{u} \approx \sum_{j=1}^N \left(\mathbf{u} \cdot \hat{\mathbf{e}}_j\right) \hat{\mathbf{e}}_j. } \tag{47}$$

Let us summarize these results.

THEOREM 4.9.5

Best Approximation

Let $\mathbf{u}$ be any vector in a normed inner product vector space $\mathcal{S}$ with the norm $\|\mathbf{u}\| = \sqrt{\mathbf{u} \cdot \mathbf{u}}$, and let $\{\hat{\mathbf{e}}_1, \ldots, \hat{\mathbf{e}}_N\}$ be an ON set in $\mathcal{S}$. Then the best approximation (43) is obtained when the c_j's are given by $c_j = \mathbf{u} \cdot \hat{\mathbf{e}}_j$, as indicated in (47).

---

[1] This fact may not be obvious due to the minus sign in front of the last summation. But remember that the right-hand side of (46) is equal to $\|\mathbf{E}\|^2$, and surely $\|\mathbf{E}\|^2 \geq 0$.

EXAMPLE 7
Let S be $\mathbb{R}^2$, $N = 1$, $\hat{\mathbf{e}}_1 = \frac{1}{13}(12, 5)$, and $\mathbf{u} = (1, 1)$, as shown in Fig. 2. Find
the best approximation $\mathbf{u} \approx c_1\hat{\mathbf{e}}_1$, that is, the best approximation of $\mathbf{u}$ in span $\{\hat{\mathbf{e}}_1\}$
(which is the line L). Theorem 4.9.5 gives $c_1 = \mathbf{u}\cdot\hat{\mathbf{e}}_1 = 17/13$, and hence the best
approximation

$$\mathbf{u} \approx \frac{17}{13}\hat{\mathbf{e}}_1, \tag{48}$$

which is the vector **OA** in Fig. 2.

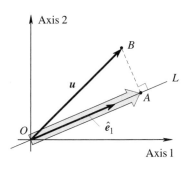

FIGURE 2
Best approximation
of $\mathbf{u}$ in span $\{\hat{\mathbf{e}}_1\}$.

COMMENT.   Observe from the figure that the best approximation **OA** is the *orthogonal projection* of $\mathbf{u}$ onto span $\{\mathbf{e}_1\}$, which orthogonality is verified by the calculation

$$\mathbf{AB}\cdot\hat{\mathbf{e}}_1 = (\mathbf{u} - \mathbf{OA})\cdot\hat{\mathbf{e}}_1 = (\mathbf{u} - \frac{17}{13}\hat{\mathbf{e}}_1)\cdot\hat{\mathbf{e}}_1 = \frac{17}{13} - \frac{17}{13} = 0.$$

That result makes sense geometrically since the norm of the error vector **E** is the
distance from the tip of $c_1\hat{\mathbf{e}}_1$ (on L) to the tip of $\mathbf{u}$. That distance is the shortest when
the line BA is perpendicular to L.   ∎

EXAMPLE 8
Let S be $\mathbb{R}^3$, let $N = 2$ with $\hat{\mathbf{e}}_1 = (1, 0, 0)$ and $\hat{\mathbf{e}}_1 = (0, 1, 0)$, and let $\mathbf{u} = (a, b, c)$,
as shown in Fig. 3. Computing the coefficients in (47) as $\mathbf{u}\cdot\hat{\mathbf{e}}_1 = a$ and $\mathbf{u}\cdot\hat{\mathbf{e}}_2 = b$,
(47) becomes

$$\mathbf{u} \approx a\hat{\mathbf{e}}_1 + b\hat{\mathbf{e}}_2. \tag{49}$$

The latter is an equality if $c = 0$. That is, (7) is an equality if $\mathbf{u}$ happens to lie in
span $\{\hat{\mathbf{e}}_1, \hat{\mathbf{e}}_2\}$, but if $c \neq 0$ then the best approximation $a\hat{\mathbf{e}}_1 + b\hat{\mathbf{e}}_2$ to $\mathbf{u}$ is the orthogonal
projection of $\mathbf{u}$ onto span $\{\hat{\mathbf{e}}_1, \hat{\mathbf{e}}_2\}$.   ∎

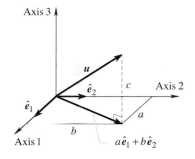

FIGURE 3
Best approximation
of $\mathbf{u}$ in span $\{\hat{\mathbf{e}}_1, \hat{\mathbf{e}}_2\}$.

In Examples 7 and 8, S was $\mathbb{R}^2$ and $\mathbb{R}^3$, respectively, so we were able to draw
useful pictures. In each case we discovered that the best approximation of $\mathbf{u}$ in the
subspace T of S spanned by $\hat{\mathbf{e}}_1, \ldots, \hat{\mathbf{e}}_N$ was the orthogonal projection of $\mathbf{u}$ onto T.
Is that result true in all cases? That is, is the error vector **E** necessarily orthogonal
to T? Since the error vector is

$$\mathbf{E} = \mathbf{u} - \sum_{j=1}^{N} (\mathbf{u}\cdot\hat{\mathbf{e}}_j)\,\hat{\mathbf{e}}_j, \tag{50}$$

we have

$$\mathbf{E}\cdot\hat{\mathbf{e}}_k = \left[\mathbf{u} - \sum_{j=1}^{N} (\mathbf{u}\cdot\hat{\mathbf{e}}_j)\,\hat{\mathbf{e}}_j\right]\cdot\hat{\mathbf{e}}_k$$
$$= \mathbf{u}\cdot\hat{\mathbf{e}}_k - (\mathbf{u}\cdot\hat{\mathbf{e}}_k)(1) = 0 \tag{51}$$

for each $k = 1, 2, \ldots, N$, where the second equality follows from the fact that $\hat{\mathbf{e}}_j \cdot \hat{\mathbf{e}}_k = 0$ if $j \neq k$ and 1 if $j = k$. Since $\mathbf{E}$ is orthogonal to every one of the $\hat{\mathbf{e}}_j$'s, it is therefore orthogonal to every vector in $\mathcal{T}$. In that sense we say that the right-hand side of (47) is the **orthogonal projection of u onto** $\mathcal{T}$, and we denote it as $\text{proj}_{\mathcal{T}} \mathbf{u}$:

$$\boxed{\text{proj}_{\mathcal{T}} \mathbf{u} \equiv \sum_{j=1}^{N} \left(\mathbf{u} \cdot \hat{\mathbf{e}}_j \right) \hat{\mathbf{e}}_j.} \tag{52}$$

The idea that the best approximation of $\mathbf{u}$ in $\mathcal{T}$ is the orthogonal projection of $\mathbf{u}$ onto $\mathcal{T}$ lends a welcome geometrical interpretation to the problem of best approximation. In fact, let us rephrase Theorem 4.9.5 in terms of orthogonal projection.

THEOREM 4.9.6
Best Approximation by Orthogonal Projection
Let $\mathbf{u}$ be any vector in a normed inner product vector space $\mathcal{S}$ with the norm $\|\mathbf{u}\| = \sqrt{\mathbf{u} \cdot \mathbf{u}}$, and let $\{\hat{\mathbf{e}}_1, \ldots, \hat{\mathbf{e}}_N\}$ be an ON set in $\mathcal{S}$. Denote the subspace span$\{\hat{\mathbf{e}}_1, \ldots, \hat{\mathbf{e}}_N\}$ of $\mathcal{S}$ as $\mathcal{T}$. Then the best approximation of $\mathbf{u}$ in $\mathcal{T}$ (i.e., of the form $c_1 \hat{\mathbf{e}}_1 + \cdots + c_N \hat{\mathbf{e}}_N$) is given by the orthogonal projection of $\mathbf{u}$ onto $\mathcal{T}$, namely, by $\text{proj}_{\mathcal{T}} \mathbf{u}$.

Closure. This section is about the **expansion** of vectors, in a given vector space $\mathcal{S}$, in terms of a set of base vectors. A set of vectors $\{\mathbf{e}_1, \ldots, \mathbf{e}_k\}$ in $\mathcal{S}$ is a *basis* for $\mathcal{S}$ if each vector $\mathbf{u}$ in $\mathcal{S}$ *can* be expanded as a *unique* linear combination of the $\mathbf{e}_j$'s. We showed (Theorem 4.9.1) that $\{\mathbf{e}_1, \ldots, \mathbf{e}_k\}$ is indeed a basis for $\mathcal{S}$ if and only if it spans $\mathcal{S}$ (so each $\mathbf{u}$ *can* be expanded) and is LI (so the expansion is *unique*). The number of vectors in any basis for $\mathcal{S}$ is called the *dimension* of $\mathcal{S}$. For instance, $\mathbb{R}^n$ admits the *standard basis* (19), comprised of n vectors, so $\mathbb{R}^n$ is n-dimensional. And the greatest number of LI vectors in a set $\{\mathbf{u}_1, \ldots, \mathbf{u}_k\}$ is the dimension of their span.

We found that the expansion process (i.e., the determination of the expansion coefficients) can be laborious if there are many base vectors, but is simple if the basis is orthogonal or ON, in which case the expansions are given by (25) or (26), respectively. You should remember those two formulas and be able to derive them as well.

In view of the advantage of orthogonal bases over ones that are merely LI, we showed how to "trade in" a LI set for an orthogonal one, by the *Gram-Schmidt orthogonalization procedure*. We also showed how to do the best we can if we don't have a full basis. We found that if $\mathbf{u}$ is a vector in $\mathcal{S}$ and $\{\hat{\mathbf{e}}_1, \ldots \hat{\mathbf{e}}_N\}$ is an ON set in $\mathcal{S}$, then the *best approximation* of $\mathbf{u}$, as a linear combination of the $\hat{\mathbf{e}}_j$'s, is given by the orthogonal projection of $\mathbf{u}$ onto span$\{\hat{\mathbf{e}}_1, \ldots \hat{\mathbf{e}}_N\}$.

EXERCISES 4.9

1. Show whether the following is a basis.
   (a) $(1, 0), (1, 1), (1, 2)$ for $\mathbb{R}^2$
   (b) $(3, 2), (-1, -5)$ for $\mathbb{R}^2$
   (c) $(1, 1)$ for $\mathbb{R}^2$
   (d) $(2, 0, 1), (5, -1, 2), (1, -1, 0)$ for $\mathbb{R}^3$
   (e) $(5, -1, 2), (2, 0, 1), (1, -1, 1)$ for $\mathbb{R}^3$
   (f) $(2, 1, 0, 6), (7, -1, -2, 3), (4, 3, 2, 1)$ for $\mathbb{R}^4$
   (g) $(4, 3, -2, 1), (5, 0, 0, 0), (2, 1, -3, 0), (1, 2, 4, 5)$
       for $\mathbb{R}^4$
   (h) $(4, 2, 0, 0), (1, 2, 3, 0), (5, -2, 3, 1), (0, -6, 0, 1)$
       for $\mathbb{R}^4$
   (i) $(1, 0, 0, 0), (1, 1, 0, 0), (1, 1, 1, 0), (1, 1, 1, 1)$ for $\mathbb{R}^4$
   (j) $(3, 0, 0, 1), (2, 0, 0, 1), (1, 3, 5, 6), (4, -2, 1, 3)$ for $\mathbb{R}^4$
   (k) $(1, 3, -1, 2), (1, 2, 4, 3), (2, 5, 3, 5), (3, 7, 7, 8)$
       for $\mathbb{R}^4$
   (l) $(2, 3, 5, 0), (1, -1, 2, 3), (4, 1, 2, 3), (5, 4, 1, 0),$
       $(1, 2, 4, 6)$ for $\mathbb{R}^4$
   (m) $(1, 0, 0, 0), (0, 1, 0, 0), (0, 0, 1, 0), (0, 0, 0, 1),$
       $(0, 0, 0, 0)$ for $\mathbb{R}^4$
   (n) $(1, 0, 0, 0), (0, 1, 0, 0), (0, 0, 1, 0), (0, 0, 0, 0)$ for $\mathbb{R}^4$
   (o) $(1, 1, 2), (4, -2, -1)$ for span $\{(2, -4, -5),$
       $(1, -1, -1)\}$
   (p) $(1, 1, 2), (4, -2, -1)$ for span $\{(3, -5, -6), (1, 2, 1)\}$
   (q) $(1, 1, 1), (1, -1, 2)$ for span $\{(2, 4, 1), (1, 7, 2)\}$
   (r) $(1, 2, 3), (1, 0, 4)$ for span $\{(3, 2, 0), (1, 1, -1)\}$

2. Expand each vector $\mathbf{u}$ in terms of the orthogonal basis
   $\{\mathbf{e}_1, \mathbf{e}_2, \mathbf{e}_3\}$ of $\mathbb{R}^3$, where $\mathbf{e}_1 = (2, 1, 3)$, $\mathbf{e}_2 = (1, -2, 0)$,
   $\mathbf{e}_3 = (6, 3, -5)$.
   (a) $\mathbf{u} = (9, -2, 4)$ (b) $\mathbf{u} = (1, 0, 0)$
   (c) $\mathbf{u} = (0, 1, 5)$ (d) $\mathbf{u} = (3, -1, 1)$
   (e) $\mathbf{u} = (0, 5, 0)$ (f) $\mathbf{u} = (1, 2, 3)$

3. (a)–(f) Expand each of the $\mathbf{u}$ vectors in Exercise 2 in terms
   of the ON basis $\{\hat{\mathbf{e}}_1, \hat{\mathbf{e}}_2, \hat{\mathbf{e}}_3\}$ of $\mathbb{R}^3$, where $\hat{\mathbf{e}}_1, \hat{\mathbf{e}}_2, \hat{\mathbf{e}}_3$ are normalized versions of $\mathbf{e}_1, \mathbf{e}_2, \mathbf{e}_3$ given in Exercise 2.

4. Expand each vector $\mathbf{u}$ in terms of the orthogonal basis
   $\{\mathbf{e}_1, \ldots, \mathbf{e}_4\}$ of $\mathbb{R}^4$, where $\mathbf{e}_1 = (2, 0, -1, -5)$, $\mathbf{e}_2 = (2, 0, -1, 1)$, $\mathbf{e}_3 = (0, 1, 0, 0)$, $\mathbf{e}_4 = (1, 0, 2, 0)$.
   (a) $\mathbf{u} = (1, 0, 0, 0)$ (b) $\mathbf{u} = (0, 6, 0, 0)$
   (c) $\mathbf{u} = (2, 5, 1, -3)$ (d) $\mathbf{u} = (4, 3, -2, 0)$
   (e) $\mathbf{u} = (1, 2, 0, 5)$ (f) $\mathbf{u} = (2, -7, 4, 1)$
   (g) $\mathbf{u} = (0, 0, 0, 9)$ (h) $\mathbf{u} = (2, 3, -2, 1)$
   (i) $\mathbf{u} = (0, 0, 5, 0)$ (j) $\mathbf{u} = (1, 1, 1, 1)$

5. Verify that the $\{\mathbf{e}_1, \ldots, \mathbf{e}_4\}$ vectors given in Example 4 are
   a basis for $\mathbb{R}^4$. Also, solve (28) by Gauss elimination and

verify that the α_j's thus obtained agree with those given
(more easily) by (25).

6. If $\{\mathbf{e}_1, \ldots, \mathbf{e}_k\}$ is an orthogonal set in a vector space $\mathcal{S}$, is
   it necessarily a basis
   (a) for $\mathcal{S}$?   (b)   for span $\{\mathbf{e}_1, \ldots, \mathbf{e}_k\}$?

7. (*Zero vector space*) Use Theorem 4.9.1 to show that a zero
   vector space has no basis.

8. Let $\mathbf{u}_1 = (1, 0, 0)$, $\mathbf{u}_2 = (0, 1, 0)$, $\mathbf{u}_3 = (0, 0, 1)$, $\mathbf{u}_4 = (1, 1, 0)$, $\mathbf{u}_5 = (0, 1, 1)$, $\mathbf{u}_6 = (1, 1, 1)$, and $\mathbf{u}_7 = (3, 2, 2)$.
   Evaluate each of the following.
   (a) dim [span $\{\mathbf{u}_1\}$]
   (b) dim [span $\{\mathbf{u}_1, \mathbf{u}_2\}$]
   (c) dim [span $\{\mathbf{u}_1, \mathbf{u}_2, \mathbf{u}_3\}$]
   (d) dim [span $\{\mathbf{u}_1, \mathbf{u}_2, \mathbf{u}_3, \mathbf{u}_4\}$]
   (e) dim [span $\{\mathbf{u}_1, \mathbf{u}_2, \mathbf{u}_4\}$]
   (f) dim [span $\{\mathbf{u}_1, \mathbf{u}_4, \mathbf{u}_5\}$]
   (g) dim [span $\{\mathbf{u}_5, \mathbf{u}_6, \mathbf{u}_7\}$]
   (h) dim [span $\{\mathbf{u}_1, \mathbf{u}_5, \mathbf{u}_6, \mathbf{u}_7\}$]

9. Let $\mathbf{u}_1 = (1, 0, 0, 0)$, $\mathbf{u}_2 = (1, 1, 0, 0)$, $\mathbf{u}_3 = (1, 1, 1, 0)$,
   $\mathbf{u}_4 = (1, 1, 1, 1)$, $\mathbf{u}_5 = (0, 0, 0, 1)$, $\mathbf{u}_6 = (3, 3, 3, 3)$. Evaluate each of the following.
   (a) dim [span $\{\mathbf{u}_1, \mathbf{u}_3, \mathbf{u}_5\}$]
   (b) dim [span $\{\mathbf{u}_1, \mathbf{u}_4, \mathbf{u}_6\}$]
   (c) dim [span $\{\mathbf{u}_2, \mathbf{u}_4, \mathbf{u}_6\}$]
   (d) dim [span $\{\mathbf{u}_5\}$]
   (e) dim [span $\{\mathbf{u}_3, \mathbf{u}_4\}$]
   (f) dim [span $\{\mathbf{u}_3, \mathbf{u}_4, \mathbf{u}_5, \mathbf{u}_6\}$]
   (g) dim [span $\{\mathbf{u}_4, \mathbf{u}_6\}$]
   (h) dim [span $\{\mathbf{u}_2, \mathbf{u}_3, \mathbf{u}_4, \mathbf{u}_5, \mathbf{u}_6\}$]

10. (a)–(g) Determine the dimension of the solution space in
    Exercise 5 of Section 4.8.

11. (*Gram-Schmidt orthogonalization process*) Given k LI
    vectors $\mathbf{v}_1, \ldots, \mathbf{v}_k$, it is possible to obtain from them k ON
    vectors, say $\hat{\mathbf{w}}_1, \ldots, \hat{\mathbf{w}}_k$, in span $\{\mathbf{v}_1, \ldots, \mathbf{v}_k\}$ by the Gram-
    Schmidt process, by taking $\mathbf{w}_1$ equal to $\mathbf{v}_1$, taking $\mathbf{w}_2$ equal
    to a suitable linear combination of $\mathbf{v}_1, \mathbf{v}_2$, taking $\mathbf{w}_3$ equal
    to a suitable linear combination of $\mathbf{v}_1, \mathbf{v}_2, \mathbf{v}_3$, and so on,
    and then normalizing the results. The resulting ON set is

as follows:

$$\hat{\mathbf{w}}_1 = \frac{\mathbf{v}_1}{\|\mathbf{v}_1\|},$$

$$\hat{\mathbf{w}}_2 = \frac{\mathbf{v}_2 - (\mathbf{v}_2 \cdot \hat{\mathbf{w}}_1)\hat{\mathbf{w}}_1}{\|\mathbf{v}_2 - (\mathbf{v}_2 \cdot \hat{\mathbf{w}}_1)\hat{\mathbf{w}}_1\|},$$

$$\vdots \qquad\qquad (11.1)$$

$$\hat{\mathbf{w}}_j = \frac{\mathbf{v}_j - \sum_{i=1}^{j-1}(\mathbf{v}_j \cdot \hat{\mathbf{w}}_i)\hat{\mathbf{w}}_i}{\left\| \mathbf{v}_j - \sum_{i=1}^{j-1}(\mathbf{v}_j \cdot \hat{\mathbf{w}}_i)\hat{\mathbf{w}}_i \right\|} \quad \text{through } j = k.$$

We now state the problem: Derive the above formulas for $\hat{\mathbf{w}}_1$, $\hat{\mathbf{w}}_2$, and $\hat{\mathbf{w}}_3$.

12. In each case use the Gram-Schmidt formula (11.1) in Exercise 11 to obtain an ON set from the given LI set.

(**a**) $(4, 0)$, $(2, 1)$

(**b**) $(1, -2)$, $(3, 4)$

(**c**) $(1, 0, 0)$, $(1, 1, 0)$, $(1, 1, 1)$

(**d**) $(1, 1, 0)$, $(2, -1, 1)$, $(1, 0, 3)$

(**e**) $(1, 1, 1)$, $(2, 0, -1)$

(**f**) $(1, 1, 1)$, $(1, 0, 1)$, $(1, 1, 0)$

(**g**) $(2, 1, 1, 0)$, $(1, 5, -1, 2)$

(**h**) $(6, -1, 1, 2, 1)$, $(2, 3, -1, 1, 4)$

13. (*The dual or reciprocal vectors*) For definiteness, consider our vector space $\mathcal{S}$ to be $\mathbb{R}^n$.

(**a**) If $\{\hat{\mathbf{e}}_1, \ldots, \hat{\mathbf{e}}_n\}$ is an ON basis for $\mathbb{R}^n$, and $\mathbf{u}$ is in $\mathbb{R}^n$, then by dotting $\hat{\mathbf{e}}_i$ into both sides of the equation $\mathbf{u} = \sum_{j=1}^{n} \alpha_j \hat{\mathbf{e}}_j$ we find that $\alpha_i = \mathbf{u} \cdot \hat{\mathbf{e}}_i$ so that the expansion of $\mathbf{u}$ in terms of the given basis is

$$\mathbf{u} = \sum_{j=1}^{n} (\mathbf{u} \cdot \hat{\mathbf{e}}_j) \hat{\mathbf{e}}_j. \qquad (13.1)$$

If, instead, we have a basis $\{\mathbf{e}_1, \ldots, \mathbf{e}_n\}$ which is *not* ON, then, as noted in the text, the expansion process is not so simple. However, suppose that we can find a set $\{\mathbf{e}_1^*, \ldots, \mathbf{e}_n^*\}$ such that

$$\mathbf{e}_i \cdot \mathbf{e}_j^* = \begin{cases} 1, & i = j \\ 0, & i \neq j. \end{cases} \qquad (13.2)$$

Then show that dotting $\mathbf{e}_i^*$ into $\mathbf{u} = \sum_{j=1}^{n} \alpha_j \mathbf{e}_j$ gives $\alpha_i = \mathbf{u} \cdot \mathbf{e}_i^*$ so that

$$\mathbf{u} = \sum_{j=1}^{n} (\mathbf{u} \cdot \mathbf{e}_j^*) \mathbf{e}_j. \qquad (13.3)$$

The set $\{\mathbf{e}_1^*, \ldots, \mathbf{e}_n^*\}$ is called the **dual** or **reciprocal set** corresponding to the original set $\{\mathbf{e}_1, \ldots, \mathbf{e}_n\}$. NOTE: We will see in the last exercise in Section 5.6 that the dual set exists, is unique, and is itself a basis for $\mathbb{R}^n$, the so-called **dual** or **reciprocal basis**.

(**b**) Given the basis $\mathbf{e}_1 = (1, 0)$, $\mathbf{e}_2 = (1, 1)$ for $\mathbb{R}^2$, use equation (13.2) to determine the dual vectors $\mathbf{e}_1^*$, $\mathbf{e}_2^*$. Then use equation (13.3) to expand $\mathbf{u} = (3, 1)$. Sketch $\mathbf{e}_1, \mathbf{e}_2, \mathbf{e}_1^*, \mathbf{e}_2^*, \mathbf{u}$ to scale, and verify the expansion graphically, that is, by means of the parallelogram rule of vector addition.

(**c**) Repeat part (b), for $\mathbf{e}_1 = (2, 1)$, $\mathbf{e}_2 = (0, 2)$, $\mathbf{u} = (-3, 2)$,

(**d**) Repeat part (b), for $\mathbf{e}_1 = (-1, 1)$, $\mathbf{e}_2 = (2, 1)$, $\mathbf{u} = (0, 4)$,

(**e**) Given the basis $\mathbf{e}_1 = (1, 0, 0)$, $\mathbf{e}_2 = (1, 1, 0)$, $\mathbf{e}_3 = (1, 1, 1)$ for $\mathbb{R}^3$, use equation (13.2) to determine the dual vectors $\mathbf{e}_1^*$, $\mathbf{e}_2^*$, $\mathbf{e}_3^*$. Then use equation (13.3) to expand each of the vectors $\mathbf{u} = (4, -1, 5)$, $\mathbf{v} = (0, 0, 2)$, $\mathbf{w} = (5, -2, 3)$. Be sure to see that the dual vectors get computed once and for all, for a given basis $\{\mathbf{e}_1, \ldots, \mathbf{e}_n\}$; once we've got them, expansions are readily carried out by (13.3).

(**f**) Show that if the $\{\mathbf{e}_1, \ldots, \mathbf{e}_n\}$ basis does happen to be ON, then the dual vectors coalesce with the $\mathbf{e}_j$'s, i.e., $\mathbf{e}_j^* = \mathbf{e}_j$ for $j = 1, 2, \ldots, n$.

14. We concluded from (46) that the best choice for the c_j's is $c_j = \alpha_j = \mathbf{u} \cdot \hat{\mathbf{e}}_j$. Show that this same result is obtained from (44) by setting $\partial \|\mathbf{E}\|^2 / \partial c_j = 0$, and prove that the extremum thus obtained is a minimum.

15. Let $\mathcal{S}$ be $\mathbb{R}^5$, and let $N = 3$ with $\hat{\mathbf{e}}_1 = \frac{1}{\sqrt{5}}(1, 0, 2, 0, 0)$, $\hat{\mathbf{e}}_2 = \frac{1}{\sqrt{6}}(2, 0, -1, 0, 1)$, $\hat{\mathbf{e}}_3 = (0, 0, 0, 1, 0)$. Find the best approximation to the given $\mathbf{u}$ vector within span $\{\hat{\mathbf{e}}_1, \hat{\mathbf{e}}_2, \hat{\mathbf{e}}_3\}$, and the norm of the error vector.

(**a**) $(3, -2, 0, 0, 5)$ (**b**) $(0, 0, 0, 2, 1)$

(**c**) $(3, 0, 1, 4, 1)$ (**d**) $(1, 1, 0, 1, 1)$

(**e**) $(0, 2, 0, 0, 0)$ (**f**) $(1, 0, -3, 3, 1)$

16. Let $\mathcal{S}$ be $\mathbb{R}^4$, and let

$$\hat{\mathbf{e}}_1 = \frac{1}{\sqrt{3}}(1, 1, 0, -1), \quad \hat{\mathbf{e}}_2 = \frac{1}{\sqrt{3}}(1, -1, -1, 0),$$

$$\hat{\mathbf{e}}_3 = \frac{1}{\sqrt{3}}(1, 0, 1, 1), \quad \hat{\mathbf{e}}_4 = \frac{1}{\sqrt{3}}(0, 1, -1, 1).$$

Find the best approximation to $\mathbf{u} = (4, -2, 1, 6)$ within span $\{\hat{\mathbf{e}}_1\}$, span $\{\hat{\mathbf{e}}_1, \hat{\mathbf{e}}_2\}$, span $\{\hat{\mathbf{e}}_1, \hat{\mathbf{e}}_2, \hat{\mathbf{e}}_3\}$, and span $\{\hat{\mathbf{e}}_1, \hat{\mathbf{e}}_2, \hat{\mathbf{e}}_3, \hat{\mathbf{e}}_4\}$, and in each case compute the norm of the error vector, $\|\mathbf{E}\|$.

17. (*Bessel inequality*) Beginning with (46), derive the **Bessel inequality**

$$\sum_{j=1}^{N} \left(\mathbf{u} \cdot \hat{\mathbf{e}}_j \right)^2 \le \|\mathbf{u}\|^2 . \tag{17.1}$$

Notice that if **u** happens to be in span $\{\hat{\mathbf{e}}_1, \ldots, \hat{\mathbf{e}}_k\}$, or if dim $S = N$, then (17.1) becomes an equality. In two and three dimensions that equality is actually the Pythagorean theorem, and in more than three dimensions it amounts to an abstract extension of that theorem.

18. (*Proof of Theorem 4.9.2*) We proved that if S admits a basis consisting of k vectors then S is k-dimensional. Prove the converse, that if S is k-dimensional then a basis for it must consist of k vectors.

CHAPTER 4 REVIEW

We began with the two-dimensional and three-dimensional "arrow vector" concept that was probably already familiar to you from an introductory course in physics in which the vectors denoted forces, velocities, and so on. For such vectors, vector addition ($\mathbf{u} + \mathbf{v}$), scalar multiplication ($\alpha\mathbf{u}$), a zero vector (**0**), a negative inverse $[-\mathbf{u} = (-1)\mathbf{u}]$, a norm ($\|\mathbf{u}\|$), a dot product

$$\mathbf{u} \cdot \mathbf{v} = \|\mathbf{u}\| \, \|\mathbf{v}\| \cos\theta, \tag{1}$$

and an angle $\theta = \cos^{-1} \left(\dfrac{\mathbf{u} \cdot \mathbf{v}}{\|\mathbf{u}\| \, \|\mathbf{v}\|} \right)$ between **u** and **v** are all defined.

From there, we generalized to abstract n-space, where $\mathbf{u} = (u_1, \ldots, u_n)$, by defining vector addition and so on in such a way that they agree with the corresponding arrow vector definitions when $n = 2$ and $n = 3$. For instance,

$$\mathbf{u} \cdot \mathbf{v} = \sum_{j=1}^{n} u_j v_j, \tag{2}$$

$$\|\mathbf{u}\| = \sqrt{\mathbf{u} \cdot \mathbf{u}} = \sqrt{\sum_{j=1}^{n} u_j^2}, \tag{3}$$

and

$$\theta = \cos^{-1} \frac{\mathbf{u} \cdot \mathbf{v}}{\|\mathbf{u}\| \, \|\mathbf{v}\|}. \tag{4}$$

From these definitions, we derived various *properties* such as

$$\mathbf{u} + \mathbf{v} = \mathbf{v} + \mathbf{u}, \qquad \text{(commutative)} \tag{5}$$
$$(\mathbf{u} + \mathbf{v}) + \mathbf{w} = \mathbf{u} + (\mathbf{v} + \mathbf{w}), \qquad \text{(associative)} \tag{6}$$

and so on, along with the following properties of the dot product and norm.

Dot Product.

Commutative:	$\mathbf{u} \cdot \mathbf{v} = \mathbf{v} \cdot \mathbf{u},$	(7a)
Nonnegative:	$\mathbf{u} \cdot \mathbf{u} > 0 \quad \text{for all } \mathbf{u} \ne \mathbf{0}$	
	$= 0 \quad \text{for } \mathbf{u} = \mathbf{0},$	(7b)

$$\textit{Linear}: \qquad (\alpha\mathbf{u} + \beta\mathbf{v})\cdot\mathbf{w} = \alpha(\mathbf{u}\cdot\mathbf{w}) + \beta(\mathbf{v}\cdot\mathbf{w}), \tag{7c}$$

Norm.

$$\textit{Scaling}: \qquad \|\alpha\mathbf{u}\| = |\alpha|\,\|\mathbf{u}\|, \tag{8a}$$

$$\textit{Nonnegative}: \qquad \|\mathbf{u}\| > 0 \quad \text{for all } \mathbf{u} \neq \mathbf{0} \tag{8b}$$

$$= 0 \quad \text{for } \mathbf{u} = \mathbf{0},$$

$$\textit{Triangle Inequality}: \qquad \|\mathbf{u} + \mathbf{v}\| \leq \|\mathbf{u}\| + \|\mathbf{v}\|. \tag{8c}$$

To complete the extension to generalized vector space, we reversed the cart and the horse by elevating these various properties to the level of axioms or requirements. That is, we let the fundamental objects, the vectors, be whatever we choose them to be, and then define addition and scalar multiplication operations, a zero vector, a negative inverse, a dot or "inner" product (if we wish), and a norm (if we wish), so that those axioms are satisfied. Our chief interest, in this text, is in the vector space $\mathbb{R}^n$ and in subspaces thereof.

We also introduced the concept of *span* and *linear dependence*, primarily so that we could develop the idea of the *expansion* of a given vector in a vector space $\mathcal{S}$ in terms of a set of *base vectors* for $\mathcal{S}$. We defined a set of vectors $\{\mathbf{e}_1, \ldots, \mathbf{e}_k\}$ to be a *basis* for $\mathcal{S}$ if each vector $\mathbf{u}$ in $\mathcal{S}$ can be expressed ("expanded") uniquely in the form $\mathbf{u} = \alpha_1\mathbf{e}_1 + \cdots + \alpha_k\mathbf{e}_k$, and proved that a set $\{\mathbf{e}_1, \ldots, \mathbf{e}_k\}$ is a basis for $\mathcal{S}$ if and only if it spans $\mathcal{S}$ and is LI (linearly independent). In particular, orthogonal bases are especially convenient because of the ease with which one can compute the expansion coefficients α_j. The result is

$$\mathbf{u} = \left(\frac{\mathbf{u}\cdot\mathbf{e}_1}{\mathbf{e}_1\cdot\mathbf{e}_1}\right)\mathbf{e}_1 + \cdots + \left(\frac{\mathbf{u}\cdot\mathbf{e}_k}{\mathbf{e}_k\cdot\mathbf{e}_k}\right)\mathbf{e}_k \tag{9}$$

if the basis is orthogonal, and

$$\mathbf{u} = \left(\mathbf{u}\cdot\hat{\mathbf{e}}_1\right)\hat{\mathbf{e}}_1 + \cdots + \left(\mathbf{u}\cdot\hat{\mathbf{e}}_k\right)\hat{\mathbf{e}}_k \tag{10}$$

if it is ON (orthonormal); (9) and (10) should be understood and remembered.

Finally, we studied the question of the *best approximation* of a given vector $\mathbf{u}$ in a vector space $\mathcal{S}$ in terms of an ON set $\{\hat{\mathbf{e}}_1, \ldots, \hat{\mathbf{e}}_N\}$ that falls short of being a basis for $\mathcal{S}$. We showed that the best approximation (i.e., the one that minimizes the norm of the error vector) is

$$\mathbf{u} \approx \sum_{j=1}^{N} \left(\mathbf{u}\cdot\hat{\mathbf{e}}_j\right)\hat{\mathbf{e}}_j \tag{11}$$

which, in geometrical language, is the *orthogonal projection of* $\mathbf{u}$ *onto the span of* $\hat{\mathbf{e}}_1, \ldots, \hat{\mathbf{e}}_N$.

n	Computing Time
5	0.0003 sec
10	10 sec
15	4×10^6 sec ≈ 40 days
20	7×10^{12} sec $\approx 210,000$ years
25	4×10^{19} sec $\approx 10^{12}$ years

MATRICES AND LINEAR ALGEBRAIC EQUATIONS

5.1 INTRODUCTION

We have already met matrices in Section 4.5.4, but they were introduced there only as a notational convenience for the implementation of Gauss elimination and Gauss-Jordan reduction. In the present chapter we focus on matrix theory itself, which theory will enable us to obtain additional important results regarding the solution of systems of linear algebraic equations.

One way to view matrix theory is to think in terms of a parallel with function theory. In our mathematical training, we first study numbers—the points on a real number axis. Then we study functions, which are mappings, or transformations, from one real axis to another. For instance, $f(x) = x^2$ maps the point $x = 3$, say, on an x axis to the point $f = 9$ on an f axis. Just as functions act upon numbers, we shall see that matrices act upon vectors and are mappings, not from one real axis to another but from one vector space to another. Having studied vectors, in Chapter 4, we can now turn our attention to matrices.

Historically, matrix theory did not become a part of undergraduate engineering science curricula until around 1960, when digital computers became widely available in academia.

5.2 MATRICES AND MATRIX ALGEBRA

5.2.1 Introduction.
A **matrix** is a rectangular array of quantities that are called the **elements** of the matrix. Normally, the elements will be real numbers, although they may occasionally be other objects such as differential operators or even matrices. For the present, let us consider the elements to be *real numbers*.

Specifically, any matrix **A** may be expressed as

$$
\mathbf{A} = \begin{bmatrix}
a_{11} & a_{12} & \cdots & a_{1n} \\
a_{21} & a_{22} & \cdots & a_{2n} \\
\vdots & \vdots & & \vdots \\
a_{m1} & a_{m2} & \cdots & a_{mn}
\end{bmatrix},
\tag{1}
$$

where the brackets (or, in some texts, parentheses) are used to emphasize that the entire array is to be regarded as a single entity. A horizontal line of elements is called a **row**, and a vertical line is called a **column**. Counting rows from the top and columns from the left, then

$$
a_{21} \quad a_{22} \quad \cdots \quad a_{2n} \qquad \text{and} \qquad
\begin{array}{c}
a_{13} \\
a_{23} \\
\vdots \\
a_{m3},
\end{array}
$$

say, are the second row and third column, respectively. Thus, we call the first subscript on a_{ij} the *row index* and the second subscript the *column index*.

We usually use boldface capital letters to denote matrices and lightface lowercase letters to denote their elements. The matrix **A** in (1) is seen to have m rows and n columns and is therefore said to be $m \times n$ (read "m by n"); we shall refer to this as the **form** of **A**. In some applications m and/or n may be infinite, but here we shall consider only matrices of finite size: $1 \leq m < \infty$, $1 \leq n < \infty$. Furthermore, m and n may, but need not, be equal. If **A** is small we may wish to dispense with the subscript notation for the elements. For example, if $m = n = 2$, we may prefer

$$
\mathbf{A} = \begin{bmatrix} a & b \\ c & d \end{bmatrix}
\quad \text{to} \quad
\mathbf{A} = \begin{bmatrix} a_{11} & a_{12} \\ a_{21} & a_{22} \end{bmatrix},
\tag{2}
$$

but if **A** is large this becomes inconvenient. The double-subscript notation employed in (1) is especially convenient for digital computer calculations.

In view of the subscript notation in (1), one also writes

$$
\mathbf{A} = \{a_{ij}\}
\tag{3}
$$

for short, where a_{ij} is called the ij **element** and $i = 1, \ldots, m$ and $j = 1, \ldots, n$. Some authors write $a_{i,j}$ in place of a_{ij} to avoid ambiguity—for example, to prevent us from reading a_{21} as a-sub-twenty-one, but we will omit the commas except when such ambiguity is not easily resolved from the context.

EXAMPLE 1
The matrices

$$
\mathbf{A} = \begin{bmatrix} 3 & -1 \\ 0 & 2 \\ 7 & 5 \end{bmatrix},
\quad
\mathbf{B} = \begin{bmatrix} 8 & -6 & 0 \\ 1 & 3 & 27 \\ 2 & 9 & 4 \end{bmatrix},
\quad \text{and} \quad
\mathbf{C} = [8, -7, 0, 4, 3]
$$

are 3×2, 3×3, and 1×5, respectively. If we denote $\mathbf{B} = \{b_{ij}\}$, then $b_{11} = 8$, $b_{12} = -6$, $b_{32} = 9$, and so on. **C** happens to be a single row and it seems best to separate the elements by commas, but the commas are not essential. ∎

Two matrices are said to be **equal** if they are of the same form and if their corresponding elements are equal. For instance, none of the matrices above are equal, but if $\mathbf{D} = [8, -7, 0, 4, 3]$ then, $\mathbf{D} = \mathbf{C}$.

One may be tempted to identify $\mathbf{C}$, above, as a 5-tuple *vector* rather than as a 1×5 *matrix*. That would be a bit premature since vectors are not merely objects; they have rules for vector addition and scalar multiplication defined, whereas our matrices are, thus far, just mathematical "objects." In fact, our next step is to define some arithmetic operations for matrices so that they may be manipulated in useful ways. For vectors we defined two arithmetic operations, vector addition and scalar multiplication. For matrices we define three: matrix addition, scalar multiplication, *and* the multiplication of matrices. (In contrast, we did *not* define the multiplication of vectors, such as $\mathbf{uv}$.)

5.2.2 Matrix addition.   If $\mathbf{A} = \{a_{ij}\}$ and $\mathbf{B} = \{b_{ij}\}$ are any two matrices of the same form, say $m \times n$, then their sum $\mathbf{A} + \mathbf{B}$ is defined as

$$\boxed{\mathbf{A} + \mathbf{B} \equiv \{a_{ij} + b_{ij}\}} \tag{4}$$

and is itself an $m \times n$ matrix. If $\mathbf{A}$ and $\mathbf{B}$ are of the same form, they are said to be *conformable for addition*; if they are not of the same form, then $\mathbf{A} + \mathbf{B}$ is *not defined*.

E X A M P L E 2
If

$$\mathbf{A} = \begin{bmatrix} 2 & 0 & -6 \\ 1 & 3 & 4 \end{bmatrix}, \quad \mathbf{B} = \begin{bmatrix} -1 & 2 & 0 \\ 15 & 6 & 3 \end{bmatrix}, \quad \text{and} \quad \mathbf{C} = \begin{bmatrix} 4 & 2 \\ -1 & 0 \end{bmatrix}, \tag{5}$$

then,

$$\mathbf{A} + \mathbf{B} = \begin{bmatrix} 1 & 2 & -6 \\ 16 & 9 & 7 \end{bmatrix}, \tag{6}$$

but $\mathbf{A} + \mathbf{C}$ and $\mathbf{B} + \mathbf{C}$ are not defined since $\mathbf{A}$ and $\mathbf{B}$ are 2×3 while $\mathbf{C}$ is 2×2. ∎

5.2.3 Scalar multiplication.   If $\mathbf{A} = \{a_{ij}\}$ is any $m \times n$ matrix and c is any scalar, their product is defined as

$$\boxed{c\mathbf{A} \equiv \{ca_{ij}\},} \tag{7}$$

and is itself an $m \times n$ matrix; we do not distinguish between $c\mathbf{A}$ and $\mathbf{A}c$. Furthermore, we denote

$$-\mathbf{A} \equiv (-1)\mathbf{A}. \tag{8}$$

In place of $\mathbf{A} + (-\mathbf{B})$, we simply write $\mathbf{A} - \mathbf{B}$, and call it the *difference* of $\mathbf{A}$ and $\mathbf{B}$, or $\mathbf{A}$ *minus* $\mathbf{B}$.

EXAMPLE 3

If **A** and **C** are the matrices in Example 1, then

$$3\mathbf{A} = \begin{bmatrix} 9 & -3 \\ 0 & 6 \\ 21 & 15 \end{bmatrix} \quad \text{and} \quad -\mathbf{C} = [-8, 7, 0, -4, -3]. \qquad \blacksquare$$

We shall list the important properties of matrix addition and scalar multiplication in a moment, but first let us define the so-called **zero matrix 0** to be any $m \times n$ matrix all the elements of which are zero. For example,

$$\begin{bmatrix} 0 & 0 & 0 \\ 0 & 0 & 0 \end{bmatrix} = \mathbf{0} \quad \text{and} \quad \begin{bmatrix} 0 \\ 0 \\ 0 \end{bmatrix} = \mathbf{0},$$

the first being 2×3, the second being 3×1.

THEOREM 5.2.1

Properties of Matrix Addition and Scalar Multiplication

If **A**, **B**, and **C** are $m \times n$ matrices, **0** is an $m \times n$ zero matrix, and α, β are any scalars, then

$$\mathbf{A} + \mathbf{B} = \mathbf{B} + \mathbf{A}, \qquad \text{(commutative)} \qquad \text{(9a)}$$
$$(\mathbf{A} + \mathbf{B}) + \mathbf{C} = \mathbf{A} + (\mathbf{B} + \mathbf{C}), \qquad \text{(associative)} \qquad \text{(9b)}$$
$$\mathbf{A} + \mathbf{0} = \mathbf{A}, \qquad \text{(9c)}$$
$$\mathbf{A} + (-\mathbf{A}) = \mathbf{0}, \qquad \text{(9d)}$$
$$\alpha(\beta\mathbf{A}) = (\alpha\beta)\mathbf{A}, \qquad \text{(associative)} \qquad \text{(9e)}$$
$$(\alpha + \beta)\mathbf{A} = \alpha\mathbf{A} + \beta\mathbf{A}, \qquad \text{(distributive)} \qquad \text{(9f)}$$
$$\alpha(\mathbf{A} + \mathbf{B}) = \alpha\mathbf{A} + \alpha\mathbf{B}, \qquad \text{(distributive)} \qquad \text{(9g)}$$
$$1\mathbf{A} = \mathbf{A}, \qquad \text{(9h)}$$
$$0\mathbf{A} = \mathbf{0}, \qquad \text{(9i)}$$
$$\alpha\mathbf{0} = \mathbf{0}. \qquad \text{(9j)}$$

The proofs of (9a)–(9j) follow from the foregoing definitions and are left for the exercises.

There are no surprises in (9); the usual rules of arithmetic are seen to apply. For the special case where **A** consists of a single row (or column), we see that the definitions of addition and scalar multiplication above are identical to those introduced in Section 4.3 for n-tuple vectors. Thus, we may properly refer to the matrices

$$\mathbf{A} = [a_{11}, \ldots, a_{1n}] \quad \text{and} \quad \mathbf{A} = \begin{bmatrix} a_{11} \\ \vdots \\ a_{n1} \end{bmatrix} \qquad (10)$$

as n-dimensional **row** and **column vectors**, respectively, the first formatted as a row, and the second as a column.

5.2.4 Matrix multiplication.   Judging from the rather natural way in which matrix addition and scalar multiplication are defined, by (4) and (7), one might well expect the multiplication of two matrices $\mathbf{A} = \{a_{ij}\}$ and $\mathbf{B} = \{b_{ij}\}$ to be defined only if $\mathbf{A}$ and $\mathbf{B}$ are of the same form, with the definition $\mathbf{AB} \equiv \{a_{ij}b_{ij}\}$. In fact, this definition will **not** be adopted. Instead, the standard definition of matrix multiplication is the one suggested by Cayley[1] and called the **Cayley product**: If $\mathbf{A} = \{a_{ij}\}$ is any $m \times n$ matrix and $\mathbf{B} = \{b_{ij}\}$ is any $n \times p$ matrix (so that the number n of columns of $\mathbf{A}$ is equal to the number n of rows of $\mathbf{B}$), then the product $\mathbf{AB}$ is defined as

$$\mathbf{AB} \equiv \left\{ \sum_{k=1}^{n} a_{ik}b_{kj} \right\}; \qquad (1 \le i \le m, \ 1 \le j \le p) \tag{11}$$

that is, if we denote $\mathbf{AB} \equiv \mathbf{C} = \{c_{ij}\}$, then

$$c_{ij} = \sum_{k=1}^{n} a_{ik}b_{kj}. \tag{12}$$

If the number of columns of $\mathbf{A}$ is equal to the number of rows of $\mathbf{B}$, then $\mathbf{A}$ and $\mathbf{B}$ are said to be *conformable for multiplication*; if not, the product $\mathbf{AB}$ is *not defined*. NOTE: The relative forms of $\mathbf{A}$, $\mathbf{B}$, and their product $\mathbf{C}$ are important and, as stated above, are as follows:

$$\begin{array}{ccccc} & & & & (13) \\ \mathbf{A} & \text{times} & \mathbf{B} & = & \mathbf{C}. \\ m \times n & & n \times p & & m \times p \end{array}$$

$$\text{equal}$$

The Cayley definition (11) may seem obscure and confusing so let us first illustrate how to implement it, and then let us explain why we adopt (11) rather than the more obvious and simpler definition $\mathbf{AB} \equiv \{a_{ij}b_{ij}\}$.

E X A M P L E   4
Suppose that

$$\mathbf{A} = \begin{bmatrix} 2 & 0 & -5 \\ 1 & 3 & 2 \\ 4 & 1 & -1 \\ 0 & 2 & 7 \end{bmatrix} \quad \text{and} \quad \mathbf{B} = \begin{bmatrix} 5 & 1 \\ -2 & 3 \\ 1 & 0 \end{bmatrix}.$$

Then $\mathbf{A}$ is 4×3 and $\mathbf{B}$ is 3×2. Since the number of columns of $\mathbf{A}$ is the same as the number of rows of $\mathbf{B}$ (namely, 3), the product $\mathbf{AB}$ is defined and, according to (13), will be 4×2. According to the definition (12),

---

[1]*Arthur Cayley* (1821–1895) produced around 200 papers in a 15-year period during which he was engaged in the practice of law. In 1863, he accepted a professorship of mathematics at Cambridge.

(14)

$$\mathbf{AB} = \begin{bmatrix} 2 & 0 & -5 \\ 1 & 3 & 2 \\ 4 & 1 & -1 \\ 0 & 2 & 7 \end{bmatrix} \begin{bmatrix} 5 & 1 \\ -2 & 3 \\ 1 & 0 \end{bmatrix} = \begin{bmatrix} 5 & 2 \\ 1 & 10 \\ 17 & 7 \\ 3 & 6 \end{bmatrix} = \mathbf{C}.$$

$$4 \times 3 \qquad\qquad 3 \times 2 \qquad\qquad 4 \times 2$$

To compute c_{32}, for example, (12) gives

$$c_{32} = \sum_{k=1}^{3} a_{3k} b_{k2} = (4)(1) + (1)(3) + (-1)(0) = 7,$$

3rd row   2nd column
of $\mathbf{A}$ of $\mathbf{B}$

that is, $a_{31}b_{12} + a_{32}b_{22} + a_{33}b_{32}$, as indicated by the arrows in (14). One more:

$$c_{11} = \sum_{k=1}^{3} a_{1k} b_{k1} = (2)(5) + (0)(-2) + (-5)(1) = 5.$$

We move across the rows of the first matrix and down the columns of the second.

COMMENT.   Observe that c_{32} is the *dot product* of the third row of $\mathbf{A}$, considered as a 3-tuple vector, with the second column of $\mathbf{B}$, also considered as a 3-tuple vector. More generally, if $\mathbf{AB} = \mathbf{C} = \{c_{ij}\}$, then c_{ij} is the dot product of the ith row of $\mathbf{A}$ with the jth column of $\mathbf{B}$. Thus, the number of elements in the rows of $\mathbf{A}$ (namely, the number of columns in $\mathbf{A}$) must equal the number of elements in the columns of $\mathbf{B}$ (namely, the number of rows of $\mathbf{B}$).   ■

EXAMPLE 5
Two more examples:

$$\overset{4 \times 3}{\begin{bmatrix} 5 & 0 & -1 \\ 2 & 3 & 4 \\ 1 & 0 & 6 \\ 0 & 0 & 1 \end{bmatrix}} \overset{3 \times 1}{\begin{bmatrix} 3 \\ 2 \\ 5 \end{bmatrix}} = \overset{4 \times 1}{\begin{bmatrix} (5)(3) + (0)(2) + (-1)(5) \\ (2)(3) + (3)(2) + (4)(5) \\ (1)(3) + (0)(2) + (6)(5) \\ (0)(3) + (0)(2) + (1)(5) \end{bmatrix}} = \begin{bmatrix} 10 \\ 32 \\ 33 \\ 5 \end{bmatrix},$$

and

$$\overset{2 \times 2}{\begin{bmatrix} -3 & 1 \\ 10 & 2 \end{bmatrix}} \overset{2 \times 2}{\begin{bmatrix} 1 & 0 \\ 2 & 4 \end{bmatrix}} = \overset{2 \times 2}{\begin{bmatrix} (-3)(1) + (1)(2) & (-3)(0) + (1)(4) \\ (10)(1) + (2)(2) & (10)(0) + (2)(4) \end{bmatrix}}$$

$$= \begin{bmatrix} -1 & 4 \\ 14 & 8 \end{bmatrix}.$$

■

It is extremely important to see that *matrix multiplication is not, in general, commutative; that is,*

$$\boxed{AB \neq BA,} \tag{15}$$

except in exceptional cases. For suppose that $A = m \times n$ and $B = n \times p$ (by the equal signs we mean that A is $m \times n$ and B is $n \times p$) so that AB is defined. However, $BA = (n \times p)(m \times n)$ is not even defined, let alone equal to AB, unless $p = m$. Assuming that that is the case, that $p = m$,

$$BA = (n \times m)(m \times n) = n \times n, \tag{16a}$$

whereas

$$AB = (m \times n)(n \times m) = m \times m. \tag{16b}$$

Comparing (16a) with (16b), we see that we must also have $m = n$ if AB and BA are to be of the same form and hence *possibly* equal. Thus, a necessary condition for AB to equal BA (i.e., for A and B to commute under multiplication) is that A and B both be $n \times n$.

EXAMPLE 6

If

$$A = \begin{bmatrix} 2 & 3 \\ 3 & 0 \end{bmatrix}, \quad B = \begin{bmatrix} -2 & 3 \\ 3 & -4 \end{bmatrix}, \quad C = \begin{bmatrix} 2 & 3 \\ 1 & 0 \end{bmatrix},$$

we find that A and B commute ($AB = BA$), but A and C do not ($AC \neq CA$), nor do B and C. ■

It follows from Example 6 that the condition that A and B must both be $n \times n$, for A and B to commute, is *necessary but not sufficient*. In view of the importance of which factor is first and which is second in a matrix product AB, we sometimes say that B is **pre-multiplied** by A, and A is **post-multiplied** by B so as to leave no doubt as to which factor is first and which is second.

5.2.5 Motivation for Cayley definition.   The lack of commutativity, in general, is a major setback so we must wonder why Cayley's definition has been adopted rather than the much simpler one that comes to mind, $AB \equiv \{a_{ij}b_{ij}\}$, which would surely yield commutativity since $BA = \{b_{ij}a_{ij}\} = \{a_{ij}b_{ij}\} = AB$ (the second equality following from the commutativity of the multiplication of ordinary numbers).   A sufficiently compelling reason to use Cayley's definition involves the application of matrix notation to systems of linear algebraic equations for it turns out that, with Cayley's definition of multiplication, the system of m linear algebraic equations

$$\begin{aligned} a_{11}x_1 + a_{12}x_2 + \cdots + a_{1n}x_n &= c_1, \\ a_{21}x_1 + a_{22}x_2 + \cdots + a_{2n}x_n &= c_2, \\ &\vdots \\ a_{m1}x_1 + a_{m2}x_2 + \cdots + a_{mn}x_n &= c_m \end{aligned} \tag{17}$$

in the n unknowns $x_1, \ldots, x_n$ is equivalent to the single compact matrix equation

$$\boxed{\mathbf{Ax} = \mathbf{c},} \tag{18}$$

where

$$\mathbf{A} = \begin{bmatrix} a_{11} & a_{12} & \cdots & a_{1n} \\ a_{21} & a_{22} & \cdots & a_{2n} \\ \vdots & \vdots & & \vdots \\ a_{m1} & a_{m2} & \cdots & a_{mn} \end{bmatrix}, \quad \mathbf{x} = \begin{bmatrix} x_1 \\ x_2 \\ \vdots \\ x_n \end{bmatrix}, \quad \text{and} \quad \mathbf{c} = \begin{bmatrix} c_1 \\ c_2 \\ \vdots \\ c_m \end{bmatrix}. \tag{19}$$

$\mathbf{A}$ is called the **coefficient matrix**, and $\mathbf{x}$ is the unknown; that is, its components are the unknowns $x_1, \ldots, x_n$. To verify the claimed equivalence work out the product $\mathbf{Ax}$, using the Cayley definition of multiplication, and set the result equal to $\mathbf{c}$. That step gives

$$\begin{bmatrix} a_{11}x_1 + \cdots + a_{1n}x_n \\ a_{21}x_1 + \cdots + a_{2n}x_n \\ \vdots \\ a_{m1}x_1 + \cdots + a_{mn}x_n \end{bmatrix} = \begin{bmatrix} c_1 \\ c_2 \\ \vdots \\ c_m \end{bmatrix}. \tag{20}$$
$$\quad\quad\quad m \times 1 \qquad\qquad\qquad m \times 1$$

These two $m \times 1$ matrices (or m-dimensional column vectors) will be equal if and only if each of their corresponding m elements (or components) are equal. Equating those corresponding elements gives us back the m scalar equations (17), so (17) and (18) are equivalent, as claimed. This important result is of course a consequence of the Cayley definition (11) and is a strong reason for adopting that definition of matrix multiplication.

5.2.6 Matrix multiplication, continued.   Any $n \times n$ matrix $\mathbf{A} = \{a_{ij}\}$ is said to be **square**, and of **order** n,[1] and the elements $a_{11}, a_{22}, \ldots, a_{nn}$ are said to lie on the **main diagonal** of $\mathbf{A}$—that is, the diagonal from the upper left corner to the lower right corner. Notice that for us to be able to multiply any matrix $\mathbf{A}$ with itself, $\mathbf{A}$ must be square. For suppose that $\mathbf{A}$ is $m \times n$; then we have

$$\begin{array}{cc} \mathbf{A} & \mathbf{A} \\ m \times \underline{n} & \underline{m} \times n \end{array}$$

and we need n (the number of columns in the first matrix) to equal m (the number of rows in the second) for the multiplication to be defined. If $\mathbf{A}$ *is* square and p is any positive integer, we define

$$\underbrace{\mathbf{AA}\cdots\mathbf{A}}_{p \text{ factors}} \equiv \mathbf{A}^p. \tag{21}$$

---

[1]Thus, we distinguish between form and order: The **form** of an $m \times n$ matrix is $m \times n$, the **order** of an $n \times n$ matrix is n.

The familiar laws of exponents,

$$\mathbf{A}^p \mathbf{A}^q = \mathbf{A}^{p+q}, \qquad \left(\mathbf{A}^p\right)^q = \mathbf{A}^{pq} \tag{22}$$

follow for any positive integers p and q.

If, in particular, the only nonzero elements of a square matrix lie on the main diagonal, then $\mathbf{A}$ is said to be a **diagonal matrix**. For example,

$$\mathbf{A} = \begin{bmatrix} 3 & 0 \\ 0 & -5 \end{bmatrix} \quad \text{and} \quad \mathbf{B} = \begin{bmatrix} 7 & 0 & 0 \\ 0 & -2 & 0 \\ 0 & 0 & 0 \end{bmatrix}$$

are diagonal, and any diagonal matrix of order n can be denoted as

$$\mathbf{D} = \begin{bmatrix} d_{11} & 0 & \cdots & 0 \\ 0 & d_{22} & & \\ \vdots & & \ddots & \vdots \\ 0 & & \cdots & d_{nn} \end{bmatrix}, \tag{23}$$

in which all of the off-diagonal elements are zero;[1] that is, $d_{ij} = 0$ if $i \neq j$. It is left for the exercises to show that

$$\mathbf{D}^p = \begin{bmatrix} d_{11}^p & 0 & \cdots & 0 \\ 0 & d_{22}^p & & \\ \vdots & & \ddots & \vdots \\ 0 & & \cdots & d_{nn}^p \end{bmatrix}, \tag{24}$$

for any positive integer p.

Furthermore, if $d_{11} = d_{22} = \cdots = d_{nn} = 1$, then $\mathbf{D}$ is called the **identity matrix I**. Thus,

$$\mathbf{I} = \begin{bmatrix} 1 & 0 & \cdots & 0 \\ 0 & 1 & & \\ \vdots & & \ddots & \vdots \\ 0 & & \cdots & 1 \end{bmatrix} = \left\{\delta_{ij}\right\}, \tag{25}$$

where δ_{ij} is the **Kronecker delta** symbol defined in Section 4.4.5, namely,

$$\delta_{ij} = \begin{cases} 1 & \text{if } i = j, \\ 0 & \text{if } i \neq j. \end{cases} \tag{26}$$

It is sometimes convenient to include a subscript n to indicate the order of $\mathbf{I}$. For example,

$$\mathbf{I}_2 = \begin{bmatrix} 1 & 0 \\ 0 & 1 \end{bmatrix} \quad \text{and} \quad \mathbf{I}_3 = \begin{bmatrix} 1 & 0 & 0 \\ 0 & 1 & 0 \\ 0 & 0 & 1 \end{bmatrix}.$$

---

[1]Of course, if the diagonal elements are zero as well, then it would be more informative to call the matrix the zero matrix $\mathbf{0}$.

The key property of the identity matrix is that if $\mathbf{A}$ is any square matrix of the same order as $\mathbf{I}$, then

$$\boxed{\mathbf{IA} = \mathbf{A} \qquad \text{and} \qquad \mathbf{AI} = \mathbf{A},} \tag{27}$$

proof of which is left for the exercises. In other words, $\mathbf{I}$ is the matrix analog of the number 1 in scalar arithmetic. From (27) we see that $\mathbf{IA} = \mathbf{AI}$ so that one case in which commutativity *does* hold is when one of the matrices is the identity matrix $\mathbf{I}$. We urge you to convince yourself of the truth of (27) by working out $\mathbf{IA}$ and $\mathbf{AI}$ for the case where $\mathbf{A}$ is an arbitrary 3×3 matrix $\{a_{ij}\}$, say.

Finally, it is convenient to extend our definition of $\mathbf{A}^p$ [recall (21)] to the case where $p = 0$. If $\mathbf{A}$ is any $n \times n$ matrix, we define

$$\mathbf{A}^0 \equiv \mathbf{I}, \tag{28}$$

where $\mathbf{I}$ is an $n \times n$ identity matrix.

Perhaps we should take a moment to mention that whereas the identity matrix $\mathbf{I}$ is necessarily square, the zero matrix $\mathbf{0}$ is simply $m \times n$, not necessarily square. It is readily verified that

$$\mathbf{0A} = \mathbf{0}, \tag{29a}$$
$$\mathbf{A0} = \mathbf{0} \tag{29b}$$

for any matrix $\mathbf{A}$ that is conformable for the indicated multiplication. In (29a), suppose that the $\mathbf{0}$ on the left is $m \times n$ and that $\mathbf{A}$ is $n \times p$. Then the $\mathbf{0}$ on the right is $m \times p$; that is, it is not necessarily of the same form as the one on the left. Similarly in (29b).

In view of the general failure of commutativity stated in (15), we may well wonder if any other familiar arithmetic rules fail to hold for the multiplication of matrices. The answer is yes; the following rules for real numbers a, b, c do *not* carry over to matrices:

1. $ab = ba$ (commutative).
2. If $ab = ac$ and $a \neq 0$, then $b = c$ (cancelation rule).
3. If $ab = 0$, then $a = 0$ and/or $b = 0$.
4. If $a^2 = 1$, then $a = +1$ or -1.

To add emphasis, we state these results as a theorem.

THEOREM 5.2.2
"Exceptional" Properties of Matrix Multiplication

(i) $\mathbf{AB} \neq \mathbf{BA}$ in general.
(ii) Even if $\mathbf{A} \neq \mathbf{0}$, $\mathbf{AB} = \mathbf{AC}$ does not imply that $\mathbf{B} = \mathbf{C}$.
(iii) $\mathbf{AB} = \mathbf{0}$ does not imply that $\mathbf{A} = \mathbf{0}$ and/or $\mathbf{B} = \mathbf{0}$.
(iv) $\mathbf{A}^2 = \mathbf{I}$ does not imply that $\mathbf{A} = +\mathbf{I}$ or $-\mathbf{I}$.

The first of these has already been discussed, and the others are discussed in the exercises. Theorem 5.2.2 notwithstanding, several important properties do carry over from the multiplication of real numbers to the multiplication of matrices:

THEOREM 5.2.3

"Ordinary" Properties of Matrix Multiplication

If α, β are scalars, and the matrices **A**, **B**, **C** are suitably conformable, then

$$(\alpha\mathbf{A})\mathbf{B} = \mathbf{A}(\alpha\mathbf{B}) = \alpha(\mathbf{AB}), \qquad \text{(associative)} \qquad (30\text{a})$$

$$\mathbf{A}(\mathbf{BC}) = (\mathbf{AB})\mathbf{C}, \qquad \text{(associative)} \qquad (30\text{b})$$

$$(\mathbf{A} + \mathbf{B})\mathbf{C} = \mathbf{AC} + \mathbf{BC}, \qquad \text{(distributive)} \qquad (30\text{c})$$

$$\mathbf{C}(\mathbf{A} + \mathbf{B}) = \mathbf{CA} + \mathbf{CB}, \qquad \text{(distributive)} \qquad (30\text{d})$$

$$\mathbf{A}(\alpha\mathbf{B} + \beta\mathbf{C}) = \alpha\mathbf{AB} + \beta\mathbf{AC}. \qquad \text{(linear)} \qquad (30\text{e})$$

Proof is left for the exercises.

5.2.7 Partitioning.   We will find it useful to introduce the **partitioning** of matrices. The idea is that any matrix **A** (larger than 1×1) may be partitioned into a number of smaller matrices called **blocks** by vertical lines that extend from top to bottom, and horizontal lines that extend from left to right.

E X A M P L E   7
For instance,

$$\mathbf{A} = \begin{bmatrix} 2 & 0 & -3 \\ 5 & 2 & 7 \\ 1 & 3 & 0 \\ 0 & 4 & 6 \end{bmatrix} = \left[\begin{array}{cc|c} 2 & 0 & -3 \\ 5 & 2 & 7 \\ \hline 1 & 3 & 0 \\ 0 & 4 & 6 \end{array} \right] = \begin{bmatrix} \mathbf{A}_{11} & \mathbf{A}_{12} \\ \mathbf{A}_{21} & \mathbf{A}_{22} \\ \mathbf{A}_{31} & \mathbf{A}_{32} \end{bmatrix}, \qquad (31)$$

where the blocks are

$$\mathbf{A}_{11} = \begin{bmatrix} 2 & 0 \\ 5 & 2 \end{bmatrix}, \quad \mathbf{A}_{12} = \begin{bmatrix} -3 \\ 7 \end{bmatrix}, \quad \mathbf{A}_{21} = [1, 3] \quad \mathbf{A}_{22} = [0],$$

and so on. Clearly, the partition is not unique. In the present example we could also have set

$$\mathbf{A} = \begin{bmatrix} 2 & 0 & -3 \\ 5 & 2 & 7 \\ 1 & 3 & 0 \\ 0 & 4 & 6 \end{bmatrix} = \left[\begin{array}{cc|c} 2 & 0 & -3 \\ 5 & 2 & 7 \\ 1 & 3 & 0 \\ 0 & 4 & 6 \end{array} \right] = [\mathbf{A}_{11}, \mathbf{A}_{12}], \qquad (32)$$

for instance. ∎

While the matrices used here as illustrations are kept small for convenience, those encountered in modern applications may be quite large, for example 600×800. Even with modern computers such large matrices create special computational problems, and it is often advantageous to work instead with a number of smaller matrices through the use of partitioning. Such advantages might well prove illusory, however, were it not for the fact that the usual matrix arithmetic *can* be carried out with partitioned matrices.

Specifically, if $\mathbf{A}$ and $\mathbf{B}$ are partitioned as

$$\mathbf{A} = \begin{bmatrix} \mathbf{A}_{11} & \mathbf{A}_{12} & \cdots & \mathbf{A}_{1n} \\ \vdots & \vdots & & \vdots \\ \mathbf{A}_{m1} & \mathbf{A}_{m2} & \cdots & \mathbf{A}_{mn} \end{bmatrix} \quad \text{and} \quad \mathbf{B} = \begin{bmatrix} \mathbf{B}_{11} & \mathbf{B}_{12} & \cdots & \mathbf{B}_{1q} \\ \vdots & \vdots & & \vdots \\ \mathbf{B}_{p1} & \mathbf{B}_{p2} & \cdots & \mathbf{B}_{pq} \end{bmatrix}, \tag{33}$$

then

$$\alpha\mathbf{A} = \begin{bmatrix} \alpha\mathbf{A}_{11} & \alpha\mathbf{A}_{12} & \cdots & \alpha\mathbf{A}_{1n} \\ \vdots & \vdots & & \vdots \\ \alpha\mathbf{A}_{m1} & \alpha\mathbf{A}_{m2} & \cdots & \alpha\mathbf{A}_{mn} \end{bmatrix}; \tag{34}$$

if $m = p$ and $n = q$ and each $\mathbf{A}_{ij}$ block is of the same form as the corresponding $\mathbf{B}_{ij}$ block, then

$$\mathbf{A} + \mathbf{B} = \begin{bmatrix} \mathbf{A}_{11} + \mathbf{B}_{11} & \mathbf{A}_{12} + \mathbf{B}_{12} & \cdots & \mathbf{A}_{1n} + \mathbf{B}_{1n} \\ \vdots & \vdots & & \vdots \\ \mathbf{A}_{m1} + \mathbf{B}_{m1} & \mathbf{A}_{m2} + \mathbf{B}_{m2} & \cdots & \mathbf{A}_{mn} + \mathbf{B}_{mn} \end{bmatrix}; \tag{35}$$

and if $n = p$ and we denote $\mathbf{A}\mathbf{B} = \mathbf{C}$, then

$$\mathbf{C}_{ij} = \sum_{k=1}^{n} \mathbf{A}_{ik}\mathbf{B}_{kj}, \tag{36}$$

provided that the number of columns in each $\mathbf{A}_{ik}$ is the same as the number of rows in the corresponding $\mathbf{B}_{kj}$, so that the products in (36) are defined.

Verification of these three claims, (34) to (36), is left for the exercises.

EXAMPLE 8

If

$$\mathbf{A} = \left[\begin{array}{cc|c} 2 & 4 & 1 \\ -1 & 3 & 0 \\ \hline 5 & 4 & 6 \end{array} \right] = \begin{bmatrix} \mathbf{A}_{11} & \mathbf{A}_{12} \\ \mathbf{A}_{21} & \mathbf{A}_{22} \end{bmatrix}, \tag{37}$$

and

$$\mathbf{B} = \left[\begin{array}{cc|c} 0 & 1 & 3 \\ 2 & -4 & -1 \\ \hline 5 & 8 & 2 \end{array} \right] = \begin{bmatrix} \mathbf{B}_{11} & \mathbf{B}_{12} \\ \mathbf{B}_{21} & \mathbf{B}_{22} \end{bmatrix}, \tag{38}$$

then

$$\mathbf{A}\mathbf{B} = \begin{bmatrix} \mathbf{A}_{11}\mathbf{B}_{11} + \mathbf{A}_{12}\mathbf{B}_{21} & \mathbf{A}_{11}\mathbf{B}_{12} + \mathbf{A}_{12}\mathbf{B}_{22} \\ \mathbf{A}_{21}\mathbf{B}_{11} + \mathbf{A}_{22}\mathbf{B}_{21} & \mathbf{A}_{21}\mathbf{B}_{12} + \mathbf{A}_{22}\mathbf{B}_{22} \end{bmatrix}.$$

Working out the elements,

$$\mathbf{A}_{11}\mathbf{B}_{11} + \mathbf{A}_{12}\mathbf{B}_{21} = \begin{bmatrix} 2 & 4 \\ -1 & 3 \end{bmatrix}\begin{bmatrix} 0 & 1 \\ 2 & -4 \end{bmatrix} + \begin{bmatrix} 1 \\ 0 \end{bmatrix}[5, 8]$$

$$= \begin{bmatrix} 8 & -14 \\ 6 & -13 \end{bmatrix} + \begin{bmatrix} 5 & 8 \\ 0 & 0 \end{bmatrix} = \begin{bmatrix} 13 & -6 \\ 6 & -13 \end{bmatrix},$$

$$\mathbf{A}_{11}\mathbf{B}_{12} + \mathbf{A}_{12}\mathbf{B}_{22} = \begin{bmatrix} 2 & 4 \\ -1 & 3 \end{bmatrix} \begin{bmatrix} 3 \\ -1 \end{bmatrix} + \begin{bmatrix} 1 \\ 0 \end{bmatrix} [2]$$

$$= \begin{bmatrix} 2 \\ -6 \end{bmatrix} + \begin{bmatrix} 2 \\ 0 \end{bmatrix} = \begin{bmatrix} 4 \\ -6 \end{bmatrix},$$

$$\mathbf{A}_{21}\mathbf{B}_{11} + \mathbf{A}_{22}\mathbf{B}_{21} = [5, 4] \begin{bmatrix} 0 & 1 \\ 2 & -4 \end{bmatrix} + [6][5, 8]$$

$$= [8, -11] + [30, 48] = [38, 37],$$

and

$$\mathbf{A}_{21}\mathbf{B}_{12} + \mathbf{A}_{22}\mathbf{B}_{22} = [5, 4] \begin{bmatrix} 3 \\ -1 \end{bmatrix} + [6][2] = [23],$$

so

$$\mathbf{AB} = \left[\begin{array}{cc|c} 13 & -6 & 4 \\ 6 & -13 & -6 \\ \hline 38 & 37 & 23 \end{array} \right] = \begin{bmatrix} 13 & -6 & 4 \\ 6 & -13 & -6 \\ 38 & 37 & 23 \end{bmatrix}, \tag{39}$$

which is the same result as obtained by the multiplication of the unpartitioned matrices $\mathbf{A}$ and $\mathbf{B}$.

COMMENT 1.   By no means do we claim that partitioning made the preceding calculation easier; our aim was simply to illustrate the idea of partitioning.

COMMENT 2.   The partition

$$\mathbf{B} = \left[\begin{array}{c|cc} 0 & 1 & 3 \\ \hline 2 & -4 & -1 \\ 5 & 8 & 2 \end{array} \right] = \begin{bmatrix} \mathbf{B}_{11} & \mathbf{B}_{12} \\ \mathbf{B}_{21} & \mathbf{B}_{22} \end{bmatrix}, \tag{40}$$

in place of (38), would also be conformable with (37) for the multiplication $\mathbf{AB}$ and would lead to the same result, (39). On the other hand, neither of the partitions

$$\mathbf{B} = \left[\begin{array}{c|cc} 0 & 1 & 3 \\ \hline 2 & -4 & -1 \\ 5 & 8 & 2 \end{array} \right] \quad \text{or} \quad \mathbf{B} = \left[\begin{array}{cc|c} 0 & 1 & 3 \\ 2 & -4 & -1 \\ 5 & 8 & 2 \end{array} \right] \tag{41}$$

would be conformable with (37) for working out the product $\mathbf{AB}$. For example, the term $\mathbf{A}_{11}\mathbf{B}_{11}$ would not be defined then since $\mathbf{A}_{11}$ has two columns whereas the $\mathbf{B}_{11}$'s in (41) have only one row. ■

It is often useful, as we shall see in Chapter 9, to partition matrices into rows or into columns. Specifically, consider the following three cases, and let $\mathbf{r}_j$ and $\mathbf{c}_j$ denote the jth row and jth column, respectively.

First, let $\mathbf{A}$ be $m \times n$ and let $\mathbf{B}$ be $n \times p$. Then, partitioning $\mathbf{A}$ into rows and $\mathbf{B}$

into columns, it follows from the definition of matrix multiplication that

$$
\mathbf{AB} = \left[\begin{array}{ccc} \underline{a_{11}} & \cdots & a_{1n} \\ a_{21} & \cdots & a_{2n} \\ & \vdots & \\ a_{m1} & \cdots & a_{mn} \end{array}\right] \left[\begin{array}{c|c|c|c} b_{11} & b_{12} & & b_{1p} \\ \vdots & \vdots & \cdots & \vdots \\ b_{n1} & b_{n2} & & b_{np} \end{array}\right]
$$

$$
= \left[\begin{array}{c} \mathbf{r}_1 \\ \vdots \\ \mathbf{r}_m \end{array}\right] \left[\mathbf{c}_1, \ldots, \mathbf{c}_p\right] = \left[\begin{array}{ccc} \mathbf{r}_1 \cdot \mathbf{c}_1 & \cdots & \mathbf{r}_1 \cdot \mathbf{c}_p \\ \vdots & & \vdots \\ \mathbf{r}_m \cdot \mathbf{c}_1 & \cdots & \mathbf{r}_m \cdot \mathbf{c}_p \end{array}\right]. \tag{42}
$$

That is, the i, j element of $\mathbf{AB}$ is the ith row of $\mathbf{A}$ (i.e., $\mathbf{r}_i$), dotted with the jth column of $\mathbf{B}$ (i.e., $\mathbf{c}_j$). This dot product result is a simple way to think about matrix multiplication and was mentioned first in the comment at the end of Example 4.

Second, if instead we partition $\mathbf{B}$ into columns but leave $\mathbf{A}$ intact, then

$$
\mathbf{AB} = \mathbf{A}\left[\mathbf{c}_1, \ldots, \mathbf{c}_p\right] = \left[\mathbf{Ac}_1, \ldots, \mathbf{Ac}_p\right]. \tag{43}
$$

That is, the jth column of $\mathbf{AB}$ is $\mathbf{A}$ times the jth column of $\mathbf{B}$.

Third, if $\mathbf{B}$ is merely a column vector $\mathbf{x}$ then if we partition $\mathbf{A}$ into columns we obtain

$$
\mathbf{Ax} = [\mathbf{c}_1, \ldots, \mathbf{c}_n] \left[\begin{array}{c} x_1 \\ \vdots \\ x_n \end{array}\right] = x_1\mathbf{c}_1 + \cdots + x_n\mathbf{c}_n, \tag{44}
$$

from which we see that $\mathbf{Ax}$ is within the span of the column vectors of $\mathbf{A}$.

5.2.8 Matrices as operators.   Finally, we can come back to the claim made in the introduction to this chapter (Section 5.1), namely, that matrices represent mappings from one vector space to another. For let $\mathbf{x}$ be an $n \times 1$ matrix, which is the same as an n-tuple vector, a vector in n-space (arranged in column format). And let $\mathbf{A}$ be a given $m \times n$ matrix. Then the product $\mathbf{Ax}$ is an $m \times 1$ matrix, namely, a vector in m-space (arranged in column format). Thus, an $m \times n$ matrix represents a mapping from n-space to m-space: the "input vector $\mathbf{x}$" is from n-space and the "output vector $\mathbf{Ax}$" is in m-space.

Closure.   We defined matrices and three arithmetic operations on matrices: addition, multiplication by a scalar, and multiplication. Subtraction is accounted for by addition and multiplication by a scalar: $\mathbf{A} - \mathbf{B} = \mathbf{A} + (-1)\mathbf{B}$. But no division operation is defined for matrices, just as it was not defined for vectors.

It is emphasized that multiplication is not commutative (i.e., $\mathbf{AB} \neq \mathbf{BA}$ in general). This "failure" and several others are listed in Theorem 5.2.2. These shortcomings of the Cayley definition of matrix multiplication are more than offset by the fact that it permits us to express a system of m linear algebraic equations in the n unknowns $x_1, \ldots, x_n$ in compact matrix form as $\mathbf{Ax} = \mathbf{c}$.

Finally, an $m \times n$ matrix represents a mapping: from n-space to m-space, that is, from $\mathbb{R}^n$ to $\mathbb{R}^m$.

Maple.   As mentioned in Section 4.5, the *Maple* system contains many linear algebra commands within the linalg package, among which **evalm** is especially useful. For instance, to evaluate $(\mathbf{AB})^2\mathbf{P}^3 - 5\mathbf{Q}$, where

$$\mathbf{A} = \begin{bmatrix} 1 & 2 \\ 3 & 0 \end{bmatrix}, \quad \mathbf{B} = \begin{bmatrix} 1 & -1 \\ 0 & 2 \end{bmatrix}, \quad \mathbf{P} = \begin{bmatrix} 1 & 1 \\ 1 & 1 \end{bmatrix}, \quad \mathbf{Q} = \begin{bmatrix} 1 & -1 \\ 2 & 3 \end{bmatrix},$$

first enter

```
with(linalg):
```

and return. Then enter

```
A:=array([[1,2],[3,0]]):
```

and return. (If you wish the **A** matrix to be printed, type a semicolon in place of the final colon.) Similarly, for

```
B:=array([[1,-1],[0,2]]):
P:=array([[1,1],[1,1]]):
Q:=array([[1,-1],[2,3]]):
```

Then enter

```
evalm((A&*B)^2 &*P^3-5*Q);
```

and return. The printed output is the result

$$\begin{bmatrix} 11 & 21 \\ 38 & 33 \end{bmatrix}$$

Note that matrix multiplication of **A** and **B** is denoted as A&*B (not $\mathbf{A} * \mathbf{B}$), exponentiation to a positive integer power is denoted with ^, and multiplication of a matrix by a scalar is denoted by *. If we now want $((\mathbf{AB})^2\mathbf{P}^3 - 5\mathbf{Q})^4$, we can use a percent sign to carry the matrix $\begin{bmatrix} 11 & 21 \\ 38 & 33 \end{bmatrix}$ forward. Thus, enter

```
evalm(%^4);
```

and return. The result is

$$\begin{bmatrix} 2389489 & 2592744 \\ 4691632 & 5105697 \end{bmatrix}$$

Alternative to the array format indicated above, we can use a matrix format. For instance,

```
A:=matrix(2,2,[1,2,3,0]):
```

is equivalent to the array format shown above, where the "2,2" denotes that **A** is a 2×2 matrix.

EXERCISES 5.2

1. Given the matrices

$$A = \begin{bmatrix} 0 & 3 \\ 2 & -5 \\ 1 & 10 \end{bmatrix}, \quad B = \begin{bmatrix} 5 & -1 \\ 0 & 2 \end{bmatrix},$$

$$x = \begin{bmatrix} 4 \\ 3 \end{bmatrix}, \quad y = [-1, 2],$$

work out whichever of the products AB, BA, Ax, xA, Bx, xB, yB, A^2, B^2, x^2, xy, and yx are defined.

2. Let A be 6×4, B be 4×4, C be 4×3, D and E be 3×1. Determine which of the following are defined, and for those that are, give the form of the resulting matrix.

(**a**) A^{10}
(**b**) B^{10}
(**c**) ABC
(**d**) $ABCD$
(**e**) $ACBD$
(**f**) $CD + E$
(**g**) $C(2D - E)$
(**h**) $AB + AC$
(**i**) $BC + CB$
(**j**) $3BA - 5CD$

3. Let

$$A = \begin{bmatrix} 1 & 0 & 0 \\ 0 & 2 & 0 \\ 0 & 0 & 3 \end{bmatrix} \quad \text{and} \quad B = \begin{bmatrix} 1 & 2 & 3 \\ 4 & 5 & 6 \\ 7 & 8 & 9 \end{bmatrix}.$$

Evaluate the products AB and BA.

4. Suppose that A is $n \times n$, x is $n \times 1$, and c is a scalar. Can we re-express $Ax = cx$ as $(A - c)x = 0$? Explain.

5. If A and B are square matrices of the same order, are the following correct? Explain.

(**a**) $(A + B)^2 = A^2 + 2AB + B^2$
(**b**) $(A + B)(A - B) = A^2 - B^2$
(**c**) $(AB)^2 = A^2 B^2$
(**d**) $(AB)^3 = A^3 B^3$

6. (**a**) If p is a positive integer, does A need to be square for A^p to be defined? Explain.

(**b**) Let A be $m \times n$ and B be $p \times q$. What restrictions, if any, must be satisfied by m, n, p, q if $(AB)^2$ is to exist (i.e., be defined)?

7. Expand (and simplify, if possible) each of the following [e.g., the "expanded" version of $(A + B)C$ would be $AC + BC$], assuming that all of the matrices are suitably conformable. Justify each step by citing the relevant equation number in Theorem 5.2.1 or 5.2.3.

(**a**) $(2A + B)(A + 2B)$
(**b**) $(A + B)C(D + E)$
(**c**) $(A + B)^3$
(**d**) $(A - 3I)(2A + I)$

8. Given $A = \begin{bmatrix} 1 & 1 \\ 1 & 1 \end{bmatrix}$, $B = \begin{bmatrix} 1 & 1 \\ 0 & 1 \end{bmatrix}$, $C = \begin{bmatrix} 0 & 3 \\ 0 & 0 \end{bmatrix}$,

and $D = \begin{bmatrix} 0 & 5 & 7 \\ 0 & 0 & 8 \\ 0 & 0 & 0 \end{bmatrix}$, evaluate each of the following.

(**a**) A^{100}
(**b**) B^{100}
(**c**) C^{100}
(**d**) D^{100}
(**e**) $(ABC)^3$
(**f**) $(CBA)^3$
(**g**) $B^4 C^4$ and $(BC)^4$
(**h**) $C^3 B^3$ and $(CB)^3$

9. If

$$S = \begin{bmatrix} k & 0 & \cdots & 0 \\ 0 & k & & \\ \vdots & & \ddots & \vdots \\ 0 & & \cdots & k \end{bmatrix}$$

is $n \times n$, and A is any $n \times n$ matrix, show that

$$AS = SA = kA.$$

What can be said if, instead, A is $m \times n$ ($m \neq n$)?

10. If, for any given vector $x = \begin{bmatrix} x_1 \\ \vdots \\ x_4 \end{bmatrix}$, the product Ax is the column vector given below, find A.

(**a**) $\begin{bmatrix} x_1 - 3x_4 \end{bmatrix}$
(**b**) $\begin{bmatrix} x_2 + x_3 - x_4 \\ x_1 + 5x_3 \end{bmatrix}$
(**c**) $\begin{bmatrix} x_1 + x_2 \\ x_2 + x_3 \\ x_3 + x_4 \end{bmatrix}$
(**d**) $\begin{bmatrix} 2x_1 - x_3 - x_4 \\ -2x_1 + x_2 \\ x_2 + x_4 \end{bmatrix}$
(**e**) $\begin{bmatrix} x_4 \\ x_3 \\ x_2 \\ x_1 \end{bmatrix}$
(**f**) $\begin{bmatrix} x_1 + 3x_4 \\ x_2 - x_4 \\ x_3 + x_4 \\ x_4 \\ x_3 - 2x_1 \end{bmatrix}$
(**g**) $\begin{bmatrix} x_1 \\ x_4 \end{bmatrix}$
(**h**) $\begin{bmatrix} x_1 + x_2 \\ x_3 + x_4 \end{bmatrix}$

11. Make up a specific pair of matrices, A and B, both nonzero, such that $AB = 0$, where

(**a**) A is 2×2 and B is 2×2
(**b**) A is 5×2 and B is 2×2
(**c**) A is 1×2 and B is 2×4
(**d**) A is 4×3 and B is 3×2

12. Given the partitioned matrices A and B, below, carry out the products A^2 and AB for those cases in which the partitioning is suitable, i.e., conformable. If the partitioning is not suitable, explain why it is not.

(a) $\mathbf{A} = \begin{bmatrix} 2 & 0 & -1 \\ 1 & -1 & 0 \\ 5 & 2 & 4 \end{bmatrix}$, $\mathbf{B} = \begin{bmatrix} 3 & 6 & 2 \\ 0 & 1 & 0 \\ 3 & -4 & 7 \end{bmatrix}$

(b) $\mathbf{A} = \begin{bmatrix} 2 & 0 & -1 \\ 1 & -1 & 0 \\ 5 & 2 & 4 \end{bmatrix}$, $\mathbf{B} = \begin{bmatrix} 3 & 6 & 2 \\ 0 & 1 & 0 \\ 3 & -4 & 7 \end{bmatrix}$

(c) $\mathbf{A} = \begin{bmatrix} 2 & 0 & -1 \\ 1 & -1 & 0 \\ 5 & 2 & 4 \end{bmatrix}$, $\mathbf{B} = \begin{bmatrix} 3 & 6 & 2 \\ 0 & 1 & 0 \\ 3 & -4 & 7 \end{bmatrix}$

13. Use the definitions of scalar multiplication of matrices, the addition of matrices, and the multiplication of matrices to prove (34)–(36).

14. (a) If two unpartitioned matrices are not conformable for addition, can they be rendered conformable by suitable partitioning? Explain.
    (b) Same as part (a), but for multiplication.

15. If there is some positive integer p such that $\mathbf{A}^p = \mathbf{0}$, then $\mathbf{A}$ is said to be **nilpotent** (mnemonic: "potentially nil"). A square matrix $\mathbf{A} = \{a_{ij}\}$ such that $a_{ij} = 0$ for all $i > j$ is said to be an **upper triangular matrix**; if $a_{ij} = 0$ for all $i < j$, then $\mathbf{A}$ is a **lower triangular matrix**. A matrix is said to be a **triangular** if it is either upper triangular or lower triangular.
    (a) Every upper triangular matrix with null main diagonal (so that $a_{ij} = 0$ for all $i \geq j$) is nilpotent. Verify this result for second-order, third-order, and fourth-order matrices.
    (b) In fact, show that *every* upper triangular matrix (of finite order) with null main diagonal is nilpotent. HINT: Use partitioning and induction.
    (c) Is every lower diagonal matrix with null main diagonal nilpotent? Explain.
    (d) If $\mathbf{A}^p = \mathbf{0}$, show that $(\mathbf{I}+\mathbf{A}+\mathbf{A}^2+\cdots+\mathbf{A}^{p-1})(\mathbf{I}-\mathbf{A}) = (\mathbf{I} - \mathbf{A})(\mathbf{I} + \mathbf{A} + \mathbf{A}^2 + \cdots + \mathbf{A}^{p-1}) = \mathbf{I}$.

16. If $\mathbf{A}^2 = \mathbf{I}$, then $\mathbf{A}$ is called **involutory.**
    (a) Show, using Theorems 5.2.1 to 5.2.3 and (27), that $\mathbf{A}$ is involutory if and only if

    $$(\mathbf{I} - \mathbf{A})(\mathbf{I} + \mathbf{A}) = \mathbf{0}.$$

    (b) Give an example of an involutory matrix other than $\mathbf{I}$ and $-\mathbf{I}$. Thus, observe that $\mathbf{A}^2 = \mathbf{I}$ does *not* imply that $\mathbf{A} = \pm\mathbf{I}$.

17. (a) Prove (9a) to (9c).
    (b) Prove (9d) to (9f).
    (c) Prove (9g) to (9i).

18. In Theorem 5.2.2, prove
    (a) (i)   (b) (ii)   (c) (iii)   (d) (iv)

19. (a) Verify (24).
    (b) Verify (27).
    (c) Verify (30a) and (30b).
    (d) Verify (30c) and (30d).
    (e) Show that (30e) follows from (30a)–(30d).

20. Show that the most general matrix that commutes with
    $\begin{bmatrix} 1 & 2 \\ 3 & 4 \end{bmatrix}$ is $\begin{bmatrix} \alpha - \beta & 2\beta/3 \\ \beta & \alpha \end{bmatrix}$, where α and β are arbitrary.

21. Given $\mathbf{A}$, find the most general matrix $\mathbf{B}$ such that $\mathbf{AB} = \mathbf{0}$.
    (a) $\mathbf{A} = \begin{bmatrix} 1 & 2 \\ 3 & 4 \end{bmatrix}$ (b) $\mathbf{A} = \begin{bmatrix} 2 & 3 \\ 0 & 0 \end{bmatrix}$
    (c) $\mathbf{A} = \begin{bmatrix} 0 & 0 \\ 5 & 0 \end{bmatrix}$ (d) $\mathbf{A} = \begin{bmatrix} 2 & 3 \\ 4 & 6 \end{bmatrix}$

22. (*Transition probability matrix*) Beginning with $2 apiece, professors A and B proceed to match coins at $1 per match. There arise five possible states:

    $$\begin{array}{ccccc} S_1 & S_2 & S_3 & S_4 & S_5 \\ 04 & 13 & 22 & 31 & 40 \end{array}$$

    In state S_2, for example, A has $1 and B has $3. If either player is bankrupted (A is bankrupt in state S_1, B in state S_5), the game is over. Let $p_{ij}^{(n)}$ be the n-step transition probability, i.e., the probability of changing from state S_i to state S_j in n matches. For $n = 1$ we see that

    $$\mathbf{P}^{(1)} = \{p_{ij}^{(1)}\} = \begin{bmatrix} 1 & 0 & 0 & 0 & 0 \\ \frac{1}{2} & 0 & \frac{1}{2} & 0 & 0 \\ 0 & \frac{1}{2} & 0 & \frac{1}{2} & 0 \\ 0 & 0 & \frac{1}{2} & 0 & \frac{1}{2} \\ 0 & 0 & 0 & 0 & 1 \end{bmatrix}.$$

    For example, beginning in S_2, say, one match will necessarily move us to S_1 or S_3, with 50% probability in each case; thus $p_{21}^{(1)} = p_{23}^{(1)} = \frac{1}{2}$. Precisely $1 will change hands so that $p_{22}^{(1)} = p_{24}^{(1)} = p_{25}^{(1)} = 0$. Further, once in S_1 (or S_5) we remain there (according to the rules), so $p_{11}^{(1)} = 1$, $p_{12}^{(1)} = p_{13}^{(1)} = p_{14}^{(1)} = p_{15}^{(1)} = 0$. Show, by any convincing arguments or discussion, that

    $$p_{ij}^{(2)} = \sum_{k=1}^{5} p_{ik}^{(1)} p_{kj}^{(1)}, \quad p_{ij}^{(3)} = \sum_{k=1}^{5} p_{ik}^{(2)} p_{kj}^{(1)}, \quad \text{etc.}$$

    or, in matrix notation, $\mathbf{P}^{(2)} = [\mathbf{P}^{(1)}]^2$, $\mathbf{P}^{(3)} = [\mathbf{P}^{(1)}]^3$, etc. Use this result to determine $\mathbf{P}^{(2)}$ and $\mathbf{P}^{(3)}$. What is the probability that A is bankrupt after (at most) three matches if

A starts with $2? $3? $1? NOTE: $\mathbf{P}^{(1)}$ is an example of a **Markov matrix**. We meet Markov matrices again in Chapter 9.

23. Let

$$\mathbf{A} = \begin{bmatrix} 2 & -1 \\ 3 & 0 \\ 1 & 4 \end{bmatrix}, \qquad \mathbf{B} = \begin{bmatrix} 5 & 3 & 25 \\ 2 & 0.1 & -6 \end{bmatrix},$$

$$\mathbf{C} = \begin{bmatrix} 9 & 1 & -1 \\ 2 & 0 & 7 \\ 0 & 4 & 6 \end{bmatrix}, \qquad \mathbf{F} = \begin{bmatrix} 6 & 5 \\ 4 & 1 \end{bmatrix}.$$

NOTE: The letters D, I, and O are "protected," in *Maple*,

for other purposes. The problem: use *Maple* to evaluate

(a) $(\mathbf{AB})^3 + 5\mathbf{C}^2$ **(b)** $6\mathbf{AB} - 9\mathbf{C}$

(c) $6\mathbf{AFB} - 2\mathbf{C}^3$ **(d)** $4\mathbf{BAF}$

(e) $(\mathbf{BCAF})^4$ **(f)** $2\mathbf{CA} + 37.3\mathbf{A}$

(g) $(\mathbf{CAB})^2$ **(h)** $0.73\mathbf{BA} + 1.6\mathbf{F}^6$

(i) $\mathbf{F}^3\mathbf{BA} - 6\mathbf{F}^4$ **(j)** $(\mathbf{FBC}^2\mathbf{A})^3$

24. If $\mathbf{A}$ is $m \times n$ and $\mathbf{B}$ is $n \times p$, show that the product matrix $\mathbf{AB}$ is a mapping from $\mathbb{R}^p$ to $\mathbb{R}^m$.

25. (*Matrices as elements of vector space*) Show that the set of all $m \times n$ matrices, subject to the definitions (4) and (7) of matrix addition and scalar multiplication, constitute a vector space (Section 4.8). Assume that all scalars are real.

5.3   THE TRANSPOSE MATRIX

We continue the development of Section 5.2 by introducing the "transpose" of a matrix. Given any $m \times n$ matrix $\mathbf{A}$, we define the **transpose** of $\mathbf{A}$, denoted as $\mathbf{A}^T$ and read as "A-transpose," as

$$\mathbf{A}^T = \begin{bmatrix} a_{11} & a_{12} & \cdots & a_{1n} \\ a_{21} & a_{22} & \cdots & a_{2n} \\ \vdots & \vdots & & \vdots \\ a_{m1} & a_{m2} & \cdots & a_{mn} \end{bmatrix}^T \equiv \begin{bmatrix} a_{11} & a_{21} & \cdots & a_{m1} \\ a_{12} & a_{22} & \cdots & a_{m2} \\ \vdots & \vdots & & \vdots \\ a_{1n} & a_{2n} & \cdots & a_{mn} \end{bmatrix}. \qquad (1)$$

That is, the first row of $\mathbf{A}$ becomes the first column of $\mathbf{A}^T$, the second row of $\mathbf{A}$ becomes the second column of $\mathbf{A}^T$, and so on. Equivalently, the first column of $\mathbf{A}$ becomes the first row of $\mathbf{A}^T$, and so on. That is, if we denote the i, j elements of $\mathbf{A}$ and $\mathbf{A}^T$ as a_{ij} and a_{ij}^T, respectively, then

$$a_{ij}^T = a_{ji}. \qquad (2)$$

Be clear that $\mathbf{A}^T$ is not $\mathbf{A}$ to the Tth power; it is the transpose of $\mathbf{A}$.

EXAMPLE 1

If

$$\mathbf{A} = \begin{bmatrix} 2 & 0 & 1 \\ -1 & 3 & 5 \\ 4 & 6 & 7 \end{bmatrix}, \quad \mathbf{B} = \begin{bmatrix} 2 \\ 6 \\ 7 \end{bmatrix}, \quad \text{and} \quad \mathbf{C} = [1, -8, 9],$$

then

$$\mathbf{A}^T = \begin{bmatrix} 2 & -1 & 4 \\ 0 & 3 & 6 \\ 1 & 5 & 7 \end{bmatrix}, \quad \mathbf{B}^T = [2, 6, 7], \quad \text{and} \quad \mathbf{C}^T = \begin{bmatrix} 1 \\ -8 \\ 9 \end{bmatrix}. \qquad \blacksquare$$

THEOREM 5.3.1
Properties of the Transpose

$$\left(\mathbf{A}^{\mathsf{T}}\right)^{\mathsf{T}} = \mathbf{A}, \tag{3a}$$

$$(\mathbf{A} + \mathbf{B})^{\mathsf{T}} = \mathbf{A}^{\mathsf{T}} + \mathbf{B}^{\mathsf{T}}, \tag{3b}$$

$$(\alpha\mathbf{A})^{\mathsf{T}} = \alpha\mathbf{A}^{\mathsf{T}}, \tag{3c}$$

$$\boxed{(\mathbf{AB})^{\mathsf{T}} = \mathbf{B}^{\mathsf{T}}\mathbf{A}^{\mathsf{T}},} \tag{3d}$$

where it is assumed in (3b) that $\mathbf{A}$ and $\mathbf{B}$ are conformable for addition, and in (3d) that they are conformable for multiplication.

Proof.   Proof of (3a)–(3c) is left for the exercises. To prove (3d), let $\mathbf{AB} \equiv \mathbf{C} = \{c_{ij}\}$. By the definition of matrix multiplication,

$$c_{ij} = \sum_{k=1}^{n} a_{ik}b_{kj}. \tag{4}$$

Thus,

$$c_{ij}^{\mathsf{T}} = c_{ji} = \sum_{k=1}^{n} a_{jk}b_{ki} = \sum_{k=1}^{n} b_{ki}a_{jk} = \sum_{k=1}^{n} b_{ik}^{\mathsf{T}}a_{kj}^{\mathsf{T}}. \tag{5}$$

Having returned, at the end of (5), to the pattern $(\ )_{ij} = \sum(\ )_{ik}(\ )_{kj}$ as in (4), we can conclude from (5) that

$$\mathbf{C}^{\mathsf{T}} = \mathbf{B}^{\mathsf{T}}\mathbf{A}^{\mathsf{T}} \quad \text{or} \quad (\mathbf{AB})^{\mathsf{T}} = \mathbf{B}^{\mathsf{T}}\mathbf{A}^{\mathsf{T}},$$

as was to be proved. Understand that the third equality in (5) is *not* equivalent to the matrix statement $\mathbf{AB} = \mathbf{BA}$ which, we recall from Section 5.2, is generally untrue. It is simply the scalar statement that $a_{jk}b_{ki} = b_{ki}a_{jk}$, which is true because the multiplication of scalars is commutative [e.g., $(2)(3) = (3)(2)$]. ∎

The striking feature of (3d) is the reversal in the order: $(\mathbf{AB})^{\mathsf{T}}$ on the left, $\mathbf{B}^{\mathsf{T}}\mathbf{A}^{\mathsf{T}}$ on the right. Notice how (3d) checks "dimensionally":

$$[(m \times n)(n \times p)]^{\mathsf{T}} = (n \times p)^{\mathsf{T}}(m \times n)^{\mathsf{T}}$$
$$(m \times p)^{\mathsf{T}} = (p \times n)(n \times m) \tag{6}$$
$$(p \times m) = (p \times m).$$

Naturally, (6) does not *prove* (3d), but it provides a useful check, just as we check the *physical* units (such as force, mass, length, and time) of an equation to be sure that they are consistent. We've framed (3d) because it will be useful to us and, unlike (3a)–(3c), it is not an obvious result.

EXAMPLE 2

Let us verify (3d) for the case where $\mathbf{A} = \begin{bmatrix} 4 & 2 & -5 \\ 0 & 1 & 3 \end{bmatrix}$ and $\mathbf{B} = \begin{bmatrix} 6 \\ 0 \\ -1 \end{bmatrix}$. Then

$$\mathbf{AB} = \begin{bmatrix} 4 & 2 & -5 \\ 0 & 1 & 3 \end{bmatrix} \begin{bmatrix} 6 \\ 0 \\ -1 \end{bmatrix} = \begin{bmatrix} 29 \\ -3 \end{bmatrix}, \quad \text{so} \quad (\mathbf{AB})^{\mathrm{T}} = [29, -3]$$

and

$$\mathbf{B}^{\mathrm{T}}\mathbf{A}^{\mathrm{T}} = [6, 0, -1] \begin{bmatrix} 4 & 0 \\ 2 & 1 \\ -5 & 3 \end{bmatrix} = [29, -3],$$

in agreement with (3d). ∎

Furthermore, it follows from (3d) that

$$(\mathbf{ABC})^{\mathrm{T}} = \mathbf{C}^{\mathrm{T}}\mathbf{B}^{\mathrm{T}}\mathbf{A}^{\mathrm{T}}, \qquad (\mathbf{ABCD})^{\mathrm{T}} = \mathbf{D}^{\mathrm{T}}\mathbf{C}^{\mathrm{T}}\mathbf{B}^{\mathrm{T}}\mathbf{A}^{\mathrm{T}}, \tag{7}$$

and so on (Exercise 2).

Using lowercase boldface letters for column vectors from now on, let

$$\mathbf{x} = \begin{bmatrix} x_1 \\ \vdots \\ x_n \end{bmatrix} \quad \text{and} \quad \mathbf{y} = \begin{bmatrix} y_1 \\ \vdots \\ y_n \end{bmatrix}.$$

Then the standard dot product

$$\mathbf{x} \cdot \mathbf{y} = x_1 y_1 + x_2 y_2 + \cdots + x_n y_n = \sum_{j=1}^{n} x_j y_j$$

can be expressed compactly, in matrix language, as

$$\boxed{\mathbf{x} \cdot \mathbf{y} = \mathbf{x}^{\mathrm{T}}\mathbf{y}} \tag{8}$$

or, equivalently, as $\mathbf{y}^{\mathrm{T}}\mathbf{x}$, although not as $\mathbf{x}\mathbf{y}^{\mathrm{T}}$ or $\mathbf{y}\mathbf{x}^{\mathrm{T}}$, which expressions represent $n \times n$ matrices!

EXAMPLE 3

If $\mathbf{x} = \begin{bmatrix} 3 \\ 1 \end{bmatrix}$ and $\mathbf{y} = \begin{bmatrix} 5 \\ 2 \end{bmatrix}$, say, then

$$\mathbf{x} \cdot \mathbf{y} = \mathbf{x}^{\mathrm{T}}\mathbf{y} = [3, 1] \begin{bmatrix} 5 \\ 2 \end{bmatrix} = (3)(5) + (1)(2) = 17,$$

or

$$\mathbf{x} \cdot \mathbf{y} = \mathbf{y}^{\mathrm{T}}\mathbf{x} = [5, 2] \begin{bmatrix} 3 \\ 1 \end{bmatrix} = (5)(3) + (2)(1) = 17,$$

whereas

$$\mathbf{x}\mathbf{y}^{\mathrm{T}} = \begin{bmatrix} 3 \\ 1 \end{bmatrix} [5, 2] = \begin{bmatrix} 15 & 6 \\ 5 & 2 \end{bmatrix} \neq \mathbf{x} \cdot \mathbf{y},$$

and

$$\mathbf{y}\mathbf{x}^{\mathrm{T}} = \begin{bmatrix} 5 \\ 2 \end{bmatrix} [3, 1] = \begin{bmatrix} 15 & 5 \\ 6 & 2 \end{bmatrix} \neq \mathbf{x} \cdot \mathbf{y}. \qquad \blacksquare$$

Finally, two more definitions: if

$$\mathbf{A}^{\mathrm{T}} = \mathbf{A} \qquad (9)$$

we say that $\mathbf{A}$ is **symmetric**, and if

$$\mathbf{A}^{\mathrm{T}} = -\mathbf{A}, \qquad (10)$$

we say that it is **skew-symmetric** (or antisymmetric). For either of these properties to apply $\mathbf{A}$ must be square, since otherwise $\mathbf{A}^{\mathrm{T}}$ and $\mathbf{A}$ would be of different form. And for $\mathbf{A}$ to be skew-symmetric all of its diagonal elements must be zero, since (10) implies that $a_{ji} = -a_{ij}$ or, if we set $i = j$, $a_{ii} = -a_{ii}$; thus $2a_{ii} = 0$, so $a_{ii} = 0$ for each i.

EXAMPLE 4
The matrices
$$\begin{bmatrix} 1 & 3 & -2 \\ 3 & 0 & 5 \\ -2 & 5 & 4 \end{bmatrix} \quad \text{and} \quad \begin{bmatrix} 0 & 2 & 1 \\ -2 & 0 & -4 \\ -1 & 4 & 0 \end{bmatrix}$$
are symmetric and skew-symmetric, respectively. $\qquad \blacksquare$

If, in applications, the elements in the matrices were dictated by chance, then purely symmetric or skew-symmetric matrices would be rare. However, we shall see that symmetric matrices arise frequently, and that their symmetry is often a consequence of fundamental physical principles, rather than chance.

Closure.   The key points in this section are the defining of the transpose of any matrix, and the results $(\mathbf{AB})^{\mathrm{T}} = \mathbf{B}^{\mathrm{T}}\mathbf{A}^{\mathrm{T}}$ and $\mathbf{x} \cdot \mathbf{y} = \mathbf{x}^{\mathrm{T}}\mathbf{y}$ (or $\mathbf{y}^{\mathrm{T}}\mathbf{x}$). Note that the transpose notation is sometimes used to save space. For instance, it takes less vertical space on the page to write $\mathbf{x}^{\mathrm{T}} = [7, 2]$ than $\mathbf{x} = \begin{bmatrix} 7 \\ 2 \end{bmatrix}$.

Maple.   The relevant *Maple* command, to take the transpose of a matrix $\mathbf{A}$ is **transpose**, within the linalg package. For instance, to obtain the transpose of

$$\mathbf{A} = \begin{bmatrix} 1 & 2 & 3 \\ 4 & 5 & 6 \end{bmatrix} \text{ enter}$$

$$\texttt{with(linalg):}$$

to access the transpose command. Enter

$$A:=array([[1,2,3],[4,5,6]]);$$

and return, then enter

$$transpose(A);$$

and return. The output is

$$\begin{bmatrix} 1 & 4 \\ 2 & 5 \\ 3 & 6 \end{bmatrix}$$

EXERCISES 5.3

1. (**a**) If $\mathbf{x} = [3, -3]^T$ and $\mathbf{y} = [1, 2]^T$, work out $\mathbf{x}^T\mathbf{y}$ and $\mathbf{xy}^T$.

(**b**) If $\mathbf{x} = [4, -1, 0]^T$ and $\mathbf{y} = [1, 2, 3]^T$, work out $\mathbf{x}^T\mathbf{y}$ and $\mathbf{xy}^T$.

(**c**) If $\mathbf{x} = [0, 4, -2, 1]^T$ and $\mathbf{y} = [3, 0, 1, -2]^T$, work out $\mathbf{x}^T\mathbf{y}$ and $\mathbf{xy}^T$.

2. Show that (7) follows from (3d).

3. Recall that in general $\mathbf{AB} \neq \mathbf{BA}$, and that a *necessary* (but not sufficient) condition for equality to hold is that both $\mathbf{A}$ and $\mathbf{B}$ be square and of the same order. Perhaps a *sufficient* condition is that $\mathbf{A}$ and $\mathbf{B}$ both be of the same order and *symmetric*. Prove or disprove this hypothesis.

4. Verify $(\mathbf{ABC})^T = \mathbf{C}^T\mathbf{B}^T\mathbf{A}^T$ directly, for

(**a**) $\mathbf{A} = \begin{bmatrix} 5 & -2 \\ 0 & 1 \\ 1 & 3 \end{bmatrix}$, $\mathbf{B} = \begin{bmatrix} 0 \\ 7 \end{bmatrix}$, $\mathbf{C} = [3, 1, 2, 9]$

(**b**) $\mathbf{A} = \begin{bmatrix} 1 & 2 \\ 5 & 4 \end{bmatrix}$, $\mathbf{B} = \begin{bmatrix} 0 & 0 \\ 1 & 1 \end{bmatrix}$,

$\mathbf{C} = \begin{bmatrix} 4 & -1 & 5 \\ 0 & 3 & 0 \end{bmatrix}$

(**c**) $\mathbf{A} = [5, 3, 0]$, $\mathbf{B} = \begin{bmatrix} -2 & 1 \\ 6 & 4 \\ 5 & 3 \end{bmatrix}$, $\mathbf{C} = \begin{bmatrix} 4 \\ 8 \end{bmatrix}$

(**d**) $\mathbf{A} = \begin{bmatrix} 1 & 0 & 2 \\ 0 & 1 & 1 \end{bmatrix}$, $\mathbf{B} = \begin{bmatrix} 3 & 4 \\ 1 & 2 \\ 0 & 1 \end{bmatrix}$, $\mathbf{C} = \begin{bmatrix} 6 \\ 0 \end{bmatrix}$

5. Prove the properties (3a), (3b), and (3c).

6. Even if a (square) matrix is neither symmetric nor skew-symmetric it can be decomposed as the sum of a symmetric matrix and a skew-symmetric matrix. Specifically, writing

$$\mathbf{A} = \frac{1}{2}(\mathbf{A} + \mathbf{A}^T) + \frac{1}{2}(\mathbf{A} - \mathbf{A}^T) \tag{6.1}$$

$$\equiv \quad \mathbf{A}_1 \quad + \quad \mathbf{A}_2,$$

show that $\mathbf{A}_1$ is symmetric and $\mathbf{A}_2$ is skew-symmetric.

7. Decompose the given matrix as the sum of two matrices, one symmetric and one skew symmetric, as indicated in Exercise 6.

(**a**) $\begin{bmatrix} 3 & 2 \\ 1 & -5 \end{bmatrix}$ (**b**) $\begin{bmatrix} 1 & 2 \\ 3 & 4 \end{bmatrix}$

(**c**) $\begin{bmatrix} 1 & 0 \\ 2 & 0 \end{bmatrix}$ (**d**) $\begin{bmatrix} 8 & -2 \\ 4 & 0 \end{bmatrix}$

(**e**) $\begin{bmatrix} 9 & 8 & 7 \\ 6 & 5 & 4 \\ 3 & 2 & 1 \end{bmatrix}$ (**f**) $\begin{bmatrix} 1 & 0 & 1 \\ 2 & -1 & 0 \\ 3 & 0 & 6 \end{bmatrix}$

8. (*Quadratic forms*) The quadratic function $ax^2 + by^2 + cxy$ is said to be a **quadratic form** in x and y, $ax^2 + by^2 + cz^2 + dxy + exz + fyz$ is a quadratic form in x, y, z, and so on. Every quadratic form in the n variables $x_1, \ldots, x_n$ can be expressed in matrix notation as $\mathbf{x}^T\mathbf{Ax}$, where $\mathbf{A}$ is a symmetric $n \times n$ matrix. For each given quadratic form, determine the $\mathbf{A}$ matrix.

(**a**) $6x_1^2 + x_2^2 - 8x_1x_2$ in x_1, x_2

(**b**) $x_1^2 - 3x_2^2 + 6x_1x_2$ in x_1, x_2

(**c**) $4x_1^2 + x_2^2 - x_3^2 + 8x_1x_2 + 3x_1x_3 - 2x_2x_3$ in x_1, x_2, x_3

(**d**) $x_1^2 - 4x_3^2 + 2x_1x_3 - 10x_2x_3$ in x_1, x_2, x_3

(**e**) $3x_2^2 + x_3^2 - 6x_1x_3$ in x_1, x_2, x_3

(**f**) $x_1^2 + x_2^2 + x_4^2 + x_1x_3 + x_1x_4 + x_2x_3$ in x_1, x_2, x_3, x_4

9. Show that if $\mathbf{A}$ is an $m \times n$ matrix, then $\mathbf{AA}^T$ is symmetric. HINT: There is often an inclination to work out a problem like this using "brute force," i.e., by actually writing out the $\mathbf{A}$ and $\mathbf{A}^T$ matrices, multiplying them, and examining the resulting matrix to see if it is symmetric. Whenever possible, we advise against such an approach. In this problem, for example, we wish to show that $\mathbf{C}^T = \mathbf{C}$, where $\mathbf{C}$ is short for $\mathbf{AA}^T$; i.e., we wish to show that $(\mathbf{AA}^T)^T = \mathbf{AA}^T$, and this can be done (in one or two short lines) using the properties stated in Theorem 5.3.1.

<u>10</u>. Prove that *the product* **AB** *need not be symmetric, even if* **A** *and* **B** *are both symmetric and of the same order.*

11. Let

$$\mathbf{A} = \begin{bmatrix} 4 & 1 & 2 \\ 0 & 5 & 7 \end{bmatrix}, \quad \mathbf{B} = \begin{bmatrix} 1 & -4 & 2 \\ 8 & 1 & 4 \end{bmatrix}.$$

Use *Maple* to evaluate

(a) $\mathbf{AB}^\mathrm{T}$ (b) $\mathbf{BA}^\mathrm{T}$

(c) $(\mathbf{AB}^\mathrm{T})^5$ (d) $(\mathbf{BA}^\mathrm{T})^\mathrm{T}$

(e) $(\mathbf{B}^\mathrm{T}\mathbf{A})^8$ (f) $2\mathbf{A}^\mathrm{T} - 7.3\mathbf{B}^\mathrm{T}$

(g) $(\mathbf{A}^\mathrm{T})^\mathrm{T}$ (h) $(\mathbf{B}^\mathrm{T})^\mathrm{T}$

5.4 DETERMINANTS

In this section we introduce a scalar quantity associated with every square matrix, the so-called "determinant" of the matrix. We denote the **determinant** of an $n \times n$ matrix $\mathbf{A} = \{a_{ij}\}$ as

$$\det\mathbf{A} = \begin{vmatrix} a_{11} & a_{12} & \cdots & a_{1n} \\ a_{21} & a_{22} & \cdots & a_{2n} \\ \vdots & \vdots & & \vdots \\ a_{n1} & a_{n2} & \cdots & a_{nn} \end{vmatrix}, \tag{1}$$

that is, with straight line braces instead of square brackets. Determinants will be prominent in Chapter 6 in connection with the linear dependence or independence of sets of functions, especially with sets of solutions of linear homogeneous differential equations. They also play a key role in the theory of systems of linear algebraic equations, as discussed in Section 5.7.

The determinant of an $n \times n$ matrix $\mathbf{A} = \{a_{ij}\}$ is defined by the **cofactor expansion**

$$\det\mathbf{A} \equiv \sum_{j=1}^{n} a_{jk} A_{jk} = \sum_{k=1}^{n} a_{jk} A_{jk}, \tag{2}$$

where *the summation is carried out on j for any fixed value of k* $(1 \le k \le n)$ *or on k for any fixed value of j* $(1 \le j \le n)$; A_{jk} is called the **cofactor** of the a_{jk} element and is defined as

$$A_{jk} \equiv (-1)^{j+k} M_{jk}, \tag{3}$$

where M_{jk} is called the **minor** of a_{jk}, namely, the determinant of the $(n-1) \times (n-1)$ matrix that survives when the row and column containing a_{jk} (the jth row and the kth column) are struck out. Probably, (2) and (3) will appear confusing, but they should be clear by the end of Example 1.

For example, if

$$\mathbf{A} = \begin{bmatrix} 4 & 7 & -2 \\ 0 & 3 & 2 \\ 1 & 5 & 6 \end{bmatrix},$$

then the minor determinants M_{11}, M_{12}, and M_{23}, say, are

$$M_{11} = \begin{vmatrix} 3 & 2 \\ 5 & 6 \end{vmatrix}, \quad M_{12} = \begin{vmatrix} 0 & 2 \\ 1 & 6 \end{vmatrix}, \quad \text{and} \quad M_{23} = \begin{vmatrix} 4 & 7 \\ 1 & 5 \end{vmatrix}.$$

For instance, M_{23} is the determinant of the 2×2 matrix that survives when the row and column containing $a_{23} = 2$ (namely, the second row and the third column) are deleted from $\mathbf{A}$.

Thus, if $\mathbf{A}$ is $n \times n$, then the right-hand side of (2) is a linear combination of n determinants, each of which is $(n-1) \times (n-1)$. Each of these, in turn, may be expressed as a linear combination of $(n-2) \times (n-2)$ determinants, and so on, until we have a (perhaps large) number of 1×1 determinants. Thus, the definition (2) is logically incomplete until we define a 1×1 determinant, which we do as follows:

$$\det \begin{bmatrix} a_{11} \end{bmatrix} = \begin{vmatrix} a_{11} \end{vmatrix} \equiv a_{11}. \tag{4}$$

That is, the determinant of a 1×1 matrix $\begin{bmatrix} a_{11} \end{bmatrix}$ is simply a_{11} itself. CAUTION: In the present context, the braces around a_{11}, in the middle member of (4), denote determinant, not absolute value. For instance, $\det[-6] = |-6| = -6$.

Using (2), (3), and (4), let us work out the determinant of any 2×2 matrix. Recalling that we can sum on j with k fixed, or vice versa, let us sum on k with $j = 1$, say. Then

$$\det \begin{bmatrix} a_{11} & a_{12} \\ a_{21} & a_{22} \end{bmatrix} = \begin{vmatrix} a_{11} & a_{12} \\ a_{21} & a_{22} \end{vmatrix} = \sum_{k=1}^{2} a_{1k} A_{1k}$$

$$= a_{11}A_{11} + a_{12}A_{12} = a_{11}(-1)^{1+1}M_{11} + a_{12}(-1)^{1+2}M_{12}$$

$$= a_{11}(+1)|a_{22}| + a_{12}(-1)|a_{21}|$$

$$= a_{11}a_{22} - a_{12}a_{21}.$$

Thus, for a 2×2 matrix,

$$\det \begin{bmatrix} a_{11} & a_{12} \\ a_{21} & a_{22} \end{bmatrix} = a_{11}a_{22} - a_{12}a_{21}, \tag{5}$$

which result may be familiar to you from earlier studies. Observe that the $(-1)^{j+k}$ in (3) is simply $+1$ if $j + k$ is even, and -1 if $j + k$ is odd.

EXAMPLE 1
Let

$$\mathbf{A} = \begin{bmatrix} 0 & 2 & -1 \\ 4 & 3 & 5 \\ 2 & 0 & -4 \end{bmatrix}. \tag{6}$$

Using (2) with $j = 1$, say,

$$\det \mathbf{A} = \sum_{k=1}^{3} a_{1k} A_{1k} = a_{11}A_{11} + a_{12}A_{12} + a_{13}A_{13}$$

$$= a_{11}(+1)M_{11} + a_{12}(-1)M_{12} + a_{13}(+1)M_{13}$$
$$= a_{11}M_{11} - a_{12}M_{12} + a_{13}M_{13}$$
$$= (0)\begin{vmatrix} 3 & 5 \\ 0 & -4 \end{vmatrix} - (2)\begin{vmatrix} 4 & 5 \\ 2 & -4 \end{vmatrix} + (-1)\begin{vmatrix} 4 & 3 \\ 2 & 0 \end{vmatrix}$$
$$= 0 - (2)(-16 - 10) + (-1)(0 - 6) = 58, \tag{7}$$

where the next to last equality follows from (5). This particular choice ($j = 1$, summing on k) is said to be the *cofactor expansion about the first row* since the $a_{11}A_{11} + a_{12}A_{12} + a_{13}A_{13}$ in (7) is the sum of the first-row elements a_{11}, a_{12}, a_{13} multiplied by their cofactors.

According to (2), we can expand about *any* row or column and the result is always the same. The truth of that claim is by no means obvious, and we will not prove it. Rather, we merely illustrate that the same answer is indeed obtained, in this example, if we expand about other rows or columns.

Expansion about second row [i.e., set $j = 2$ in (2) and sum on k]:

$$\det \mathbf{A} = \sum_{k=1}^{3} a_{2k} A_{2k} = a_{21} A_{21} + a_{22} A_{22} + a_{23} A_{23}$$

$$= a_{21}(-1)M_{21} + a_{22}(+1)M_{22} + a_{23}(-1)M_{23}$$

$$= -(4)\begin{vmatrix} 2 & -1 \\ 0 & -4 \end{vmatrix} + (3)\begin{vmatrix} 0 & -1 \\ 2 & -4 \end{vmatrix} - (5)\begin{vmatrix} 0 & 2 \\ 2 & 0 \end{vmatrix}$$

$$= -(4)(-8) + (3)(2) - (5)(-4) = 58$$

again.

Expansion about third column ($k = 3$, sum on j):

$$\det \mathbf{A} = \sum_{j=1}^{3} a_{j3} A_{j3} = a_{13} A_{13} + a_{23} A_{23} + a_{33} A_{33}$$

$$= a_{13}(+1)M_{13} + a_{23}(-1)M_{23} + a_{33}(+1)M_{33}$$

$$= (-1)\begin{vmatrix} 4 & 3 \\ 2 & 0 \end{vmatrix} - (5)\begin{vmatrix} 0 & 2 \\ 2 & 0 \end{vmatrix} + (-4)\begin{vmatrix} 0 & 2 \\ 4 & 3 \end{vmatrix}$$

$$= (-1)(-6) - (5)(-4) + (-4)(-8) = 58$$

once more. ∎

Since we may expand about any row or column, it is convenient in hand calculations to choose the row or column with the most zeros in it since those terms in the expansion drop out.

Notice carefully that for large n the cofactor expansion process is exceedingly laborious. Even if $n = 10$, say, which is still quite modest, (2) gives a linear combination of ten 9×9 determinants. In turn, each of these ten is evaluated as a linear combination of nine 8×8 determinants, and, so on! Let us see just how serious this predicament is. For estimating purposes, let us count each multiplication, addition, and subtraction as one "calculation." It can be shown (Exercise 17a) that the number of calculations $N(n)$ required in the evaluation (by cofactor expansion) of an $n \times n$ determinant is [1]

$$N(n) \sim e\,n! \tag{8}$$

as $n \to \infty$, where $e \approx 2.718$ is the base of the natural logarithm, and $n!$ is n factorial. If each calculation takes approximately one microsecond, then some time estimates are as follows.

n	Computing Time
5	0.0003 sec
10	10 sec
15	4×10^6 sec ≈ 40 days
20	7×10^{12} sec $\approx 210{,}000$ years
25	4×10^{19} sec $\approx 10^{12}$ years

It is interesting that faster computers offer no hope. For instance, even a computer that is a million times as fast would still take around 10^6 years to evaluate a 25×25 determinant, and scientific calculations can easily involve determinants that are 250×250 or larger.

It is tempting to conclude that "determinants are worthless," but let us see if we can come up with a more efficient algorithm than the cofactor expansion. A logical starting point is to first determine the various properties of determinants so that we can use them to design a better algorithm.

First, we need to introduce the idea of a "triangular" matrix. A square matrix $\mathbf{A} = \{a_{ij}\}$ is said to be **upper triangular** if $a_{ij} = 0$ for all $i > j$ and **lower triangular** if $a_{ij} = 0$ for all $i < j$. That is,

$$\begin{bmatrix} a_{11} & a_{12} & \cdots & a_{1n} \\ 0 & a_{22} & & \\ \vdots & & \ddots & \vdots \\ 0 & & \cdots & a_{nn} \end{bmatrix} \quad \text{and} \quad \begin{bmatrix} a_{11} & 0 & \cdots & 0 \\ a_{21} & a_{22} & & \\ \vdots & & \ddots & \vdots \\ a_{n1} & & \cdots & a_{nn} \end{bmatrix} \tag{9}$$

are upper triangular and lower triangular, respectively. If a matrix is upper triangular or lower triangular it is said to be **triangular**.

Here are the three properties that we will need.

PROPERTIES OF DETERMINANTS

D1. If any row (or column) of $\mathbf{A}$ is modified by adding α times the corresponding elements of another row (or column) to it, yielding a new matrix $\mathbf{B}$, then

---

[1] Isn't (8) a remarkable formula? Namely, how does e, the base of the natural logarithm, find its way into this story?

det $\mathbf{B} = \det \mathbf{A}$.

Symbolically: $\mathbf{r}_j \to \mathbf{r}_j + \alpha \mathbf{r}_k$

D2. If any two rows (or columns) of $\mathbf{A}$ are interchanged, yielding a new matrix $\mathbf{B}$, then $\det \mathbf{B} = -\det \mathbf{A}$.

Symbolically: $\mathbf{r}_j \leftrightarrow \mathbf{r}_k$

D3. If $\mathbf{A}$ is triangular, then $\det \mathbf{A}$ is simply the product of the diagonal elements, $\det \mathbf{A} = a_{11}a_{22} \cdots a_{nn}$.

Of these, D3 is easily proved. For consider the general upper triangular matrix in (9). Doing a cofactor expansion about the first column gives a_{11} times an $(n-1) \times (n-1)$ minor determinant, which is again upper triangular. Expanding the latter about its first column gives a_{22} times an $(n-2) \times (n-2)$ minor determinant, which is again upper triangular. Repeating the process leads to $\det \mathbf{A} = a_{11}a_{22} \cdots a_{nn}$. Similarly for the general lower triangular matrix in (9), except that in that case we expand about the first row, repeatedly, rather than the first column.

Let us illustrate how to use the properties D1–D3 instead of the cofactor expansion.

E X A M P L E   2

Consider the $\mathbf{A}$ matrix of Example 1 again.

$$\det \mathbf{A} = \begin{vmatrix} 0 & 2 & -1 \\ 4 & 3 & 5 \\ 2 & 0 & -4 \end{vmatrix} = -\begin{vmatrix} 2 & 0 & -4 \\ 4 & 3 & 5 \\ 0 & 2 & -1 \end{vmatrix}$$

$$= -\begin{vmatrix} 2 & 0 & -4 \\ 0 & 3 & 13 \\ 0 & 2 & -1 \end{vmatrix} = -\begin{vmatrix} 2 & 0 & -4 \\ 0 & 3 & 13 \\ 0 & 0 & -\frac{29}{3} \end{vmatrix} = -(2)(3)\left(-\frac{29}{3}\right) = 58,$$

as obtained in Example 1. In the second equality we interchanged the first and third rows ($\mathbf{r}_1 \leftrightarrow \mathbf{r}_3$), thereby changing the sign of the determinant (D2) so we compensated by putting the minus sign out in front. In the third equality we modified the second row by adding -2 times the first row to it ($\mathbf{r}_2 \to \mathbf{r}_2 - 2\mathbf{r}_1$), which step left the determinant unchanged (D1). In the fourth equality we modified the third row by adding $-\frac{2}{3}$ times the second row to it ($\mathbf{r}_3 \to \mathbf{r}_3 - \frac{2}{3}\mathbf{r}_2$), which step left the determinant unchanged (D1). Since those steps produced a triangular matrix, we could then use D3. ■

The point, then, is to use some combination of D1 and D2 steps to reduce the determinant to triangular form, in which case it is evaluated easily by D3. Of course the method is quite similar to Gauss elimination, described in Section 4.5. For instance, compare D1 and D2 with the first and third elementary equation operations listed in Section 4.5. For reference purposes, we will call the method illustrated in Example 2 the method of **triangularization**.

It is hard to tell, from the 3×3 calculation in Example 2, whether the method is more efficient than the cofactor expansion. However, in Exercise 17b it is shown that using triangularization the number of calculations $N(n)$ is

$$N(n) \sim \frac{2n^3}{3} \tag{10}$$

as $n \to \infty$. Again assuming one microsecond per calculation, (10) gives a computing time of around 0.005 second for $n = 20$ and 0.01 second for $n = 25$, compared with 210,000 years and 10^{12} years, respectively! [Comparing (8) and (10), we can see how much faster $n!$ grows than n^3, as $n \to \infty$. Factorials are powerful indeed.]

The upshot is that except for small hand calculations we should avoid the cofactor expansion, and should use triangularization instead.

Although properties D1–D3 suffice for the efficient calculation of determinants, other properties are sometimes useful as well, and are listed below.

ADDITIONAL PROPERTIES OF DETERMINANTS

D4. If all the elements of any row or column are zero, then $\det \mathbf{A} = 0$.

D5. If any two rows or columns are proportional to each other, then $\det \mathbf{A} = 0$.

D6. If any row (column) is a linear combination of other rows (columns), then $\det \mathbf{A} = 0$.

D7. If all the elements of any row or column are scaled by α, yielding a new matrix $\mathbf{B}$, then $\det \mathbf{B} = \alpha \det \mathbf{A}$.

D8. $\det(\alpha \mathbf{A}) = \alpha^n \det \mathbf{A}$.

D9. If any one row (or column) $\mathbf{a}$ of $\mathbf{A}$ is separated as $\mathbf{a} = \mathbf{b} + \mathbf{c}$, then

$$\det \mathbf{A}\big|_{\mathbf{a}} = \det \mathbf{A}\big|_{\mathbf{b}} + \det \mathbf{A}\big|_{\mathbf{c}},$$

where $\mathbf{A}\big|_{\mathbf{a}}$ denotes the $\mathbf{A}$ matrix with $\mathbf{a}$ intact, $\mathbf{A}\big|_{\mathbf{b}}$ denotes the $\mathbf{A}$ matrix with $\mathbf{b}$ in place of $\mathbf{a}$, and similarly for $\mathbf{A}\big|_{\mathbf{c}}$. For example,

$$\begin{vmatrix} 6+2 & -3+1 & 5+4 \\ 3 & 0 & 2 \\ 1 & -6 & 7 \end{vmatrix} = \begin{vmatrix} 6 & -3 & 5 \\ 3 & 0 & 2 \\ 1 & -6 & 7 \end{vmatrix} + \begin{vmatrix} 2 & 1 & 4 \\ 3 & 0 & 2 \\ 1 & -6 & 7 \end{vmatrix}.$$

D10. The determinant of $\mathbf{A}$ and its transpose are equal,

$$\det(\mathbf{A}^T) = \det \mathbf{A}.$$

D11. In general,

$$\det(\mathbf{A} + \mathbf{B}) \neq \det \mathbf{A} + \det \mathbf{B}.$$

D12. The determinant of a product equals the product of their determinants,

$$\boxed{\det(\mathbf{AB}) = (\det \mathbf{A})(\det \mathbf{B}).} \tag{11}$$

These properties are not independent of each other. For example, D5 follows from D1 and D4, and D4 follows from D6. Keep in mind that det( ) is *not linear*. That is, if α, β are scalars and $\mathbf{A}$, $\mathbf{B}$ are $n \times n$ matrices, then

$$\det(\alpha \mathbf{A} + \beta \mathbf{B}) \neq \alpha \det \mathbf{A} + \beta \det \mathbf{B}, \tag{12}$$

in general. For instance, if $\beta = 0$ is $\det(\alpha \mathbf{A}) = \alpha \det \mathbf{A}$? No, according to D8 it is $\alpha^n \det \mathbf{A}$. Or, with $\alpha = \beta = 1$, is $\det(\mathbf{A} + \mathbf{B}) = \det \mathbf{A} + \det \mathbf{B}$? Not in general, according to D11. This result may come as a surprise since we are studying "linear algebra." The truth of D12 is also surprising in view of the complexity of the matrix multiplication $\mathbf{AB}$ and the determinant calculation on the left-hand side, and the simplicity of the outcome expressed on the right-hand side. This result was proved by Cauchy in 1815.[1]

Closure.   Every $n \times n$ matrix $\mathbf{A}$ has a scalar value associated with it called its determinant and denoted as $\det \mathbf{A}$. Though $\det \mathbf{A}$ is defined, traditionally, by the cofactor expansion (2), we find that that formula is useless, computationally, unless n is quite small. Thus, we studied various properties of the determinant and put forward a computational algorithm called triangularization, based upon properties D1–D3, that is astonishingly efficient compared to the cofactor expansion.

Maple.   One can evaluate determinants using the **det** command. For instance, to evaluate the determinant of the matrix $\mathbf{A}$ given by (6), enter

```
with(linalg):
```

to access the det command. Then enter

```
det(array([[0,2,-1],[4,3,5],[2,0,-4]]));
```

and return. The output is 58. Alternatively, the sequence

```
with(linalg):
A:=array([[0,2,-1],[4,3,5],[2,0,-4]]):
det(A);
```

gives the same result.

---

[1]*Augustin-Louis Cauchy* (1789–1857) is among the great mathematicians. Unlike his contemporary Gauss, who published little of his work, Cauchy published more than 700 papers. Among the subjects on which he worked were determinants, ordinary and partial differential equations, complex variable theory, and the wave theory of light.

EXERCISES 5.4

1. In (5) we evaluated the determinant of a general 2×2 matrix using a cofactor expansion about the first row. Evaluate it again, using a cofactor expansion about the second row instead, then about the first column, and then about the second column, showing that the answer is the same in each case.

2. Evaluate each, using a cofactor expansion about the first and last rows, and also about the last column.

(a) $\begin{vmatrix} 1 & 2 & 3 \\ 3 & 2 & 1 \\ 1 & 1 & 1 \end{vmatrix}$ (b) $\begin{vmatrix} 2 & -3 & 0 \\ 1 & 4 & 2 \\ -6 & 1 & 5 \end{vmatrix}$

(c) $\begin{vmatrix} -4 & 1 & 0 \\ 3 & 2 & 0 \\ 1 & 5 & 7 \end{vmatrix}$ (d) $\begin{vmatrix} 3 & 3 & 12 \\ 0 & 6 & -1 \\ 4 & 0 & 0 \end{vmatrix}$

(e) $\begin{vmatrix} 1 & 2 & 3 \\ 2 & 3 & 4 \\ 3 & 4 & 5 \end{vmatrix}$ (f) $\begin{vmatrix} -5 & 2 & 1 & 0 \\ 4 & 0 & 3 & 0 \\ 2 & 4 & 7 & 1 \\ 0 & 0 & -2 & 6 \end{vmatrix}$

(g) $\begin{vmatrix} 2 & 0 & 1 & 0 \\ 0 & 3 & 1 & -1 \\ 0 & 4 & 5 & 0 \\ 1 & 2 & 3 & 6 \end{vmatrix}$ (h) $\begin{vmatrix} 0 & 1 & 2 & 0 \\ 3 & -1 & 1 & 4 \\ 5 & 6 & -7 & 1 \\ 0 & 2 & 1 & 0 \end{vmatrix}$

(i) $\begin{vmatrix} a & 0 & 0 & 0 \\ 0 & b & c & 0 \\ 0 & d & e & 0 \\ 0 & 0 & 0 & f \end{vmatrix}$ (j) $\begin{vmatrix} a & b & c & 0 \\ d & e & f & 0 \\ g & h & i & 0 \\ 0 & 0 & 0 & k \end{vmatrix}$

3. (a)–(j) Same as Exercise 2, but expanding about the second row, and about the first column.

4. (a)–(h) Same as Exercise 2, but using the method of triangularization.

5. (a)–(h) Same as Exercise 2, but using *Maple*.

6. Evaluate, by hand, showing your steps or logic. You may use any of the properties D1–D12.

(a) $\begin{vmatrix} 1 & 2 & 3 & 4 \\ 2 & 3 & 4 & 5 \\ 3 & 4 & 5 & 6 \\ 0 & 1 & -3 & 5 \end{vmatrix}$ (b) $\begin{vmatrix} 1 & 2 & 3 & 4 \\ 5 & 6 & 7 & 8 \\ 9 & 10 & 11 & 12 \\ 13 & 14 & 15 & 0 \end{vmatrix}$

(c) $\begin{vmatrix} 0 & 0 & a \\ 0 & b & c \\ d & e & f \end{vmatrix}$ (d) $\begin{vmatrix} a & b & c \\ d & e & 0 \\ f & 0 & 0 \end{vmatrix}$

7. A mnemonic device often put forward for evaluating 2×2 and 3×3 determinants is as shown below.

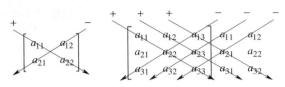

In other words, the determinants are the sums of the indicated products, with each product carrying the indicated sign. For example, in the 2×2 case this device gives

$$\begin{vmatrix} a_{11} & a_{12} \\ a_{21} & a_{22} \end{vmatrix} = +(a_{11}a_{22}) - (a_{21}a_{12}),$$

which does agree with (5). We now state the problem: Write out the mnemonic result for the 3×3 case, and verify (by cofactor expansion) that it is correct. CAUTION: This device does *not* hold, in general, for $n \times n$ determinants if $n \geq 4$.

8. Let an $n \times n$ matrix $\mathbf{A} = \{a_{ij}\}$ be diagonal. Show that

$$\det \mathbf{A} = a_{11}a_{22} \cdots a_{nn}. \qquad (8.1)$$

9. (a) Suppose that an $n \times n$ matrix $\mathbf{A}$ can be partitioned into the **block-diagonal** form

$$\mathbf{A} = \begin{bmatrix} \boxed{\mathbf{A}_1} & \mathbf{0} & \cdots & \mathbf{0} \\ \mathbf{0} & \boxed{\mathbf{A}_2} & \cdots & \\ \vdots & & \ddots & \vdots \\ \mathbf{0} & & \cdots & \boxed{\mathbf{A}_m} \end{bmatrix},$$

where $\mathbf{A}_1, \ldots, \mathbf{A}_m$ are all square, although not necessarily all of the same order. Show that

$$\det \mathbf{A} = (\det \mathbf{A}_1)(\det \mathbf{A}_2) \cdots (\det \mathbf{A}_m). \qquad (9.1)$$

This result may be regarded as a generalization of (8.1), above, wherein $\mathbf{A}_1, \ldots, \mathbf{A}_m$ were all 1×1's.

(b) Does (9.1) still hold if the elements above the m blocks are nonzero? Explain.

(c) Does (9.1) still hold if the elements below the m blocks are nonzero? Explain.

(d) To which determinants, in Exercise 2, can (9.1) be applied? In each of those cases use (9.1) to evaluate $\det \mathbf{A}$.

10. Deduce, from property D12, that if $\mathbf{A}_1, \ldots, \mathbf{A}_k$ are $n \times n$ matrices, then

$$\det(\mathbf{A}_1\mathbf{A}_2 \cdots \mathbf{A}_k) = (\det \mathbf{A}_1)(\det \mathbf{A}_2) \cdots (\det \mathbf{A}_k). \qquad (10.1)$$

11. **(a)** Derive the property D7. HINT: Write out the cofactor expansion about the row or column in question.

   (b) Then, show that D8 follows from D7.

12. Prove property D6, using *any* of the other listed properties.

13. Prove property D9. HINT: Write out the cofactor expansion about the row (or column) **a**.

14. It can be shown that the equations

$$a_0 x^2 + a_1 x + a_2 = 0, \quad (a_0 \neq 0)$$
$$b_0 x^2 + b_1 x + b_2 = 0 \quad (b_0 \neq 0)$$

have a common root if and only if

$$\begin{vmatrix} a_0 & a_1 & a_2 & 0 \\ 0 & a_0 & a_1 & a_2 \\ b_0 & b_1 & b_2 & 0 \\ 0 & b_0 & b_1 & b_2 \end{vmatrix} = 0.$$

[Similar results for two algebraic equations of degrees m and n, say, were put forward by *James Joseph Sylvester* (1814–1897).] Use this result to determine whether or not the following equation pairs have any common roots.

(a) $3x^2 + 2x - 5 = 0$ **(b)** $2x^2 + x - 15 = 0$
$\qquad 3x^2 + 3x - 2 = 0$ $\qquad\qquad\quad x^2 + 10x + 21 = 0$

15. **(a)** Suppose that the elements a_{ij} of an $n \times n$ matrix **A** are differentiable functions of some parameter t. Regarding $\det \mathbf{A}$ as a function of the n^2 variables $a_{11}, a_{12}, \ldots, a_{nn}$, show that

$$\frac{\partial}{\partial a_{ij}} (\det \mathbf{A}) = A_{ij}, \qquad (15.1)$$

where A_{ij} is the cofactor of a_{ij}. Then use (15.1) and chain differentiation to show that

$$\frac{d}{dt}(\det \mathbf{A}) = \sum_{i=1}^{n} \sum_{j=1}^{n} A_{ij} \frac{da_{ij}}{dt}, \qquad (15.2)$$

a formula first given by *Carl Gustav Jacob Jacobi* (1804–1851) in 1841. By the $\sum\sum$ notation we mean

$$\sum_{i=1}^{n} \sum_{j=1}^{n} c_{ij} \equiv \sum_{i=1}^{n} \left(\sum_{j=1}^{n} c_{ij} \right).$$

For example, if $n = 2$, then

$$\sum_{i=1}^{2} \sum_{j=1}^{2} c_{ij} = \sum_{i=1}^{2} \left(\sum_{j=1}^{2} c_{ij} \right) = \sum_{i=1}^{2} (c_{i1} + c_{i2})$$

$$= c_{11} + c_{21} + c_{12} + c_{22}.$$

Observe that (15.2) is equivalent to the statement

$$\frac{d}{dt}(\det \mathbf{A}) = \begin{vmatrix} \dfrac{da_{11}}{dt} & \cdots & \dfrac{da_{1n}}{dt} \\ a_{21} & \cdots & a_{2n} \\ \vdots & & \vdots \\ a_{n1} & \cdots & a_{nn} \end{vmatrix}$$

$$+ \cdots + \begin{vmatrix} a_{11} & \cdots & a_{1n} \\ \vdots & & \vdots \\ a_{n-1,1} & \cdots & a_{n-1,n} \\ \dfrac{da_{n1}}{dt} & \cdots & \dfrac{da_{nn}}{dt} \end{vmatrix}. \qquad (15.3)$$

   (b) Thus, evaluate $\dfrac{d}{dt}(\det \mathbf{A})$ if

$$\mathbf{A} = \begin{bmatrix} t^2 & t & 2 \\ 0 & 3t & 1 \\ 4 & 0 & \sin t \end{bmatrix},$$

and check your result by working out $\det \mathbf{A}$ and taking its t derivative.

16. (*Vandermonde determinant*) The determinant

$$\begin{vmatrix} 1 & \cdots & 1 \\ \lambda_1 & \cdots & \lambda_n \\ \vdots & & \vdots \\ \lambda_1^{n-1} & \cdots & \lambda_n^{n-1} \end{vmatrix}, \qquad (16.1)$$

is known as a **Vandermonde determinant**. It can be shown that it equals $(-1)^{n(n-1)/2} \Pi$, where Π denotes the product of all factors $\lambda_j - \lambda_k$ with $j < k \ (\leq n)$. For example, if $n = 3$ then $\Pi = (\lambda_1 - \lambda_2)(\lambda_1 - \lambda_3)(\lambda_2 - \lambda_3)$. The key property of any $n \times n$ Vandermonde determinant, which can be seen from this result, is that *it is nonzero if and only if all of the λ_j's are distinct.* The problem that we pose is for you to verify that the determinant is equal to $(-1)^{n(n-1)/2} \Pi$, as claimed above, simply by working out the determinant, but only for the cases where $n = 2$ and 3.

17. **(a)** Derive the remarkable equation (8). HINT: Show that $N(n) = nN(n-1) + 2n - 1$ for $n \geq 2$. Using the subscript notation $N(n) = P_n$,

$$P_n - nP_{n-1} = 2n - 1, \qquad (n \geq 2) \qquad (17.1)$$

with the initial condition $P_2 = 3$ (two multiplications and one subtraction). Seeking a solution P_n of (17.1) in the form $n!Q_n$, show that Q_n satisfies the equation

$$Q_n - Q_{n-1} = \frac{2n - 1}{n!} \qquad (17.2)$$

with initial condition $Q_2 = 3/2$, which admits the solution

$$Q_n = \frac{3}{2} + 2\sum_{j=2}^{n-1}\frac{1}{j!} - \sum_{j=3}^{n}\frac{1}{j!}. \tag{17.3}$$

[You need merely verify (17.3); you don't need to derive it.] Finally, show from (17.3) that $Q_n \sim e$ as $n \to \infty$ so that $N(n) \sim en!$ as $n \to \infty$.

(b) Derive (10). HINT: $n(n-1) + (n-1)(n-2) + \cdots +$

$(2)(1) = n(n^2 - 1)/3$.

NOTE: (17.1) and (17.2) are examples of linear **difference equations**, the theory of which is highly analogous to the theory of linear differential equations. In that light we can better appreciate the substitution $P_n = n!Q_n$, it changed the nonconstant-coefficient equation (17.1) to the constant-coefficient equation (17.2).

5.5 THE RANK OF A MATRIX

With determinants defined, we can now introduce one more concept, the "rank" of a matrix, which concept will enable us to obtain important results regarding the linear independence of vectors, and also, in Section 5.7, regarding the existence and uniqueness of solutions of the linear equation $\mathbf{Ax} = \mathbf{c}$.

First, we say that any matrix obtained from a given $m \times n$ matrix $\mathbf{A}$ by deleting at most $m - 1$ rows and at most $n - 1$ columns from $\mathbf{A}$ is a **submatrix** of $\mathbf{A}$. To illustrate, the 2×3 matrix

$$\mathbf{A} = \begin{bmatrix} a_{11} & a_{12} & a_{13} \\ a_{21} & a_{22} & a_{23} \end{bmatrix}$$

has 21 submatrices: one 2×3 ($\mathbf{A}$ itself), three 2×2's, two 1×3's, three 2×1's, six 1×2's, and six 1×1's. For instance,

$$\begin{bmatrix} a_{11} & a_{12} & a_{13} \\ a_{21} & a_{22} & a_{23} \end{bmatrix}, \begin{bmatrix} a_{11} & a_{12} \\ a_{21} & a_{22} \end{bmatrix}, \begin{bmatrix} a_{11} & a_{13} \\ a_{21} & a_{23} \end{bmatrix}, \begin{bmatrix} a_{13} \\ a_{23} \end{bmatrix}, [a_{21}, a_{23}],$$

and $[a_{13}]$ are submatrices of $\mathbf{A}$.

Then the rank of a matrix is defined as follows.

Definition 5.5.1   *Rank*
A matrix $\mathbf{A}$, not necessarily square, is of **rank** r, or $r(\mathbf{A})$, if it contains at least one $r \times r$ submatrix with nonzero determinant but no square submatrix larger than $r \times r$ with nonzero determinant. A matrix is of rank 0 if it is a zero matrix.

EXAMPLE 1
Let

$$\mathbf{A} = \begin{bmatrix} 2 & -1 & 1 & 0 \\ 0 & 3 & 3 & 6 \\ 1 & 4 & 5 & 9 \end{bmatrix}. \tag{1}$$

Certainly, r is at most 3 in this case since the largest possible square submatrix of $\mathbf{A}$ is 3×3. (More generally, if $\mathbf{A}$ is $m \times n$, then r is at most equal to the smaller of

m and n.) However, upon calculation, we find that all four of the 3×3 submatrices have zero determinant so that r is at most 2. In fact, there are a number of 2×2 submatrices with nonzero determinant such as

$$\begin{vmatrix} 2 & -1 \\ 0 & 3 \end{vmatrix} = 6 \neq 0,$$

but even if there were only one such 2×2 submatrix that would still be all we need to conclude that $r(\mathbf{A}) = 2$. ∎

EXAMPLE 2
The ranks of

$$\mathbf{A} = \begin{bmatrix} 5 \\ 6 \\ 0 \end{bmatrix}, \quad \mathbf{B} = \begin{bmatrix} 5 & -2 \\ 6 & 3 \\ 0 & 1 \end{bmatrix}, \quad \mathbf{C} = \begin{bmatrix} 5 & -2 & 0 \\ 6 & 3 & 0 \\ 0 & 1 & 0 \end{bmatrix}, \quad \mathbf{D} = \begin{bmatrix} 3 & 0 & 1 & 0 \\ 0 & 0 & 1 & 0 \\ 9 & 0 & 3 & 0 \end{bmatrix}$$

are $1, 2, 2$, and 2, respectively. In $\mathbf{D}$, for example, every 3×3 submatrix contains a column of zeros and hence has a vanishing determinant, but the 2×2 submatrix $\begin{bmatrix} 3 & 1 \\ 0 & 1 \end{bmatrix}$ has a nonvanishing determinant, so $r(\mathbf{D}) = 2$. There is another 2×2 submatrix of $\mathbf{D}$ with nonvanishing determinant (Do you see it?), but all we need is one such matrix to establish that the rank is 2. ∎

We may regard the rows of an $m \times n$ matrix $\mathbf{A} = \{a_{ij}\}$ as n-dimensional vectors, which we call the **row vectors** of $\mathbf{A}$ and which we denote as $\mathbf{r}_1, \ldots, \mathbf{r}_m$. (That is, we *partition* $\mathbf{A}$ into rows.) Similarly, the columns are m-dimensional vectors, which we call the **column vectors** of $\mathbf{A}$ and which we denote as $\mathbf{c}_1, \ldots, \mathbf{c}_n$. Further, we call the vector spaces span $\{\mathbf{r}_1, \ldots, \mathbf{r}_m\}$ and span $\{\mathbf{c}_1, \ldots, \mathbf{c}_n\}$ the **row** and **column spaces** of $\mathbf{A}$, respectively. From the definition of dimension, the dimensions of the row and column spaces are equal to the number of LI row vectors and the number of LI column vectors, respectively.

We can now state our main result, a connection between the rank of a matrix and the linear independence of a set of vectors.

THEOREM 5.5.1
Rank and Linear Independence
For any matrix $\mathbf{A}$, the number of LI row vectors is equal to the number of LI column vectors and these, in turn, equal the rank of $\mathbf{A}$.

In other words, the dimension of the row space equals the dimension of the column space equals the rank $r(\mathbf{A})$.

Thus, if we wish to determine how many vectors are in a given vector set $\{\mathbf{u}_1, \ldots, \mathbf{u}_k\}$ are LI we can form a matrix $\mathbf{A}$ with $\mathbf{u}_1, \ldots, \mathbf{u}_k$ as the rows (or columns) and then determine the rank of $\mathbf{A}$.

EXAMPLE 3
How many LI vectors are contained in $\{\mathbf{u}_1, \mathbf{u}_2, \mathbf{u}_3, \mathbf{u}_4\}$, where

$$\mathbf{u}_1 = \begin{bmatrix} 2 \\ 0 \\ 1 \end{bmatrix}, \quad \mathbf{u}_2 = \begin{bmatrix} -1 \\ 3 \\ 4 \end{bmatrix}, \quad \mathbf{u}_3 = \begin{bmatrix} 1 \\ 3 \\ 5 \end{bmatrix}, \quad \mathbf{u}_4 = \begin{bmatrix} 0 \\ 6 \\ 9 \end{bmatrix}?$$

If we construct a matrix having these vectors as columns, then we have the $\mathbf{A}$ matrix given in (1). As we determined in Example 1, $r(\mathbf{A}) = 2$. Hence, there are two LI vectors in $\{\mathbf{u}_1, \mathbf{u}_2, \mathbf{u}_3, \mathbf{u}_4\}$ or, put differently, $\dim[\text{span}\{\mathbf{u}_1, \mathbf{u}_2, \mathbf{u}_3, \mathbf{u}_4\}] = 2$.

COMMENT. The ordering of the columns is immaterial. For instance, we could make $\mathbf{u}_1$ the third column, $\mathbf{u}_2$ the first, $\mathbf{u}_3$ the fourth, and $\mathbf{u}_4$ the second because rank depends upon the zeroness or nonzeroness of determinants whereas the interchanging of rows or columns merely changes the *sign* of a determinant.   ■

Although the concept of the rank of a matrix is simple, it appears that the calculation of the rank can be quite laborious, though not in Examples 1 and 2 simply because those matrices were small. But suppose $\mathbf{A}$ is large, say 6×9. Then $\mathbf{A}$ will contain 84 6×6 submatrices, 756 5×5 submatrices, 1,890 4×4 submatrices, 1,680 3×3 submatrices, 540 2×2 submatrices, and 54 1×1 submatrices.[1] At best, if the rank happens to be 6 and the first 6×6 submatrix that we choose happens to have a nonzero determinant, then the calculation is not so tedious. But if the rank happens to be 3, say, then, to discover that, we need to evaluate a great many determinants: 84 6×6's, 756 5×5's, 1,890 4×4's and somewhere between 1 and 1,680 3×3's. Thus, the computational problem is serious, and it will be critical that we develop a procedure (i.e., an algorithm), for the evaluation of the rank of a matrix, that is tractable even for large matrices.

With that purpose in mind, we recall the **elementary row operations** defined in Section 4.5.4:

1. addition of a multiple of one row to another
   symbolically: $\mathbf{r}_j \to \mathbf{r}_j + \alpha\mathbf{r}_k$

2. multiplication of a row by a nonzero constant
   symbolically: $\mathbf{r}_j \to \alpha\mathbf{r}_j$

3. interchange of two rows
   symbolically: $\mathbf{r}_j \leftrightarrow \mathbf{r}_k$

---

[1] Where do we get the numbers 84, 756, ... from? It is known from the theory of probability and statistics that the number of **combinations** of p things taken q at a time (without regard to order) is

$$C(p, q) = \frac{p!}{q!(p - q)!}.$$

Thus, the number of 6×6 submatrices in a 6×9 matrix is $C(9, 6) = 9!/(6!\,3!) = 84$. The number of 5×5 submatrices in a 6×9 matrix is the product $C(9, 5)C(6, 5) = (126)(6) = 756$, the number of 4×4's is $C(9, 4)C(6, 4) = (126)(15) = 1,890$, and so on. The general formula is given in the exercises.

as well as our definition that two matrices are **row equivalent** if one can be obtained from the other by finitely many elementary row operations. The following theorem will now provide an efficient means of calculating the rank of a matrix.

THEOREM 5.5.2

Elementary Row Operations and Rank
Row equivalent matrices have the same rank. That is, elementary row operations do not alter the rank of a matrix.

Proof.   If matrices $\mathbf{A}$ and $\mathbf{B}$ are row equivalent, then $\mathbf{B}$ can be obtained from $\mathbf{A}$ by a finite number of elementary row operations. It follows that each row vector of $\mathbf{B}$ must be a linear combination of the row vectors of $\mathbf{A}$ so the row space of $\mathbf{B}$ must be a subspace of the row space of $\mathbf{A}$. Similarly, the row space of $\mathbf{A}$ must be a subspace of the row space of $\mathbf{B}$. Thus, the row space of $\mathbf{A}$ is identical to the row space of $\mathbf{B}$, and hence the dimension of the row space of $\mathbf{A}$ (which, according to Theorem 5.5.1, is the rank of $\mathbf{B}$) must equal the dimension of the row space of $\mathbf{B}$ (which, according to Theorem 5.5.1, is the rank of $\mathbf{B}$).   ∎

Thus, the idea is that we can use elementary row operations to simplify the matrix to a point where the rank (which is unchanged by the elementary row operations) is apparent by inspection.

E X A M P L E   4
Let

$$\mathbf{A} = \begin{bmatrix} 2 & 1 & -3 & 4 \\ 2 & 4 & -2 & 5 \\ 0 & 3 & 1 & 3 \\ 2 & 1 & -3 & -2 \end{bmatrix}. \tag{2}$$

Using elementary row operations,

$$\mathbf{A} \to \begin{bmatrix} 2 & 1 & -3 & 4 \\ 0 & 3 & 1 & 1 \\ 0 & 3 & 1 & 3 \\ 0 & 0 & 0 & -6 \end{bmatrix} \to \begin{bmatrix} 2 & 1 & -3 & 4 \\ 0 & 3 & 1 & 1 \\ 0 & 0 & 0 & 2 \\ 0 & 0 & 0 & -6 \end{bmatrix} \to \begin{bmatrix} 2 & 1 & -3 & 4 \\ 0 & 3 & 1 & 1 \\ 0 & 0 & 0 & 2 \\ 0 & 0 & 0 & 0 \end{bmatrix}, \tag{3}$$

where the operations were as follows: in the first step $\mathbf{r}_2 \to \mathbf{r}_2 + (-1)\mathbf{r}_1$ and $\mathbf{r}_4 \to \mathbf{r}_4 + (-1)\mathbf{r}_1$; in the second step $\mathbf{r}_3 \to \mathbf{r}_3 + (-1)\mathbf{r}_2$; and in the final step $\mathbf{r}_4 \to \mathbf{r}_4 + 3\mathbf{r}_3$. Clearly, the rank of the final matrix is 3 because (deleting the fourth row and third column)

$$\begin{vmatrix} 2 & 1 & 4 \\ 0 & 3 & 1 \\ 0 & 0 & 2 \end{vmatrix} = (2)(3)(2) = 12 \neq 0.$$

Thus, by Theorem 5.5.2, $r(\mathbf{A}) = 3$.   ■

The idea, then, is to reduce a given matrix $\mathbf{A}$ to reduced row-echelon form by means of elementary row operations.[1] It can be seen that, in that form, the nonzero rows are LI. In Example 4, for instance, a variety of conclusions follow from (3): $r(\mathbf{A}) = 3$, the number of LI vectors among the rows of $\mathbf{A}$ is 3, the dimension of the row space of $\mathbf{A}$ is 3, and a basis for that row space is given by the vectors $[2, 1, -3, 4]$, $[0, 3, 1, 1]$, and $[0, 0, 0, 2]$.

Let us close with an interesting application to stoichiometry.

EXAMPLE 5   *Application of Rank to Stoichiometry*.
To model the combustion of gasoline in an automobile engine, one can begin by writing down a list of well over 100 simultaneous chemical reactions involving various hydrocarbons, oxygen, nitrogen, and so on. In turn, these reactions can be modeled by ODE's governing the amount of each chemical species as a function of time, and one can solve the resulting system of ODE's by methods such as those discussed later, in Chapters 10 and 12. It is easy to appreciate that solving around 100 coupled ODE's is a difficult undertaking. Thus, it is important to shorten the list of reactions insofar as possible, and we can do this by eliminating ones that are redundant. For instance, if $A+B \rightarrow C$ and $A+C \rightarrow D$, then a third statement, $2A+B+C \rightarrow C+D$, surely is redundant in that it is implied by the first two.

To illustrate the shortening process, consider the burning of a mixture of CO, H_2, and CH_4 in a furnace, producing CO, CO_2, and H_2O.[2] Writing all possible reactions that the author can think of gives the list

$$CO + \frac{1}{2}O_2 \rightarrow CO_2, \tag{4a}$$

$$H_2 + \frac{1}{2}O_2 \rightarrow H_2O, \tag{4b}$$

$$CH_4 + \frac{3}{2}O_2 \rightarrow CO + 2H_2O, \tag{4c}$$

$$CH_4 + 2O_2 \rightarrow CO_2 + 2H_2O, \tag{4d}$$

where (4c) and (4d) represent the partial and complete combustion of CH_4, respectively. How many of these reactions are independent? It is convenient to re-express

---

[1]Actually, we didn't complete the reduction to reduced row-echelon form in (3), we merely carried out the reduction to a point where we could determine the rank of the final matrix by inspection.

[2]This example is discussed by Ben Noble in his book *Applications of Undergraduate Mathematics in Engineering* (New York: Macmillan, 1967). In turn, he notes that the problem was contributed by John Mahoney, Department of Chemical Engineering, West Virginia University, Morgantown, WV.

them symbolically in the equation format

$$CO + \tfrac{1}{2}O_2 - CO_2 = 0,$$
$$H_2 + \tfrac{1}{2}O_2 - H_2O = 0,$$
$$CH_4 + \tfrac{3}{2}O_2 - CO - 2H_2O = 0,$$
$$CH_4 + 2O_2 - CO_2 - 2H_2O = 0,$$

(5)

where the elements of the coefficient matrix

$$
\mathbf{A} =
\begin{array}{c}
\text{CO}\ \text{O}_2\ \ \text{CO}_2\ \text{H}_2\ \text{H}_2\text{O}\ \text{CH}_4 \\
\begin{bmatrix}
1 & \tfrac{1}{2} & -1 & 0 & 0 & 0 \\
0 & \tfrac{1}{2} & 0 & 1 & -1 & 0 \\
-1 & \tfrac{3}{2} & 0 & 0 & -2 & 1 \\
0 & 2 & -1 & 0 & -2 & 1
\end{bmatrix}
\end{array}
$$

(6)

are known as stoichiometric coefficients. To determine a minimum set of independent reactions we reduce $\mathbf{A}$ by elementary row operations and obtain

$$
\begin{array}{c}
\text{CO}\ \text{O}_2\ \ \text{CO}_2\ \ \text{H}_2\ \text{H}_2\text{O}\ \text{CH}_4 \\
\begin{bmatrix}
1 & \tfrac{1}{2} & -1 & 0 & 0 & 0 \\
0 & 1 & 0 & 2 & -2 & 0 \\
0 & 0 & -1 & -4 & 2 & 1 \\
0 & 0 & 0 & 0 & 0 & 0
\end{bmatrix}
\end{array},
$$

(7)

the rank of which is three. Thus, there are three independent reactions such as the list

$$CO + \tfrac{1}{2}O_2 - CO_2 = 0,$$
$$O_2 + 2H_2 - 2H_2O = 0,$$
$$-CO_2 - 4H_2 + 2H_2O + CH_4 = 0,$$

(8)

implied by (7). That is, $CO + \tfrac{1}{2}O_2 \rightarrow CO_2$, and so on.   ∎

Closure.   The rank $r(\mathbf{A})$ is defined as the order of the largest nonvanishing determinant within $\mathbf{A}$. Because the rank of a matrix is unaffected by elementary row operations, we can determine the rank of a given matrix efficiently by reducing it to row-echelon form, in which form the rank can be seen by inspection. Principal applications of the concept of rank include the calculation of the number of LI vectors (n-tuple vectors, that is) within a given set, and the theory of the existence and uniqueness of solutions of systems of linear algebraic equations, which will be discussed in Section 5.7.

Maple. We can evaluate the rank of a given matrix using the **rank** command. For instance, to evaluate $r(\mathbf{A})$ where the rows of $\mathbf{A}$ are [1, 2, 3, 4], [2, 4, 6, 8], and [1, 1, 1, 1], respectively, enter

```
with(linalg):
```

to access the rank command. Then enter

```
rank(array([[1,2,3,4,],[2,4,6,8],[1,1,1,1]]));
```

and return. The output is 2. Alternatively, the sequence

```
A:=array([[1,2,3,4,],[2,4,6,8],[1,1,1,1]]):
rank(A);
```

gives the same result.

EXERCISES 5.5

1. Determine the rank, the number of LI rows vectors, and the number of LI column vectors for the given matrix.

(**a**) [0, 0, 2, 0]

(**b**) [1, 2, 3]

(**c**) $\begin{bmatrix} 5 & 7 \\ 4 & 9 \end{bmatrix}$

(**d**) $\begin{bmatrix} 4 & 8 & 0 \\ 3 & 6 & 0 \end{bmatrix}$

(**e**) $\begin{bmatrix} 1 & 2 & 3 \\ 4 & 5 & 6 \\ 7 & 8 & 9 \end{bmatrix}$

(**f**) $\begin{bmatrix} 3 & 2 & 1 \\ -1 & 1 & 4 \\ 1 & 4 & 9 \end{bmatrix}$

(**g**) $\begin{bmatrix} 5 & 0 & 0 \\ 3 & 0 & 4 \\ 2 & 0 & 0 \end{bmatrix}$

(**h**) $\begin{bmatrix} 1 & 3 & 2 \\ 2 & 6 & 4 \\ 3 & 9 & 6 \end{bmatrix}$

(**i**) $\begin{bmatrix} 6 & 5 & 2 & 0 \\ 0 & 2 & 3 & 0 \\ 0 & 0 & 0 & 6 \end{bmatrix}$

(**j**) $\begin{bmatrix} 1 & 0 & 3 & 0 \\ 0 & 1 & 0 & 3 \\ 1 & 0 & 3 & 1 \end{bmatrix}$

(**k**) $\begin{bmatrix} 1 & 0 & 2 \\ 2 & 1 & -1 \\ 0 & 1 & 3 \\ 1 & 2 & 4 \end{bmatrix}$

(**l**) $\begin{bmatrix} 0 & 4 & -1 & 1 \\ 1 & 1 & 5 & -1 \\ 1 & 5 & 4 & 0 \\ 2 & 6 & 9 & -1 \end{bmatrix}$

(**m**) $\begin{bmatrix} 3 & 1 & -4 & 4 \\ 0 & 2 & 2 & 9 \\ 2 & 2 & 1 & 0 \\ -1 & 3 & 7 & 5 \end{bmatrix}$

(**n**) $\begin{bmatrix} 1 & 2 & 3 & 4 \\ 5 & -1 & 0 & 5 \\ 6 & -1 & 2 & 8 \\ 7 & -1 & 0 & 7 \\ 8 & -1 & -3 & 5 \end{bmatrix}$

2. (**a**)–(**n**) Use *Maple* to determine the rank of the matrix given in the corresponding part of Exercise 1.

3. If two matrices of the same form have the same rank, need they be row equivalent? Prove or disprove.

4. Show, by carrying out suitable row operations, that the following matrices are row equivalent.

(**a**) $\begin{bmatrix} 1 & 3 & -3 & 0 \\ 2 & 1 & 0 & 4 \end{bmatrix}$ and $\begin{bmatrix} 4 & 2 & 0 & 8 \\ 3 & 4 & -3 & 4 \end{bmatrix}$

(**b**) $\begin{bmatrix} 2 & 1 & 0 \\ 2 & 2 & 3 \\ 4 & 4 & 9 \\ 4 & 4 & 6 \end{bmatrix}$ and $\begin{bmatrix} 2 & 1 & 0 \\ 0 & 1 & 3 \\ 0 & 0 & 3 \\ 0 & 0 & 0 \end{bmatrix}$

(**c**) $\begin{bmatrix} 5 & 2 \\ 0 & -3 \\ 1 & 4 \end{bmatrix}$ and $\begin{bmatrix} 9 & 2 \\ -6 & -3 \\ 9 & 4 \end{bmatrix}$

5. Show that the following matrices are *not* row equivalent.

(**a**) $\begin{bmatrix} 1 & 2 \\ 3 & 4 \end{bmatrix}$ and $\begin{bmatrix} 1 & 2 \\ 2 & 4 \end{bmatrix}$

(**b**) $\begin{bmatrix} 0 & 2 & 0 \\ 0 & 0 & 4 \\ 6 & 0 & 0 \end{bmatrix}$ and $\begin{bmatrix} 3 & -1 & 5 \\ 1 & 5 & -5 \\ 2 & 2 & 0 \end{bmatrix}$

6. Exercises 4 and 5 are simple enough to be solvable by inspection. More generally, inspection may not suffice. Put forward a systematic procedure for determining whether or not two given matrices are row equivalent, and apply that procedure to the given matrices.

(**a**) $\begin{bmatrix} 1 & 2 & 3 & -1 \\ 2 & 4 & -1 & 1 \\ 0 & 5 & 6 & 3 \\ 4 & -2 & 0 & 6 \end{bmatrix}$ and $\begin{bmatrix} 1 & -1 & 2 & 1 \\ 3 & 0 & 5 & 1 \\ 2 & 2 & 1 & 3 \\ 3 & 1 & 3 & 4 \end{bmatrix}$

(b) $\begin{bmatrix} 4 & -1 & 2 & 3 \\ 0 & 1 & 1 & 2 \\ -1 & 0 & 2 & 0 \end{bmatrix}$ and $\begin{bmatrix} 3 & 0 & 5 & 5 \\ 6 & 0 & -1 & 5 \\ 2 & 0 & 7 & 5 \end{bmatrix}$

7. Determine whether the following set of vectors is LI or LD by computing the rank of a suitable matrix and invoking the relevant theorem.

(a) $[2, 0, 1, -1]$, $[0, 3, 0, 3]$, $[4, 3, 2, 1]$

(b) $[4, 1, 2]$, $[2, 2, 1]$, $[2, -1, 1]$, $[4, 7, 2]$, $[0, 1, 0]$

(c) $[1, 3, 2, 4, 5]$, $[2, 3, 1, 5, 4]$, $[4, 5, 3, 1, 2]$

(d) $[2, 1, 1]$, $[4, 2, 2]$, $[0, 1, 2]$, $[1, 0, 0]$

(e) $[1, -2, 0, 1]$, $[0, 1, 1, 2]$, $[1, 0, 2, 5]$, $[2, -7, -3, -4]$

8. Prove that $r(\mathbf{A}^T) = r(\mathbf{A})$ for every $m \times n$ matrix using any results given in this section.

9. The property $\det(\mathbf{AB}) = (\det \mathbf{A})(\det \mathbf{B})$, of determinants, where $\mathbf{A}$ and $\mathbf{B}$ are both $n \times n$, might seem to imply that $r(\mathbf{AB}) = r(\mathbf{A})r(\mathbf{B})$. Is the latter true? Prove or disprove.

10. Is $r(\mathbf{A} + \mathbf{B}) = r(\mathbf{A}) + r(\mathbf{B})$ true? Prove or disprove.

11. Prove that if $\mathbf{A}$ is $m \times n$ and $\mathbf{B}$ is $n \times p$, then $r(\mathbf{AB}) \leq n$. HINT: Partition $\mathbf{B}$ into rows, and write

$$\mathbf{AB} = \begin{bmatrix} a_{11} & \cdots & a_{1n} \\ \vdots & & \vdots \\ a_{m1} & \cdots & a_{mn} \end{bmatrix} \begin{bmatrix} \mathbf{r}_1 \\ \vdots \\ \mathbf{r}_n \end{bmatrix}$$

$$= \begin{bmatrix} a_{11}\mathbf{r}_1 + \cdots + a_{1n}\mathbf{r}_n \\ \vdots \\ a_{m1}\mathbf{r}_1 + \cdots + a_{mn}\mathbf{r}_n \end{bmatrix}.$$

12. (*Stoichiometry*) Determine how many of the following reactions are independent, and give such a set of independent reactions.

(a)
$$H_2 + O_2 \rightleftharpoons 2OH,$$
$$H_2 + \tfrac{1}{2}O_2 \rightleftharpoons H_2O,$$
$$H + OH \rightleftharpoons H_2 + O,$$
$$H_2 \rightleftharpoons 2H,$$
$$O_2 \rightleftharpoons 2O$$

(b)
$$H_2 + Cl_2 \rightleftharpoons 2HCl,$$
$$Cl + H_2 \rightleftharpoons HCl + H,$$
$$H_2 \rightleftharpoons 2H,$$
$$Cl_2 \rightleftharpoons 2Cl$$

(c)
$$O_2 + CH_3 \rightarrow CH_3OO,$$
$$CH_4 + CH_3OO \rightarrow CH_3 + CH_3OOH,$$
$$CH_3OOH \rightarrow CO + 2H_2 + O,$$
$$CH_4 + O \rightarrow CH_3 + OH,$$
$$CH_4 + O_2 \rightarrow CH_3OOH$$

13. Recall, from the first footnote of this section, that $C(p, q) = p!/[q!(p - q)!]$.

(a) Show that the number of $k \times k$ submatrices within an $m \times n$ matrix is $C(m, k)C(n, k)$.

(b) Give a formula for the number of $j \times k$ submatrices within an $m \times n$ matrix.

5.6 INVERSE MATRIX, CRAMER'S RULE, AND FACTORIZATION

There exist important methods for solving a linear system $\mathbf{Ax} = \mathbf{c}$ besides Gauss elimination. In this section we study three: the inverse matrix method, Cramer's rule, and LU factorization.

5.6.1 Inverse matrix. Having introduced matrix notation so that a system of linear algebraic equations can be expressed compactly as

$$\mathbf{Ax} = \mathbf{c}, \tag{1}$$

the form of (1) itself suggests other solution strategies. For how would we solve the simple scalar equation $3x = 12$? We could divide both sides by 3 and obtain $x = \frac{12}{3} = 4$. However, if we try to carry that idea over to the matrix case (1), we obtain $\mathbf{x} = \dfrac{\mathbf{c}}{\mathbf{A}}$, and need to know how to divide one matrix into another. However, matrix division has not been (and will not be) defined. Alternatively, we can solve $3x = 12$ by *multiplying* both sides by $\frac{1}{3}$, for that step gives $\frac{1}{3}3x = \frac{1}{3}12$, or $1x = 4$, and hence $x = 4$. That idea does carry over to (1) because matrix multiplication *is* defined.

The idea, then, is to seek a matrix "$\mathbf{A}^{-1}$" having the property that $\mathbf{A}^{-1}\mathbf{A} = \mathbf{I}$ for then

$$\mathbf{A}^{-1}\mathbf{A}\mathbf{x} = \mathbf{A}^{-1}\mathbf{c} \tag{2}$$

becomes

$$\mathbf{I}\mathbf{x} = \mathbf{A}^{-1}\mathbf{c}, \tag{3}$$

and since $\mathbf{I}\mathbf{x} = \mathbf{x}$ we have, from (3), the solution

$$\mathbf{x} = \mathbf{A}^{-1}\mathbf{c} \tag{4}$$

of (1). Note that $\mathbf{A}^{-1}$ does not mean $1/\mathbf{A}$ or $\mathbf{I}/\mathbf{A}$; it is simply the name of the matrix having the property

$$\mathbf{A}^{-1}\mathbf{A} = \mathbf{I}, \tag{5}$$

if one exists, just as $\mathbf{A}^{\mathrm{T}}$ means the transpose of $\mathbf{A}$, not $\mathbf{A}$ to the Tth power. We call $\mathbf{A}^{-1}$ the **inverse** of $\mathbf{A}$, or "$\mathbf{A}$-inverse" for brevity.

If we review the steps (1)–(4), it certainly appears that if, given $\mathbf{A}$, we can find a matrix $\mathbf{A}^{-1}$ satisfying (5), then (4) gives the solution to (1). But be careful: When we premultiply both sides of (1) by the matrix $\mathbf{A}^{-1}$ do we necessarily obtain an equivalent system [i.e., one having the same solution set as (1)]? That is, does (4), which follows from (2), necessarily satisfy (1)? Recall from Theorem 5.2.2 that "even if $\mathbf{A} \neq \mathbf{0}$, $\mathbf{AB} = \mathbf{AC}$ does not imply that $\mathbf{B} = \mathbf{C}$." Thus, (2) does not imply that $\mathbf{A}\mathbf{x} = \mathbf{c}$. Let us put (4) into (1) and see if it works. That step gives

$$\mathbf{A}\mathbf{A}^{-1}\mathbf{c} = \mathbf{c} \tag{6}$$

so if

$$\mathbf{A}\mathbf{A}^{-1} = \mathbf{I} \tag{7}$$

then (6) becomes an identity and (4) does indeed satisfy (1). It is tempting to think that (7) follows from (5) so that all is well, but remember that in general matrix multiplication is not commutative.

Our plan is as follows. We will show that if $\mathbf{A}$ is square and $\det \mathbf{A} \neq 0$, then we can indeed find a unique matrix $\mathbf{A}^{-1}$ satisfying both (5) *and* (7). Further, we will show that not only does (4) satisfy (1), but that it is the *unique* solution of (1).

THEOREM 5.6.1

Existence and Uniqueness of Inverse Matrix

Let $\mathbf{A} = \{a_{ij}\}$ be $n \times n$ and suppose that det $\mathbf{A} \neq 0$.

(i) Then there exists a unique matrix $\mathbf{A}^{-1}$, called the **inverse** of $\mathbf{A}$, such that

$$\boxed{\mathbf{A}^{-1}\mathbf{A} = \mathbf{A}\mathbf{A}^{-1} = \mathbf{I}.} \tag{8}$$

(ii) Let α_{ij} denote the i, j element of $\mathbf{A}^{-1}$. Then

$$\alpha_{ij} = \frac{A_{ji}}{\det \mathbf{A}}, \tag{9}$$

so

$$\boxed{\mathbf{A}^{-1} = \{\alpha_{ij}\} = \left\{ \frac{A_{ji}}{\det \mathbf{A}} \right\} = \frac{1}{\det \mathbf{A}} \begin{bmatrix} A_{11} & A_{21} & \cdots & A_{n1} \\ A_{12} & A_{22} & \cdots & A_{n2} \\ \vdots & & & \vdots \\ A_{1n} & A_{2n} & \cdots & A_{nn} \end{bmatrix},} \tag{10}$$

where A_{ji} denotes the cofactor of the a_{ji} element of $\mathbf{A}$.

Proof.   Recall that the cofactor A_{ji} was defined [by (3) in Section 5.4] as $(-1)^{j+i}$ times the minor determinant M_{ji} associated with the j, i location in $\mathbf{A}$, where M_{ji} is the $(n-1) \times (n-1)$ determinant obtained by striking out the row and column through the a_{ji} element in the determinant of $\mathbf{A}$.

Beginning our proof, we seek to determine a matrix $\mathbf{A}^{-1}$ satisfying (5). Our starting point is the cofactor expansion from Section 5.4,

$$\det \mathbf{A} = \sum a_{jk} A_{jk}, \tag{11}$$

where A_{jk} is the cofactor of the a_{jk} element, and the sum is either on j (for any fixed value of k) or on k (for any fixed value of j). Let us take the sum to be on j. Observe that

$$\sum_{j} a_{jk} A_{ji} = \begin{cases} \det \mathbf{A} & \text{if } i = k, \\ 0 & \text{if } i \neq k \end{cases} \tag{12}$$

since if $i = k$, then (11) applies, and if $i \neq k$, then the left-hand side of (12) is again a cofactor expansion, this time an expansion about the ith column—but with the ith column replaced by the kth column (because the A_{ji}'s are multiplied by the a_{jk}'s instead of the a_{ji}'s); thus, it is a determinant containing two identical columns and according to property D5 in Section 5.4, it must therefore be zero. Rearranging (12) by dividing through by det $\mathbf{A}$ (which is permissible since we have assumed that $\det \mathbf{A} \neq 0$) and using the Kronecker delta notation,[1]

$$\sum_j \left(\frac{A_{ji}}{\det \mathbf{A}} \right) a_{jk} = \delta_{ik}. \tag{13}$$

This scalar statement (which holds for $1 \leq i \leq n$ and $1 \leq k \leq n$) is equivalent, according to the definition of matrix multiplication,[2] to the matrix equation (5).

Next, we show that (7) holds as well. Again recalling the definition of matrix multiplication,[3] we have

$$\mathbf{A}\mathbf{A}^{-1} = \left\{ \sum_k a_{ik}\alpha_{kj} \right\} = \left\{ \sum_k a_{ik} \frac{A_{jk}}{\det \mathbf{A}} \right\} = \frac{1}{\det \mathbf{A}} \left\{ \sum_k a_{ik} A_{jk} \right\}. \tag{14}$$

Now,

$$\sum_k a_{ik} A_{jk} = \begin{cases} \det \mathbf{A}, & i = j \\ 0, & i \neq j \end{cases} \tag{15}$$

because if $i = j$ then (11) applies, and if $i \neq j$ then the left-hand side of (15) is again a cofactor expansion, this time an expansion about the jth row—but with the jth row replaced by the ith row. Thus, it is a determinant containing two identical rows and, according to property D5, it must be zero. Hence (14) becomes

$$\mathbf{A}\mathbf{A}^{-1} = \{\delta_{ij}\} = \mathbf{I}, \tag{16}$$

as claimed.

Finally, let us prove the uniqueness of $\mathbf{A}^{-1}$. Let matrices $\mathbf{B}$ and $\mathbf{C}$ both be inverses of $\mathbf{A}$, so that $\mathbf{B}\mathbf{A} = \mathbf{I}$ and $\mathbf{C}\mathbf{A} = \mathbf{I}$. Subtracting these gives $\mathbf{B}\mathbf{A} - \mathbf{C}\mathbf{A} = \mathbf{0}$, or $(\mathbf{B}-\mathbf{C})\mathbf{A} = \mathbf{0}$. Post-multiplying this last equation by $\mathbf{A}^{-1}$ [which is given by (10) and which exists because the $\det \mathbf{A}$ in the denominator has been assumed to be nonzero] we have $(\mathbf{B} - \mathbf{C})\mathbf{A}\mathbf{A}^{-1} = \mathbf{0}\mathbf{A}^{-1}$, or $(\mathbf{B} - \mathbf{C})\mathbf{I} = \mathbf{0}$, or $\mathbf{B} - \mathbf{C} = 0$. Thus, $\mathbf{B} = \mathbf{C}$ so the inverse of $\mathbf{A}$ is indeed unique.   ∎

The square-bracketed matrix in (10) is called the **adjoint of A** and is denoted as adj $\mathbf{A}$ so

$$\boxed{\mathbf{A}^{-1} = \frac{1}{\det \mathbf{A}} \text{ adj } \mathbf{A}.} \tag{17}$$

THEOREM 5.6.2

Nonexistence of Inverse Matrix
Let $\mathbf{A}$ be $n \times n$ and suppose that $\det \mathbf{A} = 0$. Then the inverse of $\mathbf{A}$ does not exist.

---

[1] Defined in Section 5.2.6, δ_{ik} is simply 1 if $i = k$ and 0 if $i \neq k$. Thus, a matrix $\{\delta_{ik}\}$ is an identity matrix $\mathbf{I}$.

[2] Recall that if $\mathbf{B}\mathbf{C} = \mathbf{D}$, then $d_{ij} = \sum_k b_{ik} c_{kj}$ or, what is equivalent, $d_{ik} = \sum_j b_{ij} c_{jk}$.

[3] Ibid.

Proof.   Suppose that $\mathbf{A}^{-1}$ exists. Then it follows from $\mathbf{A}^{-1}\mathbf{A} = \mathbf{I}$ that $\det(\mathbf{A}^{-1}\mathbf{A}) = \det \mathbf{I} = 1$. But by property D12 in Section 5.4, $\det\left(\mathbf{A}^{-1}\mathbf{A}\right) = \left(\det \mathbf{A}^{-1}\right)(\det \mathbf{A})$, so

$$\left(\det \mathbf{A}^{-1}\right)(\det \mathbf{A}) = 1. \tag{18}$$

But (18) cannot be satisfied if $\det \mathbf{A} = 0$, so $\mathbf{A}^{-1}$ does not exist if $\det \mathbf{A} = \mathbf{0}$.   ∎

The upshot, then, is that if $\det \mathbf{A} \neq 0$, then $\mathbf{A}^{-1}$ exists and is given by (10). In that case we say that $\mathbf{A}$ is **invertible**. If $\det \mathbf{A} = 0$, then $\mathbf{A}^{-1}$ does not exist and we say that $\mathbf{A}$ is **singular**.[1] Observe that if $\mathbf{A}$ is invertible then, since $\mathbf{A}^{-1}\mathbf{A} = \mathbf{A}\mathbf{A}^{-1}$, $\mathbf{A}$ and $\mathbf{A}^{-1}$ commute.

Furthermore, recall from Theorem 5.5.1 that $\det \mathbf{A} \neq 0$ if and only if the column vectors (equivalently, the row vectors) of $\mathbf{A}$ are LI. And $\det \mathbf{A} \neq 0$ if and only if the rank of $\mathbf{A}$ is n, so we have this equivalence:

THEOREM 5.6.3

Necessary and Sufficient Conditions for Invertibility
Equivalent necessary and sufficient conditions for the invertibility of an $n \times n$ matrix $\mathbf{A}$ are: $\det \mathbf{A} \neq 0$; the rank of $\mathbf{A}$ is n; the column vectors (equivalently, the row vectors) of $\mathbf{A}$ are LI.

E X A M P L E   1
Determine the inverse of

$$\mathbf{A} = \begin{bmatrix} 3 & 2 & -1 \\ 0 & 1 & 4 \\ 1 & 5 & -2 \end{bmatrix}, \tag{19}$$

if it exists. It does exist because $\det \mathbf{A} = -57 \neq 0$, and is given by (10). Since adj $\mathbf{A}$ is the transpose of the cofactor matrix, we have

$$\text{adj } \mathbf{A} = \begin{bmatrix} \begin{vmatrix} 1 & 4 \\ 5 & -2 \end{vmatrix} & -\begin{vmatrix} 0 & 4 \\ 1 & -2 \end{vmatrix} & \begin{vmatrix} 0 & 1 \\ 1 & 5 \end{vmatrix} \\ -\begin{vmatrix} 2 & -1 \\ 5 & -2 \end{vmatrix} & \begin{vmatrix} 3 & -1 \\ 1 & -2 \end{vmatrix} & -\begin{vmatrix} 3 & 2 \\ 1 & 5 \end{vmatrix} \\ \begin{vmatrix} 2 & -1 \\ 1 & 4 \end{vmatrix} & -\begin{vmatrix} 3 & -1 \\ 0 & 4 \end{vmatrix} & \begin{vmatrix} 3 & 2 \\ 0 & 1 \end{vmatrix} \end{bmatrix}^T$$

$$= \begin{bmatrix} -22 & 4 & -1 \\ -1 & -5 & -13 \\ 9 & -12 & 3 \end{bmatrix}^T = \begin{bmatrix} -22 & -1 & 9 \\ 4 & -5 & -12 \\ -1 & -13 & 3 \end{bmatrix} \tag{20}$$

---

[1]There exist interesting generalizations of the notion of the inverse matrix for matrices that are not strictly invertible (perhaps not even square). Such are the *Moore-Penrose generalized inverse* and the *pseudoinverse*. See, for example, Gilbert Strang, *Linear Algebra and Its Applications* (New York: Academic Press, 1976), Chap. 3.

so

$$\mathbf{A}^{-1} = \frac{1}{\det \mathbf{A}} \; \text{adj}\, \mathbf{A} = -\frac{1}{57} \begin{bmatrix} -22 & -1 & 9 \\ 4 & -5 & -12 \\ -1 & -13 & 3 \end{bmatrix} = \begin{bmatrix} \frac{22}{57} & \frac{1}{57} & -\frac{3}{19} \\ -\frac{4}{57} & \frac{5}{57} & \frac{4}{19} \\ \frac{1}{57} & \frac{13}{57} & -\frac{1}{19} \end{bmatrix}.$$

(21)

As a check, it is readily verified from (21) and (19) that $\mathbf{A}^{-1}\mathbf{A} = \mathbf{A}\mathbf{A}^{-1} = \mathbf{I}$.

COMMENT. Be careful not to forget the transpose step in (20), which follows from (9), for $\mathbf{A}^{-1} = \{\alpha_{ij}\} = \{A_{ji}\}/\det \mathbf{A} = \{A_{ij}^{\mathsf{T}}\}/\det \mathbf{A}$. ∎

THEOREM 5.6.4
Solution of $\mathbf{Ax} = \mathbf{c}$
If $\mathbf{A}$ is $n \times n$ and $\det \mathbf{A} \neq 0$, then $\mathbf{Ax} = \mathbf{c}$ admits the unique solution

$$\boxed{\mathbf{x} = \mathbf{A}^{-1}\mathbf{c}.}$$

(22)

Proof. Putting (22) into $\mathbf{Ax} = \mathbf{c}$ gives $\mathbf{AA}^{-1}\mathbf{c} = \mathbf{c}$, or $\mathbf{Ic} = \mathbf{c}$, which is the identity $\mathbf{c} = \mathbf{c}$. To prove the uniqueness of the solution, let vectors $\mathbf{u}$ and $\mathbf{v}$ both be solutions of $\mathbf{Ax} = \mathbf{c}$, so that $\mathbf{Au} = \mathbf{c}$ and $\mathbf{Av} = \mathbf{c}$. Subtracting these gives $\mathbf{Au} - \mathbf{Av} = \mathbf{0}$, or $\mathbf{A}(\mathbf{u}-\mathbf{v}) = \mathbf{0}$. Pre-multiplying this last equation by $\mathbf{A}^{-1}$ gives $\mathbf{A}^{-1}\mathbf{A}(\mathbf{u}-\mathbf{v}) = \mathbf{A}^{-1}\mathbf{0}$, or $\mathbf{I}(\mathbf{u} - \mathbf{v}) = \mathbf{0}$, or $\mathbf{u} - \mathbf{v} = \mathbf{0}$. Thus, $\mathbf{u} = \mathbf{v}$ so the solution of $\mathbf{Ax} = \mathbf{c}$ is indeed unique. ∎

EXAMPLE 2
Use (22) to solve

$$\begin{bmatrix} 3 & 2 & -1 \\ 0 & 1 & 4 \\ 1 & 5 & -2 \end{bmatrix} \begin{bmatrix} x_1 \\ x_2 \\ x_3 \end{bmatrix} = \begin{bmatrix} 6 \\ 5 \\ 1 \end{bmatrix}.$$

(23)

Understand that for any linear system $\mathbf{Ax} = \mathbf{c}$ we can *always* use Gauss elimination (or Gauss-Jordan reduction), whether or not $\mathbf{A}$ is invertible—or even square. But if $\mathbf{A}$ *is* square and $\det \mathbf{A} \neq 0$ then, alternatively, we can solve by the inverse matrix method, by which we mean evaluating $\mathbf{A}^{-1}$ and then computing $\mathbf{x}$ as $\mathbf{A}^{-1}\mathbf{c}$.

In the present example $\det \mathbf{A} = -57 \neq 0$ so $\mathbf{A}$ is indeed invertible. In fact we found $\mathbf{A}^{-1}$ in Example 1 so, using (21), we have the unique solution

$$\mathbf{x} = \mathbf{A}^{-1}\mathbf{c} = \begin{bmatrix} \frac{22}{57} & \frac{1}{57} & -\frac{3}{19} \\ -\frac{4}{57} & \frac{5}{57} & \frac{4}{19} \\ \frac{1}{57} & \frac{13}{57} & -\frac{1}{19} \end{bmatrix} \begin{bmatrix} 6 \\ 5 \\ 1 \end{bmatrix} = \begin{bmatrix} 128/57 \\ 13/57 \\ 68/57 \end{bmatrix}$$

(24)

of (23). ∎

Let us close this Section 5.6.1 with the following summary observation. The simple equation $ax = c$ admits a unique solution if and only if $a \neq 0$. The way that result generalizes to the higher-dimensional case of n equations in n unknowns, $\mathbf{Ax} = \mathbf{c}$, is that $\mathbf{Ax} = \mathbf{c}$ admits a unique solution if and only if $\det \mathbf{A} \neq 0$. That is, the condition is $\det \mathbf{A} \neq 0$, not $\mathbf{A} \neq \mathbf{0}$. Thus, think of a, in the condition $a \neq 0$ for the equation $ax = c$, as the determinant of the 1×1 coefficient matrix $[a]$.

5.6.2 Evaluation of inverse matrix by elementary row operations.   Equation (10) gives $\mathbf{A}^{-1}$ in terms of $\det \mathbf{A}$ and n^2 cofactors, each of which is ± 1 times an $(n-1) \times (n-1)$ minor determinant. Alternatively, we can bypass (10) and determine $\mathbf{A}^{-1}$ efficiently using elementary row operations, as follows.

Consider a linear system

$$\mathbf{Ax} = \mathbf{c}, \tag{25}$$

where $\mathbf{A}$ is $n \times n$ and is assumed to be invertible. On the one hand we know that (25) admits the unique solution $\mathbf{x} = \mathbf{A}^{-1}\mathbf{c}$, and on the other hand we know that we can solve (25) efficiently using elementary operations. Thus, it seems reasonable that we should, likewise, be able to evaluate $\mathbf{A}^{-1}$ using elementary row operations. Indeed we can, and the idea is as follows. Beginning with $\mathbf{Ax} = \mathbf{c}$ or, equivalently, $\mathbf{Ax} = \mathbf{Ic}$, elementary row operations produce the solution $\mathbf{x} = \mathbf{A}^{-1}\mathbf{c}$ or, equivalently, $\mathbf{Ix} = \mathbf{A}^{-1}\mathbf{c}$. Symbolically, then, the sequence of elementary row operations effects the following transformation:

$$\begin{array}{c} \mathbf{Ax} = \mathbf{Ic} \\ \downarrow \\ \mathbf{Ix} = \mathbf{A}^{-1}\mathbf{c}. \end{array} \tag{26}$$

That is, at the same time that the row operations are transforming $\mathbf{A}$ to $\mathbf{I}$ (on the left) they are also transforming $\mathbf{I}$ to $\mathbf{A}^{-1}$ (on the right). Thus, we can skip $\mathbf{x}$ and $\mathbf{c}$ altogether, put $\mathbf{A}$ and $\mathbf{I}$ "side by side" as an augmented matrix $\mathbf{A}|\mathbf{I}$, and carry out elementary row operations on $\mathbf{A}|\mathbf{I}$ so as to reduce $\mathbf{A}$, on the left, to $\mathbf{I}$. When that has been accomplished, the matrix on the right will be $\mathbf{A}^{-1}$. Symbolically,

$$\boxed{\mathbf{A}|\mathbf{I} \to \mathbf{I}|\mathbf{A}^{-1}.} \tag{27}$$

E X A M P L E   3
To illustrate, let us find the inverse of

$$\mathbf{A} = \begin{bmatrix} 1 & 3 & 0 \\ -2 & 3 & 1 \\ 0 & 1 & 1 \end{bmatrix}. \tag{28}$$

Then

$$\mathbf{A}|\mathbf{I} = \left[\begin{array}{ccc|ccc} 1 & 3 & 0 & 1 & 0 & 0 \\ -2 & 3 & 1 & 0 & 1 & 0 \\ 0 & 1 & 1 & 0 & 0 & 1 \end{array} \right] \to \left[\begin{array}{ccc|ccc} 1 & 3 & 0 & 1 & 0 & 0 \\ 0 & 9 & 1 & 2 & 1 & 0 \\ 0 & 1 & 1 & 0 & 0 & 1 \end{array} \right]$$

$$\rightarrow \begin{bmatrix} 1 & 3 & 0 & | & 1 & 0 & 0 \\ 0 & 9 & 1 & | & 2 & 1 & 0 \\ 0 & 0 & \frac{8}{9} & | & -\frac{2}{9} & -\frac{1}{9} & 1 \end{bmatrix} \rightarrow \begin{bmatrix} 1 & 3 & 0 & | & 1 & 0 & 0 \\ 0 & 9 & 1 & | & 2 & 1 & 0 \\ 0 & 0 & 1 & | & -\frac{1}{4} & -\frac{1}{8} & \frac{9}{8} \end{bmatrix}$$

$$\rightarrow \begin{bmatrix} 1 & 3 & 0 & | & 1 & 0 & 0 \\ 0 & 9 & 0 & | & \frac{9}{4} & \frac{9}{8} & -\frac{9}{8} \\ 0 & 0 & 1 & | & -\frac{1}{4} & -\frac{1}{8} & \frac{9}{8} \end{bmatrix} \rightarrow \begin{bmatrix} 1 & 3 & 0 & | & 1 & 0 & 0 \\ 0 & 1 & 0 & | & \frac{1}{4} & \frac{1}{8} & -\frac{1}{8} \\ 0 & 0 & 1 & | & -\frac{1}{4} & -\frac{1}{8} & \frac{9}{8} \end{bmatrix}$$

$$\rightarrow \begin{bmatrix} 1 & 0 & 0 & | & \frac{1}{4} & -\frac{3}{8} & \frac{3}{8} \\ 0 & 1 & 0 & | & \frac{1}{4} & \frac{1}{8} & -\frac{1}{8} \\ 0 & 0 & 1 & | & -\frac{1}{4} & -\frac{1}{8} & \frac{9}{8} \end{bmatrix} = \mathbf{I}|\mathbf{A}^{-1} \tag{29}$$

so

$$\mathbf{A}^{-1} = \begin{bmatrix} \frac{1}{4} & -\frac{3}{8} & \frac{3}{8} \\ \frac{1}{4} & \frac{1}{8} & -\frac{1}{8} \\ -\frac{1}{4} & -\frac{1}{8} & \frac{9}{8} \end{bmatrix}. \tag{30}$$

COMMENT.   Understand that although the method (27) was inspired by consideration of the problem $\mathbf{Ax} = \mathbf{c}$, the method (27) is free of any trace of the original problem $\mathbf{Ax} = \mathbf{c}$. For instance, in this example we were given an $\mathbf{A}$ matrix and were asked to invert it; there was no $\mathbf{Ax} = \mathbf{c}$ problem.   ■

Given an $n \times n$ matrix $\mathbf{A}$, can (27) *fail* to work, that is, to produce $\mathbf{A}^{-1}$? Yes indeed. It had better fail if $\det \mathbf{A}$ happens to be zero, for then $\mathbf{A}$ is not invertible! The way that failure would show up is that the elementary row operations would produce one or more rows of zeros on the left so that $\mathbf{A}$ could not be converted to $\mathbf{I}$. But if $\det \mathbf{A} \neq 0$, then (27) will always work.

Finally, here are some useful properties of inverses.

THEOREM 5.6.5
Properties of Inverses

1. If $\mathbf{A}$ and $\mathbf{B}$ are of the same order and invertible then $\mathbf{AB}$ is too, and

$$\boxed{(\mathbf{AB})^{-1} = \mathbf{B}^{-1}\mathbf{A}^{-1}.} \tag{31}$$

2. If $\mathbf{A}$ is invertible, then

$$\boxed{\left(\mathbf{A}^{\mathrm{T}}\right)^{-1} = \left(\mathbf{A}^{-1}\right)^{\mathrm{T}}} \tag{32}$$

and

$$\det\left(\mathbf{A}^{-1}\right) = \frac{1}{\det \mathbf{A}}. \tag{33}$$

3. If $\mathbf{A}$ is invertible, then $\left(\mathbf{A}^{-1}\right)^{-1} = \mathbf{A}$ and $(\mathbf{A}^m)^n = \mathbf{A}^{mn}$ for any (positive, negative, or zero) integers m and n.

4. If $\mathbf{A}$ is invertible, then $\mathbf{AB} = \mathbf{AC}$ implies that $\mathbf{B} = \mathbf{C}$, $\mathbf{BA} = \mathbf{CA}$ implies that $\mathbf{B} = \mathbf{C}$, $\mathbf{AB} = \mathbf{0}$ implies that $\mathbf{B} = \mathbf{0}$, and $\mathbf{BA} = \mathbf{0}$ implies that $\mathbf{B} = \mathbf{0}$.

Proof. Let us prove the framed results (31) and (32) and leave the remaining proofs as exercises. First (31): Since $\mathbf{A}$ and $\mathbf{B}$ are invertible, by assumption, $\det \mathbf{A} \neq 0$ and $\det \mathbf{B} \neq 0$. Thus, $\det(\mathbf{AB}) = (\det \mathbf{A})(\det \mathbf{B}) \neq 0$ so $\mathbf{AB}$ is invertible too. Let us denote $(\mathbf{AB})^{-1}$ as $\mathbf{C}$. Then $\mathbf{ABC} = \mathbf{I}$, $\mathbf{A}^{-1}\mathbf{ABC} = \mathbf{A}^{-1}\mathbf{I}$, $\mathbf{BC} = \mathbf{A}^{-1}$, $\mathbf{B}^{-1}\mathbf{BC} = \mathbf{B}^{-1}\mathbf{A}^{-1}$; hence, $\mathbf{C} = \mathbf{B}^{-1}\mathbf{A}^{-1}$, as claimed. As a mnemonic device, note the resemblance of (31) to the transpose formula $(\mathbf{AB})^{\mathrm{T}} = \mathbf{B}^{\mathrm{T}}\mathbf{A}^{\mathrm{T}}$.

To prove (32), begin with $\left(\mathbf{AA}^{-1}\right)^{\mathrm{T}} = \mathbf{I}^{\mathrm{T}} = \mathbf{I}$. But $\left(\mathbf{AA}^{-1}\right)^{\mathrm{T}} = \left(\mathbf{A}^{-1}\right)^{\mathrm{T}}\mathbf{A}^{\mathrm{T}}$. Hence, $\left(\mathbf{A}^{-1}\right)^{\mathrm{T}} = \left(\mathbf{A}^{\mathrm{T}}\right)^{-1}$. ∎

5.6.3 Application to a mass-spring system. To illustrate a number of these ideas with a physical application, and to emphasize the connections between the mathematics and the physics, consider the arrangement of masses and springs shown in Fig. 1. The three masses are in static equilibrium under the action of prescribed applied forces f_1, f_2, f_3, and the k's denote the known stiffnesses of the various springs. For instance, k_{12} denotes the stiffness of the spring connecting mass number 1 and mass number 2. The problem that we pose is to determine the resulting displacements x_1, x_2, x_3.

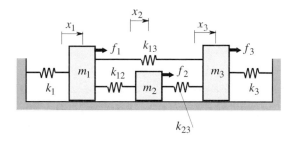

FIGURE 1   Mass-spring system.

We assume that the horizontal surface is frictionless and that the springs are neither stretched nor compressed if $x_1 = x_2 = x_3 = 0$.

The relevant physics is Newton's second law of motion and Hooke's law for each of the springs, which were discussed in Section 2.4.1. At this point, it is useful to make a concrete assumption on x_1, x_2, x_3. Specifically, let us suppose that

$$x_1 > x_2 > x_3 > 0, \tag{34}$$

as assumed in Fig. 2 (which figure, in the study of mechanics, is known as a *free-body diagram*).

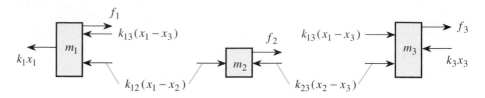

FIGURE 2   Free-body diagrams of the masses.

Consider the mass m_1. Since $x_1 > 0$, the left-hand spring is stretched so it exerts a force to the left on m_1, equal to $k_1 x_1$ according to Hooke's law, as shown in Fig. 2. Since $x_1 > x_2$, the spring of stiffness k_{12} is compressed by the amount $x_1 - x_2$, so it exerts a force $k_{12}(x_1 - x_2)$ to the left on m_1. The spring of stiffness k_{13} exerts a force $k_{13}(x_1 - x_3)$ to the left on m_1, and similarly for each of the other two masses. With the help of the information given in Fig. 2, Newton's law for each of the three masses gives the equations of motion

$$m_1 x_1'' = f_1 - k_1 x_1 - k_{12}(x_1 - x_2) - k_{13}(x_1 - x_3), \tag{35a}$$

$$m_2 x_2'' = f_2 + k_{12}(x_1 - x_2) - k_{23}(x_2 - x_3), \tag{35b}$$

$$m_3 x_3'' = f_3 + k_{13}(x_1 - x_3) + k_{23}(x_2 - x_3) - k_3 x_3, \tag{35c}$$

where primes denote differentiation with respect to the time t. Since the system is in static equilibrium by assumption, $x_1'' = x_2'' = x_3'' = 0$, and (35) becomes, in matrix form,

$$\begin{bmatrix} (k_1 + k_{12} + k_{13}) & -k_{12} & -k_{13} \\ -k_{12} & (k_{12} + k_{23}) & -k_{23} \\ -k_{13} & -k_{23} & (k_{13} + k_{23} + k_3) \end{bmatrix} \begin{bmatrix} x_1 \\ x_2 \\ x_3 \end{bmatrix} = \begin{bmatrix} f_1 \\ f_2 \\ f_3 \end{bmatrix} \tag{36}$$

or

$$\mathbf{Kx} = \mathbf{f}, \tag{37}$$

where we will call $\mathbf{K}$ the *stiffness matrix* for this system. We see that (37) is a matrix generalization of the simple Hooke's law $f = kx$ for a single spring.

Before continuing, let us explain that our assumption (34) was only for definiteness; the equations (35) are insensitive to any such assumption. For instance, suppose that

$$x_3 > x_2 > x_1 > 0 \tag{38}$$

instead. Then the free-body diagram for mass m_1 would be as shown in Fig. 3 and, in place of (35a), we would obtain

$$m_1 x_1'' = f_1 - k_1 x_1 + k_{12}(x_2 - x_1) + k_{13}(x_3 - x_1). \tag{39}$$

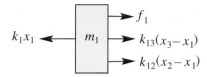

FIGURE 3   If we assume (38) instead.

But we see that (39) is really the same as (35a). Similarly for (35b) and (35c).

We now pose this question: Is there a unique solution $\mathbf{x} = \mathbf{K}^{-1}\mathbf{f}$? Experience and physical intuition probably tell us there is. In fact, we might well wonder "How could there *fail* to be unique displacements x_1, x_2, x_3 resulting from the application of the forces f_1, f_2, f_3?"

Mathematically, everything hinges upon $\det \mathbf{K}$. If $\det \mathbf{K} \neq 0$ there is a unique solution for $\mathbf{x}$, and if $\det \mathbf{K} = 0$ then there is either no solution or an infinity of solutions. With five parameters within $\mathbf{K}$ (namely, $k_1, k_{12}, k_{13}, k_{23}, k_3$) it is hard to imagine that we cannot have $\det \mathbf{K} = 0$ for some choice(s) of those parameters. Let us see. When we work out the determinant of $\mathbf{K}$, we find that

$$\det \mathbf{K} = k_1 \left(k_{12}k_{13} + k_{12}k_{23} + k_{12}k_3 + k_{23}k_{13} + k_{23}k_3 \right)$$
$$+ k_3 \left(k_{12}k_{23} + k_{12}k_{13} + k_{13}k_{23} \right). \tag{40}$$

Since each sign is positive, and the k's are positive, we see that $\det \mathbf{K} \neq 0$ so there is indeed a unique solution for $\mathbf{x}$, namely, $\mathbf{x} = \mathbf{K}^{-1}\mathbf{f}$.

However, suppose we degrade the system by removing one or more springs. We can see from (40) that even if we set any one k value equal to zero (i.e., remove that spring), $\det \mathbf{K}$ is still positive. If we are willing to remove *two* springs, then we can obtain $\det \mathbf{K} = 0$ in either of two ways: by setting $k_1 = k_3 = 0$ or by setting $k_{12} = k_{23} = 0$. Let us consider the former, and leave the latter for the exercises.

With $k_1 = k_3 = 0$, $\mathbf{K}$ is singular (noninvertible) and (37) admits either no solution or an infinity of them. Which is it, and how is such a result to be understood physically? Setting $k_1 = k_3 = 0$, (36) reduces to

$$\begin{bmatrix} (k_{12} + k_{13}) & -k_{12} & -k_{13} \\ -k_{12} & (k_{12} + k_{23}) & -k_{23} \\ -k_{13} & -k_{23} & (k_{13} + k_{23}) \end{bmatrix} \begin{bmatrix} x_1 \\ x_2 \\ x_3 \end{bmatrix} = \begin{bmatrix} f_1 \\ f_2 \\ f_3 \end{bmatrix} \tag{41}$$

and, in augmented matrix form, Gauss elimination gives

$$\begin{bmatrix} 1 & \dfrac{k_{23}}{k_{13}} & -1 - \dfrac{k_{23}}{k_{13}} & -\dfrac{f_3}{k_{13}} \\ 0 & 1 + \dfrac{k_{23}}{k_{12}} + \dfrac{k_{23}}{k_{13}} & -1 - \dfrac{k_{23}}{k_{12}} - \dfrac{k_{23}}{k_{13}} & \dfrac{f_2}{k_{12}} - \dfrac{f_3}{k_{13}} \\ 0 & 0 & 0 & f_1 + f_2 + f_3 \end{bmatrix}, \tag{42}$$

which result reveals two possibilities.

(i) If $f_1 + f_2 + f_3 \neq 0$, then there is *no* solution. That mathematical result makes perfect sense physically, because with the end springs removed $f_1 + f_2 + f_3$ is the net lateral force on the three-mass system, and if that net force is nonzero, then the system cannot be in static equilibrium, as was assumed when we set $x_1'' = x_2'' = x_3'' = 0$ in (35)!

(ii) If $f_1 + f_2 + f_3 = 0$, then we see from (42) that there is an *infinity* of solutions, of the form

$$x_3 = \alpha, \quad x_2 = \alpha + \text{etc.}, \quad x_1 = \alpha + \text{etc.},$$

where α is an arbitrary constant and the two etc.'s involve the f's and k's. Since the arbitrary additive constant α is the same for x_1, x_2, and x_3, the solution is unique to within an *arbitrary translation* α. Again that result makes sense physically because with $k_1 = k_3 = 0$ there are no end springs to restrain the three-mass system laterally, so the entire arrangement "doesn't care" if we pick it up and translate it to the left or right by any amount α.

Let us make one more important point. Observe from (36) that the **K** matrix is *symmetric*. Yet the system (Fig. 1) is not *physically* symmetric; that is, in general, $k_1 \neq k_3$ and $k_{12} \neq k_{23}$. Thus, the mathematical symmetry is somewhat unexpected and mysterious. In fact, we state without proof that for *any* number of masses interconnected with springs the resulting **K** matrix will be symmetric.

There is a striking consequence of the symmetry of **K**, which we now explain. Property (32) gives $(\mathbf{K}^{-1})^{\mathrm{T}} = (\mathbf{K}^{\mathrm{T}})^{-1} = \mathbf{K}^{-1}$ since $\mathbf{K}^{\mathrm{T}} = \mathbf{K}$. Let us denote $\mathbf{K} = \{\beta_{ij}\}$, say, and let us compare the displacement x_3 of m_3 due to a unit load $f_1 = 1$ on m_1 (with $f_2 = f_3 = 0$) with the displacement x_1 of m_1 due to a unit load $f_3 = 1$ on m_3 (with $f_1 = f_2 = 0$). In the first case,

$$\begin{bmatrix} x_1 \\ x_2 \\ x_3 \end{bmatrix} = \begin{bmatrix} \beta_{11} & \beta_{12} & \beta_{13} \\ \beta_{21} & \beta_{22} & \beta_{23} \\ \beta_{31} & \beta_{32} & \beta_{33} \end{bmatrix} \begin{bmatrix} 1 \\ 0 \\ 0 \end{bmatrix}$$

gives $x_3 = \beta_{31}$ and, in the second case,

$$\begin{bmatrix} x_1 \\ x_2 \\ x_3 \end{bmatrix} = \begin{bmatrix} \beta_{11} & \beta_{12} & \beta_{13} \\ \beta_{21} & \beta_{22} & \beta_{23} \\ \beta_{31} & \beta_{32} & \beta_{33} \end{bmatrix} \begin{bmatrix} 0 \\ 0 \\ 1 \end{bmatrix}$$

gives $x_1 = \beta_{13}$. But these are the same ($\beta_{31} = \beta_{13}$) because **K** is symmetric. In this manner we find that

$$\left(\begin{array}{c} \text{displacement } x_j \text{ of mass } m_j \\ \text{due to unit load on mass } m_k \end{array} \right) = \left(\begin{array}{c} \text{displacement } x_k \text{ of mass } m_k \\ \text{due to unit load on mass } m_j \end{array} \right). \tag{43}$$

The latter "reciprocity" result can be generalized so as to apply to any linear elastic system and is known as *Maxwell reciprocity*.[1]

---

[1] By a linear elastic system is meant one obeying Hooke's law. For a statement and proof of Maxwell's reciprocity theorem see, for instance, Den Hartog's *Advanced Strength of Materials* (New York: McGraw-Hill, 1952).

There is an electrical analog of the mechanical system shown in Fig. 1, a circuit containing resistors and voltage sources (such as batteries), and discussion of that case is left for the exercises.

In this Section 5.6.3 we have tried to emphasize the deep and intimate connection that exists between the mathematics and the physics.

5.6.4 Cramer's rule.   We have seen that if $\mathbf{A}$ is $n \times n$ and $\det \mathbf{A} \neq 0$, then $\mathbf{Ax} = \mathbf{c}$ has the unique solution

$$\mathbf{x} = \mathbf{A}^{-1}\mathbf{c}. \tag{44}$$

To focus on the individual components of $\mathbf{x}$, rather than the entire $\mathbf{x}$ vector, let us write out (44):

$$\begin{bmatrix} x_1 \\ \vdots \\ x_n \end{bmatrix} = \begin{bmatrix} \alpha_{11} & \alpha_{12} & \cdots & \alpha_{1n} \\ \vdots & \vdots & & \vdots \\ \alpha_{n1} & \alpha_{n2} & \cdots & \alpha_{nn} \end{bmatrix} \begin{bmatrix} c_1 \\ \vdots \\ c_n \end{bmatrix} = \begin{bmatrix} \sum_j \alpha_{1j} c_j \\ \vdots \\ \sum_j \alpha_{nj} c_j \end{bmatrix} \tag{45}$$

where we use the notation $\mathbf{A}^{-1} = \{\alpha_{ij}\}$ introduced in (10). Equating the ith component on the left with the ith component on the right, we have the scalar statement

$$x_i = \sum_j \alpha_{ij} c_j \tag{46}$$

for any desired i $(1 \leq i \leq n)$. Or, recalling α_{ij} from (10),

$$x_i = \sum_j \left(\frac{A_{ji}}{\det \mathbf{A}} \right) c_j = \frac{\sum_j A_{ji} c_j}{\det \mathbf{A}}. \tag{47}$$

Now, if the numerator on the right-hand side were $\sum_j A_{ji} a_{ji}$, instead, it would be recognizable as the determinant of $\mathbf{A}$, namely, the cofactor expansion about the ith column.[1] But the a_{ji}'s are replaced, in (47), by the c_j's, so the numerator of (47) amounts to a determinant but not the determinant of $\mathbf{A}$; rather, it is the determinant of the $\mathbf{A}$ matrix with its ith column replaced by the column of c_j's (or the $\mathbf{c}$ vector if you like).

The result, known as **Cramer's rule**, after *Gabriel Cramer* (1704–1752), is as follows.

THEOREM 5.6.6

Cramer's Rule

If $\mathbf{Ax} = \mathbf{c}$ where $\mathbf{A}$ is invertible, then each component x_i of $\mathbf{x}$ may be computed as the ratio of two determinants; the denominator is $\det \mathbf{A}$, and the numerator is also the determinant of the $\mathbf{A}$ matrix but with its ith column replaced by $\mathbf{c}$.

---

[1] If this claim is not clear see the last part of Example 1 of Section 5.4, the expansion about the third column.

EXAMPLE 4

Let us solve the system

$$\begin{bmatrix} 1 & 3 & 0 \\ -2 & 3 & 1 \\ 0 & 1 & 1 \end{bmatrix} \begin{bmatrix} x_1 \\ x_2 \\ x_3 \end{bmatrix} = \begin{bmatrix} 5 \\ 1 \\ -2 \end{bmatrix} \tag{48}$$

for x_1 and x_2, say, using Cramer's rule. In this case $\det \mathbf{A} = 8 \neq 0$ so the method is, first of all, applicable. Thus,

$$x_1 = \frac{\begin{vmatrix} \mathbf{5} & 3 & 0 \\ \mathbf{1} & 3 & 1 \\ \mathbf{-2} & 1 & 1 \end{vmatrix}}{\begin{vmatrix} 1 & 3 & 0 \\ -2 & 3 & 1 \\ 0 & 1 & 1 \end{vmatrix}} = \frac{1}{8} \quad \text{and} \quad x_2 = \frac{\begin{vmatrix} 1 & \mathbf{5} & 0 \\ -2 & \mathbf{1} & 1 \\ 0 & \mathbf{-2} & 1 \end{vmatrix}}{\begin{vmatrix} 1 & 3 & 0 \\ -2 & 3 & 1 \\ 0 & 1 & 1 \end{vmatrix}} = \frac{13}{8}, \tag{49}$$

where we have printed the "replacement columns" as boldface for emphasis. ∎

Cramer's rule, like the inverse matrix solution (44) from which it comes, has the advantage of being an explicit *formula*, rather than a *method*. It is also useful in that it permits us to focus on any single component of $\mathbf{x}$ without having to compute the entire $\mathbf{x}$ vector.

5.6.5 LU factorization. (Optional). This final subsection is not really about the inverse matrix, or about the inverse matrix method of solving $\mathbf{Ax} = \mathbf{c}$. Rather, it is about an alternative method of solution that is based upon the factorization of an $n \times n$ matrix $\mathbf{A}$ as a lower triangular matrix $\mathbf{L}$ times an upper triangular matrix $\mathbf{U}$:

$$\mathbf{A} = \mathbf{LU} = \begin{bmatrix} l_{11} & 0 & 0 \\ l_{21} & l_{22} & 0 \\ l_{31} & l_{32} & l_{33} \end{bmatrix} \begin{bmatrix} u_{11} & u_{12} & u_{13} \\ 0 & u_{22} & u_{23} \\ 0 & 0 & u_{33} \end{bmatrix}, \tag{50}$$

where we have taken $n = 3$ simply to illustrate. If we carry out the multiplication on the right and equate the nine elements of $\mathbf{LU}$ to the corresponding elements of $\mathbf{A}$ we obtain nine equations in the twelve unknown l_{ij}'s and u_{ij}'s. Since we have more unknowns than equations, there is some flexibility in implementing the idea. Hence there are various versions of LU factorization.

According to **Doolittle's method** we can set each $l_{jj} = 1$ in $\mathbf{L}$ (i.e., the three diagonal elements) and solve uniquely for the remaining l_{ij}'s and the u_{ij}'s. With $\mathbf{L}$ and $\mathbf{U}$ determined, we then solve $\mathbf{Ax} = \mathbf{LUx} = \mathbf{c}$ by setting $\mathbf{Ux} = \mathbf{y}$, so that $\mathbf{L(Ux)} = \mathbf{c}$ breaks into the two problems

$$\mathbf{Ly} = \mathbf{c}, \tag{51a}$$

$$\mathbf{Ux} = \mathbf{y}, \tag{51b}$$

each of which is simple because $\mathbf{L}$ and $\mathbf{U}$ are triangular. We solve (51a) for $\mathbf{y}$, put that $\mathbf{y}$ into (51b), and then solve (51b) for $\mathbf{x}$. Let us illustrate the procedure.

EXAMPLE 5

To solve

$$\begin{bmatrix} 2 & -3 & 3 \\ 6 & -8 & 7 \\ -2 & 6 & -1 \end{bmatrix} \begin{bmatrix} x_1 \\ x_2 \\ x_3 \end{bmatrix} = \begin{bmatrix} -2 \\ -3 \\ 3 \end{bmatrix} \tag{52}$$

by the Doolittle LU factorization method, we first need to determine **L** and **U** by equating

$$\begin{bmatrix} 2 & -3 & 3 \\ 6 & -8 & 7 \\ -2 & 6 & -1 \end{bmatrix} = \begin{bmatrix} 1 & 0 & 0 \\ l_{21} & 1 & 0 \\ l_{31} & l_{32} & 1 \end{bmatrix} \begin{bmatrix} u_{11} & u_{12} & u_{13} \\ 0 & u_{22} & u_{23} \\ 0 & 0 & u_{33} \end{bmatrix}$$

$$= \begin{bmatrix} \underline{u_{11}} & \underline{u_{12}} & \underline{u_{13}} \\ \underline{l_{21}}u_{11} & l_{21}u_{12} + \underline{u_{22}} & l_{21}u_{13} + \underline{u_{23}} \\ \underline{l_{31}}u_{11} & l_{31}u_{12} + \underline{l_{32}}u_{22} & l_{31}u_{13} + l_{32}u_{23} + \underline{u_{33}} \end{bmatrix}. \tag{53}$$

Matching $a_{11}, a_{12}, a_{13}, a_{21}, \ldots, a_{32}, a_{33}$ [i.e., the elements of the 3×3 matrix in (52)] to the corresponding terms on the right gives a sequence of equations for $u_{11}, u_{12}, u_{13}, l_{21}, u_{22}, u_{23}, l_{31}, l_{32},$ and u_{33} (i.e., the underlined entries) in turn:

$$\begin{aligned}
u_{11} &= 2 \\
u_{12} &= -3 \\
u_{13} &= 3 \\
l_{21} &= 6/u_{11} = 6/2 = 3 \\
u_{22} &= -8 - l_{21}u_{12} = -8 - (3)(-3) = 1 \\
u_{23} &= 7 - l_{21}u_{13} = 7 - (3)(3) = -2 \\
l_{31} &= -2/u_{11} = -2/2 = -1 \\
l_{32} &= (6 - l_{31}u_{12})/u_{22} = [6 - (-1)(-3)]/1 = 3 \\
u_{33} &= -1 - l_{31}u_{13} - l_{32}u_{23} = -1 - (-1)(3) - (3)(-2) = 8.
\end{aligned} \tag{54}$$

Then (51a) becomes

$$\begin{bmatrix} 1 & 0 & 0 \\ 3 & 1 & 0 \\ -1 & 3 & 1 \end{bmatrix} \begin{bmatrix} y_1 \\ y_2 \\ y_3 \end{bmatrix} = \begin{bmatrix} -2 \\ -3 \\ 3 \end{bmatrix},$$

which gives $\mathbf{y} = [-2, 3, -8]^{\mathrm{T}}$. Finally, (51b) becomes

$$\begin{bmatrix} 2 & -3 & 3 \\ 0 & 1 & -2 \\ 0 & 0 & 8 \end{bmatrix} \begin{bmatrix} x_1 \\ x_2 \\ x_3 \end{bmatrix} = \begin{bmatrix} -2 \\ -3 \\ 3 \end{bmatrix},$$

which gives the final solution $\mathbf{x} = [2, 1, -1]^{\mathrm{T}}$. ∎

The beauty of the method is that the l_{ij}'s and u_{ij}'s are found not by solving simultaneous coupled equations, but by solving a sequence of linear equations in only one unknown [as illustrated in (54)]. With $\mathbf{L}$ and $\mathbf{U}$ thus determined, the solution of (51a) and (51b) is likewise simple since $\mathbf{L}$ and $\mathbf{U}$ are triangular. In fact, the method is approximately twice as fast as Gauss-Jordan elimination.

We can always successfully factor an $n \times n$ matrix $\mathbf{A}$ as a lower triangular matrix $\mathbf{L}$ times an upper triangular matrix $\mathbf{U}$, whether or not $\det \mathbf{A} \neq 0$. However, if $\det \mathbf{A} = 0$, then (51) will yield either no solution for $\mathbf{x}$ or a nonunique solution.

Closure.   The inverse of a matrix $\mathbf{A}$, denoted as $\mathbf{A}^{-1}$, exists if and only if $\mathbf{A}$ is square $(n \times n)$ and $\det \mathbf{A} \neq 0$. If it exists it is given uniquely as

$$\mathbf{A}^{-1} = \frac{1}{\det \mathbf{A}} \operatorname{adj} \mathbf{A}, \tag{55}$$

where the matrix $\operatorname{adj} \mathbf{A}$ is the adjoint of $\mathbf{A}$ (the transpose of the cofactor matrix), and is such that

$$\mathbf{A}^{-1}\mathbf{A} = \mathbf{A}\mathbf{A}^{-1} = \mathbf{I}. \tag{56}$$

The case where $\mathbf{A}$ is not invertible (i.e., is singular) is the exceptional case; in the generic case a given $n \times n$ matrix is invertible. That is, if we make up an $n \times n$ matrix at random, it is only by extreme coincidence that its determinant will be zero. Besides (56), several useful properties of inverses were given in Theorem 5.6.5.

If $\mathbf{A}$ is invertible, then $\mathbf{A}\mathbf{x} = \mathbf{c}$ admits the unique solution

$$\mathbf{x} = \mathbf{A}^{-1}\mathbf{c} = \frac{1}{\det \mathbf{A}} (\operatorname{adj} \mathbf{A}) \, \mathbf{c}, \tag{57}$$

which result enabled us to derive Cramer's rule, whereby each component of $\mathbf{x}$ is expressed as the ratio of $n \times n$ determinants.

Notice that the equation $ax = c$ has a unique solution if and only if $a \neq 0$. For $\mathbf{A}\mathbf{x} = \mathbf{c}$, where $\mathbf{A}$ is $n \times n$, that condition generalizes *not* to $\mathbf{A} \neq \mathbf{0}$ but to $\det \mathbf{A} \neq 0$. That is, if $\mathbf{A}$ is $n \times n$ then $\mathbf{A}\mathbf{x} = \mathbf{c}$ has a unique solution if and only if $\det \mathbf{A} \neq 0$, which result is one of the most important results in all of linear algebra.

In contrast with Gauss-Jordan reduction and LU factorization, which are solution *methods*, (57) and Cramer's rule are explicit *formulas* for the solution (when a unique solution does exist).

Finally, we urge you to be careful with the sequencing of matrices because of the general absence of commutativity under multiplication. For instance $\mathbf{A}\mathbf{x} = \mathbf{c}$ implies $\mathbf{x} = \mathbf{A}^{-1}\mathbf{c}$ (if $\mathbf{A}$ is invertible), NOT $\mathbf{x} = \mathbf{c}\mathbf{A}^{-1}$. Indeed, the product $\mathbf{c}\mathbf{A}^{-1}$ is not even defined since $\mathbf{c}$ is $n \times 1$ and $\mathbf{A}^{-1}$ is $n \times n$.

Maple.   The relevant *Maple* command is **inverse**, within the linalg package. For instance, to invert

$$\mathbf{A} = \begin{bmatrix} 1 & 2 \\ 3 & 4 \end{bmatrix},$$

enter

```
with(linalg):
```

and return. Then the steps

    A:=array([[1,2],[3,4]]):

and

    inverse(A);

give the result

$$\begin{bmatrix} -2 & 1 \\ \frac{3}{2} & -\frac{1}{2} \end{bmatrix}$$

EXERCISES 5.6

1. Use (10) to evaluate the inverse matrix. If the matrix is not invertible, state that.

(a) $\begin{bmatrix} a & b \\ c & d \end{bmatrix}$, where $ad - bc \neq 0$

(b) $\begin{bmatrix} 5 & 4 \\ 3 & 2 \end{bmatrix}$

(c) $\begin{bmatrix} 0 & 1 \\ 1 & 0 \end{bmatrix}$

(d) $\begin{bmatrix} 2 & 1 \\ 2 & -1 \end{bmatrix}$

(e) $\begin{bmatrix} 3 & 0 \\ 0 & 4 \end{bmatrix}$

(f) $\begin{bmatrix} 1 & 2 & 1 \\ 2 & 1 & 3 \\ 0 & 3 & -1 \end{bmatrix}$

(g) $\begin{bmatrix} 0 & 1 & 0 \\ 2 & 0 & 5 \\ 0 & 0 & 3 \end{bmatrix}$

(h) $\begin{bmatrix} 1 & 1 & 1 \\ 0 & 1 & 1 \\ 0 & 0 & 1 \end{bmatrix}$

(i) $\begin{bmatrix} 0 & 0 & 2 \\ 0 & -1 & 0 \\ 1 & 0 & 0 \end{bmatrix}$

(j) $\begin{bmatrix} 3 & 1 & -2 \\ 1 & 2 & 1 \\ 4 & 3 & -1 \end{bmatrix}$

(k) $\begin{bmatrix} 7 & 1 & 3 \\ 2 & -1 & 1 \\ 0 & 1 & 4 \end{bmatrix}$

(l) $\begin{bmatrix} 1 & 0 & 2 \\ 0 & 3 & 0 \\ 4 & 0 & 5 \end{bmatrix}$

(m) $\begin{bmatrix} 2 & 3 & -5 \\ 12 & 1 & 0 \\ 3 & 4 & 1 \end{bmatrix}$

(n) $\begin{bmatrix} 2 & 0 & 0 & 0 \\ 0 & 1 & 3 & 0 \\ 0 & 0 & 1 & 0 \\ 0 & 0 & 4 & 1 \end{bmatrix}$

(o) $\begin{bmatrix} 2 & 3 & 0 & 0 \\ 1 & 1 & 0 & 0 \\ 0 & 0 & 1 & 0 \\ 0 & 0 & 4 & 1 \end{bmatrix}$

(p) $\begin{bmatrix} 7 & 1 & 3 & 0 \\ 2 & -1 & 1 & 0 \\ 0 & 1 & 4 & 0 \\ 0 & 0 & 0 & 1 \end{bmatrix}$

(q) $\begin{bmatrix} \cos\theta & -\sin\theta \\ \sin\theta & \cos\theta \end{bmatrix}$

(r) $\begin{bmatrix} \cos\theta & -\sin\theta & 0 \\ \sin\theta & \cos\theta & 0 \\ 0 & 0 & 1 \end{bmatrix}$

(s) $\begin{bmatrix} \cos\theta & 0 & -\sin\theta \\ 0 & 1 & 0 \\ \sin\theta & 0 & \cos\theta \end{bmatrix}$

2. (c)–(o) Evaluate the inverse of the matrix given in the corresponding part of Exercise 1 using elementary row operations, as we did in Example 3.

3. (c)–(o) Evaluate the inverse of the matrix given in the corresponding part of Exercise 1 using *Maple*.

4. (*Block-diagonal matrices*) If an $n \times n$ matrix **A** can be partitioned as

$$\mathbf{A} = \begin{bmatrix} \mathbf{A}_1 & \mathbf{0} & \cdots & \mathbf{0} \\ \mathbf{0} & \mathbf{A}_2 & & \\ \vdots & & \ddots & \vdots \\ \mathbf{0} & & \cdots & \mathbf{A}_k \end{bmatrix} \quad (k \leq n)$$

it is said to be **block diagonal** All of the $\mathbf{A}_j$ submatrices need to be square, although not necessarily of equal order, with their main diagonals coinciding with the main diagonal of **A**. For instance,

$$\mathbf{A} = \begin{bmatrix} 2 & 1 & 0 & 0 & 0 & 0 \\ 1 & 2 & 0 & 0 & 0 & 0 \\ 0 & 0 & -1 & 0 & 3 & 0 \\ 0 & 0 & 2 & 5 & 1 & 0 \\ 0 & 0 & 3 & 1 & 0 & 0 \\ 0 & 0 & 0 & 0 & 0 & 4 \end{bmatrix} = \begin{bmatrix} \mathbf{A}_1 & \mathbf{0} & \mathbf{0} \\ \mathbf{0} & \mathbf{A}_2 & \mathbf{0} \\ \mathbf{0} & \mathbf{0} & \mathbf{A}_3 \end{bmatrix}$$

(4.1)

is block diagonal. Such matrices exhibit essentially the same simple features as diagonal matrices.

(a) Show that **A** is invertible if and only if $\mathbf{A}_1, \ldots, \mathbf{A}_k$ are. HINT: Recall equation (9.1) in Exercise 9, Section 5.4.

(b) Assuming that $\mathbf{A}_1, \ldots, \mathbf{A}_k$ are invertible, verify that

$$\mathbf{A}^{-1} = \begin{bmatrix} \mathbf{A}_1^{-1} & \mathbf{0} & \cdots & \mathbf{0} \\ \mathbf{0} & \mathbf{A}_2^{-1} & & \\ \vdots & & \ddots & \vdots \\ \mathbf{0} & & \cdots & \mathbf{A}_k^{-1} \end{bmatrix}.$$

(c) Use the latter result to evaluate A^{-1}, where A is given in (4.1).

5. Solve for x_1 and x_2 by Cramer's rule.

(a) $\begin{aligned} x_1 + 4x_2 &= 0 \\ 3x_1 - x_2 &= 6 \end{aligned}$

(b) $\begin{aligned} ax_1 + bx_2 &= c \\ dx_1 + ex_2 &= f \end{aligned}$

(c) $\begin{aligned} x_1 - 2x_2 + x_3 &= 4 \\ 2x_1 + 3x_2 + x_3 &= -7 \\ 4x_1 + x_2 + 2x_3 &= 0 \end{aligned}$

(d) $\begin{aligned} x_1 + 2x_2 + 3x_3 &= 9 \\ x_1 + 4x_2 &= 6 \\ x_1 - 5x_3 &= 2 \end{aligned}$

(e) $\begin{aligned} 2x_1 + x_2 &= 1 \\ x_1 + 2x_2 + x_3 &= 0 \\ x_2 + 2x_3 + x_4 &= 0 \\ x_3 + 2x_4 &= 0 \end{aligned}$

(f) $\begin{aligned} x_1 + x_2 + x_3 &= 1 \\ x_1 + 2x_2 + 3x_3 &= 0 \\ x_1 - x_2 + 4x_3 &= 0 \end{aligned}$

6. (a) Given a certain 3×3 matrix A, we find its inverse to be

$$A^{-1} = \begin{bmatrix} 1 & 2 & 0 \\ 0 & 1 & 0 \\ 3 & 0 & 0 \end{bmatrix}.$$

Can that result be correct? Explain.

(b) Same as (a), for

$$A^{-1} = \begin{bmatrix} 1 & 1 & 1 \\ 2 & 2 & 2 \\ 3 & 0 & 0 \end{bmatrix}.$$

7. If A^{-1} is the given matrix, find A.

(a) $\begin{bmatrix} 3 & -1 \\ 3 & -2 \end{bmatrix}$ **(b)** $\begin{bmatrix} 1 & 2 \\ 3 & 4 \end{bmatrix}$

(c) $\begin{bmatrix} 1 & 2 & 1 \\ 0 & 3 & 1 \\ 1 & 1 & 0 \end{bmatrix}$ **(d)** $\begin{bmatrix} 2 & 1 & 1 \\ 4 & 1 & 1 \\ 6 & 2 & 1 \end{bmatrix}$

8. Suppose that $Ax = c$, where A is 3×3 and x and c are 3×1, and that to the c vectors

$$c = \begin{bmatrix} 1 \\ 0 \\ 0 \end{bmatrix}, \quad \begin{bmatrix} 0 \\ 1 \\ 0 \end{bmatrix}, \quad \begin{bmatrix} 0 \\ 0 \\ 1 \end{bmatrix}$$

there correspond unique solutions

$$x = \begin{bmatrix} 2 \\ 5 \\ -1 \end{bmatrix}, \quad \begin{bmatrix} 1 \\ 1 \\ 2 \end{bmatrix}, \quad \begin{bmatrix} 3 \\ 0 \\ 4 \end{bmatrix},$$

respectively. Then what is the solution x corresponding to $c = [4, 3, -1]^T$? Is it unique? Explain.

9. Suppose that Gauss elimination gives the solution of a linear system $Ax = c$ as $x = x_0 + \alpha_1 x_1 + \alpha_2 x_2$, where A is 6×6 and α_1 and α_2 are arbitrary. Is A invertible? Explain.

10. (*Nilpotent matrices*) If there is some positive integer p such that $A^p = 0$, then A said to be **nilpotent** (i.e., potentially nil).

(a) Show that a nilpotent matrix is necessarily singular.

(b) If A is nilpotent, with $A^p = 0$,

show that

$$(I - A)^{-1} = I + A + A^2 + \cdots + A^{p-1}. \qquad (10.1)$$

11. First, read Exercise 10. Use (10.1) to find the inverse of the given matrix. HINT: You will need to identify the A matrix in (10.1).

(a) $\begin{bmatrix} 1 & 2 & 3 \\ 0 & 1 & 8 \\ 0 & 0 & 1 \end{bmatrix}$ **(b)** $\begin{bmatrix} 1 & 0 & 0 \\ -3 & 1 & 0 \\ 2 & 7 & 1 \end{bmatrix}$

(c) $\begin{bmatrix} 1 & 5 & 1 & 0 \\ 0 & 1 & 2 & 7 \\ 0 & 0 & 1 & 3 \\ 0 & 0 & 0 & 1 \end{bmatrix}$ **(d)** $\begin{bmatrix} 1 & 0 & 0 & 0 \\ -4 & 1 & 0 & 0 \\ 0 & 3 & 1 & 0 \\ 2 & 1 & 6 & 1 \end{bmatrix}$

12. First, read Exercise 10. Use (10.1) to find the inverse of the given matrix

(a) $\begin{bmatrix} 2 & 8 & 10 \\ 0 & 3 & 12 \\ 0 & 0 & 4 \end{bmatrix}$ HINT: $\begin{bmatrix} 2 & 8 & 10 \\ 0 & 3 & 12 \\ 0 & 0 & 4 \end{bmatrix}^{-1}$

$$= \left[\begin{bmatrix} 2 & 0 & 0 \\ 0 & 3 & 0 \\ 0 & 0 & 4 \end{bmatrix} \begin{bmatrix} 1 & 4 & 5 \\ 0 & 1 & 4 \\ 0 & 0 & 1 \end{bmatrix} \right]^{-1}$$

$$= \begin{bmatrix} 1 & 4 & 5 \\ 0 & 1 & 4 \\ 0 & 0 & 1 \end{bmatrix}^{-1} \begin{bmatrix} \frac{1}{2} & 0 & 0 \\ 0 & \frac{1}{3} & 0 \\ 0 & 0 & \frac{1}{4} \end{bmatrix}.$$

(b) $\begin{bmatrix} 3 & 0 & 0 \\ 4 & -2 & 0 \\ 10 & 0 & 2 \end{bmatrix}$ **(c)** $\begin{bmatrix} 2 & 0 & 0 \\ 4 & 2 & 0 \\ 1 & 6 & 2 \end{bmatrix}$

13. (*Ill-conditioned systems*) Consider the system

$$\begin{bmatrix} 1 & \frac{1}{2} & \frac{1}{3} \\ \frac{1}{2} & \frac{1}{3} & \frac{1}{4} \\ \frac{1}{3} & \frac{1}{4} & \frac{1}{5} \end{bmatrix} \begin{bmatrix} x_1 \\ x_2 \\ x_3 \end{bmatrix} = \begin{bmatrix} 1 \\ 2 \\ 2 \end{bmatrix} \quad \text{or} \quad Ax = c.$$

$$(13.1)$$

Here A is a third-order **Hilbert matrix**, named after *David Hilbert* (1862–1943).

(a) Evaluate A^{-1}, by any analytical means that you wish, and show that $x = A^{-1}c = [-3, -12, 30]^T$.

(b) To simulate the effects of roundoff error, consider in place of (13.1) the rounded off system

$$\begin{bmatrix} 1 & 0.5 & 0.33 \\ 0.5 & 0.33 & 0.25 \\ 0.33 & 0.25 & 0.2 \end{bmatrix} \begin{bmatrix} x_1 \\ x_2 \\ x_3 \end{bmatrix} = \begin{bmatrix} 1 \\ 2 \\ 2 \end{bmatrix}. \quad (13.2)$$

Solving (13.2) by any means you wish (computer software being the easiest), show that the solution of (13.2) is $\mathbf{x} \approx [11.1, -84.1, 96.8]^T$. NOTE: In this example we see that only a slight error in $\mathbf{A}$ leads to a disproportionately large error in the solution. Hence, (13.1) is said to be **ill-conditioned**. (Ill-conditioned systems were also mentioned in Exercise 19, Section 4.5.) In applications it is important to know if a given system is ill-conditioned so that steps can be taken to obtain a sufficiently accurate solution. According to one criterion in the literature, an $n \times n$ matrix $\mathbf{A}$ may be considered as ill-conditioned if

$$\frac{|\det \mathbf{A}|}{\sqrt{\sum_{i=1}^{n} \sum_{j=1}^{n} a_{ij}^2}} \ll 1. \quad (13.3)$$

For the Hilbert matrix in (13.1), the left-hand side of (13.3) is 0.00033, which is indeed much smaller than unity.

(c) In place of (13.2), use the more accurate rounded off system

$$\begin{bmatrix} 1 & 0.5 & 0.333 \\ 0.5 & 0.333 & 0.25 \\ 0.333 & 0.25 & 0.2 \end{bmatrix} \begin{bmatrix} x_1 \\ x_2 \\ x_3 \end{bmatrix} = \begin{bmatrix} 1 \\ 2 \\ 2 \end{bmatrix}$$
$$(13.4)$$

and see how much closer the solution of (13.4) comes to the exact solution $\mathbf{x} = [-3, -12, 30]^T$ of (13.1). Also, compute the left-hand side of (13.3) for the $\mathbf{A}$ matrix in (13.4).

14. Prove property 3 in Theorem 5.6.5.

15. Prove property 4 in Theorem 5.6.5.

16. In Section 5.6.3 we stated that if we are willing to remove two springs then we can have $\det \mathbf{K} = 0$ either by setting $k_1 = k_3 = 0$ or by setting $k_{12} = k_{23} = 0$. We discussed the former choice, $k_1 = k_3 = 0$, both mathematically and physically as well. Do the same for the latter choice, $k_{12} = k_{23} = 0$.

17. (*A dc circuit*) Application of Kirchhoff's laws to the circuit shown in Example 2 of Section 4.5.1 produced the five equations within equation (5) on the currents i_1, i_2, i_3. Of these equations, the second is the same as the first, and the

fifth is the fourth minus the third. Thus, deleting those two redundant equations leaves the system

$$\begin{aligned} i_1 - i_2 - i_3 &= 0 \\ R_2 i_2 - R_3 i_3 &= 0 \qquad (17.1) \\ R_1 i_1 + R_2 i_2 \quad\quad &= E. \end{aligned}$$

(a) Show that the determinant of the coefficient matrix in (17.1) is necessarily nonzero, so that the system $\mathbf{Ri} = \mathbf{e}$ given by (17.1) necessarily admits the unique solution $\mathbf{i} = \mathbf{R}^{-1}\mathbf{e}$. NOTE: $R_1 > 0$, $R_2 > 0$, and $R_3 > 0$.

(b) Solve for $\mathbf{i}$ by the inverse matrix method. Also, solve for i_1, i_2, i_3 by Cramer's rule and verify that the results are the same.

(c) Suppose, instead, that we allow one or more of the resistances to be zero so that $R_1 \geq 0$, $R_2 \geq 0$, $R_3 \geq 0$. Show that if any two, or all three, of the resistances are zero, then the determinant *does* vanish so that equations (17.1) admit either no solution or an infinity of solutions. For each of these four "singular" cases determine whether there is no solution or an infinity of solutions by applying Gauss elimination. Explain the *physical significance* of each of these results insofar as possible.

18. (*Circuit analog*) The electrical circuit analog of the mass-spring system in Fig. 1 is shown in the figure below.

(a) Applying Kirchhoff's voltage law, show that

$$\begin{bmatrix} R_1 + R_{12} + R_{13} & -R_{12} & -R_{13} \\ -R_{12} & R_{12} + R_{23} & -R_{23} \\ -R_{13} & -R_{23} & R_{13} + R_{23} + R_3 \end{bmatrix} \begin{bmatrix} I_1 \\ I_2 \\ I_3 \end{bmatrix}$$
$$= \begin{bmatrix} E_1 \\ E_2 \\ E_3 \end{bmatrix} \quad (18.1)$$

which is the analog of (36) under the correspondence $R_{ij} \leftrightarrow k_{ij}$, $I_j \leftrightarrow x_j$, $E_j \leftrightarrow f_j$.

(b) Discuss the existence and uniqueness of solutions of (18.1) in the same way that we did that for the mass-spring system in Section 5.6.3, including a reciprocity result analogous to (43).

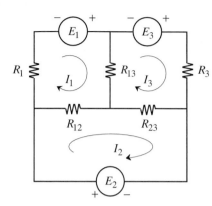

$$\text{(c)} \begin{bmatrix} 2 & 5 & 1 \\ 2 & 8 & 0 \\ 8 & 2 & 2 \end{bmatrix} \begin{bmatrix} x_1 \\ x_2 \\ x_3 \end{bmatrix} = \begin{bmatrix} 0 \\ -7 \\ 10 \end{bmatrix}$$

$$\text{(d)} \begin{bmatrix} 1 & 3 & -1 \\ 2 & 2 & 0 \\ 3 & 1 & 1 \end{bmatrix} \begin{bmatrix} x_1 \\ x_2 \\ x_3 \end{bmatrix} = \begin{bmatrix} 0 \\ 6 \\ 12 \end{bmatrix}$$

20. (*Singular case*, $\det \mathbf{A} = 0$) To see how the Doolittle LU-factorization method works when $\mathbf{A}$ is singular, apply the method to each of the following systems.

$$\text{(a)} \begin{bmatrix} 1 & 3 \\ 2 & 6 \end{bmatrix} \begin{bmatrix} x_1 \\ x_2 \end{bmatrix} = \begin{bmatrix} 4 \\ 3 \end{bmatrix}$$

$$\text{(b)} \begin{bmatrix} 1 & 3 \\ 2 & 6 \end{bmatrix} \begin{bmatrix} x_1 \\ x_2 \end{bmatrix} = \begin{bmatrix} 4 \\ 8 \end{bmatrix}$$

19. (*Regular case*, $\det \mathbf{A} \neq 0$) Solve by the Doolittle LU factorization method.

$$\text{(a)} \begin{bmatrix} 2 & 3 \\ 8 & -1 \end{bmatrix} \begin{bmatrix} x_1 \\ x_2 \end{bmatrix} = \begin{bmatrix} -4 \\ 10 \end{bmatrix}$$

$$\text{(b)} \begin{bmatrix} 2 & -1 \\ 2 & 1 \end{bmatrix} \begin{bmatrix} x_1 \\ x_2 \end{bmatrix} = \begin{bmatrix} 7 \\ 13 \end{bmatrix}$$

21. (*Dual or reciprocal set*) In Exercise 13 of Section 4.9 we introduced the concept of a set of dual or reciprocal vectors $\{\mathbf{e}_1^*, \dots, \mathbf{e}_n^*\}$ corresponding to a basis $\{\mathbf{e}_1, \dots, \mathbf{e}_n\}$ that is not necessarily orthogonal, or ON. Having learned more about the solution of systems of linear algebraic equations, we can now return to that exercise and prove the claim made in part (a) therein. Specifically, prove that the dual set exists, is unique, and is itself a basis for $\mathbb{R}^n$.

5.7 EXISTENCE AND UNIQUENESS FOR THE SYSTEM Ax = c

5.7.1 Existence and uniqueness in terms of rank. Linear algebra has important and diverse applications but a central theme is always the fundamental problem of m linear algebraic equations in n unknowns,

$$\mathbf{Ax} = \mathbf{c}. \tag{1}$$

In Section 4.5 we showed how to solve any such system by the systematic and efficient method of Gauss elimination, and in Section 5.6.1 we considered the special case where $m = n$ and $\det \mathbf{A} \neq 0$ (i.e., the case where $\mathbf{A}$ is invertible, not singular) and showed how to find the inverse matrix $\mathbf{A}^{-1}$ and hence the unique solution $\mathbf{x} = \mathbf{A}^{-1}\mathbf{c}$ of (1).

In the present section we return to the questions of existence and uniqueness for the system (1) and answer those questions in terms of the rank of two matrices, the coefficient matrix $\mathbf{A}$ and the augmented matrix $\mathbf{A}|\mathbf{c}$. In doing so, it will help to have a concrete example before us.

EXAMPLE 1

Consider the system $\mathbf{Ax} = \mathbf{c}$ given by

$$
\begin{aligned}
x_1 - x_2 + x_3 + 3x_4 + 2x_6 &= 4, \\
x_1 + 3x_3 + 3x_4 - x_5 + 6x_6 &= 3, \\
2x_1 - x_2 + 2x_3 + x_4 - x_5 + 7x_6 &= 9, \\
x_1 + 5x_3 + 8x_4 - x_5 + 7x_6 &= 1.
\end{aligned}
\tag{2}
$$

Carrying out Gauss elimination by applying elementary row operations to the augmented matrix

$$
\mathbf{A}|\mathbf{c} =
\begin{bmatrix}
1 & -1 & 1 & 3 & 0 & 2 & 4 \\
1 & 0 & 3 & 3 & -1 & 6 & 3 \\
2 & -1 & 2 & 1 & -1 & 7 & 9 \\
1 & 0 & 5 & 8 & -1 & 7 & 1
\end{bmatrix}
\tag{3}
$$

gives

$$
\begin{bmatrix}
1 & -1 & 1 & 3 & 0 & 2 & 4 \\
0 & 1 & 2 & 0 & -1 & 4 & -1 \\
0 & 0 & 2 & 5 & 0 & 1 & -2 \\
0 & 0 & 0 & 0 & 0 & 0 & 0
\end{bmatrix}
\tag{4}
$$

and hence the three-parameter family of solutions

$$
\begin{aligned}
x_6 = \alpha_1, \quad x_5 = \alpha_2, \quad x_4 = \alpha_3, \quad x_3 = -1 - \tfrac{1}{2}\alpha_1 - \tfrac{5}{2}\alpha_3, \\
x_2 = 1 - 3\alpha_1 + \alpha_2 + 5\alpha_3, \quad x_1 = 6 - \tfrac{9}{2}\alpha_1 + \alpha_2 + \tfrac{9}{2}\alpha_3,
\end{aligned}
\tag{5}
$$

where the parameters $\alpha_1, \alpha_2, \alpha_3$ are arbitrary.

It is illuminating to express (5) in vector form as

$$
\mathbf{x} =
\begin{bmatrix}
6 - \tfrac{9}{2}\alpha_1 + \alpha_2 + \tfrac{9}{2}\alpha_3 \\
1 - 3\alpha_1 + \alpha_2 + 5\alpha_3 \\
-1 - \tfrac{1}{2}\alpha_1 - \tfrac{5}{2}\alpha_3 \\
\alpha_3 \\
\alpha_2 \\
\alpha_1
\end{bmatrix}
=
\begin{bmatrix}
6 \\ 1 \\ -1 \\ 0 \\ 0 \\ 0
\end{bmatrix}
+ \alpha_1
\begin{bmatrix}
-\tfrac{9}{2} \\ -3 \\ -\tfrac{1}{2} \\ 0 \\ 0 \\ 1
\end{bmatrix}
+ \alpha_2
\begin{bmatrix}
1 \\ 1 \\ 0 \\ 0 \\ 1 \\ 0
\end{bmatrix}
+ \alpha_3
\begin{bmatrix}
\tfrac{9}{2} \\ 5 \\ -\tfrac{5}{2} \\ 1 \\ 0 \\ 0
\end{bmatrix}
$$

$$
= \mathbf{x}_p + \alpha_1 \mathbf{x}_1 + \alpha_2 \mathbf{x}_2 + \alpha_3 \mathbf{x}_3.
\tag{6}
$$

To understand the significance of $\mathbf{x}_p, \mathbf{x}_1, \mathbf{x}_2$, and $\mathbf{x}_3$ in (6), let us put (6) into $\mathbf{Ax} = \mathbf{c}$:

$$
\mathbf{A}\left(\mathbf{x}_p + \alpha_1 \mathbf{x}_1 + \alpha_2 \mathbf{x}_2 + \alpha_3 \mathbf{x}_3\right) = \mathbf{c},
\tag{7}
$$

hence,

$$
\mathbf{Ax}_p + \alpha_1 \mathbf{Ax}_1 + \alpha_2 \mathbf{Ax}_2 + \alpha_3 \mathbf{Ax}_3 = \mathbf{c}.
\tag{8}
$$

Since (6) is a solution for *any* α's, we can set $\alpha_1 = \alpha_2 = \alpha_3 = 0$ in (8) and thereby learn that $\mathbf{Ax}_p = \mathbf{c}$. Thus, $\mathbf{x}_p$ is itself a solution of $\mathbf{Ax} = \mathbf{c}$, a *particular solution* (i.e., one specific solution). Therefore we can cancel the $\mathbf{A}\mathbf{x}_p$ and $\mathbf{c}$ terms in (8) and obtain

$$\alpha_1 \mathbf{A}\mathbf{x}_1 + \alpha_2 \mathbf{A}\mathbf{x}_2 + \alpha_3 \mathbf{A}\mathbf{x}_3 = \mathbf{0}, \tag{9}$$

which holds for any choice of the α's. The choice $\alpha_1 = 1, \alpha_2 = \alpha_3 = 0$ in (9) reveals that $\mathbf{A}\mathbf{x}_1 = \mathbf{0}$; similarly, $\alpha_2 = 1, \alpha_1 = \alpha_3 = 0$ reveals that $\mathbf{A}\mathbf{x}_2 = \mathbf{0}$ and $\alpha_3 = 1, \alpha_1 = \alpha_2 = 0$ reveals that $\mathbf{A}\mathbf{x}_3 = \mathbf{0}$, so each of $\mathbf{x}_1, \mathbf{x}_2, \mathbf{x}_3$ is a *homogeneous solution* of $\mathbf{A}\mathbf{x} = \mathbf{c}$, namely a solution of the homogeneous version $\mathbf{A}\mathbf{x} = \mathbf{0}$.

Observe further that $\mathbf{x}_1, \mathbf{x}_2, \mathbf{x}_3$ are LI, for the rank of

$$[\mathbf{x}_1, \mathbf{x}_2, \mathbf{x}_3] = \begin{bmatrix} -\frac{9}{2} & 1 & \frac{9}{2} \\ -3 & 1 & 5 \\ -\frac{1}{2} & 0 & -\frac{5}{2} \\ 0 & 0 & 1 \\ 0 & 1 & 0 \\ 1 & 0 & 0 \end{bmatrix} \tag{10}$$

is 3, as is readily seen from the bottom three rows which surely have a nonzero determinant. ∎

Generalizing the results of Example 1, suppose that a system

$$\mathbf{A}\mathbf{x} = \mathbf{c}, \tag{11}$$

where $\mathbf{A}$ is $m \times n$, is consistent and has an N-parameter family of solutions

$$\mathbf{x} = \mathbf{x}_p + \alpha_1 \mathbf{x}_1 + \cdots + \alpha_N \mathbf{x}_N. \tag{12}$$

Then $\mathbf{x}_p$ is necessarily a **particular solution** (i.e., one specific solution of $\mathbf{A}\mathbf{x} = \mathbf{c}$), and $\mathbf{x}_1, \ldots, \mathbf{x}_N$ are necessarily LI **homogeneous solutions** (i.e., solutions of the homogeneous equation $\mathbf{A}\mathbf{x} = \mathbf{0}$). Recall from Section 4.8.2 that span $\{\mathbf{x}_1, \ldots, \mathbf{x}_N\}$ is the **solution space** of the homogeneous equation $\mathbf{A}\mathbf{x} = \mathbf{0}$, or the **null space** (or **kernel**) of $\mathbf{A}$ because it is the set of $\mathbf{x}$ vectors that are "nullified" by $\mathbf{A}$ (since $\mathbf{A}\mathbf{x} = \mathbf{0}$). The dimension of that null space, N, is called the **nullity** of $\mathbf{A}$.

Observe that if $\{\mathbf{c}_1, \ldots, \mathbf{c}_n\}$ denote the columns of $\mathbf{A}$, then [recalling (44) in Section 5.2] $\mathbf{A}\mathbf{x} = \mathbf{c}$ can be expressed as

$$x_1 \mathbf{c}_1 + x_2 \mathbf{c}_2 + \cdots + x_n \mathbf{c}_n = \mathbf{c}, \tag{13}$$

from which we can see that (11) is consistent (i.e., has one or more solutions) if and only if $\mathbf{c}$ happens to be in the column space of $\mathbf{A}$ [namely, span $\{\mathbf{c}_1, \ldots, \mathbf{c}_n\}$]. Or, expressed in terms of rank, (11) is consistent if and only if the rank of the augmented matrix $\mathbf{A}|\mathbf{c}$ equals the rank of the coefficient matrix $\mathbf{A}$: $r(\mathbf{A}|\mathbf{c}) = r(\mathbf{A})$.

In Example 1, for instance, we see from (4) that $r(\mathbf{A}|\mathbf{c}) = 3$ and (by covering up the last column, which is $\mathbf{c}$) that $r(\mathbf{A}) = 3$ as well. Thus, $r(\mathbf{A}|\mathbf{c}) = r(\mathbf{A})$ and, sure enough, (2) is consistent, solutions being given by (5). However, if we modify (2) by changing the underlined 1 to a 2, say, then the underlined 0 in (4) becomes a 1. In that case $r(\mathbf{A}|\mathbf{c}) = 4$ and $r(\mathbf{A}) = 3$ are unequal and, sure enough, there is no solution because the bottom row of (4) would then be equivalent to $0x_1 + \cdots + 0x_6 = 1$, which cannot be satisfied by any combination of x_j's.

Finally, what can be said about the value of N in (12)? In Example 1, for instance, we see from (4) that $N = 3$ for there are three arbitrary x_j values (as seen from the third row), and that that value of N arises as the difference between the number of unknowns $n = 6$ and the rank $r(\mathbf{A}) = r(\mathbf{A}|\mathbf{c}) = 3$.

These results suggest the following important generalization.

THEOREM 5.7.1

Existence and Uniqueness for $\mathbf{Ax} = \mathbf{c}$
Consider the linear system

$$\mathbf{Ax} = \mathbf{c}, \tag{14}$$

where $\mathbf{A}$ is $m \times n$. There is

1. no solution if and only if $r(\mathbf{A}|\mathbf{c}) \neq r(\mathbf{A})$,
2. a unique solution $\mathbf{x} = \mathbf{x}_p$, say, if and only if $r(\mathbf{A}|\mathbf{c}) = r(\mathbf{A}) = n$,
3. an N-parameter family of solutions of the form

$$\boxed{\mathbf{x} = \mathbf{x}_p + \alpha_1\mathbf{x}_1 + \cdots + \alpha_N\mathbf{x}_N} \tag{15}$$

if and only if $r(\mathbf{A}|\mathbf{c}) = r(\mathbf{A}) \equiv r$ is less than n. Here, $\mathbf{x}_p$ is a particular solution of $\mathbf{Ax} = \mathbf{c}$ (i.e., $\mathbf{Ax}_p = \mathbf{c}$); $\mathbf{x}_1, \ldots, \mathbf{x}_N$ are LI solutions of the homogeneous equation $\mathbf{Ax} = \mathbf{0}$; the parameters $\alpha_1, \ldots, \alpha_N$ are arbitrary constants; and $N = n - r$ is the nullity of $\mathbf{A}$.

The essential ideas required to prove these three results were developed above, and the proofs are left for the exercises.

Naturally, the **homogeneous** system

$$\mathbf{Ax} = \mathbf{0} \tag{16}$$

is but a special case of (14), hence, it is already covered by Theorem 5.7.1. However, it is such an important case that it deserves special attention. In (16), the augmented matrix $r(\mathbf{A}|\mathbf{c})$ is the $\mathbf{A}$ matrix augmented by a column of zeros so it is surely true that $r(\mathbf{A}|\mathbf{c}) = r(\mathbf{A})$ and, according to Theorem 5.7.1, it must be true that (16) is consistent. That result is no great surprise since (16) always admits the trivial solution $\mathbf{x} = \mathbf{0}$. Hence, the significant question about (16) is not whether or not it is consistent, but whether $\mathbf{x} = \mathbf{0}$ is the *only* solution. That is, does (16) admit *non*trivial solutions as well? That question is answered by parts 1 and 2 of Theorem 5.7.1 so we can state the following more specialized results.

THEOREM 5.7.2

Homogeneous Case Where **A** *is* $m \times n$

If **A** is $m \times n$, then

$$\mathbf{Ax} = \mathbf{0} \tag{17}$$

1. is consistent,
2. admits the trivial solution $\mathbf{x} = \mathbf{0}$,
3. admits the unique solution $\mathbf{x} = \mathbf{0}$ if and only if $r(\mathbf{A}) = n$,
4. admits an N-parameter family of nontrivial solutions

$$\boxed{\mathbf{x} = \alpha_1 \mathbf{x}_1 + \cdots + \alpha_N \mathbf{x}_N,} \tag{18}$$

in addition to the trivial solution, if and only if $r(\mathbf{A}) \equiv r < n$. Here, $\mathbf{x}_1, \ldots, \mathbf{x}_N$ are linearly independent solutions of $\mathbf{Ax} = \mathbf{0}$; the parameters $\alpha_1, \ldots, \alpha_N$ are arbitrary constants; and $N = n - r$ is the nullity of **A**.

THEOREM 5.7.3

Homogeneous Case Where **A** *is* $n \times n$

If **A** is $n \times n$, then

$$\mathbf{Ax} = \mathbf{0} \tag{19}$$

admits nontrivial solutions, besides the trivial solution $\mathbf{x} = \mathbf{0}$, if and only if $\det \mathbf{A} = 0$.

Theorem 5.7.3 will be crucial when we study the "eigenvalue problem" in Chapter 9.

As a final example, consider an interesting application of these concepts to "dimensional analysis."

5.7.2 Application to dimensional analysis. (Optional). Consider a rectangular flat plate (i.e., a flat rectangular wing, or "airfoil") in steady motion through otherwise undisturbed air as shown in Fig. 1; V is the flight speed, θ is the incidence or angle of attack of the airfoil, A is the chord length, and B is the span (the dimension normal to the paper). Equivalently, it is experimentally more convenient to keep the airfoil fixed (in a wind tunnel) and to blow air past it at a speed V. Imagine that our object is to conduct an experimental determination of the lift force ℓ generated on the airfoil, that is, to experimentally determine the functional dependence of ℓ on the various relevant quantities. What quantities *are* relevant? Surely A, B, θ, V are important, as well as the air density ρ (for instance we expect a much greater lift in water than in air, and no lift at all in a vacuum). A reasonable list of the relevant variables is given in Table 1.

Other variables come to mind, such as the ambient temperature, but if we expect ℓ to be only weakly dependent on them we can leave them out.

Major difficulties are now apparent. If ℓ depends upon the seven variables listed in Table 1, and we measure ℓ for five different values of each variable, then we need

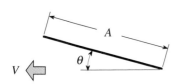

FIGURE 1
Flat plate airfoil.

Table 1. Relevant variables.

Variable	Symbol	Fundamental Units
Chord	A	L
Span	B	L
Incidence	θ	$M^0 L^0 T^0$
Flight velocity	V	LT^{-1}
Velocity of sound in air	V_0	LT^{-1}
Air density	ρ	ML^{-3}
Absolute viscosity	μ	$ML^{-1}T^{-1}$
Lift	ℓ	MLT^{-2}

to conduct $5^7 = 78,125$ experimental runs, and then present the results (using graphs, tables, or whatever) in a user-friendly way. Furthermore, whereas some variables are easily varied (such as A, B, θ, V) others are not (such as ρ and μ). The principal object of the following "dimensional analysis" is to reduce the number of variables as much as possible.

To begin, we express each variable in terms of the fundamental units M (mass), L (length), and T (time), in the right-hand column. We do not need to include F (force) as a fundamental unit because Newton's second law, $f = ma$, gives the dimensional relation $F = MLT^{-2}$. Also, observe that θ is dimensionless so we've written its units as $M^0 L^0 T^0$.[1]

Next, we seek all possible dimensionless products of the form

$$A^a B^b \theta^c V^d V_0^e \rho^f \mu^g \ell^h. \tag{20}$$

That is, we seek the exponents $a, \ldots, h$ such that

$$(L)^a (L)^b \left(M^0 L^0 T^0\right)^c \left(LT^{-1}\right)^d \left(LT^{-1}\right)^e \left(ML^{-3}\right)^f \left(ML^{-1}T^{-1}\right)^g \left(MLT^{-2}\right)^h$$
$$= M^0 L^0 T^0. \tag{21}$$

Equating exponents of L, T, M on both sides, we see that $a, \ldots, h$ must satisfy the homogeneous linear system

$$\begin{aligned} a + b + \quad d + e - 3f - g + \ h &= 0, \\ -d - e \quad\quad - g - 2h &= 0, \\ f + g + \ h &= 0. \end{aligned} \tag{22}$$

Applying Theorem 5.7.1 we can see that $n = 8$ and, by inspection of (22), that $r = 3$, so there will be a 5-parameter family of solutions. In fact, solving (22) by

---

[1]Recall that angle is defined by the formula $s = r\theta$, where s is the arc length of a circular arc of radius r, subtended by an angle θ measured in radians. Thus, $\theta = s/r = $ length / length $=$ dimensionless.

Gauss elimination gives the five-parameter family of solutions

$$
\begin{bmatrix} a \\ b \\ c \\ d \\ e \\ f \\ g \\ h \end{bmatrix}
= \alpha_1 \begin{bmatrix} -2 \\ 0 \\ 0 \\ -2 \\ 0 \\ -1 \\ 0 \\ 1 \end{bmatrix}
+ \alpha_2 \begin{bmatrix} -1 \\ 0 \\ 0 \\ -1 \\ 0 \\ -1 \\ 1 \\ 0 \end{bmatrix}
+ \alpha_3 \begin{bmatrix} 0 \\ 0 \\ 0 \\ -1 \\ 1 \\ 0 \\ 0 \\ 0 \end{bmatrix}
+ \alpha_4 \begin{bmatrix} 0 \\ 0 \\ 1 \\ 0 \\ 0 \\ 0 \\ 0 \\ 0 \end{bmatrix}
+ \alpha_5 \begin{bmatrix} -1 \\ 1 \\ 0 \\ 0 \\ 0 \\ 0 \\ 0 \\ 0 \end{bmatrix}, \tag{23}
$$

where $\alpha_1, \dots, \alpha_5$ are arbitrary constants. With $\alpha_1 = 1$ and $\alpha_2 = \cdots = \alpha_5 = 0$, say, (23) gives $a = -2$, $b = c = 0$, etc., and hence the nondimensional parameter $A^{-2}B^0\theta^0V^{-2}V_0^0\rho^{-1}\mu^0\ell^1$, namely, the nondimensional lift $\ell/(\rho V^2 A^2)$.[1] Similarly, setting $\alpha_2 = -1$ and the other α_j's $= 0$ gives the nondimensional parameter $\rho A V/\mu$, well known in fluid mechanics as the *Reynolds number* and denoted as Re;[2] setting $\alpha_3 = -1$ and the other α_j's $= 0$ gives the *Mach number* V/V_0, denoted as $\mathcal{M}$; setting $\alpha_4 = 1$ and the other α_j's $= 0$ gives the *incidence* θ (which was nondimensional to begin with); and setting $\alpha_5 = 1$ and the other α_j's $= 0$ gives B/A, known as the *aspect ratio* and denoted as $A\!R$.

The upshot is that rather than seek a functional relationship on the eight variables listed in the table, we can seek a relationship on the five nondimensional variables $[\ell/(\rho V^2 A^2),\ \text{Re},\ \mathcal{M},\ \theta,\ A\!R]$. Or, singling out the nondimensional lift, we can express that one in terms of the other four,

$$
\frac{\ell}{\rho V^2 A^2} = f\,(\text{Re},\, \mathcal{M},\, \theta,\, A\!R), \tag{24}
$$

and determine f experimentally by measuring $\ell/(\rho V^2 A^2)$ for various combinations of Re, $\mathcal{M}$, θ, and $A\!R$ values. Thus, we have been successful in reducing the number of variables from seven (A, B, θ, V, V_0, ρ, and μ) to four (Re, $\mathcal{M}$, θ, $A\!R$).

COMMENT 1.   A fluid mechanicist could probably simplify the problem even further. For example, it is known (from the governing equations of fluid mechanics) that the effect of the Mach number $\mathcal{M}$ will be negligible if $\mathcal{M}^2 \ll 1$. Thus, if the flight speed V is less than 200 miles per hour, say, then $\mathcal{M}$ can, to a good approximation, be dropped from (24), in which case f reduces to a function of three variables instead of four.[3]

---

[1] In fluid mechanics one would probably change this to $\ell/(\rho V^2 AB)$ (corresponding to $\alpha_1 = 1, \alpha_2 = \alpha_3 = \alpha_4 = 0, \alpha_5 = -1$), or to $\ell/(\frac{1}{2}\rho V^2 AB)$ because $\frac{1}{2}\rho V^2$ has physical significance as the stagnation pressure, and AB is the area of the airfoil. In that case the "reference force" $\frac{1}{2}\rho V^2 AB$, to which the lift force ℓ is being compared, has the clear physical significance of being the stagnation pressure times the area of the plate.

[2] Of course it is more natural to set $\alpha_2 = +1$ than -1, but $\alpha_2 = +1$ gives the nondimensional parameter $\mu/\rho AV$ which is legitimate but less desirable because it is $1/\text{Re}$ instead of Re.

[3] At ground level, the speed of sound is 762 mph, so if $V = 200$ mph then $\mathcal{M}^2 = (200/762)^2 = 0.069$, which is indeed small compared to 1.

COMMENT 2.   We mentioned, above, the practical difficulty in carrying out the experiment for a range of values of the fluid density. For instance, we could use air and water, but their densities are widely different and we would need both wind tunnel and water tunnel facilities. In the right-hand side of (24), however, ρ shows up only within the Reynolds number Re $= \rho AV/\mu$, which can be varied conveniently by varying the wind speed V in the wind tunnel.

COMMENT 3.   Of course, (23) gives an *infinite* number of nondimensional parameters. However, there are only five independent ones, such as the ones named above. For instance, we could choose $\alpha_3 = \alpha_5 = 1$ and $\alpha_1 = \alpha_2 = \alpha_4 = 0$, but the resulting nondimensional parameter, $V_0 B/(V A)$, is merely the aspect ratio divided by the Mach number.

Closure.   The chief result of this section is Theorem 5.7.1, which should be understood and remembered. This theorem and Theorem 5.5.2 [which relates the number of LI vectors in a set of vectors to the rank of the matrix having these vectors as its columns (or rows)] establish rank as one of the important concepts of linear algebra.

EXERCISES 5.7

1. Consider the problem **Ax** = **c**, where **A** is the given $m \times n$ matrix. In each case let **c** be the m-dimensional vector $[1, 1, \ldots, 1]^T$. Use Theorem 5.7.1 and suitable rank calculations to determine whether or not the system is consistent. If consistent, determine whether it admits a unique solution or an N-parameter family of solutions. If the latter, determine N. Do not solve the system; merely use the concept of rank and Theorem 5.7.1.

   (a) $[0, 0, 2, 0]$

   (b) $[1, 2, 3]$

   (c) $\begin{bmatrix} 5 & 7 \\ 4 & 9 \end{bmatrix}$

   (d) $\begin{bmatrix} 4 & 8 & 0 \\ 3 & 6 & 0 \end{bmatrix}$

   (e) $\begin{bmatrix} 1 & 2 & 3 \\ 4 & 5 & 6 \\ 7 & 8 & 9 \end{bmatrix}$

   (f) $\begin{bmatrix} 3 & 2 & 1 \\ -1 & 1 & 4 \\ 1 & 4 & 9 \end{bmatrix}$

   (g) $\begin{bmatrix} 5 & 0 & 0 \\ 3 & 0 & 4 \\ 2 & 0 & 0 \end{bmatrix}$

   (h) $\begin{bmatrix} 1 & 3 & 2 \\ 2 & 6 & 4 \\ 3 & 9 & 6 \end{bmatrix}$

   (i) $\begin{bmatrix} 6 & 5 & 2 & 0 \\ 0 & 2 & 3 & 0 \\ 0 & 0 & 0 & 6 \end{bmatrix}$

   (j) $\begin{bmatrix} 1 & 0 & 3 & 0 \\ 0 & 1 & 0 & 3 \\ 1 & 0 & 3 & 1 \end{bmatrix}$

   (k) $\begin{bmatrix} 1 & 0 & 2 \\ 2 & 1 & -1 \\ 0 & 1 & 3 \\ 1 & 2 & 4 \end{bmatrix}$

   (l) $\begin{bmatrix} 0 & 4 & -1 & 1 \\ 1 & 1 & 5 & -1 \\ 1 & 5 & 4 & 0 \\ 2 & 6 & 9 & -1 \end{bmatrix}$

   (m) $\begin{bmatrix} 3 & 1 & -4 & 4 \\ 0 & 2 & 2 & 9 \\ 2 & 2 & 1 & 0 \\ -1 & 3 & 7 & 5 \end{bmatrix}$

   (n) $\begin{bmatrix} 1 & 2 & 3 & 4 \\ 5 & -1 & 0 & 5 \\ 6 & -1 & 2 & 8 \\ 7 & -1 & 0 & 7 \\ 8 & -1 & -3 & 5 \end{bmatrix}$

2. **(a)** Below (12), we stated that $\mathbf{x}_1, \ldots, \mathbf{x}_N$ are necessarily LI. Prove that claim. HINT: Pattern your proof after the discussion in Example 1.

   (b) Show that $\mathbf{x}_p$ cannot be in the span of $\mathbf{x}_1, \ldots, \mathbf{x}_N$.

3. Although we made a case for the truth of Theorem 5.7.1, we did not provide a detailed proof.

   (a) Prove part 1.   **(b)** Prove part 2.   **(c)** Prove part 3.

4. This exercise refers to Example 1 and the discussion following that example. For definiteness, let **A** be 3×3.

   (a) Suppose that **Ax** = **c** admits a one-parameter family of solutions

   $$\mathbf{x} = \mathbf{x}_p + \alpha_1 \mathbf{x}_1. \qquad (4.1)$$

   Explain, with the help of a labeled sketch, the geometrical significance of $\mathbf{x}_p$, $\mathbf{x}_1$, and $\mathbf{x}_p + \alpha_1 \mathbf{x}_1$.

   (b) Suppose that **Ax** = **c** admits a two-parameter family of solutions

   $$\mathbf{x} = \mathbf{x}_p + \alpha_1 \mathbf{x}_1 + \alpha_2 \mathbf{x}_2. \qquad (4.2)$$

   Explain, with the help of a labeled sketch, the geometrical significance of $\mathbf{x}_p$ and $\mathbf{x}_p + \alpha_1 \mathbf{x}_1 + \alpha_2 \mathbf{x}_2$.

5. (*Dimensional analysis*) In studying the drag force on a sphere moving beneath a water surface, the tabulated vari-

ables are deemed relevant. Proceeding along the same lines as in Section 5.7.2, obtain the following relevant dimensionless parameters: the dimensionless drag force $D/(\rho V^2 R^2)$, the *Reynolds number* $\text{Re} = \rho RV/\mu$, the *Froude number* $V^2/(Rg)$, and the two length ratios λ/R and d/R.

Variable	Symbol	Fundamental Units
Radius of sphere	R	L
Depth below water surface	d	L
Velocity of sphere	V	LT^{-1}
Water density	ρ	ML^{-3}
Absolute viscosity	μ	$ML^{-1}T^{-1}$
Gravity	g	LT^{-2}
Wavelength of free surface waves	λ	L
Drag force	D	MLT^{-2}

5.8   VECTOR TRANSFORMATION. (OPTIONAL)

5.8.1 Transformations and linearity. Recall that a real-valued function f of a real variable x is a rule that assigns a uniquely determined value $f(x)$ to each specified value x, as illustrated in Fig. 1. Thus, f is a transformation, or mapping, from points on an x axis to points on an f axis, and we view x as the "input" and $f(x)$ as the "output." [A more effective graphical display of f, called the *graph* of f, can be obtained if, following *Descartes* (1596–1650), we arrange the x and f axes at right angles to each other and plot the set of points $(x, f(x))$, as illustrated in Fig. 2.]

In this section we reconsider vectors and matrices from this transformation point of view. Specifically, we consider vector-valued functions $\mathbf{F}$ of a vector variable $\mathbf{x}$. That is, the "input" is now a vector $\mathbf{x}$ from some vector space V, and the function $\mathbf{F}$ assigns a uniquely determined "output" vector $\mathbf{F}(\mathbf{x})$ in some vector space W. We call $\mathbf{F}$ a **transformation**, **mapping**, or **operator**, from V into W, and denote it as

$$\mathbf{F} : V \to W.$$

We call V the **domain of F** and W the **range of definition of F**. W may, but need not, be identical to V.

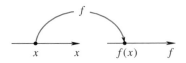

FIGURE 1
Function f as a transformation.

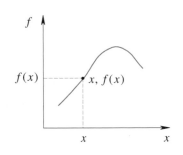

FIGURE 2
The graph of f.

E X A M P L E   1
To illustrate, consider the transformation $\mathbf{F} : \mathbb{R}^4 \to \mathbb{R}^3$ defined by

$$\mathbf{F}(\mathbf{x}) = \begin{bmatrix} x_1 - 2x_2 + x_4 \\ x_1 + x_3 \\ x_1 + 2x_2 + 2x_3 - x_4 \end{bmatrix}. \tag{1}$$

Here V is $\mathbb{R}^4$, W is $\mathbb{R}^3$, and the input vector $\mathbf{x}$ and the output vector $\mathbf{F}(\mathbf{x})$ are

$$\mathbf{x} = \begin{bmatrix} x_1 \\ x_2 \\ x_3 \\ x_4 \end{bmatrix} \text{ in } \mathbb{R}^4, \quad \text{and} \quad \mathbf{F}(\mathbf{x}) = \begin{bmatrix} x_1 - 2x_2 + x_4 \\ x_1 + x_3 \\ x_1 + 2x_2 + 2x_3 - x_4 \end{bmatrix} \text{ in } \mathbb{R}^3. \tag{2}$$

For example, if $\mathbf{x} = [2, 3, 6, -1]^T$, then $\mathbf{F}(\mathbf{x}) = [-5, 8, 21]^T$. ∎

We say that the vector $\mathbf{F}(\mathbf{x})$ in W is the **image** of the vector $\mathbf{x}$ in V under the transformation $\mathbf{F}$ and that, going in the other direction, $\mathbf{x}$ is the **inverse image** of $\mathbf{F}(\mathbf{x})$.

Since V and W are vector spaces, each must contain a zero vector. We will denote these zero vectors as $\mathbf{0}_V$ and $\mathbf{0}_W$, respectively. Finally, we define the image of V (i.e., the set of all vectors in V) in W as the **range** R of $\mathbf{F}$, and we define the inverse image of $\mathbf{0}_W$ in V as the **kernel** K or **null space** of $\mathbf{F}$. That is, the kernel K is the part of V that maps to the zero vector $\mathbf{0}_W$ in W; it is the set of all solutions of $\mathbf{F}(\mathbf{x}) = \mathbf{0}$.[1]

EXAMPLE 2

Let us find the range and kernel of the transformation $\mathbf{F}$ given in Example 1. First, the range. The range of $\mathbf{F}$ is the set of all vectors $\mathbf{c}$ in W for which the equation $\mathbf{F}(\mathbf{x}) = \mathbf{c}$ is consistent, that is, has at least one solution $\mathbf{x}$ in V. In the present case $\mathbf{F}(\mathbf{x}) = \mathbf{c}$ is

$$\begin{bmatrix} x_1 - 2x_2 + x_4 \\ x_1 + x_3 \\ x_1 + 2x_2 + 2x_3 - x_4 \end{bmatrix} = \begin{bmatrix} c_1 \\ c_2 \\ c_3 \end{bmatrix} \tag{3}$$

or, in scalar form,

$$\begin{aligned} x_1 - 2x_2 \qquad\quad + x_4 &= c_1, \\ x_1 \qquad\quad + x_3 \qquad &= c_2, \\ x_1 + 2x_2 + 2x_3 - x_4 &= c_3. \end{aligned} \tag{4}$$

Applying elementary operations to (4), we obtain the equivalent system

$$\begin{aligned} x_1 - 2x_2 \qquad\quad + x_4 &= c_1, \\ 2x_2 + x_3 - x_4 &= c_2 - c_1, \\ 0 &= c_3 - 2c_2 + c_1. \end{aligned} \tag{5a,b,c}$$

This system is consistent if and only if $\mathbf{c}$ lies in the plane (through the origin) defined by $c_3 - 2c_2 + c_1 = 0$. That plane is a two-dimensional subspace of $\mathbb{R}^3$ and is the range R of $\mathbf{F}$.

Turning to the kernel K of $\mathbf{F}$, that is, the inverse image of $[0, 0, 0]^{\mathrm{T}}$ in $\mathbb{R}^4$, we need merely set $c_1 = c_2 = c_3 = 0$ in (5) and solve (5) for $\mathbf{x}$. That solution gives

$$\mathbf{x} = \alpha_1 \begin{bmatrix} 0 \\ 1 \\ 0 \\ 2 \end{bmatrix} + \alpha_2 \begin{bmatrix} -2 \\ -1 \\ 2 \\ 0 \end{bmatrix} \equiv \alpha_1 \mathbf{x}_1 + \alpha_2 \mathbf{x}_2,$$

where α_1, α_2 are arbitrary constants. Then $K = \text{span}\,\{\mathbf{x}_1, \mathbf{x}_2\}$ and $\dim K = 2$. These results are summarized schematically in Fig. 3.   ∎

---

[1]Beware, the terminology adopted here may not be identical to that in other texts. Some authors call a transformation or mapping an "operator" only if $W = V$. Sometimes W, which we can think of as the entire target space, is called the "range." In that case, what we call the range R is probably called the "image" of $\mathbf{F}$. Further, W is sometimes called the "codomain" of V, and sometimes it is given no name at all.

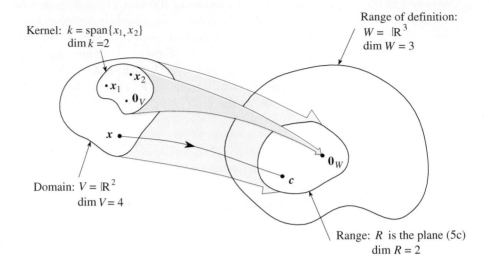

Kernel: $k = \text{span}\{x_1, x_2\}$
dim $k = 2$

Range of definition:
$W = \mathbb{R}^3$
dim $W = 3$

Domain: $V = \mathbb{R}^2$
dim $V = 4$

Range: R is the plane (5c)
dim $R = 2$

FIGURE 3   The transformation **F**.

Just as linearity, or the absence of it, is crucial in the theory of ordinary differential equations, it is likewise crucial here. We distinguish transformations as linear or nonlinear as follows.

Definition 5.8.1   *Linear Transformation*
We say that $\mathbf{F} : V \to W$ is a **linear** transformation if

$$\mathbf{F}(\alpha \mathbf{u} + \beta \mathbf{v}) = \alpha\, \mathbf{F}(\mathbf{u}) + \beta\, \mathbf{F}(\mathbf{v}) \tag{6}$$

for every choice of vectors $\mathbf{u}$, $\mathbf{v}$ in V, and for every choice of scalars α, β; otherwise, $\mathbf{F}$ is said to be **nonlinear**.

EXAMPLE 3
To illustrate, consider the transformation $\mathbf{F} : \mathbb{R}^3 \to \mathbb{R}^2$, where

$$\mathbf{F}(\mathbf{x}) = \begin{bmatrix} x_1 - 2x_2 + x_3 \\ x_2 + 5x_3 \end{bmatrix} \tag{7}$$

and x_1, x_2, and x_3 are the first, second, and third components, respectively, of the input vector $\mathbf{x}$. Let us see if this $\mathbf{F}$ is linear.

$$\mathbf{F}(\alpha \mathbf{u} + \beta \mathbf{v}) = \mathbf{F}\big((\alpha u_1 + \beta v_1, \alpha u_2 + \beta v_2, \alpha u_3 + \beta v_3)\big)$$

so we can identify x_1, x_2, x_3 in the left side of (7) as $\alpha u_1 + \beta v_1$, $\alpha u_2 + \beta v_2$, $\alpha u_3 + \beta v_3$, respectively. Thus (7) gives

$$\mathbf{F}(\alpha\mathbf{u} + \beta\mathbf{v}) = \begin{bmatrix} (\alpha u_1 + \beta v_1) - 2(\alpha u_2 + \beta v_2) + (\alpha u_3 + \beta v_3) \\ (\alpha u_2 + \beta v_2) + 5(\alpha u_3 + \beta v_3) \end{bmatrix}$$

$$= \alpha \begin{bmatrix} u_1 - 2u_2 + u_3 \\ u_2 + 5u_3 \end{bmatrix} + \beta \begin{bmatrix} v_1 - 2v_2 + v_3 \\ v_2 + 5v_3 \end{bmatrix}$$

$$= \alpha\,\mathbf{F}(\mathbf{u}) + \beta\,\mathbf{F}(\mathbf{v}) \tag{8}$$

for every choice of $\mathbf{u}$, $\mathbf{v}$, α, β, so $\mathbf{F}$ is indeed linear. Notice that the key step in (8), the second equality, follows from the definitions of the addition and scalar multiplication of matrices.   ■

E X A M P L E   4
Consider $\mathbf{F} : \mathbb{R}^2 \to \mathbb{R}^2$, where

$$\mathbf{F}(\mathbf{x}) = \begin{bmatrix} x_1^2 \\ x_1 + 2x_2 \end{bmatrix}. \tag{9}$$

Then

$$\mathbf{F}(\alpha\mathbf{u} + \beta\mathbf{v})$$

$$= \begin{bmatrix} (\alpha u_1 + \beta v_1)^2 \\ (\alpha u_1 + \beta v_1) + 2(\alpha u_2 + \beta v_2) \end{bmatrix}$$

$$= \alpha \begin{bmatrix} u_1^2 \\ u_1 + 2u_2 \end{bmatrix} + \beta \begin{bmatrix} v_1^2 \\ v_1 + 2v_2 \end{bmatrix} + \begin{bmatrix} (\alpha^2 - \alpha)u_1^2 + (\beta^2 - \beta)v_1^2 + 2\alpha\beta u_1 v_1 \\ 0 \end{bmatrix}$$

$$= \alpha\,\mathbf{F}(\mathbf{u}) + \beta\,\mathbf{F}(\mathbf{v}) + \text{deviation}, \tag{10}$$

where the "deviation"

$$\mathbf{F}(\alpha\mathbf{u} + \beta\mathbf{v}) - \alpha\,\mathbf{F}(\mathbf{u}) - \beta\,\mathbf{F}(\mathbf{v}) = \begin{bmatrix} (\alpha^2 - \alpha)u_1^2 + (\beta^2 - \beta)v_1^2 + 2\alpha\beta u_1 v_1 \\ 0 \end{bmatrix} \tag{11}$$

is obviously not zero for all choices of α, β, $\mathbf{u}$, $\mathbf{v}$. For instance, if $\alpha = \beta = u_1 = v_1 = u_2 = v_2 = 1$, then the deviation vector is $[2, 0]^{\mathrm{T}}$. Thus, (6) does not hold for all choices of α, β, $\mathbf{u}$, $\mathbf{v}$, so $\mathbf{F}$ given by (9) is nonlinear.   ■

If $\mathbf{F}$ is linear, then besides (6) we have

$$\mathbf{F}(\alpha\mathbf{u} + \beta\mathbf{v} + \gamma\mathbf{w}) = \mathbf{F}(\alpha\mathbf{u} + 1\,(\beta\mathbf{v} + \gamma\mathbf{w}))$$

$$= \alpha\,\mathbf{F}(\mathbf{u}) + 1\mathbf{F}(\beta\mathbf{v} + \gamma\mathbf{w}) = \alpha\,\mathbf{F}(\mathbf{u}) + \beta\,\mathbf{F}(\mathbf{v}) + \gamma\,\mathbf{F}(\mathbf{w}),$$

where we have used (6) in each of the last two equalities; or, more generally,

$$\boxed{\mathbf{F}(\alpha_1\mathbf{u}_1 + \cdots + \alpha_n\mathbf{u}_n) = \alpha_1\mathbf{F}(\mathbf{u}_1) + \cdots + \alpha_n\mathbf{F}(\mathbf{u}_n).} \tag{12}$$

Observe that (7) can be expressed in matrix notation as

$$\mathbf{F}(\mathbf{x}) = \begin{bmatrix} 1 & -2 & 1 \\ 0 & 1 & 5 \end{bmatrix} \begin{bmatrix} x_1 \\ x_2 \\ x_3 \end{bmatrix}. \tag{13}$$

That is, the action of $\mathbf{F}$ on $\mathbf{x}$ is equivalent to multiplication by $\mathbf{A}$, where $\mathbf{A}$ is the 2×3 matrix in (13),

$$\mathbf{F}(\mathbf{x}) = \mathbf{Ax}. \tag{14}$$

For that reason we call $\mathbf{F}$ a **matrix transformation**. Note that we do not say that "the transformation $\mathbf{F}$ is the matrix $\mathbf{A}$." Rather, we say that $\mathbf{F}$ is the transformation *multiplication by* $\mathbf{A}$.

Thus, we see that the linear transformation (7) happens to be expressible as a matrix transformation. In fact, linear transformations and matrix transformations are equivalent: Every linear transformation from $\mathbb{R}^n$ to $\mathbb{R}^m$ is a matrix transformation and every matrix transformation is a linear transformation.

THEOREM 5.8.1

Matrix Transformation

A transformation $\mathbf{F} : \mathbb{R}^n \to \mathbb{R}^m$ is linear if and only if it is a matrix transformation.

Proof.   First, we show that if $\mathbf{F}$ is a matrix transformation [i.e., if there is an $m \times n$ matrix $\mathbf{A}$ such that $\mathbf{F}(\mathbf{x}) = \mathbf{Ax}$ for each $\mathbf{x}$ in $\mathbb{R}^n$], then $\mathbf{F}$ is linear. That part is easy since $\mathbf{F}(\alpha\mathbf{u} + \beta\mathbf{v}) = \mathbf{A}(\alpha\mathbf{u} + \beta\mathbf{v}) = \alpha\mathbf{Au} + \beta\mathbf{Av} = \alpha\mathbf{F}(\mathbf{u}) + \beta\mathbf{F}(\mathbf{v})$ for all $\mathbf{u}, \mathbf{v}$ in $\mathbb{R}^n$ and for all scalars α, β. To prove the inverse, let $\{\mathbf{i}_1, \dots, \mathbf{i}_n\}$ and $\{\mathbf{j}_1, \dots, \mathbf{j}_m\}$ be bases for $\mathbb{R}^n$ and $\mathbb{R}^m$, respectively. Then we may express any $\mathbf{x}$ in $\mathbb{R}^n$ as

$$\mathbf{x} = \sum_{j=1}^{n} x_j \mathbf{i}_j,$$

so

$$\mathbf{F}(\mathbf{x}) = \mathbf{F}\left(\sum_{j=1}^{n} x_j \mathbf{i}_j\right) = \sum_{j=1}^{n} x_j \mathbf{F}(\mathbf{i}_j)$$

by (12). Since each $\mathbf{F}(\mathbf{i}_j)$ is in $\mathbb{R}^m$, it can be expressed as

$$\mathbf{F}(\mathbf{i}_j) = \sum_{k=1}^{m} a_{kj} \mathbf{j}_k$$

for some set of coefficients a_{kj}, so

$$\mathbf{F}(\mathbf{x}) = \sum_{j=1}^{n} x_j \sum_{k=1}^{m} a_{kj} \mathbf{j}_k = \sum_{k=1}^{m} \left(\sum_{j=1}^{n} a_{kj} x_j\right) \mathbf{j}_k. \tag{15a}$$

But we also have

$$\mathbf{F}(\mathbf{x}) = \mathbf{y} = \sum_{k=1}^{m} y_k \mathbf{j}_k \tag{15b}$$

for any $\mathbf{y}$ in the range of $\mathbf{F}$, and it follows from (15a) and (15b), and the linear independence of the $\mathbf{j}_k$'s, that

$$y_k = \sum_{j=1}^{n} a_{kj} x_j$$

or $\mathbf{y} = \mathbf{Ax}$, where $\mathbf{A} = \{a_{kj}\}$ is $m \times n$. Thus, $\mathbf{F}(\mathbf{x}) = \mathbf{Ax}$, as was to be proved. ∎

Now that we see that every linear transformation $\mathbf{F}$ is a matrix transformation, with matrix $\mathbf{A}$, say, we can see that *the range R of $\mathbf{F}$ is simply the column space of $\mathbf{A}$, so the dimension of R is the rank of $\mathbf{A}$*:

$$\boxed{\dim R = r(\mathbf{A}).} \tag{16}$$

We now introduce some additional terminology. First, recall that $\mathbf{F}$ is understood to be single valued. That is, to each vector $\mathbf{x}$ in the domain V of $\mathbf{F}$ there corresponds a unique image $\mathbf{F}(\mathbf{x})$ in the range R of $\mathbf{F}$. If, in addition, to each vector in R there corresponds a unique inverse image in V, then $\mathbf{F}$ is said to be **one-to-one**. Notice that we do not say "to each vector in W there corresponds a unique inverse image in V" since the range R may not be all of the range of definition W, in which case those vectors that are in W but not in R have no inverse image at all. If R does turn out to be all of W, then $\mathbf{F}$ is said to be **onto**; that is, $\mathbf{F}$ maps V "onto" W rather than "into" W. Finally, if $\mathbf{F}$ is *both onto and one-to-one*, it is said to be **invertible** for then every vector in W has a unique image in V. This inverse transformation, from W onto V, is called the **inverse of $\mathbf{F}$** and is denoted as $\mathbf{F}^{-1}$.

What is the point of the terms onto, one-to-one, and invertible? Onto has to do with the *existence* of solutions: if $\mathbf{F}$ maps V onto W then there exists a solution $\mathbf{x}$ of $\mathbf{F}(\mathbf{x}) = \mathbf{c}$ for every $\mathbf{c}$ in W. One-to-one has to do with the *uniqueness* of solutions: if $\mathbf{F}$ is one-to-one, then the solution of $\mathbf{F}(\mathbf{x}) = \mathbf{c}$ is unique. However, if $\mathbf{F}$ is onto then the solutions that exist may not be unique, and if $\mathbf{F}$ is one-to-one then a solution may (for any given $\mathbf{c}$) not exist. But if $\mathbf{F}$ is invertible, then we have both existence and uniqueness: for every $\mathbf{c}$ in W, $\mathbf{F}(\mathbf{x}) = \mathbf{c}$ has a solution $\mathbf{x}$ in V and that solution is unique.

EXAMPLE 5

Consider the matrix transformation $\mathbf{F}$ in Example 2. There, $\mathbf{F}(\mathbf{x}) = \mathbf{Ax}$ with

$$\mathbf{A} = \begin{bmatrix} 1 & -2 & 0 & 1 \\ 1 & 0 & 1 & 0 \\ 1 & 2 & 2 & -1 \end{bmatrix}. \tag{17}$$

From (5c) we see that R is only the two-dimensional subspace of W (W is $\mathbb{R}^3$) consisting of the plane $c_3 - 2c_2 + c_1 = 0$ so that $\mathbf{F}$ is not *onto*, which result was illustrated schematically in Fig. 3, where R was shown to be only a part of W. Hence

F is not invertible. Furthermore, if **c** is in R [i.e., if (5c) is satisfied], then (5) yields a nonunique solution for **x**, so **F** is not one-to-one. Summarizing, this **F** is neither onto nor one-to-one and is not invertible.

COMMENT.   That $\dim R = 2$ follows also from (16) since $r(\mathbf{A}) = 2$ for the **A** given by (17).   ∎

E X A M P L E 6
Consider the matrix transformation $\mathbf{F} : \mathbb{R}^2 \to \mathbb{R}^3$ associated with the matrix

$$\mathbf{A} = \begin{bmatrix} 1 & -2 \\ 1 & 1 \\ 2 & -1 \end{bmatrix}, \tag{18}$$

that is, $\mathbf{F}(\mathbf{x}) = \mathbf{A}\mathbf{x}$ where **A** is given by (18). Applying elementary operations to the system $\mathbf{A}\mathbf{x} = \mathbf{c}$, namely, to

$$\begin{bmatrix} 1 & -2 \\ 1 & 1 \\ 2 & -1 \end{bmatrix} \begin{bmatrix} x_1 \\ x_2 \end{bmatrix} = \begin{bmatrix} c_1 \\ c_2 \\ c_3 \end{bmatrix}, \tag{19}$$

we obtain the equivalent system

$$\begin{aligned} x_1 - 2x_2 &= c_1, \\ 3x_2 &= c_2 - c_1, \\ 0 &= c_3 - c_2 - c_1, \end{aligned} \tag{20a,b,c}$$

from which it is seen that **F** is not onto [because the range R is only the two-dimensional plane (20c) within $\mathbb{R}^3$], although it is one-to-one [because the solution of (20) is unique if (20c) is satisfied]. Thus, this **F** is not invertible.   ∎

E X A M P L E 7
Consider the matrix transformation $\mathbf{F} : \mathbb{R}^3 \to \mathbb{R}^2$ associated with the matrix

$$\mathbf{A} = \begin{bmatrix} 1 & 1 & 2 \\ 2 & 0 & -3 \end{bmatrix}. \tag{21}$$

Applying elementary operations to the system $\mathbf{A}\mathbf{x} = \mathbf{c}$, namely, to

$$\begin{bmatrix} 1 & 1 & 2 \\ 2 & 0 & -3 \end{bmatrix} \begin{bmatrix} x_1 \\ x_2 \\ x_3 \end{bmatrix} = \begin{bmatrix} c_1 \\ c_2 \end{bmatrix}, \tag{22}$$

gives the equivalent system

$$\begin{aligned} x_1 + x_2 + 2x_3 &= c_1, \\ 2x_2 + 7x_3 &= 2c_1 - c_2, \end{aligned} \tag{23a,b}$$

from which it is seen that **F** is onto [because (23) is consistent for every **c** in $\mathbb{R}^2$], although not one-to-one [because, given any **c** in $\mathbb{R}^2$, the solution of (23) is nonunique]. Thus, this **F** is not invertible.   ∎

EXAMPLE 8

Consider the matrix transformation $\mathbf{F} : \mathbb{R}^3 \to \mathbb{R}^3$ associated with the matrix

$$\mathbf{A} = \begin{bmatrix} 2 & -1 & 1 \\ 0 & 3 & 1 \\ 1 & 1 & 0 \end{bmatrix}. \tag{24}$$

Applying elementary operations to the system $\mathbf{Ax} = \mathbf{c}$, we obtain

$$\begin{aligned} 2x_1 - x_2 + x_3 &= c_1, \\ 3x_2 + x_3 &= c_2, \\ -2x_3 &= 2c_3 - c_2 - c_1, \end{aligned} \tag{25a,b,c}$$

from which we see that $\mathbf{F}$ is both onto and one-to-one, and is therefore invertible. The inverse operator $\mathbf{F}^{-1}$ is the matrix operator associated with the matrix

$$\begin{bmatrix} \frac{1}{6} & -\frac{1}{6} & \frac{2}{3} \\ -\frac{1}{6} & \frac{1}{6} & \frac{1}{3} \\ \frac{1}{2} & \frac{1}{2} & -1 \end{bmatrix}, \tag{26}$$

which is the inverse of the $\mathbf{A}$ matrix, $\mathbf{A}^{-1}$. ∎

5.8.2 Reviewing the significance of linearity.   The subject title Linear Algebra accurately announces the important role of linearity. Indeed, the problem

$$\mathbf{Ax} = \mathbf{c} \tag{27}$$

is a central theme of linear algebra, and we have seen that the theory of the existence and uniqueness of solutions of (27) depends crucially on the linearity of $\mathbf{A}$, namely, the property that

$$\mathbf{A}(\beta \mathbf{u} + \gamma \mathbf{v}) = \beta \mathbf{Au} + \gamma \mathbf{Av} \tag{28}$$

for any scalars β, γ and for any vectors $\mathbf{u}$, $\mathbf{v}$ in the range of definition of $\mathbf{A}$, which is $\mathbb{R}^n$ if $\mathbf{A}$ is $m \times n$. Recall that it follows from (28) that

$$\mathbf{A}(\beta_1 \mathbf{u}_1 + \cdots + \beta_k \mathbf{u}_k) = \beta_1 \mathbf{Au}_1 + \cdots + \beta_k \mathbf{Au}_k \tag{29}$$

for *any* finite $k \geq 1$.

Suppose that (27) admits an N-parameter family of solutions

$$\mathbf{x} = \mathbf{x}_p + \alpha_1 \mathbf{x}_1 + \cdots + \alpha_N \mathbf{x}_N, \tag{30}$$

as can be found by Gauss elimination (see, for example, Example 1 of Section 5.7), where $\mathbf{Ax}_p = \mathbf{c}$ and $\mathbf{Ax}_j = \mathbf{0}$ for each $j = 1, \ldots, N$. The latter "works" because

$$\begin{aligned} \mathbf{A}(\mathbf{x}_p + \alpha_1 \mathbf{x}_1 + \cdots + \alpha_N \mathbf{x}_N) &= \mathbf{Ax}_p + \alpha_1 \mathbf{Ax}_1 + \cdots + \alpha_N \mathbf{Ax}_N \\ &= \mathbf{c} + \alpha_1 \mathbf{0} + \cdots + \alpha_N \mathbf{0} \\ &= \mathbf{c}, \end{aligned}$$

where the first equality follows from the linearity of $\mathbf{A}$.

Closure.   This section developed the idea of a matrix as a transformation, or mapping, from one vector space to another, which idea was introduced in Section 5.1 and discussed briefly in Section 5.2.8.

Besides establishing the concept, together with the relevant mathematical terminology, a key result is given in Theorem 5.8.1, that a transformation $\mathbf{F} : \mathbb{R}^n \to \mathbb{R}^m$ is linear if and only if it is a matrix transformation.

EXERCISES 5.8

1. In general, the effect of a transformation on the input vector varies from one input vector to another. For example, let $\mathbf{F} : \mathbb{R}^2 \to \mathbb{R}^2$ be a matrix transformation $\mathbf{F}(\mathbf{x}) = \mathbf{Ax}$, where $\mathbf{A} = \begin{bmatrix} 3 & 2 \\ 0 & 1 \end{bmatrix}$. In 2-space the "effect" of a transformation on a nonzero vector $\mathbf{x}$ amounts to the resulting rotation in the plane, and the dilation (i.e., $\|\mathbf{Ax}\| / \|\mathbf{x}\|$). For the transformation $\mathbf{F}$ given above, show that these effects are as follows for the given input vectors, and notice that the effect of $\mathbf{F}$ varies from one $\mathbf{x}$ to another. Count counterclockwise rotation as positive.

	Input	Effect of $\mathbf{F}$
(a)	$\mathbf{x} = \begin{bmatrix} 1 \\ 0 \end{bmatrix}$	rotation $= 0$ radians dilation $= 3$
(b)	$\mathbf{x} = \begin{bmatrix} 0 \\ 1 \end{bmatrix}$	rotation ≈ -1.1 radians dilation $= \sqrt{5}$
(c)	$\mathbf{c} = \begin{bmatrix} 1 \\ -1 \end{bmatrix}$	rotation $= 0$ radians dilation $= 1$

2. Determine whether $\mathbf{F}$ is linear or nonlinear by determining whether or not the deviation $\mathbf{F}(\alpha\,\mathbf{u} + \beta\,\mathbf{v}) - \alpha\mathbf{F}(\mathbf{u}) - \beta\mathbf{F}(\mathbf{u})$ is necessarily zero.

(a) $\mathbf{F} : \mathbb{R}^2 \to \mathbb{R}^2$, $\mathbf{F}(\mathbf{x}) = \begin{bmatrix} x_1^2 \\ x_1 + x_2 \end{bmatrix}$

(b) $\mathbf{F} : \mathbb{R}^2 \to \mathbb{R}^2$, $\mathbf{F}(\mathbf{x}) = \begin{bmatrix} 3x_1 \\ x_1 + x_2 \end{bmatrix}$

(c) $\mathbf{F} : \mathbb{R}^3 \to \mathbb{R}^2$, $\mathbf{F}(\mathbf{x}) = \begin{bmatrix} x_1 x_2 \\ x_3 \end{bmatrix}$

(d) $\mathbf{F} : \mathbb{R}^2 \to \mathbb{R}^1$, $\mathbf{F}(\mathbf{x}) = \begin{bmatrix} 3x_2 \end{bmatrix}$

(e) $\mathbf{F} : \mathbb{R}^2 \to \mathbb{R}^2$, $\mathbf{F}(\mathbf{x}) = \begin{bmatrix} \sin x_2 \\ 4x_1 \end{bmatrix}$

(f) $\mathbf{F} : \mathbb{R}^2 \to \mathbb{R}^2$, $\mathbf{F}(\mathbf{x}) = \begin{bmatrix} x_1 + 1 \\ x_2 + 1 \end{bmatrix}$

3. (*Identity operator*) We say that $\mathbf{I} : V \to V$ is an **identity operator** if $\mathbf{I}(\mathbf{x}) = \mathbf{x}$ for each $\mathbf{x}$ in V.

(a) Suppose that $\mathbf{F} : V \to V$ is linear and that $\{\mathbf{v}_1, \ldots, \mathbf{v}_n\}$ is a basis for V. Show that if $\mathbf{F}(\mathbf{v}_1) = \mathbf{v}_1, \ldots, \mathbf{F}(\mathbf{v}_n) = \mathbf{v}_n$, then $\mathbf{F} = \mathbf{I}$.

(b) Determine the matrix $\mathbf{A}$ corresponding to the identity operator $\mathbf{I} : \mathbb{R}^n \to \mathbb{R}^n$, i.e., such that $\mathbf{I}(\mathbf{x}) = \mathbf{Ax}$.

4. (*Zero transformation*) We say that $\mathbf{\Phi} : V \to W$ is a **zero transformation** if $\mathbf{\Phi}(\mathbf{x}) = \mathbf{0}$ for all $\mathbf{x}$'s in V.

(a) Suppose that $\mathbf{F} : V \to W$ is linear and that $\{\mathbf{v}_1, \ldots, \mathbf{v}_n\}$ is a basis for V. Show that if $\mathbf{F}(\mathbf{v}_1) = \mathbf{0}, \ldots, \mathbf{F}(\mathbf{v}_n) = \mathbf{0}$, then $\mathbf{F} = \mathbf{\Phi}$.

(b) Determine the matrix $\mathbf{A}$ corresponding to the zero transformation $\mathbf{\Phi} : \mathbb{R}^n \to \mathbb{R}^m$, i.e., such that $\mathbf{\Phi}(\mathbf{x}) = \mathbf{Ax}$.

5. In each case $\mathbf{F} : \mathbb{R}^n \to \mathbb{R}^m$ is the matrix transformation corresponding to the given $m \times n$ matrix $\mathbf{A}$. Determine $\dim R$, $\dim K$, and $\dim V$. Is $\mathbf{F}$ onto? One-to-one? Invertible? Explain. Put forward any basis for K and any basis for R. NOTE: We state without proof that the relation

$$\boxed{\dim R + \dim K = \dim V} \qquad (5.1)$$

holds for every matrix transformation $\mathbf{F} : \mathbb{R}^n \to \mathbb{R}^m$. For the cases that follow, show that your values of $\dim R$, $\dim K$, and $\dim V$ satisfy (5.1).

(a) $\begin{bmatrix} 2 & 1 & 1 \\ 1 & 1 & 1 \end{bmatrix}$ (b) $\begin{bmatrix} 2 & 1 & 1 \\ 1 & 1 & 1 \\ 4 & 3 & 3 \end{bmatrix}$

(c) $\begin{bmatrix} 2 & 1 & 1 \\ 1 & 1 & 1 \\ 1 & 1 & 2 \end{bmatrix}$ (d) $\begin{bmatrix} 4 & 1 \\ 3 & 2 \\ 0 & -1 \end{bmatrix}$

(e) $\begin{bmatrix} 1 & 2 & 3 \\ 0 & 4 & 5 \\ 0 & 0 & 6 \end{bmatrix}$ (f) $\begin{bmatrix} 0 & 0 & 6 & 0 \\ 0 & -2 & 0 & 0 \\ 1 & 0 & 0 & 0 \end{bmatrix}$

$$\textbf{(g)}\quad \begin{bmatrix} 2 & -1 & 1 \\ 0 & 0 & 3 \\ 1 & 1 & 1 \\ 1 & 2 & 3 \end{bmatrix} \qquad \textbf{(h)}\quad \begin{bmatrix} 1 & 4 \\ 2 & 1 \\ 0 & 1 \\ 0 & 3 \end{bmatrix}$$

6. Make up any example of a matrix transformation $\mathbf{F}$: $\mathbb{R}^n \to \mathbb{R}^m$ that is one-to-one but not onto, one that is onto but not one-to-one, one that is neither one-to-one nor onto, and one that is both one-to-one and onto. If such an example is impossible, explain why that is so.

   (a) $n = 2, m = 2$ **(b)** $n = 3, m = 2$

   (c) $n = 2, m = 3$ **(d)** $n = 3, m = 3$

7. Prove that if $\mathbf{F} : V \to W$ is a linear transformation then

   (a) the kernel K of $\mathbf{F}$ is a subspace of V

   (b) the range R of $\mathbf{F}$ is a subspace of W

8. (*Projection operators*) Let $\mathbf{F} : \mathbb{R}^3 \to \mathbb{R}^3$ be the transformation

$$\mathbf{F}(\mathbf{x}) = (\mathbf{x}\cdot\hat{\mathbf{v}})\hat{\mathbf{v}}, \qquad (8.1)$$

where $\hat{\mathbf{v}} = [v_1, v_2, v_3]^T$ is a prescribed unit vector. In geometric terms, $\mathbf{F}(\mathbf{x})$ is the vector orthogonal projection of $\mathbf{x}$ onto the line of action of $\hat{\mathbf{v}}$, as illustrated below. Hence, $\mathbf{F}$ in (8.l) is known as a **projection operator**. We now state the problem: Show that $\mathbf{F}$ is linear, so that one can express $\mathbf{F}(\mathbf{x})$ as $\mathbf{A}\mathbf{x}$. Then determine the nine elements of the $\mathbf{A}$ matrix. (They will depend on v_1, v_2, v_3.)

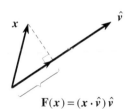

$$\mathbf{F}(x) = (x \cdot \hat{v})\,\hat{v}$$

9. (*More about projection operators*) Let $\mathbf{F} : \mathbb{R}^3 \to \mathbb{R}^3$ be the transformation

$$\mathbf{F}(\mathbf{x}) = (\mathbf{x}\cdot\hat{\mathbf{v}}_1)\hat{\mathbf{v}}_1 + (\mathbf{x}\cdot\hat{\mathbf{v}}_2)\hat{\mathbf{v}}_2, \qquad (9.1)$$

where $\hat{\mathbf{v}}_1, \hat{\mathbf{v}}_2$ are prescribed ON vectors. Then $\mathbf{F}(\mathbf{x})$ is the vector orthogonal projection of $\mathbf{x}$ onto the *plane* spanned by $\{\hat{\mathbf{v}}_1, \hat{\mathbf{v}}_2\}$.

   (a) Given that $\hat{\mathbf{v}}_1 = [1, 0, 0]^T$ and $\hat{\mathbf{v}}_2 = [0, 1, 0]^T$, work out $\mathbf{F}(\mathbf{x})$ for $\mathbf{x} = [2, 3, 4]^T$, and draw an informative, labeled picture, analogous to the one shown in Exercise 8.

   (b) Show that $\mathbf{F}$ is linear so that one can express $\mathbf{F}(\mathbf{x})$ as $\mathbf{A}\mathbf{x}$. Then determine the nine elements of the $\mathbf{A}$ matrix, in terms of the components v_{11}, v_{12}, v_{13} of $\hat{\mathbf{v}}_1$ and v_{21}, v_{22}, v_{23} of $\hat{\mathbf{v}}_2$.

10. Show that $\mathbf{F}(\alpha\mathbf{u} + \beta\mathbf{v}) = \alpha\mathbf{F}(\mathbf{u}) + \beta\mathbf{F}(\mathbf{v})$, in Definition 5.8.1, is equivalent to the two conditions $\mathbf{F}(\mathbf{u} + \mathbf{v}) = \mathbf{F}(\mathbf{u}) + \mathbf{F}(\mathbf{v})$ and $\mathbf{F}(\alpha\mathbf{u}) = \alpha\mathbf{F}(\mathbf{u})$.

11. (*Reflection about a line*) Let $\mathbf{F} : \mathbb{R}^2 \to \mathbb{R}^2$ reflect any given vector $\mathbf{x}$ about the line L, as shown in the accompanying figure.

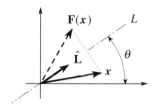

   (a) Show that $\mathbf{F}(\mathbf{x}) = \mathbf{x} + 2[(\mathbf{x}\cdot\hat{\mathbf{L}})\hat{\mathbf{L}} - \mathbf{x}]$.

   (b) Show that $\mathbf{F}$ is linear and determine the matrix $\mathbf{A}$ such that $\mathbf{F}(\mathbf{x}) = \mathbf{A}\mathbf{x}$.

   (c) Work out $\mathbf{A}^2$ and show that $\mathbf{A}^2 = \mathbf{I}$ (defined in Exercise 3). Why should it have been obvious, without working out $\mathbf{A}^2$, that $\mathbf{A}^2 = \mathbf{I}$?

12. (*Linear combination and composition*) If $\mathbf{F}$ and $\mathbf{G}$ are transformations from V into W, then we define the **linear combination** of $\mathbf{F}$ and $\mathbf{G}$, $(\alpha\mathbf{F} + \beta\mathbf{G}) : V \to W$, by

$$(\alpha\mathbf{F} + \beta\mathbf{G})(\mathbf{x}) \equiv \alpha\mathbf{F}(\mathbf{x}) + \beta\mathbf{G}(\mathbf{x}) \qquad (12.1)$$

for all $\mathbf{x}$ in V. Given transformations $\mathbf{F} : U \to V$ and $\mathbf{G} : V \to W$, we define the **composition** of $\mathbf{F}$ and $\mathbf{G}$, $(\mathbf{GF}) : U \to W$, by

$$(\mathbf{GF})(\mathbf{x}) \equiv \mathbf{G}(\mathbf{F}(\mathbf{x})) \qquad (12.2)$$

for all $\mathbf{x}$ in U.

   (a) Let $\mathbf{F} : \mathbb{R}^4 \to \mathbb{R}^2$ and $\mathbf{G} : \mathbb{R}^2 \to \mathbb{R}^2$ be matrix transformations with matrices

$$\mathbf{A} = \begin{bmatrix} 1 & 5 & 3 & -2 \\ 2 & 5 & 0 & 0 \end{bmatrix} \quad \text{and} \quad \mathbf{B} = \begin{bmatrix} 2 & 0 \\ 4 & 0 \end{bmatrix},$$

   respectively. Evaluate $(\mathbf{GF})(\mathbf{x})$ for $\mathbf{x} = [3, 1, -2, 6]^T$. Find a single matrix corresponding to the composite transformation $\mathbf{GF}$.

   (b) Let $\mathbf{F} : V \to V$ be a linear operator, and define the composite transformation $\mathbf{F}^2 : V \to V$ by $\mathbf{F}^2(\mathbf{x}) \equiv \mathbf{F}(\mathbf{F}(\mathbf{x}))$. Show that $\mathbf{F}^2$ is linear, too.

13. Show that the *translation operator* $\mathbf{F}(\mathbf{x}) = \mathbf{x} + \mathbf{c}$, where $\mathbf{c}$ is a constant vector, is nonlinear.

14. (*Applications to computer graphics*) It is basic, in computer graphics, to be able to move points about, in 3-space, by combinations of translations and rotations. Translation is easily accomplished by the operator

$$F(X) = X + \Delta X, \qquad (14.1)$$

where $X = [x, y, z]^T$ is the position vector to the point and $\Delta X = [\Delta x, \Delta y, \Delta z]^T$ is the translation. However, whereas it is convenient, in the computer software, to express all translations and rotations as matrix transformations, the operator $F : \mathbb{R}^3 \to \mathbb{R}^3$ is not linear (Exercise 13), and hence not expressible as a matrix transformation (Theorem 5.8.1). To circumvent this difficulty we define $X = [x, y, z, 1]^T$, instead, where the fourth component, unity, is included for convenience. Then we can express

$$F(X) = TX$$

$$= \begin{bmatrix} 1 & 0 & 0 & \Delta x \\ 0 & 1 & 0 & \Delta y \\ 0 & 0 & 1 & \Delta z \\ 0 & 0 & 0 & 1 \end{bmatrix} \begin{bmatrix} x \\ y \\ z \\ 1 \end{bmatrix} = \begin{bmatrix} x + \Delta x \\ y + \Delta y \\ z + \Delta z \\ 1 \end{bmatrix},$$

$$(14.2)$$

which, if we pay attention to only the first three components, effects the translation by means of multiplication by T, where T is the 4×4 matrix in (14.2).

(a) Show that

$$F(X) = R_z X = \begin{bmatrix} c_z & -s_z & 0 & 0 \\ s_z & c_z & 0 & 0 \\ 0 & 0 & 1 & 0 \\ 0 & 0 & 0 & 1 \end{bmatrix} \begin{bmatrix} x \\ y \\ z \\ 1 \end{bmatrix}$$

$$(14.3)$$

effects a rotation about the z axis through an angle θ_z, taken according to the right-hand rule, where c_z, s_z are shorthand for $\cos\theta_z, \sin\theta_z$, respectively. HINT: Letting $x = r\cos\theta, y = r\sin\theta$, show that

$$F(X) = R_z X = \begin{bmatrix} r\cos(\theta + \theta_z) \\ r\sin(\theta + \theta_z) \\ z \\ 1 \end{bmatrix}.$$

NOTE: Similarly, rotations about the x axis through an angle θ_x, and about the y axis through an angle θ_y, are effected by

$$F(X) = R_y X = \begin{bmatrix} c_y & 0 & -s_y & 0 \\ 0 & 1 & 0 & 0 \\ s_y & 0 & c_y & 0 \\ 0 & 0 & 0 & 1 \end{bmatrix} \begin{bmatrix} x \\ y \\ z \\ 1 \end{bmatrix}$$

$$(14.4)$$

and

$$F(X) = R_x X = \begin{bmatrix} 1 & 0 & 0 & 0 \\ 0 & c_x & -s_x & 0 \\ 0 & s_x & c_x & 0 \\ 0 & 0 & 0 & 1 \end{bmatrix} \begin{bmatrix} x \\ y \\ z \\ 1 \end{bmatrix},$$

$$(14.5)$$

respectively, where c_x, s_x, c_y, s_y denote $\cos\theta_x, \sin\theta_x, \cos\theta_y, \sin\theta_y$, respectively.

(b) Show that a rotation about the z axis, followed by a translation, is effected by the composite transformation (see Exercise 12)

$$F(X) = TR_z X = \begin{bmatrix} c_z & -s_z & 0 & \Delta x \\ s_z & c_z & 0 & \Delta y \\ 0 & 0 & 1 & \Delta z \\ 0 & 0 & 0 & 1 \end{bmatrix} \begin{bmatrix} x \\ y \\ z \\ 1 \end{bmatrix}.$$

Does the order of the operations matter? That is, is $TR_z = R_z T$?

(c) Compute $F(X) = TR_x R_y R_z X$ for $X = [1, 1, 0, 1]^T$, $\theta_z = -\pi/4$, $\theta_y = \pi/2$, $\theta_x = \pi$, $\Delta x = 2$, $\Delta y = 1$, $\Delta z = -3$. Verify the result by drawing the coordinate axes, identifying the initial point X, and then carrying out each rotation and translation graphically in a neat sketch.

(d) Let the point and eraser of a pencil be located, initially, by $X_p = [0, 1, 1, 1]^T$ and $X_e = [0, 1, 0, 1]^T$, respectively. Locate the point and the eraser following the composite transformation

$$F(X) = TR_x R_y R_z X,$$

where $\theta_x = 0.2$, $\theta_y = 0.3$, $\theta_z = -0.6$, $\Delta x = 1$, $\Delta y = 3$, $\Delta z = 1$. In a neat sketch, show the pencil in its initial and final configurations, and verify that its length has remained the same.

(e) Repeat part (d), with $\theta_x = \pi/2$, $\theta_y = 0$, $\theta_z = -\pi$, $\Delta x = \Delta y = \Delta z = 1$.

(f) Repeat part (d), with $\theta_x = -\pi/2$, $\theta_y = \theta_z = \pi/2$, $\Delta x = \Delta y = 1$, $\Delta z = 0$.

CHAPTER 5 REVIEW

The following review is limited to a number of isolated results and formulas that should be both understood and memorized.

Matrix multiplication:

$$\mathbf{AB} \neq \mathbf{BA} \qquad \text{(multiplication not commutative in general)}$$

Transpose:

$$(\mathbf{AB})^{\mathrm{T}} = \mathbf{B}^{\mathrm{T}}\mathbf{A}^{\mathrm{T}}$$

Determinants:

$$\det(\alpha\mathbf{A} + \beta\mathbf{B}) \neq \alpha \det\mathbf{A} + \beta \det\mathbf{B} \qquad \text{(det not linear in general)}$$
$$\det(\mathbf{AB}) = (\det\mathbf{A})(\det\mathbf{B})$$

Rank:

$$r(\mathbf{A}) = \text{number of LI columns in } \mathbf{A}$$
$$= \text{number of LI rows in } \mathbf{A}$$

Inverse matrix, $\mathbf{A}^{-1}$:

$\mathbf{A}^{-1}$ exists, and is unique, if and only if $\det\mathbf{A} \neq 0$.

$$\mathbf{A}^{-1} = \{\alpha_{ij}\} = \left\{\frac{A_{ji}}{\det\mathbf{A}}\right\}$$
$$\mathbf{A}^{-1}\mathbf{A} = \mathbf{A}\mathbf{A}^{-1} = \mathbf{I}$$
$$(\mathbf{AB})^{-1} = \mathbf{B}^{-1}\mathbf{A}^{-1}$$
$$(\mathbf{A}^{-1})^{\mathrm{T}} = (\mathbf{A}^{\mathrm{T}})^{-1}$$

Systems of linear algebraic equations, $\mathbf{Ax} = \mathbf{c}$ (where A is $m \times n$):

Inconsistent:
$$\text{No solution if } \quad r(\mathbf{A}|\mathbf{c}) \neq r(\mathbf{A})$$

Consistent:

$$\text{Unique solution if } \quad r(\mathbf{A}|\mathbf{c}) = r(\mathbf{A}) = n$$
$$(n - r)\text{-parameter family of solutions if } \quad r(\mathbf{A}|\mathbf{c}) = r(\mathbf{A}) \equiv r < n$$

The case where $m = n$:

$$\text{Unique solution } \left(\mathbf{x} = \mathbf{A}^{-1}\mathbf{c}\right) \text{ if and only if } \quad \det\mathbf{A} \neq 0 \quad [\text{i.e., } r(\mathbf{A}) = n]$$

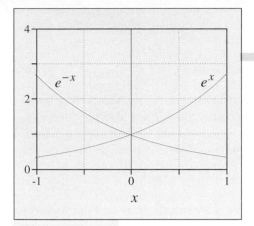

6

LINEAR DIFFERENTIAL EQUATIONS OF SECOND ORDER AND HIGHER

6.1 INTRODUCTION

We break our study of differential equations into two parts: first-order equations first, then equations of second order and higher. Further, first-order linear and nonlinear equations were treated separately in Chapters 2 and 3, respectively. Likewise for higher-order equations we treat the linear case first, in the present chapter, and non-linear equations later, in Chapter 11.

This chapter is devoted to the linear nth-order equation

$$\frac{d^n y}{dx^n} + p_1(x)\frac{d^{n-1}y}{dx^{n-1}} + \cdots + p_{n-1}(x)\frac{dy}{dx} + p_n(x)y = f(x). \tag{1}$$

To motivate our study of the differential equation (1) consider the analogous problem in linear algebra, the system

$$\mathbf{A}\mathbf{x} = \mathbf{c} \tag{2}$$

of m linear algebraic equations in n unknowns. There, $\mathbf{A}$ is a linear transformation from the vector space $\mathbb{R}^n$ to the vector space $\mathbb{R}^m$ where by $\mathbf{A}$ being linear we mean that

$$\mathbf{A}(\alpha\mathbf{u} + \beta\mathbf{v}) = \alpha\mathbf{A}\mathbf{u} + \beta\mathbf{A}\mathbf{v} \tag{3}$$

for any scalars α, β and any vectors $\mathbf{u}$, $\mathbf{v}$ in $\mathbb{R}^n$.

As to the existence and uniqueness of solutions to (2) there are three possibilities: (2) is inconsistent (i.e., has no solution), consistent with a unique solution, or consistent with a nonunique solution. In the case of a nonunique solution Theorem 5.7.1 tells us that the solution is of the form

$$\mathbf{x} = \mathbf{x}_p + \alpha_1 \mathbf{x}_1 + \cdots + \alpha_N \mathbf{x}_N \tag{4}$$

where $1 \leq N \leq n$; the α's are arbitrary constants; $\mathbf{x}_1, \ldots, \mathbf{x}_N$ are homogeneous solutions of (2) [i.e., solutions of the homogeneous version $\mathbf{A}\mathbf{x} = \mathbf{0}$ of (2)]; and $\mathbf{x}_p$ is any particular solution of (2). Since the $\alpha_1 \mathbf{x}_1 + \cdots + \alpha_N \mathbf{x}_N$ part of (4) is a general solution of the homogeneous equation $\mathbf{A}\mathbf{x} = \mathbf{0}$ let us denote it as $\mathbf{x}_h$, so (4) becomes

$$\mathbf{x} = \mathbf{x}_p + \mathbf{x}_h, \tag{5}$$

where $\mathbf{x}_p$ is any one solution of $\mathbf{A}\mathbf{x} = \mathbf{c}$ and $\mathbf{x}_h$ is a general solution of the "homogenized" version $\mathbf{A}\mathbf{x} = \mathbf{0}$. If you studied the optional Section 5.8 you will recall that the solution space of $\mathbf{A}\mathbf{x} = \mathbf{0}$, namely, span$\{\mathbf{x}_1, \ldots, \mathbf{x}_N\}$, is called the kernel of $\mathbf{A}$.

To see the analogy between (1) and (2), and therefore to see the connection between this chapter and Chapters 4 and 5, let us write (1) in operator form as

$$L[y] = f, \tag{6}$$

where L is the nth-order differential operator

$$L = \frac{d^n}{dx^n} + p_1(x)\frac{d^{n-1}}{dx^{n-1}} + \cdots + p_{n-1}(x)\frac{d}{dx} + p_n(x) \tag{7}$$

and $y(x)$ is defined on an x interval I. Just as the matrix $\mathbf{A}$ is a transformation or mapping from one vector space ($\mathbb{R}^n$), known as the domain of $\mathbf{A}$ and denoted as V, to another ($\mathbb{R}^m$), known as the range of definition of $\mathbf{A}$ and denoted as W, the differential operator L is also a transformation from one vector space (its domain V) to another vector space (its range of definition W). Let the domain V of L be the set of all functions that are n times continuously differentiable (i.e., functions that are n times differentiable, with their nth derivative being continuous) on I, and let the $p_j(x)$ coefficients in L be continuous on I. Then $L[y]$ will be continuous too, so the range of definition W of L will be the set of all functions that are continuous on I. Thus, whereas $\mathbf{A}$ is a transformation from n-tuple space to m-tuple space, L is a transformation from one *function space* to another. Like the matrix $\mathbf{A}$ in (2), L is linear; that is,

$$L[\alpha u(x) + \beta v(x)] = \alpha L[u(x)] + \beta L[v(x)] \tag{8}$$

for all constants α, β and all functions u and v in the domain V of L.

We saw in Section 5.7 that the form of the general solution to (2), as expressed in (4), follows from the linearity of $\mathbf{A}$. Likewise in this chapter we will find that the general solution of (1) is of the same form,

$$y(x) = y_p(x) + C_1 y_1(x) + \cdots + C_n y_n(x)$$
$$\equiv y_p(x) + y_h(x), \tag{9}$$

where $y_p(x)$ is a particular solution of $L[y] = f$; $C_1, \ldots, C_n$ are arbitrary constants; $y_1, \ldots, y_n$ are LI solutions of $L[y] = 0$; and $y_h(x) = C_1 y_1(x) + \cdots + C_n y_n(x)$ is a general solution of the "homogenized" equation $L[y] = 0$.

Thus, we will find much in common between the theory of the linear differential equation (1), in this chapter, and the theory of the linear algebraic equation (2) in Chapters 4 and 5, because both (1) and (2) are linear! At the same time we will find some differences. For instance, if $\mathbf{A}$ in (2) is 3×3 then N in (4) can turn out to be 3, 2, or 1 (or even 0 in the sense that we could have a unique solution $\mathbf{x} = \mathbf{x}_p$), but if L in (1) is of order n then the number of LI homogeneous solutions in (9) is necessarily n, no more no less. Another difference is that the Gauss elimination solution of (2) leads to the solution (4); that is, it gives both $\mathbf{x}_p$ and $\mathbf{x}_h$. However, we will find that to obtain the general solution (9) of (1) we will solve for the homogeneous and particular solutions separately and by different strategies.

An outline of this chapter is as follows.

Section 6.2: Covers the complex plane and the functions that will be needed—exponential, trigonometric, and hyperbolic.

Section 6.3: Covers the linear dependence and linear independence of sets of functions. This concept is needed because (9) is the desired general solution of (1) provided that the functions $y_1(x), \ldots, y_n(x)$ are indeed LI.

Section 6.4: Develops the theory for the homogeneous equation $L[y] = 0$ but does not show how to *find* the n LI homogeneous solutions $y_1(x), \ldots, y_n(x)$ that are needed in (9). Finding such solutions can be difficult so in the two sections to follow we limit our attention to two special cases for which solutions are readily found.

Section 6.5: Shows how to find homogeneous solutions for the important special case where L is of constant-coefficient type, that is, where all the $p_j(x)$ coefficients in (1) are constants.

Section 6.6: Shows how to find homogeneous solutions for another special case, namely, where L is of Cauchy-Euler type.

Section 6.7: Covers the particular solution $y_p(x)$ and gives two solution methods— the methods of undetermined coefficients and variation of parameters.

6.2 THE COMPLEX PLANE AND THE EXPONENTIAL, TRIGONOMETRIC, AND HYPERBOLIC FUNCTIONS

6.2.1 Complex numbers and the complex plane. Historically, complex numbers were created several hundred years ago within the context of the theory of equations. For if one allowed only real numbers, then equations such as $x^2 + 1 = 0$ and $x^2 + 2x + 4 = 0$ had no solution. Thus, in a step that was slow to gain acceptance, a broader number system was devised so that the equations given above, and indeed every polynomial equation, possess solutions within that number system.[1] Those new numbers, eventually named **complex numbers**, were of the form $a + ib$ where a and b are real and where i satisfies the equation $i^2 = -1$, or $i = \sqrt{-1}$. It is important to understand that the plus sign in $a + ib$ does not denote addition; rather, $a + ib$ is a single number, not the sum of a and ib.

The (real) numbers a, b are called the **real part** and **imaginary part** of $a + ib$, respectively, and denoted by Re and Im:

$$\text{Re}(a + ib) \equiv a, \qquad \text{Im}(a + ib) \equiv b \quad (\text{not } ib). \tag{1}$$

For instance, $\text{Re}(2 + 3i) = 2$ and $\text{Im}(2 - 3i) = -3$. We do not distinguish between $a + ib$ and $a + bi$, and we generally write $a + i0$ as a, and $0 + ib$ as ib, for brevity. The former complex number is said to be *purely real* and the latter is said to be *purely imaginary*. Finally, two complex numbers are said to be *equal* if their real and imaginary parts, respectively, are equal; that is,

$$a_1 + ib_1 = a_2 + ib_2 \tag{2}$$

holds if and only if $a_1 = a_2$ and $b_1 = b_2$.

Beyond defining complex numbers, we will need an algebra for their manipulation. The idea will be to stay as close as possible to the rules of ordinary arithmetic (i.e., governing real numbers). For example, the rules of ordinary arithmetic would seem to dictate that

$$(a_1 + ib_1) + (a_2 + ib_2) = (a_1 + a_2) + i(b_1 + b_2)$$

and that

$$\begin{aligned} (a_1 + ib_1)(a_2 + ib_2) &= a_1 a_2 + i a_1 b_2 + i b_1 a_2 + i^2 b_1 b_2 \\ &= (a_1 a_2 - b_1 b_2) + i(a_1 b_2 + b_1 a_2). \end{aligned}$$

But be sure to see that these equalities have no logical support since, as noted above, $a_1 + ib_1$ is a single complex number, not a_1 *plus* ib_1; similarly for $a_2 + ib_2$. Rather, we have merely been trying to motivate definitions for the addition and multiplication of complex numbers. Accordingly, we now *define*

$$(a_1 + ib_1) + (a_2 + ib_2) \equiv (a_1 + a_2) + i(b_1 + b_2) \tag{3}$$

and

$$(a_1 + ib_1)(a_2 + ib_2) \equiv (a_1 a_2 - b_1 b_2) + i(a_1 b_2 + b_1 a_2). \tag{4}$$

For instance, $(3 - 2i) + (5 + i) = 8 - i$ and $(3 - 2i)(5 + i) = 17 - 7i$.

---

[1]Perhaps the reluctance to accept complex numbers can be better appreciated if we mention that, before complex numbers, there had even been reluctance to accept *negative* numbers. After all, how could one have a negative amount of something. That was before the invention of the credit card.

We often denote complex numbers by a single letter, usually z, for brevity. Thus, we write $z = a + ib$. With the definitions given above, it is readily verified that the familiar rules of algebra hold for complex numbers. For instance, if we denote any three complex numbers as z_1, z_2, and z_3, then

$$z_1 + z_2 = z_2 + z_1, \qquad\qquad z_1 z_2 = z_2 z_1 \qquad \text{(commutative)} \qquad (5)$$

$$(z_1 + z_2) + z_3 = z_1 + (z_2 + z_3), \quad (z_1 z_2) z_3 = z_1 (z_2 z_3) \qquad \text{(associative)} \qquad (6)$$

$$z_1 (z_2 + z_3) = z_1 z_2 + z_1 z_3 \qquad\qquad\qquad \text{(distributive)}. \qquad (7)$$

Further, there are zero and unit complex numbers since, with $z = a + ib$, it follows from (3) that

$$z + (0 + i0) = (a + 0) + i(b + 0) = a + ib = z,$$

and it follows from (4) that

$$(1 + i0)z = (a - 0) + i(b + 0) = a + ib = z$$

for all z. For brevity, we write the zero complex number $0 + i0$ as 0 and the unit complex number $1 + i0$ as 1. Finally, $-z \equiv (-1)z$, and the *subtraction* of complex numbers is defined, in terms of the already-defined operations of addition and multiplication, by $z_1 - z_2 \equiv z_1 + (-z_2) = z_1 + (-1)z_2$.

Just as it is standard to represent a real number, graphically, as a point on a real axis, it is standard to represent a complex number z as a point in a Cartesian a, b plane. Since Cartesian axes are more usually denoted by x and y rather than a and b, let us write $z = x + iy$ rather than $a + ib$ and represent z as a point in a so-called **complex z plane**, as shown in Fig. 1.[1] The x axis is called the **real axis**, and the y axis is called the **imaginary axis**.[2]

Observe from Fig. 2 that the addition of complex numbers defined by (3) satisfies the parallelogram law for the addition of vectors so it is possible, and is often convenient, to think of complex numbers as *vectors*. That is, a z vector is the vector from the origin to the point z.

The distance from the origin to the point z (i.e., the "length of the z vector") is called the **modulus** of z and, by analogy with the absolute value of a real number, is denoted as $|z|$, or as mod(z). Then, from the Pythagorean theorem (Fig. 1),

$$|z| = \sqrt{x^2 + y^2}. \qquad (8)$$

For example, $|2 - i| = \sqrt{5}$ and $|3i| = 3$. It is easy to show (Exercise 2) that

$$|z_1 z_2| = |z_1| \, |z_2|; \qquad (9)$$

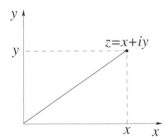

FIGURE 1
Complex z plane.

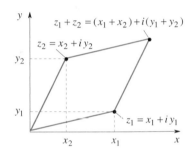

FIGURE 2
Addition in the z plane.

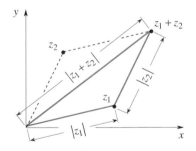

FIGURE 3
Triangle inequality.

---

[1]The complex plane is sometimes called the *Argand diagram* after the French mathematician *Jean Robert Argand* (1768–1822), who was one of the first to propose such representation of complex numbers. A bookkeeper by profession, Argand was a self-taught mathematician.

[2]One might be concerned that we have labeled the point on the y axis as "y." Shouldn't it be "iy?" The answer is that either the real or the complex label is correct, depending on whether we are regarding the point as a point on the real y axis or as a point in the complex z plane.

that is, the modulus of a product equals the product of the moduli of the factors. Further (Fig. 3), the inequality

$$|z_1 + z_2| \leq |z_1| + |z_2| \tag{10}$$

follows from the Euclidean proposition that the length of any one side of a triangle is less than or equal to the sum of the lengths of the other two sides. Hence (10) is known as the **triangle inequality**.

Understand that the complex numbers are not *ordered* as real numbers are. For example, whereas $-6 < 2$ and $10 > 7$, analogous statements such as $z < 0$, $z > 3$, and $1 + 3i > 1 + 2i$ are not meaningful. Of course, statements such as $\text{Re}\, z < 6$, $\text{Im}\, z < 6$, $|z| > 4$, and $|1 - i| < |3 - i|$ *do* make sense because $\text{Re}\, z$, $\text{Im}\, z$, and $|z|$ are real numbers.

Besides $z = x + iy$, it is useful to define the **complex conjugate** of z as

$$\bar{z} \equiv x - iy. \tag{11}$$

Thus, if $z_1 = 8 + 3i$ and $z_2 = -4i$, then $\bar{z}_1 = 8 - 3i$ and $\bar{z}_2 = 4i$. It is readily shown (Exercise 3) that

$$\overline{z_1 + z_2} = \bar{z}_1 + \bar{z}_2, \tag{12a}$$

$$\overline{z_1 z_2} = \bar{z}_1 \bar{z}_2, \tag{12b}$$

and that

$$|z| = \sqrt{z\bar{z}}. \tag{13}$$

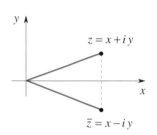

FIGURE 4
Complex conjugate.

Graphically, $\bar{z}$ is simply the reflection of z about the real axis (Fig. 4).

The complex conjugate is useful in defining the *division* of two complex numbers. Division is defined as the inverse of multiplication. That is, the quotient $z = z_1/z_2$ is the complex number $z = x + iy$ such that $z_2 z = z_1$. Writing out the latter as

$$(x_2 + iy_2)(x + iy) = x_1 + iy_1, \tag{14}$$

expanding the left-hand side and equating real and imaginary parts on the left-hand and right-hand sides gives

$$x = \frac{x_1 x_2 + y_1 y_2}{x_2^2 + y_2^2}, \qquad y = \frac{x_2 y_1 - x_1 y_2}{x_2^2 + y_2^2}. \tag{15}$$

Practically speaking, however, it is simpler to evaluate the ratio z_1/z_2 by multiplying numerator and denominator by the complex conjugate of the denominator,

$$z = \frac{z_1}{z_2} = \frac{z_1}{z_2} \frac{\bar{z}_2}{\bar{z}_2} \tag{16}$$

because the denominator $z_2 \bar{z}_2 = |z|^2$ is now real. Thus, writing out (16), we find that

$$z = \frac{z_1}{z_2} = \frac{z_1 \bar{z}_2}{z_2 \bar{z}_2} = \frac{(x_1 + iy_1)(x_2 - iy_2)}{x_2^2 + y_2^2}$$

$$= \frac{x_1 x_2 + y_1 y_2}{x_2^2 + y_2^2} + i \frac{x_2 y_1 - x_1 y_2}{x_2^2 + y_2^2}, \tag{17}$$

which result agrees with (15).

EXAMPLE 1

$$\frac{2+i}{3-4i} = \frac{2+i}{3-4i}\frac{3+4i}{3+4i} = \frac{(6-4)+(8+3)i}{9+16} = \frac{2}{25} + \frac{11}{25}i,$$

which result can be checked by showing that $3 - 4i$ times $\frac{2}{25} + \frac{11}{25}i$ gives $2 + i$.   ∎

6.2.2 Euler's formula and review of the circular and hyperbolic functions.   Having introduced complex numbers and rules for their algebraic manipulation, we can now introduce complex valued algebraic *functions* of z, such as $f(z) = 3z^2 + 2$ and $g(z) = (z + 2i)/(z^4 + 1)$. To evaluate $f(z) = 3z^2 + 2$ at $z = 2 - i$, for instance, write

$$f(2-i) = 3(2-i)^2 + 2 = 3(2-i)(2-i) + 2$$
$$= 3(4 - 4i - 1) + 2 = 11 - 12i.$$

The complex function of special interest to us, in this book, will be the exponential function e^z.

Consider the exponential function e^z. Understand first that we cannot "figure out" how to evaluate e^{x+iy} from our knowledge of the function e^x where x is real. That is, e^{x+iy} is a new object and its values are a matter of definition. To motivate that definition, let us proceed tentatively as follows:

$$e^z = e^{x+iy} = e^x e^{iy}$$

$$= e^x \left[1 + iy + \frac{(iy)^2}{2!} + \frac{(iy)^3}{3!} + \frac{(iy)^4}{4!} + \cdots \right]$$

$$= e^x \left[\left(1 - \frac{y^2}{2!} + \frac{y^4}{4!} - \cdots \right) + i \left(y - \frac{y^3}{3!} + \frac{y^5}{5!} - \cdots \right) \right]. \quad (18)$$

Recognizing the final two series as the Taylor series representations of $\cos y$ and $\sin y$, respectively, we have

$$\boxed{e^{x+iy} = e^x (\cos y + i \sin y).} \quad (19)$$

We say that (19) defines e^z (namely e^{x+iy}) since it gives e^{x+iy} in the standard Cartesian form $a + ib$, where the real part a is $e^x \cos y$ and the imaginary part b is $e^x \sin y$. Observe carefully that we cannot defend certain steps in (18). Specifically, the second equality seems to be the familiar formula $e^{a+b} = e^a e^b$, but the latter is for real numbers a and b whereas the exponent iy is not real. Likewise, the third equality rests upon the Taylor series formula $e^u = 1 + u + \frac{u^2}{2!} + \cdots$ that is derived in the calculus for the case where u is real, but iy is not real. The point to understand, then, is that the steps in (18) are only informal. Trying to stay as close to real-variable theory as possible, we arrive at (19). Once (19) is obtained, we can discard (18) and take (19) as our (i.e., Euler's) *definition* of e^{x+iy}.

As a special case of (19), let $x = 0$. Then (19) becomes

$$e^{iy} = \cos y + i \sin y, \tag{20a}$$

which is known as **Euler's formula**, after the great Swiss mathematician *Leonhard Euler* (1707–1783), whose many contributions to mathematics included the systematic development of the theory of linear constant-coefficient differential equations. In this text it will be convenient to refer to (20a) and the more general formula (19) collectively as Euler's formula. For instance, (20a) gives $e^{\pi i} = \cos \pi + i \sin \pi = -1 + 0i = -1$, and $e^{2-3i} = e^2 (\cos 3 - i \sin 3) = 7.39(-0.990 - 0.141i) = -7.32 - 1.04i$. Since (20a) holds for all y, it must hold also with y changed to $-y$:

$$e^{-iy} = \cos(-y) + i \sin(-y),$$

and since $\cos(-y) = \cos y$ and $\sin(-y) = -\sin y$ (since the graphs of $\cos y$ and $\sin y$ are symmetric and antisymmetric, respectively, about, $y = 0$), it follows that

$$e^{-iy} = \cos y - i \sin y. \tag{20b}$$

Conversely, we can express $\cos y$ and $\sin y$ as linear combinations of the complex exponentials e^{iy} and e^{-iy}: adding (20a) and (20b) and subtracting them gives $\cos y = \left(e^{iy} + e^{-iy}\right)/2$ and $\sin y = \left(e^{iy} - e^{-iy}\right)/(2i)$, respectively. Let us frame these formulas for emphasis and reference:

$$\boxed{\begin{aligned} e^{iy} &= \cos y + i \sin y, \\ e^{-iy} &= \cos y - i \sin y \end{aligned}} \tag{21a,b}$$

and

$$\boxed{\begin{aligned} \cos y &= \frac{e^{iy} + e^{-iy}}{2}, \\ \sin y &= \frac{e^{iy} - e^{-iy}}{2i}. \end{aligned}} \tag{22a,b}$$

Observe that all four of these formulas come from the single formula (20a). Of course there is nothing essential about the name of the variable in these formulas. For instance, $e^{ix} = \cos x + i \sin x$, $e^{i\theta} = \cos \theta + i \sin \theta$, and so on.

There are analogous formulas relating the hyperbolic cosine and hyperbolic sine to real exponentials, for if we recall the definitions

$$\boxed{\begin{aligned} \cosh y &= \frac{e^y + e^{-y}}{2}, \\ \sinh y &= \frac{e^y - e^{-y}}{2} \end{aligned}} \tag{23a,b}$$

of the hyperbolic cosine and hyperbolic sine, we find, by addition and subtraction of these formulas, that

$$\boxed{\begin{aligned} e^y &= \cosh y + \sinh y, \\ e^{-y} &= \cosh y - \sinh y. \end{aligned}} \tag{24a,b}$$

Compare (21) with (24), and (22) with (23). The familiar graphs of $\cosh x$, $\sinh x$, e^x, and e^{-x} are given in Fig. 5.

For reference, some useful properties of the cosine, sine, hyperbolic cosine, and hyperbolic sine functions are as follows:

$$\cos^2 y + \sin^2 y = 1, \tag{25}$$
$$\cosh^2 y - \sinh^2 y = 1, \tag{26}$$

as well as

$$\sin(A + B) = \sin A \cos B + \sin B \cos A, \tag{27}$$
$$\cos(A + B) = \cos A \cos B - \sin A \sin B, \tag{28}$$
$$\sinh(A + B) = \sinh A \cosh B + \sinh B \cosh A, \tag{29}$$
$$\cosh(A + B) = \cosh A \cosh B + \sinh A \sinh B, \tag{30}$$

and the derivative formulas

$$\frac{d}{dx}\cos x = -\sin x, \qquad \frac{d}{dx}\sin x = \cos x,$$
$$\frac{d}{dx}\cosh x = \sinh x, \qquad \frac{d}{dx}\sinh x = \cosh x. \tag{31}$$

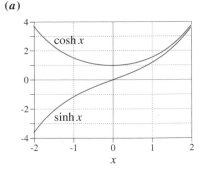

(a)

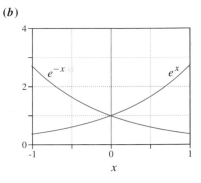

(b)

FIGURE 5
$\cosh x$, $\sinh x$, e^x, and e^{-x}.

We shall be interested specifically in the function e^{rx} and its derivatives with respect to x, where r is a constant that may be complex, say $r = a + ib$. We know from the calculus that

$$\boxed{\frac{d}{dx}e^{rx} = re^{rx}} \tag{32}$$

when r is a real constant. Does (32) hold when $r = a + ib$ is complex? To answer that question, use Euler's formula (19) to express

$$e^{rx} = e^{(a+ib)x} = e^{ax}(\cos bx + i\sin bx).$$

Then,

$$\frac{d}{dx}e^{rx} = \frac{d}{dx}\left[e^{ax}(\cos bx + i\sin bx)\right]$$
$$= \frac{d}{dx}\left(e^{ax}\cos bx\right) + i\frac{d}{dx}\left(e^{ax}\sin bx\right)$$
$$= \left(ae^{ax}\cos bx - be^{ax}\sin bx\right) + i\left(ae^{ax}\sin bx + be^{ax}\cos bx\right)$$
$$= e^{ax}(a + ib)(\cos bx + i\sin bx)$$
$$= re^{ax}(\cos bx + i\sin bx) = re^{rx},$$

so the familiar formula (32) does hold even for complex r.

There is one more fact about the exponential function e^z that we will need, namely, that e^z cannot be zero for any choice of z, for

$$\left|e^z\right| = \left|e^{x+iy}\right| = \left|e^x(\cos y + i\sin y)\right|$$

$$= |e^x| |\cos y + i \sin y| = e^x |\cos y + i \sin y|$$

$$= e^x \sqrt{\cos^2 y + \sin^2 y} = e^x (1) = e^x.$$

The fourth equality follows from the fact that the real exponential is everywhere positive. Finally, we know that $e^x > 0$ for all x, so $|e^z| > 0$ for all z, so $e^z \neq 0$ for all z, as claimed. Thus, we say that e^z *has no zeros*, no values of z for which $e^z = 0$.

Closure. We have introduced the complex number system, defined the arithmetic operations of addition, subtraction, multiplication, and division, and have defined the complex valued exponential function e^z by

$$e^z = e^{x+iy} \equiv e^x (\cos y + i \sin y). \tag{33}$$

The special case $x = 0$ gives

$$e^{iy} = \cos y + i \sin y. \tag{34}$$

From the single formula (34) follow the four relations (21a,b) and (22a,b) between the complex exponentials and the cosine and sine functions, and from (33) follows the fact that

$$\frac{d}{dx} e^{rx} = r e^{rx}$$

even if r is complex. We also showed that the exponential function e^z has no zeros; that is, it is not equal to zero anywhere in the complex z plane. Besides the exponential function we could have defined many other useful functions, such as $\sin z$, $\cos z$, $\sinh z$, $\cosh z$, $\log z$, and so on, but have not because they will not be needed in this text.

Maple. The *Maple* designation of i is I. Here are several illustrative calculations in *Maple*. To evaluate $1/(1 + i)$ (namely, to express it in the Cartesian form $a + bi$) enter

$$1/(1 + I);$$

and obtain

$$\frac{1}{2} - \frac{1}{2} I$$

To evaluate the latter in floating point form use the **evalf** command, which *eval*uates an expression in floating point form. Thus, enter

$$\texttt{evalf(\%);}$$

where the % serves to input the preceding output, and obtain

$$.5000000000 - .5000000000 I$$

The default number of digits is 10 but that number can be increased or decreased using a digits option in evalf(expression,digits). For instance, if we enter

$$\texttt{evalf(\%,3);}$$

rather than `evalf(%)`, we obtain

$$.500 - .500 I$$

EXERCISES 6.2

1. Using the definitions (3) and (4), verify
   (a) the commutative properties (5)
   (b) the associative properties (6)
   (c) the distributive property (7)

2. Verify (9).

3. Verify (12a), (12b), and (13).

4. Show that
   (a) $|z^3| = |z|^3$
   (b) $|z^n| = |z|^n$ and $|1/z^n| = 1/|z|^n$ for $n = 1, 2, \ldots$
   (c) $|z_1 z_2 z_3| = |z_1||z_2||z_3|$
   (d) $|z_1 z_2 \cdots z_n| = |z_1||z_2|\cdots|z_n|$ for $n = 1, 2, \ldots$
   (e) $|z_1 + z_2 + z_3| \leq |z_1| + |z_2| + |z_3|$
   (f) $|z_1 + z_2 + \cdots + z_n| \leq |z_1| + |z_2| + \cdots + |z_n|$
      for $n = 1, 2, \ldots$

5. Show that
   (a) $\overline{\overline{z}} = z$
   (b) $\overline{\left(\dfrac{1}{z}\right)} = \dfrac{1}{\overline{z}}$
   (c) z is real if and only if $z = \overline{z}$
   (d) $\overline{z^3} = \overline{z}^3$
   (e) $\overline{z^n} = \overline{z}^n$ for $n = 1, 2, \ldots$

6. Show that if $z = x + iy$, then $x = (z + \overline{z})/2$ and $y = (z - \overline{z})/2i$.

7. Show that if $z_1 z_2 = 0$, then at least one of the two factors must be zero.

8. Evaluate each of the following. That is, express each in standard Cartesian form $x + iy$.
   (a) $(2 - i)^3$
   (b) $\dfrac{1}{1 - 2i}$
   (c) $\dfrac{i}{2 + 5i}$
   (d) $\dfrac{1 + i}{1 - i}$
   (e) $\left(\dfrac{1 + i}{2 - i}\right)^3$
   (f) $\dfrac{1}{(1 + i)^3}$
   (g) $\text{Re}\,\dfrac{2 + 3i}{4 + 5i}$
   (h) $\text{Im}\,(1 + i)^3$
   (i) $\left(\text{Re}\,\dfrac{1}{1 + i}\right)^3$
   (j) $\text{Im}\,\dfrac{a + ib}{c + id}$
   (k) $\text{Re}\,\dfrac{1}{1 - i}$
   (l) $\text{Im}\,\dfrac{1}{(2 + i)^3}$

9. Evaluate
   (a) $\left|\dfrac{1 - i}{1 + i}\right|$
   (b) $\left|\dfrac{(2 - i)^3}{(1 + 3i)^2}\right|$
   (c) $\left|(1 - 2i)^2 + (1 + i)^2\right|$
   (d) $\left|\left(\dfrac{1 + i}{1 + 2i}\right)^2\right|$

10. Verify the triangle inequality (10) for each of the following cases by working out the left-hand and right-hand sides.
    (a) $z_1 = 2 + 3i,\ z_2 = 4 - i$

    (b) $z_1 = 1 + i,\ z_2 = 7i$
    (c) $z_1 = 5,\ z_2 = 4i$
    (d) $z_1 = 3 + 4i,\ z_2 = 2 + i$
    (e) $z_1 = 1 + i,\ z_2 = 1 - i$

11. Evaluate each in Cartesian form
    (a) $e^{2 + \pi i}$
    (b) $e^{1 - i}$
    (c) $e^{\pi i/4}$
    (d) $e^{-\pi i/3}$
    (e) $e^{(1 + i)^2}$
    (f) $e^{(1 - i)^3}$

12. (a)–(f) Evaluate the expression in the corresponding part of Exercise 11 using *Maple*. NOTE: The *Maple* designations for π and e^z are Pi and exp(z), respectively.

13. Show whether or not $|e^z| = e^{|z|}$.

14. Show that $e^z = 1$ if and only if $z = 2n\pi i$, where n is any integer.

15. Consider this reasoning: $\left|e^{iz}\right| = |\cos z + i \sin z| = \sqrt{\cos^2 z + \sin^2 z} = 1$ for all z, yet $z = -2i$, say, gives $\left|e^{iz}\right| = e^2 \neq 1$. Explain the flaw(s) in that reasoning.

16. Using definitions (19) and (4), show that
$$e^{z_1} e^{z_2} = e^{z_1 + z_2}, \tag{16.1}$$
just as $e^a e^b = e^{a+b}$ holds if a and b are real.

17. As noted in the closure, we have not defined other useful functions, such as $\sin z$, $\cos z$, $\sinh z$, $\cosh z$, and so on, because they will not be needed in this text. Let us do so in this exercise. Since (22a,b) give
$$\cos y = \frac{e^{iy} + e^{-iy}}{2}, \qquad \sin y = \frac{e^{iy} - e^{-iy}}{2i}$$
for any real argument y, we *define*
$$\cos z \equiv \frac{e^{iz} + e^{-iz}}{2} \tag{17.1}$$
and
$$\sin z \equiv \frac{e^{iz} - e^{-iz}}{2i}. \tag{17.2}$$
The right-hand sides are meaningful by virtue of (19). For example, $e^{iz} = e^{i(x+iy)} = e^{-y+ix} = e^{-y}(\cos x + i \sin x)$. Similarly, motivated by (23a,b), we *define*
$$\cosh z \equiv \frac{e^z + e^{-z}}{2} \tag{17.3}$$
and
$$\sinh z \equiv \frac{e^z - e^{-z}}{2}. \tag{17.4}$$

(a) Use (17.1) and (17.2) to show that

$$\cos z = \cos x \cosh y - i \sin x \sinh y \qquad (17.5)$$

and that

$$\sin z = \sin x \cosh y + i \cos x \sinh y. \qquad (17.6)$$

(b) Use (17.3) and (17.4) to show that

$$\cosh z = \cosh x \cos y + i \sinh x \sin y \qquad (17.7)$$

and

$$\sinh z = \sinh x \cos y + i \cosh x \sin y. \qquad (17.8)$$

(c) Derive the following "connections" between the circular and hyperbolic functions:

$$\cos(iz) = \cosh z \qquad (17.9)$$
$$\sin(iz) = i \sinh z, \qquad (17.10)$$
$$\cosh(iz) = \cos z, \qquad (17.11)$$
$$\sinh(iz) = i \sin z. \qquad (17.12)$$

6.3   LINEAR DEPENDENCE AND LINEAR INDEPENDENCE OF FUNCTIONS

Asked how many different paints he had, a painter replied five: red, blue, green, yellow, and purple. However, it could be argued that the count was inflated since only three (e.g., red, blue, and yellow) are independent: the green can be obtained from a certain proportion of the blue and the yellow, and the purple can be obtained from the red and the blue. Similarly, in studying linear differential equations, we will need to determine how many "different," or "independent," functions are contained within a given set of functions. The concept is the same for functions as it is for vectors. Since the linear dependence and linear independence of vectors was discussed in Section 4.7, our extension to the case of functions will be brief, and proofs will be omitted since most mimic the proofs of corresponding theorems in Section 4.7.

Definition 6.3.1   *Linear Dependence and Linear Independence of Functions*
A set of functions $\{u_1, \ldots, u_n\}$ of an independent variable x, say, is **linearly dependent** on an interval I if at least one of them can be expressed as a linear combination of the others on I. If none can be so expressed, then the set is **linearly independent**.[1]

If we do not specify the interval I, then it will be understood to be the entire x axis. NOTE: As in Chapters 4 and 5, we continue to abbreviate the terms linearly dependent and linearly independent as **LD** and **LI**, respectively, with the reminder that this notation is not standard outside of this text.

E X A M P L E   1
The set $\{x^2, e^x, e^{-x}, \sinh x\}$ is seen to be LD (linearly dependent) because we can express $\sinh x$ as a linear combination of the others:

$$\sinh x = \frac{e^x - e^{-x}}{2} = \frac{1}{2}e^x - \frac{1}{2}e^{-x} + 0 \, x^2. \qquad (1)$$

---

[2]Usually, $n > 1$. If $n = 1$ the set $\{u_1\}$ is defined to be linearly dependent if u_1 is the zero function $u_1(x) = 0$ and linearly independent if it is not the zero function.

In fact, we could express e^x as a linear combination of the others too, since solving (1) for e^x gives $e^x = 2\sinh x + e^{-x} + 0\,x^2$. Likewise, we could express $e^{-x} = e^x - 2\sinh x + 0\,x^2$. We cannot express x^2 as a linear combination of the others [since we cannot solve (1) for x^2], but the set is LD nonetheless, because we need only to be able to express "at least one" member as a linear combination of the others. ∎

The foregoing example was simple enough to be worked by inspection. In more complicated cases, the following theorem provides a test for determining whether a given set of functions is LD or LI.

THEOREM 6.3.1

Test for Linear Dependence/Independence
A finite set of functions $\{u_1, \ldots, u_n\}$ is LD on an interval I if and only if there exist scalars α_j, not all zero, such that

$$\alpha_1 u_1(x) + \alpha_2 u_2(x) + \cdots + \alpha_n u_n(x) = 0 \qquad (2)$$

identically on I. If (2) is true only if all the α's are zero, then the set is LI on I.

EXAMPLE 2
To determine if the set $\{1, x, x^2\}$ is LD or LI (on $-\infty < x < \infty$) using Theorem 6.3.1, write equation (2),

$$\alpha_1 + \alpha_2 x + \alpha_3 x^2 = 0, \qquad (3)$$

and see if the truth of (3) requires all the α's to be zero. Since (3) is to hold for *all* x's in the interval, let us write it for $x = 0, 1, 2$, say, to generate three equations on the three α's:

$$\alpha_1 = 0,$$
$$\alpha_1 + \alpha_2 + \alpha_3 = 0, \qquad (4)$$
$$\alpha_1 + 2\alpha_2 + 4\alpha_3 = 0.$$

Solution of (4) gives only the trivial solution $\alpha_1 = \alpha_2 = \alpha_3 = 0$, so the set is LI.

In fact, (3) amounts to an *infinite* number of linear algebraic equations on the three α's since there is no limit to the number of x values that could be chosen in (3). However, three different x values sufficed to show that all the α's must be zero. ∎

Instead of writing out (2) for n specific x values, to generate n equations on $\alpha_1, \ldots, \alpha_n$, it is more common to write (2) and its first $n - 1$ derivatives (assuming, of course, that $u_1, \ldots, u_n$ are $n - 1$ times differentiable on I),

$$\alpha_1 u_1(x) + \cdots + \alpha_n u_n(x) = 0,$$
$$\alpha_1 u_1'(x) + \cdots + \alpha_n u_n'(x) = 0,$$
$$\vdots \qquad\qquad (5)$$
$$\alpha_1 u_1^{(n-1)}(x) + \cdots + \alpha_n u_n^{(n-1)}(x) = 0.$$

Let us denote the determinant of the coefficients as

$$W\left[u_1, \ldots, u_n\right](x) = \begin{vmatrix} u_1(x) & \cdots & u_n(x) \\ u_1'(x) & \cdots & u_n'(x) \\ \vdots & & \vdots \\ u_1^{(n-1)}(x) & \cdots & u_n^{(n-1)}(x) \end{vmatrix}, \tag{6}$$

which is known as the **Wronskian determinant** of $u_1, \ldots, u_n$, or simply the **Wronskian** of $u_1, \ldots, u_n$, after the Polish mathematician *Josef M. H. Wronski* (1778–1853). The Wronskian W is itself a function of x.

Observe that (5) is a system of n homogeneous linear algebraic equations in the n unknowns $\alpha_1, \ldots, \alpha_n$. Surely they admit the trivial solution $\alpha_1 = \alpha_2 = \cdots = \alpha_n = 0$, but the question is: Does (5) admit any *non*trivial solutions as well? Suppose there is a value of x in I, say x_0, such that the determinant of the coefficient matrix is nonzero there, $W[u_1, \ldots, u_n](x_0) \neq 0$. If we set x in (5) equal to x_0, then (5) admits *only* the trivial solution $\alpha_1 = \cdots = \alpha_n = 0$, so the set is LI on I. You may be wondering: What about other values of x? Might the α's be nonzero for other choices of x? But realize that the α's are constants, not functions of x. If, by choosing $x = x_0$ in (5), we learn that the α's are zero, then they are zero for *all* values of x in I.

Thus, we have this useful condition for linear independence.

THEOREM 6.3.2
Wronskian Condition for Linear Independence
If, for a set of functions $\{u_1, \ldots, u_n\}$ of x having derivatives through order $n - 1$ on an interval I, $W[u_1, \ldots, u_n](x)$ is nonzero anywhere on I, then the set is LI on I.

Be careful not to read into Theorem 6.3.2 an inverse, namely, that if $W[u_1, \ldots, u_n](x)$ *is* identically zero on I (which we sometimes emphasize by writing $W \equiv 0$), then the set is LD on I. In fact, the latter is not true, as shown by the following example.

EXAMPLE 3
Consider the set $\{u_1, u_2\}$, where

$$u_1(x) = \begin{cases} x^2, & x \leq 0 \\ 0, & x \geq 0, \end{cases} \qquad u_2(x) = \begin{cases} 0, & x \leq 0 \\ x^2, & x \geq 0. \end{cases} \tag{7}$$

These are displayed in Fig. 1. Then (2) becomes

$$\alpha_1 x^2 + \alpha_2(0) = 0 \qquad \text{for } x \leq 0$$
$$\alpha_1(0) + \alpha_2 x^2 = 0 \qquad \text{for } x \geq 0.$$

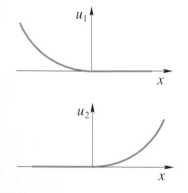

FIGURE 1
u_1 and u_2 in Example 3.

The first implies that $\alpha_1 = 0$, and the second implies that $\alpha_2 = 0$. Hence $\{u_1, u_2\}$ is LI. Yet, $W[u_1, u_2](x) = \begin{vmatrix} x^2 & 0 \\ 2x & 0 \end{vmatrix} = 0$ on $x \leq 0$, and $W[u_1, u_2](x) = \begin{vmatrix} 0 & x^2 \\ 0 & 2x \end{vmatrix} = 0$ on $x \geq 0$, so $W[u_1, u_2](x) \equiv 0$ for all x. ∎

The upshot is that if W is nonzero anywhere on I then the set is LI on I, but if W is identically zero on I then we learn nothing: the set could be LD or LI.

However, our interest in linear dependence and independence, in this chapter, is not going to be in connection with sets of randomly chosen functions, but with sets of functions which have in common that they are solutions of a given linear homogeneous differential equation. In that case, it can be shown that the inverse of Theorem 6.3.2 *is* true: that is, if $W \equiv 0$, then the set is LD. Thus, for that case we have the following stronger theorem, which we will rely on extensively.

THEOREM 6.3.3
A Necessary and Sufficient Condition for Linear Dependence
If $u_1, \ldots, u_n$ are solutions of an nth-order linear homogeneous differential equation

$$\frac{d^n y}{dx^n} + p_1(x)\frac{d^{n-1}y}{dx^{n-1}} + \cdots + p_{n-1}(x)\frac{dy}{dx} + p_n(x)y = 0, \tag{8}$$

where the coefficients $p_j(x)$ are continuous on an interval I, then $W[u_1, \ldots, u_n](x) \equiv 0$ on I is both necessary and sufficient for the linear dependence of the set $\{u_1, \ldots, u_n\}$ on I.

EXAMPLE 4
It is readily verified that each of the functions $1, e^x, e^{-x}$ satisfies the equation $y''' - y' = 0$. Since their Wronskian is

$$W\left[1, e^x, e^{-x}\right](x) = \begin{vmatrix} 1 & e^x & e^{-x} \\ 0 & e^x & -e^{-x} \\ 0 & e^x & e^{-x} \end{vmatrix} = 2 \neq 0,$$

it follows from Theorem 6.3.2 that the set $\{1, e^x, e^{-x}\}$ is LI. Another set of solutions of $y''' - y' = 0$ is $e^x, e^{-x}, \cosh x$. Their Wronskian is

$$W\left[e^x, e^{-x}, \cosh x\right](x) = \begin{vmatrix} e^x & e^{-x} & \cosh x \\ e^x & -e^{-x} & \sinh x \\ e^x & e^{-x} & \cosh x \end{vmatrix} = 0,$$

so the set $\{e^x, e^{-x}, \cosh x\}$ is LD. This result is no surprise because we know that $\cosh x$ is a linear combination of e^x and e^{-x}. ∎

There is an interesting formula for the Wronskian. Namely, subject to the conditions of Theorem 6.3.3 it can be shown (Exercise 5) that

$$W(x) = W(\xi)e^{-\int_\xi^x p_1(t)\,dt}, \tag{9}$$

where ξ is any point in the x interval, where p_1 is the coefficient of the next-to-highest derivative in (8), and where we have written $W[u_1, \ldots, u_n](x)$ as $W(x)$, and $W[u_1, \ldots, u_n](\xi)$ as $W(\xi)$, for brevity; t does not denote time, it is merely a dummy integration variable. Due to the French mathematician *Joseph Liouville* (1809–1882) and known as **Liouville's formula**, (9) shows that under the conditions of Theorem 6.3.3 *the Wronskian is either everywhere zero or everywhere nonzero on I*, for the exponential function is positive for *all* finite values of its argument and the constant $W(\xi)$ is either zero or not. Incidentally, in Example 4, $p_1(x)$ was zero so (9) became $W(x) = W(\xi)e^0 = W(\xi) = $ constant. That is why both Wronskians in that example happened to be constants.

Finally, we cite the following three simple and useful results which are the function analogs of the vector Theorems 4.7.2 – 4.7.4.

> **THEOREM 6.3.4**
> *Linear Dependence/Independence of Two Functions*
> A set of two functions, $\{u_1, u_2\}$, is LD if and only if one is expressible as a scalar multiple of the other.

> **THEOREM 6.3.5**
> *Linear Dependence of Sets Containing the Zero Function*
> If a set of functions $\{u_1, \ldots, u_n\}$ contains the zero function [that is, $u_j(x) = 0$ for some j], then the set is LD.

> **THEOREM 6.3.6**
> *Equating Coefficients*
> Let $\{u_1, \ldots, u_n\}$ be LI on an interval I. Then, for
>
> $$a_1 u_1(x) + \cdots + a_n u_n(x) = b_1 u_1(x) + \cdots + b_n u_n(x)$$
>
> to hold on I, it is necessary and sufficient that $a_j = b_j$ for each $j = 1, \ldots, n$. That is, the coefficients of corresponding terms on the left-hand and right-hand sides must match.

E X A M P L E   5

The set $\{x, \sin x\}$ is LI on $-\infty < x < \infty$ according to Theorem 6.3.4 because x is not expressible as a constant times $\sin x$, nor is $\sin x$ expressible as a constant times x. ∎

EXAMPLE 6

We've seen that $\{1, e^x, e^{-x}\}$ is LI on $-\infty < x < \infty$. Thus, if we meet the equation

$$a + be^x + ce^{-x} = 6 - 2e^{-x}, \tag{10}$$

then it follows from Theorem 6.3.6 that we must have $a = 6$, $b = 0$, $c = -2$, for if we rewrite (10) as

$$(a - 6)(1) + be^x + (c + 2)e^{-x} = 0,$$

then it follows from the linear independence of $1, e^x, e^{-x}$ that $a - 6 = 0$, $b = 0$, $c + 2 = 0$; that is, $a = 6$, $b = 0$, $c = -2$. ∎

Closure.   We have introduced the concept of linear dependence and linear independence as preliminary to our development of the theory of linear differential equations of second order and higher, which follows next. Following the definitions of these terms, we gave three theorems for the testing of a given set of functions to determine if they are LI or LD. Of these, Theorem 6.3.3 will be most useful to us in the sections to follow because it applies to sets of functions that arise as solutions of a given differential equation.

In case you have trouble remembering which of the conditions $W = 0$ and $W \neq 0$ corresponds to linear dependence and which to linear independence, think of it this way. If we randomly make up a determinant, the chances are that its value is nonzero; that is the generic case. Likewise, if we randomly select a set of functions out of the set of all possible functions, the generic case is for them to be unrelated—namely, LI. The generic cases go together ($W \neq 0$ corresponding to linear independence) and the nongeneric cases go together ($W = 0$ corresponding to linear dependence).

EXERCISES 6.3

1. **(a)** Can a set be neither LD nor LI? Explain.
   (b) Can a set be both LD and LI? Explain.

2. Show that the following sets are LD by expressing one of the functions as a linear combination of the others.
   (a) $\{1, x + 2, 3x - 5\}$
   (b) $\{x^2, x^2 + x, x^2 + x + 1, x - 1\}$
   (c) $\{x^4 - x^2, x^4 - x^2 + 1, x^4 - x^2 - 1\}$
   (d) $\{e^x, e^{2x}, \sinh x, \cosh x\}$
   (e) $\{\sinh 3x, e^x, e^{3x}, e^{5x}, e^{-3x}\}$
   (f) $\{e^x, e^{2x}, xe^x, (7x - 2)e^x\}$
   (g) $\{0, x, x^3\}$
   (h) $\{x, 2x, x^2\}$

3. Show whether the given set is LD or LI. HINT: In one of the problems it will be useful to recall the Vandermonde determinant.

   (a) $\{1, x, x^2, \ldots, x^n\}$
   (b) $\{e^{a_1 x}, e^{a_2 x}, \ldots, e^{a_n x}\}$
   (c) $\{1, 1 + x, 1 + x^2\}$
   (d) $\{e^x, e^{2x}\}$
   (e) $\{\sin x, \cos x, \sinh x\}$
   (f) $\{x, x^2\}$
   (g) $\{1, \sin 3x\}$
   (h) $\{1 - x, 1 + x, x^3\}$
   (i) $\{1 - x, 2 + x, 4\}$
   (j) $\{x, e^x, \cos x\}$

4. Verify that each of the given functions is a solution of the given differential equation, and then use Theorem 6.3.3 to determine if the set is LD or LI. As a check, use Theorem 6.3.4 if that theorem applies.

   (a) $y''' - 6y'' + 11y' - 6y = 0$, $\{e^x, e^{2x}, e^{3x}\}$
   (b) $y'' + 4y = 0$, $\{\sin 2x, \cos 2x\}$
   (c) $y''' - 6y'' + 9y' - 4y = 0$, $\{e^x, xe^x, e^{4x}\}$
   (d) $y''' - 6y'' + 9y' - 4y = 0$, $\{e^x, xe^x, (1 - x)e^x\}$
   (e) $y''' - y'' - 2y' = 0$, $\{1, e^{-x}, e^{2x}\}$
   (f) $y'''' - 5y'' + 4y = 0$, $\{e^x, e^{-x}, e^{2x}, e^{-2x}\}$

(g) $x^2 y'' - 3xy' + 3y = 0$, $\{x, x^3\}$,   on $x > 0$

(h) $x^2 y'' - 3xy' + 4y = 0$, $\{x^2, x^2 \ln x\}$,   on $x > 0$

(i) $y'' - 4y' + 4y = 0$, $\{e^{2x}, xe^{2x}\}$

(j) $x^3 y''' + xy' - y = 0$, $\{x, x \ln x, x(\ln x)^2\}$, on $x > 0$

5. (*Liouville's formula*)

(a) Derive Liouville's formula, (9), for the special case where $n = 2$, by writing out $W'(x)$, showing that

$$W'(x) = -p_1(x)W(x), \qquad (5.1)$$

and integrating the latter to obtain (9).

(b) Derive (9) for the general case (i.e., where n need not equal 2), by using equation (15.3) in Exercise 15 of Section 5.4. Show that each of the n determinants on the right-hand side of that equation, except the nth one, has two identical rows and hence vanishes, so that

$$W'(x) = \begin{vmatrix} u_1(x) & \cdots & u_n(x) \\ \vdots & & \vdots \\ u_1^{(n-2)}(x) & \cdots & u_n^{(n-2)}(x) \\ u_1^{(n)}(x) & \cdots & u_n^{(n)}(x) \end{vmatrix}. \qquad (5.2)$$

In the last row, substitute

$$u^{(n)}(x) = -p_1(x)u^{(n-1)}(x) - \cdots - p_n(x)u(x)$$

from (8), again omit vanishing determinants, and again obtain (5.1) and hence the solution (9).

6. (a) Prove Theorem 6.3.4.

(b) Prove Theorem 6.3.5.

(c) Prove Theorem 6.3.6.

7. If u_1 and u_2 are LI, u_1 and u_3 are LI, and u_2 and u_3 are LI, does it follow that $\{u_1, u_2, u_3\}$ is LI? Prove or disprove. HINT: If a proposition is false it can be disproved by a single counterexample, but if it is true then a single example does not suffice as proof.

8. Verify that x^2 and x^3 are solutions of $x^2 y'' - 4xy' + 6y = 0$ on $-\infty < x < \infty$. Also verify, from Theorem 6.3.4, that they are LI on that interval. Does the fact that their Wronskian $W[x^2, x^3](x) = x^4$ vanishes at $x = 0$, together with their linear independence on $-\infty < x < \infty$ violate Theorem 6.3.3? Explain.

6.4 HOMOGENEOUS EQUATION; GENERAL SOLUTION

6.4.1 General solution and solution to initial value problem.   We studied the first-order linear homogeneous equation

$$y' + p(x)y = 0 \qquad (1)$$

in Chapter 2, where $p(x)$ is continuous on the x interval of interest, I, and found the solution to be

$$y(x) = Ce^{-\int p(x)\,dx}, \qquad (2)$$

where C is an arbitrary constant. If we append to (1) an initial condition $y(a) = b$, where a is a point in I, then we obtain, from (2),

$$y(x) = be^{-\int_a^x p(\xi)\,d\xi}, \qquad (3)$$

as was shown in Section 2.2.

The solution (2) is really a family of solutions because of the arbitrary constant C. Since (2) contains *all* solutions of (1), we called it a general solution of (1). In contrast, (3) was only one member of that family, so we called it a particular solution.

Now we turn to the nth-order linear equation

$$\frac{d^n y}{dx^n} + p_1(x)\frac{d^{n-1}y}{dx^{n-1}} + \cdots + p_{n-1}(x)\frac{dy}{dx} + p_n(x)y = 0, \tag{4}$$

and once again we are interested in general and particular solutions. By a **general solution** of (4) on an interval I, we mean a family of solutions that contains every solution of (4) on that interval, and by a **particular solution** of (4) we mean any one member of that family of solutions.

We begin with a fundamental existence and uniqueness theorem.[1]

THEOREM 6.4.1

Existence and Uniqueness for Initial Value Problem

If $p_1(x), \ldots, p_n(x)$ are continuous on an open interval I, then the initial value problem consisting of the differential equation

$$\frac{d^n y}{dx^n} + p_1(x)\frac{d^{n-1}y}{dx^{n-1}} + \cdots + p_{n-1}(x)\frac{dy}{dx} + p_n(x)y = 0, \tag{5a}$$

together with initial conditions

$$y(a) = b_1, \quad y'(a) = b_2, \quad \ldots, \quad y^{(n-1)}(a) = b_n, \tag{5b}$$

where the initial point a is in I, has a solution on I, and that solution is unique.

Let us leave the initial value problem (5) now, and turn our attention to determining the form of the general solution of the nth-order linear homogeneous equation (4). We begin by re-expressing (4) more compactly as

$$L[y] = 0, \tag{6}$$

where

$$L = \frac{d^n}{dx^n} + p_1(x)\frac{d^{n-1}}{dx^{n-1}} + \cdots + p_{n-1}(x)\frac{d}{dx} + p_n(x) \tag{7}$$

is called an nth-order **differential operator** and

$$\begin{aligned}
L[y] &= \left(\frac{d^n}{dx^n} + p_1(x)\frac{d^{n-1}}{dx^{n-1}} + \cdots + p_{n-1}(x)\frac{d}{dx} + p_n(x) \right)[y] \\
&\equiv \frac{d^n}{dx^n}y(x) + p_1(x)\frac{d^{n-1}}{dx^{n-1}}y(x) + \cdots + p_n(x)y(x) \tag{8}
\end{aligned}$$

defines the action of L on the function y. $L[y]$ is itself a function of x, with values $L[y](x)$. For instance, if

$$L[y] = y'' + 2xy' - e^x y$$

---

[1] For a more complete sequence of theorems, and their proofs, we refer the interested reader to the little book by J. C. Burkill, *The Theory of Ordinary Differential Equations* (Edinburgh: Oliver and Boyd, 1956) or to William E. Boyce and Richard C. DiPrima, *Elementary Differential Equations and Boundary Value Problems*, 5th ed. (New York: Wiley, 1992).

and $y(x)$ is x^2, then

$$L[y](x) = (x^2)'' + 2x(x^2)' - e^x x^2 = 2 + 4x^2 - x^2 e^x.$$

The key property of the operator L defined by (8) is that

$$\boxed{L[\alpha u + \beta v] = \alpha L[u] + \beta L[v]} \qquad (9)$$

for any (n-times differentiable) functions u, v and for any constants α, β. Let us verify (9) for the case where L is of second order:

$$
\begin{aligned}
L[\alpha u + \beta v] &= \left(\frac{d^2}{dx^2} + p_1 \frac{d}{dx} + p_2 \right) (\alpha u + \beta v) \\
&= (\alpha u + \beta v)'' + p_1 (\alpha u + \beta v)' + p_2 (\alpha u + \beta v) \\
&= \alpha u'' + \beta v'' + p_1 \alpha u' + p_1 \beta v' + p_2 \alpha u + p_2 \beta v \\
&= \alpha \left(u'' + p_1 u' + p_2 u \right) + \beta \left(v'' + p_1 v' + p_2 v \right) \\
&= \alpha L[u] + \beta L[v],
\end{aligned}
\qquad (10)
$$

as claimed in (9). Similarly for $n = 1$ and for $n \geq 3$.

Recall that the differential equation (4) is classified as linear. Correspondingly, the operator L given by (8) is said to be a **linear differential operator**. The key and defining feature of a linear differential operator is the linearity property (9), which will be of great importance to us.

In fact, (9) holds not just for two functions u and v, but for any finite number of functions, say $u_1, \ldots, u_k$. That is,

$$\boxed{L[\alpha_1 u_1 + \cdots + \alpha_k u_k] = \alpha_1 L[u_1] + \cdots + \alpha_k L[u_k]} \qquad (11)$$

for any functions $u_1, \ldots, u_k$ and for any constants $\alpha_1, \ldots, \alpha_k$. (Of course it should be understood, whether we say so explicitly or not, that $u_1, \ldots, u_k$ must be n times differentiable since they are being operated on by the nth-order differential operator L.) To prove (11) we apply (9) step by step. For instance, if $k = 3$ we have

$$
\begin{aligned}
L[\alpha_1 u_1 + \alpha_2 u_2 + \alpha_3 u_3] &= L[\alpha_1 u_1 + 1(\alpha_2 u_2 + \alpha_3 u_3)] \\
&= \alpha_1 L[u_1] + 1L[\alpha_2 u_2 + \alpha_3 u_3] \qquad \text{from (9)} \\
&= \alpha_1 L[u_1] + \alpha_2 L[u_2] + \alpha_3 L[u_3] \quad \text{from (9) again.}
\end{aligned}
$$

From the linearity property (11) we have the following superposition principle:

THEOREM 6.4.2

Superposition of Solutions of $y^{(n)} + p_1 y^{(n-1)} + \cdots + p_n y_n = 0$
If $y_1(x), \ldots, y_k(x)$, are solutions of the homogeneous and linear equation (4), then $C_1 y_1(x) + \cdots + C_k y_k(x)$ is a solution too, for any constants $C_1, \ldots, C_k$.

Proof. By $y_1, \ldots, y_k$ being solutions of (4), we mean that $L[y_1] = 0, \ldots, L[y_k] = 0$. Then it follows from (11) that

$$L[C_1 y_1 + \cdots + C_k y_k] = C_1 L[y_1] + \cdots + C_k L[y_k]$$
$$= C_1(0) + \cdots + C_k(0)$$
$$= 0.$$

∎

E X A M P L E 1   *Superposition*.
It is readily verified by direct substitution that $y_1 = e^{3x}$ and $y_2 = e^{-3x}$ are solutions of the equation

$$y'' - 9y = 0. \tag{12}$$

(We are not yet concerned with how to *find* such solutions; we will come to that in the next section.) Then, according to Theorem 6.4.2, $y = C_1 e^{3x} + C_2 e^{-3x}$ is also a solution, for any constants C_1 and C_2, as can be verified by substituting it into (12). ∎

To illustrate that the theorem does not apply to nonlinear or nonhomogeneous equations, we offer two counter-examples:

E X A M P L E 2   *A Nonlinear Equation*.
It can be verified that $y_1 = 1$ and $y_2 = x^2$ are solutions of the nonlinear equation $x^3 y'' - yy' = 0$, yet their linear combination $4 + 3x^2$ is not. ∎

E X A M P L E 3   *A Nonhomogeneous Equation*.
It can be verified that $y_1 = 4e^{3x} - 2$ and $y_2 = e^{3x} - 2$ are solutions of the nonhomogeneous equation $y'' - 9y = 18$, yet their sum $5e^{3x} - 4$ is not. ∎

We can now prove the following major result.

THEOREM 6.4.3

General Solution of $y^{(n)} + p_1 y^{(n-1)} + \cdots + p_n y = 0$
Let $p_1(x), \ldots, p_n(x)$ be continuous on an open interval I. Then the nth-order linear homogeneous differential equation (4) admits exactly n LI solutions; that is, at least n and no more than n. Further, if $y_1(x), \ldots, y_n(x)$ is such a set of LI solutions on I, then a general solution of (4), on I, is

$$y(x) = C_1 y_1(x) + \cdots + C_n y_n(x), \tag{13}$$

where $C_1, \ldots, C_n$ are arbitrary constants.

Proof. To show that (4) has no more than n LI solutions, suppose that $y_1(x), \ldots, y_m(x)$ are solutions of (4), where $m > n$. Let ξ be some point in I. The n linear algebraic equations

$$c_1 y_1(\xi) + \cdots + c_m y_m(\xi) = 0$$

$$\vdots \tag{14}$$

$$c_1 y_1^{(n-1)}(\xi) + \cdots + c_m y_m^{(n-1)}(\xi) = 0$$

in the m unknown c_j's have nontrivial solutions because $m > n$. (Do you see why?) Choosing such a nontrivial set of c_j's, define

$$v(x) \equiv c_1 y_1(x) + \cdots + c_m y_m(x), \tag{15}$$

and observe first that

$$L[v] = L[c_1 y_1 + \cdots + c_m y_m]$$
$$= c_1 L[y_1] + \cdots + c_m L[y_m] = c_1(0) + \cdots + c_m(0) = 0, \tag{16}$$

where L is the differential operator in (4). Second, observe from the definition (15) that the left-hand sides of (14) are $v(\xi)$, $v'(\xi)$, ..., $v^{(n-1)}(\xi)$, so that $v(\xi) = v'(\xi) = \cdots = v^{(n-1)}(\xi) = 0$. One function $v(x)$ that satisfies $L[v] = 0$ and $v(\xi) = \cdots = v^{(n-1)}(\xi) = 0$ is simply $v(x) = 0$. By the uniqueness part of Theorem 6.4.1 it is the *only* such function, so $v(x) = 0$. Recalling that the c_j's in $v(x) = c_1 y_1(x) + \cdots + c_m y_m(x) = 0$ are not all zero, it follows that $y_1(x), \ldots, y_m(x)$ must be LD. Thus, (4) cannot have more than n LI solutions.

To show that there are indeed n LI solutions of (4), let us put forward n such solutions. According to the existence part of Theorem 6.4.1, there must be solutions $y_1(x), \ldots, y_n(x)$ of (4) satisfying the initial conditions

$$y_1(a) = \alpha_{11}, \quad y_1'(a) = \alpha_{12}, \quad \cdots, \quad y_1^{(n-1)}(a) = \alpha_{1n},$$

$$\vdots \qquad\qquad \vdots \qquad\qquad\qquad \vdots \tag{17}$$

$$y_n(a) = \alpha_{n1}, \quad y_n'(a) = \alpha_{n2}, \quad \cdots, \quad y_n^{(n-1)}(a) = \alpha_{nn},$$

where a is any chosen point in I and the α's are any chosen numbers such that their determinant is nonzero. (For instance, one such set of α's is given by $\alpha_{ij} = \delta_{ij}$ where δ_{ij} is the Kronecker delta symbol.) Then, according to Theorem 6.3.3, $y_1(x), \ldots, y_n(x)$ must be LI since their Wronskian is nonzero at $x = a$. Thus, there are indeed n LI solutions of (4).

Finally, every solution of (4) must be expressible as a linear combination of those n LI solutions, as in (13), for otherwise there would be more than n LI solutions of (4), which we have already proved is impossible.   ∎

Any such set of n LI solutions is called a **basis**, or **fundamental set**, of solutions of the differential equation.

EXAMPLE 4
Suppose we begin writing solutions of the equation $y'' - 9y = 0$ (from Example 1): $e^{3x}, 5e^{3x}, e^{-3x}, 2e^{3x} + e^{-3x}, \sinh 3x, \cosh 3x, e^{3x} - 4\cosh 3x$, and so on. (That each is a solution is easily verified.) From among these we can indeed find

two that are LI, but no more than two. For instance, $\{e^{3x}, e^{-3x}\}$, $\{e^{3x}, 2e^{3x} + e^{-3x}\}$, $\{e^{3x}, \sinh 3x\}$, $\{\sinh 3x, \cosh 3x\}$, $\{\sinh 3x, e^{-3x}\}$ are bases, so the general solution can be expressed in any of these ways:

$$y(x) = C_1 e^{3x} + C_2 e^{-3x}, \tag{18a}$$

$$y(x) = C_1 e^{3x} + C_2 \left(2e^{3x} + e^{-3x}\right), \tag{18b}$$

$$y(x) = C_1 e^{3x} + C_2 \sinh 3x, \tag{18c}$$

$$y(x) = C_1 \sinh 3x + C_2 \cosh 3x, \tag{18d}$$

and so on. Each of these is a general solution of $y'' - 9y = 0$, and *they are equivalent.* For instance, (18a) implies (18d) since

$$y(x) = C_1 e^{3x} + C_2 e^{-3x}$$
$$= C_1 (\cosh 3x + \sinh 3x) + C_2 (\cosh 3x - \sinh 3x)$$
$$= (C_1 + C_2) \cosh 3x + (C_1 - C_2) \sinh 3x$$
$$= C_1' \cosh 3x + C_2' \sinh 3x,$$

where C_1', C_2' are arbitrary constants (Exercise 13).   ■

EXAMPLE 5

Solve the initial value problem

$$y''' + y' = 0 \tag{19a}$$

$$y(0) = 3, \quad y'(0) = 5, \quad y''(0) = -4, \tag{19b}$$

given that $\cos x$, $\sin x$, and 1 are LI solutions of (19a). Then a general solution of (19a) is

$$y(x) = C_1 \cos x + C_2 \sin x + C_3. \tag{20}$$

Imposing (19b) on (20)gives

$$y(0) = 3 = C_1 + C_3,$$
$$y'(0) = 5 = C_2,$$
$$y''(0) = -4 = -C_1,$$

which equations admit the unique solution $C_1 = 4$, $C_2 = 5$, $C_3 = -1$. Thus,

$$y(x) = 4 \cos x + 5 \sin x - 1$$

is the unique solution to the initial value problem (19).   ■

6.4.2 Boundary value problems.   Remember that the existence and uniqueness theorem, Theorem 6.4.1, is for *initial* value problems. Though most of our interest in this text is in initial value problems, one also encounters problems of boundary value type, where conditions are specified at *two* points, normally the ends of the interval I of interest. Not only are boundary value problems not covered by Theorem 6.4.1, *boundary value problems need not have unique solutions.* In fact, they may have *no solution, a unique solution, or a nonunique solution,* as shown by the following example.

EXAMPLE 6 *Boundary Value Problem*.
It is readily verified that the differential equation

$$y'' + y = 0 \tag{21}$$

admits a general solution

$$y(x) = C_1 \cos x + C_2 \sin x. \tag{22}$$

Consider three different sets of boundary values.

Case 1: $y(0) = 2, y(\pi) = 1$. Then

$$y(0) = 2 = C_1 + 0,$$
$$y(\pi) = 1 = -C_1 + 0,$$

which has no solution for C_1, C_2, so the boundary value problem has *no* solution for $y(x)$.

Case 2: $y(0) = 2, y(\pi/2) = 3$. Then

$$y(0) = 2 = C_1 + 0,$$
$$y(\pi/2) = 3 = 0 + C_2,$$

which gives the unique solution $C_1 = 2, C_2 = 3$. Thus, the boundary value problem has the *unique* solution $y(x) = 2 \cos x + 3 \sin x$.

Case 3: $y(0) = 2, y(\pi) = -2$. Then

$$y(0) = 2 = C_1 + 0,$$
$$y(\pi) = -2 = -C_1 + 0,$$

so $C_1 = 2$, and C_2 is arbitrary, and the boundary value problem has the *nonunique* solution (indeed, the infinity of solutions) $y(x) = 2 \cos x + C_2 \sin x$, where C_2 is an arbitrary constant.   ■

Closure.   In this section we have considered the nth-order linear homogeneous differential equation $L[y] = 0$. The principal result was that a general solution

$$y(x) = C_1 y_1(x) + \cdots + C_n y_n(x)$$

can be built up by the superposition of n LI solutions $y_1(x), \ldots, y_n(x)$ thanks to the linearity of L. Any such set of n LI solutions $\{y_1, \ldots, y_n\}$ of $L[y] = 0$ is called a basis of solutions (or basis, for brevity) for that equation.

For the initial value problem $L[y] = 0$ together with the initial conditions (5b), we found that a solution exists and is unique, but for the boundary value version we found that a solution need not exist, it may exist and be unique, or there may exist a nonunique solution.

Theorems 6.4.1 and 6.4.3 will be especially important to us.

EXERCISES 6.4

NOTE: If not specified, the interval I is understood to be the entire x axis.

1. Show whether or not each of the following is a general solution to the equation given.

   (a) $y'' - 3y' + 2y = 0$; $C_1 e^x + C_2 e^{2x}$

   (b) $y'' - 3y' + 2y = 0$; $C_1(e^x - e^{2x}) + C_2 e^x$

   (c) $y'' - y' - 2y = 0$; $C_1(e^{-x} + e^{2x})$

   (d) $y'' - y' - 2y = 0$; $C_1 e^{-x} + C_2 e^{2x}$

   (e) $y''' + 4y' = 0$; $C_1 \cos 2x + C_2 \sin 2x$

   (f) $y''' + 4y' = 0$; $C_1 + C_2 \cos 2x + C_2 \sin 2x$

   (g) $y''' - 2y'' + y' = 0$; $\left(C_1 + C_2 x + C_3 x^2\right) e^x$

   (h) $y''' - 2y'' + y' = 0$; $(C_1 + C_2 x) e^x + C_3$

   (i) $y''' - y'' - y' + y = 0$; $C_1 e^x + C_2 e^{-x} + C_3 x e^x$

   (j) $y''' = 0$; $C_1 + C_2 x + C_3 x^2 + C_4 x^3$

   (k) $y'''' - y'' = 0$; $C_1 + C_2 x + C_3 e^x + C_4 e^{-x}$

2. Show whether or not each of the following is a basis for the given equation.

   (a) $y'' - 9y = 0$; $\left\{e^{3x}, \cosh 3x, \sinh 3x\right\}$

   (b) $y'' - 9y = 0$; $\left\{e^{3x}, \cosh 3x\right\}$

   (c) $y'' - y = 0$; $\{\sinh 3x, 2\cosh 3x\}$

   (d) $y''' - 3y'' + 3y' - y = 0$; $\left\{e^x, x e^x, x^2 e^x\right\}$

   (e) $y''' - 3y'' = 0$; $\left\{1, x, e^{3x}\right\}$

   (f) $y'''' + 2y'' + y = 0$; $\{\cos x, \sin x, x \cos x, x \sin x\}$

3. Are the following general solutions of $x^2 y'' + x y' - 4y = 0$ on $0 < x < \infty$? On $\infty < x < \infty$? Explain.

   (a) $C_1 x^2$ (b) $C_1 x^2 + C_2 x^{-2}$

   (c) $C_1(x^2 + x^{-2}) + C_2(x^2 - x^{-2})$

4. Are the following bases for the equation $x^2 y'' - x y' + y = 0$ on $0 < x < \infty$? On $-\infty < x < 0$? On $-\infty < x < \infty$? On $6 < x < 10$? Explain.

   (a) $\left\{x, x^2\right\}$

   (b) $\left\{e^x, e^{-x}\right\}$

   (c) $\{x, x \ln|x|\}$

   (d) $\{x + x \ln|x|, x - x \ln|x|\}$

5. Show whether or not the following is a general solution of
   $y^{(vii)} - 4y^{(vi)} - 14y^{(v)} + 56y^{(iv)} + 49y''' - 196y'' - 36y' + 144y = 0$.

   (a) $C_1 e^x + C_2 e^{-x} + C_3 e^{2x} + C_4 e^{-2x} + C_5 e^{3x} + C_6 e^{-3x}$

   (b) $C_1 e^x + C_2 e^{-x} + C_3 e^{2x} + C_4 e^{-2x} + C_5 e^{3x} + C_6 e^{-3x} + C_7 \sinh x + C_8 \cosh 2x$

6. Show that each of the functions $y_1 = 3x^2 - x$ and $y_2 = x^2 - x$ is a solution of the equation $x^2 y'' - 2y = 2x$. Is the linear combination $C_1 y_1 + C_2 y_2$ a solution as well, for all choices of the constants C_1 and C_2? Explain.

7. Does the problem stated have a unique solution? No solution? A nonunique solution? Explain.

   (a) $y'' + 2y' + 3y = 0$; $y(0) = 5$, $y'(0) = -1$

   (b) $y'' + 2y' + 3y = 0$; $y(3) = 2$, $y'(3) = 37$

   (c) $y'' + x y' - y = 0$; $y(3) = y'(3) = 0$

   (d) $x y''' + x y' - y = 0$; $y(-1) = y'(-1) = 0$, $y''(-1) = 4$

   (e) $x^2 y'' - y' - y = 0$; $y(6) = 0$, $y'(6) = 1$

   (f) $(\sin x) y'''' + x y''' = 0$; $y(2) = y'(2) = y''(2) = 0$, $y'''(2) = -9$

8. Verify that (22) is indeed a general solution of (21).

9. Consider the boundary value problem consisting of the differential equation $y'' + y = 0$ plus the boundary conditions given. Does the problem have any solutions? If so, find them. Is the solution unique? HINT: A general solution of the differential equation is $y = C_1 \cos x + C_2 \sin x$.

   (a) $y(0) = 0$, $y(2) = 0$

   (b) $y(0) = 0$, $y(2\pi) = -3$

   (c) $y(1) = 1$, $y(2) = 2$

   (d) $y'(0) = 0$, $y(5) = 1$

   (e) $y'(0) = 0$, $y'(\pi) = 0$

   (f) $y'(0) = 0$, $y'(6\pi) = 0$

   (g) $y'(0) = 0$, $y'(2\pi) = 38$

10. Consider the boundary value problem consisting of the differential equation $y'''' + 2y'' + y = 0$ plus the boundary conditions given. Does the problem have any solutions? If so, find them. Is the solution unique? HINT: A general solution of the differential equation is $y = (C_1 + C_2 x) \cos x + (C_3 + C_4 x) \sin x$.

    (a) $y(0) = y'(0) = 0$, $y(\pi) = 0$, $y'(\pi) = 2$

    (b) $y(0) = y'(0) = y''(0) = 0$, $y(\pi) = 1$

    (c) $y(0) = y''(0) = 0$, $y(\pi) = y''(\pi) = 0$

    (d) $y(0) = y''(0) = 0$, $y(\pi) = y''(\pi) = 3$

11. Prove that the linearity property (9) is equivalent to the two properties

    $$L[u + v] = L[u] + L[v], \tag{11.1a}$$

    $$L[\alpha u] = \alpha L[u]. \tag{11.1b}$$

    That is, show that the truth of (9) implies the truth of (11.1), and conversely.

12. We showed that (11) holds for the case $k = 3$, but did not prove it in general. Here, we ask you to prove (11) for any integer $k \geq 1$. HINT: Use **mathematical induction**, whereby a proposition $P(k)$, for $k \geq 1$, is proved by first showing that it holds for $k = 1$, and then showing that if it

holds for k then it must also hold for $k + 1$. In the present example, the proposition $P(k)$ is the equation (11).

13. (*Example 4, Continued*)

    (a) Verify that each of (18a) through (18d) is a general solution of $y'' - 9y = 0$.

    (b) It seems reasonable that if C_1, C_2 are arbitrary constants, and if we call

$$C_1 + C_2 = C_1' \quad \text{and} \quad C_1 - C_2 = C_2', \qquad (13.1)$$

then C_1', C_2' are arbitrary too, as we claimed at the end of Example 4. Actually, for that claim to be true we need to be able to show that corresponding to any chosen values C_1', C_2' the equations (13.1) on C_1, C_2 are consistent—that is, that they admit one or more solutions for C_1, C_2. Show that (13.1) is indeed consistent.

    (c) Show that if, instead, we had $C_1 + C_2 = C_1'$ and $2C_1 + 2C_2 = C_2'$, where C_1, C_2 are arbitrary constants, then it does *not* follow that C_1', C_2' are arbitrary too.

6.5 HOMOGENEOUS EQUATIONS WITH CONSTANT COEFFICIENTS

Knowing that the general solution of an nth-order homogeneous differential equation is an arbitrary linear combination of n LI (linearly independent) solutions, the question is: How do we find those solutions? That question will occupy us for the remainder of this chapter. In this section we consider the constant-coefficient case,

$$\frac{d^n y}{dx^n} + a_1 \frac{d^{n-1} y}{dx^{n-1}} + \cdots + a_{n-1} \frac{dy}{dx} + a_n y = 0; \qquad (1)$$

that is, where the $a_j(x)$ coefficients are merely constants. This case is said to be "elementary" in the sense that the solutions will always be found among the **elementary functions** (such as powers of x, trigonometric functions, exponentials, and logarithms), but it is also elementary in the sense that it is the simplest case: nonconstant-coefficient equations are generally much harder, and nonlinear equations are much harder still.

    Fortunately, the constant-coefficient case is not only the simplest, it is also of great importance in science and engineering. For instance, the equations

$$mx'' + cx' + kx = 0$$

and

$$Li'' + Ri' + \frac{1}{C}i = 0,$$

governing mechanical and electrical oscillators, where primes denote derivatives with respect to the independent variable t, are both of the type (1) because m, c, k, and L, R, C are constants; they do not vary with t.

6.5.1 Exponential solutions. To guide our search for solutions of (1), it is sensible to begin with the simplest case, $n = 1$:

$$\frac{dy}{dx} + a_1 y = 0, \qquad (2)$$

the general solution of which is

$$y(x) = Ce^{-a_1 x}, \tag{3}$$

where C is an arbitrary constant. One can derive (3) by noticing that (2) is a first-order linear equation and using the general solution developed in Section 2.2, or by using the fact that (2) is separable.

Since (3) is of exponential form, it is natural to wonder if higher-order equations admit exponential solutions too. Consider the second-order equation

$$y'' + a_1 y' + a_2 y = 0, \tag{4}$$

where a_1 and a_2 are real numbers, and let us seek a solution in the exponential form

$$y(x) = e^{rx}, \tag{5}$$

where the constant r is not yet known. If (5) is to be a solution of (4) then, by putting (5) into (4), we find that[1]

$$r^2 e^{rx} + a_1 r e^{rx} + a_2 e^{rx} = 0, \tag{6}$$

or

$$\left(r^2 + a_1 r + a_2 \right) e^{rx} = 0. \tag{7}$$

For (5) to be a solution of (4) on some interval I, we need (7) to be satisfied on I. We recall from Section 6.2 that the exponential function e^z is not zero for *any* value of z, real or complex, so the e^{rx} factor in (7) is nonzero and can be canceled. Thus, if (5) is to satisfy (4) we need r to satisfy the quadratic equation

$$r^2 + a_1 r + a_2 = 0. \tag{8}$$

This equation and its left-hand side are called the **characteristic equation** and **characteristic polynomial**, respectively, corresponding to the differential equation (4). In general, (8) gives two distinct roots, say r_1 and r_2, which can be found from the quadratic formula as

$$r = \frac{-a_1 \pm \sqrt{a_1^2 - 4a_2}}{2}.$$

Thus, our choice of the exponential form (5) has been successful. Indeed, we have found *two* solutions of that form, $e^{r_1 x}$ and $e^{r_2 x}$. (The nongeneric case of repeated roots, which occurs if $a_1^2 - 4a_2$ vanishes, is discussed separately, below.) Next, from Theorem 6.4.2 it follows [thanks to the linearity of (4)] that if $e^{r_1 x}$ and $e^{r_2 x}$ are solutions of (4) then so is any linear combination of them,

$$y(x) = C_1 e^{r_1 x} + C_2 e^{r_2 x}. \tag{9}$$

Theorem 6.4.3 guarantees that (9) is a general solution of (4) if $e^{r_1 x}$ and $e^{r_2 x}$ are LI on I, and Theorem 6.3.4 tells us that they are indeed LI since neither one is expressible as a scalar multiple of the other. Thus, by seeking solutions in the form (5) we were successful in finding the general solution (9) of (4).

---

[1] We know that $d(e^{rx})/dx = re^{rx}$ if the constant r is real, but what if r turns out to be complex? This point was anticipated in Section 6.2, where we showed that the latter formula holds even if r is complex.

EXAMPLE 1

For the equation

$$y'' - y' - 6y = 0, \tag{10}$$

the characteristic equation is $r^2 - r - 6 = 0$, with roots $r = -2, 3$, so

$$y(x) = C_1 e^{-2x} + C_2 e^{3x} \tag{11}$$

is a general solution of (10).   ∎

EXAMPLE 2

For the equation

$$y'' - 9y = 0, \tag{12}$$

the characteristic equation is $r^2 - 9 = 0$, with roots $r = \pm 3$, so a general solution of (12) is

$$y(x) = C_1 e^{3x} + C_2 e^{-3x}. \tag{13}$$

COMMENT 1.   As discussed in Example 4 of Section 6.4, an infinite number of forms of the general solution to (12) are equivalent to (13), such as

$$y(x) = C_1 \cosh 3x + C_2 \sinh 3x, \tag{14}$$

$$y(x) = C_1 \sinh 3x + C_2 \left(5e^{-3x} - 2 \cosh 3x \right), \tag{15}$$

$$y(x) = C_1 \left(e^{3x} + 4 \sinh 3x \right) + C_2 \left(\cosh 3x - \sqrt{\pi} \sinh 3x \right), \tag{16}$$

and so on. Of these, one would normally choose either (13) or (14). What is wrong with (15) and (16)? Nothing, but whereas e^{3x} and e^{-3x} make a "handsome couple," and $\cosh 3x$ and $\sinh 3x$ do too, the choices in (15) and (16) seem ugly and purposeless.

COMMENT 2.   If (13) and (14) are equivalent, does it matter whether we choose one or the other? No, since they are equivalent. However, one may be slightly more convenient than the other insofar as the application of initial or boundary conditions.

For instance, suppose we append to (12) the initial conditions $y(0) = 4$, $y'(0) = -5$. Applying these to (13) gives

$$\begin{aligned} y(0) &= 4 = C_1 + C_2, \\ y'(0) &= -5 = 3C_1 - 3C_2, \end{aligned} \tag{17}$$

so $C_1 = 7/6$, $C_2 = 17/6$, and $y(x) = (7e^{3x} + 17e^{-3x})/6$ (Fig. 1). Applying these initial conditions to (14), instead, gives

$$\begin{aligned} y(0) &= 4 = C_1, \\ y'(0) &= -5 = 3C_2, \end{aligned} \tag{18}$$

so $C_1 = 4$, $C_2 = -5/3$, and $y(x) = 4 \cosh 3x - (5/3) \sinh 3x$. Whereas our final results are equivalent, we see that (18) was simpler than (17), the simplicity resulting from the fact that $\cosh 0 = 1$ and $\sinh 0 = 0$. Thus, the hyperbolic functions are a bit more convenient than the exponential functions when conditions are given at $x = 0$.   ∎

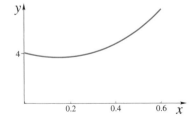

FIGURE 1

The particular solution $y(x) = (7e^{3x} + 17e^{-3x})/6$ of (12), with the initial conditions $y(0) = 4$, $y'(0) = -5$.

EXAMPLE 3

For

$$y'' + 9y = 0 \tag{19}$$

the characteristic equation is $r^2 + 9 = 0$, with roots $r = \pm 3i$, so a general solution of (19) is

$$y(x) = C_1 e^{i3x} + C_2 e^{-i3x}. \tag{20}$$

COMMENT 1.   Just as the general solution of $y'' - 9y = 0$ was expressible in terms of the real exponentials e^{3x}, e^{-3x} or the hyperbolic functions $\cosh 3x$, $\sinh 3x$, the general solution of (19) is expressible in terms of the complex exponentials e^{i3x}, e^{-i3x} or in terms of the circular functions $\cos 3x$, $\sin 3x$, for we can use Euler's formula to re-express (20) as

$$\begin{aligned} y(x) &= C_1 \left(\cos 3x + i \sin 3x \right) + C_2 \left(\cos 3x - i \sin 3x \right) \\ &= (C_1 + C_2) \cos 3x + i \, (C_1 - C_2) \sin 3x. \end{aligned} \tag{21}$$

Since C_1 and C_2 are arbitrary constants, we can simplify this result by letting $C_1 + C_2$ be a new constant A, and letting $i(C_1 - C_2)$ be a new constant B, so we have, from (21), the form

$$y(x) = A \cos 3x + B \sin 3x, \tag{22}$$

where A, B are arbitrary constants. As in Example 1, we note that (20) and (22) are but two out of an infinite number of equivalent forms.

COMMENT 2.   You may be concerned that if $y(x)$ is a physical quantity such as the displacement of a mass or the current in an electrical circuit then it should be real, whereas the right side of (20) seems to be complex. To explore this point, let us solve a complete problem, the differential equation (19) plus a representative set of initial conditions, say $y(0) = 7$, $y'(0) = 3$, and see if the final answer is real or not. Imposing these initial conditions on (20) gives

$$\begin{aligned} y(0) &= 7 = C_1 + C_2, \\ y'(0) &= 3 = i3C_1 - i3C_2, \end{aligned}$$

so $C_1 = (7 - i)/2$ and $C_2 = (7 + i)/2$. Putting these values into (20), we see from (21) that $y(x) = (1/2)[(7 - i) + (7 + i)] \cos 3x + (i/2)[(7 - i) - (7 + i)] \sin 3x = 7 \cos 3x + \sin 3x$ (Fig. 2), which is indeed real. Put differently, if the differential equation and initial conditions represent some physical system, then the mathematics "knows all about" the physics; it is built in, and we need not be anxious. ∎

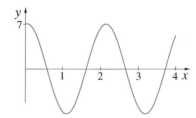

FIGURE 2
The particular solution
$y(x) = 7 \cos 3x + \sin 3x$ of (19),
with the initial conditions
$y(0) = 7$, $y'(0) = 3$.

EXAMPLE 4

The equation

$$y'' + 4y' + 7y = 0 \tag{23}$$

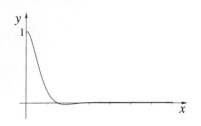

FIGURE 3
The particular solution
$y(x) = e^{-2x}(\cos \sqrt{3}x +$
$\frac{2}{\sqrt{3}} \sin \sqrt{3}x)$ of (23), with the
initial conditions $y(0) = 1$,
$y'(0) = 0$.

has the characteristic equation $r^2 + 4r + 7 = 0$, with distinct roots $r = -2 \pm i\sqrt{3}$, so a general solution of (23) is

$$
\begin{aligned}
y(x) &= C_1 e^{(-2+i\sqrt{3})x} + C_2 e^{(-2-i\sqrt{3})x} \\
&= e^{-2x}\left(C_1 e^{i\sqrt{3}x} + C_2 e^{-i\sqrt{3}x} \right) \\
&= e^{-2x}\left(A \cos \sqrt{3}x + B \sin \sqrt{3}x \right).
\end{aligned} \tag{24}
$$

That is, first we factor out the common factor e^{-2x}, then we re-express the complex exponentials in terms of the circular functions.

If we impose initial conditions $y(0) = 1$, $y'(0) = 0$, say, we find that $A = 1$ and $B = 2/\sqrt{3}$, so $y(x) = e^{-2x}\left(\cos \sqrt{3}x + \frac{2}{\sqrt{3}} \sin \sqrt{3}x \right)$ (Fig. 3). According to Theorem 6.4.1, that solution is unique.   ∎

Examples 1–4 are representative of the four possible cases for second-order equations having distinct roots of the characteristic equation: If the roots are both real then the solution is expressible as a linear combination of two real exponentials (Example 1); if the roots are real and equal in magnitude and opposite in sign, then the solution is expressible either as exponentials or as a hyperbolic cosine and a hyperbolic sine (Example 2); if the roots are not both real then they will be complex conjugates. If those complex conjugates are purely imaginary, then the solution is expressible as a linear combination of two complex exponentials or as a cosine and a sine (Example 3); if they are not purely imaginary, then the solution is expressible as a real exponential times a linear combination of complex exponentials or a cosine and a sine (Example 4).

6.5.2 Higher-order equations ($n > 2$).   Turning to higher-order equations ($n > 2$), our attention focuses on the characteristic equation

$$
r^n + a_1 r^{n-1} + \cdots + a_{n-1}r + a_n = 0. \tag{25}
$$

If $n = 1$, then (25) becomes $r + a_1 = 0$ which, of course, has the root $r = -a_1$ on the real axis. If $n = 2$, then (25) becomes $r^2 + a_1 r + a_2 = 0$. If the roots are complex (and both a_1 and a_2 are real) they will necessarily occur as a complex conjugate pair, as in Example 4.

One might wonder if a further extension of the number system, beyond the complex plane, is required to assure the existence of solutions to the nth-degree polynomial equation (25) for $n \geq 3$. However, it turns out that the complex plane continues to suffice. The characteristic equation (25) admits n roots, which need not be distinct and which need not be real, but if there are complex roots then they necessarily occur in complex conjugate pairs [assuming that all of the a_j coefficients in (1), and hence in (25), are real].

Just as we can solve (25) explicitly by the quadratic formula if $n = 2$, analogous formulas exist for the cases $n = 3$ and $n = 4$ and are given in mathematical handbooks. In this text, however, if we cannot solve (25) readily then we will rely on *Maple*.

In this subsection we limit attention to the case where there are n *distinct* roots of (25), which we denote as $r_1, r_2, \ldots, r_n$. Then each of the exponentials $e^{r_1 x}, \ldots, e^{r_n x}$ is a solution of (1) and, by Theorem 6.4.3,

$$y(x) = C_1 e^{r_1 x} + \cdots + C_n e^{r_n x} \tag{26}$$

is a general solution of (1) if and only if the set of exponentials is LI. The following theorem assures us that they are indeed LI.

THEOREM 6.5.1
Linear Independence of a Set of Exponentials
Let $r_1, \ldots, r_n$ be any numbers, real or complex. The set $\{e^{r_1 x}, \ldots, e^{r_n x}\}$ is LI (on any given interval I) if and only if the r's are distinct.

Proof. Recall from Theorem 6.3.2 that if the Wronskian determinant

$$W\left[e^{r_1 x}, \ldots, e^{r_n x}\right](x) = \begin{vmatrix} e^{r_1 x} & \cdots & e^{r_n x} \\ r_1 e^{r_1 x} & \cdots & r_n e^{r_n x} \\ \vdots & & \vdots \\ r_1^{n-1} e^{r_1 x} & \cdots & r_n^{n-1} e^{r_n x} \end{vmatrix} \tag{27}$$

is not identically zero on I, then the set is LI on I. According to the property D7 of determinants (in Section 5.4), we can factor $e^{r_1 x}$ out of the first column, $e^{r_2 x}$ out of the second, and so on, so that we can re-express W as

$$W\left[e^{r_1 x}, \ldots, e^{r_n x}\right](x) = e^{(r_1 + \cdots + r_n)x} \begin{vmatrix} 1 & \cdots & 1 \\ r_1 & \cdots & r_n \\ \vdots & & \vdots \\ r_1^{n-1} & \cdots & r_n^{n-1} \end{vmatrix}. \tag{28}$$

The exponential function on the right-hand side is nonzero. Further, the determinant in (28) is of Vandermonde type (see the exercises in Section 5.4), so it is nonzero if the r's are distinct, as we have assumed. Thus, W is nonzero, so the given set is LI.

Conversely, if the r's are not distinct, then surely the set is LD because at least two of its members are identical. ∎

Consider the following examples.

EXAMPLE 5
The equation

$$y''' - 8y' + 8y = 0 \tag{29}$$

has the characteristic equation $r^3 - 8r + 8 = 0$. Trial and error reveals that $r = 2$ is one root. Hence we can factor $r^3 - 8r + 8$ as $(r - 2)p(r)$, where $p(r)$ is a quadratic function of r. To find $p(r)$ we divide $r - 2$ into $r^3 - 8r + 8$ by long division and obtain

$p(r) = r^2 + 2r - 4$ which, in turn, can be factored as $[r - (-1 + \sqrt{5})][r - (-1 - \sqrt{5})]$. Thus, r equals 2 and $-1 \pm \sqrt{5}$, so

$$\begin{aligned} y(x) &= C_1 e^{2x} + C_2 e^{(-1+\sqrt{5})x} + C_3 e^{(-1-\sqrt{5})x} \\ &= C_1 e^{2x} + e^{-x} \left(C_2 e^{\sqrt{5}x} + C_3 e^{-\sqrt{5}x} \right) \end{aligned} \tag{30}$$

is a general solution of (29). Of course, equivalent to the $C_2 e^{\sqrt{5}x} + C_3^{-\sqrt{5}x}$ part we could write $C_4 \cosh(\sqrt{5}x) + C_5 \sinh(\sqrt{5}x)$. ∎

EXAMPLE 6
The equation

$$y''' - y = 0 \tag{31}$$

has the characteristic equation $r^3 - 1 = 0$, which surely has the root $r = 1$. Thus, $r^3 - 1$ is $r - 1$ times a quadratic function of r, which function can be found, as above, by long division. In that manner we obtain

$$(r - 1)\left(r^2 + r + 1\right) = 0,$$

so r equals 1 and $(-1 \pm \sqrt{3}\,i)/2$. Hence

$$\begin{aligned} y(x) &= C_1 e^x + C_2 e^{(-1+\sqrt{3}i)x/2} + C_3 e^{(-1-\sqrt{3}i)x/2} \\ &= C_1 e^x + e^{-x/2} \left(C_2 e^{i\sqrt{3}x/2} + C_3 e^{-i\sqrt{3}x/2} \right) \\ &= C_1 e^x + e^{-x/2} \left(C_4 \cos \frac{\sqrt{3}}{2}x + C_5 \sin \frac{\sqrt{3}}{2}x \right), \end{aligned} \tag{32}$$

where C_1, C_4, C_5 are arbitrary constants. ∎

EXAMPLE 7
The equation

$$y^{(v)} - 7y''' + 12y' = 0 \tag{33}$$

has the characteristic equation $r^5 - 7r^3 + 12r = 0$ or, $r(r^4 - 7r^2 + 12) = 0$. The r factor gives the root $r = 0$. The quartic factor is actually a quadratic in r^2, with roots 4 and 3, so the quadratic formula gives $r^2 = 4$ and $r^2 = 3$. Thus, r equals 0, $\pm 2, \pm\sqrt{3}$, so

$$y(x) = C_1 + C_2 e^{2x} + C_3 e^{-2x} + C_4 e^{\sqrt{3}x} + C_5 e^{-\sqrt{3}x} \tag{34}$$

is a general solution of (33). ∎

6.5.3 Repeated roots and reduction of order.   Thus far we have considered only the generic case, where the nth-order characteristic equation (25) admits n distinct roots $r_1, \ldots, r_n$. To complete our discussion, we need to consider the case where one or more roots is repeated. We say that a root r_j of (25) is **repeated** if (25) contains the factor $r - r_j$ more than once. More specifically, we say that r_j is a **root of order k** if (25) contains the factor $r - r_j$ k times. For instance, if the characteristic equation for some given sixth-order equation can be factored as $(r + 2)(r - 5)^3(r - 1)^2 = 0$, then the roots $r = 5$ and $r = 1$ are repeated; $r = 5$ is a root of order 3 and $r = 1$ is a root of order 2. In that case

$$y(x) = C_1 e^{-2x} + C_2 e^{5x} + C_3 e^x$$

is a solution for any constants C_1, C_2, C_3, but the latter falls short of being a *general* solution of the sixth-order differential equation since it is not a linear combination of six LI solutions (Theorem 6.4.3). The problem, in such a case of repeated roots, is how to find the missing solutions. Evidently, they will not be of the form e^{rx}, for if they were then we would have found them when we sought $y(x)$ in that form.

We will use a simple example to show how to obtain such "missing solutions," and will then state the general result as a theorem.

E X A M P L E   8   *Reduction of Order*.
The equation

$$y'' + 2y' + y = 0 \tag{35}$$

has the characteristic equation $r^2 + 2r + 1 = (r + 1)^2 = 0$, so $r = -1$ is a root of order 2. Thus, we have the solution Ae^{-x} but are missing a second linearly independent solution, which is needed if we are to obtain a general solution of (35).

To find the missing solution, we use Lagrange's method of **reduction of order**, which works as follows. Suppose that we know one solution, say $y_1(x)$, of a given linear homogeneous differential equation, and we seek one or more other linearly independent solutions. If $y_1(x)$ is a solution then, of course, so is $Ay_1(x)$, where A is an arbitrary constant. According to the method of reduction of order, we let A vary and seek $y(x)$ in the form $y(x) = A(x)y_1(x)$. Putting that form into the given differential equation results in another differential equation on the unknown $A(x)$, but that equation inevitably will be simpler than the original differential equation on y, as we shall see.

In the present example, $y_1(x)$ is e^{-x}, so to find the missing solution we seek

$$y(x) = A(x)e^{-x}. \tag{36}$$

From (36), $y' = (A' - A)e^{-x}$ and $y'' = (A'' - 2A' + A)e^{-x}$, and putting these expressions into (36) gives

$$\left(A'' - 2A' + A + 2A' - 2A + A\right)e^{-x} = 0, \tag{37}$$

so that $A(x)$ must satisfy the second-order differential equation obtained by equating the coefficient of e^{-x} in (37) to zero, namely, $A'' - 2A' + A + 2A' - 2A + A = 0$.

The cancelation of the three A terms in that equation is not a coincidence, for if $A(x)$ were a constant [in which case the A' and A'' terms in (37) would drop out] then the terms on the left-hand side of (37) would have to cancel to zero because Ae^{-x} is a solution of the original homogeneous differential equation if A is a constant. Thanks to that (inevitable) cancelation, the differential equation governing $A(x)$ will be of the form

$$A'' + \alpha A' = 0, \tag{38}$$

for some constant α, and this second-order equation can be reduced to the first-order equation $v' + \alpha v = 0$ by setting $A' = v$; hence the name reduction of order for the method. In fact, not only do the A terms cancel, as they must, the A' terms happen to cancel as well, so in place of (38) we have the even simpler equation

$$A'' = 0 \tag{39}$$

on $A(x)$. Integration gives $A(x) = C_1 + C_2 x$, so that (36) becomes

$$y(x) = C_1 e^{-x} + C_2 x e^{-x}. \tag{40}$$

The $C_1 e^{-x}$ term merely reproduces that which was already known (recall the second sentence of this example), and the $C_2 x e^{-x}$ term is the desired missing solution. Since the two are LI, (40) is a general solution of (35). ∎

Similarly, suppose we have an eighth-order equation, the characteristic equation of which can be factored as $(r - 2)^3 (r + 1)^4 (r + 5)$, say, so that 2 is a root of order 3 and -1 is a root of order 4. If we take the solution Ae^{2x} associated with the root $r = 2$, and apply reduction of order by seeking y in the form $A(x)e^{2x}$, then we obtain $A''' = 0$ and $A(x) = C_1 + C_2 x + C_3 x^2$ and hence the "string" of solutions $C_1 e^{2x}, C_2 x e^{2x}, C_3 x^2 e^{2x}$ coming from the root $r = 2$. Likewise, if we take the solution Ae^{-x} associated with the root $r = -1$, and apply reduction of order, we obtain $A(x) = C_4 + C_5 x + C_6 x^2 + C_7 x^3$ and hence the string of solutions $C_4 e^{-x}, C_5 x e^{-x}, C_6 x^2 e^{-x}, C_7 x^3 e^{-x}$ coming from the root $r = -1$, so that we have a general solution

$$y(x) = \left(C_1 + C_2 x + C_3 x^2 \right) e^{2x}$$
$$+ \left(C_4 + C_5 x + C_6 x^2 + C_7 x^3 \right) e^{-x} + C_8 e^{-5x} \tag{41}$$

of the original differential equation. [To verify that this is indeed a general solution one would need to show that the eight solutions contained within (41) are LI, as could be done by working out the Wronskian W and showing that $W \neq 0$.]

EXAMPLE 9　*A Small Complication*.
For

$$y'''' - y'' = 0 \tag{42}$$

the characteristic equation $r^4 - r^2 = 0$ gives $r = 0, 0, 1, -1$ and hence the solution $y(x) = A + Be^x + Ce^{-x}$. The latter falls short of being a general solution of (42) because the repeated root $r = 0$ gave the single solution A. To find the missing solution by reduction of order we could vary the parameter A and seek $y(x) = A(x)$, but this time there can be no gain in that step since it merely amounts to a name change, from $y(x)$ to $A(x)$. This situation will always occur when the repeated root is zero, but in that case we can always achieve a reduction of order more directly. In the case of (42) we can set $y'' = p$. Then the fourth-order equation (42) is reduced to the second-order equation $p'' - p = 0$ with general solution

$$p(x) = Ae^x + Be^{-x}.$$

But $y'' = p$, so

$$y'(x) = \int p(x)\,dx = Ae^x - Be^{-x} + C.$$

Integrating again,

$$y(x) = \int \left(Ae^x - Be^{-x} + C\right)dx = Ae^x + Be^{-x} + Cx + D \qquad (43)$$

is a general solution of (42). Observe that the pattern is the same as before: The repeated root $r = 0$ contributes the two terms $Cx + D$ to the general solution (43); that is, it contributes a "string" of solutions of the form $(C_1 + C_2 x)e^{0x}$.   ■

E X A M P L E   10
The equation

$$y'''' + 5y''' = 0 \qquad (44)$$

has the characteristic equation $r^4 + 5r^3 = 0$ with roots $r = 0, 0, 0, -5$. Hence, a general solution of (44) is

$$y(x) = (C_1 + C_2 x + C_3 x^2)e^{0x} + C_4 e^{-5x}$$
$$= C_1 + C_2 x + C_3 x^2 + C_4 e^{-5x}.$$ ■

We organize these results as a theorem:

THEOREM 6.5.2
Repeated Roots of Characteristic Equation
If r_1 is a root of order k, of the characteristic equation (25), then $e^{r_1 x}, xe^{r_1 x}, \ldots, x^{k-1}e^{r_1 x}$ are k LI solutions of the differential equation (1).

Proof.   Denote (1) in operator form as $L[y] = 0$, where

$$L = \frac{d^{(n)}}{dx^{(n)}} + a_1 \frac{d^{(n-1)}}{dx^{(n-1)}} + \cdots + a_{n-1}\frac{d}{dx} + a_n. \qquad (45)$$

Then

$$L\left[e^{rx}\right] = \left(r^n + a_1 r^{n-1} + \cdots + a_{n-1}r + a_n\right)e^{rx}$$

or,

$$L\left[e^{rx}\right] = (r - r_1)^k\, p(r)e^{rx}, \tag{46}$$

where $p(r)$ is a polynomial in r, of degree $n - k$. Since (46) holds for all r, we can set $r = r_1$ in that formula. Doing so, the right-hand side of (46) vanishes, so that $L\left[e^{r_1 x}\right] = 0$ and hence $e^{r_1 x}$ is a solution of $L[y] = 0$. Our object, now, is to show that $xe^{r_1 x}, \ldots, x^{k-1}e^{r_1 x}$ are solutions as well if $k > 1$.

To proceed, differentiate (46) with respect to r (r, not x):

$$\frac{d}{dr}L\left[e^{rx}\right] = k\,(r - r_1)^{k-1}\, p(r)e^{rx} + (r - r_1)^k\,\frac{d}{dr}\left(p(r)e^{rx}\right). \tag{47}$$

The left-hand side of (47) calls for e^{rx} to be differentiated first with respect to x according to the operator L defined in (45) and then with respect to r. Since we can interchange the order of these differentiations, we can express the left-hand side as $L\left[\frac{d}{dr}e^{rx}\right]$, that is, as $L\left[xe^{rx}\right]$. Thus, one differentiation of (46) with respect to r gives

$$L\left[xe^{rx}\right] = k\,(r - r_1)^{k-1}\, p(r)e^{rx} + (r - r_1)^k\,\frac{d}{dr}\left(p(r)e^{rx}\right). \tag{48}$$

Setting $r = r_1$ in (48) gives $L\left[xe^{r_1 x}\right] = 0$. Hence, not only is $e^{r_1 x}$ a solution of $L[y] = 0$, so is $xe^{r_1 x}$. Repeated differentiation with respect to r reveals that $x^2 e^{r_1 x}, \ldots, x^{k-1}e^{r_1 x}$ are solutions as well, as was to be proved.

Verifying the linear independence of the solutions $e^{r_1 x}, xe^{r_1 x}, \ldots, x^{k-1}e^{r_1 x}$ is left for the exercises.   ∎

E X A M P L E   11
As a final example, consider the equation

$$y'''' - 8y''' + 26y'' - 40y' + 25y = 0 \tag{49}$$

with characteristic equation $r^4 - 8r^3 + 26r^2 - 40r + 25 = 0$ and repeated complex roots $r = 2 + i,\ 2 + i,\ 2 - i,\ 2 - i$. It follows that

$$
\begin{aligned}
y(x) &= (C_1 + C_2 x)\, e^{(2+i)x} + (C_3 + C_4 x)\, e^{(2-i)x} \\
&= e^{2x}\left[\left(C_1 e^{ix} + C_3 e^{-ix}\right) + x\left(C_2 e^{ix} + C_4 e^{-ix}\right)\right] \\
&= e^{2x}\left[(A\cos x + B\sin x) + x\,(C\cos x + D\sin x)\right]
\end{aligned} \tag{50}
$$

is a general solution of (49).   ∎

6.5.4 Stability.   An important consideration in applications, especially feedback control systems, is whether or not a system is "stable." Normally, stability has to do with the behavior of a system over *time*, so let us change the name of the independent variable from x to t in (1):

$$\frac{d^n y}{dt^n} + a_1 \frac{d^{n-1}y}{dt^{n-1}} + \cdots + a_{n-1}\frac{dy}{dt} + a_n y = 0, \tag{51}$$

and let us denote the general solution of (51) as $y(t) = C_1 y_1(t) + \cdots + C_n y_n(t)$. We say that the system described by (51) (be it mechanical, electrical, economic, or whatever) is **stable** if all of its solutions are bounded—that is, if there exists a constant M_j for each solution $y_j(t)$ such that $\left| y_j(t) \right| < M_j$ for all $t \geq 0$. If the system is not stable, then it is **unstable**.

THEOREM 6.5.3

Stability

For the system described by (51) to be stable, it is necessary and sufficient that the characteristic equation of (51) have no roots to the right of the imaginary axis in the complex plane and that any roots on the imaginary axis be nonrepeated.

Proof. Let $r = a + ib$ be any nonrepeated root of the characteristic equation. Such a root will contribute a solution $e^{(a+ib)t} = e^{at}(\cos bt + i \sin bt)$. Since the magnitude (modulus, to be more precise) of a complex number $x + iy$ is defined as $|x + iy| = \sqrt{x^2 + y^2}$, and the magnitude of the product of real or complex numbers is the product of their magnitudes, we see that

$$|e^{(a+ib)t}| = |e^{at}(\cos bt + i \sin bt)|$$
$$= |e^{at}||\cos bt + i \sin bt|$$
$$= e^{at}\sqrt{\cos^2 bt + \sin^2 bt} = e^{at}$$

so that solution will be bounded if and only if $a \leq 0$, that is, if r does not lie to the right of the imaginary axis.

Next, let $r = a + ib$ be a repeated root of order k, with $a \neq 0$. Such a root will contribute solutions of the form $t^p e^{(a+ib)t} = t^p e^{at}(\cos bt + i \sin bt)$, for $p = 0, \ldots, k - 1$, with magnitude $t^p e^{at}$. Surely the latter grows unboundedly if $a > 0$ because both factors do, but its behavior is less obvious if $a < 0$ since then the t^p factor grows and the e^{at} decays. To see which one "wins," one can rewrite the product as t^p / e^{-at} and then apply l'Hôpital's rule p times. Doing so, one finds that the ratio tends to zero as $t \to \infty$. [Recall that l'Hôpital's rule applies to indeterminate forms of the type $0/0$ or ∞/∞, not $(\infty)(0)$; that is why we first rewrote $t^p e^{at}$ in the form t^p / e^{-at}.] The upshot is that such solutions are bounded if $r = a + ib$ lies in the left half plane ($a < 0$), and unbounded if it lies in the right half plane ($a > 0$). If r lies *on* the imaginary axis ($a = 0$), then $\left| t^p e^{(a+ib)t} \right| = \left| t^p e^{ibt} \right| = |t^p(\cos bt + i \sin bt)| = t^p$, which grows unboundedly. Our conclusion is that all solutions are bounded if and only if no roots lie to the right of the imaginary axis and no repeated roots lie on the imaginary axis, as was to be proved. ∎

One is often interested in being able to determine whether the system is stable or not without actually evaluating the n roots of the characteristic equation (25). There are theorems that provide information about stability based directly upon the a_j coefficients in (25). One such theorem is stated below. Another, the Routh-Hurwitz criterion, is given in the exercises.

> **THEOREM 6.5.4**
>
> *A Sufficient Condition for Instability*
>
> If the coefficients in (25) are real and of mixed sign (i.e., there is at least one positive and at least one negative), then there is at least one root r with $\text{Re}\, r > 0$, so the system is unstable.

EXAMPLE 12

For instance, if a system is governed by a differential equation $y''' - 2y'' + y' + 5y = 0$, it follows immediately from Theorem 6.5.4 that that system is unstable. But be careful: if the differential equation were $y''' + 2y'' + y' + 5y = 0$, it would *not* follow from the theorem that the system is stable. In that case the theorem would simply give no information. ∎

Closure. In this section we limited our attention to linear homogeneous differential equations with constant coefficients, a case of great importance in applications. Seeking solutions in exponential form, we found the characteristic equation to be central. According to the fundamental theorem of algebra, such equations always have least one root, so we are guaranteed of finding at least one exponential solution of such a differential equation. If the n roots $r_1, \ldots, r_n$ are distinct, then each root r_j contributes a solution $e^{r_j x}$, and their superposition gives a general solution of (1) in the form

$$y(x) = C_1 e^{r_1 x} + \cdots + C_n e^{r_n x}. \tag{52}$$

If any root r_j is repeated, say of order k, then it contributes not only the solution $e^{r_j x}$, but the "string" of k LI solutions $e^{r_j x}, x e^{r_j x}, \ldots, x^{k-1} e^{r_j x}$ to the general solution. Thus, in the generic case of distinct roots, the general solution of (1) is of the form (52); in the nongeneric case of repeated roots, the solution also contains one or more terms which are powers of x times exponentials.

The solution process for linear constant-coefficient homogeneous equations is strikingly simple, with the only difficulty being algebraic—the need to find the roots of the characteristic equation. The reason for this simplicity is that most of the work was done simply in deciding to look for solutions in the right place, within the set of exponential functions. Also, observe that although in a fundamental sense the solving of a differential equation in some way involves integration, the methods discussed in this section required no integrations, in contrast to most of the methods of solution of first-order equations in Chapters 2 and 3.

In the final section we introduced the concept of stability, and in Theorem 6.5.3 we related the stability of the physical system to the location of the roots of the characteristic equation in the complex plane.

Maple. To solve characteristic equations we can use the **solve** command, which can give analytical solutions of equations up to fourth degree, even if one or more coefficients are unspecified and are treated as parameters. For instance, to solve the quadratic equation $ax^2 + bx + c = 0$, the command

```
solve(a*x^2+b*x+c=0,x);
```

gives the roots

$$\frac{-b + \sqrt{b^2 - 4ac}}{2a}, \quad \frac{-b - \sqrt{b^2 - 4ac}}{2a}$$

The ", x" in the command is needed to clarify which letter is the unknown. For the equation $x^4 + x^3 + x^2 + x - 6 = 0$ the command

```
solve(x^4+x^3+x^2+x-6=0);
```

gives the roots

$$2, \quad -1, \quad -1 + \sqrt{2}I, \quad -1 - \sqrt{2}I$$

For higher-degree equations solve may fail. For instance, for $x^5 + x^4 + x^3 + x^2 + x + 1 = 0$ the solve command gives the real root -1 and two pairs of complex conjugate roots, but if we change the $+1$ in the equation to a -1 then the command fails, giving no roots. In that case we can use instead the **fsolve** command, which seeks roots in floating point form. We find that the command

```
fsolve(x^5+x^4+x^3+x^2+x-1=0);
```

gives the single root

$$.5086603916$$

Evidently, the four missing roots are complex. To obtain *all* roots use the command

```
fsolve(x^5+x^4+x^3+x^2+x-1=0,x,complex);
```

and obtain the five roots

$$.5086603916, \quad -1.011836827 - .6839585956I, \quad -1.011836827 + .6839585956I,$$
$$.2575066316 - 1.1187903141I, \quad .2575066316 + 1.1187903141I$$

Turning from the characteristic equation to the differential equation, to obtain a general solution of $y''' - 9y' = 0$ use the command

```
dsolve(diff(y(x),x,x,x)-9*diff(y(x),x)=0,y(x));
```

and to solve that equation subject to the initial conditions $y(0) = 5$, $y'(0) = 2$, $y''(0) = -4$, use the command

```
dsolve({diff(y(x),x,x,x)-9*diff(y(x),x)=0,y(0)=5,
D(y)(0)=2,D(D(y))(0)=-4},y(x));
```

In place of diff($y(x), x, x, x$) we could use diff($y(x), x\$3$) for brevity.

If we wish to go directly from an initial problem to a plot of the solution, we can use the **phaseportrait** command that was used first in Section 2.2. For instance, to obtain a plot of $y(x)$ versus x over $0 \le x \le 1$, for the initial value problem cited above, we can use the commands

```
with(DEtools):
phaseportrait(diff(y(x),x,x,x)-9*diff(y(x),x)=0,
y(x),x=0..1,[[y(0)=5,D(y)(0)=2,D(D(y))(0)=-4]],
stepsize=.05,arrows=NONE);
```

EXERCISES 6.5

1. Theorem 6.5.2 states that $e^{r_1 x}, x e^{r_1 x}, \ldots, x^{k-1} e^{r_1 x}$ are LI. Prove that claim.

2. (*Nonrepeated roots*) Find a general solution of each of the following equations, and a particular solution satisfying the given conditions, if such conditions are given.

(**a**) $y'' + 5y' = 0$

(**b**) $y'' - y' = 0$

(**c**) $y'' + y' = 0; \quad y(0) = 3, \quad y'(0) = 0$

(**d**) $y'' - 3y' + 2y = 0; \quad y(1) = 1, \quad y'(1) = 0$

(**e**) $y'' - 4y' - 5y = 0; \quad y(1) = 1, \quad y'(1) = 0$

(**f**) $y'' + y' - 12y = 0; \quad y(-1) = 2, \quad y'(-1) = 5$

(**g**) $y'' - 4y' + 5y = 0; \quad y(0) = 2, \quad y'(0) = 5$

(**h**) $y'' - 2y' + 3y = 0; \quad y(0) = 4, \quad y'(0) = -1$

(**i**) $y'' - 2y' + 2y = 0; \quad y(0) = 0, \quad y'(0) = -5$

(**j**) $y'' + 2y' + 3y = 0; \quad y(0) = 0, \quad y'(0) = 3$

(**k**) $y''' + 3y' - 4y = 0; \quad y(0) = 0, \quad y'(0) = 0, \quad y''(0) = 6$

(**l**) $y''' - y'' + 2y' = 0; \quad y(0) = 1, \quad y'(0) = 0, \quad y''(0) = 0$

(**m**) $y''' + y'' - 2y = 0$

(**n**) $y''' - y = 0$

(**o**) $y'''' - y = 0$

(**p**) $y'''' - 2y'' - 3y = 0$

(**q**) $y'''' + 6y'' + 8y = 0$

(**r**) $y'''' + 7y'' + 12y = 0$

(**s**) $y''''' - 2y''' - y'' + 2y' = 0$

3. (**a**)–(**s**) Solve the corresponding problem in Exercise 2 using *Maple*.

4. (*Repeated roots*) Find a general solution of each of the following equations, and a particular solution satisfying the given conditions, if such conditions are given.

(**a**) $y'' = 0; \quad y(-3) = 5, \quad y'(-3) = -1$

(**b**) $y'' + 6y' + 9y = 0; \quad y(1) = e, \quad y'(1) = -2$

(**c**) $y''' = 0; \quad y(0) = 3, \quad y'(0) = -5, \quad y''(0) = 1$

(**d**) $y''' + 5y'' = 0; \quad y(0) = 1, \quad y'(0) = 0, \quad y''(0) = 0$

(**e**) $y''' + 3y'' + 3y' + y = 0$

(**f**) $y''' - 3y'' + 3y' - y = 0$

(**g**) $y''' - y'' - y' + y = 0$

(**h**) $y''''' + 3y''' = 0$

(**i**) $y'''' + y''' + y'' = 0$

(**j**) $y'''' + 8y'' + 16y = 0$

(**k**) $y^{(vi)} = 0; \quad y(0) = y'(0) = y''(0) = y'''(0) = y^{(iv)}(0) = 0, \quad y^{(v)}(0) = 3$

5. (**a**)–(**k**) Solve the corresponding problem in Exercise 4 using *Maple*.

6. If the roots of the characteristic equation are as follows, then find the original differential equation and also a general solution of it:

(**a**) $2, 6$

(**b**) $2i, -2i$

(**c**) $4 - 2i, 4 + 2i$

(**d**) $-2, 3, 5$

(**e**) $2, 3, -1$

(**f**) $1, 1, -2$

(**g**) $4, 4, 4, i, -i$

(**h**) $1, -1, 2 + i, 2 - i$

(**i**) $0, 0, 0, 0, 7, 9$

(**j**) $1 + i, 1 + i, 1 - i, 1 - i$

(**k**) $0, 1, 2, 3, 4$

(**l**) $0, 1, -1, 2, -2$

7. (*Complex a_j's*) Find a general solution of each of the following equations. NOTE: Normally, the a_j coefficients in (1) are real, but the results of this section hold even if they are not (except for Theorem 6.5.4, which requires that the coefficients be real). However, *be aware that if the a_j coefficients are not all real, then complex roots do not necessarily occur in complex conjugate pairs.* For instance, $r^2 + 2ir + 1 = 0$ has the roots $r = (\sqrt{2} - 1)i, -(\sqrt{2} + 1)i$.

(**a**) $y'' - 2iy' + y = 0$ (**b**) $y'' - 3iy' - 2y = 0$

(**c**) $y'' + iy' - y = 0$ (**d**) $y'' - 2iy' - y = 0$

(**e**) $y'' - iy = 0$ HINT: Verify, and use, the fact that $\sqrt{i} = \pm(1 + i)/\sqrt{2}$.

(**f**) $y''' + 4iy'' - y' = 0$

(**g**) $y''' + iy' = 0$ HINT: Verify, and use, the fact that $\sqrt{-i} = \pm(1 - i)/\sqrt{2}$.

8. (**a**)–(**h**) Solve the corresponding problem in Exercise 7 using *Maple*.

9. Use *Maple* to obtain the roots of the given characteristic equation, state whether the system is stable or unstable, and explain why. If Theorem 6.5.4 applies, then show that your results are consistent with the predictions of that theorem.

(**a**) $r^3 - 3r^2 + 26r - 2 = 0$

(**b**) $r^3 + 3r^2 + 2r + 2 = 0$

(**c**) $r^4 + r^3 + 3r^2 + 2r + 2 = 0$

(**d**) $r^4 + r^3 + 5r^2 + r + 4 = 0$

(**e**) $r^6 + r^5 + 5r^4 + 2r^3 - r^2 + r + 3 = 0$

(**f**) $r^6 + 9r^5 + 5r^4 + 2r^3 + 7r^2 + r + 3 = 0$

(**g**) $r^6 + r^5 + 5r^4 + 4r^3 + 4r^2 + 8r + 4 = 0$

(**h**) $r^6 + r^5 + 5r^4 + 2r^3 + 7r^2 + r + 3 = 0$

(**i**) $r^8 - r^6 + r^5 + 5r^4 + 2r^3 + 7r^2 + r + 3 = 0$

(**j**) $r^8 + r^7 + r^6 + r^5 + 5r^4 + 21r^3 + 7r^2 + r + 3 = 0$

10. Given the following characteristic equation, where α is a real but unspecified constant (i.e., it is a "parameter"), use one or more inequalities to define the range of values of α

for which the system is stable, and also the range of values of α for which the system is unstable. Do not use *Maple*.

(a) $r^2 + \alpha r + 1 = 0$ (b) $r^2 + \alpha r - 1 = 0$

(c) $r^2 + r + \alpha = 0$ (d) $r^2 - r + \alpha = 0$

11. The same as Exercise 10, but this time you may use *Maple* if you wish.

(a) $r^3 - 3(\alpha + 1)r^2 + (3\alpha^2 + 6\alpha + 2)r - \alpha(\alpha^2 + 3\alpha + 2) = 0$

(b) $r^4 + r^3 + 5(\alpha + 1)r^2 - (\alpha + 3)r + 6r + 1 = 0$

(c) $r^3 + \alpha^3 r^2 + r - \alpha^2 - 4\alpha - 3 = 0$

12. (*Routh-Hurwitz criterion*) According to the **Routh-Hurwitz criterion**, necessary and sufficient conditions for all the roots of the characteristic equation of (51) to have negative real parts are that $a_j > 0$ for each $j = 1, \ldots, n$, and that $\Delta_j > 0$ for each $j = 1, \ldots, n$, where

$$\Delta_j = \begin{vmatrix} a_1 & 1 & 0 & 0 & 0 & \cdots & 0 \\ a_3 & a_2 & a_1 & 1 & 0 & \cdots & 0 \\ a_5 & a_4 & a_3 & a_2 & a_1 & \cdots & \\ \vdots & & & & & & \vdots \\ a_{2j-1} & a_{2j-2} & a_{2j-3} & a_{2j-4} & & \cdots & a_j \end{vmatrix}.$$

Zeros are entered for any a_k's that are called for where $k > n$. For example, if $n = 3$, then

$$\Delta_1 = a_1, \quad \Delta_2 = \begin{vmatrix} a_1 & 1 \\ a_3 & a_2 \end{vmatrix}, \quad \Delta_3 = \begin{vmatrix} a_1 & 1 & 0 \\ a_3 & a_2 & a_1 \\ 0 & 0 & a_3 \end{vmatrix};$$

expanding Δ_3 about the third row yields the simplification $\Delta_3 = a_3 \Delta_2$. Here is the problem: Use the Routh-Hurwitz criterion to determine whether or not the all roots have negative real parts.

(a) $r^4 + 6r^3 + 5r^2 + 4r + 1 = 0$

(b) $r^4 + 2r^3 + 7r^2 + 4r + 8 = 0$

(c) $r^4 + 2r^3 + 5r^2 + 8r + 12 = 0$

(d) $r^4 + r^3 + 4r + 8 = 0$

(e) $r^5 + r^4 + r^3 + r^2 + r + 1 = 0$

(f) $r^5 + r^4 + r^3 + r^2 + r + 8 = 0$

(g) $r^5 + 2r^4 + 3r^3 + 4r^2 + 5r + 6 = 0$

6.6 HOMOGENEOUS EQUATIONS WITH NONCONSTANT COEFFICIENTS; CAUCHY-EULER EQUATION

We return to the linear nth-order homogeneous equation

$$a_0(x)\frac{d^n y}{dx^n} + a_1(x)\frac{d^{n-1} y}{dx^{n-1}} + \cdots + a_{n-1}(x)\frac{dy}{dx} + a_n(x)y = 0, \tag{1}$$

but this time we allow the $a_j(x)$ coefficients to be nonconstant. The theory of the homogeneous equation, given in Section 6.4, holds whether the coefficients are constant or not. However, the task of finding solutions is generally much more difficult if the coefficients in (1) are not all constants. Only in special cases are we able to find solutions in terms of the elementary functions and in closed form (as opposed, say, to infinite series). This section is devoted to the most important of those special cases, equations of Cauchy-Euler type. In other cases we generally give up on finding closed-form solutions. Instead, we might seek solutions in the form of infinite series (which is the subject of Chapter 8) or we might pursue a numerical solution (which is the subject of Chapter 12).

6.6.1 Cauchy-Euler equation. If (1) is of the special form

$$x^n\frac{d^n y}{dx^n} + c_1 x^{n-1}\frac{d^{n-1} y}{dx^{n-1}} + \cdots + c_{n-1}x\frac{dy}{dx} + c_n y = 0, \tag{2}$$

where the c_j's are constants, it is called a **Cauchy-Euler equation** or, by some authors, an **equidimensional equation**. For instance,

$$x^2 y'' - 5xy' + 2y = 0$$

and

$$x^3 y''' + xy' = 0 \qquad \text{or, equivalently,} \qquad x^2 y''' + y' = 0$$

are Cauchy-Euler equations of orders 2 and 3, respectively. Most important to us will be equations of second order, namely,

$$\boxed{x^2 y'' + c_1 xy' + c_2 y = 0,} \tag{3}$$

so we treat that case first. *We will consider (3) on the interval* $0 < x < \infty$.

Suppose we try to solve (3) by seeking y in the form $y = e^{rx}$, where r is a yet-to-be-determined constant. After all, that form proved spectacularly successful for constant-coefficient equations (in Section 6.5). Let us try. If $y = e^{rx}$ then (if r is a constant) $y' = re^{rx}$, and $y'' = r^2 e^{rx}$, so (3) becomes

$$r^2 x^2 e^{rx} + rc_1 x e^{rx} + c_2 e^{rx} = 0 \tag{4}$$

or, canceling the nonzero e^{rx}'s

$$r^2 x^2 + rc_1 x + c_2 = 0. \tag{5}$$

We can solve (5) for r by the quadratic formula but we find that the roots r are functions of x, which result contradicts our assumption that r is a constant. Thus, *whereas the form* $y = e^{rx}$ *works for constant-coefficient equations it does **not** work for Cauchy-Euler equations.* Perhaps a different form will work. To motivate that choice let us back up and look at the *first*-order Cauchy-Euler equation

$$xy' + c_1 y = 0 \tag{6}$$

because it is readily solved by separation of variables, its general solution being

$$y(x) = Ax^{-c_1}. \tag{7}$$

From the form of (7), we wonder whether, in place of e^{rx}, we might seek solutions of Cauchy-Euler equations as powers of x:

$$\boxed{y(x) = x^r,} \tag{8}$$

where r is a yet-to-be-determined constant. Then $y' = rx^{r-1}$ and $y'' = r(r-1)x^{r-2}$, so (3) becomes

$$r(r-1)x^r + rc_1 x^r + c_2 x^r = 0 \tag{9}$$

or, canceling the nonzero x^r's,

$$r^2 + (c_1 - 1)r + c_2 = 0, \tag{10}$$

which equation can be solved for r by the quadratic formula. Thus, the form (8) works for Cauchy-Euler equations (of any order, not merely second order) just as surely as e^{rx} works for constant-coefficient equations. The secret of (8)'s success is this. If $y = x^r$ then the third term in (3) is an x^r term; likewise for the second term because the derivative knocks the exponent of x down by one but then multiplying by x builds it back up by one; likewise for the first term because the double derivative knocks the exponent down by two but multiplying by x^2 builds it back up by two. Thus, all three terms in (9) are of the same type, x^r, so, by suitable choice of r they can be made to cancel to zero. That is, if $y = x^r$ then the three terms in (9) are LD (linearly dependent), as are the $n + 1$ terms in the Cauchy-Euler equation (1) of order n. [Similarly for constant-coefficient equations, if we take $y = e^{rx}$ then all terms in the equation are of the same type, e^{rx}.]

The quadratic equation (10) admits three possibilities:

Distinct real roots.   If (10) gives distinct real roots $r = r_1, r_2$, then (3) admits a general solution

$$y(x) = Ax^{r_1} + Bx^{r_2}. \tag{11}$$

EXAMPLE 1
To solve the Cauchy-Euler equation

$$x^2 y'' - 2xy' - 10y = 0, \tag{12}$$

seek $y = x^r$. That form gives $r^2 - 3r - 10 = 0$ (Do you agree?), with roots $r = -2$ and 5, so a general solution of (12) is

$$y(x) = \frac{A}{x^2} + Bx^5. \qquad\blacksquare$$

Repeated real roots. If (10) gives repeated real roots $r = r_1, r_1$, then we have

$$y(x) = Ax^{r_1} + B(?). \tag{13}$$

That is, we are missing a second LI solution, which is needed if we are to obtain a general solution of (3). Evidently, the missing solution is not of the form x^r or we would have found it when we sought $y(x) = x^r$.

To find the missing solution we can use Lagrange's method of **reduction of order**, as we did in Section 6.5.3 for constant-coefficient differential equations with repeated roots of the characteristic equation. That is, we let A vary and seek

$$y(x) = A(x)x^{r_1}. \tag{14}$$

For brevity, let us illustrate with an example and then state the general result.

EXAMPLE 2
To solve

$$x^2 y'' + 7xy' + 9y = 0, \tag{15}$$

seek $y = x^r$. That form gives $r^2 + 6r + 9 = 0$ with the repeated root $r = -3, -3$. Thus we have found the single solution

$$y(x) = Ax^{-3}. \tag{16}$$

To find the missing solution, according to Lagrange's method, we seek

$$y(x) = A(x)x^{-3}. \tag{17}$$

Putting the latter into (15) gives the differential equation

$$A''x^{-1} - 6A'x^{-2} + 12Ax^{-3} + 7A'x^{-2} - 21Ax^{-3} + 9Ax^{-3} = 0 \tag{18}$$

or, upon simplification,

$$xA'' + A' = 0. \tag{19}$$

Next, set $A' \equiv p$, say, to reduce the order:

$$x\frac{dp}{dx} + p = 0,$$

which gives $p = D/x$, where D is an arbitrary constant. But p is A', so $A' = D/x$, integration of which gives $A(x) = D \ln x + C$, where C is also an arbitrary constant. Then (17) gives the general solution of (15) as

$$y(x) = (C + D \ln x)x^{-3}. \tag{20}$$

The Cx^{-3} term is merely the solution (16) that we already knew about, so the "missing solution" is seen from (20) to be the $D(\ln x)x^{-3}$ term. No wonder we didn't find it when we sought $y(x) = x^r$, because it is not of that form.

COMMENT.   Observe that the success of Lagrange's method lies in the absence of an A term in (19), so that we can reduce the order of (19) by setting $A' = p$. That cancelation was no accident. It *had* to occur because if $A(x)$ were a constant then the A'' and A' terms in (18) would be zero and the remaining terms would have to cancel to zero because (16) is indeed a solution of the homogeneous differential equation (15).   ∎

In fact, the same result is obtained for the general case $x^2y'' + c_1xy' + c_2y = 0$: if, seeking $y = x^r$, we obtain repeated real roots $r = r_1, r_1$, then the general solution is

$$\boxed{y(x) = (C + D \ln x)x^{r_1}.} \tag{21}$$

Details for the general case, namely, the nth-order equation (2), follow the lines indicated in Example 2 and are left for the exercises.

Complex conjugate roots.   If all of the c_j's in (2) and (3) are real, then any complex conjugate roots r will necessarily occur in complex conjugate pairs. Supposing that the roots of (10) are $\alpha \pm i\beta$, we can write a general solution of (3) as

$$y(x) = Ax^{\alpha+i\beta} + Bx^{\alpha-i\beta}$$

$$= x^\alpha \left(A x^{i\beta} + B x^{-i\beta} \right). \tag{22}$$

What to do with the unfamiliar $x^{i\beta}$ and $x^{-i\beta}$ terms? Since we normally prefer real solution forms, let us use the identity $u = e^{\ln u}$ to re-express (22) as

$$y(x) = x^\alpha \left(A e^{\ln x^{i\beta}} + B e^{\ln x^{-i\beta}} \right) = x^\alpha \left(A e^{i\beta \ln x} + B e^{-i\beta \ln x} \right)$$

$$= x^\alpha \{ A[\cos(\beta \ln x) + i \sin(\beta \ln x)] + B[\cos(\beta \ln x) - i \sin(\beta \ln x)] \}$$

$$= x^\alpha \left[(A + B) \cos(\beta \ln x) + i(A - B) \sin(\beta \ln x) \right]. \tag{23}$$

Or, letting $A + B \equiv C$ and $i(A - B) \equiv D$,

$$\boxed{y(x) = x^\alpha \left[C \cos(\beta \ln x) + D \sin(\beta \ln x) \right].} \tag{24}$$

EXAMPLE 3
To solve

$$x^2 y'' - 2xy' + 4y = 0, \tag{25}$$

seek $y = x^r$. That form gives $r^2 - 3r + 4 = 0$, so $r = \frac{3}{2} \pm i \frac{\sqrt{7}}{2}$. Hence

$$y(x) = A x^{3/2 + i\sqrt{7}/2} + B x^{3/2 - i\sqrt{7}/2} = x^{3/2} \left(A x^{i\sqrt{7}/2} + B x^{-i\sqrt{7}/2} \right)$$

$$= x^{3/2} \left(A e^{i \frac{\sqrt{7}}{2} \ln x} + B e^{-i \frac{\sqrt{7}}{2} \ln x} \right)$$

$$= x^{3/2} \left[C \cos \left(\frac{\sqrt{7}}{2} \ln x \right) + D \sin \left(\frac{\sqrt{7}}{2} \ln x \right) \right]. \qquad \blacksquare$$

Although the trigonometric and logarithmic functions are familiar elementary functions, the combinations $\cos(\beta \ln x)$ and $\sin(\beta \ln x)$ in (24) may be unfamiliar to you. To visualize the graph of $\cos(\beta \ln x)$, say, let us rewrite $\cos(\beta \ln x)$ as

$$\cos \left[\left(\beta \frac{\ln x}{x} \right) x \right] = \cos \omega x, \text{ because the form } \cos \omega x \text{ is familiar to us. But the}$$

frequency ω is not a constant, it varies with x:

$$\omega(x) = \beta \frac{\ln x}{x}. \tag{26}$$

As $x \to 0$, $\ln x \to -\infty$ so $\omega(x) \to$ infinity ($+\infty$ if $\beta < 0$, $-\infty$ if $\beta > 0$). And as $x \to \infty$, $\ln x / x \to \infty / \infty$ is indeterminate, but l'Hôpital's rule shows that $\ln x / x$, and hence ω, tends to zero as $x \to \infty$. Thus, the $\cos(\beta \ln x)$ term represents an oscillation between $+1$ and -1, but with a frequency that varies with x, the frequency becoming infinite as $x \to 0$ and zero as $x \to \infty$. With $\beta = 10$, say, a computer plot of $\cos(\beta \ln x)$ is given in Fig. 1a.

Since the $\cos(10 \ln x)$ in (24) is multiplied by x^α we have also plotted x^α, for representative values of α, in Fig. 1b. Finally, the products of x^α and $\cos(\beta \ln x)$, for those α's and for $\beta = 10$, are shown in Fig. 2.

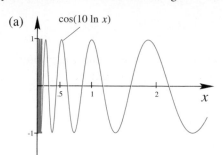

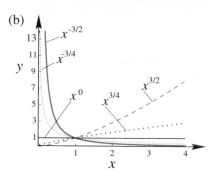

FIGURE 1   Getting ready to plot
$x^\alpha \cos(\beta \ln x)$.

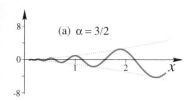

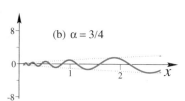

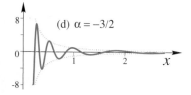

FIGURE 2
Graphs of the function
$x^\alpha \cos(\beta \ln x)$ for   representative
values of α and β: $\alpha = 3/2$,
$3/4, -3/4, -3/2$ and $\beta = 10$.

6.6.2 Conversion to equation with constant coefficients. (Optional).   It turns out that *the change of variables from x to t, say, according to*

$$\boxed{x = e^t} \tag{27}$$

reduces any Cauchy-Euler equation (2) to an equation with constant coefficients. Let us verify this claim for the general second-order case

$$x^2 y'' + c_1 x y' + c_2 y = 0, \tag{28}$$

leaving higher-order equations for the exercises.
    By chain differentiation,

$$\frac{d(\ )}{dx} = \frac{d(\ )}{dt}\frac{dt}{dx} = \frac{d(\ )}{dt}\left(\frac{dx}{dt}\right)^{-1} = e^{-t}\frac{d(\ )}{dt}. \tag{29}$$

Thus, (28) can be re-expressed in terms of t as

$$e^{2t}\left(e^{-t}\frac{d}{dt}\right)\left(e^{-t}\frac{d}{dt}\right)Y + c_1 e^t\left(e^{-t}\frac{d}{dt}\right)Y + c_2 Y = 0, \tag{30}$$

where we denote $y(x(t))$ as $Y(t)$.[1] Simplifying,

$$e^t\frac{d}{dt}\left(e^{-t}\frac{dY}{dt}\right) + c_1\frac{dY}{dt} + c_2 Y = 0,$$

$$e^t\left(e^{-t}\frac{d^2Y}{dt^2} - e^{-t}\frac{dY}{dt}\right) + c_1\frac{dY}{dt} + c_2 Y = 0,$$

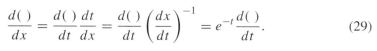

[1]Why do we need the new letter Y? Let us illustrate with an example. Let $y(x)$ be x^2 and let $x = 3t^4$, say. Then $y(x(t)) = x(t)^2 = (3t^4)^2 = 9t^8 \equiv Y(t)$. To compare the functional forms of y and Y let us use "arg" as their arguments instead of the different letters x and t. Then $y(\text{arg}) = (\text{arg})^2$, whereas $Y(\text{arg}) = 9(\text{arg})^8$. Hence, the function Y is not the same function as y, so it should be assigned a different name.

or, finally,

$$Y'' + (c_1 - 1)Y' + c_2 Y = 0. \tag{31}$$

Thus, we can solve the constant-coefficient equation (31) by the methods of Section 6.5 and, finally, return from t to x according to (27). To illustrate, let us re-solve Examples 1 and 2 in this manner.

EXAMPLE 4

To solve (12) by this method set $x = e^t$ and obtain [from (31), with $c_1 = -2$ and $c_2 = -10$]

$$Y'' - 3Y' - 10Y = 0. \tag{32}$$

Seeking $Y(t) = e^{rt}$, (32) gives $r^2 - 3r - 10 = 0$, so $r = -2, 5$. Hence,

$$Y(t) = Ae^{-2t} + Be^{5t}$$
$$= A\left(e^t\right)^{-2} + B\left(e^t\right)^5 = Ax^{-2} + Bx^5,$$

so

$$y(x) = \frac{A}{x^2} + Bx^5,$$

as obtained in Example 1.   ■

EXAMPLE 5

To solve (15) by this method set $x = e^t$ and obtain [from (31), with $c_1 = 7$ and $c_2 = 9$]

$$Y'' + 6Y' + 9Y = 0. \tag{33}$$

Seeking $Y(t) = e^{rt}$, (33) gives $r^2 + 6r + 9 = 0$ with repeated roots $r = -3, -3$. Hence,

$$Y(t) = (A + Bt)e^{-3t} = (A + Bt)(e^t)^{-3}$$
$$= (A + B\ln x)x^{-3},$$

so

$$y(x) = (A + B\ln x)x^{-3},$$

as obtained in Example 2.   ■

We are not proposing that this approach [namely, using (27) to convert the Cauchy-Euler equation to a constant-coefficient equation] is "better" than seeking solutions of the original Cauchy-Euler equation in the form x^r. Rather, our aim here was to show the close relationship between Cauchy-Euler equations and constant-coefficient equations.

Closure.   The Cauchy-Euler equation (1) is the simplest nonconstant coefficient differential equation and admits closed-form solutions in terms of elementary functions: powers of x and $\ln x$. The exponential form $y(x) = e^{rx}$ does *not* work for Cauchy-Euler equations. Rather, the appropriate form is $y(x) = x^r$. We limited attention to the second-order case $x^2y'' + c_1xy' + c_2y = 0$, and found that r is determined from the quadratic equation $r^2 + (c_1 - 1)r + c_2 = 0$:

If the roots are real and distinct ($r = r_1, r_2$) the general solution is

$$y(x) = Ax^{r_1} + Bx^{r_2}.$$

If they are real and repeated ($r = r_1, r_1$) the general solution is

$$y(x) = (A + B \ln x)x^{r_1}.$$

And if they are complex conjugates ($r = \alpha \pm i\beta$) the general solution is

$$y(x) = x^\alpha [A \cos(\beta \ln x) + B \sin(\beta \ln x)].$$

In the optional Section 6.6.2 we showed that the change of variables $x = e^t$ reduces the Cauchy-Euler equation $x^2y'' + c_1xy' + c_2y = 0$ to the constant-coefficient equation $Y'' + (c_1 - 1)Y + c_2Y = 0$, solution of which is simple by using the methods of Section 6.5. Be sure to understand the steps (28)–(31) involving chain differentiation.

We will find, in Chapter 8, that the Cauchy-Euler equation will play a prominent role when we develop a series solution method for more difficult nonconstant-coefficient differential equations (such as the Bessel and Legendre equations).

Beyond our striking success with the Cauchy-Euler equation, other successes for nonconstant-coefficient equations are few and far between. For instance, we might be able to obtain one solution by inspection and others, from it, by reduction of order but such successes are exceptional. Thus, other lines of approach will be needed for nonconstant-coefficient equations, and they are developed in Chapters 8 and 12.

EXERCISES 6.6

1. Derive a general solution for the given Cauchy-Euler equation by seeking $y(x) = x^r$. In addition, find the particular solution corresponding to the initial conditions. Recall that we are considering $0 < x < \infty$.

(**a**) $xy' + y = 0$; $y(3) = 4$

(**b**) $xy' - y = 0$; $y(2) = 5$

(**c**) $xy'' + y' = 0$; $y(1) = 50$, $y'(1) = 10$

(**d**) $xy'' - 4y' = 0$; $y(1) = 0$, $y'(1) = 3$

(**e**) $x^2y'' + xy' - 9y = 0$; $y(2) = 1$, $y'(2) = 2$

(**f**) $x^2y'' + xy' + y = 0$; $y(1) = 1$, $y'(1) = 0$

(**g**) $x^2y'' + 3xy' + 2y = 0$; $y(1) = 0$, $y'(1) = 2$

(**h**) $x^2y'' - 2y = 0$; $y(5) = 3$, $y'(5) = 0$

(**i**) $4x^2y'' + 5y = 0$; $y(1) = 0$, $y'(1) = 1$

(**j**) $x^2y'' + xy' + 4y = 0$; $y(1) = 0$, $y'(1) = -1$

(**k**) $x^2y'' + 2xy' - 2y = 0$; $y(3) = 2$, $y'(3) = 2$

(**l**) $x^2y'' - 2xy' + 2y = 0$; $y(6) = 1$, $y'(6) = 0$

(**m**) $x^2y'' - 3xy' + 3y = 0$; $y(1) = 0$, $y'(1) = 75$

2. (**a**)-(**m**) For the corresponding part of Exercise 1, use *Maple* to obtain both the general solution and the particular solution satisfying the given initial conditions. Also, use *Maple* to obtain a plot of the particular solution.

3. (**a**) Verify that the two solutions in (11) are LI.

(**b**) Verify that the two solutions in (21) are LI.

(**c**) Verify that the two solutions in (24) are LI.

4. Using Lagrange's method of reduction of order, along the lines of Example 2, show that if (10) has repeated roots $r = r_1, r_1$ then (3) has the general solution $y(x) = (A + B \ln x)x^{r_1}$.

5. (*Legendre's equation by reduction of order*) The equation

$$(1 - x^2)y'' - 2xy' + 2y = 0 \qquad (-1 < x < 1) \quad (5.1)$$

is a special case of **Legendre's equation**, named after the French mathematician *Adrien Marie Legendre* (1752–1833). By inspection, one solution is $y_1(x) = x$. Using reduction of order, seek $y(x) = A(x)x$ and thus derive the general solution,

$$y(x) = C_1 x + C_2 \left(\frac{x}{2} \ln \frac{1+x}{1-x} - 1 \right). \quad (5.2)$$

NOTE: Besides the Cauchy-Euler equation, (5.1) is an exceptional nonconstant-coefficient equation in that it admits a closed-form solution in terms of elementary functions.

6. (*Reduction of order*)

(a) By inspection, one solution of

$$y'' + xy' - y = 0 \quad (6.1)$$

is $y_1(x) = x$. Using reduction of order, seek $y(x) = A(x)x$ and thus derive the general solution

$$y(x) = Ax + B \left(e^{-x^2/2} + x \int_0^x e^{-t^2/2} dt \right). \quad (6.2)$$

(b) The integral in (6.2) is nonelementary; that is, it cannot be evaluated in closed form in terms of elementary functions. However, there exists a nonelementary function erf(x) defined by

$$\text{erf}(x) \equiv \frac{2}{\sqrt{\pi}} \int_0^x e^{-t^2} dt \quad (6.3)$$

that is tabulated and is called the **error function**. Its graph is shown here. The $2/\sqrt{\pi}$ factor is included so as to "normalize" the function so that erf(x) $\rightarrow$ 1 as $x \rightarrow \infty$.

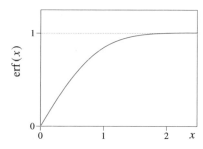

Here is the problem: Using (6.3), show that (6.2) can be re-expressed as

$$y(x) = Ax + B \left[e^{-x^2/2} + \sqrt{\frac{\pi}{2}} x \, \text{erf}\left(\frac{x}{\sqrt{2}}\right) \right]. \quad (6.4)$$

7. (*Electric potential*) The electric potential Φ (i.e., the voltage) within an annular region such as a long metal pipe of inner radius r_1 and outer radius r_2, satisfies the differential equation

$$r \frac{d^2\Phi}{dr^2} + \frac{d\Phi}{dr} = 0. \qquad (r_1 < r < r_2)$$

Solve for $\Phi(r)$ subject to these boundary conditions:

(a) $\Phi(r_1) = \Phi_1, \quad \Phi(r_2) = \Phi_2$

(b) $\dfrac{d\Phi}{dr}(r_1) = 0, \quad \Phi(r_2) = \Phi_2$

8. (*Steady-state temperature distribution*) The steady-state temperature distribution u within a hollow sphere, of inner radius r_1 and outer radius r_2, is governed by the differential equation

$$r \frac{d^2u}{dr^2} + 2\frac{du}{dr} = 0.$$

Solve for $u(r)$ subject to these boundary conditions:

(a) $u(r_1) = u_1, \quad u(r_2) = u_2$

(b) $\dfrac{du}{dr}(r_1) = 3, \quad u(r_2) = 0$

(c) $u(r_1) = u_1, \quad \dfrac{du}{dr}(r_2) = 0$

9. (*Higher-order Cauchy-Euler equations*) The solution form $y(x) = x^r$ works just as well for Cauchy-Euler equations of order higher than two. Use that method to find the general solution of these equations. If you encounter repeated roots for r you may use the results stated in Exercise 11(b), without proving them.

(a) $x^3 y''' - 3x^2 y'' + 6xy' - 6y = 0$
(b) $x^3 y''' + x^2 y'' - 2xy' + 2y = 0$
(c) $x^3 y''' + 2x^2 y'' - 4xy' + 4y = 0$
(d) $xy''' + 3y'' = 0$
(e) $x^2 y''' - 3xy'' + 4y' = 0$
(f) $x^3 y''' - 4x^2 y'' - 10xy' + 4y = 0$
(g) $x^2 y''' + 3xy'' + y' = 0$
(h) $x^3 y''' - 3x^2 y'' + 7xy' - 8y = 0$
(i) $x^3 y''' - 4x^2 y'' + 11xy' - 15y = 0$
(j) $x^2 y''' - 3xy'' + 8y' = 0$

EXERCISES FOR OPTIONAL SECTION 6.6.2

10. Determine the general solution by making the change of variables $x = e^t$ and then solving the resulting constant-coefficient equation.

(a) $x^2 y'' + xy' - 3y = 0$ **(b)** $x^2 y'' + xy' - 4y = 0$
(c) $x^2 y'' + xy' + 4y = 0$ **(d)** $x^2 y'' + 3xy' + y = 0$
(e) $x^2 y'' + xy' - 9y = 0$ **(f)** $x^2 y'' + y = 0$
(g) $x^2 y'' + 2xy' - 2y = 0$ **(h)** $4x^2 y'' - y = 0$

11. *Repeated roots*

(a) Show that if we solve $x^2 y'' + c_1 x y' + c_2 y = 0$ by seeking $y(x) = x^r$ and obtain the repeated roots $r = r_1, r_1$, then a general solution is given by (21). HINT: You can use variation of parameters and seek $y(x) = A(x) x^{r_1}$ (unless $r_1 = 0$, in which case you can solve by reduction of order, letting $y' = p$), but it is simpler to use the change of variables $x = e^t$ so that you can work with the constant-coefficient equation (31) on $Y(t)$.

(b) Show that if we solve $x^3 y''' + c_1 x^2 y'' + c_2 x y' + c_3 y = 0$ by seeking $y(x) = x^r$ and obtain roots $r = r_1, r_1, r_2$ ($r_2 \neq r_1$), then

$$y(x) = (A + B \ln x) x^{r_1} + C x^{r_2}, \qquad (11.1)$$

and if the roots are $r = r_1, r_1, r_1$, then

$$y(x) = \left[A + B \ln x + C (\ln x)^2 \right] x^{r_1}. \qquad (11.2)$$

6.7 SOLUTION OF NONHOMOGENEOUS EQUATION

Thus far, for differential equations of second-order and higher, we have studied only the homogeneous equation $L[y] = 0$. In this section we turn to the *non*homogeneous case

$$L[y] = f(x), \qquad (1)$$

where L is an nth-order linear differential operator. That is, this time we include a nonzero forcing function $f(x)$.

Before proceeding with solution techniques, let us reiterate that the function $f(x)$ (that is, what's left on the right-hand side when the terms involving y and its derivatives are put on the left-hand side) is, essentially, a **forcing function**, and we will call it that, in this text, since that is often its physical significance.

For instance, we have already met (Section 2.4.1) the equation

$$m \frac{d^2 x}{dt^2} + c \frac{dx}{dt} + k x = F(t) \qquad (2)$$

governing the displacement $x(t)$ of a mechanical oscillator. Here, the forcing function is the applied force $F(t)$. For the analogous electrical oscillator (Section 2.4.2), governed by the equations

$$L \frac{d^2 i}{dt^2} + R \frac{di}{dt} + \frac{1}{C} i = \frac{dE(t)}{dt} \qquad (3)$$

on the current $i(t)$, and

$$L \frac{d^2 Q}{dt^2} + R \frac{dQ}{dt} + \frac{1}{C} Q = E(t) \qquad (4)$$

on the charge $Q(t)$ on the capacitor, the forcing functions are the time derivative of the applied voltage $E(t)$, and the applied voltage $E(t)$, respectively.

As one more example, we give (without derivation) the differential equation

$$EI \frac{d^4 y}{dx^4} + k y = w(x) \qquad (5)$$

governing the deflection $y(x)$ of a beam that rests upon an elastic foundation, under a load distribution $w(x)$ (i.e., load per unit x length), as sketched in Fig. 1. E, I, and k are known physical constants: E is the Young's modulus of the beam material, I is the inertia of the beam's cross section, and k is the spring stiffness per unit length (force per unit length) of the foundation. Thus, in this case the forcing function is $w(x)$, the applied load distribution. [Derivation of (5) involves the so-called Euler beam theory and is part of a first course in solid mechanics.]

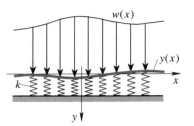

FIGURE 1
Beam on elastic foundation.

6.7.1 General solution.   Remember that the general solution of the linear homogeneous equation $L[y] = 0$ is a family of solutions that contains every solution of that equation, over the interval of interest. Likewise, by a general solution of the linear nonhomogeneous equation $L[y] = f$ we mean a family of solutions that contains every solution of that equation, over the interval of interest.

Like virtually all of the concepts and methods developed in this chapter, the concepts that follow *rest upon the assumed linearity* of the differential equation (1), in particular, upon the fact that if L is linear then

$$L\left[\alpha u(x) + \beta v(x)\right] = \alpha L[u(x)] + \beta L[v(x)] \tag{6}$$

for any two functions u, v (n-times differentiable if L is an nth-order operator) and any constants α, β. Indeed, recall the analogous result for any finite number of functions:

$$L\left[\alpha_1 u_1(x) + \cdots + \alpha_k u_k(x)\right] = \alpha_1 L[u_1(x)] + \cdots + \alpha_k L[u_k(x)] \tag{7}$$

for any constants $\alpha_1, \ldots, \alpha_k$ and any functions $u_1, \ldots, u_k$.

To begin, we suppose that $y_h(x)$ is a general solution of the homogeneous version of (1), $L[y_h(x)] = 0$, and that $y_p(x)$ is any particular solution of (1): $L[y_p(x)] = f(x)$. That is, $y_p(x)$ is any function which, when put into the left-hand side of (1), gives $f(x)$. We will refer to $y_h(x)$ and $y_p(x)$ as **homogeneous** and **particular solutions** of (1), respectively.

THEOREM 6.7.1
General Solution of $L = [y] = f$
If $y_h(x)$ is a general solution of $L[y] = 0$ and $y_p(x)$ is a solution of $L[y] = f$, on an x interval I, then a general solution of $L[y] = f$ on I is

$$\boxed{y(x) = y_h(x) + y_p(x).} \tag{8}$$

Proof.   That (8) satisfies (1) follows from the linearity of (1):

$$L\left[y_h(x) + y_p(x)\right] = L\left[y_h(x)\right] + L\left[y_p(x)\right]$$
$$= 0 + f(x) = f(x),$$

where the first equality follows from (6), with $\alpha = \beta = 1$ and u, v equal to y_h and y_p, respectively.

To see that it is a general solution, let y be any solution of (1). Again using the linearity of L, we have

$$L\left[y - y_p\right] = L\left[y\right] - L\left[y_p\right] = f - f = 0,$$

so that the most general $y - y_p$ is a linear combination of a fundamental set of solutions of the homogeneous version of (1), namely y_h. Hence $y = y_h + y_p$ is a general solution of (1).   ∎

Thus, to solve the nonhomogeneous equation (1) we need to augment the homogeneous solution $y_h(x)$ by adding to it any particular solution $y_p(x)$.

Often, in applications, $f(x)$ is not a single term but a linear combination of terms: $f(x) = f_1(x) + \cdots + f_k(x)$. In the equation $L[y] = 5x^2 - 2\sin x + 6$, for instance, we can identify $f_1(x) = 5x^2$, $f_2(x) = -2\sin x$, and $f_3(x) = 6$. In that case, the particular solution due to f is the sum of the particular solutions to $f_1, f_2, \ldots, f_k$.

THEOREM 6.7.2

General Solution of $L[y] = f_1(x) + \cdots + f_k(x)$

If $y_h(x)$ is a general solution of $L[y] = 0$ on an interval I, and $y_{p1}(x), \ldots, y_{pk}(x)$ are particular solutions of $L[y] = f_1(x), \ldots, L[y] = f_k(x)$ on I, respectively, then a general solution of $L[y] = f_1(x) + \cdots + f_k(x)$ on I is

$$y(x) = y_h(x) + y_{p1}(x) + \cdots + y_{pk}(x). \tag{9}$$

Proof.   That (9) satisfies (1) follows from (7), with all the α's in (7) equal to 1:

$$
\begin{aligned}
L\left[y_h + y_{p1} + \cdots + y_{pk}\right] &= L\left[y_h\right] + L\left[y_{p1}\right] + \cdots + L\left[y_{pk}\right] \\
&= 0 + f_1 + \cdots + f_k \\
&= f_1 + \cdots + f_k, \tag{10}
\end{aligned}
$$

as was to be verified.

To see that (9) is a *general* solution of (1), let y be any solution of (1). Then

$$
\begin{aligned}
L\left[y - y_{p1} - \cdots - y_{pk}\right] &= L\left[y\right] - L\left[y_{p1}\right] - \cdots - L\left[y_{pk}\right] \\
&= f - f_1 - \cdots - f_k = 0,
\end{aligned}
$$

so the most general $y - y_{p1} - \cdots - y_{pk}$ is a general solution y_h of the homogeneous version of (1).   ∎

This result is a **superposition principle**. It tells us that the response or output y_p to a superposition of inputs (the forcing functions $f_1, \ldots, f_k$) is the superposition of their individual outputs $(y_{p1}, \ldots, y_{pk})$.[1]

The upshot is that to solve a nonhomogeneous equation we need to find both the homogeneous solution y_h and a particular solution y_p. Having already developed methods for determining y_h—at least for certain types of equations—we now need to present methods for determining particular solutions y_p, and that is the subject of Sections 6.7.2–6.7.5 that follow.

---

[1] This principle was introduced first in Section 3.6, for linear first-order equations.

6.7.2 Undetermined coefficients. The method of undetermined coefficients is a procedure for determining a particular solution to the linear equation

$$L[y] = f(x)$$
$$= f_1(x) + \cdots + f_k(x), \tag{11}$$

subject to two conditions:

(i) Besides being linear, L is of constant-coefficient type.

(ii) Repeated differentiation of each $f_j(x)$ term in (11) produces only a finite number of LI (linearly independent) terms.

To explain the latter condition, consider $f_j(x) = 2xe^{-x}$. The sequence consisting of this term and its successive derivatives is

$$2xe^{-x} \longrightarrow \left\{ 2xe^{-x},\ 2e^{-x} - 2xe^{-x},\ -4e^{-x} + 2xe^{-x},\ \ldots \right\},$$

and we can see that this sequence contains only the two LI functions e^{-x} and xe^{-x}. Thus, $f_j(x) = 2xe^{-x}$ satisfies condition (ii).

As a second example, consider $f_j(x) = x^2$. This term generates the sequence

$$x^2 \longrightarrow \left\{ x^2, 2x, 2, 0, 0, \ldots \right\},$$

which contains only the three LI functions x^2, x, and 1. Thus, $f_j(x) = x^2$ satisfies condition (ii).

The term $f_j(x) = 1/x$, however, generates the sequence

$$1/x \longrightarrow \left\{ 1/x, -1/x^2, 2/x^3, -6/x^4, \ldots \right\},$$

which contains an infinite number of LI terms ($1/x, 1/x^2, 1/x^3, \ldots$). Thus, $f_j(x) = 1/x$ does not satisfy condition (ii).

If the term $f_j(x)$ does satisfy condition (ii), then we will call the finite set of LI terms generated by it, through repeated differentiation, the **family generated by** $f_j(x)$. (That terminology is for convenience here and is not standard.) Thus, the family generated by $2xe^{-x}$ is comprised of e^{-x} and xe^{-x}, and the family generated by $3x^2$ is comprised of x^2, x, and 1.

Let us now illustrate the method of undetermined coefficients.

EXAMPLE 1

Consider the differential equation

$$y'''' - y'' = 3x^2 - \sin 2x. \tag{12}$$

(It may seem odd that we begin with a fourth-order equation rather than a second-order equation, for instance, but this higher-order equation will permit a better view of the method.) First, we see that L is linear, with constant coefficients, so condition (i) is satisfied. Next, we identify $f_1(x)$, $f_2(x)$, and their generated sequences as

$$f_1(x) = 3x^2 \longrightarrow \left\{3x^2, 6x, 6, 0, 0, \ldots\right\}, \tag{13a}$$

$$f_2(x) = -\sin 2x \longrightarrow \{-\sin 2x, -2\cos 2x, 4\sin 2x, \ldots\}. \tag{13b}$$

Thus, f_1 and f_2 do generate the finite families

$$f_1(x) = 3x^2 \longrightarrow \left\{x^2, x, 1\right\}, \tag{14a}$$

$$f_2(x) = -\sin 2x \longrightarrow \{\sin 2x, \cos 2x\}, \tag{14b}$$

so condition (ii) is satisfied.

To find a particular solution y_{p1} corresponding to f_1, tentatively seek it as a linear combination of the terms in (14a):

$$y_{p1}(x) = Ax^2 + Bx + C, \tag{15}$$

where A, B, C are the so-called **undetermined coefficients**. Next, we write down the homogeneous solution of (12),

$$y_h(x) = C_1 + C_2 x + C_3 e^x + C_4 e^{-x}, \tag{16}$$

and check each term in y_{p1} [i.e., in (15)] for any possible duplication with terms in y_h. Doing so, we find that the Bx and C terms in (15) duplicate (to within constant scale factors) the $C_2 x$ and C_1 terms, respectively, in (16). The method then calls for multiplying the entire family involved in the duplication by the lowest positive integer power of x needed to eliminate all such duplication. Thus, we revise (15) as

$$y_{p1}(x) = x\left(Ax^2 + Bx + C\right) = Ax^3 + Bx^2 + Cx, \tag{17}$$

but find that the Cx term in (17) is still "in violation" in that it duplicates the $C_2 x$ term in (16). Multiplying by another x, try

$$y_{p1}(x) = x^2\left(Ax^2 + Bx + C\right) = Ax^4 + Bx^3 + Cx^2. \tag{18}$$

This time we are done, since all duplication has now been eliminated.

Next, we put the final revised form (18) into the equation $y'''' - y'' = 3x^2$ [i.e., $L[y] = f_1(x)$] and obtain

$$24A - 12Ax^2 - 6Bx - 2C = 3x^2. \tag{19}$$

Finally, equating coefficients of like terms gives

$$\begin{array}{rl} x^2: & -12A = 3 \\ x: & -6B = 0 \\ 1: & 24A - 2C = 0, \end{array} \tag{20}$$

so that $A = -1/4$, $B = 0$, $C = -3$. Thus

$$y_{p1}(x) = -\frac{1}{4}x^4 - 3x^2. \qquad (21)$$

Next, we need to find y_{p2} corresponding to f_2. To do so, we seek it as a linear combination of the terms in (14b):

$$y_{p2}(x) = D \sin 2x + E \cos 2x. \qquad (22)$$

Checking each term in (22) for duplication with terms in y_h, we see that there is no such duplication. Thus, we accept the form (22), put it into the equation $y'''' - y'' = -\sin 2x$ [i.e., $L[y] = f_2(x)$], and obtain

$$20D \sin 2x + 20E \cos 2x = -\sin 2x. \qquad (23)$$

Equating coefficients of like terms gives $20D = -1$, and $20E = 0$, so that $D = -1/20$ and $E = 0$. Thus,

$$y_{p2}(x) = -\frac{1}{20} \sin 2x. \qquad (24)$$

Finally, a general solution of (12) is, according to Theorem 6.8.2,

$$y(x) = y_h(x) + y_p(x) = y_h(x) + y_{p1}(x) + y_{p2}(x),$$

namely,

$$y(x) = C_1 + C_2 x + C_3 e^x + C_4 e^{-x} - \frac{1}{4}x^4 - 3x^2 - \frac{1}{20} \sin 2x. \qquad (25)$$

COMMENT 1. We obtained (20) by "equating coefficients of like terms" in (19). That step amounted to using the concept of linear independence—namely, noting that $1, x, x^2$ are LI (on any given x interval) and then using Theorem 6.3.6. Alternatively, we could have rewritten (19) as

$$(24A - 2C)1 + (-6B)x + (-12A - 3)x^2 = 0 \qquad (26)$$

and used the linear independence of $1, x, x^2$ to infer that the coefficient of each term must be zero, which step once again gives (20).

COMMENT 2. The key point in the analysis is that the system (20), consisting of three linear algebraic equations in the three unknowns A, B, C, is consistent. Similarly for the system $20D = -1$, $20E = 0$ governing D and E. The guarantee provided by the method of undetermined coefficients is that the resulting system of linear algebraic equations on the unknown coefficients will indeed be consistent, so that we can successfully solve for the undetermined coefficients. What would have happened if we had used the form (15) for $y_p(x)$ instead of the revised form (18)— that is, if we had not adhered to the prescribed procedure? We would have obtained, in place of (20), the equations

$$
\begin{array}{rl}
x^2: & 0 = 3 \\
x: & 0 = 0 \\
1: & -2A = 0,
\end{array}
$$

which are *in*consistent because the "equation" $0 = 3$ cannot be satisfied by any choice of A, B, C. That is, (15) would not have worked.   ∎

Let us summarize.

STEPS IN THE METHOD OF UNDETERMINED COEFFICIENTS:

1. Obtain the general solution $y_h(x)$ of the homogeneous equation [(16) in Example 1].

2. Verify that condition (i) is satisfied (namely that L is linear, with constant coefficients).

3. Identify the $f_j(x)$'s and verify that each one satisfies condition (ii) [namely, that repeated differentiation of $f_j(x)$ produces only a finite number of LI terms].

4. Determine the finite family corresponding to each $f_j(x)$ [(14a,b) in Example 1].

5. Seek $y_{p1}(x)$, tentatively, as a linear combination of the terms in the family corresponding to $f_1(x)$ [(15) in Example 1].

6. Examine the terms in $y_{p1}(x)$ for possible duplication of terms, or linear combinations of terms, in $y_h(x)$. If such duplication is found, multiply the entire family group by the lowest positive integer power of x necessary to remove all such duplication [(18) in Example 1].

7. Substitute the final version of the form assumed for $y_{p1}(x)$ into the left-hand side of the equation $L[y] = f_1$, and equate coefficients of like terms.

8. Solve the resulting system of linear algebraic equations for the undetermined coefficients. That step completes our determination of $y_{p1}(x)$.

9. Repeat steps 5–8 for $y_{p2}(x), \ldots, y_{pk}(x)$, in turn.

10. Then the general solution of $L[y] = f_1 + \cdots + f_k$ is given, according to Theorem 6.8.2, by $y(x) = y_h(x) + y_p(x) = y_h(x) + y_{p1}(x) + \cdots + y_{pk}(x)$.

E X A M P L E 2
As a final example, consider

$$y'' - 9y = 4 + 5\sinh 3x, \tag{27}$$

which is indeed linear and of constant-coefficient type. Since

$$f_1(x) = 4 \quad \longrightarrow \quad \{4, 0, 0, \ldots\},$$
$$f_2(x) = 5\sinh 3x \quad \longrightarrow \quad \{5\sinh 3x, 15\cosh 3x, 45\sinh 3x, \ldots\},$$

we see that these terms generate the finite families

$$f_1(x) = 4 \quad \longrightarrow \quad \{1\},$$

$$f_2(x) = 5 \sinh 3x \quad \longrightarrow \quad \{\sinh 3x, \cosh 3x\},$$

so we tentatively seek

$$y_{p1}(x) = A. \tag{28}$$

Since

$$y_h(x) = C_1 e^{3x} + C_2 e^{-3x}, \tag{29}$$

there is no duplication between the term in (28) and those in (29). Putting (28) into $y'' - 9y = 4$ gives $-9A = 4$, so $A = -4/9$. Thus,

$$y_{p1}(x) = -\frac{4}{9}. \tag{30}$$

Next, we tentatively seek

$$y_{p2}(x) = B \sinh 3x + C \cosh 3x. \tag{31}$$

At first glance, it appears that there is no duplication between any of the terms in (31) and those in (29). However, since the $\sinh 3x$ and $\cosh 3x$ are linear combinations of e^{3x} and e^{-3x}, we do indeed have duplication. In other words, each of the $\sinh 3x$ and $\cosh 3x$ terms are solutions of the homogeneous equation. Thus, we need to multiply the right-hand side of (31) by x and revise y_{p2} as

$$y_{p2}(x) = x \left(B \sinh 3x + C \cosh 3x \right). \tag{32}$$

Now that y_{p2} is in a satisfactory form, we put that form into $y'' - 9y = 5 \sinh 3x$ [i.e., $L[y] = f_2(x)$] and obtain the equation

$$(3C + 3C) \sinh 3x + (3B + 3B) \cosh 3x$$
$$+ (9B - 9B)x \sinh 3x + (9C - 9C)x \cosh 3x = 5 \sinh 3x.$$

Equating coefficients of like terms gives $B = 0$ and $C = 5/6$, so

$$y_{p2}(x) = \frac{5}{6}x \cosh 3x. \tag{33}$$

It follows then, from Theorem 6.8.2, that a general solution of (27) is

$$y(x) = C_1 e^{3x} + C_2 e^{-3x} - \frac{4}{9} + \frac{5}{6}x \cosh 3x. \tag{34}$$

Naturally, your final result can (and should) be checked by direct substitution into the original differential equation.

COMMENT.   Suppose that in addition to the differential equation (27), initial conditions $y(0) = 0$, $y'(0) = 2$ are specified. Imposing these conditions on (34) gives $C_1 = 5/12$, $C_2 = 1/36$, and hence the particular solution

$$y(x) = \frac{5}{12}e^{3x} + \frac{1}{36}e^{-3x} - \frac{4}{9} + \frac{5}{6}x \cosh 3x. \tag{35}$$

Do not be concerned that we call (35) a particular solution even though each of the two exponential terms in (35) is a homogeneous solution, because if we put (35) into the left-hand side of (27) it does give the right-hand side of (27); thus, it is a particular solution of (27).   ∎

As a word of caution, suppose the differential equation is

$$y'' - 3y' + 2y = 2 \sinh x,$$

with homogeneous solution $y_h(x) = C_1 e^x + C_2 e^{2x}$. Observe that $2 \sinh x = e^x - e^{-x}$ contains an e^x term, which corresponds to one of the homogeneous solutions. To bring this duplication into the light, we should re-express the differential equation as

$$y'' - 3y' + 2y = e^x - e^{-x}$$

before beginning the method of undetermined coefficients. Then, the particular solution due to $f_1(x) = e^x$ will be $y_{p1}(x) = Axe^x$ and the particular solution due to $f_2(x) = e^{-x}$ will be $y_{p2}(x) = Be^{-x}$. We find that $A = -1$ and $B = -1/6$ so the general solution is

$$y(x) = C_1 e^x + C_2 e^{2x} - xe^x - \frac{1}{6} e^{-x}.$$

We close this discussion of the method of undetermined coefficients by re-considering condition (ii), that repeated differentiation of each term in the forcing function must produce only a finite number of LI terms. How broad is the class of functions that satisfy that condition? If a forcing function f satisfies that condition, then it must be true that numbers α_j exist, not all of them zero, such that

$$a_0 f^{(N)} + a_1 f^{(N-1)} + \cdots + a_{N-1} f' + a_N f = 0 \tag{36}$$

over the x interval under consideration, for some finite positive integer N. From our discussion of the solution of such constant-coefficient equations we know that solutions f of (36) must be of the form $Cx^m e^{(\alpha + i\beta)x}$, or a linear combination of such terms. Such functions are so common in applications that condition (ii) is not as restrictive as it might seem.

6.7.3 Complex function method. (Optional).   If the forcing function happens to be a sine or a cosine then it is possible to simplify the method of undetermined coefficients by considering the forcing function to be a single complex exponential.

For instance, consider the task of finding a particular solution $x_p(t)$ of the differential equation

$$L[x] = F \cos \omega t, \tag{37}$$

where L is a linear differential operator with constant coefficients, and F and ω are constants. According to the method of undetermined coefficients, we can obtain $x_p(t)$ in the form $C_1 \cos \omega t + C_2 \sin \omega t$ [provided that $\cos \omega t$ and $\sin \omega t$ are not homogenous solutions of (37)]. However, it is simpler to solve the modified problem

$$L[v] = F e^{i\omega t} \tag{38}$$

for a particular solution $v_p(t)$, and to then obtain $x_p(t)$ from $v_p(t)$ as

$$x_p(t) = \operatorname{Re} v_p(t). \tag{39}$$

The reason the v problem is simpler is that by the method of undetermined coefficients a particular solution $v_p(t)$ of (38) can be found in the form $Ae^{i\omega t}$, which form is simpler than the *two*-term form $C_1 \cos \omega t + C_2 \sin \omega t$ needed for equation (37).

To verify the truth of (39), break all quantities into real and imaginary parts. Thus, write (38) as

$$L[\operatorname{Re} v + i \operatorname{Im} v] = F \cos \omega t + i F \sin \omega t. \tag{40}$$

Since L is linear, by assumption, $L[\operatorname{Re} v + i \operatorname{Im} v] = L[\operatorname{Re} v] + i L[\operatorname{Im} v]$. Putting the latter expression into (40) and equating real and imaginary parts on the left-hand and right-hand sides gives

$$L[\operatorname{Re} v] = F \cos \omega t \tag{41a}$$

and

$$L[\operatorname{Im} v] = F \sin \omega t. \tag{41b}$$

Comparing (41a) with (37), the truth of (39) follows. [If the right-hand side of (37) were $F \sin \omega t$ instead, then in place of (39) we would use $x_p(t) = \operatorname{Im} v(t)$.]

E X A M P L E 3
If the applied voltage is $E(t) = E_0 \sin \omega t$, then the current $i(t)$ in the electrical circuit shown in Fig. 2 is governed by the differential equation

$$Li'' + Ri' + \frac{1}{C} i = \frac{dE(t)}{dt}$$
$$= \omega E_0 \cos \omega t, \tag{42}$$

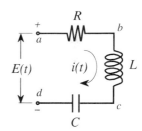

FIGURE 2
RLC circuit.

where L, R, C, E_0, ω are constants, and the inductance L is not to be confused with the operator L in (37)–(41). According to the **complex function method** described above, to find a particular solution $i_p(t)$ of (42), consider instead the simpler equation

$$Lv'' + Rv' + \frac{1}{C} v = \omega E_0 e^{i\omega t} \tag{43}$$

and seek

$$v_p(t) = Ae^{i\omega t}. \tag{44}$$

Putting (44) into (43) gives

$$\left(-L\omega^2 + i R\omega + \frac{1}{C}\right) Ae^{i\omega t} = \omega E_0 e^{i\omega t}. \tag{45}$$

Thus, $A = \omega E_0 / \left(-L\omega^2 + i R\omega + \dfrac{1}{C}\right)$ so

$$v_p(t) = \frac{\omega E_0 C}{(1 - LC\omega^2) + i RC\omega} e^{i\omega t}, \tag{46}$$

and $i_p(t) = \operatorname{Re} v_p(t)$. Thus,

$$i_p(t) = \omega E_0 C \operatorname{Re} \left(\frac{\cos \omega t + i \sin \omega t}{(1 - LC\omega^2) + i RC\omega} \frac{(1 - LC\omega^2) - i RC\omega}{(1 - LC\omega^2) - i RC\omega}\right)$$
$$= \frac{\omega E_0 C}{(1 - LC\omega^2)^2 + R^2 C^2 \omega^2} \left[(1 - LC\omega^2) \cos \omega t + RC\omega \sin \omega t\right]. \tag{47}$$

COMMENT 1. If the inductor and capacitor were removed from the circuit, then the parentheses in (45) would contain only $iR\omega$, where R is the resistance. With the inductor and capacitor present, we can express the terms within those parentheses as

$$-L\omega^2 + iR\omega + \frac{1}{C} = i\omega\left[R + i\left(L\omega - \frac{1}{\omega C}\right)\right] \tag{48}$$

so we can think of

$$Z \equiv R + i\left(L\omega - \frac{1}{\omega C}\right) \tag{49}$$

as a sort of equivalent or generalized resistance. In electrical engineering terminology, Z is called the **complex impedance** of the circuit.

COMMENT 2. Note that because of the Ri' term in (42) the homogeneous solution of (42) will inevitably tend to zero as $t \to \infty$. Thus, (47) is not only a particular solution, it is the **steady-state solution**. The complex function method is commonly used, in engineering, as a convenient method for obtaining steady-state responses.

∎

6.7.4 Variation of parameters. Although easy to apply, the method of undetermined coefficients is limited by the two conditions cited above—that L be of constant-coefficient type, and that repeated differentiation of each $f_j(x)$ forcing term produces only a finite number of LI terms.

The method of variation of parameters, due to Lagrange, is more powerful in that it is not subject to those restrictions. As with automobile engines, we can expect more power to come at a higher price and, as we shall see, Lagrange's method is indeed the more difficult to apply.

In fact, we have already presented the method, in Section 2.3.3, for the general linear first-order equation

$$y' + p(x)y = q(x), \tag{50}$$

and we urge you to review that discussion. The idea was to seek a particular solution y_p by varying the parameter A (i.e., the constant of integration) in the homogeneous solution

$$y_h(x) = Ae^{-\int p(x)\,dx}.$$

Thus, we sought

$$y_p(x) = A(x)e^{-\int p(x)\,dx},$$

put that form into (50) and solved for $A(x)$.

Likewise, if an nth-order linear differential equation $L[y] = f$ has a homogeneous solution

$$y_h(x) = C_1 y_1(x) + \cdots + C_n y_n(x), \tag{51}$$

then according to the method of variation of parameters we seek a particular solution in the form

$$y_p(x) = C_1(x)y_1(x) + \cdots + C_n(x)y_n(x); \tag{52}$$

that is, we "vary the parameters" $C_1, \ldots, C_n$ in (51).

Let us demonstrate the procedure for the linear second-order equation

$$L[y] = y'' + p_1(x)y' + p_2(x)y = f(x), \tag{53}$$

where the coefficients $p_1(x)$ and $p_2(x)$ are assumed to be continuous on the x interval of interest, say I. We suppose that

$$y_h(x) = C_1 y_1(x) + C_2 y_2(x) \tag{54}$$

is a known general solution of the homogeneous equation, on I, and we seek

$$y_p(x) = C_1(x)y_1(x) + C_2(x)y_2(x). \tag{55}$$

Needing y_p' and y_p'', to substitute into (53), we differentiate (55):

$$y_p' = C_1 y_1' + C_2 y_2' + \underline{C_1' y_1 + C_2' y_2}. \tag{56}$$

Looking ahead, y_p'' will include $C_1, C_2, C_1', C_2', C_1'', C_2''$ terms, so that (53) will become a nonhomogeneous second-order differential equation in $C_1(x)$ and $C_2(x)$, which can hardly be expected to be simpler than the original equation (53)! However, it will be only one equation in the two unknowns C_1, C_2, so we are free to impose another condition on C_1, C_2 to complete, and simplify, the system.

An especially convenient condition to impose will be

$$C_1' y_1 + C_2' y_2 = 0, \tag{57}$$

for this condition will knock out the underlined C_1', C_2' terms in (56), so that y_p'' will contain only *first*-order derivatives of C_1 and C_2. Then (56) reduces to $y_p' = C_1 y_1' + C_2 y_2'$, so

$$y_p'' = C_1 y_1'' + C_2 y_2'' + C_1' y_1' + C_2' y_2', \tag{58}$$

and (53) becomes

$$C_1 \left(y_1'' + p_1 y_1' + p_2 y_1 \right) + C_2 \left(y_2'' + p_1 y_2' + p_2 y_2 \right) + C_1' y_1' + C_2' y_2' = f. \tag{59}$$

The two parenthetic groups vanish by virtue of y_1 and y_2 being solutions of the homogeneous equation $L[y] = 0$, so (59) simplifies to $C_1' y_1' + C_2' y_2' = f$. That result, together with (57), gives us the equations

$$\begin{aligned} y_1 C_1' + y_2 C_2' &= 0, \\ y_1' C_1' + y_2' C_2' &= f \end{aligned} \tag{60}$$

on C_1', C_2'. We recognize the determinant of the coefficients as the Wronskian of y_1 and y_2,

$$W[y_1, y_2](x) = \begin{vmatrix} y_1(x) & y_2(x) \\ y_1'(x) & y_2'(x) \end{vmatrix}, \tag{61}$$

and the latter is necessarily nonzero on I by Theorem 6.3.3 because y_1 and y_2 are LI solutions of $L[y] = 0$ by assumption. Therefore, (60) admits a unique solution for C_1', C_2', which can be found by Cramer's rule as

$$C_1'(x) = \frac{\begin{vmatrix} 0 & y_2 \\ f & y_2' \end{vmatrix}}{\begin{vmatrix} y_1 & y_2 \\ y_1' & y_2' \end{vmatrix}} \equiv \frac{W_1(x)}{W(x)}, \quad C_2'(x) = \frac{\begin{vmatrix} y_1 & 0 \\ y_1' & f \end{vmatrix}}{\begin{vmatrix} y_1 & y_2 \\ y_1' & y_2' \end{vmatrix}} \equiv \frac{W_2(x)}{W(x)}, \quad (62)$$

where W_1, W_2 simply denote the determinants in the numerators. Integrating these equations and putting the results into (55) gives

$$y_p(x) = \left[\int \frac{W_1(x)}{W(x)} \, dx \right] y_1(x) + \left[\int \frac{W_2(x)}{W(x)} \, dx \right] y_2(x). \quad (63)$$

Note carefully that the arbitrary integration constants from the two indefinite integrals in (63) can be omitted because they would contribute terms to $y_p(x)$ that are constants times $y_1(x)$ and $y_2(x)$, which terms are already present in the homogeneous solution (54).

EXAMPLE 4
To solve

$$y'' - 4y = 8e^{2x}, \quad (64)$$

we note the general solution

$$y_h(x) = C_1 e^{2x} + C_2 e^{-2x} \quad (65)$$

of the homogeneous equation, so that we may take $y_1(x) = e^{2x}$, $y_2(x) = e^{-2x}$. Then $W(x) = y_1 y_2' - y_1' y_2 = -4$, $W_1(x) = -f(x)y_2(x) = -8e^{2x}e^{-2x} = -8$, and $W_2(x) = f(x)y_1(x) = 8e^{2x}e^{2x} = 8e^{4x}$, so (63) gives

$$y_p(x) = \left(\int 2 \, dx \right) e^{2x} + \left(\int -2e^{4x} \, dx \right) e^{-2x}$$

$$= (2x)e^{2x} + \left(-\frac{e^{4x}}{2} \right) e^{-2x} = 2xe^{2x} - \frac{e^{2x}}{2}. \quad (66)$$

We can drop the $-e^{2x}/2$ term in the right side of (66) since it is a homogeneous solution [and can be absorbed in the $C_1 e^{2x}$ term in y_h]. Thus, we write

$$y_p(x) = 2xe^{2x}. \quad (67)$$

Finally,

$$y(x) = y_h(x) + y_p(x) = C_1 e^{2x} + C_2 e^{-2x} + 2xe^{2x} \quad (68)$$

gives a general solution of (64). ∎

The method of variation of parameters applies to higher-order equations as well but we leave that discussion to the exercises.

Closure.   In this section we considered the linear *non*homogeneous nth-order differential equation

$$L[y] = f(x), \tag{69}$$

that is, where the forcing function $f(x)$ is not zero. Naturally, the names of the variables are not significant here; for instance, the independent variable might be x or t or whatever, and similarly for the dependent variable.

We showed that a general solution of the nonhomogeneous equation (69) is of the form

$$y(x) = y_h(x) + y_p(x), \tag{70}$$

where

$$y_h(x) = C_1 y_1(x) + \cdots + C_n y_n(x) \tag{71}$$

is a general solution of the homogeneous version of (69), $L[y] = 0$, and where $y_p(x)$ is any particular solution of the full equation (69). Having already studied the homogeneous equation earlier in this chapter, we focused on two methods for determining $y_p(x)$: the methods of undetermined coefficients and variation of parameters.

The simpler of these two methods, and hence the one that we recommend (whenever it applies), is undetermined coefficients. The latter is applicable if the linear operator L is of *constant-coefficient* type and if, by repeated differentiation, each term in $f(x)$ generates only a *finite* family of LI functions. Variation of parameters is more powerful since it does not require the latter two conditions. In fact, to apply variation of parameters we do not even need $f(x)$ to be specified; the answer can be obtained as an integral with $f(x)$ as part of the integrand (Exercise 6).

And in the optional Section 6.7.3 we showed how the method of undetermined coefficients can be simplified if the forcing function is a sine or cosine (which cases are frequent in applications), by changing the forcing function to a complex exponential instead.

EXERCISES 6.7

1. Show whether or not the given forcing function satisfies condition (ii), below equation (11). If so, give a finite family of LI functions generated by it.

   (**a**) $x^2 \cos x$ (**b**) $\cos x \sinh 2x$

   (**c**) $\ln x$ (**d**) $x^2 \ln x$

   (**e**) $\sin x / x$ (**f**) e^{x^2}

   (**g**) e^{9x} (**h**) $(x-1)/(x+2)$

   (**i**) $\tan x$ (**j**) $e^x \cos 3x$

   (**k**) $x^3 e^{-x}$ (**l**) $\cos x \cos 2x$

   (**m**) $x \sin x$ (**n**) $4x^2 \cos x$

2. Obtain a general solution using the method of undetermined coefficients.

   (**a**) $y' - 3y = xe^{2x} + 6$

   (**b**) $y' + y = x^4 + 2x$

   (**c**) $y' + 2y = 3e^{2x} + 4 \sin x$

   (**d**) $y' - 3y = xe^{3x} + 4$

   (**e**) $y' + y = 5 - e^{-x}$

   (**f**) $y' - y = x^2 e^x$

   (**g**) $y'' - y' = 5 \sin 2x$

   (**h**) $y'' + y' = 4xe^x + 3 \sin x$

   (**i**) $y'' + y = 3 \sin 2x - 5$

   (**j**) $y'' + y' - 2y = x^3 - e^{-x}$

(k) $y'' + y = 6\cos x + 2$
(l) $y'' + 2y' = x^2 + 4e^{2x}$
(m) $y'' - 2y' + y = x^2 e^x$
(n) $y'' - 4y = 24\cosh 2x$
(o) $y'' - y' = 2xe^x$
(p) $y''' - y' = 25\cos 2x$
(q) $y''' - y'' = 6x$
(r) $y'''' + y'' - 2y = 3x^2 - 1$
(s) $y'''' - y = 5x^3$

3. **(a)–(s)** Use *Maple* to solve the corresponding problem in Exercise 2.

4. Obtain a general solution using the method of variation of parameters.
(a) $y' + 2y = 4e^{2x}$
(b) $y' - y = xe^x + 1$
(c) $xy' - y = x^3$
(d) $xy' + y = 1/x \quad (x > 0)$
(e) $x^3 y' + x^2 y = 1 \quad (x > 0)$
(f) $y'' - y = 8x$
(g) $y'' - y = 8e^x$
(h) $y'' - 2y' + y = 6x^2$
(i) $y'' - 2y' + y = 2e^x$
(j) $y'' + y = 4\sin x$
(k) $y'' + 4y' + 4y = 2e^{-2x}$
(l) $6y'' - 5y' + y = x^2$
(m) $x^3 y'' + x^2 y' - 4xy = 1 \quad (x > 0)$
(n) $x^2 y'' - xy' - 3y = 4x \quad (x < 0)$

5. **(a)–(n)** Use *Maple* to solve the corresponding problem in Exercise 4.

6. We cannot use the method of undetermined coefficients for the equation
$$y'' + \omega^2 y = f(x), \qquad (6.1)$$
where ω is a prescribed constant, since $f(x)$ is not specified (so that we cannot tell what form to assume for y_p). Nevertheless, we can use the method of variation of parameters.

(a) Doing so, show that a general solution of (6.1) is
$$y(x) = C_1 \sin \omega x + C_2 \cos \omega x$$
$$+ \frac{\sin \omega x}{\omega} \int^x \cos \omega \xi \, f(\xi) \, d\xi$$
$$- \frac{\cos \omega x}{\omega} \int^x \sin \omega \xi \, f(\xi) \, d\xi. \qquad (6.2)$$

(b) Explain why the following definite integral form is equivalent to (6.2):
$$y(x) = C_1 \sin \omega x + C_2 \cos \omega x$$

$$+ \frac{\sin \omega x}{\omega} \int_a^x \cos \omega \xi \, f(\xi) \, d\xi$$
$$- \frac{\cos \omega x}{\omega} \int_a^x \sin \omega \xi \, f(\xi) \, d\xi. \qquad (6.3)$$
where the lower limit a is arbitrary.

(c) Complete the solution if y is subjected to the initial conditions $y(0) = 5$, $y'(0) = 2$. HINT: You should find that the form (6.3) is more convenient than (6.2), especially if you choose a to be the initial x, namely, 0.

7. (*Variation of parameters for higher-order equations*) To show how to apply the method of variation of parameters to higher-order equations consider the third-order equation
$$y''' + p_1(x)y'' + p_2(x)y' + p_3(x)y = f(x). \qquad (7.1)$$
Since the homogeneous solution of (7.1) is of the form
$$y_h(x) = C_1 y_1(x) + \cdots + C_3 y_3(x), \qquad (7.2)$$
seek $y_p(x)$, according to the method of variation of parameters, as
$$y_p(x) = C_1(x)y_1(x) + \cdots + C_3(x)y_3(x). \qquad (7.3)$$
To put (7.3) into (7.1) you will need to differentiate (7.3) three times. Proceeding essentially along the same lines as for the second-order equation (53), set the sum of the $C_j'(x)$ terms in the expression for $y_p'(x)$ equal to zero. Then work out $y_p''(x)$ and set the sum of the $C_j'(x)$ terms equal to zero. Those steps give two equations on the three unknowns C_1', C_2', C_3'. The third equation is obtained by putting your expressions for y_p, y_p', y_p'' and y_p''' into (7.1). Solving those three equations for C_1', C_2', C_3' by Cramer's rule, show that
$$C_1'(x) = \frac{W_1(x)}{W(x)}, \quad \ldots, \quad C_3'(x) = \frac{W_3(x)}{W(x)}, \qquad (7.4)$$
where W is the Wronskian of y_1, y_2, y_3 and where W_j is identical to W but with its jth column replaced by the column $[0, 0, f]^T$. Finally, integrating (7.4) show that
$$y_p(x) = \left[\int \frac{W_1(x)}{W(x)} \, dx \right] y_1(x) + \cdots + \left[\int \frac{W_3(x)}{W(x)} \, dx \right] y_3(x). \qquad (7.5)$$

8. With the help of (7.5) in Exercise 7, obtain a general solution of the given differential equation.
(a) $y''' - 2y'' - y' + 2y = 24e^{3x}$
(b) $y''' + 3y'' - y' - 3y = 36x - 108$
(c) $x^3 y''' - 3x^2 y'' + 6xy' - 6y = 4x \quad (x > 0)$
(d) $x^3 y''' + x^2 y'' - 2xy' + 2y = 2/x \quad (x > 0)$

CHAPTER 6 REVIEW

A differential equation is far more tractable, insofar as analytical solution is concerned, if it is linear than if it is nonlinear. We could see that even in Chapter 2, where we were able to derive a general solution of the general first-order linear equation $y' + p(x)y = q(x)$ but had success with nonlinear equations only in special cases. In fact, for linear equations of any order (first, second, or higher) a number of important results can be found.

The most important is that for an nth-order *linear* differential equation $L[y] = f(x)$ a general solution is expressible as the sum of any particular solution $y_p(x)$ of the full equation $L[y] = f$ and a general solution $y_h(x)$ of the homogeneous equation $L[y] = 0$:

$$\boxed{y(x) = y_p(x) + y_h(x).}$$

In turn, $y_h(x)$ is expressible as an arbitrary linear combination of any n LI (linearly independent) solutions of $L[y] = 0$:

$$\boxed{y_h(x) = C_1 y_1(x) + \cdots + C_n y_n(x).}$$

Thus, we needed to introduce the concept of linear independence early, and theorems are provided in Section 6.3 for testing a given set of functions to see if they are LI or not. We then showed how to find the solutions $y_1(x), \ldots, y_n(x)$ for two important cases: for constant-coefficient equations and for Cauchy-Euler equations.

For *constant-coefficient equations* the idea is to seek $y_h(x)$ in the exponential form

$$\boxed{y(x) = e^{rx}.}$$

Putting that form into $L[y] = 0$ gives an nth-degree polynomial equation on r called the *characteristic equation*. Each nonrepeated root r_j contributes a solution $e^{r_j x}$, and each repeated root r_j of order k contributes k solutions $e^{r_j x}, xe^{r_j x}, \ldots, x^{k-1}e^{r_j x}$.

For *Cauchy-Euler equations* the exponential form $y(x) = e^{rx}$ does *not* work. Rather, the idea is to seek $y_h(x)$ in the power form

$$\boxed{y(x) = x^r.}$$

Each nonrepeated root r_j contributes a solution x^{r_j}, and each repeated root r_j of order k contributes k solutions $x^{r_j}, (\ln x)x^{r_j}, \ldots, (\ln x)^{k-1}x^{r_j}$.

Two different methods were put forward for finding particular solutions, the method of *undetermined coefficients* and Lagrange's method of *variation of parameters*. Undetermined coefficients is easier to apply but is subject to the conditions that

  (i) besides being linear, L must be of constant-coefficient type, and

  (ii) repeated differentiation of each term in f must produce only a finite number of LI terms.

Variation of parameters, on the other hand, merely requires L to be linear. If L is of second order then, according to the method, we vary the parameters (i.e., the arbitrary constants in y_h) C_1 and C_2, and seek $y_p(x) = C_1(x)y_1(x) + C_2(x)y_2(x)$. Putting that form into the given differential equation gives one equation on the $C_j(x)$'s. That condition is augmented by an additional condition that is designed to preclude the presence of derivatives of the $C_j(x)$'s that are of order higher than first.

Besides this review of Chapter 6 we urge you to reread Section 6.1, which gives an overview of the chapter.

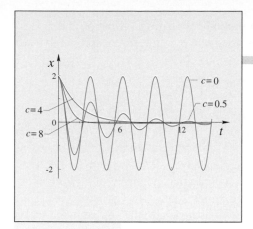

APPLICATIONS OF LINEAR CONSTANT-COEFFICIENT EQUATIONS

7.1 INTRODUCTION

Having devoted the preceding chapter to the theory and solution of equations of second order and higher, in this chapter we study a number of physical applications. We devote the most space to the linear harmonic oscillator because that application is of great importance in a variety of disciplines. Mechanical oscillators are of fundamental importance to the mechanical engineer in connection with vibrations and control. Similarly, electrical oscillators are important to the electrical engineer interested in circuit theory, and chemical and biological oscillators are important to chemical engineers and biologists. Further, we will use the harmonic oscillator as a vehicle for introducing the concept of the phase plane, in Section 7.3.

Finally, in Section 7.5 we study additional applications: pollution in a stream and Euler beam theory. These applications provide some diversity, for whereas the harmonic oscillator is most naturally an initial value problem, these additional applications are of boundary value type.

7.2 LINEAR HARMONIC OSCILLATOR; FREE OSCILLATION

In Section 2.3 we discussed the modeling of the mechanical oscillator shown again here in Fig. 1. Neglecting air resistance, the block of mass m is subjected to a restoring force due to the spring, a "drag" force due to the friction between the block and the lubricated table top, and an applied force $f(t)$. (By a restoring force, we mean

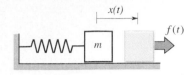

FIGURE 1
Mechanical oscillator.

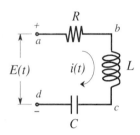

FIGURE 2
Electrical oscillator;
RLC circuit.

(a)

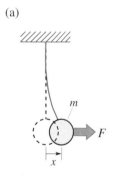

(b)

FIGURE 3
Equivalent systems.

that the force opposes the stretch or compression in the spring.) Most of that discussion focused on the modeling of the spring force and friction force, and we derived the approximate equation of motion

$$mx'' + cx' + kx = f(t),$$ (1)

where c is the *damping coefficient* and k is the *spring stiffness*.

In Section 2.4 we found that an equation of the same form as (1) governs the current in an electrical circuit comprised of a single loop containing a resistor, an inductor, a capacitor, and a voltage source $E(t)$. Hence we call that system an electrical oscillator (Fig. 2). It should not be surprising to learn that examples of biological oscillators abound because many biological activities, such as respiration, the heart beat, menstruation, and the daily sleep-wake cycle, are oscillatory phenomena.

For definiteness, we will use the mechanical oscillator of Fig. 1 and equation (1) as the vehicle for our consideration of oscillators. Realize that the actual physical system does not necessarily look like the one shown in Fig. 1. For instance, consider a beam cantilevered downward from the ceiling with a mass m at its end, as shown in Fig. 3a. We assume the mass of the beam to be negligible compared to m. The lateral displacement x of the mass due to a static lateral force F is found (either empirically or theoretically using Euler beam theory) to be proportional to the force F, so that insofar as the lateral motion of the mass is concerned the beam might just as well be replaced by an equivalent spring, as shown in Fig. 3b. Thus, the cantilever beam and mass system of Fig. 3a is equivalent to the spring and mass mechanical oscillator shown in Fig. 1.

In this section we consider the case where $f(t) = 0$ in (1); this is the so-called *unforced* or *free* oscillation. We also consider the initial displacement and velocity to be specified as $x(0) = x_0$ and $x'(0) = x_0'$, respectively, so we have the initial value problem

$$mx'' + cx' + kx = 0, \quad (0 \le t < \infty)$$ (2a)
$$x(0) = x_0, \quad x'(0) = x_0'.$$ (2b)

According to Theorem 6.4.1 the solution $x(t)$ to (2) does exist and is unique. To find it, we seek $x(t) = e^{rt}$ and obtain the characteristic equation $mr^2 + cr + k = 0$, with roots

$$r = \frac{-c \pm \sqrt{c^2 - 4mk}}{2m}.$$ (3)

Undamped case (c = 0).   Consider first the case where there is no damping.[1] Then $c = 0$, so $r = \pm i\sqrt{k/m}$ and we have as the general solution of (2a)

$$x(t) = Ae^{i\omega t} + Be^{-i\omega t},$$ (4)

where $\omega = \sqrt{k/m}$ is called the **natural frequency** of the system, in rad/sec. Or, equivalent to (4),

$$x(t) = C \cos \omega t + D \sin \omega t.$$ (5)

---

[1] It is difficult to imagine ever having $c = 0$. Rather, by $c = 0$ we mean that the damping is so small as to be negligible.

In fact, besides (3) and (4) there is another form of the general solution that is often useful,

$$x(t) = E \sin(\omega t + \phi). \tag{6}$$

To show the equivalence of (5) and (6) we use the trigonometric identity $\sin(A+B) = \sin A \cos B + \sin B \cos A$. Then

$$E \sin(\omega t + \phi) = E \sin\phi \cos\omega t + E \cos\phi \sin\omega t,$$

the right-hand side of which is identical to $C \cos\omega t + D \sin\omega t$ if

$$C = E \sin\phi, \tag{7a}$$

$$D = E \cos\phi. \tag{7b}$$

Squaring and adding equations (7), and also dividing one by the other, gives

$$\boxed{E = \sqrt{C^2 + D^2} \quad \text{and} \quad \phi = \arctan\frac{C}{D},} \tag{8a,b}$$

respectively, as the connection between the equivalent forms (5) and (6). We will understand the square root as the positive one; for example, $\sqrt{9} = +3$. Still, the arctan is multi-valued and we need to be careful, as we now illustrate.

EXAMPLE 1

Express $5 \cos 2t - 3 \sin 2t$ in the form (6). Then $\omega = 2$, $C = 5$, $D = -3$, (8a) gives $E = \sqrt{25 + 9} = \sqrt{34}$, and (8b) gives

$$\phi = \arctan\left(\frac{+5}{-3}\right) \tag{9}$$

which, using a hand-held calculator, gives $\phi = -1.0304$ rad. Thus, it appears that

$$5 \cos 2t - 3 \sin 2t = \sqrt{34} \sin(2t - 1.0304). \tag{10}$$

However, if we test (10) by letting $t = 2$, say, we find that the left side gives -0.9978 whereas the right side gives 0.9978. Thus, our computed value $\phi = -1.0304$ is incorrect and the reason is this: The angle $\phi = \arctan\dfrac{C}{D}$ (Fig. 4a) is in the second quadrant if $C > 0$ and $D < 0$, as is true in this case since $C = +5$ and $D = -3$ (Fig. 4b). However, our calculator cannot distinguish between $\arctan\left(\frac{+5}{-3}\right)$ and $\arctan\left(\frac{-5}{+3}\right)$ and it is programmed to always give values in the interval $-\pi/2 < \phi < \pi/2$. Thus, the calculator interprets our $\arctan\left(-\frac{5}{3}\right)$ as $\arctan\left(\frac{-5}{+3}\right)$ and gives the angle $\phi_1 = -1.0304$ in Fig. 4b.

To obtain the correct ϕ (which, in this example, is in the second quadrant) we can express it as the difference of the angles in Fig. 4c, namely,

$$\phi = \pi - 1.0304 = 2.111 \text{ rad.}$$

Thus,

$$5 \cos 2t - 3 \sin 2t = \sqrt{34} \sin(2t + 2.111),$$

as depicted in Fig. 5. ∎

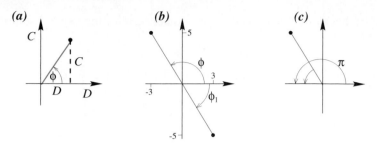

FIGURE 4   Choosing the correct quadrant for ϕ.

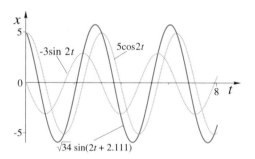

FIGURE 5
$5\cos 2t - 3\sin 2t = \sqrt{34}\sin(2t + 2.111)$.

To generalize the discussion in Example 1, let us call the value of the arctangent that is in the interval $-\pi/2 < \phi < \pi/2$ the **principal value** of the arctangent and let us denote it as Arctan( ), that is, with a capital A. Then we have these cases for the calculation of ϕ in (6):[1]

(i) $D > 0$ (1st or 4th quadrant)

$$\phi = \text{Arctan}\,\frac{C}{D} \tag{11a}$$

(ii) $D < 0$ (2nd or 3rd quadrant)

$$\phi = \text{Arctan}\,\frac{C}{D} + \pi. \tag{11b}$$

An advantage of (6) over (5) is that whereas C and D in (5) have no obvious physical significance, E, ω, and ϕ in (6) are these important quantities: the **amplitude**, **frequency**, and **phase angle** of the vibration, respectively (Fig. 6a).

---

[1]In (i) and (ii), below, we do not consider the cases where C or D is zero because these cases are simple. If $C = 0$, then (5) is already in the form (6), with $\phi = 0$. If $D = 0$, then $\phi = +\pi/2$ if $C > 0$ and $\phi = -\pi/2$ if $C < 0$.

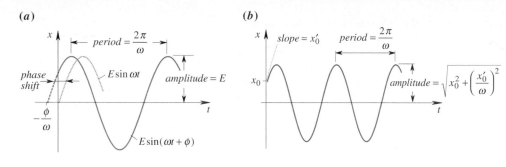

FIGURE 6   (a) Graphical significance of E, ω, ϕ.
(b) A representative undamped free oscillation.

However, an advantage of (5) over (6) is that the application of initial conditions is especially simple using (5). Specifically, the application of the initial conditions (2b) to (5) gives

$$x(0) = x_0 = C, \tag{12a}$$
$$x'(0) = x_0' = \omega D, \tag{12b}$$

so $C = x_0$, $D = x_0'/\omega$ and the solution of (2) is

$$x(t) = x_0 \cos \omega t + \frac{x_0'}{\omega} \sin \omega t, \tag{13}$$

a plot of which is given in Fig. 6b for representative initial conditions x_0 and x_0'.

Consider these results from a physical point of view. Does it seem reasonable that the frequency $\omega = \sqrt{k/m}$ should increase with the spring stiffness k and decrease with the mass m? Yes, experience tells us that the stiffer the spring the higher the frequency, and it seems reasonable that the greater the mass the more sluggish the vibration, hence, the lower the frequency. The following result, however, is less intuitive, that the frequency is a constant and does not depend upon the amplitude of the vibration, for intuition probably tells us that as we increase amplitude $\sqrt{x_0^2 + (x_0'/\omega)^2}$ (by increasing the initial displacement x_0 and/or the initial velocity x_0') the period $2\pi/\omega$ should increase, in which case the frequency ω should decrease. Yet the frequency $\omega = \sqrt{k/m}$ is a constant. The frequency being independent of the amplitude is a consequence of our assumption that the spring force is linear, that is, is governed by *Hooke's law* $F_s = kx$. For larger amplitudes we can expect the deviation of $F_s(x)$ from kx to come into play (as in Fig. 4 of Section 2.4). Then our differential equation $mx'' + F_s(x) = 0$ is *nonlinear* and the frequency *will*, in that case, be amplitude dependent. We explore this point in Chapter 11.

Before moving on to the case of damped vibrations (i.e., where $c \neq 0$), let us collect, in one place, three different equivalent forms for the general solution of

$$mx'' + kx = 0 \text{ or } x'' + \omega^2 x = 0 \text{ where } \omega = \sqrt{k/m}:$$

$$x(t) = \begin{cases} Ae^{i\omega t} + Be^{-i\omega t}, \\ C\cos\omega t + D\sin\omega t, \\ E\sin(\omega t + \phi). \end{cases} \tag{14}$$

That equivalence is important and should be understood and remembered.

Underdamped case ($c < c_{cr}$).   Now suppose there is some damping, $c > 0$, but let us first suppose that $c^2 < 4mk$ so that the square root in (3) is still imaginary. Then (3) gives

$$\begin{aligned} r &= -\frac{c}{2m} \pm \sqrt{\left(\frac{c}{2m}\right)^2 - \frac{k}{m}} = -\frac{c}{2m} \pm i\sqrt{\frac{k}{m} - \left(\frac{c}{2m}\right)^2} \\ &= -\frac{c}{2m} \pm i\sqrt{\omega^2 - \left(\frac{c}{2m}\right)^2}, \end{aligned} \tag{15}$$

where we continue to denote $\omega = \sqrt{k/m}$ as the "natural frequency." Thus, a general solution of (2a) is

$$x(t) = e^{-(c/2m)t}\left[A\cos\sqrt{\omega^2 - \left(\frac{c}{2m}\right)^2}\, t + B\sin\sqrt{\omega^2 - \left(\frac{c}{2m}\right)^2}\, t \right], \tag{16}$$

where the constants A and B can be determined from the initial conditions (2b). Observe from (16) that the damping c has two effects. First, it introduces the $e^{-(c/2m)t}$ factor, which causes the oscillation to "damp out" as $t \to \infty$, as illustrated in Fig. 7 for representative values of m, c, k, x_0, and x_0'; that is, the amplitude of the oscillation tends to zero as $t \to \infty$. We call the dotted curves the "envelope" of the graph of $x(t)$. Second, it reduces the frequency from the natural frequency ω to $\sqrt{\omega^2 - (c/2m)^2}$; that is, it makes the system more sluggish, as seems intuitively reasonable. (It might appear, by comparing Figs. 6b and 7, that the damping has increased the frequency, but that is only because we have compressed the t scale in Fig. 7.)

Critical damping ($c = c_{cr}$).   From (3) we see that as we continue to increase c we reach a critical value

$$c_{cr} = \sqrt{4mk}, \tag{17}$$

for which the square root becomes zero. In that case we have the repeated roots $r = -c/2m, -c/2m$, and hence the general solution

$$x(t) = (A + Bt)\, e^{-(c/2m)t}. \tag{18}$$

Thus, the frequency has been reduced so much, by the damping, that it has become zero and there is no oscillation at all. A representative case is shown in Fig. 8.

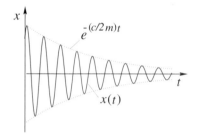

FIGURE 7
A representative underdamped free oscillation.

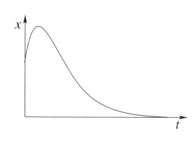

FIGURE 8
A representative critically damped case.

Overdamped case ($c > c_{cr}$).   Finally, if we increase the damping beyond c_{cr}, then $c^2 - 4mk$ in (3) becomes positive and (3) gives two distinct real roots, each negative since the $\sqrt{c^2 - 4mk}$ is smaller than c. We can express the general solution of (2a) as

$$x(t) = e^{-(c/2m)t}\left[A\cosh\sqrt{\left(\frac{c}{2m}\right)^2 - \omega^2}\, t + B\sinh\sqrt{\left(\frac{c}{2m}\right)^2 - \omega^2}\, t \right] \quad (19)$$

or we could have used exponentials in place of the cosh and the sinh.

EXAMPLE 2

To contrast these four cases (undamped, underdamped, critically damped, and over-damped) in one example, let $m = 1$, $k = 4$ (so $\omega = 2$ and $c_{cr} = 4$), $x_0 = 2$, and $x_0' = 0$. Let us find the solution $x(t)$ and plot it for these cases: $c = 0$ (no damping), $c = 0.5$ (underdamped), $c = 4$ (critically damped), and $c = 8$ (overdamped). Applying the initial conditions $x(0) = 2$ and $x'(0) = 0$ to (5) for $c = 0$, to (16) for $c = 0.5$, to (18) for $c = 4$, and to (19) for $c = 8$, we obtain these results (Exercise 5).

$c = 0$:

$$x(t) = 2\cos 2t, \quad (20a)$$

$c = 0.5$:

$$x(t) = e^{-0.25t}\left[2\cos\left(\frac{\sqrt{63}}{4}t\right) + \frac{2}{\sqrt{63}}\sin\left(\frac{\sqrt{63}}{4}t\right) \right], \quad (20b)$$

$c = c_{cr} = 4$:

$$x(t) = (2 + 4t)e^{-2t}, \quad (20c)$$

$c = 8$:

$$x(t) = e^{-4t}\left[2\cosh\left(2\sqrt{3}\,t\right) + \frac{4}{\sqrt{3}}\sinh\left(2\sqrt{3}\,t\right) \right], \quad (20d)$$

which are plotted in Fig. 9. Observe that as c increases from 0 to 0.5 the frequency decreases only slightly, from 2 to $\sqrt{63}/4 = 1.984$, but as c increases to 4 the frequency decreases to zero and the oscillation disappears entirely.   ■

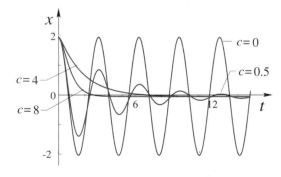

FIGURE 9   Varying the damping: (20a,b,c,d).

Periodicity.   Of the four solutions graphed in Fig. 9 we say that those corresponding to $c = 0$ and $c = 0.5$ are oscillatory and that those corresponding to $c = 4$ and $c = 8$ are nonoscillatory. Of the oscillatory solutions, the one for $c = 0.5$ damps out and the one for $c = 0$ does not. That undamped oscillation is one example of what we call a periodic function. We say that $f(t)$ is **periodic**, with **period** T if

$$\boxed{f(t + T) = f(t)} \tag{21}$$

for all $t \geq 0$ [or, if f is defined on $-\infty < t < \infty$, not just on $t \geq 0$, then periodicity would require (21) to hold for all $-\infty < t < \infty$].

For example, the function f shown in Fig. 10 is seen to be periodic with period $T = 4$ for if, starting at any point t, we make a step of 4 units to the right then we come back to the same function value as when we started. In other words, if we take any segment of length 4, such as ABC, and reproduce it repeatedly to the right, then we generate the rest of the graph. As additional examples, $100 \sin t$ is periodic with period 2π, and $25 \sin 3t$ and $50 \sin(3t + \phi)$ are periodic with period $2\pi/3$.

Notice that if f is periodic with period T then it is also periodic with period $2T, 3T, 4T, \ldots$, as well. For instance, $f(t + 2T) = f((t + T) + T) = f(t + T) = f(t)$, the second and third equalities following from (21), so that if f is periodic with period T then it is also periodic with period $2T$. Of all these possible periods, if there exists a *smallest* one, then that period is called the **fundamental period**.[1] For instance, f in Fig. 10 is periodic with period 4, 8, 12, $\ldots$, so its fundamental period is 4. For brevity, one usually omits the adjective "fundamental" and simply says that f is periodic with period 4.

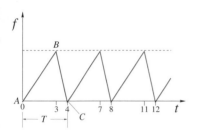

FIGURE 10
Periodic function f.

Closure.   Considering the free vibration of a mechanical oscillator governed by $mx'' + cx' + kx = 0$, we began with the undamped case ($c = 0$), in which case the motion is purely **harmonic**, that is, it is of the form $C \cos \omega t + D \sin \omega t$ [or either of the other forms given in (14)], where $\omega = \sqrt{k/m}$ is the natural frequency. If we increase c but keep it less than the critical value $c_{cr} = \sqrt{4mk}$ we find, from (16), that the frequency is reduced from ω to $\sqrt{\omega^2 - (c/2m)^2}$ and the amplitude is damped to zero (as $t \to \infty$) by the $e^{-(c/2m)t}$ factor. As $c \to c_{cr}$ the frequency tends to zero and the motion decays without oscillation. As c is increased further, the decay rate increases further.

The key formulas are boxed, and Example 2 and its Fig. 9 provide a summary of these results for a specific system.

---

[1]Observe that if $f(t)$ is a constant, such as $f(t) = 50$, then it is periodic and *every* $T > 0$ is a period. Thus, there exists no smallest period, so in this case f does not have a fundamental period.

EXERCISES 7.2

1. (*Undamped case*) Consider the initial value problem (2) with $c = 0$. In each case obtain the solution in each of the three forms given in (14).

   (a) $m = 1, k = 1, x_0 = 4, x_0' = 0$
   (b) $m = 1, k = 1, x_0 = 4, x_0' = 3$
   (c) $m = 1, k = 4, x_0 = 5, x_0' = -2$
   (d) $m = 2, k = 4, x_0 = -3, x_0' = 5$
   (e) $m = 2, k = 4, x_0 = 0, x_0' = 5$
   (f) $m = 1, k = 9, x_0 = -4, x_0' = -2$
   (g) $m = 4, k = 1, x_0 = -3, x_0' = -3$
   (h) $m = 4, k = 1, x_0 = -3, x_0' = 3$
   (i) $m = 4, k = 9, x_0 = 2, x_0' = 6$
   (j) $m = 4, k = 9, x_0 = 2, x_0' = -6$

2. (a)–(j) Generate a computer plot of the solution to the corresponding part of Exercise 1. Provide any key labeling.

3. (a)–(j) (*Damped case*) This time let $c = 1$ and take for m, k, x_0, and x_0' the values given in the corresponding part of Exercise 1. Solve for $x(t)$.

4. (a)–(j) (*Damped case*) This time let $c = 4$ and take for m, k, x_0, and x_0' the values given in the corresponding part of Exercise 1. Solve for $x(t)$.

5. Derive the solutions (20a,b,c,d) in Example 2.

6. We emphasized the equivalence of the solution forms (4), (5), and (6), and discussed the equations (8a,b) that relate C and D in (5) to E and ϕ in (6). Of course, we could have used the cosine in place of the sine, and expressed

$$x(t) = G \cos(\omega t - \psi) \qquad (6.1)$$

instead, with the minus sign for convenience. Derive formulas analogous to (8a,b), expressing G and ψ in terms of C and D.

7. We mentioned in the text that the oscillation ceases altogether when c is increased to c_{cr} or beyond. Let us make that statement more precise: for $c \geq c_{cr}$ the graph of $x(t)$ has at most one "flat spot" (on $0 \leq t < \infty$), that is, where $x' = 0$.

   (a) Prove that claim.
   (b) Make up a case (i.e., give numerical values of m, c, k, x_0, x_0') where there is *no* flat spot on $0 \leq t < \infty$.
   (c) Make up a case where there is one flat spot on $0 \leq t < \infty$.

8. (*Logarithmic decrement*) For the underdamped case ($c < c_{cr}$), let x_n and x_{n+1} denote any two successive maxima of $x(t)$.

   (a) Show that the ratio $r_n = x_n/x_{n+1}$ is a constant, say r;

that is, $x_1/x_2 = x_2/x_3 = \cdots = r$.

   (b) Further, show that the natural logarithm of r, called the **logarithmic decrement** δ, is given by

$$\delta = \frac{c}{m} \frac{\pi}{\sqrt{\omega^2 - (c/2m)^2}}.$$

9. (*Correction for the mass of the spring*) Recall that our model of the mechanical oscillator neglects the effect of the mass of the spring on the grounds that it is sufficiently small compared to that of the mass m. In this exercise we seek to improve our model so as to account, if only approximately, for the mass of the spring. In doing so, we consider the undamped case, for which the equation of motion is $mx'' + kx = 0$.

   (a) Multiplying that equation by dx and integrating, derive the "first integral"

$$\frac{1}{2}m(x')^2 + \frac{1}{2}kx^2 = C, \qquad (9.1)$$

which states that the total energy, the kinetic energy of the mass plus the potential energy of the spring, is a constant.

   (b) Let the mass of the spring be m_s. Suppose that the velocity of the elements within the spring at any time t varies linearly from 0 at the fixed end to $x'(t)$ at its attachment to the mass m. Show, subject to that assumption, that the kinetic energy in the spring is $\frac{1}{6}m_s(x')^2(t)$. Improving (9.1) to the form

$$\frac{1}{2}\left(m + \frac{1}{3}m_s\right)x'^2 + \frac{1}{2}kx^2 = C, \qquad (9.2)$$

obtain, by differentiation with respect to t, the improved equation of motion

$$\left(m + \frac{1}{3}m_s\right)x'' + kx = 0. \qquad (9.3)$$

Thus, as a correction, to take into account the mass of the spring, we merely replace the mass m in $mx'' + kx = 0$ by an "effective mass" $m + \frac{1}{3}m_s$, which incorporates the effect of the mass of the spring. NOTE: This analysis is approximate in that it assumes a linear velocity distribution within the spring, whereas that distribution itself needs to be determined, that determination involving the solution of a certain partial differential equation.

(c) In obtaining an effective mass of the form $m + \alpha m_s$, why is it reasonable that α turns out to be less than 1?

10. (*Piston oscillator*) Let a piston of mass m be placed at the midpoint of a closed cylinder of cross-sectional area A and length $2L$, as sketched. Assume that the pressure p on either side of the piston satisfies Boyle's law (namely, that the pressure times the volume is constant), and let p_0 be the pressure on both sides when $x = 0$.

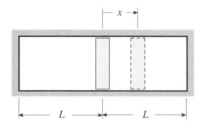

(a) If the piston is disturbed from its equilibrium position $x = 0$, show that the governing equation of motion is

$$mx'' + 2p_0 AL \frac{x}{L^2 - x^2} = 0. \qquad (10.1)$$

(b) Is (10.1) linear or nonlinear? Explain.

(c) Expand the $x/(L^2 - x^2)$ term in a Taylor series about $x = 0$, up to the third-order term. Keeping only the leading term, derive the linearized version

$$mx'' + \frac{2p_0 A}{L} x = 0 \qquad (10.2)$$

of (10.1), which is restricted to the case of small oscillations—that is, where the amplitude of oscillation is small compared to L.

(d) From (10.2), determine the frequency of oscillation, in cycles per second.

(e) Is the resulting linearized model equivalent to the vibration of a mass/spring system, with an equivalent spring stiffness of $k_{eq} = 2p_0 A/L$? Explain.

11. (*Oscillating platform*) A uniform horizontal platform of mass m is supported by counter-rotating cylinders a distance L apart (see figure). The friction force f exerted on the platform by each cylinder is proportional to the normal force N between the platform and the cylinder, with constant of proportionality (coefficient of sliding friction) μ: $f = \mu N$. Show that if the cylinder is disturbed from its equilibrium position ($x = 0$), then it will undergo a lateral oscillation of frequency $\omega = \sqrt{2\mu g/L}$ rad/sec, where g is

the acceleration due to gravity. HINT: Derive the equation of motion governing the lateral displacement x of the midpoint of the platform relative to a point midway between the cylinders.

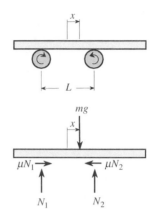

12. (*Tilted pendulum*) Consider a rod of length L with a point mass m at its end, where the mass of the rod is negligible compared to m. The rod is welded at a right angle to another, which rotates without friction about an axis that is tilted by an angle of α with respect to the vertical (see figure). Let θ denote the angle of rotation of the pendulum, with respect to its equilibrium position (where m is at its lowest possible point, namely, in the plane of the paper).

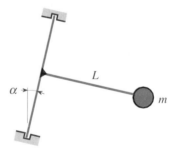

(a) Derive the governing equation of motion

$$\theta'' + \frac{g}{L} \sin \alpha \sin \theta = 0. \qquad (12.1)$$

HINT: Write down an equation of conservation of energy (kinetic plus potential energy equal a constant), and differentiate it with respect to the time t.

(b) What is the frequency of small amplitude oscillations, in rad/sec? In cycles/sec?

7.3 LINEAR HARMONIC OSCILLATOR; PHASE PLANE

In this section we continue to study the free oscillation of a linear harmonic oscillator (Fig. 1), governed by

$$mx'' + cx' + kx = 0, \tag{1}$$

but this time our method is different and involves what we call the phase plane.

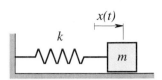

FIGURE 1
Simple harmonic mechanical oscillator.

Undamped case   **(c = 0).** Consider first the undamped case,

$$mx'' + kx = 0. \tag{2}$$

Of course, we can readily solve (2) and obtain the general solution $x(t) = C_1 \cos \omega t + C_2 \sin \omega t$, where $\omega = \sqrt{k/m}$ is the natural frequency, or, equivalently, $x(t) = A \sin(\omega t + \phi)$, where A and ϕ are the amplitude and phase angle, respectively. To present this result graphically one can plot x versus t, as we did in Section 7.2, but here we proceed differently.

We begin by defining the velocity x' as y, say. Then x'' is y', so the single second-order equation (2) is equivalent to the following system of two first-order equations:

$$\frac{dx}{dt} = y, \tag{3a}$$

$$\frac{dy}{dt} = -\frac{k}{m}x, \tag{3b}$$

the first serving to define y and the second following from (2). Next, we divide (3b) by (3a), obtaining

$$\frac{dy}{dx} = -\frac{k}{m}\frac{x}{y} \qquad \text{or} \qquad my\,dy + kx\,dx = 0, \tag{4}$$

integration of which gives

$$\frac{1}{2}my^2 + \frac{1}{2}kx^2 = C. \tag{5}$$

Since y is $x'(t)$, (5) is actually a first-order differential equation on $x(t)$. If we solve it for y we obtain

$$y = x'(t) = \sqrt{\frac{2}{m}C - \frac{k}{m}x^2} \tag{6}$$

which is variable separable and can be solved accordingly for $x(t)$. The result, of course, will be to obtain $x(t) = C_1 \cos \omega t + C_2 \sin \omega t$ or $x(t) = A \sin(\omega t + \phi)$ once again.

Instead, let us take (5) as our final result and plot the one-parameter family of ellipses that it defines (Fig. 2), the parameter being the integration constant C. In this example C happens to have a clear physical significance; it is the total energy (the kinetic energy $mx'^2/2$ of the mass plus the potential energy $kx^2/2$ of the spring). $C = 0$ gives the "point ellipse" $x = y = 0$ which represents the case where the

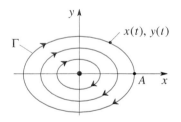

FIGURE 2
Phase portrait of (1).

mass is at rest in equilibrium, and as we increase C we obtain larger ellipses, which correspond to larger amplitude motions.

Thus, we are not plotting the displacement x versus the time t in an x, t plane, but the velocity y versus the displacement x. We call the x, y plane the **phase plane**. Each integral curve in the x, y plane represents a possible motion of the mass, and each point on the curve represents an instantaneous state of the mass (the horizontal coordinate being the displacement and the vertical coordinate being the velocity). Observe that the time t enters only as a parameter, through the parametric representation $x = x(t)$, $y = y(t)$. So we can visualize the **representative point** $x(t)$, $y(t)$ as moving along a given curve as suggested by the arrows in Fig. 2. The direction of the arrows is implied by the fact that $y = dx/dt$, so that $y > 0$ implies that $x(t)$ is increasing and $y < 0$ implies that $x(t)$ is decreasing. Thus, in the upper half plane the arrows are to the right, and in the lower half plane they are to the left. One generally calls the integral curves **phase trajectories**, or simply **trajectories**, to suggest the idea of movement of the representative point. A display of a number of such trajectories in the phase plane is called a **phase portrait** of the original differential equation, in this case (2). There is a trajectory through each point of the phase plane, so if we showed *all* possible trajectories we would simply have a black picture; the idea is to plot enough trajectories to establish the key features of the phase portrait.

It is interesting to think of the phase portrait as a two-dimensional **flow**, in the x, y phase plane, where dx/dt and dy/dt are the x and y velocity components, respectively, of the point under consideration and dy/dx is the slope of the tangent line through that point (Fig. 3).

What are the advantages of presenting results in the form of a phase portrait, rather than as traditional plots of $x(t)$ versus t? One advantage of the phase portrait is that it requires only a "first integral" of the original second-order equation such as equation (5) in this example, and sometimes we can obtain the first integral even when the original differential equation is nonlinear and too difficult to solve for $x(t)$. For instance, let us complicate (2) by supposing that the spring force is not given by the linear function $F_s = kx$, but by the nonlinear function $F_s = ax + bx^3$, and suppose that $a > 0$ and $b > 0$ so that the spring is a "hard" spring: it grows stiffer as x increases in magnitude (Fig. 4), as does a typical rubber band. If we take $a = b = m$, say, for definiteness and for simplicity, then in place of (2) we have the difficult *nonlinear* equation

$$x'' + x + x^3 = 0. \tag{7}$$

Proceeding as before, we re-express (7) as the system

$$x' = y, \tag{8a}$$

$$y' = -x - x^3. \tag{8b}$$

Division gives

$$\frac{dy}{dx} = -\frac{x + x^3}{y} \qquad \text{or} \qquad y\,dy + (x + x^3)dx = 0, \tag{9}$$

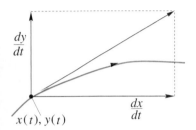

FIGURE 3
Flow velocity.

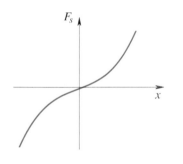

FIGURE 4
Hard spring.

which readily yields the first integral

$$\frac{1}{2}y^2 + \frac{1}{2}x^2 + \frac{1}{4}x^4 = C. \qquad (10)$$

If we plot these curves for various values of C we obtain the phase portrait shown in Fig. 5.

To repeat, one advantage of the phase portrait presentation is that it requires only a first integral. In the present case (7) was nonlinear due to the x^3 term, yet its first integral (10) was readily obtained. Nevertheless, even if the resulting first-order equation can *not* be solved analytically the phase plane approach is still valuable, as we shall see in Chapter 11 where we study the method in more detail.

A second attractive feature of the phase portrait is its compactness. For instance, observe that the single phase trajectory Γ in Fig. 2 corresponds to an entire family of oscillations of amplitude A, several of which are shown in Fig. 6 since any point on Γ can be designated as the initial point ($t = 0$): If the initial point on Γ is $(x(0), y(0)) = (A, 0)$, then we get the curve #1 in Fig. 6; if the initial point on Γ is a bit counterclockwise of $(A, 0)$ then we get the curve #2; and so on. Passing from the x, t plane to the x, y plane, the infinite family of curves suggested in Fig. 6 collapse onto the single trajectory Γ in Fig. 2.

Damped case (c > 0).   Now consider the damped case,

$$mx'' + cx' + kx = 0. \qquad (11)$$

Proceeding as we did for (2), we write, in place of (11), the equivalent system of first-order equations

$$\frac{dx}{dt} = y, \qquad (12a)$$

$$\frac{dy}{dt} = -\frac{k}{m}x - \frac{c}{m}y \qquad (12b)$$

and then divide (12b) by (12a) to obtain the equation

$$\frac{dy}{dx} = -\frac{1}{m}\frac{kx + cy}{y} \qquad (13)$$

in which the t variable does not appear. Equation (13) can be solved analytically, but for the purpose of this discussion it will be better to leave those steps for the exercises and to merely show the results obtained when we solve (12a,b) using *Maple*. For definiteness, let us set $m = 1$ and $k = 4$.

As representative of the underdamped case, set $c = 0.5$ (whereas $c_{cr} = \sqrt{4mk} = 4$). Then the phase portrait corresponding to the representative initial conditions $x(0) = -2$, $y(0) = 2$ and $x(0) = -2.5$, $y(0) = 2.5$ is shown in Fig. 7. To visualize the solution $x(t)$ from the phase portrait, visualize the **representative point** $x(t)$, $y(t)$ (shown by the heavy dot in Fig. 7) moving along the phase trajectory (clockwise) as t increases; $x(t)$ is simply the projection of that point onto the x axis. Thus, we see that as the representative point moves in the diminishing spiral the

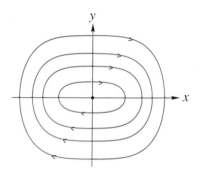

FIGURE 5
Phase portrait of (7) (hard spring).

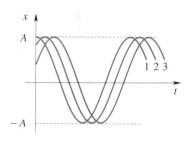

FIGURE 6
Solutions $x(t)$ corresponding to the trajectory Γ in Fig. 2.

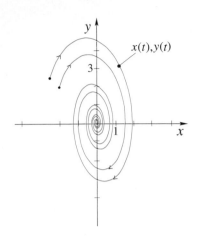

FIGURE 7
Phase portrait for underdamped case: $m = 1, k = 4, c = 0.5$.

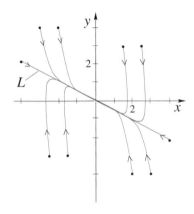

FIGURE 8
Phase portrait for overdamped case: $m = 1$, $k = 4, c = 8$.

x projection undergoes a diminishing oscillation. In fact, the $x(t)$ versus t plot of this damped oscillation was shown in Fig. 9 of Section 7.2 [for the initial condition $x(0) = 2$, $y(0) = 0$]. As $t \to \infty$ the representative point in Fig. 7 converges to the origin, the steady-state solution being the rest state $x(t) = 0$.

As representative of the overdamped case, set $c = 8$ (whereas $c_{cr} = 4$). Then the phase portrait corresponding to ten different initial conditions is shown in Fig. 8. It is striking that, no matter where the initial point is, the representative point is "attracted" toward the line L and approaches the origin and the line L as $t \to \infty$. This and other interesting features will be explored in the exercises. Here, our objective has merely been to introduce the phase plane concept.

If we compare Figs. 2, 5, 7, and 8, we see that in Figs. 2 and 5 the trajectories are closed "orbits," and in Figs. 7 and 8 they are not. Those *closed orbits* in the x, y phase plane correspond to *periodic motions* in the t, x plane. For instance, the orbit Γ in Fig. 2 corresponds to the family of periodic motions shown in Fig. 6, depending upon which point on Γ is taken as the starting point (i.e., the initial condition).

Closure.   Beginning with the linear oscillator equation (2), our first step was to re-express that single second-order equation as the equivalent pair of first-order equations (3). At that stage x and y were functions of the time t, but by dividing (3b) by (3a) we were able to eliminate t by cancelation of the dt's, and to obtain the first-order equation (4) relating x and y. Integration of (4) gave the family of ellipses (5), which define the solution curves, or "phase trajectories," in the x, y "phase plane." Thus, the result is not the usual plot of x versus t but of y versus x. Although the time t has been suppressed and does not appear explicitly, it is present in the sense that the representative point x, y can be thought of as moving along the phase trajectory, in time, according to $x = x(t)$ and $y = y(t)$.

In the present examples we began with a second-order equation and rewrote it as an equivalent pair of first-order equations by setting $x' = y$. In other applications we might actually start with two first-order equations. For example, consider the equations

$$x' = \mu(1 - y)x, \tag{14a}$$

$$y' = -\nu(1 - x)y, \tag{14b}$$

used by Volterra to model the populations of two coexisting species, one a predator and the other its prey. Here, $x(t)$ and $y(t)$ are the populations of the prey and the predator, respectively, t is the time, and μ and ν are known constants. Thus, in this application x and y represent populations rather than displacement and velocity. Nevertheless, we can proceed as before. Dividing (14b) by (14a) eliminates the explicit t dependence and gives the equation

$$\frac{dy}{dx} = -\frac{\nu}{\mu}\frac{(1 - x)y}{(1 - y)x} \tag{15}$$

governing the x, y phase plane.

In this section our purpose was merely to introduce the phase plane concept. We will return to the phase plane in Chapter 11, where its power will be revealed—chiefly through its application to *nonlinear* equations.

Maple.   The **phaseportrait** command is convenient for generating phase trajectories. For instance, to generate Fig. 7 we used these commands:

```
with(DEtools):
phaseportrait({diff(x(t),t)=y(t),diff(y(t),t)=-.5*y(t)-4*x(t)},
{x(t),y(t)},t=0..15,{[x(0)=-2,y(0)=2],[x(0)=-2.5,y(0)=2.5]},
stepsize=.05,x=-4..4,y=-5..5,scene=[x,y],arrows=NONE);
```

The phaseportrait command was explained at the end of Section 2.2, but here we have adapted it to a case where there are two differential equations rather than one.

The stepsize option was discussed at the end of Section 2.4. The idea is to reduce the stepsize if the default stepsize is too large. How can we tell if that is the case? Try running the above commands but omitting the stepsize = .05 option. Then *Maple* will use the default stepsize, which will be one twentieth of the t interval, $(15-0)/20 = 0.75$. You will find that the plotted phase trajectories are crudely given by a jagged series of straight lines. Only by reducing the stepsize below around 0.1 do we obtain smooth trajectories. In other words, if we keep reducing the stepsize we find that the trajectories become smoother and smoother. No further smoothing can be observed (at least, to the accuracy of the plot) once the stepsize is reduced below around 0.1. As a rule of thumb we suggest that for the applications in this text it is best to impose a stepsize of 0.05 or less. A better understanding of the role of the stepsize must await Chapter 12, in which we study the numerical integration of differential equations.

We have also used the scene = $[x, y]$ option which results in the horizontal and vertical axes being x and y, respectively. Actually, scene = $[x, y]$ is the default, so we could have omitted it. But observe that if we wanted to plot x versus t, say, we could have obtained that plot by using the option scene = $[t, x]$.

If we desire only the direction field we could try writing arrows=LINE and entering no initial points, but we would obtain an error message reminding us that we must include at least one initial point. In this example we could use the single initial point $[x(0) = 0, y(0) = 0]$ because that initial point gives rise to only the "point trajectory" consisting of the single point at the origin.

EXERCISES 7.3

1. Solve (6) by separation of variables and thus obtain the solution $x(t) = A \sin(\omega t + \phi)$, where $\omega = \sqrt{k/m}$ and A, ϕ are arbitrary constants.

2. Consider the equation $mx'' + cx' + kx = 0$. Given the values of m, c, k, the initial conditions $x(0), x'(0)$, and the t interval of interest, use the *Maple* phaseportrait command to generate the corresponding trajectories in the x, y phase plane. Omit the direction field by specifying arrows=NONE, but do add arrowheads by hand to show the direction of movement along the trajectories.

   (**a**) $m = 1, k = 4, c = 8$; $(x(0), x'(0)) = (-3, 4), (-2, 4),$ $(1.5, 3),$ $(2.5, 3),$ $(-2.5, -3),$ $(-1.5, -3),$ $(2, -4),$ $(3, -4),$ $(-4, 2.1436),$ $(4, -2.1436);$ $0 \le t \le 15.$ NOTE: This case should produce Fig. 8.

   (**b**) $m = 1, k = 4, c = 0$; $(x(0), x'(0)) = (0, 1), (0, 2),$ $(0, 3), (0, 4); 0 \le t \le 5$

   (**c**) $m = 1, k = 1, c = 0.1$; $(x(0), x'(0)) = (0, 5);$ $0 \le t \le 40$

(d) Same as (c) but with $c = 0.3$

(e) Same as (c) but with $c = 0.5$

(f) Same as (c) but with $c = 2$

(g) Same as (c) but with $c = 4$

(h) $m = 4$, $k = 1$, $c = 0.5$; $(x(0), x'(0)) = (3, 3)$; $0 \leq t \leq 50$

(i) $m = 4$, $k = 1$, $c = 5$; $(x(0), x'(0)) = (-6, 1.5)$, $(3, -3)$, $(0, -3)$, $(-3, -3)$, $(6, -1.5)$, $(-3, 3)$, $(0, 3)$, $(3, 3)$; $0 \leq t \leq 10$

(j) $m = 1$, $k = 2$, $c = 4$; $(x(0), x'(0)) = (-4, 2.344)$, $(4, -2.344)$, $(-0.879, 3)$, $(0.879, -3)$, $(-2, 3)$, $(2, -3)$, $(0.8, 3)$, $(-0.8, -3)$, $(2.4, 3)$, $(-2.4, -3)$; $0 \leq t \leq 10$

3. (*The role of t*) Consider $x'' + 9x = 0$ with the initial conditions $x(0) = 0$, $x'(0) = 5$. Give a labeled sketch of the phase trajectory over the t interval $0 \leq t \leq T$, for the given value of T.

(a) $T = \pi/12$ **(b)** $T = \pi/6$ **(c)** $T = \pi/4$

(d) $T = \pi/3$ **(e)** $T = 2\pi/3$ **(f)** $T = 5$

4. (a) Show that the phase portrait of $mx'' + cx' + kx = 0$ includes straight line trajectories through the origin if and only if $c > c_{cr}$. Show that there are two such trajectories, namely,

$$y = \alpha x; \qquad \alpha = (-c \pm \sqrt{c^2 - 4mk})/2m. \quad (4.1)$$

(b) Accordingly, there should be two such trajectories in Fig. 8, whereas the figure shows only one. The second one is missing from the figure simply because we did not include any initial points on that line. For the case depicted in Fig. 8 determine the equations of those two lines (one of which should be the line L in Fig. 8).

(c) Using *Maple*, obtain the phase portrait shown in Fig. 8, but this time include the second straight line trajectory. NOTE: The initial points used to generate Fig. 8 are listed in Exercise 2(a).

(d) What happens to the two straight line trajectories given by (4.1) as c decreases and approaches c_{cr}?

5. (*Phase speed*) Let $s(t)$ denote the arc length along a given phase trajectory, from an initial point to the representative point $(x(t), y(t))$. Then the speed of the representative point in the phase plane is

$$\frac{ds}{dt} = \frac{\sqrt{(dx)^2 + (dy)^2}}{dt} = \sqrt{[x'(t)]^2 + [y'(t)]^2}. \quad (5.1)$$

We call s' (i.e., ds/dt) the **phase speed**. It turns out that there exists a simple geometrical interpretation of the phase speed s', as follows. Construct a perpendicular to the trajectory, at $(x(t), y(t))$, and let a be the distance from

$(x(t), y(t))$ to the point where that perpendicular line intersects the x axis, as indicated in the figure.

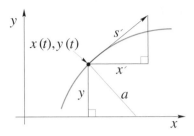

(a) Using similar triangles, show that the phase speed is simply

$$\boxed{s' = a,} \quad (5.2)$$

provided that $x' = y$. We did use $x' = y$ to break equations (1) and (7) down into pairs of first-order equations, but pointed out in the Closure that x' is not always equal to y, as occurs, for instance, in the Volterra equations (14a,b). In such cases (5.2) does not apply.

(b) For (3a,b) show that if $m = k$ then the phase speed is a constant on each trajectory, namely, $\sqrt{2E/m}$, where E is the total energy (i.e., kinetic plus potential).

(c) Consider a rectangular phase trajectory $ABCDA$, where the corner points have the x, y coordinates $A = (-1, 1)$, $B = (3, 1)$, $C = (3, -1)$, $D = (-1, -1)$. Using (5.2), plot the graph of $x(t)$ versus t, from $t = 0$ through $t = 20$, if the representative point P is at A at $t = 0$.

(d) Consider a phase trajectory ABC consisting of straight line segments from $A = (-1, 0)$ to $B = (0, 1)$ to $C = (1, 0)$ with P at B at $t = 0$. Using (5.2), sketch the graph of $x(t)$ versus t over $-\infty < t < \infty$. Also, give $x(t)$ analytically over $-\infty < t < 0$ and $0 < t < \infty$.

(e) Consider a straight line phase trajectory from $A = (0, 5)$ to $B = (10, -5)$. Using (5.2), sketch the graph of $x(t)$ versus t over $0 < t < \infty$, if P is at A at $t = 0$.

(f) Same as (e), but with P at B at $t = 0$.

6. (*The nonlinear case (7)*) Recall that for the undamped linear oscillator equation $mx'' + kx = 0$ the phase trajectories are the ellipses given by (5) and shown in Fig. 2, with closed orbits in the x, y plane corresponding to periodic motions in the t, x plane. As noted in Section 7.2, the frequency $\sqrt{k/m}$ is a constant (since it depends only upon m and k) and does not vary with the amplitude of the motion. Turning to the nonlinear equation (7), we see from the orbital trajectories in Fig. 5 that it too gives oscillatory motions, but we have no reason to expect that its frequency will be a constant. As a numerical experiment, solve (8) using the *Maple* phaseportrait command but use the option scene $= [t, x]$ to obtain plots of $x(t)$ versus t. Do so for the initial conditions $x(0) = 0.1, 0.3, 0.6, 1.0, 1.5, 2, 3, 5, 10$, with $x'(0) = 0$ in each case. From each plot determine the period of the motion and plot the period versus $x(0)$ (which will be the amplitude of the oscillation). Discuss your results.

7.4 LINEAR HARMONIC OSCILLATOR; FORCED OSCILLATION

7.4.1 Response to harmonic excitation.
The free oscillation of the harmonic oscillator (Fig. 1) was studied in Section 7.2. Now we turn to the case of forced oscillations, governed by the equation

$$mx'' + cx' + kx = f(t), \tag{1}$$

where a forcing function $f(t)$ is prescribed and where we wish to solve (1) for the resulting displacement history $x(t)$. In practical applications the case of most interest is the case of "harmonic" excitation, namely, when $f(t)$ is sinusoidal. When we say sinusoidal we mean give or take a phase shift, so $f(t)$ could be a pure sine, a pure cosine, or a sine or cosine with a phase shift. As representative of these cases, let us consider

$$f(t) = F_0 \cos \Omega t, \tag{2}$$

say, with prescribed amplitude F_0 and prescribed frequency Ω, so (1) becomes

$$\boxed{mx'' + cx' + kx = F_0 \cos \Omega t.} \tag{3}$$

Undamped case $c = 0$. To begin, consider the undamped case ($c = 0$),

$$mx'' + kx = F_0 \cos \Omega t. \tag{4}$$

To illustrate how (4) might arise, suppose that we do not apply a force directly to the mass, but only indirectly by vibrating the "wall," as shown in Fig. 2. If $x(t)$ is the displacement of the mass m relative to the wall (i.e., the point P), and $X(t)$ is the absolute displacement of the wall, then the total absolute displacement of the mass m is $x(t) + X(t)$. Hence, application of Newton's second law to the mass m gives

$$m \frac{d^2}{dt^2} [x(t) + X(t)] = -kx. \tag{5}$$

If $X(t) = A \cos \Omega t$, say, then (5) gives

$$mx'' + kx = mA\Omega^2 \cos \Omega t \tag{6}$$

which is the same as (4), where F_0 happens to be $mA\Omega^2$.

For instance, an automobile is, essentially, a mass resting on the springs of its suspension system. When it is driven over a rippled road bed the effect is the same as in Fig. 2, although vertically rather than horizontally.

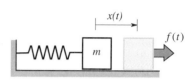

FIGURE 1
Mechanical oscillator.

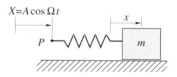

FIGURE 2
Applying a force by vibrating the wall.

The homogeneous solution of (4) is

$$x_h(t) = A \cos \omega t + B \sin \omega t, \tag{7}$$

where $\omega = \sqrt{k/m}$ is the natural frequency (i.e., the frequency of the free oscillation), and the forcing function $F_0 \cos \Omega t$ generates the family $\{\cos \Omega t, \sin \Omega t\}$. Thus, to find a particular solution of (4) by the method of undetermined coefficients, seek

$$x_p(t) = C \cos \Omega t + D \sin \Omega t. \tag{8}$$

Two cases present themselves. In the generic case, the driving frequency Ω is different from the natural frequency ω, so the terms in (8) do not duplicate any of those in (7) and we can accept (8) without modification. In the exceptional or "singular" case where Ω is equal to ω, the terms in (8) repeat those in (7), so we need to modify (8) by multiplying the right side of (8) by t. For reasons that will become clear below, these cases are known as *nonresonance* and *resonance*, respectively.

Nonresonant oscillation, $\Omega \neq \omega$.   Putting (8) into the left side of (4) gives

$$\left(\omega^2 - \Omega^2\right) C \cos \Omega t + \left(\omega^2 - \Omega^2\right) D \sin \Omega t = \frac{F_0}{m} \cos \Omega t. \tag{9}$$

It follows from (9), by equating the coefficients of $\cos \Omega t$ and $\sin \Omega t$ on the left and right sides, that $C = (F_0/m)/(\omega^2 - \Omega^2)$ and $D = 0$. Thus

$$x_p(t) = \frac{F_0/m}{\omega^2 - \Omega^2} \cos \Omega t, \tag{10}$$

so a general solution of (6) is

$$\begin{aligned} x(t) &= x_h(t) + x_p(t) \\ &= A \cos \omega t + B \sin \omega t + \frac{F_0/m}{\omega^2 - \Omega^2} \cos \Omega t. \end{aligned} \tag{11}$$

In a sense we are done, and if we wish to impose any initial conditions $x(0)$ and $x'(0)$, then we could use those conditions to evaluate the constants A and B in (11). Then, for any desired numerical values of m, k, F_0, and Ω we could plot $x(t)$ versus t and see what the solution looks like. However, in science and engineering one is interested not only in obtaining answers, but also in understanding phenomena, so we ask: How can we extract, from (11), an understanding of the phenomenon? To answer that question, let us first rewrite (11) in the equivalent form

$$x(t) = E \sin (\omega t + \phi) + \frac{F_0/m}{\omega^2 - \Omega^2} \cos \Omega t \tag{12}$$

since then we can see it more clearly as a superposition of two harmonic solutions of different amplitude, frequency, and phase.

The homogeneous solution $E \sin (\omega t + \phi)$ in (12), the "free vibration," was already discussed in Section 7.2. Thus, we turn to the particular solution, or "forced response," given by (10), which is the last term in (12). It is natural to regard m and k

(and hence ω) as fixed, and F_0 and Ω as controllable quantities or "parameters." It is no surprise that the response (10) is merely proportional to F_0, for it follows from the linearity of the differential operator in (4) that the response to the forcing function input will be proportional to that input. We also see, from (10), that the response is at the same frequency as the forcing function, Ω.

More interesting is the variation of the amplitude $(F_0/m)/(\omega^2 - \Omega^2)$ with Ω, which is plotted in Fig. 3. The change in sign, as Ω increases through ω, is awkward since it prevents us from interpreting the plotted quantity as a pure amplitude, for an amplitude cannot be negative since it is an absolute magnitude. However, we can re-express (10) in the equivalent form

$$x_p(t) = \frac{F_0/m}{\left|\omega^2 - \Omega^2\right|} \cos\left(\Omega t + \Phi\right), \tag{13}$$

where the phase angle Φ is 0 for $\Omega < \omega$ and π for $\Omega > \omega$. Let us verify the equivalence: For $\Omega < \omega$ the $\omega^2 - \Omega^2$ in (13) is positive so we can drop the absolute magnitude; further, $\cos(\Omega t + \Phi) = \cos(\Omega t + 0) = \cos\Omega t$, so for $\Omega < \omega$ the right-hand side of (13) does agree with the right-hand side of (10). For $\Omega > \omega$ we can express the $|\omega^2 - \Omega^2|$ as $\Omega^2 - \omega^2$ and the $\cos(\Omega t + \Phi)$ as $\cos(\Omega t + \pi) = \cos\Omega t \cos\pi - \sin\Omega t \sin\pi = -\cos\Omega t$ so, once again, we see that the right-hand side of (13) does agree with the right-hand side of (10).

The upshot is that $(F_0/m)/|\omega^2 - \Omega^2|$ *is* the amplitude (because it is positive), and that Φ is the phase angle. *Both* are functions of the "driving frequency" Ω, so it is of interest to plot both, as we have done in Fig. 4.

From Fig. 4a, observe that as the driving frequency Ω increases and approaches the natural frequency ω the amplitude increases without bound and tends to infinity![1] Further, as Ω increases beyond ω the amplitude decreases, and as $\Omega \to \infty$ the amplitude tends to zero. Finally, we see from Fig. 4b that the response [i.e., $x_p(t)$] is *in-phase* (i.e., $\Phi = 0$) with the forcing function if $\Omega < \omega$, but is 180° *out-of-phase* with it if $\Omega > \omega$. This is a striking result, for imagine varying Ω very slowly, beginning with a value less than ω. Then the response $x_p(t)$ is in-phase with the force. However, when Ω increases from being slightly less than ω to being slightly greater than ω the response will undergo a change so that it is then opposing (i.e., 180° out-of-phase) the force for all $\Omega > \omega$.

Resonant oscillation, $\Omega = \omega$. For the special case $\Omega = \omega$ (that is, where we force the system precisely at its natural frequency), the terms in (8) duplicate those in (7) so, according to the method of undetermined coefficients, we need to revise x_p as

$$x_p(t) = t\left(C \cos\omega t + D \sin\omega t\right). \tag{14}$$

Since the duplication has thereby been removed, we accept (14). Putting that form

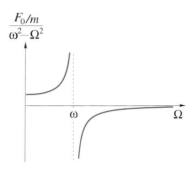

FIGURE 3
The coefficient of $\cos\Omega t$ in (12).

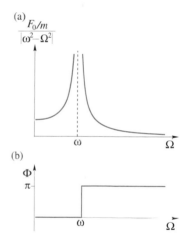

FIGURE 4
Amplitude response and phase response curves.

---

[1]Of course, we should keep in mind that right *at* $\Omega = \omega$ the particular solution (10) is not valid. That special case, the case of resonance, will be treated below.

into (4), we find that $C = 0$ and $D = F_0/(2m\omega)$, so

$$x_p(t) = \frac{F_0}{2m\omega}t\sin\omega t, \qquad (15)$$

which is plotted versus t in Fig. 5.

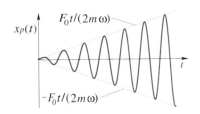

FIGURE 5
Resonant oscillation.

In this special case the response is not a harmonic oscillation but a harmonic function times t, which factor causes the magnitude to tend to infinity as $t \to \infty$. This unbounded growth of the amplitude is known as **resonance**. Of course, the magnitude does not grow unboundedly in a real application since the mathematical model of the system (the governing differential equation) will become inaccurate for sufficiently large amplitudes, parts will break, and so on.

Resonance is sometimes welcome and sometimes unwelcome. We will have more to say about this below, after we have considered the more realistic case in which there is some damping present ($c > 0$).

Damped case. We now consider the harmonically driven oscillator with a cx' damping term included ($c > 0$):

$$mx'' + cx' + kx = F_0\cos\Omega t. \qquad (16)$$

Recall from Section 7.2 that the homogeneous solution is

$$x_h(t) = \begin{cases} e^{-\frac{c}{2m}t}\left[A\cos\sqrt{\omega^2 - \left(\frac{c}{2m}\right)^2}\,t + B\sin\sqrt{\omega^2 - \left(\frac{c}{2m}\right)^2}\,t\right] \\ e^{-\frac{c}{2m}t}(A + Bt) \\ e^{-\frac{c}{2m}t}\left[A\cosh\sqrt{\left(\frac{c}{2m}\right)^2 - \omega^2}\,t + B\sinh\sqrt{\left(\frac{c}{2m}\right)^2 - \omega^2}\,t\right] \end{cases} \qquad (17)$$

for the underdamped ($c < c_{cr}$), critically damped ($c = c_{cr}$), and overdamped ($c > c_{cr}$) cases, respectively, and where $\omega = \sqrt{k/m}$ and $c_{cr} = \sqrt{4mk}$.

This time, when we write

$$x_p(t) = C\cos\Omega t + D\sin\Omega t, \qquad (18)$$

according to the method of undetermined coefficients, there is no duplication between terms in (18) and (17), even if $\Omega = \omega$, so we can accept (18) without modification; we don't need to treat the case $\Omega = \omega$ separately because (18) applies for *all* values of Ω. If we put (18) into (16) and equate coefficients of the $\cos\Omega t$ terms on both sides of the equation, and also equate the coefficients of the $\sin\Omega t$ terms, we can solve for C and D. The result (Exercise 3a) is that

$$x_p(t) = \frac{(F_0/m)\left(\omega^2 - \Omega^2\right)}{\left(\omega^2 - \Omega^2\right)^2 + (c\,\Omega/m)^2}\cos\Omega t$$

$$+ \frac{F_0 c\,\Omega/m^2}{\left(\omega^2 - \Omega^2\right)^2 + (c\,\Omega/m)^2}\sin\Omega t, \qquad (19)$$

or (Exercise 3b), equivalently,[1]

$$x_p(t) = E \cos(\Omega t - \Phi), \tag{20a}$$

where the amplitude E and phase Φ are

$$E = \frac{F_0/m}{\sqrt{\left(\omega^2 - \Omega^2\right)^2 + (c\,\Omega/m)^2}}, \tag{20b}$$

$$\Phi = \arctan \frac{c\,\Omega/m}{\omega^2 - \Omega^2}, \tag{20c}$$

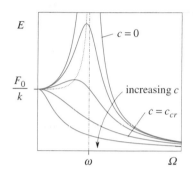

with the arctan understood to lie between 0 and π (Exercise 3c).

As for the undamped case, we have great interest in the amplitude and frequency response curves, namely, the graphs of the amplitude E, and the phase Φ versus the driving frequency Ω. The former is given in Fig. 6 for various values of the damping coefficient c.

From Fig. 6 we see that true resonance (i.e., an infinite response amplitude when Ω equals the natural frequency ω) is possible only in the case of no damping ($c = 0$), which case is an idealization since in reality there is inevitably some damping present. Analytically we see the same thing: (20b) shows that the amplitude E can become infinite only if $c = 0$, and that occurs only for $\Omega = \omega$. However, for $c > 0$ there is still a peaking of the amplitude, even if that peak is now finite, at a driving frequency Ω that decreases from ω as c increases and is 0 for all $c \geq c_{cr}$; the location of that peak is shown in Fig. 6 by the dotted curve. That peak magnitude diminishes from ∞ to F_0/k as c is increased from 0 to c_{cr}, and remains F_0/k for all $c > c_{cr}$.

What is the significance of the F_0/k intercept on the E axis? For $\Omega = 0$ the differential equation becomes $mx'' + cx' + kx = F_0$, and the method of undetermined coefficients gives $x_p(t) = \text{constant} = F_0/k$, which is merely the static deflection of the mass under the steady force F_0.

Even if true resonance is possible only for the undamped case ($c = 0$), the term resonance is often used, nevertheless, to refer to the dramatic peaking of the amplitude response curves if c is small (compared to c_{cr}).

The general solution, of course, is the sum

$$\begin{aligned} x(t) &= x_h(t) + x_p(t) \\ &= x_h(t) + E \cos(\Omega t - \Phi), \end{aligned} \tag{21}$$

where E and Φ are given by (20b,c) and $x_h(t)$ is given by the appropriate right-hand side of (17) according to whether the system is underdamped, critically damped, or overdamped. If we impose initial conditions $x(0)$ and $x'(0)$ on (21), then we can solve for the integration constants A and B within $x_h(t)$.

FIGURE 6
Amplitude response curves.

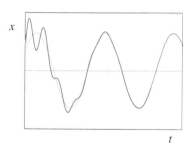

FIGURE 7
A representative response $x(t)$ (solid); approach to the steady-state oscillation $x_p(t)$ (dotted).

---

[1] We could use either cosine or sine in (20a) but it seems more natural to choose the cosine because the forcing function in (16) is a cosine. Then the phase Φ of the response will be *relative to* the forcing function $F_0 \cos \Omega t$.

Note that the $x_h(t)$ part of the solution inevitably tends to zero as $t \to \infty$ because of the $\exp(-ct/2m)$ factor, no matter how small c is, as long as $c > 0$. Thus, we call $x_h(t)$ in (21) the **transient** part of the solution and we call $x_p(t)$ the **steady-state** part since $x(t) \to E \cos(\Omega t - \Phi)$ as $t \to \infty$. The transient part depends upon the initial conditions, whereas the steady-state part does not. A representative underdamped case is shown in Fig. 7, where we see the approach to the steady-state oscillation $x_p(t)$ (the light dotted curve) as t increases.

7.4.2 Beats. (Optional).

There is a lingering question to address. For the undamped case ($c = 0$) the response to the harmonic forcing function is given by (13) and the amplitude and phase response curves are plotted in Fig. 4. These results are valid for all $0 \le \Omega < \infty$ *except for $\Omega = \omega$*. In that case we have resonance and the solution grows with t, as seen in (15) and in Fig. 5. Does it not seem strange that this *single* value of Ω must be treated separately? If this resonant behavior occurs only at $\Omega = \omega$ then one could argue that it is of only academic interest because in practice we cannot make Ω *exactly* equal to ω anyway. But if such behavior occurs throughout some small neighborhood of ω then the resonance phenomenon is indeed of practical interest.

To address this question let us examine the solution $x(t)$ as Ω *approaches* ω. If we adopt the simple initial conditions $x(0) = 0$ and $x'(0) = 0$, for definiteness, we can evaluate A and B in (11), and obtain

$$x(t) = -\frac{F_0/m}{\omega^2 - \Omega^2}(\cos \omega t - \cos \Omega t), \tag{22}$$

or, recalling the trigonometric identity $\cos A - \cos B = 2 \sin \frac{B+A}{2} \sin \frac{B-A}{2}$,

$$x(t) = \frac{2F_0/m}{\omega^2 - \Omega^2} \sin\left(\frac{\omega + \Omega}{2}\right)t \; \sin\left(\frac{\omega - \Omega}{2}\right)t. \tag{23}$$

Now, suppose that Ω is close to (but not equal to) the natural frequency ω. Then the frequency of the second sinusoid in (23) is very small compared to that of the first, so the $\sin\left(\frac{\omega - \Omega}{2}\right)t$ factor amounts, essentially, to a slow "amplitude modulation" of the relatively high frequency $\sin\left(\frac{\omega + \Omega}{2}\right)t$ factor. This phenomenon is known as **beats**, and is seen in Fig. 8.

In each case we chose $2F_0/m = 1$ and $\omega = 1$, for definiteness, and plotted t from 0 to 75 and x from -15 to 15. In Fig. 8a Ω is not close to ω (i.e., 1) so the amplitude $(2F_0/m)/(\omega^2 - \Omega^2) = 1/(1 - \Omega^2) = 1/(1 - 0.04) = 1.04$ is relatively small and there is no discernible beat pattern. But as $\Omega \to \omega$ (i.e., 1), in parts (b), (c), and (d) of the figure, two things are observed. First, the amplitude $(2F_0/m)/(\omega^2 - \Omega^2)$ increases because the denominator $\omega^2 - \Omega^2$ becomes small. Second, the beat phenomenon becomes well established, as seen in Figs. 8b, 8c, and 8d, and as we have emphasized by plotting the "envelope" $\pm \sin\left(\frac{\omega - \Omega}{2}t\right)$ as dotted in each case. The trend of these responses as $\Omega \to \omega$ (i.e., as $\Omega \to 1$) to the resonant oscillation depicted in Fig. 5 is evident.

(a) $\Omega = 0.2$

x

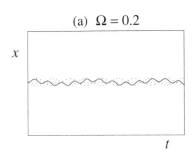

t

(b) $\Omega = 0.7$

x

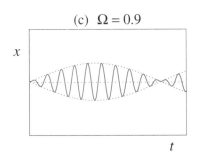

t

(c) $\Omega = 0.9$

x

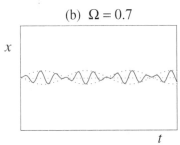

t

(d) $\Omega = 0.96$

x

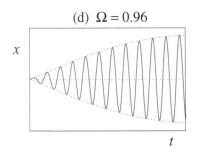

t

FIGURE 8
Beats, and approach
to resonance.

We can now see that the resonance phenomenon at $\Omega = \omega$ is not an isolated behavior but is a limiting case as $\Omega \to \omega$. That is, resonance (Fig. 4) is actually a limit of the sequence shown in Fig. 8, as $\Omega \to \omega$. Rather than depend only on these suggestive graphical results, we can proceed analytically as well. Specifically, we can take the limit of the response (22) as $\Omega \to \omega$. That limit is indeterminate (namely, 0/0) but, with the help of l'Hôpital's rule, we do obtain (15)!

Closure.   We have considered the forced vibration of a harmonic oscillator, namely, a system governed by the differential equation $mx'' + cx' + kx = f(t)$ for the case of the harmonic excitation $f(t) = F_0 \cos \Omega t$. Thus, besides a homogeneous solution we needed to find a particular solution, and that was done by the method of undetermined coefficients. We could have used the complex function method (Section 6.7.3) instead, but have left that method for the exercises. The particular solution is important physically, since even an infinitesimal amount of damping will cause the homogeneous solution to tend to zero as $t \to \infty$, so that the particular solution becomes the steady-state response. To understand the physical significance of that response we studied the amplitude and phase response curves and discussed the phenomena of resonance and beats. Our discussion has been limited to the case of harmonic excitation, whereas in applications $f(t)$ need not be harmonic. However, that case is important enough to deserve this special section. When we study the Laplace transform method in Chapter 13, we will be able to return to problems such as $mx'' + cx' + kx = f(t)$ and use the convenient Laplace transform methodology to obtain solutions for more general forcing functions.

Besides harmonic excitation being important in its own right, there is another reason for its prominence in books on engineering mathematics. Namely, in many applications the forcing functions are *periodic* functions of t, such as the "saw tooth" function shown in Fig. 9. It turns out that such functions can be represented as infinite series of cosine and/or sine functions, of various amplitudes and frequencies. The idea, then, is that the response to the periodic forcing function $f(t)$ can be found (thanks to the linearity of the differential equation) by summing the responses due to each term in the series. Such series are called **Fourier series.**[1]

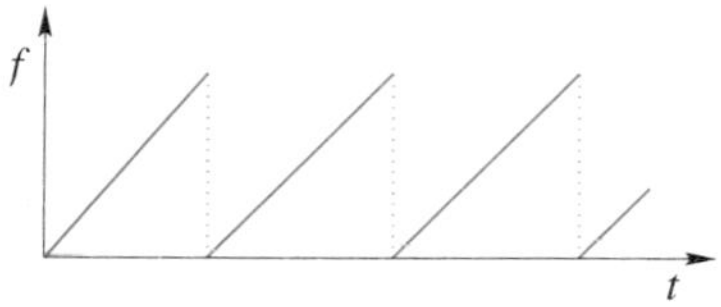

FIGURE 9
Example of a periodic forcing function.

EXERCISES 7.4

1. Applying the initial conditions $x(0) = 0$ and $x'(0) = 0$ to (11), derive (22).

2. Derive (15) from (14).

3. (a) Derive (19).

 (b) Derive (20a,b,c).

 (c) Explain why Φ in (20c) must lie between 0 and π.

Show that

$$\Phi = \begin{cases} \text{Arctan} \dfrac{c\,\Omega/m}{\omega^2 - \Omega^2} & \text{if} \quad \Omega < \omega \\[2ex] \text{Arctan} \dfrac{c\,\Omega/m}{\omega^2 - \Omega^2} + \pi & \text{if} \quad \Omega > \omega \end{cases}$$

(3.1)

where the capitalized Arctan is the principal value defined in Section 7.2.

---

[1] See, for example, M. Greenberg, *Advanced Engineering Mathematics*, 2nd ed. (Upper Saddle River, NJ: Prentice Hall, 1998), Chap. 17.

4. In Fig. 7 we show the approach of a representative response curve (solid) to the steady-state oscillation (dotted), for an underdamped system. To explore the role of damping, obtain computer plots of the response $x(t)$ versus t, over $0 \le t \le 100$, for the representative system

$$x'' + cx' + x = \cos 2t; \qquad x(0) = 3, \quad x'(0) = 4$$

for three cases: light damping ($c = 0.1$), critical damping ($c = 2$), and heavy damping ($c = 20$). How does the speed of approach to steady state vary with c?

5. Show that taking the limit of the response (22) as $\Omega \to \omega$, with the help of l'Hôpital's rule, does give (15), as claimed three paragraphs below (23).

6. For the mechanical oscillator governed by the differential equation $mx'' + cx' + kx = F(t)$, obtain computer plots of the amplitude response curves [E versus Ω, where E is given by (20b)], for the case where $F(t) = 25 \sin \Omega t$, for these six values of the damping coefficient: $c = 0, 0.25c_{cr}, 0.5c_{cr}, c_{cr}, 2c_{cr}, 4c_{cr}$, where

  (**a**) $m = 1, k = 1$ (**b**) $m = 2, k = 5$
  (**c**) $m = 2, k = 10$ (**d**) $m = 4, k = 2$

7. (*Complex function method*) Use the complex function method (described in Section 6.7.3) to obtain a particular solution to the given equation.

  (**a**) $mx'' + cx' + kx = F_0 \cos \Omega t$
  (**b**) $mx'' + cx' + kx = F_0 \sin \Omega t$
  (**c**) $x' + 3x = 5 \cos 2t$
  (**d**) $x' - x = 4 \sin 3t$
  (**e**) $x'' - x' + x = \cos 2t$
  (**f**) $x'' + 5x' + x = 3 \sin 4t$
  (**g**) $x'' - 2x' + x = 6 \cos 5t$
  (**h**) $x''' + x'' + x' + x = 3 \sin t$
  (**i**) $x'''' + x' + x = 3 \cos t$
  (**j**) $x'''' + 2x''' + 4x = 9 \sin 6t$

8. (*Electrical circuit*) Recall from Section 2.4.2 that the equations governing the current $i(t)$ in the circuit shown, and the charge $Q(t)$ on the capacitor are

$$L\frac{d^2i}{dt^2} + R\frac{di}{dt} + \frac{1}{C}i = \frac{dE(t)}{dt}, \qquad (8.1)$$

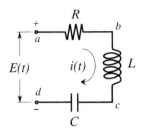

and

$$L\frac{d^2Q}{dt^2} + R\frac{dQ}{dt} + \frac{1}{C}Q = E(t), \qquad (8.2)$$

respectively, where L, R, C, E, i, and Q are measured in henrys, ohms, farads, volts, amperes, and coulombs, respectively.

  (**a**) Let $L = 2$, $R = 4$, and $C = 0.05$. Solve for $Q(t)$ subject to the initial conditions $Q(0) = Q'(0) = 0$, where $E(t) = 100$. Identify the steady-state solution. Obtain a computer plot of the solution for $Q(t)$ over a sufficiently long time period to clearly show the approach of Q to its steady state. (Naturally, all plots should be suitably labeled.)

  (**b**) Same as (a), but for $E(t) = 10e^{-t}$.

  (**c**) Same as (a), but for $E(t) = 10\left(1 - e^{-t}\right)$.

  (**d**) Same as (a), but for $E(t) = 50\left(1 + e^{-0.5t}\right)$.

7.5   ADDITIONAL APPLICATIONS

The harmonic oscillator problems in Sections 7.2 and 7.4 involved a second-order differential equation on the interval $0 \le t < \infty$, and initial conditions at $t = 0$. In this final section we consider two additional applications, one concerning the distribution of a pollutant in a stream and one concerning the deflection of a loaded beam. Both are boundary value problems and the beam problem involves a fourth-order differential equation, so these problems complement the coverage in Sections 7.2–7.4.

7.5.1 Pollution in a stream.

Suppose that a certain pollutant is discharged into a stream at a steady rate, along its bank, and suppose that this situation has persisted for a long enough time for the concentration of that pollutant in the stream to attain a steady-state spatial distribution. Let U be the stream velocity (meters/sec), let $q(x)$ be the pollutant input to the stream along the banks (grams per second per meter), and let $c(x)$ be the pollution concentration in the stream (grams/meter^3) that we seek to determine. A schematic perspective aerial view is shown in Fig. 1, where x is the coordinate measured downstream from some reference point. Of course the stream might meander, in which case x would be measured as arc length along the stream, but there is no harm in straightening the stream out for simplicity in the schematic. Let us also suppose that both the velocity U and the cross-sectional area of the stream A (meters^2) are constant and that c is constant over each cross section of the stream; that is, it does not vary across the stream or over the depth so it is, approximately, a function of x only.

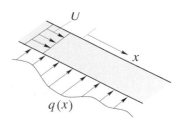

FIGURE 1
Aerial (schematic) view of the river.

To determine $c(x)$, first we derive a differential equation that it must satisfy, then we determine suitable boundary conditions and, finally, we solve for $c(x)$. To derive the governing differential equation we follow a procedure that is typical in engineering: *We consider, as a "control volume," a typical element of the system and then apply to that control volume the relevant physical laws.* Accordingly, we consider an element of the stream between x and $x + \Delta x$ and apply the principal that the rate of increase of pollutant within the element equals the rate at which it enters minus the rate at which it leaves minus the rate at which it disappears by virtue of chemical decay. But the rate of increase must be zero since the system is, by assumption, in steady state. Hence, the foregoing principal reduces to this:

$$\text{Rate in} = \text{Rate out} + \text{Rate of decay.} \tag{1}$$

To understand the "Rate in" and "Rate out" terms in (1) we ask this question: What are the *transport* mechanisms whereby pollutant moves about? There are two: **diffusion** and **convection**. To explain these two mechanisms, first consider the water to be stationary ($U = 0$) and imagine the pollutant to be black ink. If we insert a drop of ink into the water we know that it will spread in all directions. That transport is by diffusion. If, instead, the stream is flowing ($U > 0$) then the ink will not only spread by diffusion, it will also be swept downstream by convection.

According to **Fick's law of diffusion**, the transport (i.e., the flux) of material (the pollutant in this case) across any cross section of the river will be proportional to both the area A (of course, the more area the more flux) and the "gradient" dc/dx at that location, namely, the slope of the graph of $c(x)$. In a similar manner, heat flows by diffusion from a point A to a nearby point B because it is hotter at A than at B, that is, because of the temperature gradient. Whether the diffusion of heat or the diffusion of mass, the process of diffusion is driven by gradients. Thus, Fick's law takes the form

$$\text{flux} = -kA\frac{dc}{dx}, \tag{2}$$

where k is a (positive) constant of proportionality known as the *diffusivity* of the material. For instance, one expects that the diffusivity of ink in molasses will be less than its diffusivity in water. To understand the minus sign in (2), note that the flux is taken to be positive if it is in the $+x$ direction and negative if it is in the $-x$ direction. If $dc/dx > 0$ then the flux will be to the left, hence the minus sign in (2); or, if $dc/dx < 0$ then the flux will be to the right and hence positive.

Let us show the various fluxes in Fig. 2, and then discuss them. The fluxes $-kAc'(x)$ at x and $-kAc'(x + \Delta x)$ at $x + \Delta x$, by diffusion, follow from (2). The flux $q(x)\Delta x$ is due to the inflow along the bank, the Δx being needed because $q(x)$ is the flux per unit x-length. Note that all terms in Fig. 2 have units of mass per unit time (grams/sec).

Next, consider the two AUc terms. In time Δt the influx of pollutant across the face x will [with physical units (meters, seconds, and grams) shown in parentheses] be

$$A(m^2) \times U\left(\frac{m}{s}\right) \times c\left(\frac{g}{m^3}\right) \times \Delta t(s) = AUc\Delta t \text{ grams} \tag{3}$$

so AUc will be the flux in mass per unit time (grams/sec).

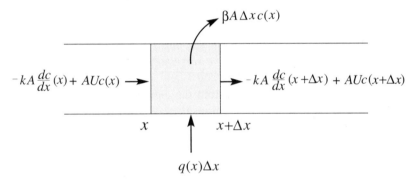

FIGURE 2   Transport of pollutant into and out of the control volume.

Finally, let us suppose that the pollutant participates in chemical reactions so that there is a disappearance of pollutant, within the control volume, at a rate that is proportional to the amount of pollutant within that volume, namely, $A\Delta xc(x)$ grams. If the constant of proportionality is β (which is a chemical rate constant, with units of 1/sec), then the disappearance of pollutant in the control volume is at the rate $\beta A\Delta xc(x)$ grams/sec.

Thus, (1) becomes

$$-kAc'(x)+AUc(x)+q(x)\Delta x = -kAc'(x+\Delta x)+AUc(x+\Delta x)+\beta A\Delta xc(x), \tag{4}$$

or,

$$kA[c'(x + \Delta x) - c'(x)] - AU[c(x + \Delta x) - c(x)] - \beta A\Delta xc(x) = -q(x)\Delta x. \tag{5}$$

Dividing (5) by $A\Delta x$ and taking the limit as $\Delta x \to 0$ then gives

$$\boxed{kc''(x) - Uc'(x) - \beta c(x) = -\frac{q(x)}{A}} \tag{6}$$

as the differential equation governing $c(x)$. On the surface, (6) resembles the forced harmonic oscillator equation studied in the foregoing sections (with x as independent variable in place of t), but a major difference is that the signs of the c' and c terms in (6) are negative.

EXAMPLE 1

To pose a specific case let $U = 0$, so the stream is not flowing, and let $q(x) = QH(x)$, where Q is a constant and $H(x)$ is the Heaviside function (Fig. 3); that is, the input of pollutant along the bank is zero for $x < 0$ and Q grams per second per meter for $x > 0$. Then (6) becomes

$$kc'' - \beta c = -\frac{Q}{A}H(x). \tag{7}$$

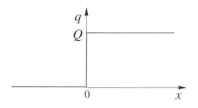

FIGURE 3
$q(x) = QH(x)$.

Since (7) is of second order there will be two arbitrary constants in the homogeneous solution, so it will be appropriate to require two boundary or initial conditions on $c(x)$. Since the x interval is $-\infty < x < \infty$ it seems reasonable to have one boundary condition at $x = -\infty$ and one at $x = +\infty$.

Understand that in most examples in this text we are given the differential equation and any initial or boundary conditions, and are asked to solve for the unknown function. In the present example, however, we are also addressing the *modeling* phase, in which we develop the differential equation and the boundary conditions as well.

Since the boundary conditions at $\pm\infty$ are not yet apparent, let us begin the solution of (7), for when the general solution is in hand we will understand the problem more fully and will be in a better position to specify appropriate boundary conditions. The homogeneous solution of (7) is readily found, but how do we find a particular solution? The method of undetermined coefficients does not apply because $H(x)$ does not generate a finite number of linearly independent derivatives; in fact, $H(x)$ is not even differentiable because it is discontinuous. The method of variation of parameters does apply but is less convenient than the method we will adopt.

Specifically, it seems best in this example to solve by breaking the x domain into two parts, $-\infty < x < 0$ and $0 < x < \infty$, because then we can replace the single equation (7), which has a complicated forcing function, by two equations with simple forcing functions:

$$c'' - a^2 c = 0 \quad \text{on} \quad -\infty < x < 0, \tag{8a}$$

$$c'' - a^2 c = -b \quad \text{on} \quad 0 < x < \infty, \tag{8b}$$

where, for notational simplicity, we have set

$$a \equiv \sqrt{\beta/k} \quad \text{and} \quad b \equiv Q/kA. \tag{9}$$

A particular solution to (8b) is simply b/a^2, so the general solution to (8) is

$$c(x) = \begin{cases} De^{ax} + Ee^{-ax}, & x < 0 \\ Fe^{ax} + Ge^{-ax} + \dfrac{b}{a^2}, & x > 0. \end{cases} \tag{10a,b}$$

Note carefully that the arbitrary constants D and E in the homogeneous solution for $x < 0$ are independent of the arbitrary constants F and G in the homogeneous solution for $x > 0$ so we have used different letters for them. Note also that now we have four arbitrary constants so we need four boundary conditions rather than two. That result makes sense because now we have two second-order equations, one on $x < 0$ and one on $x > 0$.

Observe in (10a) that the $e^{-ax} \to \infty$ as $x \to -\infty$, and in (10b) that the $e^{ax} \to \infty$ as $x \to +\infty$. Correspondingly, if $E \neq 0$ then $c(x)$ will blow up (i.e., tend to infinity in magnitude) as $x \to -\infty$, and if $F \neq 0$ then $c(x)$ will blow up as $x \to +\infty$. From the nature of the physical problem it seems unrealistic that $c(x)$ should blow up, either at $+\infty$ or at $-\infty$. Thus, let us now prescribe these two boundary conditions:

$$c(x) \text{ bounded as } x \to -\infty, \tag{11a}$$

$$c(x) \text{ bounded as } x \to +\infty. \tag{11b}$$

Such boundary conditions are called **boundedness conditions**. This is the first example, in this book, where we have had boundedness boundary conditions rather than specific numerical values.

By applying (11a) we learn that $E = 0$, and by applying (11b) we learn that $F = 0$. Thus, (10) reduces to

$$c(x) = \begin{cases} De^{ax}, & x < 0 \\ Ge^{-ax} + \dfrac{b}{a^2}, & x > 0. \end{cases} \tag{12a,b}$$

We still need two more conditions (so we can solve for D and E) and these are obtained by suitably blending the two solutions (i.e., for $x < 0$ and for $x > 0$) at $x = 0$. Specifically, suppose we define the left-hand solution [i.e., $c(x)$ given by (12a)] as $c_L(x)$ and the right-hand solution [given by (12b)] as $c_R(x)$. As our two blending, or "matching," conditions we require that these functions and their slopes agree at $x = 0$:[1]

$$c_L(0) = c_R(0), \tag{13a}$$

$$c_L'(0) = c_R'(0), \tag{13b}$$

so that the graph of $c(x)$ is both continuous and smooth at $x = 0$. These conditions appear reasonable, but deserve explanation. To avoid getting sidetracked right now let us defer that discussion to Comment 2, below.

Putting (12) into (13a) and (13b) gives the equations

$$D = G + \frac{b}{a^2}, \tag{14a}$$

$$aD = -aG, \tag{14b}$$

---

[1]Strictly speaking, we should express (13a) as $\lim_{x \to 0} c_L(x) = \lim_{x \to 0} c_R(x)$ since (12a) applies for $x < 0$ and (12b) applies for $x > 0$. Similarly for (13b).

so $D = b/2a^2$ and $G = -b/2a^2$. Putting these values into (12), and recalling the definitions of a and b given by (9), gives the solution as

$$c(x) = \begin{cases} \dfrac{Q}{2\beta A} e^{\sqrt{\beta/k}\,x}, & x \le 0 \\[2mm] \dfrac{Q}{2\beta A}(2 - e^{-\sqrt{\beta/k}\,x}), & x \ge 0. \end{cases} \qquad (15\text{a,b})$$

which is shown in Fig. 4 by the solid curve.

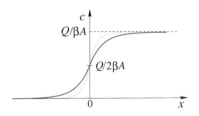

FIGURE 4
Graph of the solution (15).

COMMENT 1. This result appears to be quite reasonable for these reasons. First, if diffusion is driven by gradients, as noted above, then the effects of diffusion should be most pronounced near $x = 0$ where the input $q(x) = QH(x)$ has a "vertical slope" and hence an infinite gradient, and we see from Fig. 4 that most of the variation in $c(x)$ does indeed occur near $x = 0$. Far upstream and far downstream we can neglect the c'' diffusion term in (7) and write (7) as $-\beta c(x) \approx -(Q A)H(x)$; that is, $c(x) \approx 0$ far upstream and $c(x) \approx Q/\beta A$ far downstream, which results are corroborated in Fig. 4. It also makes sense that the asymptotic value $c(x) \sim Q/\beta A$ (as $x \to \infty$) increases with the input Q and decreases with the decay rate β and the area A. Finally, observe the $\sqrt{\beta/k}$ in the exponent in (15). As we increase $\sqrt{\beta/k}$ the x interval over which c suffers rapid change narrows (dotted curve in Fig. 4), as occurs if we decrease k or increase β. Does that result make sense? Yes, for k is the diffusivity, so if we decrease k then we decrease the diffusion as is evident either from Fick's law (2) or by observing that if k decreases then the role of the kc'' diffusion term in (7) decreases too. Also, if β increases then pollutant is disappearing (i.e., it is being converted by chemical reactions to other forms) more rapidly; hence, it is available, to diffuse, for a shorter time.

COMMENT 2. How can we justify our choice of the matching conditions (13a,b)? After all, the forcing function $-(QA)H(x)$ in (7) is discontinuous at $x = 0$, so should not $c(x)$, or perhaps $c'(x)$, be discontinuous there too? That is, if the "input" $c(x)$ is discontinuous should not the "output" $c(x)$ be discontinuous as well? This point was raised earlier, following Example 4 in Section 2.4.1, in connection with a first-order differential equation with a discontinuous forcing function. The idea emphasized there was that integration is a smoothing process. Since (7) is a second-order differential equation its solution involves, in effect, two integrations. As illustrated in Fig. 14 of Section 2.4.1, if we begin with the discontinuous forcing function $H(x)$ in (7) then the first integration will give a function that is continuous but with a kink (i.e., a jump in slope) at $x = 0$, and the second integration will give a function that is not only continuous but also smooth (i.e., with a continuous slope) at $x = 0$. The foregoing justification for the blending conditions (13a,b) was informal, not rigorous, so let us reassure you that the same solution (15) is obtained using variation of parameters, which method does not involve any matching conditions because the method retains the full x interval and does not break it into two parts. Solution by that method is left for the exercises.

COMMENT 3.   The boundedness conditions (11) are the first such conditions in this book. Application of these conditions to the solution (10) was simple: The e^{-ax} in (10a) blows up as $x \to -\infty$ so we set $E = 0$, and the e^{ax} in (10b) blows up as $x \to +\infty$ so we set $F = 0$. What if we had used $\sinh ax$ and $\cosh ax$ in place of e^{ax} and e^{-ax}? Then in place of (10) we would have obtained

$$c(x) = \begin{cases} H \sinh ax + I \cosh ax, & x < 0 \\[2mm] J \sinh ax + K \cosh ax + \dfrac{b}{a^2}, & x > 0. \end{cases} \qquad (16\text{a,b})$$

which form is equivalent to (10). In this case the application of the boundedness conditions is less straightforward because *both* $\sinh ax$ and $\cosh ax$ blow up as $x \to \pm\infty$. Do we therefore need to set both H and I to zero and similarly for J and K? No. Consider $x < 0$. Then

$$c(x) = H \sinh ax + I \cosh ax$$
$$= \frac{H}{2}\left(e^{ax} - e^{-ax}\right) + \frac{I}{2}\left(e^{ax} + e^{-ax}\right)$$
$$= \frac{H+I}{2}e^{ax} - \frac{H-I}{2}e^{-ax}.$$

Since $e^{ax} \to 0$ as $x \to -\infty$ that term is acceptable. But $e^{-ax} \to \infty$ as $x \to -\infty$ so we need to set $(H - I)/2 = 0$; that is, we need to choose $H = I$, in which case (16a) reduces to $c(x) = He^{ax}$, just as we obtained in (12a). Similarly for $x > 0$.

In summary, we obtain the same final result whether we use the exponentials or the sinh and the cosh, but the exponentials are more convenient if we have boundedness conditions—because we can see immediately which exponential is bounded and which is not. ∎

7.5.2 Euler beam theory.   By a beam an engineer means a stiff slender horizontal structural member that supports a load distribution normal to itself. Typically it is supported at both ends but it might be free at one end, as is a diving board. The deflection $y(x)$ is of considerable interest and it is shown in a course on solid mechanics that $y(x)$ is governed by the fourth-order differential equation

$$EIy'''' = -w(x), \qquad (17)$$

where $w(x)$ is the load (in force per unit length), the constant E is *Young's modulus* for the beam material, and I is the area moment of inertia of the cross-sectional area about an axis that is normal to the plane of bending and through the centroid of the area (Fig. 5). In any case, EI is a known constant.

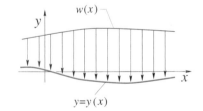

FIGURE 5
A portion of a loaded beam.

EXAMPLE 2   *Uniformly Loaded Cantilever Beam.*
Consider the loaded cantilever beam shown in Fig. 6, and let the load be uniform so that $w(x) = \text{constant} \equiv w_0$. Then

$$EIy'''' = -w_0 \quad (0 \le x \le L) \qquad (18\text{a})$$

$$y(0) = 0, \quad y'(0) = 0, \quad y''(L) = 0, \quad y'''(L) = 0, \qquad (18b)$$

which is a boundary value problem because conditions are given at both ends. The significance of the boundary conditions is as follows: $y(0) = 0$ simply by virtue of our chosen placement of the origin of the x, y coordinate system; $y'(0) = 0$ because the beam is cantilevered out of the wall so its slope at $x = 0$ is zero because our x axis is normal to the wall; and $y''(L) = 0$ and $y'''(L) = 0$ because (in the language of engineering mechanics) the moment and shear force, respectively, applied at the end of the beam are zero. The differential equation (18a) can be solved by direct integration. Integrating it repeatedly with respect to x gives

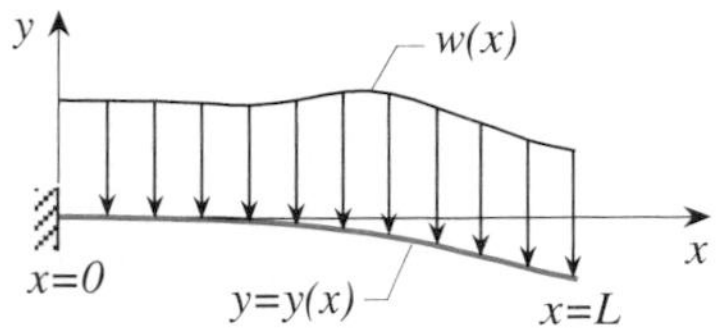

FIGURE 6
Loaded cantilever beam.

$$EIy''' = -w_0 x + A, \qquad (19a)$$

$$EIy'' = -\frac{w_0}{2}x^2 + Ax + B, \qquad (19b)$$

$$EIy' = -\frac{w_0}{6}x^3 + \frac{A}{2}x^2 + Bx + C, \qquad (19c)$$

$$EIy = -\frac{w_0}{24}x^4 + \frac{A}{6}x^3 + \frac{B}{2}x^2 + Cx + D. \qquad (19d)$$

Then, the boundary conditions (18b) give

$$0 = D,$$
$$0 = C,$$
$$0 = -\frac{w_0}{2}L^2 + AL + B,$$
$$0 = -w_0 L + A.$$

Solving these for A, B, C, D, (19d) gives the solution

$$y(x) = -\frac{w_0}{24EI}\left(x^4 - 4Lx^3 + 6L^2x^2\right). \qquad (20)$$

∎

EXAMPLE 3 *Semi-Infinite Beam on Elastic Foundation.*
This time let the beam be semi-infinite, extending over $0 \le x < \infty$, under a uniform load $w(x) = $ constant $= w_0$ again, but suppose that the deflection is resisted by a distributed spring, like an infinitely long "mattress," having a stiffness k (force per unit deflection, per unit x length), as depicted in Fig. 7. And let the beam be supported by a frictionless pin at the left end, as shown.

In railroad engineering the "mattress" models the track bed, which consists of gravel and soil and which behaves as a distributed spring. As in Example 2, we wish to find the deflection $y(x)$.

In this case, besides the downward load $w(x)$ there is an upward spring force $ky(x)$ per unit x length, so (17) becomes $EIy'''' = -w(x) - ky(x)$, $-ky$ because if y is negative the spring force is positive (upward) and if y is positive the spring force is negative (downward). As boundary conditions we have $y(0) = 0$ (zero displacement) and $y''(0) = 0$ (zero applied moment at the left end), and we will also require that

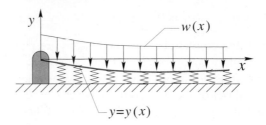

FIGURE 7   Semi-infinite beam on elastic
foundation.

$y(x)$ be bounded as $x \to \infty$, as seems reasonable. Thus, we have the following boundary value problem:

$$y'''' + \beta^4 y = -W \qquad (0 \le x < \infty) \tag{21a}$$

$$y(0) = 0, \quad y''(0) = 0, \quad y \text{ bounded as } x \to \infty, \tag{21b}$$

where we have set $k/EI \equiv \beta^4$ (the exponent 4 being included for convenience because we will be taking the one-fourth root of that quantity in the analysis to follow) and $w_0/EI \equiv W$.

To find the homogeneous solution of (21a) seek $y_h(x) = e^{rx}$ and obtain the characteristic equation

$$r^4 + \beta^4 = 0. \tag{22}$$

Solving (22) for r is equivalent to finding the four one-fourth roots

$$r = \left(-\beta^4\right)^{1/4}. \tag{23}$$

One can do that using complex variable theory, but our introduction to that subject in Section 6.2 did not go far enough to show how to do that. However, if we use *Maple*, we obtain these four complex roots:

$$r = \frac{\beta}{\sqrt{2}}(1+i), \ \frac{\beta}{\sqrt{2}}(-1+i), \ \frac{\beta}{\sqrt{2}}(-1-i), \ \frac{\beta}{\sqrt{2}}(1-i). \tag{24}$$

Thus,

$$\begin{aligned}
y_h(x) &= Ae^{(1+i)\frac{\beta}{\sqrt{2}}x} + Be^{(-1+i)\frac{\beta}{\sqrt{2}}x} + Ce^{(-1-i)\frac{\beta}{\sqrt{2}}x} + De^{(1-i)\frac{\beta}{\sqrt{2}}x} \\
&= e^{\frac{\beta}{\sqrt{2}}x}\left(Ae^{i\frac{\beta}{\sqrt{2}}x} + De^{-i\frac{\beta}{\sqrt{2}}x}\right) + e^{\frac{-\beta}{\sqrt{2}}x}\left(Be^{i\frac{\beta}{\sqrt{2}}x} + Ce^{-i\frac{\beta}{\sqrt{2}}x}\right) \\
&= e^{\frac{\beta}{\sqrt{2}}x}\left(E\cos\frac{\beta}{\sqrt{2}}x + F\sin\frac{\beta x}{\sqrt{2}}\right) + e^{\frac{-\beta}{\sqrt{2}}x}\left(G\cos\frac{\beta x}{\sqrt{2}} + H\sin\frac{\beta x}{\sqrt{2}}\right).
\end{aligned} \tag{25}$$

Turning to the particular solution of (21a), we find by the method of undetermined coefficients or simply by inspection that

$$y_p(x) = -W/\beta^4,$$

so

$$y(x) = y_h(x) + y_p(x)$$

$$= e^{\beta x/\sqrt{2}} \left(E \cos \frac{\beta x}{\sqrt{2}} + F \sin \frac{\beta x}{\sqrt{2}} \right)$$

$$+ e^{-\beta x/\sqrt{2}} \left(G \cos \frac{\beta x}{\sqrt{2}} + H \sin \frac{\beta x}{\sqrt{2}} \right) - \frac{W}{\beta^4}. \tag{26}$$

The boundedness condition immediately implies that $E = 0$ and $F = 0$ so (26) reduces to

$$y(x) = e^{-\beta x/\sqrt{2}} \left(G \cos \frac{\beta x}{\sqrt{2}} + H \sin \frac{\beta x}{\sqrt{2}} \right) - \frac{W}{\beta^4}. \tag{27}$$

Then,

$$y(0) = 0 = G - \frac{W}{\beta^4},$$

$$y''(0) = 0 = -\beta^2 H,$$

so $G = W/\beta^4$ and $H = 0$. Thus, we have

$$y(x) = -\frac{W}{\beta^4} \left(1 - e^{-\beta x/\sqrt{2}} \cos \frac{\beta x}{\sqrt{2}} \right). \tag{28}$$

COMMENT 1.   It does seem reasonable that $y(x)$ should tend to a constant as $x \to \infty$. In that case the y'''' in (21a) tends to zero as $x \to \infty$ so (21a) gives $y(x) \sim -W/\beta^4$ as $x \to \infty$, which result agrees with (28). Further, the exponential approach to that value seems reasonable, but the cosinusoidal oscillations are not so obvious from a physical point of view. We have plotted (28) in Fig. 8.

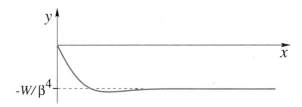

FIGURE 8   The deflection $y(x)$ given by (28).

COMMENT 2.   You might have been puzzled that (21b) contains only three boundary conditions whereas (21a) is of fourth order. However, the boundedness condition is equivalent to two boundary conditions in the sense that its application results in the evaluation of *two* of the arbitrary constants (E and F). ∎

Closure. Prior to this section, most examples consisted of second-order equations with initial conditions. Therefore, in this section we have tried to provide some balance by emphasizing boundary value problems and including an application involving a fourth-order equation as well. In doing so we have also met, for the first time in this book, boundedness conditions.

Maple. Asked to solve $r^4 + \beta^4 = 0$, *Maple* gives no response—because it knows nothing about the nature of β. Since we know β is real and positive we can write $r = \beta(-1)^{1/4}$, in place of (23) or, equivalently, we can solve $r^4 + 1 = 0$ and multiply the roots of that equation by β. Then the relevant *Maple* command is

```
solve(r^4+1=0,r);
```

EXERCISES 7.5

1. Solve (6) subject to the condition that $c(x)$ is bounded as $x \to \pm\infty$, where $k = 1$, $\beta = 4$, $A = 1$ and
  (a) $U = 0$, $q(x) = 10$
  (b) $U = 5$, $q(x) = 10$
  (c) $U = -5$, $q(x) = 10$
  (d) $U = 0$, $q(x) = 1 + \sin x$
  (e) $U = 3$, $q(x) = 1 + \sin x$
  (f) $U = -3$, $q(x) = 1 + \sin x$
  (g) $U = 0$, $q(x) = 50 \cos^2 x$
  (h) $U = 3$, $q(x) = 5 + 10 \cos x$

2. Let the beam in Example 2 be cantilevered at *both* ends, so

$$EIy'''' = -w_0 \quad (0 \le x \le L)$$
$$y(0) = 0, \ y'(0) = 0, \ y(L) = 0, \ y'(L) = 0.$$

Solve for $y(x)$ and give a labeled sketch of the solution. Show that the maximum deflection of the beam is

$$y\left(\frac{L}{2}\right) = -\frac{w_0 L^4}{384EI}. \tag{2.1}$$

3. Let the beam in Example 2 be "simply supported" at both ends. That is, imagine it supported on knife edges at both ends, so

$$EIy'''' = -w_0 \quad (0 \le x \le L)$$
$$y(0) = 0, \ y''(0) = 0, \ y(L) = 0, \ y''(L) = 0,$$

because both the deflection and the applied bending moment are zero at both ends. Solve for $y(x)$ and give a labeled sketch of the solution. Show that the maximum deflection of the beam is

$$y\left(\frac{L}{2}\right) = -\frac{5w_0 L^4}{384EI}. \tag{3.1}$$

[Notice that the maximum deflection (3.1) for the simply supported beam is substantially greater than the maximum deflection (2.1) of the beam cantilevered at both ends, which result seems quite reasonable.]

4. Verify that if we raise the roots given by (24) to the fourth power we do obtain $-\beta^4$.

5. In Example 3 we considered the semi-infinite beam problem (21). Here, consider an *infinite* beam $(-\infty < x < \infty)$ on an elastic foundation, governed by the boundary value problem

$$EIy'''' + ky = -w(x),$$

where y is to be bounded as $x \to \pm\infty$. Solve for $y(x)$, for the case where
  (a) $w(x) = \text{constant} = w_0$
  (b) $w(x) = w_0(1 + \cos x)$
  (c) $w(x) = w_0 \cos^2 x$

6. Solve Exercise 2 for $y(x)$ using *Maple*.

7. Solve Exercise 3 for $y(x)$ using *Maple*.

CHAPTER 7 REVIEW

After studying the theory and solution of linear constant coefficient equations in Chapter 6, we devoted Chapter 7 to important and representative applications of such equations. Of these, the most important case was the linear harmonic oscillator

$$mx'' + cx' + kx = f(t), \tag{1}$$

which is one of the most prominent differential equations in science and engineering. The latter was the subject of Sections 7.2–7.4.

In Section 7.2 we considered the free vibration, where $f(t) = 0$, and studied the effect of damping on the solution. For the undamped case ($c = 0$) we found the solution to be a harmonic oscillation, namely, of the form $A \cos \omega t + B \sin \omega t$ or, equivalently, $C \sin(\omega t + \phi)$, where $\omega = \sqrt{k/m}$ is the "natural frequency" and ϕ is a phase angle. If damping is present but c is less than the critical value $c_{cr} = \sqrt{4mk}$ the system is said to be underdamped. In that case we found that

$$x(t) = e^{-(c/2m)t} \left[A \cos \sqrt{\omega^2 - \left(\frac{c}{2m}\right)^2}\, t + B \sin \sqrt{\omega^2 - \left(\frac{c}{2m}\right)^2}\, t \right], \qquad (2)$$

so that the frequency is less than ω, and there now appears an exponential factor $e^{-(c/2m)t}$ that causes the oscillation to "damp out" (whereby its amplitude tends to zero as $t \to \infty$). As c is increased the rate of exponential decay increases and the frequency continues to decrease until, when c reaches its critical value $c_{cr} = \sqrt{4mk}$, the frequency in (2) becomes zero. As c is increased beyond c_{cr} the square roots in (2) become imaginary and the cosine and sine terms become hyperbolic cosine and hyperbolic sine, respectively, for it follows from the definitions of the circular and hyperbolic functions (Exercise 17c of Section 6.2) that

$$\cos iz = \cosh z, \quad \sin iz = i \sinh z.$$

That is, the system becomes so sluggish that the oscillation disappears entirely.

In Section 7.3 we continued to study the free vibration, given by (1) with $f(t) = 0$, but this time we considered not the traditional x, t plane but the x, x' "phase plane." This concept will be used again in Chapter 11.

In Section 7.4 we turned to the forced oscillation and limited our attention to the most important case, that of harmonic excitation. Specifically, we studied the case

$$mx'' + cx' + kx = F_0 \cos \Omega t. \qquad (3)$$

Beginning with the undamped case ($c = 0$), we solved (3) and plotted the amplitude-response and phase-response curves. We found that as the driving frequency Ω approaches the natural frequency ω a "beat" phenomenon develops, that leads to resonance when $\Omega = \omega$. The most notable result was that very large amplitude oscillations can result if the system is forced near its natural frequency.

Finally, in Section 7.5 we presented applications to pollution in a stream and to the deflection of a loaded beam. The latter problem involved a fourth-order equation, and both the former and the latter were of boundary value type. Hence these applications complemented the harmonic oscillator applications in Sections 7.2–7.4 since all of those involved second-order equations and were of initial value type.

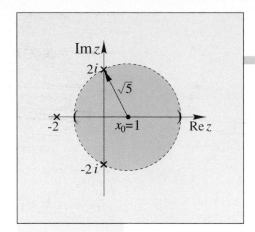

8

POWER SERIES SOLUTIONS

8.1 INTRODUCTION

In this chapter we continue to restrict ourselves to linear equations of second order and higher. We've shown how to solve constant-coefficient equations and Cauchy-Euler equations, so in this chapter we turn to nonconstant-coefficient equations that are not of Cauchy-Euler type, such as

$$y'' + x^3 y' + \left(2 - e^x\right) y = 0.$$

We will limit our attention to second-order equations, this case being the most common in applications, but the methods we develop can be applied to higher-order equations as well.[1]

Nonconstant-coefficient equations that are not of Cauchy-Euler type are generally so difficult that we give up on the hope of finding solutions in the form of finite linear combinations of elementary functions and seek, instead, solutions in the form of *infinite series*.

To illustrate the series solution method that will be developed in subsequent sections, consider the simple problem

$$\frac{dy}{dx} + y = 0. \tag{1}$$

Of course (1) is readily solved, a general solution being

$$y(x) = A e^{-x}, \tag{2}$$

where A is an arbitrary constant. Nevertheless, let us ignore (2) and seek a solution to (1) in the form of an infinite series,

---

[1] There is no need to use these methods for *first*-order equations because we can solve the linear first-order equation in closed form; see (14) in Section 2.3.

$$y(x) = \sum_{0}^{\infty} a_n x^n = a_0 + a_1 x + a_2 x^2 + \cdots , \tag{3}$$

where the constants $a_0, a_1, \ldots$ are to be determined so that (3) satisfies (1). From (3),

$$\frac{dy}{dx} = \frac{d}{dx}\left(a_0 + a_1 x + a_2 x^2 + \cdots\right) = a_1 + 2a_2 x + 3a_3 x^2 + \cdots \tag{4}$$

so putting (3) and (4) into (1) gives

$$\left(a_1 + 2a_2 x + 3a_3 x^2 + \cdots\right) + \left(a_0 + a_1 x + a_2 x^2 + \cdots\right) = 0, \tag{5}$$

or, rearranging terms,

$$(a_1 + a_0) + (2a_2 + a_1)\, x + (3a_3 + a_2)\, x^2 + \cdots = 0. \tag{6}$$

If we realize that the right side of (6) is really $0 + 0x + 0x^2 + \cdots$, then, by equating coefficients of like powers of x on both sides of (6), we obtain $a_1 + a_0 = 0$, $2a_2 + a_1 = 0$, $3a_3 + a_2 = 0$, and so on. Solving these gives

$$\begin{aligned}
a_1 &= -a_0, \\
a_2 &= -a_1/2 = -(-a_0)/2 = a_0/2, \\
a_3 &= -a_2/3 = -(a_0/2)/3 = -a_0/6,
\end{aligned} \tag{7}$$

and so on, where a_0 remains arbitrary. Thus, we have

$$y(x) = a_0 \left(1 - x + \frac{1}{2}x^2 - \frac{1}{6}x^3 + \cdots\right), \tag{8}$$

as the general solution to (1). If we remember the Taylor series formula

$$e^{-x} = 1 - x + \frac{1}{2!}x^2 - \frac{1}{3!}x^3 + \cdots , \tag{9}$$

then we can identify (8) as

$$y(x) = a_0 e^{-x} \tag{10}$$

which is identical to (2) (with a_0 in place of A).

In this simple example we were able to recognize the series obtained in (8) as e^{-x}. Typically, however, when we solve a *nonconstant*-coefficient differential equation by the series method we are not so fortunate and must leave the solution in series form.

As simple as the above steps appear, there are several points of rigor to address if we are to have confidence in the final result (8):

(i) In (4) we differentiated an infinite series term by term. That is, instead of summing the series $a_0 + a_1 x + \cdots$ and *then* differentiating the sum function we differentiated the individual terms and *then* summed the result. That is, we interchanged the order of the differentiation and the summation and wrote

$$\frac{d}{dx} \sum a_n x^n = \sum \frac{d}{dx} \left(a_n x^n \right). \tag{11}$$

Surely (11) is true if the sum is over a *finite* number of terms, but if it is an infinite series then it is possible that reversing the order of the summation and the differentiation might give different results.[1]

(ii) Re-expressing (5) in the form of (6) is based on a supposition that we can add series term by term:

$$\sum A_n + \sum B_n = \sum (A_n + B_n). \tag{12}$$

Again, that step looks reasonable, but is it necessarily correct?

(iii) Finally, inferring (7) from (6) is based upon a supposition that if

$$\sum A_n x^n = \sum B_n x^n \tag{13}$$

for all x in some interval of interest, then it must be true that $A_n = B_n$ for each n. Though reasonable, does it really follow that for the sums to be the same the corresponding individual terms need to be the same?

---

[1]This is a subtle point, so let us illustrate with another example. Given the function $f(x, y) = (x^2 - y^2)/(x^2 + y^2)$, what is the limit of f as x and y tend to zero? The answer depends upon the limit sequence: if we first let $x \to 0$ and then let $y \to 0$ then

$$\lim_{y \to 0} \left(\lim_{x \to 0} \frac{x^2 - y^2}{x^2 + y^2} \right) = \lim_{y \to 0} (-1) = -1,$$

whereas if we first let $y \to 0$ and then let $x \to 0$ then

$$\lim_{x \to 0} \left(\lim_{y \to 0} \frac{x^2 - y^2}{x^2 + y^2} \right) = \lim_{x \to 0} (+1) = +1.$$

Generalizing this example, we can say that *interchanging the order of two limit operations does not necessarily lead to the same result.* In (11) we do indeed have two limit operations, for an infinite series is defined (in Section 8.2.1) as the limit of the sequence of partial sums,

$$\sum_{1}^{\infty} b_n \equiv \lim_{N \to \infty} \sum_{1}^{N} b_n,$$

and a derivative is defined as a limit of difference quotients,

$$\frac{df}{dx} \equiv \lim_{\Delta x \to 0} \frac{f(x + \Delta x) - f(x)}{\Delta x}.$$

Thus, there are some technical questions we need to address, and we do that in the next section. Our approach in deriving (8) was heuristic not rigorous, since we did not attend to the issues mentioned above. We can sidestep the several questions of rigor that arose in deriving the series (8) if, instead, we verify, a posteriori, that (8) does satisfy the given differential equation (1). However, that procedure begs exactly the same questions: termwise differentiation of the series, termwise addition of series, and equating the coefficients of like powers of x on both sides of the equation.

8.2 POWER SERIES

8.2.1 Review of series.   Whereas a finite sum,

$$\sum_{k=1}^{N} a_k = a_1 + a_2 + \cdots + a_N, \tag{1}$$

is well-defined thanks to the commutative and associative laws of addition, an infinite sum, or **infinite series**,

$$\sum_{k=1}^{\infty} a_k = a_1 + a_2 + a_3 + \cdots, \tag{2}$$

is not. For example, is the series $\sum_{1}^{\infty}(-1)^{k-1} = 1 - 1 + 1 - 1 + \cdots$ equal to $(1 - 1) + (1 - 1) + \cdots = 0 + 0 + \cdots = 0$? Is it (by grouping differently) $1 - (1 - 1) - (1 - 1) - \cdots = 1 - 0 - 0 - \cdots = 1$? In fact, besides grouping the numbers in different ways we could rearrange their order as well. The point, then, is that (2) is not self-explanatory, it needs to be defined; we need to decide, or be told, how to do the calculation. To give the traditional definition of (2), we first define the sequence of **partial sums** of the series (2) as

$$s_1 = a_1, \qquad s_2 = a_1 + a_2, \qquad s_3 = a_1 + a_2 + a_3, \tag{3}$$

and so on:

$$s_n = \sum_{k=1}^{n} a_k, \tag{4}$$

where a_k is called the kth **term** of the series. If the limit of the sequence s_n exists, as $n \to \infty$, and equals some number s, then we say that the series (2) is **convergent**, and that it **converges to** s; otherwise it is **divergent**. That is, an infinite series is defined as the limit (if that limit exists) of its sequence of partial sums:

$$\sum_{k=1}^{\infty} a_k \equiv \lim_{n \to \infty} \sum_{k=1}^{n} a_k = \lim_{n \to \infty} s_n = s. \tag{5}$$

That definition, known as **ordinary convergence**, is not the only one possible. For instance, another definition, due to *Cesàro*, is discussed in the exercises. However, ordinary convergence is the traditional definition and is the one that is understood unless specifically stated otherwise.

Recall from the calculus that by $\lim_{n\to\infty} s_n = s$, in (5), we mean that to each number $\epsilon > 0$, no matter how small, there exists an integer N such that $|s - s_n| < \epsilon$ for all $n > N$. (Logically, the words "no matter how small" are unnecessary, but we include them for emphasis.) In general, the smaller the chosen ϵ, the larger the N that is needed, so that N is a function of ϵ.

The significance of the limit concept cannot be overstated, for in mathematics it is often as limits of "old things" that we introduce "new things." For instance, the derivative is introduced as the limit of a difference quotient, the Riemann integral is introduced as the limit of a sequence of Riemann sums, infinite series are introduced as limits of sequences of partial sums, and so on.

To illustrate the definition of convergence given above, consider two simple examples. The series $1 + 1 + 1 + \cdots$ diverges because $s_n = n$ fails to approach a limit as $n \to \infty$. However, for a series to diverge its partial sums need not grow unboundedly. For instance, the series $1 - 1 + 1 - 1 + \cdots$, mentioned above, diverges because its sequence of partial sums (namely, $1, 0, 1, 0, 1, \ldots$) fails to approach a limit. Of course, determining whether a series is convergent or divergent is usually more difficult than for these examples. Ideally, one would like a theorem that gives both necessary and sufficient conditions for convergence. We do have such a theorem:

THEOREM 8.2.1

Cauchy Convergence Theorem
An infinite series is convergent if and only if its sequence of partial sums s_n is a **Cauchy sequence**—that is, if to each $\epsilon > 0$ (no matter how small) there corresponds an integer $N(\epsilon)$ such that $|s_m - s_n| < \epsilon$ for all m and n greater than N.

Unfortunately, this theorem is difficult to apply, so one develops (in the calculus) an array of theorems (i.e., tests for convergence/divergence), each of which is more specialized (and hence less powerful) than the Cauchy convergence theorem, but easier to apply. For instance, if in Theorem 8.2.1 we set $m = n - 1$, then the stated condition becomes: to each $\epsilon > 0$ (no matter how small) there corresponds an integer $N(\epsilon)$ such that $|s_m - s_n| = |s_{n-1} - s_n| = |s_n - s_{n-1}| = |a_n| < \epsilon$ for all $n > N$. The latter is equivalent to saying that $a_n \to 0$ as $n \to \infty$. Thus, we have the specialized but readily applied theorem that *for the infinite series $\sum a_n$ to converge, it is necessary (but not sufficient) that $a_n \to 0$ as $n \to \infty$.* From this theorem it follows immediately that the series $1 + 1 + 1 + \cdots$ and $1 - 1 + 1 - 1 + \cdots$, cited above, both diverge because in each case the nth term does not tend to zero as $n \to \infty$.

8.2.2 Power series.   Let us now focus on the specific needs of this chapter, **power series**—that is, series of the form

$$\sum_0^\infty a_n(x - x_0)^n = a_0 + a_1(x - x_0) + a_2(x - x_0)^2 + \cdots, \qquad (6)$$

where the a_n's are numbers called the **coefficients** of the series, x is a variable, and x_0 is a fixed point called the **center** of the series. We say that the expansion is "about the point x_0." There also exist complex series, but here we restrict all variables and constants to be real. Notice that the quantity $(x - x_0)^n$ on the left side of (6) is the indeterminate form 0^0 when $n = 0$ and $x = x_0$; that form must be interpreted as 1 if the leading term of the series is to be a_0, as desired.

The terms in (6) are now functions of x rather than numbers; consequently, the series may converge at some points on the x axis and diverge at others. At the very least (6) converges at $x = x_0$ since then it reduces to the single term a_0.

THEOREM 8.2.2

Interval of Convergence of Power Series

The power series (6) converges at $x = x_0$. If it converges at other points as well, then those points necessarily comprise an interval $|x - x_0| < R$ centered at x_0 and, possibly, one or both endpoints of that interval (Fig. 1), where R can be determined from either of the formulas

$$R = \frac{1}{\lim\limits_{n\to\infty} \left| \dfrac{a_{n+1}}{a_n} \right|} \quad \text{or} \quad R = \frac{1}{\lim\limits_{n\to\infty} \sqrt[n]{|a_n|}}, \tag{7a,b}$$

if the limits in the denominators exist and are nonzero. If the limits in (7a,b) are zero, then (6) converges for all x (i.e., for every finite x, no matter how large), and we say that "$R = \infty$." If the limits fail to exist by virtue of being infinite, then $R = 0$ and (6) converges only at x_0.

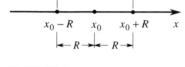

FIGURE 1
Interval of convergence of power series.

We call $|x - x_0| < R$ the **interval of convergence**, and R the **radius of convergence**. If a power series converges to a function f on some interval, we say that it **represents** f on that interval, and we call f its **sum function**.

EXAMPLE 1
Consider $\sum_0^\infty n!\, x^n$, so $a_n = n!$ and $x_0 = 0$. Then (7a) is simpler to apply than (7b), and gives

$$R = 1/ \lim_{n\to\infty} \frac{(n+1)!}{n!} = 1/ \lim_{n\to\infty} (n+1) = 1/\infty = 0,$$

so the series converges only at $x = x_0 = 0$. ∎

EXAMPLE 2
Consider $\sum_0^\infty (-1)^n \,[(x+5)/2]^n$. Then $a_n = (-1)^n/2^n$, $x_0 = -5$, and (7a) gives

$$R = 1/ \lim_{n\to\infty} \left| \frac{(-1)^{n+1}}{2^{n+1}} \frac{2^n}{(-1)^n} \right| = 1/ \lim_{n\to\infty} \frac{1}{2} = 1/\left(\frac{1}{2}\right) = 2,$$

so the series converges in $|x + 5| < 2$ and diverges in $|x + 5| > 2$. For $|x + 5| = 2$ ($x = -7, -3$) the theorem gives no information. However, we see that for $x = -7$ and -3 the terms do not tend to zero as $n \to \infty$, so the series diverges at the endpoints $x = -7$ and -3. ∎

EXAMPLE 3

Consider $\displaystyle\sum_{4}^{\infty} \frac{(x-1)^n}{(n+1)^n}$. Then $a_n = (n+1)^{-n}$, $x_0 = 1$, and (7b) gives

$$R = 1/\lim_{n\to\infty} \sqrt[n]{(n+1)^{-n}} = 1/\lim_{n\to\infty} \frac{1}{n+1} = 1/0 = \infty,$$

so the series converges for all x; that is, the interval of convergence is $|x-1| < \infty$. ∎

EXAMPLE 4

Consider the power series

$$1 + \frac{(x-3)^2}{5} + \frac{(x-3)^4}{5^2} + \cdots = \sum_{0}^{\infty} \frac{1}{5^n}(x-3)^{2n}. \qquad (8)$$

This series is not of the form (6) because the powers of $x-3$ are in steps of 2. However, if we set $X = (x-3)^2$, then we have the standard form $\displaystyle\sum_{0}^{\infty} \frac{1}{5^n} X^n$, with $a_n = 1/5^n$ and $\displaystyle\lim_{n\to\infty} \left| \frac{a_{n+1}}{a_n} \right| = \lim_{n\to\infty} \left| \frac{5^n}{5^{n+1}} \right| = \frac{1}{5}$. Thus, (7a) gives $R = 5$, and the series converges in $|X| < 5$ (i.e., in $|x-3| < \sqrt{5}$), and diverges in $|X| > 5$ (i.e., in $|x-3| > \sqrt{5}$). ∎

Recall from our introductory example, in Section 8.1, that several questions arise regarding the manipulation of power series. The following theorem answers those questions and will be needed when we apply the power series method.

THEOREM 8.2.3

Manipulation of Power Series

(a) *Termwise differentiation (or integration) permissible.* A power series may be differentiated (or integrated) termwise (i.e., term by term) within its interval of convergence I. The series that results has the same interval of convergence I and represents the derivative (or integral) of the sum function of the original series.

(b) *Termwise addition (or subtraction or multiplication) permissible.* Two power series (about the same point x_0) may be added (or subtracted or multiplied) termwise within their common interval of convergence I. The series that results has the same interval of convergence I and represents the sum (or difference or product) of their two sum functions.

(c) *If two power series are equal, then their corresponding coefficients must be equal.* That is, for

$$\sum_{0}^{\infty} a_n(x-x_0)^n = \sum_{0}^{\infty} b_n(x-x_0)^n \qquad (9)$$

to hold in some common interval of convergence, it must be true that $a_n = b_n$ for each n. In particular, if

$$\sum_0^\infty a_n(x - x_0)^n = 0 \tag{10}$$

in some interval, then each a_n must be zero.

Part (a) means that if $f(x) = \sum_0^\infty a_n(x - x_0)^n$ within I, then

$$f'(x) = \frac{d}{dx} \sum_0^\infty a_n(x - x_0)^n = \sum_0^\infty \frac{d}{dx} \left[a_n(x - x_0)^n \right] = \sum_1^\infty n a_n(x - x_0)^{n-1} \tag{11}$$

and

$$\int_a^b f(x)\, dx = \int_a^b \sum_0^\infty a_n(x - x_0)^n\, dx$$

$$= \sum_0^\infty a_n \int_a^b (x - x_0)^n\, dx$$

$$= \sum_0^\infty a_n \frac{(b - x_0)^{n+1} - (a - x_0)^{n+1}}{n + 1} \tag{12}$$

within I, where a, b are any two points within I.

Part (b) means that if $f(x) = \sum_0^\infty a_n(x - x_0)^n$ and $g(x) = \sum_0^\infty b_n(x - x_0)^n$ on I, then

$$f(x) \pm g(x) = \sum_0^\infty (a_n \pm b_n)(x - x_0)^n, \tag{13}$$

and, with $z = x - x_0$ for brevity,

$$f(x)g(x) = \left(\sum_0^\infty a_n z^n \right) \left(\sum_0^\infty b_n z^n \right)$$

$$= (a_0 + a_1 z + \cdots)(b_0 + b_1 z + \cdots)$$

$$= a_0 \left(b_0 + b_1 z + b_2 z^2 + \cdots \right) + a_1 z \left(b_0 + b_1 z + b_2 z^2 + \cdots \right)$$

$$+ a_2 z^2 \left(b_0 + b_1 z + b_2 z^2 + \cdots \right) + \cdots$$

$$= a_0 b_0 + (a_0 b_1 + a_1 b_0)\, z + (a_0 b_2 + a_1 b_1 + a_2 b_0) z^2 + \cdots$$

$$- \sum_0^\infty (a_0 b_n + a_1 b_{n-1} + \cdots + a_n b_0)\, z^n \tag{14}$$

within I. The series on the right-hand side of (14) is known as the **Cauchy product** of the two series. Of course, if the two convergence intervals have different radii, then the common interval means the smaller of the two.

In summary, we see that convergent power series can be manipulated in essentially the same way as if they were finite-degree polynomials.

8.2.3 Taylor series.   The last items to address, before coming to the power series method of solution of differential equations, are Taylor series and analyticity. Recall from the calculus that the **Taylor series** of a given function $f(x)$ about a chosen point x_0, which we denote here as TS $f|_{x_0}$, is defined as the infinite series

$$
\begin{aligned}
\text{TS } f|_{x_0} &= f(x_0) + \frac{f'(x_0)}{1!}(x - x_0) + \frac{f''(x_0)}{2!}(x - x_0)^2 + \cdots \\
&= \sum_0^\infty \frac{f^{(n)}(x_0)}{n!}(x - x_0)^n,
\end{aligned}
\tag{15}
$$

where $0! = 1$. The purpose of Taylor series is to represent the given function f,[1] so the fundamental question is: Does it? Does the Taylor series really converge to $f(x)$ on some x interval, in which case we can write, in place of (15),

$$
f(x) = \sum_0^\infty \frac{f^{(n)}(x_0)}{n!}(x - x_0)^n?
\tag{16}
$$

For that to be the case three conditions must be met:

(i) First, f must *have* a Taylor series (15) about that point. Namely, f must be infinitely differentiable at x_0 so that all of the coefficients $f^{(n)}(x_0)/n!$ in (15) exist.

(ii) Second, the resulting series in (15) must *converge* in some interval $|x - x_0| < R$, where $R > 0$.

(iii) Third, the sum of the Taylor series must *equal* f in the interval, so that the Taylor series represents f on that interval—which, after all, is our objective.

The third condition might seem strange, for how could the Taylor series of $f(x)$ converge, but to something other than $f(x)$? Let us see.

EXAMPLE 5
Consider the Taylor series of

$$
f(x) = \begin{cases} e^{-1/x^2}, & x \neq 0 \\ 0, & x = 0 \end{cases}
\tag{17}
$$

---

[1] An expression is said to *represent* f on an x interval I if it gives the value $f(x)$ at each x in I.

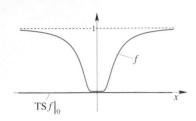

FIGURE 2
Graph of f in Example 5.

about $x = 0$.[1] (See Fig. 2.) It can be shown (Exercise 6) that $f'(0) = f''(0) = \cdots = 0$, so that

$$\text{TS } f|_0 = 0 + 0x + 0x^2 + \cdots. \tag{18}$$

Surely the latter series converges for all x, but its sum function is identically zero, so it is not equal to the function $f(x)$ defined in (17). Of course the sum function and $f(x)$ do agree at the point of expansion, $x = 0$, as will always be true, but that is not good enough, and we conclude that the Taylor series of f, about $x = 0$, does not represent f on *any* nonzero interval I even though it converges for all x!

COMMENT.   The graphs of f and TS $f|_0$ are so close near the origin (Fig. 2) that it *looks* like they coincide, but they do not; $\exp(-1/x^2)$ is nonzero for all $x \neq 0$.   ∎

However, such examples rarely occur in applications. Typically, if f is infinitely differentiable at some point x_0 so that it *has* a Taylor series about x_0, then that Taylor series will converge in some nonzero interval $|x - x_0| < R$, and its sum function will indeed be $f(x)$ over that interval. Thus, typically, if f is infinitely differentiable at x_0, then its Taylor series will *represent* it over some nonzero interval centered at x_0. For instance, such is the case for each of the following Taylor series (about $x = 0$), which we recall from the calculus:

$$e^x = 1 + \frac{1}{1!}x + \frac{1}{2!}x^2 + \frac{1}{3!}x^3 + \cdots, \qquad (|x| < \infty) \tag{19}$$

$$\sin x = x - \frac{1}{3!}x^3 + \frac{1}{5!}x^5 - \cdots, \qquad (|x| < \infty) \tag{20}$$

$$\cos x = 1 - \frac{1}{2!}x^2 + \frac{1}{4!}x^4 - \cdots, \qquad (|x| < \infty) \tag{21}$$

$$\frac{1}{1 - x} = 1 + x + x^2 + x^3 + \cdots. \qquad (|x| < 1) \tag{22}$$

The Taylor series in (22) is well known as the **geometric series**.

EXAMPLE 6
To obtain a more graphic understanding of Taylor series consider the Taylor series of e^{-x} about $x = 0$,

$$e^{-x} = 1 - \frac{1}{1!}x + \frac{1}{2!}x^2 - \frac{1}{3!}x^3 + \cdots, \tag{23}$$

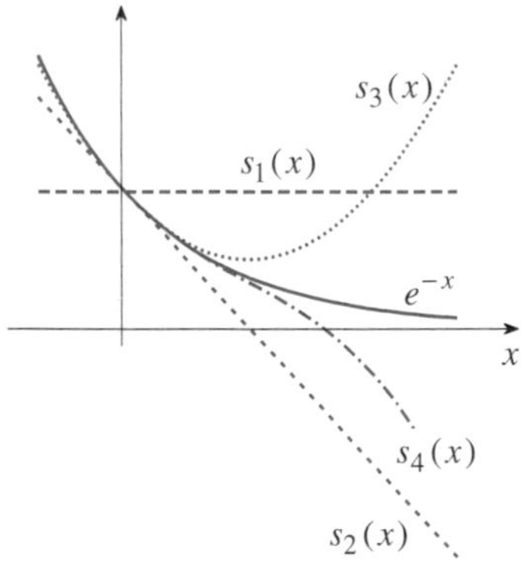

obtained from (15) or, more readily, by replacing each x in (19) by $-x$. Specifically, let us examine the convergence of the right-hand side of (23) to e^{-x}, graphically.

The partial sums of the series are $s_1(x) = 1$, $s_2(x) = 1 - x$, $s_3(x) = 1 - x + x^2/2$, and so on, and the first few are shown in Fig. 3. The first partial sum $s_1(x)$ matches the value of e^{-x} at $x = 0$ (i.e., the point of expansion); $s_2(x)$ matches

FIGURE 3
Convergence of the sequence of partial sums to e^{-x}.

[1]Why do we need to define $f(0) = 0$ separately? Does not $e^{-1/x^2} = e^{-1/0} = e^{-\infty} = 0$ at $x = 0$? No, $e^{-1/0}$ is simply undefined because $1/0$ is undefined. Of course we could define $f(0)$ to be any value we wish; we chose 0 in this example so that the resulting function would be continuous at $x = 0$.

both the value of e^{-x} and its slope at $x = 0$; $s_3(x)$ matches the value of e^{-x}, its slope, and its second derivative at $x = 0$; and so on. It's true that for each fixed n the discrepancy between $s_n(x)$ and e^{-x} becomes infinite as $x \to \infty$, but that fact is irrelevant insofar as convergence is concerned because by the convergence of $s_n(x)$ to e^{-x} we mean that $s_n(x)$ tends to e^{-x}, at a *fixed value of x*, as $n \to \infty$.

Just as a rodeo rider endeavors to stay on the horse, we can imagine the polynomials $s_1(x)$, $s_2(x)$, ... endeavoring to "stay on the function": $s_2(x)$ does a better job than $s_1(x)$, $s_3(x)$ does a better job than $s_2(x)$, and so on, but eventually , as x increases, they all "bite the dust." ∎

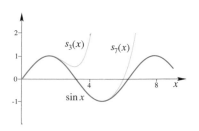

FIGURE 4
Representative partial sums in (20).

We have presented a similar plot for the Taylor series of $\sin x$, given by (20), in Fig. 4.

What is the connection between power series and Taylor series? By virtue of its form, every Taylor series is a power series. The inverse, that every convergent power series is a Taylor series, is not as obvious; it is the Taylor series of its sum function (i.e., the function to which the series converges).

THEOREM 8.2.4

Power Series and Taylor Series
Every Taylor series is a power series. Further, if a power series converges on some x interval, then it is the Taylor series of its sum function on that interval.

Proof. To prove the latter, suppose a power series $\sum_{n=0}^{\infty} a_n(x - x_0)^n$ converges to $f(x)$ on an interval I, so

$$f(x) = \sum_{n=0}^{\infty} a_n(x - x_0)^n = a_0 + a_1(x - x_0) + a_2(x - x_0)^2 + \cdots \quad (24)$$

on I. Putting $x = x_0$ in (24) gives $f(x_0) = a_0$. Next, differentiate (24) termwise [which is permissible according to Theorem 8.2.3(a), the resulting power series also being convergent on I], obtaining

$$f'(x) = \sum_{n=1}^{\infty} n a_n(x - x_0)^{n-1} = a_1 + 2a_2(x - x_0) + \cdots , \quad (25)$$

from which it follows, by setting $x = x_0$, that $f'(x_0) = a_1$. If we proceed in this manner, we find that $f(x_0) = a_0$, $f'(x_0) = a_1$, $f''(x_0)/2 = a_2, \ldots, f^{(n)}(x_0)/n! = a_n$, so the power series (24) is indeed the Taylor series of f. ∎

8.2.4 Analyticity. Finally, we introduce the concept of analyticity, which will be needed when we study power series solutions of differential equations in the remainder of this chapter.

A function f is said to be **analytic** *at a given point x_0 if its Taylor series about x_0 represents f on some nonzero interval $|x - x_0| < R$.* If f is not analytic at x_0 then it is said to be **singular** there.

Observe that there is a subtle difficulty here. We know how to test a given Taylor series for convergence since a Taylor series is a power series and power series convergence is addressed in Theorem 8.2.2. But how can we determine if the sum function is the same as the original function f? To fully understand analyticity one needs to study a more advanced topic known as complex variable theory.[1] However, the cases where the Taylor series of f converges, but not to f, are rather pathological and will not occur in the present chapter, so it will suffice to understand analyticity at x_0 to correspond to the convergence of the Taylor series in some nonzero interval about x_0. In fact, it is also exceptional for f to have a Taylor series about a point (i.e., be infinitely differentiable at that point) and to have that Taylor series fail to converge in some nonzero interval about x_0. Thus, *as a rule of thumb, we will test a function for analyticity at a given point simply by seeing if it is infinitely differentiable at that point.*

Most functions encountered in applications are analytic for all x, or for all x with the exception of one or more points called **singular points** of f. (Of course, the points are not singular, the function is.) For instance, polynomial functions, $\sin x$, $\cos x$, e^x, and e^{-x} are analytic for all x. On the other hand, $f(x) = 1/(x-1)$ is analytic for all x except $x = 1$, where f and all of its derivatives are undefined; they fail to exist. The function $f(x) = \tan x = \sin x / \cos x$ is analytic for all x except $x = n\pi/2$ ($n = \pm1, \pm3, \ldots$), where it is undefined because $\cos x$ vanishes in the denominator.

The function

$$f(x) = |x|^{5/2} \tag{26}$$

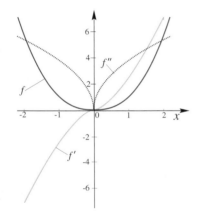

FIGURE 5
f, f', f'' for $f(x)$
given by (26).

is less obvious. To avoid the bothersome absolute value signs let it suffice to discuss (26) only for $x \geq 0$ (although we've plotted it in Fig. 5 for $x < 0$ as well), in which case $f(x) = x^{5/2}$. Then $f'(x) = \frac{5}{2}x^{3/2}$, $f''(x) = \frac{15}{4}x^{1/2}$, $f'''(x) = \frac{15}{8}x^{-1/2}$, $f''''(x) = -\frac{15}{16}x^{-3/2}$, and so on. We see from this list that $f(x)$ is infinitely differentiable for every $x > 0$, but that at $x = 0$ it is only twice differentiable since x to a negative power is undefined at $x = 0$. That is, although we cannot see it, visually, from a graph of f (Fig. 5), there is a "latent" problem at $x = 0$, which surfaces only through repeated differentiation. Thus, $f(x) = |x|^{5/2}$ is analytic for all x except for $x = 0$, at which point it is singular.

Closure. We have reviewed infinite series, power series, and Taylor series. The connection between power series and Taylor series is indicated in Theorem 8.2.4: every Taylor series is a power series, and every power series is the Taylor series of its sum function within its interval of convergence. We defined f as being analytic at x_0 if it admits a Taylor series TS $f|_{x_0}$ that converges to $f(x)$ on some nonzero interval $|x - x_0| < R$. Typically, functions are analytic for all x or for all x except for one or more singular points. For instance, $\sin x$, e^x, and $1/(1+x^2)$ are analytic for all x but $1/(1-x)$ is analytic for all $x \neq 1$ and is singular at $x = 1$.

---

[1]See, for example, Chapter 21 of M. D. Greenberg, *Advanced Engineering Mathematics*, 2nd ed. (Upper Saddle River, NJ: Prentice Hall, 1998).

Of special importance, in Sections 8.3 and 8.4, will be Theorem 8.2.3 on the manipulation of power series, and the concepts of radius of convergence and analyticity.

Maple.   We can obtain Taylor series by using the **taylor** command. For instance, to obtain the Taylor series of e^{-x} about $x = 1$, up to and including terms of order 3, use the command

```
taylor(exp(-x),x=1,4);
```

The result is

$$e^{-1} - e^{-1}(x-1) + \frac{1}{2}e^{-1}(x-1)^2 - \frac{1}{6}e^{-1}(x-1)^3 + O((x-1)^4)$$

where the $O((x-1)^4)$ signifies all of the subsequent terms, which are of fourth order and higher.

EXERCISES 8.2

1. Use (7a) or (7b) to determine the radius of convergence R of the given power series, and its interval of convergence.

(**a**) $\displaystyle\sum_0^\infty nx^n$

(**b**) $\displaystyle\sum_0^\infty (-1)^n n^{1000} x^n$

(**c**) $\displaystyle\sum_5^\infty e^n x^n$

(**d**) $\displaystyle\sum_0^{1000} n! \, x^n$

(**e**) $\displaystyle\sum_0^\infty \left(\frac{x+3}{2}\right)^n$

(**f**) $\displaystyle\sum_2^\infty (n-1)^3(x-5)^n$

(**g**) $\displaystyle\sum_1^\infty \frac{n^{50}}{n!}(x+7)^n$

(**h**) $\displaystyle\sum_2^\infty (\ln n)^{n+1}(x-2)^n$

(**i**) $\displaystyle\sum_3^\infty \frac{(-1)^n}{4^n}(x+2)^{3n}$

(**j**) $\displaystyle\sum_0^\infty (-1)^n \frac{n}{2^n}(x-5)^{2n}$

(**k**) $\displaystyle\sum_0^\infty \frac{n^6}{3^n+n}(x+4)^{8n+1}$

(**l**) $\displaystyle\sum_0^\infty \frac{(-1)^{n+1}}{4^n+1}(x-3)^{2n+1}$

2. Work out the Taylor series of the given function, about the given point x_0, and use (7a) or (7b) to determine its radius of convergence.

(**a**) e^x, $\quad x_0 = 1$

(**b**) e^{-x}, $\quad x_0 = -2$

(**c**) $\sin x$, $\quad x_0 = \pi$

(**d**) $\sin x$, $\quad x_0 = \pi/2$

(**e**) $\cos x$, $\quad x_0 = \pi/2$

(**f**) $\cos(x-2)$, $\quad x_0 = 2$

(**g**) $1/x$, $\quad x_0 = 4$

(**h**) $\ln x$, $\quad x_0 = 1$

(**i**) x^2, $\quad x_0 = 3$

(**j**) $2x^3 - 4$, $\quad x_0 = 0$

(**k**) $1/x^2$, $\quad x_0 = 3$

(**l**) $\dfrac{1}{1-x^{10}}$, $\quad x_0 = 0$

(**m**) $1/x^2$, $\quad x_0 = 3$

(**n**) $\sin(3x^{10})$, $\quad x_0 = 0$

3. Use *Maple* to obtain the first six nonzero terms in the Taylor series expansion of the given function f, about the given point x_0, and obtain a computer plot of f and the partial sums $s_2(x)$, $s_4(x)$, and $s_6(x)$ over the given interval I.

(**a**) $f(x) = e^{-x}$, $\quad x_0 = 0$, $\quad I : 0 < x < 4$

(**b**) $f(x) = \sin x$, $\quad x_0 = 0$, $\quad I : 0 < x < 10$

(**c**) $f(x) = \ln x$, $\quad x_0 = 1$, $\quad I : 0 < x < 2$

(**d**) $f(x) = 1/(1-x)$, $\quad x_0 = 0$, $\quad I : -1 < x < 1$

(**e**) $f(x) = 1/x$, $\quad x_0 = 2$, $\quad I : 0 < x < 4$

(**f**) $f(x) = 1/(1+x^2)$, $\quad x_0 = 0$, $\quad I : -1 < x < 1$

(**g**) $f(x) = 4/(4+x+x^2)$, $\quad x_0 = 0$, $\quad I : -1.3 < x < 0.36$

4. (*Geometric series*)

(**a**) Show that

$$\boxed{\frac{1}{1-x} = 1 + x + x^2 + \cdots + x^{n-1} + \frac{x^n}{1-x}} \qquad (4.1)$$

is an *identity* for all $x \neq 1$ and any positive integer n, by multiplying through by $1 - x$ (which is nonzero since $x \neq 1$) and simplifying.

(b) The identity (4.1) can be used to study the Taylor series known as the **geometric series** $\sum_{k=0}^{\infty} x^k$ since, according to (4.1), its partial sum $s_n(x)$ is

$$s_n(x) = \sum_{k=0}^{n-1} x^k = \frac{1 - x^n}{1 - x}. \qquad (x \neq 1) \qquad (4.2)$$

Show, from (4.2), that the sequence $s_n(x)$ converges, as $n \to \infty$, for $|x| < 1$, and diverges for $|x| > 1$, and that

$$\boxed{\frac{1}{1 - x} = \sum_{0}^{\infty} x^n, \qquad (|x| < 1)} \qquad (4.3)$$

as was given in (22).

(c) Determine, by any means, the convergence or divergence of the geometric series for the points at the ends of the interval of convergence, $x = \pm 1$.

NOTE: The formula (4.2) is striking because it reduces $s_n(x)$ to the *closed form* $(1 - x^n)/(1 - x)$, direct examination of which gives not only the interval of convergence but also the sum function $1/(1 - x)$. It is rare that one can reduce $s_n(x)$ to closed form.

5. (a) Derive the Taylor series of $1/(x - 1)$ about $x = 4$ using the Taylor series formula (15).

(b) Show that the same result is obtained (more readily) by writing

$$\frac{1}{x - 1} = \frac{1}{3 + (x - 4)} = \frac{1}{3} \frac{1}{1 + \dfrac{x - 4}{3}} \qquad (5.1)$$

and using the geometric series formula (4.3) from Exercise 4, with "x"$= -(x - 4)/3$. Further, deduce the

x interval of convergence of the result from the convergence condition "$|x| < 1$" in (4.3).

6. In Example 5 we stated that $f'(0) = f''(0) = \cdots = 0$. Verify that

(a) $f'(0) = 0$ **(b)** $f''(0) = 0$ **(c)** $f'''(0) = 0$

HINT: It will be necessary to fall back on the limit-of-the-difference-quotient definition of derivative.

7. (*Cesàro summability*) Although (5) gives the usual definition of infinite series, it is not the only possible one nor the only one used. For example, according to **Cesàro summability**, which is especially useful in the theory of Fourier series, one defines

$$\sum_{1}^{\infty} a_n \equiv \lim_{N \to \infty} \frac{s_1 + s_2 + \cdots + s_N}{N}, \qquad (7.1)$$

that is, the limit of the arithmetic means of the partial sums. It can be shown that if a series converges to s according to "ordinary convergence" [equation (5)], then it will also converge to the same value in the Cesàro sense. Yet, there are series that diverge in the ordinary sense but that converge in the Cesàro sense. Show that for the geometric series (see Exercise 4),

$$\frac{s_1 + s_2 + \cdots + s_N}{N} = \frac{1}{1 - x} - \frac{x}{N} \frac{1 - x^N}{(1 - x)^2} \qquad (7.2)$$

for all $x \neq 1$, and use that result to show that the Cesàro definition gives divergence for all $|x| > 1$ and for $x = 1$, and convergence for $|x| < 1$, as does ordinary convergence, but that for $x = -1$ it gives convergence to $1/2$, whereas according to ordinary convergence the series diverges for $x = -1$.

8.3 POWER SERIES SOLUTIONS

8.3.1 Theorem and applications.   As a theoretical basis for seeking power series solutions, the following fundamental theorem often suffices.

THEOREM 8.3.1

Power Series Solution

If $p(x)$ and $q(x)$ are analytic at x_0, then every solution of

$$y'' + p(x)y' + q(x)y = 0 \qquad (1)$$

is too, and can therefore be found in the form

$$y(x) = a_0 + a_1(x - x_0) + a_2(x - x_0)^2 + \cdots = \sum_0^\infty a_n(x - x_0)^n. \qquad (2)$$

Further, the radius of convergence of every solution (2) is at least as large as the smaller of the radii of convergence of TS $p|_{x_0}$ and TS $q|_{x_0}$.

The point is that if $p(x)$ and $q(x)$ are not constants and if (1) is not of Cauchy-Euler type, then it is very unlikely that we can find closed-form solutions. Turning to power series, the question is: Does (1) *admit* power series solutions, namely, in the form given by (2)? Theorem 8.3.1 assures us that it does—if both $p(x)$ and $q(x)$ are analytic at x_0, where x_0 is any point that we choose, about which to expand. If initial conditions are prescribed at some point then that point makes an especially convenient choice for x_0 because then most of the terms in (2) drop out when we apply the initial condition.

Further, the theorem gives information about the interval of validity of the power series solution (2).

Although we will not prove Theorem 8.3.1, we can at least suggest that its claim is plausible, as follows. Since $p(x)$ and $q(x)$ are, by assumption, analytic at the chosen point x_0, they can be represented, in some neighborhood of x_0, by their Taylor series about x_0, so we can write (1) as

$$y'' + \left[p(x_0) + p'(x_0)(x - x_0) + \cdots \right] y' + \left[q(x_0) + q'(x_0)(x - x_0) + \cdots \right] y = 0. \quad (3)$$

Locally, near x_0, if we keep only the leading terms of those series we can approximate (3) as

$$y'' + p(x_0)y' + q(x_0)y = 0. \qquad (4)$$

Since (4) is of constant-coefficient type every solution of (4) is either an exponential or x times an exponential, and is therefore analytic and representable in the form (2), as claimed. Again, this argument is intended only as heuristic.

In many applications $p(x)$ and $q(x)$ are **rational functions**, that is, one polynomial in x divided by another. In that case we can easily determine the radii of convergence of TS $p|_{x_0}$ and TS $q|_{x_0}$. Let

$$p(x) = \frac{N(x)}{D(x)} \qquad (5)$$

denote any rational function, where the numerator and denominator polynomials are $N(x)$ and $D(x)$, respectively, and where any common factors have been canceled. And let the zeros of $D(x)$ be the points $z_1, \ldots, z_k$ in the complex plane; that is, they are the solutions of $D(x) = 0$. We say that $z_1, \ldots, z_k$ are **singular points** of p. Then it is known from complex variable theory[1] that TS $p|_{x_0}$ will converge in $|x - x_0| < R$, where R is the distance from the point x_0 to the nearest singular point of p. Similarly for q.

---

[1] Ibid, Chapter 24.

(a)

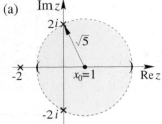

(b)

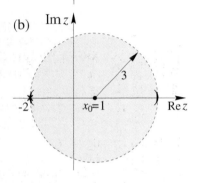

FIGURE 1
Intervals of convergence of
TS $p|_{x_0}$ and TS $q|_{x_0}$, respectively.

EXAMPLE 1
Given the differential equation

$$(x + 2)\left(x^2 + 4\right) y'' + \left(x^2 + 1\right) y' - \left(x^2 + 4\right) y = 0, \tag{6}$$

we first divide by $(x + 2)\left(x^2 + 4\right)$ to obtain the form (1):

$$y'' + \frac{x^2 + 1}{(x + 2)\left(x^2 + 4\right)} y' - \frac{1}{x + 2} y = 0. \tag{7}$$

Thus, we identify

$$p(x) = \frac{x^2 + 1}{(x + 2)\left(x^2 + 4\right)} \quad \text{and} \quad q(x) = -\frac{1}{x + 2} \tag{8}$$

so p has singular points at -2 and at $\pm 2i$, and q has a singular point only at -2. If we choose $x_0 = 1$, say, then we see from Fig. 1a that TS $p|_{x_0}$ converges in $|x - 1| < \sqrt{5}$ (i.e., in $1 - \sqrt{5} < x < 1 + \sqrt{5}$) and from Fig. 1b that TS $q|_{x_0}$ converges in $|x - 1| < 3$ (i.e., in $-2 < x < 4$). Then Theorem 8.3.1 assures us that two LI (linearly independent) solutions to (7) can be found, about $x_0 = 1$, with radii of convergence at least as large as $\sqrt{5}$, $\sqrt{5}$ being the smaller of the two radii of convergence. ∎

In the examples to follow we show how to find such power series solutions.

EXAMPLE 2
Solve

$$y'' + y = 0 \tag{9}$$

by the power series method. Of course, (9) is elementary. We know the solution and do not need the power series method to find it. Let us use this equation nevertheless, as a first example, to illustrate the method.

We can choose the point of expansion x_0 in (2) as any point at which both $p(x)$ and $q(x)$ are analytic. In the present example, $p(x) = 0$ and $q(x) = 1$ are analytic for all x, so we can choose the point of expansion x_0 to be whatever we like. Let $x_0 = 0$, for simplicity. Then Theorem 8.3.1 assures us that all solutions of (9) can be found in the form

$$y(x) = \sum_0^\infty a_n x^n, \tag{10}$$

which series will have infinite radii of convergence. Within that (infinite) interval of convergence we have, from Theorem 8.2.3(a),[1]

$$y'(x) = \sum_1^\infty n a_n x^{n-1}, \tag{11}$$

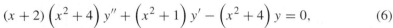

[1]Whether or not we leave the lower summation limit in (11) as 0, as in (10), doesn't matter since the coefficient na_n in (11) is 0 for $n = 0$ anyhow. The same point is true for (12).

$$y''(x) = \sum_{2}^{\infty} n(n-1)a_n x^{n-2}. \tag{12}$$

Putting (10) and (12) into (9) gives

$$\sum_{n=2}^{\infty} n(n-1)a_n x^{n-2} + \sum_{n=0}^{\infty} a_n x^n = 0. \tag{13}$$

We would like to combine the two sums, but to do that we need the exponents of x to be the same, whereas they are $n - 2$ and n. To have the same exponents, let us set $n - 2 = m$ in the first sum, just as one might make a change of variables in an integral. In the integral case, one would then need to change the integration limits, if necessary, consistent with the change of variables; here, we need to do that with the summation limits. With $n - 2 = m$, $n = \infty$ corresponds to $m = \infty$ and $n = 2$ corresponds to $m = 0$ so (13) becomes

$$\sum_{m=0}^{\infty} (m+2)(m+1)a_{m+2} x^m + \sum_{n=0}^{\infty} a_n x^n = 0. \tag{14}$$

Next, and we shall explain this step in a moment, let $m = n$ in the first sum in (14):

$$\sum_{n=0}^{\infty} (n+2)(n+1)a_{n+2} x^n + \sum_{n=0}^{\infty} a_n x^n = 0 \tag{15}$$

or, with the help of Theorem 8.2.3(b),

$$\sum_{n=0}^{\infty} \left[(n+2)(n+1)a_{n+2} + a_n\right] x^n = 0. \tag{16}$$

Finally, it follows from Theorem 8.2.3(c) that each coefficient in (16) must be zero:

$$(n+2)(n+1)a_{n+2} + a_n = 0. \qquad (n = 0, 1, 2, \ldots) \tag{17}$$

Before using (17), let us explain our setting $m = n$ in (14) since that step might seem to contradict the preceding change of variables $n - 2 = m$. The point to appreciate is that m in the first sum in (14) is a *dummy index* just as t is a *dummy variable* in $\int_0^1 t^2 \, dt$. Just as $\int_0^1 t^2 \, dt = \int_0^1 r^2 \, dr = \int_0^1 x^2 \, dx = \cdots = \frac{1}{3}$, the sums in (14) are insensitive to whether the dummy index is m or n; that is,

$$\sum_{m=0}^{\infty} (m+2)(m+1)a_{m+2} x^m = 2a_2 + 6a_3 x + 12a_4 x^2 + \cdots,$$

and

$$\sum_{n=0}^{\infty} (n+2)(n+1)a_{n+2} x^n = 2a_2 + 6a_3 x + 12a_4 x^2 + \cdots$$

are identical, even though the summation indices are different.

Equation (17) is known as a **recursion** (or **recurrence**) **formula** on the unknown coefficients since it gives us the nth coefficient in terms of preceding ones. Specifically,

$$a_{n+2} = -\frac{1}{(n+2)(n+1)}a_n, \qquad (n = 0, 1, 2, \ldots) \qquad (18)$$

so that

$$n = 0: \quad a_2 = -\frac{1}{(2)(1)}a_0,$$

$$n = 1: \quad a_3 = -\frac{1}{(3)(2)}a_1,$$

$$n = 2: \quad a_4 = -\frac{1}{(4)(3)}a_2 = \frac{1}{(4)(3)(2)(1)}a_0 = \frac{1}{4!}a_0, \qquad (19)$$

$$n = 3: \quad a_5 = -\frac{1}{(5)(4)}a_3 = \frac{1}{5!}a_1,$$

and so on, where a_0 and a_1 remain arbitrary. Putting these results into (10) gives

$$y(x) = a_0 + a_1 x - \frac{1}{2!}a_0 x^2 - \frac{1}{3!}a_1 x^3 + \frac{1}{4!}a_0 x^4 + \frac{1}{5!}a_1 x^5 + \cdots$$

$$= a_0\left(1 - \frac{1}{2!}x^2 + \frac{1}{4!}x^4 - \cdots\right) + a_1\left(x - \frac{1}{3!}x^3 + \frac{1}{5!}x^5 - \cdots\right) \qquad (20)$$

or

$$y(x) = a_0 y_1(x) + a_1 y_2(x), \qquad (21)$$

where $y_1(x)$ and $y_2(x)$ are the series within the first and second pairs of parentheses in (20), respectively. From their series, we recognize $y_1(x)$ as $\cos x$ and $y_2(x)$ as $\sin x$ but, in general, we can't expect to identify the power series in terms of elementary functions. Thus, let us continue to call the two series "$y_1(x)$" and "$y_2(x)$."

It is not necessary to check the series for convergence because Theorem 8.3.1 guarantees that they will converge for all x. We should, however, check to see if y_1, y_2 are LI. To do so, it suffices to evaluate the Wronskian $W[y_1, y_2](x)$ at a single point, say $x = 0$:

$$W[y_1, y_2](x)|_{x=0} = \begin{vmatrix} y_1(0) & y_2(0) \\ y_1'(0) & y_2'(0) \end{vmatrix} = \begin{vmatrix} 1 & 0 \\ 0 & 1 \end{vmatrix} = 1, \qquad (22)$$

which is nonzero. It follows from that result and Theorem 6.3.3 that y_1, y_2 are LI on the entire x axis, so (21) is a general solution of (9) for all x. Actually, since there are only two functions it would have been simpler to apply Theorem 6.3.4: y_1, y_2 are LI because neither one is a constant multiple of the other.

COMMENT 1. To evaluate $y_1(x)$ or $y_2(x)$ at a given x, we need to add enough terms of the series to achieve the desired accuracy. For small values of x (i.e., for x's that are close to the point of expansion x_0, which in this case is 0) just a few terms may suffice. For example, the first four terms of $y_1(x)$ give $y_1(0.5) = 0.877582$, whereas the exact value is 0.877583 (to six decimal places). As x increases, more and more terms are needed for comparable accuracy. The situation is depicted graphically in Fig. 2, where we plot the partial sums s_3 and s_6, along with the sum function $y_1(x)$ (i.e., $\cos x$). Observe that the larger n is, the broader the x interval is over which the n-term approximation s_n stays close to the sum function. However, whereas $y_1(x)$ is oscillatory and remains between -1 and $+1$, $s_n(x)$ is a polynomial, and therefore it eventually tends to $+\infty$ or $-\infty$ as x increases ($-\infty$ if n is even and $+\infty$ if n is odd). Observe that if we do need to add a great many terms, then it is important to have an expression for the general term in the series. In this example it is not hard to establish that

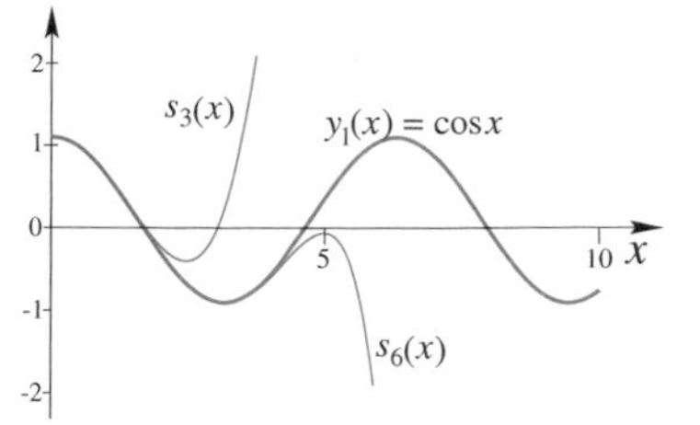

FIGURE 2
Partial sums of $y_1(x)$, compared with the sum function $y_1(x) = \cos x$.

$$y_1(x) = \sum_0^\infty (-1)^n \frac{x^{2n}}{(2n)!}, \qquad y_2(x) = \sum_0^\infty (-1)^n \frac{x^{2n+1}}{(2n+1)!}. \qquad (23)$$

COMMENT 2. We posed (9) without initial conditions. Suppose that we wish to impose the initial conditions $y(0) = 4$ and $y'(0) = -1$. Then, from (21),

$$\begin{aligned} y(0) &= \quad 4 = a_0 y_1(0) + a_1 y_2(0), \\ y'(0) &= -1 = a_0 y_1'(0) + a_1 y_2'(0). \end{aligned} \qquad (24)$$

From the series representations of y_1 and y_2 in (20), we see that $y_1(0) = 1$, $y_2(0) = 0$, $y_1'(0) = 0$, and $y_2'(0) = 1$, so we can solve (24) for a_0 and a_1: $a_0 - 4$ and $a_1 = -1$, hence the desired particular solution is $y(x) = 4y_1(x) - y_2(x)$, on $-\infty < x < \infty$. ∎

EXAMPLE 3
Solve the initial value problem

$$xy'' + y' + xy = 0, \qquad y(3) = 5, \quad y'(3) = 0 \qquad (25)$$

on the interval $3 \le x < \infty$. To get (25) into the standard form $y'' + p(x)y' + q(x)y = 0$, we divide by x (which is permissible since $x \ne 0$ on the interval of interest):

$$y'' + \frac{1}{x}y' + y = 0, \qquad (26)$$

so $p(x) = 1/x$ and $q(x) = 1$. These are analytic for all x except $x = 0$, where $p(x)$ is undefined. In particular, they are analytic at the initial point $x = 3$, so let us choose $x_0 = 3$ and seek $y(x)$ in the form

$$y(x) = \sum_0^\infty a_n(x-3)^n. \qquad (27)$$

To proceed we can use either (25) or (26). Since we are expanding each term in the differential equation about $x = 3$, we need to expand the coefficient x of y'' and of y if we use (25), or the coefficient $1/x$ if we use (26). The former is simpler since

$$x = 3 + (x - 3) \tag{28}$$

is merely a two-term Taylor series, whereas (Exercise 1)

$$\frac{1}{x} = \frac{1}{3} \sum_0^\infty \frac{(-1)^n}{3^n} (x - 3)^n \tag{29}$$

is an infinite series. Thus, let us use (25) rather than (26). Putting (27) and its derivatives and (28) into (25) gives

$$[3 + (x - 3)] \sum_2^\infty n(n - 1)a_n(x - 3)^{n-2} + \sum_1^\infty na_n(x - 3)^{n-1}$$

$$+ [3 + (x - 3)] \sum_0^\infty a_n(x - 3)^n = 0 \tag{30}$$

or, absorbing the $3 + (x - 3)$ terms into the series that they multiply and setting $t = x - 3$ for compactness,

$$\sum_2^\infty 3n(n - 1)a_n t^{n-2} + \sum_2^\infty n(n - 1)a_n t^{n-1}$$

$$+ \sum_1^\infty na_n t^{n-1} + \sum_0^\infty 3a_n t^n + \sum_0^\infty a_n t^{n+1} = 0. \tag{31}$$

To adjust all t exponents to n, let $n - 2 = m$ in the first sum, $n - 1 = m$ in the second and third, and $n + 1 = m$ in the last:

$$\sum_0^\infty 3(m + 2)(m + 1)a_{m+2} t^m + \sum_1^\infty (m + 1)m a_{m+1} t^m$$

$$+ \sum_0^\infty (m + 1)a_{m+1} t^m + \sum_0^\infty 3a_n t^n + \sum_1^\infty a_{m-1} t^m = 0. \tag{32}$$

Next, we change all of the m indices to n. Then we have t^n in each sum, but we cannot yet combine the five sums because the lower summation limits are not all the same; three are 0 and two are 1. We can handle that problem as follows. The lower limit in the second sum can be changed from 1 to 0 because the zeroth term is zero anyhow (due to the m factor). And the lower limit in the last sum can be changed to 0 if we agree that the a_{-1}, that occurs in the zeroth term, be zero by definition. Then (32) becomes

$$\sum_0^\infty 3(n + 2)(n + 1)a_{n+2} t^n + \sum_0^\infty (n + 1)n a_{n+1} t^n$$

$$+ \sum_{0}^{\infty}(n+1)a_{n+1}t^n + \sum_{0}^{\infty}3a_nt^n + \sum_{0}^{\infty}a_{n-1}t^n = 0 \qquad (33)$$

or

$$\sum_{0}^{\infty}\left[3(n+2)(n+1)a_{n+2} + (n+1)^2a_{n+1} + 3a_n + a_{n-1}\right]t^n = 0, \qquad (34)$$

with $a_{-1} \equiv 0$.

Setting the square-bracketed coefficients to zero for each n [Theorem 8.2.3(c)] then gives the recursion formula

$$3(n+2)(n+1)a_{n+2} + (n+1)^2a_{n+1} + 3a_n + a_{n-1} = 0$$

or

$$a_{n+2} = -\frac{n+1}{3(n+2)}a_{n+1} - \frac{1}{(n+2)(n+1)}a_n - \frac{1}{3(n+2)(n+1)}a_{n-1} \qquad (35)$$

for $n = 0, 1, 2, \ldots$. Thus,

$$n = 0: \quad a_2 = -\frac{1}{6}a_1 - \frac{1}{2}a_0 - \frac{1}{6}a_{-1} = -\frac{1}{6}a_1 - \frac{1}{2}a_0$$

$$n = 1: \quad a_3 = -\frac{2}{9}a_2 - \frac{1}{6}a_1 - \frac{1}{18}a_0$$

$$= -\frac{2}{9}\left(-\frac{1}{6}a_1 - \frac{1}{2}a_0\right) - \frac{1}{6}a_1 - \frac{1}{18}a_0 = -\frac{7}{54}a_1 + \frac{1}{18}a_0 \qquad (36)$$

$$n = 2: \quad a_4 = -\frac{1}{4}a_3 - \frac{1}{12}a_2 - \frac{1}{36}a_1$$

$$= -\frac{1}{4}\left(-\frac{7}{54}a_1 + \frac{1}{18}a_0\right) - \frac{1}{12}\left(-\frac{1}{6}a_1 - \frac{1}{2}a_0\right) - \frac{1}{36}a_1$$

$$= \frac{1}{54}a_1 + \frac{1}{36}a_0,$$

and so on, where a_0 and a_1 remain arbitrary. Putting these expressions for the a_n's back into (27) then gives

$$y(x) = a_0 + a_1(x-3) + \left(-\frac{1}{6}a_1 - \frac{1}{2}a_0\right)(x-3)^2 + \left(-\frac{7}{54}a_1 + \frac{1}{18}a_0\right)(x-3)^3$$

$$+ \left(\frac{1}{54}a_1 + \frac{1}{36}a_0\right)(x-3)^4 + \cdots$$

$$= a_0\left[1 - \frac{1}{2}(x-3)^2 + \frac{1}{18}(x-3)^3 + \frac{1}{36}(x-3)^4 + \cdots\right]$$

$$+ a_1\left[(x-3) - \frac{1}{6}(x-3)^2 - \frac{7}{54}(x-3)^3 + \frac{1}{54}(x-3)^4 + \cdots\right]$$

$$= a_0 y_1(x) + a_1 y_2(x), \qquad (37)$$

where $y_1(x)$, $y_2(x)$ are the functions represented by the bracketed series. To test y_1, y_2 for linear independence it is simplest to use Theorem 6.3.4: y_1, y_2 are LI because neither one is a constant multiple of the other. Thus, $y(x) = a_0 y_1(x) + a_1 y_2(x)$ is a general solution of $xy'' + y' + xy = 0$.

It is simple to impose the initial conditions because the expansions are about the initial point $x = 3$:

$$y(3) = 5 = a_0 y_1(3) + a_1 y_2(3) = a_0(1) + a_1(0),$$
$$y'(3) = 0 = a_0 y_1'(3) + a_1 y_2'(3) = a_0(0) + a_1(1),$$

(38)

so $a_0 = 5$ and $a_1 = 0$, hence

$$y(x) = 5y_1(x) = 5\left[1 - \frac{1}{2}(x-3)^2 + \frac{1}{18}(x-3)^3 + \frac{1}{36}(x-3)^4 + \cdots\right]$$

(39)

is the desired solution to the initial value problem (25).

COMMENT 1. Recall that Theorem 8.3.1 guaranteed that the power series solution would have a radius of convergence R at least as large as 3, namely, the distance from the center of the expansion ($x_0 = 3$) to the singularity in $1/x$ at $x = 0$. For comparison, let us determine R directly from our results. In this example it is difficult to obtain a general expression for a_n. (Indeed, we didn't attempt to; we were content to develop the first several terms of the series, knowing that we could obtain as many more as we desire, from the recursion formula.) Can we obtain R without an explicit expression for a_n? Yes, we can use the recursion formula (35), which reveals that $a_{n+2} \sim -\frac{1}{3}a_{n+1}$ as $n \to \infty$ or, equivalently, $a_{n+1} \sim -\frac{1}{3}a_n$. Then, from (7a) in Section 8.2,

$$R = \frac{1}{\lim_{n\to\infty}\left|\dfrac{a_{n+1}}{a_n}\right|} = \frac{1}{\lim_{n\to\infty}\left|-\dfrac{1}{3}\right|} = \frac{1}{\dfrac{1}{3}} = 3.$$

(40)

Thus, if we were hoping to obtain the solution over the entire interval $3 \le x < \infty$ we are disappointed to find that the power series converges only over $0 < x < 6$, and hence only over the $3 \le x < 6$ part of the problem domain. Does this result mean that the solution is singular at $x = 6$ and cannot be continued beyond that point, or that it doesn't exist beyond $x = 6$? No, the convergence is simply being limited by the singularity in $p(x) = 1/x$ at $x = 0$, which lies outside of the problem domain $3 \le x < \infty$.

COMMENT 2. We have plotted the two LI solutions,

$$y_1(x) = 1 - \frac{1}{2}(x-3)^2 + \frac{1}{18}(x-3)^3 + \frac{1}{36}(x-3)^4 + \cdots,$$

(41)

$$y_2(x) = (x-3) - \frac{1}{6}(x-3)^2 - \frac{7}{54}(x-3)^3 + \frac{1}{54}(x-3)^4 + \cdots,$$

(42)

but only over the $3 < x < 6$ part of the problem domain since both series diverge for $x > 6$ (Fig. 3).

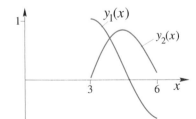

FIGURE 3
The LI solutions $y_1(x)$ and $y_2(x)$ given by (41) and (42).

COMMENT 3.   Actually, the equation (25),

$$xy'' + y' + xy = 0, \tag{43}$$

is well known; it is a Bessel equation. More generally the **Bessel equation of order ν** is

$$xy'' + xy' + (x^2 - \nu^2)y = 0, \tag{44}$$

where ν is a constant, so (43) is a Bessel equation of order zero. The equation (44) was studied systematically first by *Friedrich Wilhelm Bessel* (1784–1846) who was director of the astronomical observatory at Königsberg, in connection with the motion of the planets. We will return to this equation in Section 8.4.   ∎

8.3.2 Legendre's equation. (Optional).   The differential equation

$$\left(1 - x^2\right) y'' - 2xy' + \lambda y = 0, \tag{45}$$

where λ is a constant, is known as **Legendre's equation**, after the French mathematician *Adrien-Marie Legendre* (1752–1833). The latter is prominent in *potential theory*, namely, in problems governed by the **Laplace partial differential equation**.[1] In such applications the x interval of interest in (45) is $-1 < x < 1$ and one is interested in power series solutions of (45) about $x = 0$,

$$y(x) = \sum_{k=0}^{\infty} a_k x^k. \tag{46}$$

Since

$$p(x) = -\frac{2x}{1 - x^2} \quad \text{and} \quad q(x) = \frac{\lambda}{1 - x^2}$$

have singular points at $x = \pm 1$ (on the real x axis), it follows from Theorem 8.3.1 that we can find two LI solutions in the form (46), and that those power series will converge on the interval $-1 < x < 1$ at the least.

This time leaving the detailed steps for the exercises, we state that if we put (46) into (45) and proceed as in Examples 2 and 3 we obtain the recursion formula

$$a_{k+2} = \frac{k(k + 1) - \lambda}{(k + 1)(k + 2)} a_k. \quad (k = 0, 1, 2, \ldots) \tag{47}$$

Setting $k = 0, 1, 2, \ldots$, in turn, reveals that a_0 and a_1 are arbitrary, and that subsequent a_k's can be expressed alternately in terms of a_0 and a_1:

$$a_2 = -\tfrac{\lambda}{2}a_0,$$

$$a_3 = \tfrac{2-\lambda}{(2)(3)}a_1 = \tfrac{2-\lambda}{6}a_1, \tag{48}$$

$$a_4 = \tfrac{(2)(3)-\lambda}{(3)(4)}a_2 = \tfrac{6-\lambda}{12}\left(-\tfrac{\lambda}{2}a_0\right) = -\tfrac{(6-\lambda)\lambda}{24}a_0,$$

---

[1] See, for instance, M. D. Greenberg, *Advanced Engineering Mathematics*, 2nd ed. (Upper Saddle River, NJ: Prentice Hall, 1998), Section 20.3.3.

and so on, and we have the general solution of (45):

$$y(x) = a_0 \left[1 - \frac{\lambda}{2}x^2 - \frac{(6-\lambda)\lambda}{24}x^4 - \frac{(20-\lambda)(6-\lambda)\lambda}{720}x^6 - \cdots \right]$$

$$+ a_1 \left[x + \frac{2-\lambda}{6}x^3 + \frac{(12-\lambda)(2-\lambda)}{120}x^5 + \cdots \right]$$

$$= a_0 y_1(x) + a_1 y_2(x) \tag{49}$$

To determine the radii of convergence of the two series in (49) we can use the recursion formula (47) and Theorem 8.2.2, provided that we realize that the a_{k+2} on the left side of (47) is really the next coefficient after a_k, the "a_{k+1}" in Theorem 8.2.2, since every other term in each series is missing. Thus, (47) gives

$$\lim_{k \to \infty} \left| \frac{"a_{k+1}"}{a_k} \right| = \lim_{k \to \infty} \left| \frac{k(k+1)-\lambda}{(k+1)(k+2)} \right| = 1, \tag{50}$$

and it follows from Theorem 8.2.2 that $R = 1$, so each of the two series converges on $-1 < x < 1$.

In physical applications of the Legendre equation, such as finding the steady-state temperature distribution within a sphere subjected to a known temperature distribution on its surface, one needs solutions to (45) that are *bounded* on $-1 \le x \le 1$.[1] Thus, let us add boundedness boundary conditions and consider the boundary value problem

$$(1 - x^2)y'' - 2xy' + \lambda y = 0 \qquad (-1 \le x \le 1) \tag{51a}$$

$$y(-1), \ y(+1) \ \text{finite}, \tag{51b}$$

with the finiteness conditions added to force the solutions to be bounded on $-1 \le x \le 1$.

Recall that λ in (45) is a parameter. If we specify a value for λ, such as $\lambda = 3.172$, to pick a value at random, then we find that $y_1(x)$ tends to $-\infty$ as $x \to \pm 1$, and that $y_2(x)$ tends to $+\infty$ as $x \to -1$ and to $-\infty$ as $x \to +1$. Thus, no matter how we choose a_0 and a_1 in (49) we cannot obtain nontrivial solutions to (51).

However, if we set $\lambda = 6$, say, then (46) becomes

$$y(x) = a_0 \left(1 - 3x^2 \right) + a_1 \left(x - \frac{2}{3}x^3 - \frac{1}{5}x^5 - \cdots \right). \tag{52}$$

That is, the first series *terminates* because of the $6 - \lambda$ factor in all terms after the second, leaving the finite-degree polynomial $1 - 3x^2$, which of *course* is bounded on $-1 \le x \le 1$. The nonterminating series in (49) and (52) is unbounded, but we can set $a_1 = 0$ to eliminate it, thus obtaining the bounded solutions

$$y(x) = a_0 \left(1 - 3x^2 \right) \tag{53}$$

of (51) (for the case where $\lambda = 6$), where a_0 remains arbitrary.

---

[1] $F(x)$ being **bounded** on an interval I means that there exists a finite constant M, such that $|F(x)| \le M$, for all x in I. If $F(x)$ is not bounded then it is **unbounded**.

Table 1. The first five Legendre polynomials.

n	$\lambda = n(n+1)$	Polynomial Solution	Legendre Polynomial $P_n(x)$
0	0	1	$P_0(x) = 1$
1	2	x	$P_1(x) = x$
2	6	$1 - 3x^2$	$P_2(x) = \frac{1}{2}(3x^2 - 1)$
3	12	$x - \frac{5}{3}x^3$	$P_3(x) = \frac{1}{2}(5x^3 - 3x)$
4	20	$1 - 10x^2 + \frac{35}{3}x^4$	$P_4(x) = \frac{1}{8}(35x^4 - 30x^2 + 3)$

In fact, we can see from (47) that such termination will occur whenever λ is such that

$$\lambda = n(n+1) \tag{54}$$

for any integer $n = 0, 1, 2, \ldots$, for then one of the two series will terminate at $k = n$. For example, if $n = 2$ and $\lambda = 2(2+1) = 6$, then the $6 - \lambda$ factor in every term after the second, in the even-powered series, causes all those terms to vanish, so the series terminates as $1 - 3x^2$. Similarly, if $\lambda = n(n+1)$ where n is odd then the odd-powered series will terminate at $k = n$. For example, if $n = 3$ and $\lambda = 3(3+1) = 12$, then the odd-powered series terminates as $x - \frac{5}{3}x^3$. The nonterminating series in (52) is unbounded, but we can set $a_0 = 0$ to eliminate it, thus obtaining the bounded solution

$$y(x) = a_1\left(x - \frac{5}{3}x^3\right) \tag{55}$$

of (51) (for the case where $\lambda = 12$), where a_1 remains arbitrary.

The first five such λ's and their corresponding polynomial solutions of (51) are shown in the second and third columns of Table 1. These polynomial solutions can be scaled by any desired numerical factor [for instance, recall that a_0 was arbitrary in (53) and a_1 was arbitrary in (55)]. Scaling them to equal unity at $x = 1$, by convention, they become the so-called **Legendre polynomials**. Thus, the Legendre polynomial $P_n(x)$ is a polynomial solution of the Legendre equation

$$(1 - x^2)y'' - 2xy' + n(n+1)y = 0, \tag{56}$$

scaled so that $P_n(1) = 1$. In fact, it can be shown that they are given explicitly by the formula

$$P_n(x) = \frac{1}{2^n n!}\frac{d^n}{dx^n}\left[(x^2 - 1)^n\right], \tag{57}$$

which result is known as **Rodrigues's formula**. The first several are given in Table 1 and plotted in Fig. 4.

Various properties of these important functions are included among the exercises.

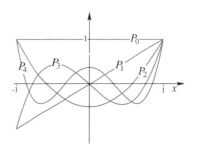

FIGURE 4
The first five Legendre polynomials.

Closure. The focus of this section is Theorem 8.3.1, which assures us that if $p(x)$ and $q(x)$ are analytic at a point x_0 then every solution of the equation $y'' + p(x)y' + q(x)y = 0$ can be found in the form of a power series expansion about x_0. Practically speaking, that means that two LI solutions, and hence a general solution, of the equation can be found by assuming a solution form

$$y(x) = \sum_{0}^{\infty} a_n (x - x_0)^n . \tag{58}$$

Putting (58) into the equation and equating coefficients of powers of $x - x_0$ on the two sides of the equation, using Theorem 8.2.3(a,b,c), we can solve for the a_n's, recursively, in terms of two of them that remain arbitrary.

Further, Theorem 8.3.1 tells us about the radii of convergence of our series solutions: they will be at least as large as the smaller of the radii of convergence of the Taylor series of $p(x)$ and $q(x)$ about the point x_0. In the best case both of those radii are infinite, for then both power series solutions will converge for all x. That case occurred in Example 2 and would occur for the equations

$$y'' + e^x y' - (\cos x) y = 0$$

and

$$y'' + x^3 y = 0,$$

for example, no matter what point is chosen for x_0, because e^x, $-\cos x$, and x^3 are analytic for all x. However, and this is a serious drawback, if one or both of the Taylor series of $p(x)$ and $q(x)$, about the chosen point x_0 (which will probably be the initial point if initial conditions are given), have finite radii of convergence, the smaller of the two being R, say, then the power series solutions will probably converge only on $|x - x_0| < R$, which interval might be only a part of the problem domain. This difficulty is discussed further in the optional Section 8.4.

Finally, in Section 8.3.2 we considered the well known Legendre equation (45). Expanding about $x = 0$, we obtained two LI power series solutions which converged on $-1 < x < 1$. For special values of the parameter λ we were able to obtain finite-degree polynomial solutions, the famous Legendre polynomials $P_n(x)$, that are bounded on $-1 \le x \le 1$.

Maple. The **dsolve** command can be used to obtain power series solutions by using the type = series option. For instance, to solve the problem

$$xy'' + y' + xy = 0; \qquad y(3) = 5, \; y'(3) = 0 \tag{59}$$

of Example 3, use the command

```
dsolve({x*diff(y(x),x,x)+diff(y(x),x)+x*y(x)=0,
y(3)=5,D(y)(3)=0},y(x),type=series);
```

and obtain

$$y(x) = 5 - \frac{5}{2}(x-3)^2 + \frac{5}{18}(x-3)^3 + \frac{5}{36}(x-3)^4 - \frac{1}{108}(x-3)^5 + O\left((x-3)^6\right).$$

This command does not yield the general term of the series but it does permit us to obtain as many terms as we like. For instance, if we desire the solution of (59) up to (but not including) terms of order eight (the default being six, as obtained above), enter

```
Order:=8:
dsolve({x*diff(y(x),x,x)+diff(y(x),x)+x*y(x)=0,
y(3)=5,D(y)(3)=0},y(x),type=series);
```

and obtain

$$y(x) = 5 - \frac{5}{2}(x-3)^2 + \frac{5}{18}(x-3)^3 + \frac{5}{36}(x-3)^4 - \frac{1}{108}(x-3)^5$$
$$- \frac{5}{972}(x-3)^6 + \frac{1}{1701}(x-3)^7 + O\left((x-3)^8\right)$$

Finally, suppose we desire a general solution to an equation, say $y'' + y = 0$, in the form of power series expansions about $x_0 = 2$, say, where no initial conditions are given. If we use the command

```
dsolve({diff(y(x),x,x)+y(x)=0,y(2)=a,D(y)(2)=b},y(x),
type=series);
```

we obtain

$$y(x) = a + b(x-2) - \frac{1}{2}a(x-2)^2 - \frac{1}{6}b(x-2)^3 + \frac{1}{24}a(x-2)^4$$
$$+ \frac{1}{120}b(x-2)^5 + O\left((x-2)^6\right).$$

That is,

$$y(x) = a\left[1 - \frac{1}{2}(x-2)^2 + \frac{1}{24}(x-2)^4 + O\left((x-2)^6\right)\right]$$
$$+ b\left[(x-2) - \frac{1}{6}(x-2)^3 + \frac{1}{120}(x-2)^5 + O\left((x-2)^6\right)\right], \tag{60}$$

where the "dummy initial conditions" a and b serve as the two arbitrary constants in the general solution (60). The purpose of including the dummy initial conditions $y(2) = a$ and $D(y)(2) = b$ was to tell the computer that the point of expansion is $x_0 = 2$, for if we omit them and use the command

```
dsolve(diff(y(x),x,x)+y(x)=0,y(x),type=series);
```

then we would obtain, as the default case, expansions about $x_0 = 0$.

Finally, to check Rodrigues's formula (57) against the entries in Table 1 let $n = 4$, say. The *Maple* commands

```
diff((x^2-1)^4,x,x,x,x)/(2^4*factorial(4));
simplify(%);
```

give $(35x^4 - 30x^2 + 3)/8$, as in Table 1.

EXERCISES 8.3

1. Derive the Taylor expansion (29).

2. Obtain the general solution by seeking $y(x)$ in the form of a power series about the designated point x_0, giving at least the first three nonvanishing terms of each series. Show that your results agree with the general solution obtained, alternatively (and more simply), by seeking a solution in the form $y(x) = e^{rx}$.

   (a) $y'' - 9y = 0$, $x_0 = 0$

   (b) $y'' - 9y = 0$, $x_0 = 2$

   (c) $y'' + y' = 0$, $x_0 = 0$

   (d) $y'' - y' = 0$, $x_0 = 3$

   (e) $y'' + y = 0$, $x_0 = -5$

   (f) $y'' - 3y' + 2y = 0$, $x_0 = 0$

3. In Exercise 2 we kept the differential equation simple (namely, of constant-coefficient type) so that you could also solve it analytically and in closed form, for comparison with your series solution. In this exercise we turn to nonconstant-coefficient equations which can be solved by the methods of this section but which do not admit simple closed form solutions of the type e^{rx}. In each case obtain a general solution by seeking $y(x)$ in the form of a power series about the designated point x_0, giving the first four nonvanishing terms of each series and determining the radii of convergence.

   (a) $y'' + xy = 0$, $x_0 = 0$

   (b) $y'' + xy = 0$, $x_0 = 1$

   (c) $y'' - xy = 0$, $x_0 = 0$

   (d) $y'' + xy' + y = 0$, $x_0 = 0$

   (e) $y'' + 3xy' - y = 0$, $x_0 = -2$

   (f) $y'' + e^{-x}y = 0$, $x_0 = 0$

   (g) $y'' - e^{-x}y = 0$, $x_0 = 2$

   (h) $y'' - xy' + 4y = 0$, $x_0 = 0$

   (i) $y'' - xy' + y = 0$, $x_0 = 5$

4. (a)–(i) Use *Maple* to solve the corresponding problem in Exercise 3. You need not determine the radius of convergence.

5. Consider the solution of the given differential equation in the form of power series about the given point x_0. On what x interval will those power series converge? HINT: You can determine the interval of convergence directly, from an examination of $p(x)$ and $q(x)$, without actually working out the power series solutions.

   (a) $y'' - 3y' + 6y = 0$; $x_0 = 0$

   (b) $y'' + 2y' + 7y = 0$; $x_0 = 6$

   (c) $y'' + 2y' + 3xy = 0$; $x_0 = -2$

   (d) $xy'' + y' + xy = 0$; $x_0 = 5$

   (e) $xy'' + y' - y = 0$; $x_0 = -10$

   (f) $(x + 2)y'' - (x + 2)y' + y = 0$; $x_0 = 0$

   (g) $(x^2 - 1)y'' + (x + 1)y' - (x - 1)y = 0$; $x_0 = 1/3$

   (h) $(x^2 - 1)y'' - (x + 1)y' + (x - 1)y = 0$; $x_0 = 4$

   (i) $(x^2 + 2x + 2)y'' + x^2y' - (2x + 7)y = 0$; $x_0 = 0$

   (j) $(x^2 + 2x + 5)y'' - y' + (x - 1)y = 0$; $x_0 = 1$

   (k) $(4x^2 + 1)y'' + 3xy' + y = 0$; $x_0 = 5$

   (l) $(x^2 - 4x)y'' + y' + y = 0$; $x_0 = 3$

   (m) $(x^2 + 4x)y'' + y' - y = 0$; $x_0 = 3$

   (n) $(x^3 + 1)y'' - xy' + y = 0$; $x_0 = 1$

   (o) $(x^4 - 16)y'' + x^3y' + xy = 0$; $x_0 = 3$

   (p) $(x^4 - 16)y'' + x^3y' + xy = 0$; $x_0 = -6$

6. (*Nonhomogeneous equations*)

   (a) Proceeding informally [since Theorem 8.3.1 is for the *homogeneous* equation (1)], find the general solution to the *nonhomogeneous* equation

   $$y'' + xy = e^x \qquad (6.1)$$

   by seeking y in the power series form

   $$y(x) = \sum_0^\infty a_n x^n, \qquad (6.2)$$

   and thus show that the general solution to (6.1) is

   $$y(x) = \left(\frac{1}{2!}x^2 + \frac{1}{3!}x^3 + \frac{1}{4!}x^4 - \frac{2}{5!}x^5 - \frac{3}{6!}x^6\right.$$
   $$\left. - \frac{4}{7!}x^7 - \cdots\right) + a_0\left(1 - \frac{1}{3!}x^3 + \frac{4}{6!}x^6 - \cdots\right)$$
   $$+ a_1\left(x - \frac{2}{4!}x^4 + \frac{10}{7!}x^7 - \cdots\right),$$
   $$(6.3)$$

   through terms of seventh order. The first series on the right-hand side of (6.3) is a particular solution of (6.1) and the other two series are LI homogeneous solutions.

   (b) Solve (6.1) by *Maple* instead, in series form, and verify that the *Maple* result agrees with (6.3).

EXERCISES FOR OPTIONAL SECTION 8.3.2

7. Use Rodrigues's formula (57) to work out $P_0(x)$, $P_1(x)$, $P_2(x)$, and $P_3(x)$.

8. (*Generating function*) If we regard $(1 - 2xr + r^2)^{-1/2}$ as a function of r, with x fixed, and expand it in a Taylor series

about $r = 0$ then the coefficients will, of course, be functions of x. In fact, it turns out that those coefficients are the Legendre polynomials $P_n(x)$:

$$\frac{1}{\sqrt{1 - 2xr + r^2}} = \sum_0^\infty P_n(x)r^n. \quad (|x| \leq 1, \; |r| < 1)$$

(8.1)

Thus, $(1 - 2xr + r^2)^{-1/2}$ is known as the **generating function** for the P_n's; (8.1) is the source of much information about the Legendre polynomials. In this exercise we ask you to verify (8.1). Specifically, expand the left-hand side in a Taylor series in r, about $r = 0$, through r^3, say, and verify that the coefficients of $r^0, \ldots, r^3$, are indeed $P_0(x), \ldots, P_3(x)$, respectively.

9. By changing x to $-x$ on both sides of (8.1), show that

$$P_n(-x) = (-1)^n P_n(x), \quad (9.1)$$

so that if the integer n is even then $P_n(-x) = P_n(x)$ so the graph of $P_n(x)$ is symmetric about $x = 0$, whereas if n is odd then $P_n(-x) = -P_n(x)$ so the graph of $P_n(x)$ is antisymmetric about $x = 0$, as seen in Fig. 4 for $n = 0, 1, 2, 3, 4$.

10. (*Recursion formula*) By taking $\partial/\partial r$ of (8.1), derive the recursion formula

$$nP_n(x) = (2n - 1)x P_{n-1}(x) - (n - 1)P_{n-2}(x) \quad (10.1)$$

for $n = 2, 3, \ldots$, and using the entries in Table 1, verify (10.1) for $n = 2$ and for $n = 3$.

11. (*Orthogonality relation*) A particularly important property of Legendre polynomials is the "**orthogonality relation**"

$$\int_{-1}^1 P_j(x) P_k(x) \, dx = 0 \quad \text{if } j \neq k. \quad (11.1)$$

(a) Prove (11.1). HINT: Show that the Legendre equation

$$\left(1 - x^2\right) y'' - 2xy' + n(n + 1)y = 0 \quad (11.2)$$

can be expressed as

$$\left[\left(1 - x^2\right) y'\right]' + n(n + 1)y = 0. \quad (11.3)$$

Considering $\int_{-1}^1 \left[(1 - x^2)P_j'\right]' P_k \, dx$, show, by integrating by parts, that

$$\int_{-1}^1 \left[(1 - x^2)P_j'\right]' P_k \, dx = \int_{-1}^1 P_j \left[(1 - x^2)P_k'\right]' dx.$$

(11.4)

Then, since P_j and P_k are solutions of the Legendre equation (11.2) for $n = j$ and for $n = k$, respectively, show that (11.4) can be re-expressed as

$$-j(j+1) \int_{-1}^1 P_j P_k \, dx = -k(k+1) \int_{-1}^1 P_j P_k \, dx \quad (12.5)$$

Finally, show that (11.1) follows from (11.5) because $j \neq k$ (by assumption).
(b) Verify (11.1) by working out the integral for these cases: $j = 1$ and $k = 2$, $j = 1$ and $k = 3$, and $j = 2$ and $k = 3$, say.

12. (a) As a companion to (12.1), for the case where $j = k$, derive the formula

$$\int_{-1}^1 [P_n(x)]^2 \, dx = \frac{2}{2n + 1} \quad (12.1)$$

for $n = 0, 1, 2, \ldots$. HINT: Squaring (8.1) and integrating,

$$\int_{-1}^1 \frac{dx}{1 - 2rx + r^2}$$
$$= \int_{-1}^1 \sum_{m=0}^\infty r^m P_m(x) \sum_{n=0}^\infty r^n P_n(x) \, dx. \quad (12.2)$$

Integrating the left side, and using the orthogonality relation (11.1) to simplify the right side, obtain

$$\frac{1}{r} \ln\left(\frac{1 + r}{1 - r}\right) = \sum_{n=0}^\infty \left\{\int_{-1}^1 [P_n(x)]^2 \, dx\right\} r^{2n}. \quad (12.3)$$

Finally, expanding the left-hand side in a Taylor series in r, show that (12.1) follows.

(b) Verify (12.1), by working out the integral, for the cases $n = 0, 1,$ and 2.

13. (*Integral representation of P_n*) It can be shown that

$$P_n(x) = \frac{1}{\pi} \int_0^\pi \left(x + \sqrt{x^2 - 1} \, \cos t\right)^n dt \quad (13.1)$$

for $n = 0, 1, 2, \ldots$, which is called **Laplace's integral form** for $P_n(x)$. Here, we ask you to verify (13.1) for the cases $n = 0, 1,$ and 2, by working out the integral for those cases.

14. In the paragraph preceding (52) we stated that $y_1(x) \to -\infty$ as $x \to \pm 1$ and that $y_2(x) \to +\infty$ as $x \to -1$ and to $-\infty$ as $x \to +1$. We don't ask you to prove those claims but only to use *Maple* to obtain evidence to support them. HINT: Consider using either phaseportrait to obtain a plot of y versus x or dsolve with the value = array option (Section 3.2) to obtain specific values of $y(x)$.

8.4 INTRODUCTION TO THE SINGULAR CASE (OPTIONAL)

If, to solve

$$y'' + p(x)y' + q(x)y = 0, \tag{1}$$

we seek $y(x)$ in power series form, then Theorem 8.3.1 assures success if both $p(x)$ and $q(x)$ are analytic at the point of expansion x_0; that case is called the **regular** case and x_0 is called a **regular point** of (1). If $p(x)$ and/or $q(x)$ are *not* analytic at the chosen point x_0, that case is called the **singular** case and x_0 is called a **singular point** of (1). In that case the theorem offers no information one way or the other. In practice one can avoid the singular case by choosing x_0 to be a regular point, yet sometimes it is more fruitful to expand about a singular point than to avoid it.

In that event Theorem 8.3.1 is of no help, but we may be able to use a modified series approach known as the **method of Frobenius**.[1] Although that method is beyond our present scope we will, in this section, explain the main ideas and apply them to the Bessel equation that we met in Section 8.3.

8.4.1 Bessel's equation.   We will use the Bessel equation of order zero,

$$xy'' + y' + xy = 0, \quad (0 < x < \infty) \tag{2}$$

as the focal point of our discussion. Dividing by x, to put (2) into the standard form (1), we see that $p(x) = 1/x$ and $q(x) = 1$, so the only singular point of (2) is $x = 0$. When we studied (2) in Example 3 of Section 8.3 we gave initial conditions at $x = 3$, which led us to choose $x_0 = 3$. Hence, that analysis involved the regular case. Here, however, let us "jump in" right *at* the singular point $x = 0$ and seek a power series solution about that point,

$$y(x) = \sum_{0}^{\infty} a_n x^n. \tag{3}$$

Although Theorem 8.3.1 does not apply, let us try (3) and see what happens.

Putting (3) into (2) gives (after some cancelation)

$$\sum_{0}^{\infty} n^2 a_n x^{n-1} + \sum_{0}^{\infty} a_n x^{n+1} = 0. \tag{4}$$

To obtain the same exponents on the x's set $n - 1 = m + 1$ (i.e., $n = m + 2$) in the first sum, so

$$\sum_{-2}^{\infty} (m + 2)^2 a_{m+2} x^{m+1} + \sum_{0}^{\infty} a_n x^{n+1} = 0 \tag{5}$$

---

[1] See, for instance, M. D. Greenberg, *Advanced Engineering Mathematics*, 2nd ed. (Upper Saddle River, NJ: Prentice Hall, 1998), Section 4.3.

or, since m is just a dummy summation index that can be changed to n (or j or k or whatever),

$$\sum_{-2}^{\infty}(n+2)^2 a_{n+2}x^{n+1} + \sum_{0}^{\infty} a_n x^{n+1} = 0. \qquad (6)$$

Now we have the same exponents on the x's but the summation limits differ. But, if we define $a_{-1} = a_{-2} \equiv 0$ we can change the lower limit of the second sum to -2 and pull the two sums together:

$$\sum_{-2}^{\infty} \left[(n+2)^2 a_{n+2} + a_n \right] x^{n+1} = 0. \qquad (7)$$

It follows from (7) that

$$(n+2)^2 a_{n+2} + a_n = 0 \qquad (8)$$

for each $n = -2, -1, 0, 1, \ldots$. Writing out the latter recursion formula for $n = -2, -1, 0, \ldots$ gives $a_1 = a_3 = a_5 = \cdots = 0$, and it gives $a_2, a_4, a_6, \ldots$ in terms of a_0. Putting these results into (3) we have

$$y(x) = a_0 \left(1 - \frac{1}{2^2}x^2 + \frac{1}{4^2 2^2}x^4 - \frac{1}{6^2 4^2 2^2}x^6 + \cdots \right)$$

$$= a_0 \sum_{0}^{\infty} \frac{(-1)^n}{2^{2n}(n!)^2} x^{2n} \equiv a_0 y_1(x). \qquad (9)$$

The solution $y_1(x)$ that we have found is called $J_0(x)$, the **Bessel function of the first kind and of order zero.** [Recall that the "order" of the Bessel equation is the ν in equation (44) of Section 8.3. In the present case, equation (2) above, $\nu = 0$.] The latter is plotted in Fig. 1 and is seen to resemble a damped cosine wave.

FIGURE 1
Bessel function of the first kind and order zero.

The bad news is that the expansion (3) about the singular point $x = 0$ led to only one solution, namely,

$$\boxed{ J_0(x) = 1 - \frac{1}{2^2}x^2 + \frac{1}{4^2 2^2}x^4 - \cdots = \sum_{0}^{\infty} \frac{(-1)^n}{2^{2n}(n!)^2} x^{2n}, } \qquad (10)$$

so we are missing a second LI solution. The good news is that the series in (10) converges with an *infinite* radius of convergence (proof of which is left for the exercises), on $-\infty < x < \infty$, whereas when we expanded about the regular point $x_0 = 3$ (in Example 3 of Section 8.3) both power series solutions converged only on $0 < x < 6$, the limitation being due to the presence of the singular point of $p(x) = 1/x$ at $x = 0$.

Evidently, the missing second solution to (2), say $y_2(x)$, is not analytic at $x = 0$ or else we would have found it when we sought $y(x)$ in the power series form (3). How can we find $y_2(x)$? Recall that when we know one solution we can use it to determine other solutions by means of Lagrange's method of **reduction of order**, which method was introduced in Example 8 of Section 6.5 and used again in Section

6.6. According to that method, we seek $y_2(x)$ by varying the parameter a_0 in (9). That is, seek

$$y_2(x) = A(x)y_1(x) = A(x)J_0(x). \tag{11}$$

Putting (11) into (2) gives the differential equation

$$(x J_0) A'' + \left(J_0 + 2x J_0'\right) A' + \left[x J_0'' + J_0' + x J_0\right] A = 0 \tag{12}$$

on $A(x)$. The square-bracketed terms cancel because $J_0(x)$ is a solution of (2), so if we define

$$A'(x) \equiv s(x), \tag{13}$$

say, then (12) reduces to the *first*-order equation

$$x J_0 s' + \left(J_0 + 2x J_0'\right) s = 0 \tag{14}$$

on $s(x)$. Equivalently,

$$\frac{ds}{s} + \left(2\frac{J_0'}{J_0} + \frac{1}{x}\right) dx = 0 \tag{15}$$

or

$$\frac{ds}{s} + 2\frac{d J_0}{J_0} + \frac{dx}{x} = 0, \tag{16}$$

integration of which gives

$$\ln |s| + 2 \ln |J_0| + \ln |x| = \text{arbitrary constant}$$
$$\equiv \ln B, \tag{17}$$

where B is an arbitrary constant such that $0 < B < \infty$ [for if $0 < B < \infty$ then $-\infty < \ln B < \infty$ so the "arbitrary constant" in (17) can be designated as $\ln B$, for some positive value of B]. Thus,

$$\ln \left|s J_0^2 x\right| = \ln B$$

so

$$\left|s J_0^2 x\right| = B,$$
$$s J_0^2 x = \pm B \equiv C,$$

say, where C is an arbitrary (but nonzero) constant. Then

$$A'(x) = s(x) = \frac{C}{x J_0^2(x)}$$

so

$$A(x) = C \int \frac{1}{x J_0^2(x)} dx. \tag{18}$$

We can complete the evaluation of $A(x)$, and hence of $y_2(x)$, in either of two ways. First, we can use (18) to infer the *form* of $A(x)$. Specifically, $J_0(x)$ is analytic at $x = 0$ and is nonzero at $x = 0$, so the $1/J_0^2(x)$ factor within the integrand is analytic at $x = 0$ and can be expanded in the form $\sum_0^\infty b_n x^n$. Then

$$A(x) = C \int \left(\frac{b_0}{x} + b_1 + b_2 x + b_3 x^2 + \cdots \right) dx$$

$$= C \left(\ln x + b_1 x + \frac{b_2}{2} x^2 + \frac{b_3}{3} x^3 + \cdots \right) \tag{19}$$

and putting this expression back into (11) reveals that $y_2(x)$ is of the form

$$y_2(x) = C \left(\ln x + b_1 x + \frac{b_2}{2} x^2 + \cdots \right) J_0(x)$$

$$= C \left[(\ln x) J_0(x) + \left(b_1 x + \frac{b_2}{2} x^2 + \cdots \right) J_0(x) \right]$$

$$= C \left[(\ln x) J_0(x) + \sum_1^\infty c_n x^n \right]. \tag{20}$$

Thus, with $C = 1$, say, we could start over and seek $y_2(x)$ in the form

$$y_2(x) = (\ln x) J_0(x) + \sum_1^\infty c_n x^n. \tag{21}$$

Putting (21) into (2) will give the c_n's. We leave this approach to the exercises.

Alternatively, we could put our series (10) for $J_0(x)$ into (18) and integrate termwise. First, we need to express $1/J_0^2(x)$ in the form

$$\frac{1}{J_0^2(x)} = \sum_0^\infty b_n x^n. \tag{22}$$

To evaluate the b_n's in (22), write

$$1 = \left(\sum_0^\infty b_n x^n \right) J_0^2(x)$$

$$= \left(b_0 + b_1 x + b_2 x^2 + \cdots \right) \left(1 - \frac{1}{2^2} x^2 + \frac{1}{4^2 2^2} x^4 - \cdots \right)^2$$

$$= b_0 + b_1 x + \left(b_2 - \frac{1}{2} b_0 \right) x^2 + \left(b_3 - \frac{1}{2} b_1 \right) x^3 + \left(b_4 - \frac{1}{2} b_2 + \frac{3}{32} b_0 \right) x^4 +$$

$$+ \left(b_5 - \frac{1}{2} b_3 + \frac{3}{32} b_1 \right) x^5 + \left(b_6 - \frac{1}{2} b_4 + \frac{3}{32} b_2 - \frac{5}{5184} b_0 \right) x^6 + \cdots \tag{23}$$

and equate coefficients of each power of x on the left-hand and right-hand sides:

$$x^0: \quad 1 = b_0, \tag{24a}$$

$$x: \quad 0 = b_1, \tag{24b}$$

$$x^2: \quad 0 = b_2 - \frac{1}{2}b_0, \tag{24c}$$

$$x^3: \quad 0 = b_3 - \frac{1}{2}b_1, \tag{24d}$$

$$x^4: \quad 0 = b_4 - \frac{1}{2}b_2 + \frac{3}{32}b_0, \tag{24e}$$

$$x^5: \quad 0 = b_5 - \frac{1}{2}b_3 + \frac{3}{32}b_1, \tag{24f}$$

$$x^6: \quad 0 = b_6 - \frac{1}{2}b_4 + \frac{3}{32}b_2 - \frac{5}{576}b_0, \tag{24g}$$

and so on. Solving these gives $b_0 = 1$, $b_2 = 1/2$, $b_4 = 5/32$, $b_6 = 23/576$, ... and $b_1 = b_3 = b_5 = \cdots = 0$. Then (19) gives (with $C = 1$, say)

$$A(x) = \ln x + \frac{1}{4}x^2 + \frac{5}{128}x^4 + \frac{23}{3456}x^6 + \cdots \tag{25}$$

so

$$y_2(x) = A(x)J_0(x)$$

$$= \left(\ln x + \frac{1}{4}x^2 + \frac{5}{128}x^4 + \frac{23}{3456}x^6 + \cdots \right) \times$$

$$\left(1 - \frac{1}{4}x^2 + \frac{1}{64}x^4 - \frac{1}{2304}x^6 + \cdots \right)$$

$$= (\ln x) J_0(x) + \frac{1}{4}x^2 - \frac{3}{128}x^4 + \frac{11}{13824}x^6 + \cdots \tag{26}$$

which is called $\mathbf{Y}_0(x)$, the **Neumann function of order zero**. Thus, we have found the two LI solutions $y_1(x) = J_0(x)$ and $y_2(x) = \mathbf{Y}_0(x)$, so we can use them to form a general solution of (2). However, following *Weber*, it proves to be convenient and standard to use, in place of $\mathbf{Y}_0(x)$, a linear combination of $J_0(x)$ and $\mathbf{Y}_0(x)$, namely,

$$y_2(x) = \frac{2}{\pi} [\mathbf{Y}_0(x) + (\gamma - \ln 2)J_0(x)] \equiv Y_0(x), \tag{27}$$

where

$$Y_0(x) = \frac{2}{\pi} \left[\left(\ln \frac{x}{2} + \gamma \right) J_0(x) + \frac{x^2}{2^2} - \left(1 + \frac{1}{2} \right) \frac{x^4}{2^4(2!)^2} \right.$$

$$\left. + \left(1 + \frac{1}{2} + \frac{1}{3} \right) \frac{x^6}{2^6(3!)^2} - \cdots \right] \tag{28}$$

is Weber's **Bessel function of the second kind, of order zero**; $\gamma \approx 0.5772157$ is known as **Euler's constant**, and $Y_0(x)$ is sometimes written as $N_0(x)$. The graphs of $J_0(x)$ and $Y_0(x)$ are shown in Fig. 2. Important features are that $J_0(x)$ and $Y_0(x)$ look a bit like damped cosine and sine functions, except that $Y_0(x)$ tends to $-\infty$ as $x \to 0$. Specifically, we see from (10) and (28) that[1]

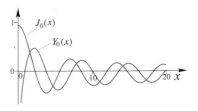

$$J_0(x) \sim 1, \tag{29a}$$

$$Y_0(x) \sim \frac{2}{\pi} \ln x \tag{29b}$$

FIGURE 2
J_0 and Y_0.

as $x \to 0$, and it can be shown that

$$J_0(x) \sim \sqrt{\frac{2}{\pi x}} \cos\left(x - \frac{\pi}{4}\right), \tag{30a}$$

$$Y_0(x) \sim \sqrt{\frac{2}{\pi x}} \sin\left(x - \frac{\pi}{4}\right) \tag{30b}$$

as $x \to \infty$. Indeed, we can see from (30) why Weber's Bessel function Y_0 is a nicer companion for J_0 than Neumann's Bessel function $\mathbf{Y}_0$, for

$$\mathbf{Y}_0(x) = \frac{\pi}{2} Y_0(x) - (\gamma - \ln 2) J_0(x)$$

$$\sim \sqrt{\frac{2}{\pi x}} \left[\frac{\pi}{2} \sin\left(x - \frac{\pi}{4}\right) - (\gamma - \ln 2) \cos\left(x - \frac{\pi}{4}\right)\right] \tag{30c}$$

as $x \to \infty$; surely (30b) makes a nicer companion for (30a) than does (30c).[2]

It might appear, from Fig. 2 and (30), that the *zeros* of J_0 and Y_0 [i.e., the roots of $J_0(x) = 0$ and $Y_0(x) = 0$] are equally spaced, but they are not; they approach an equal spacing only as $x \to \infty$. For instance, the first several zeros of J_0 are 2.405, 5.520, 8.654, 11.792, 14.931. Their differences are 3.115, 3.134, 3.138, 3.139, and these are seen to rapidly approach a constant [namely, π, the spacing between the zeros of $\cos(x - \pi/4)$ in (30a)]. The zeros of the various Bessel functions turn out to be important, and they are tabulated to many significant figures in the literature.

In summary, we have found the general solution

$$\boxed{y(x) = A J_0(x) + B Y_0(x)} \tag{31}$$

of the Bessel equation of order zero, $xy'' + y' + xy = 0$, where $J_0(x)$ is the Bessel function of order zero of the first kind, defined by (10), and $Y_0(x)$ is the Bessel function of order zero of the second kind, defined by (28).

---

[1]Note that $\ln(x/2) = \ln x - \ln 2 \sim \ln x$ as $x \to 0$.

[2]If this point is unclear consider the following more familiar example. We normally write a general solution of $y'' + y = 0$ as $y = A \cos x + B \sin x$ or as $y = C e^{ix} + D e^{-ix}$. That is, if we choose $y_1 = \cos x$ then $\sin x$ is the most natural choice for y_2, and if we choose $y_1 = e^{ix}$ then e^{-ix} is the most natural choice for y_2. In contrast, if we choose $y_1 = \cos x$, say, then $y_2 = 5 \cos x - 3 e^{ix}$ is a *legitimate* second LI solution, but it surely seems a less attractive choice than $y_2 = \sin x$.

Let us reflect upon the outcome. Since $x = 0$ is a singular point of the Bessel equation (2), Theorem 8.3.1 does not assure us that any power series solution can be found about $x = 0$, since the conditions of the theorem are not met. We were fortunate, nevertheless, to find one such power series solution, $J_0(x)$, and then we used reduction of order to find a second LI solution $Y_0(x)$. The latter is *not* of power series type since it includes a $\ln x$ term; $Y_0(x)$ is singular at $x = 0$. The effect of the singular $\ln x$ term can be seen in Fig. 2 : $Y_0(x) \to -\infty$ as $x \to 0$. In a sense, we can think of the singularity in the Bessel equation, since $p(x) = 1/x$ is singular at $x = 0$, as having been "passed on," like a computer virus, to the $Y_0(x)$ solution.

Please note that expanding about the singular point $x = 0$ brings the logarithmic singularity at $x = 0$ into explicit view. The same logarithmic singularity is contained in the solutions (41) and (42) in Section 8.3 but they cannot be seen explicitly in those solutions. Whereas the useful asymptotic formula (29b) follows easily from (28), that result cannot be seen explicitly from (41) and (42) in Section 8.3.

Finally, observe that when we expanded about the regular point $x_0 = 3$, in Section 8.3, the singularity at $x = 0$ limited the radii of convergence of the series to $R = 3$, whereas when we expand about the singular point $x = 0$ the resulting series [in (10) and (28)] have *infinite* radii of convergence!

8.4.2 Overview of the general case.   The foregoing is but one example, albeit an important one. More generally, consider *any* equation

$$y'' + p(x)y' + q(x)y = 0 \tag{32}$$

on $0 < x < \infty$, with a singular point at the left endpoint $x = 0$. (If the singular point is at x_0, where $x_0 \neq 0$, we can easily move it to the origin $t = 0$ by changing the independent variable from x to t according to $t = x - x_0$.)

Multiplying (32) through by x^2 and grouping terms by rectangular brackets, let us re-express (32) in the form

$$x^2 y'' + x\, [xp(x)]\, y' + \left[x^2 q(x) \right] y = 0. \tag{33}$$

We have supposed that $p(x)$ and $q(x)$ are not both analytic at $x = 0$, so that $x = 0$ is a singular point of (32). But, suppose that $xp(x)$ and $x^2 q(x)$ *are* analytic at $x = 0$. Then they can be expanded in Taylor series, that is, power series

$$xp(x) = p_0 + p_1 x + p_2 x^2 + \cdots, \tag{34a}$$

$$x^2 q(x) = q_0 + q_1 x + q_2 x^2 + \cdots, \tag{34b}$$

which converge in some interval about $x = 0$, so we can express (33) as

$$x^2 y'' + x\,(p_0 + p_1 x + \cdots)\, y' + (q_0 + q_1 x + \cdots)\, y = 0. \tag{35}$$

Locally, in the neighborhood of $x = 0$, we can keep just the leading terms of the series and approximate (35) by the *reduced equation*

$$\boxed{x^2 y'' + p_0 x y' + q_0 y = 0.} \tag{36}$$

The latter is a Cauchy-Euler equation, which type of equation always admits at least one solution in the form $y(x) = x^r$, where r is a root of the quadratic equation

$$r^2 + (p_0 - 1)r + q_0 = 0. \qquad (37)$$

[If p_0 and q_0 are real numbers, then the roots of (37) will be real or else complex conjugates.]

If the roots of (37) are distinct, say r_1 and r_2, then (36) admits the general solution

$$y(x) = Ax^{r_1} + Bx^{r_2}, \qquad (38)$$

and if the roots are repeated, say r_1, then (36) admits the general solution

$$y(x) = (A + B \ln x) x^{r_1}. \qquad (39)$$

These results provide the clue as to a promising form in which to seek solutions of the original equation (32), for if (36) approximates (32) near $x = 0$, then we can expect (32) to admit solutions of the form

$$y(x) = \left(a_0 + a_1 x + a_2 x^2 + \cdots\right) x^{r_1} + \left(b_0 + b_1 x + b_2 x^2 + \cdots\right) x^{r_2} \qquad (40)$$

if (37) has distinct roots r_1, r_2, and of the form

$$y(x) = [(a_0 + a_1 x + \cdots) + (b_0 + b_1 x + \cdots) \ln x] x^{r_1} \qquad (41)$$

if (37) has a repeated root r_1, because as $x \to 0$ (40) and (41) reduce to $y(x) \sim a_0 x^{r_1} + b_0 x^{r_2}$ and $y(x) \sim (a_0 + b_0 \ln x) x^{r_1}$, like (38) and (39). Away from $x = 0$ we can expect the two power series in (40) to account for the deviation of the solution from its local behavior (38), and similarly for (41) and (39).

One finds, in the Frobenius theory, that these expectations are met, except in the case where r_1 and r_2 are distinct and happen to differ by an integer, in which case the general solution may not be of the form (40).

EXAMPLE 1
Consider the equation

$$8x^2 y'' + \left(2x + 10x^2\right) y' + (1 - x) y = 0, \qquad (42)$$

for which $p(x) = \left(2x + 10x^2\right)/8x^2$ and $q(x) = (1 - x)/8x^2$. These functions are singular at $x = 0$, so $x = 0$ is a singular point of (42). Hence, Theorem 8.3.1 does not apply. However, both $xp(x) = (1 + 5x)/4$ and $x^2 q(x) = (1 - x)/8$ *are* analytic at $x = 0$, and the expansions (34a,b) are simply

$$xp(x) = \frac{1}{4} + \frac{5}{4}x + 0x^2 + 0x^3 + \cdots,$$

$$x^2 q(x) = \frac{1}{8} - \frac{1}{8}x + 0x^2 + 0x^3 + \cdots.$$

Thus $p_0 = 1/4$ and $q_0 = 1/8$, so the reduced equation (36) is

$$x^2 y'' + \frac{1}{4} x y' + \frac{1}{8} y = 0 \tag{43}$$

with the general solution

$$y(x) = A x^{1/4} + B x^{1/2}. \tag{44}$$

That is, with $p_0 = 1/4$ and $q_0 = 1/8$ equation (37) gives the roots $r = 1/2, 1/4$.

Thus, we can expect the general solution of the original equation (42) to be of the form (40), with $r_1 = 1/2$ and $r_2 = 1/4$. To determine that solution we could use the method of Frobenius, but since that method is not given in this text let us use *Maple* instead. Doing so, we obtain the general solution

$$y(x) = C_1 x^{1/2} \left(1 - \frac{2}{5} x + \frac{7}{45} x^2 - \frac{28}{585} x^3 + \cdots \right)$$

$$+ C_2 x^{1/4} \left(1 - \frac{1}{4} x + \frac{23}{224} x^2 - \frac{989}{29568} x^3 + \cdots \right), \tag{45}$$

which is indeed of the form (40), as anticipated. That is, for (40) to satisfy (42), with $r_1 = 1/2$ and $r_2 = 1/4$, it turns out that we need $a_1 = -\frac{2}{5} a_0$, $a_2 = \frac{7}{45} a_0$, ... and $b_1 = -\frac{1}{4} b_0$, $b_2 = \frac{23}{224} b_0$, ... Factoring out the a_0's and renaming a_0 as C_1, say, and factoring out the b_0's and renaming b_0 as C_2, gives the general solution (45).   ∎

E X A M P L E   2   *Bessel Equation*.
For the Bessel equation of order zero,

$$x y'' + y' + x y = 0, \tag{46}$$

which was solved earlier in this section, $p(x) = 1/x$ and $q(x) = 1$. Since $p(x)$ is singular at $x = 0$, $x = 0$ is a singular point of (46), so Theorem 8.3.1 did not apply. However, both $x p(x) = 1$ and $x^2 q(x) = x^2$ *are* analytic at $x = 0$, and the expansions (34a,b) are simply

$$x p(x) = 1 + 0x + 0x^2 + \cdots,$$

$$x^2 q(x) = 0 + 0x + 1x^2 + 0x^3 + \cdots.$$

Thus, $p_0 = 1$ and $q_0 = 0$, so the reduced equation (36) is

$$x^2 y'' + x y' = 0$$

with general solution

$$y(x) = A + B \ln x.$$

That is, with $p_0 = 1$ and $q_0 = 0$ equation (37) gives the repeated roots $r = 0, 0$. Hence, (39) applies, with $r_1 = 0$.

Thus, we can expect the general solution of (46) to be of the form (41) with $r_1 = 0$. In fact, we found that a general solution of (46) is

$$y(x) = A J_0(x) + B Y_0(x). \tag{47}$$

Recall from (10) and (28) that $J_0(x)$ is a power series and $Y_0(x)$ is a power series plus $\ln x$ times a power series. Hence the linear combination $y(x) = A J_0(x) + B Y_0(x)$ in (47) is indeed of the anticipated form (41).   ∎

Observe that the assumption underlying the foregoing (heuristic) reasoning is that even though $p(x)$ and $q(x)$ are not both analytic at $x = 0$, both $xp(x)$ and $x^2 q(x)$ *are* analytic at $x = 0$. This means, essentially, that even though $p(x)$ and/or $q(x)$ are singular, they are not "too singular"; multiplying $p(x)$ by x and $q(x)$ by x^2 is sufficient to "remove" the singular behavior. For the Bessel equation, for instance, the singular behavior of $p(x) = 1/x$ at $x = 0$ is removed when we multiply by x, for $xp(x) = 1$ is analytic there. Thus, if $x = 0$ is a singular point of (32) but both $xp(x)$ and $x^2 q(x)$ are analytic at $x = 0$, then we say that $x = 0$ is a **regular singular point** of (32). If a singular point of (32) is a regular singular point then we can be guided by the reduced equation (36) toward solutions of the form (40) or (41), depending upon the roots r_1 and r_2 of (37), as is detailed in the Frobenius theory. If it is not a regular singular point then it is called an **irregular singular point**, which case is more difficult.[1]

EXAMPLE 3
For the equation

$$xy'' - 3xy' + 2e^{-x}y = 0 \tag{48}$$

$p(x) = -3x/x = -3$ is analytic at $x = 0$ but $q(x) = 2e^{-x}/x$ is not, so $x = 0$ is a singular point of (48). Since both $xp(x) = -3x$ and $x^2 q(x) = 2xe^{-x}$ are analytic there, $x = 0$ is a *regular singular point* of (48). ∎

EXAMPLE 4
For the equation

$$x^3 y'' + 4xe^x y' + 5(\sin x)y = 0 \tag{49}$$

neither $p(x) = 4e^x/x^2$ nor $q(x) = 5(\sin x)/x^3$ are analytic at $x = 0$, so $x = 0$ is a singular point of (49). Since $xp(x) = 4e^x/x$ is singular (i.e., not analytic) at $x = 0$, $x = 0$ is an *irregular singular point* of (49), even though $x^2 q(x) = 5(\sin x)/x = 5 - 5x^2/6 + x^4/24 - \cdots$ is analytic there. ∎

Closure.   In this section we have considered the singular case, where $p(x)$ and $q(x)$ in $y'' + p(x)y' + q(x)y = 0$ are not both analytic at the chosen point of expansion (which we chose, in every case, to be at the origin, $x = 0$). In that case Theorem 8.3.1 does not apply.

As a concrete and important example we considered the Bessel equation of order zero, $xy'' + y' + xy = 0$. Without being able to use Theorem 8.3.1 we nonetheless sought a power series solution and succeeded in finding one such solution, the Bessel function of the first kind and order zero, $J_0(x)$. We knew that the missing second solution would be singular at $x = 0$, since otherwise we would have found it when we sought $y(x)$ in power series form. To obtain a second LI solution we used Lagrange's method of reduction of order and obtained the Bessel function of the second kind and

---

[1] For a more advanced treatment that covers irregular singular points, see C. M. Bender and S. A. Orszag, *Advanced Mathematical Methods for Scientists and Engineers* (New York: McGraw-Hill, 1978).

order zero, $Y_0(x)$. The latter is indeed singular at $x = 0$ due to a $\ln x$ term, and tends to $-\infty$ as $x \to 0$.

In Section 8.4.2 we presented a heuristic plan for the general singular case. There, we argued that even if $p(x)$ and $q(x)$ are not both analytic at $x = 0$, nevertheless we should be able to find a general solution, either in the form of (40) or (41), if $p(x)$ and $q(x)$ are not too singular there, namely, if both $xp(x)$ and $x^2q(x)$ are analytic there, in which case we say that $x = 0$ is a regular singular point of the differential equation. It turns out that the Frobenius theory reveals that a possible exception is the case where the roots r_1 and r_2 of (37) differ by an integer, in which case (40) may not apply.

Why do we study the singular case at all, when there is no lack of regular points about which to expand? (For the Bessel equation, for instance, $x = 0$ is the only singular point; every other point on the x axis is a regular point.) The answer is twofold. First, when we expand about the singular point the singular nature of the solution, at that point, comes into explicit view—for instance, the $\ln x$ term in the Y_0 solution of the Bessel equation. Second, when we expand about the singular point the singularity in the solution at that point does not limit the radius of convergence of the series—for instance, the series in $J_0(x)$ and $Y_0(x)$ have infinite radii of convergence.

Maple.   The *Maple* command that gave (45) was

```
dsolve(8*x^2*diff(y(x),x,x)+(2*x+10*x^2)*diff(y(x),x)
+(1-x)*y(x)=0,y(x),type=series);
```

The latter gives terms up to but not including order six. If we want more terms, up to but not including order ten, say, then precede the dsolve command with the command

```
Order:= 10:
```

NOTE: The *Maple* names for $J_0(x)$ and $Y_0(x)$ are BesselJ(0,x) and BesselY(0,x), respectively.

EXERCISES 8.4

1. We noted, following (21), that we could have put (21) into (2) and solved for the c_n's. Do that, and show that your result agrees with (26).

2. Using (10) and (28), and a hand held calculator or *Maple*, evaluate $J_0(1)$ and $Y_0(1)$ to four significant figures.

3. Show that the series in (10) converges for all x, as claimed.

4. (*Maple experiment*) We made the claim that the general solution of (32) is of the form (40) if the roots r_1, r_2 of (37) are distinct (and do not differ by an integer), and is of the form (41) if the roots are repeated. In each case determine r_1, r_2, find the general solution using the *Maple* dsolve command, and then indicate whether your results

are consistent with the foregoing claim. NOTE: In each of these equations we use coefficients that are simple polynomials. The reason is that at this time *Maple* is not giving an output when coefficients are taken to be transcendental functions (such as sines, cosines).

(a) $2xy'' + y' + 2x^2y = 0$

(b) $4xy'' + (1 + x)y' - 4x^2y = 0$

(c) $x^2y'' + (x + x^2)y' + y = 0$

(d) $8x^2y'' + (2x - x^4)y' + (1 + 3x^2)y = 0$

(e) $x^2y'' + 3x(1 - 2x^2)y' + y = 0$

(f) $x^2y'' + x(1 - x)y' + y = 0$

(g) $4x^2y'' + 4x(x-1)y' + 3y = 0$

(h) $16x^2y'' - 16x(1+x)y' + 15y = 0$

(i) $x^2y'' - xy' + (1+x)y = 0$

(j) $9x^2(1+x)y'' - xy' + 5y = 0$

(k) $x^2(1-x)y'' + xy' + y = 0$

(l) $x^2(1+5x)y'' + 3xy' + y = 0$

5. Determine all singular points, if any, and classify them as regular or irregular.

(a) $y'' - x^3y' + xy = 0$

(b) $xy'' - (\cos x)y' + 5y = 0$

(c) $(x^2 - 3)y'' - y = 0$

(d) $x(x^2 + 3)y'' + y = 0$

(e) $(x+1)^2y'' - 4y' + (x+1)y = 0$

(f) $y'' + (\ln x)y' + 2y = 0$

(g) $(x-1)(x+3)^2y'' + y' + y = 0$

(h) $xy'' + (\sin x)y' - (\cos x)y = 0$

(i) $x(x^4 + 2)y'' + y = 0$

(j) $(x^4 - 1)y'' + xy' - x^2y = 0$

(k) $(x^4 - 1)^3y'' + (x^2 - 1)^2y' - y = 0$

(l) $(x^4 - 1)^3y'' - 3(x+1)^2y' + x(x+1)y = 0$

(m) $(xy')' - 5y = 0$

(n) $\left[x^3(x-1)y'\right]' + 2y = 0$

(o) $2x^2y'' - xy' + 7y = 0$

(p) $xy'' + 4y' = 0$

(q) $x^2y'' - 3y = 0$

(r) $2x^2y'' + \sqrt{\pi}\,y = 0$

CHAPTER 8 REVIEW

Aside from the class of Cauchy-Euler equations, nonconstant coefficient linear differential equations of order greater than one are so difficult that, in general, we cannot find closed form solutions. In that case the standard options are as follows: to seek solutions in "open form" as power series, or to solve numerically. We discuss the former in this chapter and limit attention to equations of second order, that case being the most important to us; numerical solution is the subject of Chapter 12.

Theorem 8.3.1 reveals that if we seek solutions of

$$y'' + p(x)y' + q(x)y = 0 \tag{1}$$

in the power series form

$$y(x) = \sum_0^\infty a_n(x - x_0)^n \tag{2}$$

then the success of the method depends on $p(x)$ and $q(x)$ being analytic at the expansion point x_0. Specifically, if $p(x)$ and $q(x)$ are analytic at x_0 then two LI solutions of (1) [and therefore a general solution of (1)] can be found by seeking $y(x)$ in the form (2). Further, the radius of convergence of (2) will be at least as large as the smaller of the radii of convergence of the Taylor series of $p(x)$ and $q(x)$ about x_0.

Typically, $p(x)$ and $q(x)$ are analytic for all x or have at most a few singular points. For instance, the Bessel equation of order zero

$$y'' + \frac{1}{x}y' + y = 0 \tag{3}$$

has only one singular point, at $x = 0$, and the Legendre equation

$$y'' - \frac{2x}{1-x^2}y' + \frac{\lambda}{1-x^2}y = 0 \tag{4}$$

has two singular points, at $x = \pm 1$. It would appear that Theorem 8.3.1 is all we need, for it is easy to choose x_0 so as to avoid the singular points, if any, of $p(x)$ and $q(x)$. However, even if we choose x_0 to avoid any singular points of $p(x)$ and $q(x)$ that lie on the x axis, the presence of those singular points can be expected to limit the radii of convergence of our power series solutions so that those series solutions may not converge on the entire x interval of interest.

Rather than avoid a singular point that lies on the x axis, we can seek solutions *about* such a singular point, provided that we modify the form (2) so as to "build in" the kind of singular behavior in $y(x)$ that can result from the singular behavior of the coefficients $p(x)$ and/or $q(x)$ in the differential equation. This strategy is guaranteed to work if $p(x)$ and $q(x)$ are not "too singular" at the point in question, specifically, if the singular point is a *regular singular point* rather than an *irregular singular point*. In that case $y(x)$ can admit singular behavior of the form x^r, $\ln x$, or $x^r \ln x$ at $x = 0$, so the appropriate solution forms involve power series multiplied by one or more of those singular forms. This strategy is motivated in Section 8.4.2.

What is the benefit of expanding about a singular point of the differential equation? For the Bessel equation of order zero, for instance, the equation admits only one singular point, a regular singular point at $x = 0$. We showed in Section 8.4.1 that expanding about that singular point gives two LI solutions, $J_0(x)$ which is analytic and hence which is obtained in power series form, and $Y_0(x)$ which has a logarithmic singularity at $x = 0$. The advantage of this result is that both the power series expression for $J_0(x)$ and the power series part of $Y_0(x)$ converge for all x, their convergence being in no way limited by the regular singular point at $x = 0$. Further, from the expression for $Y_0(x)$ we can even see its singular behavior explicitly, namely,

$$Y_0(x) \sim \frac{2}{\pi} \ln x \tag{5}$$

as $x \to 0$. If instead we sought to avoid the singular point at $x = 0$ by expanding about $x = 4$, say, then we would have obtained two LI solutions in power series form (as guaranteed by Theorem 8.3.1), but one or both of them would have converged only on $0 < x < 8$ rather than for all x and, furthermore, we would have failed to explicitly apprehend the singular $\ln x$ behavior at $x = 0$.

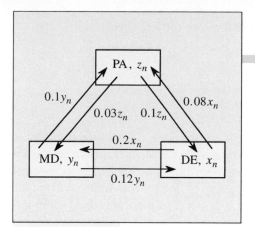

THE EIGENVALUE PROBLEM

9.1 INTRODUCTION

To continue our study of differential equations, we turn to *systems* of differential equations. However, to proceed with that discussion we will need to be familiar with a problem in linear algebra known as the eigenvalue problem, so we devote the present chapter to the eigenvalue problem before studying systems of differential equations in the next two chapters.

Specifically, we will be studying the problem

$$\mathbf{Ax} = \lambda\mathbf{x},\tag{1}$$

where $\mathbf{A}$ is a given $n \times n$ matrix, $\mathbf{x}$ is an unknown $n \times 1$ vector, and λ is an unknown scalar. If we re-express (1) as $\mathbf{Ax} = \lambda\mathbf{Ix}$ (where $\mathbf{I}$ is an $n \times n$ identity matrix) and subtract $\lambda\mathbf{Ix}$ from both sides gives, we obtain the equivalent equation[1]

$$(\mathbf{A} - \lambda\mathbf{I})\mathbf{x} = \mathbf{0},\tag{2}$$

which is a homogeneous system of n equations in the n unknown x_j's, where the coefficient matrix $\mathbf{A} - \lambda\mathbf{I}$ contains the parameter λ.

To be sure that (1) and (2) are clear, let us write them out in scalar form, for $n = 3$, for example. Then (1) is the system

$$a_{11}x_1 + a_{12}x_2 + a_{13}x_3 = \lambda x_1,$$
$$a_{21}x_1 + a_{22}x_2 + a_{23}x_3 = \lambda x_2,$$
$$a_{31}x_1 + a_{32}x_2 + a_{33}x_3 = \lambda x_3.$$

---

[1]Of course we don't need to insert the $\mathbf{I}$. We could re-express (1), correctly, as $\mathbf{Ax} - \lambda\mathbf{x} = \mathbf{0}$, but it would *not* follow from the latter that $(\mathbf{A} - \lambda)\mathbf{x} = \mathbf{0}$ because subtraction of a scalar (λ) from a matrix ($\mathbf{A}$) is not defined; hence, the need to insert $\mathbf{I}$.

Subtracting the terms on the right from those on the left gives

$$
\begin{aligned}
(a_{11} - \lambda)x_1 + \quad & a_{12}x_2 + \quad & a_{13}x_3 = 0, \\
a_{21}x_1 + (a_{22} - \lambda)x_2 + \quad & a_{23}x_3 = 0, \\
a_{31}x_1 + \quad & a_{32}x_2 + (a_{33} - \lambda)x_3 = 0,
\end{aligned}
$$

which, in matrix form, is equation (2).

We know that (2) is consistent because it is homogeneous and therefore admits the "trivial" solution $\mathbf{x} = \mathbf{0}$. However, our interest in (2) shall be in the search for *nontrivial* solutions, and we anticipate that whether or not nontrivial solutions exist will depend upon the value of λ. Thus, the problem of interest is as follows: given the $n \times n$ matrix $\mathbf{A}$, find the value(s) of λ (if any) such that (2) admits nontrivial solutions, and find those nontrivial solutions. The latter is called the **eigenvalue problem** and is the focus of this chapter. The λ's that lead to nontrivial solutions for $\mathbf{x}$ are called the **eigenvalues** (or **characteristic values** of $\mathbf{A}$), and the corresponding nontrivial solutions for $\mathbf{x}$ are called the **eigenvectors** (or **characteristic vectors** of $\mathbf{A}$).

We can interpret (1) geometrically as follows. If we multiply the input vector $\mathbf{x}$ by $\mathbf{A}$, then the output vector $\mathbf{Ax}$ is to be a scalar multiple of $\mathbf{x}$. For instance, consider the various input/output pairs illustrated in Fig. 1. The $\mathbf{x}$ vector in (a) is *not* an eigenvector of $\mathbf{A}$ because $\mathbf{Ax}$ is not a scalar multiple of $\mathbf{x}$. However, the $\mathbf{x}$ vector in (b) *is* an eigenvector of $\mathbf{A}$ because $\mathbf{x}$ and $\mathbf{Ax}$ are colinear. Measuring $\mathbf{x}$ and $\mathbf{Ax}$ we find that the scalar multiple is 2 so that the eigenvalue corresponding to the eigenvector $\mathbf{x}$ is $\lambda = 2$. Likewise, the $\mathbf{x}$ in (c) is an eigenvector, with eigenvalue $\lambda = 0$, and the $\mathbf{x}$ in (d) is also an eigenvector, with eigenvalue $\lambda = -1.5$. Of course, if $n > 3$ then we cannot display the vectors graphically, as we have in Fig. 1, except in a schematic sense.

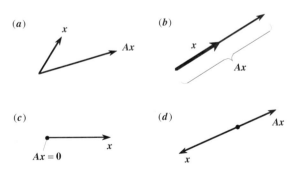

FIGURE 1　Is $\mathbf{x}$ an eigenvector of $\mathbf{A}$?

The eigenvalue problem (1) [or, equivalently, (2)] occurs in a wide variety of applications such as vibration theory, chemical kinetics, stability of equilibria, buckling of structures, convergence of iterative techniques, and systems of coupled ordinary differential equations.

9.2 SOLUTION OF THE EIGENVALUE PROBLEM

9.2.1 Solution method.   The eigenvalue problem consists of the equation

$$\mathbf{Ax} = \lambda\mathbf{x}. \tag{1}$$

To put that problem in perspective, recall that we have already studied the problem of m linear algebraic equations in n unknowns in Sections 4.6, 5.6, and 5.7. That problem was of the form

$$\mathbf{Ax} = \mathbf{c}, \tag{2}$$

where the coefficient matrix $\mathbf{A}$ was $m \times n$. In fact, if we re-express (1) in the form

$$\boxed{(\mathbf{A} - \lambda\mathbf{I})\,\mathbf{x} = \mathbf{0}} \tag{3}$$

we can see that it is really a special case of (2), where $\mathbf{c} = \mathbf{0}$ (so the equation is homogeneous), where $m = n$, and where the coefficient matrix "$\mathbf{A}$"$= \mathbf{A} - \lambda\mathbf{I}$ contains the parameter λ.

Since $\mathbf{A}$ is square ($n \times n$) so is $\mathbf{A} - \lambda\mathbf{I}$, so the latter has a determinant. Then, from Sections 5.6 and 5.7 we know that (3) will admit nontrivial solutions (in addition to the trivial solution $\mathbf{x} = \mathbf{0}$) if and only if

$$\boxed{\det(\mathbf{A} - \lambda\mathbf{I}) = 0.} \tag{4}$$

The latter is not a vector or matrix equation; it is an algebraic equation in λ known as the **characteristic equation** corresponding to the matrix $\mathbf{A}$, and its left-hand side is an nth-degree polynomial known as the **characteristic polynomial**. It is known that such an equation has precisely n roots in the complex plane. Since one or more of these roots can be repeated, we can say that there is at least one eigenvalue λ, and at most n distinct eigenvalues λ, corresponding to any given $n \times n$ matrix $\mathbf{A}$.

We continue to consider only real matrices, as in Chapter 5. However, even if $\mathbf{A}$ is real (so that the coefficients of the characteristic polynomial are too), the characteristic equation can still have complex roots; for instance, $\lambda^2 - 4\lambda + 5 = 0$ has the complex roots $\lambda = 2 + i$ and $2 - i$. This case will be deferred until Section 9.2.3.

This is not the first time we have run into the need to solve polynomial equations. In Section 6.5 we sought solutions to linear, homogeneous, constant-coefficient differential equations by seeking $y(x) = e^{\lambda r}$. Putting that solution form into the nth-order differential equation gave an nth-degree polynomial equation on r. In fact, even the terminology was the same: the equation was called the **characteristic equation** of the differential equation, and the nth-degree polynomial was called the **characteristic polynomial**. If $n = 2$ we can solve the characteristic equation by the quadratic formula. For larger n's we can, if necessary, use computer software such as the *Maple* solve or fsolve commands discussed in Sections 6.5 and 7.5. Thus, let us consider (4) to have been solved for the eigenvalues, for the moment, and let us designate them as $\lambda_1, \ldots, \lambda_k$ ($1 \le k \le n$).

Next, set $\lambda = \lambda_1$ in (3). Since $\det(\mathbf{A} - \lambda_1\mathbf{I}) = 0$, it is guaranteed that $(\mathbf{A} - \lambda_1\mathbf{I})\mathbf{x} = \mathbf{0}$ will have nontrivial solutions. We can find those solutions by Gauss elimination, and we designate them as $\mathbf{e}_1$, where the letter e is for eigenvector. The $\mathbf{e}_1$ solution space is called the **eigenspace** corresponding to the eigenvalue λ_1. Next, we set $\lambda = \lambda_2, \ldots, \lambda_k$, in turn, and repeat the process until the k eigenspaces have been found.

E X A M P L E 1

Determine all eigenvalues and eigenspaces of

$$\mathbf{A} = \begin{bmatrix} 3 & 4 \\ 2 & 1 \end{bmatrix}. \tag{5}$$

The characteristic equation is

$$\det(\mathbf{A} - \lambda \mathbf{I}) = \begin{vmatrix} 3 - \lambda & 4 \\ 2 & 1 - \lambda \end{vmatrix} = \lambda^2 - 4\lambda - 5$$

$$= (\lambda - 5)(\lambda + 1) = 0 \tag{6}$$

so the eigenvalues of $\mathbf{A}$ are $\lambda_1 = 5$ and $\lambda_2 = -1$ or vice versa; the order is immaterial.

With the eigenvalues found from (4), we find the corresponding eigenspaces from (3).

$\lambda_1 = 5$: Then $(\mathbf{A} - \lambda_1 \mathbf{I})\mathbf{x} = \mathbf{0}$ becomes

$$\begin{bmatrix} 3 - 5 & 4 \\ 2 & 1 - 5 \end{bmatrix} \begin{bmatrix} x_1 \\ x_2 \end{bmatrix} = \begin{bmatrix} -2 & 4 \\ 2 & -4 \end{bmatrix} \begin{bmatrix} x_1 \\ x_2 \end{bmatrix} = \begin{bmatrix} 0 \\ 0 \end{bmatrix}, \tag{7}$$

Gauss elimination of which gives

$$\begin{bmatrix} -2 & 4 \\ 0 & 0 \end{bmatrix} \begin{bmatrix} x_1 \\ x_2 \end{bmatrix} = \begin{bmatrix} 0 \\ 0 \end{bmatrix}. \tag{8}$$

The solution is $x_2 = \alpha$ (arbitrary) and $x_1 = 2\alpha$ so, using the letter $\mathbf{e}_1$ in place of $\mathbf{x}$ and the subscript 1 because $\mathbf{e}_1$ corresponds to the eigenvalue λ_1, we have

$$\mathbf{e}_1 = \begin{bmatrix} 2\alpha \\ \alpha \end{bmatrix} = \alpha \begin{bmatrix} 2 \\ 1 \end{bmatrix}. \tag{9}$$

Thus, the eigenspace corresponding to $\lambda = 5$ is $\text{span}\{[2, 1]^T\}$, which is a one-dimensional subspace of $\mathbb{R}^2$, namely, the line $-2x_1 + 4x_2 = 0$ through the origin; we have denoted that eigenspace as $\mathbf{S}_1$ in Fig. 2. NOTE: Here we use the transpose notation $[2, 1]^T$ simply because the equivalent column vector format uses more vertical space.

$\lambda_2 = -1$: Then $(\mathbf{A} - \lambda_1 \mathbf{I})\mathbf{x} = \mathbf{0}$ becomes

$$\begin{bmatrix} 3 + 1 & 4 \\ 2 & 1 + 1 \end{bmatrix} \begin{bmatrix} x_1 \\ x_2 \end{bmatrix} = \begin{bmatrix} 4 & 4 \\ 2 & 2 \end{bmatrix} \begin{bmatrix} x_1 \\ x_2 \end{bmatrix} = \begin{bmatrix} 0 \\ 0 \end{bmatrix}, \tag{10}$$

Gauss elimination of which gives

$$\begin{bmatrix} 4 & 4 \\ 0 & 0 \end{bmatrix} \begin{bmatrix} x_1 \\ x_2 \end{bmatrix} = \begin{bmatrix} 0 \\ 0 \end{bmatrix}. \tag{11}$$

The solution is $x_2 = \beta$ (arbitrary) and $x_1 = -\beta$ so

$$\mathbf{e}_2 = \begin{bmatrix} -\beta \\ \beta \end{bmatrix} = \beta \begin{bmatrix} -1 \\ 1 \end{bmatrix}. \tag{12}$$

Thus, the eigenspace corresponding to $\lambda_2 = -1$ is span $\{[-1, 1]^T\}$, which is a one-dimensional subspace of $\mathbb{R}^2$, namely, the line $x_1 + x_2 = 0$ through the origin, denoted as $\mathbf{S}_2$ in Fig. 1.

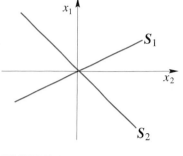

COMMENT 1. Notice that not all vectors in an eigenspace are eigenvectors because an eigenspace is a span, which includes the zero vector. The latter is not itself an eigenvector since an eigenvector is, by definition, a *non*trivial solution of $\mathbf{Ax} = \lambda\mathbf{x}$. The zero vector is included so that the eigenspace will be a vector space, for recall that a vector space necessarily includes a zero vector.

FIGURE 1
The eigenspaces $\mathbf{S}_1, \mathbf{S}_2$.

COMMENT 2. We can determine eigenvectors only to within an arbitrary (but nonzero) scale factor, such as the α in (9) and the β in (12), because if a vector $\mathbf{e}$ satisfies $\mathbf{Ae} = \lambda\mathbf{e}$ then so does any scalar multiple of $\mathbf{e}$. Note that it would be equally correct to express $\mathbf{e}_2$ as $\beta[1, -1]^T$, say, since the scaling of the vector in (12) by -1 can be absorbed by the arbitrary β. ∎

EXAMPLE 2
Determine all eigenvalues and eigenspaces of

$$\mathbf{A} = \begin{bmatrix} 2 & 2 & 1 \\ 1 & 3 & 1 \\ 1 & 2 & 2 \end{bmatrix}. \tag{13}$$

The characteristic equation is

$$\det(\mathbf{A} - \lambda\mathbf{I}) = \begin{vmatrix} 2-\lambda & 2 & 1 \\ 1 & 3-\lambda & 1 \\ 1 & 2 & 2-\lambda \end{vmatrix} = -\lambda^3 + 7\lambda^2 - 11\lambda + 5$$

$$= -(\lambda - 5)(\lambda - 1)^2 = 0 \tag{14}$$

so the eigenvalues of $\mathbf{A}$ are $\lambda_1 = 5$ and $\lambda_2 = 1$ with $\lambda_2 = 1$ called a **repeated eigenvalue**—specifically, an eigenvalue of **multiplicity** 2 because it is a double root of the characteristic equation (14).

Next, find the eigenspaces.

$\lambda_1 = 5$:   Then $(\mathbf{A} - \lambda_1\mathbf{I})\mathbf{x} = \mathbf{0}$ becomes

$$\begin{bmatrix} 2-5 & 2 & 1 \\ 1 & 3-5 & 1 \\ 1 & 2 & 2-5 \end{bmatrix} \begin{bmatrix} x_1 \\ x_2 \\ x_3 \end{bmatrix} = \begin{bmatrix} -3 & 2 & 1 \\ 1 & -2 & 1 \\ 1 & 2 & -3 \end{bmatrix} \begin{bmatrix} x_1 \\ x_2 \\ x_3 \end{bmatrix} = \begin{bmatrix} 0 \\ 0 \\ 0 \end{bmatrix},$$

$$\tag{15}$$

Gauss elimination of which gives

$$\begin{bmatrix} -3 & 2 & 1 \\ 0 & 1 & -1 \\ 0 & 0 & 0 \end{bmatrix} \begin{bmatrix} x_1 \\ x_2 \\ x_3 \end{bmatrix} = \begin{bmatrix} 0 \\ 0 \\ 0 \end{bmatrix}. \tag{16}$$

The solution is $x_3 = \alpha$ (arbitrary), $x_2 = \alpha$, $x_1 = \alpha$ so, using $\mathbf{e}$ in place of $\mathbf{x}$,

$$\mathbf{e} = \begin{bmatrix} \alpha \\ \alpha \\ \alpha \end{bmatrix} = \alpha \begin{bmatrix} 1 \\ 1 \\ 1 \end{bmatrix}. \tag{17}$$

Thus, the eigenspace corresponding to $\lambda_1 = 5$ is span $\{[1, 1, 1]^T\}$, the latter being a one-dimensional subspace of $\mathbb{R}^3$, namely, the line through the origin given by (17).

$\lambda_2 = 1$:   Then $(\mathbf{A} - \lambda_2\mathbf{I})\mathbf{x} = \mathbf{0}$ becomes

$$\begin{bmatrix} 2-1 & 2 & 1 \\ 1 & 3-1 & 1 \\ 1 & 2 & 2-1 \end{bmatrix} \begin{bmatrix} x_1 \\ x_2 \\ x_3 \end{bmatrix} = \begin{bmatrix} 1 & 2 & 1 \\ 1 & 2 & 1 \\ 1 & 2 & 1 \end{bmatrix} \begin{bmatrix} x_1 \\ x_2 \\ x_3 \end{bmatrix} = \begin{bmatrix} 0 \\ 0 \\ 0 \end{bmatrix}, \tag{18}$$

Gauss elimination of which gives

$$\begin{bmatrix} 1 & 2 & 1 \\ 0 & 0 & 0 \\ 0 & 0 & 0 \end{bmatrix} \begin{bmatrix} x_1 \\ x_2 \\ x_3 \end{bmatrix} = \begin{bmatrix} 0 \\ 0 \\ 0 \end{bmatrix}. \tag{19}$$

The solution is $x_3 = \beta$ (arbitrary), $x_2 = \gamma$ (arbitrary), $x_1 = -\beta - 2\gamma$ so

$$\mathbf{e} = \begin{bmatrix} -\beta - 2\gamma \\ \gamma \\ \beta \end{bmatrix} = \beta \begin{bmatrix} -1 \\ 0 \\ 1 \end{bmatrix} + \gamma \begin{bmatrix} -2 \\ 1 \\ 0 \end{bmatrix}. \tag{20}$$

Thus, the eigenspace corresponding to $\lambda_2 = 1$ is span $\{[-1, 0, 1]^T, [-2, 1, 0]^T\}$, the latter being a two-dimensional subspace of $\mathbb{R}^3$, namely, the plane through the origin, spanned by $[-1, 0, 1]^T$ and $[-2, 1, 0]^T$. In fact, the equation of that plane is seen, in (19), to be $x_1 + 2x_2 + x_3 = 0$.[1]

COMMENT.   In the language of Section 5.7, the rank of the $\mathbf{A} - \lambda_1\mathbf{I}$ coefficient matrix in (16) is 2 so $n - r = 3 - 2 = 1$, and (according to Theorem 5.7.1) (16) admits a one-parameter family of solutions. That is, the nullity of $\mathbf{A} - \lambda_1\mathbf{I}$ is 1 and the $\mathbf{e}_1$ eigenspace is one-dimensional. Similarly, the rank of $\mathbf{A} - \lambda_2\mathbf{I}$ in (19) is 1, so $n - r = 3 - 1 = 2$, and (19) admits a two-parameter family of solutions. That is, the nullity of $\mathbf{A} - \lambda_2\mathbf{I}$ is 2, and the $\mathbf{e}_2$ eigenspace is two-dimensional.   ∎

---

[1]The plane $ax_1 + bx_2 + cx_3 = d$ passes through the origin if and only if $d = 0$.

EXAMPLE 3
The matrix

$$\mathbf{A} = \begin{bmatrix} 1 & 0 & 1 \\ 1 & 1 & 0 \\ 0 & 0 & 1 \end{bmatrix} \qquad (21)$$

has the characteristic equation

$$\det(\mathbf{A} - \lambda\mathbf{I}) = \begin{vmatrix} 1-\lambda & 0 & 1 \\ 1 & 1-\lambda & 0 \\ 0 & 0 & 1-\lambda \end{vmatrix} = (1-\lambda)^3 = 0 \qquad (22)$$

with roots $\lambda = 1, 1, 1$. That is, $\lambda_1 = 1$ is a root of multiplicity three. To find its eigenspace, write out $(\mathbf{A} - \lambda_1\mathbf{I})\mathbf{x} = \mathbf{0}$ as

$$\begin{bmatrix} 1-1 & 0 & 1 \\ 1 & 1-1 & 0 \\ 0 & 0 & 1-1 \end{bmatrix} \begin{bmatrix} x_1 \\ x_2 \\ x_3 \end{bmatrix} = \begin{bmatrix} 0 & 0 & 1 \\ 1 & 0 & 0 \\ 0 & 0 & 0 \end{bmatrix} \begin{bmatrix} x_1 \\ x_2 \\ x_3 \end{bmatrix} = \begin{bmatrix} 0 \\ 0 \\ 0 \end{bmatrix}. \qquad (23)$$

The solution is $x_3 = 0$, $x_1 = 0$, $x_2 = \alpha$ (arbitrary) so

$$\mathbf{e} = \begin{bmatrix} 0 \\ \alpha \\ 0 \end{bmatrix} = \alpha \begin{bmatrix} 0 \\ 1 \\ 0 \end{bmatrix}. \qquad (24)$$

In Examples 1 and 2 the dimension of each eigenspace was the same as the multiplicity of the corresponding eigenvalue. Lest you infer that this equality always holds, the present example shows that such a claim would be incorrect, for $\lambda - 1$ is of multiplicity three here, yet its eigenspace is only one-dimensional. ∎

EXAMPLE 4
The matrix

$$\mathbf{A} = \begin{bmatrix} 1 & 2 & 3 & 4 \\ 0 & 0 & 0 & 0 \\ 0 & 0 & 0 & 0 \\ 0 & 0 & 0 & 0 \end{bmatrix} \qquad (25)$$

has the characteristic equation

$$\det(\mathbf{A} - \lambda\mathbf{I}) = \det \begin{bmatrix} 1-\lambda & 2 & 3 & 4 \\ 0 & -\lambda & 0 & 0 \\ 0 & 0 & \lambda & 0 \\ 0 & 0 & 0 & -\lambda \end{bmatrix} = -\lambda^3(1-\lambda) = 0 \qquad (26)$$

with roots $\lambda = 0, 0, 0, 1$. Thus, $\lambda_1 = 0$ is a root of multiplicity three. To find its eigenspace write out $(\mathbf{A} - \lambda_1 \mathbf{I}) \mathbf{x} = \mathbf{0}$ as

$$
\begin{bmatrix} 1-0 & 2 & 3 & 4 \\ 0 & 0-0 & 0 & 0 \\ 0 & 0 & 0-0 & 0 \\ 0 & 0 & 0 & 0-0 \end{bmatrix} \begin{bmatrix} x_1 \\ x_2 \\ x_3 \\ x_4 \end{bmatrix} = \begin{bmatrix} 1 & 2 & 3 & 4 \\ 0 & 0 & 0 & 0 \\ 0 & 0 & 0 & 0 \\ 0 & 0 & 0 & 0 \end{bmatrix} \begin{bmatrix} x_1 \\ x_2 \\ x_3 \\ x_4 \end{bmatrix} = \begin{bmatrix} 0 \\ 0 \\ 0 \\ 0 \end{bmatrix}.
\tag{27}
$$

The solution is $x_4 = \alpha$ (arbitrary), $x_3 = \beta$ (arbitrary), $x_2 = \gamma$ (arbitrary), $x_1 = -4\alpha - 3\beta - 2\gamma$ so the eigenspace corresponding to $\lambda_1 = 0$ is given by

$$
\mathbf{e}_1 = \begin{bmatrix} -4\alpha - 3\beta - 2\gamma \\ \gamma \\ \beta \\ \alpha \end{bmatrix} = \alpha \begin{bmatrix} -4 \\ 0 \\ 0 \\ 1 \end{bmatrix} + \beta \begin{bmatrix} -3 \\ 0 \\ 1 \\ 0 \end{bmatrix} + \gamma \begin{bmatrix} -2 \\ 1 \\ 0 \\ 0 \end{bmatrix}, \tag{28}
$$

which is a three-dimensional subspace of $\mathbb{R}^4$, namely, the "hyperplane" $x_1 + 2x_2 + 3x_3 + 4x_4 = 0$.[1]

Next, $\lambda_2 = 1$. To find its eigenspace write out $(\mathbf{A} - \lambda \mathbf{I}) \mathbf{x} = \mathbf{0}$ as

$$
\begin{bmatrix} 1-1 & 2 & 3 & 4 \\ 0 & 0-1 & 0 & 0 \\ 0 & 0 & 0-1 & 0 \\ 4 & 0 & 0 & 0-1 \end{bmatrix} \begin{bmatrix} x_1 \\ x_2 \\ x_3 \\ x_4 \end{bmatrix} = \begin{bmatrix} 0 & 2 & 3 & 4 \\ 0 & -1 & 0 & 0 \\ 0 & 0 & -1 & 0 \\ 0 & 0 & 0 & -1 \end{bmatrix} \begin{bmatrix} x_1 \\ x_2 \\ x_3 \\ x_4 \end{bmatrix} = \begin{bmatrix} 0 \\ 0 \\ 0 \\ 0 \end{bmatrix}.
\tag{29}
$$

The solution is $x_4 = 0$, $x_3 = 0$, $x_2 = 0$, $x_1 = \gamma$ (arbitrary) so the eigenspace corresponding to $\lambda_2 = 1$ is

$$
\mathbf{e}_2 = \begin{bmatrix} \gamma \\ 0 \\ 0 \\ 0 \end{bmatrix} = \gamma \begin{bmatrix} 1 \\ 0 \\ 0 \\ 0 \end{bmatrix}, \tag{30}
$$

which is a one-dimensional subspace of $\mathbb{R}^4$, a line.

COMMENT.   Be clear that although an eigenvector of a matrix $\mathbf{A}$ is, by definition, a *nontrivial* solution of $\mathbf{Ax} = \lambda\mathbf{x}$, there is nothing wrong with an eigen*value* being zero, as occurs in this example and as was also illustrated in Fig. 1c of Section 9.1. We can see from (4) that zero will indeed be among the eigenvalues of $\mathbf{A}$ if and only if $\det \mathbf{A}$ happens to be zero.   ∎

9.2.2 Application to population dynamics; Markov process.

What is the point of the eigenvalue problem? That is, from an applied viewpoint, what are its applications? The most prominent, in this text, will be in connection with the solution of

---

[1] Observe that a plane is not necessarily two-dimensional. In $\mathbb{R}^3$ it is, but in $\mathbb{R}^4$ it is three-dimensional, in $\mathbb{R}^5$ it is four-dimensional, and so on; $x_1 + 2x_2 + 3x_3 + 4x_4 = 0$ is a three-dimensional structure because, as we see from (28), it is the span of three LI vectors.

systems of differential equations, in Chapters 10 and 11. However, applications of the eigenvalue problem are plentiful and diverse. Let us illustrate with an application to population dynamics.

Suppose that there is a population exchange between Delaware, Maryland, and Pennsylvania such that, each year, 20% of Delaware's residents move to Maryland and 8% to Pennsylvania; 12% of Maryland's residents move to Delaware and 10% to Pennsylvania; 10% of Pennsylvania's residents move to Delaware and 3% to Maryland. For simplicity, let us ignore gains in population due to births and losses due to deaths—or, equivalently, suppose that these effects are nonzero but equal and opposite so as to cancel. Further, let us suppose that the three states are a closed system; that is, they exchange populations only among themselves.

If we denote the populations at the end of the nth year, in DE, MD, and PA as x_n, y_n, z_n, respectively, then (see Fig. 2)

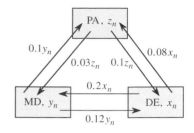

FIGURE 2
Population exchange model.

$$x_{n+1} = x_n - (0.2 + 0.08)x_n + 0.12y_n + 0.1z_n,$$
$$y_{n+1} = y_n + 0.2x_n - (0.12 + 0.1)y_n + 0.03z_n, \qquad (31)$$
$$z_{n+1} = z_n + 0.08x_n + 0.1y_n - (0.1 + 0.03)z_n,$$

or, in matrix form,

$$\begin{bmatrix} x_{n+1} \\ y_{n+1} \\ z_{n+1} \end{bmatrix} = \begin{bmatrix} 0.72 & 0.12 & 0.1 \\ 0.2 & 0.78 & 0.03 \\ 0.08 & 0.1 & 0.87 \end{bmatrix} \begin{bmatrix} x_n \\ y_n \\ z_n \end{bmatrix} \qquad (32)$$

or,

$$\mathbf{p}_{n+1} = \mathbf{A}\mathbf{p}_n, \qquad (33)$$

where $\mathbf{p}_n = [x_n, y_n, z_n]^{\mathrm{T}}$ is the "population vector" at the end of the nth year.

The first problem that we pose is to find the population $\mathbf{p}_n$ as a function of n, n being essentially a discrete time variable, given some initial population $\mathbf{p}_0$. That's simple, because (33) gives $\mathbf{p}_1 = \mathbf{A}\mathbf{p}_0$, $\mathbf{p}_2 = \mathbf{A}\mathbf{p}_1 = \mathbf{A}(\mathbf{A}\mathbf{p}_0) = \mathbf{A}^2\mathbf{p}_0$, $\mathbf{p}_3 = \mathbf{A}\mathbf{p}_2 = \mathbf{A}(\mathbf{A}^2\mathbf{p}_0) = \mathbf{A}^3\mathbf{p}_0$, and so on, so

$$\mathbf{p}_n = \mathbf{A}^n\mathbf{p}_0. \qquad (34)$$

We wonder whether $\mathbf{A}^n\mathbf{p}_0$ keeps changing as n increases, or whether it settles down and approaches an equilibrium (or steady-state) vector, say $\mathbf{P}$. If there is such an equilibrium vector then, by the definition of equilibrium, $\mathbf{p}_{n+1} = \mathbf{p}_n = \mathbf{P}$, so (33) becomes $\mathbf{P} = \mathbf{A}\mathbf{P}$. Surely, $\mathbf{P} = \mathbf{0}$ satisfies

$$\mathbf{A}\mathbf{P} = \mathbf{P} \qquad (35)$$

(of course, if we start with no people then we always have no people), but the interesting question is whether or not there exist *non*trivial $\mathbf{P}$'s. In fact, observe that (35) is an eigenvalue problem with $\lambda = 1$, so we can say that nontrivial equilibrium vectors exist if and only if 1 happens to be an eigenvalue of $\mathbf{A}$. As explained at the end of this section, we can use *Maple* to obtain the following eigenvalues and eigenspaces of $\mathbf{A}$:

$$\lambda_1 = 1, \; \mathbf{e}_1 = \alpha \begin{bmatrix} -0.56 \\ -0.62 \\ -0.82 \end{bmatrix}; \qquad \lambda_2 = 0.77, \; \mathbf{e}_2 = \beta \begin{bmatrix} -0.09 \\ -0.70 \\ 0.79 \end{bmatrix};$$

$$\lambda_3 = 0.60, \; \mathbf{e}_3 = \gamma \begin{bmatrix} -0.62 \\ 0.70 \\ -0.07 \end{bmatrix}. \tag{36}$$

(Actually, *Maple* gives the λ_j's and $\mathbf{e}_j$'s to nine and ten significant figures, respectively, but we have rounded off for brevity.)

Sure enough, $\lambda = 1$ is among the eigenvalues of $\mathbf{A}$ so there is an equilibrium population vector $\mathbf{P}$ given by the corresponding eigenvector

$$\mathbf{P} = \alpha \begin{bmatrix} 0.56 \\ 0.62 \\ 0.82 \end{bmatrix}, \tag{37}$$

where we have absorbed a factor of -1 into the scale factor α, to avoid the appearance of negative populations. If desired, we can compute α by conserving the total population: in equilibrium it is $0.56\alpha + 0.62\alpha + 0.82\alpha$, and initially it is $x_0 + y_0 + z_0$; equating the two gives $\alpha = (x_0 + y_0 + z_0)/2.0$.[1]

Stability of the equilibrium population.   Finally, it is important to determine whether or not the equilibrium is *stable* for it will be observed only if it is stable, just as marbles are found in valleys, but not on hilltops. To address the question of stability, let us use the set of LI eigenvectors $\{\mathbf{e}_1, \mathbf{e}_2, \mathbf{e}_3\}$, with any nonzero values of α, β, and γ, such as $\alpha = \beta = \gamma = 1$, as a basis for $\mathbb{R}^3$, and expand the initial vector $\mathbf{p}_0$ in terms of that basis as

$$\mathbf{p}_0 = c_1\mathbf{e}_1 + c_2\mathbf{e}_2 + c_3\mathbf{e}_3. \tag{38}$$

Then (33) gives

$$\mathbf{p}_1 = \mathbf{A}\mathbf{p}_0 = c_1\mathbf{A}\mathbf{e}_1 + c_2\mathbf{A}\mathbf{e}_2 + c_3\mathbf{A}\mathbf{e}_3$$
$$= c_1\lambda_1\mathbf{e}_1 + c_2\lambda_2\mathbf{e}_2 + c_3\lambda_3\mathbf{e}_3,$$

$$\mathbf{p}_2 = \mathbf{A}\mathbf{p}_1 = c_1\lambda_1\mathbf{A}\mathbf{e}_1 + c_2\lambda_2\mathbf{A}\mathbf{e}_2 + c_3\lambda_3\mathbf{A}\mathbf{e}_3$$
$$= c_1\lambda_1^2\mathbf{e}_1 + c_2\lambda_2^2\mathbf{e}_2 + c_3\lambda_3^2\mathbf{e}_3,$$

$$\vdots$$

$$\mathbf{p}_n = \mathbf{A}\mathbf{p}_{n-1} = c_1\lambda_1^n\mathbf{e}_1 + c_2\lambda_2^n\mathbf{e}_2 + c_3\lambda_3^n\mathbf{e}_3$$
$$= c_1\mathbf{e}_1 + c_2(0.77)^n\mathbf{e}_2 + c_3(0.60)^n\mathbf{e}_3$$
$$\sim c_1\mathbf{e}_1 \tag{39}$$

---

[1]Recall that we built into (32) the assumption that any births and deaths cancel, in number, as revealed by adding the three scalar equations in (32), for that step gives $x_{n+1} + y_{n+1} + z_{n+1} = x_n + y_n + z_n$; that is, the total population is a constant from year to year.

as $n \rightarrow \infty$ (because 0.77^n and 0.60^n both tend to zero as $n \rightarrow \infty$), provided that $c_1 \neq 0$. In fact, c_1 cannot be zero because if it were zero then (39) would give $\mathbf{p}_n \rightarrow \mathbf{0}$ as $n \rightarrow \infty$ and, since the total population is conserved, that could happen only in the uninteresting case where $\mathbf{p}_0 = \mathbf{0}$. Thus, we see from (39) that $\mathbf{p}_n$ inevitably tends to a multiple of $\mathbf{e}_1$, namely, to the equilibrium vector $\mathbf{P}$.

The upshot is that the population history is given by (34), and that $\mathbf{p}_n$ inevitably tends to a unique steady state which is some scalar multiple of $\mathbf{e}_1$, the multiple being fixed by the conservation of the total population.

COMMENT.   This example incorporates a number of linear algebra concepts: matrix multiplication in expressing (31) compactly as (33) and in deriving the solution (34) for $\mathbf{p}_n$; the eigenvalue problem in regard to the possibility of a steady-state solution $\mathbf{P}$; and bases and expansions in assessing the stability of that steady state, in (38)–(39). Consider how effective these linear algebra concepts and methods are in providing a systematic approach to solving this problem, especially in determining the stability of the steady state. The same approach applies whether the system includes only three states, as in the present example, or many states.

The matrix $\mathbf{A}$ in (32) is an example of a "Markov" matrix. An $n \times n$ matrix $\mathbf{A} = \{a_{ij}\}$ is called a **Markov (or stochastic) matrix** if $a_{ij} \geq 0$ for each i, j, and if the elements of each column sum to unity, or if the elements of each row sum to unity. [In the case of the $\mathbf{A}$ given in (32), its columns sum to unity.] It was no coincidence that $\lambda = 1$ was an eigenvalue, since $\lambda = 1$ is an eigenvalue of *every* Markov matrix.

THEOREM 9.2.1
Eigenvalues of Markov Matrix
Every Markov matrix has among its eigenvalues $\lambda = 1$.

Proof.   Let $\mathbf{A}$ be a Markov matrix. The value $\lambda = 1$ will be among the eigenvalues of $\mathbf{A}$ if and only if $\mathbf{Ax} = \mathbf{x}$ has nontrivial solutions for $\mathbf{x}$ or, equivalently, if the rows or columns of $\mathbf{A} - \mathbf{I}$ are linearly dependent. Since $\mathbf{A}$ is a Markov matrix, either the elements of each of its columns sum to unity or the elements of each of its rows sum to unity. It follows that either the elements of each of the columns of $\mathbf{A} - \mathbf{I}$ sum to zero (in which case the rows vectors in $\mathbf{A} - \mathbf{I}$ are linearly dependent) or the elements of each of the rows of $\mathbf{A} - \mathbf{I}$ sum to zero (in which case the columns vectors in $\mathbf{A} - \mathbf{I}$ are linearly dependent), or both. Thus, our claim is proved.   ∎

9.2.3 Complex eigenvalues and eigenvectors. (Optional).   Thus far in this text we have assumed that all matrix elements, vector components, and scalars are real numbers, both for simplicity and because this case is the most important to us in applications. Surely, if an $n \times n$ matrix $\mathbf{A}$ is real, then the coefficients of its nth-degree characteristic polynomial $\det(\mathbf{A} - \lambda \mathbf{I})$ are real too. Nevertheless, the characteristic equation $\det(\mathbf{A} - \lambda \mathbf{I}) = 0$ *can* have complex roots; from the theory of equations, we know that any such complex roots will occur in complex conjugate pairs. If indeed an eigenvalue is complex, then its eigenspace will be too.

EXAMPLE 5

The matrix

$$\mathbf{A} = \begin{bmatrix} 1 & -4 & 0 \\ 1 & 1 & 1 \\ 0 & 0 & 3 \end{bmatrix} \tag{40}$$

has the characteristic equation

$$\det(\mathbf{A} - \lambda\mathbf{I}) = \begin{vmatrix} 1-\lambda & -4 & 0 \\ 1 & 1-\lambda & 1 \\ 0 & 0 & 3-\lambda \end{vmatrix} = (3-\lambda)(\lambda^2 - 2\lambda + 5) = 0 \tag{41}$$

with roots $\lambda = 3, 1 + 2i, 1 - 2i$.[1] Proceeding as in Examples 1–4, we find that the eigenspace corresponding to $\lambda_1 = 3$ is

$$\mathbf{e}_1 = \alpha \begin{bmatrix} -1/2 \\ 1/4 \\ 1 \end{bmatrix} \tag{42}$$

or, scaling by 4 to eliminate fractions,

$$\mathbf{e}_1 = \alpha \begin{bmatrix} -2 \\ 1 \\ 4 \end{bmatrix}. \tag{43}$$

Next, $\lambda_2 = 1 + 2i$. To find its eigenspace write out $(\mathbf{A} - \lambda_2\mathbf{I})\mathbf{x} = \mathbf{0}$ as

$$\begin{bmatrix} 1-(1+2i) & -4 & 0 \\ 1 & 1-(1+2i) & 1 \\ 0 & 0 & 3-(1+2i) \end{bmatrix} \begin{bmatrix} x_1 \\ x_2 \\ x_3 \end{bmatrix}$$

$$= \begin{bmatrix} -2i & -4 & 0 \\ 1 & -2i & 1 \\ 0 & 0 & 2-2i \end{bmatrix} \begin{bmatrix} x_1 \\ x_2 \\ x_3 \end{bmatrix} = \begin{bmatrix} 0 \\ 0 \\ 0 \end{bmatrix}. \tag{44}$$

To solve (44) by Gauss elimination write

$$\begin{bmatrix} -2i & -4 & 0 & 0 \\ 1 & -2i & 1 & 0 \\ 0 & 0 & 2-2i & 0 \end{bmatrix} \rightarrow \begin{bmatrix} 1 & -2i & 1 & 0 \\ -2i & -4 & 0 & 0 \\ 0 & 0 & 2-2i & 0 \end{bmatrix}$$

$$\rightarrow \begin{bmatrix} 1 & -2i & 1 & 0 \\ 0 & 0 & 2i & 0 \\ 0 & 0 & 2-2i & 0 \end{bmatrix} \rightarrow \begin{bmatrix} 1 & -2i & 1 & 0 \\ 0 & 0 & 1 & 0 \\ 0 & 0 & 0 & 0 \end{bmatrix}, \tag{45}$$

where in the first step we interchanged the first two rows and in the second step we added $2i$ times the first row to the second. The solution is $x_3 = 0$, $x_2 = \beta$ (arbitrary),

---

[1] The temptation to multiply out the right-hand side of (41) as $-\lambda^3 + 5\lambda^2 - 11\lambda + 15$ is to be resisted since that step would be counterproductive; our aim is to factor the characteristic polynomial in order to solve for the eigenvalues.

$x_1 = 2i\beta$, so the eigenspace corresponding to $\lambda_2 = 1 + 2i$ is

$$\mathbf{e}_2 = \beta \begin{bmatrix} 2i \\ 1 \\ 0 \end{bmatrix}. \tag{46}$$

We will show, below, that if λ has an eigenvector $\mathbf{e}$, then $\overline{\lambda}$ has an eigenvector $\overline{\mathbf{e}}$, where the overhead bars denote complex conjugates, as usual. Thus, since $\lambda_2 + 1 + 2i$ has $\mathbf{e}_2$ given by (46), then $\lambda_3 = 1 - 2i$ has

$$\mathbf{e}_3 = \gamma \begin{bmatrix} -2i \\ 1 \\ 0 \end{bmatrix}. \tag{47}$$

[Or, one can derive (47) in the same way that we derived (46), by solving $(\mathbf{A} - \lambda_2\mathbf{I})\mathbf{x} = \mathbf{0}$ by Gauss elimination.] ∎

THEOREM 9.2.2
Eigenvectors for Complex Conjugate Eigenvalues
If λ is a complex eigenvalue of a real matrix $\mathbf{A}$, with corresponding eigenvector $\mathbf{e}$, then $\overline{\lambda}$ is also an eigenvalue of $\mathbf{A}$, with corresponding eigenvector $\overline{\mathbf{e}}$.

Proof. If λ is an eigenvalue of $\mathbf{A}$, with eigenvector $\mathbf{e}$, then $\mathbf{Ae} = \lambda\mathbf{e}$. Taking the complex conjugate of each side gives $\overline{\mathbf{Ae}} = \overline{\lambda\mathbf{e}}$. But $\overline{\mathbf{Ae}} = \overline{\mathbf{A}}\overline{\mathbf{e}} = \mathbf{A}\overline{\mathbf{e}}$ because $\mathbf{A}$ is real, and $\overline{\lambda\mathbf{e}} = \overline{\lambda}\overline{\mathbf{e}}$ [by (12b) in Section 6.2], so

$$\mathbf{A}\overline{\mathbf{e}} = \overline{\lambda}\overline{\mathbf{e}}. \tag{48}$$

Surely $\overline{\mathbf{e}} \neq \mathbf{0}$, because $\mathbf{e} \neq \mathbf{0}$, so it follows from (48) that $\overline{\lambda}$ is an eigenvalue of $\mathbf{A}$, with eigenvector $\overline{\mathbf{e}}$. ∎

The upshot is that even if $\mathbf{A}$ is real it is nevertheless possible that it will have complex eigenvalues and eigenvectors. Thus, if $\mathbf{A}$ is $n \times n$ then we are not working with the usual real vector space $\mathbb{R}^n$, but with the **complex vector space** $\mathbb{C}^n$. For recall that all scalars in the vector space $\mathbb{R}^n$ defined in Section 4.3 (namely, the scalars that multiplied vectors and the scalar components of the vectors themselves) are real. If we allow these scalars to be complex, then in place of $\mathbb{R}^n$ we have the **complex n-space** denoted here as $\mathbb{C}^n$:

$$\mathbb{C}^n = \{[a_1, \ldots, a_n]^T \mid a_1, \ldots, a_n \text{ complex numbers }\}. \tag{49}$$

The definitions $\mathbf{u} + \mathbf{v} \equiv [u_1 + v_1, \ldots, u_n + v_n]^T$, $\alpha\mathbf{u} \equiv [\alpha u_1, \ldots, \alpha u_n]^T$, $\mathbf{0} \equiv [0, \ldots, 0]^T$, and $-\mathbf{u} \equiv [-u_1, \ldots, -u_n]^T$ are the same as for $\mathbb{R}^n$, except that now the scalars may be complex numbers. From these definitions the same properties follow for $\mathbb{C}^n$ as for $\mathbb{R}^n$ ($\mathbf{u} + \mathbf{v} = \mathbf{v} + \mathbf{u}$, etc.), as in (10) in Section 4.3.[1]

---

[1]Note that in Chapter 4 we used a row format for vectors, whereas beginning in Chapter 5 we adopted a column format. Here we adhere to the column format in (49).

However, the Euclidean dot product $\mathbf{u} \cdot \mathbf{v} = u_1 v_1 + \cdots + u_n v_n$ which we adopted for real vector space is unacceptable because the resulting norm $\|\mathbf{u}\| = \sqrt{\mathbf{u} \cdot \mathbf{u}} = \sqrt{u_1^2 + \cdots + u_n^2}$ fails to display the key properties expected of a norm, in particular, the nonnegativeness condition

$$\|\mathbf{u}\| > 0 \quad \text{for all } \mathbf{u} \neq \mathbf{0},$$
$$= 0 \quad \text{for } \mathbf{u} = \mathbf{0}. \tag{50}$$

For example, if $\mathbf{u} = [2, 2i, 0, 0]^T$, then $\|\mathbf{u}\| = \sqrt{(2)^2 + (2i)^2 + (0)^2 + (0)^2} = 0$ even though $\mathbf{u} \neq \mathbf{0}$; and if $\mathbf{u} = [0, 2i, 0, 0]^T$, then $\|\mathbf{u}\| = \sqrt{0 - 4 + 0 + 0} = 2i$ is not even real so it cannot satisfy the condition $\|\mathbf{u}\| > 0$.

To avoid this problem with the norm, we adopt the modified dot product

$$\mathbf{u} \cdot \mathbf{v} \equiv u_1 \bar{v}_1 + u_2 \bar{v}_2 + \cdots + u_n \bar{v}_n = \sum_{j=1}^{n} u_j \bar{v}_j \tag{51}$$

for $\mathbb{C}^n$ or, in matrix form, $\mathbf{u} \cdot \mathbf{v} = \mathbf{u}^T \bar{\mathbf{v}}$, for then

$$\|\mathbf{u}\| \equiv \sqrt{\mathbf{u} \cdot \mathbf{u}} = \sqrt{\sum_{j=1}^{n} u_j \bar{u}_j} = \sqrt{\sum_{j=1}^{n} |u_j|^2} \tag{52}$$

does satisfy the nonnegativeness condition (50). [Recall that if $z = a + ib$, then $z\bar{z} = (a + ib)(a - ib) = a^2 + b^2 = |z|^2$.]

EXAMPLE 6
If $\mathbf{u} = [2, 3 - 5i, 0, 4i]^T$ and $\mathbf{v} = [i, 1 + i, 2 - i, 6]^T$, then

$$\|\mathbf{u}\| = \sqrt{(2)(2) + (3 - 5i)(3 + 5i) + (0)(0) + (4i)(-4i)} = \sqrt{54},$$

$$\|\mathbf{v}\| = \sqrt{(i)(-i) + (1 + i)(1 - i) + (2 - i)(2 + i) + (6)(6)} = \sqrt{44},$$

and

$$\mathbf{u} \cdot \mathbf{v} = (2)(-i) + (3 - 5i)(1 - i) + (0)(2 + i) + (4i)(6) = -2 + 14i. \quad \blacksquare$$

We will use (51) as our dot product in Section 9.3.

Finally, what is the dimension of $\mathbb{C}^n$?

THEOREM 9.2.3
Dimension of $\mathbb{C}^n$
The dimension of $\mathbb{C}^n$ is n: dim $\mathbb{C}^n = n$.

Proof.   The vectors

$$\mathbf{e}_1 = [1, 0, 0, \ldots, 0]^T,$$
$$\mathbf{e}_2 = [0, 1, 0, \ldots, 0]^T,$$
$$\vdots \tag{53}$$
$$\mathbf{e}_n = [0, \ldots, 0, 0, 1]^T$$

constitute a basis for $\mathbb{C}^n$ just as they do for $\mathbb{R}^n$ because any vector $\mathbf{u} = [u_1, \dots, u_n]^T$ in $\mathbb{C}^n$ can be expanded uniquely as $\mathbf{u} = u_1\mathbf{e}_1 + \cdots + u_n\mathbf{e}_n$. Since this basis contains n vectors, it follows from Theorem 4.9.2 that $\dim\mathbb{C} = n$. ∎

One might have expected that $\mathbb{C}^n$ is $2n$-dimensional on the grounds that there are n vector components and each, being complex, is two-dimensional. No, $\mathbb{C}^n$ is n-dimensional since the vectors given in (53) constitute a basis for $\mathbb{C}^n$, and there are n of them.

Closure. The eigenvalue problem $\mathbf{Ax} = \lambda\mathbf{x}$ is homogeneous since it is equivalent to $(\mathbf{A} - \lambda\mathbf{I})\mathbf{x} = \mathbf{0}$. The latter admits nontrivial solutions $\mathbf{x}$, besides the trivial solution $\mathbf{x} = \mathbf{0}$, if and only if λ is chosen so that $\det(\mathbf{A} - \lambda\mathbf{I})$. Such a λ is called an eigenvalue of $\mathbf{A}$ and the corresponding nontrivial vector solutions of $(\mathbf{A} - \lambda\mathbf{I})\mathbf{x} = \mathbf{0}$ are called eigenvectors corresponding to the eigenvalue λ and are denoted as $\mathbf{e}$. If, for a given eigenvalue λ, you solve $(\mathbf{A} - \lambda\mathbf{I})\mathbf{x} = \mathbf{0}$ by Gauss elimination and obtain a unique solution $\mathbf{e} = \mathbf{0}$, then your calculations are incorrect: either the eigenvalue is incorrect and/or the Gauss elimination is incorrect.

Observe that our solution strategy uncouples the calculation of the eigenvalues and the eigenvectors: First we solve the characteristic polynomial equation $\det(\mathbf{A} - \lambda\mathbf{I}) = 0$ for the λ's, and then for each λ we solve $(\mathbf{A} - \lambda\mathbf{I})\mathbf{x} = \mathbf{0}$ by Gauss elimination for the corresponding eigenvectors.

Observe that every $n \times n$ matrix $\mathbf{A}$ has at least one eigenvalue since, by the fundamental theorem of algebra, its characteristic equation must have at least one root.

Finally, be aware that even if $\mathbf{A}$ is real it can have complex eigenvalues and eigenvectors, as in Example 5. In that case the relevant vector space is $\mathbb{C}^n$ rather than $\mathbb{R}^n$.

Maple. In *Maple*, the relevant commands are **eigenvals** and **eigenvects**, both of which are in the linalg package. The command eigenvals gives just the eigenvalues, and eigenvects gives the eigenvalues, their multiplicity, and a basis for its eigenspace. For instance, let $\mathbf{A}$ be the matrix in Example 1. First, enter

```
with(linalg):
```

and return. Next, enter

```
A:=array([[2,2,1],[1,3,1],[1,2,2]]);
```

Then

```
eigenvals(A);
```

gives the eigenvalues as

```
5,1,1
```

and

```
eigenvects(A);
```

gives both the eigenvalues and the eigenvectors as

$$[5,1,\{[1,1,1]\}],[1,2,\{[-2,1,0],[-1,0,1]\}]$$

Suppose we want the eigenvalues of $\mathbf{A}^{20}$. Enter

```
A:=array([[2,2,1],[1,3,1],[1,2,2]]);
```

then

```
evalm(A^20);
```

and then

```
eigenvals(%);
```

The percentage sign inputs the preceding output, thereby saving us the trouble of entering the matrix $\mathbf{A}^{20}$ that was calculated in the preceding step.

EXERCISES 9.2

1. Find the eigenvalues and eigenspaces, as well as a basis for each eigenspace.

(**a**) $\begin{bmatrix} 0 & 0 \\ 0 & 0 \end{bmatrix}$

(**b**) $\begin{bmatrix} 1 & -3 \\ 0 & 0 \end{bmatrix}$

(**c**) $\begin{bmatrix} 1 & 2 \\ 3 & 4 \end{bmatrix}$

(**d**) $\begin{bmatrix} -3 & 2 \\ 6 & -4 \end{bmatrix}$

(**e**) $\begin{bmatrix} 0 & 0 & 0 \\ 0 & 3 & 0 \\ 0 & 0 & 0 \end{bmatrix}$

(**f**) $\begin{bmatrix} 0 & 0 & 0 \\ 0 & 0 & 0 \\ 0 & 0 & 0 \end{bmatrix}$

(**g**) $\begin{bmatrix} 2 & 0 & 0 \\ 0 & -5 & 0 \\ 0 & 0 & 4 \end{bmatrix}$

(**h**) $\begin{bmatrix} 2 & 1 & 6 \\ 0 & -5 & 3 \\ 0 & 0 & 4 \end{bmatrix}$

(**i**) $\begin{bmatrix} 2 & 0 & 0 \\ 0 & 1 & 1 \\ 0 & 1 & 1 \end{bmatrix}$

(**j**) $\begin{bmatrix} 4 & 4 & 4 \\ 4 & 4 & 4 \\ 4 & 4 & 4 \end{bmatrix}$

(**k**) $\begin{bmatrix} 1 & 0 & 2 \\ 1 & 0 & 2 \\ 1 & 0 & 2 \end{bmatrix}$

(**l**) $\begin{bmatrix} 0 & 0 & 1 \\ 0 & 0 & 1 \\ 1 & 1 & 1 \end{bmatrix}$

(**m**) $\begin{bmatrix} 1 & 0 & 1 \\ 0 & 2 & 0 \\ 4 & 0 & 4 \end{bmatrix}$

(**n**) $\begin{bmatrix} 0 & 0 & 3 \\ 0 & 0 & 1 \\ 1 & -13 & 7 \end{bmatrix}$

(**o**) $\begin{bmatrix} 1 & 0 & 1 \\ 1 & 1 & 0 \\ 0 & 0 & 1 \end{bmatrix}$

(**p**) $\begin{bmatrix} 0 & 2 & 0 \\ 0 & 3 & 0 \\ 0 & 4 & 0 \end{bmatrix}$

(**q**) $\begin{bmatrix} 2 & 0 & 0 & 0 \\ 0 & 0 & 0 & 1 \\ 0 & 0 & 0 & 1 \\ 0 & 1 & 1 & 1 \end{bmatrix}$

(**r**) $\begin{bmatrix} 1 & 1 & 1 & 1 \\ 2 & 2 & 2 & 2 \\ 3 & 3 & 3 & 3 \\ 4 & 4 & 4 & 4 \end{bmatrix}$

2. (**a**)–(**r**) Use *Maple* to find the eigenvalues and eigenspaces for the matrix in the corresponding part of Exercise 1.

3. Is the following an eigenvector of the matrix $\mathbf{A}$? Explain.

$$\mathbf{A} = \begin{bmatrix} 1 & 8 & 5 & 3 \\ 2 & 16 & 10 & 6 \\ 5 & -14 & -11 & -3 \\ -1 & -8 & -5 & -3 \end{bmatrix}.$$

(**a**) $[1, 2, -1, 3]^\mathrm{T}$ (**b**) $[1, 2, -4, -1]^\mathrm{T}$

(**c**) $[1, 2, 1, 1]^\mathrm{T}$ (**d**) $[1, 0, 1, -2]^\mathrm{T}$

(**e**) $[1, 0, 1, -1]^\mathrm{T}$ (**f**) $[1, 1, 0, -3]^\mathrm{T}$

(**g**) $[1, 2, -1, -1]^\mathrm{T}$ (**h**) $[2, 1, 0, 1]^\mathrm{T}$

(**i**) $[2, 1, 1, -5]^\mathrm{T}$

4. The given matrix has $\lambda = 2$ among its eigenvalues. Find the eigenspace corresponding to that eigenvalue.

(**a**) $\begin{bmatrix} 3 & 2 & 2 & 1 \\ 2 & 3 & 1 & 2 \\ -1 & 1 & 2 & 0 \\ 2 & 4 & 3 & 5 \end{bmatrix}$

(**b**) $\begin{bmatrix} 3 & 1 & 2 & 1 \\ -1 & 3 & 1 & 2 \\ 0 & 2 & 5 & 3 \\ 1 & 3 & 5 & 6 \end{bmatrix}$

(**c**) $\begin{bmatrix} 3 & 1 & 1 & 1 \\ 1 & 3 & 1 & 1 \\ 1 & 1 & 3 & 1 \\ 1 & 1 & 1 & 3 \end{bmatrix}$

(**d**) $\begin{bmatrix} 3 & 0 & 1 & 1 \\ 1 & 3 & 0 & 1 \\ -1 & 1 & 3 & -1 \\ 1 & 2 & 2 & 3 \end{bmatrix}$

(e) $\begin{bmatrix} 2 & 0 & 0 & 0 \\ 3 & 2 & 0 & 0 \\ 0 & 4 & 2 & 0 \\ 1 & 0 & 5 & 1 \end{bmatrix}$ (f) $\begin{bmatrix} 0 & 0 & 0 & 0 \\ 0 & 2 & 0 & 0 \\ 0 & 0 & 2 & 0 \\ 0 & 0 & 0 & 0 \end{bmatrix}$

5. How do you think we made up Exercise 4? That is, describe a process by which we can obtain various 4×4 (for instance) matrices having $\lambda = 2$ (for instance) among their eigenvalues. Illustrate by using that process to generate one such matrix.

6. It is known that the $n \times n$ tridiagonal matrix

$$\mathbf{A} = \begin{bmatrix} b & c & 0 & 0 & & \cdots & 0 \\ a & b & c & 0 & & & \vdots \\ 0 & a & b & c & & & \\ & & & \ddots & & & \vdots \\ \vdots & & & & a & b & c \\ 0 & \cdots & & \cdots & 0 & a & b \end{bmatrix}$$

has eigenvalues

$$\lambda_j = b + 2\sqrt{ac}\,\cos\frac{j\pi}{n+1} \qquad (6.1)$$

for $j = 1, 2, \ldots, n$. (**A** is called **tridiagonal** because all elements are zero except for those on the main diagonal and the two adjacent diagonals.)

(a) Verify (6.1) by calculating the eigenvalues for $n = 1$ and $n = 2$.

(b) Verify (6.1) by using *Maple* to determine the eigenvalues for $n = 1$ and 2 and 3.

(c) Verify (6.1) by using *Maple* to determine the eigenvalues for $n = 4$ and $a = 1, b = 2, c = 1$.

(d) Same as (c), for $n = 4$ and $a = 2, b = 3, c = -1$.

(e) Same as (c), for $n = 5$ and $a = 1, b = 5, c = 3$.

7. We saw in Example 1 that a given eigenvalue can have more than one LI eigenvector. Can a given eigenvector correspond to more than one eigenvalue? Explain.

8. Show that if **A** is triangular, its eigenvalues are simply the diagonal elements of **A**.

9. Show that the eigenvalues of $k\mathbf{A}$, for any scalar k, are k times those of **A**. Are the corresponding eigenspaces the same? Explain.

10. Show that the eigenvalues of $\mathbf{A}^{\mathrm{T}}$ are the same as those of **A**. Is the eigenspace corresponding to an eigenvalue λ of **A** the same as the eigenspace corresponding to the same eigenvalue λ of $\mathbf{A}^{\mathrm{T}}$? Prove or disprove.

11. Show that if λ is an eigenvalue of **A**, with a corresponding eigenvector **e**, then λ^n is an eigenvalue of $\mathbf{A}^n$, with the same eigenvector **e**, for any integer n. (Of course, if n is

negative, **A** needs to be nonsingular if $\mathbf{A}^n$ is to exist in the first place.) HINT: Pre-multiply $\mathbf{Ae} = \lambda\mathbf{e}$ by $\mathbf{A}, \mathbf{A}^2, \ldots$.

12. Use the results stated in the preceding two exercises to determine the eigenvalues and eigenspaces of $\mathbf{A}^{10}$ for each of the following **A** matrices. Check your results by working out $\mathbf{A}^{10}$ and its eigenvalues and eigenvectors.

(a) $\begin{bmatrix} 1 & 0 & 0 \\ 1 & 2 & 0 \\ 1 & 1 & 3 \end{bmatrix}$ (b) $\begin{bmatrix} 3 & 0 & 1 \\ 0 & 0 & 0 \\ 0 & 0 & 0 \end{bmatrix}$

(c) $\begin{bmatrix} 1 & 0 & 1 \\ 0 & 0 & 0 \\ 0 & 0 & -2 \end{bmatrix}$ (d) $\begin{bmatrix} 1 & 1 & 2 \\ 0 & -1 & 2 \\ 0 & 0 & 2 \end{bmatrix}$

13. For the given **A** matrix, use *Maple* to determine its eigenvalues and eigenspaces. Then, use *Maple* to obtain $\mathbf{A}^5$ and to determine its eigenvalues and eigenspaces. Then, verify the result stated in Exercise 11, for this case.

(a) $\begin{bmatrix} 2 & 2 & 0 \\ 0 & 2 & 0 \\ 2 & 0 & 2 \end{bmatrix}$ (b) $\begin{bmatrix} 1 & 1 & 1 \\ 2 & 2 & 2 \\ 3 & 3 & 3 \end{bmatrix}$

14. (*Similar matrices*)

(a) Suppose that $\mathbf{Ax} = \mathbf{y}$, where **A** is square. Setting $\mathbf{x} = \mathbf{Q}\widetilde{\mathbf{x}}$ and $\mathbf{y} = \mathbf{Q}\widetilde{\mathbf{y}}$, where **Q** is invertible, show that

$$\widetilde{\mathbf{A}}\widetilde{\mathbf{x}} = \widetilde{\mathbf{y}}, \qquad (14.1)$$

where

$$\widetilde{\mathbf{A}} = \mathbf{Q}^{-1}\mathbf{A}\mathbf{Q}, \qquad (14.2)$$

Given any invertible matrix **Q**, matrices **A** and $\widetilde{\mathbf{A}}$ related by (14.2) are said to be **similar**.

(b) (b) Show that if **A** and $\widetilde{\mathbf{A}}$ are similar, then they have the same characteristic polynomials and hence the same eigenvalues.

15. Show that if two $n \times n$ matrices **A** and **B** have the same eigenvalues $\lambda_1, \ldots, \lambda_n$ and the same n LI eigenvectors $\mathbf{e}_1, \ldots, \mathbf{e}_n$, then it must be true that $\mathbf{A} = \mathbf{B}$.

16. Can an $n \times n$ matrix have more than n LI eigenvectors? Explain.

17. (*Markov matrices*) Recall the definition of a Markov matrix, above Theorem 9.2.1, and that every Markov matrix contains $\lambda = 1$ among its eigenvalues. You may use the result stated in Exercise 9 if you need it.

(a) Find one eigenvalue of

$$\mathbf{A} = \begin{bmatrix} 8 & 10 & 12 \\ 9 & 10 & 11 \\ 10 & 10 & 10 \end{bmatrix}.$$

(b) Find one eigenvalue of

$$A = \begin{bmatrix} 3 & 1 & 2 \\ 2 & 2 & 1 \\ 0 & 2 & 2 \end{bmatrix}.$$

(c) Determine the eigenvalues and a basis for each eigenspace for the 20×20 matrix A having unity for each of its 400 elements.

(d) Determine the eigenvalues and a basis for each eigenspace for the 30×30 matrix A having 2 for each of its 900 elements.

18. (*Cayley-Hamilton theorem*) The **Cayley-Hamilton theorem** states that if the characteristic equation of any square matrix A is $\lambda^n + \alpha_1 \lambda^{n-1} + \cdots + \alpha_{n-1}\lambda + \alpha_n = 0$, then $A^n + \alpha_1 A^{n-1} + \cdots + \alpha_{n-1}A + \alpha_n I = 0$; i.e., A *satisfies its own characteristic equation.*

(a) Prove this theorem for the general 2×2 case, $A = \begin{bmatrix} a & b \\ c & d \end{bmatrix}$.

(b) If $A = \begin{bmatrix} 2 & 1 \\ 1 & 2 \end{bmatrix}$, show that $A^2 - 4A + 3I = 0$ so that

$$A^{-1} = \frac{4}{3}I - \frac{1}{3}A = \begin{bmatrix} \frac{2}{3} & -\frac{1}{3} \\ -\frac{1}{3} & \frac{2}{3} \end{bmatrix}.$$

19. (*Generalized eigenvalue problem*) If $B \neq I$, then $Ax = \lambda Bx$ is called a **generalized eigenvalue problem**. It should be easy to see that in this case the characteristic equation is $\det(A - \lambda B) = 0$, and that the eigenvectors then follow as the nontrivial solutions of $(A - \lambda B)x = 0$. Find the eigenvalues and eigenspaces in each case.

(a) $\begin{bmatrix} 5 & 1 \\ 1 & 5 \end{bmatrix} \begin{bmatrix} x_1 \\ x_2 \end{bmatrix} = \lambda \begin{bmatrix} 8 & -4 \\ -4 & 8 \end{bmatrix} \begin{bmatrix} x_1 \\ x_2 \end{bmatrix}$

(b) $\begin{bmatrix} x_1 \\ x_2 \end{bmatrix} = \lambda \begin{bmatrix} 1 & 0 \\ 1 & 1 \end{bmatrix} \begin{bmatrix} x_1 \\ x_2 \end{bmatrix}$

(c) $\begin{bmatrix} 2 & -1 \\ -1 & 2 \end{bmatrix} \begin{bmatrix} x_1 \\ x_2 \end{bmatrix} = \lambda \begin{bmatrix} 1 & 0 \\ 0 & 2 \end{bmatrix} \begin{bmatrix} x_1 \\ x_2 \end{bmatrix}$

(d) $\begin{bmatrix} x_1 \\ x_2 \\ x_3 \end{bmatrix} = \lambda \begin{bmatrix} 2 & 1 & 1 \\ 1 & 2 & 1 \\ 1 & 1 & 2 \end{bmatrix} \begin{bmatrix} x_1 \\ x_2 \\ x_3 \end{bmatrix}$

20. In seeking the eigenvalues and eigenvectors of a given matrix A, is it permissible first to simplify A by means of some elementary row operations? (That is, are the eigenvalues and eigenvectors of A invariant with respect to elementary row operations?)

21. (*Complex eigenvalues and eigenvectors*) Find the eigenvalues and eigenspaces, and a basis for each eigenspace.

(a) $\begin{bmatrix} 1 & 0 & 0 \\ 0 & 0 & -1 \\ -1 & 1 & 1 \end{bmatrix}$ **(b)** $\begin{bmatrix} 0 & 0 & 1 \\ 0 & 0 & 0 \\ -1 & 0 & 0 \end{bmatrix}$

(c) $\begin{bmatrix} 0 & 2 & 1 \\ -2 & 0 & 0 \\ -1 & 0 & 0 \end{bmatrix}$ **(d)** $\begin{bmatrix} 0 & 1 & 2 \\ -1 & 0 & 2 \\ -2 & -2 & 0 \end{bmatrix}$

22. (*Properties of the dot product for complex vector space*) Recall that whereas we use the dot product $\mathbf{u} \cdot \mathbf{v} = \mathbf{u}^T \mathbf{v}$ for $\mathbb{R}^n$, we use $\mathbf{u} \cdot \mathbf{v} = \mathbf{u}^T \overline{\mathbf{v}}$ for $\mathbb{C}^n$. Show that these properties follow from the latter definition.

(a) $\mathbf{u} \cdot \mathbf{v} = \overline{\mathbf{v} \cdot \mathbf{u}}$ (complex commutativity)

(b) $(\alpha \mathbf{u}) \cdot \mathbf{v} = \alpha (\mathbf{u} \cdot \mathbf{v})$

(c) $\mathbf{u} \cdot (\alpha \mathbf{v}) = \overline{\alpha}(\mathbf{u} \cdot \mathbf{v})$

23. Vectors in $\mathbb{R}^1, \mathbb{R}^2, \mathbb{R}^3$ can be displayed, graphically, as arrow vectors. Is the same true for $\mathbb{C}^1, \mathbb{C}^2, \mathbb{C}^3$? Explain.

9.3   THE SPECIAL CASE OF SYMMETRIC MATRICES

We continue to consider matrices that are square ($n \times n$) and real ($\overline{A} = A$) and we assume in this section that they are symmetric as well. Recall that by symmetric we mean symmetric about the main diagonal, so that $a_{ij} = a_{ji}$ and hence $A^T = A$.

We are especially interested in symmetric matrices because they occur frequently in applications. For instance, the stiffness matrix K in Section 5.6.3 was symmetric—even though the mass-spring system was not physically symmetric.

We begin with three important theorems. Proof of the second is deferred to the exercises.

THEOREM 9.3.1
Real Eigenvalues
If a real matrix $\mathbf{A}$ is symmetric ($\mathbf{A}^T = \mathbf{A}$), then all of its eigenvalues are real.

Proof. The following proof will use material from the optional Section 9.2.3 and can be omitted if you did not study that section. Specifically, since we don't yet know that the eigenvalues are real we must allow for both the eigenvalues and eigenvectors of $\mathbf{A}$ to be complex. Accordingly, we will use the dot product $\mathbf{u} \cdot \mathbf{v} = \mathbf{u}^T \overline{\mathbf{v}}$ as discussed in Section 9.2.3. From the latter definition these properties follow (Exercise 1):

$$(\alpha \mathbf{u}) \cdot \mathbf{v} = \alpha(\mathbf{u} \cdot \mathbf{v}), \tag{1a}$$

$$\mathbf{u} \cdot (\alpha \mathbf{v}) = \overline{\alpha}(\mathbf{u} \cdot \mathbf{v}). \tag{1b}$$

To begin, if λ is an eigenvalue of $\mathbf{A}$ and $\mathbf{e}$ is a corresponding eigenvector, then

$$\mathbf{A}\mathbf{e} = \lambda \mathbf{e}. \tag{2}$$

Next, we "post-dot" and "pre-dot" both sides of (2) with $\mathbf{e}$.[1] *Post*-dotting gives

$$(\mathbf{A}\mathbf{e}) \cdot \mathbf{e} = (\lambda \mathbf{e}) \cdot \mathbf{e}, \tag{3a}$$

$$(\mathbf{A}\mathbf{e})^T \overline{\mathbf{e}} = \lambda(\mathbf{e} \cdot \mathbf{e}), \tag{3b}$$

$$\mathbf{e}^T \mathbf{A}^T \overline{\mathbf{e}} = \lambda \|\mathbf{e}\|^2, \tag{3c}$$

where we've used (1a) and recalled that $(\mathbf{A}\mathbf{B})^T = \mathbf{B}^T \mathbf{A}^T$ for any matrices $\mathbf{A}$ and $\mathbf{B}$ that are conformable for multiplication. Next, *pre*-dotting (2) with $\mathbf{e}$ gives

$$\mathbf{e} \cdot (\mathbf{A}\mathbf{e}) = \mathbf{e} \cdot (\lambda \mathbf{e}), \tag{4a}$$

$$\mathbf{e}^T \overline{\mathbf{A}\mathbf{e}} = \overline{\lambda}(\mathbf{e} \cdot \mathbf{e}), \tag{4b}$$

$$\mathbf{e}^T \overline{\mathbf{A}\mathbf{e}} = \overline{\lambda} \|\mathbf{e}\|^2. \tag{4c}$$

But $\mathbf{A}^T = \mathbf{A}$ and $\overline{\mathbf{A}} = \mathbf{A}$, by assumption, so subtracting (3c) from (4c) gives

$$0 = (\overline{\lambda} - \lambda) \|\mathbf{e}\|^2. \tag{5}$$

Since $\mathbf{e}$ is an eigenvector $\|\mathbf{e}\| \neq 0$, so it follows from (5) that $\overline{\lambda} = \lambda$; hence, λ is real. ∎

THEOREM 9.3.2
Dimension of Eigenspace
If an eigenvalue λ of a real symmetric matrix $\mathbf{A}$ is of multiplicity k, then the eigenspace corresponding to λ is of dimension k.

---

[1] By post-dotting $\mathbf{u}$ into $\mathbf{v}$ we mean forming $\mathbf{v} \cdot \mathbf{u}$; by pre-dotting $\mathbf{u}$ into $\mathbf{v}$ we mean forming $\mathbf{u} \cdot \mathbf{v}$. In the real case, the order does not matter because $\mathbf{u} \cdot \mathbf{v} = \mathbf{v} \cdot \mathbf{u}$, but in the complex case the order does matter because $\mathbf{u} \cdot \mathbf{v} \neq \mathbf{v} \cdot \mathbf{u}$; rather, $\mathbf{u} \cdot \mathbf{v} = \overline{\mathbf{v} \cdot \mathbf{u}}$.

THEOREM 9.3.3
Orthogonality of Eigenvectors
If a real matrix $\mathbf{A}$ is symmetric, then eigenvectors corresponding to distinct eigenvalues are orthogonal.

Proof.   Let $\mathbf{e}_j$ and $\mathbf{e}_k$ be eigenvectors corresponding to distinct eigenvalues λ_j and λ_k, respectively. Thus,

$$\mathbf{A}\mathbf{e}_j = \lambda_j \mathbf{e}_j \quad \text{and} \quad \mathbf{A}\mathbf{e}_k = \lambda_k \mathbf{e}_k. \tag{6a,b}$$

Now that we know from Theorem 9.3.1 that the λ's are real, it follows that the $\mathbf{e}$'s are too, so we can dispense with the complex dot product ($\mathbf{u}\cdot\mathbf{v} = \mathbf{u}^{\mathrm{T}}\overline{\mathbf{v}}$) and use the real one ($\mathbf{u}\cdot\mathbf{v} = \mathbf{u}^{\mathrm{T}}\mathbf{v}$). Pre-dotting (6a) with $\mathbf{e}_k$ and post-dotting (6b) with $\mathbf{e}_j$ gives

$$
\begin{array}{c|c}
\begin{aligned}
\mathbf{e}_k\cdot(\mathbf{A}\mathbf{e}_j) &= \mathbf{e}_k\cdot(\lambda_j\mathbf{e}_j) \\
\mathbf{e}_k^{\mathrm{T}}\mathbf{A}\mathbf{e}_j &= \lambda_j\mathbf{e}_k^{\mathrm{T}}\mathbf{e}_j
\end{aligned}
&
\begin{aligned}
(\mathbf{A}\mathbf{e}_k)\cdot\mathbf{e}_j &= (\lambda_k\mathbf{e}_k)\cdot\mathbf{e}_j \\
(\mathbf{A}\mathbf{e}_k)^{\mathrm{T}}\mathbf{e}_j &= \lambda_k\mathbf{e}_k^{\mathrm{T}}\mathbf{e}_j \\
\mathbf{e}_k^{\mathrm{T}}\mathbf{A}^{\mathrm{T}}\mathbf{e}_j &= \lambda_k\mathbf{e}_k^{\mathrm{T}}\mathbf{e}_j.
\end{aligned}
\end{array}
\tag{7}
$$

But $\mathbf{A}^{\mathrm{T}} = \mathbf{A}$ by assumption so if we subtract the bottom equations on the left and right of the vertical divider, we obtain

$$0 = (\lambda_j - \lambda_k)\mathbf{e}_k^{\mathrm{T}}\mathbf{e}_j. \tag{8}$$

Finally, $\lambda_j - \lambda_k \neq 0$ since λ_j and λ_k were assumed to be distinct so it follows from (8) that $\mathbf{e}_k^{\mathrm{T}}\mathbf{e}_j = 0$. Thus, $\mathbf{e}_k\cdot\mathbf{e}_j = 0$, as claimed.   ∎

Be careful not to read inverses into theorems. For instance, Theorem 9.3.1 says that if a real matrix $\mathbf{A}$ is symmetric, then its eigenvalues are real. It does *not* say that the eigenvalues of $\mathbf{A}$ are real if and only if $\mathbf{A}$ is symmetric. For instance, in each of the four examples of Section 9.2 the λ's are real, yet none of the matrices is symmetric.

EXAMPLE 1
For the symmetric matrix

$$\mathbf{A} = \begin{bmatrix} 2 & 1 & 1 \\ 1 & 2 & 1 \\ 1 & 1 & 2 \end{bmatrix}, \tag{9}$$

we find the eigenvalues and eigenspaces

$$\lambda_1 = 4, \ \mathbf{e}_1 = \alpha\begin{bmatrix} 1 \\ 1 \\ 1 \end{bmatrix}, \qquad \lambda_2 = 1, \ \mathbf{e}_2 = \beta\begin{bmatrix} -1 \\ 0 \\ 1 \end{bmatrix} + \gamma\begin{bmatrix} -1 \\ 1 \\ 0 \end{bmatrix}, \tag{10}$$

where $\lambda_2 = 1$ is of multiplicity two [i.e., the characteristic polynomial can be factored as $-(\lambda - 4)(\lambda - 1)^2$]. Since $\mathbf{A}$ is real and symmetric, the theorems apply. In accordance with Theorem 9.3.1 the λ's are real; in accordance with Theorem 9.3.2, λ_1 is of multiplicity 1 and its eigenspace is one-dimensional (namely, span$\{[1, 1, 1]^T\}$), and λ_2 is of multiplicity 2 and its eigenspace is two-dimensional (namely, span$\{[-1, 0, 1]^T, [-1, 1, 0]^T\}$); and in accordance with Theorem 9.3.3, $\mathbf{e}_1 \cdot \mathbf{e}_2 = 0$ for all choices of α, β, γ. The eigenspace $\mathbf{e}_2$ is the plane through the origin (in 3-space) that is spanned by $[-1, 0, 1]^T$ and $[-1, 1, 0]^T$, and the eigenspace $\mathbf{e}_1$ is the line through the origin that is spanned by $[1, 1, 1]^T$ and is normal to that plane.

The vectors $[-1, 0, 1]^T$ and $[-1, 1, 0]^T$ in $\mathbf{e}_2$ are LI and a basis for $\mathbf{e}_2$, but happen not to be orthogonal. Their lack of orthogonality does not violate Theorem 9.3.3 since they come from the same λ, not from distinct λ's. Nonetheless, we can "trade those vectors in" for two within $\mathbf{e}_2$ that *are* orthogonal (in a nonunique way, in fact, for there is an infinite number of pairs of orthogonal vectors within that plane). For instance, we can choose

$$\mathbf{e}_2 = [-1, 0, 1]^T \tag{11}$$

[i.e., by setting $\beta = 1$ and $\gamma = 0$ in (10)] and seek a second vector in the plane in the from

$$\mathbf{e}_3 = \beta[-1, 0, 1]^T + \gamma[-1, 1, 0]^T \tag{12}$$

such that

$$\mathbf{e}_2 \cdot \mathbf{e}_3 = 2\beta + \gamma = 0. \tag{13}$$

Choosing $\beta = 1$, say, then (13) gives $\gamma = -2$, and (12) gives

$$\mathbf{e}_3 = [1, -2, 1]^T, \tag{14}$$

and the vectors given by (11) and (14) constitute an orthogonal basis for the eigenspace corresponding to the eigenvalue 1.

And since (with $\alpha = 1$, say)

$$\mathbf{e}_1 = [1, 1, 1]^T \tag{15}$$

is orthogonal to each of those vectors, it follows that the eigenvectors given by (11), (14), and (15) constitute an orthogonal basis for 3-space. That is, among the eigenvectors of the 3×3 matrix $\mathbf{A}$ given by (9) we can find an orthogonal basis for 3-space.

COMMENT 1.   The $\mathbf{A}$ matrix in (9) happens to be symmetric about the other diagonal as well as about the main diagonal, but that symmetry is irrelevant; by symmetry we always mean that $\mathbf{A}^T = \mathbf{A}$, which is symmetry about the *main* diagonal (from upper left to bottom right).

COMMENT 2.   The procedure that we used to obtain the orthogonal set $\{\mathbf{e}_2, \mathbf{e}_3\}$ from the LI set $\{[-1, 0, 1]^T, [-1, 1, 0]^T\}$ is essentially the Gram-Schmidt orthogonalization procedure explained in Section 4.9.4.   ∎

Generalization of the ideas contained in Example 1 yields the following theorem.

THEOREM 9.3.4

Orthogonal Basis

If an $n \times n$ real matrix $\mathbf{A}$ is symmetric, then its eigenvectors provide an orthogonal basis for n-space.

Proof. If all of $\mathbf{A}$'s n eigenvalues are distinct then, according to Theorem 9.3.3, the n eigenspaces are orthogonal (each being a one-dimensional line in n-space) and therefore provide n orthogonal vectors, which necessarily constitute a basis for n-space. What if the eigenvalues are not distinct? Suppose that all are distinct except for one, say λ, which is of multiplicity k. Then the $n - k$ eigenvectors corresponding to the other eigenvalues are orthogonal to each other and also to all vectors in the eigenspace corresponding to λ. Further, λ's eigenspace is k-dimensional (Theorem 9.3.2) and hence contains k orthogonal vectors. Altogether then we have $(n-k)+k = n$ orthogonal eigenvectors and hence an orthogonal basis for n-space. A similar argument applies if there is more than one repeated eigenvalue. ■

Closure. In this brief section we considered the special case where $\mathbf{A}$ is not only real but also symmetric. In that case four major results were shown to follow:

(i) the eigenvalues are real (Theorem 9.3.1);

(ii) if an eigenvalue is of multiplicity k, then the eigenspace corresponding to it is of dimension k (Theorem 9.3.2);

(iii) eigenvectors corresponding to distinct eigenvalues are orthogonal (Theorem 9.3.3);

(iv) the eigenvectors provide an orthogonal basis for n-space (Theorem 9.3.4).

EXERCISES 9.3

1. Let α be any complex scalar, and let $\mathbf{u}$ and $\mathbf{v}$ be any vectors in $\mathbb{C}^n$. Then use the definition $\mathbf{u} \cdot \mathbf{v} = \mathbf{u}^T \bar{\mathbf{v}}$ to derive the property stated in

(a) equation (1a)   (b) equation (1b)

2. Prove Theorem 9.3.2 for the special case where $n = 2$.

3. From the eigenvectors of the given $n \times n$ matrix obtain an orthogonal basis for $\mathbb{R}^n$.

(a) $\begin{bmatrix} 1 & 1 \\ 1 & 1 \end{bmatrix}$

(b) $\begin{bmatrix} 2 & 1 \\ 1 & 0 \end{bmatrix}$

(c) $\begin{bmatrix} 0 & 1 & 0 \\ 1 & 0 & 0 \\ 0 & 0 & 0 \end{bmatrix}$

(d) $\begin{bmatrix} 0 & 2 & 0 \\ 2 & 0 & 0 \\ 0 & 0 & -2 \end{bmatrix}$

(e) $\begin{bmatrix} 4 & 2 & 2 \\ 2 & 4 & 2 \\ 2 & 2 & 4 \end{bmatrix}$

(f) $\begin{bmatrix} 7 & 4 & -4 \\ 4 & 1 & 8 \\ -4 & 8 & 1 \end{bmatrix}$

(g) $\begin{bmatrix} 0 & 0 & 0 & 4 \\ 0 & 0 & 4 & 0 \\ 0 & 4 & 0 & 0 \\ 4 & 0 & 0 & 0 \end{bmatrix}$

(h) $\begin{bmatrix} 0 & 3 & 3 & 0 \\ 3 & 0 & 3 & 0 \\ 3 & 3 & 0 & 0 \\ 0 & 0 & 0 & -3 \end{bmatrix}$

(i) $\begin{bmatrix} 6 & 0 & 0 & 0 \\ 0 & 2 & 2 & 2 \\ 0 & 2 & 2 & 2 \\ 0 & 2 & 2 & 2 \end{bmatrix}$

(j) $\begin{bmatrix} 1 & 0 & 0 & 1 \\ 0 & 0 & 0 & 0 \\ 0 & 0 & 0 & 0 \\ 1 & 0 & 0 & 1 \end{bmatrix}$

(k) $\begin{bmatrix} 1 & 1 & 1 \\ 1 & 1 & 1 \\ 1 & 1 & 1 \\ 1 & 1 & 1 \end{bmatrix}$

(l) $\begin{bmatrix} 1 & 1 & 1 & 1 \\ 1 & 1 & 1 & 1 \\ 1 & 1 & 1 & 1 \\ 1 & 1 & 1 & 1 \end{bmatrix}$

4. (*Eigenvector expansion solution of* $\mathbf{Ax} = \mathbf{c}$) We've seen how to solve the general nonhomogeneous equation

$$\mathbf{Ax} = \mathbf{c} \qquad (4.1)$$

by Gauss elimination in Chapter 4, and by the inverse matrix method (if $\det \mathbf{A} \neq 0$) in Chapter 5. Here, we outline a different approach that involves the eigenvalue problem. Suppose that besides being real and square ($n \times n$) $\mathbf{A}$ is symmetric. To solve (4.1), first solve the associated eigenvalue problem $\mathbf{Ax} = \lambda \mathbf{x}$. Suppose that step is done and the eigenvalues are $\lambda_1, \ldots, \lambda_n$ (not necessarily distinct) with orthogonal eigenvectors $\mathbf{e}_1, \ldots, \mathbf{e}_n$. The latter constitute an orthogonal basis for n-space. To solve (4.1), expand $\mathbf{x}$ and $\mathbf{c}$ in terms of that basis:

$$\mathbf{x} = \sum_1^n a_j \mathbf{e}_j \qquad \text{and} \qquad \mathbf{c} = \sum_1^n c_j \mathbf{e}_j. \qquad (4.2)$$

Knowing $\mathbf{c}$ and the $\mathbf{e}_j$'s, we can compute the c_j's; the a_j's are our unknowns.

(a) Putting (4.2) into (4.1), show that we obtain the equation

$$\lambda_j a_j = c_j \quad (j = 1, \ldots, n) \qquad (4.3)$$

for the a_j's.

(b) Next, distinguish two cases. First, suppose that none of the λ_j's is zero. In that case $\det \mathbf{A} \neq 0$ so we know from Chapter 5 that there will be a unique solution, $\mathbf{x} = \mathbf{A}^{-1}\mathbf{c}$. Show that our eigenvector expansion method does indeed give a *unique solution*, namely,

$$\mathbf{x} = \sum_1^n \frac{c_j}{\lambda_j} \mathbf{e}_j. \qquad (4.4)$$

(c) Second, suppose that $\lambda_1 = 0$ *is* an eigenvalue of multiplicity N ($1 \leq N \leq n$). In that case $\det \mathbf{A} = 0$ so we know from Chapter 5 that there will be either no solution of (4.1), or a nonunique solution. Show that our eigenvector expansion method does indeed give the same result. Specifically, show that if $c_1, \ldots, c_N$ are not all zero then there is *no solution* to (4.1), and if $c_1 = \cdots = c_N = 0$ then there is a nonunique solution, the N-parameter family of solutions

$$\mathbf{x} = a_1 \mathbf{e}_1 + \cdots + a_N \mathbf{e}_N + \sum_{N+1}^n \frac{c_j}{\lambda_j} \mathbf{e}_j \qquad (4.5)$$

where $a_1, \ldots, a_N$ are arbitrary constants.

(d) Show that $\mathbf{e}_1, \cdots, \mathbf{e}_N$ in (4.5) are "homogeneous solutions," that is, solutions of the homogeneous version

$\mathbf{Ax} = \mathbf{0}$ of (4.1). And show that the $\sum_{N+1}^n (c_j / \lambda_j) \mathbf{e}_j$ term in (4.5) is a "particular solution," that is, a solution of the full equation $\mathbf{Ax} = \mathbf{c}$. Thus, note the similarity between the form of the solution (4.5) to the linear algebraic equation $\mathbf{Ax} = \mathbf{c}$ and of the general solution to a linear differential equation.

(e) What is the advantage of using the eigenvectors of $\mathbf{A}$ for our base vectors? After all, finding the λ_j's and $\mathbf{e}$'s may require a significant amount of computation, computation that could have been avoided if we had used an "off-the-shelf" basis such as $\{[1, 0, \ldots, 0]^T, [0, 1, 0, \ldots, 0]^T, \ldots, [0, \ldots, 0, 1]^T\}$. So why do we use the eigenvectors of $\mathbf{A}$?

5. In each case, use the eigenvector expansion method outlined in Exercise 4 to solve the problem $\mathbf{Ax} = \mathbf{c}$ for $\mathbf{x}$, if indeed a solution does exist. Then solve again, this time using Gauss elimination, and show that your two results agree.

(a) $\begin{bmatrix} 2 & 1 \\ 1 & 2 \end{bmatrix} \begin{bmatrix} x_1 \\ x_2 \end{bmatrix} = \begin{bmatrix} 3 \\ 0 \end{bmatrix}$

(b) $\begin{bmatrix} 1 & 2 \\ 2 & 0 \end{bmatrix} \begin{bmatrix} x_1 \\ x_2 \end{bmatrix} = \begin{bmatrix} 6 \\ 8 \end{bmatrix}$

(c) $\begin{bmatrix} 3 & 3 \\ 3 & 3 \end{bmatrix} \begin{bmatrix} x_1 \\ x_2 \end{bmatrix} = \begin{bmatrix} 1 \\ 0 \end{bmatrix}$

(d) $\begin{bmatrix} 3 & 3 \\ 3 & 3 \end{bmatrix} \begin{bmatrix} x_1 \\ x_2 \end{bmatrix} = \begin{bmatrix} 6 \\ 6 \end{bmatrix}$

(e) $\begin{bmatrix} 0 & 1 & 1 \\ 1 & 0 & 1 \\ 1 & 1 & 0 \end{bmatrix} \begin{bmatrix} x_1 \\ x_2 \\ x_3 \end{bmatrix} = \begin{bmatrix} 1 \\ 2 \\ 3 \end{bmatrix}$

(f) $\begin{bmatrix} 2 & 2 & 2 \\ 2 & 2 & 2 \\ 2 & 2 & 2 \end{bmatrix} \begin{bmatrix} x_1 \\ x_2 \\ x_3 \end{bmatrix} = \begin{bmatrix} 5 \\ 3 \\ 1 \end{bmatrix}$

(g) $\begin{bmatrix} 0 & 0 & 1 \\ 0 & 0 & 1 \\ 1 & 1 & 1 \end{bmatrix} \begin{bmatrix} x_1 \\ x_2 \\ x_3 \end{bmatrix} = \begin{bmatrix} 2 \\ 3 \\ 4 \end{bmatrix}$

(h) $\begin{bmatrix} 0 & 0 & 1 \\ 0 & 0 & 1 \\ 1 & 1 & 1 \end{bmatrix} \begin{bmatrix} x_1 \\ x_2 \\ x_3 \end{bmatrix} = \begin{bmatrix} 4 \\ 4 \\ 3 \end{bmatrix}$

6. (*Rayleigh's quotient*) Let $\mathbf{A}$ be a symmetric $n \times n$ matrix. Dotting any eigenvector $\mathbf{e}$ of $\mathbf{A}$ into both sides of $\mathbf{Ae} = \lambda \mathbf{e}$ and solving for λ, gives

$$\lambda = \frac{\mathbf{e} \cdot \mathbf{Ae}}{\mathbf{e} \cdot \mathbf{e}} = \frac{\mathbf{e}^T \mathbf{Ae}}{\mathbf{e}^T \mathbf{e}}. \qquad (6.1)$$

More generally, if $\mathbf{x}$ is any vector, not necessarily an eigenvector of $\mathbf{A}$, then the number

$$R(\mathbf{x}) = \frac{\mathbf{x}^T \mathbf{A} \mathbf{x}}{\mathbf{x}^T \mathbf{x}} \qquad (6.2)$$

is known as **Rayleigh's quotient**, after *Lord Rayleigh* (*John William Strutt*; 1842–1919). Imagine putting randomly chosen $\mathbf{x}$ vectors, one after another, into Rayleigh's quotient. If $\mathbf{x}$ happens to coincide with an eigenvector, then, according to (6.1), $R(\mathbf{x})$ gives the corresponding eigenvalue. In any case,

$$|R(\mathbf{x})| = \left| \frac{\mathbf{x}^T \mathbf{A} \mathbf{x}}{\mathbf{x}^T \mathbf{x}} \right| \leq |\lambda_1|, \qquad (6.3)$$

where the eigenvalues are ordered so that

$$|\lambda_1| \geq |\lambda_2| \geq \cdots \geq |\lambda_n|.$$

That is, $|R(\mathbf{x})|$ provides a **lower bound** on the magnitude of the largest eigenvalue of $\mathbf{A}$, where $\mathbf{x}$ is any vector [i.e., any nonzero vector, since $R(\mathbf{0}) = 0/0$ is undefined]. Upper and lower bounds on eigenvalues are sometimes useful, and Rayleigh's quotient is used in Exercise 7.

(a) Prove the inequality in (6.3). HINT: Since $\mathbf{A}$ is symmetric, it has n orthogonal eigenvectors $\mathbf{e}_1, \ldots, \mathbf{e}_n$, corresponding to the eigenvalues $\lambda_1, \ldots, \lambda_n$. Expand $\mathbf{x}$ in terms of the orthogonal basis $\{\mathbf{e}_1, \ldots, \mathbf{e}_n\}$:

$$\mathbf{x} = \sum_{j=1}^{n} a_j \mathbf{e}_j. \qquad (6.4)$$

(b) Verify (6.3) for the matrix

$$\mathbf{A} = \begin{bmatrix} 2 & 0 & 0 \\ 0 & -1 & -2 \\ 0 & -2 & -1 \end{bmatrix} \qquad (6.5)$$

by taking $\mathbf{x} = [1, 0, 0]^T, [0, 1, 0]^T, [0, 0, 1]^T, [1, 2, 3]^T,$ $[3, -1, 4]^T,$ and $[0.98, 0, 0.01]^T,$ say, verifying that $|R(\mathbf{x})| \leq |\lambda_1|$ in each case.

7. (*The power method*) There exists a simple *iterative* procedure for calculating eigenvalues and eigenvectors, which is known as the **power method**. To begin, select any nonzero vector $\mathbf{x}^{(0)}$ and then compute $\mathbf{x}^{(1)} \equiv \mathbf{A}\mathbf{x}^{(0)}, \mathbf{x}^{(2)} \equiv \mathbf{A}\mathbf{x}^{(1)},$ and so on. That is,

$$\mathbf{x}^{(k+1)} \equiv \mathbf{A}\mathbf{x}^{(k)} \quad (k = 0, 1, 2, \ldots). \qquad (7.1)$$

Before analyzing the situation, let us apply (7.1) and see what happens. Let

$$\mathbf{A} = \begin{bmatrix} 1 & 1 & 1 \\ 1 & 0 & 0 \\ 1 & 0 & 0 \end{bmatrix}, \qquad (7.2)$$

for example. Choosing $\mathbf{x}^{(0)} = [1, 0, 0]^T$, say, successive application of (7.1) gives

$$\begin{array}{cccccc} \mathbf{x}^{(0)} & \mathbf{x}^{(1)} & \mathbf{x}^{(2)} & \mathbf{x}^{(3)} & \mathbf{x}^{(4)} & \mathbf{x}^{(5)} \end{array}$$
$$\begin{bmatrix} 1 & 1 & 1 \\ 1 & 0 & 0 \\ 1 & 0 & 0 \end{bmatrix} \begin{bmatrix} 1 \\ 0 \\ 0 \end{bmatrix} \begin{bmatrix} 1 \\ 1 \\ 1 \end{bmatrix} \begin{bmatrix} 3 \\ 1 \\ 1 \end{bmatrix} \begin{bmatrix} 5 \\ 3 \\ 3 \end{bmatrix} \begin{bmatrix} 11 \\ 5 \\ 5 \end{bmatrix} \begin{bmatrix} 21 \\ 11 \\ 11 \end{bmatrix},$$

$$\qquad (7.3)$$

and so on. Now, observe, from the very nature of the eigenvalue problem $\mathbf{A}\mathbf{x} = \lambda\mathbf{x}$, that if $\mathbf{x}^{(0)}$ were an eigenvector of $\mathbf{A}$, then $\mathbf{A}\mathbf{x}^{(0)}$ would be some scalar multiple of $\mathbf{x}^{(0)}$. But $\mathbf{A}\mathbf{x}^{(0)} \equiv \mathbf{x}^{(1)}$, and $\mathbf{x}^{(1)}$ is seen from (7.3) *not* to be a scalar multiple of $\mathbf{x}^{(0)}$. Hence, $\mathbf{x}^{(0)}$ is not an eigenvector of $\mathbf{A}$. Similarly, $\mathbf{x}^{(1)}$ is not an eigenvector because $\mathbf{x}^{(2)}$ is not a multiple of $\mathbf{x}^{(1)}$, and so on. Nevertheless, we see that with each successive step $\mathbf{x}^{(k+1)}$ draws closer and closer to being a multiple of $\mathbf{x}^{(k)}$ so that the sequence $\mathbf{x}^{(k)}$ is evidently *approaching* an eigenvector of $\mathbf{A}$. In fact, $\mathbf{x}^{(5)}$ is very close to being a multiple of $\mathbf{x}^{(4)}$ so that there is an eigenvector

$$\mathbf{e} \approx \begin{bmatrix} 21 \\ 11 \\ 11 \end{bmatrix}. \qquad (7.4)$$

What is the corresponding λ; is $\lambda \approx 21/11$? $\lambda \approx 11/5$? An average of the two? We state without proof that one does well to use the Rayleigh quotient from Exercise 6:

$$\lambda \approx \frac{\mathbf{x}^{(4)T} \mathbf{A} \mathbf{x}^{(4)}}{\mathbf{x}^{(4)T} \mathbf{x}^{(4)}} = \frac{\mathbf{x}^{(4)T} \mathbf{x}^{(5)}}{\mathbf{x}^{(4)T} \mathbf{x}^{(4)}} = \frac{341}{171} = 1.994. \qquad (7.5)$$

How do (7.4) and (7.5) compare with exact values? The eigenvalues and eigenvectors of $\mathbf{A}$ are

$$\lambda_1 = 2, \qquad \mathbf{e}_1 = \begin{bmatrix} 2 \\ 1 \\ 1 \end{bmatrix},$$

$$\lambda_2 = -1, \qquad \mathbf{e}_2 = \begin{bmatrix} 1 \\ -1 \\ -1 \end{bmatrix},$$

$$\lambda_3 = 0, \qquad \mathbf{e}_3 = \begin{bmatrix} 0 \\ 1 \\ -1 \end{bmatrix} \qquad (7.6)$$

so the iteration (7.3) is evidently converging to $\mathbf{e}_1$; of course $\mathbf{e}_1$, $\mathbf{e}_2$, and $\mathbf{e}_3$ can be scaled by arbitrary nonzero constants. [It is striking that whereas (7.4) is accurate to only around one part in 20, (7.5) is accurate to around one part in 200. This enhancement of accuracy, using Rayleigh's quotient to determine λ, is not a coincidence and can be explained theoretically. See, for instance, Stephen H. Crandall, *Engineering Analysis* (New York: McGraw-Hill, 1956), Chap. 2.]

To see what is going on, suppose that $\mathbf{A}$ is a symmetric matrix of order n (although symmetry is more than we need; it would suffice for $\mathbf{A}$ to have n LI eigenvectors) and let its eigenvalues be ordered so that

$$|\lambda_1| \geq |\lambda_2| \geq \cdots \geq |\lambda_n|$$

as in (7.6). Since $\mathbf{A}$ is symmetric, its eigenvectors $\mathbf{e}_1, \ldots, \mathbf{e}_n$ provide an orthogonal basis for n-space. Hence, our initial vector $\mathbf{x}^{(0)}$ must be expressible as

$$\mathbf{x}^{(0)} = \sum_{j=1}^{n} a_j \mathbf{e}_j. \tag{7.7}$$

Of course, we cannot *compute* the a_j's since we do not know the $\mathbf{e}_j$'s yet, but that is no problem; it is the *form* of (7.7) that is important here. It follows from (7.1) and (7.7) that

$$\mathbf{x}^{(k)} = \sum_{1}^{n} a_j \lambda_j^k \mathbf{e}_j$$

$$= \lambda_1^k \left[a_1 \mathbf{e}_1 + a_2 \left(\frac{\lambda_2}{\lambda_1}\right)^k \mathbf{e}_2 + \cdots + a_n \left(\frac{\lambda_n}{\lambda_1}\right)^k \mathbf{e}_n \right]. \tag{7.8}$$

If λ_1 is in fact *dominant*, i.e., $|\lambda_1| > |\lambda_2| \geq \cdots \geq |\lambda_n|$, then $(\lambda_2/\lambda_1)^k, \ldots, (\lambda_n/\lambda_1)^k$ all tend to zero as $k \to \infty$, so that $\mathbf{x}^{(k)} \sim \lambda_1^k a_1 \mathbf{e}_1$ as $k \to \infty$ (provided that $a_1 \neq 0$, i.e., provided that $\mathbf{x}^{(0)}$ does not happen to be orthogonal to $\mathbf{e}_1$). Eigenvectors can be scaled arbitrarily so the $\lambda_1^k a_1$ factor is of little interest; the point is that $\mathbf{x}^{(k)}$ converges to $\mathbf{e}_1$, the eigenvector corresponding to the dominant eigenvalue. That is precisely what was found in the preceding illustration, wherein the dominant eigenvalue was $\lambda_1 = 2$. NOTE: As in the problem on population dynamics (Section 9.2.2) and in the eigenvector expansion method (Exercise 4, above), we see *that the most convenient basis to use is the basis provided by the $\mathbf{A}$ matrix itself.*

(a) Show that (7.8) follows from (7.1) and (7.7).

(b) Determine the dominant eigenvalue and corresponding eigenvector by the power method for

$$\mathbf{A} = \begin{bmatrix} 0 & 0 & 1 \\ 0 & 0 & 1 \\ 1 & 1 & 1 \end{bmatrix}.$$

Also, evaluate the eigenvalues and eigenvectors exactly, either by hand or by using *Maple*. NOTE: Observe that even though your iteration converges, you cannot be certain that the eigenvalue obtained is the *dominant* one, λ_1, since your chosen $\mathbf{x}^{(0)}$ may (without your knowing it) be orthogonal to $\mathbf{e}_1$, as mentioned in the sentence following (7.8). Hence, we recommend that you carry out the iteration three times, once with $\mathbf{x}^{(0)} = [1, 0, 0]^T$, once with $\mathbf{x}^{(0)} = [0, 1, 0]^T$, and once with $\mathbf{x}^{(0)} = [0, 0, 1]^T$ since there is no way that all *three* of these $\mathbf{x}^{(0)}$'s can be orthogonal to $\mathbf{e}_1$. (Do you see why this is so?) Go as far as $\mathbf{x}^{(5)}$ in each case, and use the Rayleigh quotient to estimate λ, as we did in (7.5).

8. The same as Exercise 7(b), for the given matrix.

(a) $\begin{bmatrix} 2 & 1 & -1 \\ 1 & 4 & 3 \\ -1 & 3 & 4 \end{bmatrix}$ **(b)** $\begin{bmatrix} 2 & 1 & 1 \\ 1 & 2 & 1 \\ 1 & 1 & 2 \end{bmatrix}$

(c) $\begin{bmatrix} 1 & 0 & 1 \\ 0 & 1 & 0 \\ 1 & 0 & 1 \end{bmatrix}$ **(d)** $\begin{bmatrix} 1 & 0 & 1 \\ 0 & 3 & 0 \\ 1 & 0 & 1 \end{bmatrix}$

(e) $\begin{bmatrix} 0 & 1 & -1 \\ 1 & 3 & 4 \\ -1 & 4 & 3 \end{bmatrix}$ **(f)** $\begin{bmatrix} 4 & 1 & 3 \\ 1 & 0 & -1 \\ 3 & -1 & 4 \end{bmatrix}$

(g) $\begin{bmatrix} 2 & 1 & 1 & 2 \\ 0 & 1 & 1 & 0 \\ 0 & 1 & 1 & 0 \\ 2 & 1 & 1 & 2 \end{bmatrix}$

(h) $\begin{bmatrix} 1 & 0 & 0 & 1 \\ 0 & 0 & 0 & 0 \\ 0 & 0 & 0 & 0 \\ 1 & 0 & 0 & 1 \end{bmatrix}$

9. To be done either by hand or using *Maple*. For the $\mathbf{A}$ matrix given in (7.2) above, work out $\mathbf{A}^4$. Apply the power method (Exercise 7) to $\mathbf{A}^4$ beginning with $\mathbf{x}^{(0)} = [1, 0, 0]^T$. Go as far as $\mathbf{x}^{(5)}$ and use the Rayleigh quotient to estimate λ. Finally, recover the eigenvalue and the eigenvector of $\mathbf{A}$ from those of $\mathbf{A}^4$. (HINT: See Exercise 11 in Section 9.2.) Explain why the iteration converges much more rapidly for $\mathbf{A}^4$ (in this exercise) than it did for $\mathbf{A}$ (in Exercise 7).

9.4 DIFFERENTIAL EQUATION BOUNDARY VALUE PROBLEMS AS EIGENVALUE PROBLEMS

Besides the eigenvalue problem $\mathbf{Ax} = \lambda \mathbf{x}$ where $\mathbf{A}$ is an $n \times n$ matrix operator and $\mathbf{x}$ is a vector in n-space, there is also a differential equation version of the eigenvalue problem, of the form $L[y] = \lambda y$ where L is a differential operator and $y(x)$ is a vector in some function space, that is, a vector space where the vectors are functions. Let us begin with a concrete application in which such an eigenvalue problem arises.

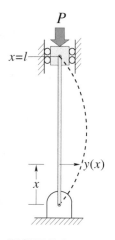

FIGURE 1
Column buckling.

9.4.1 The buckling of a column.   By a column we mean a stiff slender vertical structural member that supports a vertical load, denoted as P in Fig. 1. It would be natural to expect that the column, which is straight when $P = 0$, will bow out [according to $y = y(x)$] more and more as the load P is increased. Remarkably, it is found that the column remains straight [$y(x) = 0$] as P is increased—until a critical load P_{cr} is reached, at which point it bows out and collapses under the load. That event is known as *buckling*. Since buckling amounts to a failure of the load bearing column, it is important for the structural engineer to be able to predict the critical load, P_{cr}, known as the *buckling load*.

According to Euler beam theory, the lateral deflection $y(x)$ is governed by the boundary value problem

$$EIy'' = -Py \qquad (0 \leq x \leq l) \tag{1a}$$

$$y(0) = 0, \quad y(l) = 0, \tag{1b}$$

where E is Young's modulus for the column material, I is the area moment of inertia of the cross-sectional area about an axis that is normal to the plane of bending and through the centroid of the area, and $-Py(x)$ is the moment of the force P about the point $(x, y(x))$ of the column. With $\lambda = P/EI$, let us re-express (1) as

$$y'' + \lambda y = 0 \qquad (0 \leq x \leq l) \tag{2a}$$

$$y(0) = 0, \quad y(l) = 0, \tag{2b}$$

where the constant λ is a parameter such that $0 < \lambda < \infty$; that is, E and I are fixed, but we can apply any load P that we like, such that $0 < P < \infty$. Thus, by our choice of P we can have λ be any positive value that we like.[1]

Let us solve for $y(x)$. A general solution of (2a) is

$$y(x) = A \cos \sqrt{\lambda}\, x + B \sin \sqrt{\lambda}\, x. \tag{3}$$

Then

$$y(0) = 0 = (1)A + (0)B, \tag{4a}$$

---

[1] We rule out the case $\lambda = 0$ as uninteresting since $\lambda = 0$ implies $P = 0$, in which case there is no load on the column.

$$y(l) = 0 = \left(\cos \sqrt{\lambda}\, l\right) A + \left(\sin \sqrt{\lambda}\, l\right) B. \qquad (4b)$$

Equation (4a) gives $A = 0$, and then (4b) gives

$$\left(\sin \sqrt{\lambda}\, l\right) B = 0. \qquad (5)$$

If $\sin \sqrt{\lambda}\, l \neq 0$ then (5) implies that $B = 0$ as well. In that case (3) gives the unique solution $y(x) = 0$, which we call the *trivial solution*.

Of *course* (2) admits the trivial solution $y(x) = 0$, because both the differential equation (2a) and the boundary conditions (2b) are homogeneous, so the presence of the trivial solution is no surprise. However, beyond finding the trivial solution we have learned that it is the *only* solution if $\sin \sqrt{\lambda}\, l \neq 0$.

On the other hand, suppose that P, and hence λ, is chosen so that

$$\sin \sqrt{\lambda}\, l = 0. \qquad (6)$$

In that case (4a) gives $A = 0$ but (5) reveals that B is arbitrary, so (3) gives the solution

$$y(x) = B \sin \sqrt{\lambda}\, x, \qquad (7)$$

where B is arbitrary. Since $\sin x = 0$ at $x = n\pi$ for any integer n, it follows from (6) that those special values of λ are found from $\sqrt{\lambda}\, l = n\pi$; they are

$$\lambda_n = \left(\frac{n\pi}{l}\right)^2 \qquad (8)$$

for $n = 1, 2, \ldots$.

The upshot is that if λ is not equal to one of the λ_n values given by (8) then (2) gives the unique solution $y(x) = 0$, but if λ is equal to one of the λ_n values then we have, besides the trivial solution obtained by choosing $B = 0$, the nontrivial solution

$$y(x) = B \sin \frac{n\pi x}{l} \qquad (9)$$

obtained by putting (8) into (7).

Before interpreting these findings in terms of buckling, let us review the problem (2) as an eigenvalue problem. It is indeed an eigenvalue problem because it gives the unique trivial solution $y(x) = 0$ unless λ takes on one of a certain set of special values λ_n (called **eigenvalues**), and it gives nontrivial solutions [which we call **eigenfunctions** and henceforth denote as $\phi_n(x)$] if $\lambda = \lambda_n$:

$$\text{eigenvalues:} \quad \lambda_n = \left(\frac{n\pi}{l}\right)^2, \qquad (10a)$$

$$\text{eigenfunctions:} \quad \phi_n(x) = \sin\left(\frac{n\pi x}{l}\right), \qquad (10b)$$

for $n = 1, 2, \ldots$. [We have chosen $B = 1$ in (10b), but remember that the eigenfunction $\sin(n\pi x/l)$ can be scaled by any arbitrary nonzero constant.]

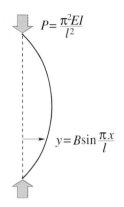

$P = \dfrac{\pi^2 EI}{l^2}$

$y = B \sin \dfrac{\pi x}{l}$

FIGURE 2
The buckling configuration.

Be clear that the solution (9) is *nonunique* since B is arbitrary; that is, we have the solutions $y(x) = 6\sin(n\pi x/l)$, $-2.73\sin(n\pi x/l)$, 0, and so on. Doesn't that result violate our fundamental existence and uniqueness theorem (Theorem 6.4.1)? No, that theorem is for *initial value* problems, whereas (2) is of *boundary value* type; this important distinction was emphasized in Section 6.4.2, which section you may wish to review. That is, if we change (2b) to the initial conditions $y(0) = 0$ and $y'(0) = 0$, say, then (2) would admit a unique solution, namely, $y(x) = 0$, for any value of λ.

Finally, let us interpret our findings physically in terms of buckling. Imagine starting out with $P = 0$ and increasing it very slowly. The solution for y is simply $y(x) = 0$ for all P's until λ reaches the first eigenvalue, $\lambda_1 = (\pi/l)^2$, at which point nontrivial solutions $y(x) = B\sin(\pi x/l)$ appear; $\lambda = \lambda_1 = (\pi/l)^2$ corresponds to $P = (\pi/l)^2 EI$, which therefore is the buckling load. The buckling shape, or configuration, or mode, is given by $y(x) = B\sin(\pi x/l)$, as shown in Fig. 2. How do we determine B? We cannot, from the foregoing analysis. The idea is that when P reaches the value $\pi^2 EI/l^2$ we can imagine the column bending according to the shape $y(x) = B\sin(\pi x/l)$ where B increases dramatically with the time t. Thus, realize that the buckling process is actually a dynamic process and to track it we would need to replace the static Euler beam theory equation with a partial differential equation, on $y(x, t)$, that incorporates Newton's second law of motion. Although our foregoing analysis fails to predict the subsequent process of collapse it does successfully predict the most important feature, the value of P at which that collapse will occur.

What about the higher eigenvalues $\lambda_2 = (2\pi/l)^2$, $\lambda_3 = (3\pi/l)^2$, ... and the corresponding buckling loads $P = (2\pi/l)^2 EI$, $(3\pi/l)^2 EI$,...? In order to reach one of these higher buckling loads, by increasing P, we must pass through the first one, $P = (\pi/l)^2 EI$. At that point buckling will occur, so we will not be able to reach the higher modes. Thus, we call the smallest buckling load the *critical* buckling load and write

$$P_{\text{cr}} = \frac{\pi^2 EI}{l^2}. \tag{11}$$

EXAMPLE 1
As a variation on the eigenvalue problem (2), suppose we change $y(0) = 0$ to $y'(0) = 0$ so

$$y'' + \lambda y = 0 \qquad (0 \le x \le l) \tag{12a}$$
$$y'(0) = 0, \quad y(l) = 0. \tag{12b}$$

Then

$$y(x) = A\cos\sqrt{\lambda}\,x + B\sin\sqrt{\lambda}\,x \tag{13}$$

and

$$y'(0) = 0 = (0)A + \left(\sqrt{\lambda}\right)B, \tag{14a}$$
$$y(l) = 0 = \left(\cos\sqrt{\lambda}\,l\right)A + \left(\sin\sqrt{\lambda}\,l\right)B \tag{14b}$$

so $B = 0$ from (14a) and

$$\left(\cos \sqrt{\lambda}\, l\right) A = 0 \tag{15}$$

from (14b). This time, for nontrivial solutions we need

$$\cos \sqrt{\lambda}\, l = 0. \tag{16}$$

Since the positive roots of $\cos x = 0$ are $x = \pi/2, 3\pi/2, 5\pi/2, \ldots$, it follows from (16) that the eigenvalues of (12) are

$$\lambda_n = \left(\frac{n\pi}{2l}\right)^2 \tag{17}$$

for $n = 1, 3, 5, \ldots$. Then, from (13) with $B = 0$, A arbitrary, and $\lambda = (n\pi/2l)^2$, we have

$$\phi_n(x) = \cos \frac{n\pi x}{2l} \tag{18}$$

as the corresponding eigenfunctions.   ■

9.4.2 Interpreting the differential equation eigenvalue problem as a matrix eigenvalue problem. (Optional).   Let us reconsider the eigenvalue problem (2),

$$y'' + \lambda y = 0 \qquad (0 \le x \le l) \tag{19a}$$
$$y(0) = 0, \quad y(l) = 0, \tag{19b}$$

but this time let us seek an approximate numerical solution—even though we found an exact solution analytically. We do so in order to relate (19) to a matrix eigenvalue problem and to thereby better understand the differential equation version of the eigenvalue problem.

We will *discretize* the problem (19) by seeking $y(x)$ not over the entire interval $0 \le x \le l$ but only at discrete points called *nodal points*. Specifically, we divide l into N equal parts, of length $\Delta x = l/N$, and denote y at $j\Delta x$ as y_j for $j = 0, 1, 2, \ldots, N$, as illustrated in Fig. 3 for $N = 4$. The end values, y_0 and y_4, are zero by virtue of the boundary conditions (19b).

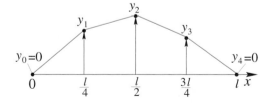

FIGURE 3   Discrete model.

Next, we need a discrete version of the y'' term in (19a). To obtain that we use Taylor series. Expanding $y(x)$ about x first with a positive step $+\Delta x$ and then with a negative step $-\Delta x$ gives

$$y(x + \Delta x) = y(x) + y'(x)\Delta x + \frac{y''(x)}{2!}(\Delta x)^2 + \frac{y'''(x)}{3!}(\Delta x)^3 + O\left((\Delta x)^4\right),$$
(20a)

$$y(x - \Delta x) = y(x) - y'(x)\Delta x + \frac{y''(x)}{2!}(\Delta x)^2 - \frac{y'''(x)}{3!}(\Delta x)^3 + O\left((\Delta x)^4\right),$$
(20b)

where the $O\left((\Delta x)^4\right)$ notation is shorthand to indicate that the remaining terms are of order $(\Delta x)^4$ and higher. Adding (20a) and (20b) we can cancel the y' terms and obtain

$$y''(x) = \frac{y(x + \Delta x) - 2y(x) + y(x - \Delta x)}{(\Delta x)^2} + O\left((\Delta x)^2\right)$$
(21)

or, neglecting the $O\left((\Delta x)^2\right)$ term,

$$y''(x) \approx \frac{y(x + \Delta x) - 2y(x) + y(x - \Delta x)}{(\Delta x)^2}.$$
(22)

In the limit as $\Delta x \to 0$ (22) tends to an equality because the $O\left((\Delta x)^2\right)$ error tends to zero. However, we will not take the limit as $\Delta x \to 0$; we will simply choose a small Δx (i.e., small compared to l) and accept the error that results.

Thus, in terms of our discrete notation, (22) gives

$$y_j''(x) = \frac{y_{j+1} - 2y_j + y_{j-1}}{(\Delta x)^2},$$
(23)

where we use an equal sign but understand that (23) is not exact. Then (19a) becomes

$$\frac{y_{j+1} - 2y_j + y_{j-1}}{(\Delta x)^2} + \lambda y_j = 0$$
(24)

for $j = 1, 2, \ldots, N - 1$. Let it suffice, for purposes of this example, to choose $N = 4$, as in Fig. 3, even though $\Delta x = l/4$ is not very small compared to l. Then, writing out (24) for $j = 1, 2, 3$ gives

$$\frac{y_2 - 2y_1 + y_0}{(\Delta x)^2} + \lambda y_1 = 0,$$

$$\frac{y_3 - 2y_2 + y_1}{(\Delta x)^2} + \lambda y_2 = 0,$$

$$\frac{y_4 - 2y_3 + y_2}{(\Delta x)^2} + \lambda y_3 = 0.$$

Putting $y_0 = y_4 = 0$, according to (19b), and rearranging terms gives

$$\begin{bmatrix} 2 & -1 & 0 \\ -1 & 2 & -1 \\ 0 & -1 & 2 \end{bmatrix} \begin{bmatrix} y_1 \\ y_2 \\ y_3 \end{bmatrix} = \lambda(\Delta x)^2 \begin{bmatrix} y_1 \\ y_2 \\ y_3 \end{bmatrix},$$
(25)

which is an eigenvalue problem, with $\lambda(\Delta x)^2 \equiv \Lambda$, say, as the eigenvalue. The eigenpairs are found (by hand or by *Maple*) to be

$$\Lambda_1 = 2 - \sqrt{2}, \quad \mathbf{e}_1 = [1, \sqrt{2}, 1]^{\mathrm{T}},$$
$$\Lambda_2 = 2, \quad \mathbf{e}_2 = [1, 0, -1]^{\mathrm{T}},$$
$$\Lambda_3 = 2 + \sqrt{2}, \quad \mathbf{e}_3 = [1, -\sqrt{2}, 1]^{\mathrm{T}}$$

or, since $\lambda = \Lambda/(\Delta x)^2 = \Lambda/(l/4)^2 = 16\Lambda/l^2$,

$$\lambda_1 = 9.37/l^2, \quad \mathbf{e}_1 = [1, 1.41, 1]^{\mathrm{T}},$$
$$\lambda_2 = 32/l^2, \quad \mathbf{e}_2 = [1, 0, -1]^{\mathrm{T}}, \tag{26}$$
$$\lambda_3 = 54.6/l^2, \quad \mathbf{e}_3 = [1, -1.41, 1]^{\mathrm{T}}.$$

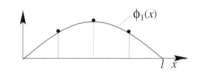

These are an approximation of the first three eigenpairs of the exact problem (19), the differential equation version, which are

$$\lambda_1 = \pi^2/l^2 = 9.87/l^2, \quad \phi_1(x) = \sin(\pi x/l),$$
$$\lambda_2 = 4\pi^2/l^2 = 39.5/l^2, \quad \phi_2(x) = \sin(2\pi x/l), \tag{27}$$
$$\lambda_3 = 9\pi^2/l^2 = 88.8/l^2, \quad \phi_3(x) = (3\pi x/l).$$

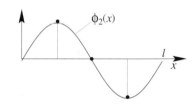

We can compare the eigenvalues in (26) with those in (27) by inspection. In order to compare the (discrete) eigenvectors in (26) with the (continuous) eigenfunctions in (27) we have plotted them in Fig. 4, where we have arbitrarily scaled the eigenvectors in (26) so as to fit the sinusoids in (27) as well as possible; such scaling is permissible because eigenvectors are determined only to within an arbitrary (nonzero) scaling anyway. If we were to repeat the calculation with $N = 10$, say, we would obtain an approximation of the first 10 eigenpairs. In the limit as $N \to \infty$ the eigenvalues and eigenvectors of the discrete problem will converge to the corresponding eigenvalues and eigenfunctions of the continuous problem (19). That convergence is explored heuristically in the exercises.

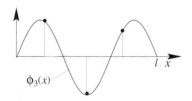

FIGURE 4
Comparison of the eigenvectors in (26) (dots) with the eigenfunctions in (27) (solid).

9.4.3 The General case; Sturm-Liouville theory. (Optional). The differential equation eigenvalue problems (2) and (12) are but two examples of the following more general eigenvalue problem:

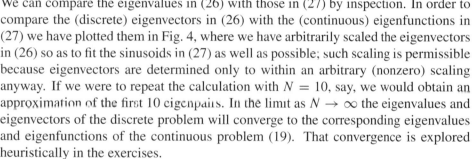

$$p(x)y'' + p'(x)y' + q(x)y + \lambda w(x)y = 0 \qquad (a \leq x \leq b)$$
$$\alpha y(a) + \beta y'(a) = 0, \tag{28a,b,c}$$
$$\gamma y(b) + \delta y'(b) = 0,$$

which is well known as the **Sturm-Liouville problem.**[1]

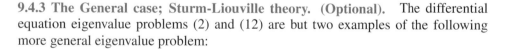

---

[1] After *Charles Sturm* (1803–1855) and *Joseph Liouville* (1809–1882), professors of mechanics at the Sorbonne and mathematics at the Collège de France, respectively. Note: Liouville, pronounced Lee-oo-vil, not Louisville.

For instance, (12) is a Sturm-Liouville problem with $p(x) = 1$, $q(x) = 0$, $w(x) = 1$, $a = 0$, $b = l$, $\alpha = \delta = 0$, and $\beta = \gamma = 1$.

While discussion of the Sturm-Liouville theory is beyond the scope of this book, a brief overview of it will be helpful in putting Sections 9.4.1 and 9.4.2 in perspective and also in emphasizing the importance of *function spaces* in applied mathematics. First, to make (28) look like the matrix eigenvalue problem $\mathbf{A}\mathbf{x} = \lambda\mathbf{x}$ we re-express it as

$$L[y] = \lambda y \tag{29}$$

where the differential operator

$$L = -\frac{1}{w}\left(p\frac{d^2}{dx^2} + p'\frac{d}{dx} + q\right) \tag{30}$$

acts on the function space of functions that have continuous second derivatives on $[a, b]$ and that satisfy the homogeneous boundary conditions (28b,c). For that space we define the inner product of any functions u and v in the space as

$$\langle u(x), v(x)\rangle = \int_a^b u(x)v(x)w(x)\,dx \tag{31}$$

where the weight function $w(x)$ in (31) is the same $w(x)$ as in (28a). (These matters were introduced in Section 4.8, which you might wish to review.)

Suppose that a and b are finite, that p, p', q, w are continuous on $[a, b]$, and that $p(x) > 0$ and $w(x) > 0$ on $[a, b]$. Then the Sturm-Liouville theory gives several important results. Let λ_n and ϕ_n denote any eigenvalue and corresponding eigenfunction of the Sturm-Liouville problem (28). Then:

(a) There are an infinite number of eigenvalues. They are *real* and can be ordered so that $\lambda_1 < \lambda_2 < \lambda_3 < \cdots$, where $\lambda_n \to \infty$ as $n \to \infty$.

(b) Eigenfunctions corresponding to distinct eigenvalues are *orthogonal*. That is, if $\lambda_j \neq \lambda_k$, then $\langle \phi_j, \phi_k \rangle = 0$.

(c) The infinite set of orthogonal eigenfunctions constitutes an *orthogonal basis* for the infinite-dimensional space.[1]

Observe the striking similarity between these results and the analogous results for real symmetric matrices. Specifically, compare items (a), (b), (c) with Theorems 9.3.1, 9.3.3, and 9.3.4, respectively. Roughly speaking, we can think of the Sturm-Liouville eigenvalue problem as an infinite-dimensional function space version of the matrix eigenvalue problem $\mathbf{A}\mathbf{x} = \lambda\mathbf{x}$ where $\mathbf{A}$ is real and symmetric. That connection can be seen in the example studied in Section 9.4.2. However, we should stress that the Sturm-Liouville theory is deeper than the corresponding matrix theory in Section 9.4.2, especially with respect to item (c). Specifically, in the matrix case if

---

[1]Stronger statements can be made. See, for example, M. Greenberg, *Advanced Engineering Mathematics*, 2nd ed. (Upper Saddle River, NJ: Prentice Hall, 1998), Sections 17.7 and 17.8.

the $n \times n$ matrix gives n orthogonal eigenvectors (as it will if it is real and symmetric), then surely those vectors will constitute a basis for the n-space. In the Sturm-Liouville case we obtain an infinite set of orthogonal eigenfunctions so it would be natural to feel assured that they must constitute a basis for the infinite-dimensional function space. However, suppose they *are* a basis. If we remove three of them, say, then the remaining set will not be a basis; it will not be "complete," even though it still contains an infinite number of vectors. That is, in an infinite-dimensional space an infinite set of LI (or even orthogonal) vectors might still fall short of being a basis. Furthermore, expansions will be infinite series so there will be subtle convergence issues to address.

Let us close with an illustration of items (a) and (b), above.

EXAMPLE 2

The boundary value problem (2) is of Sturm-Liouville type and the eigenvalues and eigenfunctions are given by (10a) and (10b), respectively. In accord with item (a), there is an infinite number of λ_n's, they are real, and they do tend to infinity as $n \to \infty$. To check item (b) we need to first know what the weight function w is for the inner product (31). Comparing (2a) with (28a) we see that $w(x) = 1$. Then

$$\langle \phi_j(x), \phi_k(x) \rangle = \int_0^l \sin \frac{j\pi x}{l} \sin \frac{k\pi x}{l} \, dx. \tag{32}$$

Using the trigonometric identity $\sin A \sin B = [\cos(A - B) - \cos(A + B)]/2$ to evaluate the integral, we do find that $\langle \phi_j(x), \phi_k(x) \rangle = 0$ if $j \neq k$, in accord with item (b). We reserve discussion of item (c) for the exercises.   ∎

Closure.   In this section we see that the eigenvalue problem is not peculiar to matrix operators; it occurs for differential operators as well. One difference between the two cases is that there are no "boundary conditions" in the matrix case, as there are in the case of differential operators. Differential equation eigenvalue problems are always of *boundary* value type so that nonunique solutions can arise, for if there were a unique solution it would be the trivial solution and there would be no eigenfunctions. Solution of a differential equation eigenvalue problem does not require special techniques; we merely obtain a general solution of the differential equation and then impose the homogeneous boundary conditions. To obtain nontrivial solutions we will need λ to satisfy a certain *characteristic equation*, just as in the matrix case. For instance, (6) was the characteristic equation corresponding to the eigenvalue problem (2).

As a physical application we studied the buckling of columns and found that the smallest eigenvalue gave the critical buckling load and that the corresponding eigenfunction gave the buckling mode shape.

It is most striking that eigenvalue problems of this type generate infinite sets of orthogonal eigenfunctions that constitute bases for the expansion of functions over that interval. This concept, established by the Sturm-Liouville theory, proves to be invaluable in the study of partial differential equations.

EXERCISES 9.4

1. For the differential equation $y'' + \lambda y = 0$ and the given boundary conditions, find all eigenvalues in the interval $0 \leq \lambda < \infty$. HINT: As we did in the text, you will need to apply the boundary conditions to the general solution $y(x) = A \cos \sqrt{\lambda} x + B \sin \sqrt{\lambda} x$. However, notice that the latter is not the general solution if $\lambda = 0$. Rather,

$$y(x) = \begin{cases} A \cos \sqrt{\lambda} x + B \sin \sqrt{\lambda} x, & \lambda \neq 0 \\ C + Dx, & \lambda = 0. \end{cases}$$
(1.1a,b)

Thus, use (1.1b) to see if $\lambda = 0$ is an eigenvalue, and use (1.1a) for $0 < \lambda < \infty$. In the buckling example in the text we did not consider the case $\lambda = 0$ because if $\lambda = 0$ then $P = 0$, since $\lambda = P/EI$, and if the load P is zero then surely the column will remain straight. However, in other cases $\lambda = 0$ might be an eigenvalue; it *is* in at least one of the following exercises:

(a) $y'(0) = 0, \ y(l) = 0$
(b) $y(0) = 0, \ y(l) = 0$
(c) $y'(0) = 0, \ y'(l) = 0$
(d) $y'(0) = 0, \ y'(l) = 0$

2. Consider the eigenvalue problem

$$y'' + 2y' + \lambda y = 0; \quad y(0) = 0, \ y(\pi) = 0. \quad (2.1)$$

(a) Derive the general solution

$$y(x) = e^{-x} \left(A \cos \sqrt{\lambda - 1} x + B \sin \sqrt{\lambda - 1} x \right).$$
(2.2)

(b) Applying the boundary conditions to (2.2), derive these eigenvalues and eigenfunctions:

$$\lambda_n = 1 + n^2, \quad \phi_n(x) = e^{-x} \sin nx \quad (2.3)$$

for $n = 1, 2, \ldots$.

(c) Observe that $\lambda = 1$ is not among the eigenvalues listed in (2.3). However, observe further that if $\lambda = 1$ then (2.2) is not a general solution of the given differential equation because the $e^{-x} \sin \sqrt{\lambda - 1} x$ solution drops out since $\sin 0 = 0$. Thus, we must treat the $\lambda = 1$ case separately. Show that if $\lambda = 1$ then a general solution is

$$y(x) = (C + Dx) e^{-x}. \quad (2.4)$$

Applying the boundary conditions to (2.4), show that $\lambda = 1$ is not an eigenvalue.
NOTE: Thus, we suggest this procedure in solving

eigenvalue problems: *Obtain a general solution, such as (2.2). Then examine that solution to see if there are any exceptional values of λ for which that solution degenerates so as not to be a general solution. Treat those exceptional λ's separately.* Thus, in this exercise the first step is to derive

$$y(x) =$$
$$\begin{cases} e^{-x} \left(A \cos \sqrt{\lambda - 1} x + B \sin \sqrt{\lambda - 1} x \right), & \lambda \neq 1 \\ (C + Dx) e^{-x}, & \lambda = 1. \end{cases}$$
(2.5a,b)

Then apply the boundary conditions to (2.5a) and, separately, to (2.5b). Note that in Exercise 1 $\lambda = 0$ was exceptional; in this exercise $\lambda = 1$ is exceptional.

(d) Show that alternative to the solution form (2.5a) are the forms

$$y(x) = e^{-x} \left(E \cosh \sqrt{1 - \lambda} x + F \sinh \sqrt{1 - \lambda} x \right)$$
(2.6)

and

$$y(x) = e^{-x} \left(G e^{\sqrt{1 - \lambda} x} + H e^{-\sqrt{1 - \lambda} x} \right). \quad (2.7)$$

Since equations (2.5a), (2.6), and (2.7) are equivalent, they will all lead to the result (2.3). Here we ask you to use (2.6), instead of (2.5a), and obtain (2.3) from it. HINT: You will need to use formulas given in Exercise 17(c) of Section 6.2. You will find that although (2.5a) and (2.6) are equivalent, (2.5a) is the simpler to use. Given such a choice in other examples, how can you tell which solution form to use? As a rule of thumb, expect eigenfunctions to be oscillatory. In the present exercise that rule would lead us to choose (2.5a) over (2.6) or (2.7).

(e) Observe that (2.1) is not of the form (28a) because the coefficient of y' is not the derivative of the coefficient of y''. However, if we multiply (2.1) by e^{2x} then we have

$$e^{2x} y'' + 2e^{2x} y' + \lambda e^{2x} y = 0 \quad (2.8)$$

which *is* of the form (28a), with $p(x) = e^{2x}, q(x) = 0$, and $w(x) = e^{2x}$. Thus, the weight function to be used is $w(x) = e^{2x}$. With that weight function verify, by carrying out the integration in (31), that the eigenfunctions in (2.3) are indeed orthogonal.

3. Consider the eigenvalue problem

$$x^2 y'' + xy' + \lambda y = 0; \quad y(1) = 0, \ y(a) = 0, \quad (3.1)$$

where $a > 1$.

(a) First, derive the general solution

$$y(x) =$$

$$\begin{cases} A \cos\left(\sqrt{\lambda}\ \ln x\right) + B \sin\left(\sqrt{\lambda}\ \ln x\right), & \lambda \neq 0 \\ C + D \ln x, & \lambda = 0. \end{cases}$$

$$(3.2a,b)$$

(b) Using (3.2b), show that $\lambda = 0$ is not an eigenvalue of (3.1).

(c) Turning to $\lambda \neq 0$, apply the boundary conditions to (3.2a) and derive the results

$$\lambda_n = \frac{n^2 \pi^2}{(\ln a)^2}, \qquad \phi_n(x) = \sin\left(n\pi \frac{\ln x}{\ln a}\right), \quad (3.3)$$

for $n = 1, 2, \ldots$.

(d) Observe that (3.1) is not of the form (28a) because the coefficient x is not the derivative of the coefficient x^2. However, if we multiply (3.1) by $1/x$ then we have

$$xy'' + y' + \lambda \frac{1}{x} y = 0, \quad (3.4)$$

which *is* of the form (28a), with $p(x) = x$, $q(x) = 0$, and $w(x) = 1/x$. Thus, we identify the weight function to be used in (31) as $w(x) = 1/x$. With that weight function verify, by carrying out the integration in (31), that the eigenfunctions are orthogonal, namely, that

$$\langle \phi_j, \phi_k \rangle = \int_1^a \sin\left(j\pi \frac{\ln x}{\ln a}\right) \sin\left(k\pi \frac{\ln x}{\ln a}\right) \frac{1}{x}\, dx$$

$$= 0 \quad (3.5)$$

for $j \neq k$. HINT: Use the substitution $t = \ln x$.

4. (*Buckling of linearly tapered column*) Consider a column of circular cross section, the radius of which varies linearly with x. It extends over $a < x < b$, as shown below, and is loaded by a vertical force P. Then the inertia I in (1a) is not a constant; it is $I(x) = I_0(x/b)^4$, where I_0 is its value at $x = b$. The eigenvalue problem governing buckling is

$$x^4 y'' + \lambda y = 0; \quad y(a) = 0, \ y(b) = 0, \quad (4.1)$$

where $\lambda = b^4 P / E I_0$.

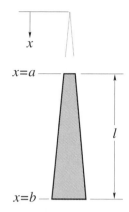

(a) Verify that a general solution is

$$y(x) =$$

$$\begin{cases} x\left[A \cos\left(\dfrac{\sqrt{\lambda}}{x}\right) + B \sin\left(\dfrac{\sqrt{\lambda}}{x}\right)\right], & \lambda \neq 0 \\ C + Dx, & \lambda = 0. \end{cases}$$

$$(4.2a,b)$$

You may verify (4.2a) without deriving it.

(b) Applying the boundary conditions, show that the eigenvalues and eigenfunctions are

$$\lambda_n = \left(\frac{n\pi ab}{l}\right)^2, \qquad \phi_n(x) = x \sin\left[n\pi \frac{b}{l}\left(1 - \frac{a}{x}\right)\right]$$

$$(4.3)$$

for $n = 1, 2, \ldots$, and that the buckling load is $P_{\text{cr}} = \left(\dfrac{\pi a}{bl}\right)^2 E I_0$, where $l = b - a$. As a partial check on this result, show that it agrees with the formula (11) if the column is uniform rather than tapered. HINT: Keeping $b - a$ equal to l and keeping the inertia at $x = b$ equal to I_0, let $b \to \infty$.

5. Consider the eigenvalue problem

$$y'' + \lambda y = 0; \quad y(0) = 0, \ y(1) + y'(1) = 0. \quad (5.1)$$

(a) Beginning with the general solution

$$y(x) = \begin{cases} A \cos\sqrt{\lambda}\, x + B \sin\sqrt{\lambda}\, x, & \lambda \neq 0 \\ C + Dx, & \lambda = 0, \end{cases}$$

$$(5.2a,b)$$

first show that $\lambda = 0$ is not an eigenvalue.

(b) Show that the eigenvalues are found as the roots of the equation

$$\sin\sqrt{\lambda} + \lambda \cos\sqrt{\lambda} = 0 \quad (5.3)$$

and that the corresponding eigenfunctions are

$$\phi_n(x) = \sin\sqrt{\lambda_n}x. \qquad (5.4)$$

(c) To explore the roots of (5.3) it is convenient to let $\sqrt{\lambda} = \Lambda$, say, to sketch the graph of $\sin\Lambda + \Lambda\cos\Lambda$, and to look for crossings of the Λ axis . However, it is simpler to write

$$\tan\Lambda = -\Lambda \qquad (5.5)$$

instead, to sketch the graphs of $\tan\Lambda$ and $-\Lambda$, and to look for intersections of those two graphs. Doing that, show that

$$\Lambda_n \sim (2n-1)\pi/2 \qquad (n=1,2,\dots) \qquad (5.6)$$

as $n \to \infty$.

(d) Use the *Maple* fsolve command to find the first four positive roots of (5.5) and observe that those values do seem to be consistent with (5.6).

(e) Verify, by integration of (31), that the eigenfunctions (5.4) are orthogonal.

6. Show that for

$$y'' + \lambda y = 0; \quad y(0) - y(1) = 0, \ y'(0) + y'(1) = 0 \quad (6.1)$$

every λ (real or complex) is an eigenvalue! Is (6.1) of the Sturm-Liouville form (28)? Explain.

7. Consider the eigenvalue problem

$$y'''' + \lambda y'' = 0; \quad y(0) = y'(0) = y(1) = y'(1) = 0. \qquad (7.1)$$

(a) Derive a general solution of the differential equation. You will see that you need to distinguish the case $\lambda = 0$ from the case $\lambda \neq 0$, as discussed in the preceding exercises.

(b) Show that $\lambda = 0$ is not an eigenvalue.

(c) Show that the eigenvalues can be found as the nonzero roots of the equation

$$2 - 2\cos\sqrt{\lambda} - \sqrt{\lambda}\sin\sqrt{\lambda} = 0. \qquad (7.2)$$

Show that if λ_n is such a root then the corresponding eigenfunction is

$$\phi_n(x) = \sin a - a + a(1 - \cos a)x$$
$$+ (a - \sin a)\cos ax + (\cos a - 1)\sin ax, \quad (7.3)$$

where a is short for $\sqrt{\lambda_n}$.

8. We stated, below (25), that the eigenvalues are found to be $\Lambda = 2 - \sqrt{2}, 2, 2 + \sqrt{2}$. Show that those values follow from equation (6.1) in Exercise 6 of Section 9.2.

9. In Section 9.4.2 we gave numerical results for the choice $N = 4$. Try a larger N: use $N = 8$. Using *Maple*, evaluate the eight eigenvalues of the discrete model and compare them with the first eight eigenvalues of the exact problem, as we did in (26) and (27) for the case where $N = 4$.

10. (*Fourier series*) We did not illustrate item (c), below (31), in the text, so let us do so in this exercise. Consider, for instance, the eigenfunctions $\phi_n(x) = \sin(n\pi x/l)$ in (10b), and let $l = \pi$ for convenience, so $\phi_n(x) = \sin nx$. To expand a given function $f(x)$ in terms of the orthogonal basis $\{\sin nx\}$, on the interval $0 \le x \le \pi$, write

$$f(x) = a_1\sin x + a_2\sin 2x + a_3\sin 3x + \cdots. \quad (10.1)$$

Such expansions, in terms of the eigenfunctions generated by a Sturm-Liouville problem, are called **Fourier series** (or "generalized" Fourier series).

(a) Mimicking our derivation of (25) in Section 4.9.3, derive the formula

$$a_j = \frac{\langle f, \sin jx\rangle}{\langle\sin jx, \sin jx\rangle} = \frac{\int_0^\pi f(x)\sin jx\,dx}{\int_0^\pi \sin^2 jx\,dx}. \quad (10.2)$$

Evaluating the integral in the denominator (using *Maple* if you wish) obtain the final result

$$a_j = \frac{2}{\pi}\int_0^\pi f(x)\sin jx\,dx. \qquad (10.3)$$

NOTE: We know that $\langle u, \alpha_1 v_1 + \cdots + \alpha_k v_k\rangle = \alpha_1\langle u, v_1\rangle + \cdots + \alpha_k\langle u, v_k\rangle$ for finite k; you may assume in this exercise that distributivity holds even for an infinite series: $\langle u, \alpha_1 v_1 + \alpha_2 v_2 + \cdots\rangle = \alpha_1\langle u, v_1\rangle + \alpha_2\langle u, v_2\rangle + \cdots$.

(b) To illustrate, let

$$f(x) = \begin{cases} x^2, & 0 \le x \le 1 \\ (\pi - x)/(\pi - 1), & 1 \le x \le \pi. \end{cases} \quad (10.4)$$

Using *Maple* to evaluate the integral in (10.3), show that (10.1) gives the expansion

$$f(x) = 0.7362\sin x + 0.1316\sin 2x$$
$$- 0.0692\sin 3x - 0.1072\sin 4x$$
$$- 0.0675\sin 5x - 0.0124\sin 6x$$
$$+ 0.0201\sin 7x + 0.0214\sin 8x$$
$$+ 0.0047\sin 9x - 0.0109\sin 10x + \cdots. \quad (10.5)$$

HINT: Here is a set of *Maple* commands to evaluate the a_j coefficients:

```
f:=x^2*(1-Heaviside(x-1))+((Pi-x)/(Pi-1))
```

```
    *(Heaviside(x-1)-Heaviside(x-Pi));
aj:=(2/Pi)*int(f*sin(j*x),x=0..Pi);
evalf(subs(j=1,aj));
```

The last command gives a_1; changing "$j = 1$" to $j = 2$ it gives a_2, and so on. Note the use of the **subs** command, which substitutes $j = 1$ in a_j.

(c) Use *Maple* to plot these together: $f(x)$ given by (10.4);

the sum of the first five terms in the right-hand side of (10.5); and the sum of the first ten terms. HINT: In the implicitplot command, use the option numpoints = 1000, say, to obtain sufficient plotting accuracy. Also, to plot (10.4) using the *Maple* implicitplot command you will need to express (10.4) in Heaviside notation.

CHAPTER 9 REVIEW

The matrix eigenvalue problem is the search for nontrivial solutions of $\mathbf{Ax} = \lambda\mathbf{x}$ or, equivalently,

$$(\mathbf{A} - \lambda\mathbf{I})\mathbf{x} = \mathbf{0}; \tag{1}$$

that is, solutions other than the trivial solution $\mathbf{x} = \mathbf{0}$. Here, $\mathbf{A}$ is necessarily square, $n \times n$. The eigenvalues are found by setting

$$\det(\mathbf{A} - \lambda\mathbf{I}) = 0, \tag{2}$$

which condition on λ guarantees the existence of nontrivial solutions of (1). Known as the characteristic equation of $\mathbf{A}$, (2) is an nth-degree polynomial equation, which always has at least one and at most n distinct roots. For each eigenvalue λ_j thus found, the solution of $(\mathbf{A} - \lambda_j\mathbf{I})\mathbf{x} = \mathbf{0}$ by Gauss elimination then gives the corresponding eigenvectors $\mathbf{e}_j$.

In more picturesque language, imagine turning a "λ knob," just as we turn the frequency knob on a radio. In the latter case, only at a discrete set of specific frequencies do we tune in to a station; in between we pick up (approximately) no signal at all. Analogously, only when λ takes on one of a discrete set of values will (1) give nontrivial solutions.

In Section 9.3 we studied the special case of symmetric matrices, which are common in applications. For such matrices several important results follow: If a real $n \times n$ matrix $\mathbf{A}$ is symmetric, then all of its eigenvalues are real, the eigenspace corresponding to an eigenvalue of multiplicity k is of dimension k, eigenvectors corresponding to distinct eigenvalues are orthogonal, and the eigenvectors of $\mathbf{A}$ provide an orthogonal basis for n-space.

In the final section, 9.4, we introduced eigenvalue problems for differential, rather than matrix, equations. The latter are of the form $L[y] = \lambda y$, where L is a differential operator, together with homogeneous boundary conditions. Thus, they are homogeneous boundary value problems. For problems of Sturm-Liouville type, we stated that there is always an infinite set of real eigenvalues, with an infinite set of orthogonal eigenfunctions. Just as the eigenvectors of a real symmetric $n \times n$ matrix constitute an orthogonal basis for the n-dimensional n-space, so do the eigenfunctions of a Sturm-Liouville problem constitute an orthogonal basis for the relevant infinite-dimensional function space of functions defined on the given x interval.

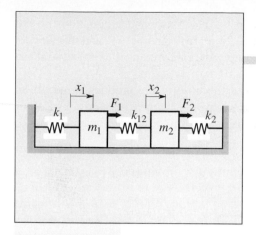

SYSTEMS OF LINEAR DIFFERENTIAL EQUATIONS

10.1 INTRODUCTION

Thus far we have considered problems in which there is only one dependent variable. However, many problems involve two or more interrelated dependent variables, so that instead of a single differential equation we have a system of simultaneous differential equations.

For instance, consider an ecological system containing two species, one a predator (such as hawks) and the other its prey (such as field mice). Denote their populations, as functions of the time t, by $y(t)$ and $x(t)$, respectively. According to the well known Lotka-Volterra model of predator-prey dynamics, the populations are governed by the system of differential equations

$$\frac{dx}{dt} = (\alpha - \beta y)x, \tag{1a}$$

$$\frac{dy}{dt} = (-\gamma + \delta x)y, \tag{1b}$$

where $\alpha, \beta, \gamma, \delta$ are empirical positive constants. In the absence of predator ($y = 0$) equation (1a) reduces to $dx/dt = \alpha x$, where α is a net birth/death rate, so the prey population exhibits exponential growth. But if the predator is present ($y > 0$) then the prey's net birth/death rate $\alpha - \beta y$ in (1a) shows the adverse effect of the presence of the predator. Similarly, in the absence of prey ($x = 0$) equation (1b) reduces to $dy/dt = -\gamma y$, so the predator population exhibits exponential decline. But if the prey is present ($x > 0$) then the predator's net birth/death rate $-\gamma + \delta x$ shows the positive effect of the availability of prey.

We say that (1a) and (1b) are **coupled** because the x and y equations do not contain only x and y, respectively; the x equation (1a) contains a y, and the y equation (1b) contains an x. The coupling is simply the mathematical expression of the relatedness of the two populations.

The system (1) happens to be nonlinear because of the product terms, yx in (1a) and xy in (1b). In this chapter we consider only linear equations. By definition, a **linear** first-order system of n equations in n unknown dependent variables $x_1(t), \ldots, x_n(t)$ is of the form

$$b_{11}(t)\frac{dx_1}{dt} + \cdots + b_{1n}(t)\frac{dx_n}{dt} + c_{11}(t)x_1 + \cdots + c_{1n}(t)x_n = g_1(t),$$

$$\vdots \tag{2}$$

$$b_{n1}(t)\frac{dx_1}{dt} + \cdots + b_{nn}(t)\frac{dx_n}{dt} + c_{n1}(t)x_1 + \cdots + c_{nn}(t)x_n = g_n(t),$$

where the $g_j(t)$'s and the coefficients $b_{jk}(t)$ and $c_{jk}(t)$ are prescribed functions of the independent variable t. The independent variable is often the time t, so we use t as a generic independent variable. Likewise, the names $x_1, \ldots, x_n$ are intended as generic. We refer to the $g_j(t)$'s as *forcing functions* and classify (2) as **homogeneous** if all of the $g_j(t)$'s are identically zero, and as **nonhomogeneous** otherwise.

We call (2) a **first-order system** because the highest derivatives in it are of first order. If the highest derivatives were of second order we would call it a **second-order system**, and so on.

In matrix form we can write (2) as

$$\begin{bmatrix} b_{11} & \cdots & b_{1n} \\ \vdots & & \vdots \\ b_{n1} & \cdots & b_{nn} \end{bmatrix} \begin{bmatrix} x_1' \\ \vdots \\ x_n' \end{bmatrix} + \begin{bmatrix} c_{11} & \cdots & c_{1n} \\ \vdots & & \vdots \\ c_{n1} & \cdots & c_{nn} \end{bmatrix} \begin{bmatrix} x_1 \\ \vdots \\ x_n \end{bmatrix} = \begin{bmatrix} g_1 \\ \vdots \\ g_n \end{bmatrix} \tag{3}$$

or

$$\mathbf{B}(t)\mathbf{x}' + \mathbf{C}(t)\mathbf{x} = \mathbf{g}(t), \tag{4}$$

where primes denote d/dt. If the $\mathbf{B}$ matrix is nonsingular (i.e., if $\det \mathbf{B} \neq 0$) over the t interval of interest, then we can compute $\mathbf{B}^{-1}$ and multiply it into (4), giving

$$\mathbf{x}' + \mathbf{B}^{-1}\mathbf{C}\mathbf{x} = \mathbf{B}^{-1}\mathbf{g} \tag{5}$$

or

$$\mathbf{x}' = -\mathbf{B}^{-1}\mathbf{C}\mathbf{x} + \mathbf{B}^{-1}\mathbf{g}. \tag{6}$$

Denoting the matrix $-\mathbf{B}^{-1}\mathbf{C}$ as $\mathbf{A}$ and the vector $\mathbf{B}^{-1}\mathbf{g}$ as $\mathbf{f}$, for brevity, (6) becomes

$$\boxed{\mathbf{x}' = \mathbf{A}(t)\mathbf{x} + \mathbf{f}(t)} \tag{7}$$

or, in scalar form,

$$\boxed{\begin{aligned} x_1' &= a_{11}(t)x_1 + \cdots + a_{n1}(t)x_n + f_1(t), \\ &\vdots \\ x_n' &= a_{n1}(t)x_1 + \cdots + a_{nn}(t)x_n + f_n(t). \end{aligned}} \tag{8}$$

In applications, linear first-order systems occur sometimes in the form (3) and sometimes in the form (8). We have framed the latter because it is the form that is assumed in most of the remainder of this chapter.

Examples. Let us begin by giving a few examples of how such systems arise in applications.

E X A M P L E 1 *RL Circuit.*

Consider the circuit shown in Fig. 1, comprised of three loops. We wish to obtain the differential equations governing the various currents in the circuit. There are two ways to proceed that are different but equivalent, and which correspond to the current labeling shown in Fig. 2a and 2b (in which we have omitted the circuit elements, for simplicity). First consider the former. If the current approaching the junction p from the "west" is designated as i_1 and the current leaving to the east is i_2, then it follows from Kirchhoff's current law (namely, that the algebraic sum of the currents approaching or leaving any point of a circuit is zero) that the current to the south must be $i_1 - i_2$. Similarly, if we designate the current leaving the junction q to the east as i_3, then the current leaving to the south must be $i_2 - i_3$. With the current approaching r from the north and east being $i_2 - i_3$ and i_3, it follows that the current leaving to the west must be i_2. Similarly, the current leaving s to the west must be i_1.

Next, apply Kirchhoff's voltage law (namely, that the algebraic sum of the voltage drops around each loop of the circuit must be zero) to each loop, recalling from Section 2.4 that the voltage drops across inductors, resistors, and capacitors (of which there are none in this particular circuit) are $L\dfrac{di}{dt}$, Ri, and $\frac{1}{C}\int i\,dt$, respectively.

For the left-hand loop that step gives $L_1\dfrac{di_1}{dt} + R_1(i_1 - i_2) - E_1(t) = 0$, where the last term (corresponding to the applied voltage E_1) is counted as negative because it amounts to a voltage rise (according to the polarity denoted by the $\pm$ signs in Fig. 1) rather than a drop. Thus, we have for the left, middle, and right loops,

$$L_1 i_1' + R_1(i_1 - i_2) = E_1(t),$$

$$L_2 i_2' + R_2(i_2 - i_3) + R_1(i_2 - i_1) = E_2(t), \qquad (9)$$

$$L_3 i_3' + R_3 i_3 + R_2(i_3 - i_2) = E_3(t),$$

respectively, or,

$$i_1' = -\frac{R_1}{L_1}i_1 + \frac{R_1}{L_1}i_2 + \frac{1}{L_1}E_1(t),$$

$$i_2' = \frac{R_1}{L_2}i_1 - \frac{R_1 + R_2}{L_2}i_2 + \frac{R_2}{L_2}i_3 + \frac{1}{L_2}E_2(t), \qquad (10)$$

$$i_3' = \frac{R_2}{L_3}i_2 - \frac{R_2 + R_3}{L_3}i_3 + \frac{1}{L_3}E_3(t),$$

where $E_1(t)$, $E_2(t)$, $E_3(t)$ are prescribed. It must be remembered that the currents do not need to flow in the directions assumed by the arrows; after all, they are the

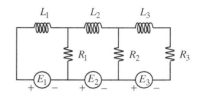

FIGURE 1
Circuit of Example 1.

(a)

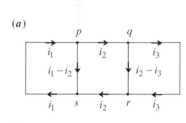

(b)

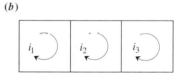

FIGURE 2
Current designations.

unknowns. If any of them turn out to be negative (at any given instant t), that merely means that they are flowing in the direction opposite to that tentatively assumed in Fig. 2a.

Alternatively, one can use the idea of "loop currents," as denoted in Fig. 2b. In that case the south-flowing currents in R_1 and R_2 (Fig. 1) are the net currents $i_1 - i_2$ and $i_2 - i_3$, respectively, just as in Fig. 2a. Either way, the result is the linear first-order system (10).   ∎

E X A M P L E   2   *LC Circuit*.

For the circuit shown in Fig. 3, the same reasoning as above gives the equations

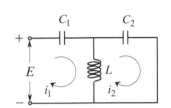

$$\frac{1}{C_1} \int i_1 \, dt + L \frac{d}{dt}(i_1 - i_2) = E(t),$$
$$\frac{1}{C_2} \int i_2 \, dt + L \frac{d}{dt}(i_2 - i_1) = 0 \tag{11}$$

FIGURE 3
LC circuit.

on the currents $i_1(t)$ and $i_2(t)$ or, differentiating to eliminate the integral signs,

$$Li_1'' - Li_2'' + \frac{1}{C_1}i_1 = E'(t),$$
$$Li_2'' - Li_1'' + \frac{1}{C_2}i_2 = 0. \tag{12}$$

Whereas (10) was a first-order system, (12) is of second order.   ∎

E X A M P L E   3   *Mass-Spring System*.

This time consider a mechanical system, shown in Fig. 4 and comprised of masses and springs. The masses rest on a frictionless table and are subjected to applied forces $F_1(t)$, $F_2(t)$, respectively. When the displacements x_1 and x_2 are zero, the springs are neither stretched nor compressed, and we seek the equations of motion of the system, that is, the differential equations governing $x_1(t)$ and $x_2(t)$.

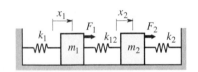

FIGURE 4
Mass-spring system.

The relevant physics consists of Newton's second law of motion and Hooke's law for each of the three springs. To proceed, it is useful to make a concrete assumption on x_1 and x_2. Specifically, suppose that at the instant t we have $x_1 > x_2 > 0$, as assumed in Fig. 5 (which figure, in the study of mechanics, is called a *free-body diagram*). Then the left spring is stretched by x_1 so it exerts a force to the left on m_1 equal (according to Hooke's law) to $k_1 x_1$. The middle spring is compressed by $x_1 - x_2$ so it exerts a force $k_{12}(x_1 - x_2)$ to the left on m_1 and to the right on m_2, and the right spring is compressed by x_2 and exerts a force $k_2 x_2$ to the left on m_2, as shown in Fig. 5. With the help of the information given in Fig. 5, Newton's second law for each of the two masses gives

$$m_1 x_1'' = -k_1 x_1 - k_{12}(x_1 - x_2) + F_1(t),$$
$$m_2 x_2'' = -k_2 x_2 + k_{12}(x_1 - x_2) + F_2(t) \tag{13}$$

as the desired equations of motion.

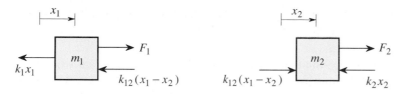

FIGURE 5   Free-body diagram of the masses if
$$x_1 > x_2 > 0.$$

COMMENT. Our assumption that $x_1 > x_2 > 0$ was only for definiteness; the resulting equations (13) are insensitive to whatever such assumption is made. For instance, suppose that we assume, instead, that $x_2 > x_1 > 0$. Then the middle spring is stretched by $x_2 - x_1$, so the free-body diagram of m_1 changes to that shown in Fig. 6, and Newton's law for m_1 gives $m_1 x_1'' = -k_1 x_1 + k_{12}(x_2 - x_1) + F_1(t)$, which is seen to be equivalent to the first of equations (13); similarly for the second of equations (13). ∎

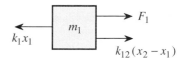

FIGURE 6
Revised free-body
diagram for m_1 if $x_2 > x_1 > 0$.

Reviewing these three examples, we see that the system (10) is of our "standard" form (8), where the $a_{jk}(t)$ coefficients in (8) happen to be constants, but both (12) and (13) are second-order systems. Nevertheless, *we can always reduce a higher-order system to a first-order system by the introduction of* **artificial** *or* **auxiliary variables**.

To reduce (12) to a first-order system we define auxiliary dependent variables "$i_3(t)$" and "$i_4(t)$" according to $i_1'(t) \equiv i_3(t)$ and $i_2'(t) \equiv i_4(t)$, for then (12) becomes

$$i_1' = i_3,$$
$$i_2' = i_4,$$
$$Li_3' - Li_4' + \frac{1}{C_1}i_1 = E'(t), \tag{14}$$
$$Li_4' - Li_3' + \frac{1}{C_2}i_2 = 0,$$

which is of the form (4) where the **B** matrix is

$$\mathbf{B} = \begin{bmatrix} 1 & 0 & 0 & 0 \\ 0 & 1 & 0 & 0 \\ 0 & 0 & L & -L \\ 0 & 0 & -L & L \end{bmatrix}. \tag{15}$$

The latter happens to be singular, so (14) cannot be re-expressed in the "standard" form (8).

Similarly, to reduce (13) to the form (8) we can introduce auxiliary variables "$x_3(t)$" and "$x_4(t)$" according to $x_1' \equiv x_3$ and $x_2' \equiv x_4$, for then (13) becomes

$$x_1' \equiv x_3, \qquad x_2' \equiv x_4,$$
$$x_3' = -\frac{k_1 + k_{12}}{m_1}x_1 + \frac{k_{12}}{m_1}x_2 + \frac{1}{m_1}F_1(t), \tag{16}$$

$$x_4' = \frac{k_{12}}{m_2}x_1 - \frac{k_2 + k_{12}}{m_2}x_2 + \frac{1}{m_2}F_2(t),$$

which is of the form (8).

The technique of auxiliary variables will also be important when we study numerical solutions, in Chapter 12. Let us illustrate the idea with two more examples.

E X A M P L E   4

To reduce

$$x''' = x + 2y', \tag{17a}$$

$$y'' = x - y \tag{17b}$$

to a first-order system, set $x' \equiv u$, $x'' = u' \equiv v$, and $y' \equiv w$. Then (17) is converted to the first-order system

$$x' = u,$$

$$u' = v,$$

$$y' = w, \tag{18}$$

$$v' = x + 2w,$$

$$w' = x - y,$$

where the first three equations serve to define the auxiliary variables u,v,w, the fourth equation follows from (17a), and the fifth follows from (17b). ∎

E X A M P L E   5

Even a single equation such as

$$x''' - tx'' + 3x = \sin t \tag{19}$$

can be reduced to a first-order system, namely, to the system

$$x' = u,$$

$$u' = v, \tag{20}$$

$$v' = -3x + tv + \sin t$$

on $x(t)$, $u(t)$, and $v(t)$. ∎

Closure.   The general first-order linear system is of the form (4), but in what follows we focus mostly on the form (7). Even if we do start with a system in the form (4), we can convert it to the form (7) if the **B** matrix is nonsingular. We illustrated, in Examples 1–3, how systems of coupled differential equations arise in applications, and showed how to introduce auxiliary dependent variables so as to reduce a higher-order system to a first-order system.

EXERCISES 10.1

1. Derive the system of differential equations governing the displacements $x_j(t)$, using the assumption that $x_1 > x_2 > x_3 > 0$. Repeat the derivation assuming instead that $x_3 > x_2 > x_1 > 0$ and again, assuming that $x_1 > x_3 > x_2 > 0$, and show that the resulting equations are the same, independent of these different assumptions.

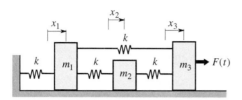

2. **(a)**, **(b)**, **(c)** Derive the system of differential equations governing the currents $i_j(t)$. State any physical laws that you use.

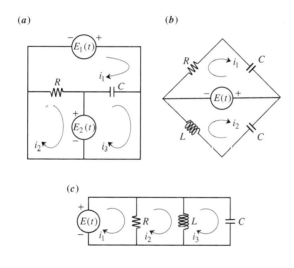

3. (*A two-compartment system*) Consider the pair of mixing tanks shown below, with a constant inflow and a constant outflow of Q gal/min. The inflow is at a constant concentration c_0 (lb/gal) of a particular solute and the tanks are stirred so that the concentrations $c_1(t)$ and $c_2(t)$ are spatially uniform. There is an exchange of liquid at the constant rates Q_{21} gal/min from tank 2 to tank 1 and Q_{12} gal/min from tank 1 to tank 2. The tanks contain V_1 and V_2 gal, respectively. Use the principle of conservation of mass to derive the governing system of equations

$$c_1' = -\frac{Q_{12}}{V_1}c_1(t) + \frac{Q_{21}}{V_1}c_2(t) + \frac{Q}{V_1}c_0, \qquad (3.1)$$

$$c_2' = \frac{Q_{12}}{V_2}c_1(t) - \left(\frac{Q_{21} + Q}{V_2}\right)c_2(t), \qquad (3.2)$$

to which it would be appropriate to append initial conditions $c_1(0) = b_1$ and $c_2(0) = b_2$, say. HINT: If necessary, see Section 2.4.4. Note that for V_1 and V_2 to be constant, as assumed, we need Q, Q_{12}, and Q_{21} to be related according to $Q_{12} = Q + Q_{21}$.

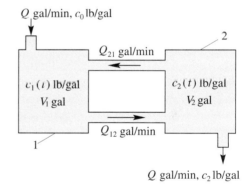

10.2 EXISTENCE, UNIQUENESS, AND GENERAL SOLUTION

We will consider the first-order linear system

$$\begin{array}{l} x_1' = a_{11}(t)x_1 + \cdots + a_{n1}(t)x_n + f_1(t), \\ \qquad \vdots \\ x_n' = a_{n1}(t)x_1 + \cdots + a_{nn}(t)x_n + f_n(t) \end{array} \qquad (1)$$

or, in matrix form,

$$\mathbf{x}' = \mathbf{A}(t)\mathbf{x} + \mathbf{f}(t). \qquad (2)$$

By a set $\{x_1(t), \ldots, x_n(t)\}$ being a **solution** of (1) on a given t interval I we mean that if these functions are substituted into (1), then each equation in (1) is reduced to an identity on I. Similarly for the matrix version (2); by a vector $\mathbf{x}(t)$ being a solution of (2) on I we mean that its scalar components $\{x_1(t), \ldots, x_n(t)\}$ are a solution of (1).

If we append to (1) initial conditions

$$x_1(a) = b_1, \ \ x_2(a) = b_2, \ \ \ldots, \ \ x_n(a) = b_n \tag{3}$$

at some point $t = a$, then we call (1) and (3), together, an **initial value problem**. We can write (3) more compactly as $\mathbf{x}(a) = \mathbf{b}$.

The theory regarding the solutions of (2) is similar to the corresponding discussion in Chapter 6 for the single nth-order equation

$$\frac{d^n y}{dx^n} + p_1(x)\frac{d^{n-1}y}{dx^{n-1}} + \cdots + p_n(x)y = f(x), \tag{4}$$

so in this section we streamline the discussion and go directly to the main results.

THEOREM 10.2.1

Existence and Uniqueness of Solutions of Initial Value Problem
Let $\mathbf{A}(t)$ and $\mathbf{f}(t)$ in (2) be continuous on a t interval I, and let a be a point in I. Then the initial value problem

$$\boxed{\mathbf{x}' = \mathbf{A}(t)\mathbf{x} + \mathbf{f}(t); \quad \mathbf{x}(a) = \mathbf{b}} \tag{5}$$

has a unique solution on I.

By the $n \times n$ matrix $\mathbf{A}(t)$ being continuous on I, we mean that every $a_{jk}(t)$ element of $\mathbf{A}(t)$ is continuous on I; similarly for $\mathbf{f}(t)$.

For the case where $n = 1$, (5) is the same (though in different notation) as the initial value problem

$$y' + p(x)y = q(x); \quad y(a) = b$$

studied in Chapter 2. Sure enough, Theorem 10.2.1 is, for the case $n = 1$, identical to Theorem 2.3.1, except that Theorem 2.3.1 goes further and actually gives the solution; however, we are not able to do that if n is greater than 1.

Theorem 10.2.1 addresses the initial value problem. If initial conditions are not prescribed then we are interested in a "general solution" of (2). Recall that for the equation (4), we say that a family of solutions

$$y(x) = C_1 y_1(x) + \cdots + C_n y_n(x) + y_p(x) \tag{6}$$

is a general solution of (4) if it contains (by suitable choice of the arbitrary constants $C_1, \ldots, C_n$) every solution of (4). Similarly, we say that a family of solutions of (2) is a **general solution** of (2) if (by suitable choice of the constants) it contains every solution of (2). First, a preliminary result:

THEOREM 10.2.2

Superposition of Solutions of $\mathbf{x}' = \mathbf{A}(t)\mathbf{x}$

If $\mathbf{x}_1(t), \ldots, \mathbf{x}_k(t)$ are solutions of the homogeneous equation $\mathbf{x}' = \mathbf{A}(t)\mathbf{x}$, then $C_1\mathbf{x}_1(t) + \cdots + C_k\mathbf{x}_k(t)$ is a solution too, for any constants $C_1, \ldots, C_k$.

Proof.   Putting $\mathbf{x} = C_1\mathbf{x}_1 + \cdots + C_k\mathbf{x}_k$ into $\mathbf{x}' = \mathbf{A}\mathbf{x}$ gives

$$\frac{d}{dt}(C_1\mathbf{x}_1 + \cdots + C_k\mathbf{x}_k) = \mathbf{A}(C_1\mathbf{x}_1 + \cdots + C_k\mathbf{x}_k)$$

or

$$C_1\mathbf{x}'_1 + \cdots + C_k\mathbf{x}'_k = C_1\mathbf{A}\mathbf{x}_1 + \cdots + C_k\mathbf{A}\mathbf{x}_k,$$

which is satisfied for any constants $C_1, \ldots, C_k$ because $\mathbf{x}'_1 = \mathbf{A}(t)\mathbf{x}_1, \ldots, \mathbf{x}'_k = \mathbf{A}(t)\mathbf{x}_k$ by assumption.   ∎

The next theorem establishes the form of the general solution to (2).

THEOREM 10.2.3

General Solution

Let $\mathbf{A}(t)$ and $\mathbf{f}(t)$ in (2) be continuous on a t interval I. Then there exist n LI solutions of the homogeneous equation $\mathbf{x}' = \mathbf{A}(t)\mathbf{x}$ on I. Let $\mathbf{x}_1(t), \ldots, \mathbf{x}_n(t)$ be n such LI solutions, and let $\mathbf{x}_p(t)$ be any solution of the nonhomogeneous equation

$$\boxed{\mathbf{x}' = \mathbf{A}(t)\mathbf{x} + \mathbf{f}(t)} \tag{7}$$

on I. Then

$$\boxed{\mathbf{x} = C_1\mathbf{x}_1(t) + \cdots + C_n\mathbf{x}_n(t) + \mathbf{x}_p(t)} \tag{8}$$

is a general solution of (7) on I.

Any such set $\mathbf{x}_1(t), \ldots, \mathbf{x}_n(t)$ is called a **basis** or **fundamental set** of solutions of the system $\mathbf{x}' = \mathbf{A}(t)\mathbf{x}$. Observe that we have not yet shown how to *find* homogeneous solutions $\mathbf{x}_1(t), \ldots, \mathbf{x}_n(t)$ nor a particular solution $\mathbf{x}_p(t)$ in (8); how to find solutions will be the subject of subsequent sections. Observe also that we did not bother to assert that a particular solution $\mathbf{x}_p(t)$ does indeed exist, because the existence of an $\mathbf{x}_p(t)$ follows from Theorem 10.2.1. All we need to do is specify a $\mathbf{b}$ vector in (5) (any choice will do), and there follows a particular solution of (5) to use in (8).

E X A M P L E   1
Consider the system

$$\begin{aligned} x'_1 &= x_1 + 3x_2 - 4t, \\ x'_2 &= x_1 - x_2 - 8. \end{aligned} \tag{9}$$

It can be verified by substitution that each of

$$\mathbf{x}_1(t) = \begin{bmatrix} 3e^{2t} \\ e^{2t} \end{bmatrix} = \begin{bmatrix} 3 \\ 1 \end{bmatrix} e^{2t} \tag{10}$$

and

$$\mathbf{x}_2(t) = \begin{bmatrix} -e^{-2t} \\ e^{-2t} \end{bmatrix} = \begin{bmatrix} -1 \\ 1 \end{bmatrix} e^{-2t} \tag{11}$$

is a solution of the homogeneous version

$$\begin{aligned} x_1' &= x_1 + 3x_2, \\ x_2' &= x_1 - x_2 \end{aligned} \tag{12}$$

of (9), and that

$$\mathbf{x}_p(t) = \begin{bmatrix} t + 7 \\ t - 2 \end{bmatrix} \tag{13}$$

is a particular solution of the full equation (9), on $I = (-\infty, \infty)$ (i.e., on $-\infty < t < \infty$). Further,

$$\det [\mathbf{x}_1(t), \mathbf{x}_2(t)] = \begin{vmatrix} 3e^{2t} & -e^{-2t} \\ e^{2t} & e^{-2t} \end{vmatrix} = 4 \neq 0 \tag{14}$$

for all t so, by Theorem 5.5.1, $\mathbf{x}_1(t)$ and $\mathbf{x}_2(t)$ are LI on I. Thus, by Theorem 10.2.3,

$$\mathbf{x}(t) = C_1 \begin{bmatrix} 3 \\ 1 \end{bmatrix} e^{2t} + C_2 \begin{bmatrix} -1 \\ 1 \end{bmatrix} e^{-2t} + \begin{bmatrix} t + 7 \\ t - 2 \end{bmatrix} \tag{15}$$

is a general solution of (9). In scalar form,

$$\begin{aligned} x_1(t) &= 3C_1 e^{2t} - C_2 e^{-2t} + t + 7, \\ x_2(t) &= C_1 e^{2t} + C_2 e^{-2t} + t - 2. \end{aligned} \tag{16}$$

In (9) we have $a_{11}(t) = 1$, $a_{12}(t) = 3$, $f_1(t) = -4t$, $a_{21}(t) = 1$, $a_{22}(t) = -1$, and $f_2(t) = -8$, all of which are continuous on I. Thus, by Theorem 10.2.1, the initial value version of (9) will have a unique solution on I for any initial condition $\mathbf{x}(a) = \mathbf{b}$. To determine C_1 and C_2 write

$$\begin{aligned} x_1(a) &= b_1 = 3C_1 e^{2a} - C_2 e^{-2a} + a + 7, \\ x_2(a) &= b_2 = C_1 e^{2a} + C_2 e^{-2a} + a - 2 \end{aligned} \tag{17}$$

or,

$$\begin{bmatrix} 3e^{2a} & -e^{-2a} \\ e^{2a} & e^{-2a} \end{bmatrix} \begin{bmatrix} C_1 \\ C_2 \end{bmatrix} = \begin{bmatrix} b_1 - a - 7 \\ b_2 - a + 2 \end{bmatrix}. \tag{18}$$

The latter gives a unique solution for C_1 and C_2 because the determinant of the coefficient matrix is nonzero. Of course, that determinant is the same as the one in (14) (for $t = a$), which is nonzero because $\mathbf{x}_1(t)$ and $\mathbf{x}_2(t)$ are LI.

COMMENT.   You might have noticed that our verification of the linear indepen-dence of $\mathbf{x}_1$ and $\mathbf{x}_2$ was harder than necessary since for *two* vectors we can see by inspection whether or not one is a scalar multiple of the other (Theorem 4.7.2). How-ever, the method used here can be used when there are more than two vectors.   ∎

Closure.   The main results are given in Theorems 10.2.1–10.2.3. Example 1 should also be studied carefully. Thus far we have permitted $\mathbf{A}$ in $\mathbf{x}' = \mathbf{A}(t)\mathbf{x} + \mathbf{f}(t)$ to be a function of t. However, in Section 10.3 we will restrict $\mathbf{A}(t)$ to be a constant matrix so that we can solve by a method of elimination.

EXERCISES 10.2

1. Verify that $\mathbf{x}_1(t)$ and $\mathbf{x}_2(t)$ are solutions of the homogeneous system, and that $\mathbf{x}_p(t)$ is a particular solution. From them, obtain a general solution. Finally, find the solution $\mathbf{x}(t)$ satisfying the given initial conditions. Is it unique? How do you know that?

   (a) System: $x_1' = x_1 + x_2$, $x_2' = x_1 + x_2$
   Solutions:
   $$\mathbf{x}_1 = \begin{bmatrix} e^{2t} \\ e^{2t} \end{bmatrix}, \ \mathbf{x}_2 = \begin{bmatrix} 1 \\ -1 \end{bmatrix}$$
   Initial conditions: $x_1(0) = 5$, $x_2(0) = -3$

   (b) System: $x_1' = 4x_1 + 4x_2$, $x_2' = x_1 + x_2 + 125t$
   Solutions:
   $$\mathbf{x}_1 = \begin{bmatrix} 4e^{5t} \\ e^{5t} \end{bmatrix}, \ \mathbf{x}_2 = \begin{bmatrix} 1 \\ -1 \end{bmatrix},$$
   $$\mathbf{x}_p = \begin{bmatrix} -50t^2 - 20t - 4 \\ 50t^2 - 5t - 1 \end{bmatrix}$$
   Initial conditions: $x_1(1) = 0$, $x_2(1) = 0$

   (c) System: $x_1' = 4x_2 + 144(1 + t)$, $x_2' = 3x_1 + x_2$
   Solutions:
   $$\mathbf{x}_1 = \begin{bmatrix} e^{4t} \\ e^{4t} \end{bmatrix}, \ \mathbf{x}_2 = \begin{bmatrix} 4e^{-3t} \\ -3e^{-3t} \end{bmatrix},$$
   $$\mathbf{x}_p = \begin{bmatrix} 12t - 1 \\ -36t - 33 \end{bmatrix}$$
   Initial conditions: $x_1(-1) = 0$, $x_2(-1) = 5$

   (d) System: $x_1' = 4x_1 + 2x_2$, $x_2' = 2x_1 + x_2 - 12e^t$
   Solutions:
   $$\mathbf{x}_1 = \begin{bmatrix} 1 \\ -2 \end{bmatrix}, \ \mathbf{x}_2 = \begin{bmatrix} 2e^{5t} \\ e^{5t} \end{bmatrix}, \ \mathbf{x}_p = \begin{bmatrix} 6e^t \\ -9e^t \end{bmatrix}$$
   Initial conditions: $x_1(4) = 3$, $x_2(4) = 0$

2. Same as Exercise 1, but for systems of three equations.
   (a) System: $x_1' = x_1 + x_2 + x_3$,
   $x_2' = x_1 + x_2 + x_3$, $x_3' = x_1 + x_2 + x_3$
   Solutions:
   $$\mathbf{x}_1 = \begin{bmatrix} 2 \\ -1 \\ -1 \end{bmatrix}, \ \mathbf{x}_2 = \begin{bmatrix} 1 \\ -2 \\ 1 \end{bmatrix}, \ \mathbf{x}_3 = \begin{bmatrix} e^{3t} \\ e^{3t} \\ e^{3t} \end{bmatrix}$$

   Initial conditions: $x_1(0) = 15$, $x_2(0) = 21$, $x_3(0) = -12$

   (b) System: $x_1' = x_3 + 8$, $x_2' = x_3 + 8t$,
   $x_3' = x_1 + x_2 + x_3$
   Solutions:
   $$\mathbf{x}_1 = \begin{bmatrix} 1 \\ -1 \\ 0 \end{bmatrix}, \ \mathbf{x}_2 = \begin{bmatrix} e^{-t} \\ e^{-t} \\ -e^{-t} \end{bmatrix}, \ \mathbf{x}_3 = \begin{bmatrix} e^{2t} + 3 \\ e^{2t} - 3 \\ 2e^{2t} \end{bmatrix}.$$
   $$\mathbf{x}_p = \begin{bmatrix} -2t^2 + 6t - 1 \\ 2t^2 - 2t - 1 \\ -4t - 2 \end{bmatrix}$$
   Initial conditions: $x_1(0) = 0$, $x_2(0) = 0$, $x_3(0) = 120$

   (c) System: $x_1' = 3x_3 + 9e^{-t}$, $x_2' = x_3 - 9e^{-t}$,
   $x_3' = x_1 - 13x_2 + 7x_3$
   Solutions:
   $$\mathbf{x}_1 = \begin{bmatrix} 13 \\ 1 \\ 0 \end{bmatrix}, \ \mathbf{x}_2 = \begin{bmatrix} 3e^{5t} \\ e^{5t} \\ 5e^{5t} \end{bmatrix}, \ \mathbf{x}_3 = \begin{bmatrix} 3e^{2t} \\ e^{2t} \\ 2e^{2t} \end{bmatrix},$$
   $$\mathbf{x}_p = \begin{bmatrix} -30e^{-t} \\ 2e^{-t} \\ 7e^{-t} \end{bmatrix}$$
   Initial conditions: $x_1(0) = 3$, $x_2(0) = 0$, $x_3(0) = 0$

   (d) System: $x_1' = 2x_1 + x_2 + x_3$, $x_2' = x_1 + 2x_2 + x_3$,
   $x_3' = x_1 + x_2 + 2x_3$
   Solutions:
   $$\mathbf{x}_1 = \begin{bmatrix} e^{4t} \\ e^{4t} \\ e^{4t} \end{bmatrix}, \ \mathbf{x}_2 = \begin{bmatrix} 2e^t \\ -e^t \\ -e^t \end{bmatrix}, \ \mathbf{x}_3 = \begin{bmatrix} -e^t \\ 2e^t \\ -e^t \end{bmatrix}$$
   Initial conditions: $x_1(0) = 3$, $x_2(0) = 0$, $x_3(0) = 0$

3. Show that if (8) satisfies (7) where $C_1, \ldots, C_n$ are arbitrary constants, then $\mathbf{x}_p(t)$ must be a particular solution [i.e., $\mathbf{x}_p' = \mathbf{A}\mathbf{x}_p + \mathbf{f}(t)$] and each $\mathbf{x}_j(t)$ must be a homogeneous solution [i.e., $\mathbf{x}_j' = \mathbf{A}\mathbf{x}_j$ for each $j = 1, \ldots, n$].

10.3 SOLUTION BY ELIMINATION; CONSTANT-COEFFICIENT EQUATIONS

We now turn to methods of solution.

10.3.1 Elimination.   Just as we can solve a system of linear algebraic equations by a systematic elimination procedure, we can do the same with a system of linear differential equations. To do so we use the operator notation

$$
\begin{aligned}
L &= a_0(t)\frac{d^n}{dt^n} + a_1(t)\frac{d^{n-1}}{dt^{n-1}} + \cdots + a_n(t) \\
&= a_0(t)D^n + a_1(t)D^{n-1} + \cdots + a_n(t)
\end{aligned}
\tag{1}
$$

that was introduced in Section 6.4.1, where D denotes d/dt, D^2 denotes d^2/dt^2, and so on. We say that L is of order n (if a_0 is not identically zero) and that it "acts" on x, or "operates" on x. Further, by $L_1 L_2[x]$ we mean $L_1[L_2[x]]$; that is, first the operator immediately to the left of x acts on x, then the operator to the left of that acts on the result. Two operators, say L_1 and L_2, are said to be equal if $L_1[x] = L_2[x]$ for all functions $x(t)$ (that are sufficiently differentiable for L_1 and L_2 to act on them), in which case we write $L_1 = L_2$.

In general, differential operators do not commute; that is, in general

$$
\boxed{L_1 L_2 \neq L_2 L_1.}
\tag{2}
$$

For instance, if $L_1 = D$ and $L_2 = tD$ then

$$
L_1 L_2[x] = (D)(tD)x = D(tx') = tx'' + x',
$$

whereas

$$
L_2 L_1[x] = (tD)(D)x = tDx' = tx''.
$$

The result (2) should not be surprising since we've already seen in Chapter 5 that, in general, matrix operators don't commute either, namely, $\mathbf{AB} \neq \mathbf{BA}$.

However, *differential operators do commute if their a_j coefficients [in (1)] are constants*. To illustrate, let $L_1 = 2D - 1$ and $L_2 = D^2 + 3$. Then

$$
(2D - 1)(D^2 + 3)x = (2D - 1)(x'' + 3x) = 2x''' - x'' + 6x' - 3x
$$

and

$$
(D^2 + 3)(2D - 1)x = (D^2 + 3)(2x' - x) = 2x''' - x' + 6x' - 3x
$$

are identical for all (twice differentiable) functions $x(t)$, so in this case $L_1 L_2 = L_2 L_1$.

The method of elimination that we present in this section requires commutativity, so we restrict this section to differential equations with *constant coefficients*.

Let us begin with an example.

EXAMPLE 1

Consider the system

$$x' = x + 3y - 4t, \tag{3a}$$
$$y' = x - y - 8, \tag{3b}$$

which is coupled by virtue of the $3y$ term in (3a) and the x term in (3b).[1] If not for the presence of those terms we could merely solve (3a) and (3b) independently, the first for $x(t)$ and the second for $y(t)$. To solve by elimination we first re-express the system in operator form as

$$(D - 1)x - 3y = -4t, \tag{4a}$$
$$-x + (D + 1)y = -8, \tag{4b}$$

or

$$L_1[x] + L_2[y] = -4t, \tag{5a}$$
$$L_3[x] + L_4[y] = -8, \tag{5b}$$

where $L_1 = D - 1$, $L_2 = -3$, $L_3 = -1$, and $L_4 = D + 1$. Next, operate on (5a) with L_3 and on (5b) with L_1, giving

$$L_3 L_1[x] + L_3 L_2[y] = L_3[-4t], \tag{6a}$$
$$L_1 L_3[x] + L_1 L_4[y] = L_1[-8], \tag{6b}$$

where we have used the linearity of L_3 in writing $L_3[L_1[x] + L_2[y]]$ as $L_3 L_1[x] + L_3 L_2[y]$ and the linearity of L_1 in writing $L_1[L_3[x] + L_4[y]]$ as $L_1 L_3[x] + L_1 L_4[y]$. Then, subtracting (6a) from (6b) gives the equation

$$(L_1 L_4 - L_3 L_2)[y] = L_1[-8] - L_3[-4t] \tag{7}$$

on y alone! Note that we have used the commutativity of L_1 and L_3 in using the $-L_3 L_1[x]$ term to cancel the $L_1 L_3[x]$ term.

At this point we can return to non-operator form:

$$L_1 L_4 - L_3 L_2 = (D - 1)(D + 1) - (-1)(-3) = D^2 - 4,$$

$L_1[-8] = (D - 1)(-8) = 8$, and $L_3[-4t] = (-1)(-4t) = 4t$, so (7) becomes

$$y'' - 4y = 8 - 4t, \tag{8}$$

which admits the general solution

$$y(t) = C_1 e^{2t} + C_2 e^{-2t} + t - 2. \tag{9}$$

---

[1] Here we favor the simpler x, y notation over the more formal subscripted dependent variable notation used in Section 10.2.

To find $x(t)$ we can proceed in the same manner. This time, operate on (5a) with L_4, on (5b) with L_2, subtract one from the other, and obtain the equation

$$(L_4 L_1 - L_2 L_3)[x] = L_4[-4t] - L_2[-8] \tag{10}$$

on x alone. Or in non-operator form,

$$x'' - 4x = -4t - 28, \tag{11}$$

with general solution

$$x(t) = C_3 e^{2t} + C_4 e^{-2t} + t + 7. \tag{12}$$

It is tempting to say that (9) and (12) give a general solution of (3). However, Theorem 10.2.3 indicates that a general solution of a system of two first-order equations contains only two arbitrary constants. It must turn out that C_1, C_2, C_3, C_4 are not mutually independent. After all, x and y are related through (3). In fact, putting (9) and (12) into (3a) gives, after rearrangement of terms,

$$(C_3 - 3C_1)e^{2t} - (3C_4 + 3C_2)e^{-2t} = 0, \tag{13}$$

and the linear independence of e^{2t} and e^{-2t} requires that $C_3 - 3C_1 = 0$ and $3C_4 + 3C_2 = 0$ so $C_3 = 3C_1$ and $C_4 = -C_2$. [Putting (9) and (12) into (3b) gives the same result.]

Thus, a general solution of (3) is

$$x(t) = 3C_1 e^{2t} - C_2 e^{-2t} + t + 7, \tag{14a}$$

$$y(t) = C_1 e^{2t} + C_2 e^{-2t} + t - 2. \tag{14b}$$

COMMENT 1.   Suppose we append initial conditions, say $x(0) = 0$ and $y(0) = 5$. Then (14a,b) give

$$x(0) = 0 = 3C_1 - C_2 + 7,$$
$$y(0) = 5 = C_1 + C_2 - 2.$$

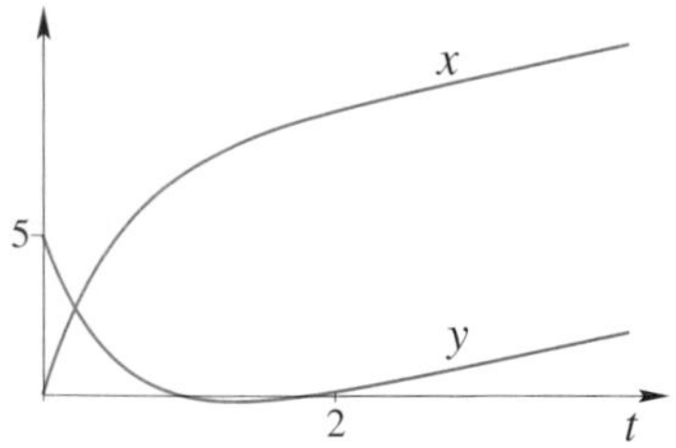

Thus, $C_1 = 0$ and $C_2 = 7$, so

$$x(t) = -7e^{-2t} + t + 7, \tag{15a}$$

$$y(t) = 7e^{-2t} + t - 2, \tag{15b}$$

and these are plotted in Fig. 1.

FIGURE 1
The solution (15).

COMMENT 2.   In this example, once we obtain the solution (9) for $y(t)$ we can solve for $x(t)$ more easily by substituting (9) into (3b) for $y(t)$ and solving by algebra for $x(t)$, which step does produce the same solution as before, as given by (14a). However, the availability of this simpler approach is fortuitous here, so we used instead a systematic approach that always works.

COMMENT 3.   Let us re-examine this example from a vector space point of view. First, write (4) as

$$\begin{bmatrix} D-1 & -3 \\ -1 & D+1 \end{bmatrix} \begin{bmatrix} x(t) \\ y(t) \end{bmatrix} = \begin{bmatrix} -4t \\ -8 \end{bmatrix} \tag{16}$$

or, more compactly, as

$$\mathbf{L}[\mathbf{x}] = \mathbf{f}(t), \tag{17}$$

where

$$\mathbf{L} = \begin{bmatrix} D-1 & -3 \\ -1 & D+1 \end{bmatrix}, \quad \mathbf{x}(t) = \begin{bmatrix} x(t) \\ y(t) \end{bmatrix}, \quad \mathbf{f}(t) = \begin{bmatrix} -4t \\ -8 \end{bmatrix}.$$

We can think of $\mathbf{L}$ as a *matrix differential operator* because it is both a matrix and an operator. Let its domain and range of definition be the vector space consisting of the set of two-dimensional column vectors whose elements are continuously differentiable over $0 \le t < \infty$, say. We have found that, in vector form, a general solution of (17) is

$$\mathbf{x}(t) = \begin{bmatrix} x(t) \\ y(t) \end{bmatrix} = C_1 \begin{bmatrix} 3e^{2t} \\ e^{2t} \end{bmatrix} + C_2 \begin{bmatrix} -e^{-2t} \\ e^{-2t} \end{bmatrix} + \begin{bmatrix} t+7 \\ t-2 \end{bmatrix}$$

$$= C_1 \mathbf{x}_1(t) + C_2 \mathbf{x}_2(t) + \mathbf{x}_p(t). \tag{18}$$

From our discussion in Section 10.2, we know that $\mathbf{x}_p$ is a particular solution of (17), and that $\mathbf{x}_1$ and $\mathbf{x}_2$ are homogeneous solutions of (17); $\mathbf{x}_1$ and $\mathbf{x}_2$ satisfy $\mathbf{L}[\mathbf{x}] = \mathbf{0}$. Thus, $\mathbf{x}_1$ and $\mathbf{x}_2$ constitute a basis for the kernel or null space of $\mathbf{L}$, the null space being two-dimensional since there are two vectors in a basis for it. ∎

A review of the steps in the elimination process reveals that the operators $L_1, \dots, L_4$ might just as well have been constants the way we have manipulated them. In fact, a useful way to organize the procedure is to use Cramer's rule (Section 5.6.4). For instance, if we have two differential equations

$$\boxed{\begin{aligned} L_1[x] + L_2[y] &= f_1(t), \\ L_3[x] + L_4[y] &= f_2(t), \end{aligned}} \tag{19a,b}$$

we can, heuristically, use Cramer's rule to write

$$x = \frac{\begin{vmatrix} f_1 & L_2 \\ f_2 & L_4 \end{vmatrix}}{\begin{vmatrix} L_1 & L_2 \\ L_3 & L_4 \end{vmatrix}} = \frac{L_4[f_1] - L_2[f_2]}{L_1 L_4 - L_2 L_3}, \tag{20a}$$

$$y = \frac{\begin{vmatrix} L_1 & f_1 \\ L_3 & f_2 \end{vmatrix}}{\begin{vmatrix} L_1 & L_2 \\ L_3 & L_4 \end{vmatrix}} = \frac{L_1[f_2] - L_3[f_1]}{L_1 L_4 - L_2 L_3}. \tag{20b}$$

Having obtained (20a,b) heuristically (namely, treating the L_j differential operators as though they were merely numbers) we must interpret the f_1 times L_4 term in the numerator determinant in (20a), for instance, as $L_4[f_1]$ not f_1L_4. Similarly, the division by an operator on the right-hand sides of (20a,b) is not defined, so we need to put the $L_1L_4 - L_2L_3$ back up on the left-hand side, where it came from. That step gives

$$\begin{vmatrix} L_1 & L_2 \\ L_3 & L_4 \end{vmatrix} x = \begin{vmatrix} f_1 & L_2 \\ f_2 & L_4 \end{vmatrix},$$

$$\begin{vmatrix} L_1 & L_2 \\ L_3 & L_4 \end{vmatrix} y = \begin{vmatrix} L_1 & f_1 \\ L_3 & f_2 \end{vmatrix}$$

or,

$$\boxed{\begin{aligned} (L_1L_4 - L_2L_3)\,[x] &= L_4[f_1] - L_2[f_2], \\ (L_1L_4 - L_2L_3)\,[y] &= L_1[f_2] - L_3[f_1], \end{aligned}} \qquad \text{(21a,b)}$$

which equations correspond to (10) and (7), respectively, in Example 1 (since $L_1L_4 = L_4L_1$ and $L_2L_3 = L_3L_2$). We will refer to (19)–(21) as the *Cramer's rule version of the elimination method*. The latter is readily applied and is not limited to systems of two equations in two unknowns.

EXAMPLE 2
To solidify the foregoing ideas, let us solve the system

$$x' = x + 4y, \qquad \text{(22a)}$$

$$y' = x + y - 9t. \qquad \text{(22b)}$$

First, express (22) in operator form as

$$(D - 1)x - 4y = 0, \qquad \text{(23a)}$$

$$-x + (D - 1)y = -9t. \qquad \text{(23b)}$$

Then $L_1 = D - 1$, $L_2 = -4$, $L_3 = -1$, $L_4 = D - 1$, $f_1 = 0$, and $f_2 = -9t$, so (21) becomes

$$\left(D^2 - 2D - 3\right)x = -36t, \qquad \text{(24a)}$$

$$\left(D^2 - 2D - 3\right)y = -9 + 9t, \qquad \text{(24b)}$$

with general solutions

$$x(t) = C_1 e^{3t} + C_2 e^{-t} - 8 + 12t, \qquad \text{(25a)}$$

$$y(t) = C_3 e^{3t} + C_4 e^{-t} + 5 - 3t. \qquad \text{(25b)}$$

Finally, put (25) into either (22a) or (22b) to determine any algebraic relations among C_1, C_2, C_3, and C_4. That step gives

$$(2C_1 - 4C_3)\, e^{3t} - (2C_2 + 4C_4)\, e^{-t} = 0, \qquad \text{(26)}$$

so, by the linear independence of e^{3t} and e^{-t}, $C_3 = C_1/2$ and $C_4 = -C_2/2$. Putting these results into (25) gives

$$x(t) = C_1 e^{3t} + C_2 e^{-t} - 8 + 12t, \tag{27a}$$

$$y(t) = \frac{1}{2} C_1 e^{3t} - \frac{1}{2} C_2 e^{-t} + 5 - 3t. \tag{27b}$$

as a general solution of (22).   ■

   Although the system must be in the form $\mathbf{x}' = \mathbf{Ax} + \mathbf{f}$ to apply Theorems 10.2.1 or 10.2.3, it does not need to be in that form to proceed with our method of solution by elimination.

EXAMPLE 3
For instance,

$$x'' + y' + 6y = 4e^t, \tag{28a}$$

$$x' + y' + 2y = 0 \tag{28b}$$

is not of the form $\mathbf{x}' = \mathbf{Ax} + \mathbf{f}$. We can reduce it to that form by introducing an auxiliary variable, but we don't need to; we can solve it by elimination, as in Examples 1 and 2. First, express (28) as

$$D^2 x + (D + 6)y = 4e^t, \tag{29a}$$

$$Dx + (D + 2)y = 0. \tag{29b}$$

Then $L_1 = D^2$, and $L_2 = D + 6$, so $L_3 = D$, $L_4 = D + 2$, $f_1 = 4e^t$, and $f_2 = 0$, so (21) gives

$$\left(D^3 + D^2 - 6D \right) x = 12e^t, \tag{30a}$$

$$\left(D^3 + D^2 - 6D \right) y = -4e^t, \tag{30b}$$

with general solutions

$$x(t) = C_1 + C_2 e^{2t} + C_3 e^{-3t} - 3e^t, \tag{31a}$$

$$y(t) = C_4 + C_5 e^{2t} + C_6 e^{-3t} + e^t. \tag{31b}$$

Finally, if we put (31) into (28a) or (28b) we find that $C_4 = 0$, $C_5 = -C_2/2$, and $C_6 = -3C_3$, so (31) gives

$$x(t) = C_1 + C_2 e^{2t} + C_3 e^{-3t} - 3e^t, \tag{32a}$$

$$y(t) = -\frac{1}{2} C_2 e^{2t} - 3C_3 e^{-3t} + e^t \tag{32b}$$

as a general solution of (28).

COMMENT.   Observe that if we introduce an auxiliary dependent variable (as we explained in Section 10.1; refer to Example 4) we can change (29) to a system of three first-order equations. Therefore, according to Theorem 10.2.3 a general solution of (29) must contain exactly three arbitrary constants. These constants are the constants C_1, C_2, C_3 in (32). In other words, (29a) is a second-order equation and (29b) is a first-order equation so there will be $2 + 1 = 3$ arbitrary constants. To generalize further, *the number of arbitrary constants in a general solution of a system of linear differential equations equals the sum of the orders of the differential equations in the system.* Be sure to keep this point in mind.   ■

Let us consider an important physical application.

10.3.2 Application to a coupled oscillator.   Understand that the applications that we discuss in this text are intended to represent types or classes of problems, rather than being specific and hence of limited value. The following example is important for it is representative of coupled oscillators of any type, be they mechanical, electrical, biological, or whatever.

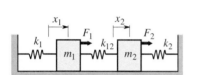

FIGURE 2
Two-mass oscillator.

The problem we consider is that of the free and forced oscillations of the two-mass oscillator shown in Fig. 2. In Example 3 of Section 10.1 we derived the governing equations of motion, namely,

$$m_1 x_1'' + (k_1 + k_{12}) x_1 - k_{12} x_2 = F_1(t),$$
$$m_2 x_2'' - k_{12} x_1 + (k_2 + k_{12}) x_2 = F_2(t). \tag{33}$$

Let us set $m_1 = m_2 = k_1 = k_{12} = k_2 = 1$, for simplicity, and let $F_1(t) = F \sin \Omega t$ and $F_2(t) = 0$. Then (33) becomes

$$\left(D^2 + 2\right) x_1 - x_2 = F \sin \Omega t, \tag{34a}$$

$$-x_1 + \left(D^2 + 2\right) x_2 = 0 \tag{34b}$$

and our usual process of elimination [using (21) with $L_1 = D^2 + 2$, $L_2 = -1$, $L_3 = -1$, $L_4 = D^2 + 2$, $f_1 = F \sin \Omega t$, $f_2 = 0$] gives

$$\left(D^4 + 4D^2 + 3\right) x_1 = \left(2 - \Omega^2\right) F \sin \Omega t, \tag{35a}$$

$$\left(D^4 + 4D^2 + 3\right) x_2 = F \sin \Omega t. \tag{35b}$$

General solutions of (35a,b) are found to be

$$x_1(t) = C_1 \cos t + C_2 \sin t + C_3 \cos \sqrt{3}t + C_4 \sin \sqrt{3}t$$
$$+ \frac{\left(2 - \Omega^2\right) F}{\left(\Omega^2 - 1\right)\left(\Omega^2 - 3\right)} \sin \Omega t, \tag{36a}$$

$$x_2(t) = C_5 \cos t + C_6 \sin t + C_7 \cos \sqrt{3}t + C_8 \sin \sqrt{3}t$$
$$+ \frac{F}{\left(\Omega^2 - 1\right)\left(\Omega^2 - 3\right)} \sin \Omega t. \tag{36b}$$

Putting (36) into either (34a) or (34b) and using the linear independence of $\cos t$, $\sin t$, $\cos \sqrt{3}t$, and $\sin \sqrt{3}t$ gives the relations $C_5 = C_1$, $C_6 = C_2$, $C_7 = -C_3$, $C_8 = -C_4$. Then (36) becomes

$$x_1(t) = C_1 \cos t + C_2 \sin t + C_3 \cos \sqrt{3}t + C_4 \sin \sqrt{3}t$$

$$+ \frac{(2 - \Omega^2) F}{(\Omega^2 - 1)(\Omega^2 - 3)} \sin \Omega t, \tag{37a}$$

$$x_2(t) = C_1 \cos t + C_2 \sin t - C_3 \cos \sqrt{3}t - C_4 \sin \sqrt{3}t$$

$$+ \frac{F}{(\Omega^2 - 1)(\Omega^2 - 3)} \sin \Omega t. \tag{37b}$$

Recall from equation (14) of Section 7.2 that we can write $C \cos \omega t + D \sin \omega t$ in the equivalent form $E \sin(\omega t + \phi)$, which form is more illuminating physically since E, ω, ϕ bear the physical significance of amplitude, frequency, and phase, respectively. Thus, although this step is not essential, let us re-express (37) as

$$x_1(t) = G \sin(t + \phi) + H \sin(\sqrt{3}t + \psi) + \frac{(2 - \Omega^2) F}{(\Omega^2 - 1)(\Omega^2 - 3)} \sin \Omega t, \tag{38a}$$

$$x_2(t) = G \sin(t + \phi) - H \sin(\sqrt{3}t + \psi) + \frac{F}{(\Omega^2 - 1)(\Omega^2 - 3)} \sin \Omega t, \tag{38b}$$

where G, H, ϕ, ψ are constants. If we prescribe four initial conditions, which we denote as

$$x_1(0), \quad x_1'(0), \quad x_2(0), \quad x_2'(0), \tag{39}$$

then we can determine C_1, C_2, C_3, C_4 if we use (37), or G, H, ϕ, ψ if we use (38).

If $F = 0$, so there is no forcing, then only the $\sin(t + \phi)$ and $\sin(\sqrt{3}t + \psi)$ terms survive in (38). Mathematically, these give the homogeneous solution to (34); physically, they represent the *free vibration*. The $\sin \Omega t$ terms in (38) give a particular solution to (34) and, physically, represent the *forced vibration*. The general solution (38) is the sum of the two. For clarity, it will be best to discuss the free and forced vibrations separately.

Free vibration.   In vector form,

$$\begin{bmatrix} x_1(t) \\ x_2(t) \end{bmatrix} = G \begin{bmatrix} \sin(t + \phi) \\ \sin(t + \phi) \end{bmatrix} + H \begin{bmatrix} \sin(\sqrt{3}t + \psi) \\ -\sin(\sqrt{3}t + \psi) \end{bmatrix} \tag{40}$$

shows the general solution as a superposition of two distinct **modes**, defined by the two vectors on the right-hand side. These are called the **low mode** and the **high mode** because the first is at the frequency 1 rad/sec and the second is at the higher frequency $\sqrt{3}$ rad/sec. In the low mode the two masses move with the same amplitude G, with the same frequency 1, and with the same phase ϕ; they are *in-phase*. In the high mode they move with the same amplitude H and with the same frequency $\sqrt{3}$, but they are 180° *out-of-phase* due to the minus sign in the second element of the vector. Or, since $\sin(A + \pi) = \sin A \cos \pi + \sin \pi \cos A = -\sin A$, we can re-express the

$-\sin(\sqrt{3}t+\psi)$ as $\sin(\sqrt{3}t+\psi+\pi)$ to see the phase difference of π between the x_1 and x_2 motions. The modal frequencies are called the **natural frequencies** of the system.

To apply initial conditions it is a bit easier to use the "$C\cos\omega t + D\sin\omega t$" form rather than the equivalent "$E\sin(\omega t + \phi)$" form, namely,

$$x_1(t) = C_1\cos t + C_2\sin t + C_3\cos\sqrt{3}t + C_4\sin\sqrt{3}t, \qquad (41a)$$

$$x_2(t) = C_1\cos t + C_2\sin t - C_3\cos\sqrt{3}t - C_4\sin\sqrt{3}t. \qquad (41b)$$

It seems clear that to obtain a purely low mode motion we need $x_2(0) = x_1(0)$ and $x_2'(0) = x_1'(0)$ in (39). For instance, let $x_1(0) = x_2(0) = a$ and let $x_1'(0) = x_2'(0) = 0$. Then

$$\begin{aligned}
x_1(0) = a = & \ C_1 + C_3, \\
x_2(0) = a = & \ C_1 - C_3, \\
x_1'(0) = 0 = & \ C_2 + \sqrt{3}C_4, \\
x_2'(0) = 0 = & \ C_2 - \sqrt{3}C_4
\end{aligned} \qquad (42)$$

give $C_1 = a$ and $C_2 = C_3 = C_4 = 0$, so

$$\begin{bmatrix} x_1(t) \\ x_2(t) \end{bmatrix} = a \begin{bmatrix} \cos t \\ \cos t \end{bmatrix}, \qquad (43)$$

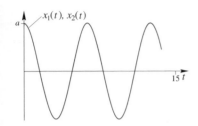

FIGURE 3
The low-mode response (43).

which is indeed purely a low mode motion (Fig. 3).

And it seems clear that to obtain a purely high mode motion we need $x_2(0) = -x_1(0)$ and $x_2'(0) = -x_1'(0)$. For instance, if $x_1(0) = +b$ and $x_2(0) = -b$, $x_1'(0) = x_2'(0) = 0$, then we find that $C_1 = C_2 = C_4 = 0$ and $C_3 = a$, so (Fig. 4)

$$\begin{bmatrix} x_1(t) \\ x_2(t) \end{bmatrix} = b \begin{bmatrix} \cos\sqrt{3}t \\ -\cos\sqrt{3}t \end{bmatrix}. \qquad (44)$$

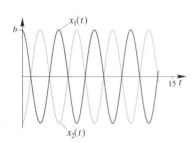

FIGURE 4
The high-mode response (44).

Otherwise, the motion will be a mix of the two modes. For instance, if $x_1(0) = c$ and $x_2(0) = x_1'(0) = x_2'(0) = 0$, then we obtain

$$\begin{bmatrix} x_1(t) \\ x_2(t) \end{bmatrix} = \frac{c}{2} \begin{bmatrix} \cos t \\ \cos t \end{bmatrix} + \frac{c}{2} \begin{bmatrix} \cos\sqrt{3}t \\ -\cos\sqrt{3}t \end{bmatrix}, \qquad (45)$$

which is displayed in Fig. 5. Observe that the pure-mode motions are "clean and simple" (e.g., they are easily described in words) and that the mixed-mode motion is not.

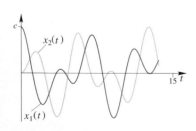

FIGURE 5
The mixed-mode response (45).

Forced vibration.   The particular solution or forced vibration is given in scalar form as

$$x_1(t) = \frac{(2-\Omega^2)F}{(\Omega^2-1)(\Omega^2-3)}\sin\Omega t \equiv I_1(\Omega)\sin\Omega t, \qquad (46a)$$

$$x_2(t) = \frac{F}{(\Omega^2-1)(\Omega^2-3)}\sin\Omega t \equiv I_2(\Omega)\sin\Omega t. \qquad (46b)$$

Of course this is only part of the solution since we need to add to (46) the homogeneous solution. However, recall that we did not include any damping terms (proportional to x_1' and to x_2') in (33). If we did, we would find that—even for an arbitrarily small amount of damping—the homogeneous solution tends to zero as $t \to \infty$, leaving just the particular solution. Thus, we can think of the particular solution (46) as the steady-state response in the presence of an "infinitesimal" amount of damping. The idea is essentially the same as for the single-mass oscillator; see the paragraph below equation (21) in Section 7.4.

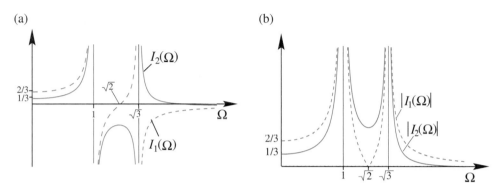

(a)   (b)

FIGURE 6   The amplitude response curves.

We have plotted the coefficients $I_1(\Omega)$ and $I_2(\Omega)$ in (46) in Fig. 6a. To think of these as the amplitudes of the forced vibration we need to take their absolute magnitudes, $|I_1(\Omega)|$ and $|I_2(\Omega)|$, and to account for any negative segments of the graphs in Fig. 6a by introducing phase angles. That is, in place of (46) write

$$x_1(t) = |I_1(\Omega)| \sin\left[\Omega t + \phi_1(\Omega)\right], \tag{47a}$$

$$x_2(t) = |I_2(\Omega)| \sin\left[\Omega t + \phi_2(\Omega)\right]. \tag{47b}$$

The amplitudes $|I_1(\Omega)|$ and $|I_2(\Omega)|$ are shown in Fig. 6b and the phase angles $\phi_1(\Omega)$ and $\phi_2(\Omega)$ in Fig. 7.

From Fig. 6b we see that the vibrational amplitude of each mass tends to infinity as the driving frequency Ω tends to either of the natural frequencies 1 and $\sqrt{3}$.

COMMENT 1.   In this example there were two masses and we obtained two modes of free vibration, hence two distinct natural frequencies. If there were three masses (Fig. 8) there would be three modes of free vibration, hence three distinct natural frequencies, and so on.

COMMENT 2.   Consider the two-loop electrical circuit shown in Fig. 9. If we use the expressions for the voltage drops in the various circuit elements, given in Fig. 15 of Section 2.4, and apply Kirchhoff's voltage law to each of the two loops, we obtain these equations:

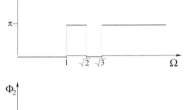

FIGURE 7
The phase angles in (47).

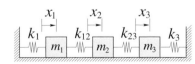

FIGURE 8
A three-mass oscillator.

$$E_1(t) - \frac{1}{C_1} \int i_1 \, dt - \frac{1}{C_{12}} \int (i_1 - i_2) \, dt - L_1 \frac{di_1}{dt} = 0,$$

$$E_2(t) - L_2 \frac{di_2}{dt} - \frac{1}{C_{12}} \int (i_2 - i_1) \, dt - \frac{1}{C_2} \int i_2 \, dt = 0.$$

Differentiating each with respect to t and rearranging terms gives

$$L_1 i_1'' + \left(\frac{1}{C_1} + \frac{1}{C_{12}} \right) i_1 - \frac{1}{C_{12}} i_2 = E_1'(t),$$

$$L_2 i_2'' - \frac{1}{C_{12}} i_1 + \left(\frac{1}{C_2} + \frac{1}{C_{12}} \right) i_2 = E_2'(t), \tag{48}$$

which is of exactly the same form as (33), with the equivalence

$$x(t) \leftrightarrow i(t), \quad m \leftrightarrow L, \quad k \leftrightarrow \frac{1}{C}, \quad F(t) \leftrightarrow E'(t). \tag{49}$$

If we change letters, according to (49), the entire preceding analysis applies to the electrical oscillator of Fig. 9.

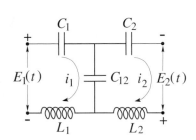

FIGURE 9
Two-loop electrical oscillator.

10.3.3 The singular case. (Optional).   Remember that a system $\mathbf{Ax} = \mathbf{c}$ of n linear algebraic equations in n unknowns is singular if $\det \mathbf{A} = 0$. In that case the system admits either no solution or an infinity of solutions. It should not be surprising that systems of linear differential equations can exhibit similar behavior. Specifically, recall from Section 10.1 that a first-order system might arise in the form

$$\mathbf{Bx'} + \mathbf{Cx} = \mathbf{g}. \tag{50}$$

If $\mathbf{B}$ is nonsingular then we can multiply (50) by $\mathbf{B}^{-1}$, convert it to the form

$$\mathbf{x'} = \mathbf{Ax} + \mathbf{f}, \tag{51}$$

and [if $\mathbf{A}(t)$ and $\mathbf{f}(t)$ are continuous functions of t on the t interval of interest] apply the existence and uniqueness Theorem 10.2.1. However, if $\mathbf{B}$ is singular then the conversion of (50) to the form (51) cannot be accomplished and we say that the system (50) is **singular**. In that case Theorem 10.2.1 does not apply and a solution to the initial value problem

$$\mathbf{Bx'} + \mathbf{Cx} = \mathbf{g}; \quad \mathbf{x}(a) = \mathbf{b}$$

may fail to exist, and if it does exist it may fail to be unique, as we will illustrate in the next two examples.

E X A M P L E 4
The system

$$x' + y' + x = t, \tag{52a}$$

$$2x' + 2y' - y = 0 \tag{52b}$$

is singular because the coefficient matrix

$$\mathbf{B} = \begin{bmatrix} 1 & 1 \\ 2 & 2 \end{bmatrix}$$

of the x' and y' terms is singular. Nevertheless, the elimination method described above still works and gives the general solution

$$x(t) = C_1 e^t + t - 1, \tag{53a}$$
$$y(t) = -2C_1 e^t + 2. \tag{53b}$$

The reason there is only one arbitrary constant in (53) instead of two is as follows: (52a) gives $x' + y' = t - x$ and putting that result into (52b) gives the relation

$$y = 2t - 2x. \tag{54}$$

Putting (54) into either (52a) or (52b), for y, gives a *first*-order equation on x, with general solution (53a). Finally, (53a) and (54) give (53b).

If we append initial conditions $x(0) = x_0$ and $y(0) = y_0$, say, to (52), then there are two initial conditions to satisfy but only one arbitrary constant is available in (53). Applying those initial conditions to (53), you will find that there is *no solution* for C_1, unless y_0 and x_0 happen to satisfy the relation $y_0 = -2x_0$ implied by (54), in which case we obtain the *unique* solution $x(t) = (x_0 + 1)e^t + t - 1$ and $y(t) = -2(x_0 + 1)e^t + 2$. Thus, the upshot is that, depending on the initial conditions $x(0) = x_0$ and $y(0) = y_0$, (52) has either no solution or a unique solution. ■

EXAMPLE 5
Likewise, the system

$$x' - y' = 0, \quad x(0) = x_0, \tag{55a}$$
$$x' - y' = 0, \quad y(0) = y_0 \tag{55b}$$

is singular because

$$\mathbf{B} = \begin{bmatrix} 1 & -1 \\ 1 & -1 \end{bmatrix}$$

is singular. In this case, however, there is an *infinity of solutions*, namely,

$$x(t) = x_0 \frac{\phi(t)}{\phi(0)}, \tag{56a}$$

$$y(t) = x_0 \frac{\phi(t)}{\phi(0)} + y_0 - x_0, \tag{56b}$$

where $\phi(t)$ is any differentiable function of t with $\phi(0) \neq 0$, as can be verified by substituting (56) into (55). ■

Closure.   Consider a system of linear constant-coefficient differential equations of first order, and let the number of equations n be the same as the number of unknowns. Let $n = 2$: To solve by elimination, first write the system in operator form as

$$L_1[x] + L_2[y] = f_1(t), \tag{57a}$$

$$L_3[x] + L_4[y] = f_2(t). \tag{57b}$$

To solve (57) we can use elimination to obtain differential equations on x alone and on y alone, but it is simpler to accomplish that result by using Cramer's rule in a heuristic manner (namely, treating the L_j differential operators as though they were numbers), obtaining

$$\begin{vmatrix} L_1 & L_2 \\ L_3 & L_4 \end{vmatrix} x = \begin{vmatrix} f_1 & L_2 \\ f_2 & L_4 \end{vmatrix}, \quad \begin{vmatrix} L_1 & L_2 \\ L_3 & L_4 \end{vmatrix} y = \begin{vmatrix} L_1 & f_1 \\ L_3 & f_2 \end{vmatrix}, \tag{58a,b}$$

provided that we interpret quantities such as $f_1 L_4$ [in (58a)] as $L_4[f_1]$; that is, the operators act on the functions rather than the functions multiplying the operators. Obtain general solutions of (58a) and (58b), then put those solutions back into (57a) and (57b) to determine any relations on the arbitrary constants. Then apply initial conditions if there are any. Similarly, for $n = 3$

$$L_1[x] + L_2[y] + L_3[z] = f_1(t),$$

$$L_4[x] + L_5[y] + L_6[z] = f_2(t),$$

$$L_7[x] + L_8[y] + L_9[z] = f_3(t),$$

and so on. The method works even if the system is of higher order than first. CAUTION: Be sure to be clear, in a given problem, what the independent and dependent variables are. For example, in Example 1 x and y are functions of t; y is not a function of x.

Maple.   The *Maple* commands that gave Fig. 1, for instance, were these:

```
with(plots):
implicitplot({x=-7*exp(-2*t)+t+7,x=7*exp(-2*t)+t-2},
t=0..4,x=-3..12);
```

To find a general solution to (3), using *Maple*, use the command

```
dsolve({diff(x(t),t)=x(t)+3*y(t)-4*t,diff(y(t),t)=
x(t)-y(t)-8},{x(t),y(t)});
```

To include initial conditions $x(0) = 0$ and $y(0) = 5$, say, use

```
dsolve({diff(x(t),t)=x(t)+3*y(t)-4*t,diff(y(t),t)=
x(t)-y(t)-8,x(0)=0,y(0)=5},{x(t),y(t)});
```

instead.

Alternatively, one can first define the two equations and then call them in the dsolve command. That is, enter

```
deq1:=diff(x(t),t)=x(t)+3*y(t)-4*t:
deq2:=diff(y(t),t)=x(t)-y(t)-8:
```

The colon at the end of each line indicates that it is a definition, not a command. Commands are followed by semicolons. Now, enter the dsolve command:

```
dsolve({deq1,deq2},x(0)=0,y(0)=5},{x(t),y(t)});
```

and return. The result is

$$\{x(t) = -7e^{-2t} + t + 7, \ y(t) = 7e^{-2t} + t - 2\}$$

EXERCISES 10.3

1. Obtain a general solution by the method of elimination, either step-by-step or using the Cramer's rule shortcut. In all cases the independent variable is t.

(a) $(D-1)x + Dy = 0$
$(D+1)x + (2D+2)y = 0$

(b) $(D-1)x + 2Dy = 0$
$(D+1)x + 4Dy = 0$

(c) $Dx + (D-1)y = 5$
$2(D+1)x + (D+1)y = 0$

(d) $x' + y' = y + t$
$x' - 3y' = -x + 2$

(e) $x' = \sin t - y$
$y' = -9x + 4$

(f) $x' = x - 8y$
$y' = -x - y - 3t^2$

(g) $x' = 2x + 6y - t + 7$
$y' = 2x - 2y$

(h) $2x' + y' + x + y = t^2 - 1$
$x' + y' + x + y = 0$

(i) $x' + y' + x - y = e^t$
$x' + 2y' + 2x - 2y = 1 - t$

(j) $x'' = x - 4y$
$y'' = -2x - y$

(k) $x'' = x - 2y$
$y'' = 2x - 4y$

(l) $x'' + y'' = x$
$3x'' - y'' = -x + 6$

(m) $x' = x + y + z + 6$
$y' = x + y + z$
$z' = x + y + z$

(n) $x' = z + e^t$
$y' = z - e^t$
$z' = x + y + z$

(o) $x' = x + y + z + 5$
$y' = 2x + 2y + 2z$
$z' = 3x + 3y + 3z$

2. **(a)** – **(s)** Find the general solution to the corresponding problem in Exercise 1 using *Maple*. Then make up any set of initial conditions (not all zero if the system is homogeneous) and obtain the particular solution, again using *Maple*.

3. (*Chemical kinetics*) Two substances, with concentrations $x(t)$ and $y(t)$, react to form a third substance, with concentration $z(t)$. The reaction is governed by the system $x' + \alpha x = 0$, $z' = \beta y$ and $x + y + z = \gamma$, where α, β, γ are known positive constants. Solve for $x(t)$, $y(t)$, $z(t)$, subject to the initial conditions $z(0) = z'(0) = 0$ for these cases:

(a) $\alpha \neq \beta$

(b) $\alpha = \beta$ HINT: It is simplest to take the limit of your answer to part (a), as $\beta \to \alpha$ (or as $\alpha \to \beta$).

4. (*Motion of a charged mass*) Consider a particle of mass m, carrying an electrical charge q, and moving in a uniform magnetic field of strength B. The field is in the positive z direction. The equations of motion of the particle are

$$\begin{aligned} mx'' &= qBy', \\ my'' &= -qBx', \\ mz'' &= 0, \end{aligned} \tag{4.1}$$

where $x(t), y(t), z(t)$ are the x, y, z displacements as a function of the time t.

(a) Find a general solution of (4.1) for $x(t)$, $y(t)$, $z(t)$.

(b) Show that by a suitable choice of initial conditions the motion can be a circle in the x, y plane, of any desired radius R and centered at any desired point x_0, y_0. Propose such a set of initial conditions.

(c) Besides a circular motion in a constant z plane, are any other types of motion possible? Explain.

5. (a) Derive (35) from (34).

(b) Derive the general solution (36) of (35).

6. In each case derive a general solution of (33), analogous to (37), and give the natural frequencies. If irrational numbers arise, use decimal form for them, to three significant figures.

(a) $m_1 = 2$, $m_2 = 1$, $k_1 = 1$, $k_2 = 1$, $k_{12} = 1$, $F_1(t) = F \sin \Omega t$, $F_2(t) = 0$

(b) Same as (a), but with $m_1 = 1$, $F_1(t) = F_1 \sin \Omega t$, $F_2(t) = F_2 \sin \Omega t$

(c) Same as (a), but with $m_1 = 1$, $F_1(t) = 0$, $F_2(t) = F \sin \Omega t$

7. (*Beats*) Suppose in (33) that $m_1 = m_2 = k_{12} = 1$, $k_1 = k_2 = 20$ so that the coupling of m_1 and m_2 is weak since k_{12} is much smaller than k_1 and k_2. (In the limit, if k_{12} were zero, there would be no coupling at all, and the motions of m_1 and m_2 would be entirely independent.) Supposing further that $x_1(0) = 1$, $x_2(0) = x_1'(0) = x_2'(0) = 0$, proceed as before and show that for the free vibration [i.e., for the case where $F_1(t) = F_2(t) = 0$]

$$x_1(t) = \tfrac{1}{2} \left(\cos \sqrt{20}\, t + \cos \sqrt{22}\, t \right),$$
$$x_2(t) = \tfrac{1}{2} \left(\cos \sqrt{20}\, t - \cos \sqrt{22}\, t \right). \tag{7.1}$$

Next, use the trigonometric identities

$$\cos A + \cos B = 2 \cos \tfrac{A+B}{2} \cos \tfrac{A-B}{2},$$
$$\cos A - \cos B = -2 \sin \tfrac{A+B}{2} \sin \tfrac{A-B}{2} \tag{7.2}$$

to show that

$$x_1(t) = \cos 4.58t \cos 0.11t,$$
$$x_2(t) = \sin 4.58t \sin 0.11t. \tag{7.3}$$

Use (7.3) to sketch $x_1(t)$ and $x_2(t)$ versus t in separate graphs, one below the other, labeling key values. Observe the *slow transfer of energy back and forth* between m_1 and m_2. In vibration theory this important phenomenon is known as **beats**.

8. It is interesting, in Section 10.3.2, that by adjusting the driving frequency Ω we can get the first mass to be stationary even though there is a force $F_1 \sin \Omega t$ applied to it. Specifically, we see in Fig. 6b that the amplitude $|I_1(\Omega)| = 0$ if $\Omega = \sqrt{2}$. Explain, in simple physical terms, why this result does not violate Newton's second law.

9. (*Singular systems*) Obtain a general solution by the method of elimination or using Cramer's rule. Further, obtain the particular solution(s) satisfying the given initial conditions. If there is no solution state that.

(a) $x' - y' + x = 3$, $x(0) = 1$
         $2x' - 2y' + y = 0$, $y(0) = 2$

(b) $x' + y' + x = e^t$, $x(0) = 1$
         $x' + y' + y = 2$, $y(0) = 2$

(c) $x' + 2y' - x = 0$, $x(1) = 6$
         $2x' + 4y' - 3y = 0$, $y(1) = 4$

(d) $x' + z' + x = t$, $x(0) = 0$
         $y' + z' + x = 0$, $y(0) = 0$
         $x' + y' + 2z' - y = 0$, $z(0) = 0$

(e) $x' + z' = x + z - t$, $x(0) = 0$
         $y' - z' = 4y - z$, $y(0) = 0$
         $x' + y' = x + 4y$, $z(0) = 1$

(f) $x' + y' + x = \sin t$, $x(0) = 5$
         $x' + y' + y = \sin t$, $y(0) = 5$
         $x' + y' + z = \sin t$, $z(0) = 5$

10.4 SOLUTION OF HOMOGENEOUS SYSTEMS AS EIGENVALUE PROBLEMS

In Section 10.3 we showed how to use elimination to convert a system of n coupled constant-coefficient differential equations on n dependent variables to an *uncoupled* system of equations. Alternatively, if the system of linear differential equations has constant coefficients and is homogeneous, both of which we assume in this section, then we can solve by assuming a solution in exponential form and then solving the eigenvalue problem that results.

EXAMPLE 1

Consider the system

$$x' = x + 4y,$$
$$y' = x + y \tag{1}$$

on $x(t)$ and $y(t)$. Since (1) is linear with constant coefficients and homogeneous, we can find exponential solutions. Thus, let us seek x, y in the form

$$x(t) = q_1 e^{rt},$$
$$y(t) = q_2 e^{rt}, \tag{2}$$

where the constants q_1, q_2, and r are to be determined. Put (2) into (1):

$$rq_1 e^{rt} = q_1 e^{rt} + 4q_2 e^{rt},$$
$$rq_2 e^{rt} = q_1 e^{rt} + q_2 e^{rt}. \tag{3}$$

Next, cancel the e^{rt}'s (because they are nonzero) and express the result in matrix form, so

$$\begin{bmatrix} 1 & 4 \\ 1 & 1 \end{bmatrix} \begin{bmatrix} q_1 \\ q_2 \end{bmatrix} = r \begin{bmatrix} q_1 \\ q_2 \end{bmatrix}$$

or

$$\mathbf{Aq} = r\mathbf{q}. \tag{4}$$

We reject the obvious trivial solution $\mathbf{q} = \mathbf{0}$ of (4) because if $q_1 = q_2 = 0$ in (2) then (2) merely gives the trivial solution $x(t) = y(t) = 0$ of (1), whereas we seek a *general* solution of (1). Thus, we need to choose r so that (4) gives nontrivial solutions as well; hence (4) is an eigenvalue problem where the eigenvalue "λ" is r.

The eigenvalues and eigenvectors of $\mathbf{A}$ are readily found to be

$$\lambda_1 = 3, \ \mathbf{e}_1 = \alpha \begin{bmatrix} 2 \\ 1 \end{bmatrix}; \qquad \lambda_2 = -1, \ \mathbf{e}_2 = \beta \begin{bmatrix} -2 \\ 1 \end{bmatrix}. \tag{5}$$

Denoting $\mathbf{x}(t) = [x(t), y(t)]^T$, each "eigenpair" in (5) gives a solution in the form (2), where r is the eigenvalue and the q_j's are given by the eigenvector. Thus we have found the solutions

$$\mathbf{x}(t) = \alpha \begin{bmatrix} 2 \\ 1 \end{bmatrix} e^{3t} \quad \text{and} \quad \mathbf{x}(t) = \beta \begin{bmatrix} -2 \\ 1 \end{bmatrix} e^{-t}. \tag{6}$$

By the linearity of (1), we can superimpose these solutions (recall Theorems 10.2.2 and 10.2.3) and obtain the general solution

$$\mathbf{x}(t) = \alpha \begin{bmatrix} 2 \\ 1 \end{bmatrix} e^{3t} + \beta \begin{bmatrix} -2 \\ 1 \end{bmatrix} e^{-t} \tag{7}$$

or, in scalar form,

$$x(t) = 2\alpha e^{3t} - 2\beta e^{-t},$$
$$y(t) = \alpha e^{3t} + \beta e^{-t}, \tag{8}$$

where α and β are arbitrary constants.

COMMENT.   Since we use q_1 and q_2 in (2), it would be natural to wonder why we don't also use "r_1" and "r_2." If we try that we obtain, in place of (3), we have

$$r_1 q_1 e^{r_1 t} = q_1 e^{r_1 t} + 4 q_2 e^{r_2 t}, \tag{9a}$$

$$r_2 q_2 e^{r_2 t} = q_1 e^{r_1 t} + q_2 e^{r_2 t}. \tag{9b}$$

If $r_2 \neq r_1$ then $e^{r_1 t}$ and $e^{r_2 t}$ are linearly independent, so it follows from (9) that

$$r_1 q_1 = q_1, \quad 0 = 4 q_2, \quad r_2 q_2 = q_2, \quad 0 = q_1. \tag{10}$$

These equations require that $q_1 = q_2 = 0$ so only the trivial solution is obtained. Only by having $r_2 = r_1 \equiv r$ can we obtain nontrivial solutions for $\mathbf{q}$, and hence for $\mathbf{x}(t)$.   ∎

EXAMPLE 2

To solve

$$\begin{aligned}
x' &= x + z, \\
y' &= -x - y, \\
z' &= 2x + 2z,
\end{aligned} \tag{11}$$

seek

$$x(t) = q_1 e^{rt}, \quad y(t) = q_2 e^{rt}, \quad z(t) = q_3 e^{rt}. \tag{12}$$

That step leads to the eigenvalue problem

$$\begin{bmatrix} 1 & 0 & 1 \\ -1 & -1 & 0 \\ 2 & 0 & 2 \end{bmatrix} \begin{bmatrix} q_1 \\ q_2 \\ q_3 \end{bmatrix} = r \begin{bmatrix} q_1 \\ q_2 \\ q_3 \end{bmatrix} \tag{13}$$

with the "eigenpairs" (i.e., eigenvalues and corresponding eigenvectors)

$$\lambda_1 = 0, \; \mathbf{e}_1 = \alpha \begin{bmatrix} -1 \\ 1 \\ 1 \end{bmatrix}; \quad \lambda_2 = -1, \; \mathbf{e}_2 = \beta \begin{bmatrix} 0 \\ 1 \\ 0 \end{bmatrix}; \quad \lambda_3 = 3, \; \mathbf{e}_3 = \gamma \begin{bmatrix} 4 \\ -1 \\ 8 \end{bmatrix}. \tag{14}$$

Thus, we have

$$\mathbf{x}(t) = \alpha \begin{bmatrix} -1 \\ 1 \\ 1 \end{bmatrix} + \beta \begin{bmatrix} 0 \\ 1 \\ 0 \end{bmatrix} e^{-t} + \gamma \begin{bmatrix} 4 \\ -1 \\ 8 \end{bmatrix} e^{3t} \tag{15}$$

or, in scalar form,

$$\begin{aligned}
x(t) &= -\alpha + 4\gamma e^{3t}, \\
y(t) &= \alpha + \beta e^{-t} - \gamma e^{3t}, \\
z(t) &= \alpha + 8\gamma e^{3t}
\end{aligned} \tag{16}$$

as the desired general solution of (11).   ∎

Thus, the solution procedure is this: To solve

$$\boxed{\mathbf{x}' = \mathbf{A}\mathbf{x},} \tag{17}$$

where the elements of the $n \times n$ matrix $\mathbf{A}$ are constants, seek

$$\boxed{\mathbf{x}(t) = \mathbf{q}e^{rt}.} \tag{18}$$

Put (18) into (17), cancel the e^{rt}'s, and obtain the eigenvalue problem

$$\boxed{\mathbf{A}\mathbf{q} = r\mathbf{q}.} \tag{19}$$

Solve (19) for the eigenvalues r and their corresponding eigenvectors $\mathbf{q}$. Each eigen-pair gives a solution of the form (18), as in (6). Finally, since the original system $\mathbf{x}' = \mathbf{A}\mathbf{x}$ was linear we can superimpose those solutions, as we did in (7).

But be careful; *that solution might fall short of being a general solution of (17),* as we demonstrate in the next example.

EXAMPLE 3
Consider the system

$$\begin{aligned} x' &= y, \\ y' &= -9x + 6y. \end{aligned} \tag{20}$$

Seeking a solution in the form (2) gives the eigenvalue problem

$$\begin{bmatrix} 0 & 1 \\ -9 & 6 \end{bmatrix} \begin{bmatrix} q_1 \\ q_2 \end{bmatrix} = r \begin{bmatrix} q_1 \\ q_2 \end{bmatrix}. \tag{21}$$

The latter has the repeated eigenvalue $r = 3, 3$ with one-dimensional eigenspace $\mathbf{e} = \alpha[1, 3]^{\mathrm{T}}$. These give

$$\mathbf{x}(t) = \alpha \begin{bmatrix} e^{3t} \\ 3e^{3t} \end{bmatrix} = \alpha \begin{bmatrix} 1 \\ 3 \end{bmatrix} e^{3t}, \tag{22}$$

but the latter falls short of being a general solution of (20) since, according to Theorem 10.2.3, there should be two LI solutions and two arbitrary constants. If, instead, we solve (20) by elimination we do obtain the general solution

$$\mathbf{x}(t) = C_1 \begin{bmatrix} e^{3t} \\ 3e^{3t} \end{bmatrix} + C_2 \begin{bmatrix} te^{3t} \\ (1 + 3t)e^{3t} \end{bmatrix}. \tag{23}$$

The first solution on the right-hand side of (23) is the same as that which was given in (22), and the second is the solution that was missing in (22). The reason it was missing, in (22), is that it includes te^{3t} terms, whereas (2) allowed for solutions only of the form e^{rt}.   ∎

EXAMPLE 4

The system

$$x' = 2x + 2y + z,$$
$$y' = x + 3y + z, \tag{24}$$
$$z' = x + 2y + 2z$$

also leads to a repeated eigenvalue: $r = 5, 1, 1$. The eigenpairs are

$$\lambda_1 = 5, \ \mathbf{e}_1 = \alpha \begin{bmatrix} 1 \\ 1 \\ 1 \end{bmatrix}; \qquad \lambda_2 = 1, \ \mathbf{e}_2 = \beta \begin{bmatrix} -1 \\ 0 \\ 1 \end{bmatrix} + \gamma \begin{bmatrix} -2 \\ 1 \\ 0 \end{bmatrix} \tag{25}$$

so

$$\mathbf{x}(t) = \alpha \begin{bmatrix} 1 \\ 1 \\ 1 \end{bmatrix} e^{5t} + \left(\beta \begin{bmatrix} -1 \\ 0 \\ 1 \end{bmatrix} + \gamma \begin{bmatrix} -2 \\ 1 \\ 0 \end{bmatrix} \right) e^{t}$$

$$= \alpha \begin{bmatrix} 1 \\ 1 \\ 1 \end{bmatrix} e^{5t} + \beta \begin{bmatrix} -1 \\ 0 \\ 1 \end{bmatrix} e^{t} + \gamma \begin{bmatrix} -2 \\ 1 \\ 0 \end{bmatrix} e^{t}, \tag{26}$$

which *is* a general solution of (24). [We leave it for you to verify that the three solutions in (26) are linearly independent.] ∎

From Examples 3 and 4 we see that multiple eigenvalues may, but need not, cause a defect, in the sense that seeking $\mathbf{x} = \mathbf{q}e^{rt}$ does not lead to a general solution of the system $\mathbf{x}' = \mathbf{A}\mathbf{x}$. In Example 4 the repeated eigenvalue was of multiplicity two and the dimension of the corresponding eigenspace was also two, so we were led to the general solution of (24). In Example 3, however, the eigenvalue of multiplicity two had an eigenspace that was only of dimension one. Thus, we say that the eigenvalue $r = 3$ in Example 3 was **defective**. More generally, if an eigenvalue of multiplicity k has an eigenspace of dimension l where $l < k$, then that eigenvalue is said to be of **defect** $k - l$ since there will result a shortage in the solution of $\mathbf{x}' = \mathbf{A}\mathbf{x}$, by $k - l$ LI solutions. In that case a general solution can nevertheless be found by modifying the solution form $\mathbf{x} = \mathbf{q}e^{rt}$, as outlined in the exercises, or by using the method of elimination (Section 10.3) instead.

EXAMPLE 5   *Coupled Oscillator*.

As our final example, let us return to the coupled mechanical oscillator (Fig. 1) that was the subject of Section 10.3.2. With $m_1 = m_2 = k_1 = k_{12} = k_2 = 1$ again, the free vibration is governed by the system

$$x_1'' + 2x_1 - x_2 = 0, \tag{27a}$$
$$x_2'' - x_1 + 2x_2 = 0. \tag{27b}$$

Seek

$$x_1(t) = q_1 e^{rt},$$
$$x_2(t) = q_2 e^{rt}, \tag{28}$$

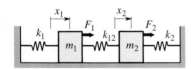

FIGURE 1
Two-mass oscillator.

put the latter into (27), cancel the e^{rt}'s, and obtain the eigenvalue problem

$$\begin{bmatrix} -2 & 1 \\ 1 & -2 \end{bmatrix} \begin{bmatrix} q_1 \\ q_2 \end{bmatrix} = r^2 \begin{bmatrix} q_1 \\ q_2 \end{bmatrix} \tag{29}$$

where "λ" is r^2. The eigenpairs are readily found to be

$$\lambda_1 = r_1^2 = -1, \ \mathbf{e}_1 = \alpha \begin{bmatrix} 1 \\ 1 \end{bmatrix}; \quad \lambda_2 = r_2^2 = -3, \ \mathbf{e}_2 = \beta \begin{bmatrix} 1 \\ -1 \end{bmatrix}. \tag{30}$$

However, realize that each of these eigenpairs gives two solutions since $r_1^2 = -1$ gives both $r_1 = +i$ and $r_1 = -i$; similarly for r_2. Put these results into (28) and use superposition:

$$\mathbf{x}(t) = C_1 \begin{bmatrix} 1 \\ 1 \end{bmatrix} e^{it} + C_2 \begin{bmatrix} 1 \\ 1 \end{bmatrix} e^{-it} + C_3 \begin{bmatrix} 1 \\ -1 \end{bmatrix} e^{i\sqrt{3}t} + C_4 \begin{bmatrix} 1 \\ -1 \end{bmatrix} e^{-i\sqrt{3}t}. \tag{31}$$

Alternatively, we can express the complex exponentials in terms of cosines and sines as follows:

$$\mathbf{x}(t) = \begin{bmatrix} C_1 e^{it} + C_2 e^{-it} \\ C_1 e^{it} + C_2 e^{-it} \end{bmatrix} + \begin{bmatrix} C_3 e^{i\sqrt{3}t} + C_4 e^{-i\sqrt{3}t} \\ -C_3 e^{i\sqrt{3}t} - C_4 e^{-i\sqrt{3}t} \end{bmatrix}$$

$$= \begin{bmatrix} C_5 \cos t + C_6 \sin t \\ C_5 \cos t + C_6 \sin t \end{bmatrix} + \begin{bmatrix} C_7 \cos \sqrt{3}t + C_8 \sin \sqrt{3}t \\ -C_7 \cos \sqrt{3}t - C_8 \sin \sqrt{3}t \end{bmatrix} \tag{32}$$

so

$$\mathbf{x}(t) = C_5 \begin{bmatrix} 1 \\ 1 \end{bmatrix} \cos t + C_6 \begin{bmatrix} 1 \\ 1 \end{bmatrix} \sin t + C_7 \begin{bmatrix} 1 \\ -1 \end{bmatrix} \cos \sqrt{3}t + C_8 \begin{bmatrix} 1 \\ -1 \end{bmatrix} \sin \sqrt{3}t. \tag{33}$$

Or, if we prefer the $\sin(\omega t + \phi)$ form we can obtain that from (32) as

$$\mathbf{x}(t) = \begin{bmatrix} G \sin(t + \phi) \\ G \sin(t + \phi) \end{bmatrix} + \begin{bmatrix} H \sin(\sqrt{3}t + \psi) \\ -H \sin(\sqrt{3}t + \psi) \end{bmatrix}$$

$$= G \begin{bmatrix} 1 \\ 1 \end{bmatrix} \sin(t + \phi) + H \begin{bmatrix} 1 \\ -1 \end{bmatrix} \sin(\sqrt{3}t + \psi). \tag{34}$$

According to the terminology introduced in Section 10.3, the first term on the right-hand side of (34) is called the low mode, that is, the low frequency mode, and the second term is the high mode, the high frequency mode.

COMMENT 1. They are also called **eigenmodes** because each mode is determined by an eigenpair; the vibrational frequency is dictated by the eigenvalue and is called the **eigenfrequency**, and the "mode shape" is dictated by the eigenvector.[1] Further, they are called **normal modes**, or **orthogonal modes**, because the eigenvectors

---

[1]Actually, the frequencies are not the eigenvalues, they are the square roots of the absolute values of the eigenvalues.

$[1, 1]^T$ and $[1, -1]^T$ are orthogonal, which orthogonality results from the symmetry of the matrix in (29). Be clear that the motions are not physically perpendicular to each other; on the contrary, they are colinear. Rather, the orthogonality is in the sense that the dot product of the modal vectors is zero. In a physical sense, the significance of the orthogonality is that the two modes are—roughly speaking—about as different from each other as we can imagine. If this is not clear, try mimicking each of the two modal motions using your fists as the two masses.

COMMENT 2. An interesting application of this type of analysis can be found in courses on statistical mechanics, where a molecule of a perfect gas such as CO_2 is modeled by point masses (representing the atoms) connected by linear springs (representing the interatomic electrical forces). From the eigenfrequencies and eigenmodes one can determine the specific heat and entropy of the gas, thereby relating the microscale behavior of a single molecule to the macroscopic behavior of the gas. ∎

Closure. Given a linear homogeneous system

$$\mathbf{x}' = \mathbf{Ax}, \tag{35}$$

where the $n \times n$ matrix $\mathbf{A}$ is constant, we seek a solution in the form

$$\mathbf{x}(t) = \mathbf{q}e^{rt}. \tag{36}$$

Putting (36) into (35) gives the eigenvalue problem

$$\mathbf{Aq} = r\mathbf{q}. \tag{37}$$

Each $r, \mathbf{q}$ eigenpair gives a solution of the form (36) and, since (35) is linear, these can be superimposed (Theorem 10.2.2). If the n eigenvalues of $\mathbf{A}$ are distinct then our superimposed solution is a general solution of (35). The same will be true even if there are repeated eigenvalues, provided that in each case the dimension l of the corresponding eigenspace equals the multiplicity k of the eigenvalue (as occurred here in Example 4). If, instead, $l < k$ for a repeated eigenvalue, then we say that the eigenvalue is of defect $k - l$ and there will result a shortage in the solution of $\mathbf{x}' = \mathbf{Ax}$ by $k - l$ LI solutions (as occurred in Example 3). Nevertheless, a general solution of $\mathbf{x}' = \mathbf{Ax}$ can be found in that case by suitably modifying the solution form $\mathbf{x} = \mathbf{q}e^{rt}$, as is discussed in the exercises, or by using the method of elimination instead.

EXERCISES 10.4

1. Solve, as we did in Examples 1–5, by seeking $\mathbf{x}(t)$ in the form $\mathbf{q}e^{rt}$. State whether or not the solution that you obtain is a general solution. Primes denote d/dt. Use *Maple* to evaluate the eigenvalues and eigenvectors if you cannot solve for them by hand.

(a) $\quad \begin{aligned} x' &= x + 2y \\ y' &= 3x + 6y \end{aligned}$

(b) $\quad \begin{aligned} x' &= x + y \\ y' &= x + y \end{aligned}$

(c) $\quad \begin{aligned} x' &= 12x + 3y \\ y' &= -20x - 5y \end{aligned}$

(d) $\quad \begin{aligned} x' &= 3x + y \\ y' &= -3x - y \end{aligned}$

(e) $x'' = 2x + y$
$y'' = 9x + 2y$

(f) $x'' = x + y$
$y'' = x + y$

(g) $x'' = 20x + 9y$
$y'' = -35x - 16y$

(h) $x'' = 3x + y$
$y'' = -3x - y$

(i) $x' = 4x + y$
$y' = -5x + z$
$z' = x - y - z$

(j) $x' = y - z$
$y' = -5x + 4y + z$
$z' = x - z$

(k) $x' = 9x + 10y + 9z$
$y' = -6x - 5y - 3z$
$z' = x - z$

(l) $x' = 5x + 2y$
$y' = -6x - 2y$
$z' = x + y$

(m) $x' = 2x - y$
$y' = y - z$
$z' = -x + y$

(n) $x' = 2x + z$
$y' = y + z$
$z' = -x$

(o) $x'' = -x + y$
$y'' = -x + z$
$z'' = x - 2z$

(p) $x'' = y$
$y'' = -2x - 2y + z$
$z'' = x + y - z$

2. First, solve the equation for the general solution "as usual," that is, by the methods studied in Chapter 6. Then convert the equation to a first-order system by introducing auxiliary variables, and solve that system by the method explained in this section. State whether or not the latter method gives you a general solution for $x(t)$.

(a) $x'' + 3x' + 2x = 0$
(b) $x'' + x' - 2x = 0$
(c) $x'' + 4x' + 4x = 0$
(d) $x'' - 2x' + x = 0$
(e) $x'' + 3x' = 0$
(f) $x''' + 3x'' - 4x' = 0$

3. Consider a mass-spring system like the one shown in Section 10.3.2, but with five masses and six springs. If all the masses and spring stiffnesses are 1, say, then the equations of motion in matrix form are

$$
\begin{bmatrix} x_1'' \\ x_2'' \\ x_3'' \\ x_4'' \\ x_5'' \end{bmatrix} + \begin{bmatrix} 2 & -1 & 0 & 0 & 0 \\ -1 & 2 & -1 & 0 & 0 \\ 0 & -1 & 2 & -1 & 0 \\ 0 & 0 & -1 & 2 & -1 \\ 0 & 0 & 0 & -1 & 2 \end{bmatrix} \begin{bmatrix} x_1 \\ x_2 \\ x_3 \\ x_4 \\ x_5 \end{bmatrix} = \begin{bmatrix} 0 \\ 0 \\ 0 \\ 0 \\ 0 \end{bmatrix}
$$

(3.1)

or $\mathbf{x}'' + \mathbf{A}\mathbf{x} = \mathbf{0}$. Observe that the $\mathbf{A}$ matrix is **tridiagonal** (i.e., all elements are zero except for the main diagonal and the two neighboring diagonals). Physically, this result corresponds to **nearest-neighbor coupling** whereby each mass feels only its immediate neighbors. Nearest-neighbor coupling occurs in other systems as well. For instance, in modeling single-lane traffic flow each driver accelerates or decelerates according to the motion of the cars immediately ahead and immediately behind, so the resulting coupled differential equations exhibit nearest-neighbor coupling. The problem that we pose is for you to use *Maple* to determine the natural frequencies and mode shapes (eigenvectors).

4. (*Lumped parameter model of vibrating string*) Let us consider the plane vibration of a taut string such as a guitar string. If the tension is τ, the mass per unit length is ρ, and the lateral displacement is $y(x, t)$, then the motion $y(x, t)$ is governed by the well known wave equation, namely, the partial differential equation ("PDE")

$$
\tau \frac{\partial^2 y}{\partial x^2} = \rho \frac{\partial^2 y}{\partial t^2}.
$$
(4.1)

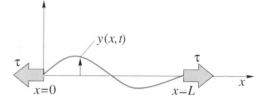

(a) Derive (4.1). HINT: A free body diagram of a typical element of the string, between x and $x + \Delta x$, is shown below, where we have neglected the gravitational force on the element compared to the effects of the tension τ.

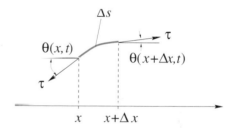

(After all, doesn't a guitar sound the same if it is held horizontally or vertically?) Applying Newton's second law in the y direction, obtain

$$
\rho \Delta s \frac{\partial^2 y(x + \alpha \Delta x, t)}{\partial t^2}
$$
$$
= \tau [\sin \theta(x + \Delta x, t) - \sin \theta(x, t)], \quad (4.2)
$$

where Δs is the arc length of the element and $x + \alpha \Delta x$ is the x location of the mass center of the element. Next, make the classical assumption that the motion is small; that is, $|y| \ll L$ and $|\theta| \ll 1$. In that case $\Delta s \approx \Delta x$ and, since θ is small,

$$
\sin \theta = \theta - \frac{1}{3!}\theta^2 + \frac{1}{5!}\theta^5 - \cdots \approx \theta \quad (4.3)
$$

and

$$
\tan \theta = \theta + \frac{1}{3}\theta^3 + \frac{2}{15}\theta^5 + \cdots \approx \theta \quad (4.4)
$$

so we can replace

$$\sin\theta \approx \tan\theta = \frac{\partial y}{\partial x}, \tag{4.5}$$

which step enables us to eliminate the temporary variable θ, in (4.2), in favor of y. With these approximations in (4.2), let $\Delta x \to 0$ and obtain (4.1). NOTE: Physically, (4.1) is a statement of Newton's second law, that "$F = ma$." To understand the significance of the $\tau \partial^2 y/\partial x^2$ term, recall from the calculus that the *curvature* κ and the *radius of curvature* R at each point of a plane curve $y(x)$ are related according to

$$\kappa = \frac{1}{R} = \frac{y''(x)}{\left[1 + y'(x)^2\right]^{3/2}}. \tag{4.6}$$

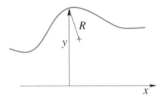

Since $|y'| \ll 1$ by assumption, (4.6) gives $\kappa \approx y''(x)$. Thus, the vertical force $\tau \partial^2 y/\partial x^2$ in (4.1) is seen to be the interaction of (specifically, the product of) the tension and the curvature, which result makes sense since we see from the second figure that the vertical force is due to the presence of both the tension and the curvature, for without curvature the two tension forces would cancel.

(b) Suppose that the string is tied so that

$$y(0, t) = 0, \quad y(L, t) = 0 \tag{4.7}$$

and suppose we initiate its motion by pulling it into some initial shape

$$y(x, 0) = f(x), \tag{4.8}$$

holding it still so

$$\frac{\partial y}{\partial t}(x, 0) = 0, \tag{4.9}$$

and releasing it at $t = 0$. Show that the function

$$\sin\frac{n\pi x}{L}\cos\frac{n\pi ct}{L}, \tag{4.10}$$

where $c = \sqrt{\tau/\rho}$ and n is any positive integer, satisfies the PDE (4.1), the boundary conditions (4.7), and the initial condition (4.9), so that the linear combination

$$y(x, t) =$$
$$C_1 \sin\frac{\pi x}{L}\cos\frac{\pi ct}{L} + \cdots + C_M \sin\frac{M\pi x}{L}\cos\frac{M\pi ct}{L} \tag{4.11}$$

does too, for arbitrarily large M. Letting $M \to \infty$ we can write

$$y(x, t) = \sum_{1}^{\infty} C_n \sin\frac{n\pi x}{L}\cos\frac{n\pi ct}{L}. \tag{4.12}$$

The C_n's can be determined (using the theory of Fourier series, which is not discussed in this text) from the remaining condition (4.8),

$$f(x) = \sum_{1}^{\infty} C_n \sin\frac{n\pi x}{L},$$

but we will not address that calculation here. Rather, we observe from (4.12) that the motion $y(x, t)$ is the superposition of an *infinite* number of modes corresponding to the terms in the series (4.12). The first three modes are shown below.

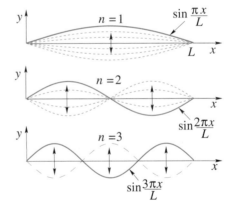

(c) In (a) and (b) we dealt with the string as a *continuous* system, which it is. Alternatively, one can model such continuous systems by *discrete* approximations known as **lumped parameter** models. In the present case we can do that by dividing the string into N equal parts (the larger N, the better) and representing each string segment by a point mass or "bead," of mass $\rho L/N$, located at its midpoint, as shown below for $N = 4$, say. Using the small angle approximation $\sin\beta \approx \tan\beta = (y_2 - y_1)/(L/4)$, and similarly for

α, apply Newton's law to the first bead and show that

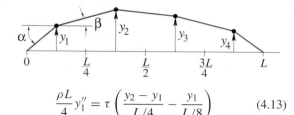

$$\frac{\rho L}{4} y_1'' = \tau \left(\frac{y_2 - y_1}{L/4} - \frac{y_1}{L/8} \right) \qquad (4.13)$$

where the y_j's are functions of t only and primes denote d/dt. Doing the same for the other three masses show that the resulting system of ODE's can be expressed as

$$\mathbf{y}'' + \mathbf{A}\mathbf{y} = \mathbf{0}, \qquad (4.14)$$

where $\mathbf{y} = [y_1(t), \dots, y_4(t)]^{\mathrm{T}}$ and

$$\mathbf{A} = \frac{16\tau}{\rho L^2} \begin{bmatrix} 3 & -1 & 0 & 0 \\ -1 & 2 & -1 & 0 \\ 0 & -1 & 2 & -1 \\ 0 & 0 & -1 & 3 \end{bmatrix}. \qquad (4.15)$$

(**d**) To solve (4.14) we could seek $\mathbf{y}(t) = \boldsymbol{\eta} e^{rt}$, say, but realizing that the motion will be oscillatory it is convenient to go right to the form

$$\mathbf{y}(t) = \boldsymbol{\eta} \sin(\omega t + \phi), \qquad (4.16)$$

where the vector $\boldsymbol{\eta}$ and the scalars ω and ϕ remain to be determined. Putting (4.16) into (4.14), derive the eigenvalue problem

$$\mathbf{B}\boldsymbol{\eta} = \lambda\boldsymbol{\eta}, \qquad (4.17)$$

where $\mathbf{B}$ is the $\mathbf{A}$ matrix in (4.15) but without the $16\tau/\rho L^2$ factor, and $\lambda = (\rho/\tau)(L/4)^2\omega^2$, so $\omega = 4\sqrt{\lambda\tau/\rho}/L = 4\sqrt{\lambda}\,c/L$. Solve (4.17) using *Maple* and show that, to three decimal places,

$$\mathbf{y}(t) =$$

$$D_1 \begin{bmatrix} 1 \\ 2.414 \\ 2.414 \\ 1 \end{bmatrix} \cos\frac{3.061 ct}{L} + D_2 \begin{bmatrix} 1 \\ 1 \\ -1 \\ -1 \end{bmatrix} \cos\frac{5.657 ct}{L}$$

$$+ D_3 \begin{bmatrix} 1 \\ -0.414 \\ -0.414 \\ 1 \end{bmatrix} \cos\frac{7.391 ct}{L} + D_4 \begin{bmatrix} 1 \\ -1 \\ 1 \\ -1 \end{bmatrix} \cos\frac{8 ct}{L},$$

$$(4.18)$$

where the D_j's are arbitrary constants. Here we have changed the $\sin(\omega t + \phi)$ in (4.16) to $\cos \omega t$

to satisfy (4.9). Comparing (4.18) with (4.12) we see that the discretized version (4.18) approximates the first four modes in (4.12). The eigenvectors [which are the bracketed column vectors in (4.18)] give discrete approximations of the mode shapes $\sin \pi x/L, \dots, \sin 4\pi x/L$ (to within scale factors since the C_j's and D_j's are arbitrary) and 3.061, 5.657, 7.391, and 8 approximate π, 2π, 3π, and 4π, respectively. Plot the mode shapes $\sin \pi x/L, \dots, \sin 4\pi x/L$ and superimpose on those plots the points indicated by the eigenvectors; you can scale one or the other to obtain what appears to be the best fit.

(**e**) Since the $\mathbf{A}$ matrix in (4.15) is symmetric the eigenvectors are mutually orthogonal, as we emphasized in Section 9.3. Thus, should not the continuous mode shapes $\sin(n\pi x/L)$ be mutually orthogonal too? Yes. Recall from Example 9 of Section 4.8 that we can use the inner product

$$\langle u(x), v(x) \rangle = \int_0^L u(x)v(x)\,dx. \qquad (4.19)$$

By carrying out the integration, show that the inner product

$$\left\langle \sin\frac{m\pi x}{L}, \sin\frac{n\pi x}{L} \right\rangle = \int_0^L \sin\frac{m\pi x}{L} \sin\frac{n\pi x}{L}\,dx$$

$$(4.20)$$

is indeed zero for any integers m and n, if $m \neq n$. HINT: $\sin A \sin B = [\cos(A - B) - \cos(A + B)]/2$.

(**f**) In parts (c) – (e) we used $N = 4$. More generally, with N beads one obtains (4.14), where $\mathbf{y} = [y_1(t), \dots, y_N(t)]^{\mathrm{T}}$ and $\mathbf{A}$ is the $N \times N$ matrix

$$\mathbf{A} = \left(\frac{N}{L}\right)^2 \frac{\tau}{\rho} \begin{bmatrix} 3 & -1 & 0 & 0 & \cdots & 0 \\ -1 & 2 & -1 & 0 & \cdots & 0 \\ 0 & -1 & 2 & -1 & \cdots & 0 \\ \vdots & & & \ddots & & \vdots \\ 0 & \cdots & & -1 & 2 & -1 \\ 0 & \cdots & & 0 & -1 & 3 \end{bmatrix}.$$

$$(4.21)$$

(You need not derive this result.) Since $\mathbf{A}$ has N eigenpairs, we see that as we increase N we pick up more and more modes, and we (correctly) expect that as $N \to \infty$ the discrete results will converge to the continuous exact results. To explore this point, rework the problem (using *Maple*) with $N = 10$ and compare just the first eigenmode with the exact one, $\sin(\pi x/L)\cos(\pi ct/L)$, insofar as the mode shape and the temporal frequency.

5. (*Dealing with a defective eigenvalue*) First, reread Example 3. There, we sought $\mathbf{x}(t) = \mathbf{q}e^{rt}$ and found the eigenvalue $r = 3$ of multiplicity two, but its eigenspace was only one-dimensional, so $r = 3$ was of defect one. To find not only the partial solution (22) but the missing solution as well, seek $\mathbf{x}(t)$ in the modified form $\mathbf{x}(t) = (\mathbf{p} + \mathbf{q}t)e^{3t}$; that is,

$$\boxed{\mathbf{x}(t) = (\mathbf{p} + \mathbf{q}t)e^{rt},} \tag{5.1}$$

where in this example $r = 3$. Putting (5.1) into the original system $\mathbf{x}' = \mathbf{A}\mathbf{x}$, show that

$$(\mathbf{A} - r\mathbf{I})\,\mathbf{q} = \mathbf{0}, \tag{5.2}$$

$$(\mathbf{A} - r\mathbf{I})\,\mathbf{p} = \mathbf{q}. \tag{5.3}$$

With $r = 3$, (5.2) simply says that $\mathbf{q}$ is the eigenvector $\alpha[1, 3]^T$ already found in Example 3. Then, putting $r = 3$ and $\mathbf{q} = \alpha[1, 3]^T$ into (5.3), solve for $\mathbf{p}$ by Gauss elimination and show that your solution (5.1) agrees with (23),

which was found by elimination and is a general solution of (20).

6. (*Exercise 5, continued*) Find a general solution of the given system as follows. First, seek $\mathbf{x}(t) = \mathbf{q}e^{rt}$. For each defective eigenvalue r, start over, seeking

$$\mathbf{x}(t) = (\mathbf{p} + \mathbf{q}t)e^{rt} \tag{6.1}$$

if r is of defect one,

$$\mathbf{x}(t) = (\boldsymbol{\sigma} + \mathbf{p}t + \mathbf{q}t^2)e^{rt} \tag{6.2}$$

if r is of defect two, and so on.

(a) $\begin{aligned} x' &= 2x + 4y \\ y' &= -x + 6y \end{aligned}$ (b) $\begin{aligned} x' &= 2x - 4y \\ y' &= x - 2y \end{aligned}$

(c) $\begin{aligned} x' &= -x + y - 3z \\ y' &= x + 2z \\ z' &= 2x - y + 4z \end{aligned}$ (d) $\begin{aligned} x' &= -4x - 4y \\ y' &= x - 8y \\ z' &= z \end{aligned}$

10.5 DIAGONALIZATION. (OPTIONAL)

One of the examples in this section will require knowledge of Taylor series of functions of two variables. We will give that necessary background first, so that we can then proceed without interruption.

10.5.1 A preliminary; Taylor's series in two variables.
The Taylor series of a real-valued function f of a single real variable x about a point $x = a$,

$$f(x) = f(a) + \frac{f'(a)}{1!}(x - a) + \frac{f''(a)}{2!}(x - a)^2 + \cdots, \tag{1}$$

is studied in the calculus and was reviewed in Section 8.2.3. From an applied point of view, (1) is valuable because it enables us to approximate a "hard function" [i.e., the given function $f(x)$] by a "simple one," namely, a finite-degree polynomial consisting of the first few terms on the right-hand side of (1); the more terms we retain and the closer we take x to the point of expansion a, the more accurate the approximation. Often, it suffices to keep just the first two terms,

$$f(x) \approx f(a) + f'(a)(x - a), \tag{2}$$

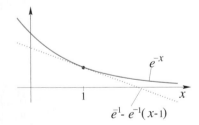

FIGURE 1
Taylor series linearization of e^{-x} about $x = 1$.

which approximation *linearizes* $f(x)$ about a, for the right side of (2) gives the tangent line to the graph of $f(x)$, as is illustrated in Fig. 1 for the case where $f(x)$ is e^{-x} and $a = 1$.

Likewise, we can expand a function of two variables $f(x, y)$ in a Taylor series about a point $(x, y) = (a, b)$ in the x, y plane. To derive the formula for the two-variable Taylor series we can expand $f(x, y)$ in one variable at a time. Expanding first in x, with y temporarily held fixed, gives

$$f(x, y) = f(a, y) + \frac{1}{1!} f_x(a, y)(x - a) + \frac{1}{2!} f_{xx}(a, y)(x - a)^2 + \cdots, \quad (3)$$

where subscripts denote partial derivatives. Then, expanding the coefficients in (3) (which are functions of y only) in y about $y = b$,

$$f(a, y) = f(a, b) + f_y(a, b)(y - b) + \frac{1}{2!} f_{yy}(a, b)(y - b)^2 + \cdots,$$

$$f_x(a, y) = f_x(a, b) + f_{xy}(a, b)(y - b) + \frac{f_{xyy}(a, b)}{2!}(y - b)^2 + \cdots, \quad (4)$$

$$f_{xx}(a, y) = f_{xx}(a, b) + f_{xxy}(a, b)(y - b) + \frac{f_{xxyy}(a, b)}{2!}(y - b)^2 + \cdots.$$

Finally, putting (4) into (3) and arranging terms in ascending order gives the **Taylor series**

$$\begin{aligned}
f(x, y) = {}& f(a, b) + \frac{1}{1!}\left[f_x(a, b)(x - a) + f_y(a, b)(y - b) \right] \\
& + \frac{1}{2!}\left[f_{xx}(a, b)(x - a)^2 + 2f_{xy}(a, b)(x - a)(y - b) \right. \\
& \left. + f_{yy}(a, b)(y - b)^2 \right] + \cdots
\end{aligned} \quad (5)$$

of $f(x, y)$ about (a, b); by the **order** of a term $(x - a)^m (y - b)^n$ we mean the sum of the exponents $m + n$. The idea is the same for functions of more than two variables.

Just as (2) gives the tangent *line* approximation of $f(x)$ about the point $(a, f(a))$ in the x, y plane, obtained by cutting off (1) after the first-order term,

$$f(x, y) \approx f(a, b) + f_x(a, b)(x - a) + f_y(a, b)(y - b) \quad (6)$$

gives the tangent *plane* approximation of $f(x, y)$ about the point $(a, b, f(a, b))$ in x, y, f space, obtained by cutting (5) off after the first-order terms.

EXAMPLE 1

The Taylor series of e^{xy} about the point $(2, 1)$ is, according to (5),

$$e^{xy} = e^2 + e^2(x - 2) + 2e^2(y - 1)$$

$$+ \frac{e^2}{2}(x - 2)^2 + 2e^2(x - 2)(y - 1) + 2e^2(y - 1)^2 + \cdots,$$

and the tangent plane approximation about $(2, 1)$ is (Fig. 2)

$$e^{xy} \approx e^2 + e^2(x - 2) + 2e^2(y - 1).$$

COMMENT. Is it clear that the latter is the equation of a plane, as claimed? If we denote e^{xy} as $f(x, y)$, then $e^2 x + 2e^2 y - f = 3e^2$ is indeed the equation of a plane in x, y, f space just as $ax + by + cz = d$ is the equation of a plane in x, y, z space. ∎

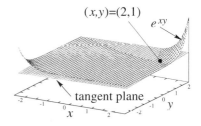

FIGURE 2
Comparison of e^{xy} with the tangent plane from its Taylor series near $(2, 1)$.

10.5.2 Diagonalization. We have seen that diagonal matrices are particularly simple to work with. For instance, the solution of the system of algebraic equations $\mathbf{Ax} = \mathbf{c}$, where $\mathbf{A}$ is $n \times n$, is generally tedious if n is large but is simple if $\mathbf{A}$ is diagonal, for then the scalar equations are not coupled. Similarly, raising $\mathbf{A}$ to the mth power is generally tedious if m is large but is simple if $\mathbf{A}$ is diagonal. Likewise, and relevant to the present chapter, the solution of a system of differential equations

$$\mathbf{x}'(t) = \mathbf{Ax}(t) \tag{7}$$

is generally tedious but is simple if $\mathbf{A}$ is diagonal, for then the scalar equations are uncoupled. Thus, we are interested here in ways to convert a given system (7) to one in which the matrix on the right-hand side is diagonal.

We begin the solution of (7) by the "method of diagonalization" by introducing a linear change of variables from the original variables $x_1(t), \ldots, x_n(t)$ to new variables $\widetilde{x}_1(t), \ldots, \widetilde{x}_n(t)$:

$$\boxed{\mathbf{x} = \mathbf{Q}\widetilde{\mathbf{x}},} \tag{8}$$

where $\mathbf{Q}$ is a constant matrix, $\mathbf{x} = [x_1, \ldots, x_n]^\mathrm{T}$, and $\widetilde{\mathbf{x}} = [\widetilde{x}_1, \ldots, \widetilde{x}_n]^\mathrm{T}$. Written out,

$$x_1(t) = q_{11}\widetilde{x}_1(t) + \cdots + q_{1n}\widetilde{x}_n(t),$$
$$\vdots \tag{9}$$
$$x_n(t) = q_{n1}\widetilde{x}_1(t) + \cdots + q_{nn}\widetilde{x}_n(t).$$

Putting (8) into (7) [and observing that $(\mathbf{Q}\widetilde{\mathbf{x}})' = \mathbf{Q}\widetilde{\mathbf{x}}'$ because the matrix $\mathbf{Q}$ is constant] gives

$$\mathbf{Q}\widetilde{\mathbf{x}}' = \mathbf{AQ}\widetilde{\mathbf{x}}. \tag{10}$$

Since the choice of $\mathbf{Q}$ is ours, we can ask $\mathbf{Q}$ to be invertible. Then, multiplying (10) by $\mathbf{Q}^{-1}$ gives $\mathbf{Q}^{-1}\mathbf{Q}\widetilde{\mathbf{x}}' = \mathbf{Q}^{-1}\mathbf{AQ}\widetilde{\mathbf{x}}$ or

$$\widetilde{\mathbf{x}}' = \mathbf{Q}^{-1}\mathbf{AQ}\widetilde{\mathbf{x}}. \tag{11}$$

Thus, given the $\mathbf{A}$ matrix the idea is to try to find a $\mathbf{Q}$ matrix such that

$$\boxed{\mathbf{Q}^{-1}\mathbf{AQ} = \mathbf{D}} \tag{12}$$

is diagonal because then (11) becomes

$$\boxed{\widetilde{\mathbf{x}}' = \mathbf{D}\widetilde{\mathbf{x}},} \tag{13}$$

and the scalar differential equations within (13) will be *uncoupled*. If there does exist such a $\mathbf{Q}$ then we say that $\mathbf{A}$ is **diagonalizable** and that $\mathbf{Q}$ **diagonalizes** $\mathbf{A}$. [By virtue of the form of (12) we say that the matrices $\mathbf{A}$ and $\mathbf{D}$ are **similar**; see Exercise 14 of Section 9.2.]

Two questions present themselves: given $\mathbf{A}$, does there exist such a $\mathbf{Q}$ and, if so, how do we find it? (There is also a question of uniqueness, but we are not especially interested in whether or not $\mathbf{Q}$ is unique; we'll be content to find *any* $\mathbf{Q}$ that diagonalizes $\mathbf{A}$.)

THEOREM 10.5.1

Diagonalization

Let $\mathbf{A}$ be $n \times n$.

1. $\mathbf{A}$ is diagonalizable if and only if it has n LI eigenvectors.

2. If $\mathbf{A}$ has n LI eigenvectors $\mathbf{e}_1, \ldots, \mathbf{e}_n$ and we make these the columns of $\mathbf{Q}$, so that

$$\boxed{\mathbf{Q} = [\mathbf{e}_1, \ldots, \mathbf{e}_n],} \tag{14}$$

then $\mathbf{Q}^{-1}\mathbf{A}\mathbf{Q} = \mathbf{D}$ is diagonal and the jth diagonal element of $\mathbf{D}$ is the jth eigenvalue of $\mathbf{A}$.

Proof.   First, by the notation $\mathbf{Q} = [\mathbf{e}_1, \ldots, \mathbf{e}_n]$ we mean that $\mathbf{Q}$ is partitioned (Section 5.2.7) into columns, the columns being the eigenvectors $\mathbf{e}_1, \ldots, \mathbf{e}_n$ of $\mathbf{A}$.

Let us prove that if $\mathbf{A}$ is diagonalizable then it must have n LI eigenvectors. If $\mathbf{A}$ is diagonalizable then there is an invertible matrix $\mathbf{Q}$ such that

$$\mathbf{Q}^{-1}\mathbf{A}\mathbf{Q} = \mathbf{D} = \begin{bmatrix} d_1 & 0 & \cdots & 0 \\ 0 & d_2 & & \vdots \\ \vdots & & \ddots & \vdots \\ 0 & & \cdots & d_n \end{bmatrix}. \tag{15}$$

Pre-multiplying both sides of (15) by $\mathbf{Q}$ gives $\mathbf{A}\mathbf{Q} = \mathbf{Q}\mathbf{D}$:

$$\mathbf{A}\mathbf{Q} = \begin{bmatrix} q_{11} & \cdots & q_{1n} \\ \vdots & & \vdots \\ q_{n1} & \cdots & q_{nn} \end{bmatrix} \begin{bmatrix} d_1 & 0 & \cdots & 0 \\ 0 & d_2 & & \vdots \\ \vdots & & \ddots & \vdots \\ 0 & & \cdots & d_n \end{bmatrix}$$

$$= \begin{bmatrix} d_1 q_{11} & \cdots & d_n q_{1n} \\ \vdots & & \vdots \\ d_1 q_{n1} & \cdots & d_n q_{nn} \end{bmatrix} = [d_1\mathbf{q}_1, \ldots, d_n\mathbf{q}_n], \tag{16}$$

where the vector $\mathbf{q}_j$ simply denotes the jth column of $\mathbf{Q}$. Alternatively,

$$\mathbf{A}\mathbf{Q} = \mathbf{A}[\mathbf{q}_1, \mathbf{q}_2, \ldots, \mathbf{q}_n] = [\mathbf{A}\mathbf{q}_1, \mathbf{A}\mathbf{q}_2, \ldots, \mathbf{A}\mathbf{q}_n] \tag{17}$$

and comparing (16) and (17) we see that

$$\mathbf{A}\mathbf{q}_1 = d_1\mathbf{q}_1, \quad \ldots, \quad \mathbf{A}\mathbf{q}_n = d_n\mathbf{q}_n. \tag{18}$$

Does (18) imply that the $\mathbf{q}_j$'s are eigenvectors of $\mathbf{A}$? Yes, provided that they are nonzero. Since we have assumed $\mathbf{A}$ to be diagonalizable, $\mathbf{Q}$ must be invertible. Hence, none of its columns $\mathbf{q}_j$ can be $\mathbf{0}$. Thus, the d_j's and $\mathbf{q}_j$'s are the eigenvalues λ_j and eigenvectors $\mathbf{e}_j$ of $\mathbf{A}$. Furthermore, the rank of $\mathbf{Q}$ must be n since $\mathbf{Q}$ is to be invertible, so (Theorem 5.6.3) its columns must be LI.

Thus far we have proved half of item 1, that if $\mathbf{A}$ is diagonalizable, then it must have n LI eigenvectors. In doing so we have also proved item 2. It remains to prove the rest of item 1, that if $\mathbf{A}$ has n LI eigenvectors then it is diagonalizable. To do so, let us take $\mathbf{Q}$ to be made up of columns which are the eigenvectors of $\mathbf{A}$, so $\mathbf{Q} = [\mathbf{e}_1, \ldots, \mathbf{e}_n]$. Then

$$\mathbf{AQ} = [\mathbf{Ae}_1, \ldots, \mathbf{Ae}_n] = [\lambda_1 \mathbf{e}_1, \ldots, \lambda_n \mathbf{e}_n] = \begin{bmatrix} \lambda_1 e_{11} & \cdots & \lambda_n e_{n1} \\ \vdots & & \vdots \\ \lambda_1 e_{1n} & \cdots & \lambda_n e_{nn} \end{bmatrix}$$

$$= \begin{bmatrix} e_{11} & \cdots & e_{n1} \\ \vdots & & \vdots \\ e_{1n} & \cdots & e_{nn} \end{bmatrix} \begin{bmatrix} \lambda_1 & 0 & \cdots & 0 \\ 0 & \lambda_2 & & \vdots \\ \vdots & & \ddots & 0 \\ 0 & \cdots & 0 & \lambda_n \end{bmatrix} = \mathbf{QD}. \tag{19}$$

Finally, $\mathbf{Q}$ is invertible since its columns are LI, so pre-multiplying (19) by $\mathbf{Q}^{-1}$ gives $\mathbf{Q}^{-1}\mathbf{AQ} = \mathbf{D}$. Hence $\mathbf{A}$ is diagonalizable, and the proof is complete.   ∎

Since the columns of $\mathbf{Q}$ are LI eigenvectors of $\mathbf{A}$, $\mathbf{Q}$ is called a **modal matrix** of $\mathbf{A}$.

Theorem 10.5.1 relates the diagonalizability of $\mathbf{A}$ to the eigenvectors of $\mathbf{A}$. With the help of Theorem 10.5.2, we will be able to relate the diagonalizability of $\mathbf{A}$ to the eigenvalues of $\mathbf{A}$ as well. When that is done we will turn to applications.

THEOREM 10.5.2

Distinct Eigenvalues, LI Eigenvectors
If an $n \times n$ matrix $\mathbf{A}$ has distinct eigenvalues $\lambda_1, \ldots, \lambda_n$ then the corresponding eigenvectors $\mathbf{e}_1, \ldots, \mathbf{e}_n$ are LI.

Proof.   We need to show that

$$c_1 \mathbf{e}_1 + c_2 \mathbf{e}_2 + \cdots + c_n \mathbf{e}_n = \mathbf{0} \tag{20}$$

holds only if $c_1 = c_2 = \cdots = c_n = 0$. Multiplying (20) by $\mathbf{A}$ and noting that $\mathbf{Ae}_j = \lambda_j \mathbf{e}_j$, we have

$$c_1 \lambda_1 \mathbf{e}_1 + c_2 \lambda_2 \mathbf{e}_2 + \cdots + c_n \lambda_n \mathbf{e}_n = \mathbf{0}. \tag{21}$$

Repeating the process gives

$$\begin{aligned} c_1 \lambda_1^2 \mathbf{e}_1 + c_2 \lambda_2^2 \mathbf{e}_2 + \cdots + c_n \lambda_n^2 \mathbf{e}_n &= \mathbf{0}, \\ &\vdots \\ c_1 \lambda_1^{n-1} \mathbf{e}_1 + c_2 \lambda_2^{n-1} \mathbf{e}_2 + \cdots + c_n \lambda_n^{n-1} \mathbf{e}_n &= \mathbf{0}. \end{aligned} \tag{22}$$

Expressing (20)–(22) in matrix form,[1]

$$
\begin{bmatrix}
1 & \cdots & 1 \\
\lambda_1 & \cdots & \lambda_n \\
\vdots & & \vdots \\
\lambda_1^{n-1} & \cdots & \lambda_n^{n-1}
\end{bmatrix}
\begin{bmatrix}
c_1 \mathbf{e}_1 \\
c_2 \mathbf{e}_2 \\
\vdots \\
c_n \mathbf{e}_n
\end{bmatrix}
=
\begin{bmatrix}
\mathbf{0} \\
\mathbf{0} \\
\vdots \\
\mathbf{0}
\end{bmatrix}.
\tag{23}
$$

The determinant of the coefficient matrix is a *Vandermonde determinant*, which (see Exercise 16 in Section 5.4) is nonzero if the λ_j's are distinct. Since the λ_j's are indeed distinct by assumption, (23) admits the unique trivial solution $c_1 \mathbf{e}_1 = \mathbf{0}$, $c_2 \mathbf{e}_2 = \mathbf{0}, \ldots, c_n \mathbf{e}_n = \mathbf{0}$, and since the $\mathbf{e}_j$'s are nonzero (because they are eigenvectors) it follows that $c_1 = c_2 = \cdots = c_n = 0$, so $\mathbf{e}_1, \ldots, \mathbf{e}_n$ are LI. ∎

From Theorems 10.5.1 and 10.5.2 we can draw the following conclusion.

THEOREM 10.5.3
Diagonalizability
If an $n \times n$ matrix has n distinct eigenvalues, then it is diagonalizable.

As usual, be careful not to read inverses into theorems when they are not stated. Specifically, Theorem 10.5.3 does *not* say that an $n \times n$ matrix is diagonalizable *if and only if* it has n distinct eigenvalues.

Consider an application.

E X A M P L E   2   *A Problem in Chemical Kinetics*.
We consider here a special class of chemical reactions known as *first-order reactions*. These reactions are governed by systems of linear, coupled, first-order ordinary differential equations. Specifically, suppose that $X_1, \ldots, X_n$ are the chemical names of n reacting species (elements or molecules), that $x_j(t)$ denotes the concentration of X_j (in suitable units) as a function of the time t, and that the *rate constant* for the conversion of X_i to X_j is the positive constant k_{ji}. For a two-component reaction, for example, denoted schematically in Fig. 3, this means that

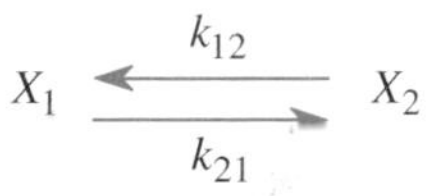

FIGURE 3
Two-component reaction.

$$
x_1' = -k_{21}x_1 + k_{12}x_2, \tag{24a}
$$
$$
x_2' = k_{21}x_1 - k_{12}x_2. \tag{24b}
$$

The first term on the right-hand side of (24a) accounts for the loss of X_1 due to the $X_1 \to X_2$ reaction; it is proportional to the concentration of X_1, namely x_1, and the constant of proportionality is the relevant rate constant k_{21}. The second term on the right-hand side of (24a) accounts for the rate of gain of X_1 due to the reverse reaction $X_2 \to X_1$. A similar accounting gives the terms in (24b).

The difficulty in solving (24), and similar systems for n-component reactions where $n > 2$, is due to the coupling. Equations (24) are coupled due to the $k_{12}x_2$ term in (24a) and the $k_{21}x_1$ term in (24b).

---

[1] Usually, the elements of our vectors are scalars, but in (23) they themselves are vectors. That is perfectly legitimate. After all, if we carry out the multiplication on the left-hand side of (23), the result does imply (20)–(22).

We could solve (24) by the method of elimination (Section 10.3) or by seeking exponential solutions and solving the resulting eigenvalue problem (Section 10.4), but let us solve, instead, by the method of diagonalization. In matrix form (24) is

$$\mathbf{x}' = \mathbf{A}\mathbf{x}, \quad \text{where} \quad \mathbf{A} = \begin{bmatrix} -k_{21} & k_{12} \\ k_{21} & -k_{12} \end{bmatrix}. \tag{25}$$

The eigenvalues and eigenvectors of $\mathbf{A}$ are readily found to be

$$\lambda_1 = 0, \quad \mathbf{e}_1 = \alpha \begin{bmatrix} k_{12} \\ k_{21} \end{bmatrix}; \qquad \lambda_2 = -(k_{12} + k_{21}), \quad \mathbf{e}_2 = \beta \begin{bmatrix} 1 \\ -1 \end{bmatrix}. \tag{26}$$

The λ_j's are distinct, because $k_{12} > 0$ and $k_{21} > 0$, so Theorem 10.5.3 guarantees that $\mathbf{A}$ is diagonalizable. Alternatively, observe that the $\mathbf{e}_j$'s are necessarily LI because for them to be LD we would need $k_{21} = -k_{12}$, which is impossible since $k_{12} > 0$ and $k_{21} > 0$. Their linear independence implies that $\mathbf{A}$ is diagonalizable, by Theorem 10.5.1.

Thus, if we set $\mathbf{x} = \mathbf{Q}\widetilde{\mathbf{x}}$, where (with $\alpha = \beta = 1$, say)

$$\mathbf{Q} = [\mathbf{e}_1, \mathbf{e}_2] = \begin{bmatrix} k_{12} & 1 \\ k_{21} & -1 \end{bmatrix}, \tag{27}$$

then the preceding analysis assures us (without having to work out the matrix product $\mathbf{Q}^{-1}\mathbf{A}\mathbf{Q}$) that

$$\widetilde{\mathbf{x}}' = \mathbf{Q}^{-1}\mathbf{A}\mathbf{Q}\widetilde{\mathbf{x}} = \mathbf{D}\widetilde{\mathbf{x}} = \begin{bmatrix} \lambda_1 & 0 \\ 0 & \lambda_2 \end{bmatrix} \widetilde{\mathbf{x}}. \tag{28}$$

Thus, we have the *uncoupled* system (which, of course, was our objective)

$$\begin{aligned} \widetilde{x}'_1 &= \lambda_1 \widetilde{x}_1, \\ \widetilde{x}'_2 &= \lambda_2 \widetilde{x}_2, \end{aligned} \tag{29}$$

the general solution of which is

$$\begin{aligned} \widetilde{x}_1 &= C_1 e^{\lambda_1 t} = C_1 e^{0t} = C_1, \\ \widetilde{x}_2 &= C_2 e^{\lambda_2 t} = C_2 e^{-(k_{12}+k_{21})t}. \end{aligned} \tag{30}$$

Finally, putting these expressions into $\mathbf{x} = \mathbf{Q}\widetilde{\mathbf{x}}$ recovers

$$\begin{bmatrix} x_1 \\ x_2 \end{bmatrix} = \begin{bmatrix} k_{12} & 1 \\ k_{21} & -1 \end{bmatrix} \begin{bmatrix} C_1 \\ C_2 e^{-(k_{12}+k_{21})t} \end{bmatrix} \tag{31}$$

or

$$\begin{aligned} x_1(t) &= C_1 k_{12} + C_2 e^{-(k_{12}+k_{21})t}, \\ x_2(t) &= C_1 k_{21} - C_2 e^{-(k_{12}+k_{21})t}. \end{aligned} \tag{32}$$

COMMENT 1. Here we have emphasized the mathematics rather than the chemistry and have assumed the rate constants to be known. A problem of importance to the chemist is the determination of those constants. Such determination normally involves a blend of the foregoing theory with suitable experiments.

COMMENT 2.   The numbering of the eigenvalues and eigenvectors is immaterial. For instance, we could just as well take $\lambda_1 = -(k_{12} + k_{21})$ and $\lambda_2 = 0$. The final result, (32), would be the same.   ■

Theorem 10.5.3 revealed that diagonalizability is the typical case, the generic case, because an nth-degree algebraic equation (namely, the characteristic equation of $\mathbf{A}$) typically has n distinct roots. Furthermore, every *symmetric* matrix is diagonalizable:

THEOREM 10.5.4
Symmetric Matrices
Every symmetric matrix is diagonalizable.

Proof.   Theorem 10.5.1 states that $\mathbf{A}$ is diagonalizable if and only if it has n LI eigenvectors, and Theorem 9.3.4 assures us that every $n \times n$ symmetric matrix has n orthogonal (and hence LI) eigenvectors.   ■

Note that the scaling of the eigenvectors in (14) is immaterial. For instance, in (27) we took $\alpha = \beta = 1$ but we could just as well have taken $\alpha = 38$ and $\beta = \sqrt{5}$, say. However, for a *symmetric matrix* $\mathbf{A}$ it is convenient to use the *normalized* eigenvectors of $\mathbf{A}$ to form its modal matrix $\mathbf{Q}$,

$$\mathbf{Q} = [\hat{\mathbf{e}}_1, \ldots, \hat{\mathbf{e}}_n], \tag{33}$$

because then

$$\mathbf{Q}^{\mathrm{T}}\mathbf{Q} = \begin{bmatrix} \hat{\mathbf{e}}_1^{\mathrm{T}} \\ \vdots \\ \hat{\mathbf{e}}_n^{\mathrm{T}} \end{bmatrix} [\hat{\mathbf{e}}_1, \ldots, \hat{\mathbf{e}}_n] = \begin{bmatrix} \hat{\mathbf{e}}_1^{\mathrm{T}}\hat{\mathbf{e}}_1 & \cdots & \hat{\mathbf{e}}_1^{\mathrm{T}}\hat{\mathbf{e}}_n \\ \vdots & & \vdots \\ \hat{\mathbf{e}}_n^{\mathrm{T}}\hat{\mathbf{e}}_1 & \cdots & \hat{\mathbf{e}}_n^{\mathrm{T}}\hat{\mathbf{e}}_n \end{bmatrix}$$

$$= \begin{bmatrix} \hat{\mathbf{e}}_1 \cdot \hat{\mathbf{e}}_1 & \cdots & \hat{\mathbf{e}}_1 \cdot \hat{\mathbf{e}}_n \\ \vdots & & \vdots \\ \hat{\mathbf{e}}_n \cdot \hat{\mathbf{e}}_1 & \cdots & \hat{\mathbf{e}}_n \cdot \hat{\mathbf{e}}_n \end{bmatrix} = \begin{bmatrix} 1 & \cdots & 0 \\ \vdots & \ddots & \vdots \\ 0 & \cdots & 1 \end{bmatrix} = \mathbf{I} \tag{34}$$

so that

$$\boxed{\mathbf{Q}^{-1} = \mathbf{Q}^{\mathrm{T}}.} \tag{35}$$

The property (35) is convenient because if we ever need the inverse of $\mathbf{Q}$, it is simply $\mathbf{Q}^{\mathrm{T}}$.

Be sure to understand each step in (34). $\mathbf{Q}$ starts out as an $n \times n$ matrix, but when we partition it into columns, as $[\hat{\mathbf{e}}_1, \ldots, \hat{\mathbf{e}}_n]$, it is then a $1 \times n$ matrix with elements $\hat{\mathbf{e}}_1, \ldots, \hat{\mathbf{e}}_n$. To form $\mathbf{Q}^{\mathrm{T}}$, we make the jth column of $\mathbf{Q}$, namely $\hat{\mathbf{e}}_j$, the jth row of $\mathbf{Q}^{\mathrm{T}}$, and to put it into row format we need to write it as $\hat{\mathbf{e}}_j^{\mathrm{T}}$ rather than $\hat{\mathbf{e}}_j$. Thus, working with the partitioned $\mathbf{Q}$ and $\mathbf{Q}^{\mathrm{T}}$ matrices, the product to the right of the first equal sign in (34) is an $n \times 1$ matrix times a $1 \times n$ matrix, which product gives the $n \times n$ matrix to the right of the second equal sign.

Understand also that (35) has nothing to do with the $\hat{\mathbf{e}}_j$'s being eigenvectors. The steps in (34) rely only on the fact that the columns of $\mathbf{Q}$ are ON. *Any* square matrix, the columns of which are ON, satisfies (35) and is called an **orthogonal matrix**.

Let us return to diagonalization. Note that if $\mathbf{A}$ is symmetric then, $\mathbf{Q}^{-1}\mathbf{A}\mathbf{Q} = \mathbf{D}$ is diagonal whether or not the columns of the modal matrix $\mathbf{Q}$ are normalized. However, let us agree (at least within this text) to always normalize them if $\mathbf{A}$ is symmetric, so as to have the property (35) if we need it.

We close with one more application.

EXAMPLE 3   *A Free-Vibration Problem*.

Consider a mass m constrained by two mechanical springs, of stiffnesses k_1 and k_2, as sketched in Fig. 4. Imagine Fig. 4 as a view looking down on the apparatus, which lies in a horizontal plane on a frictionless table. In the configuration shown, the springs are neither stretched nor compressed, and m is at rest in static equilibrium. However, if some initial displacement and/or velocity is imparted to m, some motion, no doubt vibrational, will result, and it is that motion that we wish to determine.

The first step in the formulation is to introduce a coordinate system. A reasonable choice is the Cartesian system shown in Fig. 4, with its origin at the equilibrium position of the mass m (which we regard as a "point mass").

If m is at some point x, y other than the origin, then one or both springs will be stretched or compressed and will exert forces $\mathbf{F}_1$ and $\mathbf{F}_2$ on m (Fig. 5). The magnitude of $\mathbf{F}_1$ is

$$\|\mathbf{F}_1\| = k_1 \text{ times the stretch in spring \#1}$$
$$= k_1 \left\{ \sqrt{[x-(-1)]^2 + (y-0)^2} - 1 \right\}$$
$$= k_1 \left\{ \sqrt{(x+1)^2 + y^2} - 1 \right\}. \tag{36}$$

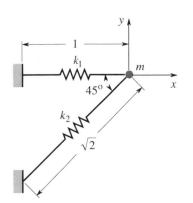

FIGURE 4
Mass-spring system;
view from above.

If we multiply this magnitude by a unit vector directed from (x, y) toward $(-1, 0)$, we will have $\mathbf{F}_1$. To obtain that unit vector, note that the vector from (x, y) to $(-1, 0)$ is $(-1-x)\hat{\mathbf{i}} + (0-y)\hat{\mathbf{j}}$, where $\hat{\mathbf{i}}, \hat{\mathbf{j}}$ are unit base vectors in the x, y directions, respectively. Normalizing that vector gives the desired unit vector

$$\hat{\mathbf{F}}_1 = -\frac{(1+x)\hat{\mathbf{i}} + y\hat{\mathbf{j}}}{\sqrt{(1+x)^2 + y^2}}, \tag{37}$$

so

$$\mathbf{F}_1 = \|\mathbf{F}_1\|\,\hat{\mathbf{F}}_1 = -k_1 \left[\frac{\sqrt{(x+1)^2 + y^2} - 1}{\sqrt{(x+1)^2 + y^2}} \right] \left[(x+1)\hat{\mathbf{i}} + y\hat{\mathbf{j}} \right]. \tag{38}$$

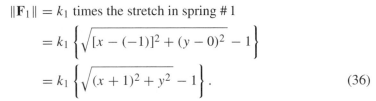

In like manner, we find that

$$\mathbf{F}_2 = \|\mathbf{F}_2\|\,\hat{\mathbf{F}}_2 = -k_2 \left[\frac{\sqrt{(x+1)^2 + (y+1)^2} - \sqrt{2}}{\sqrt{(x+1)^2 + (y+1)^2}} \right] \left[(x+1)\hat{\mathbf{i}} + (y+1)\hat{\mathbf{j}} \right]. \tag{39}$$

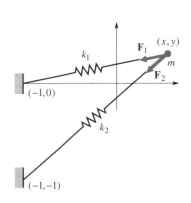

FIGURE 5
The forces on m.

According to Newton's second law,

$$mx'' = F_x \quad \text{and} \quad my'' = F_y, \tag{40}$$

where F_x is the sum of the x components of $\mathbf{F}_1$ and $\mathbf{F}_2$, and F_y is the sum of the y components of $\mathbf{F}_1$ and $\mathbf{F}_2$. Thus, the governing equations of motion are

$$mx'' = -k_1 \left[\frac{\sqrt{(x+1)^2 + y^2} - 1}{\sqrt{(x+1)^2 + y^2}} \right] (x+1)$$

$$\quad - k_2 \left[\frac{\sqrt{(x+1)^2 + (y+1)^2} - \sqrt{2}}{\sqrt{(x+1)^2 + (y+1)^2}} \right] (x+1), \tag{41a}$$

$$my'' = -k_1 \left[\frac{\sqrt{(x+1)^2 + y^2} - 1}{\sqrt{(x+1)^2 + y^2}} \right] y$$

$$\quad - k_2 \left[\frac{\sqrt{(x+1)^2 + (y+1)^2} - \sqrt{2}}{\sqrt{(x+1)^2 + (y+1)^2}} \right] (y+1). \tag{41b}$$

The latter coupled nonlinear differential equations are clearly quite intractable analytically. Two possibilities present themselves. First, if we assign numerical values to m, k_1, k_2, $x(0)$, $y(0)$, $x'(0)$, $y'(0)$, then we can generate $x(t)$ and $y(t)$ by one of the numerical methods studied in Chapter 12 (such as fourth-order Runge-Kutta integration) or by using computer software such as *Maple*.

Second, we can limit our attention to small motions, motions that remain close to the equilibrium position at the origin: $|x| \ll 1$ and $|y| \ll 1$. In that case we can simplify (41) in essentially the same way that we can simplify the nonlinear differential equation

$$x'' + \frac{g}{l} \sin x = 0 \tag{42}$$

governing the motion of a pendulum (Fig. 6): for small motions, near the equilibrium point $x = 0$,

$$\sin x = x - \frac{x^3}{3!} + \frac{x^5}{5!} - \cdots \quad \text{[Taylor series (1) with } a = 0\text{]}$$

$$\approx x \quad \quad \text{(linearized)}$$

for $|x| \ll 1$ so (42) can be approximated by the simple (linearized) equation

$$x'' + \frac{g}{l} x = 0. \tag{43}$$

We will follow the same steps for (41), but instead of using the linearized Taylor series in one variable we will use the linearized Taylor series in two variables, given by (6) with $a = b = 0$, to simplify the right-hand sides of (41a) and (41b). That step gives, as the *linearized* version of (41),

$$mx'' = -\left(k_1 + \frac{k_2}{2}\right) x - \frac{k_2}{2} y, \tag{44a}$$

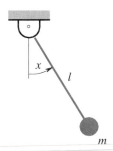

FIGURE 6
Pendulum.

$$my'' = -\frac{k_2}{2}x - \frac{k_2}{2}y \tag{44b}$$

or

$$\mathbf{x}'' + \mathbf{A}\mathbf{x} = \mathbf{0}, \tag{45}$$

where

$$\mathbf{x} = \begin{bmatrix} x(t) \\ y(t) \end{bmatrix}, \quad \mathbf{A} = \begin{bmatrix} \dfrac{2k_1 + k_2}{2m} & \dfrac{k_2}{2m} \\ \dfrac{k_2}{2m} & \dfrac{k_2}{2m} \end{bmatrix}. \tag{46}$$

Let us solve (45) by diagonalization. It is striking that $\mathbf{A}$ is symmetric even though there is no "physical" symmetry to be seen in Fig. 4. For definiteness, let us set

$$m = 1, \; k_1 = 3, \; k_2 = 4. \tag{47}$$

Then the eigenvalues and normalized eigenvectors of $\mathbf{A}$ are

$$\lambda_1 = 1, \; \hat{\mathbf{e}}_1 = \frac{1}{\sqrt{5}}\begin{bmatrix} 1 \\ -2 \end{bmatrix}; \quad \lambda_2 = 6, \; \hat{\mathbf{e}}_2 = \frac{1}{\sqrt{5}}\begin{bmatrix} 2 \\ 1 \end{bmatrix}. \tag{48}$$

With

$$\mathbf{Q} = [\hat{\mathbf{e}}_1, \hat{\mathbf{e}}_2] = \begin{bmatrix} 1/\sqrt{5} & 2/\sqrt{5} \\ -2/\sqrt{5} & 1/\sqrt{5} \end{bmatrix}, \tag{49}$$

set $\mathbf{x} = \mathbf{Q}\widetilde{\mathbf{x}}$ in (45). Thus,

$$\mathbf{Q}\widetilde{\mathbf{x}}'' + \mathbf{A}\mathbf{Q}\widetilde{\mathbf{x}} = \mathbf{0} \tag{50}$$

and hence

$$\widetilde{\mathbf{x}}'' + \mathbf{Q}^{-1}\mathbf{A}\mathbf{Q}\widetilde{\mathbf{x}} = \mathbf{0} \tag{51}$$

or

$$\widetilde{\mathbf{x}}'' + \mathbf{D}\widetilde{\mathbf{x}} = \mathbf{0}, \tag{52}$$

where

$$\mathbf{D} = \begin{bmatrix} \lambda_1 & 0 \\ 0 & \lambda_2 \end{bmatrix} = \begin{bmatrix} 1 & 0 \\ 0 & 6 \end{bmatrix}. \tag{53}$$

In scalar form, (52) gives the simple *uncoupled* equations

$$\begin{aligned} \widetilde{x}'' + \widetilde{x} &= 0, \\ \widetilde{y}'' + 6\widetilde{y} &= 0 \end{aligned} \tag{54}$$

with general solution [expressed in the $A \sin(\omega t + \phi)$ form]

$$\begin{aligned} \widetilde{x} &= A_1 \sin(t + \phi_1), \\ \widetilde{y} &= A_2 \sin(\sqrt{6}\,t + \phi_2), \end{aligned} \tag{55}$$

where the amplitudes A_1, A_2 and phase angles ϕ_1, ϕ_2 are arbitrary constants. To recover the original x, y variables, write

$$\begin{bmatrix} x \\ y \end{bmatrix} = \mathbf{Q}\widetilde{\mathbf{x}} = \begin{bmatrix} 1/\sqrt{5} & 2/\sqrt{5} \\ -2/\sqrt{5} & 1/\sqrt{5} \end{bmatrix}\begin{bmatrix} \widetilde{x} \\ \widetilde{y} \end{bmatrix}$$

$$= A_1 \begin{bmatrix} \dfrac{1}{\sqrt{5}} \sin(t + \phi_1) \\[2mm] -\dfrac{2}{\sqrt{5}} \sin(t + \phi_1) \end{bmatrix} + A_2 \begin{bmatrix} \dfrac{2}{\sqrt{5}} \sin(\sqrt{6}\,t + \phi_2) \\[2mm] \dfrac{1}{\sqrt{5}} \sin(\sqrt{6}\,t + \phi_2) \end{bmatrix}$$

$$= A_1 \hat{\mathbf{e}}_1 \sin(t + \phi_1) + A_2 \hat{\mathbf{e}}_2 \sin(\sqrt{6}\,t + \phi_2) \tag{56}$$

or, in scalar form,

$$x(t) = C_1 \sin(t + \phi_1) + 2C_2 \sin(\sqrt{6}\,t + \phi_2),$$
$$y(t) = -2C_1 \sin(t + \phi_1) + C_2 \sin(\sqrt{6}\,t + \phi_2), \tag{57}$$

where C_1 is $A_1/\sqrt{5}$ and C_2 is $A_2/\sqrt{5}$ for brevity.

COMMENT 1.   It is seen from (56) that the general solution of (45) is a linear combination of two orthogonal modes, as in Example 5 of Section 10.4. The low mode is a vibration along the $\hat{\mathbf{e}}_1$ direction, at a frequency that is the square root of λ_1, and the high mode is a vibration along the $\hat{\mathbf{e}}_2$ direction, at a frequency that is the square root of λ_2, as summarized in Fig. 7. In this example the orthogonality of the modes is "physical" since the low-mode and high-mode motions are perpendicular to each other; in Example 5 of Section 10.4, the modes were orthogonal but not physically perpendicular (see Comment 1 in that example).

COMMENT 2.   Observe that the high-frequency motion along the $\tilde{y}$ axis (Fig. 7) is along the direction of maximum stiffness of the two-spring system and the low-frequency motion along the direction of the $\tilde{x}$ axis is along the direction of minimum stiffness.

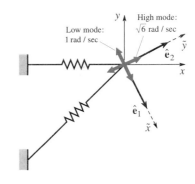

FIGURE 7
The orthogonal modes.

COMMENT 3.   Why do we show the positive $\tilde{x}$ and $\tilde{y}$ coordinate axes as being in the $\hat{\mathbf{e}}_1$ and $\hat{\mathbf{e}}_2$ directions, respectively, in Fig. 7? Because if we set $\tilde{x} = 1$ and $\tilde{y} = 0$ (as a "marker" on the positive $\tilde{x}$ axis) in

$$\begin{bmatrix} x \\ y \end{bmatrix} = \begin{bmatrix} 1/\sqrt{5} & 2/\sqrt{5} \\ -2/\sqrt{5} & 1/\sqrt{5} \end{bmatrix} \begin{bmatrix} \tilde{x} \\ \tilde{y} \end{bmatrix}$$

we get $[x, y]^T = [1/\sqrt{5}, -2/\sqrt{5}]^T = \hat{\mathbf{e}}_1$, and if we set $\tilde{x} = 0$ and $\tilde{y} = 1$ (as a "marker" on the positive $\tilde{y}$ axis) we get $[x, y]^T = [2/\sqrt{5}, 1/\sqrt{5}]^T = \hat{\mathbf{e}}_2$. In fact, the effect of the change of variables $\mathbf{x} = \mathbf{Q}\tilde{\mathbf{x}}$, where $\mathbf{Q}$ is an *orthogonal* matrix with its determinant equal to $+1$, as here, is a pure *rotation* of axes. Thus, we have the vivid visual image of the elements of the coupling matrix varying as we rotate the Cartesian coordinate system (somewhat like looking into a kaleidoscope), until the off-diagonal terms become zero and the equations uncouple. See Exercise 10.

COMMENT 4.   In principle, it would have been best to choose the $\tilde{x}$, $\tilde{y}$ coordinate system in the first place, but its orientation was not known. Thus, we chose any x, y system, to get started, and then used the method of diagonalization to find the optimal $\tilde{x}$, $\tilde{y}$ coordinate system.   ∎

Closure. From a mathematical viewpoint, this section is about finding an $n \times n$ matrix $\mathbf{Q}$, given an $n \times n$ matrix $\mathbf{A}$, such that $\mathbf{Q}^{-1}\mathbf{AQ} = \mathbf{D}$ is diagonal. We find that in the generic case $\mathbf{A}$ is diagonalizable: It is diagonalizable if and only if it has n LI eigenvectors, and it is diagonalizable if it has n distinct eigenvalues or is symmetric. $\mathbf{Q}$ can be made up of columns which are the eigenvectors of $\mathbf{A}$, and the diagonal elements of $\mathbf{D}$ are the corresponding eigenvalues. In the event that $\mathbf{A}$ is symmetric, we suggest always normalizing the eigenvectors that are the columns of $\mathbf{Q}$, so that $\mathbf{Q}$ will admit the useful property $\mathbf{Q}^{-1} = \mathbf{Q}^{\mathrm{T}}$.

From an applications standpoint, we have used diagonalization to uncouple systems of coupled differential equations, but other applications are to be found in the exercises and in subsequent sections.

It turns out that even if an $n \times n$ matrix $\mathbf{A}$ cannot be diagonalized, it can be triangularized. That is, a generalized modal matrix $\mathbf{P}$ can be found for $\mathbf{A}$ so that

$$\mathbf{P}^{-1}\mathbf{AP} = \mathbf{J} \tag{58}$$

is triangular. Called the **Jordan normal form**, or simply the **Jordan form** for $\mathbf{A}$, $\mathbf{J}$ is upper triangular, with zeros above its main diagonal except for 1's immediately above one or more diagonal elements. This case is discussed in the optional Section 10.6.

EXERCISES 10.5

1. Diagonalize each of the given $\mathbf{A}$ matrices. That is, determine matrices $\mathbf{Q}$ and $\mathbf{D}$ such that $\mathbf{Q}^{-1}\mathbf{AQ} = \mathbf{D}$ is diagonal. Also, work out $\mathbf{Q}^{-1}$ and verify that $\mathbf{Q}^{-1}\mathbf{AQ}$ is diagonal and that its diagonal elements are the eigenvalues of $\mathbf{A}$. If $\mathbf{A}$ is not diagonalizable, state that and give the reason.

(<u>a</u>) $\begin{bmatrix} 2 & -3 \\ 0 & 0 \end{bmatrix}$

(**b**) $\begin{bmatrix} 2 & 4 \\ -1 & -2 \end{bmatrix}$

(c) $\begin{bmatrix} 1 & 0 \\ 1 & 0 \end{bmatrix}$

(**d**) $\begin{bmatrix} 0 & 0 & 1 \\ 0 & 0 & 1 \\ 1 & 1 & 1 \end{bmatrix}$

(e) $\begin{bmatrix} 0 & 1 & 1 \\ 0 & 0 & 1 \\ 0 & 0 & 0 \end{bmatrix}$

(**f**) $\begin{bmatrix} 2 & 0 & 0 \\ 0 & 1 & 1 \\ 0 & 1 & 1 \end{bmatrix}$

(<u>g</u>) $\begin{bmatrix} 2 & 1 & -1 \\ 1 & 4 & 3 \\ -1 & 3 & 4 \end{bmatrix}$

(**h**) $\begin{bmatrix} 0 & 1 & 0 \\ 1 & 0 & 0 \\ 0 & 0 & 0 \end{bmatrix}$

(**i**) $\begin{bmatrix} 4 & 0 & 0 & 0 \\ 0 & 1 & 1 & 0 \\ 0 & 1 & 1 & 0 \\ 0 & 0 & 0 & 2 \end{bmatrix}$

(<u>j</u>) $\begin{bmatrix} 1 & 1 & 1 & 1 \\ 1 & 1 & 1 & 1 \\ 1 & 1 & 1 & 1 \\ 1 & 1 & 1 & 1 \end{bmatrix}$

2. Use the method of diagonalization to obtain the general solution of the given system of differential equations, where primes denote d/dt.

(<u>a</u>) $\begin{aligned} x' &= x + y \\ y' &= x + y \end{aligned}$

(**b**) $\begin{aligned} x' &= x + 4y \\ y' &= x + y \end{aligned}$

(c) $\begin{aligned} x'' &= 2x + 4y \\ y'' &= x - y \end{aligned}$

(**d**) $\begin{aligned} x' + 2x + y &= 0 \\ y' + x + 2y + z &= 0 \\ z' + y + 2z &= 0 \end{aligned}$

(e) $\begin{aligned} x' &= 4x + y + 3z \\ y' &= x - z \\ z' &= 3x - y + 4z \end{aligned}$

(**f**) $\begin{aligned} x'' &= -y + z \\ y'' &= -x - 3y - 4z \\ z'' &= x - 4y - 3z \end{aligned}$

3. Can a singular matrix be diagonalized? Explain.

4. We see from Fig. 7 that the line of action of the high mode falls between the two springs. Show that that situation holds for all possible combinations of stiffnesses k_1 and k_2.

5. Determine (as in Example 3) the natural frequencies and mode shapes for each of the systems shown below. You

need carry only three or four significant figures.   Each spring is of unit length.

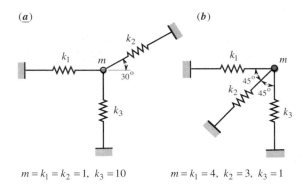

(*a*)

(*b*)

$m = k_1 = k_2 = 1, \ k_3 = 10$ $m = k_1 = 4, \ k_2 = 3, \ k_3 = 1$

6. Show why the second equality in (17) is true.

7. (*Application to exponentiation*) Diagonalization can be helpful in raising a square matrix to a large power. Specifically, show that if $\mathbf{A}$ is diagonalizable so that $\mathbf{Q}^{-1}\mathbf{A}\mathbf{Q} = \mathbf{D}$, then

$$\mathbf{A}^m = \mathbf{Q}\mathbf{D}^m\mathbf{Q}^{-1}, \qquad (7.1)$$

the advantage being that $\mathbf{D}^m$ is simply

$$\mathbf{D}^m = \begin{bmatrix} \lambda_1 & \cdots & 0 \\ \vdots & \ddots & \vdots \\ 0 & \cdots & \lambda_n \end{bmatrix}^m = \begin{bmatrix} \lambda_1^m & \cdots & 0 \\ \vdots & \ddots & \vdots \\ 0 & \cdots & \lambda_n^m \end{bmatrix}. \qquad (7.2)$$

8. Use (7.1), above, to evaluate $\mathbf{A}^{1000}$, where

(**a**) $\mathbf{A} = \begin{bmatrix} 0 & 2 & 2 \\ 2 & 0 & 2 \\ 2 & 2 & 0 \end{bmatrix}$

(**b**) $\mathbf{A} = \begin{bmatrix} 2 & 2 & 1 \\ 1 & 3 & 1 \\ 1 & 2 & 2 \end{bmatrix}$

(**c**) $\mathbf{A} = \begin{bmatrix} 1 & 0 & 1 \\ -1 & -1 & 0 \\ 2 & 0 & 2 \end{bmatrix}$

(**d**) $\mathbf{A} = \begin{bmatrix} 9 & 10 & 9 \\ -6 & -5 & -3 \\ 1 & 0 & -1 \end{bmatrix}$

9. (*Application to principal inertias and principal axes*) Two vectors of importance in studying the dynamics of a rigid body $\mathcal{B}$ are the *moment of momentum* $\mathbf{H}_p$ and the *angular velocity* $\boldsymbol{\omega}$, of $\mathcal{B}$. These are related according to $\mathbf{H}_p = \boldsymbol{\mathcal{I}}\boldsymbol{\omega}$,

where $\boldsymbol{\mathcal{I}}$ is the *inertia matrix*. Written out, we have

$$\begin{bmatrix} (H_p)_x \\ (H_p)_y \\ (H_p)_z \end{bmatrix} = \begin{bmatrix} I_{xx} & -I_{xy} & -I_{xz} \\ -I_{yx} & I_{yy} & -I_{yz} \\ -I_{zx} & -I_{zy} & I_{zz} \end{bmatrix} \begin{bmatrix} \omega_x \\ \omega_y \\ \omega_z \end{bmatrix}, \qquad (9.1)$$

where P is the origin of a Cartesian x, y, z coordinate system (see the figure), and

$$I_{xx} = \int_{\mathcal{B}} \left(y^2 + z^2 \right) dm, \quad I_{xy} = I_{yx} = \int_{\mathcal{B}} xy \, dm$$

$$I_{yy} = \int_{\mathcal{B}} \left(x^2 + z^2 \right) dm, \quad I_{xz} = I_{zx} = \int_{\mathcal{B}} xz \, dm$$

$$I_{zz} = \int_{\mathcal{B}} \left(x^2 + y^2 \right) dm, \quad I_{yz} = I_{zy} = \int_{\mathcal{B}} yz \, dm. \qquad (9.2)$$

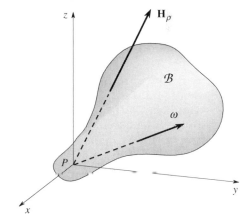

I_{xx}, I_{yy}, I_{zz} are known as the *moments of inertia* of $\mathcal{B}$ about the x, y, z axes, respectively, and I_{xy}, I_{xz}, I_{yz} are the *products of inertia* of $\mathcal{B}$; dm in (9.2) is "d(mass)." Now, the relation (9.1) and hence the subsequent dynamic analysis (which will be of no concern here) will be simplest if $\boldsymbol{\mathcal{I}}$ is diagonal, i.e., if all of the products of inertia are zero. In general, it is too difficult to see, by inspection, how to orient the coordinate axes to achieve this result. Instead, we go ahead and choose *some* x, y, z reference frame, compute the nine inertia components, and then rotate to a new Cartesian $\tilde{x}, \tilde{y}, \tilde{z}$ frame so as to diagonalize $\boldsymbol{\mathcal{I}}$. That is, if $\mathbf{x} = \mathbf{Q}\tilde{\mathbf{x}}$, where $\mathbf{x} = [x, y, z]^{\mathrm{T}}$ and $\tilde{\mathbf{x}} = [\tilde{x}, \tilde{y}, \tilde{z}]^{\mathrm{T}}$, then

$$\mathbf{H}_p = \mathbf{Q}\tilde{\mathbf{H}}_p \qquad \text{and} \qquad \boldsymbol{\omega} = \mathbf{Q}\tilde{\boldsymbol{\omega}}$$

so that $\mathbf{H}_p = \boldsymbol{\mathcal{I}}\boldsymbol{\omega}$ becomes

$$\mathbf{Q}\tilde{\mathbf{H}}_p = \boldsymbol{\mathcal{I}}\mathbf{Q}\tilde{\boldsymbol{\omega}} \quad \text{and} \quad \tilde{\mathbf{H}}_p = \left(\mathbf{Q}^{-1}\boldsymbol{\mathcal{I}}\mathbf{Q} \right)\tilde{\boldsymbol{\omega}}, \qquad (9.3)$$

where $\mathbf{Q}^{-1}\mathcal{I}\mathbf{Q} \equiv \tilde{\mathcal{I}}$ is diagonal. That such diagonalization is possible follows from the fact that $\mathcal{I}$ is *symmetric* since $I_{xy} = I_{yx}$, $I_{xz} = I_{zx}$, and $I_{yz} = I_{zy}$. $I_{\tilde{x}\tilde{x}}$, $I_{\tilde{y}\tilde{y}}$, and $I_{\tilde{z}\tilde{z}}$ are called the **principal inertias** of $\mathcal{B}$ (with respect to coordinates with origin at P), and the $\tilde{x}, \tilde{y}, \tilde{z}$ axes are called the **principal axes**. We finally state the problem: Compute the principal inertias and determine the principal axes for each of the following bodies; sketch the principal axes. In each case $\mathcal{B}$ can be assumed, for simplicity, to be infinitely thin in the z direction, with mass density σ mass units per unit area.

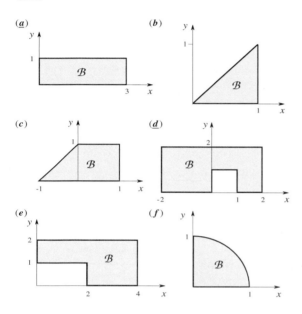

(*a*)

(*b*)

(*c*)

(*d*)

(*e*)

(*f*)

10. (*Rotation in the plane*) In Example 3 we changed variables from the original Cartesian x, y system to new $\tilde{x}$, $\tilde{y}$ coordinates according to the relation $\mathbf{x} = \mathbf{Q}\tilde{\mathbf{x}}$, where $\mathbf{x} = [x, y]^T$ and $\tilde{\mathbf{x}} = [\tilde{x}, \tilde{y}]^T$. We found that the new coordinate system, $\tilde{x}$ and $\tilde{y}$, was also Cartesian and that the change of variables $\mathbf{x} = \mathbf{Q}\tilde{\mathbf{x}}$ corresponded, geometrically, to a rotation of coordinate axes, as can be seen from Fig. 7.

(a) Show that if we take $\mathbf{Q} = [\hat{\mathbf{e}}_2, \hat{\mathbf{e}}_1]$ instead of $\mathbf{Q} = [\hat{\mathbf{e}}_1, \hat{\mathbf{e}}_2]$, which is legitimate since the numbering of the eigenvalues and eigenvectors is arbitrary, then the new $\tilde{x}, \tilde{y}$ coordinate system will be as shown below. Observe that in this case the $\tilde{x}, \tilde{y}$ system can*not* be obtained from the x, y system by a rotation, but by both a rotation (about a z axis normal to the plane) and a reflection (about the $\tilde{x}$ or $\tilde{y}$ axis).

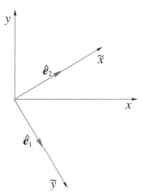

(**b**) More generally, consider any rotation (left-hand figure).

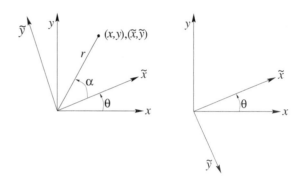

Then

$$x = r\cos(\theta + \alpha) = r\cos\theta\cos\alpha - r\sin\theta\sin\alpha,$$
$$y = r\sin(\theta + \alpha) = r\sin\theta\cos\alpha + r\cos\theta\sin\alpha$$

or, since $r\cos\alpha$ is $\tilde{x}$ and $r\sin\alpha$ is $\tilde{y}$,

$$\begin{bmatrix} x \\ y \end{bmatrix} = \begin{bmatrix} \cos\theta & -\sin\theta \\ \sin\theta & \cos\theta \end{bmatrix} \begin{bmatrix} \tilde{x} \\ \tilde{y} \end{bmatrix}. \qquad (10.1)$$

Comparing the transformation matrix in (10.1) with the $\mathbf{Q}$ matrix given by (49), determine θ (which result should agree with the θ observed in Fig. 7).

(**c**) Whereas the transformation (10.1) corresponds to the rotation shown in the left-hand figure (above), show that the transformation

$$\begin{bmatrix} x \\ y \end{bmatrix} = \begin{bmatrix} \cos\theta & \sin\theta \\ \sin\theta & -\cos\theta \end{bmatrix} \begin{bmatrix} \tilde{x} \\ \tilde{y} \end{bmatrix} \qquad (10.2)$$

corresponds to the **rotation plus reflection** shown in the right-hand figure. NOTE: Observe that the determinant of the transformation matrix is $+1$ in (10.1) and -1 in (10.2). For the $\mathbf{Q}$ matrix given by (49) $\det \mathbf{Q} = +1$, so that transformation effects a rotation only.

(d) For the transformation

$$\begin{bmatrix} x \\ y \end{bmatrix} = \begin{bmatrix} 2 & 1 \\ 1 & 1 \end{bmatrix} \begin{bmatrix} \tilde{x} \\ \tilde{y} \end{bmatrix},$$

which is **not** of the form (10.1) or (10.2), give a labeled diagram of the x, y and $\tilde{x}$, $\tilde{y}$ coordinate axes. That is, if the original x, y coordinates are Cartesian, how are the $\tilde{x}$, $\tilde{y}$ axes oriented relative to the x, y axes?

11. (*Proof of Cayley-Hamilton theorem*) Prove the Cayley-Hamilton theorem, which was stated in Exercise 18 of Section 9.2, for the case where $\mathbf{A}$ is diagonalizable. HINT: Beginning with $\mathbf{Q}^{-1}\mathbf{A}\mathbf{Q} = \mathbf{D}$, show that $\mathbf{A} = \mathbf{Q}\mathbf{D}\mathbf{Q}^{-1}$, $\mathbf{A}^2 = \mathbf{Q}\mathbf{D}^2\mathbf{Q}^{-1}$, and so on.

10.6 TRIANGULARIZATION. (OPTIONAL)

Recall that an $n \times n$ matrix $\mathbf{A}$ is diagonalizable if there exists a $\mathbf{Q}$ matrix such that $\mathbf{Q}^{-1}\mathbf{A}\mathbf{Q}$ is diagonal. For $\mathbf{A}$ to be diagonalizable, a necessary and sufficient condition is that it have n LI eigenvectors (Theorem 10.5.1), and a sufficient condition is that it have n distinct eigenvalues (Theorem 10.5.3). Typically, a given $n \times n$ matrix will be diagonalizable since, typically, an nth-degree algebraic equation (the characteristic equation of $\mathbf{A}$) will have n distinct roots (the eigenvalues of $\mathbf{A}$). It turns out that even if $\mathbf{A}$ cannot be diagonalized it can at least be triangularized, which is "almost as nice." That is, given any $n \times n$ matrix $\mathbf{A}$ there always exists a $\mathbf{P}$ matrix such that

$$\boxed{\mathbf{P}^{-1}\mathbf{A}\mathbf{P} = \mathbf{J}} \tag{1}$$

is triangular, upper triangular to be precise. We use the letter $\mathbf{J}$ because that matrix is called a **Jordan form** for $\mathbf{A}$, after *Camille Jordan* (1838–1922).[1]

Suppose that $\mathbf{A}$ is *not* diagonalizable. For that to be true $\mathbf{A}$ must have at least one repeated eigenvalue with a corresponding eigenspace that is of lower dimension than the multiplicity of that eigenvalue.

To illustrate, consider

$$\mathbf{A} = \begin{bmatrix} 2 & -1 & 2 & 0 \\ 0 & 3 & -1 & 0 \\ 0 & 1 & 1 & 0 \\ 0 & 1 & -3 & 5 \end{bmatrix}. \tag{2}$$

Calculation (Exercise 1) reveals that

$$\lambda_1 = \lambda_2 = \lambda_3 = 2, \qquad \mathbf{e}_1 = \alpha\,[1, 0, 0, 0]^{\mathrm{T}}; \tag{3a}$$

$$\lambda_4 = 5, \qquad \mathbf{e}_4 = \beta\,[0, 0, 0, 1]^{\mathrm{T}}. \tag{3b}$$

For definiteness, let us set $\alpha = 1$ and $\beta = 1$. In this case the eigenvalue of multiplicity three, $\lambda = 2$, contributes only one LI eigenvector, instead of three, so we have only two LI eigenvectors, $\mathbf{e}_1$ and $\mathbf{e}_4$, instead of the four that are needed for $\mathbf{A}$ to be diagonalizable. Thus, this $\mathbf{A}$ is not diagonalizable.

---

[1]This is not the same Jordan as in Gauss-Jordan reduction; that was Wilhelm Jordan (1842–1839).

But we can find so-called *generalized eigenvectors* $\mathbf{e}_2$ and $\mathbf{e}_3$, associated with $\lambda = 2$, such that $\mathbf{e}_1, \mathbf{e}_2, \mathbf{e}_3, \mathbf{e}_4$ are LI. Noting that $\mathbf{e}_1$ satisfies

$$(\mathbf{A} - \lambda_1 \mathbf{I})\,\mathbf{e}_1 = \mathbf{0}, \tag{4}$$

the idea is to introduce $\mathbf{e}_2$ and $\mathbf{e}_3$ so as to satisfy the conditions

$$(\mathbf{A} - \lambda_1 \mathbf{I})\,\mathbf{e}_2 = \mathbf{e}_1, \tag{5}$$
$$(\mathbf{A} - \lambda_1 \mathbf{I})\,\mathbf{e}_3 = \mathbf{e}_2. \tag{6}$$

With $\mathbf{e}_1$ known from (4), solve (5) for $\mathbf{e}_2$; then, put that $\mathbf{e}_2$ into (6) and solve (6) for $\mathbf{e}_3$.

Since we do not wish to be distracted by those calculations right now, let us suppose that $\mathbf{e}_2$ and $\mathbf{e}_3$ are in hand. We will show that $\mathbf{e}_1, \mathbf{e}_2, \mathbf{e}_3, \mathbf{e}_4$ are necessarily LI. For suppose that

$$\alpha_1 \mathbf{e}_1 + \alpha_2 \mathbf{e}_2 + \alpha_3 \mathbf{e}_3 + \alpha_4 \mathbf{e}_4 = \mathbf{0}. \tag{7}$$

Multiplying each term in (7) by $\mathbf{A} - \lambda_1 \mathbf{I}$ and recalling (4)–(6) and the fact that $\mathbf{A}\mathbf{e}_4 = \lambda_4 \mathbf{e}_4$, we obtain

$$\mathbf{0} + \alpha_2 \mathbf{e}_1 + \alpha_3 \mathbf{e}_2 + \alpha_4\,(\lambda_4 - \lambda_1)\,\mathbf{e}_4 = \mathbf{0}. \tag{8}$$

Repeating the process, multiply (8) by $\mathbf{A} - \lambda_1 \mathbf{I}$ and obtain

$$\mathbf{0} + \mathbf{0} + \alpha_3 \mathbf{e}_1 + \alpha_4\,(\lambda_4 - \lambda_1)^2\,\mathbf{e}_4 = \mathbf{0}. \tag{9}$$

Repeating the process once more gives

$$\mathbf{0} + \mathbf{0} + \mathbf{0} + \alpha_4\,(\lambda_4 - \lambda_1)^3\,\mathbf{e}_4 = \mathbf{0}. \tag{10}$$

Now $\mathbf{e}_4 \neq \mathbf{0}$ (because it is an eigenvector) and $\lambda_4 - \lambda_1 \neq 0$, so it follows from (10) that $\alpha_4 = 0$. With $\alpha_4 = 0$, (9) gives $\alpha_3 = 0$. Then (8) gives $\alpha_2 = 0$ and (7) gives $\alpha_1 = 0$. Since the α_j's are necessarily zero in (7) it follows that $\mathbf{e}_1, \ldots, \mathbf{e}_4$ are LI, as claimed.

Next, using these vectors as columns, form the matrix

$$\mathbf{P} = [\mathbf{e}_1, \mathbf{e}_2, \mathbf{e}_3, \mathbf{e}_4]. \tag{11}$$

Since its columns are the eigenvectors $\mathbf{e}_1, \mathbf{e}_4$ and the generalized eigenvectors $\mathbf{e}_2$ and $\mathbf{e}_3$, we refer to $\mathbf{P}$ as a *generalized modal matrix* associated with $\mathbf{A}$. We plan to show, next, that although $\mathbf{P}^{-1}\mathbf{A}\mathbf{P}$ is not diagonal, it is *almost* diagonal, namely, triangular.

First, observe that $\mathbf{P}^{-1}$ does exist since the columns of $\mathbf{P}$ (namely, $\mathbf{e}_1, \ldots, \mathbf{e}_4$) are LI. Let us denote the rows of $\mathbf{P}^{-1}$ as $\mathbf{r}_1, \ldots, \mathbf{r}_4$. Since $\mathbf{P}^{-1}\mathbf{P} = \mathbf{I}$, it follows that[1]

$$\mathbf{r}_i \mathbf{e}_j = \delta_{ij}, \tag{12}$$

where δ_{ij} is the Kronecker delta.

---

[1] Note that $\mathbf{r}_i \mathbf{e}_j$ in (12) is actually a dot product but we don't include a dot between the $\mathbf{r}_i$ and the $\mathbf{e}_j$ because $\mathbf{r}_i$ is in row format and $\mathbf{e}_j$ is in column format; hence, the ordinary matrix product $\mathbf{r}_i \mathbf{e}_j$ is in fact a dot product of the column vectors $\mathbf{r}_i^{\mathrm{T}}$ and $\mathbf{e}_j$.

We are now ready to show that $\mathbf{P}^{-1}\mathbf{AP}$ is almost diagonal:

$$\mathbf{P}^{-1}\mathbf{AP} = \mathbf{P}^{-1}\mathbf{A}\,[\mathbf{e}_1, \mathbf{e}_2, \mathbf{e}_3, \mathbf{e}_4]$$

$$= \mathbf{P}^{-1}\,[\mathbf{Ae}_1, \mathbf{Ae}_2, \mathbf{Ae}_3, \mathbf{Ae}_4]$$

$$= \mathbf{P}^{-1}\,[\lambda_1\mathbf{e}_1, \mathbf{e}_1 + \lambda_1\mathbf{e}_2, \mathbf{e}_2 + \lambda_1\mathbf{e}_3, \lambda_4\mathbf{e}_4]$$

$$= \begin{bmatrix} \mathbf{r}_1 \\ \mathbf{r}_2 \\ \mathbf{r}_3 \\ \mathbf{r}_4 \end{bmatrix}[\lambda_1\mathbf{e}_1, \lambda_1\mathbf{e}_2, \lambda_1\mathbf{e}_3, \lambda_4\mathbf{e}_4] + \begin{bmatrix} \mathbf{r}_1 \\ \mathbf{r}_2 \\ \mathbf{r}_3 \\ \mathbf{r}_4 \end{bmatrix}[\mathbf{0}, \mathbf{e}_1, \mathbf{e}_2, \mathbf{0}]$$

$$= \begin{bmatrix} \lambda_1 & 0 & 0 & 0 \\ 0 & \lambda_1 & 0 & 0 \\ 0 & 0 & \lambda_1 & 0 \\ 0 & 0 & 0 & \lambda_4 \end{bmatrix} + \begin{bmatrix} 0 & 1 & 0 & 0 \\ 0 & 0 & 1 & 0 \\ 0 & 0 & 0 & 0 \\ 0 & 0 & 0 & 0 \end{bmatrix} \tag{13}$$

so, in this example,

$$\mathbf{P}^{-1}\mathbf{AP} = \begin{bmatrix} \lambda_1 & 1 & 0 & 0 \\ 0 & \lambda_1 & 1 & 0 \\ 0 & 0 & \lambda_1 & 0 \\ 0 & 0 & 0 & \lambda_4 \end{bmatrix} = \begin{bmatrix} 2 & 1 & 0 & 0 \\ 0 & 2 & 1 & 0 \\ 0 & 0 & 2 & 0 \\ 0 & 0 & 0 & 5 \end{bmatrix} \equiv \mathbf{J}, \tag{14}$$

which is known as a *Jordan form* for the $\mathbf{A}$ matrix given by (2).

It remains to determine $\mathbf{e}_2$, $\mathbf{e}_3$ and hence $\mathbf{P}$. Notice that of all the steps in obtaining the Jordan form (14) the only possible weak link is in the calculation of the generalized eigenvectors $\mathbf{e}_2$ and $\mathbf{e}_3$ for, after all, (5) and (6) are nonhomogeneous equations and the coefficient matrix is singular because $\det(\mathbf{A} - \lambda_1\mathbf{I}) = 0$; thus, each of (5) and (6) will have either no solution or a nonunique solution. And if either (5) or (6) has no solution then the method fails. Let us see. Denoting $\mathbf{e}_2 = \left[x_1^{(2)}, x_2^{(2)}, x_3^{(3)}, x_4^{(2)}\right]^{\mathrm{T}}$, (5) becomes

$$\begin{bmatrix} 0 & -1 & 2 & 0 \\ 0 & 1 & -1 & 0 \\ 0 & 1 & -1 & 0 \\ 0 & 1 & -3 & 3 \end{bmatrix} \begin{bmatrix} x_1^{(2)} \\ x_2^{(2)} \\ x_3^{(2)} \\ x_4^{(2)} \end{bmatrix} = \begin{bmatrix} 1 \\ 0 \\ 0 \\ 0 \end{bmatrix}. \tag{15}$$

The latter *is* consistent and admits the one-parameter family of solutions $\mathbf{e}_2 = [\gamma, 1, 1, 2/3]^{\mathrm{T}}$ where γ is arbitrary.[1] Then (6) becomes

$$\begin{bmatrix} 0 & -1 & 2 & 0 \\ 0 & 1 & -1 & 0 \\ 0 & 1 & -1 & 0 \\ 0 & 1 & -3 & 3 \end{bmatrix} \begin{bmatrix} x_1^{(3)} \\ x_2^{(3)} \\ x_3^{(3)} \\ x_4^{(3)} \end{bmatrix} = \begin{bmatrix} \gamma \\ 1 \\ 1 \\ 2/3 \end{bmatrix}. \tag{16}$$

---

[1]Can we scale $\mathbf{e}_2$ by 3 so as to eliminate the fraction 2/3? No, eigenvectors can be scaled since they satisfy homogeneous equations, but the generalized eigenvectors $\mathbf{e}_2$ and $\mathbf{e}_3$ satisfy the *non*homogeneous equations (5) and (6).

with the solution $e_3 = [\delta, \gamma + 2, \gamma + 1, 2\gamma/3 + 5/9]^T$, where both γ and δ are arbitrary. We do not need all possible e_2's and e_3's, so let us choose $\gamma = \delta = 0$, say. Then

$$\mathbf{P} = [e_1, e_2, e_3, e_4] = \begin{bmatrix} 1 & 0 & 0 & 0 \\ 0 & 1 & 2 & 0 \\ 0 & 1 & 1 & 0 \\ 0 & 2/3 & 5/9 & 1 \end{bmatrix}. \tag{17}$$

Here, e_1 and e_4 are eigenvectors of $\mathbf{A}$, and e_2 and e_3 are generalized eigenvectors associated with e_1. We speak of e_1, e_2, e_3 as a **chain** of vectors headed by the eigenvector e_1.

EXAMPLE 1

To apply these ideas to the solution of a system of differential equations, let us solve the system

$$\mathbf{x}' = \mathbf{A}\mathbf{x}, \tag{18}$$

where $\mathbf{A}$ is given by (2). We would like to diagonalize $\mathbf{A}$, to uncouple the equations, but we cannot because $\mathbf{A}$ is not diagonalizable. Nevertheless, we can at least minimize the coupling by reducing $\mathbf{A}$ to Jordan form. Accordingly, let $\mathbf{x}(t) = \mathbf{P}\widetilde{\mathbf{x}}(t)$, where $\mathbf{P}$ is a constant 4×4 matrix. Then (18) becomes

$$\mathbf{P}\widetilde{\mathbf{x}}' = \mathbf{A}\mathbf{P}\widetilde{\mathbf{x}} \tag{19}$$

or,

$$\widetilde{\mathbf{x}}' = \mathbf{P}^{-1}\mathbf{A}\mathbf{P}\widetilde{\mathbf{x}}. \tag{20}$$

Choosing $\mathbf{P}$ according to (17), $\mathbf{P}^{-1}\mathbf{A}\mathbf{P}$ is given by (14) so (20) becomes

$$\widetilde{x}_1' = 2\widetilde{x}_1 + \widetilde{x}_2, \tag{21a}$$

$$\widetilde{x}_2' = 2\widetilde{x}_2 + \widetilde{x}_3, \tag{21b}$$

$$\widetilde{x}_3' = 2\widetilde{x}_3, \tag{21c}$$

$$\widetilde{x}_4' = 5\widetilde{x}_4. \tag{21d}$$

The latter is not uncoupled, but thanks to the triangular form of $\mathbf{J}$ we can solve (21) by starting at the bottom and working our way up, just as we solve a Gauss eliminated system of linear algebraic equations by starting at the bottom and working our way up. Carrying out those steps (Exercise 2), we obtain

$$\widetilde{x}_4(t) = Ae^{5t}, \tag{22a}$$

$$\widetilde{x}_3(t) = Be^{2t}, \tag{22b}$$

$$\widetilde{x}_2(t) = (C + Bt)e^{2t}, \tag{22c}$$

$$\widetilde{x}_1(t) = \left(D + Ct + \frac{1}{2}Bt^2\right)e^{2t}, \tag{22d}$$

in turn, where A, B, C, D are arbitrary constants. Finally, putting (22) into $\mathbf{x}(t) = \mathbf{P}\widetilde{\mathbf{x}}(t)$ gives

$$x_1(t) = \left(D + Ct + \frac{1}{2}Bt^2 \right) e^{2t}, \tag{23a}$$

$$x_2(t) = [(C + 2B) + Bt]e^{2t}, \tag{23b}$$

$$x_3(t) = [(C + B) + Bt]e^{2t}, \tag{23c}$$

$$x_4(t) = \left[\left(\frac{2}{3}C + \frac{5}{9}B \right) + \frac{2}{3}Bt \right] e^{2t} + Ae^{5t} \tag{23d}$$

as a general solution of (18).   ∎

EXAMPLE 2

Consider the system of differential equations

$$x_1' = 6x_1 - x_3, \tag{24a}$$
$$x_2' = -x_1 + 5x_2 + x_3, \tag{24b}$$
$$x_3' = x_1 + 4x_3 \tag{24c}$$

or, in matrix form, $\mathbf{x}' = \mathbf{A}\mathbf{x}$, where

$$\mathbf{A} = \begin{bmatrix} 6 & 0 & -1 \\ -1 & 5 & 1 \\ 1 & 0 & 4 \end{bmatrix}. \tag{25}$$

The eigenvalues and eigenvectors of $\mathbf{A}$ are found to be

$$\lambda_1 = \lambda_2 = \lambda_3 = 5; \quad \mathbf{e} = \begin{bmatrix} \alpha \\ \beta \\ \alpha \end{bmatrix} = \alpha \begin{bmatrix} 1 \\ 0 \\ 1 \end{bmatrix} + \beta \begin{bmatrix} 0 \\ 1 \\ 0 \end{bmatrix}. \tag{26}$$

We see that $\mathbf{A}$ is not diagonalizable because $\lambda = 5$ is of multiplicity three but its eigenspace is only two-dimensional. Nevertheless, we can triangularize $\mathbf{A}$. To do so we could let $\mathbf{e}_1 = [1, 0, 1]^T$ and $\mathbf{e}_2 = [0, 1, 0]^T$, say, and then seek $\mathbf{e}_3$ according to

$$(\mathbf{A} - \lambda_2\mathbf{I})\mathbf{e}_3 = \mathbf{e}_2. \tag{27}$$

That is, we could let $\mathbf{e}_2 = [0, 1, 0]^T$ head the two-vector chain $\mathbf{e}_2, \mathbf{e}_3$, where $\mathbf{e}_3$ is a generalized eigenvector associated with the eigenvector $\mathbf{e}_2$. However, we would find that the system (27),

$$\begin{bmatrix} 1 & 0 & -1 \\ -1 & 0 & 1 \\ 1 & 0 & -1 \end{bmatrix} \begin{bmatrix} x_1^{(3)} \\ x_2^{(3)} \\ x_3^{(3)} \end{bmatrix} = \begin{bmatrix} 0 \\ 1 \\ 0 \end{bmatrix}, \tag{28}$$

is inconsistent! Instead of choosing $\mathbf{e}_2 = [0, 1, 0]^T$, let us choose $\mathbf{e}_2 = [\alpha, \beta, \alpha]^T$ to head the chain so that we can use the arbitrary constants α and β to render (27) consistent. In that case (27) becomes

$$\begin{bmatrix} 1 & 0 & -1 \\ -1 & 0 & 1 \\ 1 & 0 & -1 \end{bmatrix} \begin{bmatrix} x_1^{(3)} \\ x_2^{(3)} \\ x_3^{(3)} \end{bmatrix} = \begin{bmatrix} \alpha \\ \beta \\ \alpha \end{bmatrix}, \tag{29}$$

which is now consistent if we choose $\beta = -\alpha$. With $\alpha = 1$ and $\beta = -1$, say, $\mathbf{e}_2$ is $[1, -1, 1]^T$ and then solving (29) gives $\mathbf{e}_3 = [1, 0, 0]^T$. Thus,

$$\mathbf{P} = [\mathbf{e}_1, \mathbf{e}_2, \mathbf{e}_3] = \begin{bmatrix} 1 & 1 & 1 \\ 0 & -1 & 0 \\ 1 & 1 & 0 \end{bmatrix}. \tag{30}$$

With $\mathbf{x} = \mathbf{P}\widetilde{\mathbf{x}}$, $\mathbf{x}' = \mathbf{A}\mathbf{x}$ becomes $\mathbf{P}\widetilde{\mathbf{x}}' = \mathbf{A}\mathbf{P}\widetilde{\mathbf{x}}$. Hence,

$$\widetilde{\mathbf{x}}' = \mathbf{P}^{-1}\mathbf{A}\mathbf{P}\widetilde{\mathbf{x}} = \mathbf{J}\widetilde{\mathbf{x}} = \begin{bmatrix} 5 & 0 & 0 \\ 0 & 5 & 1 \\ 0 & 0 & 5 \end{bmatrix} \widetilde{\mathbf{x}} \tag{31}$$

or, in scalar form,

$$\widetilde{x}_1' = 5\widetilde{x}_1, \tag{32a}$$
$$\widetilde{x}_2' = 5\widetilde{x}_2 + \widetilde{x}_3, \tag{32b}$$
$$\widetilde{x}_3' = 5\widetilde{x}_3. \tag{32c}$$

If we solve from the bottom up, $\widetilde{x}_3(t) = Ae^{5t}$, then $\widetilde{x}_2(t) = (B + At)e^{5t}$, and $\widetilde{x}_1(t) = Ce^{5t}$, where A, B, C are arbitrary constants. Finally,

$$\mathbf{x}(t) = \mathbf{P}\widetilde{\mathbf{x}}(t) = \begin{bmatrix} 1 & 1 & 1 \\ 0 & -1 & 0 \\ 1 & 1 & 0 \end{bmatrix} \begin{bmatrix} Ce^{5t} \\ (B + At)e^{5t} \\ Ae^{5t} \end{bmatrix} = \begin{bmatrix} (A + B + C + At)e^{5t} \\ -(B + At)e^{5t} \\ (B + C + At)e^{5t} \end{bmatrix}, \tag{33}$$

so

$$x_1(t) = (A + B + C + At)e^{5t},$$
$$x_2(t) = -(B + At)e^{5t}, \tag{34}$$
$$x_3(t) = (B + C + At)e^{5t},$$

is a general solution of (24).   ∎

With these examples behind us, we can now state the general procedure for triangularizing any given $n \times n$ nondiagonalizable matrix $\mathbf{A}$. Let λ be an eigenvalue of $\mathbf{A}$ of multiplicity K, and let the eigenspace corresponding to λ be of dimension k, where $k < K$. Within that eigenspace there can be found k LI eigenvectors of $\mathbf{A}$, say

e_1 through e_k. Then, **generalized eigenvectors** $e_{k+1}, \ldots, e_K$ associated with e_k can be found from the equations

$$\begin{aligned}
(\mathbf{A} - \lambda\mathbf{I})\mathbf{e}_{k+1} &= \mathbf{e}_k, \\
(\mathbf{A} - \lambda\mathbf{I})\mathbf{e}_{k+2} &= \mathbf{e}_{k+1}, \\
&\vdots \\
(\mathbf{A} - \lambda\mathbf{I})\mathbf{e}_K &= \mathbf{e}_{K-1}.
\end{aligned} \tag{35}$$

Thus, the eigenvalue λ generates the eigenvectors $e_1, \ldots, e_k$ and the generalized eigenvectors $e_{k+1}, \ldots, e_K$. We say that $e_k, e_{k+1}, \ldots, e_K$ is a **chain** of vectors headed by the eigenvector e_k.

Keeping all chain sequences intact, form a **generalized modal matrix P** as

$$\mathbf{P} = [\mathbf{e}_1, \ldots, \mathbf{e}_n], \tag{36}$$

some columns of which will be eigenvectors of $\mathbf{A}$ and some (i.e, at least one) of which will be generalized eigenvectors of $\mathbf{A}$. Then the Jordan form for $\mathbf{A}$,

$$\mathbf{P}^{-1}\mathbf{A}\mathbf{P} = \mathbf{J}, \tag{37}$$

will have the eigenvalues of $\mathbf{A}$ on its main diagonal and a 1 immediately above the diagonal element in each column that corresponds to a generalized eigenvector within $\mathbf{P}$; all other elements in $\mathbf{J}$ will be zero.

EXAMPLE 3

Suppose a 7×7 matrix $\mathbf{A}$ has eigenvalues with the multiplicities indicated in column 2 of Table 10.6.1, with the dimension of each eigenspace as indicated in column 3. Then the eigenvectors and generalized eigenvectors will be distributed as indicated in columns 4 and 5, and we set

$$\mathbf{P} = \left[\underbrace{\mathbf{e}_1}_{\text{chain}} , \underbrace{\mathbf{e}_2}_{\text{chain}} , \underbrace{\mathbf{e}_3, \mathbf{e}_4, \mathbf{e}_5}_{\text{chain}}, \underbrace{\mathbf{e}_6, \mathbf{e}_7}_{\text{chain}} \right]. \tag{38}$$

Table 10.6.1 Distribution of eigenvectors and generalized eigenvectors.

λ	Multiplicity	Dimension of Eigenspace	Eigenvectors	Generalized Eigenvectors
λ_1	1	1	$\mathbf{e}_1$	None
λ_2	4	2	$\mathbf{e}_2, \mathbf{e}_3$	$\mathbf{e}_4, \mathbf{e}_5$
λ_3	2	1	$\mathbf{e}_6$	$\mathbf{e}_7$

Then $\mathbf{J}$ will be of the form

$$\mathbf{J} = \begin{bmatrix} \lambda_1 & 0 & \cdots & & & \cdots & 0 \\ 0 & \lambda_2 & & & & & \vdots \\ \vdots & & \lambda_2 & 1 & & & \\ & & & \lambda_2 & 1 & & \\ & & & & \lambda_2 & & \\ \vdots & & & & & \lambda_3 & 1 \\ 0 & \cdots & & & \cdots & 0 & \lambda_3 \end{bmatrix}, \tag{39}$$

where all elements not shown are zero. Since the generalized eigenvectors in (38) are $\mathbf{e}_4$, $\mathbf{e}_5$, and $\mathbf{e}_7$, there are 1's above the main diagonal in columns 4, 5, and 7. ∎

Remember, *every* square matrix can be triangularized:

THEOREM 10.6.1

Triangularization
Every square matrix can be triangularized. That is, given a square matrix $\mathbf{A}$ there exists an invertible matrix $\mathbf{P}$ such that $\mathbf{P}^{-1}\mathbf{A}\mathbf{P}$ is upper triangular.

Of course, if the $n \times n$ matrix $\mathbf{A}$ has n LI eigenvectors then there will be *no* generalized eigenvectors and $\mathbf{P}^{-1}\mathbf{A}\mathbf{P}$ will be not only tridiagonal, it will be *diagonal*.

We will not prove Theorem 10.6.1, but our foregoing discussion has outlined the main ideas—except how to prove that the generalized eigenvectors exist, that is, that the equations (35) are consistent. For proofs, see Strang or Faddeev and Faddeeva.[1]

Closure. We have seen that for every $n \times n$ matrix $\mathbf{A}$ there exists an invertible matrix $\mathbf{P}$ such that $\mathbf{P}^{-1}\mathbf{A}\mathbf{P} = \mathbf{J}$ is upper triangular. To construct $\mathbf{P}$ we take its columns to be the eigenvectors and generalized eigenvectors of $\mathbf{A}$, keeping all chains intact, so $\mathbf{P} = [\mathbf{e}_1, \ldots, \mathbf{e}_n]$. The eigenvectors of $\mathbf{A}$ are found in the usual way, and each of the one or more chains of generalized eigenvectors is found by solving a system of nonhomogeneous algebraic equations of the form (35). Then the Jordan form $\mathbf{J}$ will have the eigenvalues of $\mathbf{A}$ on its main diagonal and 1's above the diagonal element in each column that corresponds to a generalized eigenvector within $\mathbf{P}$; all other elements in $\mathbf{J}$ will be zero.

Our principal application of triangularization, in this section, is to the solution of systems of differential equations of the form $\mathbf{x}' = \mathbf{A}\mathbf{x}$ where $\mathbf{A}$ is not diagonalizable, for the change of variables $\mathbf{x}(t) = \mathbf{P}\widetilde{\mathbf{x}}(t)$ gives $\mathbf{P}\widetilde{\mathbf{x}}' = \mathbf{A}\mathbf{P}\widetilde{\mathbf{x}}$ or,

$$\widetilde{\mathbf{x}}' = \mathbf{P}^{-1}\mathbf{A}\mathbf{P}\widetilde{\mathbf{x}} \tag{40}$$

---

[1]Gilbert Strang, *Linear Algebra and its Applications* (New York: Academic Press, 1976); D. Faddeev and V. Faddeeva, *Computational Methods in Linear Algebra* (San Francisco: W. H. Freeman, 1963).

or,

$$\widetilde{\mathbf{x}}' = \mathbf{J}\widetilde{\mathbf{x}}. \qquad (41)$$

Although (41) is not entirely uncoupled (because of one or more 1's above the main diagonal), it is upper triangular and can therefore be solved from the bottom up.

EXERCISES 10.6

1. Derive the eigenvalues and eigenspaces of the **A** matrix given by (2), and show that they agree with the results stated in (3).

2. Derive the general solution (22) of (21).

3. Reduce the given **A** matrix to Jordan triangular form. That is, determine **P** and **J** so that $\mathbf{P}^{-1}\mathbf{AP} = \mathbf{J}$.

   (a) $\begin{bmatrix} 3 & 1 \\ -4 & -1 \end{bmatrix}$ (b) $\begin{bmatrix} 2 & 4 \\ -1 & 6 \end{bmatrix}$

   (c) $\begin{bmatrix} 1 & 4 \\ -1 & 5 \end{bmatrix}$ (d) $\begin{bmatrix} -4 & -4 \\ 1 & -8 \end{bmatrix}$

   (e) $\begin{bmatrix} -1 & 1 & -3 \\ 1 & 0 & 2 \\ 2 & -1 & 4 \end{bmatrix}$ (f) $\begin{bmatrix} -2 & 2 & 1 \\ -5 & 4 & 2 \\ -1 & 1 & 1 \end{bmatrix}$

   (g) $\begin{bmatrix} 1 & 1 & 1 \\ 0 & 1 & 1 \\ 0 & 0 & 1 \end{bmatrix}$ (h) $\begin{bmatrix} 0 & 1 & -3 \\ 1 & 1 & 2 \\ 2 & -1 & 5 \end{bmatrix}$

   (i) $\begin{bmatrix} 1 & 0 & 0 & 0 \\ 1 & 1 & 0 & 0 \\ 1 & 1 & 1 & 0 \\ 1 & 1 & 1 & 1 \end{bmatrix}$ (j) $\begin{bmatrix} 3 & 1 & -1 & 1 \\ -1 & 1 & 1 & 0 \\ 0 & 0 & 2 & 0 \\ 0 & 0 & 0 & 2 \end{bmatrix}$

   (k) $\begin{bmatrix} 0 & 1 & 1 & 1 \\ 0 & -1 & 1 & -2 \\ 0 & 0 & -1 & 1 \\ 0 & 1 & -1 & 2 \end{bmatrix}$ (l) $\begin{bmatrix} 1 & 0 & 0 & 0 \\ 0 & 3 & 0 & -1 \\ 0 & 0 & 4 & 0 \\ 0 & 1 & 0 & 5 \end{bmatrix}$

4. (a)–(l) Use the method of triangularization to determine a general solution of $\mathbf{x}' = \mathbf{Ax}$ where **A** is given in the corresponding part of Exercise 3 and **x** denotes $[x(t), y(t)]^{\mathrm{T}}$.

5. (a)–(d) Use the method of triangularization to determine a general solution of $\mathbf{x}'' = \mathbf{Ax}$, where **A** is given in the corresponding part of Exercise 3 and **x** denotes $[x(t), y(t)]^{\mathrm{T}}$.

6. If **J** is a Jordan form for **A**, is $\alpha\mathbf{J}$ a Jordan form for $\alpha\mathbf{A}$? Prove or disprove.

7. Prove that if **J** is a Jordan form for **A**, and **A** is similar to **B**, then **J** is also a Jordan form for **B**. NOTE: The term *similar* was defined in Exercise 14 of Section 9.2.

10.7 EXPLICIT SOLUTION OF $\mathbf{x}' = \mathbf{Ax}$ AND THE MATRIX EXPONENTIAL FUNCTION (OPTIONAL)

In Section 10.5 we showed how to solve the system $\mathbf{x}' = \mathbf{Ax}$ by diagonalization if **A** *is* diagonalizable, and in Section 10.6 we showed how to solve it by triangularization if **A** is *not* diagonalizable. In those sections we emphasized the solution *method*; in this final section we use those methods to find an *explicit solution* to the initial value problem

$$x_1' = a_{11}x_1 + \cdots + a_{1n}x_n; \qquad x_1(0) = c_1$$
$$\vdots \qquad\qquad (1)$$
$$x_n' = a_{n1}x_1 + \cdots + a_{nn}x_n; \qquad x_n(0) = c_n$$

where the a_{ij}'s are constants, and to express that solution in terms of a "matrix exponential function." First, we express (1) in the compact matrix form

$$\mathbf{x}' = \mathbf{Ax}; \qquad \mathbf{x}(0) = \mathbf{c}. \qquad (2)$$

10.7.1 If A is diagonalizable.   Consider first the case where $\mathbf{A}$ is diagonalizable—that is, it has n LI eigenvectors $\mathbf{e}_1, \ldots, \mathbf{e}_n$ with corresponding eigenvalues $\lambda_1, \ldots, \lambda_n$. Setting

$$\mathbf{x}(t) = \mathbf{Q}\widetilde{\mathbf{x}}(t), \tag{3}$$

where $\mathbf{Q} = [\mathbf{e}_1, \ldots, \mathbf{e}_n]$ is a modal matrix of $\mathbf{A}$, (2) becomes

$$\mathbf{Q}\widetilde{\mathbf{x}}' = \mathbf{A}\mathbf{Q}\widetilde{\mathbf{x}}; \qquad \mathbf{Q}\widetilde{\mathbf{x}}(0) = \mathbf{c} \tag{4}$$

or, equivalently,

$$\widetilde{\mathbf{x}}' = \mathbf{Q}^{-1}\mathbf{A}\mathbf{Q}\widetilde{\mathbf{x}}; \qquad \widetilde{\mathbf{x}}(0) = \mathbf{Q}^{-1}\mathbf{c}. \tag{5}$$

We know that

$$\mathbf{Q}^{-1}\mathbf{A}\mathbf{Q} = \mathbf{D} = \begin{bmatrix} \lambda_1 & 0 & \cdots & 0 \\ 0 & \lambda_2 & & \\ \vdots & & \ddots & \vdots \\ 0 & & \cdots & \lambda_n \end{bmatrix}, \tag{6}$$

and if we denote $\mathbf{Q}^{-1}\mathbf{c}$ as $\widetilde{\mathbf{c}}$, for brevity, then

$$\boxed{\widetilde{\mathbf{x}}' = \mathbf{D}\widetilde{\mathbf{x}}; \qquad \widetilde{\mathbf{x}}(0) = \widetilde{\mathbf{c}}} \tag{7}$$

or, returning to scalar form,

$$\begin{aligned} \widetilde{x}_1' &= \lambda_1 \widetilde{x}_1; \qquad \widetilde{x}_1(0) = \widetilde{c}_1 \\ &\vdots \\ \widetilde{x}_n' &= \lambda_n \widetilde{x}_n; \qquad \widetilde{x}_n(0) = \widetilde{c}_n. \end{aligned} \tag{8}$$

These uncoupled problems are easily solved, giving

$$\begin{aligned} \widetilde{x}_1(t) &= \widetilde{c}_1 e^{\lambda_1 t}, \\ &\vdots \\ \widetilde{x}_n(t) &= \widetilde{c}_n e^{\lambda_n t}. \end{aligned} \tag{9}$$

If we define

$$\mathbf{E}(t) \equiv \begin{bmatrix} e^{\lambda_1 t} & 0 & \cdots & 0 \\ 0 & e^{\lambda_2 t} & & \\ \vdots & & \ddots & \vdots \\ 0 & & \cdots & e^{\lambda_n t} \end{bmatrix}, \tag{10}$$

then we can express (9) in matrix form as

$$\widetilde{\mathbf{x}}(t) = \mathbf{E}(t)\widetilde{\mathbf{c}}. \tag{11}$$

Note the order $\mathbf{E}(t)\mathbf{c}$ not $\mathbf{c}\mathbf{E}(t)$ in (11), which result we urge you to verify for yourself. Finally, since $\widetilde{\mathbf{x}} = \mathbf{Q}^{-1}\mathbf{x}$ and $\widetilde{\mathbf{c}} = \mathbf{Q}^{-1}\mathbf{c}$, (11) gives

$$\boxed{\mathbf{x}(t) = \mathbf{Q}\mathbf{E}(t)\mathbf{Q}^{-1}\mathbf{c}} \tag{12}$$

as the solution of (1).

EXAMPLE 1

Use (12) to solve the system

$$x' = x + 4y; \qquad x(0) = 0$$
$$y' = x + y; \qquad y(0) = 2. \tag{13}$$

Then

$$\mathbf{A} = \begin{bmatrix} 1 & 4 \\ 1 & 1 \end{bmatrix}, \quad \mathbf{c} = \begin{bmatrix} 0 \\ 2 \end{bmatrix}. \tag{14}$$

The eigenvalues and eigenvectors of $\mathbf{A}$ are found to be

$$\lambda_1 = 3, \ \mathbf{e}_1 = \alpha \begin{bmatrix} 2 \\ 1 \end{bmatrix}; \qquad \lambda_2 = -1, \ \mathbf{e}_1 = \beta \begin{bmatrix} -2 \\ 1 \end{bmatrix}. \tag{15}$$

Since these $\mathbf{e}_j$'s are LI, $\mathbf{A}$ is diagonalizable so (12) applies. With

$$\mathbf{Q} = \begin{bmatrix} 2 & -2 \\ 1 & 1 \end{bmatrix}, \quad \mathbf{Q}^{-1} = \begin{bmatrix} \frac{1}{4} & \frac{1}{2} \\ -\frac{1}{4} & \frac{1}{2} \end{bmatrix}, \tag{16}$$

(12) gives

$$\mathbf{x}(t) = \begin{bmatrix} 2 & -2 \\ 1 & 1 \end{bmatrix} \begin{bmatrix} e^{3t} & 0 \\ 0 & e^{-t} \end{bmatrix} \begin{bmatrix} \frac{1}{4} & \frac{1}{2} \\ -\frac{1}{4} & \frac{1}{2} \end{bmatrix} \begin{bmatrix} 0 \\ 2 \end{bmatrix}$$

$$= \begin{bmatrix} 2e^{3t} - 2e^{-t} \\ e^{3t} + e^{-t} \end{bmatrix} \tag{17}$$

so

$$x(t) = 2e^{3t} - 2e^{-t},$$

$$y(t) = e^{3t} + e^{-t} \tag{18}$$

is the desired solution of (13).   ∎

10.7.2 Exponential matrix function.

It will be interesting to consider the matrix product $\mathbf{Q}\mathbf{E}(t)\mathbf{Q}^{-1}$ in (12) a bit further. Recalling the definitions

$$\mathbf{D} = \begin{bmatrix} \lambda_1 & \cdots & 0 \\ \vdots & \ddots & \vdots \\ 0 & \cdots & \lambda_n \end{bmatrix}, \quad \mathbf{E} = \begin{bmatrix} e^{\lambda_1 t} & \cdots & 0 \\ \vdots & \ddots & \vdots \\ 0 & \cdots & e^{\lambda_n t} \end{bmatrix}$$

of $\mathbf{D}$ and $\mathbf{E}$, and the familiar Taylor series

$$e^x = 1 + x + \frac{x^2}{2!} + \frac{x^3}{3!} + \cdots, \tag{19}$$

we have

$$\mathbf{QE}(t)\mathbf{Q}^{-1} = \mathbf{Q} \begin{bmatrix} 1 + \lambda_1 t + \lambda_1^2 \frac{t^2}{2!} + \cdots & \cdots & 0 \\ \vdots & \ddots & \vdots \\ 0 & \cdots & 1 + \lambda_n t + \lambda_n^2 \frac{t^2}{2!} + \cdots \end{bmatrix} \mathbf{Q}^{-1}$$

$$= \mathbf{Q} \left[\mathbf{I} + \mathbf{D}t + \mathbf{D}^2 \frac{t^2}{2!} + \cdots \right] \mathbf{Q}^{-1}. \tag{20}$$

But

$$\mathbf{D} = \mathbf{Q}^{-1}\mathbf{A}\mathbf{Q},$$
$$\mathbf{D}^2 = \mathbf{Q}^{-1}\mathbf{A}\mathbf{Q}\mathbf{Q}^{-1}\mathbf{A}\mathbf{Q} = \mathbf{Q}^{-1}\mathbf{A}^2\mathbf{Q}, \tag{21}$$
$$\mathbf{D}^3 = \mathbf{D}^2\mathbf{D} = \mathbf{Q}^{-1}\mathbf{A}^2\mathbf{Q}\mathbf{Q}^{-1}\mathbf{A}\mathbf{Q} = \mathbf{Q}^{-1}\mathbf{A}^3\mathbf{Q},$$

and so on, so (20) becomes

$$\mathbf{QE}(t)\mathbf{Q}^{-1} = \mathbf{Q} \left[\mathbf{I} + \mathbf{Q}^{-1}\mathbf{A}\mathbf{Q}t + \mathbf{Q}^{-1}\mathbf{A}^2\mathbf{Q}\frac{t^2}{2!} + \cdots \right] \mathbf{Q}^{-1}$$

$$= \mathbf{I} + \mathbf{A}t + \mathbf{A}^2\frac{t^2}{2!} + \cdots. \tag{22}$$

If, by analogy with (19), we define the latter series as $e^{\mathbf{A}t}$, then (22) gives

$$\mathbf{QE}(t)\mathbf{Q}^{-1} = e^{\mathbf{A}t}, \tag{23}$$

so (12) can be expressed, alternatively, in the form

$$\boxed{\mathbf{x}(t) = e^{\mathbf{A}t}\mathbf{c}.} \tag{24}$$

We have derived (24) for two reasons. First, suppose that $n = 1$ in (1) so that, discarding subscripts, (1) is simply

$$x' = ax; \qquad x(0) = c, \tag{25}$$

the solution of which is[1]

$$x(t) = ce^{at}. \tag{26}$$

Unlike (12), the solution form (24) is striking in that it can be seen to be the matrix generalization of (26). Computationally, however, it is simpler to use (12) than (24). That is, it is simpler to compute $\mathbf{QE}(t)\mathbf{Q}^{-1}$ than the infinite series $\mathbf{I} + \mathbf{A}t + \mathbf{A}^2 t^2/2! + \cdots$.

---

[1]Whether we write ce^{at} or $e^{at}c$ in (26) does not matter, but in (24) we *must* write $e^{\mathbf{A}t}\mathbf{c}$ not $\mathbf{c}e^{\mathbf{A}t}$.

Second, our derivation of (24) has led us naturally to the exponential matrix function, say $e^{\mathbf{A}}$, defined by[1]

$$e^{\mathbf{A}} \equiv \mathbf{I} + \mathbf{A} + \frac{1}{2!}\mathbf{A}^2 + \frac{1}{3!}\mathbf{A}^3 + \cdots . \tag{27}$$

This step is significant because it introduces the concept of a **matrix function** $f(\mathbf{A})$. In the case $f(\mathbf{A}) = e^{\mathbf{A}}$ the input $\mathbf{A}$ is an $n \times n$ matrix and the output $f(\mathbf{A})$ is also an $n \times n$ matrix. In the case $f(\mathbf{A}) = \det \mathbf{A}$, say, the input $\mathbf{A}$ is an $n \times n$ matrix and the output $f(\mathbf{A})$ is a scalar. Of course, we've worked with other matrix functions before, such as $6\mathbf{A} - 4\mathbf{A}^3$ and $5\mathbf{A}^{-1} + \mathbf{A}^2$, but have not yet regarded them as functions. In fact, (27) suggests that additional matrix functions, such as $\sin \mathbf{A}$ and $\cos \mathbf{A}$, might fruitfully be defined by such power series, but here we will limit our attention to the exponential function $e^{\mathbf{A}}$.

10.7.3 If A is not diagonalizable.   Actually, (24) is the solution to (2) for any $n \times n$ matrix $\mathbf{A}$, whether it is diagonalizable or not, because $e^{\mathbf{A}t}$ admits the derivative

$$\frac{d}{dt}e^{\mathbf{A}t} = \mathbf{A}e^{\mathbf{A}t}, \tag{28}$$

just as $(d/dt)e^{at} = ae^{at}$ (Exercise 4). By virtue of (28), if $\mathbf{x}(t) = e^{\mathbf{A}t}\mathbf{c}$ then

$$\mathbf{x}'(t) = \mathbf{A}e^{\mathbf{A}t}\mathbf{c} = \mathbf{A}\mathbf{x}$$

so (24) does satisfy the differential equation $\mathbf{x}' = \mathbf{A}\mathbf{x}$. Furthermore, (24) satisfies the initial condition $\mathbf{x}(0) = \mathbf{c}$ because (27) shows that

$$e^{\mathbf{0}} = \mathbf{I}, \tag{29}$$

---

[1]Be aware that the right-hand side of (27) is an *infinite series of matrices*, which we have not yet defined. Mimicking the usual definition of convergence for series of scalars, we define an infinite series of matrices $\sum_{j=1}^{\infty} \mathbf{A}_j$ as the limit of the sequence of the partial sums $\mathbf{S}_N$, where $\mathbf{S}_N = \sum_{j=1}^{N} \mathbf{A}_j$. That is,

$$\sum_{j=1}^{\infty} \mathbf{A}_j \equiv \lim_{N\to\infty} \mathbf{S}_N = \lim_{N\to\infty} \sum_{j=1}^{N} \mathbf{A}_j.$$

The infinite series is said to **converge** if the limit on the right exists and to diverge if that limit does not exist. Finally, observe that $\lim_{N\to\infty} \mathbf{S}_N$ is the *limit of a sequence of matrices*, which we have not yet defined so we are not done. Let $\mathbf{C}_1, \mathbf{C}_2, \ldots$ be a sequence of $m \times n$ matrices, with $(c_{ij})_n$ as the i, j element of $\mathbf{C}_n$. We say that the sequence **converges** to a matrix $\mathbf{C} = \{c_{ij}\}$ if

$$\lim_{n\to\infty} (c_{ij})_n = c_{ij}$$

for each i, j, and we denote such convergence by writing either $\lim_{n\to\infty} \mathbf{C}_n = \mathbf{C}$ or $\mathbf{C}_n \to \mathbf{C}$ as $n \to \infty$. If the sequence does not converge, then it is said to **diverge**.

just as $e^0 = 1$. By virtue of (29), if $\mathbf{x}(t) = e^{\mathbf{A}t}\mathbf{c}$ then

$$\mathbf{x}(0) = e^0\mathbf{c} = \mathbf{I}\mathbf{c} = \mathbf{c}$$

so (24) does satisfy (2) for any square matrix $\mathbf{A}$, as claimed.

However, (24) is not computationally convenient because it requires an infinite series calculation of $e^{\mathbf{A}t}$, so let us seek a more convenient form analogous to (12). To begin, set

$$\mathbf{x}(t) = \mathbf{P}\widetilde{\mathbf{x}}(t), \tag{30}$$

where $\mathbf{P}$ is a generalized modal matrix of $\mathbf{A}$. Putting (30) into (2) gives

$$\mathbf{P}\widetilde{\mathbf{x}}' = \mathbf{A}\mathbf{P}\widetilde{\mathbf{x}}; \qquad \mathbf{P}\widetilde{\mathbf{x}}(0) = \mathbf{c} \tag{31}$$

or

$$\widetilde{\mathbf{x}}' = \mathbf{P}^{-1}\mathbf{A}\mathbf{P}\widetilde{\mathbf{x}}; \qquad \widetilde{\mathbf{x}}(0) = \mathbf{P}^{-1}\mathbf{c} \equiv \widetilde{\mathbf{c}}. \tag{32}$$

But

$$\mathbf{P}^{-1}\mathbf{A}\mathbf{P} = \mathbf{J}, \tag{33}$$

where $\mathbf{J}$ is the Jordan form for $\mathbf{A}$ corresponding to the generalized modal matrix $\mathbf{P}$, so

$$\widetilde{\mathbf{x}}' = \mathbf{J}\widetilde{\mathbf{x}}; \qquad \widetilde{\mathbf{x}}(0) = \widetilde{\mathbf{c}}. \tag{34}$$

Then, since (24) is the solution of (2) for any $n \times n$ matrix $\mathbf{A}$, it follows that

$$\widetilde{\mathbf{x}}(t) = e^{\mathbf{J}t}\widetilde{\mathbf{c}} \tag{35}$$

is the solution of (34). Finally, putting $\widetilde{\mathbf{x}} = \mathbf{P}^{-1}\mathbf{x}$ and $\widetilde{\mathbf{c}} = \mathbf{P}^{-1}\mathbf{c}$ into (35) gives

$$\mathbf{x}(t) = \mathbf{P}e^{\mathbf{J}t}\mathbf{P}^{-1}\mathbf{c}. \tag{36}$$

How do we evaluate the $e^{\mathbf{J}t}$ in (36)? Recall from Section 10.6 that $\mathbf{J} = \mathbf{D} + $ "$\boldsymbol{\delta}$" where $\mathbf{D}$ is a diagonal matrix the diagonal elements of which are the eigenvalues of $\mathbf{A}$, and where $\boldsymbol{\delta}$ is identically zero except for one or more 1's immediately above the main diagonal. Then

$$e^{\mathbf{J}t} = e^{(\mathbf{D}+\boldsymbol{\delta})t} = e^{\mathbf{D}t+\boldsymbol{\delta}t} = e^{\mathbf{D}t}e^{\boldsymbol{\delta}t}$$

$$= \left(\mathbf{I} + \mathbf{D}t + \mathbf{D}^2\frac{t^2}{2!} + \cdots\right)e^{\boldsymbol{\delta}t}$$

$$= \begin{bmatrix} 1 + \lambda_1 t + \cdots & \cdots & 0 \\ & \ddots & \vdots \\ 0 & \cdots & 1 + \lambda_n t + \cdots \end{bmatrix} e^{\boldsymbol{\delta}t}$$

$$= \begin{bmatrix} e^{\lambda_1 t} & \cdots & 0 \\ \vdots & \ddots & \vdots \\ 0 & \cdots & e^{\lambda_n t} \end{bmatrix} e^{\boldsymbol{\delta}t} = \mathbf{E}(t)e^{\boldsymbol{\delta}t}, \tag{37}$$

so (36) and (37) give the final result

$$\boxed{\mathbf{x}(t) = \mathbf{P}\mathbf{E}(t)e^{\delta t}\mathbf{P}^{-1}\mathbf{c}.} \tag{38}$$

In writing $e^{\mathbf{D}t+\delta t} = e^{\mathbf{D}t}e^{\delta t}$ in (37), we have used the fact, which we state without proof, that

$$e^{\mathbf{A}+\mathbf{B}} = e^{\mathbf{A}}e^{\mathbf{B}} \tag{39}$$

if $\mathbf{A}$ and $\mathbf{B}$ commute (i.e., if $\mathbf{AB} = \mathbf{BA}$); it can be shown that $\mathbf{D}$ and δ do commute (Exercises 5 and 6).

Finally, computing the $e^{\delta t}$ in (38) according to the infinite series

$$e^{\delta t} = \mathbf{I} + \delta t + \delta^2\frac{t^2}{2!} + \cdots \tag{40}$$

is simple because δ is nilpotent so the series (40) will terminate!

Observe that if $\mathbf{A}$ is diagonalizable then the generalized modal matrix $\mathbf{P}$ reduces to the modal matrix $\mathbf{Q}$ and $\delta = \mathbf{0}$ so $e^{\delta t} = \mathbf{I}$; then (38) reduces to (12), as it should. The advantage of (38) over (24) is that (38) is in closed form whereas (24) is in open form because it requires the evaluation of the infinite series $e^{\mathbf{A}t} = \mathbf{I} + \mathbf{A}t + \cdots$.

EXAMPLE 2

To illustrate the application of (38), consider the system (2) where

$$\mathbf{A} = \begin{bmatrix} 6 & 0 & -1 \\ -1 & 5 & 1 \\ 1 & 0 & 4 \end{bmatrix} \quad \text{and} \quad \mathbf{c} = \begin{bmatrix} 3 \\ 1 \\ 0 \end{bmatrix}. \tag{41}$$

This is the $\mathbf{A}$ matrix of Example 2 in Section 10.6, so let us bring forward from that example the results

$$\lambda_1 = \lambda_2 = \lambda_3 = 5, \quad \mathbf{P} = \begin{bmatrix} 1 & 1 & 1 \\ 0 & -1 & 0 \\ 1 & 1 & 0 \end{bmatrix}, \quad \mathbf{P}^{-1} = \begin{bmatrix} 0 & 1 & 1 \\ 0 & -1 & 0 \\ 1 & 0 & -1 \end{bmatrix} \tag{42}$$

and

$$\mathbf{J} = \begin{bmatrix} 5 & 0 & 0 \\ 0 & 5 & 1 \\ 0 & 0 & 5 \end{bmatrix} = \mathbf{D} + \delta = \begin{bmatrix} 5 & 0 & 0 \\ 0 & 5 & 0 \\ 0 & 0 & 5 \end{bmatrix} + \begin{bmatrix} 0 & 0 & 0 \\ 0 & 0 & 1 \\ 0 & 0 & 0 \end{bmatrix}. \tag{43}$$

We find that $\delta^2 = \delta^3 = \cdots = \mathbf{0}$, so

$$e^{\delta t} = \mathbf{I} + \delta t + \frac{1}{2}\delta^2 t^2 + \cdots = \mathbf{I} + \delta t = \begin{bmatrix} 1 & 0 & 0 \\ 0 & 1 & t \\ 0 & 0 & 1 \end{bmatrix}. \tag{44}$$

Hence, (38) gives

$$\mathbf{x}(t) = \begin{bmatrix} 1 & 1 & 1 \\ 0 & -1 & 0 \\ 1 & 1 & 0 \end{bmatrix} \begin{bmatrix} e^{5t} & 0 & 0 \\ 0 & e^{5t} & 0 \\ 0 & 0 & e^{5t} \end{bmatrix} \begin{bmatrix} 1 & 0 & 0 \\ 0 & 1 & t \\ 0 & 0 & 1 \end{bmatrix} \begin{bmatrix} 0 & 1 & 1 \\ 0 & -1 & 0 \\ 1 & 0 & -1 \end{bmatrix} \begin{bmatrix} 3 \\ 1 \\ 0 \end{bmatrix}$$

$$= \begin{bmatrix} (3+3t)\,e^{5t} \\ (1-3t)\,e^{5t} \\ 3t\,e^{5t} \end{bmatrix} \tag{45}$$

so

$$\begin{aligned}
x_1(t) &= (3+3t)\,e^{5t}, \\
x_2(t) &= (1-3t)\,e^{5t}, \\
x_2(t) &= 3t\,e^{5t},
\end{aligned} \tag{46}$$

is the solution.   ∎

Closure.   We have shown that the initial value problem

$$\mathbf{x}' = \mathbf{A}\mathbf{x}; \qquad \mathbf{x}(0) = \mathbf{c} \tag{47}$$

admits the solution

$$\mathbf{x}(t) = e^{\mathbf{A}t}\mathbf{c}. \tag{48}$$

where

$$e^{\mathbf{A}t} = \mathbf{I} + \mathbf{A}t + \mathbf{A}^2 \frac{t^2}{2!} + \cdots, \tag{49}$$

whether $\mathbf{A}$ is diagonalizable or not. The solution (47) is of interest because it provides the matrix generalization of the solution $x(t) = ce^{at}$ of the problem

$$x' = ax; \qquad x(0) = c$$

and because it introduces the exponential matrix function, but it is computationally inconvenient since to evaluate the $e^{\mathbf{A}t}$ matrix we must sum the series in (49).

Thus, we sought a closed form solution as well, as a computational alternative to (48). For the case where $\mathbf{A}$ is diagonalizable we obtained the result

$$\mathbf{x}(t) = \mathbf{Q}\mathbf{E}(t)\mathbf{Q}^{-1}\mathbf{c}.$$

and for the case where $\mathbf{A}$ is not diagonalizable we obtained

$$\mathbf{x}(t) = \mathbf{P}\mathbf{E}(t)e^{\delta t}\mathbf{P}^{-1}\mathbf{c}.$$

In this section $\mathbf{c}$ has been the initial condition. If no initial condition is prescribed and we seek a general solution instead, merely set $\mathbf{c} = [C_1, \ldots, C_n]^{\mathrm{T}}$ in the foregoing analysis, where the C_j's are arbitrary constants.

Maple.   To compute $e^{\mathbf{A}}$ for

$$\mathbf{A} = \begin{bmatrix} 1 & 2 \\ 3 & 4 \end{bmatrix},$$

say, use the **exponential** command

```
with(linalg):
A:=array([[1,2],[3,4]]);
exponential(A);
```

EXERCISES 10.7

1. (**A** *diagonalizable*) For the given **A** matrix solve (2) for $\mathbf{x}(t)$ using (12). In (a) and (b) use $\mathbf{c} = [3, -1]^T$ and in (c)–(f) use $\mathbf{c} = [4, 3, 0]^T$.

(a) $\begin{bmatrix} 2 & -3 \\ 0 & 0 \end{bmatrix}$ (b) $\begin{bmatrix} 1 & 0 \\ 1 & 0 \end{bmatrix}$

(c) $\begin{bmatrix} 0 & 0 & 1 \\ 0 & 0 & 1 \\ 1 & 1 & 1 \end{bmatrix}$ (d) $\begin{bmatrix} 2 & 0 & 0 \\ 0 & 1 & 1 \\ 0 & 1 & 1 \end{bmatrix}$

(e) $\begin{bmatrix} 2 & 1 & -1 \\ 1 & 4 & 3 \\ -1 & 3 & 4 \end{bmatrix}$ (f) $\begin{bmatrix} 0 & 1 & 0 \\ 1 & 0 & 0 \\ 0 & 0 & 0 \end{bmatrix}$

2. (a)–(f) Same as the corresponding part of Exercise 1, but discard the given **c** vector and find, instead, a *general* solution.

3. (**A** *not diagonalizable*) For the given **A** matrix solve (2) for $\mathbf{x}(t)$ using (38). In (a)–(d) use $\mathbf{c} = [3, 2]^T$ and in (e)–(h) use $\mathbf{c} = [4, 0, 6]^T$.

(a) $\begin{bmatrix} 3 & 1 \\ -4 & -1 \end{bmatrix}$ (b) $\begin{bmatrix} 2 & 4 \\ -1 & 6 \end{bmatrix}$

(c) $\begin{bmatrix} 1 & 4 \\ -1 & 5 \end{bmatrix}$ (d) $\begin{bmatrix} -4 & -4 \\ 1 & -8 \end{bmatrix}$

(e) $\begin{bmatrix} -1 & 1 & -3 \\ 1 & 0 & 2 \\ 2 & -1 & 4 \end{bmatrix}$ (f) $\begin{bmatrix} -2 & 2 & 1 \\ -5 & 4 & 2 \\ -1 & 1 & 1 \end{bmatrix}$

(g) $\begin{bmatrix} 1 & 1 & 1 \\ 0 & 1 & 1 \\ 0 & 0 & 1 \end{bmatrix}$ (h) $\begin{bmatrix} 0 & 1 & -3 \\ 1 & 1 & 2 \\ 2 & -1 & 5 \end{bmatrix}$

4. Show that formal termwise differentiation of $e^{\mathbf{A}t} = \mathbf{I} + \mathbf{A}t + \mathbf{A}^2 t^2/2! + \cdots$ does give $\mathbf{A}e^{\mathbf{A}t}$, as claimed in (28).

5. We stated that (39) holds if **A** and **B** commute. Using the series definition (27), work out the left-hand and right-hand sides of (39) up to and including third-order terms, say, and verify that they do agree if $\mathbf{AB} = \mathbf{BA}$.

6. Proof of the claim, made below (39), that $\mathbf{D}\delta = \delta\mathbf{D}$, would be tedious. In this exercise we outline only the main ideas of such a proof.

(a) First, consider a single Jordan "block,"

$$\mathbf{J} = \begin{bmatrix} a & 1 & 0 & \cdots & 0 \\ 0 & a & 1 & & \vdots \\ 0 & 0 & a & \ddots & 0 \\ \vdots & & \ddots & \ddots & 1 \\ 0 & \cdots & 0 & 0 & a \end{bmatrix}$$

and verify directly that $\mathbf{D}\delta = \delta\mathbf{D}$, where

$$\mathbf{D} = \begin{bmatrix} a & \cdots & 0 \\ \vdots & \ddots & \vdots \\ 0 & \cdots & a \end{bmatrix}$$

and

$$\delta = \begin{bmatrix} 0 & 1 & \cdots & 0 \\ 0 & & \ddots & \vdots \\ \vdots & & & 1 \\ 0 & \cdots & 0 & 0 \end{bmatrix}.$$

(b) For the case of multiple Jordan blocks one can use partitioning. To illustrate the idea, consider the case where

$$\mathbf{J} = \begin{bmatrix} b & 1 & 0 & 0 & 0 \\ 0 & b & 1 & 0 & 0 \\ 0 & 0 & b & 0 & 0 \\ \hline 0 & 0 & 0 & c & 1 \\ 0 & 0 & 0 & 0 & c \end{bmatrix} = \begin{bmatrix} \mathbf{J}_1 & \mathbf{0} \\ \mathbf{0} & \mathbf{J}_2 \end{bmatrix}.$$

$\mathbf{J}_1$ and $\mathbf{J}_2$ are two Jordan blocks within **J**. Then

$$\mathbf{D} = \begin{bmatrix} b & 0 & 0 & 0 & 0 \\ 0 & b & 0 & 0 & 0 \\ 0 & 0 & b & 0 & 0 \\ \hline 0 & 0 & 0 & c & 0 \\ 0 & 0 & 0 & 0 & c \end{bmatrix} = \begin{bmatrix} \mathbf{D}_1 & \mathbf{0} \\ \mathbf{0} & \mathbf{D}_2 \end{bmatrix}$$

and

$$\delta = \begin{bmatrix} 0 & 1 & 0 & 0 & 0 \\ 0 & 0 & 1 & 0 & 0 \\ 0 & 0 & 0 & 0 & 0 \\ \hline 0 & 0 & 0 & 0 & 1 \\ 0 & 0 & 0 & 0 & 0 \end{bmatrix} = \begin{bmatrix} \delta_1 & \mathbf{0} \\ \mathbf{0} & \delta_2 \end{bmatrix}$$

show, using the result established in part (a), that $\mathbf{D}\delta = \delta\mathbf{D}$.

7. Evaluate $e^{\mathbf{A}}$ for the given **A** matrix. HINT: Each **A** is diagonalizable so you can use the formula

$$e^{\mathbf{A}} = \mathbf{Q}\mathbf{E}(1)\mathbf{Q}^{-1} \qquad (7.1)$$

[i.e., equation (23) with $t = 1$.] Alternatively, you can always use the definition

$$e^{\mathbf{A}} = \mathbf{I} + \mathbf{A} + \frac{1}{2!}\mathbf{A}^2 + \frac{1}{3!}\mathbf{A}^3 + \cdots. \qquad (7.2)$$

In general the series evaluation is unwieldy, but if $\mathbf{A}$ happens to be nilpotent then the series reduces to a finite number of terms.

(a) $\begin{bmatrix} 1 & 2 \\ 0 & 3 \end{bmatrix}$ (b) $\begin{bmatrix} 2 & 2 \\ 2 & 2 \end{bmatrix}$

(c) $\begin{bmatrix} 0 & 4 \\ 9 & 0 \end{bmatrix}$ (d) $\begin{bmatrix} 2 & 1 \\ 1 & 2 \end{bmatrix}$

(e) $\begin{bmatrix} 2 & 0 \\ 7 & 3 \end{bmatrix}$ (f) $\begin{bmatrix} 0 & 8 \\ 0 & 0 \end{bmatrix}$

(g) $\begin{bmatrix} 0 & 0 & 1 \\ 0 & 0 & 0 \\ 1 & 0 & 0 \end{bmatrix}$ (h) $\begin{bmatrix} 0 & 0 & 1 \\ 0 & 1 & 0 \\ 1 & 0 & 0 \end{bmatrix}$

8. (a) – (h) Evaluate $e^{\mathbf{A}}$ and $e^{\mathbf{A}t}$ using *Maple*, where $\mathbf{A}$ is given in the corresponding part of Exercise 7.

9. Whereas (2) was homogeneous, consider the nonhomogeneous initial value problem

$$\mathbf{x}' = \mathbf{A}\mathbf{x} + \mathbf{f}(t); \qquad \mathbf{x}(t_0) = \mathbf{c} \qquad (9.1)$$

and suppose that $\mathbf{A}$ is diagonalizable. Derive the solution

$$\mathbf{x}(t) = \mathbf{Q}\mathbf{E}(t - t_0)\mathbf{Q}^{-1}\mathbf{c} + \int_{t_0}^{t} \mathbf{Q}\mathbf{E}(t - \tau)\mathbf{Q}^{-1}\mathbf{f}(\tau)d\tau. \quad (9.2)$$

NOTE: An analogous result can be derived for the case where $\mathbf{A}$ is not diagonalizable, but we will not consider that case.

10. Use (9.2) to solve the problem

$$\begin{array}{ll} x_1' = x_2 + 2t - 3; & x_1(0) = 3 \\ x_2' = \quad x_1 - t^2; & x_2(0) = 4. \end{array}$$

CHAPTER 10 REVIEW

This chapter is about solving systems of linear coupled differential equations. It suffices to consider first-order systems (i.e., in which the highest derivatives are of first order) because a higher-order system can be reduced to a first-order system by introducing one or more auxiliary dependent variables (Section 10.1).

The fundamental existence and uniqueness theorem (Theorem 10.2.1) assures the existence and uniqueness of a solution to the initial value problem

$$\mathbf{x}' = \mathbf{A}(t)\mathbf{x} + \mathbf{f}(t); \qquad \mathbf{x}(a) = \mathbf{b} \qquad (1)$$

if the $n \times n$ matrix $\mathbf{A}(t)$ and the $n \times 1$ vector $\mathbf{f}(t)$ are continuous on the t interval of interest. If there is no initial condition the general solution is of the form

$$\mathbf{x}(t) = C_1\mathbf{x}_1(t) + \cdots + C_n\mathbf{x}_n(t) + \mathbf{x}_p(t), \qquad (2)$$

where the $\mathbf{x}_j$'s are LI solutions of the homogeneous equation $\mathbf{x}' = \mathbf{A}(t)\mathbf{x}$, $\mathbf{x}_p(t)$ is any particular solution of the full equation $\mathbf{x}' = \mathbf{A}(t)\mathbf{x} + \mathbf{f}(t)$, the C_j's are arbitrary constants, and n is the number of differential equations in the system.

If $n = 1$ we can solve $x' = A(t)x + f(t)$ explicitly; indeed, we did so in Chapter 2 [where we used the notation $y' + p(x)y = q(x)$ instead]. But if $n > 1$ then, in general,

$$\mathbf{x}' = \mathbf{A}(t)\mathbf{x} + \mathbf{f}(t) \qquad (3)$$

is too difficult to solve explicitly (although it can be solved numerically, as will be discussed in Chapter 12) unless the coefficient matrix $\mathbf{A}$ is constant. Thus, in Sections 10.3–10.7 we restricted $\mathbf{A}$ to be constant.

In Section 10.3 we developed a powerful solution method known as elimination, patterned after the Gauss elimination of systems of linear algebraic equations, which converted the system of coupled differential equations to uncoupled equations of higher order on each of the unknowns, which equations can be solved by the methods presented in Chapter 6. Alternatively, we found it convenient to use Cramer's rule in place of the elimination procedure. An important application to a coupled mechanical oscillator was given in Section 10.3.2.

In Sections 10.4–10.7 we considered the homogeneous version,

$$\mathbf{x}' = \mathbf{Ax}. \tag{4}$$

In Section 10.4 we sought solutions of (4) in the form

$$\mathbf{x}(t) = \mathbf{q}e^{rt}. \tag{5}$$

Putting (5) into (4) leads to an eigenvalue problem

$$\mathbf{Aq} = r\mathbf{q} \tag{6}$$

with each eigenpair giving a solution of the form (5). Superposition of those solutions gives a general solution of (4) if the n eigenvalues are distinct. If any eigenvalue of $\mathbf{A}$ is repeated, say of multiplicity k, then it is said to be *defective* if the dimension of its eigenspace is less than k. In that case the solution obtained by this method falls short of being a general solution of (4). As a physical application we considered the coupled mechanical oscillator of Section 10.3.2 once again and identified the free oscillation as the superposition of *orthogonal modes*.

Still another alternative solution method was considered in Section 10.5, diagonalization. An $n \times n$ matrix $\mathbf{A}$ is *diagonalizable* if there exists a $\mathbf{Q}$ matrix such that

$$\mathbf{Q}^{-1}\mathbf{AQ} = \mathbf{D} \tag{7}$$

is diagonal. For $\mathbf{A}$ to be diagonalizable it is necessary and sufficient that it have n LI eigenvectors, and it is sufficient (but not necessary) that it have n distinct eigenvalues. If $\mathbf{A}$ *is* diagonalizable then a *modal matrix* $\mathbf{Q}$ can be formed by letting its columns be the eigenvectors of $\mathbf{A}$, in which case the diagonal elements of $\mathbf{D}$ will be the corresponding eigenvalues of $\mathbf{A}$. To use (7) we make a change of variables $\mathbf{x}(t) = \mathbf{Q}\widetilde{\mathbf{x}}(t)$ in (4) and obtain

$$\widetilde{\mathbf{x}}' = \mathbf{Q}^{-1}\mathbf{AQ}\widetilde{\mathbf{x}} \tag{8}$$

or

$$\widetilde{\mathbf{x}}' = \mathbf{D}\widetilde{\mathbf{x}}, \tag{9}$$

which system is now uncoupled (because $\mathbf{D}$ is diagonal) and is readily solved for $\widetilde{\mathbf{x}}$. With $\widetilde{\mathbf{x}}$ found, the solution of (4) is then obtained as $\mathbf{x}(t) = \mathbf{Q}\widetilde{\mathbf{x}}(t)$.

The exceptional case, where $\mathbf{A}$ is *not* diagonalizable, is considered in Section 10.6. There, we showed how to construct a *generalized modal matrix* $\mathbf{P}$ so that

$$\mathbf{P}^{-1}\mathbf{AP} = \mathbf{J}, \tag{10}$$

known as the *Jordan form* for $\mathbf{A}$, is upper triangular. Like $\mathbf{D}$ in (7), $\mathbf{J}$ has the eigenvalues of $\mathbf{A}$ as its diagonal elements, but it will have one or more 1's immediately above its main diagonal. With $\mathbf{P}$ determined, the change of variables $\mathbf{x}(t) = \mathbf{P}\widetilde{\mathbf{x}}(t)$ reduces (4) to

$$\widetilde{\mathbf{x}}' = \mathbf{J}\widetilde{\mathbf{x}}, \tag{11}$$

which system can be solved from the bottom up because $\mathbf{J}$ is upper triangular.

Finally, in Section 10.7 we used the methods of diagonalization (if $\mathbf{A}$ is diagonalizable) and triangularization (if $\mathbf{A}$ is not diagonalizable) to obtain explicit solutions to the problem

$$\mathbf{x}' = \mathbf{A}\mathbf{x}; \qquad \mathbf{x}(0) = \mathbf{c} \tag{12}$$

and showed that in either case the solution can be expressed in terms of the *matrix exponential function* as

$$\mathbf{x}(t) = e^{\mathbf{A}t}\mathbf{c},$$

where

$$e^{\mathbf{A}} = \mathbf{I} + \mathbf{A} + \frac{1}{2!}\mathbf{A}^2 + \frac{1}{3!}\mathbf{A}^3 + \cdots . \tag{13}$$

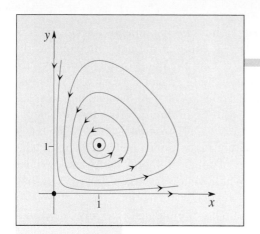

11

QUALITATIVE METHODS;
THE PHASE PLANE

11.1 INTRODUCTION

We did learn how to solve certain types of nonlinear first-order differential equations in Section 3.2 (namely, nonlinear equations that are separable) and in Section 3.4 (namely, nonlinear equations that are exact or can be made exact by an integrating factor), but all the rest of our development thus far has been restricted to linear differential equations. For instance, our standard technique of building up a general solution of a differential equation as a linear combination of homogeneous solutions plus a particular solution (i.e., using superposition) applies only if the differential equation $L[y] = f(x)$ is linear.

In this chapter and the next we discuss methods that apply even if the differential equation is nonlinear. The two methods, the phase plane method in this chapter and the method of numerical integration in Chapter 12, complement each other in that phase plane analysis is essentially qualitative, combining topological and geometric ideas, whereas numerical integration is entirely quantitative.

Historically, interest in nonlinear differential equations is as old as the subject of differential equations itself, which dates back to Newton, but little progress was made until the late 1880s when the great mathematician and astronomer *Henri Poincaré* (1854–1912) took up a systematic study of the subject in connection with celestial mechanics. Realizing that nonlinear equations are rarely solvable analytically, and not having the benefit of computers to generate solutions numerically, he sidestepped the search for solutions altogether and instead sought to answer fundamental questions about the qualitative and topological nature of solutions of nonlinear differential equations without actually finding them.

Though Poincaré's work was motivated primarily by problems of celestial mechanics, the subject began to attract broader attention during and following World War II, especially in connection with nonlinear control theory. In the postwar years, interest was stimulated further by the publication in English of N. Minorsky's *Nonlinear Mechanics* (Ann Arbor, MI: J. W. Edwards) in 1947. With that and other books, such as A. Andronov and C. Chaikin's *Theory of Oscillations* (Princeton: Princeton University Press, 1949) and J. J. Stoker's *Nonlinear Vibrations* (New York: Interscience, 1950) available as texts, the subject appeared in university curricula by the end of the 1950's. With that base, and the availability of digital computers by then, the subject of nonlinear dynamics, generally known now as *dynamical systems*, has blossomed into one of the most active research areas, with applications well beyond celestial mechanics and engineering, for example to biological systems, the social sciences, economics, and chemistry.

The introductory Section 7.3 on the phase plane is a prerequisite for this chapter.

11.2 THE PHASE PLANE

In Section 7.3 we introduced the phase plane as part of our study of the free vibration of the linear harmonic oscillator. In the present chapter we shift our emphasis from the harmonic oscillator application to the phase plane method itself.

11.2.1 The phase plane method.   The method applies to systems of the form

$$\frac{dx}{dt} = P(x, y),$$
$$\frac{dy}{dt} = Q(x, y),$$

(1a,b)

where the independent variable t is usually the time, and where the physical significance of the dependent variables x and y will vary from one application to another, as we will see in Examples 1 and 2.

E X A M P L E   1   *Lotka-Volterra Population Dynamics*.
Consider an ecological system consisting of two species, one a predator and the other its prey. According to the Lotka-Volterra model, the populations $y(t)$ of the predator and $x(t)$ of the prey are governed by the system

$$\frac{dx}{dt} = (\alpha - \beta y)x,$$

(2a)

$$\frac{dy}{dt} = (-\gamma + \delta x)y,$$

(2b)

where $\alpha, \beta, \gamma, \delta$ are nonnegative empirical constants.   In this case $P(x, y)$ is $(\alpha - \beta y)x$ and $Q(x, y)$ is $(-\gamma + \delta x)y$, and $x(t)$ and $y(t)$ are populations.   ∎

E X A M P L E 2 *Damped Mechanical Oscillator*.
Consider the equation

$$mx'' + cx' + kx = 0 \tag{3}$$

governing the free vibration of a damped mechanical oscillator; m, c, k are the mass, damping coefficient, and spring stiffness, respectively. Although (3) is not of the form (1), it can be reduced to that form by introducing an auxiliary or "artificial" dependent variable y according to $x' \equiv y$, for then (3) is equivalent to the system

$$x' = y, \tag{4a}$$

$$y' = -\frac{k}{m}x - \frac{c}{m}y, \tag{4b}$$

which *is* of the form (1). In this example x is displacement and $y = x'$ is velocity. ∎

The system (1) need *not* be linear; for instance, (4) happens to be linear but (2) is not (because of the xy terms). In fact, we will see that the phase plane method is most valuable in the difficult case when the system is nonlinear.

Since we restrict P and Q in (1) to be functions of x and y but *not* to depend explicitly on t, we say that the system (1) is **autonomous**. Since (1) is autonomous we can eliminate t entirely by dividing (1b) by (1a), as was discussed in Section 7.3, obtaining the differential equation

$$\boxed{\frac{dy}{dx} = \frac{Q(x, y)}{P(x, y)}} \tag{5}$$

on x and y.

If we are able to solve (5) we obtain a family of relations on x and y, a family because of the arbitrary constant of integration. That family defines the phase trajectories (or, more simply, the **trajectories**) in the x, y **phase plane**. A plot of numerous representative trajectories is known as a **phase portrait** of (1).

Thus, instead of solving the coupled first-order system (1) for $x(t)$ and $y(t)$ and presenting the results as graphs of x and y versus t, as we did in Chapter 10, according to the phase plane method we solve the single first-order equation (5) for the trajectories and plot them in the x, y phase plane. The phase plane method does not depend upon our being able to solve (5) analytically. We will see that even if we are not able to solve (5) we can still develop the phase portrait using a combination of "critical point" analysis and *Maple* (or other software).

11.2.2 The nonlinear pendulum. To illustrate the ideas already outlined and to motivate the importance of singular point analysis (which follows in Section 11.2.3), consider the free oscillation of a pendulum in the absence of friction and air resistance (Fig. 1).

To derive its equation of motion, recall from physics that for a rigid body undergoing pure rotation about an axis the inertia of the body about that axis times the body's angular acceleration is equal to the torque applied about that axis. For the

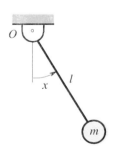

FIGURE 1
Pendulum.

pendulum, the pivot axis is at the pin O, the inertia of the mass m about O is ml^2, the angular acceleration is $x''(t)$ [since $x(t)$ is the angular displacement], and the torque about O due to the downward gravitational force mg is $-mgl \sin x$. Thus, the equation of motion is $ml^2 x'' = -mgl \sin x$ or

$$x'' + \frac{g}{l} \sin x = 0, \tag{6}$$

which is nonlinear because of the $\sin x$ term.[1] For definiteness, suppose that $g/l = 1$ so (6) becomes

$$\boxed{x'' + \sin x = 0.} \tag{7}$$

Following the program outlined above, we re-express (7) as the equivalent system

$$x' = y, \tag{8a}$$
$$y' = -\sin x. \tag{8b}$$

Next, division of (8b) by (8a) gives

$$\frac{dy}{dx} = -\frac{\sin x}{y}. \tag{9}$$

The latter is separable and the step

$$\int y \, dy = -\int \sin x \, dx \tag{10}$$

gives

$$\frac{1}{2} y^2 = \cos x + C \tag{11}$$

for the trajectories. In this example the equation (11) of the trajectories happens to be a statement of conservation of energy; that fact will not be used here so we relegate its discussion to the exercises.

Finally, if we plot the trajectories (11) for various values of C we obtain the phase portrait shown in Fig. 2.

How did we decide in which direction to draw the arrows in Fig. 2? In this example we see from (8a) that if $y > 0$ then $x' > 0$ (since $x' = y$), and if $y < 0$ then $x' < 0$; thus, the "flow" is rightward in the upper half plane and leftward in the lower half plane.

Besides the flow *direction* we are also interested in the flow *speed* in the phase plane. Since the velocity components of the flow are x' and y', it follows from Fig. 3 and the Pythagorean theorem that the flow speed is

$$\boxed{s' = \sqrt{x'^2 + y'^2},} \tag{12}$$

---

[1] In case that is not clear, recall that $\sin x = x - x^3/3! + x^5/5! - \cdots$; that is, $\sin x$ is not merely a constant times x.

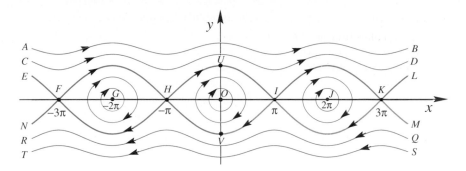

FIGURE 2   Phase portrait of $x'' + \sin x = 0$.

where primes denote d/dt and $s(t)$ is arc length along the trajectory of interest. We call s' the **phase speed** and emphasize that it is the speed of the representative point P in the x, y phase plane, not the physical speed of the mass m in three-space. From (12) and (1) we see that

$$s' = \sqrt{P^2(x, y) + Q^2(x, y)} \tag{13}$$

varies from point to point in the phase plane.

Before returning to Fig. 2 to see what it can tell us, let us also introduce the idea of an equilibrium point of the system (1). We say that (x_0, y_0) is an **equilibrium point** of (1) if

$$x(t) = x_0, \quad y(t) = y_0 \tag{14}$$

is a **point solution** of (1). That is, if the representative point $P = (x(t), y(t))$ starts at (x_0, y_0) then it remains there forever; the phase velocity is zero there.[1] Since x' and y' are zero at an equilibrium point it follows from (1) that equilibrium points (x_0, y_0), if any, are found as solutions of the simultaneous equations

$$\boxed{\begin{aligned} P(x, y) &= 0, \\ Q(x, y) &= 0. \end{aligned}} \tag{15a,b}$$

In the pendulum example (8) and (15) give

$$y = 0 \quad \text{and} \quad -\sin x = 0. \tag{16a,b}$$

The solution sets of (16a) and (16b) are the x axis and the lines $x = n\pi$ ($n = 0, \pm 1, \pm 2, \dots$), respectively, so the solution set of the system (16) is the intersection of those two sets (Fig. 4), namely, the dotted points on the x axis in Fig. 4. Thus, the points $F, G, H, O, I, J, K, \dots$ in Fig. 2 are the equilibrium points of (8).

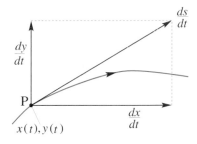

FIGURE 3
Phase speed s'.

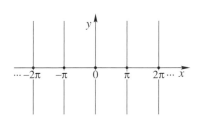

FIGURE 4
Equilibrium points of (8).

---

[1] In fluid mechanics terminology, we would say that (x_0, y_0) is a *stagnation point* of the flow defined by (1).

Of special importance in Fig. 2 is the **separatrix**, which is the collection of trajectories linking E with L and M with N, and which is depicted with slightly thicker lines for emphasis. It is called a separatrix because it separates qualitatively different flows, an interior flow consisting of closed loops and an exterior flow consisting of "open" paths on the interval $-\infty < x < \infty$. (We're not in a position to give a precise definition of the term separatrix, but the foregoing idea, that a separatrix is a trajectory or collection of trajectories that separates qualitatively different flows, should suffice for our purposes.)

What value of C in (11) gives the separatrix? At the point I, for instance, which lies on the separatrix, $x = \pi$ and $y = 0$, so (11) gives $0 = -1 + C$. Hence $C = 1$. Thus, the upper and lower branches of the separatrix are given by

$$y = \pm\sqrt{2\cos x + 2}. \tag{17}$$

For instance, $y = +2$ at U and $y = -2$ at V.

Now we can explain how we selected C values in (11) to generate Fig. 2. We (arbitrarily) chose to have five trajectories cross the positive y axis and to have five cross the negative y axis, including the upper and lower branches of the separatrix. Since $y = +2$ at U and -2 at V, we sought the trajectories through $(0, \pm2/3)$, $(0, \pm4/3)$, $(0, \pm2)$, $(0, \pm8/3)$, and $(0, \pm10/3)$. Putting $(0, \pm10/3)$, say, into (11) gives $C = 41/9$ so the implicit form

$$\frac{1}{2}y^2 = \cos x + \frac{41}{9} \tag{18}$$

or, equivalently, the explicit form

$$y = \pm\sqrt{2\cos x + \frac{82}{9}}, \tag{19}$$

yields the trajectories AB and ST. Similarly, $(0, \pm8/3)$ gives $C = 23/9$, $(0, \pm2)$ gives $C = 1$, $(0, \pm4/3)$ gives $C = -1/9$, and $(0, \pm2/3)$ gives $C = -7/9$. These C values yield all the remaining trajectories shown in Fig. 2, which you can verify using *Maple* or other software, as outlined at the end of this section.

It is also important to understand Fig. 2 in physical terms. To do so, imagine the following experiment. If the pendulum hangs straight down and is at rest ($x' = 0$), then it will remain at rest. That "motion" corresponds to a point trajectory, namely, the equilibrium point O. If, instead, with $x(0) = 0$ we impart an initial velocity $x'(0) = y(0) > 0$ to the mass by striking it with a hammer then, beginning at point W (Fig. 5) at $t = 0$ the representative point P will traverse the closed loop trajectory Γ_1 repeatedly so that the projection x (i.e., onto the x axis) undergoes a periodic motion of amplitude A; physically, the pendulum swings back and forth with amplitude A.[1] If we strike the mass harder, so as to start the motion closer to U, then there results a periodic motion Γ_2 of larger amplitude, and so on.

---

[1]Of course, to generate the motion Γ_1 we don't need to start at W, we can start at *any* point on Γ_1.

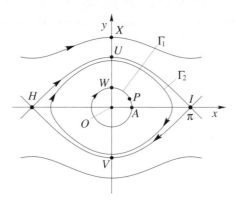

FIGURE 5   The periodic motions inside
$HUIV$.

As we impart more initial energy the resulting "orbit" not only becomes larger, it also becomes more and more distorted so as to conform to the pointed "football" shape of the separatrix $HUIVH$. As we choose orbits closer and closer to the separatrix, we find that the period of the motion tends to infinity as the orbit approaches the separatrix; by the **period** we mean the time that it takes, beginning at any point on Γ, to traverse Γ one time and return to the starting point. And if we strike the mass hard enough so as to start at the point U *on* the separatrix, then the point P will approach I as $t \to \infty$ but will not reach it in finite time.

We leave the details of these claims to the exercises, but support them here heuristically by noting that the phase speed

$$s' = \sqrt{x'^2 + y'^2} = \sqrt{y^2 + \sin^2 x} \tag{20}$$

is not only zero *at* the equilibrium point I, it tends to zero as we *approach* I. Thus, the flow very close to I is very slow, so for orbits closer and closer to I the "residence time" of P in the neighborhood of I is larger and larger. Hence the period of Γ is greater and greater as Γ approaches $HUIVH$.

If we impart still more energy so as to start at X, say, then the pendulum does slow down (i.e., $x' = y$ decreases) as the pendulum approaches the inverted position $x = \pi$, but has enough energy to go "up and over." We would not say that this motion in the phase plane is periodic because the representative point does not undergo a closed loop, as seen from CD in Fig. 2.

Do you see how we can initiate motions within the other "footballs?" To obtain an orbit around J, say, in Fig. 2, rotate the pendulum by hand counterclockwise through one revolution, then let it hang motionless. That step puts us at J. If at time $t = 0$ we then strike the mass so as to impart a nonzero value of $y(0)$ [where $-2 < y(0) < 2$ so the representative point P remains within the separatrix] we will generate an orbit within the separatrix between I and K.

11.2.3 Singular points and their stability.   In our discussion of the pendulum

(a)

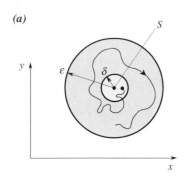

(b)

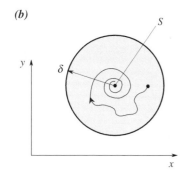

FIGURE 6
Stability and
asymptotic stability.

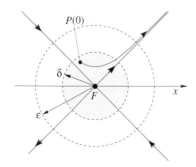

FIGURE 7
Breakout.

problem we can see the important role played by the equilibrium points and their impact on the overall flow field in the phase plane. Observe that because $P(x, y) = 0$ and $Q(x, y) = 0$ at an equilibrium point, the slope

$$\frac{dy}{dx} = \frac{Q(x, y)}{P(x, y)} = \frac{0}{0} \tag{21}$$

is undefined at such a point; put differently, a unique slope does not exist there. To illustrate that point note that *two* different slopes exist at the equilibrium points F, H, I, and K in Fig. 2 and that *none* exist at the equilibrium points G, O, and J. Thus, we say that an equilibrium point of (1) is necessarily a **singular point** since a unique value of dy/dx is not defined there. (Some authors use the term **critical point** instead.)

Physically, the concept of an equilibrium point is closely related to the concept of stability. For instance, we know that a marble is in equilibrium in a valley or on a hill, and that the former is said to be stable and the latter unstable. Similarly, we classify equilibrium points (i.e., singular points) in the phase plane as stable or unstable, as follows.

A singular point S of the autonomous system (1) is said to be **stable** *if motions (i.e., trajectories) that start out close to S remain close to S.* To make that intuitively stated definition precise, let us understand the distance between any two points $P_1 = (x_1, y_1)$ and $P_2 = (x_2, y_2)$ as the usual Euclidean distance $\sqrt{(x_1 - x_2)^2 + (y_1 - y_2)^2}$ and let $P(t) = (x(t), y(t))$ denote the representative point, as usual. Then, a singular point S of (1) is stable if, given any $\epsilon > 0$ (i.e., as small as we wish) there corresponds a $\delta > 0$ such that $P(t)$ will remain closer to S than ϵ for all $t > 0$ if it starts out closer to S than δ (Fig. 6a).[1] If S is not stable then it is **unstable**.

Further, we say that S is not only stable but **asymptotically stable** if motions that start out sufficiently close to S not only stay close to but actually approach S as $t \to \infty$. That is, S is asymptotically stable if there is a $\delta > 0$ such that the distance between $P(t)$ and S tends to zero as $t \to \infty$ if $P(t)$ starts closer to S than δ (Fig. 6b).

For instance, in Fig. 2 the singular point G is stable because the trajectories that start out close to G remain close to G, but it is not asymptotically stable because those trajectories do not approach G as $t \to \infty$. Similarly for O and J (and all the singular points on the x axis at $x = 0, \pm 2\pi, \pm 4\pi, \ldots$).

However, the singular point F is unstable—even though if $P(0)$ is on EF or on the lower branch of HF then $P(t) \to F$ as $t \to \infty$. For no matter how small we make δ, we cannot keep *all* trajectories originating within the δ-circle (Fig. 7) within a circle of radius $\epsilon = 0.5$, say, for all $t > 0$.

Of the three possible types of singular point (asymptotically stable, stable but not asymptotically stable, unstable) the latter two occur in Fig. 2 but the first (asymptotically stable) does not. That case is illustrated in Figs. 7 and 8 of Section 7.3, in each of which the origin is an asymptotically stable singular point.

---

[1]In general we can expect that the smaller we choose ϵ to be, the smaller will δ have to be. Thus, $\delta(\epsilon)$ will be a function of ϵ.

Closure.   To apply the phase plane method to an autonomous system

$$\frac{dx}{dt} = P(x, y), \tag{22a}$$

$$\frac{dy}{dt} = Q(x, y) \tag{22b}$$

we divide the second of these by the first to obtain

$$\frac{dy}{dx} = \frac{Q(x, y)}{P(x, y)}, \tag{23}$$

in which the time t is not present. Thus, rather than try to solve the coupled equations (22) for $x(t)$ and $y(t)$, which might be too difficult a task if (22) is nonlinear, we try to solve the single equation (23). We will see, in the subsequent sections, that even if we are not able to solve (23) we can still develop the phase portrait of (22) using a blend of singular point analysis and *Maple* (or other software).

To illustrate the method, we studied the nonlinear pendulum equation $x'' + \sin x = 0$ or, equivalently, the system

$$x' = y,$$
$$y' = -\sin x,$$

and found that the resulting phase portrait (Fig. 2) revealed, all in one plot, the variety of motions that are possible—periodic motions corresponding to the closed loop trajectories inside the separatrix, more energetic "up and over" motions corresponding to the trajectories above and below the separatrix, and the borderline motions corresponding to segments of the separatrix itself.

We found that the equilibrium points played an important role so we focused attention on equilibrium points and their stability in Section 11.2.3. Since, at an equilibrium point, both $P(x, y)$ and $Q(x, y)$ are zero, we can see from (23) that the slope dy/dx is not uniquely defined there. In that sense an equilibrium point is a singular point and we will subsequently call such points *singular points*.

Maple.   As we mentioned, the phase portrait in Fig. 2 can be obtained by plotting the trajectories defined implicitly by

$$\frac{y^2}{2} - \cos x = C,$$

for the values $C = -7/9, -1/9, 1, 23/9,$ and $41/9$. This can be done using the *Maple* commands

```
with(plots):
implicitplot({y^2/2-cos(x)=-7/9,y^2/2-cos(x)=-1/9,
y^2/2-cos(x)=1,y^2/2-cos(x)=23/9,y^2/2-cos(x)=41/9},
x=-11..11,y=-5..5,numpoints=2000);
```

where we find that the numpoints option is needed to get good plotting accuracy near the equilibrium points F, H, I, and K.

If the equation $dy/dx = Q(x, y)/P(x, y)$ is not readily integrated we can still obtain the phase portrait using the powerful *Maple* phaseportrait command that was discussed in Section 7.3. For instance, the commands

```
with(DEtools):
phaseportrait({diff(x(t),t)=y(t),diff(y(t),t)=-sin(x(t))},
{x(t),y(t)},t=-10..10,[[x(0)=0,y(0)=2/3],[x(0)=0,y(0)=4/3],
[x(0)=0,y(0)=2],[x(0)=0,y(0)=8/3],[x(0)=0,y(0)=-2],
[x(0)=2*Pi,y(0)=2]],x=-11..11,y=-4..4,
scene=[x,y],stepsize=.05,arrows=NONE);
```

will generate (in Fig. 2) the two orbits inside $HUIVH$, CD, IVH and the upper half of the separatrix between H and K. As mentioned, the scene $= [x, y]$ option is unnecessary since that option is the default choice.

If, instead, we desire plots of $x(t)$ versus t we can obtain them from a modified version of the foregoing phaseportrait command. Specifically, change the $t = -10..10$ to $t = 0..40$ or whatever t interval is desired, delete the $x = -11..11$, $y = -4..4$, and change the scene option to scene $= [t, x]$.

EXERCISES 11.2

1. Determine the equation of the phase trajectories for the given system, and sketch several representative trajectories. Use arrows to indicate the direction of movement along those trajectories.

(a) $x' = y$, $y' = -x$ (b) $x' = xy$, $y' = -x^2$
(c) $x' = y^2$, $y' = -xy$

2. Determine the equation of the phase trajectories and sketch enough representative trajectories to show the essential features of the phase portrait. Use arrows to indicate the direction of movement along those trajectories.

(a) $x' = y$, $y' = -y$ (b) $x' = y$, $y' = y$
(c) $x' = y$, $y' = x$ (d) $x' = y$, $y' = 9x$
(e) $x' = x$, $y' = x$ (f) $x' = x$, $y' = -4x$

3. Generate Fig. 2 using the *Maple* implicitplot command and then again using the phaseportrait command. In the latter case use the option arrows $=$ THIN to obtain thin direction field arrows (whereas we omitted them in Fig. 2). HINT: See the discussion of Fig. 2 at the end of this section.

4. (*Graphical interpretation of phase speed*) Consider the system (1) for the special case where $P(x, y) = y$. [This case is particularly important since it occurs whenever we reduce a second-order equation $x'' = f(x, x')$ to a system of two first-order equations by setting $x' \equiv y$.] From

the accompanying sketch, show that in that case the phase speed s' can be interpreted graphically as

$$\boxed{s' = a,} \qquad (4.1)$$

where a is the distance, perpendicular to s', from E to the x axis. [But remember that (4.1) holds only if $P(x, y)$ is y.]

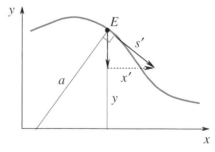

5. Let us continue our discussion from Exercise 4. Suppose $P(x, y)$ is y in (1) so that (4.1) holds, and consider a trajectory Γ that crosses the x axis at an angle other than 90°, as sketched at the left, below. Since $x' = y$, it follows that the flow directions must be as indicated by the arrows, and

that Q is necessarily an equilibrium point. Let it suffice to consider the simpler case where Γ is straight, as shown at the right. From (4.1) it follows that $s' \to 0$ as the representative point P approaches Q (along the upper branch or along the lower branch). That fact is necessary but not sufficient to conclude that P approaches Q as $t \to \infty$ but does not reach Q in finite time. The problem that we pose is this: Show that P does not reach Q in finite time. HINT: Write the equation of Γ as $y = mx + b$, say, and remember that y is x'.

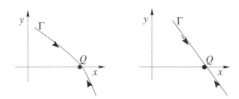

6. Consider the orbit Γ_1 in Fig. 5. We are interested in the period of that orbit as a function of the amplitude A, for $0 < A < \pi$. At one extreme, as $A \to 0$, we claim that the period corresponding to the full equation $x'' + \sin x = 0$ approaches the period corresponding to the linearized equation $x'' + x = 0$. And at the other extreme, as $A \to \pi$, we claim that the period tends to infinity. Here, we ask you to explore those two claims with calculations. Specifically, use the *Maple* phaseportrait command to solve (7) subject to the initial conditions $x(0) = A$, $y(0) = 0$, for $A = 0.1, 0.5, 1, 2, 2.5, 3, 3.1$, and to plot $x(t)$ versus t for these cases. From these plots estimate the period for each case and then plot (a simple hand plot will suffice) the period versus A, for those A values. Do your results appear to support the foregoing claims for the limiting cases $A \to 0$ and $A \to \pi$? Explain.

7. (*Lotka-Volterra equations*) Consider the Lotka-Volterra system (2), with $\alpha = \beta = \gamma = \delta = 1$ for definiteness. Since x and y are populations they cannot be negative, so we are interested only in the first quadrant of the phase plane: $0 \leq x < \infty, 0 \leq y < \infty$.

(**a**) Determine, from the differential equations, any equilibrium points.

(**b**) Use the *Maple* phaseportrait command to obtain the direction field over $0 \leq x \leq 3$ and $0 \leq y \leq 3$, say. To avoid also obtaining trajectories you can take both initial points to be at the origin; that is, $x(0) = 0$ and $y(0) = 0$, which choice will merely produce the point trajectory at the origin.

(**c**) Directly on your direction field plot sketch, by hand, the trajectories through the points $(2, 0)$, $(0, 2)$, $(1, 0.25)$, $(1, 0.5)$, and $(1, 0.8)$.

(**d**) Determine, by any means you wish, the periods of the orbits through $(1, 0.25)$, $(1, 0.5)$, and $(1, 0.8)$.

8. Just as we developed the phase portrait for the nonlinear pendulum equation

$$x'' + \sin x = 0 \qquad (8.1)$$

in Fig. 2, we ask you to do the same, even following the same steps and using *Maple* as needed, for the equation

$$x'' + x - x^3 = 0. \qquad (8.2)$$

HINT: This time you will find only three equilibrium points; be sure to identify them. There is a separatrix; be sure to include it and to label it in your phase portrait.

9. Consider the system

$$x' = ax + by, \qquad (9.1a)$$

$$y' = cx + dy, \qquad (9.1b)$$

where a, b, c, d are given constants. Show, in each case, that (9.1) has just one singular point, namely, at $x = y = 0$. Then, determine whether that singular point is stable or unstable. If it is stable, determine whether or not it is asymptotically stable. HINT: Solve (9.1) for $x(t)$ and $y(t)$ by any method studied in Chapter 10, and examine those solutions.

(**a**) $a = 0, b = 1, c = -9, d = 0$
(**b**) (b) $a = 0, b = 1, c = 4, d = 0$
(**c**) $a = 0, b = 1, c = -5, d = -2$
(**d**) (d) $a = 0, b = 1, c = -5, d = 2$
(**e**) (e) $a = 1, b = -5, c = 2, d = -1$
(**f**) (f) $a = 1, b = 1, c = -1, d = 1$
(**g**) $a = 1, b = 1, c = 1, d = 2$
(**h**) (h) $a = 1, b = 2, c = 2, d = 1$

10. (*Conservation of energy*)

(**a**) Show that (11) is a statement of conservation of energy (i.e., the kinetic energy plus the potential energy is a constant).

(**b**) More generally, suppose that Newton's second law gives

$$mx'' = F(x), \qquad (10.1)$$

that is, where the force F is an explicit function of x but not of x' or t. Show that if we multiply (10.1) by dx and integrate we obtain

$$\frac{1}{2}mx'^2 + V(x) = \text{constant} \qquad (10.2)$$

where $V(x)$ is $-\int F(x)\,dx$. The latter is a statement of conservation of energy and $V(x)$ is called the *potential energy* of the system. Further, the system (10.1) is therefore said to be *conservative*.

11.3 SINGULAR POINT ANALYSIS

11.3.1 Motivation.   Suppose we wish to study the graph of a given function $f(x)$. We know from the calculus that a useful first step is to solve $f'(x) = 0$, for the roots of that equation give the locations of any extrema of $f(x)$. Suppose x_0 is such a root. To "see what we've got" at x_0 (a maximum, a minimum, or a horizontal inflection point) we expand $f(x)$ in a Taylor series about x_0 as

$$f(x) = f(x_0) + f'(x_0)(x - x_0) + \frac{f''(x_0)}{2!}(x - x_0)^2 + \frac{f'''(x_0)}{3!}(x - x_0)^3 + \cdots$$
$$= f(x_0) + \frac{f''(x_0)}{2!}(x - x_0)^2 + \frac{f'''(x_0)}{3!}(x - x_0)^3 + \cdots. \tag{1}$$

Next, we approximate f *locally*, near x_0, by the truncated version

$$f(x) \approx f(x_0) + \frac{f''(x_0)}{2!}(x - x_0)^2 \tag{2}$$

of (1), which is a parabolic "fit" to the graph of f at x_0. If $f''(x_0) > 0$ then f has a minimum at x_0, if $f''(x_0) < 0$ it has a maximum at x_0, and in the borderline case where $f''(x_0) = 0$ we can't tell without further examination. In the latter case we must compute $f'''(x_0)$, $f''''(x_0)$, and so on, until we come to the first *non*vanishing derivative there, say $f^{(k)}(x_0)$. If k is odd then f has a horizontal inflection point at x_0, if k is even then it has a maximum if $f^{(k)}(x_0) < 0$ and a minimum if $f^{(k)}(x_0) > 0$.[1]

To illustrate, consider $f(x) = x^2 e^{-x}$. Then $f'(x) = 0$ gives $x_0 = 0$ and $x_0 = 2$, and (2) gives

$$f(x) \approx 0 + x^2 \tag{3}$$

near $x = 0$, and

$$f(x) \approx 4e^{-2} - e^{-2}(x - 2)^2 \tag{4}$$

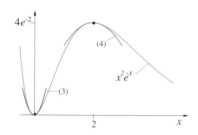

$4e^{-2}$

(4)

$x^2 e^{-x}$

(3)

2

x

FIGURE 1
Key features of $x^2 e^{-x}$.

near $x = 2$. Plotting the two local approximations (3) and (4) in Fig. 1, as well as the graph of $f(x) = x^2 e^{-x}$, we see how the global graph of f is established, to a large degree, by the local behaviors (3) and (4) near the two extrema.

Likewise, we shall see that the global phase portrait of the system

$$x' = P(x, y), \tag{5a}$$
$$y' = Q(x, y) \tag{5b}$$

is established, to a large degree, by the local flows near the various singular points of (5), that is, the equilibrium points (x_0, y_0) at which $P(x, y) = Q(x, y) = 0$. For example, recall the phase portrait of the system

$$x' = y, \tag{6a}$$
$$y' = -\sin x \tag{6b}$$

shown in Fig. 2 of Section 11.2, and observe how the global flow pattern is so much determined by the local flow patterns near the singular points at $x = n\pi$ on the x axis, a few of which are sketched here in Fig. 2; roughly speaking, the global flow pattern can be thought of as an interpolation of the singular point flows shown here in Fig. 2.

Suppose that a singular point (x_0, y_0) is known. To ascertain the local flow near (x_0, y_0) we can expand $P(x, y)$ and $Q(x, y)$ in Taylor series about (x_0, y_0) using equation (5) from Section 10.5:

$$x' = P(x_0, y_0) + P_x(x_0, y_0)(x - x_0) + P_y(x_0, y_0)(y - y_0) + \cdots, \tag{7a}$$

$$y' = Q(x_0, y_0) + Q_x(x_0, y_0)(x - x_0) + Q_y(x_0, y_0)(y - y_0) + \cdots. \tag{7b}$$

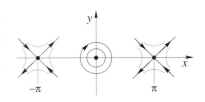

FIGURE 2
The flow near the singular points of (6).

Both leading terms are zero because (x_0, y_0) is a singular point of (5). Deleting those terms and cutting off (7a) and (7b) after the first-order terms gives the approximations

$$\boxed{\begin{aligned} x' &\approx P_x(x_0, y_0)(x - x_0) + P_y(x_0, y_0)(y - y_0), \\ y' &\approx Q_x(x_0, y_0)(x - x_0) + Q_y(x_0, y_0)(y - y_0). \end{aligned}} \tag{8a,b}$$

The idea is that whereas (5) is in general nonlinear, the local approximation (8) is linear and hence readily solved. Our hope is that the solution of (8) gives a local approximation to the solution of (5) near (x_0, y_0) somewhat as (2) gives a local approximation to f near x_0. In general that hope is indeed fulfilled, as explained at the end of this section.

11.3.2 The linearized equations and their singularities. To study the linearized equations (8), it is convenient to move the origin to (x_0, y_0) by the change of variables

$$X = x - x_0, \qquad Y = y - y_0. \tag{9}$$

If we also denote the constants $P_x(x_0, y_0)$, $P_y(x_0, y_0)$, $Q_x(x_0, y_0)$, and $Q_y(x_0, y_0)$ as a, b, c, d, respectively, then (8) becomes

$$\boxed{\begin{aligned} X' &= aX + bY, \\ Y' &= cX + dY, \end{aligned}} \tag{10a,b}$$

which system is now the focus of our attention. Singular points of (10) are found by setting

$$aX + bY = 0, \tag{11a}$$

$$cX + dY = 0, \tag{11b}$$

which system gives the unique solution $X = Y = 0$ provided that

$$\boxed{\begin{vmatrix} a & b \\ c & d \end{vmatrix} = ad - bc \neq 0.} \tag{12}$$

---

[1]Be careful; it is a common misconception that if $f'(x_0) = f''(x_0) = 0$ then f necessarily has a horizontal inflection point at x_0.

If $ab - bc$ is zero then the solutions of (11) constitute either a line through the origin or the whole X, Y plane (the latter case occurring in the uninteresting case where $a = b = c = d = 0$). We will assume that the condition (12) is satisfied so that *(10) has a singular point only at the origin* [just as the original system (5) presumably has only an isolated[1] singular point at (x_0, y_0)].

E X A M P L E   1
The system

$$x' = 2x^3 - y, \tag{13a}$$

$$y' = y^2 \sin \pi x \tag{13b}$$

has singular points at the roots of the simultaneous equations

$$P(x, y) = 2x^3 - y = 0, \tag{14a}$$

$$Q(x, y) = y^2 \sin \pi x = 0. \tag{14b}$$

These roots are at the intersection of the curve $y = 2x^3$ and the vertical lines $x = 0, \pm 1, \pm 2, \ldots$ in the x, y plane, namely, the points $(0, 0)$, $(1, 2)$, $(-1, -2)$, $(2, 16)$, $(-2, -16)$, and so on.

Consider first the singular point at $(1, 2)$, say. With $x_0 = 1$ and $y_0 = 2$, (8) gives

$$x' = (6x^2)\big|_{1,2}(x - 1) + (-1)\big|_{1,2}(y - 2) = 6(x - 1) - (y - 2),$$

$$y' = (\pi y^2 \cos \pi x)\big|_{1,2}(x - 1) + (2y \sin \pi x)\big|_{1,2}(y - 2) = -4\pi(x - 1) + 0(y - 2)$$

or, with $X = x - 1$ and $Y = y - 2$,

$$X' = 6X - Y, \tag{15a}$$

$$Y' = -4\pi X + 0Y. \tag{15b}$$

Then $ad - bc = -4\pi \neq 0$ so the origin $X = Y = 0$ ($x = 1$ and $y = 2$ in the original x, y plane) is indeed the only singular point of (15).

Next, consider the singular point at $(0, 0)$ in the x, y plane. With $x_0 = 0$ and $y_0 = 0$, (8) gives

$$x' = (6x^2)\big|_{0,0} x + (-1)\big|_{0,0} y = 0x - y,$$

$$y' = (\pi y^2 \cos \pi x)\big|_{0,0} x + (2y \sin \pi x)\big|_{0,0} y = 0x + 0y$$

or, with $X = x$ and $Y = y$,

$$X' = 0X - Y, \tag{16a}$$

$$Y' = 0X + 0Y. \tag{16b}$$

---

[1]A singular point is said to be *isolated* if the distance from it to the nearest other singular point (if there is another) is greater than δ for some number $\delta > 0$.

In this case $ad - bc = 0$ so the origin is *not* an isolated singular point of (16) [just as $(0, 0)$ is not an isolated singular point of (13)]. Rather, every point on the line $Y = 0$ (i.e., the X axis) through the origin is a singular point of (16).

Thus, whereas the singular point $(1, 2)$ of (13) will be covered by the methods that will be developed in this section (because $ad - bc \neq 0$), the singular point $(0, 0)$ of (13) will not (because $ad - bc = 0$ in that case). ∎

To proceed, let us solve (10) by seeking a solution in the form

$$\boxed{\begin{aligned} X(t) &= q_1 e^{rt}, \\ Y(t) &= q_2 e^{rt}, \end{aligned}} \tag{17a,b}$$

which method was explained in Section 10.4. Putting (17) into (10) gives, after canceling e^{rt} factors and expressing the equations in matrix form,

$$\begin{bmatrix} a & b \\ c & d \end{bmatrix} \begin{bmatrix} q_1 \\ q_2 \end{bmatrix} = r \begin{bmatrix} q_1 \\ q_2 \end{bmatrix} \tag{18}$$

or

$$\mathbf{A}\mathbf{q} = r\mathbf{q}.$$

Since we seek a general solution of (10) we are not interested in the "trivial solution" $q_1 = q_2 = 0$ of (18) because the latter merely gives $X(t) = 0$ and $Y(t) = 0$. Thus, (18) is an eigenvalue problem with "λ"$= r$. Suppose that $\mathbf{A}$ admits distinct eigenvalues λ_1 and λ_2 with corresponding eigenvectors

$$\mathbf{e}_1 = \alpha \begin{bmatrix} e_{11} \\ e_{12} \end{bmatrix} \quad \text{and} \quad \mathbf{e}_2 = \beta \begin{bmatrix} e_{21} \\ e_{22} \end{bmatrix}. \tag{19}$$

(The case where $\lambda_1 = \lambda_2$ will be considered last.) Then (10) admits the general solution

$$\begin{aligned} \begin{bmatrix} X(t) \\ Y(t) \end{bmatrix} &= \mathbf{e}_1 e^{\lambda_1 t} + \mathbf{e}_2 e^{\lambda_2 t} \\ &= \alpha \begin{bmatrix} e_{11} \\ e_{12} \end{bmatrix} e^{\lambda_1 t} + \beta \begin{bmatrix} e_{21} \\ e_{22} \end{bmatrix} e^{\lambda_2 t} \end{aligned} \tag{20}$$

or, in scalar form,

$$\boxed{\begin{aligned} X(t) &= \alpha e_{11} e^{\lambda_1 t} + \beta e_{21} e^{\lambda_2 t}, \\ Y(t) &= \alpha e_{12} e^{\lambda_1 t} + \beta e_{22} e^{\lambda_2 t}, \end{aligned}} \tag{21a,b}$$

where α, β are arbitrary constants.

The eigenvalues of $\mathbf{A}$ are found as the roots of

$$\det(\mathbf{A} - \lambda \mathbf{I}) = \begin{vmatrix} a - \lambda & b \\ c & d - \lambda \end{vmatrix} = \lambda^2 - (a + d)\lambda + (ad - bc) = 0, \tag{22}$$

namely,

$$\lambda = \frac{a + d \pm \sqrt{(a+d)^2 - 4(ad - bc)}}{2}$$

$$= \frac{a + d \pm \sqrt{(a-d)^2 + 4bc}}{2}. \tag{23}$$

In particular, we can see from (22) that $\lambda = 0$ *will not be among the eigenvalues of* **A** since if $\lambda = 0$ then (22) becomes $\det\mathbf{A} = ad - bc = 0$, which is contrary to our assumption (12).

The qualitative nature of the flow in the phase plane, defined by (21), will depend upon the roots (23). There are four types:

1. purely imaginary eigenvalues

2. complex conjugate eigenvalues

3. real eigenvalues of the same sign

4. real eigenvalues of opposite sign

Consider these cases in turn.

1. Purely imaginary eigenvalues. (CENTER)   If $a + d = 0$ and $(a - d)^2 + 4bc = 4(a^2 + bc) < 0$ in (23), then

$$\lambda = \pm i \sqrt{|a^2 + 4bc|} \equiv \pm i\omega, \tag{24}$$

say. That is, $\lambda_1 = +i\omega$ and $\lambda_2 = -i\omega$.[1] Since the $e^{\lambda_1 t}$ and $e^{\lambda_2 t}$ terms in (21) are $e^{i\omega t}$ and $e^{-i\omega t}$ we see that both the $X(t)$ and $Y(t)$ motions will be linear combinations of $\cos \omega t$ and $\sin \omega t$ and will therefore be oscillatory. We state without proof that *the resulting trajectories in the X, Y phase plane will be a family of ellipses the principal axes of which might be the X, Y axes but which, in general, will be rotated.* In this case the singular point at the origin is called a **center**. Let us illustrate.

E X A M P L E   2
As an example of (10) consider the system

$$X' = 4Y, \tag{25a}$$

$$Y' = -X. \tag{25b}$$

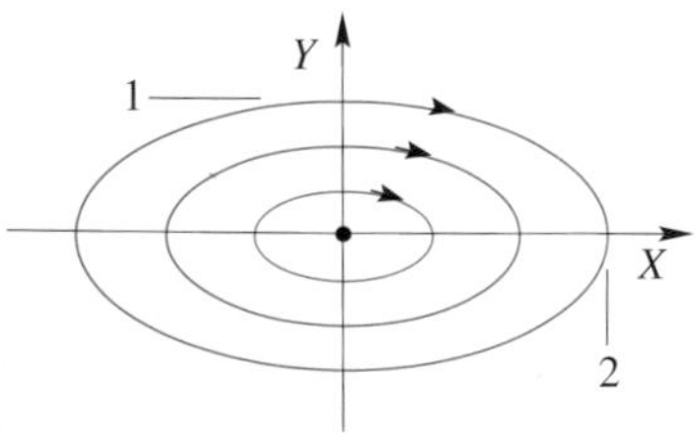

Then $a = 0$, $b = 4$, $c = -1$, and $d = 0$ so (23) gives $\lambda = \pm 2i$. Rather than use the solution form (21) it is simpler in this case to obtain the phase trajectories directly in the usual way. That is, dividing (25b) by (25a) and integrating gives

$$X^2 + 4Y^2 = C \tag{26}$$

which is the family of ellipses shown in Fig. 3.   ∎

FIGURE 3
Example 2; a center.

---

[1]Recall that the case of complex eigenvalues and eigenvectors was discussed in the optional Section 9.2.3. That section is recommended but is not essential for the present discussion.

EXAMPLE 3

Consider the system

$$X' = X - 3Y, \tag{27a}$$
$$Y' = 2X - Y. \tag{27b}$$

Then $a = 1$, $b = -3$, $c = 2$, and $d = -1$ so (23) gives $\lambda = \pm\sqrt{5}i$. Dividing (27b) by (27a) and integrating gives the trajectories

$$2X^2 - 2XY + 3Y^2 = C \tag{28}$$

which, like (26), is a family of ellipses, but with their axes rotated by 31.7°, as shown in Fig. 4 (Exercise 7).

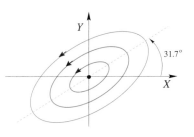

FIGURE 4
Example 3; a center.

COMMENT. In Example 2 the flow was clockwise since (25a) showed that $X' > 0$ in the upper half plane and $X' < 0$ in the lower half plane. In the present example it suffices to look along the X axis, say, by setting $Y = 0$ in (27a). Then $X' = X$ is positive for $X > 0$ and negative for $X < 0$ so the flow is counterclockwise.[1] You might be wondering if this logic suffices, for might not the flow be counterclockwise over one part of the ellipse and clockwise over another part of it, for instance as shown in Fig. 5? No, for that occurrence would require the presence of equilibrium (i.e., singular) points at A and B whereas we are assured by (10) that the only singular point of (27) is at the origin. ∎

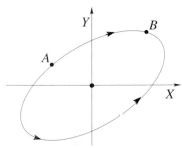

FIGURE 5
Is this possible?

Observe that a center is stable, though not asymptotically stable since (in intuitive language) points initially close to the origin remain close to it but do not approach it as $t \to \infty$.

2. Complex conjugate eigenvalues. (SPIRAL)   If a, b, c, d are real numbers such that the square root in (23) is imaginary and $a + d \neq 0$, then we can see from (21) that each of $X(t)$ and $Y(t)$ is a linear combination of terms of the form

$$e^{(a+d)t/2}e^{\pm i\sqrt{4(ad-bc)-(a+d)^2}\,t/2}$$

so each of $X(t)$ and $Y(t)$ is of the form $e^{(a+d)t/2}$ times a linear combination of $\cos\left[\sqrt{4(ad-bc)-(a+d)^2}\,t/2\right]$ and $\sin\left[\sqrt{4(ad-bc)-(a+d)^2}\,t/2\right]$. Thus, the motion is an oscillation that grows as $e^{(a+d)t/2}$ if $a + d > 0$ or decays as $e^{(a+d)t/2}$ if $a + d < 0$. Accordingly, the trajectories either spiral out away from the origin or in toward the origin so the singularity is called a **spiral** (or **focus**). If $a + d < 0$ the spiral will be *asymptotically stable*; if $a + d > 0$ it will be *unstable*.

---

[1] If you use the *Maple* phaseportrait command to generate the phase portrait and do not include the option arrows = NONE, then besides the trajectories the plot will also include the direction field with barbed arrows, so there will be no question as to the flow direction.

(a)

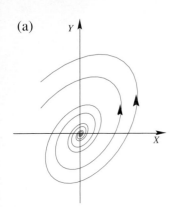

(b)

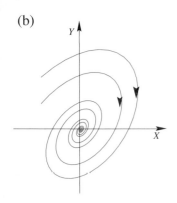

FIGURE 6
Unstable and stable
spirals.

EXAMPLE 4
Consider the system

$$X' = 2X - 4Y, \tag{29a}$$
$$Y' = 3X - Y. \tag{29b}$$

Then $a = 2$, $b = -4$, $c = 3$, and $d = -1$ so (23) gives $\lambda = \frac{1}{2} \pm i\frac{\sqrt{39}}{2}$. We could divide (29b) by (29a) and integrate the resulting differential equation but the steps would be tedious. Let it suffice to use the *Maple* phaseportrait command, which gives the representative trajectories shown in Fig. 6a.

COMMENT.   If we reverse all signs on the right-hand side of (29), so that the system becomes

$$X' = -2X + 4Y, \tag{30a}$$
$$Y' = -3X + Y, \tag{30b}$$

then the ratio $dY/dX = (-3X+Y)/(-2X+4Y) = (3X-Y)/(2X-4Y)$ is the same as for (29) so the trajectories of (30) are the same as for (28) but the flow direction is reversed; that is, the arrows are reversed so the singularity is an asymptotically stable spiral (Fig. 6b). ∎

3. Real eigenvalues of the same sign. (NODE) Suppose first that the λ roots are both *negative* and, for definiteness, let us number them so that $\lambda_2 < \lambda_1 < 0$. Then in the limit as $t \to \infty$, (21) gives

$$\begin{aligned}
X(t) &= \alpha e_{11}e^{\lambda_1 t} + \beta e_{21}e^{\lambda_2 t} \\
&= e^{\lambda_1 t}\left[\alpha e_{11} + \beta e_{21}e^{-(\lambda_1 - \lambda_2)t}\right] \sim \alpha e_{11}e^{\lambda_1 t}
\end{aligned} \tag{31a}$$

because $\exp[-(\lambda_1 - \lambda_2)t] \to 0$ as $t \to \infty$, and

$$\begin{aligned}
Y(t) &= \alpha e_{12}e^{\lambda_1 t} + \beta e_{22}e^{\lambda_2 t} \\
&= e^{\lambda_1 t}\left[\alpha e_{12} + \beta e_{22}e^{-(\lambda_1 - \lambda_2)t}\right] \sim \alpha e_{12}e^{\lambda_1 t},
\end{aligned} \tag{31b}$$

from which it follows that as $t \to \infty$ the trajectories tend to a straight line approach to the origin, where the straight line is defined by[1]

$$\frac{Y}{X} = \frac{e_{12}}{e_{11}}, \tag{32}$$

obtained by dividing (31b) by (31a) and canceling the $\alpha e^{\lambda_1 t}$'s, that is, along the line defined by the eigenvector $\mathbf{e}_1 = [e_{11}, e_{12}]^\mathrm{T}$ corresponding to the eigenvalue λ_1. There is an exception: If the initial conditions give $\alpha = 0$ then (21) is simply

$$X(t) = \beta e_{21}e^{\lambda_2 t}, \tag{33a}$$

---

[1]If $e_{11} = 0$ then $e_{12} \neq 0$ (because the eigenvector $\mathbf{e}_1$ cannot be the zero vector) so we interpret (32) as the vertical line $X = 0$.

$$Y(t) = \beta e_{22} e^{\lambda_2 t}, \tag{33b}$$

so instead of approaching the origin along the line defined by (32) the representative point approaches the origin along the line defined by

$$\frac{Y}{X} = \frac{e_{22}}{e_{21}}, \tag{34}$$

that is, along the line defined by the eigenvector $\mathbf{e}_2 = [e_{21}, e_{22}]^{\mathrm{T}}$ corresponding to the eigenvalue λ_2.

EXAMPLE 5
Consider the system

$$X' = -4X + 3Y, \tag{35a}$$
$$Y' = X - 2Y. \tag{35b}$$

Then $a = -4$, $b = 3$, $c = 1$, and $d = -2$ so (23) gives $\lambda_1 = -1$ and $\lambda_2 = -5$. Representative trajectories are shown in Fig. 7.

Observe from the figure that there are two straight line paths of approach to the origin, along the directions of $\mathbf{e}_1$ and $\mathbf{e}_2$, which we call **eigendirections**, in accordance with (32) and (34). The approach along the $\mathbf{e}_2$ eigendirection is relatively fast (with X and Y proportional to e^{-5t}) and the approach along the $\mathbf{e}_1$ eigendirection is relatively slow (with X and Y proportional to e^{-t}) so we call these lines the *fast manifold* and the *slow manifold*, respectively.

What if we start at a point such as B not on one of those manifolds? Let us think of the $\mathbf{e}_1$ and $\mathbf{e}_2$ lines as an oblique coordinate system (Fig. 8). According to (20) the coordinates OC and OD of B along those axes are proportional to $e^{\lambda_1 t}$ (i.e., e^{-t}) and $e^{\lambda_2 t}$ (i.e., e^{-5t}), respectively. Thus, OD diminishes rapidly and OC diminishes relatively slowly, as we can see in Fig. 7.

COMMENT.   To create the phase portrait in Fig. 7 we first established the two straight line trajectories. With those lines in place we were then able to assign eight initial points so as to "fill out" the flow field in a somewhat uniform manner.   ∎

Suppose instead that the λ roots are both *positive* and let us number them so that $0 < \lambda_1 < \lambda_2$. In that case the asymptotic results (31) still hold, but this time as $t \to -\infty$ instead of as $t \to +\infty$. For instance, if we modify (35) by reversing the four signs on the right-hand sides then we will have $\lambda_1 = +1$ and $\lambda_2 = +5$. The flow will be the same as in Fig. 7 but with the arrows reversed.

EXAMPLE 6
Consider the system

$$X' = 2X + Y, \tag{36a}$$
$$Y' = X + 2Y. \tag{36b}$$

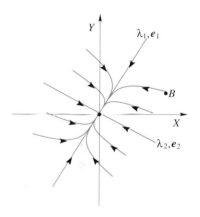

FIGURE 7
Example 5;
asymptotically stable node.

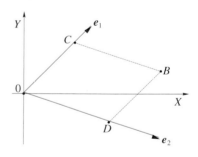

FIGURE 8
The oblique coordinate system.

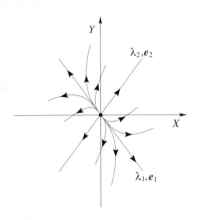

FIGURE 9
Example 6; unstable node.

Then $\lambda_1 = +1$, $\lambda_2 = +3$, and representative trajectories are shown in Fig. 9. In this case the slow and fast manifolds are orthogonal because the **A** matrix in (36) happens to be symmetric.[1]   ■

This type of singularity is known as a **node** and is *asymptotically stable* if both λ's are negative (Fig. 7) and *unstable* if both λ's are positive (Fig. 9).

What happens if two real eigenvalues are *equal*? The nature of the flow then depends upon whether the corresponding eigenspace is one-dimensional or two-dimensional. If it is *one-dimensional* then the fast and slow manifolds merge into a single manifold of attraction (if $\lambda_1 = \lambda_2 < 0$) or repulsion (if $\lambda_1 = \lambda_2 > 0$) defined by the eigenvector, as we now illustrate.

EXAMPLE 7
Consider the system

$$X' = -X + Y, \tag{37a}$$
$$Y' = -X - 3Y. \tag{37b}$$

Then $a = -1$, $b = 1$, $c = -1$, and $d = -3$ so $\lambda_1 = \lambda_2 = -2$ with the one-dimensional eigenspace $\mathbf{e}_1 = \alpha[1, -1]^{\mathrm{T}}$. Thus, instead of the general solution (20) we obtain only the partial solution

$$\left[\begin{array}{c} X(t) \\ Y(t) \end{array} \right] = \mathbf{e}_1 e^{\lambda_1 t} = \alpha \left[\begin{array}{c} 1 \\ -1 \end{array} \right] e^{-2t}. \tag{38}$$

[One way to see that (38) falls short of being a general solution is to observe that (38) contains only one arbitrary constant, whereas a general solution of (37) must contain two arbitrary constants.] To obtain a general solution we could solve (37) by the method of elimination (Section 10.3), by the method of triangularization (Section 10.6), or by the method of Section 10.7.3. By any of these methods we obtain the general solution

$$\left[\begin{array}{c} X(t) \\ Y(t) \end{array} \right] = \alpha \left[\begin{array}{c} 1 \\ -1 \end{array} \right] e^{-2t} + \beta \left[\begin{array}{c} t \\ -1 - t \end{array} \right] e^{-2t}. \tag{39}$$

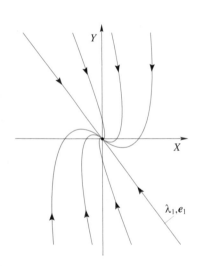

FIGURE 10
Example 7; repeated eigenvalue with one-dimensional eigenspace.

As $t \to \infty$ the te^{-2t} terms dominate the e^{-2t} terms so

$$\left[\begin{array}{c} X(t) \\ Y(t) \end{array} \right] \sim \beta \left[\begin{array}{c} 1 \\ -1 \end{array} \right] te^{-2t}. \tag{40}$$

That is, the flow approaches the line $\beta[1, -1]^{\mathrm{T}}$ (i.e., $Y = -X$) corresponding to the eigenvector $\mathbf{e}_1$ as it approaches the origin,[2] as can also be seen in Fig. 10. Because of the merging of the two eigendirections we call this singularity a **degenerate node**.

---

[1] The slow and fast manifolds don't look orthogonal in Fig. 9, so it must be that the X and Y scales used in the plot were not quite the same.

[2] It does indeed approach the origin since $te^{-2t} \to 0$ in (40) as $t \to \infty$.

COMMENT. This node is asymptotically stable. If we wish to illustrate an unstable node for the case of a repeated eigenvalue with a one-dimensional eigenspace, without making up an entirely new example, it will suffice to reverse the four signs on the right-hand sides of (37). Then we obtain $\lambda_1 = \lambda_2 = +2$, $\mathbf{e}_1 = \alpha[1, -1]^T$, and the same flow field as in Fig. 10 but with the arrows reversed. ∎

Since **A** in (18) is only 2×2, the only way a repeated eigenvalue λ_1 can have a *two-dimensional eigenspace* is if $\mathbf{A} - \lambda_1 \mathbf{I}$ is of rank zero, that is, if $\mathbf{A} - \lambda_1 \mathbf{I}$ is the zero matrix, that is, if

$$\mathbf{A} = \lambda_1 \mathbf{I} = \begin{bmatrix} \lambda_1 & 0 \\ 0 & \lambda_1 \end{bmatrix}, \tag{41}$$

that is, if $\mathbf{A} = \begin{bmatrix} a & b \\ c & d \end{bmatrix}$ is diagonal with $a = d$ and $b = c = 0$. Then our system (9) must be of the simple form

$$X' = aX, \tag{42a}$$
$$Y' = aY, \tag{42b}$$

with general solution

$$X(t) = C_1 e^{at}, \tag{43a}$$
$$Y(t) = C_2 e^{at}, \tag{43b}$$

where C_1 and C_2 are arbitrary constants. In terms of eigenvalues and eigenvectors, we have $\lambda_1 = \lambda_2 = a$ with the two-dimensional eigenspace

$$\mathbf{e} = \begin{bmatrix} \alpha \\ \beta \end{bmatrix} = \alpha \begin{bmatrix} 1 \\ 0 \end{bmatrix} + \beta \begin{bmatrix} 0 \\ 1 \end{bmatrix}. \tag{44}$$

Thus,

$$\begin{bmatrix} X(t) \\ Y(t) \end{bmatrix} = \left(\alpha \begin{bmatrix} 1 \\ 0 \end{bmatrix} + \beta \begin{bmatrix} 0 \\ 1 \end{bmatrix} \right) e^{at} = \begin{bmatrix} \alpha \\ \beta \end{bmatrix} e^{at}, \tag{45}$$

which result is equivalent to (43). From (43) we see that $Y/X = C_2/C_1$ is a constant, and since C_1 and C_2 are arbitrary it follows that every ray through the origin is a trajectory (Fig. 11). Such nodes are called **stars**.

4. Real eigenvalues of opposite sign. (SADDLE) As the last of the four cases suppose that the eigenvalues are real and of opposite sign, with $\lambda_1 < 0 < \lambda_2$. Recall the general solution (20),

$$\begin{bmatrix} X(t) \\ Y(t) \end{bmatrix} = \alpha \begin{bmatrix} e_{11} \\ e_{12} \end{bmatrix} e^{\lambda_1 t} + \beta \begin{bmatrix} e_{21} \\ e_{22} \end{bmatrix} e^{\lambda_2 t}. \tag{46}$$

If the initial conditions give $\beta = 0$ then the solution $\alpha[e_{11}, e_{12}]^T e^{\lambda_1 t}$ amounts to the straight line trajectory $Y/X = e_{12}/e_{11}$ (which is the Y axis if $e_{11} = 0$) along which the flow is toward the origin (since $\lambda_1 < 0$). And if the initial conditions give $\alpha = 0$ then the solution $\beta[e_{21}, e_{22}]^T e^{\lambda_2 t}$ amounts to the straight line trajectory $Y/X = e_{22}/e_{21}$ (which is the Y axis if $e_{21} = 0$) along which the flow is away from the origin (since $\lambda_2 > 0$).

(a)

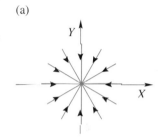

(b)

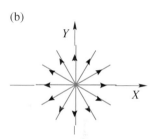

FIGURE 11
Repeated eigenvalue with two-dimensional eigenspace; stable and unstable stars.

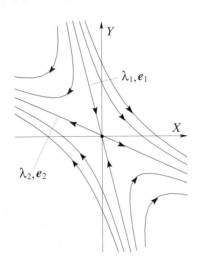

FIGURE 12
Example 8; saddle.

EXAMPLE 8
Consider the system

$$X' = 5X + 3Y, \tag{47a}$$
$$Y' = -3X - 5Y. \tag{47b}$$

Then $\lambda_1 = -4$, $\mathbf{e}_1 = \alpha[1, -3]^T$ and $\lambda_2 = +4$, $\mathbf{e}_2 = \beta[3, -1]^T$. Representative trajectories are shown in Fig. 12. ∎

This type of singularity is called a **saddle** and is always *unstable*.[1]

Having completed our discussion of the four types of singularity of the linear system (10) (namely, centers, spirals, nodes, and saddles), let us summarize the results in Table 11.3.1.

Table 11.3.1   The singularities of $\mathbf{X}' = \mathbf{AX}$.

Eigenvalues of $\mathbf{A}$	Singularity Type	Stability
Purely imaginary	CENTER	Stable
Complex conjugates	SPIRAL	Stable if $a + d < 0$
		Unstable if $a + d > 0$
Real of same sign	NODE	Stable if λ's < 0
		Unstable if λ's > 0
Real of opposite sign	SADDLE	Unstable

11.3.3 Singularities of nonlinear systems.   Recall the plan that was laid out in Section 11.3.1, to study a given singular point (x_0, y_0) of a nonlinear system

$$x' = P(x, y), \tag{48a}$$
$$y' = Q(x, y) \tag{48b}$$

by linearizing the Taylor series of P and Q about that point and considering the approximate equations

$$x' = P_x(x_0, y_0)(x - x_0) + P_y(x_0, y_0)(y - y_0), \tag{49a}$$
$$y' = Q_x(x_0, y_0)(x - x_0) + Q_y(x_0, y_0)(y - y_0), \tag{49b}$$

the latter being readily solved because they are linear. Thus, letting $X = x - x_0$, $Y = y - y_0$, $a = P_x(x_0, y_0)$, $b = P_y(x_0, y_0)$, $c = Q_x(x_0, y_0)$, and $d = Q_y(x_0, y_0)$, we studied the linear system

$$X' = aX + bY, \tag{50a}$$

---

[1] It is true that if the representative point starts on the $\mathbf{e}_1$ manifold then it approaches the origin as $t \to \infty$, but if it starts at any point not on that manifold then it moves arbitrarily far from the origin as $t \to \infty$.

$$Y' = cX + dY. \tag{50b}$$

Keep in mind that we are trying to ascertain the local behavior of the flow corresponding to the original nonlinear system (48) near the singular point (x_0, y_0) by studying the simpler linearized version (50). That plan begs this question: *Is the nature of the flow corresponding to the nonlinear system (48) faithfully captured, near the singular point, by its linearized version (50)?* We state without proof that the answer is yes, subject to the following qualifications:

(i) If the λ's are purely imaginary (i.e., $\lambda = \pm i\omega$ where ω is real) then the linearized version (50) has a center but the nonlinear system (48) can have a center OR a stable or unstable spiral.

(ii) If $\lambda_1 = \lambda_2 < 0$ then the linearized version (50) has a stable node but the nonlinear system (48) can have a stable node OR a stable spiral.

(iii) If $\lambda_1 = \lambda_2 > 0$ then the linearized version (50) can have an unstable node OR an unstable spiral.

The reason for these possible discrepancies is that these are "borderline" cases. In case (i), for instance, observe from (23) that to obtain purely imaginary λ's we need to have $a + d = 0$; thus, we need d to be *exactly* equal to $-a$ in (50). Heuristically, we can imagine the neglected higher-order terms in the Taylor series of $P(x, y)$ and $Q(x, y)$ being equivalent to a slight perturbation of the coefficients a, b, c, and d in (50). For instance, if the a, b, c, d values in (50) give $\lambda = 0 \pm 3i$, say, which values correspond to a center, then slightly perturbed values of a, b, c, d might lead to $\lambda = -0.04 \pm 3.02i$ or $\lambda = +0.01 \pm 3.03i$, say, these values corresponding not to centers but to stable and unstable spirals, respectively. Similarly for cases (ii) and (iii) because $\lambda_1 = \lambda_2$ requires that $(a - d)^2 + 4bc$ must be exactly equal to zero.

EXAMPLE 9   *Nonlinear Oscillator with Cubic Damping.*
Recall that the free vibrations of the damped harmonic oscillator are governed by the linear equation $mx'' + px' + kx = 0$, where $x(t)$ is displacement and m, p, k are the mass, damping coefficient, and spring stiffness, respectively.[1] Suppose that instead of linear damping (i.e., proportional to x') there is nonlinear damping that is proportional to x'^3 so that $x(t)$ is governed instead by the nonlinear equation

$$mx'' + px'^3 + kx = 0. \tag{51}$$

For definiteness, let $m = p = k = 1$. Then, express (51) as the nonlinear system

$$x' = y, \tag{52a}$$

$$y' = -x - y^3. \tag{52b}$$

---

[1]Previously we used c for the damping coefficient. Here we use p instead, to avoid confusing this parameter with the c in (10).

Setting $P(x, y) = y = 0$ and $Q(x, y) = -x - y^3 = 0$ reveals that there is only one singular point, namely, at $x = y = 0$. Observe that the right-hand sides of (52) are already in the form of Taylor series about $(0, 0)$:

$$P(x, y) = y = 0x + 1y + 0x^2 + 0xy + 0y^2 + \cdots,$$

$$Q(x, y) = -x - y^3$$
$$= -1x + 0y + 0x^2 + 0xy + 0y^2$$
$$\quad + 0x^3 + 0x^2y + 0xy^2 - 1y^3 + 0x^4 + \cdots,$$

so the linearized form of (52), about the singular point $(0, 0)$, is

$$X' = Y, \tag{53a}$$
$$Y' = -X, \tag{53b}$$

where $X = x - 0$ and $Y = y - 0$. Then $a = 0$, $b = 1$, $c = -1$, and $d = 0$ so $\lambda = \pm i$. Hence, the linearized version (53) of (52) has a center at the origin. Does the nonlinear system (52) likewise have a center at the origin? Item (i) above indicates that the singularity of (52) at the origin could be a center *or* a spiral. To see which it is let us use the *Maple* phaseportrait command to plot a typical trajectory near the origin. That result (Fig. 13) indicates that (52) has a stable spiral at the origin.   ■

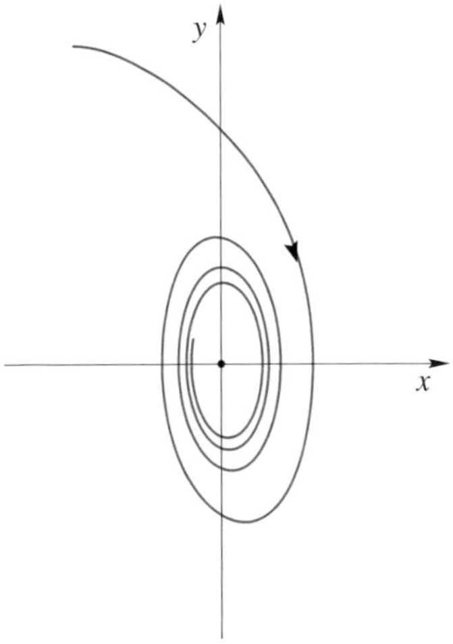

FIGURE 13
The nonlinear system (52) with cubic damping.

Closure.   In this section we have been interested in approximating a nonlinear system (48) in the neighborhood of a singular point by linearizing it about that point, that is, expanding $P(x, y)$ and $Q(x, y)$ in Taylor series about that point and cutting off those series after their first-order terms. Solving the resulting linear system (50) we distinguished four qualitatively distinct types, depending upon the eigenvalues of the matrix $\mathbf{A} = \begin{bmatrix} a & b \\ c & d \end{bmatrix}$: the center (if the λ's are purely imaginary), the spiral (if they are complex conjugates), the node (if they are real and of the same sign), and the saddle (if they are real and of opposite sign). Of these, centers are always stable and saddles are always unstable, but spirals and nodes can be either stable (namely, asymptotically stable) or unstable.

Finally, we stated without proof that the linearized system does indeed faithfully capture the singularity type of the original nonlinear system—except possibly in three cases: when the λ's are purely complex, and when they are real with $\lambda_1 = \lambda_2 < 0$ or with $\lambda_1 = \lambda_2 > 0$.

Maple.   The *Maple* commands used to generate Fig. 6a were as follows:

```
with(DEtools):
phaseportrait({diff(x(t),t)=2*x(t)-4*y(t),diff(y(t),t)=
3*x(t)-y(t)},{x(t),y(t)},t=-10..0,{[x(0)=-2,y(0)=1],
[x(0)=-2,y(0)=2]},stepsize=.05,x=-3..4,y=-3..4,arrows=NONE);
```

EXERCISES 11.3

1. Recall from Sections 2.4.1 and 7.2 that the differential equation governing the displacement $x(t)$ of a linear mechanical oscillator is of the form

$$mx'' + px' + kx = 0. \tag{1.1}$$

Re-express (1.1) equivalently as the system

$$x' = y \tag{1.2a}$$

$$y' = -\frac{k}{m}x - \frac{p}{m}y. \tag{1.2b}$$

(a) Considering the behavior as we vary p over $0 \le p < \infty$, show that the system has a center for $p = 0$, a stable spiral for $0 < p < \sqrt{4km}$ (where $\sqrt{4km}$ is called the *critical* damping coefficient), and a stable node for $p \ge \sqrt{4km}$. (Here we consider $m > 0$ and $p > 0$.)

(b) If we could somehow have $k < 0$ then (with $m > 0$ and $p > 0$) show that the motion is a saddle no matter how large we make p. (That is, if $k < 0$ then we cannot stabilize the singular point at the origin of the x, y phase plane no matter how much we increase the damping coefficient p.)

2. Given $a, b, c,$ and d in (10), classify the singularity of (10) at $X = Y = 0$. Is it stable or unstable? If stable, is it asymptotically stable?

(a) $a = 1, b = 2, c = 3, d = 4$
(b) $a = 1, b = 4, c = 1, d = -1$
(c) $a = 2, b = 0, c = 0, d = 4$
(d) $a = 3, b = 0, c = 1, d = -3$
(e) $a = -3, b = 1, c = 0, d = -3$
(f) $a = 2, b = 4, c = -4, d = -2$
(g) $a = 0, b = 4, c = -1, d = 0$
(h) $a = 1, b = -2, c = 2, d = 1$
(i) $a = -1, b = -2, c = 2, d = -1$
(j) $a = 1, b = 1, c = 0, d = 1$
(k) $a = 5, b = -1, c = 2, d = 4$
(l) $a = 3, b = 1, c = -1, d = 1$

3. (a) – (l) Use *Maple* to generate the phase portrait of the system defined in the corresponding part of Exercise 2. Include enough trajectories, and any key trajectories, to clarify the flow pattern and use arrows, as we have, to show the flow direction.

4. Determine all singular points, if any, and classify each insofar as possible.

(a) $x' = y, \quad y' = 1 - x^4$
(b) $x' = 1 - y^2, \quad y' = 1 - x$

(c) $x' = y, \quad y' = (1 - x^2)/(1 + x^2)$
(d) $x' = x - y, \quad y' = \sin(x + y)$
(e) $x' = (1 - x^2)y, \quad y' = -x - 2y$
(f) $x' = (1 - x^2)y, \quad y' = -x + 2y$
(g) $x' = -2x - y, \quad y' = x + x^3$
(h) $x' = -2x - y, \quad y' = \sin x$
(i) $x' = ye^x - 1, \quad y' = y - x - 1$
(j) $x' = x^2 - 2y, \quad y' = 2x - y$
(k) $x' = x^2 - y^2, \quad y' = x^2 + y - 2$
(l) $x' = y, \quad y' = -3\sin x$
(m) $x' = x + 2y, \quad y' = -x - \sin y$
(n) $x' = (x^6 + 1)y, \quad y' = x^2 - 4$

5. We stated the solution (39) of (37) without derivation. Derive it.

6. In the footnote in Example 7 we stated that $te^{-2t} \to 0$ as $t \to \infty$. Prove that.

7. In Example 3 we claimed that the trajectories

$$2X^2 - 2XY + 3Y^2 = C \tag{7.1}$$

arc ellipses with their axes rotated by $31.7°$, as shown in Fig. 4. Prove that claim by introducing a rotated coordinate system as follows.

(a) If $\bar{X}, \bar{Y}$ is a Cartesian coordinate system that is rotated counterclockwise through an angle α, as shown here, then show that the X, Y and $\bar{X}, \bar{Y}$ coordinates of any given point are related according to

$$X = \bar{X}\cos\alpha - \bar{Y}\sin\alpha, \tag{7.2a}$$

$$Y = \bar{X}\sin\alpha + \bar{Y}\cos\alpha. \tag{7.2b}$$

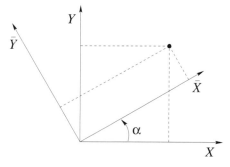

(b) Putting (7.2) into (27), choose α so that the result is of the form

$$\bar{X}' = \beta\bar{Y}, \qquad \bar{Y}' = -\gamma\bar{X}. \tag{7.3}$$

You should find that both constants β, γ are positive or negative so that the integration of $d\bar{Y}/d\bar{X} = -\gamma\bar{X}/\beta\bar{Y}$ gives a family of ellipses. Show that these steps give $\alpha = 31.7°$ and another value as well. Explain the significance of the other α.

(c) Alternatively, and more simply, show that $\alpha = 31.7°$ by putting $X = r\cos\theta$ and $Y = r\sin\theta$ into (7.1) and setting $dr/d\theta = 0$.

8. Prove that the linear system (10) can have one, two, or an infinite number of straight-line trajectories through the origin, but never a finite number greater than two. HINT: Set $Y = \kappa X$ in (10a) and (10b).

9. (*More about Example 9*) Let us compare oscillators having linear damping with those having cubic damping, governed by the equations $x'' + x' + x = 0$ and $x'' + x'^3 + x = 0$. Equivalently, consider the systems

$$x' = y, \quad y' = -x - y \tag{9.1}$$

and

$$x' = y, \quad y' = -x - y^3, \tag{9.2}$$

respectively. Use the *Maple* phaseportrait command to plot the trajectories, both for (9.1) and for (9.2), with the initial point $x(0) = -1$, $y(0) = 1$, say. Does the representative point approach the origin faster for the case of cubic damping or for the case of linear damping? Explain why that is so.

10. (*More about Example 9*) Doing whatever you need to, determine the type of the singular point at the origin. (In each case reduce the second-order ODE to a system of first-order ODE's by letting $x' = y$.)

(a) $x'' - x'^3 + x = 0$ (b) $x'' + x'^3 - x = 0$
(c) $x'' + x'^2 + x = 0$ (d) $x'' - x'^2 + x = 0$

11. (*Higher-order singularities*) Recall that the center, spiral, node, and saddle are the *only* types of singularity of the linear system

$$x' = ax + by, \tag{11.1a}$$
$$y' = cx + dy, \tag{11.1b}$$

assuming that the system is "not too degenerate," namely, assuming that $ad - bc \neq 0$. They are called the **elementary singularities**. (Surely, whether we use X, Y or x, y as the variables is immaterial.)

(a) To illustrate the case of nonelementary singularities consider the equation $x'' + x^2 = 0$ or, equivalently, the system

$$x' = y, \tag{11.2a}$$
$$y' = -x^2 \tag{11.2b}$$

[in which case $ad - bc = (0)(0) - (1)(0) = 0$]. Nevertheless, (11.2) is simple enough so that $dy/dx = -x^2/y$ can be integrated and the trajectories obtained analytically. Do that and then sketch the flow in the neighborhood of the origin, using *Maple* to assist you if necessary. Be sure to determine the equation of any trajectories that pass through the singular point itself (namely, the origin).

(b) Same as (a), for $x'' - x^2 = 0$.
(c) Same as (a), for $x'' + x^3 = 0$.
(d) Same as (a), for $x'' - x^3 = 0$.

11.4 ADDITIONAL APPLICATIONS

11.4.1 Applications. Basing our discussion on three examples, let us use the ideas developed in Section 11.3 to study nonlinear systems in the phase plane.

E X A M P L E 1   *Lotka-Volterra Population Dynamics*.
The Lotka-Volterra model of predator-prey population dynamics was introduced in Section 10.1. With $\alpha = \beta = \delta = \gamma = 1$, say, the governing equations are

$$x' = (1 - y)x, \tag{1a}$$
$$y' = (-1 + x)y, \tag{1b}$$

where $x(t)$, $y(t)$ are the populations of the prey and predator, respectively. Setting $(1 - y)x = 0$ and $(-1 + x)y = 0$ we find two singular (i.e., equilibrium) points, $(0, 0)$ and $(1, 1)$.

Consider first $(0, 0)$. Expanding the right-hand sides of (1) about $(0, 0)$, linearizing, and letting $X = x - 0$ and $Y = y - 0$ gives[1]

$$X' = 1X + 0Y, \tag{2a}$$
$$Y' = 0X - 1Y. \tag{2b}$$

The eigenvalues and eigenvectors of the coefficient matrix are $\lambda_1 = -1$, $\mathbf{e}_1 = \alpha[0, 1]^T$, and $\lambda_2 = 1$, $\mathbf{e}_2 = \beta[1, 0]^T$. Since the λ's are of opposite sign the singularity in (1) at $x = y = 0$ ($X = Y = 0$) is a saddle, as shown in Fig. 1. Of course, since X and Y are populations we are interested only in the first quadrant ($X \geq 0$, $Y \geq 0$) of this plot.

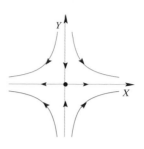

FIGURE 1
Saddle at $x = y = 0$.

Next, consider the singularity at $(1, 1)$. Expanding the right-hand sides of (1) about $(1, 1)$, linearizing, and letting $X = x - 1$ and $Y = y - 1$ gives[2]

$$X' = 0X - 1Y, \tag{3a}$$
$$Y' = 1X - 0Y. \tag{3b}$$

We find that $\lambda = \pm i$ so the singularity of (1) at $x = 1$, $y = 1$ is a center, as shown in Fig. 2. (After studying Section 11.3 you should be able to fill in the gaps in this more concise discussion, such as obtaining the equations of the trajectories in Figs. 1 and 2, the direction of the flow arrows, and so on.)

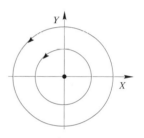

FIGURE 2
Center at $x = y = 1$.

If we place these two local flows into the x, y plane we obtain the picture shown in Fig. 3. To "fill in the gaps" we can integrate the equation

$$\frac{dy}{dx} = \frac{(-1 + x)y}{(1 - y)x} \tag{4}$$

obtained by dividing (1b) by (1a), obtain the equations of the trajectories (in implicit form), and plot any number of them by computer. However, applications are typically complicated enough so that we are not able to solve $dy/dx = Q(x, y)/P(x, y)$ analytically. Thus, our discussion here will presume the use of the *Maple* phaseportrait command or equivalent software.

To proceed in that manner we need to decide on a set of initial points. For instance, we could specify a number of initial points on the line connecting $(1, 0)$ with $(1, 1)$. In addition, it will be important to show the two trajectories through the singular point at the origin, the one along the y axis and the other along the x axis. To obtain the former we can choose an initial point at $x = 0$, $y = 0.01$, say, and to obtain the latter we can choose an initial point at $x = 0.01$, $y = 0$, say. It is important to choose the points very close to the origin since the saddle flow shown at the origin in Fig. 3 is expected to be correct only locally; the saddle flow and the exact flow become identical only in the limit as the singular point is approached. The resulting phase portrait is shown in Fig. 4.

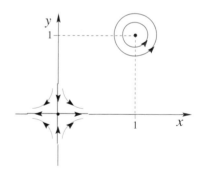

FIGURE 3
Local clues.

---

[1]Of course we don't really need to change letters from x, y to X, Y if $X = x$ and $Y = y$, but we prefer to because the capitalized notation reminds us that we are looking at the locally linearized equations rather than at the original nonlinear equations.

[2]For example, $P(x, y) = (1 - y)x \approx P(1, 1) + P_x(1, 1)(x - 1) + P_y(1, 1)(y - 1) = 0 + 0(x - 1) - 1(y - 1) = 0X - 1Y$ gives (3a); similarly for (3b).

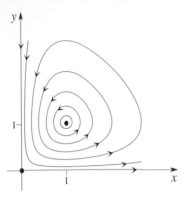

FIGURE 4
Phase portrait for (1).

COMMENT 1.   To illustrate our claim that the linearized flow (i.e., the saddle flow) and the exact flow become identical in the limit as the singular point is approached, we have plotted both the saddle flow (dotted lines) and the exact flow (solid lines) in Fig. 5. Observe that the two sets of curves coincide more and more closely as we approach the origin (i.e., the point about which we expanded and linearized). In this example the exact and linearized trajectories through the origin (namely, the x and y axes) happen to coincide over their entire length.

COMMENT 2.   Observe from Fig. 4 that the motions in the phase plane are periodic motions about the point $(1, 1)$ (except for the trajectories consisting of the positive y and positive x axes). To emphasize this result, and to complement the phase plane plot in Fig. 4, we have plotted x and y versus t in Fig. 6, corresponding to two different initial conditions: $x(0) = 1, y(0) = 0.75$ and $x(0) = 1, y(0) = 0.25$. These correspond to a small almost circular orbit around $(1, 1)$ and a larger more distorted orbit. ∎

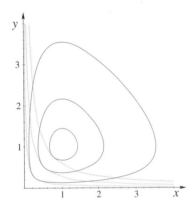

FIGURE 5
Blending of linearized
and exact flows as singularity
(at $x = y = 0$) is approached.

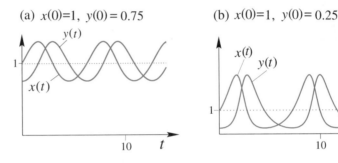

(a) $x(0)=1,\ y(0)=0.75$ (b) $x(0)=1,\ y(0)=0.25$

FIGURE 6 x and y versus t.

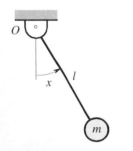

FIGURE 7
Pendulum.

EXAMPLE 2   Nonlinear Pendulum.
We have already studied the equation

$$x'' + \frac{g}{l} \sin x = 0 \tag{5}$$

governing the motion of a pendulum (Fig. 7) in the absence of friction or air resistance in Section 7.2. At that time we had not yet studied singular points so let us return to that problem now, this time paying careful attention to the singular points. Also, this time let us include the effects of air resistance. If the air resistance is proportional to the velocity lx' of the mass m, say clx' where c is a constant, then it gives an additional torque $-cl^2x'$ about O, so the equation of motion is $ml^2x'' = -mgl \sin x - cl^2x'$ or $x'' + px' + (g/l) \sin x = 0$ where $p \equiv c/m$. As in Section 11.2, let us take $g/l = 1$ for definiteness. Then

$$x'' + px' + \sin x = 0, \tag{6}$$

where p is a known constant and $p \geq 0$.

To employ the phase plane approach we first re-express (6) in the equivalent form

$$x' = y, \tag{7a}$$
$$y' = -\sin x - py. \tag{7b}$$

Setting the right-hand sides equal to zero gives the infinite set of singular points $x = n\pi$, $y = 0$ ($n = 0, \pm 1, \pm 2, \ldots$). Expanding $\sin x$ about $x = n\pi$ gives

$$\sin x = \sin n\pi + (\cos n\pi)(x - n\pi) + \cdots$$
$$= 0 + (-1)^n (x - n\pi) + \cdots$$

since $\sin n\pi = 0$ and $\cos n\pi = (-1)^n$ (namely, $+1$ if $n = 0, \pm 2, \pm 4, \ldots$ and -1 if $n = \pm 1, \pm 3, \ldots$). Thus, setting $X = x - n\pi$ and $Y = y$ and linearizing about $(n\pi, 0)$, we obtain from (7)

$$X' = 0X + 1Y, \tag{8a}$$
$$Y' = -(-1)^n X - pY. \tag{8b}$$

With $a = 0$, $b = 1$, $c = -(-1)^n$, and $d = -p$ we obtain

$$\lambda = \frac{-p \pm \sqrt{p^2 - 4(-1)^n}}{2} = \begin{cases} \dfrac{-p \pm \sqrt{p^2 - 4}}{2}, & n \text{ even} \\ \dfrac{-p \pm \sqrt{p^2 + 4}}{2}, & n \text{ odd.} \end{cases} \tag{9}$$

If $p = 0$, then (9) gives

$$\lambda = \begin{cases} \pm i, & n \text{ even} \\ \pm 1, & n \text{ odd} \end{cases} \tag{10}$$

so the singular points $(n\pi, 0)$ are centers (or possibly spirals[1]) for $n = 0, \pm 2, \pm 4, \ldots$ and saddles for $n = \pm 1, \pm 3, \ldots$. That case (namely, with no damping) was already covered in Section 11.2 and the phase portrait in Fig. 2 therein reveals that the singularities at $(n\pi, 0)$ for $n = 0, \pm 2, \pm 4, \ldots$ are centers, not spirals.

Here, let us consider the case where $p > 0$. In particular, consider p's such that $0 < p < 2$. Then from (9) we see that the singular points are stable spirals if n is even and saddles if n is odd. The phase portrait is shown in Fig. 8 for the representative case $p = 0.5$. The lightly shaded region (not including the boundaries AB and CD) is called the **basin of attraction** for the stable focus at $(2\pi, 0)$, the basin of attraction of an asymptotically stable singular point S being the set of all initial points P_0 such that the representative point "$P(t)$" tends to S as $t \to \infty$ if $P(0) = P_0$. Similarly, each of the other stable foci has its own basin of attraction.

So much for $0 < p < 2$. If we increase p beyond the critical damping coefficient $p_{cr} = 2$, so that $p > 2$, then we see from (9) that the stable spirals at $(n\pi, 0)$, where n is even, give way to stable nodes at those points, and those points are approached without oscillation. This case is left to the exercises.

---

[1] Recall item (i) from Section 11.3.3.

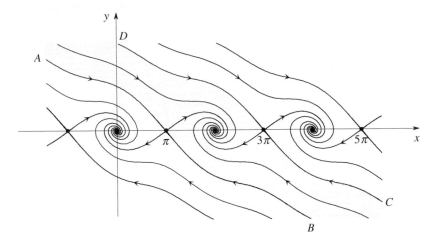

FIGURE 8   Phase portrait of (6) for $p = 0.5$.

COMMENT 1.   The trajectories that approach and leave the saddles along the eigendirections in Fig. 8 are of particular interest. For instance, the ones between A and B and between C and D together define the basin of attraction for the stable spiral at $(2\pi, 0)$. To obtain the trajectories through the saddle at $(\pi, 0)$, for example, using the *Maple* phaseportrait command or other such software, we need to identify initial points on each of those four trajectories. Of course the point $(\pi, 0)$ itself won't do because it is an equilibrium point; if we start there we remain there. Rather, we obtain the eigenvalues $\lambda_1 = -1.2808$ and $\lambda_2 = 0.7808$ of (8) (with $n = 1$ and $p = 0.5$) and their corresponding eigenvectors $\mathbf{e}_1 = \alpha[-0.7808, 1]^{\mathrm{T}}$ and $\mathbf{e}_2 = \beta[1.2808, 1]^{\mathrm{T}}$. Of the two eigenvalues, λ_1 gives the stable manifold and λ_2 the unstable manifold. For the trajectory from A to $(\pi, 0)$, for example, we can use $\mathbf{e}_1$ with a very small α such as $\alpha = 0.001$ to obtain an initial point $x(0) = \pi + 0.001(-0.7808) = 3.14081$ and $y(0) = 0 + 0.001(1) = 0.001$. For the trajectory from B to $(\pi, 0)$ we can use $\mathbf{e}_1$ with $\alpha = -0.001$, say, and obtain $x(0) = \pi - 0.001(-0.7808) = 3.14237$ and $y(0) = 0 - 0.001(1) = -0.001$. For the trajectory from $(\pi, 0)$ to $(2\pi, 0)$ we can use $\mathbf{e}_2$ with $\beta = 0.001$, say, and obtain $x(0) = \pi + 0.001(1.2808) = 3.14287$ and $y(0) = 0 + 0.001(1) = 0.001$, and similarly for the trajectory from $(\pi, 0)$ to $(0, 0)$, but with $\beta = -0.001$ instead. Additional details regarding the generation of Fig. 8 are given in the *Maple* subsection at the end of this section.

COMMENT 2.   It turns out that the nonlinear pendulum equation is also prominent in connection with a superconducting device known as a *Josephson junction*. For discussion of the Josephson junction within the context of nonlinear dynamics, we recommend the book *Nonlinear Dynamics and Chaos* (Reading, MA: Addison-Wesley, 1994) by Steven H. Strogatz.   ■

E X A M P L E 3   *Van Der Pol Equation*.

The differential equation

$$x'' - \epsilon(1 - x^2)x' + x = 0, \qquad (\epsilon > 0) \qquad (11)$$

which is nonlinear because of the x^2x' term, was studied by *Balthasar van der Pol* (1889–1959), first in connection with current oscillations in a certain vacuum tube circuit and then in connection with the modeling of the beating of the human heart.[1] Known as the **van der Pol equation**, it is among the most important nonlinear differential equations in applied mathematics. We assume that the parameter ϵ is positive, as is usually the case in applications.

To study (11) in the phase plane we first re-express it, equivalently, as the system

$$x' = y, \qquad (12a)$$
$$y' = -x + \epsilon(1 - x^2)y, \qquad (12b)$$

which has only one singular point, at $(0, 0)$. To determine the type of that singular point we linearize (12) about $(0, 0)$, obtaining

$$x' = y, \qquad (13a)$$
$$y' = -x + \epsilon y. \qquad (13b)$$

With $a = 0, b = 1, c = -1$, and $d = \epsilon$ we obtain

$$\lambda = \frac{\epsilon \pm \sqrt{\epsilon^2 - 4}}{2} \qquad (14)$$

so the singularity at the origin $(0, 0)$ is an *unstable spiral* if $\epsilon < 2$ and an *unstable node* if $\epsilon > 2$. Realize that (13) is equivalent to the single ODE $x'' - \epsilon x' + x = 0$, which is the familiar (from Chapter 7) equation of a damped harmonic oscillator, but with *negative damping*.

Near the origin in the x, y phase plane the flow is accurately described by the linearized system (13) and is shown (for $\epsilon < 2$, say) in Fig. 9a. As the motion increases in magnitude, the neglected nonlinear term $\epsilon x^2 x'$ ceases to be negligible and we wonder how the trajectory shown in Fig. 9a continues to develop as t increases. To determine that development we will need the full equations (12) rather than the version (13) that was linearized about the origin. Before turning to *Maple* and plotting the results let us think about what to expect. Similar to (11), recall the equation

$$mx'' + cx' + kx = 0 \qquad (15)$$

governing a linear mechanical mass/spring/damping system. Comparing (11) with (15), we can think of m and k as 1 and $c = -\epsilon(1 - x^2)$ as a sort of damping coefficient, at least roughly speaking since it is not merely a constant but varies with x.

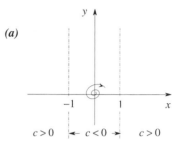

(a)

$c > 0$ $\leftarrow$ $c < 0$ $\rightarrow$ $c > 0$

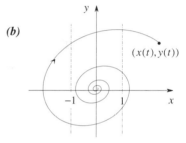

(b)

$(x(t), y(t))$

FIGURE 9
The unstable spiral at $(0, 0)$, for $\epsilon < 2$.

---

[1]B. van der Pol, "On Relaxation Oscillations," *Philosophical Magazine*, Vol. 2, (1926), pp. 978–992, and B. van der Pol and J. van der Mark, "The Heartbeat As a Relaxation Oscillation, and An Electrical Model of the Heart," *Philosophical Magazine*, Vol.6, (1928), pp. 763–775.

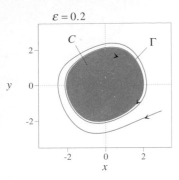

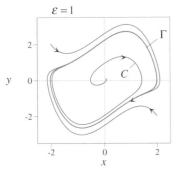

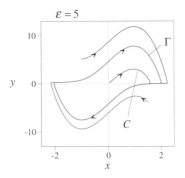

FIGURE 10
The van der Pol limit
cycle, for $\epsilon = 0.2$, 1, and 5.

Since the damping coefficient $c = -\epsilon(1 - x^2)$ in (11) is negative in the vertical strip $|x| < 1$ (between the dashed lines in Fig. 9a) we expect the spiral to continue to grow. Eventually the spiral will break out of the $|x| < 1$ strip (Fig. 9b). As the representative point $(x(t), y(t))$ spends more and more time outside that strip, where $c = -\epsilon(1 - x^2) > 0$, the effect of the positive damping in $|x| > 1$ will increase, relative to the the effect of the negative damping in $|x| < 1$, so it is natural to wonder if the trajectory might approach some limiting closed orbit as $t \to \infty$, over which the effects of the positive and negative damping are in balance.

Computer plots (using the *Maple* phaseportrait command) are shown in Fig. 10 for the representative cases $\epsilon = 0.2$, 1, and 5. As noted in connection with (14), if $\epsilon < 2$ then the singular point at the origin is an unstable spiral, and if $\epsilon > 2$ then it is an unstable node, as is confirmed in Fig. 10. In each case we see from Fig. 10 that there is indeed a closed orbit around the origin, labeled Γ, that is approached asymptotically as $t \to \infty$, both by trajectories initiated outside of Γ and by trajectories initiated inside of Γ. Such a closed orbit is called a **limit cycle**. It can be shown analytically that as $\epsilon \to 0$ the van der Pol limit cycle tends to a circle of radius 2 centered at the origin. In Fig. 10 we see that for $\epsilon = 0.2$ the limit cycle Γ is indeed approximately a circle of radius 2. As ϵ increases Γ becomes more and more distorted. We say that the van der Pol limit cycle is **stable** because trajectories *approach* it, both from the inside and from the outside. A stable limit cycle is also called an **attractor**.[1]

The stable limit cycle Γ is the most important feature of the van der Pol equation because it represents the steady-state oscillation, the place where all trajectories (except the point trajectory at the origin) "end up" as $t \to \infty$. Besides the phase plane plots in Fig. 10 we've also plotted $x(t)$ versus t in Fig. 11 corresponding to the trajectories labeled C in Fig. 10, each of which starts close to, but not at, the origin. Observe in Fig. 11 the relatively slow approach to the steady-state limit cycle oscillation for $\epsilon = 0.2$ and the more rapid approach as ϵ is increased. For instance, if $\epsilon = 0.2$ steady state is reached (approximately) by $t = 50$ and for $\epsilon = 1$ it is reached (approximately) by $t = 20$. It is striking that as ϵ becomes large the steady-state limit cycle oscillation becomes rather "herky jerky," even becoming discontinuous in the limit as $\epsilon \to \infty$; this limit is called a **relaxation oscillation**. Relaxation oscillations are commonly observed in natural phenomena. Examples mentioned in the paper by van der Pol and van der Mark include the "singing" of wires in a cross wind, the scratching noise of a knife on a plate, the periodic occurrence of epidemics and economic crises, the sleeping of flowers, menstruation, and the beating of the heart.   ∎

Observe carefully that the *limit cycle phenomenon is possible only for nonlinear systems.* The simplest way to see that is to observe that the general linear autonomous

---

[1] Similarly, a stable spiral or stable node is called a **point attractor**.

system[1]

$$x' = ax + by, \tag{16a}$$

$$y' = cx + dy \tag{16b}$$

can be solved analytically. We've already done that in Section 11.3 and found only centers, spirals, nodes, and saddles, no limit cycles.

Besides the van der Pol example, other examples of differential equations exhibiting limit cycles are given in the exercises. In other cases a limit cycle can be **unstable** (**repelling**) in that other trajectories wind away from it, or **semistable** in exceptional cases, in that trajectories wind toward it from the interior and away from it from the exterior, or vice versa.

11.4.2 Bifurcations. (Optional).

As we have stressed, our concern in this chapter is somewhat qualitative. For instance, knowing that a certain singular point is a saddle rather than a center, spiral, or node, is of greater interest than knowing whether one of the eigenvalues is 4.72 or 4.73. Of great importance then is the concept of bifurcations. That is, systems generally include one or more physical parameters [such as the constant p in (7)]. As those parameters are varied continuously one expects the system behavior (i.e., its phase portrait) to vary continuously as well. For instance, if we vary p in Example 2 from 0.5 to 0.52, say, then the phase portrait "deforms" slightly but, qualitatively, nothing dramatic happens. In other cases there may exist certain critical values of the parameters such that the system's qualitative behavior changes abruptly and dramatically as a parameter passes through such a critical value or **bifurcation** point. Let us illustrate with one example.

EXAMPLE 4   *Saddle-Node Bifurcation.*

The nonlinear system[2]

$$x' = -rx + y, \tag{17a}$$

$$y' = \frac{x^2}{1 + x^2} - y \tag{17b}$$

arises in molecular biology, where $x(t)$ and $y(t)$ are proportional to protein and messenger RNA concentrations, and r is a positive empirical constant, or parameter, associated with the "death rate" of protein in the absence of the messenger RNA [for if $y = 0$, then (17a) gives exponential decay of x, with rate constant r]. However, we will not need to understand the biology to understand this example.

The singular points of (17) correspond to intersection points of $y = rx$ and $y = x^2/(1+x^2)$, as shown (solid curves) in Fig. 12. Equating these gives $x = y = 0$

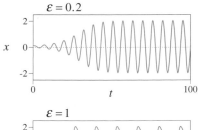
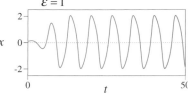
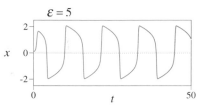
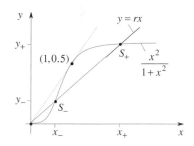

FIGURE 11

$x(t)$ versus t for the van der Pol equation (11).

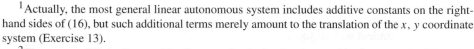

FIGURE 12

Determining the singular points of (17).

---

[1] Actually, the most general linear autonomous system includes additive constants on the right-hand sides of (16), but such additional terms merely amount to the translation of the x, y coordinate system (Exercise 13).

[2] The system (17) is discussed by Strogatz (in the book referenced in Comment 2 of Example 2) in his Example 8.1.1 and by J. S. Griffith [*Mathematical Neurobiology* (New York: Academic Press, 1971)]. In (17) we have reduced the number of parameters from two to one, for simplicity.

and also the two distinct roots

$$x_\pm = \frac{1 \pm \sqrt{1 - 4r^2}}{2r}, \qquad y_\pm = rx_\pm = \frac{1 \pm \sqrt{1 - 4r^2}}{2}, \qquad (18)$$

provided that $r < 1/2$. Thus, the critical slope of $y = rx$ is $r = 1/2$. If $r < 1/2$ we obtain the two intersections $S_+ \equiv (x_+, y_+)$ and $S_- \equiv (x_-, y_-)$, if $r = 1/2$ (dashed line in Fig. 12) these coalesce at $(1, 0.5)$, and if $r > 1/2$ they disappear and we have only the singular point at the origin.

Let us study the three singular points, for $r < 1/2$. First $(0, 0)$: We can see from (17) by inspection or by Taylor series expansion, that the linearized equations are

$$x' = -rx + y, \qquad y' = -y \qquad (19)$$

so $a = -r, b = 1, c = 0$, and $d = -1$. Thus, (23) in Section 11.3 gives $\lambda = -r$ and -1. Since both are negative, the singular point $(0, 0)$ is a stable node.

In similar fashion (which calculations we leave to Exercise 12), we find that the singularity at S_- is a saddle, and that the singularity at S_+ is an unstable node. As r is increased, S_- and at S_+ approach each other along the curve $y = x^2/(1 + x^2)$. When $r = 1/2$ they merge and form a singularity of some other type, and when r is increased beyond $1/2$ the singularity disappears altogether, leaving only the node at the origin. The bifurcation that occurs at $r = 1/2$ is an example of a **saddle-node bifurcation**. ∎

Closure.   In this section we used three examples, the Lotka-Volterra equations, the damped nonlinear pendulum, and the van der Pol equation to illustrate how to attack a nonlinear system

$$x' = P(x, y), \qquad (20a)$$
$$y' = Q(x, y) \qquad (20b)$$

to develop its phase portrait. Let us summarize the steps:

(i) First, set $P(x, y) = 0$ and $Q(x, y) = 0$ and solve these for x and y. Solutions give the coordinates of any singular points in the x, y phase plane.

(ii) For each isolated singular point, say (x_0, y_0), expand $P(x, y)$ and $Q(x, y)$ in Taylor series about that point and cut off after the first-order terms, thus "linearizing" (20) about that singular point. Doing so, and setting $X = x - x_0$ and $Y = y - y_0$, say, gives a linear system

$$X' = aX + bY, \qquad (21a)$$
$$Y' = cX + dY. \qquad (21b)$$

Solve for the eigenvalues and eigenvectors of the coefficient matrix

$$\mathbf{A} = \begin{bmatrix} a & b \\ c & d \end{bmatrix}$$

in (21). The eigenvalues reveal the singularity type (namely, center, spiral, node, or saddle, provided that $\det\mathbf{A} = ad - bc \neq 0$) and its stability, except for uncertainties in borderline cases [see (i)–(iii) in Section 11.3.3]. If the singular point is a saddle or a node then the eigenvectors of $\mathbf{A}$ give the straight-line trajectories through the singular point, the so-called eigendirections.

(iii) With that information obtained for each isolated singular point we can begin to sketch the local flows in the x, y phase plane in the neighborhood of each singular point (see, for instance, Fig. 3).

(iv) From that sketch (even a hand sketch will suffice) we can begin to apprehend the overall or global flow and can more judiciously choose a set of initial points to use in the computer generation of representative trajectories and key trajectories. The latter include sepatrices and any trajectories that intersect singular points, namely, four for every saddle and either four or two for every node (except if the node is a star).

We also met, in this section, the idea of limit cycles, which can occur only for a nonlinear system. The classic example of an equation with a limit cycle solution is the van der Pol equation which was the subject of Example 3. That limit cycle solution is said to be a **self-excited oscillation** because even the slightest disturbance from the equilibrium point at the origin results in a motion that grows and inevitably approaches the periodic limit cycle motion as $t \to \infty$ (Figs. 10 and 11).

Finally, we briefly introduced the idea of bifurcations in Example 4, namely, the dramatic qualitative change in the behavior of a system as parameters are varied continuously through critical values.

Maple.   First, some comments about the *Maple* aspects of Example 1. In generating Fig. 4 by the phaseportrait command, how do we choose a t interval that will be large enough to generate the complete closed orbits? To estimate the t interval needed, consider the orbits very close to the singular point $(1, 1)$. There, the linearized equations are given by (3), which give $X'' = -Y' = -X$. Hence $X'' + X = 0$ so $X = C_1 \cos t + C_2 \sin t$ with a period of 2π. Thus, we might try a t interval such as $0 \leq t \leq 10$ or 15, say. We can always run the command again with a larger t interval if we find that the plotted trajectories (which should be closed loops) are incomplete.

To obtain the two plots of $x(t)$ versus t in Fig. 6 we used these commands:

```
with(DEtools):
phaseportrait({diff(x(t),t)=(1-y(t))*x(t),diff(y(t),t)=
(-1+x(t))*y(t)},{x(t),y(t)},t=0..14,{[x(0)=1,y(0)=.75],
[x(0)=1,y(0)=.25]}, stepsize=.05,scene=[t,x]);
```

Changing "scene $= [t, x]$" to scene $= [t, y]$ then gave the plots of $y(t)$ versus t.

Turning to Example 2, we needed to generate 20 trajectories to obtain the phase portrait in Fig. 8, four for each of the four saddles and four in-between trajectories to add a bit more detail. For brevity, let us give the commands that generate just the four trajectories through the saddle at $(\pi, 0)$:

```
with(DEtools):
phaseportrait}({diff(x(t),t)=y(t),diff(y(t),t)=-.5*y(t)
-sin(x(t))},{x(t),y(t)},t=-10..30,{[x(0)=3.14081,y(0)=.001],
[x(0)=3.14287,y(0)=.001],[x(0)=3.14031,y(0)=-.001],
[x(0)=3.14237,y(0)=-.001]},stepsize=.05,x=-4..17,y=-6..6,
arrows=NONE);
```

where the four initial points were discussed in Comment 1 and where the t interval $-10 \leq t \leq 30$ was determined by trial and error.

EXERCISES 11.4

1. Use the *Maple* phaseportrait command to generate the phase portrait in Fig. 4, using the initial points $(1, 0.1)$, $(1, 0.3)$, $(1, 0.5)$, $(1, 0.7)$, and $(1, 0.9)$, say. Use the plotting range $0 \leq x \leq 4$ and $0 \leq y \leq 4$, say. If you include the barbed direction field arrows then they will show the flow direction. If not (i.e., if you use the option arrows = NONE, as we often do in this text), then you will need to add arrows to your trajectories by hand.

2. In Exercise 1 we asked you to generate Fig. 4 using the *Maple* phaseportrait command. Actually, we can solve (4) analytically.

    (a) Solve (4) by separation of variables and obtain the solution, in implicit form, as

    $$xy = Ce^{x+y} \qquad (2.1)$$

    where C is an arbitrary constant such that $C \geq 0$. NOTE: Isn't it remarkable that (2.1) gives closed loops which even tend to circles as the point $(1, 1)$ is approached?

    (b) From (2.1), determine C values corresponding to the initial points $(1, 0.1)$, $(1, 0.3)$, $(1, 0.5)$, $(1, 0.7)$, and $(1, 0.9)$ used in Exercise 1. Then, for each of those C values, use the *Maple* implicitplot command to plot those trajectories over the plotting range $0 \leq x \leq 4$ and $0 \leq y \leq 4$, as in Exercise 1.

3. Use the *Maple* phaseportrait command to generate Fig. 8 (without the shading of course) for the ranges $-4 \leq x \leq 17$ and $-5 \leq y \leq 5$. Sixteen of the 22 trajectories spring from or terminate at the four saddles, and the other six are chosen to fill in the gaps.

4. Repeat Exercise 3 but this time use $p = 3$. According to (9) the singularities on the x axis at $x = 0, \pm 2\pi, \pm 4\pi, \ldots$ are now stable nodes instead of saddles. To include trajectories that approach these nodes along their eigendirections you will need to determine those eigendirections (i.e.,

eigenvectors).

5. Use the *Maple* phaseportrait command to generate phase portraits of the van der Pol equation, like those in Fig. 10, for $\epsilon = 0.1$, 1, and 2. It will suffice to use only two trajectories, one inside the limit cycle and one outside it. HINT: Since we don't know any points *on* the limit cycle we can't generate the limit cycle trajectory directly. But it will suffice to use a long enough time interval to allow the two trajectories to get extremely close to the limit cycle, as we did in Fig. 10.

6. (a) It is known that as $\epsilon \to 0$ the van der Pol limit cycle approaches a circle of radius 2 and its period T approaches 2π. As an approximate check of those results run the *Maple* phaseportrait command for a small ϵ such as $\epsilon = 0.01$ and use the scene = $[t, x]$ option to obtain a plot of x versus t. From that plot measure the limit cycle's amplitude and period. Do they agree reasonably well with the claimed values of 2 and 2π, respectively?

    (b) It is also known that

    $$T \sim (3 - 2 \ln 2)\epsilon \qquad (6.1)$$

    as $\epsilon \to \infty$. Proceeding as in part (a), check (6.1) for the representative case $\epsilon = 10$.

7. In Example 3 we considered only $\epsilon > 0$. How does the phase portrait change, if at all, if instead $\epsilon < 0$? Support your claims in any convincing way that you choose.

8. (a) Explain why we might well expect the equation

    $$x'' - (1 - x^2 - x'^2)x' + x = 0 \qquad (8.1)$$

    or, equivalently, the system

    $$x' = y, \qquad (8.2a)$$
    $$y' = -x + (1 - x^2 - y^2)y \qquad (8.2b)$$

to admit a limit cycle consisting of the unit circle $x^2 + y^2 = 1$ in the x, y phase plane. Use *Maple* to see if that expectation is fulfilled.

(b) Using the experience gained in part (a), design (i.e., make up) a second-order differential equation that has a stable limit cycle on $x^2 + y^2 = 1$ and an unstable one on $x^2 + y^2 = 4$.

9. (*Dynamic formulation of a buckling problem*) Consider the buckling of the mechanical system shown in the figure, and consisting of two massless rigid rods of length l pinned to a mass m and a lateral spring of stiffness k. When the spring is neither stretched nor compressed $x = 0$ and the rods are aligned vertically. As we increase the downward load P nothing happens until we reach a critical value P_{cr}, at which value x increases (to one side or the other, we can't predict which) and the system collapses.

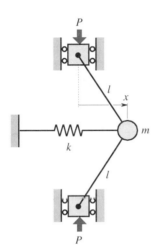

(**a**) Application of Newton's second law of motion gives

$$mx'' - \frac{2Px}{l}\left[1 - \left(\frac{x}{l}\right)^2\right]^{-1/2} + kx = 0 \quad (9.1)$$

as governing the displacement $x(t)$. With $x' = y$, show that the singularity at the origin in the x, y phase plane changes its type as P is sufficiently increased. Discuss that change of type, show why it corresponds to the onset of buckling, and use it to show that the critical buckling load is $P_{cr} = kl/2$.

(**b**) Explain what the results of part (a) have to do with bifurcation theory.

(c) Use Newton's second law to derive (9.1).

10. (*Motion of current-carrying wire*) A mutual force of attraction is exerted between parallel current-carrying wires. The infinite vertical wire shown in the figure has current I, and

the wire of length l and mass m (with leads that are perpendicular to the paper) has current i in the same direction as I. According to the Biot-Savart law of electromagnetics, the mutual force of attraction is $2Iil/(\text{separation}) = 2Iil/(a - x)$, where $x = 0$ is the position at which the spring force is zero, so the equation of motion of the restrained wire is

$$mx'' + k\left(x - \frac{r}{a - x}\right) = 0, \quad \text{where} \quad r = \frac{2Iil}{k}. \quad (10.1)$$

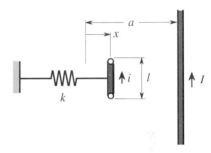

Thinking of m, k, a, and l as fixed, and the currents I and i as variable, let us study the behavior of the system in terms of the parameter r. For definiteness, let $m = k = a = 1$.

(a) With $x' = y$, identify any singularities in the x, y phase plane, determine their types, and show that their types depend upon whether r is less than, equal to, or greater than $1/4$. Suppose that $r < 1/4$. Find the equation of the phase trajectories and of the separatrix. Give a labeled sketch of the phase portrait.

(b) Let $r = 0.1$, say, and obtain a computer plot of the phase portrait.

(c) Next, consider the transitional case, where $r = 1/4$. Show that that case corresponds to the merging of the two singularities, and the forming of a single singularity of higher order (i.e., a nonelementary singularity). Give a labeled sketch of the phase portrait for that case.

(d) Let $r = 1/4$, and obtain a computer plot of the phase portrait.

(e) Next, consider the case where $r > 1/4$, and sketch the phase portrait.

(f) Let $r = 0.5$, say, and obtain a computer plot of the phase portrait.

(g) Discuss this problem from the point of view of bifurcations, insofar as the parameter r is concerned.

11. (*Hopf bifurcation*)

(**a**) Show that the nonlinear system

$$x' = \epsilon x + y - x(x^2 + y^2), \quad (11.1a)$$

$$y' = -x + \epsilon y - y(x^2 + y^2) \quad (11.1b)$$

can be simplified to

$$r' = r(\epsilon - r^2), \qquad (11.2a)$$

$$\theta' = -1 \qquad (11.2b)$$

by the change of variables $x = r\cos\theta$, $y = r\sin\theta$ from the Cartesian x, y variables to the polar r, θ variables. HINT: Putting $x = r\cos\theta$, $y = r\sin\theta$ into (11.1) gives differential equations each of which contains both r' and θ' on the left-hand side. Suitable linear combinations of those equations give (11.2a,b), respectively. We suggest that you use the shorthand $\cos\theta \equiv c$ and $\sin\theta \equiv s$ for brevity.

(b) From the form of (11.2) show that the origin in the x, y plane is a stable spiral if $\epsilon < 0$ and an unstable spiral if $\epsilon > 0$, and show that working from (11.1), instead, one obtains the same classification.

(c) Show from (11.2) that $r(t) = \sqrt{\epsilon}$ is a trajectory (if $\epsilon > 0$) and, in fact, a stable limit cycle. NOTE: Observe that zero is a bifurcation value of ϵ. As ϵ increases, a limit cycle is "born" as ϵ passes through the value zero, and its radius increases with ϵ. This is known as a **Hopf bifurcation**.

(d) Modify (11.1) so that it gives an *un*stable limit cycle at $r = \sqrt{\epsilon}$, instead.

12. (*Example 4 continued*)
(a) As a representative subcritical case, let $r = 0.3$ in

(17). Show that S_- is a saddle and determine the two eigendirections through it. Show that S_+ is a stable node and determine the two eigendirections through it. Likewise, determine the eigendirections for the stable node at $(0,0)$. Use these results to sketch the phase portrait of (17).

(b) For the critical case $r = 1/2$ show that the singularity at $S_\pm$ is nonelementary.

(c) For the representative supercritical case $r = 1$ identify and classify any singularities, and use *Maple* to generate the phase portrait of (17) in the square $0 \le x \le 1.5$, $0 \le y \le 1.5$.

13. As mentioned in the footnote preceding (16), the most general linear autonomous system is

$$x' = ax + by + e, \qquad (13.1a)$$

$$y' = cx + dy + f, \qquad (13.1b)$$

where $a, \ldots, f$ are constants. Continuing to require that $ad - bc \ne 0$, show that (13.1) can be reduced to the form (16) by a change of variables $x = \bar{x} + A$ and $y = \bar{y} + B$, that is, by a translation of coordinates to a new origin at $x = A$ and $y = B$. Evaluate A and B (in terms of $a, \ldots, f$). NOTE: Thus, just as (16) has a center, spiral, node, or saddle at $x = y = 0$, (13.1) has a center, spiral, node, or saddle at $x = A$ and $y = B$.

CHAPTER 11 REVIEW

The phase plane method applies to autonomous systems of the form

$$x' = P(x, y), \qquad (1a)$$

$$y' = Q(x, y). \qquad (1b)$$

We can also use it for a single second-order equation

$$x'' + f(x, x') = 0 \qquad (2)$$

if we first re-express (2) in the standard form (1), for instance as

$$x' = y, \qquad (3a)$$

$$y' = -f(x, y). \qquad (3b)$$

Dividing (1b) by (1a) we can eliminate t entirely and obtain the differential equation

$$\frac{dy}{dx} = \frac{Q(x, y)}{P(x, y)}. \qquad (4)$$

We call the x, y plane the *phase plane* for (1) and we call the solution curves of (4) the phase trajectories, or simplify the trajectories. Thus, rather than seeking solutions

$x(t)$ and $y(t)$ of (1) and plotting x versus t and y versus t we work in the x, y phase plane, in which t does not appear explicitly but only as the parameter in the parametric equations

$$x = x(t), \qquad y = y(t) \tag{5}$$

of a given trajectory.

Of particular importance are the singular or equilibrium points of (1), points (x_0, y_0) at which both $P(x, y) = 0$ and $Q(x, y) = 0$. If we Taylor expand (1) about (x_0, y_0) and linearize (i.e., retain terms only through first order) then, with $X = x - x_0$ and $Y = y - y_0$, we obtain a linear system of the form

$$X' = aX + bY, \tag{6a}$$

$$Y' = cX + dY \tag{6b}$$

that can be solved analytically and that reveals a local flow that is a *center*, a *spiral*, a *node*, or a *saddle* (assuming that $ad - bc \neq 0$). The flow obtained by solving the linear system (6) agrees with the flow for the original system (1) as $(x, y) \to (x_0, y_0)$, except possibly in certain borderline cases, as detailed in the text. The "local" flows obtained at the various singular points together establish the topological structure of the overall phase portrait of (1).

To obtain a detailed phase portrait we can use a combination of singular point analysis and the *Maple* phaseportrait command. Of course, if we can solve (4) analytically then we need merely plot representative trajectories using a plotting command such as the *Maple* implicitplot command but, even so, to capture any separatrix trajectories some singular point analysis will still be needed.

A great advantage of the phase plane method is that all of the various types of response, and their most salient features, are captured in a single plot, the phase portrait.

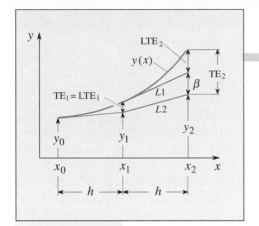

12

QUANTITATIVE METHODS; NUMERICAL SOLUTION

12.1 INTRODUCTION

Most of this book is about the underlying theory of differential equations and techniques for obtaining analytical solutions in closed form. However, many ODEs (or systems of ODEs) are too difficult to solve analytically in closed form, so additional techniques are presented in Chapters 8, 11, and 12. In Chapter 8 we showed how to solve linear second-order ODEs with nonconstant coefficients analytically but in open form, namely, in the form of infinite series. In Chapter 11 we developed an essentially qualitative and topological phase plane approach especially for nonlinear second-order equations or, equivalently, nonlinear systems of two first-order equations. And in the present chapter our approach is quantitative as we develop numerical (digital computer) solution algorithms for first-order ODEs and for systems of first-order ODEs, be they linear or not. In fact, such an algorithm lies behind the *Maple* phaseportrait command that we have used so extensively.

12.2 EULER'S METHOD

12.2.1 The algorithm. In this section and the two that follow we study the numerical solution of the first-order initial value problem

$$y' = f(x, y); \quad y(a) = b \tag{1}$$

on $y(x)$.

To motivate the first and simplest of these methods, Euler's method, consider the problem

$$y' = y + 2x - x^2; \quad y(0) = 1 \quad (0 \le x < \infty) \tag{2}$$

with the exact solution (Exercise 1)

$$y(x) = x^2 + e^x. \tag{3}$$

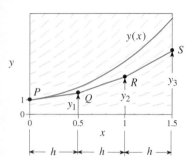

FIGURE 1
Direction field
motivation of Euler's method,
for the initial value problem (2).

Of course, in practice one wouldn't solve (2) numerically because we can solve it analytically and obtain the analytic closed form solution (3), but we will use the problem (2) as an illustration.

In Fig. 1 we display the direction field defined by $f(x, y) = y + 2x - x^2$, as well as the exact solution (3). In graphical terms, Euler's method amounts to using the direction field as a road map in developing an approximate solution to (2). Beginning at the initial point P, namely $(0, 1)$, we move in the direction dictated by the lineal element at that point. As seen from the figure, the farther we move along that line, from the starting point P to a stopping point Q, the more we expect our path to deviate from the exact solution. Thus, the idea is not to move very far. Stopping at $x = 0.5$, for the sake of illustration, we revise our direction according to the slope of the lineal element at that point Q. Moving in that new direction until $x = 1$, we revise our direction at R, and so on, moving in x increments of 0.5.

Thus, our plan is as follows. Rather than seek the function $y(x)$ that satisfies (1), we **discretize** the problem by breaking the x interval into discrete points $x_0 = a$, $x_1 = x_0 + h$, $x_2 = x_1 + h$, and so on, where h is called the **step size**. And rather than seek the function $y(x)$ we seek the discrete approximate values $y_1, y_2, \ldots$ at $x_1, x_2, \ldots$. According to the algorithm known as **Euler's method** we compute y_{n+1} as the preceding value y_n plus the slope $f(x_n, y_n)$ at (x_n, y_n) times the step size h:

$$\boxed{y_{n+1} = y_n + f(x_n, y_n)h, \qquad (n = 0, 1, 2, \ldots)} \tag{4}$$

where f is the function on the right side of the given differential equation (1), $x_0 = a$, $y_0 = b$, h is the chosen step size, and $x_n = x_0 + nh$.

Euler's method is also known as the **tangent-line method** because the first straight-line segment of the approximate solution is tangent to the exact solution $y(x)$ at P, and each subsequent segment emanating from (x_n, y_n) is tangent to the solution curve through that point.

EXAMPLE 1
For the problem (2) $f(x, y)$ is $y + 2x - x^2$, x_0 is 0, and y_0 is 1. With $h = 0.5$ Euler's method (4) gives

$$y_1 = y_0 + \left(y_0 + 2x_0 - x_0^2\right)h = 1 + (1 + 0 - 0)(0.5) = 1.5,$$

$$y_2 = y_1 + \left(y_1 + 2x_1 - x_1^2\right)h = 1.5 + (1.5 + 1 - 0.25)(0.5) = 2.625,$$

$$y_3 = y_2 + \left(y_2 + 2x_2 - x_2^2\right)h = 2.625 + (2.625 + 2 - 1)(0.5) = 4.4375,$$

and so on.   ■

Table 1. Comparison of numerical solution of (2) using Euler's method, with the exact solution (3).

x	$h = 0.5$	$h = 0.1$	$h = 0.02$	Exact $y(x)$
0	$y_0 = 1$	$y_0 = 1$	$y_0 = 1$	$y(0) = 1$
0.5	$y_1 = 1.5$	$y_5 = 1.7995$	$y_{25} = 1.8778$	$y(0.5) = 1.8987$
1.0	$y_2 = 2.625$	$y_{10} = 3.4344$	$y_{50} = 3.6578$	$y(1.0) = 3.7183$
1.5	$y_3 = 4.4375$	$y_{15} = 6.1095$	$y_{75} = 6.5975$	$y(1.5) = 6.7317$

We see that Euler's method is simple and readily implemented, even by hand calculation. (We show hand calculations to illustrate the methods. Practically speaking, however, the methods discussed in this chapter are normally programmed for digital computer calculation.)

Evidently, the greater the step size the less accurate the results, in general. For instance, the first point Q in Fig. 1 deviates more and more from the exact solution as the step size is increased—that is, as the segment PQ is extended. Conversely, we expect the approximate solution to approach the exact solution curve as h is reduced. This expectation is supported by the results shown in Table 1 for the initial value problem (2), obtained by Euler's method with step sizes of $h = 0.5, 0.1,$ and 0.02; we have included the exact solution $y(x)$, given by (3), for comparison. To keep the tabulation short we have omitted the many intermediate y_n and $y(x)$ values for the cases $h = 0.1$ and 0.02.

Scanning across each of the bottom three rows of the tabulation, we can see that the approximate solution values appear to be converging to the exact solution as $h \to 0$ (though we cannot be certain from such results no matter how small we make h), and also that the convergence is not very rapid, for even with $h = 0.02$ the computed value at $x = 1.5$ is in error by 2%. Besides tabulating the results in Table 1 we have also plotted them in Fig. 2. The results *look* accurate for $h = 0.02$, but realize that visual inspection is not very discerning, for even results with an error of around one part in 200 can look exact to the accuracy of a plot.

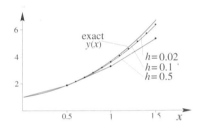

FIGURE 2
Plot of the results reported in Table 1.

12.2.2 Convergence of the Euler method. Two important theoretical questions present themselves: Does the method give convergence to the exact solution as $h \to 0$ and, if so, how fast? By a method being **convergent** we mean that *for any fixed x value in the x interval of interest the sequence of y values, obtained from (4) using smaller and smaller step size h, tends to the exact solution $y(x)$ of (1), at that point x, as $h \to 0$.*

Let us see whether the Euler method is convergent. Observe that there are two sources of error in the numerical solution. One is the tangent-line approximation upon which the method is based, and the other is the accumulation of numerical roundoff errors within the computing machine since a machine can carry only a finite number of significant figures, after which it rounds off (or chops off, depending upon the machine). In discussing convergence, one ignores the presence of such roundoff error and studies it separately. Thus, in this discussion we imagine our computer to be perfect, carrying an infinite number of significant figures.

Local truncation error.   Although we are interested in the accumulation of error after many steps have been carried out to reach a given x, it seems best to begin by investigating the error incurred *in a single step*, from x_n to x_{n+1}. We need to distinguish between the exact and the approximate solutions so let us denote the exact solution at x_{n+1} as $y(x_{n+1})$ and the approximate numerical solution at x_{n+1} as y_{n+1}. These are given by the Taylor series

$$y(x_{n+1}) = y(x_n) + y'(x_n)(x_{n+1} - x_n) + \frac{y''(x_n)}{2!}(x_{n+1} - x_n)^2 + \cdots$$

$$= y(x_n) + f(x_n, y(x_n))h + \frac{y''(x_n)}{2!}h^2 + \cdots \tag{5}$$

and the Euler algorithm

$$y_{n+1} = y_n + f(x_n, y_n)h, \tag{6}$$

respectively.

To determine the single-step error let us suppose that $y(x_n)$ and y_n are identical so that the difference

$$\boxed{\text{LTE} \equiv y(x_{n+1}) - y_{n+1}} \tag{7}$$

will truly be the error incurred in making the single step from x_n to x_{n+1}. (We will explain the name "LTE" in a moment.) By subtracting (6) from (5) we obtain

$$\text{LTE} = \frac{y''(x_n)}{2!}h^2 + \frac{y'''(x_n)}{3!}h^3 + \cdots, \tag{8}$$

where we have used the fact that $y(x_n) = y_n$ (by assumption), and hence that $f(x_n, y(x_n)) = f(x_n, y_n)$, to cancel two pairs of terms. As $h \to 0$,

$$\text{LTE} \sim \frac{y''(x_n)}{2}h^2. \tag{9}$$

We are interested in how fast LTE $\to 0$ as $h \to 0$ and (9) tells us that it tends to zero proportional to h^2 so we write, more simply than (9),

$$\boxed{\text{LTE} = O(h^2).} \tag{10}$$

In words, (10) means that the single-step error LTE tends to zero proportional to h^2; that is to say LTE $\sim Ch^2$, for some nonzero constant C, as $h \to 0$. The notation used in (10) is the so-called **big oh notation** because it it deals with *order of magnitude*. In words, read (10) as follows: LTE is big oh of h^2.

Incidentally, whereas we obtained the Euler method (4) by generalizing the graphical idea contained in Fig. 1, we could also have derived it from the Taylor series (5), for if we neglect terms of second order and higher then (5) gives

$$y(x_{n+1}) \approx y(x_n) + f(x_n, y(x_n))h$$

or, using our y_n notation for the approximate algorithmically generated values of y,

$$y_{n+1} = y_n + f(x_n, y_n)h,$$

which is the same as (4). That is, Euler's method results from truncating the Taylor series (5) after the first-order term. Since the error LTE is due to truncation of the Taylor series it is called the truncation error—more specifically, the **local truncation error** because it is the truncation error incurred in a single step; hence the symbol "LTE."

Accumulated truncation error and convergence.   Of ultimate interest, however, is the truncation error that has accumulated over *all* the preceding steps (from x_0 to x_1 to x_2, up to any given x_N) since that error is the difference between the exact solution and the computed solution at x_N. Let us denote it as

$$\boxed{\text{TE} \equiv y(x_N) - y_N} \tag{11}$$

and call it the accumulated truncation error or simply the **truncation error**.

We can estimate TE, at least insofar as its order of magnitude, as the single-step local truncation error LTE times the number of steps N. Since LTE $= O(h^2)$, this idea gives

$$\begin{aligned}
\text{TE} = O(h^2) \cdot N = O(h^2)\frac{Nh}{h} &= O(h^2)\frac{x_N - x_0}{h} \\
&= O(h)(x_N - x_0) = O(h),
\end{aligned} \tag{12}$$

where the last equality follows from the fact that $x_N - x_0$ is a fixed constant as $h \to 0$, and since the big oh notation is insensitive to scale factors, the $x_N - x_0$ factor can be absorbed by the $O(h)$. Thus,[1]

$$\boxed{\text{TE} = O(h),} \tag{13}$$

which result tells us how fast the numerical solution converges to the exact solution, at any fixed x location, as $h \to 0$. Namely, (13) tells us that TE $\sim Ch$ for some constant C that does not depend on h.

To illustrate (13) consider $x = 1.5$, say, in Table 1. According to TE $\sim Ch$, if we reduce h by a factor of five, from 0.5 to 0.1, then likewise we should reduce the error by a factor of five. We find that $(6.7317 - 4.4375)/(6.7317 - 6.1095) \approx 3.7$; we can't expect the foregoing result to be *exactly* five because (13) holds only as $h \to 0$ whereas we have used $h = 0.5$ and $h = 0.1$. If instead we use the values in Table 1 corresponding to $h = 0.1$ and $h = 0.02$ we obtain $(6.7317 - 6.1095)/(6.7317 -$

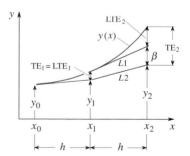

FIGURE 3
The truncation error.

---

[1]Our reasoning in writing TE $= O(h^2) \cdot N$ in (12) is that the truncation error TE is (at least insofar as order of magnitude) the sum of the N single-step errors. However, that is not quite true. We see from Fig. 3 that TE_2 (i.e., the TE at x_2) is LTE_2 (i.e., the single-step error in going from x_1 to x_2) plus β, *not* the sum of the single-step errors LTE_1 and LTE_2 since β is not identical to LTE_1. The difference between β and LTE_1 is the result of the *slightly* different slopes of the lines L_1 and L_2 acting over the *short* distance h. That difference can be shown to be a higher-order effect that does not invalidate the final result (13) provided that f is well enough behaved (e.g., if f, f_x, and f_y are all continuous on the x, y region of interest).

6.5975) ≈ 4.6, which result is indeed closer to five. Actually, we obtained the values in Table 1 using a computer and a computer introduces an additional error, due to roundoff, which has not been accounted for in our derivation of (13). Nevertheless, roundoff error is probably quite negligible in the present discussion.

In summary, (13) suggests that the Euler method (4) is convergent because the truncation error tends to zero as $h \to 0$. More generally if, for a given method, TE $= O(h^p)$ as $h \to 0$, then the method is convergent if $p > 0$ (because $h^p \to 0$ as $h \to 0$ if $p > 0$), and we say that it is **of order p**. We have the following theorem:

THEOREM 12.2.1

The Order of Euler's Method
The Euler method (4) is a first-order method.

Although convergent and easy to implement, Euler's method is usually too inaccurate for serious computation because it is only a first-order method. That is, since the accumulated truncation error is proportional to h to the first power, we need to make h extremely small if the error is to be extremely small. Why can't we do that? Why can't we merely let $h = 10^{-8}$, say? Because with $h = 10^{-8}$ it would take 10^8 steps to generate the Euler solution over a unit x interval. That number of steps might simply be impractically large in terms of computation time and expense.

Besides truncation error that we have discussed there is also machine **roundoff error**, and that error can be expected to grow with the number of calculations. Thus, as we diminish the step size h and increase the number of steps, to reduce the truncation error, we might inflict an offsetting roundoff error penalty. Roundoff error is beyond our present scope but is discussed in most books on numerical analysis.

Finally, there is an important practical question not yet addressed: How do we know how small to choose h in a given application? We will have more to say about this later, but for now let us give a simple procedure—reducing h until the results settle down to the desired accuracy. For instance, suppose we solve (2) by Euler's method using $h = 0.5$ first. Pick any fixed point x in the interval of interest, such as $x = 1.5$. The computed solution there is 4.4375. Now reduce h, say to 0.1, and run the program again. The result this time, at $x = 1.5$, is 6.1095. Since those results differ considerably, reduce h again, say to 0.02, and run the program again. Simply repeat that procedure until the solution at $x = 1.5$ settles down to the desired number of significant figures. Accept the results of the final run, and discard the others. (Of course, in a realistic application one will not have an exact solution to compare with as we did in Table 1.)

The foregoing idea is merely a rule of thumb, and is the same idea that we normally use in computing an infinite series: add more and more terms until successive partial sums agree to the desired number of significant figures.

Closure.   The Euler method is given by (4). It is easy to implement, using a handheld calculator or programming it to be run on a computer. The method is convergent but is only of first order and hence is not very accurate. Thus, it is important to develop more accurate methods, and we do that in the next section.

We also used our discussion of the Euler method to introduce the concept of the local and accumulated truncation errors LTE and TE, respectively, which are due to the approximate discretization of the problem and which have nothing to do with additional errors that enter due to machine roundoff. The LTE is the error incurred in a single step, and the TE is the accumulated error over the entire calculation. Finally, we define the method to be convergent if the truncation error TE tends to zero at any given fixed point x, as the step size h tends to zero, and to be of order p if TE $= O(h^p)$ as $h \to 0$. The Euler method is convergent and of order one.

EXERCISES 12.2

1. Derive the particular solution (3) of the initial value problem (2).

2. Use the Euler method to compute, by hand, y_1, y_2, and y_3 for the specified initial value problem using $h = 0.2$.

 (**a**) $y' = -y$; $y(0) = 1$

 (**b**) $y' = 2xy$; $y(0) = 0$

 (**c**) $y' = 3x^2y^2$; $y(0) = -1$

 (**d**) $y' = 1 + 2xy^2$; $y(1) = -2$

 (**e**) $y' = 2xe^{-y}$; $y(1) = -1$

 (**f**) $y' = x^2 - y^2$; $y(3) = 5$

 (**g**) $y' = x \sin y$; $y(0) = 0$

 (**h**) $y' = \tan(x + y)$; $y(1) = 2$

 (**i**) $y' = 5x - 2\sqrt{y}$; $y(0) = 4$

 (**j**) $y' = \sqrt{x + y}$; $y(0) = 3$

3. Using either a programmable calculator or a computer, program and run Euler's method for the initial value problem $y' = f(x, y)$, with $y(0) = 1$ and $h = 0.1$, through y_{10}. Print $y_1, \ldots, y_{10}$ and the exact solution $y(x_1), \ldots, y(x_{10})$ as well. (Six significant figures will suffice.) Evaluate TE$_{10}$ (the truncation error at x_{10}), assuming roundoff to be negligible. Use the $f(x, y)$ specified below.

 (**a**) $2x$ (**b**) $-6y^2$ (**c**) $x + y$

 (**d**) $y \sin x$ (**e**) $(y^2 + 1)/2$ (**f**) $4xe^{-y}$

 (**g**) $1 + x^2 + y$ (**h**) $-y \tan x$ (**i**) e^{x-y}

4. (**a**)–(**i**) Using either a programmable calculator or a computer, program and run Euler's method for the initial value problem $y' = f(x, y)$ (with f given in the corresponding part of Exercise 3), and print out the result at $x = 0.5$. Use $h = 0.1$, then 0.05, then 0.01, then 0.005, then 0.001, and compute the truncation error at $x = 0.5$ for each case. Is the rate of decrease of the truncation error, as h decreases, consistent with the fact that Euler's method is a first-order method? Explain.

5. Thus far we have taken the step h to be positive, and therefore developed a solution to the right of the initial point. Is

Euler's method valid if we use a negative step, $h < 0$, and hence develop a solution to the left?

6. In this section we have taken the step size h to be a constant from one step to the next. Is there any reason why we could not vary h from one step to the next?

7. (*Extrapolation*)   From (11) and (13) we have

$$y(x_N) - y_N \sim Ch \tag{7.1}$$

as $h \to 0$. If we don't actually let $h \to 0$ but merely choose a small h and compute y_N then (7.1) becomes

$$y(x_N) - y_N \approx Ch. \tag{7.2}$$

In (7.2) there are two unknowns, C and $y(x_N)$. We're not especially interested in C but we are of course interested in the exact solution $y(x_N)$. Here is the problem: Suppose we wish to find the solution $y(x)$ of (2) at $x = 0.5$. If we run the Euler method for $h = 0.1$ through $N = 5$ and for $h = 0.02$ through $N = 25$ we obtain the results $y_5 = 1.7995$ and $y_{25} = 1.8778$, respectively (Table 1). Thus (7.2) gives

$$y(0.5) - 1.7995 \approx C(0.1), \tag{7.3a}$$

$$y(0.5) - 1.8778 \approx C(0.02). \tag{7.3b}$$

Dividing (7.3a) by (7.3b) to cancel the C's, solve for $y(0.5)$ and show that the result is much more accurate than the value 1.8778 that resulted using $h = 0.02$. [For comparison, the exact solution is given by (3).] NOTE: The latter is called an **extrapolation method** since if we know how the error dies out as $h \to 0$ [namely, as given by (7.1)], and we run the method for two small but different h's, then we can "extrapolate those two results" to solve for $y(x_N)$. Though the method gives an improved result it does not yield the *exact* solution $y(x_N)$, except by coincidence, since we have used (7.2) which is only an approximate equality.

12.3 IMPROVEMENTS: RUNGE-KUTTA METHODS

Our objective in this section is to develop more accurate methods than the first-order Euler method—namely, higher-order methods. In particular, we are aiming at the widely used fourth-order Runge-Kutta method, which is an excellent general purpose differential equation solver. To bridge the gap between these two methods we begin with some general discussion about developing higher-order methods.

12.3.1 Motivation.   We saw in Section 12.2 that Euler's method could be derived by using Taylor's series and truncating after the first-order term. That procedure gave a local truncation error that was $O(h^2)$ and, accordingly, a truncation error that was $O(h)$; hence, we said that Euler's method is a first-order method. Evidently, then, one way to obtain higher-order methods is to *retain more terms* in the Taylor series

$$y(x_{n+1}) = y(x_n) + y'(x_n)h + \frac{y''(x_n)}{2!}h^2 + \frac{y'''(x_n)}{3!}h^3 + \cdots . \tag{1}$$

For instance, suppose we cut off after the h^2 term rather than the h term, obtaining from (1)

$$y(x_{n+1}) \approx y(x_n) + y'(x_n)h + \frac{y''(x_n)}{2}h^2. \tag{2}$$

Since the problem we are solving is

$$y' = f(x, y); \qquad y(a) = b, \tag{3}$$

we can replace the $y'(x_n)$ in (2) by $f(x_n, y(x_n))$. Further, the $y''(x_n)$ term in (2) can be obtained from (3), as well, by differentiating the differential equation using the chain rule:

$$y''(x) = \frac{d}{dx}y' = \frac{d}{dx}f[x, y(x)] = \frac{\partial f}{\partial x} + \frac{\partial f}{\partial y}\frac{dy}{dx} \tag{4}$$

or, since $y' = f(x, y)$,

$$y''(x) = f_x(x, y) + f_y(x, y)f(x, y). \tag{5}$$

Putting these expressions for $y'(x_n)$ and $y''(x_n)$ into (2) gives the algorithm

$$\boxed{y_{n+1} = y_n + f(x_n, y_n)h + \frac{1}{2}\left[f_x(x_n, y_n) + f_y(x_n, y_n)f(x_n, y_n)\right]h^2,} \tag{6}$$

which we call a **Taylor series method**.

In truncating (1) after the h^2 term we have incurred a local truncation error that is $O(h^3)$. Using the same heuristic reasoning as in (12) of Section 12.2, it follows that the truncation error associated with (6) is

$$\text{TE} = O(h^3) \cdot N = O(h^3)\frac{Nh}{h} = O(h^3)\frac{x_N - x_0}{h} = O(h^2), \tag{7}$$

so (6) is a second-order method.

EXAMPLE 1

To illustrate (6), consider the initial value problem

$$y' = -xy^2; \quad y(1) = 2, \qquad (1 \le x < \infty) \tag{8}$$

which is readily found (by separation of variables) to have the exact solution $y = 2/x^2$. Since $f(x, y) = -xy^2$, (6) becomes

$$y_{n+1} = y_n - x_n y_n^2 h + \frac{1}{2}\left[-y_n^2 + (-2x_n y_n)(-x_n y_n^2)\right] h^2$$

$$= y_n - x_n y_n^2 h + y_n^2 \left(x_n^2 y_n - \frac{1}{2}\right) h^2. \tag{9}$$

With $x_0 = 1$, $y_0 = 2$, and taking $h = 0.1$, say, (9) gives

$$y_1 = 2 - (1)(2^2)(0.1) + (2)^2[(1)^2(2) - 0.5](0.1)^2 = 1.66,$$

$$y_2 = 1.66 - (1.1)(1.66)^2(0.1) + (1.66)^2[(1.1)^2(1.66) - 0.5](0.1)^2 = 1.398455,$$

and so on. In Table 1 we compare the first few y_n's obtained using the second-order method (6) with the corresponding Euler results (which calculations are left for Exercise 1) and with the exact solution. The improvement in accuracy afforded by (6) relative to the Euler method is evident from those tabulated results. ■

Table 1.   Comparison of numerical solution of (8), using both the Euler and Taylor series methods, with the exact solution.

x_n	First-order Euler	Second-order Taylor series method	Exact, $y(x) = 2/x^2$
x_0	$y_0 = 2$	$y_0 = 2$	$y(1.0) = 2$
x_1	$y_1 = 1.6$	$y_1 = 1.66$	$y(1.1) = 1.65289$
x_2	$y_2 = 1.3184$	$y_2 = 1.39845$	$y(1.2) = 1.38889$
x_3	$y_3 = 1.10982$	$y_3 = 1.19338$	$y(1.3) = 1.18343$

Although the second-order convergence of (6) is an improvement over the first-order convergence of Euler's method, the attractiveness of (6) is diminished by an approximately threefold increase in the computing time per step since (6) requires three function evaluations (f, f_x, f_y) per step whereas Euler's method requires only one (f). To carry out one step of Euler's method we need to evaluate f, multiply by h, and add the result to y_n, but we can neglect the computing time for the multiplication by h and the addition of y_n on the grounds that a typical $f(x, y)$ involves many more arithmetic steps than that. Thus, as a rule of thumb, one compares the computing time per step of two methods by comparing only the number of *function evaluations per step* that they require.

In fact, the situation cited above, increasing the number of function evaluations per step, becomes more pronounced as we keep more terms of the Taylor series in (1).

Specifically, if we cut the Taylor series off after two, three, four, or five terms, then the resulting method requires one, three, six, and ten function evaluations per step, respectively (Exercise 2). Thus, let us abandon (6) and this line of approach, and let us develop higher-order methods in a different way.

12.3.2 Second-order Runge-Kutta.

Observe that the low-order Euler method $y_{n+1} = y_n + f(x_n, y_n)h$ amounts to an extrapolation of $y(x)$ away from the initial point (x_n, y_n) using the slope $f(x_n, y_n)$ at that point. Expecting an average slope to give greater accuracy, one might try the algorithm

$$y_{n+1} = y_n + \frac{1}{2}\left[f(x_n, y_n) + f(x_{n+1}, y_{n+1})\right]h \tag{10}$$

which uses an average of the slopes at the initial and final points. Unfortunately, the formula (10) does not give y_{n+1} explicitly because y_{n+1} appears not only on the left-hand side of (10) but also in the argument of $f(x_{n+1}, y_{n+1})$. Intuition tells us that we should still do well if we replace the latter y_{n+1} by an estimated value such as the Euler estimate $y_{n+1} = y_n + f(x_n, y_n)h$. Then the revised version of (10) is

$$y_{n+1} = y_n + \frac{1}{2}\left\{f(x_n, y_n) + f\left[x_{n+1}, y_n + f(x_n, y_n)h\right]\right\}h. \tag{11}$$

Intuitively, the form (11) seems like a good idea but it remains to be seen whether it really is a higher-order method, that is to say of higher order than the first-order Euler method. Rather than settle on this form too quickly, let us consider the form

$$y_{n+1} = y_n + \left\{af(x_n, y_n) + bf\left[x_n + \alpha h, y_n + \beta f(x_n, y_n)h\right]\right\}h \tag{12}$$

which is like (11) but with adjustable "design parameters" a, b, α, β included. Thus, let us consider (12) and choose a, b, α, β so as to make the order of that method as high as possible; α, β determine the second point [the first being (x_n, y_n)] in the x, y plane at which the slope is sampled, and a, b specify how the two slopes are to be weighted.

How can we make the order of the method (12) as high as possible? Recall that the way we determined the local truncation error for the Euler method was to compare the exact expression

$$y(x_{n+1}) = y(x_n) + y'(x_n)h + \frac{y''(x_n)}{2!}h^2 + \frac{y'''(x_n)}{3!}h^3 + \cdots \tag{13}$$

with the Euler method expression

$$y_{n+1} = y_n + f(x_n, y_n)h. \tag{14}$$

Assuming that the value y_n with which we begin our step is identical to the exact solution $y(x_n)$, in which case $y'(x_n) = f(x_n, y(x_n)) = f(x_n, y_n) \equiv f$ for brevity, the single-step error was

$$\text{LTE} = y(x_{n+1}) - y_{n+1}$$

$$= \left[y_n + fh + \frac{y''(x_n)}{2!}h^2 + \cdots \right] - [y_n + fh]$$

$$= \frac{y''(x_n)}{2}h^2 + \cdots = O(h^2). \tag{15}$$

We proceed in the same way for the method (12). Leaving the details to the exercises, it turns out that differencing (13) and (12) gives

$$\text{LTE} = (1 - a - b)fh + \left[\left(\frac{1}{2} - b\alpha \right) f_x + \left(\frac{1}{2} - b\beta \right) f_y f \right] h^2 + \cdots , \tag{16}$$

where the dots denote terms of third order and higher. The idea is to choose the free parameters a, b, α, β so as to eliminate as many powers of h on the right-hand side of (16) as possible, thereby maximizing the order of the local truncation error LTE. To eliminate the h term we require that

$$a + b = 1, \tag{17a}$$

and to eliminate the h^2 term we require that[1]

$$b\alpha = \frac{1}{2} \quad \text{and} \quad b\beta = \frac{1}{2}. \tag{17b}$$

The three equations (17) on the four unknowns a, b, α, β are indeed solvable and their solution is nonunique. The outcome is that any method (12), with a, b, α, β chosen so as to satisfy (17), has a local truncation error that is LTE $= O(h^3)$ [subject to mild conditions on f such as the continuity of f, f_x, and f_y so that we can justify the chain differentiation used in deriving (16)].[2]

The set of methods (12), for various combinations of a, b, α, β that satisfy (17), is known as the set of **Runge-Kutta methods of second order.**[3] That they are second-order methods follows from the fact that LTE $= O(h^3)$, together with heuristic reasoning identical to that in (7).

For instance, with $a = b = 1/2$ and $\alpha = \beta = 1$ we have

$$y_{n+1} = y_n + \frac{1}{2} \left\{ f(x_n, y_n) + f \left[x_{n+1}, y_n + f(x_n, y_n)h \right] \right\} h, \tag{18}$$

which is usually expressed in the computationally convenient form

$$y_{n+1} = y_n + \tfrac{1}{2}(k_1 + k_2),$$
$$k_1 = hf(x_n, y_n), \qquad k_2 = hf(x_{n+1}, y_n + k_1). \tag{19}$$

---

[1] Here we set the coefficients of $f_x(x_n, y_n)$ and $f_y(x_n, y_n)f(x_n, y_n)$ equal to zero separately because we need our result to hold for any function $f(x, y)$ and for each n.

[2] We have not written out the h^3 terms in (16) because their inclusion would be quite messy and because it would be found that we could not eliminate those terms (by suitable choice of a, b, α, β) anyway.

[3] The Runge-Kutta method was originated by *Carl D. Runge* (1856–1927), a German physicist and mathematician, and extended by the German aerodynamicist and mathematician *M. Wilhelm Kutta* (1867–1944). Kutta is well known for the Kutta-Joukowski formula for the lift on an airfoil, and for the "Kutta condition" of classical airfoil theory.

That is, knowing h, x_n, y_n we compute k_1. Then, knowing h, $x_{n+1} = x_n + h$, y_n, and k_1, we compute k_2. Finally, knowing y_n, k_1, and k_2 we use (19) to compute y_{n+1}.

To better understand this result, note that Euler's method gives $y_{n+1} = y_n + f(x_n, y_n)h$. If we denote that Euler estimate as y_{n+1}^{Euler}, then (18) can be expressed as

$$y_{n+1} = y_n + \frac{f(x_n, y_n) + f\left(x_{n+1}, y_{n+1}^{\text{Euler}}\right)}{2} h. \tag{20}$$

That is, we take a tentative step using Euler's method, then we average the slopes at the initial point (x_n, y_n) and at the Euler estimate of the final point $(x_{n+1}, y_{n+1}^{\text{Euler}})$, and then go back and make another Euler step, this time using the improved (average) slope. For this reason (19) is also known as the **improved Euler method**. It is the same as our intuitively derived method (11).

The improved Euler method is an example of a **predictor-corrector** method. Such a method uses a pair of formulas, a predictor formula and then a corrector formula, to carry out each step. The predictor-corrector nature of the improved Euler method can be seen in (20), which shows that a predicted value of y_{n+1} is first obtained by Euler's method; then that value is put into the corrector formula (20) to obtain the improved value y_{n+1}. If we call the predicted value v_{n+1} then the improved Euler method can be expressed in predictor-corrector form as

$$\text{Predictor}: \quad v_{n+1} = y_n + hf(x_n, y_n), \tag{21a}$$

$$\text{Corrector}: \quad y_{n+1} = y_n + \frac{f(x_n, y_n) + f(x_{n+1}, v_{n+1})}{2}. \tag{21b}$$

EXAMPLE 2

Let us proceed through the first two steps of the second-order Runge-Kutta method (19) for the same test problem as was used in Example 1,

$$y' = -xy^2; \qquad y(1) = 2, \qquad (1 \le x < \infty) \tag{22}$$

with exact solution $y = 2/x^2$. Here $f(x, y) = -xy^2$. Let $h = 0.1$.

$n = 0$:

$$k_1 = hf(x_0, y_0) = 0.1\left[-(1)(2)^2\right] = -0.4,$$

$$k_2 = hf(x_1, y_0 + k_1) = 0.1\left[-(1.1)(2 - 0.4)^2\right] = -0.2816,$$

$$y_1 = y_0 + \tfrac{1}{2}(k_1 + k_2) = 2 + 0.5(-0.4 - 0.2816) = \underline{1.6592};$$

$n = 1$:

$$k_1 = hf(x_1, y_1) = 0.1\left[-(1.1)(1.6592)^2\right] = -0.3028,$$

$$k_2 = hf(x_2, y_1 + k_1) = 0.1\left[-(1.2)(1.6592 - 0.3028)^2\right] = -0.2208,$$

$$y_2 = y_1 + \tfrac{1}{2}(k_1 + k_2) = 1.6592 + 0.5(-0.3028 - 0.2208) = \underline{1.3974},$$

compared with the values $y(x_1) = y(1.1) = \underline{1.6529}$ and $y(x_2) = y(1.2) = \underline{1.3889}$ obtained from the exact solution $y(x) = 2/x^2$.   ∎

12.3.3 Fourth-order Runge-Kutta.   This idea of a weighted average of slopes at various points in the x, y plane, with the weights and locations determined so as to maximize the order of the method, can be used to derive higher-order Runge-Kutta methods as well, although the derivations are tedious. One of the most widely used is the **fourth-order Runge-Kutta method**:

$$y_{n+1} = y_n + \tfrac{1}{6} \left(k_1 + 2k_2 + 2k_3 + k_4\right),$$

$$k_1 = hf\left(x_n, y_n\right), \qquad\qquad k_2 = hf\left(x_n + \tfrac{h}{2}, y_n + \tfrac{1}{2}k_1\right),$$

$$k_3 = hf\left(x_n + \tfrac{h}{2}, y_n + \tfrac{1}{2}k_2\right), \qquad k_4 = hf\left(x_{n+1}, y_n + k_3\right),$$

(23)

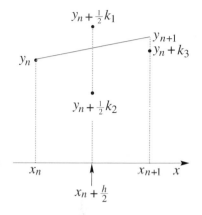

which we give without derivation. Here the effective slope used is a weighted average of the slopes at the four points (x_n, y_n), $(x_n + h/2, y_n + k_1/2)$, $(x_n + h/2, \; y_n + k_2/2)$ and $(x_{n+1}, y_n + k_3)$ in the x, y plane (Fig. 1), an *average* because the sum of the coefficients $1/6$, $2/6$, $2/6$, $1/6$ that multiply the k's is 1. Similarly, the sum of the coefficients $1/2$, $1/2$ in the second-order version (19) is 1 as well.

FIGURE 1
Computing y_{n+1} from y_n and a weighted average of the slopes at four points. (We've shown $k_1 > k_2$ but we could have $k_1 \leq k_2$.)

E X A M P L E   3

As a summary illustration, we solve a "test problem,"

$$y' = -y; \qquad y(0) = 1 \qquad (0 \leq x < \infty) \tag{24}$$

by each of the methods considered, using a step size of $h = 0.05$ and single precision arithmetic (on the computer used that amounts to carrying eight significant figures, double precision would carry 16). The results are given in Table 2, together with the exact solution $y(x) = e^{-x}$ for comparison; $0.529\mathrm{E} + 2$, for instance, means 0.529×10^2.

To illustrate the fourth-order Runge-Kutta calculation, let us go through the first two steps:

$n = 0$:

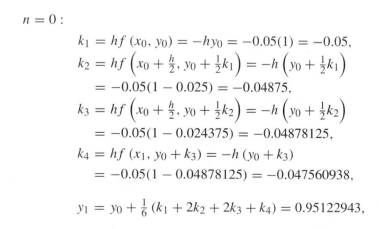

$$k_1 = hf\left(x_0, y_0\right) = -hy_0 = -0.05(1) = -0.05,$$

$$k_2 = hf\left(x_0 + \tfrac{h}{2}, y_0 + \tfrac{1}{2}k_1\right) = -h\left(y_0 + \tfrac{1}{2}k_1\right)$$

$$= -0.05(1 - 0.025) = -0.04875,$$

$$k_3 = hf\left(x_0 + \tfrac{h}{2}, y_0 + \tfrac{1}{2}k_2\right) = -h\left(y_0 + \tfrac{1}{2}k_2\right)$$

$$= -0.05(1 - 0.024375) = -0.04878125,$$

$$k_4 = hf\left(x_1, y_0 + k_3\right) = -h\left(y_0 + k_3\right)$$

$$= -0.05(1 - 0.04878125) = -0.047560938,$$

$$y_1 = y_0 + \tfrac{1}{6}\left(k_1 + 2k_2 + 2k_3 + k_4\right) = 0.95122943,$$

$$n = 1:$$

$$k_1 = hf\,(x_1, y_1) = -hy_1 = -0.05(0.95122943) = -0.04756147,$$

$$k_2 = hf\left(x_1 + \tfrac{h}{2}, y_1 + \tfrac{1}{2}k_1\right) = -h\left(y_1 + \tfrac{1}{2}k_1\right)$$

$$= -0.05(0.95122943 - 0.02378074) = -0.04637244,$$

$$k_3 = hf\left(x_1 + \tfrac{h}{2}, y_1 + \tfrac{1}{2}k_2\right) = -h\left(y_1 + \tfrac{1}{2}k_2\right)$$

$$= -0.05(0.95122943 - 0.02318622) = -0.04640216,$$

$$k_4 = hf\,(x_2, y_1 + k_3) = -h\,(y_1 + k_3)$$

$$= -0.05(0.95122943 - 0.04640216) = -0.04524136,$$

$$y_2 = y_1 + \tfrac{1}{6}\,(k_1 + 2k_2 + 2k_3 + k_4) = 0.90483743,$$

which final results do agree with the corresponding entries in Table 2. Actually, there is a discrepancy of 2 in the last digit of y_1, but an error of that size is not unreasonable in view of the fact that the machine used carried only eight significant figures.

Most striking is the excellent accuracy of the fourth-order Runge-Kutta method, with six significant figure accuracy over the entire calculation. ∎

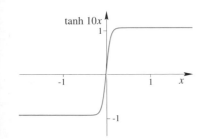

FIGURE 2

Rapid change in
$\tanh 10x$ near $x = 0$.

Of course, in real applications we do not have the exact solution to compare with the numerical results. In that case, how do we know whether or not our results are sufficiently accurate? A useful rule of thumb mentioned in Section 12.2 is to redo the entire calculation, each time with a smaller step size, until the results "settle down" to the desired number of significant digits.

Thus far we have taken h to be a constant, for simplicity, but there is no reason why it cannot be varied from one step to the next. In fact, there may be a compelling reason to do so. For instance, consider the equation $y' + y = \tanh 10x$ on $-2 \leq x \leq 2$. The function $\tanh 10x$ is almost constant, except near the origin where it varies dramatically, approximately from -1 to $+1$ (Fig. 2). Thus, we need a very fine step size h near the origin for good accuracy, but to use that h over the entire x interval would be wasteful in terms of computer time and expense.

One can devise rational schemes for varying the step size to maintain a consistent level of accuracy, and such refinements are already available within existing software. For example, the default numerical differential equation solver in *Maple* is a "fourth-fifth order Runge-Kutta-Fehlberg method" denoted as RKF45 in the literature. According to RKF45 a tentative step is made, first using a fourth-order Runge-Kutta method, and then again using a fifth-order Runge-Kutta method. If the two results agree to a specified number of significant digits, then the fifth-order result is accepted. If they agree to more than that number of significant digits, then h is increased and the next step is made. If they agree to less than that number of significant digits, then h is decreased and the step is repeated.

12.3.4 Importance of the order of the method.   Recall that for a pth-order method the truncation error is

$$\text{TE} \approx Ch^p. \tag{25}$$

Table 2. Comparison of Euler, second-order and fourth-order Runge-Kutta, and exact solutions of the initial value problem (25), with $h = 0.05$.

x	Euler	Second-order Runge-Kutta	Fourth-order Runge-Kutta	Exact, $y(x) = e^{-x}$
0.00	1.00000000 E+0	1.00000000 E+0	1.00000000 E+0	1.00000000 E+0
0.05	0.94999999 E+0	0.95125002 E+0	0.95122945 E+0	0.95122945 E+0
0.10	0.90249997 E+0	0.90487659 E+0	0.90483743 E+0	0.90483743 E+0
0.15	0.85737497 E+0	0.86076385 E+0	0.86070800 E+0	0.86070800 E+0
0.20	0.81450623 E+0	0.81880164 E+0	0.81873077 E+0	0.81873077 E+0
0.25	0.77378094 E+0	0.77888507 E+0	0.77880079 E+0	0.77880079 E+0
0.30	0.73509192 E+0	0.74091440 E+0	0.74081820 E+0	0.74081820 E+0
$\vdots$				
2.00	0.12851217 E+0	0.13545239 E+0	0.13533530 E+0	0.13533528 E+0
2.05	0.12208656 E+0	0.12884909 E+0	0.12873492 E+0	0.12873492 E+0
2.10	0.11598223 E+0	0.12256770 E+0	0.12245644 E+0	0.12245644 E+0
2.15	0.11018312 E+0	0.11659253 E+0	0.11648417 E+0	0.11648415 E+0
2.20	0.10467397 E+0	0.11090864 E+0	0.11080316 E+0	0.11080315 E+0
2.25	0.99440269 E−1	0.10550185 E+0	0.10539923 E+0	0.10539922 E+0
2.30	0.94468258 E−1	0.10035863 E+0	0.10025885 E+0	0.10025885 E+0
$\vdots$				
5.00	0.59205294 E−2	0.67525362 E−2	0.67379479 E−2	0.67379437 E−2
5.05	0.56245029 E−2	0.64233500 E−2	0.64093345 E−2	0.64093322 E−2
5.10	0.53432779 E−2	0.61102118 E−2	0.60967477 E−2	0.60967444 E−2
5.15	0.50761141 E−2	0.58123390 E−2	0.57994057 E−2	0.57994043 E−2
5.20	0.48223082 E−2	0.55289874 E−2	0.55165654 E−2	0.55165626 E−2
5.25	0.45811930 E−2	0.52594491 E−2	0.52475194 E−2	0.52475161 E−2
5.30	0.43521333 E−2	0.50030509 E−2	0.49915947 E−2	0.49915928 E−2
$\vdots$				
9.70	0.47684727 E−4	0.61541170 E−4	0.61283507 E−4	0.61283448 E−4
9.75	0.45300490 E−4	0.58541038 E−4	0.58294674 E−4	0.58294663 E−4
9.80	0.43035467 E−4	0.55687164 E−4	0.55451608 E−4	0.55451590 E−4
9.85	0.40883693 E−4	0.52972413 E−4	0.52747200 E−4	0.52747171 E−4
9.90	0.38839509 E−4	0.50390008 E−4	0.50174691 E−4	0.50174654 E−4
9.95	0.36897534 E−4	0.47933496 E−4	0.47727641 E−4	0.47727597 E−4
10.00	0.35052657 E−4	0.45596738 E−4	0.45399935 E−4	0.45399931 E−4

For this to be small we would like C to be small and p to be large, but it is important to understand *that the error (25) is much more sensitive to p than to C.* To illustrate this fact let $C = 1$, $h = 0.1$, and $p = 1$, for definiteness. If we modify the method so as to reduce C by a factor of four, from 1 to 0.25, then the truncation error will be reduced to $1/4$ of its original value. If instead we modify the method so as to increase p by a factor of four, from 1 to 4, then the truncation error will be reduced to $1/1000$ of its original value. As a concrete example consider the results in Table 2. At $x = 2$, for instance, the first-order Euler and second- and fourth-order Runge-Kutta methods are in error by approximately 5%, 0.09%, and 0.000015%, respectively. It's true that the fourth-order Runge-Kutta method will require roughly four times as much computer time as the Euler method because it requires four function evaluations per step [namely, the four f's within $k_1, \ldots, k_4$ in (23)] whereas the Euler method requires only one, but that increase in computer time is more than offset by the great increase in accuracy. Or, put differently, for a given desired accuracy we can tolerate a much larger step size h using the fourth-order Runge-Kutta method, so the number of steps (and hence the computer time) can be greatly reduced.

Since the order of the method is so important, we should know how to verify the order of whatever method we use, if only as a partial check on the programming. For instance, if we program the fourth-order Runge-Kutta method but an empirical check on the order reveals that we are getting only first-order convergence, then we can be certain that there is a programming error.

Recall that by a method being of order p we mean that at any chosen x the error behaves as Ch^p for some nonzero constant C (that does not depend on h); that is,

$$y_{\text{exact}} - y_{\text{comp}} \sim Ch^p \tag{26}$$

as $h \to 0$, where y_{exact} and y_{comp} are the exact and computed values of y at x, respectively. Suppose we wish to check the order of a given method. Select a test problem such as the one in Example 3, and use the method to compute y at any x point such as $x = 1$, for two different h's, say h_1 and h_2. Letting $y_{\text{comp}}^{(1)}$ and $y_{\text{comp}}^{(2)}$ denote the y's computed at $x = 1$ using step sizes of h_1 and h_2, respectively, we have

$$y_{\text{exact}} - y_{\text{comp}}^{(1)} \approx Ch_1^p,$$
$$y_{\text{exact}} - y_{\text{comp}}^{(2)} \approx Ch_2^p.$$

Dividing one equation by the other, to cancel the unknown C, and solving for p, gives

$$p \approx \frac{\ln\left[\dfrac{y_{\text{exact}} - y_{\text{comp}}^{(1)}}{y_{\text{exact}} - y_{\text{comp}}^{(2)}}\right]}{\ln\left[\dfrac{h_1}{h_2}\right]}. \tag{27}$$

To illustrate, let us run the test problem (24) with Euler's method, with $h_1 = 0.1$ and $h_2 = 0.05$. The results at $x = 1$ are

$$h_1 = 0.1 : \qquad y_{\text{comp}}^{(1)} = 0.348678440100,$$
$$h_2 = 0.05 : \qquad y_{\text{comp}}^{(2)} = 0.358485922409,$$

and since $y_{\text{exact}}(1) = 0.367879441171$, (27) gives $p \approx 1.03$, which is respectably close to 1. We should be able to obtain a more accurate estimate of p by using smaller h's since (27) tends to an equality only as $h \to 0$. In fact, using $h_1 = 0.05$ and $h_2 = 0.02$ gives $p \approx 1.01$. Using those same step sizes, we also obtain $p \approx 2.02$ and 4.03 for the second-order and fourth-order Runge-Kutta methods, respectively.

Why not use even smaller h's to determine p more accurately? Two difficulties arise. One is that as h is decreased and the computed solutions become more and more accurate, the $y_{\text{exact}} - y_{\text{comp}}^{(1)}$ and $y_{\text{exact}} - y_{\text{comp}}^{(2)}$ differences in (27) are known to fewer and fewer significant figures due to cancelation. This is especially true for a high-order method. The other difficulty is that (26) applies to the truncation error alone, so implicit in our use of (26) is the assumption that roundoff errors are negligible. If we make h too small, that assumption may become invalid. For both of these reasons, it is important to use extended precision for such calculations, as we have for the preceding calculations.

Closure. The Euler method is simple and readily implemented, but it is only of first order and is therefore not sufficiently accurate in many applications. Since the method follows from keeping just the first two terms on the right-hand side of (1), that is, the first two terms in the Taylor expansion of $y(x)$ about x_n, we considered obtaining higher-order methods by retaining more terms in that expansion. To obtain a fourth-order method, for instance, we would use

$$y(x_{n+1}) \approx y(x_n) + y'(x_n)h + \frac{y''(x_n)}{2!}h^2 + \frac{y'''(x_n)}{3!}h^3 + \frac{y''''(x_n)}{4!}h^4, \qquad (28)$$

where $y'(x_n), \ldots, y''''(x_n)$ are evaluated from the known function $f(x, y)$ since

$$y' = f(x, y(x)), \qquad y'' = \frac{d}{dx}f(x, y(x)) = f_x + f_y y' = f_x + f_y f,$$

and so on. We would find, in this manner, that the expressions for $y', \ldots, y''''$ involve $f, f_x, f_y, f_{xx}, f_{xy}, f_{yy}, f_{xxx}, f_{xxy}, f_{xyy}$, and f_{yyy}. Hence, this method would require 10 function evaluations per step![1]

For this reason we dropped the Taylor series approach and sought higher-order methods by using the Euler idea

$$y_{n+1} = y_n + (\text{slope})h \qquad (29)$$

again, but where the slope in (29) is a weighted average of the slopes $f(x, y)$ at a number of optimally located points in the x, y plane between $x = x_n$ and $x = x_{n+1}$. Using this approach our chief result was the fourth-order Runge-Kutta method (23) which is one of the chief algorithms of modern computation.

---

[1] We assume that $f(x, y)$ is sufficiently well behaved so that the order of differentiation does not matter; for example, $f_{xy} = f_{yx}$, $f_{xyy} = f_{yxy} = f_{yyx}$, and so on. For instance, if f_x, f_y, f_{xy}, and f_{yx} are continuous in some neighborhood of (x, y) then $f_{xy} = f_{yx}$ at (x, y). These conditions are so typically satisfied in applications, that authors of engineering and science texts generally assume the truth of $f_{xy} = f_{yx}$, $f_{xyy} = f_{yxy} = f_{yyx}$ and so on, without discussion.

Maple.   Computer-software systems such as *Maple* include numerical differential equation solvers. In *Maple* one can use the **dsolve** command together with a numeric option. The default numerical solution method is the RKF45 method mentioned above. Note that with the numeric option of dsolve one does not specify a step size h since that choice is controlled within the program and, in general, is varied from step to step to maintain a certain level of accuracy. To specify the absolute error tolerance one can use an additional option called abserr, which is formatted as abserr = Float(1,2-digits) and which means 1 times 10 to the one-digit or two-digit exponent. For instance, to solve

$$y' = -y; \quad y(0) = 1$$

for $y(x)$ with an absolute error tolerance of 1×10^{-5}, and to print the results at $x = 2, 10$, enter

```
dsolve({diff(y(x),x)=-y(x),y(0)=1},y(x),type=numeric,
value=array([2,10]),abserr=Float(1,-5));
```

and return. The printed result is

$$
\begin{bmatrix}
 & [x, y(x)] & \\
\begin{bmatrix}
2. & .1353337989 \\
10. & .00004501989224
\end{bmatrix}
\end{bmatrix}
$$

For comparison, the exact solution is $y(2) = \exp(-2) = 0.1353352832$ and $y(10) = \exp(-10) = 0.00004539992976$, respectively.

EXERCISES 12.3

1. (*Example 1*)
   (a) Derive the solution $y = 2/x^2$ of (8).
   (b) Derive the four values listed in the "Euler" column of Table 1.

2. It was stated, just above Section 12.3.2, that if we cut the Taylor series off after two, three, four, or five terms, then the resulting method requires one, three, six, and ten function evaluations per step, respectively. Prove that claim.

3. Evaluate y_1 and y_2 by hand, by the second-order and fourth-order Runge-Kutta methods, with $h = 0.02$. Obtain the exact values $y(x_1)$ and $y(x_2)$ as well, for comparison.
   (a) $y' = 3000xy^{-2}; \quad y(0) = 2$
   (b) $y' = 40xe^{-y}; \quad y(0) = 3$
   (c) $y' = x + y; \quad y(-1) = 5$
   (d) $y' = -y \tan x; \quad y(1) = -8$
   (e) $y' = (y^2 + 1)/4; \quad y(0) = 0$
   (f) $y' = -2y \sin x; \quad y(2) = 5$

4. (a)–(f) Program the second-order and fourth-order Runge-Kutta methods and run those programs to solve the initial value problem given in the corresponding part of Exercise 3 but with the initial condition $y(0) = 1$. Use $h = 0.05$. Print out all computed values of y, up to $x = 0.5$, as well as the exact solution.

5. Using the test problem

$$y' = y + 2x - x^2; \quad y(0) = 1, \qquad (5.1)$$

with the exact solution $y(x) = x^2 + e^x$, do an empirical evaluation of the order of the given method. Use (27), with $h = 0.1$ and 0.05, say. Do the evaluation at two different locations, such as $x = 1$ and $x = 2$. (The order should not depend upon x so your results at the two points should be almost identical.)
   (a) Euler's method
   (b) Second-order Runge-Kutta method

6. (*Liquid level*) Liquid is pumped into a tank of horizontal cross-sectional area A (m^2) at a rate Q (liters/sec), and is drained by a valve at its base as sketched in the figure.

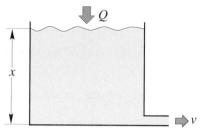

According to Bernoulli's principle, the efflux velocity $v(t)$ is approximately $\sqrt{2gx(t)}$, where g is the acceleration due to gravity. Thus, a mass balance (i.e., rate of increase of mass in tank = rate in − rate out) gives

$$Ax'(t) = Q(t) - Bv(t)$$
$$= Q(t) - B\sqrt{2gx(t)}, \quad (6.1)$$

where B is the cross-sectional area of the efflux pipe. For definiteness, suppose that $A = 1$ and $B\sqrt{2g} = 0.01$ so

$$x' = Q(t) - 0.01\sqrt{x}. \quad (6.2)$$

We wish to know the depth $x(t)$ at the end of 10 minutes ($t = 600$ sec), 20 minutes, ... , up to one hour. Program the computer solution of (6.2) by the second-order Runge-Kutta method for the following cases, and use it to solve for those x values: $x(600), x(1200), \ldots, x(3600)$. Beginning with $h = 20$, reduce h and run again, until the printed x values settle down to $\pm 10^{-4}$ or less.

(**a**) $Q(t) = 0.02; \quad x(0) = 0$

(**b**) $Q(t) = 0.02; \quad x(0) = 2$

(**c**) $Q(t) = 0.02; \quad x(0) = 4$

(**d**) $Q(t) = 0.02; \quad x(0) = 6$

(**e**) $Q(t) = 0.02\left(2 - e^{-0.004t}\right); \quad x(0) = 0$

(**f**) $Q(t) = 0.02\left(1 - e^{-0.004t}\right); \quad x(0) = 8$

(**g**) $Q(t) = 0.02t; \quad x(0) = 0$

(**h**) $Q(t) = 0.02(1 + \sin 0.1t); \quad x(0) = 0$
 NOTE: Surely, we will need h to be small compared to the period 20π of $Q(t)$.

7. (**a**)–(**h**) (*Liquid level*) Same as Exercise 6, but use fourth-order Runge-Kutta instead of second-order.

8. (**a**)–(**h**) (*Liquid level*) Same as Exercise 6, but use *Maple* for the numerical solution of the differential equation.

9. (*Liquid level*)

(**a**) For the case where $Q(t)$ is a constant, derive the general solution of (6.2) in Exercise 6 as

$$-0.01\sqrt{x} - Q\ln\left|\sqrt{x} - 100Q\right| = 0.00005t + C, \quad (9.1)$$

where C is the constant of integration.

(**b**) Evaluate C in (9.1) if $Q = 0.02$ and $x(0) = 0$. Then, solve (9.1) for $x(t)$ at $t = 600, 1200, \ldots, 3600$. NOTE: Unfortunately, (9.1) is in implicit rather than explicit form, but you can use *Maple* to solve it for x, for any given value of t.

10. (**a**) Program the fourth-order Runge-Kutta method (23) and use it to run the test problem

$$y' = y + 2x - x^2; \quad y(0) = 1$$

with the exact solution $y(x) = x^2 + e^x$, and to compute y at $x = 1$ using $h = 0.05$ and then $h = 0.02$. From those values and the known exact solution, empirically verify that the method is fourth order.

(**b**) To see what harm a programming error can cause, change the $x_n + h/2$ in the formula for k_2 to x_n, repeat the two evaluations of y at $x = 1$ using $h = 0.05$ and $h = 0.02$, and empirically determine the order of the method. Is it still a fourth-order method?

11. Derive the right-hand side of (16). HINT: Taylor expanding *all* terms in powers of h don't forget to expand the $f(x_n + \alpha h, y_n + \beta f h)$ term. You should obtain

$$f(x_n + \alpha h, y_n + \beta f h) = f + \left(\alpha f_x + \beta f_y f\right)h + \cdots.$$

12.4   APPLICATION TO SYSTEMS AND BOUNDARY VALUE PROBLEMS

The methods developed in the preceding sections are for an initial value problem with a single first-order differential equation, but what if we have a system of differential equations, a higher-order equation, or a boundary value problem? In this section we extend the Euler and fourth-order Runge-Kutta methods to these cases.

12.4.1 Systems and higher-order equations.   Consider the system of initial value problems

$$u'(x) = f(x, u, v); \qquad u(a) = u_0 \tag{1a}$$

$$v'(x) = g(x, u, v); \qquad v(a) = v_0 \tag{1b}$$

on $u(x)$ and $v(x)$. To extend Euler's method to such a system, we merely apply it to each of the problems (1a) and (1b) as follows:

$$u_{n+1} = u_n + f(x_n, u_n, v_n)h, \tag{2a}$$

$$v_{n+1} = v_n + g(x_n, u_n, v_n)h \tag{2b}$$

for $n = 0, 1, 2, \ldots$ Equations (2) are coupled [since (2a) involves v_n and (2b) involves u_n], as were equations (1), but that coupling causes no complication because the values u_n, v_n are already known from the preceding step.

E X A M P L E   1
To illustrate (2), consider the system

$$\begin{aligned} u' &= x + v; && u(0) = 0 \\ v' &= uv^2; && v(0) = 1, \end{aligned} \tag{3}$$

which is nonlinear because of the uv^2 term. We seek a numerical solution using the Euler method (2). Let $h = 0.1$, say, and let us go through the first two steps. First, $u_0 = 0$ and $v_0 = 1$ from the initial conditions. Then,

$n = 0$:

$$u_1 = u_0 + (x_0 + v_0)h = 0 + (0 + 1)(0.1) = \underline{0.1},$$

$$v_1 = v_0 + u_0 v_0^2 h = 1 + (0)(1)^2(0.1) = \underline{1}.$$

$n = 1$:

$$u_2 = u_1 + (x_1 + v_1)h = 0.1 + (0.1 + 1)(0.1) = \underline{0.21},$$

$$v_2 = v_1 + u_1 v_1^2 h = 1 + (0.1)(1)^2(0.1) = \underline{1.01},$$

and so on.   ∎

We can proceed in the same manner if the system contains more than two equations.

Next, we show how to adapt the fourth-order Runge-Kutta method to the system (1). Recall that for the single equation

$$y' = f(x, y); \qquad y(a) = y_0 \tag{4}$$

the algorithm is

$$y_{n+1} = y_n + \tfrac{1}{6}(k_1 + 2k_2 + 2k_3 + k_4),$$

$$k_1 = hf(x_n, y_n), \qquad\qquad k_2 = hf\left(x_n + \tfrac{h}{2}, y_n + \tfrac{1}{2}k_1\right), \qquad (5)$$

$$k_3 = hf\left(x_n + \tfrac{h}{2}, y_n + \tfrac{1}{2}k_2\right), \quad k_4 = hf(x_{n+1}, y_n + k_3).$$

For the system (1) it becomes

$$u_{n+1} = u_n + \frac{1}{6}(k_1 + 2k_2 + 2k_3 + k_4),$$

$$v_{n+1} = v_n + \frac{1}{6}(l_1 + 2l_2 + 2l_3 + l_4),$$

$$k_1 = hf(x_n, u_n, v_n),$$

$$l_1 = hg(x_n, u_n, v_n),$$

$$k_2 = hf\left(x_n + \frac{h}{2}, u_n + \frac{1}{2}k_1, v_n + \frac{1}{2}l_1\right), \qquad (6)$$

$$l_2 = hg\left(x_n + \frac{h}{2}, u_n + \frac{1}{2}k_1, v_n + \frac{1}{2}l_1\right),$$

$$k_3 = hf\left(x_n + \frac{h}{2}, u_n + \frac{1}{2}k_2, v_n + \frac{1}{2}l_2\right),$$

$$l_3 = hg\left(x_n + \frac{h}{2}, u_n + \frac{1}{2}k_2, v_n + \frac{1}{2}l_2\right),$$

$$k_4 = hf(x_{n+1}, u_n + k_3, v_n + l_3),$$

$$l_4 = hg(x_{n+1}, u_n + k_3, v_n + l_3),$$

and the pattern is the same for systems containing more than two equations.

EXAMPLE 2

Let us illustrate (6) using the same system as in Example 1,

$$u' = x + v; \qquad u(0) = 0$$
$$v' = uv^2; \qquad v(0) = 1. \qquad (7)$$

With $h = 0.1$, say, (6) gives

$n = 0$:

$$k_1 = h(x_0 + v_0) = (0.1)(0 + 1) = 0.1,$$

$$l_1 = hu_0v_0^2 = (0.1)(0)(1)^2 = 0,$$

$$k_2 = h\left[\left(x_0 + \frac{h}{2}\right) + \left(v_0 + \frac{l_1}{2}\right)\right]$$

$$= (0.1) \left[(0 + 0.05) + (1 + 0) \right] = 0.105,$$

$$l_2 = h \left(u_0 + \frac{k_1}{2} \right) \left(v_0 + \frac{l_1}{2} \right)^2 = (0.1)(0 + 0.05)(1 + 0)^2 = 0.005,$$

$$k_3 = h \left[\left(x_0 + \frac{h}{2} \right) + \left(v_0 + \frac{l_2}{2} \right) \right]$$

$$= (0.1) \left[(0 + 0.05) + (1 + 0.0025) \right] = 0.10525,$$

$$l_3 = h \left(u_0 + \frac{k_2}{2} \right) \left(v_0 + \frac{l_2}{2} \right)^2$$

$$= (0.1)(0 + 0.0525)(1 + 0.0025)^2 = 0.005276,$$

$$k_4 = h \left[x_1 + (v_0 + l_3) \right]$$

$$= (0.1) \left[0.1 + (1 + 0.005276) \right] = 0.110528,$$

$$l_4 = h (u_0 + k_3)(v_0 + l_3)^2$$

$$= (0.1)(0 + 0.10525)(1 + 0.005276)^2 = 0.010636,$$

$$u_1 = u_0 + \frac{1}{6}(k_1 + 2k_2 + 2k_3 + k_4)$$

$$= 0 + \frac{1}{6}(0.1 + 0.21 + 0.2105 + 0.110528) = \underline{0.105171},$$

$$v_1 = v_0 + \frac{1}{6}(l_1 + 2l_2 + 2l_3 + l_4)$$

$$= 1 + \frac{1}{6}(0 + 0.01 + 0.010552 + 0.010636) = \underline{1.005198}.$$

$n = 1:$

$$k_1 = 0.110520, \qquad l_1 = 0.010627,$$
$$k_2 = 0.116051, \qquad l_2 = 0.016382,$$
$$k_3 = 0.116339, \qquad l_3 = 0.016760,$$
$$k_4 = 0.122196, \qquad l_4 = 0.023134,$$

$$u_2 = \underline{0.221420}, \qquad v_2 = \underline{1.021872},$$

and so on for $n = 2, 3, \ldots$. We suggest that you fill in the details for the calculation of the $k_1, \ldots, v_2$ values shown above for $n = 1$.   ∎

Of course, the idea is to carry out such calculations on a computer, not by hand. The calculations shown in Examples 1 and 2 are merely intended to clarify the methods.

What about higher-order equations? The key is *to re-express an nth-order equation as an equivalent system of n first-order equations.*

EXAMPLE 3

The problem

$$y''' - xy'' + y' - 2y^3 = \sin x; \qquad y(1) = 2, \ y'(1) = 0, \ y''(1) = -3 \quad (8)$$

can be converted to an equivalent system of three first-order equations as follows. Define $y' = u$ and $y'' = v$ (hence $u' = v$). Then (8) can be re-expressed in the form

$$
\begin{aligned}
y' &= u; & y(1) &= 2 \\
u' &= v; & u(1) &= 0 \qquad\qquad \text{(9a,b,c)} \\
v' &= \sin x + 2y^3 - u + xv; & v(1) &= -3.
\end{aligned}
$$

Of the three differential equations in (9), the first two merely serve to introduce the auxiliary dependent variables u and v, and since v' is y''', the third one is a restated version of the given equation $y''' - xy'' + y' - 2y^3 = \sin x$. Equation (9a) is the y equation (because of the y' on the left-hand side), so the initial condition is on $y(1)$, namely, $y(1) = 2$, as given in (8). Equation (9b) is the u equation, so the initial condition is on $u(1)$, namely, $u(1) = y'(1) = 0$, from (8). Equation (9c) is the v equation, so the initial condition is on $v(1)$, namely, $v(1) = y''(1) = -3$.

The system (9) can now be solved by the Euler or fourth-order Runge-Kutta methods or any other such algorithm. To illustrate, let us carry out the first two steps using Euler's method, taking $h = 0.2$.

$n = 0$:

$$y_1 = y_0 + u_0 h = 2 + (0)(0.2) = \underline{2},$$

$$u_1 = u_0 + v_0 h = 0 + (-3)(0.2) = -0.6,$$

$$v_1 = v_0 + \left(\sin x_0 + 2y_0^3 - u_0 + x_0 v_0\right) h$$

$$= -3 + \left[\sin 1 + 2(2)^3 - 0 + (1)(-3)\right](0.2) = -0.231706.$$

$n = 1$:

$$y_2 = y_1 + u_1 h = 2 + (-0.6)(0.2) = \underline{1.88},$$

$$u_2 = u_1 + v_1 h = -0.6 + (-0.231706)(0.2) = -0.646341,$$

$$v_2 = v_1 + \left(\sin x_1 + 2y_1^3 - u_1 + x_1 v_1\right) h$$

$$= -0.231706 + \left[\sin 1.2 + 2(2)^3 - (-0.6) + (1.2)(-0.231706)\right](0.2)$$

$$= 3.219092,$$

and so on for $n = 2, 3, \ldots$.

COMMENT.   Observe that at each step we compute y, u, and v, yet we are not really interested in the auxiliary variables u and v. Perhaps we could compute just $y_1, y_2, \ldots$ and not the u, v values? No; equations (9) are coupled so we need to bring all three variables along together. Of course, we don't need to print or plot u and v, but we do need to compute them. ∎

EXAMPLE 4

Examples 1 and 2 involve a system of first-order equations, and Example 3 involves a single higher-order equation. As a final example, consider a combination of the two such as the initial value problem

$$u'' - 3xuv = \sin x; \quad u(0) = 4, \quad u'(0) = -1$$
$$v'' + 2u - v = 5x; \quad v(0) = 7, \quad v'(0) = 0. \tag{10}$$

The idea is the same as before. We need to recast (10) as a system of first-order initial value problems. We can do so by introducing auxiliary dependent variables w and z according to $u' = w$ and $v' = z$. Then (10) becomes

$$
\begin{aligned}
u' &= w; & u(0) &= 4 \\
w' &= \sin x + 3xuv; & w(0) &= -1 \\
v' &= z; & v(0) &= 7 \\
z' &= 5x - 2u + v; & z(0) &= 0
\end{aligned}
\tag{11}
$$

which system can now be solved by Euler's method or any other such numerical differential equation solver.   ■

12.4.2 Linear boundary value problems.   Our discussion is centered mostly upon the following example.

EXAMPLE 5

Consider the third-order boundary value problem

$$y''' - x^2 y = -x^4; \qquad y(0) = 0, \ y'(0) = 0, \ y(2) = 4. \tag{12}$$

To solve numerically, we begin by recasting (12) as the first-order system:

$$
\begin{aligned}
y' &= u; & y(0) &= 0, \quad y(2) = 4 \\
u' &= v; & u(0) &= 0 \\
v' &= x^2 y - x^4.
\end{aligned}
\tag{13a,b,c}
$$

However, we cannot apply the numerical integration techniques that we have discussed because the problem (13c) does not have an initial condition so we cannot get the algorithm started. Whereas (13c) is missing an initial condition on v, (13a) has an extra condition—the right end condition $y(2) = 4$, but that condition is of no help in a numerical integration scheme that develops a solution beginning at $x = 0$.

Nevertheless, the linearity of (12) saves the day and permits us to work with an initial value version instead. Suppose that we solve (numerically) the four initial value problems

$$L[Y_1] = 0, \qquad Y_1(0) = 1, \quad Y_1'(0) = 0, \quad Y_1''(0) = 0,$$
$$L[Y_2] = 0, \qquad Y_2(0) = 0, \quad Y_2'(0) = 1, \quad Y_2''(0) = 0,$$
$$L[Y_3] = 0, \qquad Y_3(0) = 0, \quad Y_3'(0) = 0, \quad Y_3''(0) = 1, \qquad (14)$$
$$L[Y_p] = -x^4, \quad Y_p(0) = 0, \quad Y_p'(0) = 0, \quad Y_p''(0) = 0,$$

where $L = d^3/dx^3 - x^2$ is the differential operator in (12). The nine initial conditions in the first three of these problems were chosen so that their determinant is nonzero, so Y_1, Y_2, Y_3 constitute a fundamental set of solutions (i.e., a linearly independent set of solutions) of the homogeneous equation $L[Y] = 0$. The three initial conditions on the particular solution Y_p were chosen as zero for simplicity; any values will do since any particular solution will do. Suppose we imagine that the four initial value problems in (14) have now been solved by the methods discussed above. Then Y_1, Y_2, Y_3, Y_p are known functions of x over the interval of interest $[0, 2]$, and we have the general solution

$$y(x) = C_1 Y_1(x) + C_2 Y_2(x) + C_3 Y_3(x) + Y_p(x) \qquad (15)$$

of $L[y] = -x^4$. Finally, we can evaluate the constants C_1, C_2, C_3 by imposing the boundary conditions given in (12):

$$y(0) = 0 = C_1 + 0 + 0 + 0,$$
$$y'(0) = 0 = 0 + C_2 + 0 + 0, \qquad (16)$$
$$y(2) = 4 = C_1 Y_1(2) + C_2 Y_2(2) + C_3 Y_3(2) + Y_p(2).$$

Solving (16) gives $C_1 = C_2 = 0$ and $C_3 = [4 - Y_p(2)]/Y_3(2)$, so we have the desired solution of (12) as

$$y(x) = \frac{4 - Y_p(2)}{Y_3(2)} Y_3(x) + Y_p(x). \qquad (17)$$

In fact, since $C_1 = C_2 = 0$ the functions $Y_1(x)$ and $Y_2(x)$ have dropped out, so we don't need to calculate them. All we need are $Y_3(x)$ and $Y_p(x)$, and these are found by the numerical integration of the initial value problems

$$Y_3' = U_3, \qquad Y_3(0) = 0,$$
$$U_3' = V_3, \qquad U_3(0) = 0, \qquad (18)$$
$$V_3' = x^2 Y_3, \quad V_3(0) = 1,$$

and

$$Y_p' = U_p, \qquad\qquad Y_p(0) = 0,$$
$$U_p' = V_p, \qquad\qquad U_p(0) = 0, \qquad (19)$$
$$V_p' = x^2 Y_p - x^4, \quad V_p(0) = 0,$$

respectively.

COMMENT.   Remember that whereas initial value problems have unique solutions (if the functions involved are sufficiently well behaved), boundary value problems can have no solution, a unique solution, or even an infinite number of solutions. How do these possibilities work out in this example? The clue is that (17) fails if $Y_3(2)$ turns out to be zero. The situation is seen more clearly from (16), where all the possibilities come into view. Specifically, if $Y_3(2) \neq 0$, then we can solve uniquely for C_3, and we have a unique solution, given by (17). If $Y_3(2)$ does vanish, then there are two possibilities as seen from (16): If $Y_p(2) \neq 4$, then there is no solution, and if $Y_p(2) = 4$ then there are an infinite number of solutions of (12), namely,

$$y(x) = C_3 Y_3(x) + Y_p(x), \tag{20}$$

where C_3 remains arbitrary.   ∎

We see that boundary value problems are more difficult than initial value problems. From Example 5 we see that a nonhomogeneous nth-order linear boundary value problem generally involves the solution of $n + 1$ initial value problems, although in Example 5 (in which $n = 3$) we were lucky and did not need to solve for two of the four unknowns, Y_1 and Y_2.

Nonlinear boundary value problems are more difficult still, because we cannot use the idea of finding a fundamental set of solutions plus a particular solution and thus forming a general solution as we did in Example 5, which idea is based upon linearity. One viable line of approach comes under the heading of **shooting methods**. For instance, to solve the nonlinear boundary value problem

$$y'' + \sin y = 3x; \qquad y(0) = 0, \ y(5) = 2 \tag{21}$$

we can solve the initial value problem

$$\begin{aligned} y' &= u, & y(0) &= 0 \\ u' &= 3x - \sin y, & u(0) &= u_0 \end{aligned} \tag{22}$$

iteratively. That is, we can guess at the initial condition u_0 [which is the initial slope $y'(0)$] and solve (22) for $y(x)$ and $u(x)$. Next, we compare the computed value of $y(5)$ with the boundary condition $y(5) = 2$ (which we have not yet used). If the computed value is too high, then we return to (22), reduce the value of u_0, and solve again. Comparing the new computed value of $y(5)$ with the prescribed value $y(5) = 2$, we again revise our value of u_0. If these revisions are done in a rational way, one can imagine obtaining a convergent scheme. Such a scheme is called a *shooting method* because of the obvious analogy with the shooting of a projectile such as an arrow, with the intention of having the projectile strike the ground at some distant prescribed point.

Thus, we can see the increase in difficulty as we move away from linear initial value problems. For a linear boundary value problem of order n we need to solve not one problem but $n + 1$ of them. For a nonlinear boundary value problem we need to solve an infinite sequence of them, in principle; in practice, we need to carry out only enough iterations to produce the desired accuracy.

Closure. In Section 12.4.1 we extended the Euler and fourth-order Runge-Kutta solution methods to cover systems of equations and higher-order equations. There we worked specific cases and examples, but the methods generalize nicely using vector notation. That is, the system

$$y_1'(x) = f_1(x, y_1(x), \ldots, y_n(x)); \quad y_1(a) = y_{10},$$
$$\vdots \qquad\qquad\qquad\qquad \vdots \qquad (23)$$
$$y_n'(x) = f_n(x, y_1(x), \ldots, y_n(x)); \quad y_n(a) = y_{n0},$$

can be expressed more compactly in the vector form

$$\boxed{\mathbf{y}'(x) = \mathbf{f}(x, \mathbf{y}(x)); \qquad \mathbf{y}(a) = \mathbf{y_0},} \qquad (24)$$

where the boldface letters denote n-dimensional column vectors:

$$\mathbf{y}(x) = \begin{bmatrix} y_1(x) \\ \vdots \\ y_n(x) \end{bmatrix}, \quad \mathbf{y}'(x) = \begin{bmatrix} y_1'(x) \\ \vdots \\ y_n'(x) \end{bmatrix}, \quad \mathbf{f}(x, \mathbf{y}(x)) = \begin{bmatrix} f_1(x, \mathbf{y}(x)) \\ \vdots \\ f_n(x, \mathbf{y}(x)) \end{bmatrix}, \qquad (25)$$

and where $f_j(x, \mathbf{y}(x))$ is simply a shorthand notation for $f_j(x, y_1(x), \ldots, y_n(x))$. Then the Euler algorithm corresponding to (24) is

$$\boxed{\mathbf{y_{n+1}} = \mathbf{y_n} + \mathbf{f}(x_n, \mathbf{y_n})\,h.} \qquad (26)$$

Maple. No new software is needed for the methods described in this section. For instance, we can use the *Maple* **dsolve** command with the numeric option to solve the problem

$$u' = x + v; \quad u(0) = 0$$
$$v' = -5uv; \quad v(0) = 1$$

and to print the results at $x = 1, 2$, and 3. Enter

```
dsolve({diff(u(x),x)=x+v(x),diff(v(x),x)=-5*u(x)*v(x),
u(0)=0,v(0)=1},{u(x),v(x)},type=numeric},
value=array([1,2,3]));
```

and return. The printed result is

$$\begin{bmatrix} \begin{bmatrix} & [x, u(x), v(x)] & \\ 1. & 1.032499018 & .07285274034 \\ 2. & 2.544584706 & .00001413488312 \\ 3. & 5.044585758 & -.3131498784 \times 10^{-9} \end{bmatrix} \end{bmatrix}$$

The only differences between the command above and the one given at the end of Section 12.3 is that here we have entered two differential equations, two initial conditions, and two dependent variables, and we have omitted the abserr option.

Observe that to solve a differential equation, or system of differential equations, numerically, we must first express the equations as a system of first-order equations, as illustrated in Example 4. However, to use the *Maple* dsolve command we can leave the original higher-order equations intact.

EXERCISES 12.4

1. In Example 2 we gave $k_1, l_1, k_2, l_2, k_3, l_3, k_4, l_4$ for $n = 1$, and the resulting values of u_2 and v_2, but did not show the calculations. Provide those details, as we did for the step $n = 0$.

2. As we did in Example 1, work out y_1, z_1, by hand. Use three methods: Euler, second-order Runge-Kutta, and fourth-order Runge-Kutta, and take $h = 0.2$. These problems have simple closed-form solutions, which are given in brackets. Compare your results with the exact solution.

   (a) $y' = z;$ $y(0) = 1$ $[y(x) = \cos x]$

   $z' = -y;$ $z(0) = 0$ $[z(x) = -\sin x]$

   (b) $y' = 4z;$ $y(2) = 5$ $[y(x) = 5\cos(4 - 2x)]$

   $z' = -y;$ $z(2) = 0$ $[z(x) = (5/2)\sin(4 - 2x)]$

   (c) $y' = -z^2/y;$ $y(0) = 1$ $\left[y(x) = e^{-x}\right]$

   $z' = -y;$ $z(0) = 1$ $\left[z(x) = e^{-x}\right]$

   (d) $y' = 2xz^2/y;$ $y(1) = 1$ $\left[y(x) = x^2\right]$

   $z' = y/z^2;$ $z(1) = 1$ $[z(x) = x]$

   (e) $y' = (x + y)z - 1;$ $y(1) = 1$ $[y(x) = x]$

   $z' = -yz^3;$ $z(1) = 1$ $[z(x) = 1/x]$

3. (a) – (e) First, read Exercise 2. Use *Maple* to solve the initial value problem given in the corresponding part of Exercise 2, for $y(x)$ and $z(x)$ at $x = 3, 5$, and 10.

4. (a) Just as (2) and (6) give the Euler and fourth-order Runge-Kutta algorithms for the system (1), write down the analogous Euler and second-order Runge-Kutta algorithms for the system

$$x'(t) = F(t, x, y, z),   x(a) = x_0$$

$$y'(t) = G(t, x, y, z),   y(a) = y_0   (4.1)$$

$$z'(t) = H(t, x, y, z),   z(a) = z_0.$$

   (b) Use the Euler and second-order Runge-Kutta algorithms to work out x_1, y_1, z_1 and x_2, y_2, z_2, by hand, for the case where F, G, H are $y-1, z, t+x+3(z-y+1)$, respectively, with the initial conditions $x(0) = -3$, $y(0) = 0, z(0) = 2$ using $h = 0.3$.

   (c) Same as (a), but with $x(0) = y(0) = z(0) = 0$.

   (d) Same as (a), but with $x(0) = 1, y(0) = 0, z(0) = 0$.

   (e) Same as (a), but with $x(0) = y(0) = 0, z(0) = 10$.

5. We re-expressed (8) and (10) as the equivalent systems of first-order initial value problems (9) and (11), respectively. Do the same for the problem given. You need go no further.

   (a) $mx'' + cx' + kx = f(t);$ $x(0) = x_0,$ $x'(0) = x_0'$

   (b) $Li'' + Ri' + (1/C)i = E'(t);$ $i(0) = i_0,$ $i'(0) = i_0'$

   (c) $y'' - xyy' = \sin x;$ $y(1) = 5,$ $y'(1) = -1$

   (d) $y'' + y' - 4y = 3x;$ $y(0) = 2,$ $y'(0) = 1$

   (e) $y''' - 2\sin y' = 3x;$ $y(-2) = 7,$ $y'(-2) = 4,$ $y''(-2) = 0$

   (f) $y''' + xy = \cos 2x;$ $y(1) = 3,$ $y'(1) = 2,$ $y''(1) = 0$

   (g) $x'' + 2x - 3y = 10\cos 3t;$ $x(0) = 2,$ $x'(0) = -1,$ $y'' - x + 5y = 0;$ $y(0) = 4,$ $y'(0) = 3$

   (h) $y''' + xy'z = f(x);$ $y(3) = 2,$ $y'(3) = -1,$ $y''(3) = 6z'' + y - z = g(x);$ $z(3) = 0,$ $z'(3) = 8$

   (i) $x''' - 3xz = \sin t;$ $x(1) = x'(1) = 0,$ $x''(1) = 3$ $y' + 2x + y - 5z = 0;$ $y(1) = 6$ $z' - 4xy = e^{2t};$ $z(1) = 2$

   (j) $y'''' = yz;$ $y(0) = 1,$ $y'(0) = y''(0) = 0$ $z'' = -xy + z;$ $z(0) = 5,$ $z'(0) = 1$

6. Use *Maple* to solve the given system numerically, and print out the solution for $y(x)$ and $z(x)$ at $x = 1, 2$.

   (a) $y'' - 2xz' = 5x;$ $y(0) = 2,$ $y'(0) = -1$ $z' + yz^3 = -3;$ $z(0) = 1$

   (b) $y' + 3xz = x^2;$ $y(0) = 1$ $z''' - y^2z' + z = 0;$ $z(0) = z'(0) = 2,$ $z''(0) = -1$

   (c) $y' = z' + x;$ $y(0) = 1$ $z'' + y^2z^2 = 3;$ $z(0) = -1,$ $z'(0) = 0$

   (d) $y' - z'' = x;$ $y(0) = -1$ $z''' - y^2z = 3;$ $z(0) = 1,$ $z'(0) = z''(0) = 0$

7. Complete the solution of Example 5 by using *Maple* to solve (18) for $Y_3(x)$ and (19) for $Y_p(x)$, at $x = 0.5, 1.0$, 1.5, and 2, and then using (17) to determine $y(x)$ at those points.

8. Use the method explained in Example 5 to reduce the given linear boundary value problem to a system of linear initial value problems. Then complete the solution and solve for the specified quantity, either using *Maple* or by programming any of the numerical solution methods that we have studied. If you believe that there is no solution or that it exists but is nonunique, then give your reasoning. HINT: You can specify homogeneous initial conditions for the Y_p problem, as we did in Example 5, but be aware that you do not *have* to use homogeneous conditions, and that you may be able to reduce your labor by a more optimal choice of those conditions.

   (a) $y'' - 2xy' + y = 3\sin x;$ $y(0) = 1,$ $y(2) = 3$. Determine $y(0.5), y(1)$, and $y(1.5)$.

   (b) $y'' + (\cos x)y = 0;$ $y(0) = 1,$ $y(10) = 2$. Determine $y(2)$.

(c) $y'' - [\ln(x+1)]\,y' - y = 2\sin 3x + 1$;
$y(0) = 3$, $y(2) = -1$.
Determine $y(x)$ at $x = 0.5, 1.0, 1.5$.

(d) $y'' + y' - xy = x^3$; $y(0) = 1$, $y(5) = 2$.
Determine $y(x)$ at $x = 1, 2, 3, 4$.

(e) $y''' + xy = 2x^3$; $y(1) = y'(1) = 0$, $y''(2) = -3$.
Determine $y(2)$.

(f) $y''' + xy' + y = x$; $y(1) = 2$, $y'(1) = 0.4$, $y'(5) = 3$.
Determine $y(x)$ at $x = 4, 5$.

CHAPTER 12 REVIEW

To solve a differential equation $y' = f(x, y)$ numerically we begin by discretizing the problem and working with the discrete variables x_n, y_n in place of the continuous variables $x, y(x)$. For the methods discussed in this chapter, the approximate solution is accomplished using a numerical algorithm that gives y_{n+1} in terms of one or more of the preceding values $y_n, y_{n-1}, y_{n-2}, \dots$. If y_{n+1} is given in terms of only the preceding value y_n then the method is called a single-step method; otherwise it is a multi-step method. Only single-step methods were considered in this chapter.

We are interested in methods that are convergent. That is, for any fixed x value in the x interval of interest the sequence of y values, obtained from the algorithm using smaller and smaller step size h, tends to the exact solution $y(x)$ of the initial value problem at that point x as $h \to 0$. Suppose the single-step or local truncation error LTE is of order h^p, where p is an integer. Then heuristic reasoning suggests that the accumulated truncation error TE is of order h^q where $q = p - 1$. For convergence we need $q > 0$, in which case we say that the method is of order q. For instance, the Euler method

$$y_{n+1} = y_n + f(x_n, y_n)h$$

is a first-order method ($q = 1$).

The order q of the method is extremely important because for a given step size h the truncation error at a given point x is approximately a constant times h to the q power; thus, the larger q is, the smaller the truncation error is. This fact was dramatically illustrated in Table 2 of Section 12.3.3, where the fourth-order Runge-Kutta results were correct to six or more significant figures over the entire x domain of interest whereas the first-order Euler results were correct to only one or two significant figures. Thus, while the Euler method offers the advantage of extreme simplicity, the fourth-order Runge-Kutta method is a more efficient and powerful method. That is, for a given desired level of accuracy the Runge-Kutta method can give that accuracy using a much larger step size than the Euler method.

So much for a single first-order differential equation. For a higher-order equation, or a system of equations, we noted in Section 12.4 that the first step is to introduce auxiliary dependent variables, if necessary, to obtain a system of first-order equations. Then, any of the methods studied in Section 12.2 and 12.3 can be applied. If the problem is not of initial value type then there will be at least one missing initial condition in that system so we cannot get the numerical solution started. However, if the problem is linear, then we can use the principle of superposition to generate a general solution by solving separately for a fundamental set of linearly independent homogeneous solutions as well as for a particular solution, each of which problems is of initial value type. The case of nonlinear boundary value problems was not discussed.

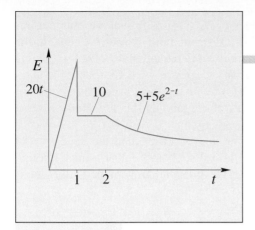

Chapter 3

LAPLACE TRANSFORM

13.1 INTRODUCTION

First, let us explain what is meant by a "transform." To do that, let us examine a transform technique that is already familiar to you—the use of the natural logarithm in numerical calculation.

While the addition of two numbers is arithmetically simple, their multiplication can be quite laborious; for example, try working out 2.761359×8.247504 by hand. Thus, given two positive numbers u and v, suppose we wish to compute their product

$$y = uv \tag{1}$$

Taking the logarithm of both sides gives $\ln y = \ln uv$. But $\ln uv = \ln u + \ln v$, so we have $\ln y = \ln u + \ln v$. Thus, whereas the original problem was one of multiplication, the problem in the "transform domain" is merely one of addition. The idea, then, is to look up $\ln u$ and $\ln v$ in a table and to add these two values. With the sum in hand, we enter the table again, this time using it in the reverse direction to find the antilog, y. (With pocket calculators and computers available logarithm tables are no longer needed, as they were in 1950, but the transform nature of the logarithm remains the same whether we use tables or not.) Of course, we don't normally call the logarithm a "transform," but it is one. In transform terminology we would say that we take the "logarithmic transform" of equation (1).

The Laplace transform works in essentially the same way, but instead of using it to evaluate products of numbers we use it to solve differential equations.[1] Namely, given a linear differential equation with constant coefficients, on an unknown function $x(t)$, if we take the Laplace transform of both sides of the equation [just as we

---

[1]The Laplace transform has other uses as well, but its chief use, and the one that is our focus in this chapter, is in the solution of differential equations.

took the logarithmic transform of both sides of equation (1)] then we obtain a linear algebraic equation on the transform $L\{x(t)\}$ of the unknown $x(t)$. Thus, in the "transform domain" we have merely a linear *algebraic* equation to solve for $L\{x(t)\}$ rather than the original *differential* equation on $x(t)$. Once we solve that linear algebraic equation for $L\{x(t)\}$, we then obtain $x(t)$ by taking the "inverse transform" of $L\{x(t)\}$; that is, we use the Laplace transform table again but this time in the reverse direction.

Observe that we used the notation $x(t)$, above, rather than the generic $y(x)$ notation used in most of this text. The reason for that choice is that in most (though not all) applications of the Laplace transform the independent variable is the time t, on the interval $0 \le t < \infty$.

13.2   DEFINITIONS AND CALCULATION OF THE TRANSFORM

Given a function $f(t)$ defined on $0 \le t < \infty$, the **Laplace transform** of $f(t)$, denoted as $L\{f(t)\}$, is defined as[1]

$$L\{f(t)\} = \int_0^\infty f(t)e^{-st}\,dt, \tag{1}$$

where the parameter s is a constant; s can be complex but in this chapter we can, and will, consider it to be real, without loss. That is, to take the Laplace transform of a given function $f(t)$ we multiply it by e^{-st} and integrate from $t = 0$ to $t = \infty$. When we evaluate the integral in (1) between the numerical limits 0 and ∞ we are left with a function of the parameter s; hence, besides the $L\{x(t)\}$ notation we will often use the notation $F(s)$ for the Laplace transform of $f(t)$, $X(s)$ for the transform of $x(t)$, and so on. Of course this notation fails if the dependent variable is already upper case. In that event we will use a standard alternative notation consisting of an overhead bar (which, in this context, does *not* mean complex conjugate); for example, $L\{Q(t)\} = \bar{Q}(s)$. Thus, we have three standard alternative notations for the Laplace transform of a given function $f(t)$:

$$L\{f(t)\}, \quad F(s), \quad \bar{f}(s).$$

Operationally speaking, the Laplace transform is a transformation or mapping from one function to another; the input function is $f(t)$ and the output function

---

[1]The Laplace transform is named after the great French applied mathematician *Pierre-Simon de Laplace* (1749–1827) who contributed chiefly to celestial mechanics but also to fluid mechanics and other branches of science. The second-order partial differential equation known as the Laplace equation is, arguably, one of the three most prominent partial differential equations of mathematical physics. However, although Laplace studied the relation (1), the Laplace transform methodology is more an outgrowth of an operational calculus developed by the British electrical engineer *Oliver Heaviside* (1850–1925). For an historical account see J. L. B. Cooper, "Heaviside and the Operational Calculus," *Math. Gazette*, Vol. 36 (1952), pp. 5–19.

is $F(s)$. By way of comparison, the logarithmic transform is also a mapping, not from one function to another but from one number to another. (A mapping from one number to another is simply a *function*, and $\ln x$ is indeed a function.)

The Laplace transform (1) is an example of an **integral transform** of the form

$$F(s) = \int_a^b f(t) K(s, t) \, dt, \tag{2}$$

where $K(t, s)$ is called the **kernel** of the transformation (not to be confused with the kernel or null space defined in Section 5.8.1).[1] One can design the transform [i.e., by choosing a, b and the kernel $K(t, s)$] so that (2) is suited to a particular class of linear differential equations; the choice $a = 0$, $b = \infty$, and $K(t, s) = e^{-st}$ gives the Laplace transform, that is tailored to *constant-coefficient* differential equations, as we shall see.

Before we illustrate (1) with examples, we must be clear about the class of functions $f(t)$ that are admissible. To address this issue we need to introduce three concepts: singular integrals, exponential order, and piecewise continuity.

First, whereas the Riemann integral (that we study in the integral calculus) has finite limits, the integral in (1) has an infinite upper limit. Thus, it is a "new object" for us, a **singular integral**,[2] and must be defined. Analogous to our definition of an infinite series [by (5) in Section 8.2] as

$$\sum_1^\infty a_n \equiv \lim_{N \to \infty} \sum_1^N a_n,$$

that is, as the limit of a sequence of finite sums, we define[3]

$$\boxed{\int_a^\infty g(t) \, dt \equiv \lim_{B \to \infty} \int_a^B g(t) \, dt,} \tag{3}$$

where a is finite. That is, the singular integral on the left is hereby defined as the limit of a sequence of regular integrals (i.e., with finite limits). If the limit in (3) exists then we say that $\int_a^\infty g(t) \, dt$ **converges**; if not, it **diverges**.

---

[1]To feel more comfortable with the function space transformation (2) it might be helpful to note the similarity between (2) and the analogous finite-dimensional transformation $\mathbf{Ax}$, in which case the input is a vector from n-space [analogous to the input $f(t)$ in (2)], $\mathbf{A}$ is an $m \times n$ matrix [analogous to $K(t, s)$ in (2)], and the output $\mathbf{Ax}$ is a vector in m-space [analogous to $F(s)$ in (2)].

[2]The term **improper integral** is also used.

[3]Frequently in mathematics we introduce a "new thing" as the limit of a sequence of "old things." For instance, with a finite sum of numbers well understood we introduce the infinite sum $\sum_1^\infty a_n$ as the limit of a sequence of partial sums $\sum_1^\infty a_n$, each of which is a finite sum; we introduce the derivative dy/dx as the limit of a sequence of difference quotients; we introduce the integral $\int_a^b f(x) \, dx$ as the limit of a sequence of partial sums; and so on. Indeed, the limit concept is the cornerstone of analysis.

EXAMPLE 1

Does the singular integral

$$\int_0^\infty e^{-\kappa t}\, dt \qquad (\kappa = \text{real constant}) \tag{4}$$

converge? Applying (3) gives

$$\int_0^\infty e^{-\kappa t}\, dt = \lim_{B\to\infty} \int_0^B e^{-\kappa t}\, dt$$

$$= \lim_{B\to\infty} \left.\frac{e^{-\kappa t}}{-\kappa}\right|_0^B = \lim_{B\to\infty} \frac{1}{\kappa}\left(1 - e^{-\kappa B}\right). \tag{5}$$

From the $e^{-\kappa B}$ term in (5) we can see that the limit exists if $\kappa > 0$ and fails to exist if $\kappa < 0$ because if $\kappa < 0$ then $e^{-\kappa B} \to \infty$ as $B \to \infty$. The case $\kappa = 0$ must be treated separately because $e^{-\kappa t}/(-\kappa)$ is undefined if $\kappa = 0$: if $\kappa = 0$, (4) gives

$$\int_0^\infty 1\, dt = \lim_{B\to\infty} \int_0^B 1\, dt = \lim_{B\to\infty} B, \tag{6}$$

which limit does not exist. Thus, (4) converges to

$$\int_0^\infty e^{-\kappa t}\, dt = \frac{1}{\kappa} \tag{7}$$

if $\kappa > 0$ and diverges if $\kappa \leq 0$.   ∎

EXAMPLE 2

Does $\int_0^\infty \sin t\, dt$ converge? From (3),

$$\int_0^\infty \sin t\, dt = \lim_{B\to\infty} \int_0^B \sin t\, dt = \lim_{B\to\infty} (1 - \cos B), \tag{8}$$

which limit does not exist because $\cos B$ keeps oscillating and does not approach a unique limit as $B \to \infty$. Thus, $\int_0^\infty \sin t\, dt$ is divergent. This case is analogous to the infinite series $1 - 1 + 1 - 1 + 1 - \cdots$ which diverges because its partial sums oscillate between the values 1 and 0 and do not approach a limit.   ∎

Next, we say that a function is of **exponential order** as $t \to \infty$ if there exist real constants K, c, and T, where K and T are positive, such that

$$\boxed{|f(t)| \leq K e^{ct}} \tag{9}$$

for all $t \geq T$. In place of the words "for all $t \geq T$" we could say that (9) holds "eventually," that is, for sufficiently large t. Thus, the set of functions of exponential order is the set of functions that do not grow faster than exponentially, which set includes the vast majority of functions of engineering interest. For instance, each of the functions $f(t) = 1$, e^{3t}, $50e^{-t}$, $-4\sin t$, t^2, $t^3 \cos 4t$ is of exponential order.

For $f(t) = 1$, say, we can satisfy (9) by choosing any $K \geq 1$, any $c \geq 0$, and any $T \geq 0$, such as $K = 4$, $c = 3$, $T = 8$.

For $f(t) = e^{3t}$ we can choose any $K \geq 1$, any $c \geq 3$, and any $T \geq 0$.

For $f(t) = 50e^{-t}$ we can choose any $K \geq 50$, any $c \geq -1$, and any $T \geq 0$.

For $f(t) = -4\sin t$ we can choose any $K \geq 4$, any $c \geq 0$, and any $T \geq 0$.

The case $f(t) = t^2$ is not as obvious. Observe first that if there is a c such that $|f(t)|/e^{ct} \to 0$ as $t \to \infty$ then surely there exist positive constants K and T such that (9) holds for all $t \geq T$, for if $|f(t)|/e^{ct} \to 0$ then for *any* positive constant K (no matter how small) there must correspond a T such that $|f(t)|/e^{ct} \leq K$ or, equivalently, $|f(t)| \leq Ke^{ct}$, for all $t \geq T$. With this idea in mind let us return to considering $f(t) = t^2$ and observe (using l'Hôpital's rule twice) that

$$\lim_{t\to\infty} \frac{t^2}{e^{ct}} = \lim_{t\to\infty} \frac{2t}{ce^{ct}} = \lim_{t\to\infty} \frac{2}{c^2 e^{ct}} = 0 \tag{10}$$

for any $c > 0$. Thus, $f(t) = t^2$ *is* of exponential order as $t \to \infty$.

Similarly, we can show that $f(t) = t^n$ is of exponential order for any $n = 1, 2, 3, \ldots$. That is, even though t^n grows rapidly as $t \to \infty$, e^{ct} grows more rapidly, no matter how small the positive constant c is, and no matter how large n is.

On the other hand, the function $f(t) = e^{t^2}$, say, is *not* of exponential order as $t \to \infty$ because

$$\lim_{t\to\infty} \frac{e^{t^2}}{e^{ct}} = \lim_{t\to\infty} e^{t^2 - ct} = \infty,$$

no matter how large c is.

Finally, we say that a function is **piecewise continuous** on $a \leq t \leq b$ if there exist a finite number of points $t_1, t_2, \ldots, t_N$ such that $f(t)$ is continuous on each open subinterval $a < t < t_1$, $t_1 < t < t_2$, $\ldots$ $t_N < t < b$, and has a finite limit as t approaches each endpoint from the interior of that subinterval.

For instance, the function $f(t)$ shown in Fig. 1 is piecewise continuous on the interval $0 \leq t \leq 4$. The values of f at the endpoints $a, t_1, t_2, \ldots, b$ are not relevant to whether or not f is piecewise continuous. Hence we have not even indicated those values in Fig. 1. For instance, the limit of f as t tends to 2 from the left exists and is 5, and the limit of f as t tends to 2 from the right exists and is 10; the value of f at $t = 2$ does not matter. Thus, piecewise continuity allows for the presence of jump discontinuities.

With the foregoing three concepts established we can now state a fundamental theorem that guarantees the convergence of the integral in (1), and hence the "existence" of the Laplace transform of a given function $f(t)$.

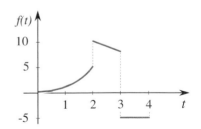

FIGURE 1
Example of piecewise continuity.

THEOREM 13.2.1

Existence of the Laplace Transform
Let $f(t)$ be

   (i) piecewise continuous on $0 \le t \le t_0$ for every $t_0 > 0$, and

   (ii) of exponential order as $t \to \infty$, so there exist real constants K, c and T, where K and T are positive, such that $|f(t)| \le Ke^{ct}$ for all $t \ge T$.

Then the Laplace transform of $f(t)$, defined by (1), exists for all $s > c$.

Proof.   We want to show that if $s > c$ then the integral in (1) converges. In

$$\int_0^\infty f(t)e^{-st}\,dt = \lim_{B \to \infty} \int_0^B f(t)e^{-st}\,dt \tag{11}$$

the integral on the right converges for each finite B because $f(t)$, and hence the integrand $f(t)e^{-st}$, is piecewise continuous by assumption. It remains to show that if $s > c$ then the limit on the right-hand side of (11) exists. We have

$$\left| \int_0^B f(t)e^{-st}\,dt \right| = \left| \int_0^T f(t)e^{-st}\,dt + \int_T^B f(t)e^{-st}\,dt \right|$$

$$\le \left| \int_0^T f(t)e^{-st}\,dt \right| + \left| \int_T^B f(t)e^{-st}\,dt \right|$$

$$\le \int_0^T \left| f(t)e^{-st} \right|\,dt + \int_T^B \left| f(t)e^{-st} \right|\,dt$$

$$= \int_0^T |f(t)|\,e^{-st}\,dt + \int_T^B |f(t)|\,e^{-st}\,dt. \tag{12}$$

Since $f(t)$ is piecewise continuous it is bounded on $0 \le t \le T$ so $|f(t)| \le M$ there, for some constant M. Further, $|f(t)| \le Ke^{ct}$ on $t \ge T$, by assumption, so (12) gives

$$\left| \int_0^B f(t)e^{-st}\,dt \right| \le \int_0^T Me^{-st}\,dt + \int_T^B Ke^{-(s-c)t}\,dt$$

$$= \frac{M}{s}\left(1 - e^{-sT}\right) + \frac{K}{s-c}\left(e^{-(s-c)T} - e^{-(s-c)B}\right). \tag{13}$$

Finally, from (3) and (13)

$$\left| \int_0^\infty f(t)e^{-st}\,dt \right| = \lim_{B \to \infty} \left| \int_0^B f(t)e^{-st}\,dt \right|$$

$$\le \frac{M}{S}\left(1 - e^{-sT}\right) + \frac{K}{s-c}e^{-(s-c)T} \tag{14}$$

if $s > c$. The bound on the right side of (14) establishes that the limit of the integral from 0 to B exists as $B \to \infty$, so the Laplace transform of $f(t)$ exists for all $s > c$.   ∎

Observe that if we let $s \to \infty$ in (14) then the right-hand side tends to zero. Hence we have the following additional result:

THEOREM 13.2.2
$F(s)$ as $s \to \infty$

If $f(t)$ satisfies the hypotheses of Theorem 13.2.1, then

$$\lim_{s \to \infty} F(s) = 0. \tag{15}$$

Though not as important as Theorem 13.2.1, Theorem 13.2.2 tells us that not every function of s is the Laplace transform of some function $f(t)$, at least not some "respectable" function $f(t)$ (i.e., a function that satisfies the hypotheses of Theorem 13.2.1). For instance, 1, s, $3s^2 + 7$, $\cos s$, and e^{2s} are not Laplace transforms of such functions because they don't tend to zero as $s \to \infty$.

Understand that Theorem 13.2.1 gives *sufficient* conditions on $f(t)$ for its transform $L\{f(t)\}$ to exist. That those conditions are not also *necessary* is shown by a single example: The function $f(t) = t^{1/2}$ does *not* satisfy those conditions since the limit of $t^{-1/2}$ as $t \to 0$ does not exist (Fig. 2), hence $t^{-1/2}$ is not piecewise continuous on $0 \le t \le B$ for *any* B. Yet the integral $L\{t^{-1/2}\}$ does exist and, we state without derivation, is

$$L\left\{ \frac{1}{\sqrt{t}} \right\} = \sqrt{\frac{\pi}{s}}. \tag{16}$$

As a final comment, notice that of the quantities K, c, and T in Theorem 13.2.1, c is particularly significant since the theorem states that the transform of $f(t)$ exits for all $s > c$.

Being assured by Theorem 13.2.1 that the transform $F(s)$ exists for a large and useful class of functions, we now proceed to illustrate the evaluation of $F(s)$ for representative elementary functions such as $f(t) = 1$, t, t^n, e^{at}, and $\sin at$, where n is a positive integer and a is a real number.

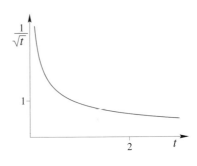

FIGURE 2
Behavior of $t^{-1/2}$.

EXAMPLE 3
If $f(t) = 1$ then the conditions of the theorem are met for any $c \ge 0$ so, according to the theorem, $F(s)$ should exist for all $s > 0$. In fact,

$$L\{1\} = F(s) = \int_0^\infty 1\, e^{-st}\, dt \tag{17}$$

is the same integral as (4) (with κ changed to s) so it follows from Example 1 that

$$\boxed{L\{1\} = \frac{1}{s}} \tag{18}$$

for all $s > 0$. ∎

EXAMPLE 4
If $f(t) = e^{at}$ then the conditions of the theorem are met for any $c \ge a$ so $F(s)$ should exist for all $s > a$. In fact,

$$L\left\{e^{at}\right\} = \int_0^\infty e^{at} e^{-st}\, dt = \lim_{B \to \infty} \int_0^B e^{-(s-a)t}\, dt$$

$$= \lim_{B \to \infty} \frac{1}{s-a} \left(1 - e^{-(s-a)B} \right). \tag{19}$$

The limit exists only if $s > a$ and gives

$$\boxed{L\left\{ e^{at} \right\} = \frac{1}{s-a}, \qquad (s > a)} \tag{20}$$

which result does agree with (18), by the way, for the special case $a = 0$.   ∎

EXAMPLE 5
If $f(t) = \sin at$, then the conditions of Theorem 13.2.1 are met for any $c \geq 0$ so $F(s)$ should exist for all $s > 0$. In fact, integrating by parts twice gives

$$F(s) = \int_0^\infty \sin at \, e^{-st} \, dt = \lim_{B \to \infty} \int_0^B \sin at \, e^{-st} \, dt$$

$$= \lim_{B \to \infty} \left[\sin at \, \frac{e^{-st}}{-s} \Big|_0^B - \int_0^B a \cos at \, \frac{e^{-st}}{-s} \, dt \right]$$

$$= \lim_{B \to \infty} \left[\left(\sin at \, \frac{e^{-st}}{-s} + \frac{a}{s} \cos at \, \frac{e^{-st}}{-s} \right) \Big|_0^B - \frac{a^2}{s^2} \int_0^B \sin at \, e^{-st} \, dt \right]$$

$$= (0 - 0) + \left(0 - \frac{a}{-s^2} \right) - \frac{a^2}{s^2} F(s), \tag{21}$$

where the limit exists if $s > 0$. The latter equation can be solved for $F(s)$, which is $L\{\sin at\}$, and gives

$$\boxed{L\{\sin at\} = \frac{a}{s^2 + a^2}. \qquad (s > 0)} \tag{22}$$

COMMENT.   An alternative approach, which requires a knowledge of the algebra of complex numbers (Section 6.2), is as follows:

$$\int_0^\infty \sin at \, e^{-st} \, dt = \int_0^\infty \left(\text{Im} \, e^{iat} \right) e^{-st} \, dt$$

$$= \text{Im} \int_0^\infty e^{-(s-ia)t} \, dt = \text{Im} \left[\lim_{B \to \infty} \frac{e^{-(s-ia)t}}{-(s-ia)} \Big|_0^B \right]$$

$$= \text{Im} \, \frac{1}{s-ia} = \text{Im} \, \frac{1}{s-ia} \frac{s+ia}{s+ia} = \frac{a}{s^2 + a^2}, \tag{23}$$

as before, where the fourth equality follows because

$$\left| e^{-(s-ia)B} \right| = \left| e^{-sB} \right| \left| e^{iaB} \right| = e^{-sB} \to 0 \tag{24}$$

as $B \to \infty$, if $s > 0$. In (24) we have used the fact that $|e^{iaB}| = |\cos aB + i \, \sin aB| = \sqrt{\cos^2 aB + \sin^2 aB} = 1$.   ∎

Since (22) is a true statement only if $s > 0$, it is important to include that qualifier explicitly in (22). Similarly, $s > a$ in (20) and $s > 0$ in (18). However, these conditions will not play an active role in this chapter or in any way limit us in applying the Laplace transform.[1] From Examples 3–5 we could begin to construct a Laplace transform table with $f(t)$ in one column and its transform $F(s)$ in another. Such a table is provided here among the endpapers. One can also obtain transforms and their inverses directly using computer software, such as *Maple*.

Tables can be used in either direction. For example, just as the transform of e^{at} is $1/(s - a)$, it is also true that the function whose transform is $1/(s - a)$ is e^{at}. Operationally, we say that

$$L\{e^{at}\} = \frac{1}{s - a} \quad \text{and} \quad L^{-1}\left\{\frac{1}{s - a}\right\} = e^{at}. \tag{25}$$

That is, $1/(s - a)$ is the Laplace transform of e^{at} and e^{at} is the **inverse Laplace transform** of $1/(s - a)$, as indicated by the L^{-1} inverse operator notation.

If we are to use our transform table in both directions then it is important that the transformation be one-to-one so that the transform $L\{f(t)\}$ is a uniquely determined function $F(s)$ and then, in the reverse direction, $L^{-1}\{F(s)\}$ is uniquely equal to the original function $f(t)$. Strictly speaking, we do not have such one-to-oneness. For instance, not only does the function $f(t) = 1$ have the transform $F(s) = 1/s$ (as found in Example 3) but so does the function

$$g(t) = \begin{cases} 1, & 0 \le t < 3, \ 3 < t < \infty \\ 2, & t = 3 \end{cases}$$

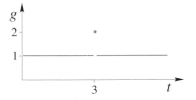

FIGURE 3
$g(t)$.

shown in Fig. 3 have the transform $G(s) = 1/s$ since the integrands in $\int_0^\infty f(t)e^{-st}\,dt$ and $\int_0^\infty g(t)e^{-st}\,dt$ differ only at the single point $t = 3$. Since there is no area under a single point (of finite height), $G(s)$ and $F(s)$ are identical: $G(s) = F(s) = 1/s$. Thus the inverse of $1/s$ is not unique. From a practical standpoint such nonunique-ness is superficial and we state without further discussion that for our purposes we can consider the Laplace transformation to be one-to-one.

Closure.   Theorem 13.2.1 guaranteed the existence of the Laplace transform of a given function $f(t)$ subject to the (sufficient but not necessary) conditions that f be piecewise continuous on $0 \le t \le t_0$ for every $t_0 > 0$ and of exponential order as $t \to \infty$. We proceeded to demonstrate the evaluation of the transforms of several simple functions by direct integration, and discussed the building up of a transform table. Regarding the use of such a table, we stated that for practical purposes we can regard the Laplace transformation as one-to-one; that is, given an $f(t)$ satisfying the

[1]Incidentally, these conditions can be determined most simply by direct inspection of the final result $F(s)$. Specifically, the c in the "$s > c$" condition in Theorem 13.2.1 is, as a rule of thumb, the real part of the most rightward point in the complex plane at which $F(s)$ is infinite. For instance, $F(s) = 1/(s - a)$ is infinite at $s = a$ so $F(s) = 1/(s - a)$ holds for $s > a$; $F(s) = a/(s^2 + a^2)$ is infinite at $s = \pm ai$ (i.e., at $0 \pm ai$) so it holds for $s > 0$; $F(s) = 1(s^2 + 8s + 25)$ is infinite at $s = -4 \pm 3i$ so it holds for $s > -4$; and so on.

conditions of Theorem 13.2.1 there is a unique transform $F(s)$ and, in the reverse direction, the inverse of $F(s)$ is (to within superficial differences) uniquely equal to the original function $f(t)$.

Maple. *Maple* can be used to obtain Laplace transforms and inverse Laplace transforms. For instance, the commands

```
with(inttrans):
laplace(exp(3*t),t,s);
```

give the transform of e^{3t} as

$$\frac{1}{s-3}$$

and the command

```
invlaplace(1/(s-3),s,t);
```

gives the inverse transform of $1/(s-3)$ as e^{3t}. Note that "inttrans" is short for integral transform, the Laplace transform being one of the most important integral transforms.

EXERCISES 13.2

1. Show whether or not the given function is of exponential order. If it is, determine a suitable set of values for K, c, and T in (9).
   (**a**) $5e^{4t}$ (**b**) $-10e^{-5t}$ (**c**) $\sinh 2t$
   (**d**) $\cosh 3t$ (**e**) $\sinh t^2$ (**f**) $e^{4t}\sin t$
   (**g**) $\cos t^3$ (**h**) t^{100} (**i**) $1/(t+2)$
   (**j**) $\cosh t^4$ (**k**) $6t + e^t\cos t$ (**l**) t^{1000}

2. If $f(t)$ is of exponential order, does it follow that df/dt is too? HINT: Consider $f(t) = \sin e^{t^2}$.

3. If $f(t)$ and $g(t)$ are each of exponential order, does it follow that $f(g(t))$ is too? HINT: Consider the case where $f(t) = e^t$ and $g(t) = t^2$.

4. Derive entry 4 of the transform table two ways:
   (**a**) using integration by parts
   (**b**) using the replacement $\cos at = \operatorname{Re} e^{iat}$ HINT: See the Comment in Example 5.

5. Derive entry 5 of the transform table.

6. Derive entry 6 of the transform table.

7. Evaluate $L\{f(t)\}$ and indicate the s interval on which the result holds. HINT: Use integration by parts, more than once if necessary. Your results should agree with entry 7 in the transform table.
   (**a**) $f(t) = t$

(**b**) $f(t) = t^2$
(**c**) $f(t) = t^n$ where n is any positive integer

8. Derive entry 9 of the transform table.

9. Derive entry 10 of the transform table.

10. Derive entry 11 of the transform table two ways:
    (**a**) by writing

    $$F(s) = \frac{1}{2i}\int_0^\infty te^{-(s-ia)t}\,dt - \frac{1}{2i}\int_0^\infty te^{-(s+ia)t}\,dt$$

    and using integration by parts
    (**b**) by differentiating both sides of the known transform

    $$\int_0^\infty \sin at\, e^{-st}\,dt = \frac{a}{s^2+a^2}$$

    (which we derived in Example 5) with respect to s, assuming the validity of the interchange

    $$\frac{d}{ds}\int_0^\infty \sin at\, e^{-st}\,dt = \int_0^\infty \frac{d}{ds}\left(\sin at\, e^{-st}\right)\,dt$$

    in the order of integration and differentiation.

11. Use the idea in Exercise 10(b) to derive
    (**a**) entry (12) from entry (4)

(b) entry (7) from entry (1)

(c) entry (13) from entry (5)

(d) entry (14) from entry (6)

(e) entry (15) from entry (2)

12. Show that $L\{e^{at}\} = 1/(s - a)$ holds even if $a = \operatorname{Re} a + i \operatorname{Im} a$ is complex, provided that $s > \operatorname{Re} a$.

13. (*Inverting as power series*) If, in inverting a given transform, we are willing to end up with a power series rather than a closed form expression we can proceed as follows. To illustrate, let us seek the inverse of $F(s) = \dfrac{1}{s - a}$. Expanding the latter in a Taylor series about $s = 0$ gives

$$F(s) = \frac{1}{s - a} = -\frac{1}{a} - \frac{1}{a^2}s - \frac{1}{a^3}s^2 - \cdots . \quad (13.1)$$

Although the individual terms are simple (merely constants times powers of s) they are not invertible because they do not tend to zero as $s \to \infty$ (Theorem 13.2.2). Thus, (13.1) is of no help. However, suppose we re-express $F(s)$ as

$$F(s) = \frac{1}{s}\frac{1}{1 - \frac{a}{s}}. \quad (13.2)$$

If we let $a/s \equiv z$, say, then when we expand

$$\frac{1}{1 - \frac{a}{s}} = \frac{1}{1 - z} = 1 + z + z^2 + \cdots$$

$$= 1 + \frac{a}{s} + \frac{a^2}{s^2} + \cdots \quad (13.3)$$

we obtain *inverse* powers of s, which *are* invertible. Thus, (13.2) and (13.3) give

$$f(t) = L^{-1}\left\{\frac{1}{s} + \frac{a}{s^2} + \frac{a^2}{s^3} + \cdots\right\}$$

$$= L^{-1}\left\{\frac{1}{s}\right\} + aL^{-1}\left\{\frac{1}{s^2}\right\} + a^2 L^{-1}\left\{\frac{1}{s^3}\right\} + \cdots$$

$$= 1 + at + \frac{1}{2!}a^2 t^2 + \frac{1}{3!}a^3 t^3 + \cdots , \quad (13.4)$$

where, in the second equality, we have assumed that the infinite series could be inverted term by term.[1] We know that the result (13.4) is correct because we know that the inverse of $1/(s - a)$ is e^{at} and the final member of (13.4) is indeed the Taylor series of e^{at} about $t = 0$. This inversion method is of interest for functions $F(s)$ that are difficult to invert

by conventional means but the drawback is that the inverse obtained is in the form of an infinite series rather than in closed form. The problem that we pose here is for you to use this method in each case and, using the transform table, verify that your resulting power series is correct. HINT: In part (a), for instance, factor an s^2 out of the denominator just as we factored an s out of the denominator in (13.2).

(a) $\dfrac{1}{s^2 + a^2}$ **(b)** $\dfrac{1}{s^2 - a^2}$

(c) $\dfrac{s}{(s^2 + 1)^2}$ **(d)** $\dfrac{s^2 - 4}{(s^2 + 4)^2}$

14. Use *Maple* to verify the given entry in the transform table. That is, show that the transform of the given $f(t)$ is the given $F(s)$ and show also that the inverse transform of the given $F(s)$ is the given $f(t)$.

(a) 1,2 **(b)** 3,4 **(c)** 5,6 **(d)** 9,10

(e) 11,12 **(f)** 13,14 **(g)** 17 **(h)** 18

(i) 7 for $n = 5$

(j) 15 for $n = 4$

15. (*Gamma function*) The integral

$$\Gamma(p) = \int_0^\infty x^{p-1} e^{-x}\, dx \qquad (p > 0) \quad (15.1)$$

is nonelementary; that is, it cannot be evaluated in closed form in terms of the so-called elementary functions. Since it arises frequently it has been given a name, the **gamma function** $\Gamma(x)$ [designated as Gamma(x) in *Maple*], and has been studied extensively. Observe that the integral is singular for two reasons: first, the upper limit is ∞ and, second, the integrand is unbounded as $x \to 0$ (if $p - 1 < 0$) because $x^{p-1} e^{-x} = x^{p-1}(1 - x + x^2/2 - \cdots) \sim x^{p-1}$ as $x \to 0$. The more negative is the exponent $p - 1$, the stronger is the "blow up" of x^{p-1} as $x \to 0$. Nevertheless, it can be shown from the theory of singular integrals that the integral in (15.1) does converge if $p - 1$ is not "too" negative, namely, if $p - 1 > -1$; i.e., if $p > 0$. Hence, that stipulation is noted in (15.1).

(a) Integrating by parts, use (15.1) to show that

$$\Gamma(p) = (p - 1)\Gamma(p - 1) \quad (p > 1) \quad (15.2)$$

NOTE: The latter *recursion formula* is the most important property of the gamma function for if we know (e.g., by numerical integration) the values of $\Gamma(p)$ over

---

[1] By the linearity of L^{-1} we know that $L^{-1}\{a_1 F_1(s) + \cdots + a_k F_k(s)\} = a_1 L^{-1}\{F_1(s)\} + \cdots + a_k L^{-1}\{F_k(s)\}$ for each finite k, no matter how large, but it does not follow from that result that $L^{-1}\{\sum_1^\infty a_j F_j(s)\} = \sum_1^\infty a_j L^{-1}\{F_j(s)\}$.

a unit interval such as $0 < x \le 1$ then we can use (15.2) to evaluate $\Gamma(p)$ for any $p > 1$. For example,

$$\Gamma(3.2) = 2.2\Gamma(2.2) = (2.2)(1.2)\Gamma(1.2)$$
$$= (2.2)(1.2)(0.2)\Gamma(0.2),$$

where $\Gamma(0.2)$ is known if, as assumed, $\Gamma(p)$ is known over $0 < p \le 1$.

(b) Show, by direct integration, that

$$\Gamma(1) = 1. \tag{15.3}$$

(c) Using (15.2) and (15.3), show that if p is a positive integer n then

$$\Gamma(n) = (n-1)! \tag{15.4}$$

where 0! is defined to be 1.

(d) Besides being able to evaluate $\Gamma(p)$ analytically at $p = 1, 2, 3, \dots$ we can evaluate it at $p = \frac{1}{2}, \frac{3}{2}, \frac{5}{2}, \dots$. In particular, show that

$$\Gamma\left(\frac{1}{2}\right) = \sqrt{\pi}. \tag{15.5}$$

HINT: By a change of variable show that

$$\Gamma\left(\frac{1}{2}\right) = 2\int_0^\infty e^{-u^2}\, du.$$

Then

$$\left[\Gamma\left(\frac{1}{2}\right)\right]^2 = 4\int_0^\infty e^{-u^2}\, du \int_0^\infty e^{-v^2}\, dv$$
$$= 4\int_0^\infty \int_0^\infty e^{-(u^2+v^2)}\, du\, dv.$$

Regarding the latter as a double integral in a Cartesian u, v plane, change from u, v to polar variables r, θ remembering that in place of the Cartesian area element $du\, dv$ the polar area element is $r\, dr\, d\theta$. The resulting double integral should be simpler to evaluate.

(e) Using (15.2) and (15.5), show that

$$\Gamma\left(\frac{3}{2}\right) = \frac{\sqrt{\pi}}{2}, \qquad \Gamma\left(\frac{5}{2}\right) = \frac{3\sqrt{\pi}}{4}. \tag{15.6}$$

NOTE: Similarly one can obtain $\Gamma(\frac{7}{2})$, $\Gamma(\frac{9}{2})$, and so on.

16. Using (15.1), show that if $p > -1$ then

$$L\left\{t^p\right\} = \frac{\Gamma(p+1)}{s^{p+1}}, \tag{16.1}$$

which is entry 8 of the transform table. If only because of (16.1) the gamma function is important in the theory and application of the Laplace transform.

13.3 PROPERTIES OF THE TRANSFORM

When we studied the integral calculus we probably evaluated a few simple integrals, such as $\int_a^b x\, dx$, directly from the definition of the Riemann integral, but mostly we learned how to evaluate integrals more indirectly, using a number of properties of integration. For instance we used the linearity property

$$\int_a^b [\alpha u(x) + \beta v(x)]\, dx = \alpha \int_a^b u(x)\, dx + \beta \int_a^b v(x)\, dx, \tag{1}$$

which holds for any constants α, β and for any functions $u(x)$, $v(x)$ if the integrals on the right-hand side are convergent. Thus, knowing that $\int_a^b 1\, dx = x|_a^b$ and $\int_a^b x\, dx = (x^2/2)|_a^b$, we were able to evaluate the integral of *any* function $\alpha + \beta x$. We also used partial fractions, integration by parts, the fundamental theorem of the integral calculus, "clever substitutions," and so on.

Our plan for the Laplace transform is the same; we evaluate a handful of transforms by direct integration and then rely on a variety of properties of the transform and the inverse transform to extend that list. In this section we cover only a few such properties—just enough to let us begin using the method to solve differential equations (in Section 13.4). Additional properties are then covered in subsequent sections and in the exercises.

We begin with the linearity property of the transform and inverse transform.

THEOREM 13.3.1

Linearity of the Transform and Inverse

(i) If $u(t)$ and $v(t)$ are any two functions such that the transforms $L\{u(t)\}$ and $L\{v(t)\}$ both exist, then

$$\boxed{L\{\alpha u(t) + \beta v(t)\} = \alpha L\{u(t)\} + \beta L\{v(t)\}}$$ (2)

for any constants α, β.

(ii) For any $U(s)$ and $V(s)$ such that the inverse transforms $L^{-1}\{U(s)\} = u(t)$ and $L^{-1}\{V(s)\} = v(t)$ exist, then

$$\boxed{\begin{aligned} L^{-1}\{\alpha U(s) + \beta V(s)\} &= \alpha L^{-1}\{U(s)\} + \beta L^{-1}\{V(s)\} \\ &= \alpha u(t) + \beta v(t) \end{aligned}}$$ (3)

for any constants α, β.

Equation (2) follows, essentially, from the corresponding linearity property of Riemann integration and then (3) follows by taking L^{-1} of (2).

EXAMPLE 1

To evaluate the transform of $6 - 5e^{4t}$, for example, we need merely know the transforms of the simpler functions 1 and e^{4t} for $L\{6 - 5e^{4t}\} = 6L\{1\} - 5L\{e^{4t}\}$. Now, $L\{1\} = 1/s$ for $s > 0$, and $L\{e^{4t}\} = 1/(s - 4)$ for $s > 4$ so (2) gives

$$L\{6 - 5e^{4t}\} = 6\frac{1}{s} - 5\frac{1}{s - 4} = \frac{s - 24}{s(s - 4)}$$

for $s > 4$.   ∎

EXAMPLE 2

Asked to evaluate the inverse of $F(s) = \dfrac{3}{s^2 + 3s - 10}$, we turn to our transformation table but do not find this $F(s)$ in the column of transforms. However, we can simplify $F(s)$ using partial fractions. Accordingly, we express

$$\frac{3}{s^2 + 3s - 10} = \frac{3}{(s + 5)(s - 2)} = \frac{A}{s + 5} + \frac{B}{s - 2}$$
$$= \frac{(A + B)s + (-2A + 5B)}{s^2 + 3s - 10}.$$ (4)

To make the latter an identity we equate the coefficients of s^1 and s^0 in the numerators of the left-hand and right-hand sides: s^1 gives $0 = A + B$, and s^0 gives $3 = -2A + 5B$

so $A = 3/7$ and $B = -3/7$. Then

$$L^{-1}\left\{\frac{3}{s^2+3s-10}\right\} = L^{-1}\left\{\frac{-3/7}{s+5}+\frac{3/7}{s-2}\right\}$$

$$= -\frac{3}{7}L^{-1}\left\{\frac{1}{s+5}\right\}+\frac{3}{7}L^{-1}\left\{\frac{1}{s-2}\right\}$$

$$= -\frac{3}{7}e^{-5t}+\frac{3}{7}e^{2t}, \tag{5}$$

where the second equality follows from (3), and the last equality follows from entry 2 in the table. ∎

If we are going to apply the Laplace transform method to differential equations, we will need to know how to take the transform of a derivative.

THEOREM 13.3.2

Transform of a Derivative

Let $f(t)$ be continuous and $f'(t)$ be piecewise continuous on $0 \le t \le t_0$ for every finite t_0, and let $f(t)$ be of exponential order as $t \to \infty$ so that there are constants K, c, T such that $|f(t)| \le Ke^{ct}$ for all $t > T$. Then $L\{f'(t)\}$ exists for all $s > c$, and

$$\boxed{L\{f'(t)\} = s\,L\{f(t)\} - f(0).} \tag{6}$$

Proof. Since $L\{f'(t)\} = \lim_{B\to\infty}\int_0^B f'(t)\,e^{-st}\,dt$, consider the integral

$$I = \int_0^B f'(t)\,e^{-st}\,dt = \int_0^{t_1} f'(t)\,e^{-st}\,dt + \cdots + \int_{t_n}^B f'(t)\,e^{-st}\,dt, \tag{7}$$

where $t_1, \ldots, t_n$ are the points, in $0 < t < B$, at which f' is discontinuous. Integrating by parts gives

$$I = f(t)\,e^{-st}\Big|_0^{t_1} + \cdots + f(t)\,e^{-st}\Big|_{t_n}^B$$

$$+ s\int_0^{t_1} f(t)\,e^{-st}\,dt + \cdots + s\int_{t_n}^B f(t)\,e^{-st}\,dt. \tag{8}$$

By virtue of the continuity of f, the boundary terms at $t_1, \ldots, t_n$ cancel in pairs so that, after recombining the integrals in (8), we have

$$I = f(B)\,e^{-sB} - f(0) + s\int_0^B f(t)\,e^{-st}\,dt. \tag{9}$$

Since f is of exponential order as $t \to \infty$ it follows that $f(B)\,e^{-sB} \to 0$ as $B \to \infty$. Thus,

$$L\{f'(t)\} = \lim_{B\to\infty}\left[f(B)\,e^{-sB} - f(0) + s\int_0^B f(t)\,e^{-st}\,dt\right]$$

$$= 0 - f(0) + s\,L\{f(t)\},$$

as was to be proved. ∎

The foregoing result can be used to obtain the transforms of higher derivatives as well. For example, if $f'(t)$ satisfies the conditions imposed on f in Theorem 13.3.2, then replacement of f by f' in (6) gives

$$L\{f''\} = s\,L\{f'\} - f'(0) = s\,[s\,L\{f\} - f(0)] - f'(0). \tag{10}$$

If, besides f', f also satisfies the conditions of Theorem 13.3.2, so that the $L\{f\}$ term on the right side of (10) exists, then

$$L\{f''\} = s^2\,L\{f\} - s\,f(0) - f'(0). \tag{11}$$

Similarly,

$$L\{f'''\} = s^3\,L\{f\} - s^2\,f(0) - s\,f'(0) - f''(0), \tag{12}$$

if f'', f', and f satisfy the conditions of Theorem 13.3.2, and so on for the transforms of higher-order derivatives.

The last of the three properties of the Laplace transform that we discuss in this section is the Laplace convolution theorem. First, we define an operation on given functions $p(t)$ and $q(t)$ that produces a new function which we denote as $p * q$ and which we call the **Laplace convolution** of p and q:

$$(p * q)(t) \equiv \int_0^t p(\tau)\,q(t - \tau)\,d\tau, \tag{13}$$

where the notation $(p * q)(t)$ is to indicate that the new function $p * q$ is, like p and q, a function of t.

EXAMPLE 3
Let $p(t) = e^t$ and $q(t) = 4e^{-3t}$. Then

$$(p * q)(t) = \int_0^t e^\tau 4\,e^{-3(t-\tau)}\,d\tau$$

$$= 4e^{-3t} \int_0^t e^{4\tau}\,d\tau = e^t - e^{-3t}.$$

■

We will establish the importance of the Laplace convolution in Theorem 13.3.3, below. For the moment merely consider the functions $p(t)$ and $q(t)$ as inputs and their convolution $(p * q)(t)$ defined by (13) as the output, being sure to realize that $p * q$ is not the product of p and q but rather their convolution. For instance, in Example 3 the ordinary multiplication of $p(t) = e^t$ and $q(t) = 4e^{-3t}$ is $p(t)q(t) = -4e^{-2t}$ whereas their convolution is $(p * q)(t) = e^t - e^{-3t}$.

Although Laplace convolution is not an ordinary product it does obey some of the rules of ordinary multiplication:

$$p * q = q * p, \qquad \text{(commutative)} \qquad \text{(14a)}$$
$$p * (q * r) = (p * q) * r, \qquad \text{(associative)} \qquad \text{(14b)}$$
$$p * (q + r) = p * q + p * r, \qquad \text{(distributive)} \qquad \text{(14c)}$$

$$p * 0 = 0. \tag{14d}$$

To prove (14a), make the change of variable $t - \tau = \mu$ from the dummy integration variable τ in (13) to the dummy integration variable μ, regarding t as fixed. Then $\tau = t - \mu$ and $d\tau = -d\mu$ so

$$(p * q)(t) = \int_0^t p(\tau) q(t - \tau) \, d\tau = \int_t^0 p(t - \mu) q(\mu) (-d\mu)$$

$$= \int_0^t q(\mu) p(t - \mu) \, d\mu = (q * p)(t). \tag{15}$$

For instance, in Example 3 we found that $(p * q)(t) = e^t - e^{-3t}$. If you work out $(q * p)(t)$ you should, according to (14a), obtain the same result: $(q * p)(t) = e^t - e^{-3t}$. Proof of (14b) is left for the exercises.

We are now ready for the Laplace convolution theorem.

THEOREM 13.3.3

Laplace Convolution Theorem

Let $p(t)$ and $q(t)$ be piecewise continuous on $0 \le t \le t_0$ for every finite t_0, and of exponential order as $t \to \infty$; that is, let there be constants K, c, T such that $|p(t)| \le Ke^{ct}$ and $|q(t)| \le Ke^{ct}$ for all $t > T$. Then

$$\boxed{L^{-1}\{P(s)Q(s)\} = (p * q)(t)} \tag{16}$$

or, equivalently,

$$\boxed{L\{(p * q)(t)\} = P(s)Q(s)} \tag{17}$$

for $s > c$.

CAUTION: It is *not* true that $L^{-1}\{P(s)Q(s)\}$ is $L^{-1}\{P(s)\}$ times $L^{-1}\{Q(s)\}$, that is, $p(t)$ times $q(t)$; rather, it is the Laplace *convolution* of p and q. Let us outline a proof of (16) and (17). Since they are equivalent it will suffice to prove just one, say (17). By definition,

$$L\{(p * q)(t)\} = L\left\{ \int_0^t p(\tau) q(t - \tau) \, d\tau \right\} = \int_0^\infty \left\{ \int_0^t p(\tau) q(t - \tau) \, d\tau \right\} e^{-st} \, dt. \tag{18}$$

The right-hand side of (18) is an iterated integral in a τ, t plane, and we can infer from the four integration limits that the region of integration is the infinite $45°$ wedge shown in Fig. 1. We recall from the calculus the equivalent notations

$$\int_c^d \left\{ \int_a^b f(x, y) \, dx \right\} dy = \int_c^d dy \int_a^b f(x, y) \, dx \tag{19}$$

for iterated integrals,[1] we change (18) from the former notation to the latter, and we

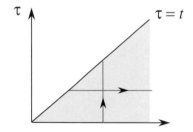

FIGURE 1
Region of integration.

---

[1] The right-hand side of (19) means this: first evaluate $\int_a^b f(x, y) \, dx$. Then take that result, which is a function of y, and integrate it on y, from $y = c$ to $y = d$.

then invert the order of the integration:

$$L\left\{\int_0^t p(\tau)\,q(t-\tau)\,d\tau\right\} = \int_0^\infty e^{-st}\,dt \int_0^t p(\tau)\,q(t-\tau)\,d\tau$$

$$= \int_0^\infty p(\tau)\,d\tau \int_\tau^\infty q(t-\tau)\,e^{-st}\,dt$$

$$= \int_0^\infty p(\tau)\,d\tau \int_0^\infty q(\mu)\,e^{-s(\mu+\tau)}\,d\mu$$

$$= \int_0^\infty p(\tau)\,e^{-s\tau}\,d\tau \int_0^\infty q(\mu)\,e^{-s\mu}\,d\mu$$

$$= P(s)Q(s), \tag{20}$$

as stated in (17). To obtain the second equality in (20) we inverted the order of integration and used Fig. 1 to deduce the new integration limits. How did we distribute the integrand $p(\tau)q(t-\tau)e^{-st}$ among the two integrals? The first integration is on t; since $q(t-\tau)$ and e^{-st} both contain t they must be included in the integrand of the t integral. However, $p(\tau)$ is a constant with respect to the t integration so we can slide it forward, into the τ integrand, and ask it to "wait there" until we come to the τ integration. The third equality followed from replacing t by μ according to $t-\tau=\mu$, regarding τ as fixed. Finally, following the fourth equal sign we see that the μ integral does not contain any τ's so what started out as an iterated integral now merely amounts to the product of two independent integrals, which are seen to be the Laplace transforms $P(s)$ and $Q(s)$.

The forgoing steps capture the key idea, inverting the order of integration, but fall short of a rigorous proof of Theorem 13.3.3 because we did not justify the inversion in the order of integration.[1]

EXAMPLE 4

In Example 2 we inverted $F(s) = \dfrac{3}{s^2+3s-10}$ using partial fractions. Let us again obtain the inverse $f(t)$, this time using the convolution theorem. Instead of breaking up $F(s)$ in an additive way by partial fractions, as

$$F(s) = -\frac{3}{7}\frac{1}{s+5} + \frac{3}{7}\frac{1}{s-2}, \tag{21}$$

we *factor* it as

$$F(s) = \frac{3}{s^3+3s-10} = \left(\frac{3}{s+5}\right)\left(\frac{1}{s-2}\right). \tag{22}$$

---

[1]Inverting the order of integration is easily justified if $p(t)$ and $q(t)$ are piecewise continuous on $0 \le t \le t_0$ for every finite t_0, as was assumed, and if the four integration limits are finite. However, one of the limits in (18) is ∞ and the infinite limit forces us to use a more delicate argument involving *uniform convergence*, which discussion is beyond our present scope. It turns out that if we assume that $p(t)$ and $q(t)$ are of exponential order as $t \to \infty$ then inverting the order of integration can be justified. For details see, for instance, R. V. Churchill, *Operational Mathematics*, 2nd ed., McGraw-Hill, NY, 1958.

Since

$$L^{-1}\left\{\frac{3}{s+5}\right\} = 3e^{-5t} \quad\text{and}\quad L^{-1}\left\{\frac{1}{s-2}\right\} = e^{2t}, \tag{23}$$

it follows from (16) that

$$L^{-1}\{F(s)\} = 3e^{-5t} * e^{2t} = 3\int_0^t e^{-5\tau} e^{2(t-\tau)}\, d\tau$$

$$= \frac{3}{7}\left(e^{2t} - e^{-5t}\right). \tag{24}$$

COMMENT. In this example we were able to choose between two possible methods of inversion. In other applications the convolution theorem may provide the *only* viable line of approach, as we will find in Example 1 of Section 13.4. ∎

Closure. The properties studied in this section—linearity, the transform of a derivative, and the convolution theorem, should be thoroughly understood. All are used in the next section, where we use the Laplace transform method to solve differential equations.

The convolution property, in particular, should be studied carefully. It is useful in both directions. If we have a transform $F(s)$ that is difficult to invert, it may be possible to factor F as $P(s)Q(s)$, where P and Q are more easily inverted. If so, then $f(t)$ is given, according to (16), as the convolution of $p(t)$ and $q(t)$. Furthermore, we may need to find the transform of an integral that is in convolution form. If so, the transform is given easily by (17); see Exercise 8.

EXERCISES 13.3

1. Find the inverse of the given transform two different ways: using partial fractions and using the convolution theorem. Cite any entries used from the transform table.
   (a) $3/[s(s+8)]$ (b) $1/(3s^2 + 5s - 2)$
   (c) $1/(s^2 - a^2)$ (d) $5/[(s+1)(3s+2)]$
   (e) $1/(s^2 + s)$ (f) $2/(2s^2 - s - 1)$

2. (a)–(f) Find the inverse of the corresponding transform in Exercise 1 using *Maple*.

3. Prove equation (14b).

4. Prove that $L\{p*q*r\} = P(s)Q(s)R(s)$ or, equivalently, that $L^{-1}\{P(s)Q(s)R(s)\} = p*q*r$. NOTE: Does $p*q*r$ mean $(p*q)*r$ or $p*(q*r)$? According to the associative property (14b) it doesn't matter; they are equal.

5. To illustrate the result stated in Exercise 4, find the inverse of $1/s^3$ as $L^{-1}\left\{\frac{1}{s^3}\right\} = L^{-1}\left\{\frac{1}{s}\frac{1}{s}\frac{1}{s}\right\} = 1*1*1$, and show that the result agrees with that given directly in the table.

6. Factoring $\dfrac{s}{(s^2+a^2)^2} = \dfrac{s}{s^2+a^2}\dfrac{1}{s^2+a^2}$, it follows from the convolution theorem and entries 3 and 4 of of the table that

   $$L^{-1}\left\{\frac{s}{(s^2+a^2)^2}\right\} = \cos at * \frac{\sin at}{a}.$$

   Evaluate this convolution and show that the result agrees with that given directly by entry 11.

7. Verify (6) and (11) directly, for each given $f(t)$, by working out the left-hand and right-hand sides and showing that they are equal. You may use the table in the endpapers to evaluate the necessary transforms.
   (a) e^{3t} (b) $e^{-4t} + 2$ (c) $t^2 + 5t - 1$
   (d) $\sinh 4t$ (e) $\cosh 3t + 5t^6$ (f) $4t^{3/2} - \cos 2t$

8. Evaluate the transform of each:
   (a) $\int_0^t e^{t-\tau} \sin 2\tau\, d\tau$ (b) $\int_0^t \cos 3(t - \tau)\, d\tau$

(c) $\int_0^t (t-\tau)^8 e^{-3\tau}\, d\tau$ **(d)** $\int_0^t \cosh 3(t-\tau)\, d\tau$

(e) $\displaystyle\int_0^t \frac{\sin \tau}{\sqrt{t-\tau}}\, d\tau$

(f) $\int_0^t \tau^{5.2} \sinh (4t-\tau)\, d\tau$

9. We emphasized that in general

$$L\{p(t)q(t)\} \neq P(s)Q(s) \tag{9.1}$$

and that in general

$$L^{-1}\{P(s)Q(s)\} \neq p(t)q(t). \tag{9.2}$$

Verify (9.1) and (9.2) for each given pair of functions.

(a) $p(t) = 1$, $q(t) = 1$ **(b)** $p(t) = 4$, $q(t) = t$
(c) $p(t) = t$, $q(t) = e^t$ **(d)** $p(t) = \sin t$, $q(t) = 4$
(e) $p(t) = t$, $q(t) = t^2$ **(f)** $p(t) = \cos t$, $q(t) = t+6$

13.4 APPLICATION TO THE SOLUTION OF DIFFERENTIAL EQUATIONS

We are ready to use the Laplace transform to solve linear constant-coefficient differential equations on the interval $0 \leq t < \infty$, with initial conditions given at $t = 0$.

EXAMPLE 1

We've already studied the important case of the harmonic oscillator—both free and forced, damped and undamped. Consider the undamped mechanical oscillator shown in Fig. 1, with a forcing function that is a constant: $f(t) = F_0$. Recall that the displacement $x(t)$ then satisfies the equation

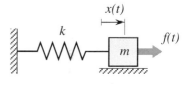

FIGURE 1
Mechanical oscillator.

$$mx'' + kx = f(t) = F_0. \tag{1}$$

Further, we assume the initial conditions $x(0)$ and $x'(0)$ are known.

To apply the Laplace transform, we transform equation (1). That is, we multiply each term in (1) by the Laplace kernel e^{-st} and integrate on t from 0 to ∞. Operationally, we use L to denote that step:

$$L\{mx'' + kx\} = L\{F_0\}. \tag{2}$$

By the linearity of L (Theorem 13.3.1), we can rewrite (2) as

$$mL\{x''(t)\} + kL\{x(t)\} = F_0\, L\{1\}. \tag{3}$$

Recalling from Section 13.3 that $L\{x''(t)\} = s^2 X(s) - sx(0) - x'(0)$, noting that $L\{x(t)\} \equiv X(s)$, and obtaining $L\{1\} = 1/s$ from the table, (3) becomes

$$m\left[s^2 X(s) - sx(0) - x'(0)\right] + kX(s) = F_0\, \frac{1}{s}. \tag{4}$$

The point to appreciate is that whereas in the "t domain" we had the linear *differential* equation (1) on $x(t)$, in the transform domain, or "s domain," we now have the linear *algebraic* equation (4) on $X(s)$. The solution now amounts to solving (4) by simple algebra for $X(s)$. Doing so gives

$$X(s) = \frac{sx(0) + x'(0)}{s^2 + \omega^2} + \frac{F_0}{ms\left(s^2 + \omega^2\right)}, \tag{5}$$

where $\omega = \sqrt{k/m}$ is the natural frequency.

With the solving for $X(s)$ completed, we now invert (5) to obtain $x(t)$:

$$x(t) = L^{-1}\left\{\frac{sx(0) + x'(0)}{s^2 + \omega^2} + \frac{F_0}{ms\left(s^2 + \omega^2\right)}\right\}$$

$$= x(0)L^{-1}\left\{\frac{s}{s^2 + \omega^2}\right\} + x'(0)L^{-1}\left\{\frac{1}{s^2 + \omega^2}\right\} + \frac{F_0}{m}L^{-1}\left\{\frac{1}{s\left(s^2 + \omega^2\right)}\right\},$$
(6)

the second equality following from the linearity of the L^{-1} operator (Theorem 13.3.1). The table gives

$$L^{-1}\left\{\frac{s}{s^2 + \omega^2}\right\} = \cos \omega t \qquad \text{and} \qquad L^{-1}\left\{\frac{1}{s^2 + \omega^2}\right\} = \frac{\sin \omega t}{\omega},$$
(7)

but the third inverse in (6) is not found in there. We could evaluate it with the help of partial fractions, but it is easier to use the convolution theorem:

$$L^{-1}\left\{\frac{1}{s\left(s^2 + \omega^2\right)}\right\} = L^{-1}\left\{\frac{1}{s}\frac{1}{s^2 + \omega^2}\right\} = L^{-1}\left\{\frac{1}{s}\right\} * L^{-1}\left\{\frac{1}{s^2 + \omega^2}\right\}$$

$$= 1 * \frac{\sin \omega t}{\omega} = \int_0^t (1)\left(\frac{\sin \omega(t - \tau)}{\omega}\right) d\tau = \frac{1 - \cos \omega t}{\omega^2},$$
(8)

so (6), (7), and (8) give the solution as

$$x(t) = x(0) \cos \omega t + \frac{x'(0)}{\omega} \sin \omega t + \frac{F_0}{k}\left(1 - \cos \omega t\right).$$
(9)

For instance, if $x(0) = x'(0) = 0$, then $x(t) = (F_0/k)(1 - \cos \omega t)$ as depicted in Fig. 2.

COMMENT 1. Does it seem correct that the constant force F_0 should cause an oscillation? Yes, for imagine rotating the apparatus 90° so that the mass hangs down. Then we can think of F_0 as the downward gravitational force on m. In static equilibrium, the mass will hang down an amount $x = F_0/k$. If we release it from $x = 0$, it will fall and then oscillate about the equilibrium position $x = F_0/k$, as shown in Fig. 2.

COMMENT 2. Recall that $f * g = g * f$, so we can write the convolution integral either as $\int_0^t f(\tau) g(t-\tau) d\tau$ or as $\int_0^t g(\tau) f(t-\tau) d\tau$; that is, we can let the argument of f be τ and the argument of g be $t - \tau$, or vice versa, whichever we choose. In (8) we chose the τ argument for 1 and the $t - \tau$ argument for $(\sin \omega t)/\omega$. Alternatively, we could have expressed the inverse in (8) as

$$\frac{\sin \omega t}{\omega} * 1 = \int_0^t \left(\frac{\sin \omega \tau}{\omega}\right)(1) d\tau = \frac{1 - \cos \omega \tau}{\omega^2},$$

which result is the same as (8). That is, it is a bit easier to put the more complicated argument $(t - \tau)$ in the simpler function (1), because 1 is simply 1 in any case.

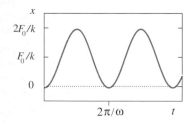

FIGURE 2
Release from rest.

COMMENT 3.   Observe that (9) is the particular solution satisfying the initial conditions $x = x(0)$ and $x' = x'(0)$ at $t = 0$. If those quantities are not prescribed, we can replace them by arbitrary constants, and then (9) amounts to the general solution of $mx'' + kx = F_0$, namely, $x(t) = A \cos \omega t + B \sin \omega t + (F_0/k)(1 - \cos \omega t)$. Thus, the method gives either a particular solution or a general solution, whichever is desired.

COMMENT 4.   If, instead of the specific forcing function $f(t) = F_0$ we allow $f(t)$ to remain unspecified, then we have, in place of (5),

$$X(s) = \frac{sx(0) + x'(0)}{s^2 + \omega^2} + \frac{F(s)}{m(s^2 + \omega^2)}, \tag{10}$$

and, in place of (9),

$$x(t) = x(0) \cos \omega t + \frac{x'(0)}{\omega} \sin \omega t + \frac{1}{m} L^{-1} \left\{ \frac{F(s)}{s^2 + \omega^2} \right\}. \tag{11}$$

Use of the convolution theorem to write

$$\begin{aligned}
L^{-1} \left\{ \frac{F(s)}{s^2 + \omega^2} \right\} &= L^{-1} \left\{ \frac{1}{s^2 + \omega^2} F(s) \right\} \\
&= L^{-1} \left\{ \frac{1}{s^2 + \omega^2} \right\} * L^{-1} \{F(s)\} = \frac{\sin \omega t}{\omega} * f(t), \tag{12}
\end{aligned}$$

and thus obtain

$$x(t) = x(0) \cos \omega t + \frac{x'(0)}{\omega} \sin \omega t + \frac{1}{m\omega} \int_0^t \sin \omega \tau \, f(t - \tau) \, d\tau \tag{13}$$

as the solution. The integral in (13) can be evaluated once $f(t)$ is specified. For instance, if $f(t) = e^{-t}$, then (13) gives

$$\begin{aligned}
x(t) &= x(0) \cos \omega t + \frac{x'(0)}{\omega} \sin \omega + \frac{1}{m\omega} \int_0^t (\sin \omega \tau) e^{-(t - \tau)} \, d\tau \\
&= x(0) \cos \omega t + \frac{x'(0)}{\omega} \sin \omega t - \frac{\omega \cos \omega t - \sin \omega t - \omega e^{-t}}{m\omega \left(1 + \omega^2\right)}. \qquad \blacksquare
\end{aligned}$$

With Example 1 completed there are general observations to make about the method. First, suppose we solve a differential equation

$$x'' + ax' + bx = f(t), \tag{14}$$

where a and b are constants, by the Laplace transform, obtaining

$$X(s) = \frac{(s + a)x(0) + x'(0)}{s^2 + as + b} + \frac{F(s)}{s^2 + as + b}. \tag{15}$$

If we invert the $F(s)/(s^2 + as + b)$ term by the convolution theorem then we convolve the inverse of $1/(s^2 + as + b)$ with the inverse of $F(s)$, namely, the original function $f(t)$. It would be wasteful to work out the transform $F(s)$ of $f(t)$ since we end up inverting $F(s)$ to obtain $f(t)$! Thus, we suggest that you simply denote the transform of $f(t)$ as $F(s)$ without actually working it out.

Second, observe how the initial conditions become incorporated when we take the transform of the derivative terms; there is no need to apply the initial conditions at the end.

Third, recall that Laplace transforms come with restrictions on s. For instance, $L\{1\} = 1/s$ for $s > 0$. However, such restrictions in no way impede the solution steps, and once we invert $X(s)$ to obtain $x(t)$ they are no longer relevant.

Finally, notice that when we use Theorem 13.3.2 to transform an x' term we don't yet know whether $x(t)$ satisfies the conditions of that theorem because we don't yet know the solution $x(t)$! Similarly for higher-order derivative terms. One viable approach is to proceed to a tentative solution $x(t)$. If at that point we verify that $x(t)$ satisfies the differential equation and the initial conditions, which is good practice in any case, then there is no need to check the validity of any assumptions used in the individual solution steps.

EXAMPLE 2
Solve the initial value problem

$$y'''' - y = 0; \qquad y(0) = 1, \ y'(0) = y''(0) = y'''(0) = 0 \tag{16}$$

for $y(x)$. That the independent and dependent variables are x and y, rather than x and t, is immaterial to the application of the Laplace transform; the transform of $y(x)$ is now $Y(s) = \int_0^\infty y(x)\, e^{-sx}\, dx$. The transform of (16) gives

$$\left[s^4 Y(s) - s^3 y(0) - s^2 y'(0) - s y''(0) - y'''(0) \right] - Y(s) = 0. \tag{17}$$

Putting the initial conditions into (17), and solving for $Y(s)$, gives

$$Y(s) = \frac{s^3}{s^4 - 1}. \tag{18}$$

To invert the latter, we can use partial fractions:

$$Y(s) = \frac{s^3}{s^4 - 1} = \frac{A}{s+1} + \frac{B}{s-1} + \frac{C}{s+i} + \frac{D}{s-i}. \tag{19}$$

Then,

$$\begin{aligned}
\frac{s^3}{s^4 - 1} &= \frac{A}{s+1} + \frac{B}{s-1} + \frac{C}{s+i} + \frac{D}{s-i} \\
&= \frac{(s-1)(s^2+1)A + (s+1)(s^2+1)B + (s-i)(s^2-1)C + (s+i)(s^2-1)D}{s^4 - 1}.
\end{aligned}$$

$$\tag{20}$$

Equating coefficients of like powers of s in the numerators gives the linear equations

$$
\begin{aligned}
s^3: &\quad 1 = A + B + C + D, \\
s^2: &\quad 0 = -A + B - iC + iD, \\
s: &\quad 0 = A + B - C - D, \\
1: &\quad 0 = -A + B + iC - iD,
\end{aligned}
$$

solution of which (for instance by Gauss elimination) gives $A = B = C = D = 1/4$. Substitution of these values into (19) and inversion gives

$$
\begin{aligned}
y(x) &= \frac{1}{4}e^{-x} + \frac{1}{4}e^{x} + \frac{1}{4}e^{-ix} + \frac{1}{4}e^{ix} \\
&= \frac{1}{2}\left(\cosh x + \cos x\right)
\end{aligned} \tag{21}
$$

as the desired solution. (See Exercise 4.) ∎

E X A M P L E   3
Solve the first-order initial-value problem

$$
x' + x = q(t); \qquad x(0) = x_0 \tag{22}
$$

for $x(t)$, where $q(t)$ is a prescribed forcing function. Application of the Laplace transform gives

$$
X(s) = \frac{x_0}{s+1} + \frac{Q(s)}{s+1} \tag{23}
$$

and hence the solution

$$
\begin{aligned}
x(t) &= x_0 e^{-t} + \int_0^t e^{-(t-\tau)} Q(\tau)\, d\tau \\
&= e^{-t}\left[x_0 + \int_0^t Q(\tau) e^{\tau}\, d\tau \right].
\end{aligned}
$$
∎

Closure.   The Laplace transform provides us with a systematic solution procedure for linear constant-coefficient differential equations. The power of the method is that it converts a linear differential equation in the t domain to a linear algebraic equation in the s domain. Initial conditions become "built into" the solution so they do not need to be applied at the end.

Thus far we have used the linearity of the transform and of the inverse transform, the transform of derivatives, the convolution theorem, partial fractions, and our transform table. In subsequent sections additional properties are discussed and applied. In particular, we will consider problems with discontinuous forcing functions, which case is common in applications. The importance of the Laplace transform in engineering mathematics is largely due to its convenient application to such problems.

On the other hand, the method does not apply to nonlinear differential equations and is only of occasional help with linear equations having nonconstant coefficients. For brevity, we limit our discussion of the nonconstant-coefficient case to the exercises.

EXERCISES 13.4

1. Use the Laplace transform to find a general solution, or the particular solution if initial conditions are given.

   (a) $x' + 2x = 4t^2$

   (b) $3x' + x = 6e^{2t}$; $x(0) = 0$

   (c) $x' - 6x = e^{-t}$; $x(0) = 4$

   (d) $x'' = 6t$; $x(0) = 2,\ x'(0) = -1$

   (e) $x'' + 5x' = 10$

   (f) $x'' - x' = 1 + t + t^2$

   (g) $x'' - 3x' + 2x = 0$; $x(0) = 3,\ x'(0) = 1$

   (h) $x'' - 4x' - 5x = 2 + e^{-t}$; $x(0) = x'(0) = 0$

   (i) $x'' - x' - 12x = t$; $x(0) = -1,\ x'(0) = 0$

   (j) $x'' + 6x' + 9x = 1$; $x(0) = 0,\ x'(0) = -2$

   (k) $x'' - 2x' + 2x = -2t$; $x(0) = 0,\ x'(0) = -5$

   (l) $x'' - 2x' + 3x = 5$; $x(0) = 1,\ x'(0) = -1$

   (m) $x''' - x'' + 2x' = t^2$; $x(0) = 1,\ x'(0) = x''(0) = 0$

   (n) $x''' + x'' - 2x' = 1 + e^t$; $x(0) = x'(0) = x''(0) = 0$

   (o) $x''' + 5x'' = t^4$; $x(0) = x'(0) = 0,\ x''(0) = 1$

   (p) $x''' - x'' - x' + x = 0$; $x(0) = 2,\ x'(0) = x''(0) = 0$

   (q) $x^{(iv)} = 2 \sin t$

   (r) $x^{(iv)} + 3x''' = 0$; $x(0) = x'(0) = 0,$
       $x''(0) = x'''(0) = 3$

   (s) $x^{(iv)} - x = 1$; $x(0) = x'(0) = x''(0) = 0,$
       $x'''(0) = 4$

2. Solve by Laplace transform.

   (a) $x'' + x = \sin t$; $x(0) = 0,\ x'(0) = 4$

   (b) $x'' + x = \cos t$; $x(0) = 0,\ x'(0) = 4$

   (c) $x'' - x = 4e^t$; $x(0) = 0,\ x'(0) = 2$

   (d) $x'' - 4x = 8e^{-2t}$; $x(0) = 0,\ x'(0) = 10$

3. (a) Show that for a constant-coefficient linear homogeneous differential equation of order n, the Laplace transform $X(s)$ of the solution $x(t)$ is necessarily of the form

   $$X(s) = \frac{p(s)}{q(s)}, \qquad (3.1)$$

   where $q(s)$ and $p(s)$ are polynomials in s, with q of degree n and p of degree less than n.

   (b) Show that if $q(s) = 0$ has n distinct roots $s_1, \ldots s_n$,

then

$$x(t) = \sum_{k=1}^{n} \frac{p(s_k)}{q'(s_k)} e^{s_k t}. \qquad (3.2)$$

4. Show that the form

$$Y(s) = \frac{s^3}{s^4 - 1} = \frac{Es + F}{s^2 - 1} + \frac{Gs + H}{s^2 + 1} \qquad (4.1)$$

is equivalent to (19). Using (4.1), evaluate E, F, G, H. Then invert (4.1) and show that your result agrees with (21).

5. (*Variable-coefficient equation*) Consider the problem

$$tx'' + x' + tx = 0 \qquad (0 \le t < \infty) \qquad (5.1)$$
$$x(0) = 1,\quad x'(0) = 0,$$

where our interest lies in seeing whether or not we can solve (5.1) by the Laplace transform method even though the differential equation has nonconstant coefficients.

   (a) Take the Laplace transform of the differential equation in (5.1). Note that the transforms of $t\,x''(t)$ and $t\,x(t)$,

$$L\{t\,x''(t)\} = \int_0^\infty t\,x''\,e^{-st}\,dt,$$

$$L\{t\,x(t)\} = \int_0^\infty t\,x\,e^{-st}\,dt,$$

present a difficulty in that we cannot express them in terms of $X(s)$ the way we can express $L\{x'(t)\} = sX(s) - x(0)$ and $L\{x''(t)\} = s^2 X(s) - sx(0) - x'(0)$. Nevertheless, these terms can be handled as follows. Observe that

$$L\{t\,x''(t)\} = \int_0^\infty t\,x''\,e^{-st}\,dt$$

$$= -\int_0^\infty \frac{d}{ds}\left(x''e^{-st}\right)dt$$

$$= -\frac{d}{ds}\int_0^\infty x''e^{-st}\,dt$$

$$= -\frac{d}{ds}\left[s^2 X(s) - sx(0) - x'(0)\right]$$

$$= -\frac{d}{ds}\left[s^2 X(s) - s\right], \tag{5.2}$$

if we assume that the unknown $x(t)$ is sufficiently well behaved for the third equality (where we have interchanged the order of two limit processes, the s differentiation and the t integration) to be justified. Handling the $L\{t\,x(t)\}$ term in the same way, show that application of the Laplace transform to (5.1) leads to the equation

$$(s^2 + 1)\frac{dX}{ds} + sX = 0 \tag{5.3}$$

on $X(s)$. Note that whereas the Laplace transform method reduces constant-coefficient differential equations to linear algebraic equations on $X(s)$, here the nonconstant coefficients result in the equation on $X(s)$ being itself a linear *differential* equation! However, (5.3) is readily solved. Doing so, show that

$$X(s) = \frac{C}{\sqrt{s^2 + 1}}. \tag{5.4}$$

(b) From our transform table, we find the inverse as $x(t) = C\,J_0(t)$, where J_0 is the Bessel function of the first kind, of order zero. (Bessel functions were discussed in Section 8.4.) Applying the initial condition once again gives $x(0) = 1 = C\,J_0(0) = (C)(1)$, so $C = 1$, and the desired solution of (5.1) is $x(t) = J_0(t)$. Here, however, we ask you to proceed as though you don't know

about Bessel functions; i.e., you may not use entry 18 in the transform table. Rather, re-express (5.4) as

$$X(s) = \frac{C}{s\sqrt{1 + (1/s^2)}} = \frac{C}{s}\left(1 - \frac{1}{2}\frac{1}{s^2} + \cdots\right), \tag{5.5}$$

where the last equality amounts to the Taylor expansion of $1/\sqrt{1+r}$ in r, about $r = 0$, where $r = 1/s^2$. Carry that expansion further; invert the resulting series term by term (assuming that that step is valid) and thus show that

$$x(t) = C\left[1 - \frac{t^2}{2^2} + \frac{1}{(2!)^2}\frac{t^4}{2^4} - \frac{1}{(3!)^2}\frac{t^6}{2^6} + \cdots\right].$$

Setting $x(0) = 1$ gives $C = 1$, and the result is that we have obtained the solution in power series form. NOTE: Of course, that power series is the Taylor series of the Bessel function $J_0(t)$. Observe that rather than pulling an s out of the square root, as we did in (5.5), and then expanding $1/\sqrt{1 + (1/s^2)}$ in powers of $1/s^2$, we could have expanded (4.4) directly in powers of s as $X(s) = C(1 - \frac{1}{2}s^2 + \cdots)$. However, positive powers of s are not invertible (recall Theorem 13.2.2) so this form is of no use. Rather, the trick was to pull an s out of the square root so that we could expand $1/\sqrt{1 + (1/s^2)}$ in powers of $1/s^2$ since *negative* powers of s *are* invertible. This idea was first put forward in Exercise 13 of Section 13.2.

13.5 DISCONTINUOUS FORCING FUNCTIONS; HEAVISIDE STEP FUNCTION

We have thus far avoided functions with discontinuities. In applications, however, systems are often subjected to discontinuous forcing functions. For instance, a circuit might be subjected to an applied voltage that is held constant at 12 volts for a minute and then shut off (i.e., reduced to zero for all subsequent time). In this section we study systems with forcing functions that are discontinuous, although we still assume that they are piecewise continuous on $0 \le t < A$ for every A and of exponential order as $t \to \infty$, so that they are Laplace transformable.

Our basic building block for piecewise continuous functions will be the Heaviside step function

$$H(t) = \begin{cases} 0, & t < 0 \\ 1, & t > 0 \end{cases} \tag{1}$$

that was discussed in Section 2.4.5, which brief section we urge you to review before continuing.

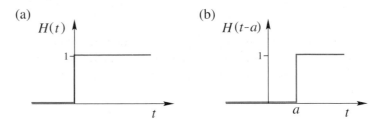

FIGURE 1   Heaviside step function.

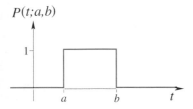

FIGURE 2
Unit rectangular pulse
function.

Since $H(t)$ is a unit step at $t = 0$ (Fig. 1a), $H(t - a)$ is a unit step shifted to $t = a$ (Fig. 1b). We showed in Section 2.4.5 that piecewise continuous functions can be expressed in terms of the unit rectangular pulse function (Fig. 2)

$$P(t; a, b) = H(t - a) - H(t - b). \qquad (2)$$

There is one more idea to introduce before we proceed, time-delayed functions. First, consider the combination

$$H(t) f(t) = \begin{cases} 0, & t < 0 \\ f(t), & t > 0 . \end{cases} \qquad (3)$$

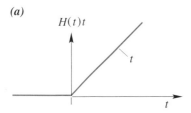

Graphically, $H(t) f(t)$ is the function $f(t)$ "cut off" for $t < 0$. To illustrate, $H(t)t$ and $H(t) \sin t$ are as shown in Fig. 3. Second, note that a function $g(t - a)$ has the same graph as $g(t)$, but shifted rightward by a, as illustrated in Figs. 4a and 4b for the functions $2t - 4$ and $\sin t$, respectively.

Putting these ideas together, we will call the combination $H(t - a) f(t - a)$ a **time-delayed function**; its graph is the same as that of $f(t)$ but shifted to the right by a, and "turned off" for all $t < a$, as illustrated in Fig. 5 for $H(t - a) \sin(t - a)$.

Turning to applications, consider the R, L, C circuit shown in Fig. 6.[1] We saw in Section 2.4.2 that the differential equation governing the charge $Q(t)$ on the capacitor is

$$L Q'' + R Q' + \frac{1}{C} Q = E(t). \qquad (4)$$

Taking the Laplace transform of (4) gives

$$L \left[s^2 \bar{Q}(s) - s Q(0) - Q'(0) \right] + R \left[s \bar{Q}(s) - Q(0) \right] + \frac{1}{C} \bar{Q}(s) = \bar{E}(s) \qquad (5)$$

where we are using the overhead bar notation $L\{Q(t)\} = \bar{Q}(s)$ and $L\{E(t)\} = \bar{E}(s)$.

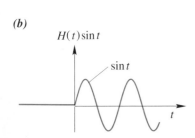

FIGURE 3
Examples of $H(t) f(t)$.

---

[1] Keep in mind that we are using the same letter L for both the inductance and the Laplace transform operator.

Solving (5) for $\bar{Q}(s)$,

$$\bar{Q}(s) = \frac{(Ls + R)Q(0) + LQ'(0)}{Ls^2 + Rs + \frac{1}{C}} + \frac{1}{Ls^2 + Rs + \frac{1}{C}}\bar{E}(s). \qquad (6)$$

Of the two terms on the right-hand side of (6) our interest, in this section, is in the second because we shall consider discontinuous forcing functions $E(t)$. To focus attention on that part, consider the initial conditions to be $Q(0) = 0$ and $Q'(0) = 0$. Then (6) becomes

$$\bar{Q}(s) = \frac{1}{Ls^2 + Rs + \frac{1}{C}}\bar{E}(s) \qquad (7)$$

which is of the form $\bar{Q}(s) = \bar{G}(s)\bar{E}(s)$. In the language of *control theory*,

$$\bar{G}(s) = \frac{1}{Ls^2 + Rs + \frac{1}{C}} \qquad (8)$$

is called the **transfer function** of the system. More generally, we have the form $\bar{O}(s) = \bar{G}(s)\bar{I}(s)$, where $\bar{O}(s)$ is the transform of the output function, $\bar{I}(s)$ is the transform of the input function, and $\bar{G}(s)$ (which is the ratio of the two corresponding to the case of zero initial conditions) is the transfer function.

Inverting (7) gives the convolution

$$Q(t) = L^{-1}\left\{\frac{1}{Ls^2 + Rs + \frac{1}{C}}\right\} * E(t). \qquad (9)$$

As we shall see in the examples to follow, if $E(t)$ is piecewise continuous and includes Heaviside functions then the convolution integral on the right-hand side of (9) will include one or more integrals of the form

$$I = \int_0^t f(\tau)H(\tau - a)\,d\tau, \qquad (a > 0) \qquad (10)$$

so let us take a moment to discuss (10) before considering examples. If $t < a$ then the integrand is zero over the entire integration interval (because of the Heaviside function) so $I = 0$. And if $t > a$ then I becomes $\int_a^t f(\tau)\,d\tau$ because the integrand is zero over $0 < \tau < a$ and $f(\tau)$ over $a < \tau < t$ (again because of the Heaviside function). Thus,

$$I(t) = \int_0^t f(\tau)H(\tau - a)\,d\tau = \begin{cases} 0, & t < a \\ \int_a^t f(\tau)\,d\tau, & t > a \end{cases} \qquad (11)$$

which can be expressed compactly as

$$\boxed{\int_0^t f(\tau)H(\tau - a)\,d\tau = H(t - a)\int_a^t f(\tau)\,d\tau,} \qquad (12)$$

where $a > 0$.

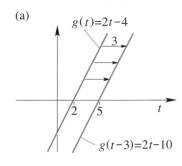

(a)

$g(t)=2t-4$

$g(t-3)=2t-10$

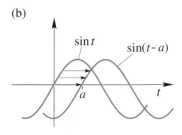

(b)

$\sin t$ $\sin(t-a)$

FIGURE 4
Examples of t shift.

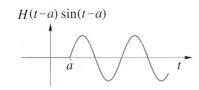

$H(t-a)\sin(t-a)$

FIGURE 5
Example of time decay.

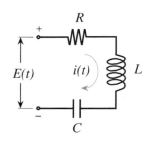

FIGURE 6
RLC circuit.

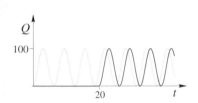

FIGURE 7
$E(t)$ in (14).

FIGURE 8
The solution (16).

EXAMPLE 1   *LC Circuit*.

As a first example let us take $R = 0$ (so that the circuit in Fig. 6 is an "LC" circuit") and, for simplicity, let $L = C = 1$. Consider the initial value problem

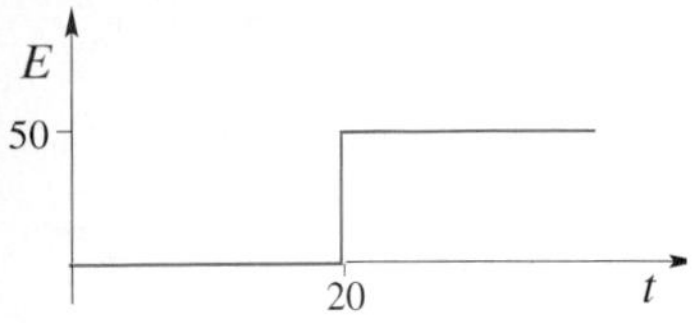

$$Q'' + Q = E(t); \qquad Q(0) = 0, \quad Q'(0) = 0, \tag{13}$$

where $E(t)$ is the piecewise continuous function shown in Fig. 7, namely,

$$E(t) = 50H(t - 20). \tag{14}$$

We can proceed right to (9), with $R = 0$ and $L = C = 1$:

$$Q(t) = L^{-1} \left\{ \frac{1}{s^2 + 1} \right\} * E(t) = \sin t * E(t)$$

$$= 50 \int_0^t \sin(t - \tau) H(\tau - 20) \, d\tau. \tag{15}$$

With the help of (12), (15) gives the solution[1]

$$Q(t) = 50H(t - 20) \int_{20}^t \sin(t - \tau) \, d\tau$$

$$= 50H(t - 20) \int_{t-20}^0 \sin\mu(-d\mu) \qquad (\mu = t - \tau)$$

$$= 50H(t - 20) \left[1 - \cos(t - 20) \right], \tag{16}$$

shown as the solid curve in Fig. 8. Because (16) is of the form $H(t - a)f(t - a)$ it is of time-delay type; it is really the function $50(1 - \cos t)$ (shown as dotted in Fig. 8) with a time delay of 20. This result is not surprising for the input $E(t)$ is delayed (Fig. 7); hence, so is the output (Fig. 8).   ■

Observe that in deriving the solution (9) of equation (4) we did not actually take the transform of $E(t) = 50H(t - 20)$; we just called its transform $\bar{E}(s)$ anticipating that $\bar{E}(s)$ would revert to $E(s)$ when we would invert $\bar{Q}(s)$ by the convolution theorem. If, however, we do not use the convolution theorem then we *do* need to work out $\bar{E}(s)$. To evaluate $\bar{E}(s)$ we need some results that we now derive.

First, observe that

$$L\{H(t - a)\} = \int_0^\infty H(t - a) e^{-st} \, dt = \int_a^\infty e^{-st} \, dt = \frac{e^{-as}}{s},$$

so the Laplace transform of $H(t - a)$ is

$$\boxed{L\{H(t - a)\} = \frac{e^{-as}}{s}.} \tag{17}$$

---

[1] Equation (12) holds even if f is a function of t as well as τ, as occurs in (15), because τ is the integration variable and t is a constant.

Further,

$$L\{H(t-a)f(t-a)\} = \int_0^\infty H(t-a)\,f(t-a)\,e^{-st}\,dt$$

$$= \int_a^\infty f(t-a)\,e^{-st}\,dt = \int_0^\infty f(\tau)\,e^{-s(\tau+a)}\,d\tau$$

$$= e^{-as}\int_0^\infty f(\tau)\,e^{-s\tau}\,d\tau = e^{-as}F(s), \tag{18}$$

where the third equality follows from the change of variables $t - a = \tau$. Thus,

$$\boxed{L\{H(t-a)f(t-a)\} = e^{-as}F(s)} \tag{19a}$$

or, equivalently,

$$\boxed{L^{-1}\{e^{-as}F(s)\} = H(t-a)f(t-a)} \tag{19b}$$

for any Laplace-transformable function $f(t)$.

To illustrate the use of (17) and (19) let us return to equation (7) and complete the steps using (17) and (19). Transforming (14) using (17) gives $\bar{E}(s) = 50e^{-20s}/s$ so (7) becomes

$$\bar{Q}(s) = 50\frac{e^{-20s}}{s(s^2+1)}. \tag{20}$$

To invert the latter use partial fractions to express

$$\frac{1}{s(s^2+1)} = \frac{A}{s} + \frac{Bs+C}{s^2+1} = \frac{As^2+A+Bs^2+Cs}{s(s^2+1)}. \tag{21}$$

Equating coefficients of s^2, s, and 1 in the numerator of the left-hand and right-hand sides gives $A + B = 0$, $C = 0$ and $A = 1$, respectively, so $A = 1$, $B = -1$, and $C = 0$. Then

$$L^{-1}\left\{\frac{1}{s(s^2+1)}\right\} = L^{-1}\left\{\frac{1}{s} - \frac{s}{s^2+1}\right\}$$

$$= L^{-1}\left\{\frac{1}{s}\right\} - L^{-1}\left\{\frac{s}{s^2+1}\right\} = 1 - \cos t. \tag{22}$$

Finally, if we apply (19b) to (20) with $a = 20$ and $F(s) = 50/[s(s^2+1)]$ then, with the help of (22), we do obtain (16) once again.

Let us work one more example.

EXAMPLE 2   RC Circuit.
We return to the circuit shown in Fig. 6 but this time we let $L = 0$ so that we have an "RC circuit." Let $R = C = 1$, and consider the initial value problem

$$Q' + Q = E(t); \qquad Q(0) = 0, \tag{23}$$

where $E(t)$ is the piecewise continuous function shown in Fig. 9, namely,

$$E(t) = [H(t-0) - H(t-1)]\,20t + [H(t-1) - H(t-2)]\,10$$

$$+ [H(t-2) - H(t-\infty)]\left(5 + 5e^{2-t}\right)$$

$$= 20t + H(t-1)(10 - 20t) - H(t-2)\left(5 - 5e^{2-t}\right) \quad (24)$$

since $H(t-0) = 1$ and $H(t-\infty) = 0$.

Transforming (23) gives $s\,\bar{Q}(s) + \bar{Q}(s) = \bar{E}(s)$ so

$$\bar{Q}(s) = \frac{1}{s+1}\bar{E}(s),$$

$$Q(t) = L^{-1}\left\{\frac{1}{s+1}\right\} * E(t) = e^{-t} * E(t)$$

$$= \int_0^t e^{-(t-\tau)}\left[20\tau + (10 - 20\tau)H(\tau-1) - (5 - 5e^{2-\tau})H(\tau-2)\right] d\tau. \quad (25)$$

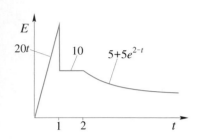

FIGURE 9
$E(t)$ in (24).

Finally, with the help of (12), (25) gives the solution

$$Q(t) = e^{-t}\left[\int_0^t 20\tau e^\tau\, d\tau + H(t-1)\int_1^t (10 - 20\tau)\,e^\tau\, d\tau\right.$$

$$\left. -H(t-2)\int_2^t \left(5 - 5e^{2-\tau}\right) e^\tau\, d\tau\right]$$

$$= 20\left(t - 1 + e^{-t}\right) + 10H(t-1)\left(3 - 2t - e^{1-t}\right)$$

$$-5H(t-2)\left[1 + (1-t)e^{2-t}\right], \quad (26)$$

which is plotted in Fig. 10.

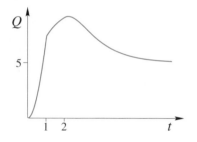

FIGURE 10
The solution (26).

COMMENT.   It is impressive how the Heaviside notation and the Laplace transform method have enabled us to solve for $Q(t)$ on the entire t domain $(0 \le t < \infty)$ at once. In contrast, if we solve by the methods of Chapters 2 and 3 we need to break (23) into three separate problems:

$$0 \le t \le 1: \qquad Q' + Q = 20t; \qquad Q(0) = 0$$

$$1 \le t \le 2: \qquad Q' + Q = 10; \qquad Q(1) = ? \qquad (27\text{a,b,c})$$

$$2 \le t < \infty: \qquad Q' + Q = 5 + 5e^{2-t}; \quad Q(2) = ?.$$

First, we solve (27a) for $Q(t)$ on $0 \le t \le 1$. The final value $Q(1)$ computed from that solution then serves as the initial condition for the next problem (27b). With that initial condition, we then solve (27b) for $Q(t)$ on $1 \le t \le 2$. The final value $Q(2)$ computed from that solution then serves as the initial condition for the final problem (27c). With that initial condition, we then solve for $Q(t)$ on $2 \le t < \infty$. This approach is more tedious than the Laplace transform approach that led to (26). ∎

If we review Examples 1 and 2 we see from Figs. 8 and 10 that in each case the output $Q(t)$ is continuous even though the input is discontinuous. To understand this interesting result consider the simpler problem

$$Q''(t) = H(t - a); \qquad Q(0) = 0, \quad Q'(0) = 0, \tag{28}$$

simpler because it can be solved by direct integration:

$$\int_0^t Q''(\tau)\, d\tau = \int_0^t H(\tau - a)\, d\tau, \tag{29a}$$

$$Q'(t) - Q'(0) = H(t - a) \int_a^t d\tau, \tag{29b}$$

where we have used (12) for the right-hand side. Thus,

$$Q'(t) = H(t - a)(t - a). \tag{30}$$

Integrating again,

$$\int_0^t Q'(\tau)\, d\tau = \int_0^t H(\tau - a)(\tau - a)\, d\tau, \tag{31a}$$

$$Q(t) - Q(0) = H(t - a) \int_a^t (\tau - a)\, d\tau, \tag{31b}$$

so

$$Q(t) = H(t - a)\frac{(t - a)^2}{2}. \tag{32}$$

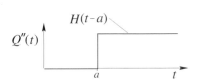

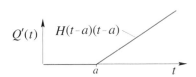

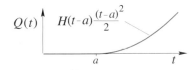

FIGURE 11
The smoothing nature of integration.

We have plotted $Q''(t)$, $Q'(t)$, and $Q(t)$ [as given by (28), (30), and (32)] in Fig. 11. What we see is that *integration is a smoothing process*: $Q''(t) = H(t - a)$ has a jump discontinuity at $t = a$ but its integral $Q'(t) = H(t - a)(t - a)$ is continuous at $t = a$, although it does suffer a "kink" there; integrating one more time, the result $Q(t) = H(t - a)(t - a)^2/2$ is not only continuous it is also *smooth* (i.e., differentiable) at $t = a$.

The same result was found in Example 1: Whereas the forcing function $50H(t - 20)$ was discontinuous at $t = 20$ the solution $Q(t)$ was both continuous and smooth there. It's true that we didn't solve (13) by direct integration as we solved (28) in (28)–(32) but, whether we see the integration process explicitly or not, *in effect* solving a second-order differential equation amounts to a process of double integration.

Similarly in Example 2. There the input $E(t)$ was discontinuous at $t = 1$ yet the output $Q(t)$ was continuous there. However, it did have a "kink" there because the differential equation was only of first order so there was, in effect, only one integration rather than two.

Closure.   We have relied heavily on the Heaviside step function $H(t - a)$, the unit rectangular pulse function $P(t; a, b)$, and time-delayed functions $H(t - a)f(t - a)$. If the forcing function is piecewise continuous then it can be expressed using rectangular pulse functions. Solution by use of the Laplace transform and the convolution theorem gives an integral with one or more Heaviside step functions in its integrand. Such terms can be integrated using (12). Alternatively, if we choose not to use the convolution theorem for the inversion then (19b) is of help in carrying out the inversion.

EXERCISES 13.5

1. Use the Laplace transform to solve

$$x' + 2x = f(t); \qquad x(0) = 0 \qquad (1.1)$$

where $f(t)$ is
(**a**) 100 on $0 < t < 5$, 0 on $5 < t < \infty$
(**b**) 100 on $0 < t < 5$, 50 on $5 < t < 10$, 0 on $10 < t < \infty$
(**c**) $20t$ on $0 < t < 1$, 20 on $1 < t < \infty$
(**d**) 0 on $0 < t < 2$, $t - 2$ on $2 < t < \infty$
(**e**) $\sin t$ on $0 < t < \pi/2$, 0 on $\pi/2 < t < \infty$
(**f**) $\sin t$ on $0 < t < 4\pi$, 0 on $4\pi < t < \infty$
(**g**) 0 on $0 < t < 1$, 10 on $1 < t < 2$, 20 on $2 < t < \infty$
(**h**) 0 on $0 < t < 5$, t on $5 < t < 10$, 5 on $10 < t < \infty$

2. (**a**)–(**h**) Same as Exercise 1 but with (1.1) changed to

$$x'' + 4x = f(t); \qquad x(0) = 0, \quad x'(0) = 0.$$

3. (**a**)–(**h**) Same as Exercise 1 but with (1.1) changed to

$$x'' - x = f(t); \qquad x(0) = 0, \quad x'(0) = 10.$$

4. (**a**)–(**h**) Same as Exercise 1 but with (1.1) changed to

$$x'' + x' = f(t); \qquad x(0) = 0, \quad x'(0) = 0.$$

5. (**a**)–(**h**) Solve the corresponding problem in Exercise 1 using *Maple* and also obtain a computer plot of the solution, similar to our Figs. 8 and 10.

6. (**a**)–(**h**) Solve the corresponding problem in Exercise 4 using *Maple* and also obtain a computer plot of the solution, similar to our Figs. 8 and 10.

7. Evaluate $L\{tH(t-1)\}$ three ways:
(**a**) by direct integration, noting that

$$\int_0^\infty tH(t-1)e^{-st}\,dt = \int_1^\infty te^{-st}\,dt$$

(**b**) by re-expressing $tH(t-1)$ as $[1 + (t-1)]H(t-1)$ and using (17) and (19). NOTE: $1 + (t-1)$ is actually the Taylor series of the function t about $t = 1$, which idea is useful because that series terminates rather than being an infinite series.
(**c**) using *Maple*

8. Evaluate $L\{t^2 H(t-1)\}$ three ways:
(**a**) by direct integration
(**b**) by re-expressing the t^2 as $1 + 2(t-1) + (t-1)^2$ and using (17) and (19)
(**c**) using *Maple*

9. Evaluate $L\{e^{-2t}H(t-3)\}$ three ways:
(**a**) by direct integration
(**b**) by re-expressing $e^{-2t} = e^{-2[3+(t-3)]} = e^{-6}e^{-2(t-3)}$ and using (19)
(**c**) using *Maple*

10. Evaluate $L\{\sin 2t\, H(t-3)\}$ two ways:
(**a**) by re-expressing $\sin 2t = \sin 2[3 + (t-3)] = \sin[6 + 2(t-3)] = \sin 6\cos 2(t-3) + \cos 6\sin 2(t-3)$ and using (19)
(**b**) using *Maple*

11. Evaluate any way you like, but without using *Maple*
(**a**) $L\{H(t-1)H(t-2)\}$ (**b**) $L\{H(t-1)H(1-t)\}$

12. Use (19b) and the transform table to invert each of the following:

(**a**) $\dfrac{1}{s^2}e^{-2s}$

(**b**) $\dfrac{1}{s-3}e^{-s}$

(**c**) $\dfrac{1}{s^2+1}e^{-s}$

(**d**) $\dfrac{s-1}{s^3}e^{-s}$

(**e**) $\dfrac{1}{s^2-9}e^{-3s}$

(**f**) $\dfrac{1}{s^2}\left(e^{-s} - e^{-2s}\right)$

13. Use *Maple* to obtain computer plots of
(**a**) Fig. 8   (**b**) Fig. 9   (**c**) Fig. 10

13.6 ADDITIONAL PROPERTIES

In this final section we present two additional properties of the Laplace transform.

13.6.1 S-shift.   Reminiscent of the t-shift formula

$$L\{H(t-a)f(t-a)\} = e^{-as}F(s) \tag{1}$$

or, equivalently,

$$L^{-1}\{e^{-as}F(s)\} = H(t-a)f(t-a), \tag{2}$$

from Section 13.5, where $F(s)$ is $L\{f(t)\}$, there is an analogous **s-shift** formula:

THEOREM 13.6.1

s-Shift

If $L\{f(t)\} = F(s)$ exists for $s > s_0$, then for any real constant a

$$\boxed{L\{e^{-at}f(t)\} = F(s+a)} \tag{3}$$

for $s + a > s_0$ or, equivalently,

$$\boxed{L^{-1}\{F(s+a)\} = e^{-at}f(t).} \tag{4}$$

Proof.

$$L\{e^{-at}f(t)\} = \int_0^\infty e^{-at}f(t)\,e^{-st}\,dt$$

$$= \int_0^\infty f(t)\,e^{-(s+a)t}\,dt = F(s+a), \tag{5}$$

and surely if $\int_0^\infty f(t)e^{-st}\,dt$ exists for $s > s_0$ then the integral in (5) exists for $s + a > s_0$. ∎

EXAMPLE 1

Determine $L\{t^3e^{5t}\}$. From the transform table, $L\{t^3\} = 6/s^4$, so it follows from (3), with $a = -5$, that

$$L\{t^3e^{5t}\} = \frac{6}{(s-5)^4}. \tag{6}$$

■

EXAMPLE 2

Invert $\dfrac{2s+1}{s^2+2s+4}$. Surely we could use partial fractions. Alternatively, we could use (4), above, as follows:

$$L^{-1}\left\{\frac{2s+1}{s^2+2s+4}\right\} = L^{-1}\left\{\frac{2s+1}{(s+1)^2+3}\right\} = L^{-1}\left\{\frac{2(s+1)-1}{(s+1)^2+3}\right\}$$

$$= 2L^{-1}\left\{\frac{(s+1)}{(s+1)^2+3}\right\} - L^{-1}\left\{\frac{1}{(s+1)^2+3}\right\}.$$

Then,

$$L^{-1}\left\{\frac{s}{s^2+3}\right\} = \cos\sqrt{3}\,t \quad \text{and} \quad L^{-1}\left\{\frac{1}{s^2+3}\right\} = \frac{\sin\sqrt{3}\,t}{\sqrt{3}}$$

from entries 3 and 4 in the transform table, so (4) with $a = 1$ gives

$$L^{-1}\left\{\frac{2s+1}{s^2+2s+4}\right\} = 2e^{-t}\cos\sqrt{3}\,t - e^{-t}\frac{\sin\sqrt{3}\,t}{\sqrt{3}}. \tag{7}$$

That is, completing the square in the denominator gave the $s + 1$ combination. Then, to express the numerator in terms of $s+1$ we re-expressed the $2s+1$ as $2(s+1)-2+1$, or, $2(s + 1) - 1$. ∎

Additional applications of Theorem 13.6.1 can be found within our transform table for with the help of that theorem entry 2 follows from entry 1, 9 from 3, 10 from 4, and 15 from 7.

13.6.2 Periodic functions. We close this chapter by deriving a formula for the transform of a **periodic function**. Periodic functions were introduced in Section 7.2 and an example is the *sawtooth wave* shown in Fig. 1, which is periodic with period $T = 2$.

To evaluate the Laplace transform of a periodic function $f(t)$ with period T it seems a good start to break up the integral on t as

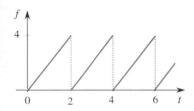

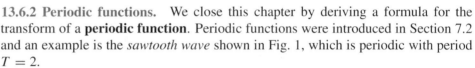

$$L\{f(t)\} = \int_0^\infty f(t)\,e^{-st}\,dt = \int_0^T f(t)\,e^{-st}\,dt + \int_T^{2T} f(t)\,e^{-st}\,dt + \cdots. \tag{8}$$

FIGURE 1
Example of a periodic
function: a sawtooth wave.

Next, let $\tau = t$ in the first integral on the right side of (8), $\tau = t - T$ in the second, $\tau = t - 2T$ in the third, and so on. Thus,

$$L\{f(t)\} = \int_0^T f(\tau)\,e^{-s\tau}\,d\tau + \int_0^T f(\tau+T)\,e^{-s(\tau+T)}\,d\tau$$

$$+ \int_0^T f(\tau+2T)\,e^{-s(\tau+2T)}\,d\tau + \cdots, \tag{9}$$

but $f(\tau + T) = f(\tau)$, $f(\tau + 2T) = f(\tau)$, and so on, because f is periodic with period T, so (9) becomes

$$L\{f(t)\} = \left(1 + e^{-sT} + e^{-2sT} + \cdots\right)\int_0^T f(\tau)\,e^{-s\tau}\,d\tau. \tag{10}$$

Unfortunately, this expression contains an infinite series. However, observe that

$$1 + e^{-sT} + e^{-2sT} + \cdots = 1 + \left(e^{-sT}\right) + \left(e^{-sT}\right)^2 + \cdots \tag{11}$$

is a geometric series $1 + z + z^2 + \cdots$ with $z = e^{-sT}$, and the latter is known to have the sum $1/(1 - z)$ if $|z| < 1$. Since $|z| = |e^{-sT}| = e^{-sT} < 1$ if $s > 0$, we can sum the parenthetic series in (10) as $1/(1 - e^{-sT})$.

Finally, if we ask that f be piecewise continuous on $0 \le t \le T$, to ensure the existence of the integral in (10), then we can state the result as follows.

THEOREM 13.6.2

Transform of Periodic Function

If f is periodic with period T on $0 \le t < \infty$ and piecewise continuous on one period, then

$$L\{f(t)\} = \frac{1}{1 - e^{-sT}} \int_0^T f(t)\, e^{-st}\, dt \tag{12}$$

for $s > 0$.

The point is that (12) requires integration only over one period rather than over $0 \le t < \infty$, and gives the transform in closed form rather than as an infinite series.

EXAMPLE 3

If f is the *sawtooth wave* shown in Fig. 1 then $T = 2$ and

$$\int_0^T f(t)\, e^{-st}\, dt = \int_0^2 2t\, e^{-st}\, dt = 2\frac{1 - (1 + 2s)e^{-2s}}{s^2}, \tag{13}$$

so the transform of f is

$$L\{f(t)\} = \frac{1}{1 - e^{-2s}} \frac{2\left[1 - (1 + 2s)e^{-2s}\right]}{s^2} = \frac{2}{s^2} - \frac{4}{s}\frac{e^{-2s}}{1 - e^{-2s}} \tag{14}$$

for $s > 0$.

A more challenging question is the reverse: What is the inverse of

$$F(s) = \frac{2}{s^2} - \frac{4}{s}\frac{e^{-2s}}{1 - e^{-2s}}, \tag{15}$$

if we have no advance knowledge of the sawtooth wave in Fig. 1, or that $f(t)$ is periodic? The key is to proceed in reverse—that is, to expand the $1/(1 - e^{-2s})$ in a geometric series in powers of e^{-2s}. Thus,

$$F(s) = \frac{2}{s^2} - \frac{4}{s}e^{-2s}\left(1 + e^{-2s} + e^{-4s} + \cdots\right)$$

$$= \frac{2}{s^2} - \frac{4}{s}\left(e^{-2s} + e^{-4s} + e^{-6s} + \cdots\right). \tag{16}$$

Assuming that the series can be inverted term by term,

$$f(t) = 2t - 4[H(t - 2) + H(t - 4) + H(t - 6) + \cdots]. \tag{17}$$

The first few partial sums,

$$f_1(t) = 2t,$$
$$f_2(t) = 2t - 4H(t - 2),$$
$$f_3(t) = 2t - 4H(t - 2) - 4H(t - 4)$$

are sketched in Fig. 2, and we can see that the infinite series (17) gives the periodic sawtooth wave shown in Fig. 1.

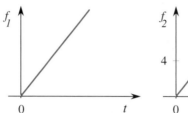
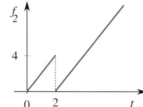
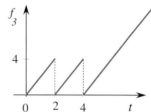

FIGURE 2   Partial sums of (17).

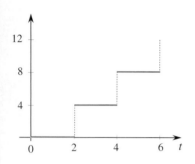

FIGURE 3
The staircase
$4[H(t - 2) + H(t - 4) + \cdots]$.

COMMENT.   Observe that the presence of $1 - e^{-sT}$ in the denominator of a transform does not suffice to imply that the inverse is a periodic function. For example, the inverse of $4e^{-2s}/[s(1 - e^{-2s})]$, in (15), is the nonperiodic "staircase" shown in Fig. 3.   ∎

E X A M P L E   4   *Differential Equation With Periodic Forcing Function*.
Solve the initial value problem

$$x' + x = f(t); \qquad x(0) = x_0, \tag{18}$$

where $f(t)$ is the periodic *square wave* shown in Fig. 4, with period $T = 2$. Transforming (18) gives

$$X(s) = \frac{x_0}{s + 1} + \frac{1}{s + 1} F(s) \tag{19}$$

where, according to (12),

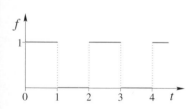

FIGURE 4
The square wave in (18).

$$F(s) = \frac{1}{1 - e^{-2s}} \int_0^2 f(t) e^{-st} \, dt$$

$$= \frac{1}{1 - e^{-2s}} \int_0^1 e^{-st} \, dt = \frac{1}{1 - e^{-2s}} \frac{1 - e^{-s}}{s}. \tag{20}$$

Thus,

$$X(s) = \frac{x_0}{s + 1} + \frac{1}{s(s + 1)} \frac{1 - e^{-s}}{1 - e^{-2s}}. \tag{21}$$

To invert (21) use the geometric series $1/(1 - z) = 1 + z + z^2 + \cdots$ with $z = e^{-2s}$ to expand the $1/(1 - e^{-2s})$ factor:

$$X(s) = \frac{x_0}{s + 1} + \frac{1}{s(s + 1)} \left(1 - e^{-s}\right) \left(1 + e^{-2s} + e^{-4s} + e^{-6s} + \cdots\right)$$

$$= \frac{x_0}{s + 1} + \frac{1}{s(s + 1)} \left(1 - e^{-s} + e^{-2s} - e^{-3s} + \cdots\right). \tag{22}$$

Next, partial fractions gives

$$L^{-1}\left\{\frac{1}{s(s + 1)}\right\} = L^{-1}\left\{\frac{1}{s} - \frac{1}{s + 1}\right\} = 1 - e^{-t}. \tag{23}$$

Finally, (23) and the t-shift property (entry 25 of the table) give, as the inverse of (22),

$$x(t) = x_0 e^{-t} + \left(1 - e^{-t}\right) - H(t - 1)\left(1 - e^{-(t-1)}\right) + H(t - 2)\left(1 - e^{-(t-2)}\right)$$

$$- H(t - 3)\left(1 - e^{-(t-3)}\right) + \cdots, \tag{24}$$

the graph of which is shown in Fig. 5 for the case where $x_0 = 0$.

Alternatively, rather than work out the transform of the square wave, which we did in (20), we could have omitted that step and used the convolution theorem to invert (19) as

$$x(t) = x_0 e^{-t} + e^{-t} * f(t). \tag{25}$$

Putting

$$f(t) = H(t) - H(t - 1) + H(t - 2) - H(t - 3) + \cdots \tag{26}$$

into (25) gives

$$x(t) = x_0 e^{-t} + \int_0^t e^{-(t-\tau)} \left[H(\tau) - H(\tau - 1) + H(\tau - 2) - \cdots\right] d\tau$$

$$= x_0 e^{-t} + e^{-t} \int_0^t e^{\tau} \left[H(\tau) - H(\tau - 1) + H(\tau - 2) - \cdots\right] d\tau.$$

Finally, we can use (12) of Section 13.5 to obtain

$$x(t) = x_0 e^{-t} + e^{-t} \left[H(t) \int_0^t e^{\tau} d\tau - H(t - 1) \int_1^t e^{\tau} d\tau + H(t - 2) \int_2^t e^{\tau} d\tau - \cdots\right]$$

$$= x_0 e^{-t} + e^{-t} \left[H(t)\left(e^t - 1\right) - H(t - 1)\left(e^t - e^1\right) + H(t - 2)\left(e^t - e^2\right) - \cdots\right]$$

$$= x_0 e^{-t} + H(t)\left(1 - e^{-t}\right) - H(t - 1)\left(1 - e^{-(t-1)}\right) + \cdots, \tag{27}$$

which is the same as (24).

COMMENT.   Typically, when we sum an infinite series we need to add more and more terms until the partial sums settle down to within the desired accuracy. It is interesting that (24) gives an exact result with only a finite number of terms. For instance, if we are interested in the interval $0 \leq t < 7$, then we need sum only the terms up to and including the $H(t - 7)$ term because $H(t - 7)$, $H(t - 8)$, ... are all zero for $t < 7$. ∎

FIGURE 5
The response (24).

Closure.   The chief results in this section are (3), (4), and (12), which, along with (1) and (2), are included for convenience as entries in our transform table.

EXERCISES 13.6

1. Use (3) and the transform table to evaluate the Laplace transform of each.

(a) $5e^{-3t} H(t-2)$

(b) $e^{2t} H(t-3)$

2. Use (4) and the transform table to invert each.

(a) $\dfrac{s}{s^2 + 2s + 3}$

(b) $\dfrac{1}{s^2 - 2s + 5}$

(c) $\dfrac{3s}{(s+4)^2}$

(d) $\dfrac{s^2 - 1}{(s-2)^3}$

(e) $\dfrac{1}{\sqrt{s+6}}$

(f) $\dfrac{s}{\sqrt{s-2}}$

(g) $\dfrac{s+1}{(s-5)^{3/2}}$

(h) $\dfrac{s^2 + s + 1}{(s+1)^4}$

3. Evaluate the transform of the sawtooth wave shown in Fig. 1 if the period is 3 instead of 2; i.e., change the 2,

4, 6 in the figure to 3, 6, 9, respectively.

4. Determine the Laplace transform of the function $f(t)$ that is periodic and defined on one period as follows.

(a) $\sin t, \quad 0 \le t \le \pi$

(b) $\begin{cases} 1, & 0 \le t < 2 \\ 0, & 2 \le t < 3 \end{cases}$

(c) $\sin 2t, \quad 0 \le t \le \pi$

(d) $e^{-t}, \quad 0 \le t < 2$

(e) $\begin{cases} t, & 0 \le t < 1 \\ 0, & 1 \le t < 2 \end{cases}$

(f) $\begin{cases} t, & 0 \le t < 1 \\ t-2, & 1 \le t < 2 \end{cases}$

5. Solve (18) where, in place of the square wave, $f(t)$ is the sawtooth wave in Fig. 1.

6. Generate a computer plot of the solution to Exercise 5. Use the time interval $0 \le t \le 15$, and let $x_0 = 0$. Is the solution periodic? Explain.

CHAPTER 13 REVIEW

The Laplace transform has a variety of uses, but its chief application is in the solution of linear ordinary and partial differential equations. In this chapter our focus was on its use in solving linear ordinary differential equations with constant coefficients. The power of the method is that it reduces such a differential equation, homogeneous or not, to a linear algebraic one. The hardest part, the inversion, is often accomplished with the help of tables, a number of theorems, and computer software. Initial conditions that are given become built in, in the process of transforming the differential equation, so they need not be applied at the end.

Chief properties of the transform and its inverse, given in Section 13.3, are:

Linearity of the transform and its inverse

$$L\{\alpha u(t) + \beta v(t)\} = \alpha L\{u(t)\} + \beta L\{v(t)\}, \tag{1}$$

$$L^{-1}\{\alpha U(s) + \beta V(s)\} = \alpha L^{-1}\{U(s)\} + \beta L^{-1}\{V(s)\}, \tag{2}$$

Transform of derivatives

$$L\{f'\} = sF(s) - f(0), \tag{3}$$

$$\vdots$$

$$L\left\{f^{(n)}\right\} = s^n F(s) - s^{n-1} f(0) - s^{n-2} f'(0)$$

$$- \cdots - sf^{(n-2)}(0) - f^{(n-1)}(0) \tag{4}$$

Convolution Theorem

$$L\{(f * g)(t)\} = F(s)G(s) \tag{5}$$

or

$$L^{-1}\{F(s)G(s)\} = (f * g)(t), \tag{6}$$

where

$$(f * g)(t) = \int_0^t f(\tau)\,g(t - \tau)\,d\tau \tag{7}$$

is the Laplace convolution of f and g.

In Section 13.5 we introduced the Heaviside function $H(t)$, defined as

$$H(t) = \begin{cases} 0, & t < 0 \\ 1, & t > 0 \end{cases}$$

which admits the useful property

$$\int_0^t f(\tau)H(\tau - a)\,d\tau = H(t - a)\int_a^t f(\tau)\,d\tau \tag{8}$$

where $a > 0$. The Heaviside function is used when the forcing function is discontinuous or is given piecewise.

Additional properties were the t-shift formula

$$L\{H(t - a)f(t - a)\} = e^{-as}F(s) \tag{9a}$$

or

$$L^{-1}\{e^{-as}F(s)\} - H(t - a)f(t - a), \tag{9b}$$

the s-shift formula

$$L\{e^{-at}f(t)\} = F(s + a) \tag{10a}$$

or

$$L^{-1}\{F(s + a)\} = e^{-at}f(t), \tag{10b}$$

and the expression

$$F(s) = \frac{1}{1 - e^{-sT}}\int_0^T f(t)\,e^{-st}\,dt \tag{11}$$

for the transform of a periodic function f of period T.

Appendix

Review of Partial Fraction Expansions

Generally, one meets the method of partial fraction expansions in the integral calculus, where the method is used to express a difficult integral as a linear combination of simpler ones. For example,

$$\int \frac{dx}{x^2 + 4x + 3} = \int \left(\frac{1}{2}\frac{1}{x+1} - \frac{1}{2}\frac{1}{x+3} \right) dx$$
$$= \frac{1}{2} \int \frac{dx}{x+1} - \frac{1}{2} \int \frac{dx}{x+3}$$
$$= \frac{1}{2} \ln |x+1| - \frac{1}{2} \ln |x+3| + \text{constant.}$$

In this text we use the method primarily to help us to invert Laplace transforms such as the Laplace transform $F(s) = 1/(s^2 + 3s + 2)$. For convenient reference, this appendix contains a review of the method.

Let $p(x)$ and $q(x)$ be finite-degree polynomials in x, of degree P and Q, respectively. Then

$$f(x) = \frac{p(x)}{q(x)} \tag{1}$$

is called a **rational function** of x. Let P be less than Q. [If $P \geq Q$, then we can, by the long division of q into p, express f as a polynomial of degree $P - Q$ plus a rational function $r(x)/q(x)$, where the degree R of $r(x)$ *is* less than degree Q of $q(x)$. For instance, long division gives

$$\frac{x^5 + 6x^2 - 5x + 6}{x^3 - x^2 + x + 1} = x^2 + x + \frac{4x^2 - 6x + 6}{x^3 - x^2 + x + 1}.$$

Whereas the method of partial fractions cannot be applied to the rational function on the left (because $P = 5$ is not less than $Q = 3$), it can be applied to the one on the right (because $P = 2$ *is* less than $Q = 3$).]

Distinct roots. Let $q(x)$ in (1) have the distinct roots $x_1, \ldots, x_Q$. Then f admits the **partial fraction expansion**

$$f(x) = \frac{p(x)}{q(x)} = \frac{a_1}{x - x_1} + \frac{a_2}{x - x_2} + \cdots + \frac{a_Q}{x - x_Q}, \tag{2}$$

where the a_j's are constants. One way to determine the a_j's is to recombine the terms on the right-hand side over a common denominator [namely, $q(x)$] and require its numerator to be identical to $p(x)$.

EXAMPLE 1
Expand $f(x) = (x - 1)/(x^2 + 5x + 6)$ in partial fractions. Since $x^2 + 5x + 6 = (x + 2)(x + 3)$, we can expand f as

$$f(x) = \frac{x - 1}{x^2 + 5x + 6} = \frac{a_1}{x + 2} + \frac{a_2}{x + 3}.$$

To determine a_1 and a_2, rewrite the latter equation as

$$\frac{x - 1}{x^2 + 5x + 6} = \frac{(a_1 + a_2)x + (3a_1 + 2a_2)}{x^2 + 5x + 6}.$$

For the numerators to be identical we need

$$x^0 : \quad -1 = 3a_1 + 2a_2,$$
$$x^1 : \quad 1 = a_1 + a_2.$$

Solving these equations gives $a_1 = -3$ and $a_2 = 4$ so

$$\frac{x - 1}{x^2 + 5x + 6} = -\frac{3}{x + 2} + \frac{4}{x + 3}. \qquad \blacksquare$$

Repeated roots. If any root x_j of $q(x)$ is of multiplicity k [i.e., $(x - x_j)^k$ is a factor of $q(x)$ but $(x - x_j)^{k+1}$ is not], then the jth term on the right-hand side of (2) must be modified to the form

$$\frac{a_{j1}}{x - x_j} + \frac{a_{j2}}{(x - x_j)^2} + \cdots + \frac{a_{jk}}{(x - x_j)^k} \qquad (3)$$

or, equivalently,

$$\frac{b_{j0} + b_{j1}x + \cdots + b_{j,k-1}x^{k-1}}{(x - x_j)^k}.$$

To solve for $a_{j1}, \ldots, a_{jk}$ in (3), we can recombine terms over a common denominator [namely, $q(x)$] and equate coefficients of powers of x in the numerator (since powers of x are linearly independent functions of x) as we did in Example 1.

EXAMPLE 2
To expand $(4x^2 + 5)/[(x - 2)^3(x + 3)]$ in partial fractions, write

$$\frac{4x^2 + 5}{(x - 2)^3(x + 3)} = \left[\frac{a}{x - 2} + \frac{b}{(x - 2)^2} + \frac{c}{(x - 2)^3} \right] + \frac{d}{x + 3}$$
$$= \frac{\left[a(x - 2)^2 + b(x - 2) + c \right](x + 3) + d(x - 2)^3}{(x - 2)^3(x + 3)}$$

$$= [(a + d)x^3 + (-a + b - 6d)x^2 + (-8a + b + c + 12d)x$$
$$+ (12a - 6b + 3c - 8d)]/[(x - 2)^3(x + 3)],$$

where the notation a, b, c, d will be simpler than using subscripted a_{jk}'s. Thus,

$$\begin{aligned}
x^0 : \quad 5 &= 12a - 6b + 3c - 8d, \\
x^1 : \quad 0 &= -8a + b + c + 12d, \\
x^2 : \quad 4 &= -a + b - 6d, \\
x^3 : \quad 0 &= a + d,
\end{aligned}$$

with solution $a = 41/125$, $b = 59/25$, $c = 21/5$, $d = -41/125$ so

$$\frac{4x^2 + 5}{(x - 2)^3(x + 3)} = \frac{41}{125}\frac{1}{x - 2} + \frac{59}{25}\frac{1}{(x - 2)^2} + \frac{21}{5}\frac{1}{(x - 2)^3} - \frac{41}{125}\frac{1}{x + 3}. \quad \blacksquare$$

Partial fraction expansions can also be carried out using computer software. For example, the relevant *Maple* command is **convert**, and the commands

```
convert((x-1)/(x^2+5*x+6),parfrac,x);
```

and

```
convert((4*x^2+5)/((x-2)^3*(x+3)),parfrac,x);
```

give the results that we obtained in Examples 1 and 2, respectively.

Answers to Selected Exercises

NOTE: The selected exercises are those with underlined exercise numbers.

CHAPTER 1

Section 1.2

1. (**a**) 1st order. y_1: $y_1'^2 - 4y_1 = (2x)^2 - 4x^2 = 0$ so $y_1(x) = x^2$ is a solution (for all x). y_2: $(4x^2) - 4(2x^2) = 8x^2 \neq 0$ so y_2 is not. y_3: $(-e^{-x})^2 - 4e^{-x} = e^{-2x} - 4e^{-x} \neq 0$ so y_3 is not. (**c**) 2nd order. No, yes, yes (**e**) 2nd order. Yes, yes (**g**) 3rd order. Put $y_1(x)$ into the differential equation and find that $y_1''' - 6y_1'' + 12y_1' - 8y_1 = 32 - 16x \neq 32 - 10x$ so y_1 is not a solution. (**i**) 2nd order. No, no, yes

2. (**a**) $A = 2$ (**c**) $A = -e^{24}$ (**e**) $A = 99$

3. (**a**) $A = 0$, $B = 2$ (**c**) $A = 1$, $B = -1/2$ (**e**) $A = 1/2$, $B = -1/2$

4. (**a**) Linear (**c**) Linear (**e**) Linear (**g**) Nonlinear (**i**) Nonlinear (**k**) Linear

5. (**a**) $r = -3$ (**c**) None (**e**) $r = 0, \pm 1$ (**g**) $r = \pm 1, \pm \sqrt{5}$

CHAPTER 2

Section 2.2

1. (**a**) $p(x) = -6x^2$, so (11) gives $y(x) = Ae^{-\int(-6x^2)\,dx} = Ae^{2x^3}$. Or, $\int dy/y = \int 6x^2\,dx$, $\ln|y| = 2x^3 + C$, $|y| = e^{C+2x^3} = e^C e^{2x^3} = Be^{2x^3}$, $y(x) = \pm Be^{2x^3} = Ae^{2x^3}$ where A is an arbitrary constant (**c**) $y(x) = Ae^{\sin x}$ (**e**) $y(x) = Ax$ (**g**) $y(x) = A(1 + 2x + x^2)$ (**i**) $y(x) = A\exp[-2\exp(-x)]$ (**k**) $y(x) = Ae^{-1/x}$ (**m**) $y(x) = Ae^{-x}/x$ (**o**) $y(x) = A/\cos x$

2. (**a**) $y(x) = 2\exp[2(x^3 - 1)]$. Since $p(x) = -6x^2$ is continuous on $-\infty < x < \infty$, Theorem 2.2.1 guarantees that a unique solution will exist on $-\infty < x < \infty$. Sure enough, we see that $y(x) = 2\exp[2(x^3 - 1)]$ and its derivative both exist and satisfy the differential equation for all x. (**c**) $y(x) = 2\exp(\sin x - \sin 1)$ on $-\infty < x < \infty$ (**e**) $y(x) = 2x$. Since $p(x) = -1/x$ is continuous on $-\infty < x < 0$ and on $0 < x < \infty$, and the initial point is at $x = 1$, Theorem 2.2.1 guarantees existence and uniqueness only on $0 < x < \infty$. Yet our solution $y(x) = 2x$ is found to exist on $-\infty < x < \infty$.

That this interval is larger that the one guaranteed by the theorem is not a contradiction of the theorem. Even if we removed the words "at least" the theorem would still be true and would not deny the possibility of existence and uniqueness over a broader interval. (**g**) $y(x) = (1 + 2x + x^2)/2$ on $-\infty < x < \infty$ (**i**) $y(x) = 2e^{-2[\exp(-x)-\exp(-1)]}$ on $-\infty < x < \infty$ (**k**) $y(x) = 2e^{(x-1)/x}$ on $0 < x < \infty$. Note that $p(x) = -1/x^2$ is continuous on $-\infty < x < 0$ and $0 < x < \infty$. Since the initial point is at $x = 1$ the relevant interval is $0 < x < \infty$. Thus, Theorem 2.2.1 guarantees existence and uniqueness at least on $0 < x < \infty$. In fact, if we examine the solution $y(x)$ we see that neither $y(x)$ nor $y'(x)$ exist at $x = 0$ so the interval of existence and uniqueness is exactly that: $0 < x < \infty$. (**m**) $y(x) = 2e^{(1-x)}/x$ on $0 < x < \infty$ (**o**) $y(x) = 2\cos 1/\cos x$ on $-\pi/2 < x < \pi/2$

3. (**a**) $y(x) = 5\exp(2x^3 + 54)$ on $-\infty < x < \infty$ (**k**) $y(x) = 5\exp\left(-\frac{1}{x} - \frac{1}{3}\right)$ on $-\infty < x < 0$ (**m**) $y(x) = -15e^{-(x+3)}/x$ on $-\infty < x < 0$ (**o**) $y(x) = 5\cos 3/\cos x$ on $-\pi/2 < x < \pi/2$

5. (**a**) `dsolve(diff(y(x),x)=6*x^2*y(x),y(x));` and then
`dsolve({diff(y(x),x)=6*x^2*y(x),y(5)=4},y(x));`

6. (**a**) Use
```
with(DEtools):
dfieldplot(diff(y(x),x)=6*x^2*y(x),y(x),
x=-2..2,y=-2..8,arrows=LINE);
```

7. (**b**) $y(x) = 3x^2$ on $-\infty < x < \infty$. Does not contradict the theorem. See solution to Exercise 2(e). (**d**) $y(0) = 5$. More generally, $y(0) = a$ where $a \neq 0$.

Section 2.3

1. (**a**) $y(x) = (A + 3x)e^x$ (**d**) $y(x) = 2\cos 2x + \sin 2x + Ae^x$ (**g**) $y(x) = Ax^2 + x^3$ (**j**) $x(y) = Ae^{6y} - 2e^y$ (**m**) $x(t) = At + t^5$

3. (**a**) $y(x) = 2x^2 - 2/x$ on $0 < x < \infty$ (**c**) $y(x) = 2x^2$ on $-\infty < x < \infty$ (**e**) $y(x) = 2x^2 + 69/x$ on $-\infty < x < 0$

4. (a) $y(x) = (x + 3 - 20/x^2)/3$ on $0 < x < \infty$ **(c)** $y(x) = (x+3+1/x^2)/3$ on $-\infty < x < 0$ **(e)** $y(x) = (x+3-4/x^2)/3$ on $-\infty < x < 0$

5. (a) $dx/dy - x = 3e^y$, so $x(y) = 3ye^y + Ae^y$ **(c)** $dx/dy + (1/y)x = 6y$, so $x(y) = 2y^2 + A/y$

6. (a) Use

```
with(DEtools):
phaseportrait(diff(y(x),x)=2+(2*x-y(x))^3,
y(x),x=0..3,{[2,1]},stepsize=.05,y=-4..4,
arrows=LINE);
```

From the resulting direction field it appears that there may be one or more straight-line solution curves. To explore that possibility set $y = mx + b$ in the ODE; we find that $y = 2x$ is indeed a solution. In fact, if we replot with $x = 0..30$ and $y = -4..60$, say, it appears that the solution curve through P has $y = 2x$ as an asymptote.   **(c)** The phase portrait suggests that the lines $y \approx \pm 1.73$ are solution curves. In fact, if we put $y = \text{constant} = C$ into the ODE we find that $y = \pm\sqrt{3}$ are solution curves. Can you now see what the other solution curves look like? Try sketching a few more, through $[-1, 0]$, $[1, 0]$, $[-1, 1.8]$, $[-1, -1.8]$, $[1, 1.8]$, and $[1, -1.8]$, for instance. **(e)** If we use stepsize= 0.05, say, we find that the solution curve through $[-3, 3]$ rises as x increases but has a suspicious kink when y reaches $y = -1$. We can try reducing the stepsize more and more to get the solution curve to "settle down" but it does not. The difficulty is that y' is infinite at $y = -1$ (and at $y = +1$). Further, the direction field suggests that the solution curve through $[-3, 3]$ would "like to" bend to the left as y increases through -1, but it cannot because $y(x)$ must be single valued. Thus, it appears that the solution through $[-3, 3]$ *does not exist* to the right of the x point at which y reaches -1. If this idea is unclear consider the example $y' = -x/y$ with $y(0) = -2$, say. It is easily verified that the solution is the circular arc $y(x) = -\sqrt{4 - x^2}$, but only over $-2 < x < 2$.

8. (a) $y(x) = 1/(Ae^{-4x} - 1)$ **(d)** $y(x) = (Ae^{-x/2} + x - 2)^{2/3}$ **(g)** $y(x) = -\ln|Ax + B|$

10. (a) $u' - 4u = -1$, $u(x) = (Ae^{4x} + 1)/4$, $y(x) = -4/(Be^{-4x}+1)$ **(d)** $u'+3u = -e^{-x}$, $u(x) = Ae^{-3x} - e^{-x}/2$, $y(x) = 2/(Be^x + e^{-x})$ **(g)** Use $Y(x) = +2$ (or -2). Then $u'+4u = -1$, $u(x) = (Ae^{-4x}-1)/4$, $y(x) = 2(B+e^{4x})/(B-e^{-4x})$

11. (b) $x = -3/2 + A/p^2$, $y = 2A/p$, where A is an arbitrary constant. In this case we can eliminate p and obtain $x = -3/2 + By^2$ where B is an arbitrary (but nonzero) constant.   **(c)** $x = Ae^{1/p}/p^2$, $y = Ae^{1/p}(1+p)/p$, where A is an arbitrary positive constant   **(f)** $f(p) = p$ gives $p^2 - 3p + 2 = 0$ so $P_0 = 1$ and $P_0 = 2$. With $g(p) = e^p$, (11.4) gives the particular solutions $y = 1x + g(1) = x + e$ for $P_0 = 1$ and $y = 2x + g(2) = 2x + e^2$ for $P_0 = 2$.   **(g)** $f(p) = p$ gives

$e^p = p$ which has no roots (at least no real roots; we will not consider the possibility of complex roots)

Section 2.4

3. (a) $x(t) = x_0 e^{-kt/c}$ for $t < t_1$ and $x(t) = x_0 e^{-kt/c} + (F_0/k)[1 - e^{-k(t-t_1)/c}]$ for $t > t_1$.   NOTE: If you wish to check this result using *Maple*, first express $F(t) = F_0 H(t-t_1)$. Then use

```
dsolve({diff(x(t),t)+(k/c)*x(t)
=(1/c)*Fsub0*Heaviside(t-tsub1),x(0)=xsub0},x(t)));
```

The result contains $H(-t_1)$ terms but these are zero because $H(t) = 0$ for all $t < 0$. Then the result reduces to $x(t) = x_0 e^{-kt/c} + (F_0/k)H(t-t_1)[1-e^{-k(t-t_1)/c}]$, which agrees with the piecewise result stated above.

4. (a) Use

```
dsolve({diff(x(t),t)+2*x(t)=5*exp(-2*t),
x(0)=10},x(t));
with(DEtools):
phaseportrait(diff(x(t),t)+2*x(t)=5*exp(-2*t),
x(t),t=0..20,{[0,10]},stepsize=.05,arrows=NONE);
```

If you omit the stepsize option you will find that the default stepsize is not sufficiently small to give a smooth graph for $x(t)$.   **(g)** Use $F(t) = 25[H(t - 5) - H(t - 10)]$. Also, note the CAUTION that precedes these exercises. A stepsize of .05 gives an error message, a stepsize of .03, for instance, does not.

5. $i(1) = 0.0318958$ and $i(10) = -0.0041710$. To check this by *Maple* do

```
dsolve({10*diff(i(t),t)+2*i(t)=diff(t*exp(-t),t),
i(0)=0},i(t),type=numeric,value=array([1,10]));
```

8. $i(1) = 1.989893$, $i(5) = 2.000000$, $i(10) = 0.000000$. If you check your analytical result using the *Maple* command

```
dsolve({diff(i(t),t)+5*i(t)
=10*(1-Heaviside(t-5)),i(0)=0.5},i(t),
type=numeric,value=array([1,5,10]));
```

you will obtain an error message associated with the discontinuity in $H(t - 5)$ at $t = 5$. If you change the 5 in array $([1, 5, 10])$ to 4.999999999 or 5.000000001, say, these commands give $i(4.999999999) = i(5.000000001) = 2.000000$.

10. $m(t) = m_0 e^{-kt}$ so $8 = 10e^{-60k}$ gives $k = 0.00372$. Then $2 = 10e^{-0.00372t}$ gives $t = 433$ years and $0.1 = 10e^{-0.00372t}$ gives $t = 1238$ years.

14. $c(t) = c_1(1 + e^{-Qt/v})$, $c(t) \to c_1$ as $t \to \infty$.

15. For $0 \le t \le 2$ use (20) in Section 2.3 with $a = 0$ and $b = 0$: $c(t) = e^{-4t}\left(\int_0^t e^{4s}4\,ds + 0\right) = 1 - e^{-4t}$ so $c(0.5) = 0.864665$ and $c(2) = 1 - e^{-8} = 0.999665$. For

$t > 2$ it is convenient to use (20) again, this time with $a = 2$ and $b = c(2) = 1 - e^{-8}$. That gives

$$c(t) = e^{-2(t-2)} \left(\int_2^t e^{2(s-2)} 2 \, ds + 1 - e^{-8} \right) = 1 - e^{-2t-4}$$

so $c(3) = 0.999955$ and $c(\infty) = 1$. As a check, let us use *Maple* to compute $c(0.5), c(2), c(3)$, and $c(\infty)$. The command

```
dsolve({diff(c(t),t)+(4-2*Heaviside(t-2))*c(t)=4
-2*Heaviside(t-2),c(0)=0},c(t),
type=numeric,value=array([.5,2.000001,3,10000]));
```

gives values that agree with those given above. Note that if we use 2 in the array then *Maple* gives an error message due to being asked to compute at $t = 2$, where $H(t - 2)$ is discontinuous. Thus, use a point *close* to 2, such as 2.000001. Also, it suffices to use 10000 here, in place of infinity.

17. (*a*) $v(t) = \frac{mg}{c}(1 - e^{-ct/m})$, terminal velocity $= mg/c$, as can be seen either by letting $t \to \infty$ in the solution or by setting $v' = 0$ in the differential equation.

18. (*a*) 0.230 inches

19. (*b*) $u(t) = 70 + 130e^{-0.016705t}$, $u = 100$ at $t = 87.78$ minutes   (*c*) $T = 2.20$ hours

20. (*b*) $S(1)/S_0 = 1.050000, 1.051162, 1.051246, 1.051267, 1.051271$, respectively. (Not much difference, is there?)

22. (*a*) $c(x) = 0$ for $x < 0$ and $(Q/A\beta)(1 - e^{-\beta x/U})$ for $x > 0$

CHAPTER 3

Section 3.2

1. (*a*) $\int e^y \, dy = \int 3x^2 \, dx$, $e^y = x^3 + A$, $y = \ln(x^3 + A)$, and $y(0) = 0 = \ln A$ gives $A = 1$ so $y(x) = \ln(x^3 + 1)$. The x interval of validity was not asked for, but we can see that it is $-1 < x < \infty$ because $\ln(x^3 + 1) \to -\infty$ as $x \to -1$. (*d*) $y(x) = \tan(x + A)$, $y(x) = \tan(x + \pi/4)$ (*g*) $y(x) = 3/(Ae^{-3x} - 1)$, $y(x) = 12/(e^{-3x} - 4)$ (*j*) $y(x) = A\sqrt{x}$, $y(x) = -\sqrt{x/3}$ (*m*) In implicit form $y + e^y = x^2 + A$ and $y(0) = 0$ gives $A = 1$ so $y + e^y = x^2 + 1$; can't solve explicitly for y.   (*p*) In implicit form $ye^y = e^x + A$ and $y(0) = 0$ gives $A = -1$ so $ye^y = e^x - 1$; can't solve explicitly for y.

2. (*a*) The command

```
dsolve({diff(y(x),x)-3*x^2*exp(-y(x))=0,y(0)=0},
y(x));
```

gives $y(x) = \ln(x^3 + 1)$. To plot the latter we could use a plotting command such as

```
with(plots):
implicitplot(y=ln(x^3+1),x=-1..5,y=-5..5);
```

or we could use

```
with(DEtools):
phaseportrait(diff(y(x),x)-3*x^2*exp(-y(x))=0,
y(x),x=-1..5,{[0,0]},stepsize=.05,y=-5..5,
arrows=NONE);
```

The latter gives an error message associated with the "blowup" of $y(x)$ at $x = -1$. Thus, change $x = -1..5$ to $x = -.999..5$, say, to stay away from $x = -1$. Then it works.

4. (*a*) No   (*b*) Yes, of degree 0

5. (*a*) $y' = 6y/x$ and $y' - x^2/y^2$ are separable, $y' = (6y/x) + 4$ and $y' = \sin(y/x)$ are not.

6. (*a*) $2x + y = A(x - y)^4$ (*c*) $y = x/(A - 2\ln|x|)$ (*e*) $y + \sqrt{x^2 + y^2} = Ax^2$

7. (*b*) $2x + y = 5 + A(x - y - 1)^4$ (*d*) $4x - 8y = 5\ln|4x + 8y - 7| + A$

Section 3.3

1. (*a*) Both $f(x, y) = 2xy$ and $\partial f/\partial y = 2x$ are continuous in the whole x, y plane so Theorem 3.3.1 assures us of the existence of a unique solution through any given initial point, but it does not predict how broad will be the x interval over which that unique solution exists. [However, since $y' = 2xy$ happens to be linear we can use the more informative Theorem 2.3.1, which tells us, since $p(x) = -2x$ and $q(x) = 0$ are continuous for all x, that a unique solution will exist on $-\infty < x < \infty$.] General solution is $y(x) = A \exp(x^2)$. $y(0) = 0$ gives $y(x) = 0$, $y(0) = 2$ gives $y(x) = 2 \exp(x^2)$, $y(0) = -1$ gives $y(x) = -\exp(x^2)$, and $y(1) = -5$ gives $y(x) = -5 \exp(x^2 - 1)$.   (*c*) $f(x, y) = x/y$ and $\partial f/\partial y = -x/y^2$ are continuous everywhere except on $y = 0$. Thus, Theorem 3.3.1 guarantees a unique solution through any initial point *not* on the x axis, though it doesn't predict how broad the interval will be on which that solution exists and is unique. If the initial point is *on* the x axis the theorem gives no information; i.e., in that case there may exist a unique solution, a nonunique solution, or no solution. The general solution is $y^2 = x^2 + A$ which gives right-hand and left-hand hyperbolas in $-x < y < x$ and upper and lower hyperbolas in $-y < x < y$. (Sketch them, as well as the solutions $y = +x$ and $y = -x$.) $y(0) = 2$ gives unique solution $y = +\sqrt{x^2 + 4}$ for all x, $y(0) = 0$ gives nonunique solutions $y = +x$ and $y = -x$ for all x, $y(0) = -2$ gives unique solution $y = -\sqrt{x^2 + 4}$ for all x, $y(5) = 3$ gives unique solution $y = +\sqrt{x^2 - 16}$ on $4 < x < \infty$ (because the graph comes down with a vertical tangent at $x = 4$ so y' does not exist at $x = 4$ and $y = 0$), $y(4) = 0$ gives no solution (because the solution curve through that point has a vertical tangent), and $y(-5) = -3$ gives the unique solution $y = -\sqrt{x^2 - 16}$ on $-\infty < x < -4$.   (*e*) $f = 1/2y$ and $f_y = -1/2y^2$ are continuous everywhere in x, y plane except on $y = 0$ (the x axis) so there will exist a unique solution through every point that is not on the x axis. General solution is $x = y^2 + A$, which is a family of parabolas with vertical tangents where they cross the x axis. $y(0) = 2$ gives unique solution $y = +\sqrt{x+4}$ on $-4 < x < \infty$, $y(0) = 0$ gives no solution (since $y' = 1/2y$ doesn't exist there), $y(0) = -3$ gives unique solution $y = -\sqrt{x+9}$ on $-9 < x < \infty$, and $y(2) = 5$ gives unique solution $y = +\sqrt{x+23}$ on $-23 < x < \infty$. (g) f and f_y are continuous everywhere in x, y plane except on the lines $y = +\pi/2, -\pi/2, +3\pi/2, -3\pi/2, \ldots$, so there will exist a unique solution through every point that is not on one of those lines. General solution is $\ln|\sin y| = x + A$. $y(3) = 2$ gives $\sin y = (\sin 2)e^{x-3}$. Since $-1 < \sin y < 1$ [the cases $\sin y = -1$ and $\sin y = 1$ corresponding to the values $y = -\pi/2$ and $y = +\pi/2$, at which y' does not exist], the latter solution is valid on $-1 < (\sin 2)e^{x-3} < 1$. The left inequality is valid for all x and the right is valid for $x < \ln(e^3/\sin 2) = 3 - \ln(\sin 2) \approx 3.095$. To check these conclusions run the *Maple* commands

```
with(DEtools):
phaseportrait(diff(y(x),x)=tan(y(x)),x=-4..4,
  [[3,2]],stepsize=.05,arrows=NONE);
```

Even refining the stepsize to .01, say, we see that the solution does seem limited to $x < 3.095$. $y(3) = 0$ gives no solution, and $y(0) = 2$ gives $\sin y = (\sin 2)e^x$, which is valid on $x < 0.095$ as can again be verified using *Maple*.

2. (a) $x = y^2/4$ (d) $y = 0$

3. (a) None (b) The general solution is $y = (x + c)^{3/2}$ with envelope $y = 0$. $y = 0$ is indeed a solution (a singular solution) of $y' = 3y^{1/3}/2$, a solution that is not found within the general solution $y = (x + c)^{3/2}$.

Section 3.4

1. (a) $y = 3x + A$, $y = 3x + 6$ (d) $10 \sin 2u + e^{-5v} = A$, $10 \sin 2u + e^{-5v} = e^{10}$ (g) $x^2 - 4xz + z^2 = A$, $x^2 - 4xz + z^2 = 94$ (j) $x^3 \sin(2y) - x^2 y = A$, $x^3 \sin 2y - x^2 y = (0.5)^3 \sin 6.2 - (0.5)^2(3.1)$

2. (d) The command

```
dsolve(diff(v(u),u)=4*cos(2*u)*exp(5*v(u)),
  v(u));
```

gives the general solution, and

```
dsolve({diff(v(u),u)=4*cos(2*u)*exp(5*v(u)),
  v(0)=-2},v(u));
```

gives the particular solution. But be aware that dsolve looks for an explicit solution so it may have trouble even if a solution is readily found in implicit form—if explicit solution is not possible.

4. (a) $b = A$

5. (a) $M = 3y$ and $N = 1$ so $(M_y - N_x)/N = 3 =$ function of x alone. Hence $\sigma(x) = \exp(\int 3\,dx) = \exp(3x)$. Then

$3ye^{3x}\,dx + e^{3x}\,dy = 0$ is exact and gives $ye^{3x} = $ constant or $y = Ae^{-3x}$. In fact, $(M_y - N_x)/M = 1/y = $ function of y alone so we can find an integrating factor $\sigma(y)$ as well, where $\sigma(y) = \exp(-\int dy/y) = 1/y$. Then $3\,dx + dy/y = 0$ is exact and gives $y = Ae^{-3x}$ once again. (d) $(M_y - N_x)/M = -1 = $ function of y alone, so $\sigma(y) = e^y$ and we obtain $xe^y - y = A$. (g) $\sigma(y) = \cos y$, $2x\cos^2 y - y\cos 2y - y = A$. NOTE: $\cos 2y = 2\cos^2 y - 1$ (j) $\sigma(y) = 1/y^2$, $x^2/y + y = A$

7. (a) $(M_y - N_x)/N = (-x - y)/(2x^2 - 3xy) \neq $ function of x and $(M_y - N_x)/M = (-x - y)/(3xy - 2y^2) \neq $ function of y, so neither $\sigma(x)$ nor $\sigma(y)$ exist. Trying $\sigma = x^a y^b$, write $x^a y^b$ times the differential equation. For exactness set $\partial/\partial y$ of the coefficient of dx equal to $\partial/\partial x$ of the coefficient of dy. If we match coefficients of the $x^{a+1} y^b$ terms we obtain $3b + 3 = 2a + 4$, and if we match coefficients of the $x^a y^{b+1}$ terms we obtain $-2b - 4 = -3a - 3$. If we solve these we obtain $a = 1$ and $b = 1$. With the equation now exact we obtain the general solution $x^2 y^2(x - y) = A$.

8. (a) $\sigma = e^{-(x+y)}$ by inspection, $e^{-x} + e^{-y} = A$

9. (a) $(y - x)\,dx + (x + y)\,dy = 0$ is already exact and gives $xy - x^2/2 + y^2/2 = A$ (c) $(2xy - e^y)\,dx - x(e^y - x)\,dy = 0$ is exact; $x^2 y - xe^y = A$

10. 1

11. (b) Not necessarily. For example, $e^{xy}\,dx + e^{yx}\,dy = 0$ is not exact.

12. If we impose $y(a) = b$ on $F(x, y) = C$ we evaluate C as $F(a, b) = C$. Thus, the particular solution is $F(x, y) = F(a, b)$.

Section 3.5

1. (a) Use the obvious particular solution $N(t) = a/b$ of (4). Then change from $N(t)$ to $u(t)$ according to $N(t) = \frac{a}{b} + \frac{1}{u(t)}$. Put the latter in (4) and obtain the first-order linear equation $u' - au = b$ on u. Thus, $u(t) = Ae^{at} - b/a$. Then put the latter into $N(t) = \frac{a}{b} + \frac{1}{u(t)}$ and obtain an expression for $N(t)$ that can be shown to be equivalent to (9).

2. d/dt of $N' = aN - bN^2$ gives $N'' = aN' - 2bNN' = (a - 2bN)N' = (a - 2bN)N(a - bN)$ so N'' changes sign as N crosses $N = a/2b$.

3. If $N' = f(N)$, sketch the graph of $f(N)$ as we did in Fig. 2, and use arrows to indicate the flow directions. (a) $N = 0$ is unstable, $N = 1$ is stable, $N = 2$ is unstable. (c) $N = 0$ is unstable, $N = 1$ is stable.

5. At $t = 0$ we have $x = 0$ and $v(0) = V$. As x increases v diminishes, according to (19), until it becomes zero when x satisfies $V^2 - 2gRx/(x + R) = 0$. Then the projectile will begin its return trip. Thus, the point is that if $V^2 < 2gR$ then x

will not tend to infinity as was supposed in the problem statement.

6. (a) With $V^2 = 2gR$ the solution of (24) is given by $(x + R)^{3/2} = 3\sqrt{2g}\,Rt/2 + R^{3/2}$. For small t, and hence for small x, use Taylor series to write $(x + R)^{3/2} = R^{3/2}(1 + x/R)^{3/2} = R^{3/2}[1+(3/2)x/R+\cdots]$ so $R^{3/2}[1+(3/2)x/R] \sim 3\sqrt{2g}\,Rt/2 + R^{3/2}$ or $x \sim \sqrt{2gR}\,t = V_e t$ as $t \to 0$. For large t, and hence for large x, write $(x + R)^{3/2} \sim x^{3/2}$ and $3\sqrt{2g}\,Rt/2 + R^{2/3} \sim 3\sqrt{2g}\,Rt/2$. Hence, $x^{3/2} \sim 3\sqrt{2g}\,Rt/2$ so $x \sim (3V_e\sqrt{R}/2)^{2/3}t^{2/3}$ as $t \to \infty$.

Section 3.6

1. $yy' = f_0$ and $y(0) = y_0 > 0$ give $y(x) = \sqrt{2f_0 x + y_0^2}$; $yy' = f_0$ and $y(0) = 0$ give $y(x) = \sqrt{2f_0 x}$; $yy' = 0$ and $y(0) = y_0 > 0$ give $y(x) = y_0$. We see that the total response is *not* the sum of the individual responses: $\sqrt{2f_0 x + y_0^2} \neq \sqrt{2f_0 x} + y_0$.

2. $y(x) = \sqrt{2f_0 x}$ is *not* proportional to f_0. For example, if we double f_0 we do not double $y(x)$, if we triple f_0 we do not triple $y(x)$, and so on.

3. (a) It must be nonlinear because if it were linear the solution would, according to (17) (in which b plays the role of y_0), be of the form $y(x)$ equals a function of x plus y_0 times a function of x; $y(x) = y_0 + y_0^2 x^3$ is not of that form.

CHAPTER 4

Section 4.2

3. (a) If $\mathbf{A} + \mathbf{B} + \mathbf{C} = \mathbf{0}$ then the vectors $\mathbf{A}$, $\mathbf{B}$, $\mathbf{C}$ placed head to head form a triangle with sides of length 1, 2, 5. Since $5 > 1 + 2$ this arrangement is impossible by the stated Euclidean proposition. Thus $\mathbf{A} + \mathbf{B} + \mathbf{C}$ cannot equal $\mathbf{0}$.

5. (b) Draw a parallelogram with top and bottom horizontal. Label lower left and right corners A and B, and the upper left as C. Let D be the midpoint of the upper side and let the diagonal BC be intersected by AD at E. Write $\mathbf{AB} + \mathbf{BC} = \mathbf{AC}$, $\mathbf{BE} = \alpha\mathbf{BC}$, $\mathbf{AB} + \mathbf{BE} = \mathbf{AE}$, and $\beta\mathbf{AE} - \mathbf{AC} = \frac{1}{2}\mathbf{AB}$, or an equivalent set of relations. Elimination $\mathbf{AB}$, $\mathbf{BE}$, and $\mathbf{AE}$, say, obtain $[1 - 3/(2\beta)]\mathbf{AC} = [1 - \alpha - 1/(2\beta)]\mathbf{BC}$. Since $\mathbf{AC}$ and $\mathbf{BC}$ are not aligned it follows that $1 - 3/(2\beta) = 0$ and $1 - \alpha - 1/(2\beta) = 0$. These give $\alpha = 2/3$.

7. $\mathbf{OB} = \mathbf{OA} + \mathbf{AB} = \mathbf{OA} + \alpha\mathbf{AC} = \mathbf{OA} + \alpha(\mathbf{OC} - \mathbf{OA}) = (1 - \alpha)\mathbf{OA} + \alpha\mathbf{OC}$

11. Draw $\mathbf{u}$, $\mathbf{v}$, $\mathbf{w}$ tail to tail and let $\mathbf{w} = \alpha\mathbf{u} + \beta\mathbf{v}$ fall between $\mathbf{u}$ and $\mathbf{v}$. Denote $\|\mathbf{u}\| = u$, $\|\mathbf{v}\| = v$, $\|\mathbf{w}\| = w$, for brevity. If θ is the angle between $\mathbf{u}$ and $\mathbf{w}$ (with $0 \leq \theta \leq \pi$) and ϕ is the angle between $\mathbf{w}$ and $\mathbf{v}$ (with $0 \leq \phi \leq \pi$) then

$\cos\theta = \mathbf{u}\cdot\mathbf{w}/uw = (\alpha\mathbf{u}\cdot\mathbf{u}+\beta\mathbf{u}\cdot\mathbf{v})/uw = (\alpha u^2 +\beta\mathbf{u}\cdot\mathbf{v})/uw$ and $\cos\phi = \mathbf{v}\cdot\mathbf{w}/vw = (\alpha\mathbf{u}\cdot\mathbf{v}+\beta v^2)/vw$. Equating these and multiplying by uvw gives $\alpha u^2 v +\beta(\mathbf{u}\cdot\mathbf{v})v = \alpha(\mathbf{u}\cdot\mathbf{v})u+\beta v^2 u$, which *is* satisfied if $\alpha = v$ and $\beta = u$. Then $\cos\theta = \cos\phi$ implies that $\theta = \phi$.

Section 4.3

1. (a) $(24, -7, 23, 32)$ **(c)** $(-92, -4, -8, -24)$ **(e)** $(6, 2, -4, -4)$ **(g)** Not defined because $\mathbf{t} + 2\mathbf{u}$ is a 4-tuple whereas $3\mathbf{v}$ is a 3-tuple   **(i)** Not defined because $\mathbf{ut}$ and $\mathbf{uw}$ are not defined   **(k)** Not defined because $2\mathbf{t} + 7\mathbf{u}$ is a 4-tuple whereas -4 is a scalar   **(m)** Not defined

2. (a) $\mathbf{x} = \frac{1}{9}(3\mathbf{u} - 4\mathbf{v}) = \frac{1}{9}(-5, 9, 20, -6) = (-5/9, 1, 20/9, -2/3)$

3. (a) $5\mathbf{v} = 5(2, 0, -5, 0) = (10, 0, -25, 0)$ by (9b). Then, $\mathbf{u} - 5\mathbf{v} = (1, 3, 0, -2) - (10, 0, -25, 0) = (-9, 3, 25, -2)$ by (9e), (9d), (9b). Then $2(\mathbf{u} - 5\mathbf{v}) = 2(-9, 3, 25, -2) = (-18, 6, 50, -4)$ by (9b). The negative of $(-18, 6, 50, -4)$ is $(18, -6, -50, 4)$ by (9d) and (9b). If we add that to $3\mathbf{x} + 2(\mathbf{u} - 5\mathbf{v}) = \mathbf{w}$ we obtain $3\mathbf{x} + \mathbf{0} = (22, -3, -48, 3)$ by (10d) and (9a). Then $3\mathbf{x} = (22, -3, -48, 3)$ by (10c). Finally, $\frac{1}{3}(3\mathbf{x}) = \frac{1}{3}(22, -3, -48, 3)$ gives $\mathbf{x} = (22/3, -1, -16, 1)$ by (10e) and (9b).   NOTE: This example is to solidify understanding of the principles of vector arithmetic. Once the principles are understood the justifications can be omitted and the calculation can be streamlined, as in our writeup of part (c) to follow. **(c)** $\mathbf{u} - 4\mathbf{x} = \mathbf{0}$, $\mathbf{x} = \frac{1}{4}\mathbf{u} = (1/4, 3/4, 0, -1/2)$

4. (a) $\alpha_1(2, 1, 3) + \alpha_2(1, 2, -4) = (2\alpha_1 +\alpha_2, \alpha_1 +2\alpha_2, 3\alpha_1 - 4\alpha_2) = (0, 0, 0)$ gives the equations $2\alpha_1 +\alpha_2 = 0$, $\alpha_1 +2\alpha_2 = 0$, $3\alpha_1 - 4\alpha_2 = 0$ on α_1, α_2. By elimination these three equations give the unique solution $\alpha_1 = \alpha_2 = 0$.   **(c)** $\alpha_1 = 2$ and $\alpha_2 = 1$ **(e)** No such scalars exist.   **(g)** $\alpha_1 = 3$, $\alpha_2 = -1$

Section 4.4

1. (a) $\|\mathbf{u}\| = 5$, $\|\mathbf{v}\| = \sqrt{5}$, $\mathbf{u}\cdot\mathbf{v} = 10$, $\theta = \cos^{-1}(10/5\sqrt{5}) = 0.464$ rad $= 26.6°$ **(c)** $\|\mathbf{u}\| = \sqrt{10}$, $\|\mathbf{v}\| = 7$, $\mathbf{u}\cdot\mathbf{v} = 0$, $\theta = \cos^{-1}0 = \pi/2$ rad $= 90°$ (orthogonal)   **(e)** $\|\mathbf{u}\| = \sqrt{29}$, $\|\mathbf{v}\| = \sqrt{116}$, $\mathbf{u}\cdot\mathbf{v} = 0$, $\theta = \cos^{-1}0 = \pi/2$ rad $= 90°$ (orthogonal)   **(g)** $\|\mathbf{u}\| = \sqrt{15}$, $\|\mathbf{v}\| = \sqrt{45}$, $\mathbf{u}\cdot\mathbf{v} = -13$, $\theta = \cos^{-1}(-13/\sqrt{15}\sqrt{45}) = 2.09$ rad $= 120.0°$

2. (a) Yes, it is a vector   **(c)** Yes, it is a scalar   **(e)** Yes, it is a scalar   **(g)** No   **(i)** Yes, it is a scalar

3. Let us denote the angle between vectors $\mathbf{OP}$ and $\mathbf{OQ}$ as POQ or QOP. **(a)** $\mathbf{AB} = (3, -1) - (2, 0) = (1, -1)$. Similarly, $\mathbf{AC} = (3, 0)$, $\mathbf{BA} = (-1, 1)$, $\mathbf{BC} = (2, 1)$, $\mathbf{CA} = (-3, 0)$, $\mathbf{CB} = (-2, -1)$. Then

$$CAB = \cos^{-1}(\mathbf{AC}\cdot\mathbf{AB}/\|\mathbf{AC}\|\|\mathbf{AB}\|) = \cos^{-1}(3/3\sqrt{2})$$
$$= 45°$$

$ABC = \cos^{-1}(\mathbf{BA} \cdot \mathbf{BC}/\|\mathbf{BA}\|\|\mathbf{BC}\|) = \cos^{-1}(-1/\sqrt{2}\sqrt{5})$
$= 108.4°$

$BCA = \cos^{-1}(\mathbf{CA} \cdot \mathbf{CB}/\|\mathbf{CA}\|\|\mathbf{CB}\|) = \cos^{-1}(6/3\sqrt{5})$
$= 26.6°$

sum $= CAB + ABC + BCA = 180°$,

which is as it should be for a triangle.   NOTE: To understand these steps you should draw an $ABCA$ triangle, even generically; it doesn't need to be to scale.   (c) Proceeding as in (a), $ABC = 108.4°$, $BCD = 90°$, $CDE = 116.6°$, $DEA = 90°$, $EAB = 135°$, sum $= 540°$ which is correct (and hence a check) since the sum of the interior angles of a 5-sided polygon is 540°; more generally, for an n-sided polygon with $n \geq 3$ it is $(n - 2)(180)°$.   (e) $BCD = 90°$, $CDE = 116.6°$, $DEB = 71.6°$, $EBC = 81.9°$, sum $= 360°$

4. (a) $\hat{\mathbf{u}} = \frac{1}{5}(4, 3) = (4/5, 3/5)$, $\hat{\mathbf{v}} = \frac{1}{\sqrt{5}}(1, 2) = (1/\sqrt{5}, 2/\sqrt{5})$
(c) $\hat{\mathbf{u}} = (3/\sqrt{10}, 0, 1/\sqrt{10})$, $\hat{\mathbf{v}} = (-2/7, 3/7, 6/7)$
(e) $\hat{\mathbf{u}} = (2/\sqrt{29}, 5/\sqrt{29})$, $\hat{\mathbf{v}} = (5/\sqrt{29}, -2/\sqrt{29})$
(g) $\hat{\mathbf{u}} = (3/\sqrt{15}, 2/\sqrt{15}, 0, -1/\sqrt{15}, 1/\sqrt{15})$, $\hat{\mathbf{v}} = (-5/\sqrt{45}, 0, 0, 2/\sqrt{45}, 4/\sqrt{45})$

6. (a) $3\sqrt{3}$ (c) 1

8. (b) $\mathbf{u}_\| = (52/25, 39/25)$, $\mathbf{u}_\perp = (-27/25, 36/25)$
(d) $\mathbf{u}_\| = (0, 18/13, 27/13)$, $\mathbf{u}_\perp = (2, 21/13, -14/13)$
(f) $\mathbf{u}_\| = (0, 0, 1/2, -1, 1/2)$, $\mathbf{u}_\perp = (2, 1, -1/2, 1, 5/2)$

10. No. For example, let $\mathbf{u} = (1, 0)$, $\mathbf{v} = (0, 1)$, $\mathbf{w} = (-1, 0)$.

11. (a) Yes   (c) Yes   (e) No

Section 4.5

2. (a) eq 2 $\to$ eq 2 $-$ 5 eq 1 gives $y = -3/16$, $x = 7/16$; unique   (c) $y = \alpha$, $x = 4 - 2\alpha$; 1-parameter family
(e) eq 2 $\to$ eq 2 $-$ 2 eq 1 gives $z = \alpha$, $y = 6 + 3\alpha$, $x = 7 + 2\alpha$; 1-parameter family   (g) eq 2 $\to$ eq 2 $-$ 5 eq 1, eq 3 $\to$ eq 3 $-$ 9 eq 1, eq 3 $\to$ eq 3 $-$ 2 eq 2, eq 2 $\to$ $-\frac{1}{4}$ eq 2 gives $z = \alpha$, $y = 3 - 2\alpha$, $x = -2 + \alpha$; 1-parameter family
(i) eq 3 $\leftrightarrow$ eq 1, eq 2 $\to$ eq 2 $-$ 3 eq 1, eq 3 $\to$ eq 3 $-$ 2 eq 1, eq 4 $\to$ eq 4 $-$ 6 eq 1, eq 2 $\to$ $-\frac{1}{28}$ eq 2, eq 3 $\to$ $-\frac{1}{21}$ eq 3, eq 4 $\to$ $-\frac{1}{49}$ eq 4, eq 3 $\to$ eq 3 $-$ eq 2, eq 4 $\to$ eq 4 $-$ eq 2 gives $x_2 = -10/7$, $x_1 = 16/7$; unique.   (k) eq 2 $\to$ eq 2 $-$ 2 eq 1, eq 3 $\to$ eq 3 $-$ 3 eq 1, eq 4 $\to$ eq 4 $-$ 4 eq 1, eq 3 $\to$ eq 3 $-$ eq 2, eq 4 $\to$ eq 4 $-$ eq 2, eq 2 $\to$ $-\frac{1}{5}$ eq 2 gives $x_2 = -\frac{1}{5}$, $x_1 = \frac{7}{5}$; unique   (m) For $c = 10$ we find that the system is inconsistent; for $c = 11$ we obtain unique solution $z = 1$, $y = 0$, $x = 2$.   (o) Unique solution $x_4 = 2$, $x_3 = 1$, $x_2 = 1$, $x_1 = 2$.
(q) 1-parameter family $x_5 = \alpha$, $x_4 = 0$, $x_3 = -\alpha$, $x_2 = 0$, $x_1 = -\alpha$

3. (q) eq 1 $\leftrightarrow$ eq 2, eq 2 $\to$ eq 2 $-$ 2 eq 1, eq 3 $\to$ eq 3 $-$ eq 1, eq 4 $\to$ eq 4 $-$ 2 eq 1, eq 2 $\to$ $-$ eq 2, eq 4 $\leftrightarrow$ eq 3, eq 4 $\to$ eq 4 $-$ 2 eq 3, eq 4 $\to$ $-\frac{1}{3}$ eq 4, eq 2 $\to$ eq 2 $-$ eq 4, eq 2 $\to$ eq 2 $+$ 2 eq 3, eq 1 $\to$ eq 1 $+$ eq 3, eq 1 $\to$ eq 1 $-$ eq 2, eq 4 $\to$ eq 4 $-$ 2 eq 3, eq 4 $\to$ $-\frac{1}{3}$ eq 4, eq 2 $\to$ eq 2 $+$ eq 4, eq 2 $\to$ eq 2 $+$ 2 eq 3, eq 1 $\to$ eq 1 $+$ eq 3, eq 1 $\to$ eq 1 $-$ eq 2 gives 1-parameter family $x_5 = \alpha$, $x_4 = 0$, $x_3 = -\alpha$, $x_2 = 0$, $x_1 = -\alpha$

5. (a) $0x_1 + 0x_2 + 0x_3 + 0x_4 = 6$ (b) Not possible   (c) Not possible
NOTE: It might appear that the equation $x_3 + x_4 = 2$, for example, has a 1-parameter family of solutions but it actually has the 3-parameter family $x_4 = \alpha_1$, $x_3 = 2 - \alpha_1$, $x_2 = \alpha_2$, $x_1 = \alpha_3$.
(d) Not possible   (e) $x_1 - x_2 + x_4 = 5$ gives $x_4 = \alpha_1$, $x_3 = \alpha_2$, $x_2 = \alpha_3$, $x_1 = 5 - \alpha_1 + \alpha_3$ (f) $0x_1 + 0x_2 + 0x_3 + 0x_4 = 0$

6. (a) $x_1 + x_2 + x_3 + x_4 = 1$, $x_1 + x_2 + x_3 + x_4 = 2$ (b) Not possible   (c) Not possible   (d) $x_1 + x_2 + x_3 + x_4 = 1$, $x_2 + x_3 + x_4 = 0$ (e) $x_1 + 2x_2 - x_3 + x_4 = 5$, $0x_1 + 0x_2 + 0x_3 + 0x_4 = 0$ (f) Let all the a_{ij}'s and all the c_j's be zero in (1).

10. Can have unique solution. Can be inconsistent. Cannot have 2-parameter family of solutions. Can have a 14-parameter family of solutions [if all a_{ij}'s and c_j's are zero in (1)]. Cannot have a 16-parameter family of solutions, since at most the 14 unknowns are arbitrary.

12. (a) If we let $y_1 = x_1^2$, $y_2 = x_2^2$, $y_3 = x_3^2$, then we have a linear system in y_1, y_2, y_3. Solve by Gauss elimination and obtain $y_3 = 0$, $y_2 = 10$, $y_1 = 9$. Then $x_1 = \pm 3$, $x_2 = \pm\sqrt{10}$, $x_3 = 0$, combinations of which give four solutions.

13. (a) $\lambda = 1$ gives $y = \alpha$, $x = -\alpha$; $\lambda = 3$ gives $y = \beta$, $x = \beta$
(c) $\lambda = 0$ gives $y = \alpha$, $x = 2\alpha$; $\lambda = -7$ gives $y = \beta$, $x = \beta/4$
(e) $\lambda = 1$ gives $z = 0$, $y = 0$, $x = \alpha$; $\lambda = 2$ gives $z = \beta$, $y = \beta$, $x = 2\beta$

14. (a) One step at a time: eq 2 $\to$ eq 2 $-$ 2 eq 1 gives the system $x_1 - 2x_2 = 0$ and $0 = 0$. Then eq 1 $\to$ eq 1 $-\frac{1}{2}$ eq 2 gives $x_1 - 2x_2 = 0$ and $0 = 0$ once again.

16. (a) $i_1 = 8/3$, $i_2 = i_3 = 4/3$ (e) No solution (i.e., inconsistent). Physically, this case corresponds to a "short circuit" with $i_1 = i_3 = \infty$ (mathematically, we cannot accept ∞'s).
(f) Nonunique solution $i_3 = \alpha$, $i_2 = 4 - \alpha$, $i_1 = 4$. Physically, since $R_2 = R_3 = 0$ the current $i_1 = 4$ can split into i_2 plus i_3 in an infinite number of ways.

17. (a) Obtain $T_1 \sin\theta_1 + T_2 \sin\theta_2 = F$ (vertical) and $T_1 \cos\theta_1 - T_2 \cos\theta_2 = 0$ (horizontal). If $\theta_1 = \theta_2 = \pi/2$ these become $T_1 + T_2 = F$ and $0 = 0$ and there is a nonunique solution $T_2 = \alpha$, $T_1 = F - \alpha$. If $\theta_1 = \theta_2 = 0$ the system becomes $0 = F$ and $T_1 - T_2 = 0$ with no (finite) solution.
NOTE: In the case $\theta_1 = \theta_2 = 0$ it makes physical sense that $T_1 = T_2 = \infty$ but the latter is not acceptable mathematically for how can we argue that the equations $\infty(0) + \infty(0) = F$

and $\infty - \infty = 0$ are satisfied? However, we *can* rigorously show that $T_1 = T_2$ tends to ∞ as θ_1 and θ_2 both tend to 0. **(b)** Obtain $0.707T_1 + 0.866T_2 + 0.5T_3 = F$ (vertical) and $0.707T_1 - 0.5T_2 - 0.866T_3 = 0$.   **(c)** Put (17.4) into the two equilibrium equations and obtain

$$0.71\frac{k_1}{\sqrt{2}}(x - y) + 0.87\left(-\frac{k_2}{2}\right)(x + \sqrt{3}y)$$
$$+ 0.5\left(-\frac{k_3}{2}\right)(\sqrt{3}x + y) = F,$$
$$0.71\frac{k_1}{\sqrt{2}}(x - y) - 0.5\left(-\frac{k_2}{2}\right)(x + \sqrt{3}y)$$
$$- 0.87\left(-\frac{k_3}{2}\right)(\sqrt{3}x + y) = 0$$

or, with $k_1 = k_2 = k_3 = 100$, $-37x - 151y = F$ and $151x + 37y = 0$ with unique solution $x \approx 0.0017F$, $y \approx -0.0070F$. Then (17.4) gives $T_1 \approx 0.62F$, $T_2 \approx 0.52F$, $T_3 \approx 0.20F$.

Section 4.6

2. (a) Yes **(c)** Yes, in fact any two of these span $\mathbb{R}^2$ without any help from the other two. **(e)** Yes **(g)** No **(i)** No **(k)** Yes **(m)** No

3. (a) $(-4, 0, 1)$, $(2, 1, 0)$ HINT: Solve the equation for x_3, x_2, x_1 and express answer in vector form. But this is not the only possible solution method, nor is the answer unique. **(c)** $(-5, 0, 1)$, $(0, 1, 0)$ **(e)** $(-2, 0, 1)$, $(0, 1, 0)$ **(g)** $(1, 0, 1)$, $(-1, 1, 0)$

4. (a) Yes **(c)** No **(e)** Yes **(g)** Yes **(i)** No

5. (a) $(1/\sqrt{5}, 2/\sqrt{5})$, $(2/\sqrt{5}, -1/\sqrt{5})$ **(c)** $(1/\sqrt{2}, -1/\sqrt{2}, 0)$, $(1/\sqrt{6}, 1/\sqrt{6}, 2/\sqrt{6})$ **(e)** $(1/\sqrt{3}, 1/\sqrt{3}, 0, 1/\sqrt{3})$, $(1/\sqrt{3}, -1/\sqrt{3}, 1/\sqrt{3}, 0)$ NOTE: These answers are by no means unique.

Section 4.7

1. (a) No

2. (a) $(3, 4) = 2(1, 1) + (1, 2)$ as can be seen either by inspection or by setting $\alpha_1(1, 1) + \alpha_2(1, 2) + \alpha_3(3, 4) = (0, 0)$ and solving the two equations in the three unknown α's for $\alpha_1, \alpha_2, \alpha_3$. Those steps give $\alpha_3 = \beta$, $\alpha_2 = -\beta$, $\alpha_1 = -2\beta$ where β is arbitrary; the choice $\beta = -1$, say, gives the result stated above. **(c)** $(-3, 3) = -3(1, -1) + 0(4, 2)$

3. (a) LD: $(1, 3) - (2, 0) - (-1, 3) + 0(7, 3) = (0, 0)$ **(c)** LI by Theorem 4.7.2 **(e)** LD: $3(0, 0, 2) - 2(0, 0, 3) + 0(2, -1, 5) + 0(1, 2, 4) = (0, 0, 0)$ **(g)** LI **(i)** LD: $0(1, 3, 0) + 0(0, 1, -1) + 5(0, 0, 0) = (0, 0, 0)$ **(k)** LD: $2(1, -3, 0, 2, 1) + (-2, 6, 0, -4, -2) = (0, 0, 0, 0, 0)$ **(m)** LI by Theorem 4.7.2 **(o)** LI **(q)** LI

5. False. For example, let $\mathbf{u}_1 = (1, 0)$, $\mathbf{u}_2 = (1, 1)$, $\mathbf{u}_3 = (2, 1)$. Then $\{\mathbf{u}_1, \mathbf{u}_2\}$, $\{\mathbf{u}_1, \mathbf{u}_3\}$, $\{\mathbf{u}_2, \mathbf{u}_3\}$ are all LI by Theorem 4.7.2, but $\{\mathbf{u}_1, \mathbf{u}_2, \mathbf{u}_3\}$ is LD because $\mathbf{u}_3 = \mathbf{u}_1 + \mathbf{u}_2$.

Section 4.8

1. (a) Yes **(c)** No. For instance, (1) fails because in general $\mathbf{u} + \mathbf{v} = (u_1 - v_1, \ldots, u_n - v_n)$ and $\mathbf{v} + \mathbf{u} = (v_1 - u_1, \ldots, v_n - u_n)$ are not the same, (2) fails because in general $(\mathbf{u}+\mathbf{v})+\mathbf{w} = (u_1-v_1-w_1, \ldots, u_n-v_n-w_n)$ and $\mathbf{u}+(\mathbf{v}+\mathbf{w}) = (u_1-v_1+w_1, \ldots, u_n-v_n+w_n)$ are not the same, (4) fails because in general $\mathbf{u} + (-\mathbf{u}) = (2u_1, \ldots, 2u_n) \neq (0, \ldots, 0)$, (6) fails because in general $(\alpha+\beta)\mathbf{u} = ((\alpha+\beta)u_1, \ldots, (\alpha+\beta)u_n)$ and $\alpha\mathbf{u} + \beta\mathbf{u} = ((\alpha - \beta)u_1, \ldots, (\alpha - \beta)u_n)$ are not the same. **(e)** No, (6) is not satisfied in general.

4. No. For instance, it would not contain a zero vector.

5. (a) $\mathbf{x} = \alpha_1(-4, 0, 1) + \alpha_2(1, 1, 0)$ **(c)** $\mathbf{x} = \alpha_1(-1, 0, 1)$ **(e)** $\mathbf{x} = \alpha_1(-2, 0, 2, 0, 1) + \alpha_2(-1, 0, 3, 1, 0) + \alpha_3(1, 1, 0, 0, 0)$

6. (a) No, the set does not contain a negative inverse $-\mathbf{u}$. Also, it is not closed under scalar multiplication since $\alpha\mathbf{u}$ is not in the space if α is negative.

7. No, the set must contain a zero vector to be a vector space.

9. $\mathbf{u}+(-1)\mathbf{u} = 1\mathbf{u}+(-1)\mathbf{u}$ [by (8)] $= (1+(-1))\mathbf{u}$ [by (6)] $= 0\mathbf{u} = \mathbf{0}$ [by (16)]. Then $(-1)\mathbf{u}$ must equal $-\mathbf{u}$ by item (iii) in Definition 4.8.1.

15. (a) Yes **(b)** No. Recall that we elevate the properties (14 a,b,c) in Section 4.4 for n-space to axioms (i.e., requirements) for generalized vector space. Of these, (14a) is satisfied but (14b) and (14c) are not. For example, let $\mathbf{u} = (2, 1, 5)$ and $\mathbf{v} = (0, 3, 2)$. Then $||\mathbf{u}|| = 1$, $||\mathbf{v}|| = 0$, and $||\mathbf{u} + \mathbf{v}|| = 2$, so $||\mathbf{v}|| = 0$ even though $\mathbf{v} \neq \mathbf{0}$, in violation of (14b), and (14c) is violated as well.

17. (a) $\mathbf{u} \cdot \mathbf{v} = 0$, $||\mathbf{u}|| = \sqrt{11}$, $||\mathbf{v}|| = \sqrt{46}$: $0 \leq \sqrt{11}\sqrt{46}$ **(c)** $\mathbf{u} \cdot \mathbf{v} = 30$, $||\mathbf{u}|| = \sqrt{15}$, $||\mathbf{v}|| = \sqrt{60}$: $30 \leq \sqrt{15}\sqrt{60}$ **(e)** $\mathbf{u} \cdot \mathbf{v} = 16$, $||\mathbf{u}|| = \sqrt{367/12}$, $||\mathbf{v}|| = \sqrt{111/10}$: $16 \leq \sqrt{367/12}\sqrt{111/10}$

Section 4.9

1. (a) No. From Theorems 4.9.2 and 4.9.3 we know that a basis for $\mathbb{R}^n$ must contain n vectors so a basis for $\mathbb{R}^2$ must contain two vectors, no more and no less.   **(c)** No. Same reasoning as in (a).   **(e)** Yes **(g)** Yes **(i)** Yes **(k)** No **(m)** No. The set includes the zero vector so it can't be LI. Also, there are five vectors whereas a basis for $\mathbb{R}^4$ must have four.   **(o)** Yes. One way to proceed is to write $\alpha(1, 1, 2) + \beta(4, -2, -1) = \gamma(2, -4, -5) + \delta(1, -1, -1)$, which gives three scalar equations on $\alpha, \beta, \gamma, \delta$. Gauss elimination shows that for any given γ and δ there is a unique solution for α and β.   **(q)** No

2. (*a*) $\mathbf{u} = 2\mathbf{e}_1 + \frac{13}{5}\mathbf{e}_2 + \frac{2}{5}\mathbf{e}_3$ (*c*) $\mathbf{u} = \frac{8}{7}\mathbf{e}_1 - \frac{2}{7}\mathbf{e}_2 - \frac{11}{35}\mathbf{e}_3$

3. (*a*) $\mathbf{u} = 2\sqrt{14}\,\hat{\mathbf{e}}_1 + \frac{13}{\sqrt{5}}\hat{\mathbf{e}}_2 + \frac{2\sqrt{70}}{5}\hat{\mathbf{e}}_3$ (*c*) $\mathbf{u} = \frac{8\sqrt{14}}{7}\hat{\mathbf{e}}_1 - \frac{2}{\sqrt{5}}\hat{\mathbf{e}}_2 - \frac{11\sqrt{70}}{35}\hat{\mathbf{e}}_3$

4. (*a*) $\mathbf{u} = \frac{1}{15}\mathbf{e}_1 + \frac{1}{3}\mathbf{e}_2 + \frac{1}{5}\mathbf{e}_4$ (*c*) $\mathbf{u} = \frac{3}{5}\mathbf{e}_1 + 5\mathbf{e}_3 + \frac{4}{5}\mathbf{e}_4$
(*e*) $\mathbf{u} = -\frac{23}{30}\mathbf{e}_1 + \frac{7}{6}\mathbf{e}_2 + 2\mathbf{e}_3 + \frac{1}{5}\mathbf{e}_4$ (*g*) $\mathbf{u} = -\frac{3}{2}\mathbf{e}_1 + \frac{3}{2}\mathbf{e}_2$

5. We know from Theorems 4.9.2 and 4.9.3 that *n* LI vectors in $\mathbb{R}^n$ are a basis for $\mathbb{R}^n$, and surely a set of orthogonal vectors is LI; after all, we can think of orthogonality as the extreme case of linear independence. Nevertheless, let us show that an orthogonal set, say $\{\mathbf{e}_1, \ldots, \mathbf{e}_n\}$, is LI. Write $\alpha_1\mathbf{e}_1 + \cdots + \alpha_n\mathbf{e}_n = \mathbf{0}$. If we dot $\mathbf{e}_1$ into both sides we find that $\alpha_1 = 0$, if we dot $\mathbf{e}_2$ into both sides we find that $\alpha_2 = 0$, and so on, so $\{\mathbf{e}_1, \ldots, \mathbf{e}_n\}$ is LI. We leave the Gauss elimination of (28) to you.

6. (*a*) Yes, if and only if $\dim \mathcal{S} = k$ (*b*) A basis for $\mathcal{S}$ must span $\mathcal{S}$ and be LI. By definition $\{\mathbf{e}_1, \ldots, \mathbf{e}_k\}$ spans $\mathrm{span}\{\mathbf{e}_1, \ldots, \mathbf{e}_k\}$. And since $\{\mathbf{e}_1, \ldots, \mathbf{e}_k\}$ is orthogonal it is surely LI (see Exercise 5, above).

8. (*a*) 1 (*c*) 3 (*e*) 2 (*g*) 2

9. (*a*) 3 (*c*) 2 (*e*) 2 (*g*) 1

10. (*a*) 2 (*c*) 1 (*e*) 3

11. Following the idea outlined in Exercise 6, seek $\mathbf{w}_1 = \mathbf{v}_1$, $\mathbf{w}_2 = \mathbf{v}_2 + \alpha\mathbf{v}_1$, and $\mathbf{w}_3 = \mathbf{v}_3 + \beta\mathbf{v}_2 + \gamma\mathbf{v}_1$, and solve for α, β, γ by setting $\mathbf{w}_2 \cdot \mathbf{w}_1 = 0$, $\mathbf{w}_3 \cdot \mathbf{w}_1 = 0$, and $\mathbf{w}_3 \cdot \mathbf{w}_2 = 0$. Simple as it sounds this proves to be messy. Rather, we suggest the following. Setting $\mathbf{w}_2 \cdot \mathbf{w}_1 = 0$ solve for α and thus obtain $\mathbf{w}_2$. To find $\mathbf{w}_3$ observe that the linear combination $\beta\mathbf{v}_2 + \gamma\mathbf{v}_1$ is equivalent to a linear combination of $\mathbf{w}_2$ and $\mathbf{w}_1$. Thus, in place of $\mathbf{w}_3 = \mathbf{v}_3 + \beta\mathbf{v}_2 + \gamma\mathbf{v}_1$ seek $\mathbf{w}_3 = \mathbf{v}_3 + \beta\mathbf{w}_2 + \gamma\mathbf{w}_1$ and solve for β and γ by setting $\mathbf{w}_3 \cdot \mathbf{w}_2 = 0$ and $\mathbf{w}_3 \cdot \mathbf{w}_1 = 0$ (and using the fact that $\mathbf{w}_2 \cdot \mathbf{w}_1 = 0$). That step gives $\mathbf{w}_3 = \mathbf{v}_3 - (\mathbf{v}_3 \cdot \hat{\mathbf{w}}_1)\hat{\mathbf{w}}_1 - (\mathbf{v}_3 \cdot \hat{\mathbf{w}}_2)\hat{\mathbf{w}}_2$. Finally, normalize $\mathbf{w}_1, \mathbf{w}_2,$ and $\mathbf{w}_3$.

12. (*a*) $\hat{\mathbf{w}}_1 = (1, 0)$, $\hat{\mathbf{w}}_2 = (0, 1)$ (*c*) $\hat{\mathbf{w}}_1 = (1, 0, 0)$, $\hat{\mathbf{w}}_2 = (0, 1, 0)$, $\hat{\mathbf{w}}_3 = (0, 0, 1)$ (*e*) $\hat{\mathbf{w}}_1 = \frac{1}{\sqrt{3}}(1, 1, 1)$, $\hat{\mathbf{w}}_2 = \frac{1}{\sqrt{42}}(5, -1, -4)$ (*g*) $\hat{\mathbf{w}}_1 = \frac{1}{\sqrt{6}}(2, 1, 1, 0)$, $\hat{\mathbf{w}}_2 = \frac{1}{5}(-1, 4, -2, 2)$

13. (*b*) $\mathbf{e}_1^* = (1, -1)$, $\mathbf{e}_2^* = (0, 1)$, $\mathbf{u} = 2\mathbf{e}_1 + \mathbf{e}_2$ (*e*) $\mathbf{e}_1^* = (1, -1, 0)$, $\mathbf{e}_2^* = (0, 1, -1)$, $\mathbf{e}_3^* = (0, 0, 1)$, $\mathbf{u} = 5\mathbf{e}_1 - 6\mathbf{e}_2 + 5\mathbf{e}_3$, $\mathbf{v} = -2\mathbf{e}_2 + 2\mathbf{e}_3$, $\mathbf{w} = 7\mathbf{e}_1 - 5\mathbf{e}_2 + 3\mathbf{e}_3$

15. (*a*) $\mathbf{u} \approx (3/\sqrt{5})\hat{\mathbf{e}}_1 + (11/\sqrt{6})\hat{\mathbf{e}}_2 + 0\hat{\mathbf{e}}_3 = \left(\frac{64}{15}, 0, -\frac{19}{30}, 0, \frac{11}{6}\right)$, $\|\mathbf{E}\| = 4.0042$ (*c*) $\mathbf{u} \approx \sqrt{5}\hat{\mathbf{e}}_1 + \sqrt{6}\hat{\mathbf{e}}_2 + 4\hat{\mathbf{e}}_3$, $\|\mathbf{E}\| = 0$. Thus, $\mathbf{u}$ happens to lie in $\mathrm{span}\{\hat{\mathbf{e}}_1, \hat{\mathbf{e}}_2, \hat{\mathbf{e}}_3\}$ (*e*) $\mathbf{u} \approx \mathbf{0}$ (i.e., $\mathbf{u}$ happens to have zero projection onto $\mathrm{span}\{\hat{\mathbf{e}}_1, \hat{\mathbf{e}}_2, \hat{\mathbf{e}}_3\}$), $\|\mathbf{E}\| = 2$.

16. $\mathbf{u} \approx (-\frac{4}{3}, -\frac{4}{3}, 0, \frac{4}{3})$, $(\frac{1}{3}, -3, -\frac{5}{3}, \frac{4}{3})$, $(4, -3, 2, 5)$, $(4, -2, 1, 6)$, respectively. $\|\mathbf{E}\| = 7.19, 6.58, 1.73, 0$, the zero occurring because the full set $\{\hat{\mathbf{e}}_1, \ldots, \hat{\mathbf{e}}_4\}$ is a *basis* for $\mathbb{R}^4$.

CHAPTER 5

Section 5.2

NOTE: For compactness we will use the *Maple* array notation for matrices. For instance, array([[1, 2], [3, 4]]) denotes the matrix $\begin{bmatrix} 1 & 2 \\ 3 & 4 \end{bmatrix}$.

1. $\mathbf{AB}$ = array([[0, 6], [10, −12], [5, 19]]), $\mathbf{Ax}$ = array([[9], [−7], [34]]), $\mathbf{Bx}$ = array([[17], [6]]), $\mathbf{yB}$ = array([[−5, 5]]), $\mathbf{B}^2$ = array([[25, −7], [0, 4]]), $\mathbf{xy}$ = array([[−4, 8], [−3, 6]]), $\mathbf{yx}$ = array([[2]]). The others are not defined.

2. (*a*) Not defined (*c*) 6×3 (*e*) 6×1 (*g*) 4×1 (*i*) Not defined

4. No, because $\mathbf{A} − c$ is not defined. However, we could write $\mathbf{Ax} = c\mathbf{x}$ as $\mathbf{Ax} = c\mathbf{Ix}$ where $\mathbf{I}$ is an $n \times n$ identity matrix. Then we *can* write $(\mathbf{A} − c\mathbf{I})\mathbf{x} = \mathbf{0}$, where $\mathbf{0}$ is $n \times 1$.

5. (*a*) $(\mathbf{A} + \mathbf{B})^2 = (\mathbf{A} + \mathbf{B})(\mathbf{A} + \mathbf{B}) = \mathbf{A}^2 + \mathbf{AB} + \mathbf{BA} + \mathbf{B}^2 = \mathbf{A}^2 + 2\mathbf{AB} + \mathbf{B}^2$ only if $\mathbf{AB} = \mathbf{BA}$ which, in general, is not true.

6. (*a*) Yes

7. (*a*) $(2\mathbf{A} + \mathbf{B})(\mathbf{A} + 2\mathbf{B}) = (2\mathbf{A})(\mathbf{A} + 2\mathbf{B}) + \mathbf{B}(\mathbf{A} + 2\mathbf{B}) = (2\mathbf{A})\mathbf{A} + (2\mathbf{A})(2\mathbf{B}) + \mathbf{BA} + \mathbf{B}(2\mathbf{B}) = 2\mathbf{A}^2 + 4\mathbf{AB} + \mathbf{BA} + 2\mathbf{B}^2$, using (30c), (30d), and (30a), respectively.

8. (*a*) If we work out the first few powers we find that each element of $\mathbf{A}^k$ is 2^{k-1} (which could then be proved by induction). Thus, $\mathbf{A}^{100} = \mathrm{array}([[2^{99}, 2^{99}], [2^{99}, 2^{99}]])$. (*c*) $\mathbf{0}$ (*e*) array([[0, 27], [0, 27]]) (*g*) $\mathbf{0}$

10. (*a*) Since $\mathbf{Ax} = [x_1 − 3x_4]$ is 1×1 and $\mathbf{x}$ is 4×1 it follows that $\mathbf{A}$ is 1×4, so $\mathbf{A} = [a_{11}, a_{12}, a_{13}, a_{14}]$. If we multiply the latter into $\mathbf{x} = [x_1, \ldots, x_4]^T$ and compare the result with $[x_1 − 3x_4]$ we find that $a_{11} = 1$, $a_{12} = a_{13} = 0$, $a_{14} = -3$, so $\mathbf{A} = [1, 0, 0, -3]$. (*c*) Since $\mathbf{Ax}$ is 3×1 and $\mathbf{x}$ is 4×1 it follows that $\mathbf{A}$ is 3×4. If we multiply a 3×4 matrix $\mathbf{A} = \{a_{ij}\}$ into $\mathbf{x} = [x_1, \ldots, x_4]^T$ and compare the result with the column vector that is given, we can determine the components a_{ij} of $\mathbf{A}$. We find that $\mathbf{A} = \mathrm{array}([[1, 0, 0, 0], [0, 1, 1, 0], [0, 0, 1, 1]])$ (*e*) $\mathbf{A}$ is a 4×4 identity matrix.

11. (*a*) E.g., $\mathbf{A}$ = array([[0, 4], [0, 0]]), $\mathbf{B}$ = array([[1, 0], [0, 0]]) (*c*) E.g., $\mathbf{A}$ = array([[1, 1]]), $\mathbf{B}$ = array([[1, −1], [−2, 2], [0, 0], [5, −5]])

12. (*a*) The partitioning is suitable for the calculation of $\mathbf{A}^2$ and $\mathbf{AB}$. The results are $\mathbf{A}^2$ = array([[a, b], [c, d]]) where a = array([[$-1, -2$], [$1, 1$]]), b = array([[-6], [-1]]), c = array([[$32, 6$]]), d = array([[11]]), and $\mathbf{AB}$ = array([[e, f], [g, h]]) where e = array([[$3, 16$], [$3, 5$]]), f = array([[-3], [2]]), g = array([[$27, 16$]]), h = array([[38]]). (*c*) The partitioning is conformable for the calculation of $\mathbf{A}^2$ but not for the calculation of $\mathbf{AB}$.

14. (*a*) No

15. (*c*) Yes   (*d*) Work out the products on the left-hand and right-hand sides and use $\mathbf{A}^p = \mathbf{0}$. By cancelation, each side will reduce to $\mathbf{I}$.

16. (*b*) $\mathbf{A}$ = array([[$1, 0$], [$4, -1$]])

21. (*a*) $\mathbf{B} = \mathbf{0}$ (*c*) $\mathbf{B}$ = array([[$0, 0$], [α, β]]) where α, β are arbitrary

23. (*a*) The commands
```
with(linalg):
A:=array([[2,-1]],[3,0],[1,4]]);
B:=array([[5,3,25],[2,.1,-6]]);
C:=array([[9,1,-1],[2,0,7],[0,4,6]]);
evalm((A&*B)^3+5*C^2);
```
give the result array([[24124.0, 11054.35, 72017.0], [36967.5, 16633.5, 102507.5], [15836.5, 5453.10, 19014.5]])

Section 5.3

1. (*a*) $\mathbf{x}^T\mathbf{y} = 3 - 6 = -3$, $\mathbf{xy}^T$ = array([[$3, 6$], [$-3, -6$]])

2. (*a*) $(\mathbf{ABC})^T = (\mathbf{A}(\mathbf{BC}))^T = (\mathbf{BC})^T\mathbf{A}^T = \mathbf{C}^T\mathbf{B}^T\mathbf{A}^T$; similarly for $(\mathbf{ABCD})^T$

3. HINT: For simplicity let $\mathbf{A}$ and $\mathbf{B}$ be 2×2. If, for that case, the hypothesis proves to be true then you will not have gained much, but if it proves false then you will have disproved it.

6. $\mathbf{A}_1^T = \frac{1}{2}(\mathbf{A} + \mathbf{A}^T)^T = \frac{1}{2}(\mathbf{A}^T + \mathbf{A}^{TT}) = \frac{1}{2}(\mathbf{A}^T + \mathbf{A}) = \mathbf{A}_1$ so $\mathbf{A}_1$ is symmetric; similarly, we find that $\mathbf{A}_2^T = -\mathbf{A}_2$ so $\mathbf{A}_2$ is skew-symmetric.

7. (*a*) $\mathbf{A}_1$ = array([[$3, 3/2$], [$3/2, -5$]]) and $\mathbf{A}_2$ = array([[$0, 1/2$], [$-1/2, 0$]])

8. (*a*) array([[$6, -4$], [$-4, 1$]])   (*c*) array([[$4, 4, 3/2$], [$4, 1, -1$], [$3/2, -1, -1$]])

10. Remember, to prove a hypothesis we need to prove it in general, but to disprove it a single counterexample will suffice.

11. (*a*) Use
```
with(linalg):
A:=array([[4,1,2],[0,5,7]]);
B:=array([[1,-4,2],[8,1,4]]);
evalm(A&*transpose(B));
```

Section 5.4

2. (*a*) About row 1, det $= 1 - 4 + 3 = 0$; about row 3, det $= -4 + 8 - 4 = 0$; about column 3, det $= 3 + 1 - 4 = 0$ (*g*) About row 1, det $= 128 - 0 + 4 - 0 = 132$; about row 4, det $= 4 + 20 - 24 + 132 = 132$; about column 4, det $= 0 + 0 + 0 + 132 = 132$

6. (*a*) The steps $\mathbf{r}_2 \rightarrow \mathbf{r}_2 + (-1)\mathbf{r}_1$ followed by $\mathbf{r}_3 \rightarrow \mathbf{r}_3 + (-1)\mathbf{r}_1$ (which, according to property D1, preserve the determinant) give a matrix with two proportional rows. Hence det $= 0$ by D5.   (*c*) By cofactor expansion det $= -abd$. NOTE: The minus sign seems to violate D3 but observe that this matrix is *not* triangular since it is not zero above or below the *main* diagonal.

9. (*b*) Yes   (*c*) Yes   (*d*) In 2(i) $\mathbf{A}_1$ is 1×1, $\mathbf{A}_2$ is 2×2, and $\mathbf{A}_3$ is 1×1. Then (9.1) gives det $= (a)(be - cd)(f)$. In 2(j) $\mathbf{A}_1$ is 3×3 and $\mathbf{A}_2$ is 1×1.

10. $\det(\mathbf{A}_1\mathbf{A}_2\mathbf{A}_3) = \det(\mathbf{A}_1(\mathbf{A}_2\mathbf{A}_3))$
$$= \det(\mathbf{A}_1)(\det(\mathbf{A}_2\mathbf{A}_3))$$
$$= (\det \mathbf{A}_1)(\det \mathbf{A}_2)(\det \mathbf{A}_3);$$
similarly for k's > 3.   NOTE: As convincing as this approach may be, it does not prove (10.1) for *all values of k*. To prove it for all values you can use induction.

14. (*a*) det $= 48 \neq 0$ so there are no common roots   (*b*) det $= 0$ so there is at least one common root

15. (*b*) $9t^2 \sin t + 3t^3 \cos t - 20$

Section 5.5

1. (*a*) 1   (*c*) 2   (*e*) 2   (*g*) 2   (*i*) 3   (*k*) 3

2. (*c*) Use
```
with(linalg):
A:=array([[5,7],[4,9]]);
rank(A);
```

3. No. For example, $(1, 0)$ and $(1, 1)$ have the same form and rank but we cannot obtain one from the other by elementary row operations.

5. (*a*) They cannot be row equivalent because their ranks are different: 2 and 1, respectively.

6. HINT: Apply Gauss-Jordan reduction to each of the two matrices and compare the results. (*a*) If you do that, it will be evident from the results that the matrices are not row equivalent; they are not even of the same rank.   (*b*) No again

7. (*a*) $r = 2 < 3$; LD   (*c*) $r = 3$; LI

8. Theorem 5.5.1 says that the number of LI row vectors equals the number of LI column vectors equals the rank. Since the transpose merely exchanges row and column vectors, $r(\mathbf{A}^T) = r(\mathbf{A})$.

12. (*a*) Four, such as $H_2 \rightleftarrows 2H$, $O_2 \rightleftarrows 2O$, $OH \rightleftarrows H + O$, and $H_2O \rightleftarrows 2H + O$

Section 5.6

1. (*a*) array($[[d/e, -b/e], [-c/e, a/e]]$) where e is $ad - bc$
(*d*) array($[[1/4, 1/4], [1/2, -1/2]]$)
(*g*) array($[[0, 1/2, -5/6], [1, 0, 0], [0, 0, 1/3]]$)   (*j*) Not invertible   (*q*) array($[[\cos\theta, \sin\theta], [-\sin\theta, \cos\theta]]$)

3. (*c*) Use
```
with(linalg):
A:=array([[0,1],[1,0]]);
inverse(A);
```

5. (*a*) $x_1 = 24/13$, $x_2 = -6/13$ (*c*) $x_1 = 3$, $x_2 = -14/5$ (x_3 wasn't asked for)   (*e*) $x_1 = 4/5$, $x_2 = -3/5$

6. (*a*) HINT: Use the fact that the given $\mathbf{A}^{-1}$ matrix is not itself invertible.

7. (*a*) $\mathbf{A} = (\mathbf{A}^{-1})^{-1} = $ array($[[2/3, -1/3], [1, -1]]$)   (*c*) $\mathbf{A} = (\mathbf{A}^{-1})^{-1} = $ array($[[1/2, -1/2, 1/2], [-1/2, 1/2, 1/2], [3/2, -1/2, -3/2]]$)

8. Unique solution $\mathbf{x} = [8, 23, -2]^T$

9. No, if $\mathbf{A}$ were invertible then the solution $\mathbf{x} = \mathbf{A}^{-1}\mathbf{c}$ would be unique, whereas the solution given is not unique due to the arbitrary α's.

11. (*a*) With the given matrix equal to $\mathbf{I} - \mathbf{A}$ we identify $\mathbf{A}$ as array($[[0, -2, -3], [0, 0, -8], [0, 0, 0]]$). Then (10.1) gives the inverse as array($[[1, 2, 3], [0, 1, 8], [0, 0, 1]]$).
(*c*) Inverse $=$ array($[[2, 1, 0, 0], [1, 2, 1, 0], [0, 1, 2, 1], [0, 0, 1, 2]]$)

12. (*a*) array($[[1/2, -4/3, 11/4], [0, 1/3, -1], [0, 0, 1/4]]$)

16. With $k_{12} = k_{23} = 0$ the middle row and middle column of $\mathbf{K}$, in (36), are zero. Thus, the second scalar equation in (36) is $0x_1 + 0x_2 + 0x_3 = f_2$ so the system has no solution if $f_2 \neq 0$. This result makes sense physically because (see Fig. 1) if $k_{12} = k_{13} = 0$ and $f_2 \neq 0$ then the middle mass will accelerate rather than be in static equilibrium. On the other hand, if f_2 *is* zero then there is a nonunique solution. Specifically, x_1 and x_3 can be found uniquely from (36) but $x_2 = \alpha$ is arbitrary. Of course, if $k_{12} = k_{23} = f_2 = 0$ then the middle mass is free to sit wherever it likes!

17. (*a*) $\det \mathbf{R} = R_1 R_2 + R_1 R_3 + R_2 R_3 > 0$ (*b*) $i_1 = E(R_2 + R_3)/\det \mathbf{R}$, $i_2 = ER_3/\det \mathbf{R}$, $i_3 = ER_2/\det \mathbf{R}$

19. (*a*) $\mathbf{L} = $ array($[[1, 0], [4, 1]]$) and $\mathbf{U} = $ array($[[2, 3], [0, -13]]$). Then (51a) gives $\mathbf{y} = [-4, 26]^T$ and (51b) gives $\mathbf{x} = [1, -2]^T$.   (*c*) $\mathbf{L} = $ array($[[1, 0, 0], [1, 1, 0], [4, -6, 1]]$), $\mathbf{U} = $ array($[[2, 5, 1], [0, 3, -1], [0, 0, -8]]$). Then (51a) gives $\mathbf{y} = [0, -7, -32]^T$ and (51b) gives $\mathbf{x} = [1/2, -1, 4]^T$.

20. (*a*) $\mathbf{L} = $ array($[[1, 0], [2, 1]]$) and $\mathbf{U} = $ array($[[1, 3], [0, 0]]$). Then (51a) gives $\mathbf{y} = [4, -5]^T$ but (51b) has no solution.

Section 5.7

1. (*a*) $n = 4$, $r(\mathbf{A}) = r(\mathbf{A} \mid \mathbf{c}) = 1$, $N = n - r = 3$-parameter family of solutions   (*c*) $n = 2$, $r(\mathbf{A}) = r(\mathbf{A} \mid \mathbf{c}) = 2$, $n = r$, so unique solution   (*e*) $n = 3$, $r(\mathbf{A}) = r(\mathbf{A} \mid \mathbf{c}) = 2$, $N = 3 - 2 = 1$-parameter family of solutions   (*g*) $n = 3$, $r(\mathbf{A}) = 2$, $r(\mathbf{A} \mid \mathbf{c}) = 3$, no solution   (*i*) $n = 4$, $r(\mathbf{A}) = r(\mathbf{A} \mid \mathbf{c}) = 3$, $N = 4 - 3 = 1$-parameter family of solutions   (*k*) $n = 3$, $r(\mathbf{A}) = 3$, $r(\mathbf{A} \mid \mathbf{c}) = 4$, no solution

4. (*a*) In words, the solution set is a straight line L, say, in 3-space. $\mathbf{x}_p$ is a vector from the origin to some point on L, $\mathbf{x}_1$ is a vector *along* L, and therefore, by suitable choice of α_1, $\mathbf{x}_p + \alpha_1 \mathbf{x}_1$ is a vector from the origin to any desired point on L. As α_1 varies over $-\infty < \alpha_1 < \infty$, $\mathbf{x}_p + \alpha_1 \mathbf{x}_1$ "generates" L.

Section 5.8

2. (*a*) Nonlinear   (*c*) Nonlinear   (*e*) Nonlinear

3. (*a*) We can expand any $\mathbf{x}$ in V as $\mathbf{x} = \alpha_1 \mathbf{v}_1 + \cdots + \alpha_n \mathbf{v}_n$. Then $\mathbf{F}(\mathbf{x}) = \mathbf{F}(\alpha_1 \mathbf{v}_1 + \cdots + \alpha_n \mathbf{v}_n) = \alpha_1 \mathbf{F}(\mathbf{v}_1) + \cdots + \alpha_n \mathbf{F}(\mathbf{v}_n) = \alpha_1 \mathbf{v}_1 + \cdots + \alpha_n \mathbf{v}_n = \mathbf{x}$, so $\mathbf{F} = \mathbf{I}$.   (*b*) The $n \times n$ identity matrix.

5. (*a*) Here we have $\mathbf{F} : \mathbb{R}^3 \to \mathbb{R}^2$. Write out $\mathbf{Ax} = \mathbf{c}$ and solve by Gauss elimination. We find that the system is consistent for every $\mathbf{c}$ in $\mathbb{R}^2$ so $\dim \mathbb{R} = 2$. For $\mathbf{c} = \mathbf{0}$ the solution is $\mathbf{x} = \alpha[0, -1, 1]^T$ so $\dim K = 1$. Finally, $\dim V = \dim \mathbb{R}^3 = 3$. $\mathbf{F}$ is onto because for every $\mathbf{c}$ in W (which is $\mathbb{R}^2$) there is an $\mathbf{x}$ such that $\mathbf{F}(\mathbf{x}) = \mathbf{Ax} = \mathbf{c}$, namely, $\mathbf{x} = [c_1 - c_2, 2c_2 - c_1, 0]^T + \alpha[0, -1, 1]^T$. However, it is not one-to-one because that $\mathbf{x}$ is not unique. Since it is not one-to-one it is not invertible. Since $\mathbf{c} = \mathbf{0}$ gives $\mathbf{x} = \alpha[0, -1, 1]^T$, a basis for K is $[0, -1, 1]^T$. Finally, R is the span of the column vectors of $\mathbf{A}$, a basis for which is $\{[2, 1]^T, [1, 1]^T\}$ or, for that matter, $\{[1, 0]^T, [0, 1]^T\}$ since R is all of $\mathbb{R}^2$.   (*c*) $\dim R = 3$, $\dim K = 0$, $\dim V = 3$. $\mathbf{F}$ is onto, one-to-one, and hence invertible. K is the zero vector space so it has no basis. A basis for R is $\{[1, 0, 0]^T, [0, 1, 0]^T, [0, 0, 1]^T\}$.   (*e*) Same as for (c).   (*g*) $\dim R = 3$, $\dim K = 0$, $\dim V = 3$, $\mathbf{F}$ is not onto but is one-to-one; not invertible. K is the zero vector space so it has no basis. A basis for R is given by the three column vectors in $\mathbf{A}$.

6. (*a*) To be onto we need $\dim R = r(\mathbf{A}) = 2$, and to be one-to-one we need $r(\mathbf{A}) = 2$. Thus, $\mathbf{F}$ will be onto, one-to-one, and hence invertible, if and only if $r(\mathbf{A}) = 2$. Examples: $\mathbf{A} = $ array($[[1, 2], [0, 3]]$) is onto and one-to-one, $\mathbf{B} = $ array($[[1, 2], [0, 0]]$) is neither. Impossible to have one and not the other because both onto and one-to-one occur if and only if $r(\mathbf{A}) = 2$.   (c) $\mathbf{F}$ cannot be onto because $\dim W = \dim \mathbb{R}^3 = 3$ whereas $\dim R = r(\mathbf{A})$ is at most 2 because $\mathbf{A}$ is 3×2. $\mathbf{A} = \text{array}([[1, 2], [0, 3], [0, 0]])$ is one-to-one but not onto, $\mathbf{B} = \text{array}([[1, 2], [0, 0]])$ is neither one-to-one nor onto.

7. (a) We need not verify *all* of the vector space requirements because V is itself a vector space; it will suffice to verify that K is *closed under scalar multiplication and vector addition and that it contains a zero vector* $\mathbf{0}_V$ *and a negative inverse* $-\mathbf{u}$. Let $\mathbf{u}$ be any vector in K, so that $\mathbf{F}(\mathbf{u}) = \mathbf{0}_W$. Then K is closed under scalar multiplication because $\mathbf{F}(\alpha\mathbf{u}) = \alpha\mathbf{F}(\mathbf{u})$(by the linearity of $\mathbf{F}$) $= \alpha\mathbf{0}_W = \mathbf{0}_W$, so if $\mathbf{u}$ is in K then so is $\alpha\mathbf{u}$ for any scalar α. (In particular, the choices $\alpha = -1$ and $\alpha = 0$ show that K contains the negative inverse $-\mathbf{u}$ of $\mathbf{u}$ and the zero vector $\mathbf{0}_V$, respectively.) Also, K is closed under vector addition because if $\mathbf{u}$ and $\mathbf{v}$ are in K then $\mathbf{F}(\mathbf{u} + \mathbf{v}) = \mathbf{F}(\mathbf{u}) + \mathbf{F}(\mathbf{v})$(by the linearity of $\mathbf{F}$) $= \mathbf{0}_W + \mathbf{0}_W = \mathbf{0}_W$ so $\mathbf{u} + \mathbf{v}$ is in K too.

8. Denoting $\mathbf{A} = \{a_{ij}\}$, $a_{ij} = v_i v_j$

9. (b) For example, $a_{11} = v_{11}^2 + v_{21}^2$ and $a_{32} = v_{12}v_{13} + v_{22}v_{23}$

11. (b) For example, $a_{11} = 2L_1^2 - 1$ and $a_{12} = 2L_1L_2$

12. (a) $(\mathbf{GF})(\mathbf{x})$ $=$ $[-20, -40]^T$, matrix $=$ array$([[2, 10, 6, -4], [4, 20, 12, -8]])$

14. (d) $\mathbf{F}(\mathbf{X}_p) = [1.2439, 3.5859, 2.2637, 1]^T$ and $\mathbf{F}(\mathbf{X}_e) = [1.5394, 3.7757, 1.3275, 1]^T$

CHAPTER 6
Section 6.2

1. (a) $z_1 + z_2$ and $z_1 z_2$ are given by (3) and (4). Then $z_2 + z_1 = (a_2 + ib_2) + (a_1 + ib_1) = (a_2 + a_1) + i(b_2 + b_1)$ is the same as the right side of (3) since $a_2 + a_1 = a_1 + a_2$ and $b_2 + b_1 = b_1 + b_2$. And $z_2 z_1 = (a_2 + ib_2)(a_1 + ib_1) = (a_2a_1 - b_2b_1) + i(a_2b_1 + b_2a_1)$ is the same as the right side of (4).

2. $|z_1 z_2| = |(x_1 x_2 - y_1 y_2) + i(x_1 y_2 + x_2 y_1)| = [(x_1 x_2 - y_1 y_2)^2 + (x_1 y_2 + x_2 y_1)^2]^{1/2}$ and $|z_1||z_2| = [(x_1^2 + y_1^2)(x_2^2 + y_2^2)]^{1/2}$. If we write these out we find that they are identical.

4. For (a)–(d) use (9); for (e) and (f) use (10).

8. (a) $2 - 11i$ (d) i (g) $23/41$ (j) $(bc - ad)/(c^2 + d^2)$

9. (a) 1 (c) $\sqrt{13}$

10. (a) left $= |6 + 2i| = \sqrt{40} = 6.32$, right $= \sqrt{13} + \sqrt{17} = 7.35$

11. (a) $-e^2$; i.e., $-e^2 + 0i$ (c) $(1/\sqrt{2}) + i(1/\sqrt{2})$ (e) $\cos 2 + i \sin 2$

12. (a) exp(2+Pi*I); does not work. Use the complex eval comand, evalc. Namely, evalc(exp(2+Pi*I));

Section 6.3

1. (a) No. Definition 6.3.1 shows linear dependence and linear independence to be mutually exclusive.

2. (a) Try expressing $x + 2$ as a linear combination of 1 and $3x - 5$: $x + 2 = a(1) + b(3x - 5) = (3b)x + (a - 5b)$ so $3b = 1$ and $a - 5b = 2$. Thus, $b = 1/3, a = 11/3$, so $x + 2 = \frac{11}{3}(1) + \frac{1}{3}(3x - 5)$. (d) E.g., $\cosh x = 1(e^x) - 1(\sinh x)$

3. (a) Use Theorem 6.3.2. The Wronskian determinant of $1, x, \ldots, x^n$ is upper triangular with nonzero integers on the main diagonal. For instance, if $n = 4$ then $W(x) = (1)(1)(2)(6) = 12$ which is nonzero for all x. Then Theorem 6.3.2 tells us that the set is LI (on $-\infty < x < \infty$). (Recall from the text that if an x interval is not specified then we will understand it to be $-\infty < x < \infty$.) (c) $W(x) = 2$ so LI by Theorem 6.3.2 (on $-\infty < x < \infty$) (e) $W(x) = -2\sinh x$ so LI (on $-\infty < x < \infty$) (g) LI (on $-\infty < x < \infty$) by Theorem 6.3.4 (i) LD because $4 = \frac{4}{3}(1 - x) + \frac{4}{3}(2 + x)$

4. (a) $W(x) = 2e^{6x}$ so LI by Theorem 6.3.3 (d) $W(x) = 0$ so LD by Theorem 6.3.3 (g) $W(x) = 2x^3$ so LI by Theorem 6.3.3. Why did we specify the interval $0 < x < \infty$? Observe that $p_1(x) = -3/x$ and $p_2(x) = 3/x^2$ are not continuous on any interval containing $x = 0$ so theorem does not apply if the interval contains $x = 0$.

7. False. For instance $\{4, 1 - x, 1 + x\}$ is LD because $4 = 2(1 - x) + 2(1 + x)$ whereas $\{4, 1 - x\}$, $\{4, 1 + x\}$, and $\{1 - x, 1 + x\}$ are LI.

8. No. Theorem 6.3.3 does not apply (if I is $-\infty < x < \infty$) because $p_1(x) = -4/x$ and $p_2(x) = 6/x^2$.

Section 6.4

1. (a) $p_1(x) = -3$ and $p_2(x) = 2$ are continuous on I, e^x and e^{2x} are LI by Theorem 6.3.4 and, as is readily verified, are solutions of the ODE. Thus, yes, it is a general solution. (c) No. The ODE is of second order but only the one LI solution $e^{-x} + e^{2x}$ is given. (e) No. Need three LI solutions and only two are given. (g) No, e^x and $x^2 e^x$ are not solutions of the ODE. (i) Yes (k) Yes

2. (a) No because they are not LI: $e^{3x} = \cosh 3x + \sinh 3x$ (c) No, they are not solutions (e) Yes, they are three LI solutions

3. (a) No, it is only one LI solution (b) Yes, for $0 < x < \infty$, or for $-\infty < x < 0$ for that matter, but not on $-\infty < x < \infty$ because x^{-2} does not satisfy the ODE at $x = 0$.

4. (*a*) No, x^2 is not a solution.   (*c*) Yes on $0 < x < \infty$, yes on $-\infty < x < 0$, yes on $6 < x < 10$, but no on $-\infty < x < \infty$ (because $x \ln |x|$ is not differentiable at $x = 0$ so it cannot satisfy the ODE at $x = 0$)

5. (*a*) No, the equation is of seventh order so any general solution must be a linear combination of *seven* LI solutions.

6. Yes, but not for all choices of C_1 and C_2; only if $C_1 + C_2 = 1$.

7. (*a*) It has a unique solution (on $-\infty < x < \infty$) according to Theorem 6.4.1.   (*d*) The coefficients $p_1(x) = 1$ and $p_2(x) = -1/x$ are continuous on the interval $-\infty < x < 0$ containing the initial point $x = -1$ so there is a unique solution (on $-\infty < x < 0$).

9. (*a*) Unique solution $y(x) = 0$ (*c*) Unique solution $y(x) = [(\sin 2 - 2\sin 1)/\sin 1]\cos x + [(2\cos 1 - \cos 2)/\sin 1]\sin x$ (*e*) Nonunique solution $y(x) = C\cos x$, C arbitrary   (*g*) No solution

10. (*a*) Unique solution $y(x) = -(2/\pi)x \sin x$ (*c*) Nonunique solution $y(x) = C \sin x$, C arbitrary

Section 6.5

1. Since $e^{r_1 x} \neq 0$ for all x, we can cancel that factor in the equaion $\alpha_1 e^{r_1 x} + \alpha_2 x e^{r_1 x} + \cdots + \alpha_k x^{k-1} e^{r_1 x} = 0$. Thus, the given set is LI if and only if the set $\{1, x, \ldots, x^{k-1}\}$ is. That the latter is LI was shown in Exercise 3(a) of Section 6.3.

2. (*a*) $y(x) = C_1 + C_2 e^{-5x}$ (*d*) $y(x) = 2e^{x-1} - e^{2(x-1)}$ (*g*) $y(x) = e^{2x}(\sin x + 2\cos x)$ (*j*) $y(x) = (3/\sqrt{2})e^{-x} \sin \sqrt{2}x$ (*m*) $y(x) = C_1 e^x + e^{-x}(C_2 \cos x + C_3 \sin x)$

4. (*a*) $y(x) = C_1 + C_2 x$, $y(x) = 2 - x$ (*d*) $y(x) = C_1 + C_2 x + C_3 e^{-5x}$, $y(x) = 1$ (*g*) $y(x) = (C_1 + C_2 x)e^x + C_3 e^{-x}$ (*j*) $y(x) = (C_1 + C_2 x)\cos 2x + (C_3 + C_4 x)\sin 2x$

6. (*a*) $(r - 2)(r - 6) = r^2 - 8r + 12$ so $y'' - 8y' + 12y = 0$, $y(x) = C_1 e^{2x} + C_2 e^{6x}$ (*d*) $y''' - 6y'' - y' + 30y = 0$, $y(x) = C_1 e^{-2x} + C_2 e^{3x} + C_3 e^{5x}$ (*g*) $y'''''' - 12y''''' + 49y''' - 76y'' + 48y' - 64y = 0$, $y(x) = (C_1 + C_2 x + C_3 x^2)e^{4x} + C_4 \cos x + C_5 \sin x$ (*j*) $y'''' - 4y''' + 8y'' - 8y' + 4y = 0$, $y(x) = e^x[(C_1 + C_2 x)\cos x + (C_3 + C_4 x)\sin x]$

7. (*a*) $y(x) = C_1 e^{i(\sqrt{2}+1)x} + C_2 e^{-i(\sqrt{2}-1)x} = e^{ix}(C_1 e^{i\sqrt{2}x} + C_2 e^{-i\sqrt{2}x}) = e^{ix}(C_3 \cos \sqrt{2}x + C_4 \sin \sqrt{2}x)$ (*d*) $y(x) = (C_1 + C_2 x)e^{ix}$ (*g*) $y(x) = C_1 + C_2 e^{(1-i)x/\sqrt{2}} + C_3 e^{-(1-i)x/\sqrt{2}}$

8. (*a*) Use

```
dsolve(diff(y(x),x,x)-2*I*diff(y(x),x)+y(x)=0,
y(x));
```

   (*d*) This ODE differs from the one in part (a) only by a sign, yet this time the *Maple* command [as in (a) but with the last

sign changed from $+$ to $-$] gives the result $y(x) = C_1 e^{ix}$ which is INCORRECT. The correct answer is $(C_1 + C_2 x)e^{ix}$.

9. (*a*) `solve(r^3-3*r^2+26*r-2=0,r);` gives an extremely messy result, `fsolve(r^3-3*r^2+26*r-2=0,r);` gives only the one real root, `fsolve(r^3-3*r^2+26*r-2=0,r,complex);` gives $r = 0.0776$, $1.4612 \pm 4.8619i$, hence unstable   (*d*) $r = \pm i$, $(-1 \pm i\sqrt{15})/2$, stable   (*g*) $r = -0.7718 \pm 0.0781i$, $-0.2683 \pm 2.0451i$, $0.5401 \pm 1.1273i$, unstable

10. (*a*) $r = (-\alpha \pm \sqrt{\alpha^2 - 4})/2$, stable if $\alpha \geq 0$, unstable if $\alpha < 0$ (*c*) stable if $\alpha \geq 0$, unstable if $\alpha < 0$

11. (*a*) Stable if $\alpha \leq -2$, unstable if $\alpha > -2$

12. (*a*) $a_1 = 6$, $a_2 = 5$, $a_3 = 4$, $a_4 = 1$ are all positive so proceed to the Δ_j's.

$$\Delta_1 = a_1 = 6,$$
$$\Delta_2 = \det(\text{array}([[6, 1], [4, 5]])) = 26,$$
$$\Delta_3 = \det(\text{array}([[6, 1, 0], [4, 5, 6], [0, 1, 4]])) = 68,$$
$$\Delta_4 = \det(\text{array}([[6, 1, 0, 0], [4, 5, 6, 1], [0, 1, 4, 5], [0, 0, 0, 1]]))$$
$$= 68$$

are all positive, hence stable.   (*e*) $a_1 = a_2 = \cdots = a_5 = 1 > 0$. $\Delta_1 = 1$, $\Delta_2 = \cdots = \Delta_5 = 0$, hence unstable

Section 6.6

1. (*a*) $y(x) = C_1/x$, $y(x) = 12/x$ (*d*) $y(x) = C_1 + C_2 x^5$, $y(x) = -\frac{3}{5} + \frac{3}{5}x^5$ (*g*) $y(x) = [C_1 \sin(\ln x) + C_2 \cos(\ln x)]/x$, $y(x) = 2[\sin(\ln x)]/x$ (*j*) $y(x) = C_1 \cos(2\ln x) + C_2 \sin(2\ln x)$, $y(x) = -[\sin(2\ln x)]/2$ (*m*) $y(x) = C_1 x + C_2 x^3$, $y(x) = -\frac{75}{2}x + \frac{75}{2}x^3$

2. (*a*) For the general solution use

```
dsolve(x*diff(y(x),x)+y(x)=0,y(x));
```

and for the particular solution use

```
dsolve({x*diff(y(x),x)+y(x)=0,y(3)=4},y(x));
```

That gives $y(x) = 12/x$. To plot use

```
with(plots):
implicitplot(y=12/x,x=0..5,y=0..100,
numpoints=2000);
```

3. Use Theorem 6.3.4

4. If we solve (10) we find that if $c_2 = (c_1 - 1)^2/4$ then there is the repeated root $r = (1 - c_1)/2 \equiv r_1$. Then the ODE is $4x^2 y'' + 4c_1 xy' + (c_1 - 1)^2 y = 0$ with solution $y(x) = Ax^{(1-c_1)/2}$. To reduce the writing let $(1-c_1)/2 \equiv a$ temporarily. Then the ODE is $x^2 y'' + (1 - 2a)xy' + a^2 y = 0$ with solution $y(x) = Ax^a$. Thus, seek $y(x) = A(x)x^a$. Putting this into the ODE gives, after simplifying, $xA'' + A' = 0$ with solution $A(x) = C_1 + C_2 \ln x$. Thus, $y(x) = (C_1 + C_2 \ln x)x^{r_1}$.

6. (a) Seeking $y(x) = A(x)x$ gives the ODE $xA'' + (2 + x^2)A' = 0$. Letting $A'(x) = p(x)$ renders this a separable first-order ODE with solution $p(x) = Be^{-x^2/2}/x^2$ so $A(x) = B\int e^{-x^2/2}\,dx/x^2$. If we integrate by parts with $u = e^{-x^2/2}$ and $dv = dx/x^2$ we obtain $A(x) = B\left(-\dfrac{e^{-x^2/2}}{x} - \int e^{-x^2/2}\,dx\right)$. Since we seek any particular $A(x)$ we can change the indefinite integral to a definite integral from 0 to x. If we do so we obtain for our second solution $y(x) = A(x)x = -B(e^{-x^2/2} + x\int_0^x e^{-t^2/2}\,dt)$ which is equivalent to the second term in (6.2) since B is arbitrary.   **(b)** In the integral in (6.2) let $t/\sqrt{2} = u$.

7. (a) $\Phi(r) = [(\Phi_2 \ln r_1 - \Phi_1 \ln r_2) + (\Phi_1 - \Phi_2)\ln r]/(\ln r_1 - \ln r_2)$

8. (a) $u(r) = \dfrac{u_2 r_2 - u_1 r_1}{r_2 - r_1} + \dfrac{u_1 - u_2}{r_2 - r_1}\dfrac{r_1 r_2}{r}$

9. (a) $y(x) = C_1 x + C_2 x^2 + C_3 x^3$ **(d)** $y(x) = C_1 + C_2 x + C_3/x$ **(g)** $y(x) = C_1 + C_2 \ln x + C_3(\ln x)^2$ **(j)** $y(x) = C_1 + x^3[C_2 \cos(2\ln x) + C_3 \sin(2\ln x)]$

10. (a) $Y'' - 3Y = 0$, $Y(t) = C_1 e^{\sqrt{3}t} + C_2 e^{-\sqrt{3}t}$, $y(x) = C_1 x^{\sqrt{3}} + C_2 x^{-\sqrt{3}}$ **(d)** $Y'' + 2Y' + Y = 0$, $Y(t) = (C_1 + C_2 t)e^{-t}$, $y(x) = (C_1 + C_2 \ln x)/x$ **(g)** $Y'' + Y' - 2Y = 0$, $Y(t) = C_1 e^t + C_2 e^{-2t}$, $y(x) = C_1 x + C_2/x^2$

Section 6.7

1. (a) Yes, $x^2 \cos x \rightarrow \{x^2 \cos x, x^2 \sin x, x \cos x, x \sin x, \cos x, \sin x\}$ **(d)** No, $x^2 \ln x \rightarrow \{x^2 \ln x, x \ln x, x, \ln x, 1, 1/x, 1/x^2, \ldots\}$ **(g)** Yes, $e^{9x} \rightarrow \{e^{9x}\}$ **(j)** Yes, $e^x \cos 3x \rightarrow \{e^x \cos 3x, e^x \sin 3x\}$

2. (a) $y_h(x) = C_1 e^{3x}$, $xe^{2x} \rightarrow \{xe^{2x}, e^{2x}\}$, $6 \rightarrow \{1\}$ so we could seek $y_{p_1}(x)$ in the form $y_{p_1}(x) = Axe^{2x} + Be^{2x}$ and $y_{p_2}(x)$ in the form $y_{p_2}(x) = C$, as in the text. However, it is simpler to seek them together as $y_p(x) = Axe^{2x} + Be^{2x} + C$. If we put that into the ODE we obtain $-Axe^{2x} + (A - B)e^{2x} - 3C = xe^{2x} + 6$ so $-A = 1$, $A - B = 0$, $-3C = 6$, so $A = B = -1$, $C = -2$, and $y(x) = y_h(x) + y_p(x) = C_1 e^{3x} - xe^{2x} - e^{2x} - 2$. **(d)** $y_h(x) = C_1 e^{3x}$, $xe^{3x} \rightarrow \{xe^{3x}, e^{3x}\} \rightarrow \{x^2 e^{3x}, xe^{3x}\}$ because of the duplication with the homogeneous solution, $4 \rightarrow \{1\}$. Seeking $y_p(x) = Ax^2 e^{3x} + Bxe^{3x} + C$ gives $2Axe^{3x} + Be^{3x} - 3C = xe^{3x} + 4$ so $2A = 1$, $B = 0$, $-3C = 4$. Thus, $A = 1/2$, $B = 0$, $C = -4/3$ and $y(x) = y_h(x) + y_p(x) = C_1 e^{3x} + x^2 e^{3x}/2 - 4/3$. **(g)** Seek $y_p(x) = A \sin 2x + B \cos 2x$, $y(x) = C_1 + C_2 e^x - \frac{5}{2} + \frac{1}{2}\cos 2x - \sin 2x$ **(j)** $y(x) = C_1 e^x + C_2 e^{-2x} - \frac{15}{8} - \frac{9}{4}x - \frac{3}{4}x^2 - \frac{1}{2}x^3 + \frac{1}{2}e^{-x}$ **(m)** $y_h(x) = (C_1 + C_2 x)e^x$, $x^2 e^x \rightarrow \{x^2 e^x, xe^x, e^x\} \rightarrow \{x^4 e^x, x^3 e^x, x^2 e^x\}$. Seek $y_p(x) = Ax^4 e^x + Bx^3 e^x + Cx^2 e^x$.

Obtain $y(x) = (C_1 + C_2 x)e^x + x^4 e^x/12$ **(p)** $y(x) = C_1 + C_2 e^x + C_3 e^{-x} - \frac{5}{2}\sin 2x$

3. You've probably noticed that sometimes a trigonometric identity is needed before your results are identical to your *Maple* results. In Exercise 2(p), for instance, whereas hand calculation leads naturally to the $-\frac{5}{2}\sin 2x$ term, *Maple* gives $-5\sin x \cos x$ which is equivalent because of the identity $\sin 2x = 2 \sin x \cos x$.

4. (a) $y_h(x) = Ae^{-2x}$ so seek $y_p(x) = A(x)e^{-2x}$. If we put that into the ODE we obtain $A' = 4e^{4x}$ so $A(x) = e^{4x}$. Thus, $y_p(x) = e^{4x}e^{-2x} = e^{2x}$. $y(x) = Ae^{-2x} + e^{2x}$. **(d)** $y_h(x) = A/x$ so seek $y_p(x) = A(x)/x$. Obtain $A' = 1/x$, $A(x) = \ln x$, $y_p(x) = (\ln x)/x$. $y(x) = A/x + (\ln x)/x$. **(g)** $y_h(x) = Ae^x + Be^{-x}$ so seek $y_p(x) = A(x)e^x + B(x)e^{-x}$. Obtain $A'e^x + B'e^{-x} = 0$ and $A'e^x - B'e^{-x} = 8e^x$ so $A' = 4$ and $B' = -4e^{2x}$. Thus, $A(x) = 4x$, $B(x) = -2e^{2x}$, and $y_p(x) = 4xe^x - 2e^{2x}e^{-x} = (4x-2)e^x$. Or, we could have used (63). $y(x) = Ae^x + Be^{-x} + (4x-2)e^x = Ce^x + Be^{-x} + 4xe^x$, where we have absorbed the $-2e^x$ term into Ae^x and renamed the arbitrary constant C. **(j)** $y_h(x) = A\cos x + B\sin x$ so seek $y_p(x) = A(x)\cos x + B(x)\sin x$. Obtain $A'c + B's = 0$ and $-A's + B'c = 4s$. (NOTE: It is often useful, in hand calculations to use the shorthand c for $\cos x$ and s for $\sin x$.) Thus, $A' = -4s^2$ and $B' = 4sc$ so $A(x) = -2x + \sin 2x$, $B(x) = 2\sin^2 x$, and $y_p(x) = (-2x + \sin 2x)\cos x + 2\sin^3 x = 2(\sin x - x \cos x)$, the last step following from the identities $\sin A \cos B = [\sin(A+B) + \sin(A-B)]/2$ and $\sin^3 A = (3\sin A - \sin 3A)/4$. $y(x) = A\cos x + B\sin x + 2(\sin x - x\cos x) = A\cos x + C\sin x - 2x\cos x$. **(m)** $y_h(x) = Ax^2 + Bx^{-2}$ so seek $y_p(x) = A(x)x^2 + B(x)x^{-2}$. Obtain $A'x^2 + B'x^{-2} = 0$ and $2A'x^4 - 2B' = 1$ so $A' = 1/(4x^4)$ and $B' = -1/4$. Thus, $A(x) = -1/(12x^3)$ and $B(x) = -x/4$ so $y_p(x) = -1/(3x)$. Thus, $y(x) = Ax^2 + Bx^{-2} - 1/(3x)$.

6. (c)

$$y(x) = C_1 \sin \omega x + C_2 \cos \omega x + \frac{\sin \omega x}{\omega}\int_0^x \cos \omega \xi f(\xi)\,d\xi - \frac{\cos \omega x}{\omega}\int_0^x \sin \omega \xi f(\xi)\,d\xi$$

gives $y(0) = 5 = C_2$ and $y'(0) = 2 = \omega C_1$, so

$$y(x) = \frac{2}{\omega}\sin \omega x + 5\cos \omega x + \frac{\sin \omega x}{\omega}\int_0^x \cos \omega \xi f(\xi)\,d\xi - \frac{\cos \omega x}{\omega}\int_0^x \sin \omega \xi f(\xi)\,d\xi$$

or, $\frac{2}{\omega}\sin \omega x + 5\cos \omega x + \frac{1}{\omega}\int_0^x \sin \omega(x - \xi)f(\xi)\,d\xi$

8. (a) $y_1(x) = e^x$, $y_2 = e^{-x}$, $y_3(x) = e^{2x}$, $W_1(x) = 72e^{4x}$, $W_2(x) = -24e^{6x}$, $W_3 = -48e^{3x}$, $W(x) = -6e^{2x}$, so (7.5) gives $y_p(x) = [-6e^{2x}]e^x + [e^{4x}]e^{-x} + [8e^x]e^{2x} = 3e^{3x}$. **(c)** $y_1(x) = x$, $y_2(x) = x^2$, $y_3(x) = x^3$, $W_1(x) = 4x^2$, $W_2(x) = -8x$, $W_3(x) = 4$ [NOTE: If we divide the ODE by x^3 to normalize the coefficient of y''' we obtain $f(x) = 4/x^2$, not $4x$.], $W(x) = 2x^3$, so $y_p(x) = [2\ln x]x + [4/x]x^2 + [-1/x^2]x^3 = 2x\ln x + 3x$.

CHAPTER 7
Section 7.2

1. (a) $x(t) = 4\cos t + 0\sin t = 2e^{it} + 2e^{-it} = 4\sin(t + \pi/2)$ **(d)** $x(t) = -3\cos\sqrt{2}t + \frac{5}{\sqrt{2}}\sin\sqrt{2}t = \left(\frac{-3\sqrt{2}-5i}{2\sqrt{2}}\right)e^{i\sqrt{2}t} + \left(\frac{-3\sqrt{2}+5i}{2\sqrt{2}}\right)e^{-i\sqrt{2}t} = \sqrt{\frac{43}{2}}\sin(\sqrt{2}t - 0.7036)$ **(g)** $x(t) = -3\cos(\frac{1}{2}t) - 6\sin(\frac{1}{2}t) = \left(\frac{-3+6i}{2}\right)e^{it/2} + \left(\frac{-3-6i}{2}\right)e^{-it/2} = \sqrt{45}\sin\left(\frac{1}{2}t + 3.605\right)$ **(j)** $x(t) = 2\cos\left(\frac{3}{2}t\right) - 4\sin\left(\frac{3}{2}t\right) = (1+2i)e^{i3t/2} + (1-2i)e^{-i3t/2} = \sqrt{20}\sin\left(\frac{3}{2}t + 2.678\right)$

4. (a) $x(t) = e^{-2t}(4\cosh\sqrt{3}t + \frac{8}{\sqrt{3}}\sinh\sqrt{3}t)$ **(d)** $x(t) = e^{-t}(-3\cos t + 2\sin t)$ **(g)** $x(t) = (-3 - \frac{9}{2}t)e^{-t/2}$ **(j)** $x(t) = e^{-t/2}(2\cos\sqrt{2}t - \frac{5}{\sqrt{2}}\sin\sqrt{2}t)$

6. $G\cos\psi = C$ and $G\sin\psi = D$ so $G = \sqrt{C^2 + D^2}$ and $\psi = \arctan(D/C)$

9. If the entire spring were moving with velocity $x'(t)$ then α would be 1, but it is not; we've assumed that the velocity varies linearly from 0 at its fixed end to $x'(t)$ at its attachment to the mass.

10. (a) Newton's second law gives $mx'' = (p_1 - p_2)A$ where p_1, p_2 are the pressures on the left and right, respectively, of the piston. Boyle's law gives $p_2(L - x)A = p_1(L + x)A + p_0LA$ so $p_1 = p_0L/(L + x)$ and $p_2 = p_0L/(L - x)$. If we put these in Newton's law we obtain (10.1). **(b)** Nonlinear because of the $x/(L^2 - x^2)$ term. **(d)** Frequency $= \sqrt{\frac{2p_0A}{mL}}\frac{\text{rad}}{\text{sec}}\frac{1\,\text{cycle}}{2\pi\,\text{rad}} = \frac{1}{2\pi}\sqrt{\frac{2p_0A}{mL}}\frac{\text{cycles}}{\text{sec}}$ **(e)** Yes

11. The vertical acceleration is zero so the sum of the vertical forces is zero: $N_1 + N_2 - mg = 0$. The angular acceleration is zero so the sum of the moments about any point P is zero. With P at top of left-hand cylinder that gives $N_2L - mg(x + L/2) = 0$. If we solve we these we obtain $N_1 = mg(L - 2x)/2L$ and $N_2 = mg(L + 2x)/2L$. Then, $mx'' = $ sum of horizontal forces $= \mu N_1 - \mu N_2$ gives $mx'' + (2mg\mu/L)x = 0$ so $\omega = \sqrt{2g\mu/L}$.

12. (a) Potential energy $= -mgL\cos\theta\sin\alpha$, kinetic energy $= m(L\theta')^2/2$

Section 7.3

2. (a) Enter

```
with(DEtools):
phaseportrait({diff(x(t),t)=y(t),diff(y(t),t)=
-4*x(t)-8*y(t)},{x(t),y(t)},t=0..15,
{[x(0)=-3,y(0)=4],[x(0)=-2,y(0)=4],
[x(0)=1.5,y(0)=3],[x(0)=2.5,y(0)=3],
[x(0)=-2.5,y(0)=-3],[x(0)=-1.5,y(0)=-3],
[x(0)=2,y(0)=-4],[x(0)=3,y(0)=-4],
[x(0)=-4,y(0)=2.1436],[x(0)=4,y(0)=-2.1436]},
stepsize=.05,x=-4..4,y=-5..5,
scene=[x,y],arrows=NONE);
```

4. (a) If we put $y = \alpha x$ into (13) we obtain the equation $m\alpha^2 + c\alpha + k = 0$ with the roots (4.1). **(b)** For the m, k, c values given in the figure caption (4.1) gives $\alpha = -0.536, -7.464$. The former is included in Fig. 8; the latter (namely, $y = -7.464x$) is not. **(d)** They coalesce as $c \to c_{cr}$.

5. (b) If $m = k$ then (5) shows that the trajectories are circles $x^2 + y^2 = 2E/m$ centered at the origin. Then $s' = a = $ radius $= \sqrt{2E/m}$. **(c)** Draw the phase trajectory $ABCDA$. Since BC is vertical the phase speed is infinite along that line. Hence the representative point moves from B to C instantaneously; likewise from D to A. The graph of $x(t)$ versus t is a continuous periodic graph with straight line segments. If (t, x) denotes the coordinates in the x, t plane then the graph is as follows: $(0, -1) \to (4, 3) \to (8, -1) \to (12, 3) \to (16, -1) \to$ etc. These points correspond to the phase plane points A, B and C, D and A, B and C, D and A, etc. **(d)** On BC $(0 < t < \infty)$ $s' = a = \sqrt{2}(1 - x)$ and also $x' = s'/\sqrt{2}$ so $x' = 1 - x$, $x(t) = 1 - e^{-t}$. On AB $(-\infty < t < 0)$ $s' = a = \sqrt{2}(x + 1)$ and also $x' = s'/\sqrt{2}$ so $x' = x + 1$, $x(t) = e^t - 1$.

6. As the amplitude tends to zero the x^3 term in (7) becomes negligible compared to the x term so (7) becomes $x'' + x = 0$ with period $= 2\pi \approx 6.28$. As the amplitude is increased [by increasing $x(0)$] the x^3 term becomes important. Qualitatively, the spring becomes stiffer so we expect the frequency to increase and hence the period to decrease. Thus, you should find that the period is very close to 6.28 for the small amplitude motions and decreases as $x(0)$ is increased. For $x(0) = 10$ you should obtain a period of around 0.74.

Section 7.4

3. (c) To derive (20c) equate $E\cos(\Omega t - \Phi) = E\cos\Omega t\cos\Phi + E\sin\Omega t\sin\Phi$ to the right side of (19) or, equivalently, (18). Thus obtain $E\cos\Phi = C$ and $E\sin\Phi = D$, dividing which gives $\tan\Phi = D/C$. Noting from (19) that $D > 0$ and from (20b) that $E > 0$, it follows from $E \sin \Phi = D$ that $\sin \Phi > 0$ so $0 < \Phi < \pi$. Since $\tan \Phi = D/C$ and $D > 0$, it follows that Φ is in the first quadrant if $C > 0$ ($\Omega < \omega$) and in the second quadrant if $C < 0$ ($\Omega > \omega$). Thus, $\Phi = \text{Arctan}(D/C)$ if $\Omega < \omega$ and $\Phi = \text{Arctan}(D/C) + \pi$ if $\Omega > \omega$.

6. (a) $\omega^2 = k/m = 1$ and $c_{cr} = \sqrt{4km} = 2$ so $c = 0, 0.5, 1, 2, 4, 8$. Use

```
with(plots):
implicitplot({E=25/sqrt((w^2-1)^2+0),
E=25/sqrt((w^2-1)^2+.25*w^2),
E=25/sqrt((w^2-1)^2+w^2),E=25/sqrt((w^2-1)^2
+4*w^2),E=25/sqrt((w^2-1)^2+16*w^2),
E=25/sqrt((w^2-1)^2+64*w^2)},
w=0..5,E=0..80,numpoints=2000);
```

7. (a) $L[x] = F_0 \cos \omega t$ so consider instead $L[v] = F_0 e^{i\Omega t}$, solve for $v_p(t)$ and then obtain $x_p(t)$ as $x_p(t) = \text{Re}\, v_p(t)$. Seek $v_p(t) = A e^{i\Omega t}$ and obtain $(-m\Omega^2 + ic\Omega + k)A = F_0$. Thus,

$$x_p(t) = \text{Re}\left(\frac{F_0}{k - m\Omega^2 + ic\Omega} e^{i\Omega t} \right)$$

$$= F_0 \frac{(k - m\Omega^2)\cos \Omega t + c\Omega \sin \Omega t}{(k - m\Omega^2)^2 + c^2 \Omega^2}.$$

(d) $L[x] = 4 \sin 3t$ so consider $L[v] = 4e^{i3t}$, solve for $v_p(t)$ and then obtain $x_p(t)$ as $x_p(t) = \text{Im}\, v_p(t)$. Seek $v_p(t) = A e^{i3t}$ and obtain $(i3 - 1)A = 4$ so $x_p(t) = \text{Im}\left(\frac{4}{-1+3i} e^{i3t} \right) = -\frac{2}{5}(3 \cos 3t + \sin 3t)$.

Section 7.5

1. (a) $c(x) = D e^{2x} + E e^{-2x} + 5/2$. c bounded as $x \to +\infty$ gives $D = 0$ and c bounded as $x \to -\infty$ gives $E = 0$. Thus, $c(x) = 5/2$.   **(d)** $c(x) = \frac{1}{4} + \frac{1}{5}\sin x$ **(g)** $c'' - 4c = -50 \cos^2 x = -25(1 + \cos 2x)$, $c(x) = \frac{25}{4} + \frac{25}{8}\cos 2x$

2. $y(x) = -w_0 x^2 (L - x)^2/(24EI)$

5. (a) $y(x) = -w_0/k$ **(b)** $y(x) = -\dfrac{w_0}{k} - \dfrac{w_0}{k + EI}\cos x$

6. dsolve works with boundary conditions, just as for initial conditions. Use

```
dsolve({EI*diff(y(x),x$4)=-w,y(0)=0,D(y)(0)=0,
y(L)=0,D(y)(L)=0},y(x));
```

where w stands for w_0.

CHAPTER 8

Section 8.2

1. (a) $\lim |a_{n+1}/a_n| = \lim |(n+1)/n| = 1$ so (7a) gives $R = 1$.

Hence, convergence in $-1 < x < 1$.   **(d)** It is simplest to note that this is not really an infinite series since it terminates at $n = 1000$; thus it converges for all x. If we insist on using (7) we need to realize that $a_n = 0$ for all $n > 1000$ so we must use $a_n = 0$ in (7). Then the $a_{n+1}/a_n = 0/0$ in (7a) is indeterminate, but the $\sqrt[n]{|a_n|} = \sqrt[n]{0} = 0$ in (7b) so (7b) gives $R = 1/0 = \infty$, as was noted above. $-\infty < x < \infty$
(g) $|a_{n+1}/a_n| = [(n+1)^{50}/(n+1)!]/n^{50}n! = (1+1/n)^{50}/(n+1) \sim 1/(n+1) \to 0$ as $n \to \infty$, so $R = 1/0 = \infty$. $-\infty < x < \infty$ **(j)** $|a_{n+1}/a_n| = [(n+1)/2^{n+1}]/[n/2^n] = (1 + 1/n)/2 \sim 1/2$ as $n \to \infty$, so $R = 1/2$. Note that to determine the interval of convergence we need to re-express $(x-5)^{2n}$ as $[(x-5)^2]^n$. Thus, convergence in $|(x-5)^2| < 1/2$ or $|x - 5| < 1/\sqrt{2}$ or $5 - 1/\sqrt{2} < x < 5 + 1/\sqrt{2}$.

2. (a) $e^x = e + e(x-1) + e(x-1)^2/2! + \cdots = e \sum_0^\infty (x-1)^n/n!$ so $|a_{n+1}/a_n| = [e/(n+1)!]/[e/n!] = 1/(n+1) \to 0$ as $n \to \infty$, so $R = \infty$ **(d)** $\sin x = \sin(\pi/2) + \cos(\pi/2)(x - \pi/2) - \sin(\pi/2)(x-\pi/2)^2/2! - \cos(\pi/2)(x-\pi/2)^3/3! + \cdots = \sum_0^\infty (-1)^n (x - \pi/2)^{2n}/(2n)!$ so $|a_{n+1}/a_n| = (2n)!/(2n+2)! = 1/[(2n+1)(2n+2)] \to 0$ as $n \to \infty$, so $R = \infty$ **(g)** $1/x = 1/4 - (x - 4)^2/4^2 + (x - 4)^3/4^3 - \cdots = \sum_0^\infty (-1)^n (x - 4)^n/4^{n+1}$ so $|a_{n+1}/a_n| = 1/4$, so $R = 4$ ($0 < x < 8$)   **(j)** $2x^3 - 4 = -4 + 0x + 0x^2 + 2x^3 + 0x^4 + \cdots = -4 + 2x^3$; the latter is only a two-term series so it converges for all x ($R = \infty$)   **(m)** Since we are expanding about $x_0 = 0$ we seek an expansion in powers of x. Thus, let $x^{18} \equiv t$ and expand $1/(1 + t)$ about $t = 0$, then replace t by x^{18} and multiply the result by the x^2. $1/(1 + t) = 1 - t + t^2 - t^3 + \cdots$ for $|t| < 1$ so $x^2/(1 + x^{18}) = x^2(1 - x^{18} + x^{36} - \cdots) = x^2 - x^{20} + x^{38} - x^{56} + \cdots$. If you don't use this technique you will obtain the same result but will waste effort by generating *all* the terms $(x^0, x, x^2, x^3, \cdots)$ only to find that most of their coefficients are zero.

3. (a) `taylor(exp(-x),x=0,6);` gives $1 - x + x^2/2 - x^3/6 + x^4/24 - x^5/120$. To plot use

```
with(plots):
implicitplot({s=exp(-x),s=1-x,s=1-x+x^2/2-x^3/6,
s=1-x+x^2/2-x^3/6+x^4/24-x^5/120},x=0..4,s=-5..5);
```

4. (c) For $x = +1$ the series $1 + 1 + 1 + \cdots$ diverges because $s_n = n \to \infty$ as $n \to \infty$. For $x = -1$ the series $1 - 1 + 1 - 1 + \cdots$ diverges because the sequence $s_n = 1, 0, 1, 0, \cdots$ does not converge.

5. (a) $\dfrac{1}{x-1} = \frac{1}{3}\sum_0^\infty (-1)^n \left(\frac{x-4}{3} \right)^n$

6. (a) $f'(0) = \lim_{h \to 0} \dfrac{e^{-1/h^2} - 0}{h} = \lim_{t \to \infty} \dfrac{e^{-t}}{1/\sqrt{t}} = \lim_{t \to \infty} \dfrac{\sqrt{t}}{e^t}$. Applying l'Hôpital's rule, $\lim_{t \to \infty} \dfrac{\sqrt{t}}{e^t} = \lim_{t \to \infty} \dfrac{(1/2)t^{-1/2}}{e^t} = 0$. Similarly for $f''(0), \ldots$.

7.

$$\frac{s_1 + s_2 + \cdots + s_N}{N} = \frac{(1-x) + (1-x^2) + \cdots + (1-x^N)}{N(1-x)}$$

$$= \frac{N - (x + x^2 + \cdots + x^N)}{N(1-x)}$$

$$= \left[N - \left(\frac{1}{1-x} - \frac{x^{N+1}}{1-x} - 1 \right) \right] / [N(1-x)]$$

$$= \frac{1}{1-x} - \frac{1}{N} \frac{x - x^{N+1}}{(1-x)^2}$$

Section 8.3

2. (a) $y = \sum_0^\infty a_n x^n$ gives

$$y(x) = a_0 + a_1 x + \tfrac{9}{2} a_0 x^2 + \cdots$$

$$= a_0 \left(1 + \tfrac{9}{2} x^2 + \tfrac{81}{24} x^4 + \cdots \right)$$

$$+ a_1 \left(x + \tfrac{9}{6} x^3 + \tfrac{81}{120} x^5 + \cdots \right).$$

On the other hand, analytical solution (using $y = e^{rx}$) gives

$$y(x) = A \cosh 3x + B \sinh 3x$$

$$= A \left(1 + \tfrac{9}{2} x^2 + \tfrac{81}{24} x^4 + \cdots \right)$$

$$+ B \left(3x + \tfrac{27}{6} x^3 + \tfrac{243}{120} x^5 + \cdots \right)$$

which is equivalent to the foregoing, with $A = a_0$ and $B = a_1/3$.

3. (a) The recursion formula is $[(n+3)^2 - (n+3)]a_{n+3} + a_n = 0$ for $n = -1, 0, 1, 2, \ldots$, where $a_{-1} \equiv 0$. Thus, $y(x) = a_0(1 - x^3/6 + x^6/180 - x^9/12960 + \cdots) + a_1(x - x^4/12 + x^7/504 - x^{10}/45360 + \cdots)$. To determine the radii of convergence we can use the recursion formula: with $a_{n+3} = $ "a_{n+1}", we have

$$\lim_{n \to \infty} \left| \frac{\text{``}a_{n+1}\text{''}}{a_n} \right| = \lim_{n \to \infty} \frac{1}{(n+3)^2 - (n+3)} = 0 \text{ so,}$$

from Theorem 8.2.2, $R = \infty$. Alternatively, it is simpler to use Theorem 8.3.1 which gives $R = \infty$ because $p(x) = 0$ and $q(x) = x$ are analytic for all x. **(d)** $y(x) = a_0(1 - x^2/2 + x^4/8 - x^6/48 + \cdots) + a_1(x - x^3/3 + x^5/15 - x^7/105 + \cdots)$, $R = \infty$ **(g)** $y(x) = a_0[1 + e^{-2}(x-2)^2/2 - e^{-2}(x-2)^3/6 + (e^{-4} + e^{-2})(x-2)^4/24 + \cdots] + a_1[(x-2) + e^{-2}(x-2)^3/6 - e^{-2}(x-2)^4/12 + (e^{-4} + 3e^{-2})(x-2)^5/120 + \cdots]$, $R = \infty$

4. (a) The commands

```
Order:=6:
dsolve({diff(y(x),x,x)+x*y(x)=0,
y(0)=1,D(y)(0)=0},y(x),type=series);
```

will give the series solution that multiplies a_0. To find the series solution that multiplies a_1 use $y(0) = 0$ and $D(y)(0) = 1$ instead.

5. Use Theorem 8.3.1. **(a)** $R = \infty$ **(d)** $p(x) = 1/x$ is singular at $x = 0$ and $q(x) = 1$ is analytic for all x. Because of the singularity of $p(x)$ at $x = 0$ we can say, according to Theorem 8.3.1, that each power series solution (about $x_0 = 5$) will have $R = 5$ at least. **(g)** $p(x) = 1/(x-1)$ is singular at $x = 1$ and $q(x) = -1/(x+1)$ is singular at $x = -1$. The closer of these two singular points, to $x_0 = 1/3$, is $x = 1$. Thus $R = 2/3$ at least. **(j)** Singular points of p and q at $-1 \pm 2i$ so $R = 2\sqrt{2}$ at least. **(m)** $R = 3$ at least. **(p)** Singular points of p and q at ± 2 and at $\pm 2i$ so $R = 4$ at least.

9. Changing the sign of x in the left-hand side of (8.1) is equivalent to changing the sign of r therein. Thus, the same must be true for the right-hand side. Thus, $\sum_0^\infty P_n(-x)r^n = \sum_0^\infty P_n(x)(-r)^n$ which implies (9.1).

12. (a) Since $\ln(1 \pm x) = \pm x - x^2/2 \pm x^3/3 - x^4/4 \pm x^5/5 - \cdots$ for $|x| < 1$, we have, for the left side of (12.3), $\frac{1}{r} \ln \left(\frac{1+r}{1-r} \right) = \frac{1}{r}[\ln(1+r) - \ln(1-r)] = 2 \left(1 + \frac{r^2}{3} + \frac{r^4}{5} + \frac{r^6}{7} + \cdots \right) = \sum_0^\infty \frac{2}{2n+1} r^{2n}$ for $|r| < 1$.

Section 8.4

2. $J_0(1) = 0.7652$, $Y_0(1) = 0.08826$

3. $J_0(x) = \sum_0^\infty [(-1)^n/2^{2n}(n!)^2](x^2)^n$ so $|a_{n+1}/a_n| = 2^{2n}(n!)^2/2^{2n+2}(n+1)!^2 = 1/4(n+1)^2 \to 0$ as $n \to \infty$ so, by Theorem 8.2.2, $R = \infty$.

4. (a) $p(x) = 1/2x$ and $q(x) = x$ so $xp(x) = 1/2$ and $x^2q(x) = x^3$. Thus, from (34), $p_0 = 1/2$ and $q_0 = 0$. Then (37) is $r^2 - (1/2)r = 0$ with $r = 0$ and $1/2$. Thus, according to (40), we expect a general solution of the form $y(x) = (a_0 + a_1 x + a_2 x^2 + \cdots)\sqrt{x} + (b_0 + b_1 x + b_2 x^2 + \cdots)$. In fact, the *Maple* command

```
dsolve(2*x*diff(y(x),x,x)+diff(y(x),x)+2*x^2*
y(x)=0,y(x),type=series);
```

does give a general solution of that form.
(e) $xp(x) = 3 - 6x^2$ and $x^2q(x) = 1$ so $p_0 = 3$ and $q_0 = 1$. Thus, $r = -1, -1$ (repeated root) so we expect a general solution of the form $y(x) = [(a_0 + a_1 x + \cdots) + (b_0 + b_1 x + \cdots) \ln x]/x$. The *Maple* command

```
dsolve(x^2*diff(y(x),x,x)+3*x*(1-2*x^2)*
diff(y(x),x)+y(x)=0,y(x),type=series);
```

does confirm that expectation.

5. (a) None **(d)** $p(x) = 0$ is analytic for all x but $q(x) = 1/[x(x^2+3)]$ is singular at $x = 0$; $x^2q(x) = x/(x^2+3)$ is analytic at $x = 0$ so the singular point at $x = 0$ is regular singular point. **(g)** Regular singular point at $x = 1$, irregular singular point at $x = -3$ (**j**) Regular singular points at $x = 1$ and at $x = -1$ (**m**) Regular singular point at $x = 0$

CHAPTER 9

Section 9.2

1. (*a*) $\lambda = 0$, $\{[1, 0]^T, [0, 1]^T\}$
(*e*) $\lambda = 3$, $\{[0, 1, 0]^T\}$; $\lambda = 0$, $\{[0, 0, 1]^T, [1, 0, 0]^T\}$ (*i*) $\lambda = 0$, $\{[0, 1, -1]^T\}$; $\lambda = 2$, $\{[1, 0, 0]^T, [0, 1, 1]^T\}$ (*m*) $\lambda = 0$, $\{[1, 0, -1]^T\}$; $\lambda = 5$, $\{[1, 0, 4]^T\}$; $\lambda = 2$, $\{[0, 1, 0]^T\}$ (*q*) $\lambda = -1$, $\{[0, 1, 1, -1]^T\}$; $\lambda = 2$, $\{[1, 0, 0, 0]^T, [0, 1, 1, 2]^T\}$; $\lambda = 0$, $\{[0, 1, -1, 0]^T\}$

2. (*a*) Use
```
with(linalg):
A:=array([[0,0],[0,0]]);
eigenvects(A);
```

3. (*a*) **x** is an eigenvector of **A** if $\mathbf{Ax} = \lambda\mathbf{x}$ for some scalar λ; i.e., if $\mathbf{Ax}$ is a scalar multiple of **x**. If so, the scalar multiple is λ. Here, $\mathbf{x} = [1, 2, -1, 3]^T$ gives $\mathbf{Ax} = [21, 42, -21, -21]^T$ which is not a scalar multiple of **x**. Thus, this **x** is not an eigenvector of **A**.   (*d*) Yes, with $\lambda = 0$ (*g*) Yes, with $\lambda = 9$

4. (*a*) Solve $(\mathbf{A} - \lambda\mathbf{I})\mathbf{x} = \mathbf{0}$ by Gauss elimination. That step gives $\mathbf{x} = \alpha[1, 1, -1, -1]^T$ or span$\{[1, 1, -1, -1]^T\}$.
(*c*) span$\{[1, 0, -1, 0]^T, [1, -1, 0, 0]^T, [1, 0, 0, -1]^T\}$

7. Let an eigenvector **e** correspond to eigenvalues λ_1 and λ_2. Then $\mathbf{Ae} = \lambda_1\mathbf{e}$ and $\mathbf{Ae} = \lambda_2\mathbf{e}$. Subtracting gives $(\lambda_1 - \lambda_2)\mathbf{e} = \mathbf{0}$. Since **e** is necessarily nonzero, $\lambda_1 - \lambda_2 = 0$. Thus, $\lambda_1 = \lambda_2$ so the answer is no.

12. (*a*) **A** has $\lambda = 1$, $\mathbf{e} = \alpha[1, -1, 0]^T$; $\lambda = 2$, $\mathbf{e} = \beta[0, 1, -1]^T$; $\lambda = 3$, $\mathbf{e} = \gamma[0, 0, 1]^T$. $\mathbf{A}^{10} = $ array([[1, 0, 0], [1023, 1024, 0], [58025, 58025, 59049]]) has $\lambda = 1$, $\mathbf{e} = \delta[1, -1, 0]^T$; $\lambda = 1024 = 2^{10}$, $\mathbf{e} = \mu[0, 1, -1]^T$; $\lambda = 59049 = 3^{10}$, $\mathbf{e} = \nu[0, 0, 1]^T$.
(*c*) **A** has $\lambda = 1$, $\mathbf{e} = \alpha[1, 0, 0]^T$; $\lambda = 0$, $\mathbf{e} = \beta[0, 1, 0]^T$; $\lambda = -2$, $\mathbf{e} = \gamma[1, 0, -3]^T$. $\mathbf{A}^{10} = $ array([[1, 0, -341], [0, 0, 0], [0, 0, 1024]]) has $\lambda = 1$, $\mathbf{e} = \delta[1, 0, 0]^T$; $\lambda = 0$, $\mathbf{e} = \mu[0, 1, 0]^T$; $\lambda = 1024 = (-2)^{10}$, $\mathbf{e} = \nu[1, 0, -3]^T$.

13. (*a*) Use
```
with(linalg):
A:=array([[2,2,0],[0,2,0],[2,0,2]]);
evalm(A^10);
eigenvects(%);
```

15. If $\mathbf{Ae}_j = \lambda_j\mathbf{e}_j$ and $\mathbf{Be}_j = \lambda_j\mathbf{e}_j$ then $(\mathbf{A} - \mathbf{B})\mathbf{e}_j = \mathbf{0}$. Let $\mathbf{r}_i$ denote the ith row vector in $\mathbf{A} - \mathbf{B}$. Then $(\mathbf{A} - \mathbf{B})\mathbf{e}_j = \mathbf{0}$ implies that $\mathbf{r}_i \cdot \mathbf{e}_j = 0$ for each $j = 1, \ldots, n$. If $\mathbf{r}_i$ is nonzero then $\{\mathbf{e}_1, \ldots, \mathbf{e}_n, \mathbf{r}_i\}$ is LI (which claim we leave for you to prove).

But we can't have $n + 1$ LI vectors in n-space so it must be that $\mathbf{r}_i = \mathbf{0}$. Since that is true for each $i = 1, \ldots, n$, it follows that $\mathbf{A} - \mathbf{B} = \mathbf{0}$ or, $\mathbf{A} = \mathbf{B}$.

16. No. Do you see why it cannot?

17. (*a*) Each row sums to 30 so $(1/30)\mathbf{A}$ is a Markov matrix with $\lambda = 1$ among its eigenvalues. Thus, **A** has 30 among its eigenvalues. (The others are $\lambda = -2$ and $\lambda = 0$.)
(*c*) $(1/20)\mathbf{A}$ is a Markov matrix with $\lambda = 1$ among its eigenvalues so **A** has $\lambda = 20$ among its eigenvalues. Following the pattern of Gauss elimination we find that its corresponding eigenvector is $\mathbf{e} = \alpha[1, 1, \ldots, 1]^T$. Also, we see by inspection that another eigenvalue is $\lambda = 0$. By Gauss elimination we find that its eigenspace is span$\{[-1, 0, \ldots, 0, 1]^T, [-1, 0, \ldots, 0, 1, 0]^T, \ldots, [-1, 1, 0, \ldots, 0]^T\}$.

19. (*a*) $\lambda = 1/3$, $\mathbf{e} = \alpha[1, -1]^T$; $\lambda = 3/2$, $\mathbf{e} = \beta[1, 1]^T$.
(*c*) $\lambda = (3 + \sqrt{3})/2$, $\mathbf{e} = \alpha[1 + \sqrt{3}, -1]^T$; $\lambda = (3 - \sqrt{3})/2$, $\mathbf{e} = \beta[1 - \sqrt{3}, -1]^T$.

21. (*a*) $\lambda = 1$, $\mathbf{e} = \alpha[1, 1, -1]^T$; $\lambda = (1 - \sqrt{3}i)/2$, $\mathbf{e} = \beta[0, -2, 1 - i\sqrt{3}]^T$; $\lambda = (1 + \sqrt{3}i)/2$, $\mathbf{e} = \gamma[0, -2, 1 + i\sqrt{3}]^T$
(*c*) $\lambda = 0$, $\mathbf{e} = \alpha[0, 1, -2]^T$; $\lambda = \sqrt{5}i$, $\mathbf{e} = \beta[-\sqrt{5}i, 2, 1]^T$; $\lambda = -\sqrt{5}i$, $\mathbf{e} = \gamma[\sqrt{5}i, 2, 1]^T$

Section 9.3

3. (*a*) $\lambda = 2$, $\mathbf{e} = \alpha[1, 1]^T$ and $\lambda = 0$, $\mathbf{e} = \beta[1, -1]^T$ so the eigenvectors give the orthogonal basis $\mathbf{e}_1 = [1, 1]^T$, $\mathbf{e}_2 = [1, -1]^T$.   (*d*) $\mathbf{e}_1 = [0, 0, 1]^T$, $\mathbf{e}_2 = [1, -1, 0]^T$, $\mathbf{e}_3 = [1, 1, 0]^T$ (*g*) $\mathbf{e}_1 = [1, 0, 0, -1]^T$ and $\mathbf{e}_2 = [0, 1, -1, 0]^T$ from the eigenspace of $\lambda = -4$, and $\mathbf{e}_3 = [1, 0, 0, 1]^T$ and $\mathbf{e}_4 = [0, 1, 1, 0]^T$ from the eigenspace of $\lambda = 4$ (*j*) $\mathbf{e}_1 = [1, 0, 0, -1]^T$, $\mathbf{e}_2 = [0, 1, 0, 0]^T$, $\mathbf{e}_3 = [0, 0, 1, 0]^T$ from the eigenspace of $\lambda = 0$, and $\mathbf{e}_4 = [1, 0, 0, 1]^T$ from the eigenspace of $\lambda = 2$

4. (*d*) Since (4.5) satisfies (4.1) for any a_j's we can set all the a_j's equal to zero. That step shows that $\sum_{N+1}^n (c_j/\lambda_j)\mathbf{e}_j$ is a particular solution of (4.1). And of course the $\mathbf{e}_j$'s are homogeneous solutions of (4.1) because they are eigenvectors corresponding to $\lambda = 0$; hence, $\mathbf{Ae}_j = \mathbf{0}$ for each $j = 1, \ldots, N$.
(*e*) This point is easily missed. Suppose we use *any* basis $\mathbf{u}_1, \ldots, \mathbf{u}_n$. Expanding $\mathbf{x} = \sum_1^n a_j\mathbf{u}_j$ and $\mathbf{c} = \sum_1^n c_j\mathbf{u}_j$, (4.1) becomes $\mathbf{A}(\sum_1^n a_j\mathbf{u}_j) = \sum_1^n c_j\mathbf{u}_j$ or $\sum_1^n a_j\mathbf{Au}_j = \sum_1^n c_j\mathbf{u}_j$. Now, each $\mathbf{Au}_j$ is a vector and can be expanded as $\mathbf{Au}_j = \sum_{k=1}^n b_{jk}\mathbf{u}_k$. Thus, $\sum_{j=1}^n \sum_{k=1}^n a_j b_{jk}\mathbf{u}_k = \sum_1^n c_j\mathbf{u}_j$. To get $\mathbf{u}_j$ on both sides interchange the dummy indices j and k on the left-hand side. Thus, $\sum_{j=1}^n (\sum_{k=1}^n b_{kj}a_k)\mathbf{u}_j = \sum_1^n c_j\mathbf{u}_j$. Since the $\mathbf{u}_j$'s are LI it follows that $\sum_{k=1}^n b_{kj}a_k = c_j$ for each $j = 1, \ldots, n$. The latter is a system of n *coupled* equations for $a_1, \ldots, a_n$ whereas (4.3) is a system of n *uncoupled* equations for those unknowns. Of course this gain is not free; using

the eigenvector expansion method we need to pay the price of evaluating the eigenvalues and eigenvectors of $\mathbf{A}$.

5. (a) $\lambda_1 = 3$, $\mathbf{e}_1 = [1, 1]^T$; $\lambda_2 = 1$, $\mathbf{e}_2 = [1, -1]^T$. $c_1 = \mathbf{c} \cdot \mathbf{e}_1 / \mathbf{e}_1 \cdot \mathbf{e}_1 = 3/2$ and $c_2 = \mathbf{c} \cdot \mathbf{e}_2 / \mathbf{e}_2 \cdot \mathbf{e}_2 = 3/2$ so (4.3) gives $3a_1 = 3/2$ and $a_2 = 3/2$. Thus, $a_1 = 1/2$ and $a_2 = 3/2$ so $\mathbf{x} = (1/2)[1, 1]^T + (3/2)[1, -1]^T = [2, -1]^T$.   **(c)** $\lambda_1 = 0$, $\mathbf{e}_1 = [1, -1]^T$; $\lambda_2 = 6$, $\mathbf{e}_2 = [1, 1]^T$. $c_1 = \mathbf{c} \cdot \mathbf{e}_1 / \mathbf{e}_1 \cdot \mathbf{e}_1 = 1 \neq 0$ so no solution.   **(d)** $\lambda_1 = 0$, $\mathbf{e}_1 = [1, -1]^T$; $\lambda_2 = 6$, $\mathbf{e}_2 = [1, 1]^T$. $c_1 = 0$, $c_2 = 12/2 = 6$ so (4.3) gives $a_1 =$ arbitrary and $a_2 = 6/6 = 1$. Thus, $\mathbf{x} = a_1[1, -1]^T + [1, 1]^T$ or $[1 + a_1, 1 - a_1]^T$.   **(e)** $\lambda_1 = 2$, $\mathbf{e}_1 = [1, 1, 1]^T$; $\lambda_2 = \lambda_3 = -1$, $\mathbf{e} = \alpha[1, -1, 0]^T + \beta[1, 0, -1]^T$. Let $\mathbf{e}_2 = [1, -1, 0]^T$ and seek $\mathbf{e}_3 = \alpha[1, -1, 0]^T + \beta[1, 0, -1]^T$ so that $\mathbf{e}_3 \cdot \mathbf{e}_2 = 2\alpha + \beta = 0$. Take $\alpha = 1$ and $\beta = -2$, say, so $\mathbf{e}_3 = [-1, -1, 2]^T$. Then $c_1 = 6/3 = 2$, $c_2 = -1/2$, $c_3 = 3/6 = 1/2$ so (4.3) gives $a_1 = 2/2 = 1$, $a_2 = (-1/2)/(-1) = 1/2$, $a_3 = (1/2)/(-1) = -1/2$. Thus, $\mathbf{x} = 1\mathbf{e}_1 + (1/2)\mathbf{e}_2 + (-1/2)\mathbf{e}_3 = [2, 1, 0]^T$.

6. (b) For $\mathbf{A}$ given by (6.5), $\lambda_1 = -3$, $\mathbf{e}_1 = [0, 1, 1]^T$; $\lambda_2 = 2$, $\mathbf{e}_2 = [1, 0, 0]^T$; $\lambda_3 = 1$, $\mathbf{e}_3 = [0, 1, -1]^T$. Let $\mathbf{x}_1 = [1, 0, 0]^T$, $\mathbf{x}_2 = [0, 1, 0]^T$, $\mathbf{x}_3 = [0, 0, 1]^T$, $\mathbf{x}_4 = [1, 2, 3]^T$, $\mathbf{x}_5 = [3, -1, 4]^T$, and $\mathbf{x}_6 = [0.98, 0, 0.01]^T$. We find $R(\mathbf{x}_1) = 2$, $R(\mathbf{x}_2) = -1$, $R(\mathbf{x}_3) = -1$, $R(\mathbf{x}_4) = -2.5$, $R(\mathbf{x}_5) = 0.654$, $R(\mathbf{x}_6) = 1.969$. Note that in all cases $|R(\mathbf{x})| \leq |-3| = 3$. Also, $R(\mathbf{x}_2) = \lambda_2$ because $\mathbf{x}_2 = \mathbf{e}_2$. Finally, note that $\mathbf{x}_6$ is close to $\mathbf{e}_2$ so $R(\mathbf{x}_6)$ is close to λ_2.

7. (b) *Maple* gives $\lambda_1 = 2$, $\mathbf{e}_1 = \alpha[1, 1, 2]^T$; $\lambda_2 = -1$, $\mathbf{e}_2 = \beta[1, 1, -1]^T$; $\lambda_3 = 0$, $\mathbf{e}_3 = \gamma[1, -1, 0]^T$. $\mathbf{x}^{(0)} = [1, 0, 0]^T$ gives $\mathbf{x}^{(4)} = [3, 3, 5]^T$ and $\mathbf{x}^{(5)} = [5, 5, 11]^T$ so $\lambda \approx \mathbf{x}^{(4)} \cdot \mathbf{x}^{(5)} / \mathbf{x}^{(4)} \cdot \mathbf{x}^{(4)} = 1.98$. $\mathbf{x}^{(0)} = [0, 1, 0]^T$ and $\mathbf{x}^{(0)} = [0, 0, 1]^T$ gives the same result so we can conclude that $\lambda_1 \approx 1.98$ and $\mathbf{e}_1 \approx [5, 5, 11]^T$. These are close to the exact values noted above.

8. (a) *Maple* gives $\lambda_1 = 7$, $\mathbf{e}_1 = \alpha[0, 1, 1]^T$; $\lambda_2 = 3$, $\mathbf{e}_2 = \beta[2, 1, -1]^T$; $\lambda_3 = 0$, $\mathbf{e}_3 = \gamma[1, -1, 1]^T$. $\mathbf{x}^{(0)} = [1, 0, 0]^T$ gives $\mathbf{x}^{(1)} = [2, 1, -1]^T$ and $\mathbf{x}^{(2)} = [6, 3, -3]^T$ so we happen to converge, in only one step, to an exact eigenpair, namely, $\lambda = 3$ and $\mathbf{e} = [2, 1, -1]^T$. Next, $\mathbf{x}^{(0)} = [0, 1, 0]^T$ gives $\mathbf{x}^{(4)} = [27, 1214, 1187]^T$ and $\mathbf{x}^{(5)} = [81, 8444, 8363]^T$ so $\lambda \approx \mathbf{x}^{(4)} \cdot \mathbf{x}^{(5)} / \mathbf{x}^{(4)} \cdot \mathbf{x}^{(4)} = 6.998$. Finally, $\mathbf{x}^{(0)} = [0, 0, 1]^T$ gives $\mathbf{x}^{(4)} = [-27, 1187, 1214]^T$ and $\mathbf{x}^{(5)} = [-81, 8363, 8444]^T$ so $\lambda \approx 6.998$. Comparing these results we see that $\mathbf{x}^{(0)} = [1, 0, 0]^T$ must have been orthogonal to $\mathbf{e}_1$ (which, from the *Maple* results given above, is seen to be true), so $\lambda_1 \approx 6.998$ and $\mathbf{e}_1 \approx [-81, 8363, 8444]^T$.

9. Use these (or equivalent) *Maple* commands:

```
with(linalg):
A:=array([[1,1,1],[1,0,0],[1,0,0]]);
A4:=evalm(A^4);
```

```
x0:=array([[1,0,0]]);
x0T:=transpose(x0);
x1:=evalm(A4&*x0T);
x2:=evalm(A4&*x1T);
x3:=evalm(A4&*x2T);
x4:=evalm(A4&*x3T);
x5:=evalm(A4&*x4T);
NUM:=evalm(transpose(x4)&*x5);
DEN:=evalm(transpose(x4)&*x4):
```

Thus, the dominant eigenvalue of $\mathbf{A}^4$ is

$$\lambda \approx \text{NUM}/\text{DEN}$$
$$= 45812984491/2863311531 = 16.00000000.$$

Then the corresponding eigenvalue of $\mathbf{A}$ is

$$\lambda = (16.00000000)^{1/4} = 2.00000000.$$

The eigenvectors of $\mathbf{A}^4$ are the same as those of $\mathbf{A}$ so

$$\mathbf{e} \approx \mathbf{x}^{(5)} = [699051, 349525, 349525]^T$$

or, scaling by $1/349525$, $\mathbf{e} \approx [2.000002861, 1., 1.]^T$. These results are virtually identical to λ_1 and $\mathbf{e}_1$ given in (7.5). To see why the power method converges more rapidly for $\mathbf{A}^4$ than for $\mathbf{A}$ consider (7.8), noting that the eigenvalues of $\mathbf{A}$ and $\mathbf{A}^4$ are $2, -1, 0$ and $16, 1, 0$, respectively. We see from (7.8) that the more dominant is λ_1 the faster is the convergence and 16 does dominate 1 and 0 more than 2 dominates -1 and 0. Or, put differently, note that $\mathbf{x}^{(0)}, \mathbf{x}^{(1)}, \mathbf{x}^{(2)}, \ldots$ for $\mathbf{A}^4$ are the same as $\mathbf{x}^{(0)}, \mathbf{x}^{(4)}, \mathbf{x}^{(8)}, \ldots$ for $\mathbf{A}$.

Section 9.4

1. (a) For $\lambda = 0$, (1.1b) gives $y'(0) = 0 = D$ and $y(l) = 0 = C + Dl$. These give $C = D = 0$ so for $\lambda = 0$ we have only the trivial solution $y(x) = 0$. Hence $\lambda = 0$ is not an eigenvalue. For $\lambda \neq 0$, (1.1a) gives $y'(0) = 0 = -\sqrt{\lambda}A\sin 0 + \sqrt{\lambda}B\cos 0 = \sqrt{\lambda}B$ and $y(l) = 0 = A\cos\sqrt{\lambda}l + B\sin\sqrt{\lambda}l$ so $B = 0$ and $\cos\sqrt{\lambda}l = 0$, which gives $\sqrt{\lambda}l = n\pi/2$ ($n = 1, 3, \ldots$) with A remaining arbitrary. Thus $\lambda_n = (n\pi/2l)^2$ and $\phi_n(x) = \cos(n\pi x/2l)$ for $n = 1, 3, \ldots$. **(c)** For $\lambda = 0$, (1.1b) gives $y'(0) = 0 = D$ and $y'(l) = 0 = D$ so $D = 0$ and C is arbitrary, say $C = 1$. For $\lambda \neq 0$, (1.1a) gives $y'(0) = 0 = -\sqrt{\lambda}A\sin 0 + \sqrt{\lambda}B\cos 0 = \sqrt{\lambda}B$ and $y'(l) = 0 = -\sqrt{\lambda}A\sin\sqrt{\lambda}l + \sqrt{\lambda}B\cos\sqrt{\lambda}l$ so $B = 0$ and $\sin\sqrt{\lambda}l = 0$, which gives $\sqrt{\lambda}l = n\pi$ ($n = 1, 2, \ldots$) with A remaining arbitrary. Thus $\lambda_0 = 0$, $\phi_0(x) = 1$ and $\lambda_n = (n\pi/l)^2$, $\phi_n(x) = \cos(n\pi x/l)$ for $n = 1, 2, \ldots$.

2. (e) $\langle \phi_j(x), \phi_k(x) \rangle = \int_0^\pi (e^{-x}\sin jx)(e^{-x}\sin kx)e^{2x}\,dx$

$$= \int_0^\pi \sin jx \sin kx\,dx$$

$$= \frac{1}{2}\int_0^\pi [\cos(j - k)x - \cos(j + k)x]\,dx.$$

If $j = k$ the latter gives $\frac{1}{2} \int_0^\pi (1 - \cos 2jx)\, dx = \pi/2$, but we are considering different eigenfunctions so $j \neq k$, in which case the integral gives $\frac{1}{2} \left[\frac{\sin(j-k)x}{j-k} - \frac{\sin(j+k)x}{j+k} \right] \Big|_{x=0}^{x=\pi} = 0$. Hence $\phi_j(x)$ and $\phi_k(x)$ are orthogonal (if $j \neq k$).

3. (b) $y(1) = 0 = C + D \ln 1 = C$ and $y(a) = 0 = C + D \ln a$ give $C = 0$ and $D = 0$. Hence $y(x) = 0$ and $\lambda = 0$ is not an eigenvalue of (3.1).

5. (a) With $y(x) = C + Dx$, $y(0) = 0 = C$ and $y(1) + y'(1) = 0 = C + D + D$ give $C = D = 0$ so $y(x) = 0$ and $\lambda = 0$ is not an eigenvalue of (5.1).   **(b)** $y(0) = 0 = A$ and $y(1) + y'(1) = 0 = A \cos\sqrt{\lambda} + B \sin\sqrt{\lambda} - \sqrt{\lambda} A \sin\sqrt{\lambda} + \sqrt{\lambda} B \cos\sqrt{\lambda}$ give $A = 0$ and $B(\sin\sqrt{\lambda} + \sqrt{\lambda}\cos\sqrt{\lambda}) = 0$. We can't afford to let $B = 0$ because $A = B = 0$ will give the trivial solution $y(x) = 0$. Thus, set $\sin\sqrt{\lambda} + \sqrt{\lambda}\cos\sqrt{\lambda} = 0$. Denoting the roots as λ_n and setting $B = 1$, say, gives the eigenfunctions $\phi_n(x) = \sin\sqrt{\lambda_n}x$.   **(d)** The command `fsolve(tan(z)+z=0,z);` merely gives the root $z = 0$. To find the roots, one after the other, we need to bracket where they are. From the sketch of the graph of $\tan \Lambda = -\Lambda$ from part (c), it is evident that the first positive root lies in the interval $\pi/2 < \Lambda < \pi$, the second in $3\pi/2 < \Lambda < 2\pi$, and so on. Thus, use the commands

```
fsolve(tan(z)+z=0,z,Pi/2..Pi);
%^2;
fsolve(tan(z)+z=0,z,3*Pi/2..2*Pi);
%^2;
```

and so on. These give

$\Lambda = 2.028757838, 4.913180439, 7.978665712, 11.08553841$

and

$\lambda = 4.115858365, 24.13934203, 63.65910654, 122.8891618.$

For comparison with the asymptotic formula (5.6), $\Lambda_n \sim (2n-1)\pi/2 = 1.57, 4.71, 7.85, 11.00.$

6. Imposing the boundary conditions on $y(x) = C + Dx$ (for $\lambda = 0$) gives $D = 0$ and $C =$ arbitrary, so $\lambda = 0$ is an eigenvalue with eigenfunction $\phi(x) = 1$. Doing so for $y(x) = A\cos\sqrt{\lambda}x + B\sin\sqrt{\lambda}x$ (for $\lambda \neq 0$) gives the equations $(1 - c)A - sB = 0$ and, after canceling a factor of $\sqrt{\lambda}$ because $\lambda \neq 0$, $-sA + (1 + c)B = 0$, where c, s are short for $\cos\sqrt{\lambda}$, $\sin\sqrt{\lambda}$, respectively. For nontrivial solutions for A and B set the determinant of the coefficient matrix to zero. That step gives $0 = 0$ so it is satisfied by *all* λ's > 0. Gauss elimination gives the equivalent system $(1 - c)A - sB = 0$ and $0 = 0$ so we can solve for B in terms of A or vice versa. For instance, $B = (1 - c)A/s$ where A remains arbitrary. With $A = s$, say, $B = 1 - c$, so the eigenfunction is $\phi(x) = \sin\sqrt{\lambda}\cos\sqrt{\lambda}x + (1 - \cos\sqrt{\lambda})\sin\sqrt{\lambda}x =$ $\sin\sqrt{\lambda}x + \sin[\sqrt{\lambda}(1 - x)]$.   No, (6.1) is not of the Sturm-Liouville form (28) because its boundary conditions are not "separated." That is, (28b) is at $x = a$ and (28c) is at $x = b$, but each boundary condition in (6.1) involves values at both endpoints.

7. (a) $y(x) = A + Bx + C\cos\sqrt{\lambda}x + D\sin\sqrt{\lambda}x$ for $\lambda \neq 0$ and $y(x) = E + Fx + Gx^2 + Hx^3$ for $\lambda = 0$.   **(c)** Applying the boundary conditions to the general solution will give

$$\begin{bmatrix} 1 & 0 & 1 & 0 \\ 0 & 1 & 0 & a \\ 1 & 1 & c & s \\ 0 & 1 & -as & ac \end{bmatrix} \begin{bmatrix} A \\ B \\ C \\ D \end{bmatrix} = \mathbf{0},$$

where $c \equiv \cos\sqrt{\lambda}$, $s \equiv \sin\sqrt{\lambda}$, and $a \equiv \sqrt{\lambda}$ for brevity. For the latter to have nontrivial solutions (in addition to the trivial solution $A = B = C = D = 0$) set the determinant of the coefficient matrix to zero (Theorem 5.7.3). That step gives (7.2). Next you need to *find* those nontrivial solutions and you can do that by Gauss elimination, which gives $D =$ arbitrary, $C = (a - s)D/(c - 1)$, $B = -aD$, $A = (s - a)D/(c - 1)$. [Note that $c - 1 = \cos\sqrt{\lambda} - 1 \neq 0$ because λ's that satisfy $\cos\sqrt{\lambda} = 1$ do not satisfy (7.2).] Choosing $D = c - 1$, say, gives (7.3).

9. With $N = 8$ the approximate eigenvalues (followed by exact values in parentheses) are $\lambda_1 = 9.74/l^2$ $(9.87/l^2)$, $\lambda_2 = 37.49/l^2$ $(39.48/l^2)$, $\lambda_3 = 79.02/l^2$ $(88.83/l^2)$, $\lambda_4 = 128/l^2$ $(157.91/l^2)$, $\lambda_5 = 176.98/l^2$ $(246.74/l^2)$, $\lambda_6 = 218.51/l^2$ $(355.31/l^2)$, $\lambda_7 = 246.26/l^2$ $(483.61/l^2)$. Comparing these results for $N = 8$ with those for $N = 4$ given in (26) and (27) we see that with $N = 8$ we improve the accuracy of $\lambda_1, \ldots, \lambda_3$ and pick up first approximations of $\lambda_4, \ldots, \lambda_7$.

CHAPTER 10

Section 10.1

1. Assuming $x_1 > x_2 > x_3 > 0$ gives $m_1 x_1'' = -kx_1 - k(x_1 - x_2) - k(x_1 - x_3)$, $m_2 x_2'' = k(x_1 - x_2) - k(x_2 - x_3)$, $m_3 x_3'' = k(x_1 - x_3) + k(x_2 - x_3) + F(t)$. Assuming $x_3 > x_2 > x_1 > 0$ gives $m_1 x_1'' = -kx_1 + k(x_2 - x_1) + k(x_3 - x_1)$, $m_2 x_2'' = -k(x_2 - x_1) + k(x_3 - x_2)$, $m_3 x_3'' = -k(x_3 - x_1) - k(x_3 - x_2) + F(t)$. Similarly if $x_1 > x_3 > x_2 > 0$.

2. (a) Assuming $i_1 > i_2 > i_3 > 0$, for instance, gives $E_1(t) - (1/C)\int(i_1 - i_3)\, dt - R(i_1 - i_2) = 0$, $E_2(t) - (1/C)\int(i_3 - i_1)\, dt = 0$, $-E_2(t) - R(i_2 - i_1) = 0$. Differentiating the first two of these three equations, to eliminate the integrals, gives $R(i_1' - i_2') + (1/C)(i_1 - i_3) = E_1'(t)$, $(1/C)(i_3 - i_1) = E_2'(t)$, $R(i_2 - i_1) = -E_2(t)$.

Section 10.2

1. (a) By Theorem 10.2.3, $\mathbf{x}(t) = C_1[1, 1]^T e^{2t} + C_2[1, -1]^T$ is a general solution so $x_1(t) = C_1 e^{2t} + C_2$ and $x_2(t) = C_1 e^{2t} - C_2$. $x_1(0) = 5 = C_1 + C_2$ and $x_2(0) = -3 = C_1 - C_2$ so $C_1 = 1$ and $C_2 = 4$. Thus, $x_1(t) = e^{2t} + 4$ and $x_2(t) = e^{2t} - 4$, which solution is unique according to Theorem 10.2.1.   **(c)** General solution: $x_1(t) = C_1 e^{4t} + 4C_2 e^{-3t} + 12t - 1$, $x_2(t) = C_1 e^{4t} - 3C_2 e^{-3t} - 36t - 33$. Unique solution satisfying $x_1(-1) = 0$ and $x_2(-1) = 5$: $x_1(t) = (47/7)e^{4(t+1)} + (44/7)e^{-3(t+1)} + 12t - 1$, $x_2(t) = (47/7)e^{4(t+1)} - (33/7)e^{-3(t+1)} - 36t - 33$.

2. (a) General solution: $x_1(t) = 2C_1 + C_2 + C_3 e^{3t}$, $x_2(t) = -C_1 - 2C_2 + C_3 e^{3t}$, $x_3(t) = -C_1 + C_2 + C_3 e^{3t}$. Unique solution satisfying the initial conditions: $x_1(t) = 7 + 8e^{3t}$, $x_2(t) = 13 + 8e^{3t}$, $x_3(t) = -20 + 8e^{3t}$.   **(c)** General solution: $x_1(t) = 13C_1 + 3C_2 e^{5t} + 3C_3 e^{2t} - 30e^{-t}$, $x_2(t) = C_1 + C_2 e^{5t} + C_3 e^{2t} + 2e^{-t}$, $x_3(t) = 5C_2 e^{5t} + 2C_3 e^{2t} + 7e^{-t}$. Unique solution satisfying the initial conditions: $x_1(t) = (234/5) + (21/5)e^{5t} - 21e^{2t} - 30e^{-t}$, $x_2(t) = (18/5) + (7/5)e^{5t} - 7e^{2t} + 2e^{-t}$, $x_3(t) = 7e^{5t} - 14e^{2t} + 7e^{-t}$.

3. Since (8) satisfies (7) for any set of C_j's (by assumption), it must satisfy (7) with $C_1 = \cdots = C_n = 0$ so it follows that $\mathbf{x}'_p = \mathbf{A}\mathbf{x}_p + \mathbf{f}(t)$. Next, put (8) into (7), obtaining $C_1\mathbf{x}'_1 + \cdots + C_n\mathbf{x}'_n + \mathbf{x}'_p = C_1\mathbf{A}\mathbf{x}_1 + \cdots + C_n\mathbf{A}\mathbf{x}_n + \mathbf{A}\mathbf{x}_p + \mathbf{f}(t)$. Canceling $\mathbf{x}'_p$ with $\mathbf{A}\mathbf{x}_p + \mathbf{f}(t)$ leaves $C_1\mathbf{x}'_1 + \cdots + C_n\mathbf{x}'_n = C_1\mathbf{A}\mathbf{x}_1 + \cdots + C_n\mathbf{A}\mathbf{x}_n$. Since the C_j's are arbitrary it follows from the latter that $\mathbf{x}'_1 = \mathbf{A}\mathbf{x}_1, \ldots, \mathbf{x}'_n = \mathbf{A}\mathbf{x}_n$.

Section 10.3

1. (a) $(D^2 - D - 2)x = 0$ gives $x(t) = C_1 e^{2t} + C_2 e^{-t}$ and $(D^2 - D - 2)y = 0$ gives $y(t) = C_3 e^{2t} + C_4 e^{-t}$. Putting these into $(D - 1)x + Dy = 0$ gives $C_1 e^{2t} - 2C_2 e^{-t} + 2C_3 e^{2t} - C_4 e^{-t} = 0$. Hence $C_1 + 2C_3 = 0$ and $2C_2 + C_4 = 0$ so $C_3 = -C_1/2$ and $C_4 = -2C_2$. Thus, $x(t) = C_1 e^{2t} + C_2 e^{-t}$ and $y(t) = -(C_1/2)e^{2t} - 2C_2 e^{-t}$.   **(d)** $(4D^2 - 1)x = 1$ and $(4D^2 - 1)y = t + 1$; $x(t) = C_1 e^{t/2} + C_2 e^{-t/2} - 1$, $y(t) = C_1 e^{t/2} - (C_2/3)e^{-t/2} - 1 - t$ **(g)** $(D^2 - 16)x = 13 - 2t$ and $(D^2 - 16)y = 14 - 2t$; $x(t) = C_1 e^{4t} + C_2 e^{-4t} - 13/16 + t/8$, $y(t) = (C_1/3)e^{4t} - C_2 e^{-4t} - 7/8 + t/8$ **(j)** $(D^4 - 9)x = 0$ and $(D^4 - 9)y = 0$; $x(t) = C_1 \cosh\sqrt{3}t + C_2 \sinh\sqrt{3}t + C_3 \cos\sqrt{3}t + C_4 \sin\sqrt{3}t$, $y(t) = -(C_1/2)\cosh\sqrt{3}t - (C_2/2)\sinh\sqrt{3}t + C_3 \cos\sqrt{3}t + C_4 \sin\sqrt{3}t$ **(m)** Cramer's rule gives $x(t) = 0/(D^3 - 3D^2)$, $y(t) = 0/(D^3 - 3D^2)$, $z(t) = 0/(D^3 - 3D^2)$, so $x''' - 3x'' = 0$, $y''' - 3y'' = 0$, and $z''' - 3z'' = 0$. Solving, $x(t) = C_1 + C_2 t + C_3 e^{3t}$, $y(t) = C_4 + C_5 t + C_6 e^{3t}$, and $z(t) = C_7 + C_8 t + C_9 e^{3t}$. To express the nine constants $C_1, \ldots, C_9$ in terms of three independent constants, put the foregoing three solutions into the three original ODEs. In each of those three equations match coefficients of the LI functions 1, t, and e^{3t}. That step gives nine linear algebraic equa-

tions in $C_1, \ldots, C_9$. Solving these (by the *Maple* linsolve command, for example) gives $C_1 = -2 - \alpha - \beta$, $C_2 = 4$, $C_3 = \gamma$, $C_4 = \alpha$, $C_5 = -2$, $C_6 = \gamma$, $C_7 = \beta$, $C_8 = -2$, $C_9 = \gamma$, so $x(t) = -2 - \alpha - \beta + 4t + \gamma e^{3t}$, $y(t) = \alpha - 2t + \gamma e^{3t}$, $z(t) = \beta - 2t + \gamma e^{3t}$.

2. (m) For the general solution use

```
dsolve({diff(x(t),t)=x(t)+y(t)+z(t)+6,
diff(y(t),t)=x(t)+y(t)+z(t),
diff(z(t),t)=x(t)+y(t)+z(t)},{x(t),y(t),z(t)});
```
With $x(0) = 3$, $y(0) = 5$, $z(0) = -4$, say, use
```
dsolve({diff(x(t),t)=x(t)+y(t)+z(t)+6,
diff(y(t),t)=x(t)+y(t)+z(t),
diff(z(t),t)=x(t)+y(t)+z(t),x(0)=3,y(0)=5,
z(0)=-4},{x(t),y(t),z(t)});
```

3. (a) $z(t) = \gamma\left(\frac{\beta e^{-\alpha t} - \alpha e^{-\beta t}}{\beta - \alpha} + 1\right)$, $x(t) = \gamma e^{-\alpha t}$, $y(t) = \gamma - z(t) - x(t)$

4. (a) $x(t) = C_1 \cos\alpha t + C_2 \sin\alpha t + C_3$, $y(t) = C_2 \cos\alpha t - C_1 \sin\alpha t + C_4$, $z(t) = C_5 + C_6 t$, where $\alpha = qB/m$

6. (a) Elimination gives $(2D^4 + 6D^2 + 3)x_1 = F(2 - \Omega^2)\sin\Omega t$ with solution $x_1(t) = C_1 \cos 0.796t + C_2 \sin 0.796t + C_3 \cos 1.538t + C_4 \sin 1.538t + [F(2 - \Omega^2)/(2\Omega^4 - 6\Omega^2 + 3)]\sin\Omega t$. Putting that into the first differential equation, $2x_1'' + 2x_1 - x_2 = F\sin\Omega t$, and solving the latter for x_2 gives

$$x_2(t) = 0.732(C_1 \cos 0.796t + C_2 \sin 0.796t)$$
$$- 2.73(C_3 \cos 1.538t + C_4 \sin 1.538t)$$
$$+ [F/(2\Omega^4 - 6\Omega^2 + 3)]\sin\Omega t.$$

The natural frequencies are 0.796 rad/sec and 1.538 rad/sec.

8. $F_1 \sin\Omega t$ is not the only force applied to the mass; there are two spring forces as well. If $\Omega = \sqrt{2}$ the applied force $F_1 \sin\Omega t$ and the two spring forces cancel.

9. (a) Elimination gives $x' - x = -3$ and $y' - y = 0$ so $x(t) = C_1 e^t + 3$ and $y(t) = C_2 e^t$. Put these into the original differential equations and find that $C_2 = 2C_1$, so $x(t) = C_1 e^t + 3$ and $y(t) = 2C_1 e^t$. Thus $x(0) = C_1 + 3$ and $y(0) = 2C_1$ so there will be a unique solution if $x(0)$ and $y(0)$ happen to satisfy $y(0) = 2[x(0) - 3]$ and no solution if $y(0) \neq 2[x(0) - 3]$. Given that $x(0) = 1$ and $y(0) = 2$, $2 \neq 2[1 - 3]$ so there is no solution.   **(c)** Unique solution $x(t) = 6e^{3(t-1)/7}$, $y(t) = 4e^{3(t-1)/7}$ **(e)** Elimination gives $0x = 5 - 4t$ so there is no solution.

Section 10.4

1. (a) $\mathbf{x}(t) = \alpha[2, -1]^T + \beta[1, 3]^T e^{7t}$, where $\mathbf{x}(t)$ denotes $[x(t), y(t)]^T$; general solution   **(d)** $\mathbf{x}(t) = \alpha[1, -3]^T +$

$\beta[1, -1]^T e^{2t}$; general solution   (g) $\lambda = r^2 = -1$, $\mathbf{e} = \alpha[3, -7]^T$; $\lambda = r^2 = 5$, $\mathbf{e} = \beta[3, -5]^T$. Thus,

$$\mathbf{x}(t) = A[3, -7]^T e^{it} + B[3, -7]^T e^{-it} + C[3, -5]^T e^{\sqrt{5}t}$$
$$+ D[3, -5]^T e^{-\sqrt{5}t}$$
$$= [3, -7]^T (Ae^{it} + Be^{-it}) + [3, -5]^T (Ce^{\sqrt{5}t} + De^{-\sqrt{5}t})$$
$$= [3, -7]^T (C_1 \cos t + C_2 \sin t)$$
$$+ [3, -5]^T (C_3 \cosh \sqrt{5}t + C_4 \sinh \sqrt{5}t);$$
general solution

(j) $\mathbf{x}(t) = \alpha[2, 3, 1]^T e^t + \beta[3, 7, 1]^T e^{2t} + \gamma[1, 1, 1]^T$; general solution
(m) $\mathbf{x}(t) = \alpha[1, 1, 0]^T e^t$; not general solution   (p) $\lambda = r^2 = -1$, $\mathbf{e} = \alpha[1, -1, 1]^T$ so $r = +i$ and $-i$; $\mathbf{x}(t) = A[1, -1, 1]^T e^{it} + B[1, -1, 1]^T e^{-it} = [1, -1, 1]^T (C_1 \cos t + C_2 \sin t)$; not general solution

2. (a) By the method of Chapter 6 seek $x(t) = e^{rt}$. Find $r = -1$ and -2 and hence the general solution $x(t) = \alpha e^{-t} + \beta e^{-2t}$. Instead, although this method is longer, convert $x'' + 3x' + 2x = 0$ to the system $x' = y$, $y' = -2x - 3y$. Seeking $\mathbf{x}(t) = [x(t), y(t)]^T = \mathbf{q}e^{rt}$ obtain $A\mathbf{q} = r\mathbf{q}$ where $A = $ array$([0, 1], [-2, -3])$. Then $\lambda = r = -1$ with $\mathbf{e} = \alpha[1, -1]^T$ and $\lambda = r = -2$ with $\mathbf{e} = \beta[1, -2]^T$ so $\mathbf{x}(t) = \alpha[1, -1]^T e^{-t} + \beta[1, -2]^T e^{-2t}$, the first component of which gives $x(t) = \alpha e^{-t} + \beta e^{-2t}$ once again.

3. $\mathbf{x}(t) = [x_1(t), \ldots, x_5(t)]^T = \mathbf{q}e^{rt}$ gives $A\mathbf{q} = -r^2 \mathbf{q}$. *Maple* gives $\lambda = -r^2 = 1, 2, 3, 2 + \sqrt{3}, 2 - \sqrt{3}$ so $r = \pm i, \pm \sqrt{2}i, \pm \sqrt{3}i, \pm \sqrt{2 + \sqrt{3}}i, \pm \sqrt{2 - \sqrt{3}}i$ so the natural frequencies are $1, \sqrt{2}, \sqrt{3}, \sqrt{2 + \sqrt{3}}, \sqrt{2 - \sqrt{3}}$ and, from *Maple*, the corresponding mode shapes (eigenvectors) are $[1, 1, 0, -1, -1]^T$, $[1, 0, -1, 0, 1]^T$, $[1, -1, 0, 1, -1]^T$, $[1, -\sqrt{3}, 2, -\sqrt{3}, 1]^T$, $[1, \sqrt{3}, 2, \sqrt{3}, 1]^T$, respectively.

4. (d) For the $\mathbf{B}$ matrix *Maple* gives $\lambda = 2 - \sqrt{2}, 2, 2 + \sqrt{2}, 4$ so $\omega = 4c\sqrt{\lambda}/L = 3.061c/L, 5.657c/L, 7.391c/L, 8c/L$ and the eigenvectors give the corresponding column vectors on the right-hand side of (4.18).   **(f)** *Maple* gives $\lambda \approx 0.0979$ as the smallest eigenvalue so $\omega = 10\sqrt{\lambda}c/L \approx 3.13c/L$, which is indeed very close to the value $\omega = \pi c/L \approx 3.14c/L$ for the continuous string. Further, if we plot the corresponding eigenvector, $[1, 2.902, 4.520, 5.70, 6.32, 6.32, 5.70, 4.520, 2.902, 1]^T$, and scale it, we find the result almost coincident with $\sin \pi x/L$.

6. (a) $\lambda = 4, 4$; $\mathbf{q} = \alpha[2, 1]^T$; $\mathbf{p} = [2\beta - \alpha, \beta]^T$. Thus, $x(t) = (2\beta - \alpha + 2\alpha t)e^{4t}$, $y(t) = (\beta + \alpha t)e^{4t}$. Note that your solution may look different but may be equivalent to the one given above. For instance, *Maple* gives $x(t) = [A + (4B - 2A)t]e^{4t}$,

$y(t) = [B + (2B - A)t]e^{4t}$. To show that the solution given above and the *Maple* solution are equivalent we need to show that these four equations in the two unknowns A, B are consistent: $A = 2\beta - \alpha$, $-2A + 4B = 2\alpha$, $B = \beta$, $-A + 2B = \alpha$. Gauss elimination shows that they are.   **(c)** $\lambda = 1, 1, 1$ so use (6.2). Putting (6.2) into $\mathbf{x}' = A\mathbf{x}$ and equating coefficients of e^t, te^t, and t^2e^t gives the equations $A\mathbf{q} = \mathbf{q}$, $(A - I)\mathbf{p} = 2\mathbf{q}$, $(A - I)\sigma = \mathbf{p}$ for $\mathbf{q}$, $\mathbf{p}$, and σ. We obtain $\mathbf{q} = \alpha[1, -1, -1]^T$, $\mathbf{p} = [-\beta, 2\alpha + \beta, \beta]^T$, $\sigma = [-2\alpha - \gamma, -4\alpha - \beta + \gamma, \gamma]^T$ so $x(t) = (-2\alpha - \gamma - \beta t + \alpha t^2)e^t$, $y(t) = [-4\alpha - \beta + \gamma + (2\alpha + \beta)t - \alpha t^2]e^t$, $z(t) = (\gamma + \beta t - \alpha t^2)e^t$.

Section 10.5

1. In each case, if A is an $n \times n$ diagonalizable matrix, then $\mathbf{D} = \{d_{ij}\}$ is diagonal with $d_{jj} = \lambda_j$. **(a)** $\lambda_1 = 2$, $\mathbf{e}_1 = \alpha[1, 0]^T$; $\lambda_2 = 0$, $\mathbf{e}_2 = \beta[3, 2]^T$. With $\alpha = \beta = 1$, say, $\mathbf{Q} = $ array$([1, 3], [0, 2])$.   **(b)** There is only one LI eigenvector; hence, A is not diagonalizable.   **(d)** $\lambda_1 = 2$, $\mathbf{e}_1 = \alpha[1, 1, 2]^T$; $\lambda_2 = 0$, $\mathbf{e}_2 = \beta[1, -1, 0]^T$; $\lambda_3 = -1$, $\mathbf{e}_3 = \gamma[1, 1, -1]^T$. With $\alpha = \beta = \gamma = 1$, say, $\mathbf{Q} = $ array$([1, 1, 1], [1, -1, 1], [2, 0, -1])$.   **(g)** $\lambda_1 = 7$, $\mathbf{e}_1 = \alpha[0, 1, 1]^T$; $\lambda_2 = 3$, $\mathbf{e}_2 = \beta[2, 1, -1]^T$; $\lambda_3 = 0$, $\mathbf{e}_3 = \gamma[1, -1, 1]^T$. With $\alpha = \beta = \gamma = 1$, say, $\mathbf{Q} = $ array$([0, 2, 1], [1, 1, -1], [1, -1, 1])$.   **(j)** $\lambda_1 = 0$, $\mathbf{e}_1 = \alpha[1, 0, -1, 0]^T$; $\lambda_2 = 0$, $\mathbf{e}_2 = \beta[1, 0, 0, -1]^T$; $\lambda_3 = 0$, $\mathbf{e}_3 = \gamma[1, -1, 0, 0]^T$; $\lambda_4 = 4$, $\mathbf{e}_4 = \delta[1, 1, 1, 1]^T$. With $\alpha = \beta = \gamma = \delta = 1$, say, $\mathbf{Q} = $ array$([1, 1, 1, 1], [0, 0, -1, 1], [-1, 0, 0, 1], [0, -1, 0, 1])$.

2. (a) $\lambda_1 = 0$, $\mathbf{e}_1 = \alpha[1, -1]^T$; $\lambda_2 = 2$, $\mathbf{e}_2 = \beta[1, 1]^T$; Take $\mathbf{Q} = $ array$([1, 1], [-1, 1])$. Then $\tilde{x}' = 0\tilde{x}$ and $\tilde{y}' = 2\tilde{y}$ so $\tilde{x}(t) = C_1$ and $\tilde{y}(t) = C_2 e^{2t}$. Then $\mathbf{x}(t) = \mathbf{Q}\tilde{\mathbf{x}}(t)$ gives $x(t) = C_1 + C_2 e^{2t}$ and $y(t) = -C_1 + C_2 e^{2t}$ **(c)** $\lambda_1 = 3$, $\mathbf{e}_1 = \alpha[4, 1]^T$; $\lambda_2 = -2$, $\mathbf{e}_2 = \beta[1, -1]^T$. Take $\mathbf{Q} = $ array$([4, 1], [1, -1])$. Then $\tilde{x}'' = 3\tilde{x}$ and $\tilde{y}'' = -2\tilde{y}$ so $\tilde{x}(t) = C_1 \cosh \sqrt{3}t + C_2 \sinh \sqrt{3}t$ and $\tilde{y}(t) = C_3 \cos \sqrt{2}t + C_4 \sin \sqrt{2}t$. Then $\mathbf{x}(t) = \mathbf{Q}\tilde{\mathbf{x}}(t)$ gives $x(t) = 4C_1 \cosh \sqrt{3}t + 4C_2 \sinh \sqrt{3}t + C_3 \cos \sqrt{2}t + C_4 \sin \sqrt{2}t$ and $y(t) = C_1 \cosh \sqrt{3}t + C_2 \sinh \sqrt{3}t - C_3 \cos \sqrt{2}t - C_4 \sin \sqrt{2}t$.   **(e)** $x(t) = C_1 e^{7t} + C_2 e^{2t} + C_3 e^{-t}$, $y(t) = C_2 e^{2t} - 2C_3 e^{-t}$, $z(t) = C_1 e^{7t} - C_2 e^{2t} - C_3 e^{-t}$

5. (a) The linearized equations of motion are $mx'' = (-k_1 + k_2 c^2)x + k_2 csy$ and $my'' = k_2 csx + (k_2 s^2 - k_3)y$, where $c = \cos 30 = \sqrt{3}/2$ and $s = \sin 30 = 1/2$. Since $m = k_1 = k_2 = 1$ and $k_3 = 10$ these become $x'' = -\frac{1}{4}x + \frac{\sqrt{3}}{4}y$ and $y'' = \frac{\sqrt{3}}{4}x - \frac{39}{4}y$. Low mode: frequency $= \sqrt{0.2303} = 0.480$, mode shape $= [1, 0.0455]^T$; high mode: frequency $= \sqrt{9.770} = 3.13$, mode shape $= [1, -21.98]^T$.

6. HINT: Recall partitioning, discussed in Section 5.2.7.

8. (a) $\lambda_1 = 4$, $\mathbf{e}_1 = \alpha[1, 1, 1]^T$; $\lambda_2 = \lambda_3 = -2$, $\mathbf{e} = \beta[0, 1, -1]^T + \gamma[1, -1, 0]^T$. If we choose $\mathbf{e}_1 = [1, 1, 1]^T$, $\mathbf{e}_2 = [0, 1, -1]^T$, $\mathbf{e}_3 = [1, -1, 0]^T$ we can make these the columns of $\mathbf{Q}$, but since those columns are not ON it will not be true that $\mathbf{Q}^{-1} = \mathbf{Q}^T$ so we will need to compute $\mathbf{Q}^{-1}$. Instead, find an orthogonal basis of the eigenspace corresponding to $\lambda = -2$, such as $\mathbf{e}_2 = [0, 1, -1]^T$ and $\mathbf{e}_3 = [2, -1, -1]^T$ ($\beta = 1, \gamma = 2$). Normalizing them, take $\mathbf{Q} = $ array($[[1/\sqrt{3}, 1/\sqrt{3}, 1/\sqrt{3}], [0, 1/\sqrt{2}, -1/\sqrt{2}], [2/\sqrt{6}, -1/\sqrt{6}, -1/\sqrt{6}]]$) in which case $\mathbf{Q}^{-1} = \mathbf{Q}^T$. $\mathbf{D}^{1000} = $ array($[[4^{1000}, 0, 0], [0, 2^{1000}, 0], [0, 0, 2^{1000}]]$). Then (7.1) gives $\mathbf{A}^{1000} = $ array($[[a, b, b], [b, a, b], [b, b, a]]$) where $a = (4^{1000} + 2^{1001})/3$ and $b = (4^{1000} - 2^{1000})/3$.
(c) $\lambda_1 = 3$, $\mathbf{e}_1 = \alpha[4, -1, 8]^T$; $\lambda_2 = 0$, $\mathbf{e}_2 = \beta[1, -1, -1]^T$; $\lambda_3 = -1$, $\mathbf{e}_3 = \gamma[0, 1, 0]^T$. Choose $\mathbf{Q} = $ array($[[4, 1, 0], [-1, -1, 1], [8, -1, 0]]$). Then $\mathbf{Q}^{-1} = \frac{1}{12}$ array($[[1, 0, 1], [8, 0, -4], [9, 12, -3]]$) and $\mathbf{A}^{1000} = \frac{1}{12}$ array($[[(4)3^{1000}, 0, (4)3^{1000}], [9 - 3^{1000}, 12, -3 - 3^{1000}], [(8)3^{1000}, 0, (8)3^{1000}]]$).

9. (a) $I_{xx} = \sigma$, $I_{xy} = 9\sigma/4$, $I_{yy} = 9\sigma$, $I_{zz} = 10\sigma$, $I_{xz} = I_{yz} = 0$ so $\mathcal{I} = $ array($[[\sigma, -9\sigma/4, 0], [-9\sigma/4, 9\sigma, 0], [0, 0, 10\sigma]]$). $\lambda_1 = 9.589\sigma$, $\mathbf{e}_1 = \alpha[1, -3.818, 0]^T$; $\lambda_2 = 0.411\sigma$, $\mathbf{e}_2 = \beta[1, 0.262, 0]^T$; $\lambda_3 = 10\sigma$, $\mathbf{e}_3 = \gamma[0, 0, 1]^T$ so $I_{\tilde{x}\tilde{x}} = 9.589\sigma$, $I_{\tilde{y}\tilde{y}} = 0.411\sigma$, $I_{\tilde{z}\tilde{z}} = 10\sigma$ where $\mathbf{e}_1, \mathbf{e}_2, \mathbf{e}_3$ give the positive $\tilde{x}, \tilde{y}, \tilde{z}$ coordinate directions. **(d)** $\mathcal{I} = $ array($[[31\sigma/3, \sigma/4, 0], [\sigma/4, 31\sigma/3, 0], [0, 0, 62\sigma/3]]$). $\lambda_1 = 127\sigma/12$, $\mathbf{e}_1 = \alpha[1, 1, 0]^T$; $\lambda_2 = 121\sigma/12$, $\mathbf{e}_2 = \beta[-1, 1, 0]^T$; $\lambda_3 = 62\sigma/3$, $\mathbf{e}_3 = \gamma[0, 0, 1]^T$ so $I_{\tilde{x}\tilde{x}} = 127\sigma/12$, $I_{\tilde{y}\tilde{y}} = 121\sigma/12$, $I_{\tilde{z}\tilde{z}} = 62\sigma/3$ where $\mathbf{e}_1, \mathbf{e}_2, \mathbf{e}_3$ give the positive $\tilde{x}, \tilde{y}, \tilde{z}$ coordinate directions.

11. Let the characteristic equation of $\mathbf{A}$ be $\lambda^n + \alpha_1\lambda^{n-1} + \cdots + \alpha_n = 0$. Since $\mathbf{Q}^{-1}\mathbf{A}\mathbf{Q} = \mathbf{D}$, it follows that $\mathbf{A}\mathbf{Q} = \mathbf{Q}\mathbf{D}$ and $\mathbf{A} = \mathbf{Q}\mathbf{D}\mathbf{Q}^{-1}$. Then $\mathbf{A}^2 = \mathbf{Q}\mathbf{D}\mathbf{Q}^{-1}\mathbf{Q}\mathbf{D}\mathbf{Q}^{-1} = \mathbf{Q}\mathbf{D}\mathbf{D}\mathbf{Q}^{-1} = \mathbf{Q}\mathbf{D}^2\mathbf{Q}^{-1}$, $\mathbf{A}^3 = \mathbf{A}^2\mathbf{A} = \mathbf{Q}\mathbf{D}^2\mathbf{Q}^{-1}\mathbf{Q}\mathbf{D}\mathbf{Q}^{-1} = \mathbf{Q}\mathbf{D}^3\mathbf{Q}^{-1}$, and so on. Thus, $\mathbf{A}^n + \alpha_1\mathbf{A}^{n-1} + \cdots + \alpha^n\mathbf{I} = \mathbf{Q}(\mathbf{D}^n + \alpha_1\mathbf{D}^{n-1} + \cdots + \alpha_n\mathbf{I})\mathbf{Q}^{-1}$. Now, $\mathbf{D}^n + \alpha_1\mathbf{D}^{n-1} + \cdots + \alpha^n\mathbf{I}$ is diagonal and its jj element is $\lambda_j^n + \alpha_1\lambda_j^{n-1} + \cdots + \alpha_n$, which is zero for each j. Thus, $\mathbf{A}^n + \alpha_1\mathbf{A}^{n-1} + \cdots + \alpha_n\mathbf{I} = \mathbf{0}$, as was to be shown.

Section 10.6

3. (a) $\lambda_1 = \lambda_2 = 1$, $\mathbf{e}_1 = \alpha[1, -2]^T$. Choose $\alpha = 1$, say. Then, $(\mathbf{A} - \mathbf{I})\mathbf{e}_2 = \mathbf{e}_1$ gives $\mathbf{e}_2 = [0, 1]^T$ so $\mathbf{P} = $ array($[[1, 0], [-2, 1]]$), $\mathbf{P}^{-1} = $ array($[[1, 0], [2, 1]]$), and $\mathbf{J} = \mathbf{P}^{-1}\mathbf{A}\mathbf{P} = $ array($[[1, 1], [0, 1]]$). **(d)** $\mathbf{P} = $ array($[[2, 1], [1, 0]]$), $\mathbf{J} = $ array($[[-6, 1], [0, -6]]$) **(g)** $\mathbf{P} = $ array($[[1, 0, 0], [0, 1, -1], [0, 0, 1]]$), $\mathbf{J} = $ array($[[1, 1, 0], [0, 1, 1], [0, 0, 1]]$) **(j)** $\lambda_1 = \lambda_2 = \lambda_3 = \lambda_4 = 2$, $\mathbf{e} =$

$\alpha[1, -1, 0, 0]^T + \beta[1, 0, 1, 0]^T$. Our choice of $\mathbf{e}_1$ will not impact our calculation of $\mathbf{e}_2$, $\mathbf{e}_3$, $\mathbf{e}_4$ so let us not commit ourselves yet on $\mathbf{e}_1$, and let us take $\mathbf{e}_2 = \alpha[1, -1, 0, 0]^T + \beta[1, 0, 1, 0]^T$. Solving $(\mathbf{A} - 2\mathbf{I})\mathbf{e}_3 = \mathbf{e}_2$ requires that $\beta = 0$ and gives $\mathbf{e}_3 = [\alpha + \gamma - \delta, \delta, \gamma, 0]^T$, where α, γ, δ remain arbitrary. Since $\beta = 0$ in $\mathbf{e}_2$, let us choose $\mathbf{e}_1 = [1, -1, 0, 0]^T$. Next, $(\mathbf{A} - 2\mathbf{I})\mathbf{e}_4 = \mathbf{e}_3$ requires that $\gamma = 0$ and gives $\mathbf{e}_4 = [-\delta + \epsilon - \mu, \mu, \epsilon, \alpha]^T$, where α, δ, ϵ, μ remain arbitrary. We can't set $\alpha = 0$ because (with β already zero) that would give $\mathbf{e}_2 = 0$, but we can take $\alpha = 1$ and $\delta = \epsilon = \mu = 0$, for simplicity. Thus, $\mathbf{P} = $ array($[[1, 1, 1, 0], [0, -1, 0, 0], [1, 0, 0, 0], [0, 0, 0, 1]]$), so $\mathbf{J} = \mathbf{P}^{-1}\mathbf{A}\mathbf{P} = $ array($[[2, 0, 0, 0], [0, 2, 1, 0], [0, 0, 2, 1], [0, 0, 0, 2]]$). Observe the considerable nonuniqueness in $\mathbf{P}$ (but not in the resulting Jordan form $\mathbf{J}$) since $\alpha(\neq 0)$, δ, ϵ, μ were all arbitrary.

4. (a) In $\mathbf{x}' = \mathbf{A}\mathbf{x}$ let $\mathbf{x} = \mathbf{P}\tilde{\mathbf{x}}$. Thus, $\tilde{\mathbf{x}}' = \mathbf{P}^{-1}\mathbf{A}\mathbf{P}\tilde{\mathbf{x}} = \mathbf{J}\tilde{\mathbf{x}}$. Written out, $\tilde{x}' = \tilde{x} + \tilde{y}$ and $\tilde{y}' = \tilde{y}$. Solving these in reverse order gives $\tilde{y} = C_1 e^t$ and $\tilde{x} = (C_2 + C_1 t)e^t$. Then, putting these into $\mathbf{x} = \mathbf{P}\tilde{\mathbf{x}}$ gives $x(t) = (C_2 + C_1 t)e^t$, $y(t) = (C_1 - 2C_2 - 2C_1 t)e^t$. **(d)** $\tilde{\mathbf{x}}' = \mathbf{J}\tilde{\mathbf{x}}$ gives $\tilde{y} = C_1 e^{-6t}$ and $\tilde{x} = (C_2 + C_1 t)e^{-6t}$, then $\mathbf{x} = \mathbf{P}\tilde{\mathbf{x}}$ gives $x(t) = (2C_2 + C_1 + 2C_1 t)e^{-6t}$, $y(t) = (C_2 + C_1 t)e^{-6t}$. **(g)** $\tilde{\mathbf{x}}' = \mathbf{J}\tilde{\mathbf{x}}$ gives $\tilde{z} = C_1 e^t$, $\tilde{y} = (C_2 - C_1 + C_1 t)e^t$, $\tilde{x} = (C_3 + C_2 t + C_1 t^2/2)e^t$. Then, putting these into $\mathbf{x} = \mathbf{P}\tilde{\mathbf{x}}$ gives $x(t) = (C_3 + C_2 t + C_1 t^2/2)e^t$, $y(t) = (C_2 - C_1 + C_1 t)e^t$, $z(t) = C_1 e^t$.

5. (a) In $\mathbf{x}'' = \mathbf{A}\mathbf{x}$ let $\mathbf{x} = \mathbf{P}\tilde{\mathbf{x}}$. Thus, $\tilde{\mathbf{x}}'' = \mathbf{P}^{-1}\mathbf{A}\mathbf{P}\tilde{\mathbf{x}} = \mathbf{J}\tilde{\mathbf{x}}$. Written out, $\tilde{x}'' = \tilde{x} + \tilde{y}$ and $\tilde{y}'' = \tilde{y}$. Thus, $\tilde{y} = C_1 e^t + C_2 e^{-t}$, $\tilde{x} = (C_3 + C_1 t/2)e^t + (C_4 - C_2 t/2)e^{-t}$. Then, putting these into $\mathbf{x} = \mathbf{P}\tilde{\mathbf{x}}$ gives $x(t) = (C_3 + C_1 t/2)e^t + (C_4 - C_2 t/2)e^{-t}$, $y(t) = (C_1 - 2C_3 - C_1 t)e^t + (C_2 - 2C_4 + C_2 t)e^{-t}$

6. Not unless $\alpha = 1$. Do you see why it is not?

Section 10.7

1. (a) $\lambda_1 = 2$, $\mathbf{e}_1 = [1, 0]^T$; $\lambda_2 = 0$, $\mathbf{e}_2 = [3, 2]^T$. $\mathbf{Q} = $ array($[[1, 3], [0, 2]]$), $\mathbf{Q}^{-1} = $ array($[[1, -3/2], [0, 1/2]]$), $\mathbf{E} = $ array($[[e^{2t}, 0], [0, 1]]$), $\mathbf{c} = [3, -1]^T$ so (12) gives $x_1(t) = \frac{9}{2}e^{2t} - \frac{3}{2}$, $x_2(t) = -1$ **(c)** $x_1(t) = \frac{7}{6}e^{-t} + \frac{7}{6}e^{2t} + \frac{1}{2}$, $x_2(t) = \frac{7}{6}e^{-t} + \frac{7}{6}e^{2t} - \frac{1}{2}$, $x_3(t) = -\frac{7}{3}e^{-t} + \frac{7}{3}e^{2t}$ **(e)** $x_1(t) = \frac{11}{3}e^{3t} + \frac{1}{3}$, $x_2(t) = \frac{11}{6}e^{3t} + \frac{3}{2}e^{7t} - \frac{1}{3}$, $x_3(t) = -\frac{11}{6}e^{3t} + \frac{3}{2}e^{7t} + \frac{1}{3}$

2. (a) To obtain a general solution use $\mathbf{c} = [C_1, C_2]^T$ where C_1, C_2 are arbitrary constants; $\mathbf{Q}\mathbf{E}\mathbf{Q}^{-1}$ is the same as was found in Exercise 1(a). The result is $x_1(t) = C_1 e^{2t} + C_2\left(-\frac{3}{2}e^{2t} + \frac{3}{2}\right)$, $x_2(t) = C_2$. **(c)** $x_1(t) = (C_1 + C_2 - C_3)e^{-t}/3 + (C_1 + C_2 + 2C_3)e^{2t}/6 + (C_1 - C_2)/2$, $x_2(t) = (C_1 + C_2 - C_3)e^{-t}/3 + (C_1 + C_2 + 2C_3)e^{2t}/6 + (-C_1 + C_2)/2$, $x_3(t) = (-C_1 - C_2 + C_3)e^{-t}/3 + (C_1 + C_2 + 2C_3)e^{2t}/3$

3. (a) $\mathbf{P}$ and $\mathbf{J}$ were found in Exercise 3 of Section 10.6 and are given in the answers, above. Further, $\mathbf{E} = $

array($[[e^t, 0], [0, e^t]]$) and $\delta = $ array($[[0, 1], [0, 0]]$). Putting these into (38), we obtain $x_1(t) = (3 + 8t)e^t$, $x_2(t) = (2 - 16t)e^t$.   (d) **P** and **J** were given in the answers to Exercise 3 of Section 10.6.   Further, **E** = array($[[e^{-6t}, 0], [0, e^{-6t}]]$) and $\delta = $ array($[[0, 1], [0, 0]]$). Thus, $x_1(t) = (3 - 2t)e^{-6t}$, $x_2(t) = (2 - t)e^{-6t}$.   (g) **P** and **J** given in the answers to Exercise 3 of Section 10.6. Further, **E** = array($[[e^t, 0, 0], [0, e^t, 0], [0, 0, e^t]]$) and $\delta = $ array($[[0, 1, 0], [0, 0, 1], [0, 0, 0]]$). Thus, $x_1(t) = (4 + 6t + 3t^2)e^t$, $x_2(t) = 6te^t$, $x_3(t) = 6e^t$.

7. (a) **Q** = array($[[1, 1], [0, 1]]$), **Q**$^{-1}$ = array($[[1, -1], [0, 1]]$), and **E**(1) = array($[[e, 0], [0, e^3]]$) so $e^\mathbf{A}$ = array($[[e, e^3 - e], [0, e^3]]$).
(d) array($[[e/2+e^3/2, e^3/2-e/2], [e^3/2-e/2, e/2+e^3/2]]$)
(g) array($[[\cosh 1, 0, \sinh 1], [0, 1, 0], [\sinh 1, 0, \cosh 1]]$)

8. (a) Using
```
with(linalg):
A:=array([[1,2],[0,3]]);
exponential(A);
```
gives $e^\mathbf{A}$. For $e^{\mathbf{A}t}$ use `A:=array([[t,2*t],[0,3*t]]);` instead.

9. With **Q** a modal matrix of **A**, write $\mathbf{x} = \mathbf{Q}\tilde{\mathbf{x}}$, $\mathbf{f} = \mathbf{Q}\tilde{\mathbf{f}}$, $\mathbf{c} = \mathbf{Q}\tilde{\mathbf{c}}$, and show that the differential equation becomes $\tilde{\mathbf{x}}' = \mathbf{D}\tilde{\mathbf{x}} + \tilde{\mathbf{f}}(t)$. In scalar form, $\tilde{x}_1' = \lambda_1\tilde{x}_1 + \tilde{f}_1(t), \dots, \tilde{x}_n' = \lambda_n\tilde{x}_n + \tilde{f}_n(t)$ with initial conditions $\tilde{x}_1(t_0) = \tilde{c}_1, \dots, \tilde{x}_n(t_0) = \tilde{c}_n$. Solving these first-order linear problems, show that

$$\tilde{x}_1(t) = e^{\lambda_1(t-t_0)}[\int_{t_0}^t e^{-\lambda_1(\tau-t_0)}\tilde{f}_1(\tau)\,d\tau + \tilde{c}_1]$$

and so on. Show that these can be re-expressed in in vector form as $\tilde{\mathbf{x}}(t) = \int_{t_0}^t \mathbf{E}(t-\tau)\tilde{\mathbf{f}}(\tau)\,d\tau + \mathbf{E}(t-t_0)\tilde{\mathbf{c}}$. Finally, put $\tilde{\mathbf{x}}(t) = \mathbf{Q}^{-1}\mathbf{x}(t)$, $\tilde{\mathbf{f}}(\tau) = \mathbf{Q}^{-1}\mathbf{f}(\tau)$, and $\tilde{\mathbf{c}} = \mathbf{Q}^{-1}\mathbf{c}$.

10. $\lambda_1 = 1$, $\mathbf{e}_1 = [1, 1]^T$; $\lambda_2 = -1$, $\mathbf{e}_2 = [1, -1]^T$ so **Q** = array($[[1, 1], [1, -1]]$), **Q**$^{-1}$ = array($[[1/2, 1/2], [1/2, -1/2]]$), **E**$(t)$ = array($[[e^t, 0], [0, e^{-t}]]$). With $t_0 = 0$, the first term on the right side of (9.2) is $[(7e^t - e^{-t})/2, (7e^t + e^{-t}/2]^T$ and the integral term gives $[t^2 - 3\sinh t, 3 - 3\cosh t]^T$, so $x_1(t) = 2e^t + e^{-t} + t^2$, $x_2(t) = 2e^t - e^{-t} + 3$.

CHAPTER 11

Section 11.2

1. (a) $x^2 + y^2 = C$ (c) $x^2 + y^2 = C$

2. (a) The family of lines $y = -x + C$ with the flow direction arrows so that $x' > 0$ in $y > 0$ and $x' < 0$ in $y < 0$.
(c) The hyperbolas $x^2 - y^2 = C$.   (e) The family of lines $y = x + C$ with the arrows so that $x' > 0$ in $x > 0$ and $x' < 0$ in $x < 0$.

5. Since $m < 0$ in $y = mx + b$, let us write $y = -\kappa(x - x_Q)$ or $x' + \kappa x = \kappa x_Q$ where $\kappa > 0$. The solution $x(t) = Ae^{-\kappa t} + x_Q$ reveals that $x(t) \to x_Q$ only as $t \to \infty$ [unless $x(0) = x_Q$ of course].

6. Your result should support the claims, both of which are true. For instance, for $A = 0.2, 2, 3$, and 3.1 you should find that the period is approximately 6.3, 8.4, 16, and 21, respectively. NOTE: How are we to understand the increase in the period as the trajectory Γ_2 (Fig. 5) approaches the separatrix HUIVH? The idea is that the phase speed is very small near the two ends of Γ_2, in accordance with formula (4.1) of Exercise 4. To see this effect graphically, plot $x(t)$ versus t for Γ_2 using an initial point very close to I, such as $x(0) = 3.14$ and $y(0) = 0$. This can be done using *Maple* as follows:
```
with(DEtools):
phaseportrait({diff(x(t),t)=y(t),diff(y(t),t)=
 -sin(x(t))},[x(t),y(t)],t=0..50,[[x(0)=3.14,
 y(0)=0]],x=-4..4,scene=[t,x],stepsize=.05,
 arrows=NONE);
```

7. (a) $(0, 0)$ and $(1, 1)$

8. The separatrix is given by $2x^2 - x^4 + 2y^2 = 1$.

9. (a) The equations $ax + by = 0$ and $cx + dy = 0$ have the trivial solution $x = y = 0$. That solution is unique if the determinant of the coefficient matrix is nonzero. With $a = 0$, $b = 1, c = -9, d = 0$ that determinant is 9 so the only singular point is at the origin. The general solution of (9.1) is $y(t) = A\cos 3t + B\sin 3t$ and $x(t) = (A/3)\sin 3t - (B/3)\cos 3t$ so the singular point is stable but not asymptotically stable. (c) Asymptotically stable   (e) Stable but not asymptotically stable   (g) Unstable

Section 11.3

2. (a) Saddle; unstable (c) Node; unstable (e) Node; asymptotically stable (g) Center; stable but not asymptotically stable (i) Spiral; asymptotically stable (k) Spiral; unstable

3. (a) First, the commands
```
with(linalg):
A:=array([[1,2],[3,4]]);
eigenvects(A):
```
give $\lambda_1 \approx 5.37$, $\mathbf{e}_1 = [1, 2.186140662]^T$ and $\lambda_2 \approx -0.37$, $\mathbf{e}_2 = [1, -0.6861406620]^T$. Since $\mathbf{e}_1$ and $\mathbf{e}_2$ give the directions of the unstable and stable straight-line manifolds, respectively, we can generate those four trajectories by the commands
```
with(DEtools):
phaseportrait({diff(x(t),t)=x(t)+2*y(t),
 diff(y(t),t)=3*x(t)+4*y(t)},{x(t),y(t)},
 t=-15..15,{[x(0)=1,y(0)=2.186140662],
 [x(0)=-1,y(0)=-2.186140662],
 [x(0)=1,y(0)=-.6861406620],
```

```
[x(0)=-1,y(0)=.6861406620]},
scene=[x,y],stepsize=.05,x=-3..3,y=-3..3,
arrows=NONE);
```

However, we find that the two stable manifolds veer away "at the last minite" instead of approaching the origin. This erroneous result is due to accumulated integration error that moves the representative point away from what should be a straight-line approach to the origin. We can rectify this situation by starting very close to the origin, namely, by using $[x(0) = .01, y(0) = -.006861406620]$ and $[x(0) = -.01, y(0) = .006861406620]$, instead, as the third and fourth initial points. To fill out the picture, choose some additional initial points such as $[x(0) = 1, y(0) = 0]$, $[x(0) = 2, y(0) = 0]$, $[x(0) = -1, y(0) = 0]$, $[x(0) = -2, y(0) = 0]$, $[x(0) = 0, y(0) = 1]$, $[x(0) = 0, y(0) = 2]$, $[x(0) = 0, y(0) = -1]$, and $[x(0) = 0, y(0) = -2]$. Finally, add the flow direction arrows by hand. To illustrate how to do that consider, for instance, the trajectory that crosses the x axis at $x = 2$. At that point the differential equations give $x' = x + 2y = 2 + 0 = 2 > 0$ and $y' = 3x + 4y = 6 + 0 = 6 > 0$. Either $x' > 0$ and/or $y' > 0$ establishes the direction of the arrow. Further, points on the unstable manifold are moving away from the origin and points on the stable manifold are moving toward the origin.

4. (a) $y = 0$ and $1 - x^4 = 0$ give the two equilibrium points $(x, y) = (1, 0)$ and $(-1, 0)$. At $(1, 0)$ the linearized ODEs are $X' = 0X + 1Y$ and $Y' = -4X + 0Y$, where $X = x - 1$ and $Y = y$, so $\mathbf{A} = \text{array}([[0, 1], [-4, 0]])$. We find $\lambda = \pm 2i$. If the system were linear it would follow from $\lambda = \pm 2i$ that $(1, 0)$ is a center, but since the system is nonlinear we conclude [see (i) above Example 9] that it is a center OR a stable or unstable spiral. At $(-1, 0)$ obtain $X' = 0X + 1Y$ and $Y' = 4X + 0Y$, where $X = x + 1$ and $Y = y$, so $\mathbf{A} = \text{array}([[0, 1], [4, 0]])$. $\lambda = \pm 2$ so $(-1, 0)$ is a saddle. **(d)** Singular points at $(x, y) = (n\pi/2, n\pi/2)$ for $n = 0, \pm 1, \pm 2, \ldots$. Noting that $\cos n\pi = (-1)^n$, obtain $\mathbf{A} = \text{array}([[1, -1], [(-1)^n, (-1)^n]])$ so $\lambda = \pm\sqrt{2}$ if n is odd and $\lambda = 1 \pm i$ if n is even. Thus, $(n\pi/2, n\pi/2)$ is a saddle if n is odd and an unstable focus if n is even. **(g)** The only singular point is at $(0, 0)$. $\mathbf{A} = \text{array}([[-2, -1], [1, 0]])$ and $\lambda = -1, -1$. As in part (a) this is a borderline case. According to (ii) above Example 9, $(0, 0)$ is a stable node OR a stable spiral. (A phaseportrait plot reveals it to be a stable node.) **(j)** Stable focus at $(0, 0)$; saddle at $(4, 8)$. **(m)** Singular point only at $(0, 0)$. [You will need to convince yourself that the only (real) root of $2y = \sin y$ is $y = 0$.] $\mathbf{A} = \text{array}([[1, 2], [-1, -1]])$ and $\lambda = \pm i$ so, according to (i) above Example 9, $(0, 0)$ is a center OR a stable or unstable spiral. (A phaseportrait plot reveals it to be an unstable spiral.)

6. This gives $(\infty)(0)$. To use l'Hôpital's rule we need the form to be $0/0$ or ∞/∞. Thus, re-express te^{-2t} as t/e^{2t} and then use l'Hôpital's rule.

9. The representative point approaches the origin more slowly for the case of cubic damping. That result might seem strange since the damping term y^3 might seem more powerful than the damping term y. However, as the oscillation dies out $y \to 0$, and as $y \to 0$ the term y^3 is much smaller than y.

10. (a) $x' = y$ and $y' = -x + y^3$ gives only $(0, 0)$ as a singular point. Then $\mathbf{A} = \text{array}([[0, 1], [-1, 0]])$ so $\lambda = \pm i$, which is a borderline case: either a center or a spiral. Intuitively, if we write the ODE as $x'' - (x'^2)x' + x = 0$ and note that $x'^2 \geq 0$, then we see that the ODE is similar to a damped harmonic oscillator with negative damping. Thus, we expect an unstable spiral at $(0, 0)$. A phaseportrait plot corroborates that expectation. **(c)** The intuitive argument used in (a) fails here because if we write the ODE as $x'' - (x')x' + x = 0$ we cannot claim, about the coefficient x', that $x' \geq 0$ or that $x' \leq 0$ for the whole motion. In fact, a phaseportrait plot reveals the singularity at $(0, 0)$ to be a center.

11. (a) The only singular point is $(0, 0)$. Integrating $dy/dx = -x^2/y$ gives $y^2/2 + x^3/3 = C$. The trajectory through $(0, 0)$ is given by $y^2/2 + x^3/3 = 0$. If $x > 0$ this equation has no roots for y, but if $x < 0$ then $y = \pm\sqrt{2/3}|x|^{3/2}$ which is a rightward cusp. To obtain the phase portrait try these commands:

```
with(DEtools):
phaseportrait({diff(x(t),t)=y(t),diff(y(t),t)=
-x(t)^2},{x(t),y(t)},t=-10..10,{[x(0)=-1,
y(0)=sqrt(2/3)],[x(0)=-1,y(0)=-sqrt(2/3)],
[x(0)=-.4,y(0)=0],[x(0)=-.8,y(0)=0],
[x(0)=-.4,y(0)=0],[x(0)=-.8,y(0)=0]},
stepsize=.05,x=-2..2,y=-2..2,scene=[x,y],
arrows=NONE);
```

Section 11.4

6. (a) Using the commands

```
with(DEtools):
phaseportrait({diff(x(t),t)=y(t),diff(y(t),t)=
-x(t)+.01*(1-x(t)^2)},{x(t),y(t)},t=70..80,
{[x(0)=2,y(0)=0]},stepsize=.05,scene=[t,x],
arrows=NONE);
```

we measure one period to be $T \approx 6.4$, compared with the limiting value $T = 2\pi \approx 6.28$ as $\epsilon \to 0$. We go to $t = 80$, say, to give the trajectory a chance to closely approach the limit cycle, and we display only $70 \leq t \leq 80$, say, to get a close look at one period. **(b)** Changing .01 to 10, and $t = 70..80$ to $t = 70..100$, we measure $T \approx 19.1$, compared with the limiting value $T \sim (3 - 2\ln 2)\epsilon \approx 2.398\epsilon \approx 23.98$ for $\epsilon = 10$.

9. (a) Integrating $m\, dy/dx = (-kx + (2P/l)x[1 - (x/l)^2]^{-1/2})/y$, obtain $my^2 + kx^2 + 4Pl[1 - (x/l)^2]^{1/2} = \text{constant} = C$. To examine this in the neighborhood of the singular point at the origin, use the Taylor expansion $[1 - (x/l)^2]^{1/2} = 1 - \frac{1}{2}(x/l)^2 + \cdots \sim 1 - \frac{1}{2}(x/l)^2$ so that $my^2 + (k - 2P/l)x^2 \approx C - 4Pl \equiv C'$, say. Increasing P, beginning with $P = 0$, the latter is a family of ellipses until $P = kl/2$; for $P > kl/2$ it is a family of hyperbolas. Thus, for $P < kl/2$ the origin is a center (stable) and for $P > kl/2$ it is a saddle (unstable), so $P_{cr} = kl/2$ is the buckling load. (b) Since the singularity at $x = y = 0$ is a center for $P < kl/2$ and a saddle for $P > kl/2$, $P = kl/2$ is a bifurcation point (on a P axis).

10. (a) Singular points at $x_\pm = (1 \pm \sqrt{1 - 4r})/2$ and $y = 0$, where $0 < x_- < x_+ < 1$, for $0 < r < 1/4$. $(x_+, 0)$ is a saddle and $(x_-, 0)$ is a center. The trajectories are given by $y^2 + x^2 + 2r \ln|x - 1| = \text{constant} = C$. Of these, the separatrix is $y^2 + x^2 + 2r \ln|x - 1| = x_+^2 + 2r \ln(1 - x_+)$. Because of the presence of the infinite vertical wire the maximum displacement is $x = a = 1$. In the x, y phase plane the vertical line $x = 1$ is a trajectory, with upward flow direction. As $r \to 1/4$, $x_\pm$ merge at $x = 1/2$ and the merging center and saddle produce a higher-order singularity with no periodic motions. For $r > 1/4$ the singularity disappears altogether. Physically, for $r > 1/4$ the force of attraction is so great as to rule out the possibility of oscillation and the vertical line $x = 1$ in the phase plane is an asymptote for every trajectory. These points are covered in parts (b)–(g).

11. (a) $x = rc$ gives $x' = r'c - rs\theta' = \epsilon rc + rs - r^3 c$ and $y = rs$ gives $y' = r's + rc\theta' = -rc + \epsilon rs - r^3 s$. Adding c times the first equation to s times the second gives (11.2a). Subtracting s times the first from c times the second gives (11.2b). (b) For $r \to 0$ (11.2a) becomes $r' \sim \epsilon r$ so the flow is away from the origin (i.e., r is increasing) if $\epsilon > 0$ and toward it if $\epsilon < 0$. Further, (11.2b) implies a uniform clockwise rotation, so it seems clear that if $\epsilon > 0$ the origin is an unstable spiral and that if $\epsilon < 0$ the origin is a stable spiral. Alternatively, we can see that from (11.1) for the linearized equations are $x' = \epsilon x + y$ and $y' = -x + \epsilon y$. Thus, $\mathbf{A} = \text{array}([[\epsilon, 1], [-1, \epsilon,]])$ so $\lambda = \epsilon \pm i$. Hence, $\epsilon > 0$ gives an unstable spiral and $\epsilon < 0$ gives a stable spiral. (d) $x' = -\epsilon x + y + x(x^2 + y^2)$ and $y' = -x - \epsilon y + y(x^2 + y^2)$

CHAPTER 12
Section 12.2

2. (a) $y_1 = 0.8$, $y_2 = 0.64$, $y_3 = 0.512$ (c) $y_1 = -1$, $y_2 = -0.976$, $y_3 = -0.88455$ (e) $y_1 = 0.08731$, $y_2 = 0.52718$, $y_3 = 0.85773$ (g) $y_1 = y_2 = y_3 = 0$ [Indeed, exact solution is $y(x) = 0$.] (i) $y_1 = 3.2$, $y_2 = 2.6845$, $y_3 = 2.4291$

3. (a) $y_1 = 1$, $y_2 = 1.02$, $y_3 = 1.06$, $y_4 = 1.12$, $y_5 = 1.2$, $y_6 = 1.3$, $y_7 = 1.42$, $y_8 = 1.56$, $y_9 = 1.72$, $y_{10} = 1.9$; $y(0.1) = 1.1$, $y(0.2) = 1.04$, $y(0.3) = 1.09$, $y(0.4) = 1.16$,

$y(0.5) = 1.25$, $y(0.6) = 1.36$, $y(0.7) = 1.49$, $y(0.8) = 1.64$, $y(0.9) = 1.81$, $y(1) = 2$; $TE_{10} = 2 - 1.9 = 0.1$ (d) $y_1 = 1$, $y_2 = 1.00983$, ..., $y_{10} = 1.500527$; $y(0.1) = 1.005008$, $y(0.2) = 1.020133$, ..., $y(1) = 1.583595$; $TE_{10} = 1.583595 - 1.500527 = 0.083069$ (g) $y_1 = 1.2$, $y_2 = 1.421$, ..., $y_{10} = 4.534344$; $y(0.1) = 1.210684$, $y(0.2) = 1.445611$, ..., $y(1) = 4.873127$; $TE_{10} = 0.338784$

4. (a) For $h = 0.1, 0.05, 0.01, 0.005, 0.001$ the computed y values at $x = 0.5$ are $1.2, 1.225, 1.245, 1.2475, 1.2495$, and the exact value is $y(0.5) = 1.25$. The corresponding TEs at $x = 0.5$ are $0.05, 0.025, 0.005, 0.0025, 0.0005$, respectively. These results are consistent with the fact that Euler's method is a first-order method for then $TE \sim Ch$ as $h \to 0$, for some constant C that does not depend on h. Thus, if we reduce h by a factor of 5, say, then TE should also be reduced by approximately a factor of 5 (approximately, since $TE \sim Ch$ holds only as $h \to 0$). The TE values do exhibit this behavior (exactly, in this example). (c) For $h = 0.1, 0.05, 0.01, 0.005, 0.001$ the computed y values at $x = 0.5$ are $0.1846, 0.2222, 0.2447, 0.2474, 0.2495$, and the exact value is $y(0.5) = 0.25$. Same discussion as in (a). (e) For $h = 0.1, 0.05, 0.01, 0.005, 0.001$ the computed y values at $x = 0.5$ are $1.6312, 1.6566, 1.6796, 1.6827, 1.6852$, and the exact value is $y(0.5) = 1.6858$. Same discussion as in (a). (g) For $h = 0.1, 0.01, 0.001, 0.0001$ the computed y values at $x = 0.5$ are $2.2531, 2.3350, 2.2439, 2.3479$, and the exact value is $y(0.5) = 2.3449$. Same discussion as in (a).

5. $h < 0$ is permissible.

6. h can be varied from step to step.

Section 12.3

3. (a) RK2: For $n = 0$, $k_1 = 0$, $k_2 = 0.3$, $y_1 = \underline{2.15}$. For $n = 1$, $k_1 = 0.25960$, $k_2 = 0.41335$, $y_2 = \underline{2.4865}$. RK4: For $n = 0$, $k_1 = 0$, $k_2 = 0.15$, $k_3 = 0.13935$, $k_4 = 0.26219$, $y_1 = \underline{2.14015}$. For $n = 1$, $k_1 = 0.26199$, $k_2 = 0.34896$, $k_3 = 0.33598$, $k_4 = 0.39144$, $y_2 = \underline{2.47737}$. Exact: $y(x) = \sqrt[3]{4500x^2 + 8}$, $y(0.02) = \underline{2.13997}$, $y(0.04) = \underline{2.47712}$. (b) RK2: For $n = 0$, $k_1 = 0$, $k_2 = 0.000797$, $y_1 = \underline{3.00040}$. For $n = 1$, $k_1 = 0.000796$, $k_2 = 0.000796$, $y_2 = \underline{3.00120}$. RK4: For $n = 0$, $k_1 = 0$, $k_2 = 0.000398$, $k_3 = 0.000398$, $k_4 = 0.000796$, $y_1 = \underline{3.000398}$. For $n = 1$, $k_1 = 0.000796$, $k_2 = 0.00119$, $k_3 = 0.00119$, $k_4 = 0.00159$, $y_2 = \underline{3.00159}$. Exact: $y(x) = \ln(20x^2 + e^3)$, $y(0.02) = \underline{3.00040}$, $y(0.04) = \underline{3.00159}$. (c) RK2: For $n = 0$, $k_1 = 0.08$, $k_2 = 0.082$, $y_1 = \underline{5.081}$. For $n = 1$, $k_1 = 0.08202$, $k_2 = 0.08366$, $y_2 = \underline{5.16384}$. RK4: For $n = 0$, $k_1 = 0.08$, $k_2 = 0.081$, $k_3 = 0.08101$, $k_4 = 0.08202$, $y_1 = \underline{5.08101}$. For $n = 1$, $k_1 = 0.08202$, $k_2 = 0.08304$, $k_3 = 0.08305$, $k_4 = 0.08408$, $y_2 = \underline{5.16406}$. Exact: $y(x) = 5e^{x+1} - x - 1$, $y(-0.98) = \underline{5.08101}$, $y(-0.96) = \underline{5.16405}$.

4. For brevity, we list only the values y_1 (at $x = 0.05$) and y_{10} (at $x = 0.5$), and we use the format $y_1 = a, b, c$; $y_{10} = d, e, f$, where a and d are the second-order RK values, b and e are the fourth-order RK values, and c and f are the exact values. **(a)** $y_1 = 4.75, 2.9927, 2.3052$; $y_{10} = 10.6582, 10.4482, 10.4035$ **(c)** $y_1 = 1.052500, 1.052542, 1.052542$; $y_{10} = 1.796781, 1.797443, 1.797442$ **(e)** $y_1 = 1.025316, 1.025318, 1.025318$; $y_{10} = 1.287402, 1.287427, 1.287427$

5. (a) For $x = 1$, $y_{10} = 3.434368$ for $h = 0.1$, $y_{20} = 3.570633$ for $h = 0.05$, exact $y(1) = 3.718282$, so (27) gives $p \approx 0.943$. For $x = 2$, $y_{20} = 10.154750$ for $h = 0.1$, $y_{40} = 10.737989$ for $h = 0.05$, exact $y(2) = 11.389056$, so (27) gives $p \approx 0.923$. These p values seem consistent with Euler's method being of first order. **(b)** For $x = 1$, $y_{10} = 3.705918$ for $h = 0.1$, $y_{20} = 3.715097$ for $h = 0.05$, exact $y(1) = 3.718282$, so (27) gives $p \approx 1.957$. For $x = 2$, $y_{20} = 11.335919$ for $h = 0.1$, $y_{40} = 11.3753429$ for $h = 0.05$, exact $y(2) = 11.389056$, so (27) gives $p \approx 1.954$.

6. For brevity we list x values only at $t = 600$ and 3600. **(a)** Reducing h to 3, $x(600) = 3.3140$ and $x(3600) = 3.9996$. **(d)** Reducing h to 8, $x(600) = 4.4875$ and $x(3600) = 4.0003$.

9. (b) $C = -0.02 \ln 2$, $x(600) = 3.313877$, $x(1200) = 3.852108, \ldots, x(3600) = 3.999637$

10. (a) $h = 0.05$ gives $y_{20} = 3.718281474$, $h = 0.02$ gives $y_{50} = 3.718281819$, $y(1) = 3.718281828$, so (27) gives $p \approx 4.008$. **(b)** This time $h = 0.05$ gives 3.701259429, $h = 0.02$ gives 3.711558039, $y(1) = 3.718281828$, so (27) gives $p \approx 0.916$. Evidently, the "programming error" has knocked the order down from 4 to 1.

Section 12.4

2. (a) Exact: $y(x_1) = 0.9800666$, $z(x_1) = -0.1986693$. Euler: $y_1 = 1$, $z_1 = -0.2$. Second-order RK: $k_1 = 0$, $l_1 = -0.2$, $k_2 = -0.04$, $l_2 = -0.2$, so $y_1 = 0.98$, $z_1 = -0.2$. Fourth-order RK: $k_1 = 0$, $l_1 = -0.2$, $k_2 = -0.02$, $l_2 = -0.2$, $k_3 = -0.02$, $l_3 = -0.198$, $k_4 = -0.0396$, $l_4 = -0.196$, so $y_1 = 0.9800667$, $z_1 = -0.1986667$ **(d)** Exact: $y(x_1) = 1.44$, $z(x_1) = 1.2$. Euler: $y_1 = 1.4$, $z_1 = 1.2$. Second-order RK: $k_1 = 0.4$, $l_1 = 0.2$, $k_2 = 0.49371$, $l_2 = 0.19444$, so $y_1 = 1.4468$, $z_1 = 1.1972$. Fourth-order RK: $k_1 = 0.4$, $l_1 = 0.2$, $k_2 = 0.4436667$, $l_2 = 0.1983471$, $k_3 = 0.4350841$, $l_3 = 0.2022597$, $k_4 = 0.4834599$, $l_4 = 0.1985687$, so $y_1 = 1.4401603$, $z_1 = 1.1999637$

3. (a) Use

```
dsolve({diff(y(x),x)=z(x),diff(z(x),x)=-y(x),
y(0)=1,z(0)=0},{y(x),z(x)},type=numeric,
value=array([3,5,10]));
```

4. (a) Euler: $x_{n+1} = x_n + F(t_n, x_n, y_n, z_n)h$, $y_{n+1} = y_n + G(t_n, x_n, y_n, z_n)h$, $z_{n+1} = z_n + H(t_n, x_n, y_n, z_n)h$. Second-order RK: $k_1 = hF(t_n, x_n, y_n, z_n)$, $l_1 = hG(t_n, x_n, y_n, z_n)$, $m_1 = hH(t_n, x_n, y_n, z_n)$, $k_2 = hF(t_{n+1}, x_n + k_1, y_n + l_1, z_n + m_1)$, $l_2 = hG(t_{n+1}, x_n + k_1, y_n + l_1, z_n + m_1)$, $m_2 = hH(t_{n+1}, x_n + k_1, y_n + l_1, z_n + m_1)$, $x_{n+1} = x_n + \frac{1}{2}(k_1 + k_2)$, $y_{n+1} = y_n + \frac{1}{2}(l_1 + l_2)$, $z_{n+1} = z_n + \frac{1}{2}(m_1 + m_2)$ **(b)** $t_0 = 0$, $x_0 = -3$, $y_0 = 0$, $z_0 = 2$. **Euler:** $x_1 = \underline{-3.3}$, $y_1 = \underline{0.6}$, $z_1 = \underline{3.8}$, $x_2 = \underline{-3.42}$, $y_2 = \underline{1.74}$, $z_2 = \underline{6.68}$. RK2: $k_1 = -0.3$, $l_1 = 0.6$, $m_1 = 1.8$, $k_2 = -0.12$, $l_2 = 1.14$, $m_2 = 2.88$, $x_1 = \underline{-3.21}$, $y_1 = \underline{0.87}$, $z_1 = \underline{4.34}$. $k_1 = -0.039$, $l_1 = 1.302$, $m_1 = 3.15$, $k_2 = 0.3516$, $l_2 = 2.247$, $m_2 = 4.8915$, $x_2 = \underline{-3.0537}$, $y_2 = \underline{2.6445}$, $z_2 = \underline{8.36075}$. NOTE: The exact solutions are $x(t) = t^2 e^t - 3 - t$, $y(t) = (t^2 + 2t)e^t$, $z(t) = (t^2 + 4t + 2)e^t$.

5. (a) $x' = y$, $x(0) = x_0$; $y' = [f(t) - kx - cy]/m$, $y(0) = x_0'$ **(e)** $y' = u$, $y(-2) = 7$; $u' = v$, $u(-2) = 4$; $v' = 3x + 2 \sin u$, $v(-2) = 0$ **(i)** $x' = u$, $x(1) = 0$; $u' = v$, $u(1) = 0$; $v' = \sin t + 3xz$, $v(1) = 3$; $y' = 5z - 2x - y$, $y(1) = 6$; $z' = e^{2t} + 4xy$, $z(1) = 2$

6. (a) Use

```
dsolve({diff(y(x),x,x)-2*x*diff(z(x),x)=5*x,
diff(z(x),x)+y(x)*z(x)^3=-3,y(0)=2,D(y)(0)=-1,
z(0)=1},{y(x),z(x)},type=numeric,
value=array([1,2]));
```

7. dsolve gives $Y_3(0.5) = 0.125018599$, $Y_3(1) = 0.502382754$, $Y_p(0.5) = -0.00003720$, $Y_p(1) = -0.00476551$, so (17) gives $y(0.5) = 0.25000000$ and $y(1) = 1.00000000$. In fact, the problem was arranged to have the simple solution $y(x) = x^2$ and the foregoing numerical results agree with this exact solution.

8. (a) $L[Y_1] = 0$, $Y_1(0) = 1$, $Y_1'(0) = 0$; $L[Y_2] = 0$, $Y_2(0) = 0$, $Y_2'(0) = 1$; $L[Y_p] = 3 \sin x$, $Y_p(0) = a$, $Y_p'(0) = b$. Then $y(x) = C_1 Y_1(x) + C_2 Y_2(x) + Y_p(x)$. $y(0) = 1 = C_1 + 0 + a$ so let us choose $a = 1$ for then $C_1 = 0$ and we don't need $Y_1(x)$; b hasn't appeared so let us choose $b = 0$, say. Then $y(2) = 3 = C_2 Y_2(2) + Y_p(2)$ gives $C_2 = [3 - Y_p(2)]/Y_2(2)$ if $Y_2(2) \neq 0$. The Y_2 problem is $Y_2' = U_2$, $Y_2(0) = 0$ and $U_2' = 2xU_2 - Y_2$, $U_2(0) = 1$. The Y_p problem is $Y_p' = U_p$, $Y_p(0) = 1$ and $U_p' = 3 \sin x + 2xU_p - Y_p$, $U_p(0) = 0$. Solving each of the Y_2 and Y_p problems by dsolve gives $Y_2(0.5) = 0.5222085$, $Y_2(1) = 1.2191612$, $Y_2(1.5) = 2.6237924$, $Y_2(2) = 7.6391401$ and $Y_p(0.5) = 0.9325135$, $Y_p(1) = 0.9656913$, $Y_p(1.5) = 1.7339756$, $Y_p(2) = 6.1560374$. Thus $C_2 = -0.4131404$ and $y(x) = C_2 Y_2(x) + Y_p(x)$ gives $y(0.5) = 0.716768$, $y(1) = 0.462007$, $y(1.5) = 0.649981$. NOTE: To solve the Y_2 and Y_p problems by dsolve you do not need to first break them down into systems of first-order equations. For instance, to solve for Y_2

you can use
```
dsolve({diff(Y2(x),x,x)-2*x*diff(Y2(x),x)+Y2(x)
=0,Y2(0)=0,D(Y2)(0)=1},{Y2(x)},type=numeric,
value=array([.5,1,1.5,2]));
```

(c) $y(0.5) = 1.3850435$, $y(1) = 0.5850356$, $y(1.5) = 0.0425983$ (e) $y(2) = -6.308132$

CHAPTER 13

Section 13.2

1. (a) Yes; $K = 5$, $c = 4$ NOTE: In this exercise K and c are not unique. Here, any $K \geq 5$ and any $c \geq 4$ will do. (c) $\sinh 2t = (e^{2t} - e^{-2t})/2 < e^{2t}/2$, so we can take $K = 1/2$, $c = 2$, say (e) $\sinh t^2 = (e^{t^2} - e^{-t^2})/2 \sim e^{t^2}/2 = (1/2)e^{(t)t}$. We could take $K = 1/2$, say, but evidently there does not exist a sufficiently large c because of the parenthetic t in the final exponent.   (g) $|\cos t^3| \leq 1$ for all t so we can take $K = 1$, $c = 0$

2. No; $f(t) = \sin e^{t^2}$ is of exponential order ($K = 1$, $c = 0$) but $f'(t) = 2te^{t^2}\cos e^{t^2}$ is not.

3. No; $f(t) = e^t$ and $g(t) = t^2$ are of exponential order but $f(g(t)) = e^{t^2}$ is not.

13. (a) $t - \frac{a^2}{3!}t^3 + \frac{a^4}{5!}t^5 - \cdots$, which does agree with $(\sin at)/a$ from the table (c) $\frac{1}{2}t^2 - \frac{1}{2(3!)}t^4 + \frac{1}{2(5!)}t^6 - \cdots$, which does agree with $(t\sin t)/2$ from the table

15. (d) $[\Gamma(1/2)]^2 = 4\int_0^{\pi/2}\int_0^\infty e^{-r^2}r\,dr\,d\theta = 2\pi\int_0^\infty e^{-r^2}r\,dr = \pi$ so $\Gamma(1/2) = \sqrt{\pi}$

Section 13.3

1. (a) By partial fractions $\frac{3}{s(s+8)} = \frac{3}{8}\frac{1}{s} - \frac{3}{8}\frac{1}{s+8}$ and then entry 2 (with $a = 0$ and $a = -8$) gives the inverse as $\frac{3}{8} - \frac{3}{8}e^{-8t}$; or, we could have used entries 1 and 2. By convolution theorem, $L^{-1}\left\{\frac{3}{s(s+8)}\right\} = L^{-1}\left\{\frac{3}{s}\right\} * L^{-1}\left\{\frac{1}{s+8}\right\} = 3 * e^{-8t} = \int_0^t e^{-8\tau}3\,d\tau = \frac{3}{8} - \frac{3}{8}e^{-8t}$ again. (c) $\frac{1}{s^2-a^2} = \frac{1}{2a}\frac{1}{s-a} - \frac{1}{2a}\frac{1}{s+a}$ and then entry 2 gives inverse $= (e^{at} - e^{-at})/2a = (\sinh at)/a$. Or, $L^{-1}\left\{\frac{1}{s^2-a^2}\right\} = L^{-1}\left\{\frac{1}{s-a}\right\} * L^{-1}\left\{\frac{1}{s+a}\right\} = e^{at} * e^{-at} = \int_0^t e^{a\tau}e^{-a(t-\tau)}\,d\tau = (\sinh at)/a$ again. (e) result is $1 - e^{-t}$

2. (c) The Maple invlaplace(1/(s^2-a^2),s,t); command gives as a result $(\sin(\sqrt{-a^2}t))/\sqrt{-a^2}$ which is correct but not in friendly form. Since $\sin ix = i\sinh x$ we can reduce this form to $(\sinh at)/a$. Better is the command invlaplace(1/((s-a)*(s+a)),s,t); which gives $(e^{at} - e^{-at})/2a$.

3. $p * (q * r) = \int_0^t p(\tau)\,d\tau \int_0^{t-\tau} r(\mu)q(t - \tau - \mu)\,d\mu$. The region of integration is a triangle in the μ, τ plane with vertices at $(\mu, \tau) = (0, t)$, $(t, 0)$, and $(0, 0)$. Inverting the order of integration (permissible because we assume p, q, r to be piecewise continuous) gives $p*(q*r) = \int_0^t r(\mu)\,d\mu \int_0^{t-\mu} p(\tau)q(t - \tau - \mu)\,d\tau = r * (p * q) = (p * q) * r$.

4. How do we know if $L\{p * q * r\}$ means $L\{p * (q * r)\}$ or $L\{(p * q) * r\}$? According to (14b) it doesn't matter; they are the same. We have $L\{p*q*r\}=L\{p*(q*r)\} = L\{p\}L\{q*r\}$ by (17). But $L\{q * r\} = L\{q\}L\{r\}$ also by (17). Thus, $L\{p*q*r\} = L\{p\}L\{q\}L\{r\} = P(s)Q(s)R(s)$.

5. $1*1 = \int_0^t d\tau = t$. Then $1*(1*1) = 1*t = \int_0^t \tau\,d\tau = t^2/2$

7.   NOTE: We will use LHS and RHS for left-hand and right-hand side, for brevity, in these Answers to Selected Exercises. (a) For (6), LHS $= L\{f'\} = L\{3e^{3t}\} = \frac{3}{s-3}$ and RHS $= sL\{f\} - f(0) = \frac{s}{s-3} - 1 = \frac{3}{s-3}$ again. For (11), LHS $= L\{f''\} = L\{9e^{3t}\} = \frac{9}{s-3}$ and RHS $= s^2L\{f\} - sf(0) - f'(0) = s^2\frac{1}{s-3} - s - 3 = \frac{9}{s-3}$ again. (c) For (6) LHS $= L\{2t + 5\} = \frac{2}{s^2} + \frac{5}{s}$ and RHS $= sL\{f\} - f(0) = s\left(\frac{2}{s^3} + \frac{5}{s^2} - \frac{1}{s}\right) - (-1) = \frac{2}{s^2} + \frac{5}{s}$ again. For (11), LHS $= L\{f''\} = L\{2\} = \frac{2}{s}$ and RHS $= s^2\left(\frac{2}{s^3} + \frac{5}{s^2} - \frac{1}{s}\right) - s(-1) - (5) = \frac{2}{s}$ again.

8. In each case we can identify the integral as a convolution integral $p * q$ and then use (17). (a) $\sin 2\tau = p(\tau)$ so $p(t) = \sin 2t$, and $e^{t-\tau} = q(t - \tau)$ so $q(t) = e^t$. Then (17) gives the transform as $P(s)Q(s) = \frac{2}{s^2+4}\frac{1}{s-1} = \frac{2}{(s-1)(s^2+4)}$. (c) $e^{-3\tau} = p(\tau)$ so $p(t) = e^{-3t}$, and $(t - \tau)^8 = q(t - \tau)$ so $q(t) = t^8$. Then (17) gives the transform as $P(s)Q(s) = \frac{1}{s+3}\frac{8!}{s^9} = \frac{40320}{s^9(s+3)}$. (e) $p(t) = \sin t$ and $q(t) = t^{-1/2}$ so transform is $P(s)Q(s) = \frac{1}{s^2+1}\frac{\Gamma(1/2)}{\sqrt{s}} = \frac{1}{s^2+1}\sqrt{\frac{\pi}{s}}$.

9. (a) $L\{p(t)q(t)\} = L\{1\} = \frac{1}{s}$ whereas $P(s)Q(s) = \frac{1}{s}\frac{1}{s} = \frac{1}{s^2}$.
$L^{-1}\{P(s)Q(s)\} = L^{-1}\left\{\frac{1}{s}\frac{1}{s}\right\} = L^{-1}\left\{\frac{1}{s^2}\right\} = t$ whereas $p(t)q(t) = 1$. (c) $L\{p(t)q(t)\} = L\{te^t\} = \frac{1}{(s-1)^2}$ whereas $P(s)Q(s) = \frac{1}{s^2}\frac{1}{s-1}$.
$L^{-1}\{P(s)Q(s)\} = L^{-1}\left\{\frac{1}{s^2}\frac{1}{s-1}\right\} = -1 - t + e^t$ whereas $p(t)q(t) = te^t$.

Section 13.4

1. (a) $sX(s) - x(0) + 2X(s) = 8/s^3$, $X(s) = \frac{x(0)}{s+2} + \frac{8}{s^3(s+2)}$, $x(t) = x(0)e^{-2t} + 1 - e^{-2t} - 2t + 2t^2$ or, if we let $x(0) - 1 \equiv C$, $x(t) = Ce^{-2t} + 1 - 2t + 2t^2$

(c) $X(s) = \frac{4}{s-6} + \frac{1}{(s+1)(s-6)} = -\frac{1}{7}\frac{1}{s+1} + \frac{29}{7}\frac{1}{s-6}$, $x(t) = -\frac{1}{7}e^{-t} + \frac{29}{7}e^{6t}$ (e) $X(s) = \frac{sx(0)+x'(0)+5x(0)}{s^2+5s} + \frac{10}{s(s^2+5s)}$, $x(t) = 2t + \frac{5x(0)+x'(0)-2}{5} + \frac{2-x'(0)}{5}e^{-5t}$ or, if we let $[5x(0) + x'(0)-2]/5 \equiv A$ and $[2-x'(0)]/5 \equiv B$, $x(t) = A + Be^{-5t} + 2t$

(g) $X(s) = \frac{3s-8}{s^2-3s+2}$, $x(t) = 5e^t - 2e^{2t}$ (i) $X(s) = \frac{s^2-s^3+1}{s^2(s^2-s-12)}$, $x(t) = \frac{1}{144} - \frac{1}{12}t - \frac{37}{63}e^{-3t} - \frac{47}{112}e^{4t}$ (k) $X(s) = -\frac{5s^2+2}{s^2(s^2-2s+2)}$, $x(t) = -1 - t + (\cos t - 5\sin t)e^t$ (m) $X(s) = \frac{1}{s}$, $x(t) = 1$

(o) $X(s) = \frac{100}{s^3(s+5)}$, $x(t) = \frac{4}{5} - 4t + 10t^2 - \frac{4}{5}e^{-5t}$ (q) $X(s) = \frac{8}{s^4} + \frac{2}{s^4(s^2+1)}$, $x(t) = \frac{5}{3}t^3 - 2t + 2\sin t$ (s) $X(s) = \frac{4s+1}{s(s^4-1)}$, $x(t) = -1 + \frac{5}{4}e^t - \frac{3}{4}e^{-t} + \frac{1}{2}\cos t - 2\sin t$

2. (a) $X(s) = \frac{4}{s^2+1} + \frac{1}{(s^2+1)^2}$, $x(t) = \frac{9}{2}\sin t - \frac{1}{2}t\cos t$
(c) $X(s) = \frac{2}{s^2-1} + \frac{4}{(s-1)(s^2-1)}$, $x(t) = 2te^t$

3. (a) The Laplace transform of $x^{(n)}(t) + c_1x^{(n-1)}(t) + \cdots + c_nx(t) = 0$ is $[s^nX - s^{n-1}x(0) - s^{n-2}x'(0) - \cdots - x^{(n-1)}(0)] + c_1[s^{n-1}X - s^{n-2}x(0) - \cdots - x^{(n-2)}(0)] + \cdots + c_nX = 0$ which is of the form $(s^n + c_1s^{n-1} + \cdots + c_n)X = p(s)$ where $p(s)$ is a polynomial of degree $n-1$ (or less) in s. Thus, $X(s)$ is a polynomial of degree $n-1$ (or less) divided by a polynomial $q(s)$ of degree n. (b) Factor $q(s) = (s-s_1)(s-s_2)\cdots(s-s_n)$ then expand $X(s) = \frac{p(s)}{q(s)} = \frac{a_1}{s-s_1} + \cdots + \frac{a_n}{s-s_n}$ by partial fractions. From the Appendix $a_j = p(s_j)/q'(s_j)$ and from the transform table $L^{-1}\left\{\frac{1}{s-s_j}\right\} = e^{s_jt}$.

Section 13.5

1. (a) $x' + 2x = 100[1 - H(t-5)]$, $x(0) = 0$, so $X(s) = \frac{1}{s+2}F(s)$, $x(t) = e^{-2t} * f(t) = \int_0^t e^{-2(t-\tau)}100[1 - H(\tau - 5)]d\tau = 50(1 - e^{-2t}) - 100e^{-2t}H(t-5)\int_5^t e^{2\tau}d\tau = 50(1 - e^{-2t}) - 50H(t-5)(1 - e^{-2(t-5)})$ (c) $f(t) = 20t[1 - H(t-1)] + 20H(t-1)$ and $x(t) = 10t - 5 + 5e^{-2t} + H(t-1)(15 - 10t - 5e^{-2(t-1)})$ (e) $f(t) = [1 - H(t - \pi/2)]\sin t$ and $x(t) = \frac{2}{5}\sin t - \frac{1}{5}\cos t + \frac{1}{5}e^{-2t} + H(t - \pi/2)\left(\frac{1}{5}\cos t - \frac{2}{5}\sin t + \frac{2}{5}e^{-2(t-\pi/2)}\right)$ (g) $f(t) = 10[H(t-1) - H(t-2)] + 20H(t-2)$ and $x(t) = 5H(t-1)(1 - e^{-2(t-1)}) + 5H(t-2)(1 - e^{-2(t-2)})$

2. (a) $x(t) = 25(1 - \cos 2t) + 25H(t-5)\{\cos[2(t-5)] - 1\}$
(c) $x(t) = \frac{5}{2}\{2t - \sin 2t + H(t-1)[2 - 2t + \sin[2(t-1)]]\}$

3. (a) $x(t) = 100(\cosh t - 1) + 10\sinh t + 100H(t-5)[1 - \cosh(t-5)]$ (c) $x(t) = 30\sinh t - 20t + 20H(t-1)[t - 1 - \sinh(t-1)]$

4. (a) $x(t) = 100\{t - 1 + e^{-t} + H(t-5)[6 - t - e^{-(t-5)}]\}$
(c) $x(t) = 10(t^2 - 2t + 2 - 2e^{-t}) - 10H(t-1)[t^2 - 4t + 5 - 2e^{-(t-1)}]$

5. (a) To solve for $x(t)$ use

```
dsolve({diff(x(t),t)+2*x(t)
=100*(1-Heaviside(t-5)),x(0)=0},x(t));
```

which gives $x(t) = 50(1 - e^{-2t}) - 50H(t-5)(1 - e^{-2(t-5)})$. To plot this use the command

```
with(plots):
implicitplot(x=50*(1-exp(-2*t))-50*Heaviside(t-
5)*(1-exp(-2*(t-5))),t=0..14,x=-10..60,
numpoints=5000);
```

To obtain the plot directly, without first generating the analytical solution, use

```
with(DEtools):
phaseportrait(diff(x(t),t)+2*x(t)=100*(1-
Heaviside(t-5)),x(t),t=0..14,[[x(0)=0]],
x=-10..60,arrows=NONE, stepsize=.051);
```

If we use stepsize $= 0.05$, instead, *Maple* gives an error message about an undefined quantity, which is $H(t-5)$ at $t = 5$. To avoid the need to evaluate $H(t-5)$ at its point of discontinuity $t = 5$, we can change the stepsize to 0.051, say.

6. (a) As in the solution to 5(a), above, we can use dsolve to get the solution,

```
dsolve({diff(x(t),t,t)+diff(x(t),t)
=100*(1-Heaviside(t-5)),x(0)=0,D(x)(0)=0},x(t));
```

which gives $x(t) = 100\{t - 1 + e^{-t} + H(t-5)[6 - t - e^{-(t-5)}]\}$. To plot use

```
with(plots):
implicitplot(x=100*(t-1+exp(-t)+Heaviside(t-5)*
(6-t-exp(-t+5)),t=0..15,x=0..600,numpoints=5000);
```

To obtain the plot directly, without first generating the analytical solution, use

```
with(DEtools):
phaseportrait(diff(x(t),t,t)+diff(x(t),t)=
100*(1-Heaviside(t-5)), x(t), t=0..15,
[[x(0)=0, D(x)(0)=0]],x=0..600, arrows=NONE,
stepsize=0.051);
```

How did we know to choose the x range 0..600? We could try $x = 0..100$, say, but will find that that range captures the solution curve only to $t \approx 2$. Thus, we need to increase the x range.

7. (b) $L\{1 + (t-1)]H(t-1)\} = L\{H(t-1)\} + L\{(t-1)H(t-1)\} = e^{-s}/s + e^{-s}/s^2$ where we have used (17) with $a = 1$ and (19a) with $a = 1$ and $f(t) = t$ [so $f(t-1) = t - 1$].   (c) Use with(inttrans): then
```
laplace(t*Heaviside(t-1),t,s);
```

10. (b) $L\{e^{-2t}H(t-3)\} = L\{e^{-6}e^{-2(t-3)}H(t-3)\} = e^{-6}L\{e^{-2(t-3)}H(t-3)\} = e^{-6}e^{-3s}/(s+2)$ where we have used (19a) with $a = 3$ and $f(t) = e^{-2t}$.

11. (a) To evaluate the $\cos 2(t-3)H(t-3)$ part use (19a) with $a = 3$ and $f(t) = \cos 2t$, and to evaluate the $\sin 2(t-3)H(t-3)$ part use (19a) with $a = 3$ and $f(t) = \sin 2t$. Answer:

$(\sin 6)e^{-3s}\frac{s}{s^2+4} + (\cos 6)e^{-3s}\frac{2}{s^2+4}$.

12. (a) The key is that $H(t-1)H(t-2) = H(t-2)$, so
$L\{H(t-1)H(t-2)\} = L\{H(t-2)\} = \frac{e^{-2s}}{s}$.

13. (a) $L\{1/s^2\} = t$ so $L\{e^{-2s}/s^2\} = (t-2)H(t-2)$
(c) $\sin(t-1)H(t-1)$ **(e)** $\frac{1}{3}\sinh 3(t-3)H(t-3)$

14. (a) Use

```
with(DEtools):
phaseportrait(diff(x(t),t,t)+x(t)=50*Heaviside(t-
20),x(t),t=0..43,[[x(0)=0,D(x)(0)=0]],x=0..120,
arrows=NONE,stepsize=.051);
```

Section 13.6

1. (a) $L\{5H(t-2)\} = 5e^{-2s}/s$ so $L\{5e^{-3t}H(t-2)\} = 5e^{-2(s+3)}/(s+3)$

2. (a) $\frac{s}{s^2+2s+3} = \frac{s}{(s^2+2s+1)-1+3} = \frac{s}{(s+1)^2+2} = \frac{(s+1)-1}{(s+1)^2+2}$,
$L^{-1}\left\{\frac{s}{s^2+2}\right\} = \cos\sqrt{2}t$ and $L^{-1}\left\{\frac{1}{s^2+2}\right\} = \frac{\sin\sqrt{2}t}{\sqrt{2}}$, so
$L^{-1}\left\{\frac{(s+1)-1}{(s+1)^2+2}\right\} = e^{-t}\cos\sqrt{2}t - e^{-t}\frac{\sin\sqrt{2}t}{\sqrt{2}}$ **(c)** $\frac{3s}{(s+4)^2} =$
$\frac{3(s+4)-12}{(s+4)^2} = \frac{3}{s+4} - \frac{12}{(s+4)^2}$, so $L^{-1}\left\{\frac{3s}{(s+4)^2}\right\} = 3e^{-4t} - 12te^{-4t}$
(e) $L^{-1}\left\{\frac{1}{\sqrt{s}}\right\} = \frac{t^{-1/2}}{\Gamma(1/2)} = \frac{1}{\sqrt{\pi t}}$, so $L^{-1}\left\{\frac{1}{\sqrt{s+6}}\right\} = \frac{1}{\sqrt{\pi t}}e^{-6t}$
(g) $L^{-1}\left\{\frac{s+1}{(s-5)^{3/2}}\right\} = L^{-1}\left\{\frac{(s-5)+6}{(s-5)^{3/2}}\right\} = L^{-1}\left\{\frac{1}{\sqrt{s-5}}\right\} +$
$6L^{-1}\left\{\frac{1}{(s-5)^{3/2}}\right\}$, and $L^{-1}\left\{\frac{1}{\sqrt{s}}\right\} = \frac{1}{\sqrt{\pi t}}$ and $L^{-1}\left\{\frac{1}{s^{3/2}}\right\} =$
$2\sqrt{\frac{t}{\pi}}$, so $L^{-1}\left\{\frac{s+1}{(s-5)^{3/2}}\right\} = \frac{1}{\sqrt{\pi t}}e^{5t} + (6)(2)\sqrt{\frac{t}{\pi}}e^{5t}$

3. $F(s) = \frac{1}{1-e^{-3s}}\int_0^3 \frac{4t}{3}e^{-st}\,dt = \frac{4}{3s^2} - \frac{4}{s}\frac{e^{-3s}}{1-e^{-3s}}$

4. (a) $F(s) = \frac{1+e^{-\pi s}}{(s^2+1)(1-e^{-\pi s})}$ **(c)** $F(s) = \frac{s(1+e^{-\pi s})}{(s^2+1)(1-e^{-\pi s})}$
(e) $F(s) = \frac{1-(s+1)e^{-s}}{s^2(1-e^{-2s})}$

5. $X(s) = \frac{x_0}{s+1} + \frac{2}{(s+1)s^2} - \frac{4}{s(s+1)}e^{-2s}(1+e^{-2s}+e^{-4s}+\cdots)$,
so $x(t) = x_0e^{-t} + 2(t-1+e^{-t}) - 4[(1-e^{-(t-2)})H(t-2) + (1-e^{-(t-4)})H(t-4) + \cdots]$

6. We can use implicitplot to give a computer plot of $x(t)$. If we plot over $0 \le t \le 6$, say, then we need to keep terms up to and including the $H(t-4)$. Thus, enter

```
with(plots):
implicitplot(x=2*(t-1+exp(-t))-4*((1-exp(2-t))*
Heaviside(t-2)+(1-exp(4-t))*Heaviside(t-4)),
t=0..6,x=0..50,numpoints=5000);
```

We can see from the plot that $x(t)$ is not periodic. Thus, it is not true in general that a periodic input [namely, the forcing function $f(t)$] leads to a periodic output [namely, $x(t)$]. As a simpler version of this problem consider $x' + x = \sin t$, with $x(0) = x_0$. The input $f(t) = \sin t$ is periodic, yet the output $x(t) = \left(\frac{1}{2} + x_0\right)e^{-t} + \frac{1}{2}(\sin t - \cos t)$ is not because of the e^{-t} term — unless we choose $x_0 = -\frac{1}{2}$ to eliminate that term. Even if $x_0 \neq -\frac{1}{2}$ it is nevertheless true that the "steady state" response $\frac{1}{2}(\sin t - \cos t)$ (namely, that which remains after the e^{-t} transient has died out) is periodic. Thus, even though this response $x(t)$ is not periodic it *approaches* a periodic function as $t \to \infty$.

Index

LAPLACE TRANSFORMS

NOTE: s is regarded as real here. If it is regarded as complex, change each "$s > c$" restriction to Re $s > c$.

0. $f(t)$ $\qquad$ $F(s) = \int_0^\infty f(t)e^{-st}\,dt$

1. 1 $\qquad$ $\dfrac{1}{s}\quad (s > 0)$

2. e^{at} $\qquad$ $\dfrac{1}{s-a}\quad (s > a)$

3. $\sin at$ $\qquad$ $\dfrac{a}{s^2+a^2}\quad (s > 0)$

4. $\cos at$ $\qquad$ $\dfrac{s}{s^2+a^2}\quad (s > 0)$

5. $\sinh at$ $\qquad$ $\dfrac{a}{s^2-a^2}\quad (s > |a|)$

6. $\cosh at$ $\qquad$ $\dfrac{s}{s^2-a^2}\quad (s > |a|)$

7. $t^n\quad (n = \text{positive integer})$ $\qquad$ $\dfrac{n!}{s^{n+1}}\quad (s > 0)$

8. $t^p\quad (p > -1)$ $\qquad$ $\dfrac{\Gamma(p+1)}{s^{p+1}}\quad (s > 0)\qquad (\Gamma \text{ is the gamma function})$

9. $e^{at}\sin bt$ $\qquad$ $\dfrac{b}{(s-a)^2+b^2}\quad (s > a)$

10. $e^{at}\cos bt$ $\qquad$ $\dfrac{s-a}{(s-a)^2+b^2}\quad (s > a)$

11. $t\sin at$ $\qquad$ $\dfrac{2as}{\left(s^2+a^2\right)^2}\quad (s > 0)$

12. $t\cos at$ $\qquad$ $\dfrac{s^2-a^2}{\left(s^2+a^2\right)^2}\quad (s > 0)$

13. $t\sinh at$ $\qquad$ $\dfrac{2as}{\left(s^2-a^2\right)^2}\quad (s > a)$

14. $t\cosh at$ $\qquad$ $\dfrac{s^2+a^2}{\left(s^2-a^2\right)^2}\quad (s > a)$

15. $t^n e^{at}\quad (n = \text{positive integer})$ $\qquad$ $\dfrac{n!}{(s-a)^{n+1}}\quad (s > a)$

16. $t^p e^{at}\quad (p > -1)$ $\qquad$ $\dfrac{\Gamma(p+1)}{(s-a)^{p+1}}\quad (s > a)$

17. $H(t-a)\quad (a \geq 0)$ $\qquad$ $\dfrac{e^{-as}}{s}\quad (s > 0)$

18. $J_0(t)$ $\qquad$ $\dfrac{1}{\sqrt{s^2+1}}\quad (s > 0)$